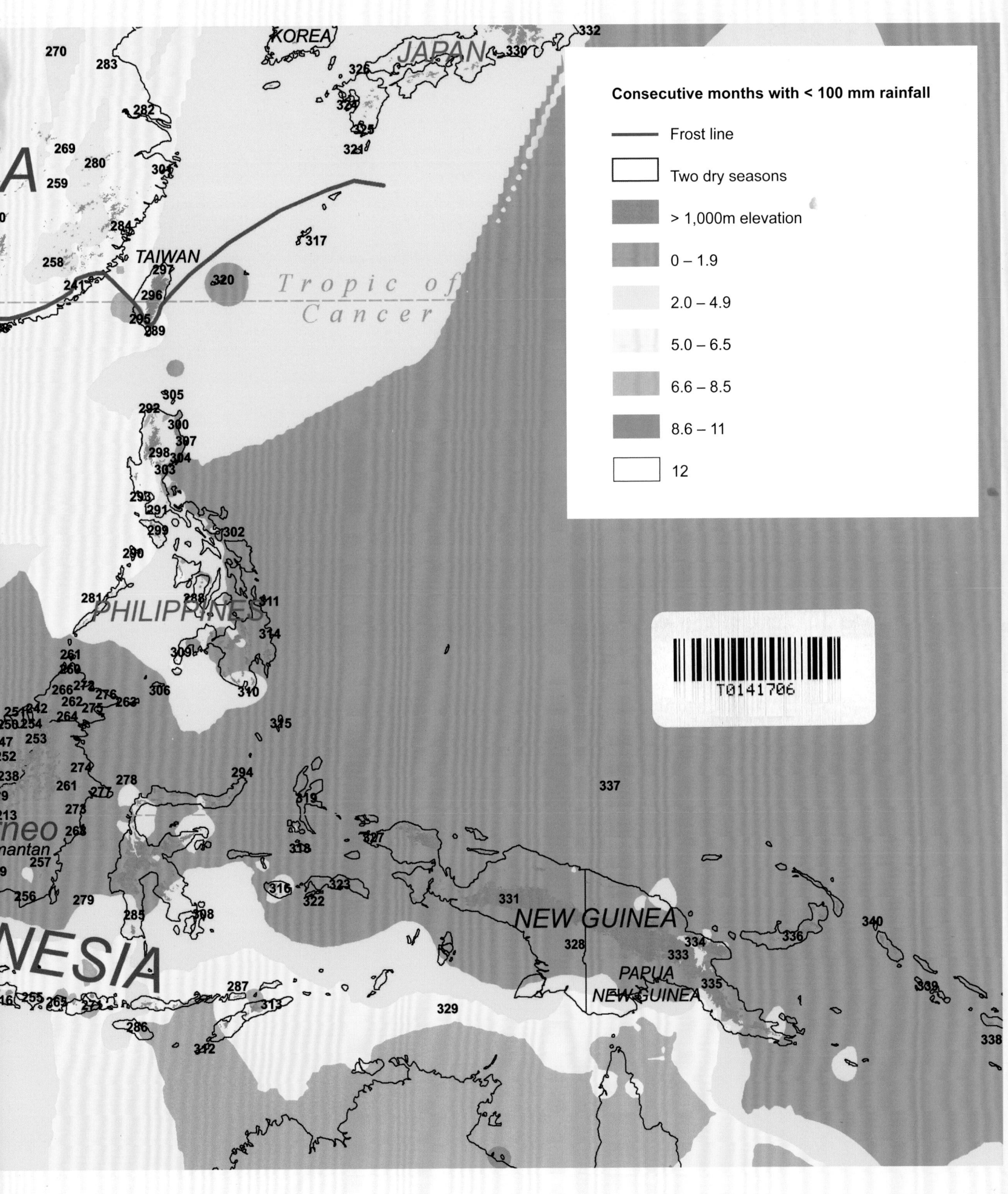

On the Forests of Tropical Asia

Tropical Asia

Lest the memory fade

PETER ASHTON

Edited and with co-authorship of the final
two chapters by Reinmar Seidler

And with much help from

Abdul Rahman Kassim, Shin-Ichiro Aiba, Robb Anderson, Simmathiri Appanah, Asah anak Kalong (Unyong), Mark Ashton, Ian Baillie, Patrick Baker, Banyeng anak Ludong, Kamaljit Bawa, Eberhard Brünig, Sarayudh Bunyavejchewin, David Burslem, Min Cao, Paul Chai P.K., Leonardo Co, Rick Condit, Gordon Congdon, H.S. Dattaraja, Priya Davidar, Stuart Davies, Chris Dick, Elisabeth Dudley, Aljos Farjon, Gan Yik Yuen, Tom Givnish, Peter Grubb, Savithri and Nimal Gunatilleke, Pamela Hall, Hasan bin Pukol, Hans Hazebroek, Michele Holbrook, Steve Hubbell, Ilias bin Pai'e, Akira Itoh, Tamiji Inoue, Bob Johns, Kwiton Jong, Jugah anak Tagi (Kudi), Koichi Kamiya, Mamoru Kanzaki, Tatuo Kira, Takeshi Kohyama, Kuswata Kartawinata, Jim LaFrankie, Ladi anak Bekas, Tim Laman, David Lee, Lee Hua Seng, Bert Leigh, Mark Leighton, Elizabeth Losos, N. Manokaran, Adrian Marshall, Bob Morley, Jan Muller, Francis Ng, Toshinori Okuda, Theo Panayoutou, Matthew Potts, Rebecca Pradhan, Mike Price, Richard Primack, Qeck Swee Peck, Shoba Nath Rai, Forbes Robertson, Steve Rogstad, Soedarsono Riswan, Shoko Sakai, Saw Leng Guan, Salleh Mohd Nor, Julia Sang, Sengelang anak Nantah, Bill Smythies, Engkik Soepadmo, Tem Smitinand, Willem Smits, Bill Smythies, Tony Start, Peter Stevens, Raman Sukumar, Supinda Bunyavanich, H.S. Suresh, Debabrata Swain, Sylvester Tan, I Goesti Made Tantra, Sean Thomas, Ranida Tourinont, Ian Turner, Jeff Vincent, Pema Wangda, Tim Whitmore, John Wyatt-Smith, Isemu Yamada, Takuo Yamakura, Yap Soon Kheong, and Zhu Hua – *but the mistakes are my own.*

Kew Publishing
Royal Botanic Gardens, Kew

The ARNOLD
ARBORETUM
of HARVARD UNIVERSITY

First published in 2014 by the Royal Botanic Gardens, Kew, in association with the Arnold Arboretumm of Harvard University.

Royal Botanic Gardens, Kew, Richmond, Surrey, TW9 3AB, UK

www.kew.org

ISBN: 978-1-84246-475-5

eISBN: 978-1-84246-516-5

Distributed on behalf of the Royal Botanic Gardens, Kew in North America by the University of Chicago Press, 1427 East 60th Street, Chicago, IL 60637, USA.

British Library Cataloguing in Publication Data
A catalogue record for this book is available from the British Library.

Copy editor: Sally Henderson
Production editor: Sharon Whitehead
Design, typesetting and page layout: Nicola Thompson, Culver Design
Production: Georgina Smith , Kew Publishing
Cover image: Hans Hazebroek

For information or to purchase all Kew titles please visit shop.kew.org/kewbooksonline or email publishing@kew.org

Kew's mission is to inspire and deliver science-based plant conservation worldwide, enhancing the quality of life.
Kew receives half of its running costs from Government through the Department for Environment, Food and Rural Affairs (Defra). All other funding needed to support Kew's vital work comes from members, foundations, donors and commercial activities including book sales.

Printed and bound in Italy by Printer Trento s.r.l.

Holder of the following quality, printing and environmental certifications: ISO 12647-2:2004, Fogra® PSO (Process Standard Offset), CERTIprint®, ISO 14001:2004, Carbon Trust Standard.

Dedicated to Asah anak Kalong (Unyong), Ladi anak
Bikas, Banyeng anak Ludong and Jugah anak Tagi (Kudi),
who taught me how to learn from the trees; and to
my love, Mary, who provided for it all to happen.

With special thanks

Without the important contributions of the following friends, this book would have been beyond my capabilities. All were essential, so I thank them in alphabetical order. Professor Mark Ashton, who generously lent his forest cottage to a demanding parent during the six falls that he was immersed in compiling this book; and who patiently introduced him to the science and art of silviculture. Dr Ian Baillie designed and prepared all graphics for the first chapter, except for those taken from Bob Morley's marvellous work. Jeff Blossom helped with satellite imagery and the climate maps at the front and back of the book. Sue Brown gently and with impressive clarity introduced me to the mysteries of lap-top computation. Bob Cook, Director of my institution, Harvard's Arnold Arboretum, gave me moral support as much as financial. Stuart Davies, Director of the Center for Tropical Forest Science, continued to exchange ideas with me throughout the book's preparation and has been unstinting in his encouragement. David Glick and Nicola Thompson masterfully prepared all other graphics. Professor Savithri Gunatilleke volunteered, with characteristic generosity without knowing the enormity of the task, to read the whole manuscript, searching for errors and omissions. Dr Pamela Hall and I have pondered and argued over rain forest issues ever since she started graduate field research in Sarawak. It is thanks to her that some statistical competence exists here; many of the graphics, there acknowledged, are hers. Dr Hans Hazebroek*, superb photographer was recommended to me by Datuk Chan Chew Lun of Borneo Natural History Publicatons. He accompanied me on two excursions, in the wet and dry seasons, to photograph trees, forests and forested landscapes of continental tropical Asia; his are the images acknowledged to *H.H.* Caroline Taylor was the first to seriously analyse my field data. Ranida Touranont McNeely initiated the indices. And Dr Reinmar Seidler who, though recognized as co-author of the last two chapters, spent more than a year in addition reorganizing the text's structure, correcting my undisciplined grammar and suggesting more economical yet felicitous means of expression.

* Hans took several thousand digital photographs; he owns the copyright. They are available in the Arnold Arboretum photographic archive for inspection.

About the author

Peter Ashton, a student of E.J.H. Corner at Cambridge University, first entered the forests of tropical Asia in 1957 as Forest Botanist in Sultan Omar of Brunei's government, when he married Mary who has been partner in his career thereafter. He spent a further five years as Forest Botanist in the Sarawak government where he pursued further forest ecological research and advised on the planning of its national park system. After serving on the botany faculty of Aberdeen University, in 1978 he was appointed Director of the Arnold Arboretum and Arnold Professor of Botany at Harvard, and later Bullard Professor of Forestry. He retired in 2000. He was the first to argue, in 1969, for niche-specificity among tree species in hyperdiverse rain forest, an hypothesis that has provided a continuing focus for him, his students and collaborators, the fruit of which is this book. A further primary interest remains the systematics and ecology of Dipterocarpaceae, dominant canopy trees and leading source of hardwood timber for quarter of a century, and their forests. Author or co-author of over 260 research papers and books, he has field experience of forests of all but three nations in the Asian tropics. He was co-founder with Stephen Hubbell of the Center for Tropical Forest Science at the Smithsonian Institution, facilitator of its large tree demography plots in Asia, doctoral advisor to students from tropical Asia who continue to extend the work, and has thereby played a major plant in establishing the long-term monitoring of tree populations and forest dynamics that will eventually test hypotheses presented here. He is a Japan Prize laureate, and recipient of the Linnean Medal for Botany, the David Fairchild Medal for Plant Exploration and, with C.V.S. and I.A.U.N. Gunatilleke and Mark Ashton, the UNESCO Sultan Qaboos Prize for Environmental Preservation. He has been President of the International Association of Botanic Gardens and Arboreta, was founding President of the U.S. Center for Plant Conservation and the Highstead Foundation, dedicated to safeguarding New England's forested landscape, is a Fellow of the American Academy of Arts and Sciences and of the Royal Society of Edinburgh, and is an Honorary Fellow of the Association for Tropical Biology and Conservation.

Contents

Botanical exploration in the Maliau
Basin, Sabah, NE Borneo (H. H.)

Prologue

I am a firm believer that without speculation there is no good and original observation…

Charles Darwin, 22 December 1857, in a letter to Alfred Russel Wallace at Ternate.

This is a book about natural history, focusing on trees and their communities. It represents, so far as I know, the first attempt to describe and compare the forests of one whole tropical region as a unity. It presents one person's experience and opinions. It is a book for all who love nature and forests, and it is a book for exploring, inquisitive minds. Its principal aim is to stimulate interest, especially among students; and to do that, I have not shunned conjecture ('speculation') in the search for testable hypotheses. In this respect, I have tried to make clear where an inference is based on field observation without rigorous test, or on the literature which I quote; but the unreferenced field observations are my own (Fig. 1).

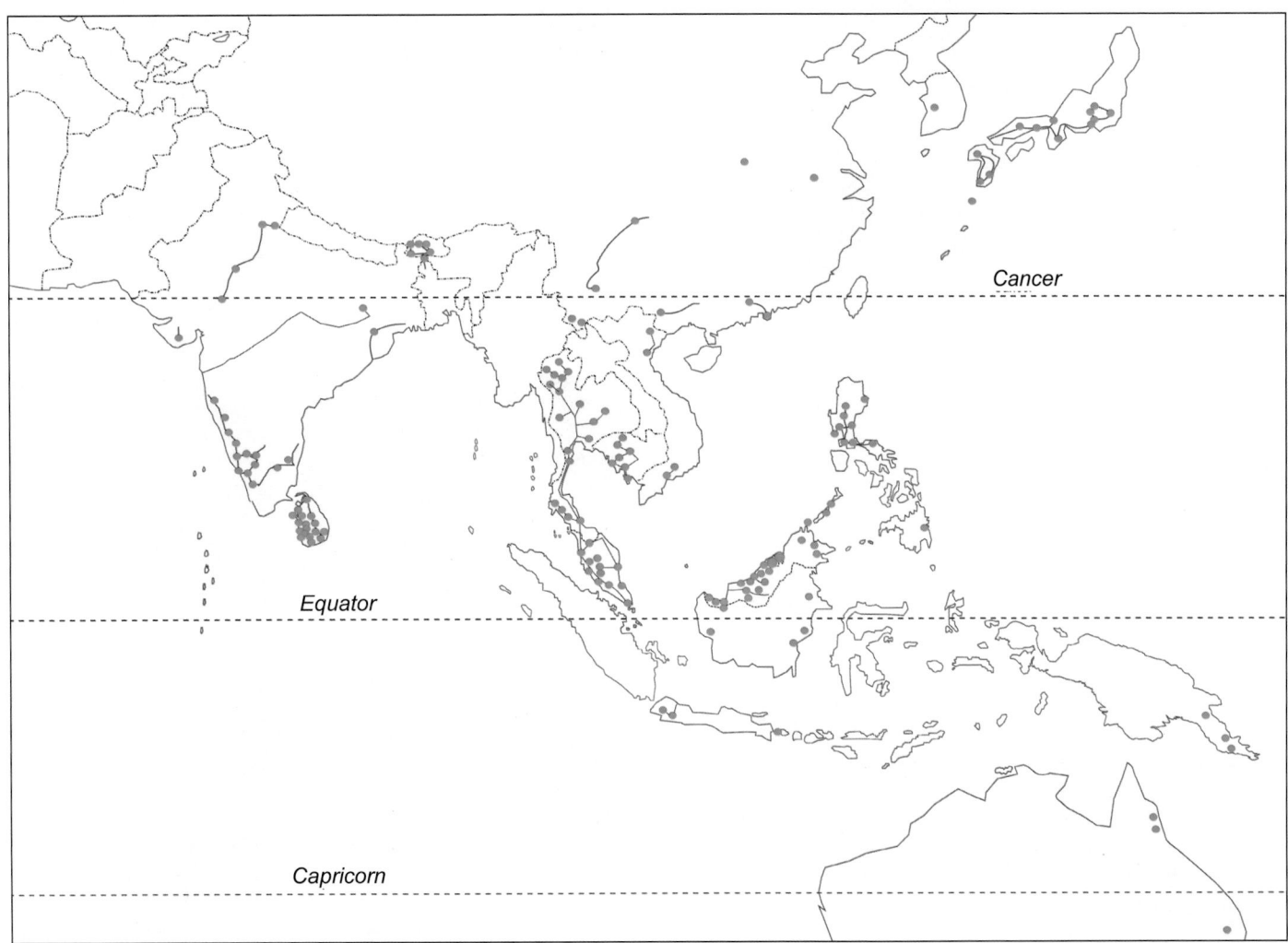

Fig. 1. Map of tropical Asia indicating the forests worked in/visited by me.

Plate 1 Some of the team (others appear throughout the text). **a,** Peter and Jugah, Semengoh, Sarawak, 2008; **b,** Asah and Mary, Kuala Belalong, Brunei, 1958; **c,** Mark (M.A. of the plates) with a student in one of his seedling shadehouses, Sinharaja, 2005; **d,** Sylvester Tan, courageous and incorruptible champion of the forest, probably the most knowledgeable student of the Borneo tree flora living today, Lambir, 2009; **e,** Sinharaja partners (left to right): Mary, Principal Investigator Savitri; naturalist Martin Wijesinghe; Peter, field co-ordinator Gunadasa; Mark. 2005; **f,** Reinmar Seidler, tireless perfectionist, Boston. (**a** P. Chai P.K.; **b** P.A.; **c** Nimal Gunatilleke; **d** S. Russo; **e** M.A.; **f** R. Seidler)

This work is not intended as a standard text, but as a personal record and interpretation, extended and provided context by the work of my former students colleagues and predecessors. I have not hesitated to refer to observations learned in conversation with my friends, who I cite. As I wrote, the text inevitably expanded, but it is still by no means comprehensive. It explores a wide range of fields relating to Asia's tropical forests – far too wide for me to claim fluency in more than a few. Inevitably, you will find misunderstandings and misinterpretations. But I stand confident of my main conclusions.

Forests have experienced radically changing values and uses in our time, changing faster than it takes to grow a tree. Our aim must be to keep future options for forests open, by not destroying opportunities through present use. That is not easy, but it is possible. It depends upon a basic understanding of the nature and functioning of these systems of unexcelled complexity. It also depends on a realistic appreciation of economics, and a time scale that takes into account the needs of all children – both ours and the forests'.

There are those who argue that taking decisive action to protect tropical forests must await a more complete understanding of how they function as ecological systems. But it takes much time to learn how long-lived organisms function, and we no longer have much time. Besides, we will never know enough. We must therefore not be ashamed of what we infer from careful observation, but have not yet rigorously tested by experiment.

I was uniquely fortunate in my first employment in the late 1950s, as Forest Botanist in H.H. the Sultan's government in Brunei. This provided me, then a graduate student, with three fruitful and happy years, more than two of which I lived in villages and under canvas. I was thus privileged to learn the trees and forests of a country which, though somewhat smaller in land area than Delaware or Wales, supports perhaps 3,000 tree species: as rich a tree flora as anywhere on Earth. Further, this was my initiation into research. Little had been written on the forests of north-west Borneo at that time, and I had but cursorily read the general literature, such as Paul Richards' *The Tropical Rain Forest*[1] and John Corner's *Wayside Trees of Malaya*.[2] I had no library and no herbarium. Instead, I shared a house with four Iban Dayak men of my own age. They had come downriver to seek work, and found an ideal calling as botanical assistants and tree-climbers. Their knowledge of the species and their uses was encyclopaedic, their camp-life humour robust and hilarious. I returned to Cambridge with an impressive knowledge of a hyperdiverse local flora, classified according to the Iban terminology. Only then, and from a truly original standpoint, did I start to consult the literature in earnest and in depth.

Such opportunities have all but gone today. Formal schooling has squeezed out traditional knowledge, and the cash economy has made it seemingly redundant. There is no longer any attraction for an eight-year-old village boy in Borneo to accompany his father up into the mountains for three months at a time in search of copal resin, living off the forest. Although it is simple to recognize a teak tree or a sal, identification of the literally thousands of tree species that may inhabit a single hill in a Borneo rainforest is another matter. Many species are known only locally, and then only by a tiny coterie of village experts, some of them traditional woodsmen and a few trained in the scientific approach. Numbers of both of these groups are shrinking by the year. Further, very few students and researchers now have the opportunity to work for prolonged periods in certain remote parts of this region; lack of regional research collaboration has been a major impediment to the advance of forest science.

Our predicament. When I first went to work in Asia, now more than 50 years ago, three quarters of Peninsular Malaysia, Sumatra and Borneo were still blanketed in primeval forest. Until one hundred years ago, the prevalence of diseases, particularly malaria, had permitted only sparse human settlement. These settlements were concentrated on the best riparian agricultural lands, leaving large swathes of forest almost completely uninhabited. Today, except for a few of the national parks and other strictly protected areas, nearly all that forest has been logged over, preventing careful insight into its ordered functioning and composition in apparent natural equilibrium. Too often the forests have been mauled by unsupervised logging and road building, the rivers silted and polluted. By the 1960s, fifty years of careful silvicultural research in the most diverse lowland forests of Peninsular Malaysia had produced the practical know-how to manage the lowland forests sustainably for timber; but forestry can never compete with land made arable for commodity crops. Therefore, the remaining lowlands of the Far Eastern tropics are now being converted to permanent agriculture at an ever-increasing rate. Additionally, as remote rural populations inexorably increase, slash and burn cultivation continues to invade the remaining forests. Remnant patches of lowland forest are increasingly confined to the uplands, on land too steep to plant, and on poor infertile soils: peat, white sands, limestone *karst*. The silviculture of these remnant forests on the poorest lands has not yet been resolved, even for timber production alone.

Meanwhile, the national parks remain as islands of forest, often in a sea of desolation where farming practices have proven unsustainable. Most are restricted to land which defies other use. Many of the parks have been logged, and almost all suffer chronic nibbling at the edges by cut-and-run operators and expanding farming communities. No forest is free of hunting today. Many have already lost their larger seed dispersers, surely affecting their future trajectory. Many forests are still hunted by those with traditional rights legislated at a time when their practices may hardly have reduced the forest resource, but since then modern fire-arms have replaced the bow and the blowpipe. How completely, and over how long a time period, will the remaining exploited forests recover, and to what extent can we hope to influence the outcome? And indeed, why bother, if massive extinction has already taken place? These must be leading questions for research; they will be addressed in the essays that follow.

Tropical rainforests are being fragmented and modified on a wider scale even than during the great periods of reduced rainfall in the tropics which coincided with the northern ice ages. But it is unhelpful to despair. Much is lost, but much yet remains. The most majestic forests that ever existed in the tropics, the mighty dipterocarp forests of Negros and of the black soils of the north-east Bornean lowlands, have all but gone now – yet the tallest trees in the tropics still survive in southeastern Sabah. So far, species extinctions have been limited, except among those large animal species that naturally occur in small numbers over vast areas. I believe that the forest can still substantially recover, but it will take time. It will require forbearance – and loss of immediate income – and often costly replanting by governments on behalf of the owners, who are the inhabitants, especially the rural population. Only then can sustainable timber production be resumed.

History has shown that forests become revalued whenever successful economic development and the accompanying urbanisation are achieved – but for the services they yield, rather than for their material products. Forests mediate the flow and quality of water, reduce erosion, and moderate weather in and around our burgeoning cities; and they provide beauty, recreation, and escape from stress. Crucially, the biological diversity of tropical rainforests constitutes the greatest library of genetic information that our world possesses. Even if that information only needs to be accessed from time to time, it will remain vitally important forever.

The challenge for a scientist. When I was a student, my research advisor, John Corner, insisted that I must first immerse myself in the rainforest for at least a year, otherwise I would be forced to rely on the conclusions of earlier workers for the choice of my research objectives. He was right. A deep experience of natural history and a sharp eye for recurring patterns are fundamental to the identification of untrampled paths of research, and the choice of the most suitable organisms by which to explore them. Few students now have those vital

opportunities, partly because career opportunities do not arise until qualifications have been won, but partly also because many forests have gone. This book aims to provide a 'second best', by describing the natural history of trees and forests in tropical Asia as I have seen them over the last fifty years. I have also presumptuously offered explanatory hypotheses of my own, and distinguished unexamined issues by a *change of font*, as challenges to future students.

Biotechnologists claim that we are on the threshold of designing and producing man-made organisms analogous to natural species, some of which will be crop plants of enhanced productivity and disease resistance. As with the 'Green Revolution' crops, such as the modern wet rice varieties, land will need to be modified to fit the needs of such crops. This will never be economically justifiable on the infertile soils and steep terrain of much of the Asian tropics, though perverse and unsustainable economic policies may appear to encourage it. In these places the economics of sustained use will continue to require the reverse approach, in which crop plants are chosen to match the varying demands of the land. The economic case for the sustainability of indigenous tree crops growing in mixture on marginal lands has never yet been invalidated. Wise management of existing natural forests will contribute substantially to this goal. It is unlikely indeed that science will ever be able to formulate ecologically and economically sustainable artificial mixed-species ecosystems of man-made organisms.

It is becoming widely known, through the media as well as through research publications, that certain tropical rainforests represent the apex of land-based biological diversity on Earth, in terms of both genetic and species diversity. They are therefore irreplaceable laboratories for testing a wide variety of evolutionary and ecological hypotheses. A general methodology for rainforest management must rely on the resolution of these hypotheses.

Primary forests and ecological near-equilibria. Why do researchers so often focus on 'primary' forest, that is, forest presumed unaffected by human activity? We assume that such forests are in the midst of a cycle of naturally-induced damage and recovery which has varied little, on a scale of centuries, since the last major change in climate 10,000 years ago. (Even that change may have been slight in the equatorial forests, as I will recount.) Only if we document change in these forests over time, therefore, can we hope to understand cycles which may govern long-term near-equilibria in species composition and dynamic processes? (Of course, evolution continues within the populations, but that process only becomes manifest in novel tree species at a scale of millions of years.) A single set of observations may provide a snapshot, but cannot set that snapshot in its context within the forest cycle. Long-term monitoring alone can do that.

A forest returning from some destructive human impact is also in a process of recovery; human disturbance varies in its nature and intensity at many scales, it cannot be precisely repeated, and is rarely documented from the beginning. Species lost from a particular area may take centuries to return. Its study therefore merely records one rung of a successional ladder whose past history is unknown and whose future trajectory cannot be ascertained. Rather, we have a better chance of gaining a thorough understanding of its successional processes through careful study of a primary forest's life cycle.

In what follows, we will see that there are certain constraints to this approach – though not that there are better alternatives. First, the canopy-tree deaths that initiate patches of forest regeneration through species succession do not occur at a constant rate in any forest. All forests experience periodic natural catastrophes. These may largely account for the losses of the big mature trees that together comprise a large proportion of the total forest biomass. Such major catastrophes may only occur in a forest at intervals of many decades. Even so, the influence of current climate and soil is continuous and measurable, that of natural catastrophe definable and amenable to monitoring. There is, therefore, no reasonable alternative to establishing sample plots in locations where observation and measurement can be continuous over time.

Again, there may no longer be any forest free of human influence. In particular, hunting reduces the populations of animal species that exert strong effects on forests through dispersing seeds and pollen and through browsing. Hunting occurs in virtually every Asian forest today. For these reasons, understanding the natural structure and function of mature tropical forests, and thereby developing best-management practices through rigorous research, becomes more difficult by the year.

From theory toward practice. The scientific method generates probabilities, not certainties. First, field observation leads to the identification of repeated patterns. The researcher must speculate on their cause, and refine these speculations into forms amenable to rigorous testing, namely hypotheses. This is the most difficult part, but necessary, since natural history must remain the wellspring of forest science. And it is the primary purpose of these essays. Carefully designed tests of hypotheses must ultimately include experiment, in which the impact of a well-defined modification of one property on a forest community or population sample is compared with an unmodified replicate. Finally, results are analysed and published after searing criticism by colleagues. Others will

thereby be challenged to pursue further tests using different approaches and different locations thus defining the limits of applicability of a hypothesis. Through this process, theory is generated.

CTFS – a network of comparable research sites. The central goal of rainforest research is to assess the extent and the ways in which natural selection operates in hyperdiverse plant communities, thereby accounting for their floristic structure; and with that to understand how to manage each forest optimally for one or more given benefits – services such as water conservation, carbon storage, biodiversity conservation or recreation, or goods such as timber or game. Botanists, perplexed by the extraordinary number of species that appear to co-occur, apparently at random, in the most diverse tropical rainforests, had long assumed that their species are ecologically complementary, that is equivalent. During my first appointment in Brunei, by laying out numerous plots at 0.6 ha^{-1} – small for the scale of their species populations – across the topography at two sites differing in soils and geology, I was able to show that rainforests are not floristically uniform. In 1969 I hypothesised that perhaps, after all, their species might be niche-specific. My subsequent research with students and colleagues, into rainforest tree-breeding systems and genetic variability added to the evidence. In 1982 I met theoretical ecologist Stephen Hubbell who had established an ambitious, 50 ha, plot, set out in Barro Colorado Island (BCI), Panama, with botanist Robin Foster; in this, trees of at least 20 cm diameter had been mapped. His plot could capture the spatial patterns of species populations. It revealed the conspicuous clumping of most species. Steve had concluded that species population patterns within the sample were due to the random exigencies of limited seed dispersal, therefore by implication ecologically complimentary. We decided to join forces, Steve with theoretical and analytical genius and myself the field naturalist with knowledge of plants, people and forests in the Far East. Thanks to the U.S. National Science Foundation, and the interest of Salleh Mohd Nor and others at The Forest Research Institute Malaysia (FRIM), we were able to replicate the Neotropical plot in Malaysia, at Pasoh Forest Reserve, Negeri Sembilan, which had already served as a site for research. The plot design had been expanded to include all trees of at least 1 cm diameter, which at Pasoh yielded a staggering 340,000 individuals. The Panama protocol was exactly replicated, including 5-yearly recensuses which continue under the leadership of FRIM. In 1986 I organised a conference in Bangkok on behalf of the U.S. State Department, to address means towards retention of biological diversity in the face of development, for the ASEAN region. A leading recommendation was the need

to establish research capabilities in biodiverse tropical forests in which long-term monitoring could be established, using strictly comparable methods and thereby achieving regional comparability. Colleagues from several countries expressed interest in the BCI concept, and a regional network was initiated. Expansion in Asia progressed rapidly. Sites were chosen which (1) represented the major lowland forest formations (although a lower montane forest plot also now exists, thanks to the efforts of Mamoru Kanzaki and Sarayudh Bunyavejchewin in Thailand); and (2) where there was the promise of continued support from institutions and colleagues. Soon Raman Sukumar at the Indian Institute of Science, and Sun I-Fang from Taiwan, joined Steve in Princeton, thereafter to establish plots beyond South-East Asia. Neotropical plots were to follow. Thus, the Center of Tropical Forest Science (CTFS) originated, to be formally established within the Smithsonian Tropical Research Institute. Participation has been coordinated on behalf of CTFS until recently by Harvard's Arnold Arboretum, further strengthening an ongoing collaboration between it and Asian plant scientists that started over a century ago.

Based substantially on BCI and Pasoh data, Steve Hubbell in 2001 published *The Unified Neutral Theory of Biodiversity and Biogeography*.[3] Hubbell demonstrated quantitatively that spatial and dynamic patterns within biodiverse communities and tree species populations occupying uniform habitats of scales up to 50 ha are compatible with predictions based on the hypothesis that they are ecologically complementary, and that the observed species richness can be as well explained by constraints on their seed dispersal alone – limiting the space that any one species can occupy at one time – as by the fine-tuning of niche specialisation through natural selection.

Were such neutral processes the sole determinant of species composition, rainforest composition would vary chaotically irrespective of habitat variation, with disastrous implications for optimising management. As I will show, that is clearly not the case, even though the clumped distribution especially of juveniles of many species must indeed be caused largely by constraints on seed dispersal. The neutral theory provides a rigorous null hypothesis against which to test claims that natural selection has led to specialisation. Resolving the relative roles of natural selection and chance in determining composition and the predictability of the dynamic cycle of forests will determine the precision with which they can be managed.

Our understanding of tropical rainforests has accelerated in the last three decades thanks to the creation of this worldwide network of lowland rainforest plots, large enough to permit analysis of the demography of most of their species, and monitored over the long term using rigorously comparable methods.

Today, the CTFS plot network has expanded beyond the tropics to more than 40 large tree demography plots in all climates with forests (those in Asia in Fig. 2), thanks to the exemplary leadership of Stuart Davies. This new programme, the Smithsonian Institution Global Earth Observatory (SIGEO), has the same monitoring protocols, but with the changed emphasis of examining the response of forests to climate change. CTFS remains as a separate entity within it.

SIGEO is the first global network of terrestrial ecosystem sample sites which allow rigorous comparison. In addition, the plot samples serve as controls for alternative land use experimentation. The network has been established through the conviction and enterprise of knowledgeable researchers in 12 nations in Asia alone. The 25 plots in tropical and temperate Asia, ranging between 16 and 50 ha each (plus a 2 ha plot in the 125 ha Bukit Timah Reserve in urban Singapore), currently contain more than three million tagged, identified and mapped trees exceeding 1cm in diameter. These represent more than 3,000 species, an estimated 12% of all tree species known from their regions to date. Half of these species include population samples large enough for rigorous demographic analysis. For the first time, the interactions between species, especially those series of co-existing congeneric species which are such a remarkable characteristic of Asian rainforest tree species diversity, can be observed over time and subjected to experiment. For the first time, variation in community and species characteristics over different climates, soils and topography can be rigorously compared. The results of this research will provide a general theoretical base for the management of tropical forests, be it for timber production or the conservation of biodiversity in forests now modified or fragmented.

Plate 2 (left) Expeditionary forest botany: **a,** Sengelang, Ladi, and a Murut shaman, dawn approaching, after our (unsuccessful) attempt to summon the spirits: Gunung Retak summit, Temburong headwaters, Brunei, March 1957; **b,** Transport in the sixties: ascending the Mujong, Sarawak; **c,** Exploring the Maliau Basin, Sabah, 2009: crossing a natural bridge. Long term research: **d,** The Royal Thai Forest Department field camp near the CTFS 50 ha plot within the Huai Kha Khaeng Wildlife Sanctuary; plot director Sarayudh Bunyavejchewin stands by. The hole in the wall was made by an investigative wild elephant; **e,** The Pasoh field station of the Malaysian Forest Research Institute in 1988, base for the 50 ha plot. Stephen Hubbell (left) by the Land Rover; **f,** Savitri and Mary assist with the layout of a seedling transplant plot, Sinharaja, 2003 (see Plate 1e). (**a, b, d, e, f** P.A.; **c,** H.H.)

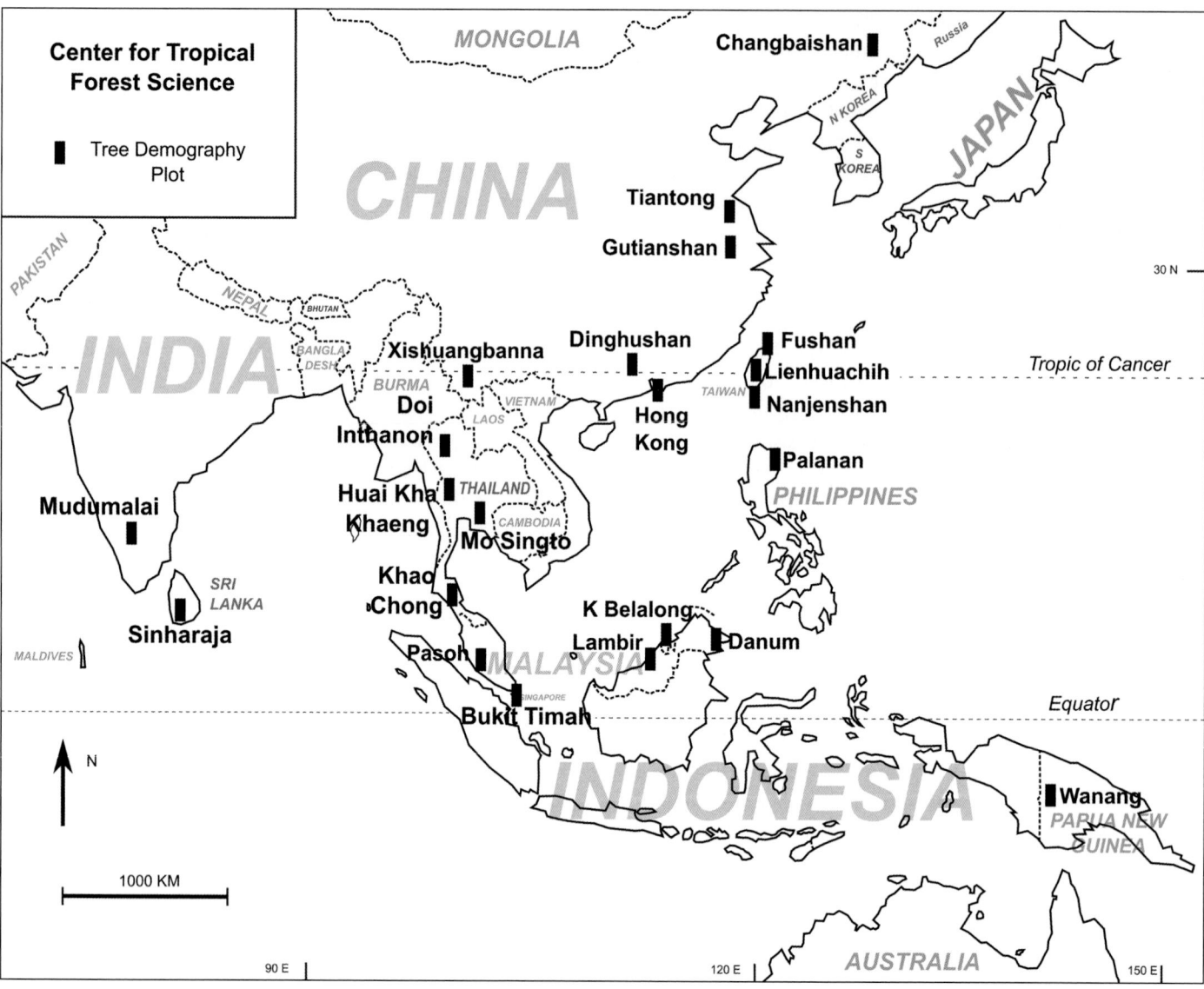

Fig. 2. The collaborations and plots of the Center for Tropical Forest Science in Asia as they were in 2012.

The aims of CTFS are ambitious and require many hands to learn and to share experience. CTFS provides the opportunity for a vast range of endeavour[4]. My essays merely set the scene, inviting you to rise to the challenge – and participate! The first permanent forest plots in tropical Asia, India and later Malaysia, were established to serve as scientific controls against which to compare results of silvicultural experiments. Although Matthew Potts has used the Pasoh plot data aimed at evaluating how management for both timber and biodiversity conservation can be optimised, none of the region's forest management researchers have as yet availed of them. Yet the opportunities are limitless.

Structure and purposes of this book

These essays are written in the conviction that only rigorous science, based on detailed field study and followed up by experiment, can achieve a sound methodology for the successful continuing management of biodiverse tropical forest in a world of economic and environmental change. They also assume that, although the forests of every region are unique in many respects, there is an underlying set of general rules which govern their form and function, whose understanding is essential because their survival must now rely on active management. It is with these practical goals in mind that hypotheses and tests, leading to the building of general theory, are presented here. And to succeed, science must be both understandable and fascinating to anyone who takes a serious interest. I write in the conviction that forest science can be communicated in plain English to anyone who wishes to listen!

This will not be a standard text. Such have already been provided by Tim Whitmore in his elegant syntheses, *Tropical Rain Forests of the Far East* (1984)[5] and *An Introduction to Tropical Rain Forests* (1998),[6] while the second edition of Paul Richards' *The Tropical Rain Forest* (1996)[7] and Ian Turner's *The Ecology of Trees in the Tropical Rain Forest* (2001)[8] provide more exhaustive reference to tropical forests and trees respectively. Instead, I here present a rather personal perspective, developed over a half-century of fieldwork in the region. And you will see that I have been immeasurably helped by the work, in Asia and elsewhere in the tropics, of those many graduate students with whom I have had the privilege of being associated, whose names I include among those on the title page. I hope to demonstrate how many questions remain unanswered, how many exciting avenues for scientific inquiry remain unexplored; and I will suggest ways in which some of them might be pursued. That is why I use the word *essays* for my chapters: observations which have led to attempts, quests, to answer questions of a lifetime.

I will provide extensive references for further reading, especially of the regional literature which is too often overlooked. I especially aim to attract students from the region who are starting out, uncertain whether to continue because the questions seem so difficult to define and to address. I have been through that, and have ended up with far too many exciting questions to answer alone. Researchers outside the region are principally interested in theoretical, generalisable aspects of those characteristics special to tropical forest. As American research has come to play the leading role in theoretical work in the Neotropics, so Japanese scientists have come to lead in the Far East. Many of the advances in theory since Tim Whitmore's book are thanks to their efforts, which include major contributions to the CTFS program as well as independent work at CTFS sites. The first priority of researchers within the region is most often to seek a better practical understanding of forest form and function. It is only with such improved understanding that the future of indigenous forests can be secured through political support for policy reform and better management. I have been overwhelmed and heartened by the increasing volume of published research emanating from some nations within the Asian tropics. The quality as well as quantity of work from India, and more recently China, reflects the growing strength of movements for sustainable forestry in these countries. This literature is still too little recognized internationally. I try here to give it the voice it is due.

A comparative approach. Traditionally, forestry has been practiced by developing empirical experience of individual forests or forest types. Prior to CTFS, few attempts had been made to draw general principles by statistically rigorous comparison at regional, let alone global, scales. This is a book about comparisons. My specific aim here, then, is to describe the forests by comparing those in different regions and habitats – with reference to the work of others elsewhere – in order to suggest explanations for their characteristics, and ways in which these suggestions can be tested by further study. I will also mention those places, known to me, where at least some of these stately groves can still be seen.

Forest research in Asia has passed on from the descriptive stage. It now focuses on deep understanding of specific cases – that is, specific samples in specific forests – from which generalisations can be inferred. But these forests differ, and conclusions derived from their study inevitably do so also. To resolve which conclusions are truly generalisable, a thorough understanding of the *differences* among these forests is required. Too few researchers in contemporary tropical Asia have the opportunity to see – and work in – countries other than their own. This is resulting in unnecessary duplication of effort, as well as deprivation of an essential tool for building hypotheses.

No comparative account of the forests throughout tropical Asia has previously been attempted, therefore no opportunity to explain how their diversity has arisen, nor how it variously functions. I have focused on their tree species and their communities, discussing other organisms only to the extent that they interact with trees. The essays derive from my field notebooks, from whose observations I planned my own more detailed studies of structure, composition and dynamics. My fieldwork has extended to most regions of the humid tropics of Asia west of Wallace's Line, Burma, Laos and Nepal excepted, with more perfunctory knowledge of Papua New Guinea and also Queensland. My experience is therefore broad, but hardly comprehensive! These accounts are restricted to inland forests. The shoreline woods, inhabited by relatively few species, are not considered, nor are the mangrove forests that have already been so ably described by Watson (1928)[9] and Tomlinson (1986).[10]

I have always concentrated my interest on one particular tree family, the Dipterocarpaceae. The dipterocarps will thus receive more reference than other tree groups in these essays. They are dominant in many lowland Asian forests, and the major source of tropical timber in the Far East. As I will show, they serve as excellent examples of many of the issues presented here, as well as the major source of tropical hardwood timber on world markets for three decades, which are ending owing to exhaustion of what had been a feasibly renewable resource.

This book makes no attempt to describe the multitude of forest communities which, characteristic of a particular climate and geology, can be recognized at the local scale. Rather, it seeks to understand forest variation in relation to habitat and history and thereby to identify questions yet unaddressed. I have focused on the forests of some countries in particular, because with the help of colleagues in them I originally sought them out as paradigmatic of their type, and now I know them best. These include areas where major indigenous forests still exist, though they are now nearly all selectively logged. They became the best examples of key forests and habitats known to me, and are almost all freely accessible to others provided appropriate clearances have been obtained. We selected many as sites for the CTFS partnership. Some areas, such as Wallacea, are more thinly treated, for my knowledge is scant and the literature on forests still sparse. The forests of Wallacea deserve a book on their own!

Aim and approach. I recognize that these essays are idiosyncratic. They address aspects of the evolutionary biology, that is, the historical and ecological biogeography of the species and the structure and dynamics of the forests in which they occur, from viewpoints gained by one person's experience. And I have not flinched, therefore, from hypothesising and sometimes speculating. That is why this book must not be read as an authoritative class text, but rather as a stimulus.

My first aim, then, is to excite the interest of a new generation of forest scientists within tropical Asia. Already there is growing interest in a few countries. Others must follow, for the need is urgent. My hope is also that these essays will encourage more research at least initially based within the forest itself, because that is the most promising means to original insights. Young scientists in the region will thereby get to the basics, which are the species and their evolutionary relationships: to merely observe the form of the trees and the structure of their forests is insufficient for gaining deep insight of their functioning. The emphasis must be on systematics rather than taxonomy, for it is the interdependence of phylogeny and ecology that will explain the mystery of hyperdiversity in rainforests, thereby facilitating its now imperative active management. One must learn to recognize at least the commoner families and species. For that, Corner's *Wayside Trees of Malaya,* which serves well for the humid tropics of all Asia, remains unsurpassed in its clarity and perceptive observation.

We botanists have too often danced on the heads of pins. Plant collectors, who have visited more forests than any other scientists, have too seldom paid attention to the details of habitat, while ecologists have often ignored

systematics; yet it is the combination of the two that yields fundamental understanding. A herbarium can be a goldmine of ecological information, but the labels of specimens gathered from jungles once remote – and too often now gone – rarely inform regarding soils, or even altitude. The floras, which are based on them, often give species unrealistic ecological amplitudes.

Reductionist and inferential science. We live in an age of reductionist science. We realise that we can only understand the mechanisms which sustain these immensely complex living systems by isolating their components, one at a time, and subjecting them to the experimental methods of hard science. But the number of such tests is potentially limitless. Key factors of interest can only be identified, tentatively, by thorough examination of the whole system, and through the identification of recurring patterns in their form and function. The purpose of these essays is to do that for one biome and biogeographical region, and from these overviews to identify key hypotheses and approaches whereby these might be tested. As for different reasons in astronomy, much of the complexity of the rainforest is not amenable to experiment. However, inference is possible through comparison of examples where only a single factor, or a single group of interdependent factors, differ. My approach is comparative, paying particular attention to those forests where most research is currently being pursued.

Content and structure. The unique feature of lowland tropical rainforest is its extraordinary species diversity. Central to any management system aimed at retaining options for future changes in priority of use must be a sound theoretical understanding of how this diversity is maintained. In part this requires knowledge of how it arose. The first essay, in the form of a book chapter, addresses the physical setting of the Asian tropics. The five essays that follow it discuss local and regional patterns of forest form, function, and history. The seventh seeks to synthesise these insights by identifying patterns of species diversity and their causes. The eighth is a reflection on the history and present status of human influences on Asian tropical forests. It provides the foundation for the final essay, in which future options for tropical forests are discussed in the context of a changing socioeconomic – as well as climatic – environment.

Other than in the first chapter, which sets the scene, each essay aims to include three main elements: (1) I describe my observations, and others that have been made in the Asian tropics, in relation to a particular topic or theme; (2) I relate these regional observations to those made in other regions of the tropics; and (3) where possible, I offer hypotheses to explore relationships among the observations which might elucidate the central topic, and suggest priorities for future research to refine these hypotheses and test them. Throughout, I focus on comparison of individual forests, their species and properties.

Recognising that the complexity of tropical forests cannot truly be understood without seeing them in person, but that few, even in those countries with the privilege of harbouring such forests, have done so, I try to remedy by generous illustration.

My primary intent. Tropical forests, the repositories of so much of the world's terrestrial biological diversity, are today understood largely through a few, often isolated, studies and locations. The vast interstices remain as yet almost unexplored through rigorous enquiry. This book will succeed if more able adventurers are inspired by the questions here posed. It is aimed at all of you who are, as I was, immensely drawn by the excitement of tropical forest biology but who are daunted by its complexity. Where to begin? What questions are most likely to yield generalisable answers, and which are answerable with minimum labour using simple, robust equipment within the funding period of a master's or doctoral dissertation? If more casual visitors with inquisitive eye also find my efforts of interest, then that is an added dividend.

Access to this book and its background information. This book is published online as well as in hard copy, so that students worldwide can readily access it at affordable cost. The photographs in it represent a small selection from the extensive archive at the Arnold Arboretum, together with many offered by friends. These, and the data accruing from current CTFS plots as well as my earlier plots in Borneo, can be accessed through the respective websites of these institutions, provided appropriate permissions are gained and acknowledgments given.

Peter Ashton (*Taji Buloh*[1]), April 2013, Tuckerton.

References

[1]Richards 1952. [2]Corner 1940, 1988. [3]Hubbell 2001. [4]Ashton *et al.* 1999. [5]Whitmore 1984a. [6]Whitmore 1990. [7]Richards 1952. [8]Turner 2001a. [9]Watson 1928. [10]Tomlinson 1986.

---

[1] A nice *double entendre* by which I am still known by my Iban forest team-mates. *Taji Buloh,* a bamboo cockspur, was the nickname of a 19th Century Iban hero who fought the first Raja of Sarawak's men; but it also implies someone who is fidgety and never satisfied, always urging his co-workers to achieve more than is reasonable.

Bukit Lambir, Sarawak, Malaysian Borneo, a cuesta
of Miocene sandstone overlying shale (H. H.)

Chapter 1

TROPICAL ASIA AS A SPECIAL CASE: THE PHYSICAL ENVIRONMENT

The patterns of natural vegetation occurring in a region can only be understood with reference to two fundamental categories of knowledge: the geological history of the region, and its current physical setting. As plant species disperse across a landscape, unsuitable climates and soils may act as barriers more impenetrable than the sea. Aspects of climate such as seasonality and mean temperatures vary continuously through time, as do sea levels; even surface soils change over the long term. Together, these physical aspects determine which species can disperse and establish new populations. The remaining factor that determines which populations will compete successfully for existence – namely, the suite of species already present – will be addressed in Chapter 3.

1. History and geology

The origins of the biotic assemblages we recognize in contemporary tropical Asia lie deep in a tumultuous geologic history. G. Evelyn Hutchinson[1] described these origins as 'composite'; as a sometimes contentious dialogue between ancient history and geology. The contemporary product of this almost infinitely slow but relentless and ongoing dialogue is a uniquely fine-grained mosaic of topographies, soils, and habitats. Yet this mosaic of substrates itself remains unfinished, hence unstable. Onto this evolving mosaic are superimposed the far more rapid and local changes wrought by water and nutrient cycling (described in Chapters 2 and 3), and finally the relatively explosive alterations currently being propelled by human activities (Chapter 8).

1.1. *History over geological time: the interplay of tectonic masses*

The Asian tropics differ dramatically from those of the other tropical continents, South America and Africa. Each of these great land masses is unique in its geological history and climate, and consequently in its landscapes and the pattern of the forests that cover them, as well as in the geography of its fauna and flora.

We live today in an epoch of reassembling continents. Like vast ice floes, the land masses of the Earth assemble, break up, and reassemble. But unlike ice floes, they do so over hundreds of millions of years rather than over a single season. Flowering plants, which dominate present day tropical floras, probably originated near the end of the Jurassic period (c.150 Ma). The fossil record confirms their existence by the early epoch of the next and final (Cretaceous) period of the Mesozoic (c.125 Ma)[2]. Their current distribution, therefore, can only have been influenced by events since that time (see also Chapter 6).

The last great global unitary continent was Pangaea. During the Permian period (c.275 Ma)[3], this immense land mass began to tear slowly open from its eastern coast towards the west, splitting Pangaea into two east-west continental blocks ranging north and south of the Equator. These blocks remained joined by the Americas in the west[4], so that as the split between them gradually deepened it created first a gulf, then a great sea, the Tethys. The result was a continent of two mighty lobes, Laurasia to the north and Gondwana to the south, joined in the west by a unitary America. Evolution began to follow separate lines on each lobe.

Almost from the beginning, coastal pieces known as terranes broke from the southern lobe, Gondwana, and began to drift northward[5]. The first chunk separated from the coast of what is now northern Australia during the early Carboniferous (c.350 Ma), when it was positioned at 25° S. This split started before the major split of Pangaea had even begun. This terrane, followed by others, drifted north as the Tethys opened, colliding and then fusing with the southeastern coast of the northern (Laurasian) lobe by the late Permian– Early Triassic (over 200 Ma). These terranes, however, parted from Gondwana far too early to have carried any early flowering plants with them. Together these chunks of the ancient Gondwanan land mass currently constitute the Tibetan Plateau, Southern China, Burma, Indo-China, the mountains running south from Yunnan along the Burmese-Thai frontier south through Peninsular Malaysia into East Sumatra known as Sinoburmalaya[6], and a triangular block with one side running down the west Borneo coast and the others tapering to Borneo's eastern coast (Fig. 1.1). Ophiolite extrusions mark the current positions of the sutures between them. The South-East Asian continental blocks moved southwards 5–7° in the

Eon	Era	Period	Epoch	Age	Start (million years ago)	
Phanerozoic	Cenozoic	Quaternary	Holocene		0.0117	Our present pluvial
			Pleistocene	Upper	0.126	
				'Ionian'	0.781	
				Calabrian	1.806	
				Gelasian	2.588	
		Neogene	Pliocene	Piacenzian	3.600	
				Zanclean	5.332	Increasing 100 my climate oscillation
			Miocene	Messinian	7.246	
				Tortonian	11.608	Collision of Africa and Laurasia; Australasia close to SE Laurasia; perhumid Sunda lowlands originate
				Serravallian	13.82	
				Langhian	15.97	
				Burdigalian	20.43	
				Aquitanian	23.03	Rise of Himalaya, India–Burma frontier range and Sunda ring of fire; Turgai epicontinental sea separates Europe and Asia
		Paleogene	Oligocene	Chattian	28.4 ± 0.1	
				Rupelian	33.9 ± 0.1	
			Eocene	Priabonian	37.2 ± 0.1	
				Bartonian	40.4 ± 0.2	
				Lutetian	48.0 ± 0.2	
				Ypresian	55.8 ± 0.2	
			Paleocene	Thanetian	58.7 ± 0.2	
				Selandian	61.1	
				Danian	65.5 ± 0.3	
	Mesozoic	Cretaceous	Upper	Maastrichtian	70.6 ± 0.6	North America distancing from Europe; Australia rifts from Gondwana
				Campanian	83.5 ± 0.7	
				Santonian	85.6 ± 0.7	
				Coniacian	88.6	
				Turonian	93.6 ± 0.7	
				Cenomanian	99.6 ± 0.9	
			Lower	Albian	112.0 ± 1.0	
				Aptian	125.0 ± 1.0	
				Barremian	130.0 ± 1.5	
				Hauterivian	133.0	Origin of flowering plants?
				Valanginian	140.2 ± 3.0	
				Berriasian	145.5 ± 4.0	
					150	India rifts from Gondwana

Senomanian — Tropical climates to high latitude in the northern hemisphere

Table 1.1 Geological ages in the era of flowering plants (age on a logarithmic scale). (From Gradstein *et al.* 2004)

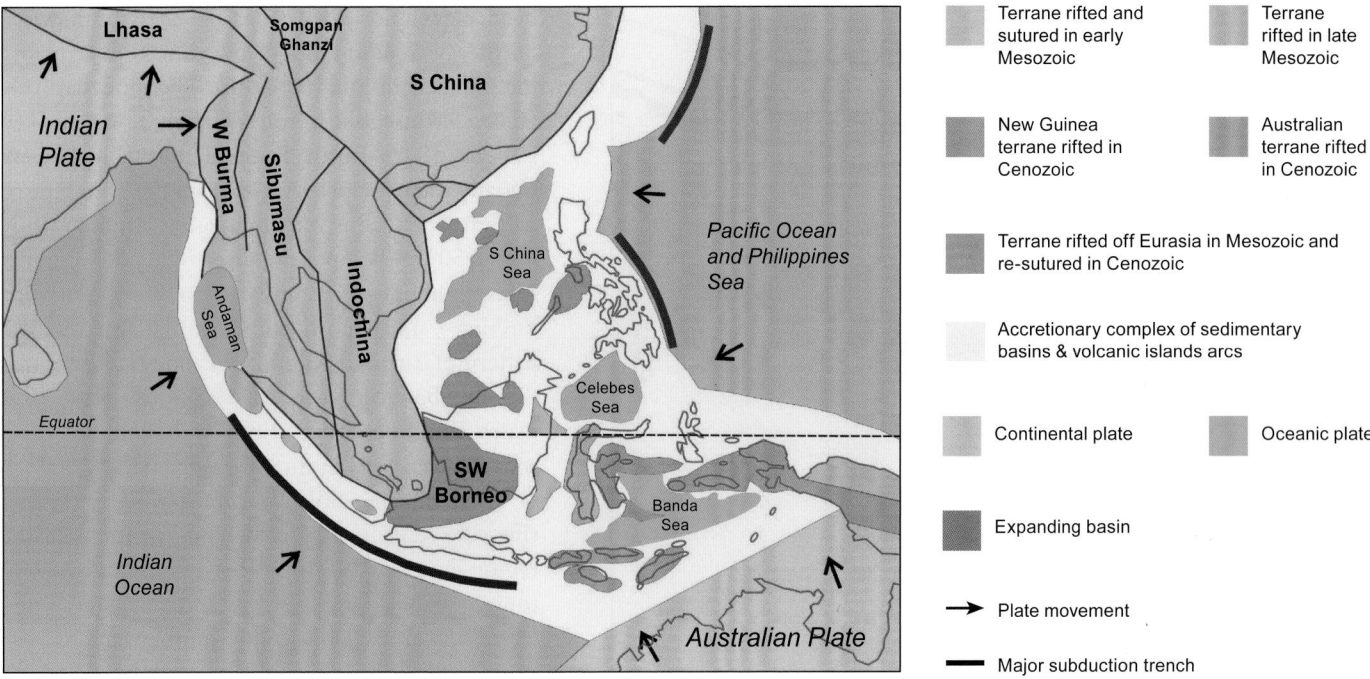

Terranes rifted off Gonwanaland, drifted northwards and sutured onto Eurasia.

Fig. 1.1 Modern tectonic features of tropical Asia; that is, the nature, location and origin of its geological building blocks: continental fragments and sediments of varying age. (I.H. Baillie del.)

Late Cretaceous to reach their present latitudes. They merged to form the currently partially submerged Sunda continental tectonic mass, which includes Peninsular Malaysia, the islands of Borneo, Java, and Sumatra, and the seas and islands between.

The first major rift of the southern (Gondwanan) lobe divided India, Australia and Madagascar from Africa and South America, though Australia did not separate from Gondwana itself at this stage. The South Asian subcontinent also originated on the northern coast of Gondwana, where it had been sandwiched between what is now the northwestern Australian coast and southeastern Africa[7]. It broke off from Australia some time during the Jurassic period (160–200 Ma) – that is, before the origins of flowering plants – then moved northward across the Tethys Sea along a course parallel to the coast of Africa but at an uncertain distance from it. By 100 Ma, it was remote from Australia. The Seychelles, a continental fragment, split off c.60 Ma and became increasingly isolated.

Then, by the end of the Cretaceous, drifting continents began to collide again. Oceanic tectonic plates consist of basaltic rocks, heavier than the granitic cores of continental plates. When an oceanic plate collides with a continental plate, it is therefore subducted beneath the continental coast, thereby raising mountains. As the huge weight of continental rocks thrusts the ocean plate edge down beneath the Earth's mantle, its basalt melts back into magma. The resulting

immense pressure is periodically released in volcanoes. The surfaces of the mountains thus formed are heavily influenced by volcanic lava and ash. Sometimes even the mantle, which is the layer of hot and semi-plastic but recognisably crystalline rock between the crust and the liquid magmatic core, is extruded. The resulting ophiolitic, ultramafic rocks often yield soils containing potentially toxic concentrations of nickel, cadmium, cobalt, chromium and copper, to which plants have adapted as a characteristic flora rich in local endemics (see below, also Chapters 3, 6).

Paleogeographers call the Indian subcontinent 'Noah's Ark' because it carried so many taxa from the southern to the northern hemisphere (Figs 1.2, 1.3). The subcontinent made first contact with the southern margin of Laurasia early in the Eocene (c.50 Ma, during our present Cenozoic Era)[8]. The collision occurred near the Equator, but since then the Indian subcontinent has continued pushing north-east owing to the spread of the Indian sea-floor plate to its south. The contact zone is now beyond the tropics, at 28–35° N, which indicates a steady northward velocity of 7.5 cm yr^{-1}.

Following the final stages of the Indian collision (c.45 Ma), the trajectory of the Indian plate shifted north-east starting in the early Oligocene epoch (c.35 Ma) and continues thus at present[9]. The Indian Gondwana Plate squeezed the sediments which had been accumulating along the shelves

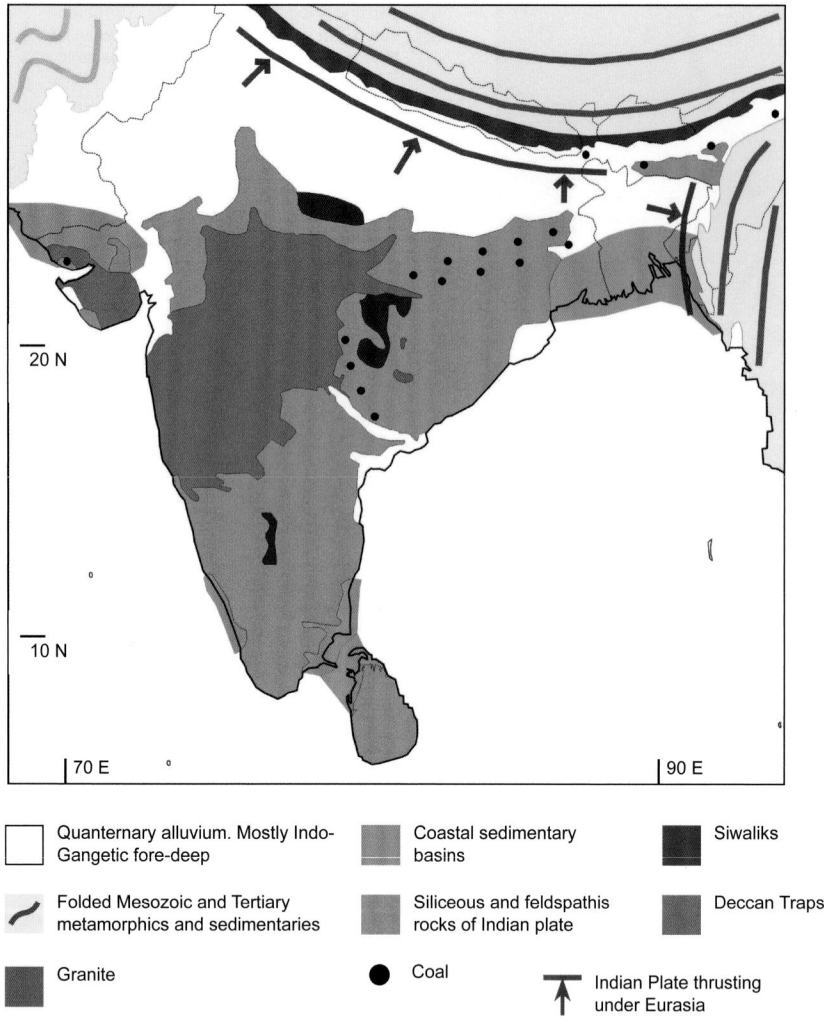

Quarternary alluvium. Mostly Indo-Gangetic fore-deep

Folded Mesozoic and Tertiary metamorphics and sedimentaries

Granite

Coastal sedimentary basins

Siliceous and feldspathis rocks of Indian plate

Coal

Siwaliks

Deccan Traps

Indian Plate thrusting under Eurasia

Fig. 1.2 The surface geology and tectonic features of South Asia, which, with climate, mediate the nature of its soils. (I.H. Baillie del.)

of the northern and southern continental blocks into huge parallel folds – mountain ranges, now the Himalaya – curving southeastwards into the Burmese Arakan Yoma (Figs 1.2, 1.3c), Andaman Islands, and the Thai-Burmese frontier ranges. The heavy basaltic ocean plate to the east of the Indian subcontinent in the Bay of Bengal subducted obliquely north-northeastwards, together with the Australasian oceanic plate to its immediate south, causing tsunamis and raising the still-active 'ring of fire' volcanoes of the Barisan Range and Java. The somewhat more northerly trajectory of the Australasian plate has caused the anti-clockwise rotation of Java and the Lesser Sundas relative to Sumatra, around the Krakatau axis. Sumatra has rotated clockwise less than 30° to approach its present position.

The Indian collision spun the bulk of northern Indo-Burma clockwise c.20°. The Sibumasu block had originally

been aligned north-east of an elongate south-west–north-east trending Indo-China block. Even before the Indian collision, Indo-Burma became massively faulted, creating shear-zones notably along the Red River and Thai-Burma Faults. The two blocks were initially rotated clockwise c.30° in the Late Jurassic to Early Cretaceous, but Sibumasu was subsequently forced 50° in the opposite direction and Peninsular Malaysia about 44°, as Indian Gondwana continued pushing north-northeastwards. The regional pattern of earthquakes and volcanoes indicates that pressure between some of the South-East Asian plates continued (Fig. 1.3b), although the main movement ceased in the Miocene epoch.[10]

Deep north–south and north-west–south-east rifts also developed across Sundaland, and the region developed a pronounced 'horst and graben' topography of parallel ridges and valleys resulting from squeezing along fault lines. Many of the valleys developed deep freshwater lakes, which then filled in with organic-rich muds. Following subsequent burial, these eventually produced most of the region's hydrocarbons. The easternmost of these rifts resulted in the separation of south-west Sulawesi from the main part of Borneo, and the formation of the Makassar Straits, creating part of the biogeographical barrier we now term Wallace's Line. Later, in the early Oligocene (c.32 Ma), a deep north-south rift opened up in the Indo-China block, creating the South China Sea and subduction zones to the east, west and south. A small oceanic plate, the Lucania Platform, now underlies the Neogene sediments south of north-west Borneo.

The Philippines comprise an arc of volcanic islands which were pushed northward by the spread of the Pacific and Philippine Sea oceanic plates arriving from the south-east during the Cenozoic.[11] They have become an orogenic zone, squeezed by subduction on both east and western flanks. Until the late Miocene, the South China Sea plate was subducted from the west at the Manila Trench. It brought with it shards of the South China block, attached at Mindoro, the Calamian Islands and northern Palawan. To this day, the West Pacific oceanic plate continues to subduct down the length of the islands along the Philippine Trench, extending southeastwards to Halmahera. By the middle Miocene epoch (15 Ma), the Philippine islands Mindoro and Palawan were approaching Sundaland, the opening of the South China Sea was arrested and these areas were joined with Borneo at about 10 Ma. Only the Zamboanga region of south-west Mindanao has an Australasian continental origin. It remained isolated all

a

Break-up of Pangaea and Gondwanaland 200 Ma

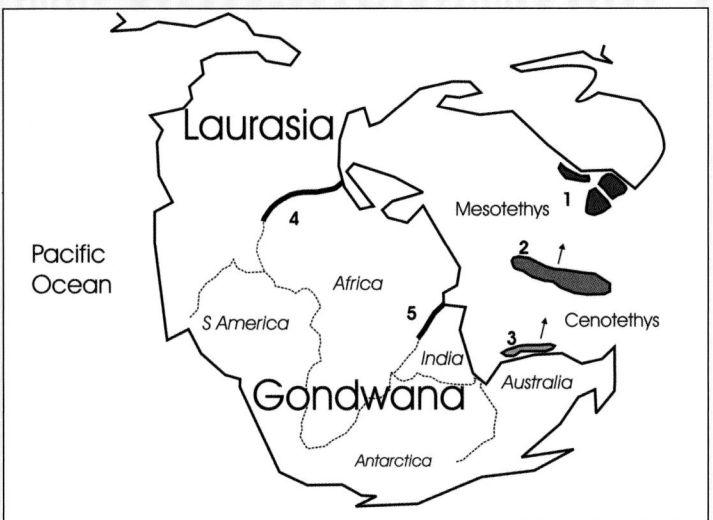

Tethys Ocean splitting supercontinent Pangaea into Laurasia in N and Gondwanaland in S

1 - 3	Terranes rifting off NW Australian section of E Gondwanaland, drifting N across Tethys and suturing onto S Laurasia from Palaeozoic (1) onwards. Large sliver in transition (2) divides Tethys.
4	Laurasia–Gondwana split continues
5	Gondwanaland beginning to fragment. India splitting from Africa

b

Break-up of Gondwanaland 100 Ma

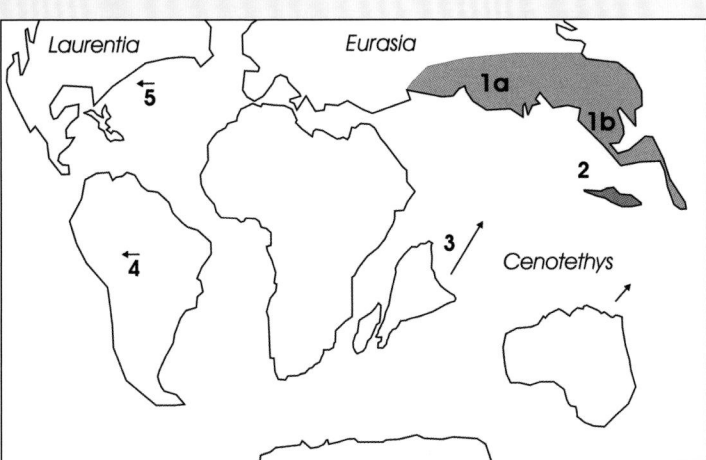

Gondwanaland fragmentation complete. Resultant macro-plates move apart

1	Eurasia extended to SE by Paleozoic suturing of terranes from NW Australian Gondwanaland to form Tibet and High Asia (1a), and in Paleo-Mesozoic to form China and Sundaland Shelf of SE Asia (1b).
2	Continuing northwards drift of younger terranes from Gondwanaland.
3	India and Madagascar rifted off SE Africa, and separating from each other.
4	Western Gondwanaland (S America) split from Africa and moving W by widening proto-S Atlantic Ocean.
5	Similarly Laurasia split by proto-N Atlantic to give Laurentia (N America) and Eurasia.

c

Migration of Greater Indian continental plate

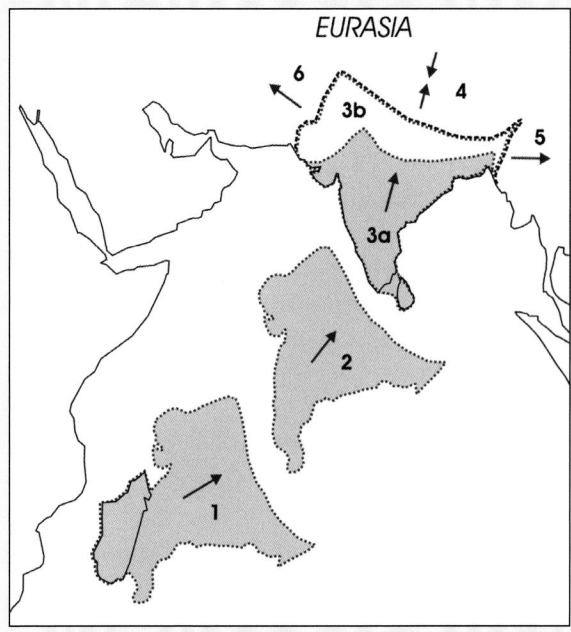

1	Jurassic (150 Ma). Greater India + Madagascar rift from Africa during break-up of Gondwanaland and drift NE.
2	Cretaceous (90 Ma). India and Madagascar separate and India drifts rapidly NNE, veering slightly anticlockwise towards N.
3	Indian plate collides with, and subducts under, Eurasian continental plate in Early Tertiary, about 55 Ma. Indian plate rotates further anticlockwise and continues to move to N, but at slower rate.
	3a is the modern outcrop of Indian plate.
	3b is where plate has been down-warped and buried by subduction and subsequent foredeep trough infill with Cenozoic Indo-Gangetic sediments.
4	Eurasian continental rocks, Paleozoic ex-Gondwanan High Asian terranes and overlying Tethyan marine sediments intensely metamorphosed and folded by collision and north-wards pressure. Eurasian crust is uplifted to give Himalayas and Tibetan Plateau and is also foreshortened, N-S.
5	Eastwards lateral pressure causes uplift, folding, faulting, metamorphism and clockwise rotation of ex-Gondwanan Sundaland terranes.
6	Similar westward pressure on Baluchistan ranges

Fig. 1.3 a. The tectonic plates that now form continental Asia as they were 200 Ma at the Triassic–Jurassic boundary, that is at least 50 my before the era of flowering plants, when Pangaea and Gondwana started to break up. **b.** The break up of Gondwana 100 Ma. **c.** The migration of the Indian continental plate. (I.H. Baillie del.)

through the Early Tertiary, not joining the archipelago until the middle or late Miocene. Taiwan, on the northern side of the opening South China Sea, remained firmly attached to the Asian continental shelf and thus at a remove from the Philippine archipelago. The continuing northward shift of the Australian Plate exerted a powerful effect on the Philippines in the Pliocene, shouldering Luzon out to the north and west, rotating the Philippine archipelago clockwise. The Philippines and Taiwan are thus closer now than at any time in the Cenozoic.

Two intermittent belts of Miocene extrusive volcanic rocks in north-western Borneo mark the impact of the last major southward thrust of the South China Sea plate. Subduction of the South China Sea beneath the south-western Philippines and Borneo-Palawan has also ceased,[12] but subduction of the Pacific Plate beneath the eastern Philippine coast continues. Similarly, both sides of the southern Philippines and Wallacea are being pushed downward, as Australia continues to raft northward. The South China Sea rift became quiescent in middle Miocene times.

By the last, Maastrichtian, age of the Cretaceous (c.75 Ma), the Australian continental plate separated from the Gondwanan mother-continent. Australia has since been drifting more or less continually northward throughout the Cenozoic era. It pulled a jumble of continental fragments along with it; these, along with a series of volcanic arcs, now comprise the eastern Indonesian islands grouped by biogeographers as Wallacea (Fig. 1.4). The Australasian-derived Gondwana fragments meet the Sibumasu South-East Asian coast at what is now Central Sulawesi. Thus, east and south-east Sulawesi originated from the Australian Gondwanan continental plate, but south-west Sulawesi shares the geology of south-west Borneo, but became separated following the formation of the Makassar Straits in the middle Eocene (Fig. 1.5).[13] About 15 Ma, the north-east volcanic arc and southern Sulawesi ophiolites amalgamated with the Bornean core of south-west Sulawesi, but it was not until the Pliocene (5 Ma) that eastern Sulawesi coalesced, with the injection of the Gondwanan Sula Spur from the east, and Sulawesi began to acquire its current form. As we will see in section 1.2.8, it is unlikely that this ophiolite provided a land connection across the Makassar Straits.

Thus, the Cretaceous period witnessed the comprehensive fragmentation of the southern Gondwana super-continent (Fig. 1.3). The Gondwanan core became isolated as the Antarctic. Africa and South America represent unitary fragments of Gondwana, which at first drifted north across the Equator together as western Gondwana. Flowering plants probably started to diversify in western Gondwana,[14] and – in contrast with the Asian tropics – must have been well advanced on these two continents before they eventually split apart and became distantly separated. They were certainly much better developed there than is suggested from the pollen record from time-equivalent sediments in South-East Asia, such as the largely Early Cretaceous Khorat succession of eastern Thailand.[15]

Therefore, although tropical Asia now lies almost entirely north of the Equator, surprisingly none of its land mass is geologically part of the original Late Jurassic northern continent of Laurasia, whose southern regions had started well within tropical climates and latitudes.

In contrast, then, to the African and the American tropics, the Asian and Australasian tropics comprise an amalgam of many separate tectonic components. They collided at different times in the Mesozoic and Cenozoic eras, both before and during the age of mammals and flowering plants. During that time, they had the potential to serve as 'Noah's Arks' by transporting the species of one continent across the seas to another. Throughout much of the Cenozoic era – which continues today – these blocks have continued to crash together, forming new amalgams and raising great mountain ranges along their colliding margins, while buckling parts of their hinterlands beneath shallow seas.

1.2. *Surface geology and geomorphology: stable tablelands set among a buckled and crushed jumble*

Though the contemporary distribution of organisms, including plants, can be substantially explained by their places of origin and past opportunities for migration, the distribution of plants and forest types is also determined by the distribution of habitats – that is, soils and climate – to which they are adapted. Geological history has determined the way in which landscapes have been created. The origin, history and composition of rocks determine their rates and processes of weathering and porosity, which generate the diversity and patterns of the soils (see section 4 below). These combine with the processes of uplift, buckling and erosion to create topography. The dispositions of continental masses, mountains and plains also have a major influence on climate.

The Gondwanan blocks which comprise the Asian tropics each have a stable core of ancient rocks and surfaces, with buckled terrain along those edges where collision continues. These stable cores, although originally continental granite, have been variously melted, eroded into sediments and crushed. In this way their rocks, albeit predominantly siliceous like granite, have been transformed into metamorphic forms or eroded and redeposited into sediments, which may themselves be metamorphosed. Many of these changes took place in Palaeozoic and early Mesozoic times well before the origins

Fig. 1.4 The surface lithology of East Asia and Malesian Australasia. Note that, whereas the sandstone and associated soils are fragmented both by geology and by the discontinuous nature of ridge-tops, the rocks yielding clay soils are continuous over great distances, the more so during periods of low sea levels (see Chapter 6). (I.H. Baillie del.)

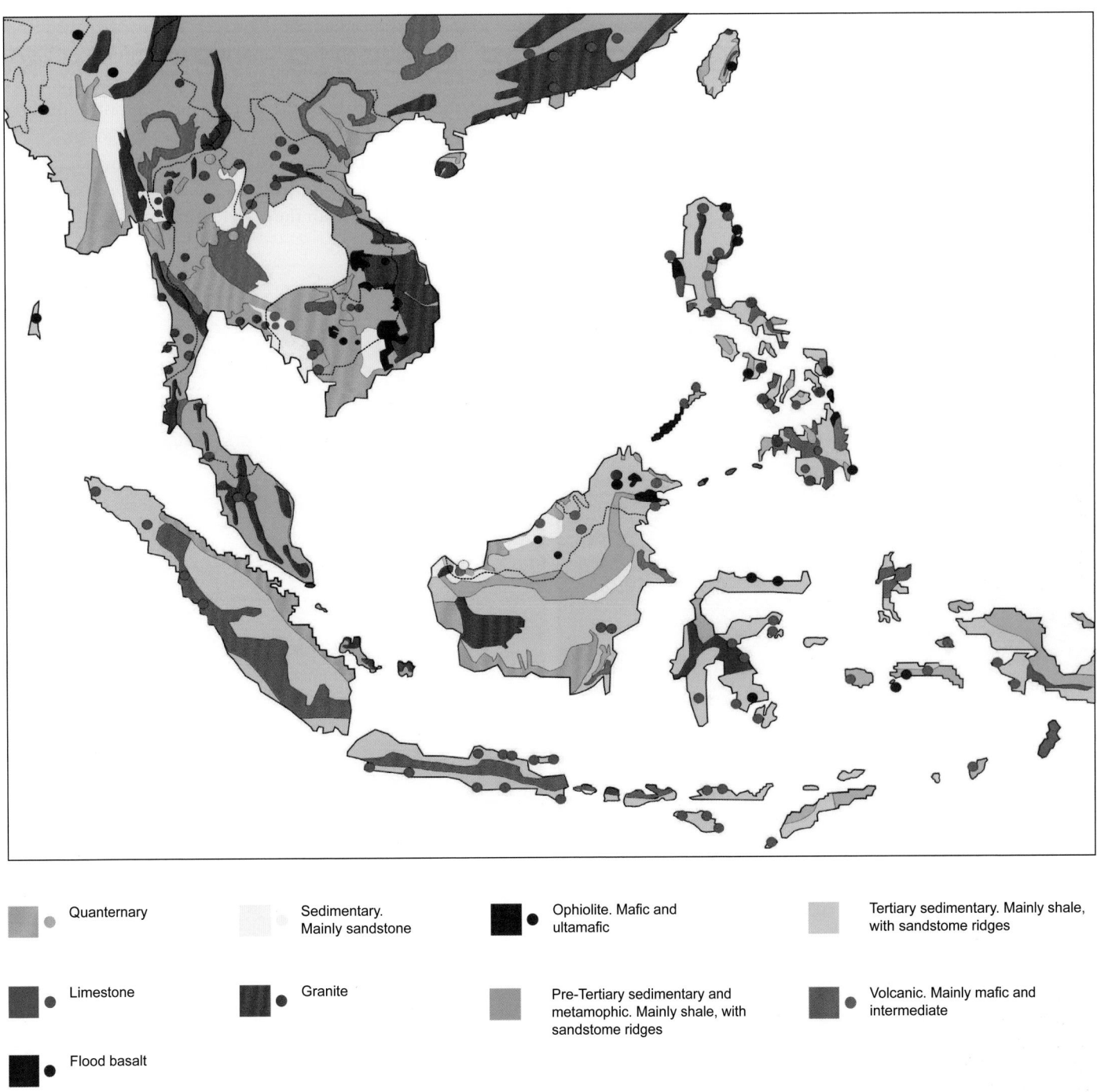

Quanternary

Sedimentary. Mainly sandstone

Ophiolite. Mafic and ultamafic

Tertiary sedimentary. Mainly shale, with sandstome ridges

Limestone

Granite

Pre-Tertiary sedimentary and metamophic. Mainly shale, with sandstome ridges

Volcanic. Mainly mafic and intermediate

Flood basalt

of our contemporary flora, but their nature greatly influences present plant distributions through their influence on soil formation.

When an oceanic plate is subducted beneath and lifts the continental margin, the resulting mountains are generally volcanic, or their surfaces heavily influenced by volcanic lava and ash. But the Himalaya was raised by an exceptional circumstance.

1.2.1. *South Asia* (Plate 1.1)

The Indian subcontinent – namely, all of South Asia south of the Himalayan summits – is by far the largest single Gondwanan continental block in Asia, and its presence has had by far the greatest climatic and plant-geographical influence on contemporary conditions. It comprises the largest expanse in tropical Asia of Precambrian and Cambrian siliceous tectonic continental rocks and associated sediments, much of which has been metamorphosed.[16] India's transit north from Gondwana was attended by a massive eruption of lava, evident now in one of the greatest extrusions of basaltic lava the world has known. This eruption occurred at the boundary between the Mesozoic and Cenozoic eras (65.5 Ma), and coincided with the extinction of the dinosaurs. Vast amounts spread and now form the Deccan Traps: horizontally bedded tablelands of the northern Western Ghats extending east into to Madhya Pradesh, and lowlands west to Gujarat (Fig. 1.2). The soils derived from them, rich in clay minerals and nutrient cations, including calcium and magnesium as well as iron and aluminium, are among the most fertile in the tropics and contrast with those over the ancient siliceous rocks to their east and south.

Its collision – obduction with another continental plate in this case – resulted in the partial submergence of the Indian continental block beneath the larger Laurasian (Figs 1.2, 1.3c).

Plate 1.1 South Asian landscapes. **a,** ancient Gondwanan continental crust: a semi-arid part of the Deccan Plateau in the lea of the Nilgiris: the Moyar Gorge, Tamil Nadu; **b,** an inselberg of metamorphosed siliceous *khondalite* rocks in the *intermediate* zone where there are two dry seasons, north-east Sri Lanka; **c,** looking down the Himalayan front range to a distant braided river issuing into the Gangetic (Brahmaputra) plain: above Phuntsholing, Bhutan; **d,** landscape at the base of the Himalaya: Buxa National Park, West Bengal, with the distant mountains behind in Bhutan. Braided river bed with grassland on siliceous shingle floodplain; low siliceous terrace of the *terai* in middle distance. (All H.H.)

This eventually led to the rise of the highest of our mountain chains, the Himalaya, probably starting toward the end of the Oligocene epoch, though they may not have attained their present heights until as late as 12 Ma. The subarctic Tibetan Plateau bulged and folded behind the Himalaya to form the so-called 'third pole', causing a massive deformation of the palaeotropical climate system known as the Indian or south-west monsoon. This has had major consequences for the migration and distribution of trees and forests in tropical Asia (see below and Chapter 6), as has the vast accumulation of sediments filling the bed of the former Tethys Sea at the southern foot of the rising mountains.

The Himalaya seems not to have reached exceptional heights before the late Oligocene. Its rise may have partially led to the compensatory upwarping of the Indian peninsular mountains, the major uplifts of which occurred during the Eocene and Miocene epochs.[17] These ranges are highest towards the south in Sri Lanka, which is part of the Indian continental shield, and in the west, such that much of the peninsula is an eastward-shelving and eroding tableland at c.900 m elevation. The narrow western coastal plain abuts an often precipitous scarp, the Western Ghats, 1,600 km long and averaging over 1,000 m elevation, with some peaks exceeding 2,500 m.

The obduction of Indian Gondwana ('Greater India') started before first surface-level contact between the southern and northern continents was made. The evidence can be found in the wide, albeit now fragmented, belt of early Eocene to Paleocene lignite fields from Gujarat east to Odisha. These were laid down in shallow saline but vegetated lagoons around the northern Gondwanan continental shelf, which may have been dotted by a string of islands upthrust by the obduction. These swampy lagoons yield the pollen, resin and other botanical evidence of the flora that had been carried north and originally derived from Africa (Chapter 6). Their alluvial sediments also indicate that the Indus-Ganges plains represent the vestige of the Tethys Sea, which has persisted as an effective ecological barrier to migration of the wet mountain floras of South Asia and the Himalaya.

The Indian and Pakistani lowlands to the north-west of the Desert of Thar (or Great Indian Desert) now experience a Mediterranean climate and do not concern us here. The Gondwana-Laurasia suture in the eastern Himalaya follows the obduction and subduction line (see section 1.1 above), running southward through Nagaland and turning sharply south-west beneath the Indo-Burmese Patkai frontier range. It then continues south, following the frontier ranges, before descending beneath the Andaman Sea. It finally continues along the deep to the west of the Sumatran outer islands and to the south of the Lesser Sundas before terminating south of Timor at the Australian continental shelf (Fig. 1.5c).

1.2.2. *Indo-Burma*

Pressure between the East Asian and Indian plates involves continental rocks, but it has also led to past subductions and obductions between the East Asian Gondwanan block fragments. Evidence for this is provided by the lines of isolated volcanoes, now extinct, behind the contact-zone mountains on the over-thrust side. Where inter-continental block movements are still active, they can produce earthquakes and periodic volcanic activity along slip-lines. Such lines of volcanoes occur east of the contact line between the Burmese and Indo-Burmalayan plates, rising between the Chindwin and Irrawaddy rivers where they culminate in the sacred mountain Popa (1,518 m). It is here that the narrow Sittang Valley, west of Pegu Yoma, represents the back arc basin compensating for the uplift caused by subduction.

Other extinct volcanoes occur west of the contact line between Indo-Burma and the Malay Peninsula and Indonesia; also along the Song-Ma fault line following the Red River valley. The easternmost, mostly granitic, Annamite chain has only one major volcanic outcrop, the Kontum plateau.

Today, the Burma continental block receives the full impact of India's continued northeastward thrust.[18] The obduction zone, where the continental plate of thrusting India continues to collide with that of Indo-Burma, is to the west of the Indo-Burmese frontier ranges and just east of the Ganges delta, thereby lifting the suture zone and exposing its ophiolite in places along the crest of the range (Fig. 1.2). Late Cretaceous flysch – that is, deep water sediments – including some limestone, is exposed at higher elevations, thrust over Eocene flysch below; while in the western foothills, a broad band of Miocene sediments and some limestone karst indicates a shallow saline depositionary environment. The main valley of the Irrawaddy and Chindwin rivers is a Cenozoic basin representing, to the west of the rivers, the northern part of the inter-arc trough which descends southward into the Andaman Sea and between Sumatra and its western outer islands. The basin is filled with Miocene to Quaternary sediments, which also compose the Pegu Yoma range along its eastern margin. Whereas sandstone predominates along much of the western slopes of these Indo-Burmese ranges, clays dominate the eastern slopes and the Pegu Yoma.

East of the Pegu Yoma and Sittang valleys, running from Yunnan south into peninsular Thailand, lies the principal north-south mountain system of Indo-Burma, the Sino-Burman Range dividing Burma from Thailand. Parallel to it in the south, but about 50 km to the east, begins the Peninsular Malaysian main range (Fig. 6.10). These ranges were uplifted in the Permian by the main Indo-Burmese orogeny (c. 220–200 Ma), arising from the collision of the East Asian blocks with Laurasia. They may have remained above sea level and within the humid tropics at least since the Early Cretaceous genesis of flowering plants, at which time the earliest sediments overlying the southern Peninsular hills were laid down. The main subsequent uplift started in the Oligocene and was followed by folding in mid-to-late Miocene times, thereby accounting for the extensive late Cenozoic sedimentation in the main valleys.

These ranges consist of Paleozoic-Mesozoic basement rocks, including the main granite exposures in Indo-Burma, flanked by extensive Paleozoic-Triassic sediments in which argillic rocks and extensive Permian-Triassic limestone predominate (Plate 1.3). Although continental deposition occurred towards the north, the extensive limestone implies shallow and sediment-free waters. This in turn implies either a relatively dry climate or the absence of a nearby source of sediments.[19] Extensive Late Jurassic to Early Cretaceous clastic turbiditic sediments are also found west of the range; extending south-east to the Khorat Plateau, these are the consequence of massive erosion, implying the severe scarification that results from heavy rainfall on seasonally dry surfaces.

Mudstones and shales, and associated karst, prevail in the mountains of north-east Burma, northern Thailand, Laos, northern Vietnam, and northwards into tropical South China flanking the Red River shear zone where karst mountains (below 2,000 m) extend far into southeastern China. This area comprises the most extensive upland region in continental tropical Asia, the ridges being mostly rounded at 700–1,300 m elevation. To the south, the predominantly Triassic continental sediments start east of the central Chao Phya valley and run south-east through southern Laos to the south-west Cambodian coast. These, and the Jurassic to Cretaceous Khorat basin sediments which overlie them, are predominantly sandstone with mudstones prevailing towards the north.

A sandstone range, exceeding 1,000 m in the west at Khao Khiaw (1,282 m) and Phnom Aural (1,813 m) in the Cardamom Mountains along the western Cambodian coast, follows the southwestern scarp of the Khorat Plateau (c.500 m), forming the south-west flank of the Mekong valley. Unlike the Annamites, this latter chain has but one granite exposure, at Khao Sai Dao (632 m) in southeastern-most Thailand. The Annamite range south to Dalat is Mesozoic granite, uplifted by westward subduction of the South China Sea plate. The range rises mostly to 700–1,200 m with peaks to 2,700 m, and extends into Laos at Luang Prabang and into northeastern Cambodia from the Kontum Basin. There is also basalt to the south of it.

This southern part of Sibumasu could therefore have served as an important cradle of regional evolution. Its ranges have served as corridors for migration of both lowland and montane humid tropical floras, whereas its drier intervening plains, continuously isolated from one another physically and also intermittently by climate, have served as barriers. The hill

ranges have similarly served as barriers to dry-climate floristic elements (Chapter 6).

1.2.3. *Sumatra and adjacent islands*
The trough, caused by downwarping inland as the coast of Sundaland became uplifted by the subduction of the Indian Ocean plate, lies mostly off the west coast, though accumulated Cenozoic sediments within it are exposed along the north-west coast. The main mountain range in this area, the Sumatran Barisan, is volcanic and still very active, representing the inner volcanic arc of the subduction zone. It is predominantly andesitic, therefore neutral though often rich in calcium. The Barisan is flanked by Permian mudstones, phyllite and shales, all of which become extensive in Jambi Province; massive karst limestone is locally present. There is also a granite belt between the volcanic rocks of the Barisan range and the overlying sediments towards the south, while the Riau Islands, Banka and Belitung are composed of granite and Triassic sediments similar to Peninsular Malaysia. The lowlands and lower massifs to the east comprise sediments accumulated in three back-arc basins. These basins began as depressions on the continental crust, but from the Eocene to Pliocene they were marine. Shale and clay are dominant.

1.2.4. *Peninsular Malaysia*
This peninsula is an ancient land mass of exceptional stability, which in the north may have started to rise above the sea surface during the Jurassic, and which has remained within 9° latitude of its present position ever since.[20] Although composed of granite from the main range westwards and of Triassic phyllites and shale to the east, its ancient rounded physiography gives the peninsula great uniformity. The east coast is characterised by very young, Holocene sandy terraces deposited at the time of highest Holocene sea levels.

1.2.5. *Borneo*
This large and lithologically diverse island has been the theatre of most forest ecological studies in the Asian tropics, therefore meriting detailed attention.

The mountains of Borneo represent the southeastern prow of the Asian tectonic continent. They are consequent to the southward thrust of the South China Sea oceanic plate and the westward thrusts of the Philippines island arc and Australian Plate. Basement Sibumasu granite is exposed mostly in the central, Muller and Schwaner ranges (reaching 2,278 m at Gunung Raya), and mountains west to the coast at Gunung Palung (1,100 m), forming the backbone of a triangle of continental Sibumasu rock terminating at the isolated eastern Kong Kemul (2,053 m). Further small granodiorite mountains occur in Sarawak west of the Lupar fault. These rocks are largely overlain at lower altitudes by Permian, Triassic and Cretaceous sediments.

Plate 1.2 a, The uniformly c.600 m high ridges of the Rejang series, viewed from the basalt extrusion of the Carapa Pila, Ulu Mujong, Balleh, Sarawak. **b,** Upper Miocene sandstone lying directly over more erodible shale, which results in the topographic differentiation of soils; Jerudong, Brunei. (All: P.A.)

At the heart of equatorial Sundaland, Borneo therefore consists of a continental core to the west with a series of sedimentary basins along its northwestern and eastern coasts, as well as continental basins in the centre and south[21] (Fig. 1.4). North-west Borneo consists of a massive sedimentary basin with scattered volcanic acid and basic outcrops, initiated in the early Cenozoic as the South China Sea opened. The core is sharply defined along its northeasterly side by the Lupar Fault, an ophiolitic suture and shear zone of Palaeogene origin, separating the Rejang series turbidite basal geosynclines from the continental core rocks and continuing south-east across the island. These include granites in the south and in some mountains to the north and west, and extensive Cretaceous to Oligocene sandstone in the north and east-centre. Two central Borneo sedimentary basins were formed to either side of this shear.

The north-west Borneo hills consist of mostly sandy Neogene sediments, resulting from erosion following the upwarping of the interior and themselves uplifted since the mid-Miocene. The collision suture of the Luconia Platform, which they overlie, is at the meeting of the Rejang turbidites with the Neogene basin sandstones to the north (Fig. 1.4). These rocks of the Balingian and Baram Delta Provinces, of sediments re-washed from the Borneo interior, are mid-Miocene to Pliocene, and predominantly moderately hard to soft sandstone. Thus, a succession of down-warping and

uplifting with buckling through the Cenozoic has left a sedimentary series. The oldest, in the south, are of argillite and hard sandstones (Plate 1.2a; Fig. 6.10), the youngest of soft sandstones and, in the north-east, shales and clays. Miocene and Pliocene Neogene basins, in which soft sandstone predominates, overlie the Luconian continental platform and comprise the coastal hills from the mouth of the Rejang to south-west Sabah (Plate 1.2.b). Pressure between some plates continues as the regional pattern of volcanoes indicates (Fig. 1.5), though the main movement ceased in the Miocene.

The Baram Line is a fault running north-west to south-east from beneath the South China Sea. It separates two groups of flysch turbidites. To its west, and south of the Neogene basins, are the ancient inland flysch turbidites of the Paleocene and Eocene Rejang series and earlier. These represent the first Cenozoic sediments deposited on the Sibumasu block. To its east are the relatively younger Paleogene to mid-Miocene flysch turbidites to the north-east and across to the east Borneo Neogene Tarakan basin. East of the Baram Line, argillic rocks dominate throughout, but with subordinate (albeit often hard) sandstone beds which crest the ridges resulting from mid-Miocene uplift and folding. Isolated karst massifs are also present.

In the north-east there was earlier southward movement of the South China Sea bed,[22] and later, collision with the Australasian continental outliers of northern Sulawesi and the eastward-trending Philippine island arc. Since the lower Miocene, these have led to complex ophiolitic (ultramafic) mantle exposures. These are associated with the Plio-Pleistocene emergence of the granite summit zone of Kinabalu (4,094 m). They are exposed in the Labuk and Segama valleys and from Kudat in the extreme north-east extending through Palawan.[23] Kinabalu, with its granite summit, is the tallest mountain between the Himalaya and New Guinea, the next tallest being the recent volcanoes Kerinci (Sumatra) at 3,805 m and Rinjani (Lombok) at 3,776 m. A further shear along the north-east Sabah coast has associated onshore and offshore Neogene sedimentary basins. The former consists of both moderately hard sandstone and of shale. Further intrusive ophiolite is exposed in the south-east Borneo Meratus Mountains.

Karst limestone belts occur locally at Bau (western Sarawak), between the Neogene sandstones and turbidites of the Baram Delta Province at Mulu, the Sangkulirang peninsula (central East Kalimantan), and more locally elsewhere. Late Miocene-Pliocene acidic to basic volcanic rocks are exposed locally between the Neogene and turbidite rocks, and within the latter.

In summary, Borneo is distinguished by an extraordinary diversity of rock substrates including, in the north-west, the most extensive exposure of sandstones and other siliceous substrates in Sundaland.

1.2.6. *The Philippines archipelago*

Palawan, composed of limestone in the north and south but otherwise largely ultramafic ophiolite, alone rests on the Sunda Shelf. There are also substantial clastic sedimentaries in northern Palawan, which are very similar to the Rejang series of Borneo. The Philippines are otherwise a volcanic island arc: isolated mountainous islands dominated throughout by basic volcanic sediments. Limestone karst exists locally in Luzon and elsewhere, while ophiolite is found especially in lands immediately to the west of the major subduction zone which borders the islands' eastern edge, and which has created the Sierra Madre ranges which follow it.

1.2.7. *Java, Bali and the eastern archipelagos*

Although Java and Bali are on the Sunda continental shelf, they share a widespread volcanic and mostly andesitic surface lithology with the Lesser Sunda Islands to their east; ash has universally influenced their soils. East of Borneo and Bali, from Sulawesi to New Guinea and from Lombok to Timor, lie archipelagos of islands with different histories, cast up by the relentless northward transit of Australia (Fig. 1.5).[24]

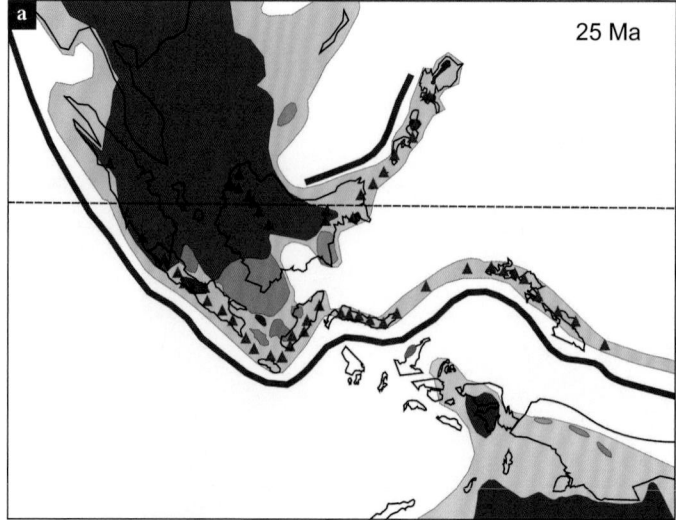

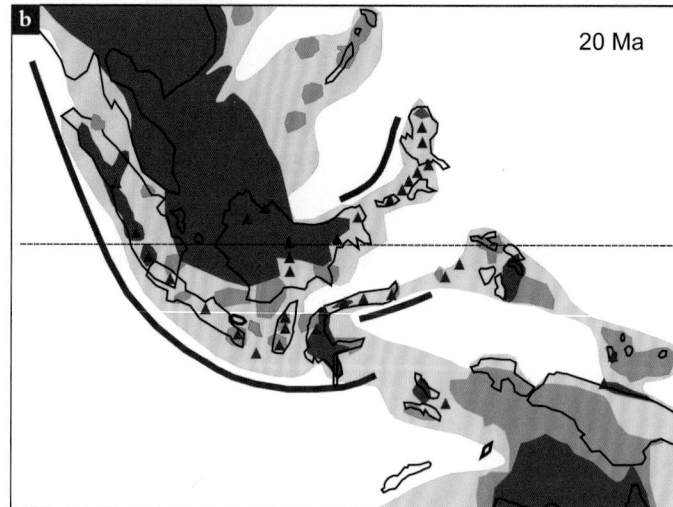

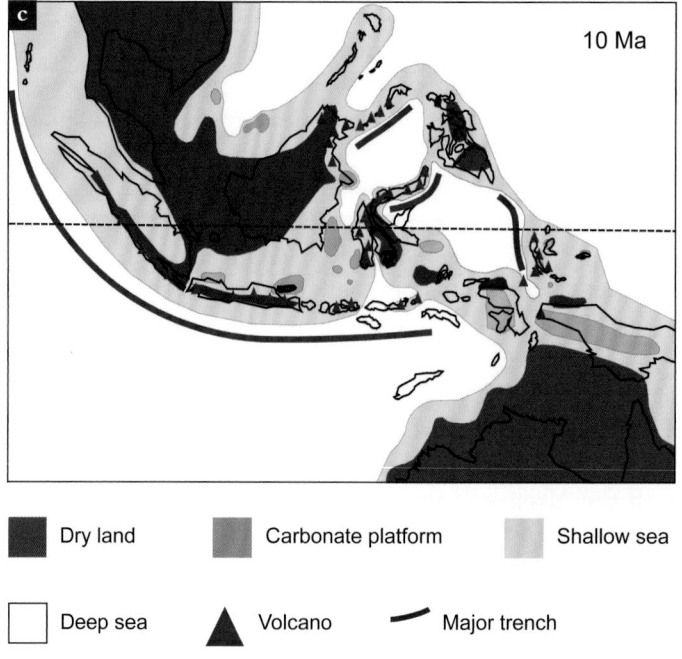

Fig. 1.5 The rafting of Australasia since the origins of perhumid lowland Sundaland; carbonate plate (active coral reefs) signal seasonal rainfall. **a**, Late Oligocene 25 Ma; **b**, Early Miocene, 20 Ma; **c**, Late Miocene, 10 Ma. (I.H. Baillie del., after Hall 1996)

■ Dry land	▨ Carbonate platform	▨ Shallow sea
□ Deep sea	▲ Volcano	— Major trench

1.2.8. *Wallacea and New Guinea*

Continental plates have remarkably sharp edges; their submarine shelves are generally narrow, dropping suddenly to the surrounding oceanic plates deep below. Such is the case with South America and Africa, and also with India including Sri Lanka. Where continental and oceanic plates collide, as they do along the southern coasts from the Andaman Islands to Timor and down the east coast of the Philippine Islands, the sharp drop at the edge of the shelf is accentuated by deep ocean trenches. A deep trough, marking the gap between the shelves of two colliding continents now but initially a rifted continental basin, follows the strait between Borneo and Sulawesi. The rifting of these Makassar Straits began in the middle Eocene.[25] Extending south and through the Sunda Islands between Bali and Lombok, it is known to biogeographers as Wallace's Line after the naturalist Alfred Russel Wallace who first documented the extraordinary change in animal species across it. The floristic and forest change across this line is a subject for Chapter 6.

Sulawesi is partly volcanic or ophiolytic. It includes a Sibumasu fragment, its southwestern arm, which separated from Borneo following middle Eocene rifting and has been separated by the Makassar strait since the Oligocene. It is the most ruggedly montane of the islands between Borneo and New Guinea.

East of Sulawesi, clay-rich sediments and karst limestone are associated with widespread volcanic rocks, including ultramafics, especially towards the east. Continental Gondwanic rocks are exposed in the Sula-Banggi islands, Sumbawa and Sumba in the Lesser Sundas, and in West Timor. Ophiolitic rocks, implying the presence of deep-sea fragments, occur in Buton and Obi, while Buru-Seram-Ambon are considered a further micro-continent detached from Australia. Most of the Wallacea islands, though, are basic volcanic. Throughout, the sandstone ridges of Borneo, and the granite ridges of Peninsular Malaysia, with the similar and distinct habitat created by them, are rare or absent in Wallacea.

New Guinea is the only eastern Malesian island large enough to support a rich, indigenous tree flora. At least in its eastern part it seems to have arisen as late as the early Miocene, through collision of the northward-thrusting Australian continent and the Pacific oceanic floor, which lifted Papua.[26] The northern part has been much folded, together with metamorphosed volcanic intrusions. It remains volcanically active, but was earlier a major depositional basin.[27] Western New Guinea arose later, beginning in middle Miocene times; but the main uplift of the New Guinea mountains in the Pliocene and Pleistocene, with massive karst limestone on their southern flanks, betrays a pre-uplift shallow sea, free of sediment. Extensive subsequent deposition created the southern foothills. New Guinea is the only island east of Wallace's Line recorded to have sandy Pleistocene raised beaches, along the foot of the western snowy mountains, which bear a unique *kerangas* forest (Chapters 3, 6).

1.2.9. *Summary*

Whereas tropical mainland Asia and Australasia comprise the ancient rocks and surfaces of stable continental blocks, the archipelago in between bears the imprint of massive collisions and upheavals, still active along the 'ring of fire', the volcanoes and earthquake zone extending from northern Sumatra eastwards to New Guinea and beyond. In the western part of this Malesian archipelago, the edge of the Asian continental shelf has been continuously uplifted and buckled as the Indian fragment from the west, Philippine island arcs from the east, and the Australasian continent from the south have crashed into it. The contrast between the volcanic, mountainous Indonesian islands fringing the Indian Ocean, with infertile, sedimentary Borneo to the north resting close to the edge of a stable continental plate, evokes that between the Andes and the Guyana Highlands of South America (though these have been stable for very much longer).

Whereas the ocean constitutes a barrier, or at least a filter, to plant migration, mighty tectonic events have provided a setting on land of enormous complexity for the evolution and intermittent spread of plants through the Asian tropics, a setting which is in stark contrast to the relative stability of the other two major tropical continents. Persistent tectonic movement has also resulted in a diversity of surface topographies and of soils far greater than in Africa (the complex and diverse Rift Valley excepted), and on a much finer spatial scale than in the South American lowlands.

Except for siliceous eastern peninsular India and southeastern Thailand (including much of Cambodia), the siliceous rocks of tropical Asia form relatively small ecological islands in a sea of clay-dominated landscapes: either in coastal basins or exposed along ridges and plateaux of harder sandstone within landscapes of predominantly fine clay-yielding sediments. Limestone karst, originating in clear seas, may also be exposed as islands associated with fine sedimentary rocks. The same is true of acid extrusions, volcanic especially rhyolite, and intrusive ultramafics. (The influence of this great diversity of habitats on Asia's forests and their composition is discussed in Chapter 3, on patterns of their tree flora and its evolution and speciation in Chapter 6, and on their species diversity in Chapter 7.)

1.3. *Among epicontinental seas*

Shallow subsidence of extensive continental surfaces beneath sea level can persist over long periods of time. Though not comparable in duration to those resulting from continental drift, such shallow seas may provide periodic barriers to migration. All three tropical continents have experienced partial submergence under shallow seas (Figs 1.6, 6.12), or freshwater deposition basins, today represented in Asia by the sedimentary basins described in the previous section.

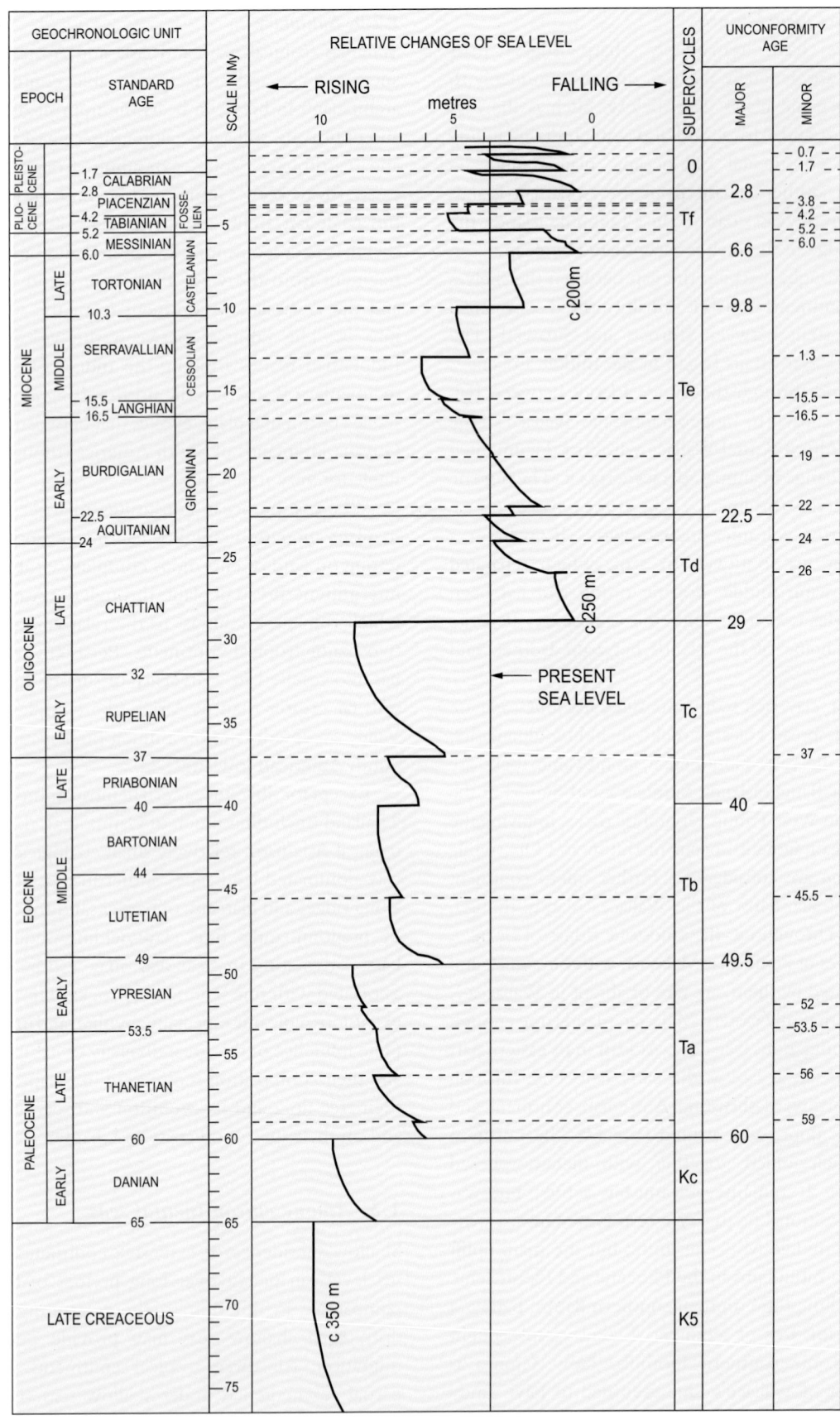

Fig. 1.6 Global cycles of sea level change through the Cenozoic. (From Hutchison 1989)

However, only present and recent epicontinental water bodies can account for discontinuities in the *current* distributions of species, unless the uplifted sediments of earlier basins yield soils inimical to the flora surrounding them.

Since the onset of the northern ice ages c.2 Ma, eustatic sea level changes have led to the periodic isolation and consolidation of the Sunda uplands by shallow seas invading and receding from the lowlands between them. During the short (c.10,000 year) intervals of high sea level such as our own Holocene epoch, sedimentation between these uplands has continued. The uplands themselves consist of previously uplifted rocks, including the sedimentary basins of former seas. The extent to which these short periods of Sunda marine fragmentation has contributed to floristic differentiation and allopatric speciation, as opposed to the influence of ecological (edaphic) fragmentation of terrestrial landscapes, is considered in Chapter 6.

Floodplains, and the continental shelf during periods of high sea level or continental downbuckling, receive sediment from surrounding uplands. The resulting soils will differ according to regional geology and the environment in which the sediments are deposited. Deposition under flowing water along rivers, in estuaries, and near the coast will be coarser, generally rich in silica sand irrespective of geology. Sediments deposited in calm water, in floodplains or offshore will be finer, rich in clay and silt, and generally richer in plant nutrient ions. Repeated uplift and sedimentation leads to re-deposition of increasingly sandy, less nutrient-rich sediments.

1.3.1. *Epicontinental seas and associated sediments of South Asia*

The South Asian Gondwanan block has hardly been inundated, the following few major areas alone meriting mention. The Ganges delta and the Bay of Bengal south as far as Sri Lanka receive sediments from the Himalaya. The variation in their coarseness is interpreted as variation over geological time in rainfall and snow regimes in the Himalaya (see section 2.3.1 below). The earliest sediments date from c.25 Ma, implying a late Oligocene-earliest Miocene origin for the high Himalaya.[28]

The Tethys Sea was closed in north-western South Asia by the collision, but the Himalayan foredeep (which later formed the Indo-Gangetic plain) long remained under shallow brackish or fresh water, receiving sediments from coasts to the north and south. The resultant Indo-Gangetic plain may therefore have continued as a physical barrier, at first marine then freshwater, to the migration of upland species. Latterly, though, it has acted solely as an ecological barrier. These will be assessed in Chapters 6 and 7.

The tropical – that is, frost-free – Himalayan slopes are confined to the Siwalik front range and below 1,500 m in the valleys leading south from the snowy summits (Chapter 4).[29]

These valleys cut through the Siwaliks, and floodwaters emerge annually into the sediment-filled foredeep of the Gangetic-Brahmaputra plain to spread vast tonnages of boulders and gravel along the broad, braided riverbeds (Plate 1.1d). The Siwaliks are comprised of schist: metamorphosed sediments, mostly siliceous, but with subordinate limestone towards the east where the Tethys persisted as a shallow epicontinental sea until the late Miocene.[30] The Siwaliks were strongly crumpled and folded by the inexorable uplift of the Himalaya.[31] They represent the erstwhile foreshores of both Gondwanan (Indian) and Laurasian coastlines, gradually rising from the late Oligocene through to the present as the Himalaya has risen. The last major uplift was c.2 Ma, in the early Pleistocene. Glaciations have left evidence in the form of side moraines ranging from as low as c.1,300 m, beneath which steep-sided, V-section valleys bear testimony to further subsequent uplift

Plate 1.3 Limestone karst jutting out of the alluvial plain of north Kedah, Peninsular Malaysia. (P. Woods)

exceeding 500 m.[32] Both the Siwaliks and the belt of riverine late Tertiary and Quaternary sediments immediately to the south are siliceous, yielding freely draining (when not flooded) but infertile soils. These form a terrace, the *terai*, bounding the northern rim of the predominantly argillic sediments of the Gangetic Plain. This edaphically diverse remnant of the southern Laurasian foreshore has likely served as a major plant migration route between the tropics of South and East Asia (Chapter 6).

Miocene sedimentary rocks mark inundation of coastal lowlands in northern Gujarat south-east of the Indus delta (the Katchchh Basin) and in the southeastern Carnatic hills (the Cauvery Basin). The Katchchh sediments are predominantly coals, shales and clays in part derived from the adjacent Deccan trap, whereas the Cauvery's are sandstone with subordinate shale. North-west Sri Lanka is also overlain with Miocene limestone thinly interbedded with sandstone and shales,[33] and more remain submerged under the Palk Strait. For c.15 My, this strait has therefore been – and currently intermittently remains – a potential barrier to species migration (Chapter 6). There is some limestone in Kerala in the western coastal plain, and some in the Siwaliks, but overall there is remarkably little limestone in South Asia, and upland soils are overwhelmingly siliceous but for the extensive Deccan trap basalts to the north-west.

1.3.2. An ancient Sunda epicontinental sea and floodplain

Sundaland periodically emerged during the Pleistocene epoch as a continent comparable in area, latitude and topography to the present northern part of South America. Eustatic sea levels are thought to have dropped by up to 120 m; perhaps only twice, although it has probably fallen below 100 m more than a dozen times in the last million years, most recently in the last ice age (Fig. 1.6).[34] The major current epicontinental sea in the Asian tropics is the southern section of the South China Sea where it overlies Sunda continental rocks. Almost a century ago, Dutch geologist G.A.M. Molengraaff described the outline of the bed of a great river, running between Java and Sumatra on the south-west and Borneo in the north-east, as far as the continental shelf between southernmost Vietnam and the southern Philippines.[35] He named the northern part the Proto-Mekong (Fig. 6.12). Some of Sundaland's mightiest modern rivers, including the Kapuas, Pahang and Rejang, flowed into it.[36]

One currently small river, Batang Lupar in north-west Borneo, follows the major Lupar fault delineating the northwestern margin of continental rocks. Its course is represented by a submarine gorge, implying that it was much larger in past eras. Its current valley and estuary, far too wide to have been carved out by the stream now flowing within it, experience the strong tidal bore characteristic of sunken estuaries and valleys large for the size of their water flow. This appears to have been an earlier course of the Kapuas, flowing toward the huge Proto-Mekong – or perhaps, prior to the uplifting of north-east Borneo and consequent lesser uplift to its west, of the Proto-Mekong itself? Its plant geographic significance will be discussed in Chapter 6.

The Proto-Mekong drained a vast area of Sundaland which had been above sea level, out onto the continental shelf during northern ice ages. This land is now below sea level partly owing to the gigantic weight of the sediments eroding out onto it from wet equatorial Sundaland since the Oligocene (32 Ma). Most of these were deposited onto the northern continental shelf of Sibumasu and the southern Sunda subplate when this area was covered by shallow seas. In large part, these sediments were disgorged from the interior Bornean mountains as the oceanic sea floor base of the South China Sea pushed south and Australasia pushed north-west, lifting these mountains. Even in times of low sea level, the vast intersecting Proto-Mekong network of rivers and their predominantly clay floodplains and deltas and young sediments of the plains would have acted as a partial barrier to many upland species, particularly those of sandy soils. The clay soil floras though, both lowland and montane, could have migrated without obstruction across the narrow upper valleys and along the slopes of the watershed divides.

Pleistocene periods of wet climate likely continued in the Sunda core through periods of moderately low sea levels coinciding with temperate glaciations.[37] During high sea levels, sedimentation continued in between the main land masses of Sundaland, whose continental heart is Peninsular Malaysia, Sumatra, Java, Bali, Borneo, Palawan and the islands strewn among them. Along the former coastlines of Quaternary Sundaland and southwestern New Guinea are numerous raised Pleistocene sand beaches, marking the highest sea levels. Some have additionally been uplifted tectonically, occasionally by as much as 15 m.[38] The middle Holocene epoch ended c.4,500 years BP with a eustatic sea level drop of c.2 m, exposing shallow sediments and causing the embayment of lagoons by dune systems down-current from river mouths.[39] Eventually, these lagoons filled with fine sediments over which peats developed (Chapter 3). These younger plains have left once-coastal dune systems isolated far inland. Of pure fine silica sand, raised white sand beaches are widespread along the coasts of Borneo and behind its coastal peat swamps. They reoccur along the east coast of Peninsular Malaysia, the western Cambodian coast at the base of the Cardamom Mountains, and the southern coast of Indonesian New Guinea.

The oceanic islands between the Asian Sunda and Australasian Sahul continental shelve (Wallacea), plus the Philippines, are dominated by shale sediments and volcanic ash and lava, although bands of raised coral limestone are widespread.

1.3.3. *Other ancient sea beds*

Another major epicontinental sea is the Arafura, currently separating New Guinea from northern Australia, whose bed lies on the Sahul continental shelf. Islands rising from this shelf include the coral Aru archipelago. The Andaman Islands are presently separated from the coastal Arakan hills of Burma by what is in principle also an epicontinental sea. In the same way Palawan – politically part of the Philippines Republic – is geologically linked to Borneo, being attached to the Sunda continental plate. But Palawan drifted into its present position relatively late (Fig. 1.5). The implications of these associations will be assessed in Chapters 6 and 7.

2. Climate

2.1. *What are the tropics?*

Geographically, the tropics are defined as that part of Earth's surface which experiences the sun vertically overhead at least once a year; that is, the region between 23.5° North and South latitude. Ecologically, these latitudes do coincide roughly with the limits of lowland tropical forests and most of their flora, but only roughly. Nowhere are these limits less precise than in continental Asia, owing to the monsoon climate and winter shelter offered by the Himalaya and southern China mountains from cold dry continental northerly winds. Here, deciduous tropical *sal* forests extend northward to 32° N in the western Central Himalaya. Mangroves (Rhizophoraceae) reach far north along the East Asian coast to Kyushu in southern Japan at nearly the same latitude. Wet seasonal dipterocarp forest reaches 28° N in North-East India and Burma, following the limits of the south-west summer monsoon (section 2.3.1 below, and Chapter 3). Specifically, the limits of these lowland forests coincide with the limits of killing frosts and ice storms in moist climates.[40]

The *climatic* Equator is the centre of the belt of tropical weather. It moves north in north-temperate summers and south in north-temperate winters, as the Earth, tilted on its axis in relation to the sun, receives solar energy differentially over its surface during its annual cycle. The tropical climatic zone consists of two separate wind circulation systems, termed Hadley Cells, north and south of the climatic Equator. Owing to the Earth's rotation, surface winds are drawn constantly from east to west. They originate at the dry margins of the tropics and pick up speed as they move diagonally towards the climatic equator, where they slow and cease. At their maximum speed over the oceans, these winds are known as the north-east and south-east trade winds, and they are the most constant winds on Earth. Where they move over water or forest, the moving air accumulates humidity. As the winds eventually slow, the increasingly warm and humid air at low altitude becomes unstable beneath cooler heavier air above. Thunderclouds form, especially where there are mountains in the path of the wind, or near the Equator where even low peaks draw the moist warm air up their slopes like giant wicks. That wet climatic-equatorial region is known as the inter-tropical convergence zone. The lowland air thus rises, cooling and losing its moisture as rain, and is returned as high-altitude dry wind to the margins of the tropical climate zone. There it descends again, now cooled and desiccated, creating the deserts. Lands to leeward of the trade winds are among the most predictably dry on Earth. In the temperate Hadley Cells immediately adjacent at higher latitudes to north and south, the prevailing winds are respectively south- and northwesterly.

The tropical margin also coincides with the change from tropical to temperate lowland soils; that is, from yellow-red soils in which particulate organic matter, visibly darkening the soil, is concentrated at the surface and hardly diffuses into the mineral horizons though carbon content generally remains high, to soils in which particulate organic matter is mixed to depth within the mineral soils. (The colour of the dark carbon compounds in the mineral horizons of tropical soils is masked by the rust-red free iron sesquioxides, mediated by the intense weathering under high temperatures.) But the poor transport of organic particles down the mineral profile of tropical soils is due to a fundamental difference in soil fauna, which again is climatically mediated (discussed in section 4.2.2.3 below and in Chapter 4).

Climate, then, is the primary mediator of the latitudinal limits of tropical forests.

2.2. *Mapping the Asian climate*

2.2.1. *Divergent mapping systems*
Climatologists have mapped the climate of tropical Asia in several ways, to emphasise influences on vegetation and agricultural production. Methods include those of Köppen,[41] who defined tropical climate classes by the mean rainfall of the driest month combined with the mean temperature of the coldest month; Holdridge,[42] whose criteria for 'life zones' were based on arbitrary divisions along gradients of mean temperature and annual rainfall combined with elevation; Gaussen,[43] who created climate maps – 'diagrams' – in which mean monthly temperature and rainfall are illustrated, and mean annual temperature and rainfall, absolute minimum temperature, and also elevation are indicated; Walter,[44] who incorporated a measure of potential evaporation into monthly estimates of temperature and rainfall indicated pictorially in a *Klimadiagramm*; and Walsh,[45] who for his climate classes combined measures of the minimum length of the dry season, mean annual rainfall and a perhumidity index which summarises the dry and wet season climates as a measure of wetness. Rory

Walsh,[46] who provided the best recent account of leading methods, used published climate diagrams as the basis for his climate maps. None of these attempts, however, has produced maps that closely correlate with the distribution of the major forest formations of tropical Asia. They are also difficult to interpret, being based on a combination of climatic variables. The closest fit was obtained by Schmidt and Ferguson,[47] who adopted a composite index of rainfall seasonality: the mean monthly rainfall of all dry months relative to that for all wet months, expressed as a percentage.

In temperate climates, regional variation is dominated by temperature, while in the tropics it is dominated by rainfall. Forests in full leaf may evaporate and transpire as much as nine-tenths the amount of water per unit area as an open lake.[48] Forests in the tropics in leaf have the capacity to transpire and evaporate 130–170 mm of rainwater per month during windy periods (1,500–2,000 mm per annum), but most release less owing to the impact on transpiration of droughts and deciduousness, and the saturated atmosphere of the wet monsoon in the seasonal tropics.[49] Under tropical temperatures, periods during which less than 100 mm of rain falls each day soon lead to water stress, constituting droughts. The number of dry days required to produce a drought varies with windiness, soil type, topography and the prevailing vegetation. I here adopt one month as an arbitrary and approximate defining threshold for drought, as it can be detected with commonly available rainfall data.

But in the tropics, much of the rain falls either in thunderstorms or in continuous periods of a day or more. Thus, much of it inevitably falls after soils are saturated, and is lost from forests as run-off into river systems. Mean monthly and annual rainfall therefore hardly correlate with the regional distribution of forest formations. More important are the regularity and length of annual dry seasons. These influence forests directly through their impact on juvenile survivorship, indirectly through their influence on the periodicity and intensity of fire (Chapters 2, 3 and 7), and directly or indirectly as triggers to flowering (Chapter 5). The bioclimate maps of Gaussen, Legris and their colleagues at the French Institute at Pondicherry and the Institute for International Vegetation Mapping at Toulouse University include a composite measure of the length and intensity of the dry season, mean annual rainfall and temperature as criteria for climate classification.[50] These maps have proven more accurate than others for lowland forest types due to their seasonality criterion. Temperature has also been factored in for montane forests, but that has proved to be only partially successful (Chapters 3, 4). But their categorisations, by mean annual rainfall, of the aseasonal wet tropics of the Far East bear little relationship with actual forest variation. The relationship between Asian tropical forests and climate are described in Chapters 2–4.

2.2.2. *The influence of the extremes*

A problem with all these attempts at climate characterisation is that the impact of occasional extreme climatic events may override that of average conditions. Walsh took account of the distribution of exceptional droughts in perhumid climates in his map of Western Malesia though their impact on primary forest remains unclear and appears – as yet - not to be great.[51] An important example with major impact on forest species composition, which has been largely ignored, is exceptional frosts at the margin of the tropics and on high mountains (Chapter 4). I prefer to relate vegetation to single measurable (and therefore more easily interpretable) variables; length of dry season, proneness to occasional extreme droughts (perhumid regions only), periodicity of fog, and frost (minimum recorded daily temperature). I will attempt to justify this procedure and support it by evidence in later chapters, especially 2, 3, 4 and 7.

Wind occurs in the tropics as prevailing winds in seasonal regions, and as hurricane-force storms in regions affected by cyclones, where their frequency affects forest structure and dynamics (Chapter 2). Occasional local windstorms, usually associated with thunderstorms, may also cause catastrophic damage and may influence composition and species richness (Chapter 3 and 7).

2.3. *The climate in tropical Asia*

I had arrived in Phnom Penh at the beginning of November, on a mission to establish a silviculture and botany capability in the Khmer forest service as part of a forest inventory project in the Cardamom and Elephant Mountains along the southern coast. It had been raining, lightly but continuously, and I was facing the prospect of prolonged fieldwork with trepidation. It was therefore with surprise and delight that, on the morning of the tenth, I stepped out of the Hotel de la Poste to my awaiting jeep into a glorious cloudless sky and the cool crisp air of early temperate autumn. Leaving town on the Khmer-Soviet Highway to Sihanoukville (now Kompong Som again) I saw, away to the south, a wall of purple thunderclouds: the retreating monsoon. Phnom Penh was to receive no more rain for three months.

The annual oscillation of the tropical climate beneath the sun draws the climatic equator with its summer rain – the monsoon – and tropical wind systems north and south with the seasons. Because the sun's rays shine vertically twice a year in the tropics (except at the margins in India and Burma), there should be two seasons of equatorial thundery humid weather, and two drier, cooler seasons. This is the case in much of Africa south of the Sahara, and in South and Central America. In Asia it holds only over a limited area of southeastern coastal India (including northern and eastern Sri Lanka) and south-east Vietnam. Drought of sufficient intensity to effect tree mortality, and therefore species distributions, occurs when there has been a sufficient number of days during which evapotranspiration has exceeded precipitation to variously affect the species.

2.3.1. *The monsoon*

Within the humid tropical lands of contemporary Asia, there is no Amazon or Congo basin densely covered in evergreen forest and recharging the rain clouds by exhaling huge quantities of water vapour. Instead, it is mainly the current and past disposition of ocean, land and topography that has determined rainfall patterns. All differed most from the present at the height of the temperate glaciations.

The climate system of tropical continental Asia is unique because it is distorted, extending far to the north beyond the geographical limit of the tropics. This is almost entirely owing to the Tibetan Plateau, specifically to its northern location and east–west disposition. This vast region of highlands north of the tropics remains cool and under constantly low pressure during the wet summer months and constantly high pressure during the snowy winter. Following the spring equinox, the inter-tropical convergence zone moves north of the Equator in the Indian Ocean. The atmosphere at the boundary layer over the sea surface absorbs moist static energy like a sponge as the sea surface warms. In April it starts to cross the coast of southernmost India. Land and air temperatures greatly increase. Fed by onshore breezes, the energy balance in the air over the land becomes greater than that over sea, radically altering atmospheric circulation. The transfer of energy to a contrasting, terrestrial environment combines synergistically with the strength of the low pressure system over Tibet. This draws the south-east trade winds with their ocean-derived humidity up and beyond the northern margin of the tropics between May and Nov. to 35° N in the upper slopes of the north-west Himalaya and, to 45° N up the western coast of the Pacific. This is the unique climate system known in English as the monsoon, the word deriving from the Arabic and Malay *musim,* meaning season.

2.3.1.1. *Two monsoons.*

The Indian and Far Eastern monsoons share the same forcing agencies, but differ in their coupled atmosphere-ocean-land dispositions[52] (Chapter 6). The Indian monsoon appears to have originated c.12 Ma; but whereas the Indian monsoon seems to have arisen consequent to the lifting of the Himalaya and Tibetan Plateau to their current height, the East Asian monsoon may have originated after the restriction of the Indian-western Pacific through-flow as the Australasian tectonic plates collided with the Sunda c.23 Ma. That led to warming of the South China Sea, which is the major source of its humid air.[53] That is also when the perhumid Sunda climate originated. Complete closure did occur during the long Pleistocene low sea level periods, when southern temperate China became drier; but the warm waters of the equatorial Proto-Mekong draining the still wet Sunda core, comparable in area to the modern Congo drainage, ensured warm seas and continuing high rainfall in north-west Borneo, and probably Peninsular Malaysia east of its Main Range.[54]

The onset of the Indian monsoon on the coast is sudden and, being controlled by the seasonal movement of the vertical sun, occurs relatively predictably 20 April–10 May. Mean wind speeds increase from c.6 to 11–14 m s^{-1} then gradually decline late June–mid-Sept.[55] Once the trades cross the geographical Equator they veer to the north-east on account of the Earth's spin (the Coriolis Force). Thus they are called the south-west monsoon from the direction of their arrival. Reaching peninsular India the monsoon winds divide. One branch chases up the Western Ghats, pouring out its rain onto their western slopes in quantities reaching an average 3 m per month on the Mahableshwar Ghats. Passing over the hills to the east and north, they exhaust what moisture remains. Thus much of continental South Asia receives much less rain and over a shorter season, which increasingly varies in intensity and time of onset (Table 1.2).

District	No. stations	Mean annual rainfall (range), millimetres	Mean length of monsoon (range), days	Mean of ranges of starting dates (full range), full date range
Garo Hills, Meghalaya	5	2248 (1841–3074)	190 (184–200)	32 (23–51) 1/4–21/5
Cannonore, Kerala	8	3017 (2310–3591)	177 (175–181)	39 (32-48) 1/5–12/6
Chandrapur, Maharashtra	17	1232 (996–1492)	111 (104–118)	40 (25-60) 10/5–4/8
Banswara, Rajasthan	15	854 (642–1075)	63 (58–75)	68 (52–91) 3/6–8/9
Mahbabnagar, Andhra	13	482 (386–600)	91 (72–110)	75 (52–101) 1/6–14/9

Table 1.2 Variability of the Indian monsoon: variation in mean annual rainfall, length of the monsoon, and date of onset: at stations within five districts at increasing distances from the point of its arrival, Cape Comorin. Cannonore is at the south-western Ghats, Garo Hills at the head of the Bay of Bengal. Climate varies regionally in the annual amount of *effective* rainfall, in the length of the monsoon, and in particular as its starting date increases with the length of the dry season gradient. (Derived from Raj 1979)

The other branch of the Indian monsoon drops even more rain onto the mountains from Sumatra north up the Bay of Bengal to the Arakan Yoma, Chittagong Hill Tracts and on to the Garo and Khasi Hills of Meghalaya and the eastern Himalaya. During its height, the monsoon precipitation comes more as driving rain than as thunderstorms, annually flooding the lower Ganges valley and drenching the hills of Odisha. Cherrapunji, in the mountains of North-East India, ranks as one of the wettest places in the world with 13 m of rain over its nine wet months, while 26 m was recorded in an exceptional year. Daily maxima can exceed 1,200 mm. Like Cherrapunji, the world's other places of extreme rainfall have a topography featuring high mountains that obstruct trade winds approaching from open ocean.

The summer-rain monsoon climate penetrates well beyond the tropics in East Asia. Far Eastern monsoons regularly reach north to the east–west barrier of the Qinling mountains in temperate north-central China, continuing up the coast into southernmost Korea and to Kyushu and north-east Honshu in Japan. Enclaves of monsoon climate can be distinguished as far north in Japan as Ibaraki Prefecture. The temperate East Asian monsoon climate can be defined as where China tea, native to warm temperate evergreen forest, can be reliably grown. The temperate monsoon climate consists of a wet tropical summer and a winter that is temperate (that is, mildly frost-prone) but relatively dry. (This will be discussed in the context of the temperate boundary of tropical lowland forests in Chapter 3, and of montane forests in Chapter 4.) The high Himalaya and northern Burmese mountains protect lands to their south, which experience a frost-free tropical climate to 32° N in the west, 28° N in the east; but frost-free lowland climates do not occur north of 24° N to their east, in South China.

During the remaining part of the year, when winter descends on the northern hemisphere, north-east trade winds dominate the continental Asian tropics. These are dry where they originate over the continent, as they do everywhere except in the Far East. There are therefore only two sharply defined seasons, respectively wet and dry.

The East Asian tropics, from eastern coastal Indo-China and the eastern coastal Philippines to New Guinea, also experience wet trade winds off the Pacific Ocean to the north-east during the northern temperate winter months Dec–March: the north-east monsoon. In the absence of a Tibetan plateau as a 'third pole', and with most land equatorial, wind speeds and maximum rainfall are less; but this monsoon is strengthened by the Miocene restriction of the Indonesian ocean current through-flow (see the beginning of this section).

2.3.1.2. *Length, intensity and variability of the south-west monsoons.* The duration of the wet monsoon over a given location generally decreases as it moves north and its humidity

is reduced by precipitation. As it ascends mountain slopes, cooling air induces further precipitation. The wet monsoon is shortest and most variable on average in the plains of Pakistan and north-west India, and in the valley of the Irrawaddy where the winds have already dropped most of their rain in the mountains of southern India or coastal Burma. It is longest in the Far East where it follows the many north–south trending valleys of north-east Burma, and the South China mountains north until halted by the east–west trending Qin Ling Shan; also along coastlines backed by mountains. Such coastlines include southwestern Sri Lanka, the west coast of peninsular India backed by the Western Ghats, and northern coastal Burma with adjacent Bangladesh, backed by the Arakan Yoma. Similarly, mountainous coasts facing the Pacific or South China Sea (including the Philippines east of the Sierra Madre, the Vietnamese coast beneath the Annamite range, eastern coastal Peninsular Malaysia, northwestern Borneo and the northern lowlands of New Guinea) all endure exceptionally wet north-east monsoons, and the most intense storms in their regions (Plate 1.4).

Plate 1.4 The north-east *landas* monsoon sweeping over Kuching, Sarawak off the South China Sea, Dec. 2006. (H.H.)

On 29 May 2004, urban Mumbai received 950 mm of rain, exceeding the mean annual rainfall over much of the rest of the Indian subcontinent. On 17 May 2003, Kudawa, at the CTFS site in Sinharaja UNESCO World Heritage Forest (Sri Lanka), received 1,400 mm in two hours during 10 days of exceptional rain. Cherrapunji received 9,300 mm in July 1861. Marudi (Sarawak, in north-west Borneo) gained 750 mm in one day and 1,200 mm in three days in Feb. 1963. Lutong, near the CTFS site at Lambir National Park, received 2,025 mm, two-thirds of the annual mean, during the month of Jan. 1963.[56]

We laid out six 0.6 ha permanent plots that March and April in the Lambir forest, where extensive landslips had followed the scarps (see Chapter 2, especially).

There may be some rain during the dry season at lower latitudes. Clouds may form by midday but precipitation is intermittent or fails to occur, according to locality and also season.

Rainfall in excess of the soil's capacity to absorb it, or the vegetation to transpire it, is ecologically superfluous except for its truncation of soils. Running off surfaces, it is lost into rivers. The length and intensity of the dry season, however, is critical in determining the nature of the vegetation and the species that can survive and compete. The Thar Desert, lying across the India-Pakistan frontier, receives no dependable annual monsoon, while the central Irrawaddy valley lies in persistent rainshadow. Indeed, it was occasional catastrophic drought in India that led to the discovery of the global El Niño Southern Oscillation pattern (see section 2.3.5 below). Where the dry season is long, the degree of annual variation in its length, and the amount of rainfall received during the intervening monsoon, become increasingly critical in defining the potential of the vegetation. In contrast, the dry season may last only 1–2 months in southern peninsular Thailand and north-west Malaysia, but it is annually dependable. However, occasional intense droughts may occur even where the wet monsoon is generally long[57] (see section 2.3.5 below). The woodlands of the driest regions of South Asia and the northern Irrawaddy plain, in contrast to the aseasonal wet regions to the south, are distinguished by markedly lower tree species diversity and a substantially different flora, shorter tree stature, and fundamentally different phenologies and plant-animal mutualisms. Trees there adapt by apportioning relatively greater biomass to roots; deciduousness and seed dormancy are obligatory (Chapters 2, 3, 5, 7).

It is those areas which experience both a wet south-west and north-east monsoon that approach double annual wet and dry seasons. As a rule, though, they receive one as a true rainy season, the other in the form of intermittent storms. Thus, south-east peninsular India and Sri Lanka receive a full north-east monsoon, but a weak south-west monsoon having already dropped most of its rain on the high mountains to their south-west. In south-east Vietnam it is the reverse.

2.3.2. *Wind storms*

Catastrophic cyclonic storms (called cyclones in the Bay of Bengal and typhoons in the Far East) occur in the wet seasonal tropics, affecting most of the coasts and coastal hills that face late summer wet monsoon winds. These regions include those from the Central Philippine islands (Fig. 1.7), from Vietnam to southern and central Japan, and down the eastern coastal hills of the Bay of Bengal and Andaman Sea from the Chittagong

Hill tracts of Bangladesh south along the Arakan Yoma and Tenasserim to peninsular Thailand. More local cyclonic storms can occur south of both belts and along the western slopes of the Western Ghats and south-west Sri Lanka. Such storms can cause landscape-scale tree-throw of varying intensity and frequency.[58] Strong local tornado-like windstorms occasionally occur in the seasonal Far East, at times causing extensive forest collapse (Chapter 2). Such storms influence forest recovery, succession and composition in multiple ways (Chapter 2, 7). They vary in frequency, affecting stands along the eastern coast of northern Luzon as frequently as once a decade, in the northern plains of Thailand no more than once or twice a century.

Even in perhumid climates, massive local turbulence or down-blasts, often preceding thunderstorms, can cause blow-downs of hundreds of hectares of forest at a time. Individual stands, though, rarely experience more than one such event in several centuries, except in peat swamps and along floodplains where rooting is shallow and the forest consequently less stable (Plate 3.9i) (see below, and Chapter 2).

2.3.3. *Temperature variability*

Singapore experiences a mean annual temperature of 27.2°C, and a mere 1°C range in mean monthly temperature, whereas the daily range is c.7°C. The minimum recorded night temperature there is 17°C. Jodhpur (at 16° N in Rajasthan) has a mean annual temperature of 26°C, but a minimum recorded temperature of -2.2°C and a maximum of 45°C. The annual range increases at higher latitudes and with the length of the dry season, increasing the intensity of summer droughts when the rain fails. Both annual and daily temperature ranges are vastly greater in desert regions where atmospheric humidity is low, exceeding that with latitude. The semi-arid and arid areas in the north-west of India and northern Pakistan coincide with incidence of winter frost (see map inside front cover). Frost occurs only occasionally to the east of the Thar Desert, but is more or less annual to its west (Chapter 3).

From about 10° N, two dry seasons, cool and hot, can be recognized. The cool dry season coincides with the north-east monsoon, the hot one with the calm, hot and humid period before the onset of the wet south-west monsoon, which is heralded by occasional pre-monsoon rains. The cool dry season lengthens and the hot shortens at increasing latitudes, until the latter ceases entirely in the moist coastal plains and hills of northernmost Vietnam and south-west Guangxi – areas that receive wet winter northeasterlies off the China Sea.

2.3.3.1. *Winter at the margins.* Minimum night-time temperatures occur during the dry winter monsoon. In inland regions from 10° N, air is sufficiently moist during the cool dry season (Oct.–March) for nightly fog, collecting in valleys and

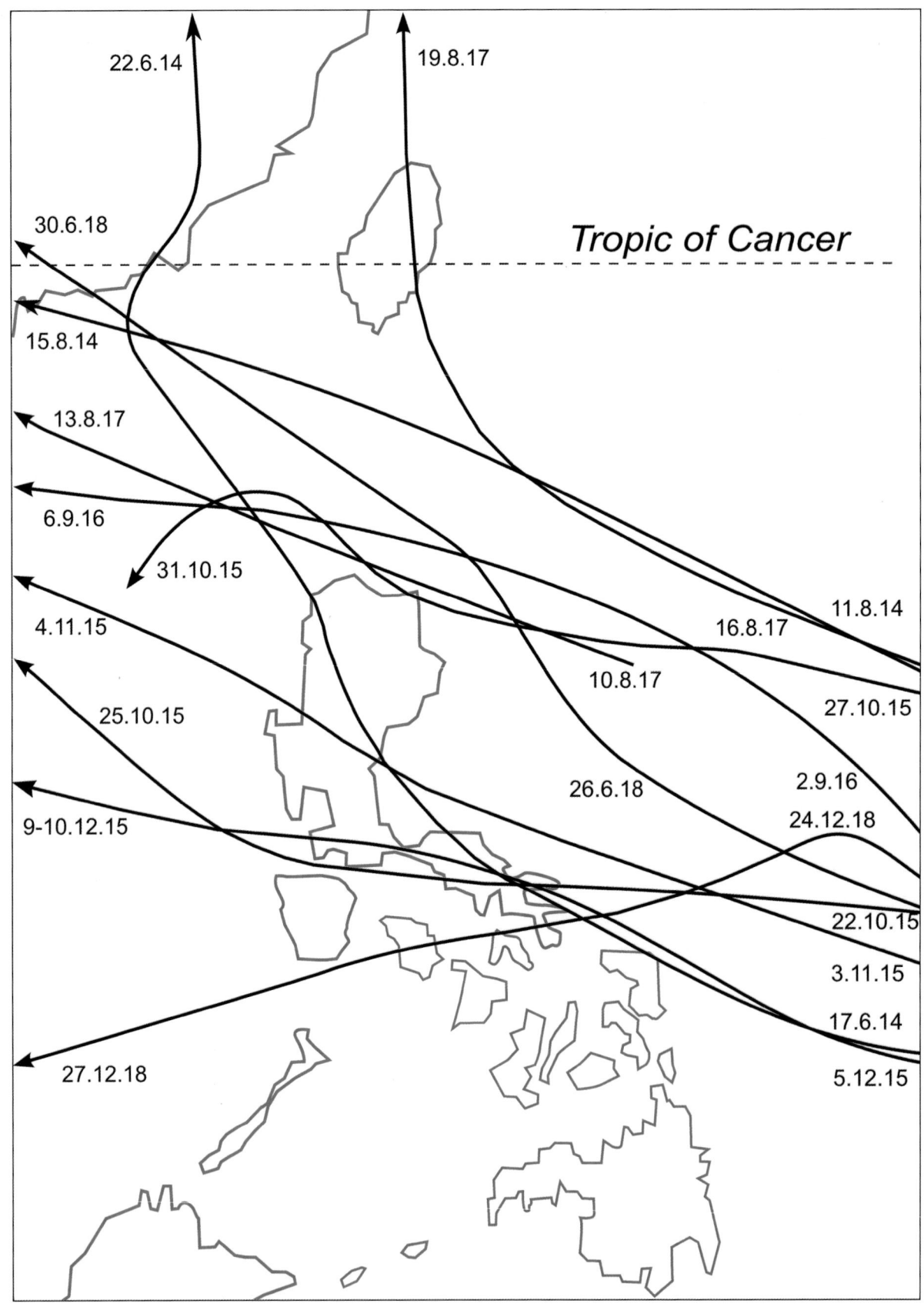

Fig. 1.7 Typhoon tracks through the Philippines over 5 years, 1914–1918. (I.H. Baillie del., from Merrill 1926)

Plate 1.5 Cloudscapes during the dry season in the seasonal Asian tropics. **a,** The dry season diurnal cloud base (the minimum altitude of winter hoar frost) at c.2,000 m on the Himalayan front ranges at Chukha, Bhutan, Nov. 2008; **b,** The haze of dry season fires in the Gangetic plain sharply cuts off at 2,000 m, (coinciding with the diurnal cloud base). Chukha, Bhutan, Nov. 2008; **c,** Looking south from Shemgang, Bhutan, 2,100 m, as the north-east dry season trade wind holds back the orographic cloud as it ascends the front range from the south. Nov. 2008; **d,** The dry season cloud base at 2,500 m, on the isolated volcano of Rinjani, 3,726 m, Lombok, 8°25' S, July 1980; **e,** November clouds high over the snows of the inner Himalaya, from Dochula Pass, 2,800 m, Bhutan, Nov. 2008. (**a–c, e** H.H.; **d** P.A.)

plains at the foot of the mountains, to condense as dew on the forest canopy (Plate 1.5). Towards the latitudinal margins of the inland lowland tropics, these night dews become copious, extending up to the frost line in mountain valleys and through most of the long winter season. Night falls initially clear as heavy cool montane downdrafts intensify, cooling the forest canopy. Dense fog forms about midnight or in the early hours, and condenses as dew in the cooling canopy.[59] Dewfall increases in the presence of breezes, implying that it is caused at least partially by impaction ('combing'), as well as condensation.[60] At higher elevations in mountain regions, the fog freezes as hoar frost. Combined with an oblique winter sun, the daily progression of temperature and humidity imposes a complex pattern of local climate in the valleys of the east-west trending regions of the Himalaya especially: nocturnal dew, morning sun and associated orographic breezes on south-facing slopes; drier colder nocturnal air causing temperature inversions, and shady moist days, on north-facing slopes where frost lines may descend up to 500 m lower in sheltered valleys.[61] In South China, frost remains confined above c.1500 m on south-facing slopes to c.23° N, but may occasionally occur lower down when cold air pools in the plains. Even Guangzhou, at 23.08° N and not far inland, occasionally experiences hard frosts, while inland frost-free south-facing pockets appear to occur far northwards (Chapter 3).

Frost defines a rough boundary between temperate and tropical lowland climates. It is therefore surprising that so little systematic observation of its occurrence or effects on vegetation, either in the lowlands or on tropical mountains, is available. In the northwestern Himalaya, winter frost occurs in the lowlands from at least 30° N 78° E northwestwards, but apparently not severely or frequently enough to entirely eliminate lowland tropical forests, which extend there to 32° N.[62] Nocturnal air at freezing temperatures appears to occur regularly above c.3,800 m on equatorial mountains. This cold heavy air descends slopes and valleys to lower altitude and 'ponds' during dry weather in wide, gentle valley bottoms. This phenomenon has been recorded as low as 1,500 m in New Guinea (c.6° S), 1,700 m in Sri Lanka (c.7° N) and the Nilgiris (10–16° N), and 800 m on Meghasani (Odisha, 21°60' N) (Plates 4.15f,g). On the other hand, such ponding has been noted only down to 2,500 m beneath the summit of Doi Inthanon (northern Thailand, 18°30' N), and to 1,500 m in the rice stubble of the foothill paddies of Bhutan (27° N), at the tropical margin but where valleys are sheltered from the cool northeasterly air by the Himalaya (Fig. 4.10). The impact of frost on vegetation is discussed in Chapter 4.

2.3.3.2. **Fire.** Fires devastating to human-modified rainforest are widespread nowadays during drought (Plate 1.5b), and have become an almost annual occurrence in some regions (Chapter 2, 8). Lightning has always been a major cause of canopy tree mortality, particularly on peaks and ridges where it occasionally initiates ground fires, and was much noted by foresters in the past[63] (Chapter 2). The *padang* savanna found at the centre of certain peat bogs bears evidence of having been initiated by fires (Chapter 3); but such fires have rarely been observed. The name of the high and periodically dry karst peak Gunung Api (Malay: fire mountain) in perhumid Sarawak speaks for itself. Direct evidence for the use of fire in cooking exists for the perhumid tropics from 45,000 BP at Niah Caves (Sarawak).[64] Fire has recently become a periodic scourge of logged and degraded forests during El Niño droughts[65] (see section 2.3.5 below) in those perhumid regions that experience droughts most intensely, but there is as yet no hard evidence that they are started naturally.

But the use of fire in the seasonal tropics by our Hominid ancestors has been inferred from 1.8 Ma, for cooking and doubtless for driving game (Chapter 2). Periodic dry season fires are endemic in regions experiencing annual dry seasons of more than five continuous dry months.[66] They may occur as low-temperature ground fires consuming the litter, or may rise through dry thickets into the canopy, thereby increasing in temperature and destructive power. These fires have a profound impact on vegetation structure and composition (Chapters 2, 3). Vegetation in turn influences fire intensity. Fire-proneness and intensity at first increase with dry season length, but both parameters reach a peak and then decline as dry biomass on the ground declines in climates with more than eight dry months each year. (Interactions between climate, fire and forest structure are discussed in Chapter 2.)

2.3.4. Perhumid Malesia: the 'land below the wind'
(Plate 1.6)

The equatorial archipelago, which extends 6,000 km from Sumatra (including Peninsular Malaysia) through the Philippines and Indonesia to New Guinea, is recognized by botanists as a unitary biogeographic region, Malesia, despite its dual tectonic origins (Chapter 6). This is because most of the land in this region – except for a few areas influenced by dry south-east trades off Australia and dry north-east trades over the west of Luzon's Sierra Madre – is under the most extensive continuously wet, that is perhumid, tropical climate in the world (map inside front cover). This climate is defined by the absence of predictable annual periods of water stress, with mean monthly rainfall in excess of expected evapotranspiration throughout the year. The pervasiveness of this perhumid climate is due to the ubiquitous open water around the Malesian Archipelago, and in particular to the presence to the east of the world's widest ocean. It is also due to their inclusion within the intertropical convergence zone throughout the year.

Plate 1.6 *The land below the wind*: Salt eroded coastal rock, Sandakan, Sabah, Malaysian Borneo, a rock that can never have experienced stormy seas. (P.A.)

Rainfall may here be due either to moist onshore winds drawn off the sea against the general flow of the trade winds by the daily heating of the land, or as orographic (that is, mountain-induced) upward cooling of local air with consequent condensation. Such a perhumid climate does not exist in lowland Africa, and is confined to a relatively small area of the upper Amazon near the Andean foothills, the Choco of the northwestern coast of South America, and also to narrow coastal regions of Central America facing the Caribbean. In both Asia and the Neotropics, perhumid climates support the most species-rich terrestrial ecosystems known (Chapter 7). They are generally associated with leached infertile soils (see sections 4.2.3.2 and 4.3.1.2 below) and with characteristic forest dynamics and species composition (Chapters 2, 3, 7).

There are two other small perhumid regions in tropical Asia. The southwestern quarter of Sri Lanka receives the brunt of both the south-west monsoon and a somewhat wet north-east monsoon owing to the mountains to its east, along with orographic rainfall during the rest of the year. The hills of central Vietnam, and the Annamite range behind them, receive a wet north-east monsoon combined with frequent orographic rain during the rest of the year, although here the coastal plains are seasonal. Each represents a unique patch in an otherwise seasonal region, and represents a potential refuge for the often hyper-rich flora of perhumid climates (Chapter 6).

2.3.5. *The El Niño Southern Oscillation*

The Pacific Ocean to its east lends Asia a greater range of yearly variation in rainfall than other wet tropical regions, with the exception of some Neotropical coasts also facing that ocean. This is due to a climatic phenomenon that influences the whole Earth's climate: the El Niño Southern Oscillation (El Niño, ENSO).[67] Regular supra-annual famines caused by the cessation of the Humboldt current down the Peruvian coast, leading to warming of the sea surface and disappearance of fish stocks, accompanied by exceptional coastal rain, were known to local fishermen since ancient times. But it was not until 1893 that a meteorologist in India, Charles Todd, observed that droughts and famines in India coincided with droughts in Australia; while Gilbert Walker, then Director-General of Indian Meteorological Services, recognized their connection with Peruvian storms across the Pacific Ocean. Walker coined the term the 'Southern Oscillation'. As with cyclonic storms originating under the influence of the trade winds, small differences between surface air and sea temperatures in the windless equatorial central-east Pacific can cause variation in the evaporative power of the air and the speed of the trade winds. Here, though, there is a coupled atmospheric and oceanic oscillation across the tropical Pacific Ocean, centred south of the Equator. In the atmosphere, this is manifested as an east–west see-saw of surface pressure and related patterns of wind, temperature, cloud and rain. In the ocean, there is an interdependent east–west flip-flop of the location and depth of warm and cool bodies of water. During El Niño events, unusual high pressure develops in the western Pacific. Trade winds weaken, and moisture off the ocean lessens. Rain moves to the central and eastern Pacific, where the weakened trades eliminate upwelling of the deep cool nutrient-rich water of the Humboldt Current. During the longer intervening La Niña periods, the easterly trades strengthen to full force, restoring upwelling of deep water in the eastern Pacific. Pressure rises in the east, and declines in the western Pacific, creating the customary high year-round rainfall.[68]

In most years, therefore, the constant flow of the easterly trade winds mediate both the currents of the tropical oceans and the concentration of rainfall over the daily warming land surfaces abutting the western Pacific Ocean. During El Niño events, rainfall around the western Pacific slackens, becoming localised and dominated by daily orographic air movement. Regions distant from coastal mountains experience droughts of varying intensity. Brush, peat, and parched grasslands become susceptible to fire (Chapter 8). Equatorial regions with mountains close to the coast and facing away from the Pacific, such as north-west Borneo and western Sumatra, are least affected. Among these are some of the most floristically rich forest communities and landscapes in the world (Chapter 7).

El Niño events may persist for more than a year.[69] The variability of the wet monsoon is also correlated: continental areas experience tropical Asia's longest droughts. El Niño affects the biology of the forest more intensely, and specifically, in tropical Asia than anywhere else (Chapters 2, 3, 5).[70]

2.4. *A day in the tropics: the diurnal march of the weather* (Plate 1.7)

Plate 1.7 The daily march of weather in perhumid Malesia. **a,** Morning cumulus clouds accumulate orographically along the curving parallel khondalite ridges of south-west Sri Lanka; **b,** Even during the north-east monsoon (landas), the cloud base inland remains above 1,000 m until the rain begins: the summit of the mountain, Matang, Sarawak is at 900 m; **c,** Orographic storm clouds, at c.1,000 m, gather over Gunung Mulu, north-east Sarawak; **d,** Valley fog in a Maliau valley, Sabah, evening after rain; **e,** Cloud hugging the Matang summit, 909 m, west Sarawak, in the early evening following rain. The summit supports upper montane forest. (**a** Swiss survey; **b–e** H.H.)

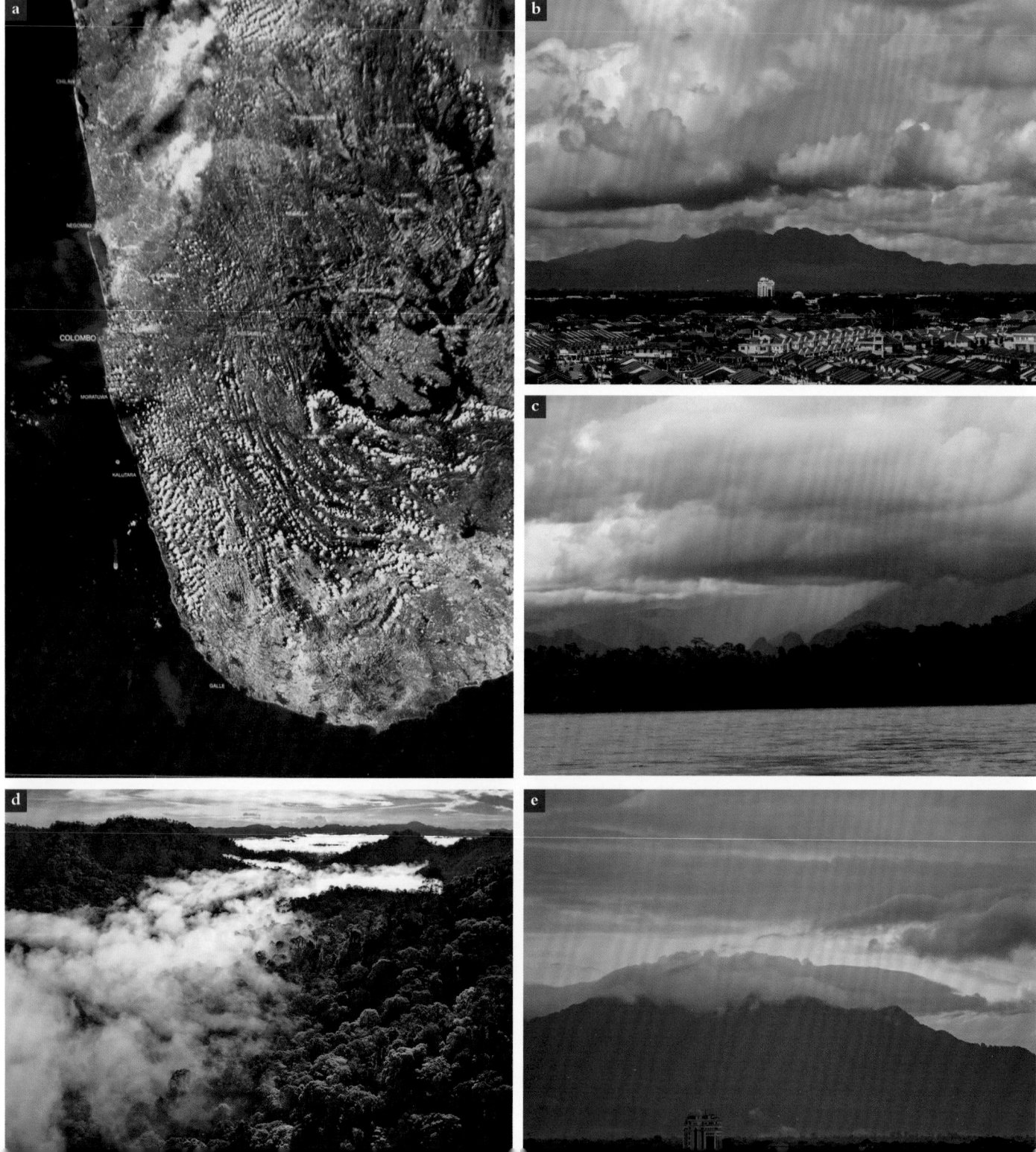

2.4.1. *Lowland weather*

Precipitation ultimately originates from the sea according to a daily pattern. During the night, land cools while ocean water retains heat, and columnar and anvil-shaped cumulonimbus clouds can consequently then be seen rising offshore. In the morning, the land heats, and by mid-morning is warmer than the sea. Humid warm air rises through the heavy cool air above, forming clouds over land. This rising warm air is at first replaced by moist air drawn inland from the relatively cooler sea, causing diurnal onshore breezes. Coastal regions of the intertropical convergence zone therefore lack prevailing winds. These areas experience more intermittent and prolonged droughts than other areas because rain most frequently falls elsewhere, day and night[71] (Fig. 1.8).

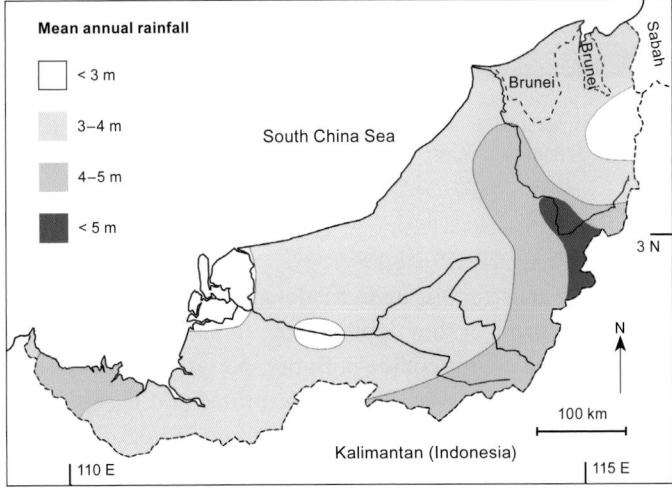

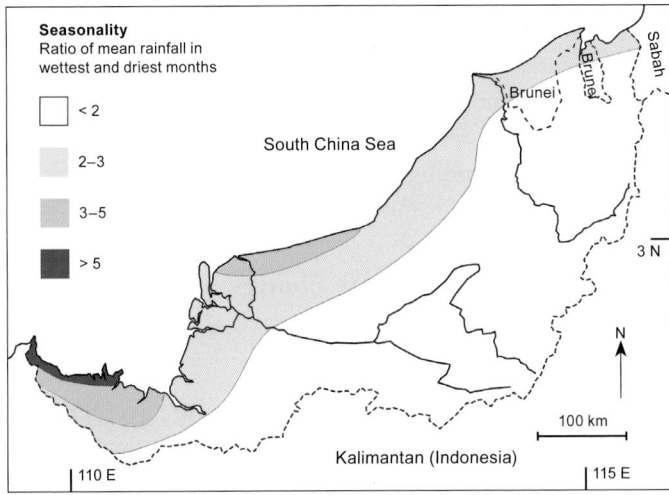

Fig. 1.8 Rainfall variation in perhumid Sarawak, north-west Borneo. Rainfall in this perhumid equatorial region is mainly orographic and diurnal (though West Sarawak experiences the north-east monsoon): coastal regions escape some diurnal precipitation owing to diurnal onshore breezes. (After Baillie 1972)

Local variations in rainfall and cloud influence forests in ways that seem to defy the climate mapper. Few careful measurements of these daily trends and their seasonal variation, especially in mountainous areas, have been made.[72] The following are personal observations, awaiting sceptical research.

The driving wet monsoon seasons aside, mornings in the humid tropics inland are typically cloudless, although fog may lie in valleys or over plains after rain the previous day. This fog, usually less than 200 m deep, rises and dissipates by 9.30am. By 10am the first small clouds appear, and cluster over high ground. These are at first shallow, but by midday they begin to spread over the plains and to billow upwards, signalling the beginning of displacement of the hot humid air of the lowlands by the cooler heavier air above. Condensation of humid lowland air, forming the flat cloud base preceding rain, occurs at the same range of altitudes in the convergence zone and during the summer wet monsoon throughout the tropics as diurnal temperatures are the same. Over the plains and smaller hills, daily clouds form a level base at c.1,200 m. This base may descend as low as c.800 m near the sea and on coastal mountains, as air rising over warming inland surfaces draws cooler moister breezes off the sea. By early afternoon, major cloud accumulations in the convergence zone have risen to heights at times exceeding 10,000 m. Precipitation starts, often accompanied by thunder. The rain is heavy, but of relatively short duration, and much of it is lost as run-off. In late afternoon it is noticeably cooler. Once the rain starts, the cloud base loses its uniformity and much of its altitude. Frequently, clouds dissipate leaving a sunny late afternoon. Trails of mist rise from the forest canopy when hot moist air, previously trapped, rises into the cooling air above.

Rain may fall continuously day and night during the wet monsoons, but it seldom continues for more than three days at a time. The most continuous rain occurs where the monsoon penetrates the perhumid zone, especially on coastal plains backed by mountains perpendicular to the rain-bearing winds. There, exceptional years can experience widespread flooding and landslips.

2.4.2. *Mountain weather*

Clouds form first over sharp summits and ridges, where ascending humid air from the lowlands follows the slopes to meet and condense overhead. In the intertropical convergence zone where prevailing winds are lacking, the latent heat of forested slopes at first prevents condensation within the canopy on those isolated peaks that penetrate the c.1,200 m lowland cloud base (Chapter 4). Fog does eventually infiltrate the forest canopy, though, as updrafts approach the ridge. The longer the upward slope, the further fog-free humid air can rise along it within the cloud base. On coastal mountains, saturated diurnal breezes off the sea drive cloud fog through

Plate 1.8 The diurnal onshore breeze on a coastal mountain: Gunung Santubong, 805 m, west Sarawak. **a,** The cloud accumulates on the summit, with a base at c.750 m, 11 am; **b,** Cloud provoked by Santubong, at 4 pm from Kuching, Sarawak; moist sea air updrafted by its sun-heated slopes, trailing inland on the diurnal on-shore breeze. (**a** P.A.; **b** H.H.)

the forest canopy to as low as 750 m (Plate 1.8). Inland, cloud penetration starts at c.1,500 m on smaller mountains, c.2,000 m on larger. On Kinabalu, diurnal fog within the forest canopy becomes pervasive at 2,200 m. On the largest mountain massifs of the Asian tropics, including the Himalaya and New Guinea mountains, the cloud base itself rises to 3,000 m or higher in the cool dry season and during droughts.

On the Himalayan front ranges of Bhutan, fog enters ridge forests in dry weather at c.1,800 m, but not until 3,000 m in the central mountains, a range of altitudes mirrored in New Guinea. On both massifs, fog may fail to penetrate the forest canopy of intervening valleys below 2,800–3,000 m. At the highest altitude, incipient cumulonimbus towers are separated by cloudless 'canyons'. Therefore, fog and drought may *both* become more frequent at subalpine and alpine elevations above 3,000 m.[73]

Following rain, fog enters the forest at lower altitudes. The mountains may retain lingering cloud at night, but are often cloud-free and cool, incurring a wide range of daily surface temperatures. In summary, equatorial tropical mountains experience a diurnal weather pattern distinguished by hot sunny mornings, moist afternoons, and cool dry nights.

In windless seasons and climates, orographic updrafts increase in speed upslope. High mountains may thus experience strong diurnal updrafts before rain. This can produce dramatic effects during the dry northeasterly monsoon, especially on coastal

mountains, when moist updrafts along sheltered coastal slopes form cloud banks as they approach the summits. Reaching the summit, these clouds collide with dry, persistent northeasterly trades from inland, above. They are turned and blown back, trailing off to the south-west. The meaning of Poporkvul, the name of an old colonial hill station on the crest of the coastal Cambodian Kamchay Mountain, is 'the clouds turn'.

At subalpine and alpine elevations above c.3,000 m, incipient cumulonimbus clouds are separated by cloudless aerial canyons. Therefore, both intermittent drought and dry season frost become increasingly frequent between long periods of fog. This is a demanding climate for plants (Chapter 4).

3. The history of climate change and stability

3.1. *Early history*

Global and regional variation in sea temperature over time can be evaluated by comparison of skeletons of the foraminiferal fauna on the ocean floor with their current distributions. As with flowering plants, these organisms become increasingly dissimilar from present taxa with increasing geological age. Regional and local terrestrial climates can be inferred from plant fossils from the Late Cretaceous onwards, but with reasonable precision only from the Oligocene, when indisputably modern taxa (at the family and genus levels) can be distinguished. The reliability of this method rests on the assumption that the climatic ranges of plant taxa have remained unchanged over geological time. This is supported by the rigidity with which current taxa, even some flowering plant families, adhere to a uniform climate range worldwide.

Bob Morley has recently produced an excellent account of the history of Cenozoic climate change in our region.[74] The Indian block and Madagascar separated early (before the origin of flowering plants) from Gondwana, from Africa and from each other. They originated in a hot and humid climate, albeit marginally tropical. The Indian block would have had to travel north through currently desert latitudes, before re-entering a probably seasonal wet tropical climate as it by-passed north-east Africa in the Late Cretaceous. It finally approached Laurasia in a very hot and wet, probably perhumid, equatorial environment early in the Cenozoic era.

The Eocene was the last period when tropical climates widely extended to 35° N or more, when wet tropical regions were more extensive than at present. The Indian plate had butted into the southern Laurasian coast, then located near the Equator. At that time, the contact zone between India and Laurasia – with the Tethys Sea retreating to the east – was still therefore at equatorial latitudes, and the Himalaya was no more than a line of hills. The climate was less seasonal than the present,[75] and the hills received rain off the Tethys and from the Pacific into which the Tethys opened. This climate changed drastically during the Cenozoic period.[76]

Since then, global climate has continued to undergo major changes. By the Paleocene of our Cenozoic era, humid tropical and near-tropical warm temperate climates had reached high latitudes in both northern and southern hemispheres. But from the beginning of the Oligocene (35 Ma), these climates receded. Seasonal tropical climates penetrated to lower latitudes, and this change would have been accelerated along the India-Laurasia suture line by India's northward thrust. Due to the Earth's rotation and hence the pattern of prevailing winds, rainfall was probably always greatest and most continuous in the regions fronting the Pacific Ocean. A corridor of perhumid climate between South Asia and the Far East is likely to have existed at the time of contact – when there was still extensive Tethys water to the east and in equatorial latitudes, and at least until the late Eocene. Evidence from sediments along the base of the Himalaya suggests that for about ten million years during the late Eocene and early Oligocene periods (c.35–20 Ma), that region was semi-arid,[77] while global climates remained hot and overall regional climates wet (see section 1.1 above). Humid tropical climates expanded again during the late Oligocene and continued until the middle Miocene thermal maximum. These climates were probably seasonal in the lands between India and the Far East, since these were by this time well north of the Equator. The great surge of the Himalaya appears to have started later, in the late Oligocene, from 24–22 Ma. The Tibetan tableland rose with it. Climate became wetter but remained seasonal. The final and highest Himalayan uplift, and with it the inception of the full monsoon climate pattern, probably occurred later (see section 2.3.1.1 above). Afterwards, global climates again cooled until the mid-Miocene (c.13 Ma), when alternating warmer and cooler periods were initiated on a 41 ka cycle correlated with a wobble in Earth's axis in relation to the sun. Finally, during the mid-Pleistocene (c.900 ka), this culminated in the present sequence of c.100 ka ice ages alternating with c.10 ka warmer periods. We currently live near the end of one of these warm periods.

The India-Burma frontier hills likely also arose during that time and possibly as late as 12 Ma.[78] They attracted rain and created a migration route for seasonal rainforest elements, but a climatic barrier to migration for deciduous forest elements between west and east, that likely has persisted to the present day. But the perhumid climates of south-west Sri Lanka and the Far East have therefore likely been isolated at least since the early Oligocene (Chapter 6).

Central and southern Europe experienced humid tropical or warm temperate climates during the Eocene (Chapter 6). However, this region was separated at the time from Asia (including India) by a broad north–south epicontinental strait, the Turgai. This strait closed at the end of the Eocene, as the Indian plate collided to its south-east. A migration route would then have opened for humid climate tropical elements between south-west Eurasia and the eastward-receding Laurasian shores of the Tethys. This may have persisted intermittently at least until the Himalayas started to rise in the late Oligocene, but is unlikely to have experienced perhumid climates throughout its length. In the absence of a low pressure zone, like the current Tibetan Plateau to its north, it is likely that rainfall would have become increasingly seasonal as the Indian block pushed north. By the Miocene, the Middle Eastern coastline was located well north of the Equator in what had become a seasonal and even semi-arid tropical climate, while the lowlands south of the northwestern Himalaya had entered temperate climes.

Warm temperate connections between East Asia and North America across the Bering Strait continued until c.14 Ma, but became intermittent. Global temperature changes appear to have been greatest at higher latitudes and were probably of reduced amplitude in equatorial regions. Equatorial temperatures substantially higher than those at present in the Eocene would likely have had an influence on rainfall patterns.

The arrival of the Gondwanan eastern Indonesian terranes, New Guinea and Australia, only occurred from the latest Oligocene onward. Pre-Miocene species immigration from northern Australasia must therefore have occurred by 'sweepstakes dispersal' across sea barriers. The modern

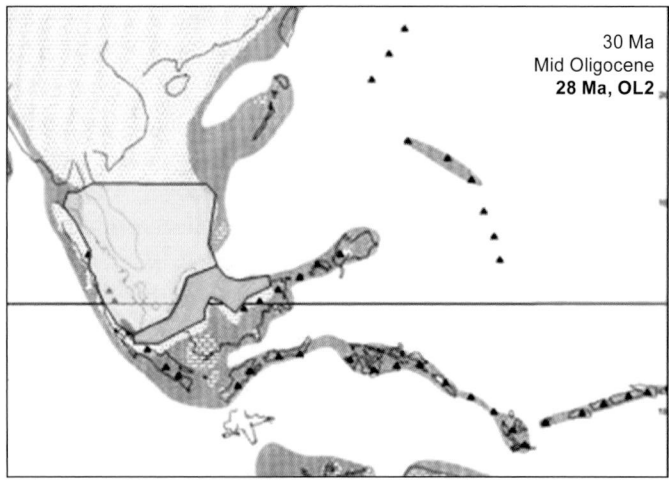

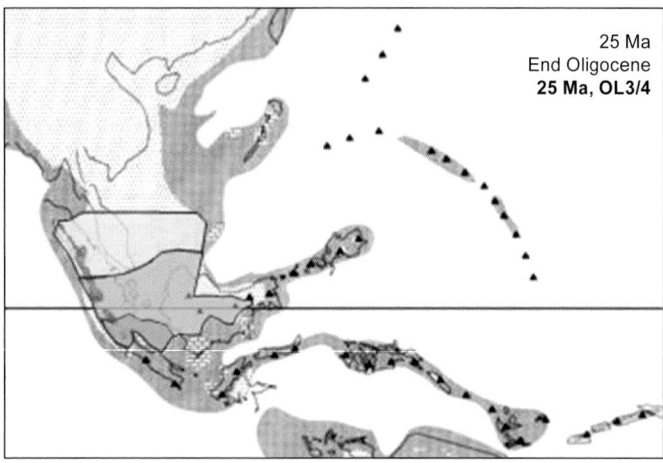

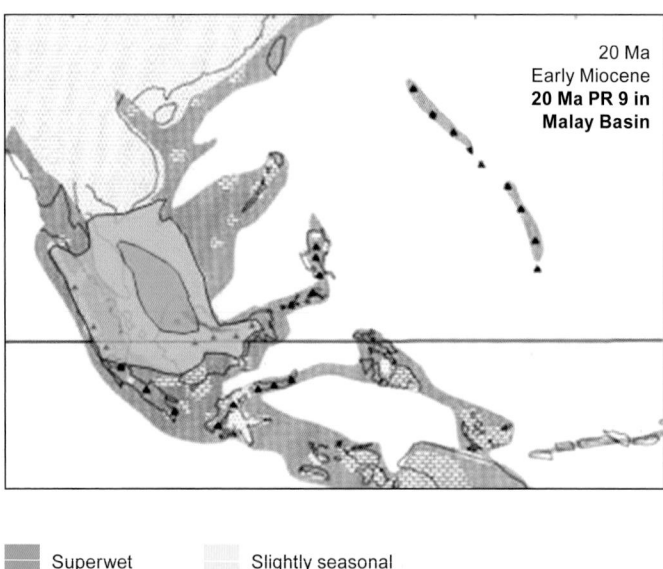

Superwet

Slightly seasonal

Everwet

Strongly seasonal

Fig. 1.9 Climate change in tropical South-East Asia in the period leading to the initiation of humid lowland Sundaland. (After Morley 1999)

distributions of flora and forest will be interpreted in the context of these events in Chapter 6.

The broad Sunda region experienced a mostly seasonal moist or wet climate during much of the Oligocene. At this time the northern part bore widespread upland areas, within which there likely remained perhumid refugia in coastal lowlands. At the end of the Oligocene, the ocean through-flow through the Malesian archipelago became constricted owing to the collision of the Australian Plate with Sunda. As a result, the Sunda climate became widely perhumid from the earliest early Miocene (c.23 Ma onward). Since that time, widespread peats formed across what is now the South China Sea and Gulf of Thailand (Fig. 1.9). Parts of the region became periodically more seasonal during the late Miocene and Pliocene, roughly as the Indian monsoon grew in intensity. This is evidenced by widespread occurrence of fossil coral reefs from that period, which require clear sediment-free water (carbonate platforms in Fig. 1.3b).

3.2. *The Pleistocene*

The biogeography of the modern flora and fauna can be best understood within the historical context of geological and climatic events over the last 10 million – and especially the last 2–3 million – years (Chapter 6).

3.2.1. *Changes in rainfall*

The monsoon was wetter 8–10 Ma than at present, but then weakened starting 2 Ma when the advent of the northern ice ages brought dramatic changes to tropical regions. Under the mountain forests of modern Bhutan lie great dunes, less of sand than of wind-blown silt. This material originates from mountain deserts at the edges of the great ice cap that covered Tibet until 12,000 BP, during the last northern ice age.[79] Tibet would at that time have been continuously white with snow, reflecting incoming solar energy and therefore a zone of constant high pressure. The monsoon would therefore have slackened or even died during the glacial periods, although the extent to which this occurred remains unclear.[80] Forests would have retreated, and deserts would likely have expanded into South Asia, though conclusive evidence is lacking as yet. Mixed dipterocarp and seasonally wet evergreen forests may have survived in South Asia only along coasts backed by mountains, where humid onshore breezes would still, as now, have been drawn upwards and drained of moisture. South Asia would thus have been an extreme example of a worldwide pattern of drying in the tropics during ice ages. The monsoon is currently weaker than it was in the early Holocene.[58]

In the Far East, seasonal climates and forests spread south as far as Kuala Lumpur, where Quaternary fossil pine pollen has been found dating from glacial maxima.[81] Kuala Lumpur lies in a lowland rainshadow between the Sumatran and peninsular mountains. During the glacial periods, much of Sundaland and New Guinea were characterised by still perhumid climates with reduced rainfall[82] or seasonal climates at times of lowest sea levels.

It has been claimed that there was a corridor of savanna across the Equator,[83] a biome probably not occurring naturally at present anywhere in Asia other than around large floodplain grasslands at the base of the Himalaya and in southern New Guinea. This claim is of major importance to palaeogeographical interpretation, not least because the time period coincides with the arrival in the region of our hominid ancestors. The evidence is further discussed in Chapter 6.

3.2.2. *Changes in global temperature and sea level*

Paleontological evidence for temperature depression in tropical Asia during the prevailing Pleistocene periods of cold northern climate and low eustatic sea levels, and particularly during the last northern ice age (100,000–12,000 BP), is of two kinds: (1) the lowest altitude of surviving glacial moraines, and (2) the presence of alpine herbaceous flora in the pollen preserved in peat cores. Both have been found as low as 2,000 m on Mount Wilhelm (Papua New Guinea), and 1,500 m in the eastern Himalaya when continuing uplift is accounted for.[84] This could therefore imply a descent of the tree line by as much as 1,700 m and therefore a reduction of mean annual temperature by up to 10°C. The current upper limit of the full lowland rainforest flora in perhumid New Guinea, Sundaland, and also the eastern Himalayan foothills is at c.1,000 m, but it is overwhelmingly concentrated below 400 m. Conservatively, therefore, any mean annual temperature drop exceeding 6°C would have led to massive extinction.

It has been claimed that, on evidence of foraminifera and other taxa, mean temperatures in the tropical lowlands were depressed by only 2–4°C, but it was likely somewhat more.[85] There are two possible explanations for this paradoxical difference in estimates of temperature depression:

1. There could have been an increase in the temperature lapse rate with elevation, from the current c.0.65°C/100 m to 1.3°C/100 m of ascent. That would represent a meteorologically impossibly high lapse rate. However, it could possibly have increased to as much as 0.75°C had the atmosphere concurrently become drier. That could indeed depress mean temperature by 6–7°C.

2. Visitors to the Franz Josef glacier on the South Island of New Zealand can see frost-sensitive podocarp forests along slopes hundreds of metres above the surface of the glacier. Moraines and peat are set in valley bottoms where night temperatures are depressed below overall means, and where inversion of mean temperature in relation to adjacent slopes

may even occur (see above). Similarly, frosts currently occasionally descend to altitudes less than 1,000 m below the tree line on equatorial mountains, killing tree regeneration and replacing forest locally with alpine grassland and a characteristic frost-hardy flora (Chapter 4, Fig. 4.15f,g). (It should be noted that pollen from the warmer adjacent slopes above would also be represented in peat cores.)

It is therefore unlikely that the overall tree line descended by as much as 1,000 m, or that the altitudinal limits of lowland forest descended more than 400 m in equatorial regions. However, the increased aridity associated with the slackening of the Indian monsoon was likely associated with more drastic temperature depression at the tropical margin in the north-east Himalayan foothills.

John Flenley has claimed that upper montane forest (mossy 'elfin' woodland, Chapter 4) disappeared during some of these periods, which would have been drier at high tropical altitudes.[86] This is consistent with its current impoverishment or absence in seasonal tropical climates (Chapter 4). Nevertheless, the persistence on Kinabalu and other major equatorial mountains of floristic elements *not* currently occurring in lower montane or subalpine forest types remains to be explained. Surely such forest types are older than the later Pleistocene (Chapters 4, 6).

4. Tropical forest soils

4.1. *Soil typology and classification*

The performance of plants depends in part on the climate to which they are adapted and within which they are competitive, and in part on the availability of water and nutrients derived from the soil in which they find themselves. The latter are themselves partly dependent on climate. They are also partly dependent on the rocky or alluvial substrate, the major source of nutrient replenishment. The mineral soil's components largely determine soil porosity, and therefore capacity for storing water while allowing oxygen – essential for the respiring roots – to penetrate. Topography also influences the soil's water economy. All these factors, but especially nutrient availability, influence vegetation litterfall rates, the amenability of litter (mostly leaves) to decomposition, and the rate and periodicity with which nutrients retained in litter are released and returned through the soil to roots. Thus, the mineral soil influences the structure, phenology and composition of the forest, while the soil is in turn influenced by the composition and phenology of its forest cover. Nutrients recycled and released by decomposing organic matter are the major nutrient source to forest plants, though the rate and composition of nutrient supplement from the substrate crucially influences and thereby differentiates forest productivity, structure and composition.

There are several soil classification systems in current use. All are designed for agricultural soils. They are of limited use for forest soils, first because forest organic matter is rapidly transformed and lost following clearance and cultivation, so that they take no account of the crucial role of litterfall and organic matter under forest conditions. Second, they focus on the nutrients available for immediate root uptake and crop productivity, with lesser consideration of the nutrient store in the substrate and mineral soil, part of which may become available during the life of a forest tree. Third, these classifications are based on readily visible variation in the appearance of the profile of the mineral soil, and processes inferred from it, whereas we are as much concerned with the appearance of the organic matter and surface horizons. For us, the conventional soil classifications over-split on mineral soil criteria and differentiate mineral profiles that are poorly correlated with observed forest variation, but over-lump changes in those observable organic matter characteristics that are correlated with forest variation.

Ian Baillie has recently provided a clear review and classification of tropical forest soils, concentrating on the wet tropics.[87] Therefore, after briefly discussing the major ways in which all tropical forest soils vary, I will focus here on soil variation in relation to the climates and surface geology of the Asian tropics.

4.2. Soils reflect landscape dynamics

Since the ultimate source of most nutrient ions (except nitrogen and most sodium) is the substrate, variation in nutrient concentrations at the landscape scale is related to variation in substrate and topography, which influence floristic composition and litter characteristics. Substrates influence soil mineral ion concentration due to their ratio of sand to clay, and the structure of the dominant clay molecules in mineral soils resulting from weathering. Inputs through rain and wind-borne dust have, until recently, been insignificant within the humid tropics of Asia owing to the wet monsoons and the lack of major deserts to the east.

The rates of many chemical processes vary with temperature. These include nutrient release during rock weathering, from mineral soil and from the decomposition of organic matter. Rates are therefore highest in the lowland tropics, and within the tropics they decline with the declines in ambient temperatures associated with increasing altitude. Chemical processes also depend on water for solution. Rates of litter decomposition, other factors being equal, are highest in perhumid climates. They may be high during the wet monsoon, but slow down or cease when soil surface horizons

dry out during the dry season, or during droughts in perhumid climates. Soils, then, are as dynamic as other parts of a forest ecosystem and, as with the forest, different components of the soil process at different rates.[88]

4.2.1. Mineral soils and pathways of erosion

The age of a soil is most often expressed in terms of the time since the mineral soil first started to establish by decomposition of a fresh substrate surface, or to stabilise from debris deposited by a previous physical process. In temperate regions this deposition is often the consequence of the most recent glaciation, so that soils are dated from the end of that event, 10,000–15,000 BP. Such temperate-zone soils are considered young. In the tropics, glaciated soils of this type are confined to the highest mountains[89] and do not occur below c.1,500 m (see section 3.2.2 above). Young mineral soils in the tropics are mostly caused by landslips on steep surfaces. These are abundant in regions of frequent earthquakes, especially where associated with tectonic uplift and buckling of sedimentary and other rocks, leading to high narrow ridges and steep slopes. They dominate slopes of the tropical Himalaya, New Guinea, and Sulawesi, and are common in Borneo due also in part to continuing uplift.

Landslips also occur where water percolates through soil and porous substrates onto a sloping impervious layer beneath (Plate 1.9a). There it may accumulate, lubricating the interface and possibly leading to slippage of the overburden. This is a frequent cause of landslips where uplift has tilted rocks into steeply inclined strata.[90] Impervious rock surfaces may be exposed on the slip itself, while a more or less deep and porous accumulation – a colluvium – may be left on the slope below, presenting two contrasting soils environments for plant establishment. This process generates steep, often narrow ridges held up by less erodible, porous strata such as sandstones. In addition, clay-rich and relatively impervious slope soils are continuously truncated, and thereby rejuvenated, by the run-off from heavy tropical rainstorms. Where they lie over bedrock in such steeply inclined landscapes, both ridge and slope soils are shallow, rarely exceeding a metre in depth and contrasting with the generally deeper colluvial accumulations below. These landscapes dominate the rugged uplands of inland Borneo, northern Thailand and north-east Burma, where sedimentary basins on submerged continental surfaces have been uplifted and buckled by tectonic activity. Neville Haile[91] estimated that the sedimentary landscapes of Borneo are eroded, on overall average, at about 50 mm per 100 years, which is consistent with the age of some charcoal discovered in a Sarawak upland soils profile dated by $C_{14.}$[92]

Where they lie more or less horizontally, porous strata such as sandstone resist surface erosion. Where they overlie impervious erodible clay-forming rocks, erosion of those layers and the consequent undercutting of the porous cap leads to breakage along the edges. Such erosion creates a landscape of cuestas or gentle dip slopes following the incline of the porous surface stratum, and steep scarps along the upper edges. These landscapes are characteristic of less steeply inclined sedimentary rocks in the humid tropics, such as in coastal northern Borneo. Soils dry out in climates with long dry seasons. Sudden saturation at the beginning of the wet monsoon can lead to total loss of soils where they overlie inclining rock surfaces. This erosion of ancient, uniformly hard rock eventually generates a rounded topography of extensive bare rock hummocks; viewed from a distance, these give the impression of sleeping elephants (Plate 1.1b). These are known as inselbergs and are widespread in regions of ancient continental rocks, such as Sri Lanka and the Eastern Ghats of peninsular India.

Such forms of ongoing erosion, whether continuous or spasmodic, have been occurring in the tropics over geological time scales. Where they are absent or uncommon, as for instance on the freely draining soils over granites of the Peninsular Malaysian Main Range, the mineral soils can accumulate as deep as 15 m and may be quite ancient (Plate 1.13a).[93] Even here, though, Pleistocene climate oscillations and the solution rates of soil minerals are such that the oldest are unlikely to exceed half a million years.

By contrast, the sedimentary landscapes of Borneo and the Siwaliks, for example, are in active process of erosion, leading to peneplanation within a few million years. Such peneplanation of young mountains is generally balanced by continuing tectonic uplift. Different regions of tropical Asia are therefore covered by mineral soils whose prevailing age varies dramatically. In consequence soil is generally shallow in steep, erodible young landscapes and deeper in old landscapes with the associated rounded 'mature' topography.[94] An exception is the geologically young, predominantly sandstone uplands arising from uplift of Miocene and later depositional basins. These occur notably along the north-west coast of Borneo. The Cenozoic sedimentary hills behind the southeastern coast of peninsular India are similar, albeit in a seasonal climate. Here, the sandstone is soft, erodible and gently dipping. Erosion occurs with the gradual shifting of ridge-top sandy soil downslope, and the consequent mixing with clays arising from argyllaceous strata below. Both sandstone and the overlying sandy soils are freely draining, and densely root-matted raw humus forms a protective blanket on top in wet climates (section 4.2.2.1 below). Deep soils and mature rounded landscapes consequently develop relatively rapidly, resisting erosion except along the sides of the incising streams. Thus, older soils may overlie the youngest rocks in this landscape.

Plate 1.9 Erosion. **a,** Landslip over clay-rich metamorphic precambrian charnokite rock following the exceptional storm of 2003, Sinharaja research station, south-west Sri Lanka; **b,** Sandy soil eroded by rainfall around a fallen leaf, isolating it on a pillar. Khao Chong National Park and CTFS plot site, peninsular Thailand. (**a** U. Goodale; **b** I.H. Baillie)

4.2.2. *Organic matter in the soil*

4.2.2.1. *Pathways of litter decomposition and their consequences in upland soils.* In the prevailing zonal yellow-red clay and sandy clay loam soils of the lowland wet tropics, rates of litter decomposition generally exceed rates of accumulation through fall of dead organic material (Plate 1.10). This condition is termed *udult*. Leaf and other fine litter is often fully decomposed in less than one year, or even during a single wet monsoon period in the seasonal climate of Asia. That explains why soil organic matter levels are relatively low, and may partially explain why depths of visible humus discoloration – betraying the presence of particulate organic matter – are generally shallow. Because soil organic matter serves as the major store for nutrients, apparent lack of it also explains the generally low fertility of tropical lowland soils. Initial comminution of fallen debris, as elsewhere in the tropics, is mostly by termites and ants, though no leaf-cutting ant species occur in Asia. Decomposition is thereafter carried out by bacteria and fungi. Bacterial decomposition prevails where organic matter pH is relatively high. In this respect, exchangeable calcium is a major mediator of soil acidity. Calcium derives from calcium carbonate in limestone, but is also an important constituent in volcanic rocks such as the neutral-reacting andesite. Though highly leachable, calcium is readily taken up by roots and may be sequestered in wood and other plant parts. Substrates relatively rich in calcium generally yield udult clay loam soils of relatively high pH, with high litter decomposition rates and fertility. Fungal decomposition proceeds at a slower rate than bacterial. Only fungal hyphae can decompose cellulose cell walls, and they can enter fallen leaves through their stomata. Fungal decomposition generates organic acids, increasing soil acidity. Termites, owing to their ability to decompose plant cell walls through the action of their fungal associates, play a dominant and ubiquitous role in lowland soils.[95]

Many species of Oomycete (mildew) fungi form symbiotic relationships with plants, penetrating and forming arbuscules within living root cells (Plate 1.11d). These symbiotic associations of fungi and roots are called vesicular arbuscular mycorrhizae. The fungus absorbs sugars from the plant host, and the plant absorbs water, phosphorus, and probably some other nutrient ions in solution from the fungus.[96] Many Basidiomycete (mushroom) fungi form ectotrophic mycorrhizal associations with tree roots, their mycelia forming

Plate 1.10 Zonal clay soils of the perhumid tropics, supporting MDF. **a,** On a hill slope over Lower Miocene Setap turbidite shale, Kuala Belalong, Brunei; grey-brown litter but no raw humus, little colour change and deep rooting in a friable soil of moderate fertility; decomposing shale from 80 cm. Jason Gathorne-Hardy indicates the deepest visible root, at 1.6 m; **b,** Similar clay loam hill slope soil albeit of lower nutrient status, over Precambrian metamorphic khondalite: Kanneliya forest, Udugama, south-west Sri Lanka; **c,** Pale yellow silty short-lattice clay soil, anoxic and with partial sesquioxide reduction (white mottling) at depth; sparse roots above, Lambir National Park, Sarawak. (**a, b** P.A.; **c** I.H. Baillie)

Plate 1.11 Leached humult yellow-red sandy soils of the perhumid tropics, supporting MDF. **a,** Soil profile over upper Miocene Belait sandstone, Lambir National Park, north-east Sarawak. Deep fine-root matted raw humus; yellow-red mineral sand horizon darkening with depth, with plentiful coarse roots; **b,** Surface litter, raw humus and root mat on soil over Belait sandstone, Lambir; **c,** Similar soil profile over siliceous Precambrian metamorphic khondalite; Sinharaja, south-west Sri Lanka; **d,** Mycorrhizal roots in the surface humic mat of shallow leached sandy clay soil on a khondalite ridge within the CTFS plot, Sinharaja UNESCO World Heritage Forest, south-west Sri Lanka; **e,** Leached humic yellow sandy soil over granite, coastal hills, Kuantan, Peninsular Malaysia. Note dense surface root mat and slender descending roots. (Digger marks on left; black and white image); **f,** Incipient podsolisation in yellow sandy soil; in the shortly seasonal coastal Triassic sandstone Cardamom Hills, Cambodia. (**a, b** C. Ziegler; **c** I.H. Baillie; **d** H.H.; **e, f** P.A.)

sheaths around swollen young roots, and their hyphae penetrating root tissues but not host cell walls. Ectotrophic mycorrhizae are ubiquitous in temperate trees but are confined to a few families in the tropics (Chapter 5).

Soils in which rates of litter decomposition are sufficiently slower than rates of litterfall for partially decomposed organic matter to accumulate, and in which particulate organic matter therefore visibly penetrates the upper mineral soil horizon, are termed *humult* (Plate 1.11). Litterfall rates seem not to vary consistently with soil acidity, but rates of decomposition decline with increasing acidity. Comminuted litter in process of decomposition accumulates between the intact litter and mineral soil where surface soils are acid, with a pH lower than c.4.2. This partially decomposed acid litter is known as raw humus or *mor*.[97] Lowland tropical soils bearing such raw humus, usually dense with fine roots of trees, are confined to climates with less than four dry months. They are found on freely draining, generally sandy mineral soils low in nutrients, especially those on coastal hills, and on high breezy ridges below the cloud base: these are prone to periodic drought which itself slows decomposition.[98] The perhumid parts of Borneo have greater areas of raw humus-bearing soils than anywhere else in the tropical world. Their moisture regime being influenced by slope, it is on convex slopes and ridges, where water deficits are most frequent, that litter decomposition is thereby most retarded; acidity increases in the accumulated surface organic layer, and carbon increases in it relative to nitrogen.[99] The mineral profile may be sandy or sandy-clay loam, yellow to coppery owing to thin coatings of iron oxide on the sand grains. These coatings can store phosphorus, thereby impeding leaching (see section 4.2.3 below). Raw humus on such soils does not exceed 5 cm depth. Alternatively, the mineral soil may be pure white silica sand, devoid of sites that can arrest leaching of nutrients. Here, rainwater, dust and leaf litter entering from trees on more fertile soils become the sole nutrient sources, and soils are consequently termed oligotrophic; yet nutrients may nevertheless accumulate in the raw humus, which can reach depths exceeding 30 cm. Soluble organic compounds — phenols and humic acids — are slowly leached from the raw humus to the base of the white sand, where they may precipitate to form a grey-black humic pan that eventually becomes impervious to water and roots (Plate 1.12a). These soils are known as tropical humic podsols (Plate 1.12).

In the wet tropics, organic matter and soil surface horizons vary in pH between c.3.5 and 7; most fall between 4.5 and 6 in udult soils, so they are somewhat acidic. Tropical forest soils in which surface pH falls below c.4.2 experience low rates of decomposition relative to litterfall because such conditions favour fungal activity over bacterial. This is exacerbated by a floristic change to tree species bearing leaves rich in lignin and with thick epidermis. These are slower to decompose, lower in nutrients, and richer in the phenolic compounds which mediate leaching of the phosphorus occluded on sesquioxides in the mineral soil.[100] The tree flora itself thereby accelerates trends within the soil towards increased acidity, leaching of nutrients, and slower rates of decomposition — trends that increase on the most freely draining soils. Some tree species thus contribute to the establishment and maintenance of conditions that are potentially less favourable to themselves[101] (Chapters 3, 7).

Two other substrates of the perhumid tropics are rich in bases yet, counter-intuitively, support soils bearing raw humus:

1. Limestone is generally free of soil on steep surfaces. In the seasonal tropics, limestone with its insoluble impurities weathers to form more or less brilliant orange-red clay-rich soils.[102] In the perhumid Sunda lands, however (particularly inner Borneo), on the high karst of New Guinea, and in wet tropical and warm temperate seasonal south-east China and tropical north-east Vietnam, the drought-prone summits support deep blankets of raw humus wherever they escape fire.

2. Ultramafic substrates of perhumid climates may yield well-structured base-rich red-brown clay soils. However, soil surfaces rich in raw humus are more common, perhaps because many of these soils and substrates, too, are coarsely granular and freely draining. Lower slopes may carry yellow soils with shallow raw humus, resembling leached humic yellow-red loam soils over granite or sedimentary rocks in profile. Elsewhere on ultramafic substrates, raw humus tends to be deeper, as in the tropical podsols (see section 4.2.3.2 below). In these cases, the mineral soil is greyish to dark grey-brown, and no humic pan forms.

The depth of raw humus reflects the ratio of the rate of litterfall to that of raw humus decomposition. In 2003 Baillie[103] observed at Lambir CTFS site (on Miocene sandstone in north-east Sarawak) (see map in Prologue) that the base of soil pits dug in 1971 had accumulated raw humus to 2 cm depth, implying an accumulation rate of c.6 mm per decade. These soils bear depths of raw humus up to c.6 cm, albeit more usually 2–3 cm, implying a minimum turnover rate of 30–90 years because the amount decomposing increases as the raw humus thickens. A lowland podsol may bear up to 30 cm raw humus, with a notional minimum turnover rate therefore of about 450 years. This time span is crucial to the processes leading to nutrient release. It may seem long, but it is short compared with the time required for a mineral soil profile to reach maturity. Nevertheless, geomorphological and edaphic evolution can sometimes interact over geological time.[104]

Plate 1.12 Lowland tropical podsols of the humid tropics, supporting *kerangas* forest.
a, Podsolisation in progress: An indurated humic B horizon at c.20 cm, now impenetrable to roots, betrays dead roots beneath it from a time before it formed. *Kerangas*-MDF ecotone, west Sarawak;
b, Giant humic podsol on a Pleistocene raised beach, Berakas Forest Reserve, Brunei, 1957. I point to the indurated humic B horizon, at 3.5 m depth over a gravel bed. Note the surface root mat, and paucity of roots in the white silica sand mineral horizon (black and white image);
c, Leaf litter with mycorrhizal roots penetrating raw humus over a humic podsol, Bako National Park, Sarawak;
d, Clear blackwater stream emanating from humic podsols, Bako National Park, Sarawak. The palm is *Eugeissona insignis*. (**a** I. Scott; **b** B.E. Smythies; **c** P.A.; **d** Mary Ashton)

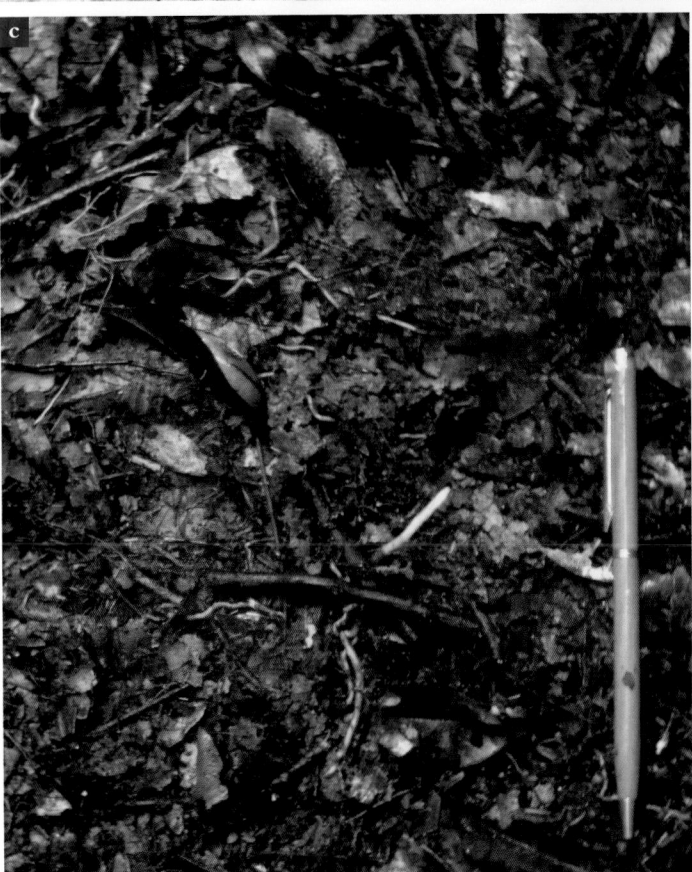

4.2.2.2. *Peat swamps.* Well-drained plains and gentle colluvial lower slopes bear soils whose root-bearing horizons are similar to (though generally deeper than) those on the steeper slopes above. Seasonal or periodic flooding from nutrient-bearing eutrophic river water restores nutrient loss and reduces acidity resulting from leaching, but anoxic subsoil conditions during waterlogging confines rooting near the surface. Beyond the reach of floods, on the extensive plains of perhumid regions, anaerobic conditions lead to increasing acidity and slowing decomposition rates. Litter accumulates and tree roots concentrate at the surface owing to anoxic conditions below. A horizon of raw humus, matted by fine feeder roots, develops above frequent flood level. Peat formation is thus initiated. Rainwater is left as the sole source of additional nutrients, and the soils become oligotrophic.

In the lowland tropics, peats only form in the absence of a dry season, since this promotes oxygenation of the surface and rapid litter decomposition. The coasts of Borneo and Sumatra, together with more limited areas in Peninsular Malaysia and south-west New Guinea, bear almost 500,000 km² of lowland tropical peat swamp (Chapter 3). (Outside Asia, such tropical peat swamps are known only on a small scale.) Although lowland tropical peat swamps resemble temperate peat swamps in their mode of formation, structure and moisture requirements, they differ greatly in their vegetation: with limited exceptions, they are densely clothed in closed evergreen broadleaved forest.

Peat swamps currently cover alluvial plains laid down in shallow coastal seas following the 2 m fall in eustatic sea level accompanying slight global cooling 4,100–3,700 BP.[105] At Marudi (north-east Sarawak), the peat has reached a depth of 15 m, implying an accumulation rate, if continuous, of up to 4 mm y^{-1}.[106] Peat forms on these coastal plains following the embayment of shallow seas by sand spits, then colonisation by mangrove, accelerating sedimentation. In due course, the innermost mangrove builds up sediments out of reach of all but the highest tides. These sediments are composed of clays and silts. Settling in still water, they form an impervious layer, restricting water flow and rapidly becoming anoxic. This leads to reduction of sesquioxides and to acidification, slowing the rate of litter breakdown. As the accumulating peat increases in area, the rate of lateral transfer of precipitation increases and peat domes up with the anoxic muck beneath towards the swamp centre. Conditions increasingly favour tree species whose litter is leathery and rich in phenolic compounds, further reducing decomposition rates and increasing surface leaching.

Peat can only accumulate where the soil surface is above the flood limits of the rivers. These bring fine sediments from the hills which then settle in their floodplains. Although the floodplains may be broader where the rivers emerge from the hills, farther out on the plain the peat gradually invades, so that the floodplains of the slow-moving, snaking rivers seldom exceed 100 m in width. The raw humus of the peat thus formed, varying little in pH between 3.7 near the fringe to 3.2 at the centre, accumulates on a dense root mat poised above the average water table height but intermittently flooded. Decomposed raw humus at its base falls, often through gaps in the soil, into an anoxic peaty soup, which includes living and partially rotted dead roots and other large litter. The inner peat swamps are fully oligotrophic, receiving nutrients solely from rain and dust to supplement whatever remains cycled from the alluvial substrate, with which roots would have lost contact more than 1,000 years ago. The water table beneath the peat domes is more variable than at the periphery. During droughts, the water table there may fall by more than a metre. The soils of peat swamps, surprisingly, therefore vary more in water stress than they do in nutrient availability. Upland peats are particularly prone in this respect.

Peat swamps up to 25 m above current coastal plains are widespread in southern Central Borneo. These originated during the last period of low sea levels (since 35,000 BP), on raised plateaux derived from former floodplains. Pollen floras indicating the presence of such upland *kerapa* peats have been recorded from the late Oligocene in the West Java Sea (Chapters 3, 6).

4.2.2.3. *Humus-rich soils of mountain and certain lowland forests.* The ratio of the rate of litterfall to that of its decomposition is also influenced by temperature, and consequently humus accumulation on average increases with altitude, because decomposition declines with temperature more than litterfall rates. But there remains much variation according to substrate and topography (Plate 1.13). The ratio has been observed to come into balance at c.1,000 m on clay loam soils over basic volcanic rocks in the Javanese mountains.[107] On such relatively base-rich clay loams, exchangeable nutrient concentrations increase as organic matter increases near the surface in the mineral soil.[108] Raw humus does not appear, nor is there an increase in leaf litter. Instead the mineral soil, observable immediately beneath the litter and rich in clay, aggregates into crumb-like particles ('friable') and becomes highly enriched with humus. Humus discoloration continues, often to 30 cm depth, and is evenly dispersed across the mineral profile (Plate 1.14a,b). Increased humus penetration was observed by Burnham above 740 m in the Cameron Highlands (Peninsular Malaysia).[109] Termite populations decline in diversity and number to levels comparable to those in warm temperate climates at this altitude. In these respects these lower montane soils are unlike lowland Asian tropical soils where humus, if present, tends to be concentrated in the mineral horizons along the channels left by dead roots or termites.

This is the soil which characterises the tea-growing country of lower montane and warm temperate climates in

Plate 1.13 Humult soils of the upper dipterocarp forest. **a,** Ancient deep soil profile over granite, 900 m, Temenggor forest, Upper Perak. I point to the substrate surface; **b,** Humult yellow-red sandy clay loam, Temenggor forest, Upper Perak, Peninsular Malaysia, with the deep root of *Anisoptera costata*. Stuart Davies as scale; **c,** Humult yellow-red sandy loam under *Dipterocarpus costatus* stand; upper dipterocarp forests, 700 m, Doi Chiang Dao, north Thailand. (**a, b** R. Harrison; **c** H.H.)

East Asia, with their evergreen oak-laurel forests (Chapter 4): well structured loams, with humus penetration deep into the mineral soil, and high cation exchange capacity indicating high fertility[110] (Plate 1.14a,b).

Humus particles can only be drawn down into the mineral soil by organisms. Collins *et al.*[111] observed on Gunung Mulu (Sarawak) that termite diversity, which is among the highest recorded in the lowlands, declines at c.1,000 m to levels comparable to warm temperate regions. They also recorded a marked increase in the biomass of earthworms, and beetle and Dipteran larvae, in the soil starting at c.800 m and peaking at 1,130 m. Since the late Professor T. Abe drew my attention to it, I have observed that lumbricine earthworm casts are omnipresent at 800 m in the lower montane *Shorea gardneri* forests of Sri Lanka, in clay-rich swales in upper dipterocarp forest at 654 m in northern Perak in Peninsular Malaysia, and

in Bhutan. On Kinabalu, lumbricine earthworms are abundant in the market gardens of Kampong Kiau (1,500 m),[112] and ascend abundantly to the last soil-filled fissures below the summit. Above c.1,800 m a giant lumbricine worm, up to 25 cm in length, also becomes locally abundant. In lowland semi-evergreen and tall deciduous forests of southern India and northern Indo-Burma, as at the CTFS site in Huai Kha Khaeng UNESCO World Heritage Forest and Wildlife Sanctuary (north-west Thailand, 15° N) (Chapter 3),[113] lumbricine worms are also often abundant as low as 600 m. Here, fire generates charcoal and increases large litterfall, that is branches and trunks, also soluble nutrients, near the soil surface.[114] Retreating deep into the soil during the dry season there, the worms reappear in their thousands after the first rains. They mate on the surface before resuming their ingestion of fine litter particles, which they transport down into the mineral profile.

Plate 1.14 Montane soils. **a,** Lower montane forest soil, humic clay loam lacking surface raw humus, abundant roots in the mineral horizon; Tia Shola, Nilgiri mountains, Western Ghats, India, 2,000 m; **b,** Lower montane humus–rich clay loam soil surface with abundant roots, Tia Shola, Nilgiri mountains, Western Ghats, India; **c,** Upper montane forest soil: shallow acid humic horizon over humus-free mineral soil, Horton Plains, Sri Lanka, 2,100 m. (**a, b** D. Mohan Das; **c** H.H.)

Although they draw humus deep into the soil, it is rapidly decomposed at lowland temperatures when soils are moist. In tropical grasslands, where abundant earthworms have been observed in Africa,[115] the mineral soil is itself dense with the short-lived fine roots of grasses and bamboo, whose decomposition may also in part explain the presence of evenly distributed humus down the soil profile. Further, these deciduous and semi-evergreen forests support large grazing and browsing mammals. Their dung likely improves the quality of substrate for earthworms, as has been shown of cows in temperate pastures.[116] Charles Darwin estimated that earthworms turned over 25 t ha⁻¹ of soil in his English garden.[117] In a northern Thai deciduous forest, earthworms were estimated to be producing 133–225 t ha⁻¹ of humus annually, equivalent to a layer of soil 9–14 mm deep![118] Soils approaching these humus characteristics are observed also in moist mesic sites on base-rich substrates

in the perhumid lowlands, and in moist swales at higher latitudes.[119] Earthworms activity, indicated by abundant casts, occurs in some lowland clay loams as at Danum (Sabah) and in clay loams over basaltic substrate in south-eastern Palawan.[120] But here, recorded organic matter levels are unexceptional.

These tropical mid-mountain soils therefore come to resemble temperate *mull* soils, termed brown earths, in which lumbricine earthworms are responsible for much of litter decomposition, and bury to depth humus thereby generated as fine organic matter. This causes the otherwise relatively impervious clay soil to become crumbly and well aerated, permitting deep rooting. The clay molecules maintain high pH, and are rich in negative ion sites which attract and hold nutrient cations thereby imbuing fertility (see section 4.2.3 below). Whereas the *climatic* margin of the tropics is defined by incidence of seasonal frost, the *edaphic* margin could be defined by the simultaneous increase of earthworms and decline of termites in zonal soils at the upper limits of the lowland forest.

Leached raw humus-bearing yellow soils and humic podsols extend on suitable substrates above 1,000 m on the Bornean mountains.[121] In forests experiencing frequent to pervasive fog, surface raw humus becomes ubiquitous on all substrates, its lower limit coinciding with that of frequent pre-rainfall fog. These soils resemble incipient temperate podsols: the mineral soil is greyish, subject to periodic waterlogging, and sometimes showing iron oxide-bearing hardpans at the base of the mineral soil horizons (Plate 1.14c). Lumbricine earthworms have not been seen in raw humus at any altitude, though earthworms up to 50 cm long were observed by Robb Anderson in lowland peat.[122]

Initially confined to ridges, acid organic soils may become widespread above 2,000 m, although their development is frequently truncated by slippage on steep slopes.[123] On base-rich valley soils, and on mountains where cloud cover is intermittent during the dry season, the mid-mountain humus-rich mull soils extend to high altitude. They become darker and shallower, enriched by humus and densely inhabited by earthworms.[124] They are said to be the prevailing soil in the upper montane valleys of New Guinea, where raw humus is apparently confined to karst summits.[125] They grade into organic muck soils on moist sites. In the subalpine zone these humic muck soils, rich in worms, are ubiquitous wherever soil has built up between rocks and on gentle surfaces, mixed with coarse sand and decomposing substrate. Sometimes, I have observed that a horizon of raw humus covers worm-dense organic muck beneath.

4.2.3. *Soil fertility*
Plants absorb a variety of essential nutrient ions in solution through their roots, including trace amounts of several elements. All nitrate anions are ultimately derived from atmospheric nitrogen, while sodium is received in precipitation from the sea in coastal forests. Nutrients otherwise arise from rock weathering. Nutrients absorbed by roots are at least partially derived from recycled organic material. Nitrate ions diffuse through soil most readily; sodium and calcium, then potassium and magnesium are readily soluble and diffusible, sulphur and phosphorus less so. Phosphorus, uniquely among the leading nutrients, is generally bound ('occluded') as phosphate anions on positive exchange sites, either on iron or on aluminum oxides, or as organic phosphates in humus. It is slow to be released in either case, but can be absorbed into roots through their mycorrhizal fungal mycelia.[126] Potassium is similarly sequestered in mica crystals. Calcium may also be immobilised as pedogenic carbonates. These form in dry climates but are rare under evergreen rainforest. However, they can be formed by biogenic accumulations as 'calcium-

hoarding' in wood or by termites. Whereas nutrients absorbed onto active surfaces are exchangeable in the short to medium term, those sequestered into the structures of mineral crystals require weathering to come back into circulation. They are released slowly, but nevertheless on time scales within the life of an old-growth tree.

The constantly high temperatures and mostly high humidity endured by tropical soils, and the consequent high rates of chemical reaction, lead to rapid leaching of nutrients unless they are either captured by roots, absorbed onto active surfaces on clay mineral and humus, or sequestered into accreting mineral grains (see above). These conditions also lead to pronounced differentiation in soil fertility among soils that differ in their capacity to hold nutrients, or in the mineralogy and erodibility of their underlying rocks.

Soils in many parts of the lowland humid tropics therefore display low concentrations of nutrients in solution; this has gained them the reputation of being among the least fertile in the world. Forest productivity may not be high in comparison to an annual herbaceous or pioneer woody succession, but slow nutrient release combined with rarity of killing droughts can achieve forest stands of great stature and biomass, if not particularly high growth rates (Chapter 2). The turnover rate of forest litter and its fate in the soil are crucial subjects generally ignored by commercial crop agronomists, because commercial agriculture produces so little litter – though it is critical for swidden farmers (Chapter 8). In tropical forests, nutrients are held overwhelmingly in the living biomass and, where it is abundant, in the dead litter and soil organic matter; but importantly, some nutrients are 'occluded' (see above).

4.2.3.1. *The problem of nutrient assay.* The availability of nutrient ions is difficult to infer by analysis of soil samples, which represent 'snapshots' and cannot reveal flow rates. Surface horizons of soils with deep accumulations of raw humus may manifest higher *volumes* of *total* nutrients than those with less raw humus or none at all. These represent recycled nutrients from organic matter. At the same time, spot estimates of *concentrations* of *available* nutrients in solution may be low and extremely variable spatially and seasonally, thanks to intense leaching or reabsorption by roots. Also, organic matter nutrient concentrations are partly determined by the species yielding them, so they vary in relation to the forest cover itself as well as in relation to the substrate and the mineral soil, from which most nutrients are ultimately derived. Under these conditions, their accurate assay requires multiple sampling. Broadly speaking, the depth of raw humus, where present, is negatively correlated with its rate of decomposition and nutrient release (see section 4.2.2.1 above). Thus it may provide a more realistic approximation of rates of nutrient release, in the least fertile soils, in solution than can spot analyses.

Tropical soils in general, though often low in available nutrient concentrations in solution, may nevertheless contain relatively high insoluble nutrient concentrations, either in organic matter or in the mineral soil itself. In the first case, they represent recycled ions within the forest cycle; in the second, they represent a gradually released subsidy from the rock substrate. The second therefore provides a more accurate spot estimate of the relative fertility of a soil for long-lived plants. Measures of *total mineral soil* nutrients have been found to correlate more closely to tropical forest composition and growth than measurements of any *available* nutrients (Chapter 3).[127] I therefore use, in this text on trees and forests, the term *fertility* in respect of the total concentration of nutrient ions in the *mineral* soil (see also section 4.2.3.2 below).

4.2.3.2. *Leaching.* Diffusible ions, more readily released from decomposing tissues in hot climates, are easily leached away during rainfall unless absorbed directly by roots. Soils in perhumid climates therefore tend to be low in nutrient ions in solution, that is 'available', especially nitrate and calcium. High concentrations of nitrogen or calcium in living leaf tissue relative to the soil indicate that trees either: (1) reabsorb these elements before leaf abscission, or (2) efficiently absorb them immediately following release in the course of organic matter decomposition. During prolonged dry seasons, absence of moisture in the soil horizons where feeder roots are concentrated prevents both nutrient uptake and leaching, so soils tend to retain exchangeable (that is, available) nutrients. Concentration of nutrients in seasonal forest topsoils is probably mostly biogenic, with some nutrients ascending by capillary action from subsoils during the first few weeks of the dry season, when leaching has ceased. Capillary action upward from the *mineral* soil is likely to be a minor contributor. For these reasons, and with litter decomposition sometimes mediated by abundant earthworms, soils of the seasonal tropics tend to be richer in available nutrients (when moist) than those of perhumid regions. In that case they are more fertile to annual crops – as well as to trees – than are soils of the lowland perhumid tropics (Fig. 1.10; Plate 1.15).

In deserts, nutrients may be concentrated in and near the topsoil, and may even crystallise on mineral surfaces. Common salt and certain other soluble crystals in desert soils are toxic, and their accumulation in seasonally irrigated fields may have been a primary cause of the decay of some irrigation-based civilisations.

Sesquioxides tend to concentrate in a single layer in soils of seasonally wet climates, where re-oxidation during dry seasons may lead to precipitation as an impervious rock-like horizon, often eventually shallow and metres thick.[128] The effective depth of soil for root penetration may thus be truncated, as in the humic 'B' horizon resulting

Plate 1.15 Forest soils of the seasonal tropics. **a,** Red clay loam soil over basalt trap, supporting short deciduous teak forest; S.N. Rai as scale. Dharwad, Deccan, Karnataka. Note deep coarse roots; **b,** Humus-rich sandy clay soil profile with abundant roots. Semi-evergreen dipterocarp forest, CTFS plot site, Huai Kha Khaeng Wildlife Sanctuary, west Thailand; **c,** Sandy clay soil; surface litter and humus-rich surface, sparse roots. Semi-evergreen dipterocarp forest, Huai Kha Khaeng; **d,** Humus-rich sandy clay loam with lumbricine worm activity in surface horizon. Semi-evergreen dipterocarp forest; Huai Kha Khaeng. (**a** P.A.; **b** C. Ziegler; **c, d** H.H.)

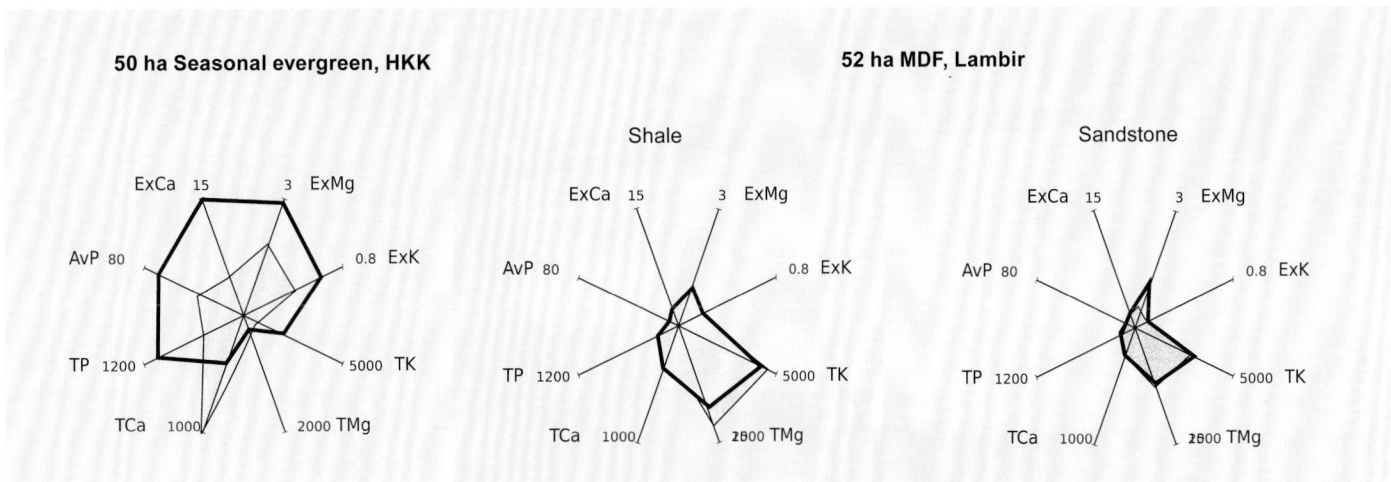

Fig. 1.10 Stoichometric roses of a sandy udult soil in semi-evergreen forest, with 5 month dry season (Huai Kha Khaeng, NW Thailand); and two in MDF and perhumid climate, on udult clay loam (left) and humult yellow sandy soil (Lambir, Sarawak, Malaysia). 90–100 cm. Exchangeable nutrient values in the top half of each rose, total values in the lower (linear scale). The seasonal climate soil notably differs, a, in its vastly greater exchangeable nutrient values near the surface; b, in higher calcium to magnesium concentrations, whereas it is the reverse in the perhumid climate soil. (I.H. Baillie)

Plate 1.16 Ferricrete and edaphically dry soils of the seasonal tropics. **a,** Deep profile in the coastal plains of the Carnatic, W. India, on a cut of the Concan railway. Deep indurated ferricrete over pale anoxic profile with reduced sesquioxides; **b,** Ferricrete, mined and cut in its living cheesy state, heaped to dry and harden into bricks. Malabar coast, Kerala, India; **c,** 12th Century Khmer bridge constructed of ferricrete. Roluos, Cambodia; **d,** Profile of dry sandy clay soil over turbidite; shallow humic coloration of mineral soil; supporting deciduous dipterocarp forest; Mae Sariang, north-west Thailand. (**a** D. Wilmore; **b, c** P.A.; **d** H.H.)

from podsolisation (see section 4.2.2.1 above). In more humid seasonal lowland climates where active sesquioxide accumulation is continuing, the iron-rich horizon assumes a cheesy texture and a pale grey-bluish hue, as in many floodplain soils of perhumid regions. Poor aeration at depth during the monsoon partially reduces the yellow-red iron oxides to bluish ferrous hydroxide. This is soluble and therefore mobile, accumulating in the plains. But exposure to the air through mining or erosion rapidly lends it a 'corned beef' appearance, then hardens it to terracotta in which form it can be used for house construction. Originally appropriately named *ferricrete*, from the classical Greek for a brick, that term has become so misused that *ferricrete* is recommended instead (Plate 1.16).

The existence of indurated ferricrete pans in currently perhumid climates has served as evidence of past seasonal rainfall climates. Unfortunately, they cannot be dated from palaeomagnetic evidence as their magnetic bearing is insufficiently stable. Further, it remains unclear whether conditions favourable for sufficient sesquioxide accumulation require open forest canopy or grassland savanna, which enhance seasonal drying and waterlogging. Such primary vegetations are rare in contemporary tropical Asia, while periodically waterlogged grasslands occur in perhumid regions (Chapters 3, 6).

Base-rich crystalline substrates such as basalt, and some clay-rich sediments, break down to form soils in which the clay molecules are predominantly large open lattices. These are responsible for freely draining soil caused by a well-developed crumb structure. These properties allow deep rooting, in contrast with short lattice clays and silty clays in which roots are concentrated in the upper part of the mineral soil (see Chapter 2). Some clay minerals can expand when wet and, though somewhat restricting permeability as in some zeolite soils of the Western Ghats, they also endow such soils with a high capacity to retain water. The clay molecules offer abundant negatively charged exchange sites which attract the abundant positively charged nutrient cations. Clay in mineral soil is therefore the source of its fertility.

In more siliceous substrates, including granites, most sedimentary schists and shale, shorter lattice clays predominate. Here the sesquioxides may keep the soil structure permeable, but where they are in low concentration and the climate continuously wet, nutrients are leached. These short lattice clays produce more compact soils with minute pore spaces and fewer exchange sites. Hydrated aluminium and iron sesquioxides, themselves hardly soluble, retain phosphorus in insoluble form. Some of the phosphorus is slowly released from the oxides which, when reduced to soluble forms, are leached along with the nutrients in the clay lattice. Mottling, with rust representing iron oxides along root channels in a pale yellow or greyish matrix, is frequent towards the base of such profiles where they become periodically waterlogged and anoxic on lower slopes in wet climates. Rooting is thereby restricted. On alluvium and other flat surfaces, such soils may be regularly or permanently waterlogged to the surface. More or less permanently inundated mineral profiles may appear grey-white, because sesquioxides are reduced to soluble colourless or bluish hydroxides.

Where leaching has been extreme and the substrate is completely lacking in nutrients, the sole remaining mineral nutrient source may be organic matter plus the minute input from rainwater. Mineral soils of pure sand, lacking clay or sesquioxide molecules, lack exchange sites other than in their organic matter. Low concentrations of nutrients result in replacement of nutrient cations on the positive charges within organic matter by ions that normally react with Al-sesquioxides, and with Al^{3+} which, where present, is usually the dominant exchangeable cation in acid rainforest soils. This leads to increased production of those soluble organic acids and phenolic compounds that are responsible for leaching.[129] This also leads to the increased relative role of fungi in humification.

In these podsols, nutrients are stored solely in the surface organic horizons, where the mat of absorbing roots is confined. In peats, nutrients are only available to roots near the surface because the waterlogged conditions below are anoxic. Peat may even accumulate on basalt plateaux at altitude, as at Ulu Arip (Tinjar, upper Rajang River, inner Sarawak, 900 m), while wet grey-brown podsolic soils lacking a humic pan occur high on the Javan volcanoes.

4.3. *Soil distribution in tropical Asia*

The regional distribution of soils is mediated by the distribution of climate and substrate, already summarised. Geographical continuity of the predominant soil types is important for interpreting the biogeography of soil-specialist species.

4.3.1. *Zonal soils*

4.3.1.1. *South Asia.* Soils were little considered by Champion in his pioneer attempt to interrelate physical environments and forests throughout South Asia and Burma.[130] His personal knowledge was mostly of India. In South Asia, substrates forming clay-rich soils and siliceous substrates yielding sandy soils form major geographic blocks (Fig. 1.2). Base-rich fertile clay loam soils are confined in South Asia to 'islands' of differing area, the most extensive of which is the Deccan trap in the north-west peninsula. There, black 'cotton' or *regur* soils, rich in calcium and exceptionally fertile owing to their dominance of large lattice montmorillonitic clays, are widespread in the strongly seasonal climates of the northern plateau. Clay-rich soils may also occur, nevertheless, on alluvium throughout siliceous regions wherever slow-moving or stationary turbid floodwater accumulates.

Two regions of mainly siliceous soils occur, separated by the plains of the Ganges and Brahmaputra: (1) those of the siliceous rocks of the eastern and southern Indian peninsula, together with northeastern Sri Lanka, and (2) the *terai* and the adjacent Siwalik sedimentary front-range of the Himalaya.

Leached, clay-rich loams of intermediate fertility, given the perhumid climate, prevail in the wet south-west of Sri Lanka,[131] and similar soils extend up the seasonally dry western flanks of the Western Ghats. These leached clay loams of South Asia are almost isolated from those of the Eastern Himalaya and Indo-Burma by surface lithology; linked only by a narrow belt running north of the Western Ghats through Rajasthan to the clay loams of the upper alluvial Gangetic Plain. They are currently also distantly isolated by climate.

4.3.1.2. *Indo-Burma and Malesia.*

The geology, topography and soils of Indo-Burma differ in many respects from those of India, including in the scale of their diversity across the landscape (sections 1 and 3 above, Chapters 3, 6). As in perhumid Sri Lanka and Sundaland, soils in seasonal Indo-Burma can be divided into relatively nutrient-rich clay and relatively nutrient-poor sandy loams. But in its fire-prone lowland forested regions, both soil types are fertile in comparison with even the most fertile upland soils in the wet tropics (Fig. 1.10). The geological marine basins, which yield clay-rich soils, are now often folded, resulting in steep-sided hills with narrow ridges. The clay loams on their slopes are generally deep and friable, but along the ridges the soils are shallow and drought-prone. Such sharp topography is rare on the ancient igneous topography of the fire-prone forested regions of India.

In Indo-Burma and throughout perhumid Sundaland, leached udult clay and sandy clay loams present a pervasive and continuous core habitat[132] (Chapter 6). The clay soils richest in bases, especially calcium, are confined to volcanic regions. Apart from the active volcanic arcs of Sumatra and Java, these are also isolated as habitat islands. The Philippines, Wallacea and New Guinea are overwhelmingly clothed with clay-rich soils, and many areas are also base-rich and fertile by the standards of the perhumid tropics. The only extensive areas of humult sandy soils in this region are found in the habitat islands created by isolated sandstone sedimentary basins of southeastern Thailand, Cambodia and northwestern Borneo. These humult sandy habitats are, however, widespread as local ecological islands where sandstone strata and siliceous granitic substrates support ridges and plateaux set in regions of fine-grained sedimentary soils.

4.3.2. *Soils restricted to specific habitats*

There was greater land continuity across the epicontinental South China Sea during the Pleistocene (Fig. 6.12), but this Sunda region was nevertheless ecologically fragmented by the broad valley of the Proto-Mekong and its tributaries. The vast alluvial plains of this valley were dominated by clay-rich soils, which would nevertheless have accumulated peat, assuming a perhumid climate – except along the narrow floodplains that must have existed. Some peat swamps are currently found inland, as in south-central Borneo and around the Kapuas lakes in West Kalimantan, and at Tasek Bera on the Pahang-Negeri Sembilan border in Peninsular Malaysia, but most are coastal and confined to the Sunda continental shelf. They are now generally limited to major landmasses, but many would have coalesced during the ebb and flow of the Pleistocene periods of low sea level, resulting in continuity of forest and flora contrasting with those of podsols and yellow sandy humic soils. Always shifting with the sea levels, peats include the youngest of terrestrial habitats in the region, although the extent to which their changing continuity has been maintained during the Pleistocene cannot be known.

Raised sand beaches are common today in Borneo, and also occur on Peninsular Malaysia's east coast, the south-west Cambodian coast and in southwestern New Guinea. These relatively small areas support deep humic tropical lowland podsols. Soils of varying texture dominate the ultramafic exposures in northwest Borneo, Palawan and elsewhere in the Philippines, likewise forming ecological islands. Granitic rocks form a continuous phalanx through the Vietnamese mountains and up western Peninsular Malaysia, where it is only broken significantly once it reaches peninsular Thailand. This substrate yields an intermediate range of soils in the humid tropics. Clays and sandy clays predominate on slopes and lowlands here, with comparable nutrient levels to the sandy clay soils found over interbedded sandstones and shales in much of lowland Borneo and eastern Sumatra.[133] Granite ridges and coastal hills provide ribbons of more freely draining soils. On them, a thin layer of acid raw humus may continue inland intermittently over hundreds of kilometres. They therefore represent potential migration routes, set within the core habitat of clay and sandy clay loams.

Limestone karst is rare in India, but pervasive in those regions of Indo-Burma dominated by clay-yielding sediments where it supports archipelagos of distinct soils and forest habitat (see above).

Ferricrete-bearing soils are widespread in tropical Asia wherever substrates yield clays rich in iron and aluminium sesquioxides. The only currently active, 'cheesy' ferrous and aluminium hydroxide accumulating soils are in the western coastal plains of India and in Burma. These have climates with a 5–7 month annual dry season, naturally supporting deciduous or semi-evergreen forest.

5. Tropical Asia: lands of forests and deserts

Like the Neotropics but in stark contrast to Africa, tropical Asia was almost entirely covered by closed forest before our ancestors arrived (Chapters 2, 3, 7). This is thanks to the strong summer low pressure zone over the Tibetan highlands, drawing enough rain to support forest and woodland over most of tropical Asia. (Though this is the case today, it may not always have been so in the more distant past). These forests are evergreen wherever there is reliable soil water. Asia has not supported a rich browsing fauna comparable to that of Africa in Pleistocene times, even where there is rain for only half the year (Chapter 5). Perhaps in part as a consequence, closed canopy deciduous forests extend to regions with ten dry months and 800 mm annual rainfall (Chapter 2). Desert is confined to the Great Indian or Thar Desert – but this is mostly frost-prone and therefore beyond the tropical margin as defined here (see cover maps); and also of course to the Arabian Peninsula, isolated from the rest of tropical Asia and devoid of forest.

Extensive lowland grasslands are only found in the floodplains of the great rivers emerging from the Himalaya, such as at Kaziranga National Park and on a smaller scale in other floodplains. Unlike the other tropical continents, tropical Asia probably has no natural upland savanna – that is, undegraded indigenous woodland with a grassy field layer in which trees are scattered, or crowns are separate from one another as in an orchard – except in the Himalayan foothills surrounding the grassy floodplains below the Terai and Duars, and the *Eucalyptus* savannas of southern New Guinea and northern Australia (see Chapter 2). The Himalayan foothill savannas occur where the mountain rivers gush out and flood the upper plains during the south-west monsoon. They are maintained in part by the intense grazing of large mammals that retreat there during the later dry season, when browse is short and upland grass has died back, and in part also by the unfavourable conditions for tree establishment. It is conceivable that hominids, such as Java man (*Pithecanthropus erectus*) whose remains 1.3 Ma are associated with a virtually treeless savanna,[134] extended (by fire) a corridor of upland savanna woodland (probably associated with naturally grassy floodplains) down the Straits of Malacca. The plant geographical evidence, however, is negative (Chapter 6). There are also grasslands above the frost line on the mountains (Chapter 4). There is nothing,

however, to compare with the plains of the Serengeti or the Brazilian *cerrado*. This important absence in tropical Asia is discussed in Chapters 2 and 5.

5.1. *Ever-changing landscapes*

The European or Japanese visitor to most parts of tropical Asia will have a sense of familiarity with much of its landscape. It is at a familiar scale; friendly, rather than massive and imposing. In this it is unlike the vast uplands, plains and deserts of Africa, North America or Australia, the great rivers and forested wildernesses of the Amazon and Congo, or the apparently limitless boreal northern taiga. The one Asian superlative, the Himalaya, indeed exceeds the Andes or Rockies in height, but differs in orientation and creates the margin of the tropics.

Asian tropical landscapes, like the landscapes of Europe, Japan or New Zealand, reflect complex local geology and diverse soils, often sudden changes in rainfall seasonality over short distances, and the diverse impacts of the last ten million years of geological history. In the seasonal tropics they also reflect continuing occupation by our ancestors, who must have massively modified the natural forest around them by the use of fire over a span of nearly two million years; and later by great civilisations with sophisticated agricultural technologies, especially irrigation, in some regions operating over more than 6,000 years. These civilisations evolved a harmony with the natural world, which – though now embattled by 'extra-tropical' philosophies – yet survives (Chapters 3, 5, 6, 8).

5.2. *Dense with the successors of ancient civilisations*

Civilisations, in the sense of great urban religious, administrative and population centres, are found elsewhere in the tropics as well as Asia, but only in Asia do their heirs remain in power today (Chapter 8). Before the industrial revolution, civilisations depended on technologies designed to intensify agriculture. In the tropics this primarily involved control of water through irrigation. While canals and conduits were developed on large scales in both continents, the irrigation methods of Central American and Andean civilisations depended essentially on raising planting beds above swamp, while in Asia the more versatile approach of ponding with bunds continues to support rice cultures in which three annual crops can be grown continuously on the same piece of land for hundreds of years. As a result, the seasonal tropics of Asia ubiquitously supported civilisations of high sophistication in the plains, where today forest remains only on exceptionally infertile soils and in riparian sites prone to extreme flooding. The surrounding hills often remained forested – sources of medicinal plants, game, and water – and venerated. Indeed, they were often under the protection of a ruler invested with divine attributes. Today they continue to be the home of minority swidden cultures (Chapter 8).

In the perhumid tropics, unpredictability of flooding for planting and dry seasons for harvesting, and the generally infertile soils have restricted intensive domestic agriculture to the base-rich volcanic loams of the foothills and plains bordering the 'ring of fire', and to the loamy humic mulls of lower montane forest habitats. Swidden cultures had remained almost entirely confined to gentle slopes and to the proximity of watercourses, their population numbers limited by disease and warfare. Uniquely, therefore, extensive forests influenced only by low levels of hunting and the collection of plant products remained throughout Asia until European influence commenced. On the continental lands of moderately low fertility in Peninsular Malaysia, eastern Sumatra and Borneo, more than 90% of the land remained under forest, hardly altered by human activity and never cultivated until the late 19th Century.

Nevertheless, with the exception of the high steep mountains of New Guinea, the Asian tropics are uniquely accessible. No substantial area is inaccessible from the sea, and the seasonal continental tropics are abundantly furnished with navigable rivers coursing through wide plains. This has made the Asian tropics particularly open to world trade, and to commercial resource exploitation. Human impacts are the subject of Chapter 8.

5.3. *The landscapes of Asia and its forests constitute a palimpsest*

The heart of my message is that tropical Asia is a region of exceptional geological and climatic diversity. Throughout the region, this physical diversity has supported forests by nature, but the character and extent of these forests have been radically altered by human activity. These activities and their effects have varied dramatically, according to the potentials of each region and site, and over historical time. The forests that remain are of unequalled spatial diversity. This diversity can only be explained by their history, and the nature and history of their land and climate over geological time. Each of these factors is unique to this region in important respects. That is the subject of the essays to follow.

References

[1]Hutchison 1989. [2]Smith & Briden 1977; Audley-Charles & Hallam 1988; Gradstein *et al.* 2004. [3]Furukawa 1997; Metcalfe 1988, 1996; Metcalfe *et al.* 1999; Gradstein *et al.* 2004. [4]Tarling 1988. [5]Searle & Treloar 1993; Metcalfe *et al.* 1999. [6]Clennell 1996. [7]Cooray 1967; Krishnan 1968. [8]Vitanage 1972. [9]Searle & Treloar 1993. [10]Hutchison 1989. [11]Hall & Blundell 1996. [12]Gansser 1983; France-Lanord *et al.* 1993; Treloar & Searle 1993. [13]Hall & Holloway 1998. [14]Raven & Axelrod 1974. [15]Morley 1999. [16]Searle & Treloar 1993. [17]Bender 1983. [18]Colenette 1964; Clennell 1996. [19]Morley 1999. [20]Tongkub 1991; Hazebroek & Tan 1993; Clennell 1996. [21]Hall 1996; Pubellier *et al.* 1996. [22]Hamilton 1979; Bergman *et al.* 1996; Hall 1996; Hall & Blundell 1996. [23]Pubellier *et al.* 1999. [24]Krishnan 1968; Gansser 1983. [25]Hall & Holloway 1998. [26]Searle & Treloar 1993. [27]Takada 1991. [28]Takada 1991. [29]France-Lanord *et al.* 1993; Treloar & Searle 1993. [30]Molengraaff & Weber 1921. [31]Voris 2000. [32]Voris 2000. [33]Cooray 1967. [34]Whitmore 1984a; Walsh 1996a. [35]Gaussen 1959, 1978; Walter *et al.* 1975. [36]Gaussen 1955; Gaussen *et al.* 1967. [37]Cannon *et al.* 2009. [38]Dykes 2000. [39]Hanebuth *et al.* 2000. [40]Corlett 2009a. [41]Köppen & Geiger 1936. [42]Holdridge 1947; Holdridge *et al.* 1971. [43]Gaussen 1955. [44]Walter & Lieth 1960; Walter *et al.* 1975. [45]Whitmore 1984a; Walsh 1992, 1996a. [46]Walsh 1992. [47]Schmidt & Ferguson 1951; Whitmore 1984a. [48]Penman 1948; Anon. 1958a. [49]Li *et al.* 2010; Kume *et al.* 2011. [50]Gaussen *et al.* 1967; Gaussen 1955, 1978. [51]Walsh 1996b. [52]Liu *et al.* 1986. [53]Sun & Wang 2005. [54]Wang *et al.* 1999. [55] Ju & Slingo 1995; Clift & Plumb 2008. [56]Anon. c.1964. [57]Walsh 1996a, b; Buckley *et al.* 2010. [58]Anon. 1891; Overpeck *et al.* 1996. [59]Liu *et al.* 2008. [60]Liu *et al.* 2004. [61]Wangda & Ohsawa 2006a, b; R. Pradhan, pers. comm. Oct. 2007. [62]Nautiyal 1934; Seth & Dabral 1957. [63]Hope 1885; McA 1885, 1888; Anon (L.M.) 1888; Lowrie 1891. [64]Higham *et al.* 2009. [65]Goldammer 1990a, b; Toma 1999. [66]Ju & Slingo 1995; Clift & Plumb 2008. [67]Walsh 1996a. [68]Ropelewski & Halpert 1989. [69]Walsh 1996a, b; Walsh & Newbery 1999, 2000. [70]Leighton & Wirawan 1986. [71]Brünig 1969, 1971; Baillie 1972. [72]Dykes 2000. [73]Lowry *et al.* 1973; Kitayama 1991. [74]Morley 1999, 2012; Morley *et al.* 2013. [75]Morley 1999. [76]Morley 1999. [77]Najman *et al.* 1993. [78]Wang et al. 2003. [79]Kuhle 1998; Baillie *et al.* 2004. [80]Shimmield *et al.* 1990; Overpeck *et al.* 1996; Wang *et al.* 1999. [81]Morley & Flenley 1987. [82]Dam 2001. [83]Morley 1999; Bird *et al.* 2005. [84]Morley & Flenley 1987; Morley 1999. [85]CLIMAP 1976; Flenley 1979. [86]Flenley 1979. [87]Baillie 1996. [88]Mohr & van Baren 1959. [89]Baillie *et al.* 2004. [90]Dykes 1994a, b, 1996, 2002. [91]Haile 1968. [92]Baillie 1975. [93]Gobbett & Hutchison 1973; Hall 1996. [94]Dykes 1994a, b; Ross & Dykes 1996. [95]Abe & Higashi 1991; Ohgushi 1997. [96]Goltapeh *et al.* 2008. [97]Baillie *et al.* 2006a. [98]Tange *et al.* 1998. [99]Hill 1972. [100]Bloomfield 1954; Inderjit & Mallik 1997. [101]Burghouts *et al.* 1998. [102]Baillie *et al.* 2005. [103]Baillie *et al.* 2006a. [104]Chadwick *et al.* 1999. [105]Fujimoto 1997; Furukawa 1997. [106]Anderson 1964; Anderson & Muller 1975. [107]Dames 1955; Andriesse 1969. [108]Ediriweera *et al.* 2008. [109]Whitmore 1984a. [110]Burnham in Whitmore 1984a. [111]Collins 1980, 1983. [112]Dawat Kemando, pers. comm. Aug. 2008. [113]Champion & Seth 1968a. [114]Nye & Greenland 1960; Ramakrishnan 1992. [115]Nye 1955. [116]Barley 1964. [117]Darwin 1882. [118]Watanabe & Sawaeng 1984. [119]Askew 1964; Baillie *et al.* 2000, 2001. [120]Baillie *et al.* 2000, 2001. [121]Askew 1964. [122]J.A.R. Anderson, pers. comm. 1966. [123]Askew 1964. [124]Askew 1964. [125]Blaker 1980. [126]Baillie 1996. [127]Baillie *et al.* 2006a. [128]Buchanan Hamilton 1807; Mohr & van Baren 1959. [129]Inderjit & Mallik 1997. [130]Champion 1936. [131]Baillie *et al.* 2006b. [132]Baillie 1996; Adzmi *et al.* 2009. [133]Adzmi *et al.* 2009. [134]R.J. & S. Morley, pers. comm. April 2012.

The forest margin seen by a roadside forest gap, Danum, Sabah (P.A.)

Chapter 2

LOWLAND FOREST FORM AND FUNCTION: RECONCILING LIGHT AND DROUGHT

Part I. Lowland forests in an aseasonal wet climate

> The observer new to the scene would perhaps first be struck by the varied yet symmetrical trunks, which rise up with perfect straightness to a great height without branching, and which being placed at a considerable distance apart, give an impression similar to that produced by the columns of some enormous building. Overhead, at a height, perhaps, of a hundred and fifty feet, is an almost unbroken canopy of foliage formed by the meeting together of these great trees and their interlacing branches …The great trees we have been hitherto describing form, however, but a portion of the forest. Beneath their lofty canopy there often exists a second forest of moderate-sized trees, whose crowns, perhaps forty or fifty feet high, do not touch the lowermost branches of those above them…Yet beneath this second set of medium-sized forest trees there is often a third undergrowth of small trees, from six to ten feet high, of dwarf palms, of tree ferns, and of gigantic herbaceous forms. Yet lower, on the surface of the ground itself, we find much variety.

> A.R. Wallace, 1891, *Natural Selection and Tropical Nature*.

I remember, back in 1964, clambering up through the hill padi *of Rumah Jelian and through the little trial plantation of oil palm which, I was confidentially informed, was soon going to be harvested for its succulent cabbages…then through a wall of secondary growth into the grandest, darkest, yet most open and vault-like forest I had seen during five years in Borneo. The sweet, fermenting aroma coming up from the soil reminded me of autumnal temperate forests of elm and ash. Even the call of the argus pheasant echoed as if in a great building.*

This was the great dipterocarp forest dominated by *paji* (*Dryobalanops lanceolata*), a camphor tree, on the lower slopes of Bukit Mersing in Sarawak, an isolated Tertiary basalt ridge 1,000 m high whose chocolate-brown spongy clay soils allow deep rooting and probably never experience a drought. We set out 30 plots, on slopes and ridges up to the summit, and the Forest Department has endeavoured to protect these northern slopes so that the growth and death of the trees there can be measured over time.

Such cathedral-like forests have always been rare in Borneo (Fig. 2.1), though they are widespread on the volcanic soils of Sumatra, the Philippines in the lee of typhoons, and perhaps, centuries ago, in West Java. The flight up the coast from Kuching to Miri nowadays reveals coastal lowlands mostly planted to oil palm. Enough forests do remain (though logged over outside of parks) to reveal that the 'pile' of the landscape surface – that is the forest canopy structure – is quite diverse. It varies with the topography. This is particularly obvious along the defiles of the narrow steep-sided parallel ridges, which are such a feature of upland Borneo. Most of these hills are of up-tipped sedimentary rock, the harder sandstone exposed along the ridges above the more erodible shale and clay slopes (Chapter 1). The narrower the ridge, the smaller and denser is its pile of the forest carpet.

On the ground, one's first experience of the celebrated lowland tropical rainforests of Asia can be disappointing. Can these trees really be so tall? For their girth is quite modest. And there is such a chaotic appearance to the canopy: has there been some recent disturbance? Few of us have the opportunity to return and carefully measure change over years. What, therefore, can reliably be inferred about the history of disturbance, about rates of growth, species' characteristics or richness, from a single snapshot of forest structure? Importantly, can one predict the response of a forest to timber harvest, the rate of growth of the surviving stock and future composition, from the pre-logging structure and composition?

Forest structure is determined by several factors, which are largely independent of one another. Without explicit knowledge of these it is hazardous to make inferences. In particular, the structure we see has a history about which we can infer little in the absence of records from plots or tagged trees, except in strongly seasonal areas where annual growth rings facilitate age estimates. Forests that have not experienced large-scale destruction consist of patches of individuals of canopy species. Each such patch represents the outcome of a chance opening in the canopy long ago, which permitted a group of juveniles to survive and grow. Foresters call these patches of trees with a shared history *stands*. (Ecologists generally use this term in a more general way, as a defined area of vegetation of moderate size; but I adopt the more narrowly defined forester's

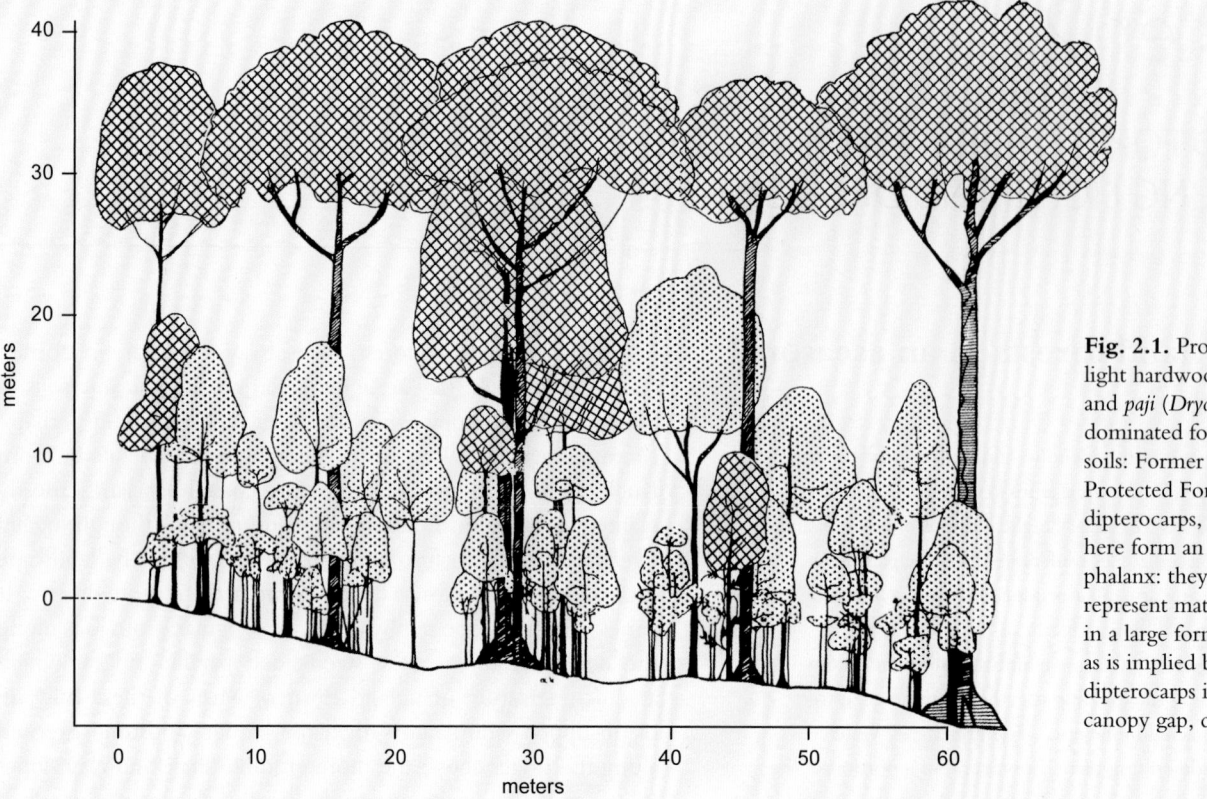

Fig. 2.1. Profile diagram of a light hardwood *meranti* (*Shorea*) and *paji* (*Dryobalanops lanceolata*)-dominated forest on clay loam soils: Former Bok-Tisam Protected Forest, Sarawak. The dipterocarps, cross-hatched, here form an almost continuous phalanx: they appear to represent maturing succession in a large former canopy gap, as is implied by the two young dipterocarps in the current canopy gap, centre.

use throughout these essays.) The stands in a forest are at various stages of recovery back into the canopy. Trees reach a maximum height and then persist, in some species for more than a century, while continuing to reproduce and to grow in girth. Therefore, most stands in old growth forest have attained full height and a semi-stable vertical structure. This stage of approaching equilibrium is called the *mature phase*. It generally occupies at least 80% of the area of a little-disturbed forest, and therefore characterises it. Mature phase structure varies with habitat conditions. It is associated with changes in species composition, abundances and richness. Structure therefore partly determines economic values and management options. Understanding the processes that engender forest structure and establish its potential for manipulation is therefore an important goal of forest science.

Forest ecologists have often concentrated their research – it might seem unreasonably – on *primary* forests, that is, forest in which all significant disturbance of canopy and regeneration can be traced to natural causes. These natural perturbations trend towards a distinct regularity on the time scale of the lifespan of a climax tree species. Their influence on forest structure and dynamics is profound. They can be studied by documenting a forest sample, waiting for the natural catastrophic event, then recording the consequences. The study of *secondary* forests, that is forests that are regenerating after human activity (ranging

from burning and cultivation to selective logging), can only be fully rewarding if a careful documentation of the forest *prior* to the disturbance has been made. Even in silvicultural research in forests subject to logging, this has too rarely been done.

1. Is the mature rainforest vertically stratified?

It is often claimed, as it was by Wallace, that the mixed dipterocarp forests of the perhumid Far East are vertically structured into three, sometimes more, fairly distinct tree strata, of the main canopy with emergent crowns forming a more or less separate stratum above, and the subcanopy trees below both[1] (Plate 2.1). These lowland rainforests include the tallest stands in the tropics. Products of a humid, windless climate, they set the standards against which I will measure other regional forest types. They lack shrubs, that is, woody plants that branch from the ground. The reproductively mature trees do appear to be vertically arrayed in either two or three strata, according to the forest. Even in mature stands, however, these strata are rarely distinct, in large part owing to the presence of juveniles growing among and between the strata that they will occupy at maturity.

The upper stratum of *emergent* canopy crowns is the most distinct. In perhumid tropical Asia, an emergent stratum of mature phase stands consists overwhelmingly of dipterocarps.

Plate 2.1 The mixed dipterocarp forest in profile. **a,** The forest margin seen by a roadside forest gap: an almost continuous emergent canopy on the skyline, but regrowth and the climber tangle in front obscure any semblance of stratification. Danum, Sabah, 2008 (P.A.); **b,** (see chapter frontispiece) MDF profile along a new road cut: the mature phase manifests distinct emergent and main canopy, and a subcanopy stratum with dense juveniles. The emergents are *Dipterocarpus crinitus*, pale crown foreground, to right of *Dryobalanops aromatica*. Labi road, Brunei, 1958. (P.A.)

These forests are termed *mixed dipterocarp* (MDF) on account of the many co-existing dipterocarp species; that differentiates them from those peat swamp and deciduous forests in which a single dipterocarp species is dominant. Dipterocarps share a common ancestry with plants such as cotton and the cacao (chocolate) tree, of the order Malvales. Several other Malvalian and also leguminous species can be found in the emergent canopy, but as scattered individuals or, notably in *Neesia* and *Coelostegia* (Bombacaeae-Durioneae), in small isolated groups. Malvalian genera with emergent representatives include most *Durio* (subgen. *Durio),* also *Pterygota,* most *Heritiera* and *Scaphium,* and a few *Sterculia* (Sterculiaceae). Scattered emergent leguminous trees also occur, including *Koompassia* (Plates 2.2a, 5.10a), the three species of which attain among the tallest stature in MDF;[2] also *Sindora* (Caesalpinoideae), *Falcataria moluccana* and several *Parkia* (Mimusoideae).

There is a further group of species which can grow as tall as these true emergents, but whose crown branches only rarely reach above the main canopy. This group includes *Triomma malaccensis* (Burseraceae), *Koordersiodendron pinnatum* and some *Mangifera* (Anacardiaceae), *Antiaris toxicaria* and some *Artocarpus,*

some *Ficus* (Plate 5.13a) and *Parartocarpus* (Moraceae), and *Azadirachta excelsa* and some *Aglaia* (Meliaceae). Other than the dipterocarps, these species occur scattered or in isolated emergent groups. They are subordinate to the emergent dipterocarp species in setting the overall emergent structure.

The emergent canopy conifers *Agathis* (Plate 3.5f), *Araucaria* and some Podocarpaceae are rare in MDF, but occasionally are abundant on freely draining or infertile sandy soils. They maintain their columnar monopodial crowns throughout maturity[3] (Box 4.2).

Finally, there is a small but prominent representation, in some MDF growing on fertile clay soils and in periodically flooded valley bottoms, of pioneer and early successional genera whose species may also become emergent. Besides *Albizia* and *Parkia* (Leguminosae-Mimusoideae) these include *Bombax anceps* (Bombacoideae-Malvaceae), *Pterocymbium* (Sterculioideae-Malvaceae), *Terminalia* (Combretaceae), *Alstonia* (Plate 2.3a) and *Dyera* (Plate 2.3b) (Apocynaceae), *Octomeles* (Plate 2.12a) and in the seasonally wet regions *Tetrameles* (Datiscaceae) (Plate 3.8e).

The emergent canopy, as described here, is unique to

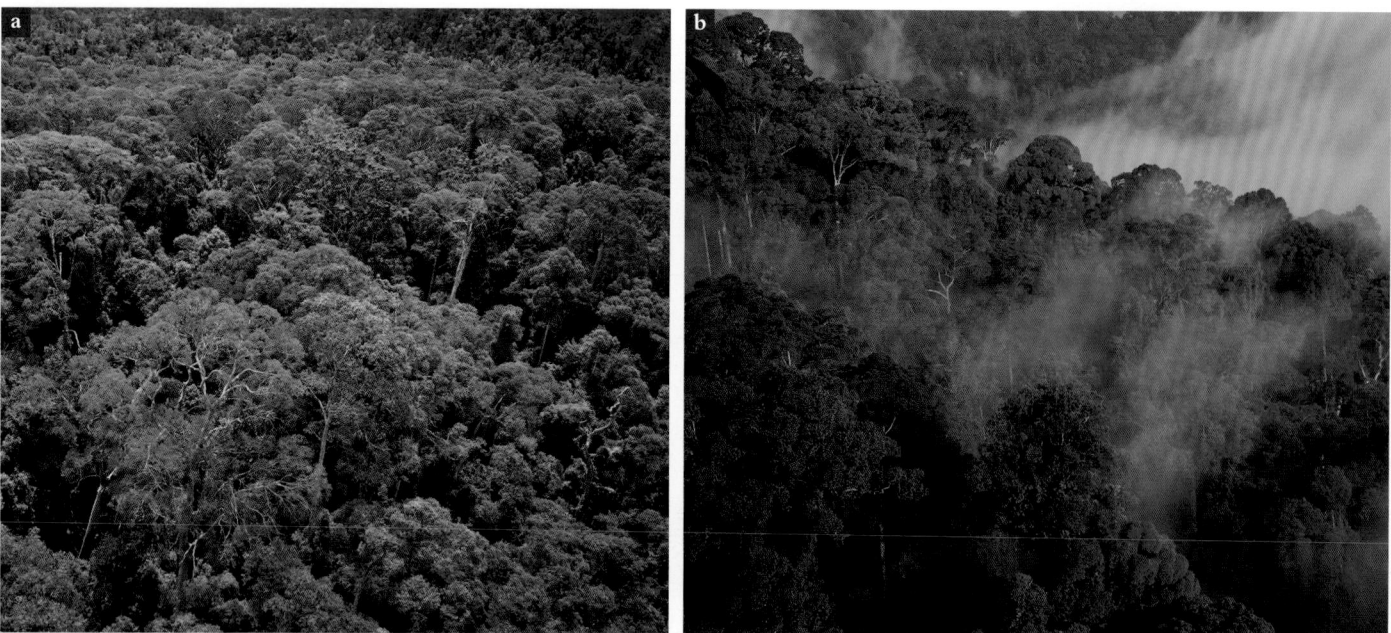

Plate 2.2 The uniformity of emergent canopy height. **a,** Remarkable canopy density, and uniformity of heights, achieved by the many species of mature emergent dipterocarps sharing friable clay loams; the copper crowns of the caesalpiniaceous *Koompassia excelsa,* which is also ectotrophic mycorrhizal, exceed the dipterocarps. East Sabah; **b,** MDF on yellow sandy soils shows similar uniformity of mature emergent height, although varying more with soil depth: Lambir National Park, Sarawak. (**a** H.H.; **b** C. Ziegler)

Plate 2.3 The layered crowns of pioneers that persist in succession to reach the emergent canopy. Such layering has the effect of densely shading without recourse to increasing the number of leaves (LAI). **a,** *Alstonia scholaris,* a pagoda tree. A pioneer, one of the most widespread of all tropical Asian trees, from Sri Lankan rainforests to Rajasthan's short deciduous forests, and from South China to New Guinea, Queensland and Melanesia; **b** *Dyera costulata.* (L.G. Saw)

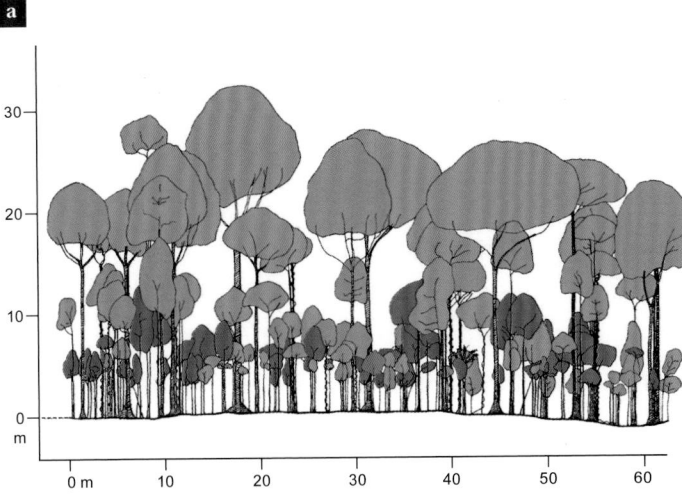

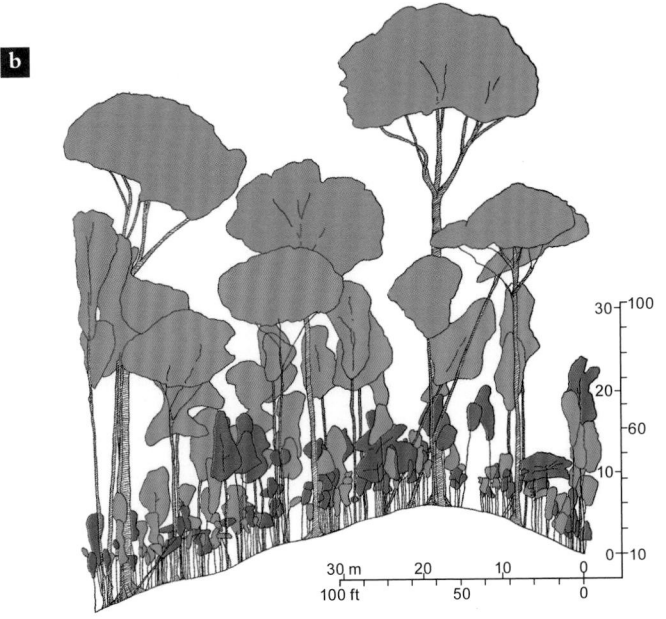

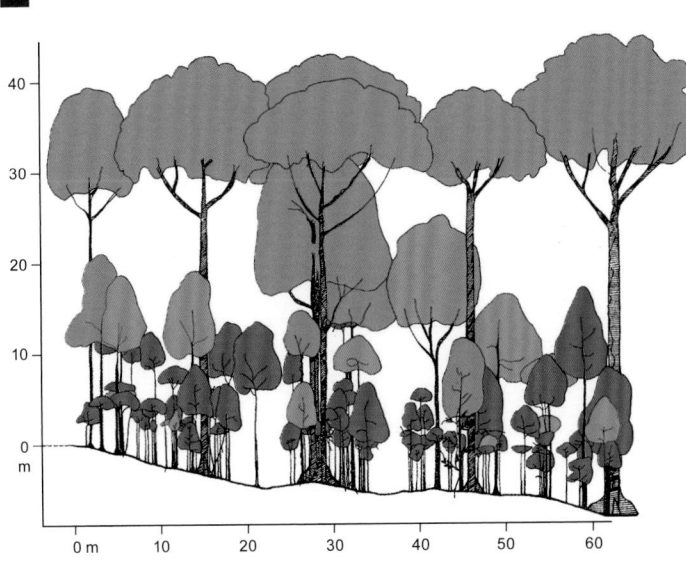

windless sites, be they in the perhumid Sunda lands or in moist sheltered valleys elsewhere. Wind and rainfall seasonality is associated with reduced overall canopy height and with contraction of the emergent canopy into the main canopy.

At the landscape scale, the emergents growing on each individual topographical element (plain, slope or ridge), and soil type trend toward a uniform height (Plate 2.2). Within homogeneous habitats, the height of the emergent canopy thus remains remarkably uniform; this phenomenon will be considered when structural variation within the landscape is discussed below. In contrast, main canopy crowns vary widely in height, and subcanopy individuals yet more so (Figs 2.1, 2.2; Plate 2.1b).

There are at least four possible explanations for these three apparent tree strata. Ecologists have long discussed whether this stratification does in fact exist, and if so, whether it is imposed by the physical environment on the trees within a forest stand irrespective of species, or whether the species themselves are genetically constrained to certain statures at maturity. Quantitative tests are difficult because the sample size required, for the emergent stratum comprised of a few large trees in particular, greatly transcends my profiles, and may stray onto different soils or topography carrying forest of different structure.

In the following sections, I consider four basic mechanisms that could generate or contribute to forest stratification: aggregation of maximum tree heights, branch architecture, crown architecture and presentation of flowers, and leaf layering.

1.1. *Stratification by maximum tree height*

It has sometimes been asserted that the perceived presence of tree stratification in tropical rainforest must be caused by an aggregation of the maximum height of tree species at maturity

Fig. 2.2. Profile diagrams, 64 x 8 m, within MDF on different soils, Sarawak. **a**, Bako National Park, shallow leached humic yellow sands; **b**, Andulau Forest Reserve, Brunei, deep humic yellow sandy loams; **c**, the former Bok-Tisam protected forest (now oil palm), udult yellow clay loams. Trees which flower in full sun as emergents, or in the main canopy, and those which flower beneath the canopy, are indicated in different colours (emergents – blue, flowering in the main canopy – green, flowering in the subcanopy – pink). Juveniles of all occur beneath the canopy. The Bako profile represents a stand recovering from major canopy mortality, with dipterocarps well represented in the subcanopy yet hardly emerging above it. The Bok-Tisam stand is in the mature phase, with an almost continuous emergent light hardwood dipterocarp canopy having likely grown as a single stand following a canopy blowdown, with sparse subcanopy and little regeneration. But the Andulau profile, with evidence of past small emergent canopy gaps now filled with main canopy species, shows evidence of intermediate canopy disturbance levels. This is discussed later in this chapter, and in Chapter 7. (D. Glick del. after P.S. Ashton 1964b, P.S. Ashton & Hall 1992, and unpubl.)

into vertical guilds, groups of species sharing a common ecological attribute. If such aggregation exists, however, the reasons for it are unclear. Woody plant species, like fish, lack genetic control of a specific height at maturity. Instead, their highest buds continue to expand but, eventually their shoot extension is halted by apparent damage. The cause has not been diagnosed in rain forest trees, but among emergents may be inability to lift water during periodic drought, because maximum heights vary with topography and soil. Theirs may be said to be an asymptotic height. As shall be seen, the maximum heights attained by co-existing emergent dipterocarps are remarkably similar within a specific habitat. Data for tree height distributions are only available for small samples because accurate height measurement is difficult and time-consuming in dense forest. These data, in which large and emergent individuals are always poorly represented, show no clear aggregation (Fig. 2.3). The vast majority of individuals in any stand or population are juveniles below their maximum potential height. This obscures

any stratification of their maxima that might be present. At the same time, different species exhibit different height growth rates. Even in the perhumid equatorial lowlands, the majority of individuals never attain their maximum height, the vast majority being first struck by some catastrophe (see section 2.2 below). Vertical stratification by maximum tree height alone is therefore improbable, the apparent uniformity of emergent canopy stature notwithstanding.

1.2. *Stratification by branch architecture*

Hallé and Oldeman,[4] in an elegant study of the patterns of branching and growth among young tropical trees, concluded that there are 24 basic architectural models that all tree species follow. These models are fundamentally defined by the life span of shoot meristems, their mode of branching and the position of their reproductive shoots.

Branching is in one of two modes: (1) There may be a

Fig. 2.3. Vertical stratification of trees >15 m tall by height, in 64 x 8 m transects in Brunei MDF: Andulau on humult yellow sandy soil; Kuala Belalong on yellow clay loams. Height in feet (0.305m) (From P.S. Ashton 1964b)

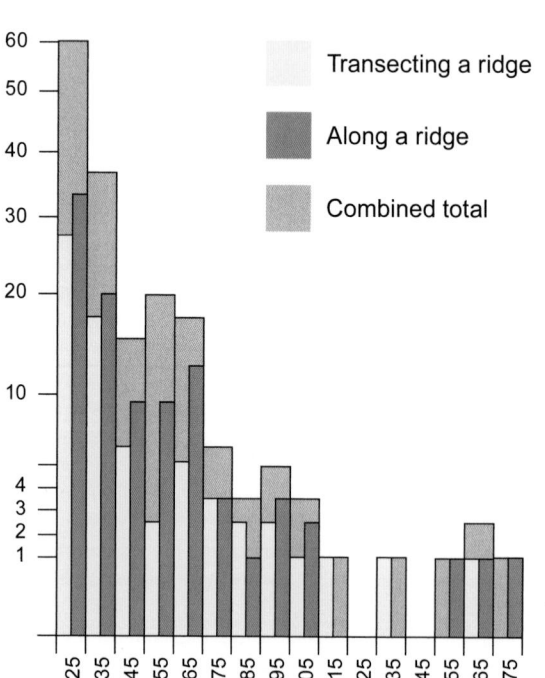

Height class histogram for two 200 x 25 ft Transects, Kuala Belalong: for trees exceeding 15 ft, 1,800 ft (c.550m) altitude.

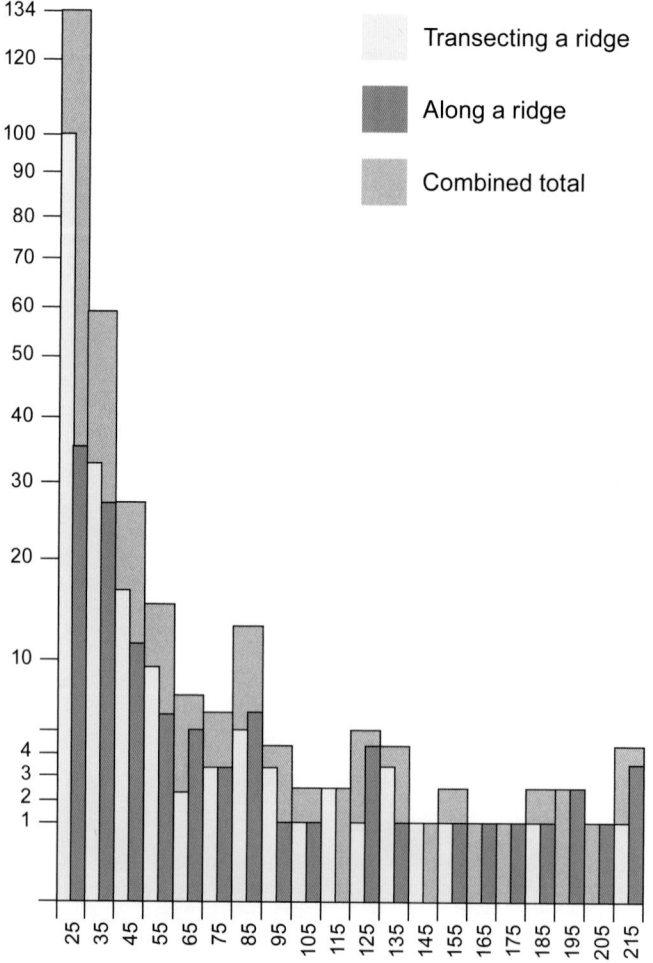

Height class histogram for two 200 x 25 ft Transects, Andulau F.R; for trees exceeding 15 ft tall.

vertical (orthotropic) leading shoot (leader) with spreading (plagiotropic) side shoots growing into the many branches. In this case, the leaves are radially arranged (spiral or in whorls) on the leader, but generally in two horizontal ranks (distichous) on the branches. (2) Branches and leader may be equivalent, orthotropic and more or less ascending, with the leader (if present) and the side branches sharing the same leaf arrangement. Growth may be continuous, with the leaves arranged at consistent more or less equal distances, or periodic with leaves aggregated towards the beginning or end of a period of shoot extension. If the reproductive shoot (the inflorescence) is terminal on the shoot, the twig is forced to branch if it is to shoot further. This sympodial branching pattern obviously has strong implications for the range of architectural forms possible for reproductively mature individuals. In a classic example, teak trees have sympodial branching, so foresters grow them close together in plantation to delay first flowering, thereby insuring a straight trunk of reasonable length. Several tree architectural 'models' are defined by how the tree responds to an inherited habit for the leader's terminal bud to abort at the end of a period of extension. Whereas many families subscribe to a single architectural model, several manifest great diversity.[5]

In Asian forests, the predominantly Malvalian species generally start with an orthotropic leader, the growth pattern of which is either intermittent ('Massart's Model') or continuous ('Rauh's'). Early on, they develop a diffuse oblong crown supported by a platform of plagiotropic branches. They grow to canopy height by often lofty extensions of a vertical leader followed by further extension of a dense layer of leaves on near-horizontal plagiotropic branches. Many Malvalian emergent species are remarkable among tropical trees in their capacity for ontogenetic architectural change. Once they break into the full sun, their branches become orthotropic (Plate 2.11), ascending with spirally arranged leaves equivalent to the leader, which eventually becomes indistinguishable from the others. The crown then becomes dome-shaped, the leaves smaller and bunched.[6] Inflorescences are in part terminal, forcing the twigs to become more branched.

The canopies of African and Neotropical forests, in contrast, are comprised substantially of leguminous trees. All of these conform to 'Troll's Model',[7] found among the short flat-topped acacias of the Serengeti Plain and also among temperate elms and beeches. Their habitats could hardly be more different, but a close examination of any of these trees reveals their architectural unity. In each of them, the leader starts vertically with spiral leaves, but gradually grows over to a horizontal position with leaves in two horizontal ranks. Then, the leader is renewed by the expansion of an axillary bud (between the base of a leaf and the twig) on the upper side of the previous shoot. This curious method of branching persists, like nearly all branching models, throughout life, although it

is obscured by bud abortion in exposed crowns. It can most easily be identified in young trees. It has proved immensely adaptable. In the rainforest understorey, the shoot may remain vertical through much of its growth and extension, so that the next 'relay' can grow vertically above it. Leguminous giants of Far Eastern forests such as *Sindora* and *Koompassia* develop among the most powerfully columnar boles in these forests!

Beneath the canopy we find a rich diversity of branching models. Among these species, optimal placement of flowers and fruits appears to take precedence over growth in height; light-capture in this challenging light climate is achieved by a variety of combined branching and juvenile leaf form adaptations. Among pioneers of mesic sites, leaves quickly form a dense umbrella above the trunk (Plate 2.7b, c). This too can be achieved by several models, but the most common are Troll's and others which form dense horizontal or drooping twigs, and those which produce orthotropic branches bearing dense rosettes of large leaves.

Temperate broadleaf trees overwhelmingly comprise only three models. With the exception of the tulip poplar, the only temperate trees which maintain a vertical leader and plagiotropic branches up to the canopy are needle-leaf conifers. All others either have orthotropic branches, or follow 'Troll's Model'.[8] The remaining models are confined to the warmer regions.

From these observations, we can infer two facts about tree architectural models. First — as is also true of several other important characteristics of tree species — models are phylogenetically conservative. They are generally (though not always) constant among related species, and frequently within genera and even families. Trees must in this case adapt within the possibilities of a set of genetic constraints imposed by evolutionary history. It is not that models are adaptively neutral, but rather that they are hardly mutable within the genome. Second, heritable architectural form cannot account for the vertical stratification of stands approaching maturity. It turns out that many basic models, such as Troll's, are sufficiently versatile to have allowed species sharing the same model (although generally coming from separate families) to evolve into different roles in mature forest structure. Also, juveniles of species of any model must pass through lower strata on their way up to their full height at maturity. Because there are always more juveniles than mature trees in any stable population, as already mentioned, juveniles will inevitably obscure any stratification by architecture which may exist among mature individuals. And finally, as trees pass through life, vicissitudes including herbivory, partial mortality in insufficient light, and drought-induced partial mortality (even in perhumid regions) are all followed by erratic recovery through reiteration of the model branching system. This, too, obscures the overall model of the individual.

1.3. Stratification through gross crown architecture and presentation of flowers

The climax species of equatorial lowland forests fall into three guilds distinguished by their crown and flowering characteristics at maturity (Fig. 2.2). These guilds form three height 'strata' at maturity, giving a characteristic form to mature stands, to the extent that they result in stratification of the forest; this is due to the characteristics of its individual species.

1.3.1. Emergent canopy flowering guild

The uppermost, emergent stratum is composed of a distinct group of species, the main branches of whose crowns at maturity arise once the leader ascends into full sun above the main canopy. Broad hemispherical crowns form, with leaves often in a single layer concentrated at the twig endings. These species present their flowering twigs in full sun. Among dipterocarps and other Malvales, heavy flowering in closed forest is confined to those individuals that have attained mature dome-shaped crowns. The emergent dipterocarps of MDF flower intensely within a well-defined season, at several-year intervals. Most other emergent taxa do likewise, albeit less consistently (Chapter 5).

1.3.2. Main canopy flowering guild

Species in this category also flower only on crown branches that experience prolonged sunlight. Few of these trees show plagiotropic branching as juveniles. They differ from emergent species in that their persisting branches at maturity arise beneath the canopy itself, and their crown branch form differs from that in immature individuals only in its degree of inclination. Flowering is less seasonally concentrated, and less confined to particular years (Chapter 5).

In Far Eastern MDF, the main canopy is the stratum richest in families, genera and species (Fig. 7.5a). Every large tree family has at least one species of the main canopy stratum. Most mature forest species in Anacardiaceae, Burseraceae, Clusiaceae, Dilleniaceae, Fagaceae, Lecythidaceae, Leguminosae, Myrtaceae, Polygalaceae, Sapindaceae and Sapotaceae are mainly of this stratum, and these families can be said to characterise it.

Compared to the emergent stratum, there are fewer pioneer or early successional genera that attain reproductive maturity in the main canopy stratum. Examples are found in *Endospermum* (Euphorbiaceae), *Jackiopsis* (Rubiaceae), some other genera in those families, and *Hibiscus* (Malvaceae).

Within a habitat, species of the main stratum vary more in maximum height than do emergents, and their leaves are less concentrated in a single layer. Therefore, they tend to convey a more mixed and less well-defined profile to the canopy as a whole.

1.3.3. Subcanopy flowering guild

This guild comprises species that reproduce in the patchily sunny conditions beneath the canopy of the forest community in which they live. Subcanopy species flower synchronously within populations, often over long periods but with little variation in intensity between years (Chapter 5). Families most abundant in the subcanopy include Euphorbiaceae (the most species-rich family in Asian evergreen forests), Rubiaceae, Melastomaceae and Labiatae (our trees formerly in Verbenaceae), as well as many Clusiaceae, Meliaceae, Sapindaceae, Myrtaceae and some other families predominant in the main canopy.

The majority apparently require *some* direct sunlight to reproduce.[9] They are diverse both in maximum height and in architecture,[10] though they undergo little architectural change during ontogeny. In the perhumid lowlands of Asia, the tree flora of the subcanopy stratum is dominated by juveniles of canopy and emergent species, especially dipterocarps. This is especially true among the smaller size classes. Partly on these grounds, subcanopy trees rarely form a distinct layer in the forest profile, leading ground observers to question whether stratification really exists. The stratum can, however, be distinguished by its distinct reproductive guild, even though this is diffuse owing to the interspecific diversity of heights at reproductive maturity. The components include a number of genera and families of monopodial habit, that is, with persisting vertical leader and plagiotropic, more or less horizontal, branches bearing distichous leaves, like a conifer. All Myristicaceae and many Clusiaceae, Ebenaceae, Rubiaceae, Aquifoliaceae and Annonaceae take this basic form. Many other species have complex branching systems and lose their straight trunks early on. They often flower at the shoot apex, forcing the trunk to branch out from lateral shoots. In other species again, including many Myrtaceae, lateral shoots expand simultaneously with the apical shoot (syllepsis), and the subsequent branches diverge from one another more or less equally. Shrubs are included in this category. They are rare in the equatorial tropics, but become more abundant in seasonal regions. Many stemless palms occur in the subcanopy (*Licuala* spp.), sometimes abundantly and almost to the exclusion of all else; also slender understorey palms of the genera *Pinanga, Nenga,* and *Eugeissona* (Box 2.1).

Whilst they subsist in the understorey, juveniles of climax canopy species have more slender trunks than do subcanopy species of the same size.[11] The former invest more in achieving height, and they apportion relatively more biomass to trunk than to branches, in anticipation of their ultimate size. These differences are substantially due to the earlier diversion of resources to reproduction in subcanopy species, which results in continuing increase in diameter without commensurate height gain. Understorey species tend to have broader crowns than canopy saplings, and a higher proportion bear plagiotropic branches with small leaves. Leaf size correlates with tree architecture: species with extending plagiotropic branches bear smaller leaves than those, mostly canopy juveniles, which extend tall leading shoots with short lateral branches.[12] Thus, there is a trade-off between height growth (enabling potential future access to much-increased sunlight) and leaf extension (to capture the energy of fleeting sun-flecks).[13]

Some species in all three vertical guilds are cauliflorous, that is they develop flowers directly from their trunks. Most cauliflorous climax trees belong to the subcanopy guild. They include Myristicaceae and Annonaceae, also many understorey species in other families. Yet there are also some cauliflorous main canopy species, such as the *cempadak* (*Artocarpus integer*, indigenous to the Far East), the *nangkah* (*A. heterophyllus* of South India, widely cultivated with enormous sausage-shaped juicy scented fruits) and the wild durians (*e.g. Durio malaccensis, D. wyatt-smithii*). Many durians are pollinated by fruit bats (Megachiroptera: Box 5.6) and bear flowers from the crown branches downwards. The flowers in the canopy are visible to these strong-sighted bats at dusk, and attract them to the trees (Chapter 5). *The pollinator of D. testudinarius, whose flowers are in part borne at the base of the trunk, remains a mystery.*

1.3.4. An overriding question

Tree architecture, reproductive structures and breeding systems (Chapter 5) are each highly diverse in every vertical guild of tropical rainforests. This diversity must reflect different evolutionary solutions among different clades, which nevertheless achieve comparable fitness. Competition between them must in this respect be relaxed. Does that imply that species are, in effect, ecologically complementary, or do they occupy different ecological niches in *other* respects? This will be a leading topic for Chapter 7. Here, I focus on the ways that forests, trees, their tissues and structures have adapted in response to the availability of light and water. One must always bear in mind that the function of a tree's different parts can only be understood within context of the whole plant, and the whole plant in relation to its habitat, the forest.[14]

1.4. Leaf layers as a source of stratification

1.4.1. Leaf geometry

The observant visitor from the temperate regions will first be amazed by the diversity of leaf size and shape in the 'jungle' along the roadside. Upon first sight of mature old growth rainforest, such a visitor may notice the unevenness of its canopy at the medium and broader scales, causing it to resemble piled morning clouds. He or she may also notice the way the canopy leaves, especially if small, are bunched towards the

ends of twigs, forming even arcs over the domed crowns. In contrast, the crowns in tropical deciduous woodlands recall in their deep leafage the north temperate deciduous woodlands, but leaf sizes are much more diverse and often surprisingly large (see section 5.7.2 below). How can this be explained?

All tropical tree species share the same biochemical pathway by which they synthesise chlorophyll through photosynthesis and the C_3 pathway by which carbon dioxide is fixed (although many epiphytes use the alternative CAM pathway[15]). It is a remarkable fact that a leaf of even the fastest-growing of plants can avail itself of no more than a third of the incoming energy from direct sunlight for photosynthesis (Fig. 2.4). The rest is either converted to heat, or reflected. The living cell contents of the leaf can endure heating only to c.45°C, and photosynthesis becomes inhibited in the leaves of many species above 35°C, a temperature commonly reached in exposed leaves on sunny days. In sun-exposed leaves, some of the carbohydrate synthesised is converted to isoprene, which, evaporated through the stomata, assists in reducing

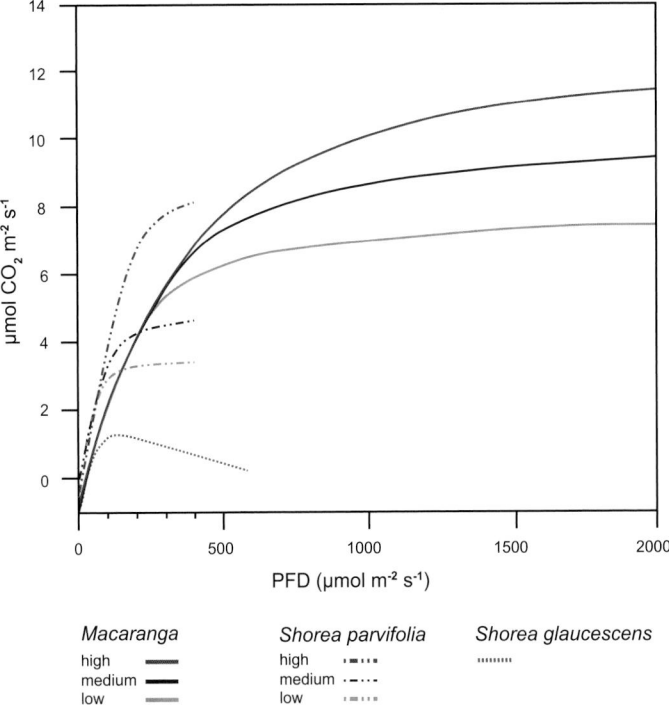

Fig. 2.4 Photosynthetic light response curves: a pioneer (*Macaranga beccariana*), a light demanding late successional (*Shorea parvifolia*) with leaves from juveniles growing in low, medium and high understorey light; and a shade tolerant climax species (*Shorea glaucescens*), which experiences photoinhibition. *Macaranga* sustains lower photosynthetic rates at low light intensities, but rates continue to increase to full sunlight radiation intensity, thereby outstripping the dipterocarps. Vertical axis: rate of photosynthesis (CO_2 fixation); horizontal axis: intensity of photosynthetically active radiation (full sunlight would have a PFD of c.2000). (*M. beccariana* after Davies *et al.* 1998, otherwise Moad 1992)

diurnal leaf temperature. Leaf temperature is mostly reduced through evaporation. and studies in tropical and other forests have shown, remarkably, that forest canopies transpire from their leaves about 90% as much water as a lake of equivalent area.[16] In equatorial regions this amounts to c.1.5 m y⁻¹, or about half the total rainfall received.

However, tropical rainfall is received in storms of short duration, much of it running over the soil directly into streams. Furthermore, droughts over several days occur at least once a year, and droughts of a month or more may occur in ENSO years (Chapter 1) at least once during the life of a climax species in the canopy. This explains why lowland rainforest canopy leaves manifest many adaptations to conserve water. These include upper surfaces which are waxy on first opening, shiny (therefore reflective) and lacking in stomata. In calm weather, leaves are cushioned in an envelope of still air, the boundary layer, in which water molecules move by diffusion rather than by convection as they do in the open air beyond. This boundary layer is kept humid by transpiration. Water vapour diffuses more slowly out of the leaf into this humid envelope, buffering the leaf against water loss.

The depth of the boundary layer declines with increasing wind speed and local convection rate. Forests of perhumid equatorial regions exist in a relatively windless environment most of the time, but the degree of unevenness of their canopies does influence local diurnal convection. Wind eddies are generated around and to some extent beneath the canopy, and these can disturb leaf boundary layers. Dense canopy foliage, on the other hand, may buffer against eddies and protect the boundary layer. Dense, smooth canopies therefore reduce water loss. In forests on drought-prone sites, where trees are otherwise drought-adapted, such a canopy structure may be regarded as an adaptation to drought. Beneath the canopy of lowland tropical evergreen forests, wind speeds sufficient to substantially reduce leaf boundary layers may be rare.

Relative to subcanopy species, a higher proportion of the leaves of emergent and other canopy species (including all dipterocarps) are heterobaric – with vascular bundle sheath extensions linking the minor veins to the epidermis.[17] The precise function of these links remains obscure, but they may serve to increase heat diffusion from the leaf surface into

the conducting tissue, promoting cooling efficiency. Many emergent (and some main canopy) species present their mature crowns with leaf undersurfaces containing stomata imbedded in fine hairs or waxy felt, thus increasing the thickness of the boundary layer. Emergent species with smaller sized, notophyll or microphyll leaves, conspicuous owing to their silvery or golden scaly or hairy undersides, predominate on high narrow ridges (Plates 2.13c, 3.2d). Most hold their leaves at an upward (or sometimes downward) inclination to incident sunlight, thereby reducing the quantity of energy received.[18] This also allows more light to penetrate between individual leaves, promoting photosynthesis in the leaves beneath. Of course, these may include leaves of neighbouring trees, which may compete with and eventually overtop the tree above. A canopy tree in direct sunlight must therefore achieve a compromise between avoiding overheating and water-loss and out-shading its competitors with a tight canopy of overlapping leaves.

1.4.2. *Leaves in the geometry of crowns and canopy*
Givnish and Vermeij[19] first predicted and defined these trade-offs. They predicted that the need to keep sun leaves cool restricts their size, shape, inclination and density (Figs 2.5, 2.10). They hypothesised that, in the more variable diffuse light and high humidity of the subcanopy, constraints on leaf size and shape should become relaxed. There, the relative cost of constructing leaves and petioles of different form and size

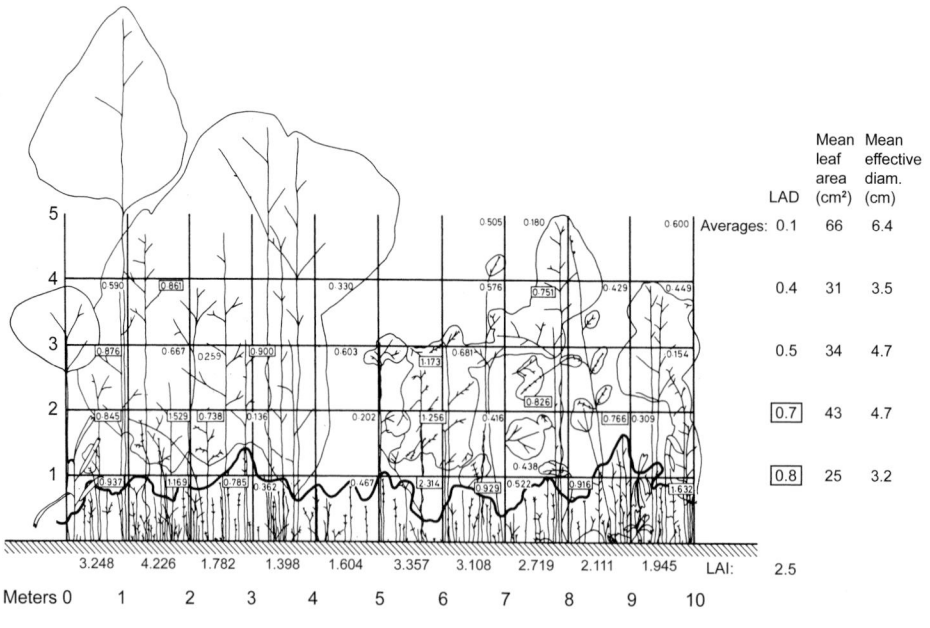

Fig. 2.5 Leaf area density (LAD) and leaf area index (LAI) (m⁻² m⁻²) of woody regeneration along a 1 m deep transect within established regeneration of climax species in a small canopy gap, Pasoh forest, Peninsular Malaysia. (From P.S. Ashton 1978)

versus the more durable twigs that bear them becomes the primary constraint. In the calm air of perhumid equatorial Asia, the tension between these conflicting needs appears to play the major part in resolving the differences in canopy structure and subcanopy density which so characterise forests in different habitats. *Their hypothesis still provides opportunities for experimental testing.*

1.4.3. *Layers of leaves*
Horn described how leaf arrangement in the crowns of temperate tree species in regenerating forests leads to their competitive replacement.[20] Each species in succession bears its leaves more densely, casting increasing shadow and running the risk of shading out their own juveniles. Such foliage stratification during succession could occur in species-rich

forest at the stand level, largely irrespective of the species participating. This stand-scale leaf stratification at the mature phase is distinctive enough to be characteristic of a forest type. Evergreen rainforests of seasonal Asia experience prevailing winds, strong on occasion, throughout the wet and sometimes also in the dry monsoon. During dry windy weather, leaf boundary layers in the canopy may be reduced to ineffectuality. Wind above c.5 m s^{-1} degrades the boundary layer and moves the leaves, scattering sun-flecks beneath the outer canopy and permitting the leaves beneath to photosynthesise more than they respire. Trees in forests subject to wind – especially pioneer species – therefore bear leaves relatively deeply within the three-dimensional space of their individual crowns; and average pioneer leaf sizes are smaller in seasonal than in perhumid regions.

Fig. 2.6 Light intensities in MDF, Sinharaja, Sri Lanka, 6°30' N latitude. Profiles of: 1. the forest; 2 *next page*. Light intensity as photosynthetically active radiation (PAR): **a**, across a large gap on a valley slope; **b**, in a single tree gap on a ridge. (After P.M.S. Ashton 1992)

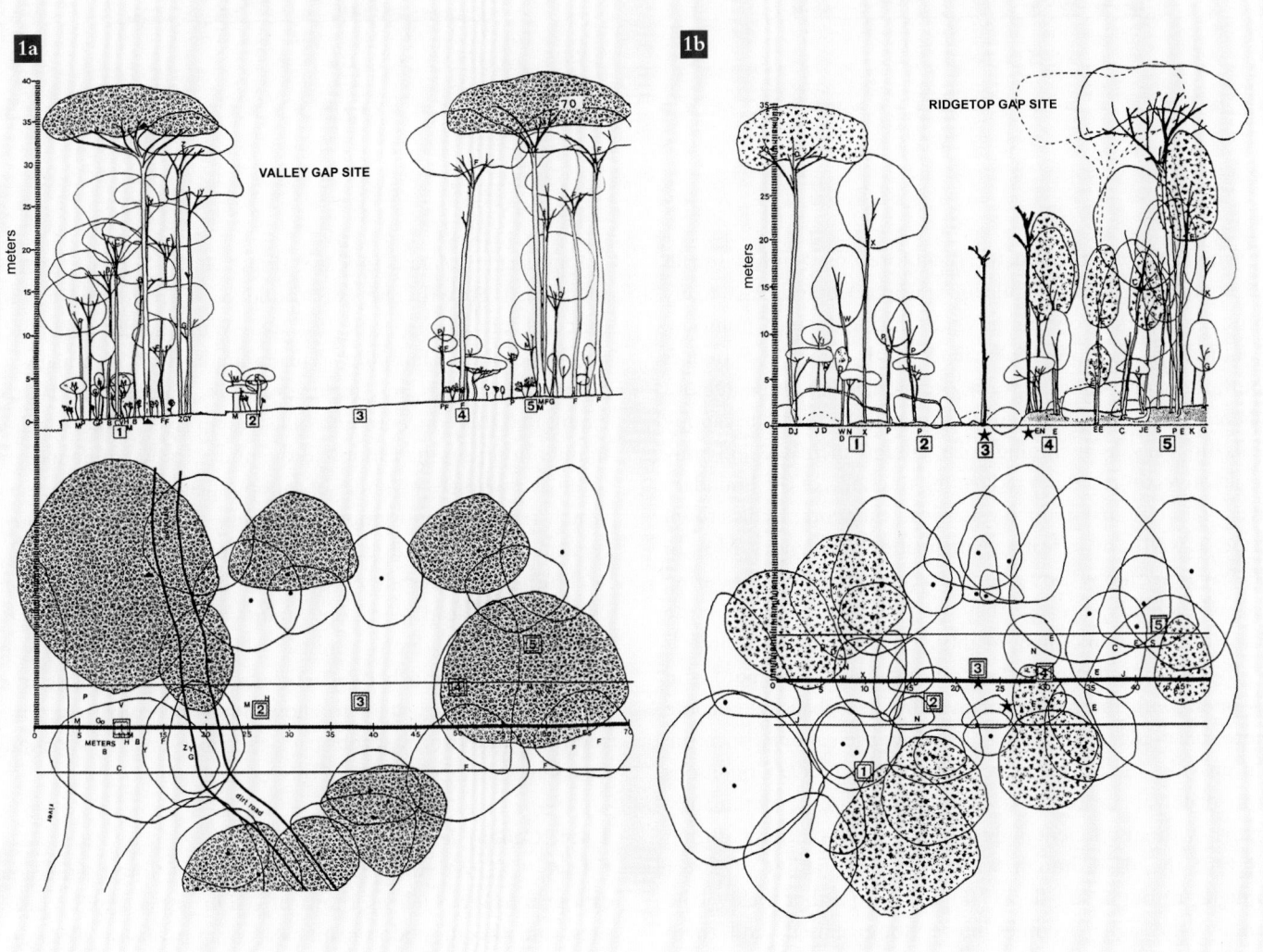

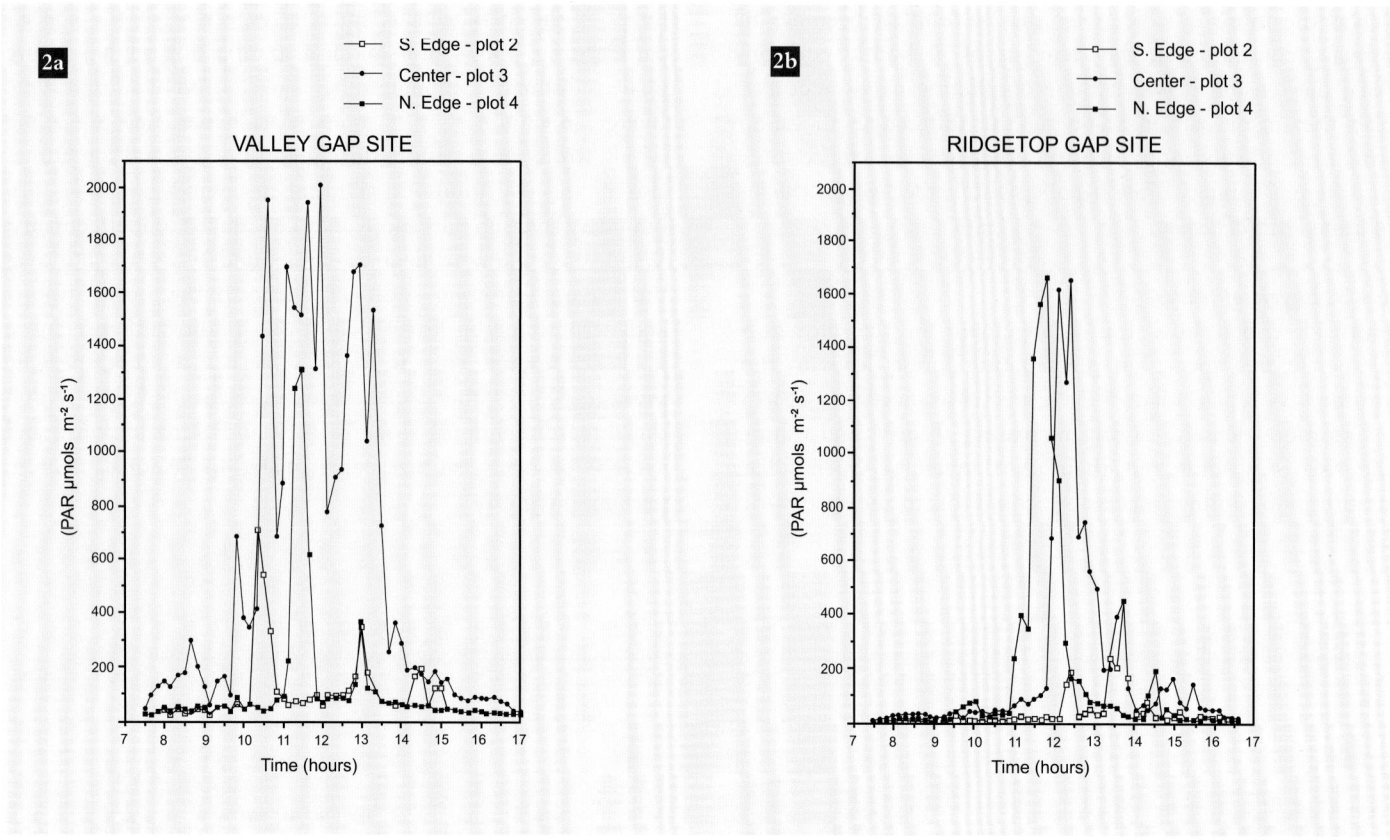

The canopy of the equatorial rainforest may endure squalls before thunderstorms, while in seasonal tropical Asia many forests experience continuous winds during the wet monsoon. Confined to perhumid climates, MDF rarely experiences continuous dry winds, except occasionally in those regions where ENSO droughts occur (Chapter 1). Instead, the canopy heats up daily as incoming solar energy increases and only local convectional eddies reduce the thickness of the boundary layer. The tree may reduce water loss, therefore, by its crown shape, as well as by its leaf size, inclination and density. Most of the leaves in canopy crowns only receive direct sunlight during part of each day, because they are otherwise shaded by the dome of their own crown, or those of adjacent emergents.[21]

In the lowland rainforest subcanopy, light flecks appear and disappear with the passage of the sun. Leaf shadows become more fuzzy-edged with distance below the canopy, since the sun provides light from a disc rather than a point. Conditions immediately beneath the canopy foliage are predominantly shady, with insufficient light for net photosynthesis during most of the day, though with occasional flashes of direct sunlight through sun-flecks (Fig. 2.6). Further below the canopy, the light becomes more homogenised, and there comes to be sufficient diffuse light for some net photosynthesis

among shade tolerant species[22] (Figs 2.6, 2.7). Here, relative (dry mass) growth rate was found to correlate more with unit leaf rate than specific leaf area. Wind hardly penetrates, and atmospheric humidity remains high. The advantages of different leaf sizes and inclinations are largely lost. To the extent that this concentration of leaves, in layers which decline in intensity from the canopy to the understorey, influences the forest as a whole – juveniles of canopy species as well as species which reach lower asymptotic heights – such stratification is regulated by the stand, not the individual species.

Horn and Terborgh describe how leaf and crown characteristics lead to canopy stratification even in maturing New Jersey deciduous forests.[23] They describe how gaps between leaves and between the crowns of the tallest trees let through enough light to allow others beneath to survive; and how understorey shrubs have therefore evolved spreading leaves and crowns to capture the residual light.

1.4.4. *Canopy leaves*
I have observed that evergreen broadleaf forest emergent canopy crowns such as those of MDF, especially in moist sites, and the monodominant peat swamp *Shorea albida* and gregarious MDF *Dryobalanops*, bear their leaves in shallower,

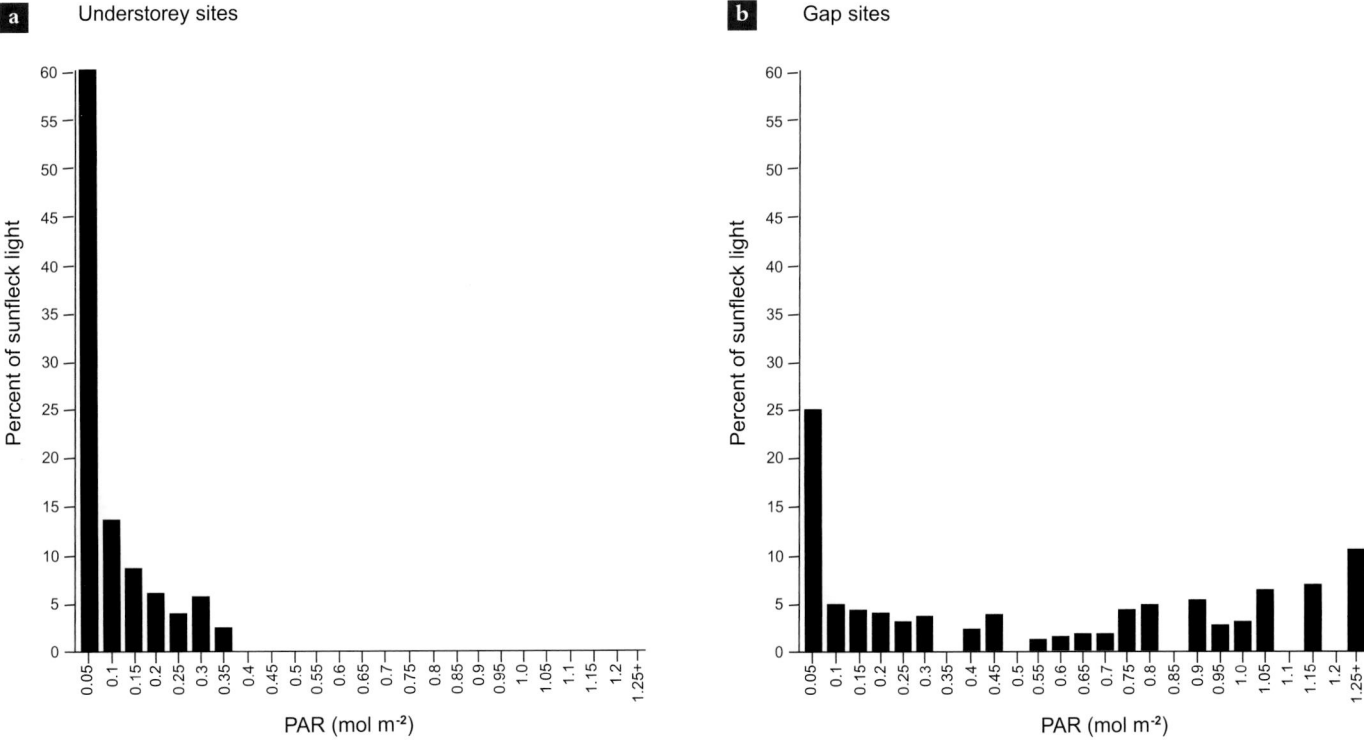

Fig. 2.7 The contribution of light from sunflecks, of different intensities of photosynthetically active radiation (PAR), in sample sites within **a**, understorey; **b**, small gaps in MDF: Sepilok forest, Sabah (From Moad 1992)

denser layers than do Asian tropical deciduous forest canopies (Plates 2.5a, 2.7e, 2.13d, 2.21). Mature canopy foliage here is more isolated from the foliage of the subcanopy beneath. In still climates, light intensity is apparently insufficient for net photosynthesis until the sharp, dark shadows cast by individual leaves (or by twigs bearing densely bunched leaves) dissipate as distance beneath increases their penumbra and intervening sun-flecks become diffuse. Once they have attained the canopy, individuals in a still climate therefore do not generally continue to carry leafy branches deep within the canopy. Subcanopy tree crowns seldom reach as high as the canopy foliage overhead, except where there is a gap between emergent crowns. It is this intense stratification of foliage near the canopy, enhanced by the stratification of branching already described, which gives the visual appearance of distinct stratification within mature lowland mixed rainforest.

Givnish[24] predicted that leaves would vary less in width (the dimension which most influences rates of evaporation in sun and low humidity) among canopy species than leaves of subcanopy species. In MDF the difference is not as great as expected. A small number of Malvalian emergent genera include species with leaves of exceptional size (Table 2.1). *Parashorea macrophylla* possesses the largest leaves among

dipterocarps, macrophylls of c.40 × 20 cm.[25] Its leaves are thin but hang in the crown with silvery undersurfaces. In bud, the leaves are folded between the veins; the folds persist as prominent corrugations when open. It is a species of clay loams on riverbanks and fresh water swamps, where water deficits are rare. However, several macrophyll species of *Dipterocarpus* and *Neesia* occur on well-drained soils of slopes and ridges. Few of these possess silvery or dense finely hairy undersurfaces, but their broad and very thick leathery leaves, borne on stout twigs, are shiny above and corrugated, thereby respectively reflecting and reducing the proportion of incident light absorbed. Their leaves are relatively widely spaced in the crown, reducing heat load but increasing demand on transpiration. Leaves of emergent dipterocarps, once fully mature, have been found to transpire relatively freely, though they do possess some stomatal control; stoamatal control is seemingly mainly controlled by

	Microphyll	**Notophyll**	**Mesophyll**	**Macrophyll**
Emergent	5	25	27	3
Main canopy	4	1	13	0

Table 2.1 Leaf size distribution among emergent versus main canopy dipterocarps at Lambir, Sarawak; N=78.

75

mesophyll assimilation.[26] Stomatal conductance in leaves of *D. sublamellatus*, an emergent at Pasoh forest with large leathery corrugated leaves, is linearly related to atmospheric saturation deficit (which was not the case with three other dipterocarp species examined[27]), implying that its leaves are able to reflect the resulting heat gain.

1.4.4.1. *Canopy repair.* Although crown repair through reiteration of new shoots from damaged branches and trunks can frequently be seen in juvenile dipterocarps and other emergent climax canopy species, such as the cultivated durian, this capacity appears generally to be lost at maturity in windless climates. This contrasts with the crowns of most pioneer and successional species in MDF, as well as nearly all broad-leaved deciduous species, which may sprout in response to both damage and exposure to increased sunlight. Among sprouting species, the crowns of adjacent trees – even of different species – may become intergrown. The capacity to resprout coppice shoots from the base of the trunk is similarly rare among mature climax MDF canopy species.

1.4.4.2. *Crown shyness*. Certain canopies manifest a pattern of leafage which, seen from below, give the mosaic-like appearance of a mangrove swamp viewed from a light aircraft: the leaves on twigs sharing the same branch are tightly packed, but there are sharp-edged open channels between branches, and these increase in width the longer the branch. The canopies taking this pattern are those of young and maturing emergent trees whose crowns are still dense and undamaged, yet whose mature orthotropic branches have developed. Sometimes even dense-crowned mature emergent trees in sheltered mesic coves, and occasionally dense wind-pruned canopies, may display this pattern. Crown shyness is a misnomer, since the pattern is due to wind-shearing of extending shoots while they are still fleshy and brittle.[28] It results in some increase in both sun-flecks and diffuse light in the subcanopy. It is confined to evergreen trees with small to mesophyll leaves in a single dense outer layer, such as dipterocarps and durians, and Fagaceae in montane forests (Chapter 4, plate 4.3d), in climates that are windy during the season of leaf flush. *Dryobalanops* species are particularly prone (Plate 2.4a), and *Shorea megistophylla* in Sri Lanka (Plate 2.4b, section 3.1.3 below).

1.4.5. *Subcanopy leaves*

Leaf sizes beneath the canopy are more variable but generally larger, presumably to enhance the likelihood of encountering sun-flecks. The need to position leaves to capture sun-flecks, constrained by the cost of constructing and maintaining their twigs and petioles, limits their size and presentation. Subcanopy leaves tend to be held horizontally at maturity, and separated loosely and haphazardly in the vertical dimension to capture light. Subcanopy individuals of many species (both juveniles of canopy trees and understorey species) do bear deep crowns in which the leaves are layered on spreading plagiotropic branches. However, there is little or no leaf stratification beneath the canopy of mixed species stands in the mature phase. Leaves on plagiotropic branches tend to be small with short petioles, the cost of support being borne by the permanent structure of the branches, while leaves of the orthotropic leading shoots are larger with longer petioles providing support and presentation to light.[29] The length of internodes and petioles increases in the relatively low red/far-red spectral ratio of the subcanopy, both among and within many species.

Shade-adapted species, and shade leaves of canopy species' seedlings have been found to have lower leaf mass and to synthesise less chlorophyll per unit leaf area. With less chlorophyll, they nevertheless photosynthesise the same quantum of carbon as do leaves of sun-adapted species.[30] This was attributed to more efficient distribution of chloroplasts in relation to incident light, in single layers towards the lower inner surfaces of shorter palisade cells on horizontal leaves (Fig. 2.8). Sun leaves both of pioneers and mature canopy species bear at least two layers of elongate palisade cells, rich

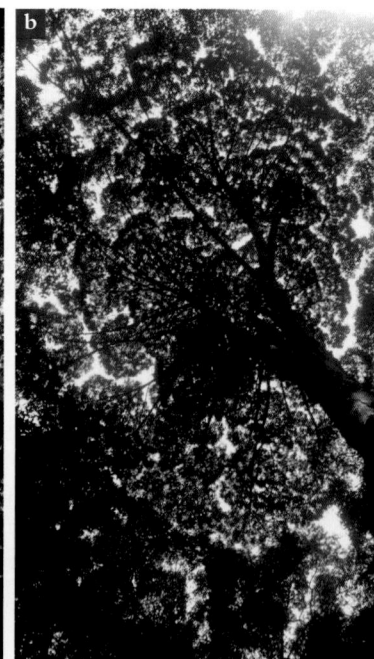

Plate 2.4 Crown shyness (see also Plate 4.3d). **a,** Canopy of the celebrated kapur (*Dryobalanops aromatica*) plantation at the Malaysian Forest Institute, Kepong, planted in the 1930s. Though the foliage is multi-layered, it is diffuse; **b,** Shyness within a single crown: *Shorea megistophylla* in its prime. Sinharaja CTFS plot, Sri Lanka. (**a** M. Moffat; **b** M.A.)

in chloroplasts yet sufficiently translucent to allow some light to penetrate the mesophyll below. Many understorey ground herbs bear leaves with a layer of purplish anthocyanin pigment within their lower (abaxial) surfaces, enhancing light capture in deep shade.[31] It is unclear why this is so uncommon among woody plants, although some treelets in Rubiaceae, and *Ardisia*, possess it.

Leaves on tropical Asian evergreen trees are rarely lobed. If lobed, they are large, and found on juveniles or understorey individuals *Heliciopsis* (Proteaceae), *Scaphium* as well as on successional *Artocarpus* and many pioneers. Shukla and Ramakrishnan[32] found that the number and depth of leaf lobes in *Artocarpus chaplasha* were inversely correlated with light intensity.

1.4.6. *Leaf ontogeny*

The size of a canopy tree species' leaf and its specific leaf area – that is, its surface area relative to its mass – is small in the first seedling leaves, when the plant is transferring dependence on seed resources to those obtained from roots and leaves. Specific leaf area enlarges to its greatest extent in the sapling, then gradually declines as the maturing tree crown experiences more direct sunlight.[33] In the western Malesian canopy species *Elateriospermum tapos* (Euphorbiaceae), Osada and colleagues[34] found that leaf area decline was related to tree height but independent of light availability, therefore not under environmental control. Leaf longevity also declined. Thomas and Ickes[35] found that the leaves of canopy species' juveniles

reduced in size as they grew taller, while those of subcanopy species generally increased. In two subcanopy *Garcinia* species, leaves increased in size until the onset of reproduction when they again declined. They attributed this to constraints on the allocation of limited assimilate in the shade, rather than to the desiccating influence of increased sunlight.

1.4.7. *Summary*

Vertical stratification of leaves occurs among mature individuals of climax species, in the mature phase (see section 2 below) of evergreen forests, owing to the height stratification of their species into emergent canopy, main canopy and subcanopy guilds, based on their floral presentation. It is most evident in the least windy climates. Leaf stratification within the mature phase of tallest forests consists of: (1) a dense but shallow, generally discontinuous emergent upper layer of fairly even height within any single homogeneous habitat; (2) a somewhat more diffuse and deeper main canopy layer of variable height beneath and between the emergent crowns, with dense overlap only at the margins of individual tree crowns; and (3) a diffuse understorey leafage, only patchily attaining the canopy where shafts of light penetrate gaps. At the scale of a few trees, leaf stratification is often obscured by individual age and performance; but leaf stratification becomes most distinct in mature stands, and is consistent throughout communities within a uniform habitat. Habitats are rarely uniform however, and stratification, though present, varies with habitat as we shall see.

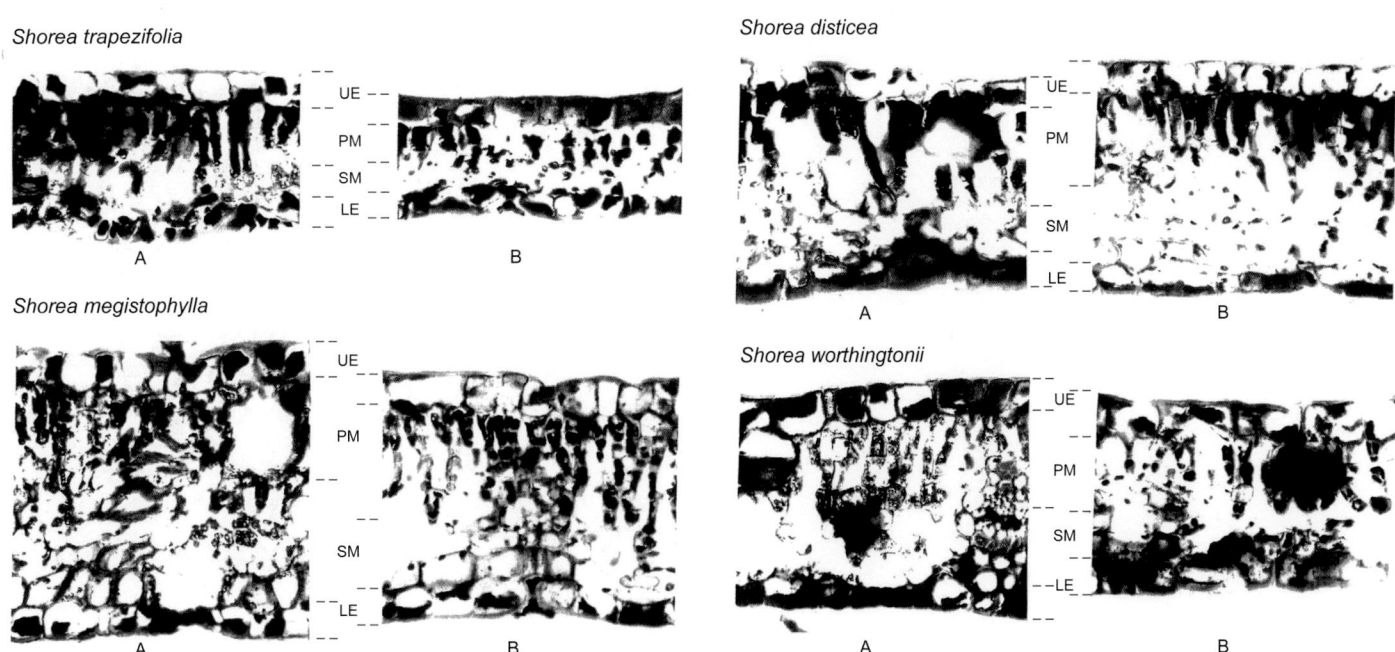

Fig. 2.8 Leaf anatomy of dipterocarp seedlings grown in: **A,** full sun; **B,** shade. Transverse sections; UE: upper epidermis; PM: palisade layer, mesophyll; SM: spongy mesophyll; LE: lower epidermis. x 15 (From P.M.S. Ashton 1990b)

Box 2.1

Palms

Canopy palms

Canopy palms constitute a special case by virtue of their exclusively primary structure, continuous growth pattern, and lack of branches; with their enormous pinnate fibrous leaves capturing sun-flecks within a wide radius, palms represent a unique approach, especially to understorey life.[36] Tree palms are locally common but rare overall in climax evergreen forests in tropical Asia. However, the fan-tail sugar palm (*Caryota*) is a common successional tree in South Asia and, especially on local volcanic soils, in East Asia. All juvenile palms must continue to present new leaves and internodes, or die. Once the palm trunks begin to grow, their leaves, unlike those of broad-leaved trees, will already have reached the size that will be retained throughout life. Water is conducted into the base of the petiole solely by tracheids, which must reduce the rate of flow when evaporative tension in the leaf is high.[37] Both petiole and blade are rigid owing to abundant sclerenchyma. The blades are held inclined upward to incident radiation when first open, declining downward before dying. The open leaf also retains corrugations originating from the folds in bud; its upper surface is often shiny and lower surface silvery-reflectant. *Ontogenetic changes in palm leaf anatomy await further study. Palms push rigid spear-like young leaves through the canopy, unfolding only after they reach full length (Plate 2.5a,b). Canopy palms must have enormous photosynthetic plasticity – which still awaits investigation.*

Understorey palms

Licuala is a species-rich genus of stemless or short-stemmed understorey palms. It produces small red berries, apparently dispersed by small mammals, and several species become gregarious in shady mesic habitats. This habit reduces opportunities for tree seedling establishment, as evidenced by their relatively low densities in these habitats[38] (Plate 2.5d) (Chapter 8).

The curious and apparently ancient genus *Eugeissona*, allied to the rattans and endemic to the Far East, consists of six species of stemless or short-stemmed palms with large woody fruit and a tendency to gregariousness.[39] Two of these, *E. tristis* (*bertam*) in Peninsular Malaysia and *E. tristis* subsp. *gracilis* (*tunjang pipit*) of Borneo, occur in the understorey of forests on freely draining sandy humult soils. Their large pinnate leaves, ascending and radiating like the feathers of a shuttlecock, leave masses of coarse litter as they die. Both living and dead leaves restrict access to soil for the establishment of tree seedlings, while reducing light and thereby favouring shade-tolerant seedlings. *Bertam* is a constant associate of the shade-tolerant *seraya* (*Shorea curtisii*) (Chapter 3, plates 2.16a, 3.2c); saplings of other species, and understorey trees of the smallest size classes are consequently few. Patchy in primary forest, *bertam* greatly expands and increases in logged forest, impeding the regeneration of potential competitors (Chapter 8).[40] In this, it recalls the New England mountain laurel, *Kalmia latifolia* (Ericaceae), which similarly impedes hardwood forest regeneration.

The lowland evergreen forests of tropical Asia are exceptionally rich in climbing Calamoid palms, in several genera (Chapter 6): the rattans.[41] It is they that make the Asian palm flora vastly richer than that of other tropical regions. They climb by means of thorny whips, which are extensions either of the midrib or of the stem-sheathing leaf base. Many also are stemless understorey plants with thin leaflets, but the majority are light-demanding vines of secondary succession.

Plate 2.5 Forest palms. **a,** A rattan (*Calamus*) rises through a narrow gap. Lambir National Park, Sarawak; **b,** *Caryota urens* emerging above the canopy. Kukulu reservoir, Sri Lanka; **c**. *Arenga undulatifolia* subcanopy palms in a small gap. Khao Chong National Park, Peninsular Thailand; **d,** Stemless understorey *Licuala* and a rattan capture residual light on the forest floor. (**a, d** S. Davies; **b, c** H.H.)

Character	Pioneer	Climax
Longevity	Short to medium	Medium to high
Ontogenetic change	Low	High
Reproduction :		
age at first flowering	Early	Late
flowering frequency	High	Low
Seeds:		
size	Usually small, epigeal, produced copiously ± continuously from early in life	Larger, mostly hypogeal, fewer, produced annually or less frequently, starting as full height is approached
quantity	Large	Small
dispersal	By wind or animals, often far	Diverse, including by gravity, mostly short
dormancy	Mostly capable	Lacking dormancy or with delayed germination
soil seed bank	Mostly abundant	Seldom abundant
Germination	Mostly reliant on the influence of direct sunlight, either on light or temperature	Involuntary
Seedlings	Require bare soil for successful establishment. Few successfully establish on raw humus in absence of fire. Cannot survive prolonged shade	Establish on a variety of soil surfaces, depending on capacity to extend tap roots through unfavorable surface materials. Variably survive shade, as a 'seedling bank'
Photosynthesis/respiration: light compensation point light saturation point optimum light conditions max. photosynthetic rate unit leaf rate in shade night respiration growth rate	Variable, relatively high High (above 400 $\mu mol\ m^{-2}\ s^{-1}$) Full sun High (<15 $\mu mol\ m^{-2}\ s^{-1}$) Low or negative Often high (0.4–1.2 $\mu mol\ CO_2\ m^{-2}\ s^{-1}$) Relatively fast	Relatively low Relatively low Variable Relatively high (3–6 $\mu mol\ m^{-2}\ s^{-1}$) Relatively high Often low (>0.2 $\mu mol\ CO_2\ m^{-2}\ s^{-1}$) Relatively slow
Height growth rate	High	Variably lower
Branching	Strongly orthotropic or strongly plagiotropic	Various
Growth periodicity	Continuous or frequent intermittent	Various
Leaf life	Short	Longer
Canopy structure	Low LAI, canopy often monolayer, adapted to maximise shade creation within the limits of prevailing soil water economy	Variable but higher LAI, adapted first to withstand all but the most extreme water stress
Herbivory	High	Lower, with more protection
Root/shoot ratio	Relatively low	Relatively high
Wood	Light, with living wood parenchyma and often no heart; resisting rotting until the tree dies, when it is rapid	Heavier, with heartwood, which is dead and protected by tannins and other chemicals which eventually degrade, exposing the heart to rotting.
Population structure	Single age cohorts	Multiple age cohorts

Table 2.2 Characteristics of pioneer and climax woody plants in MDF. (Partially after Whitmore 1984a, Ng 1986 regarding wood)

2. The lowland rainforest cycle: the life of a stand

This book's focus is on forests and stands, but a stand's composition and performance is determined in the first analysis by the suite of species of which it is composed. The ecologies of the evergreen rainforest's individual species have recently been described by Ian Turner;[42] readers wishing to understand the ecophysiology of rainforest species should refer directly to his excellent monograph.

It is important to distinguish the process of stand re-establishment overall from the attributes of the species that participate in and thereby define that process. Although re-establishment is fundamentally a continuous process, I recognise, for ease of communication, the initial phase of stand re-establishment immediately following canopy gap formation as the *gap phase* (following Whitmore). *Succession* (Whitmore's 'building phase') then follows, continuing until canopy individuals have established full height and the vertical structure of the mature stand is re-established; this then is

the *mature phase*.[43] The participating species are recognized as *pioneers* or *climax species*, and are so differentiated in the discussion that follows. Both participate at different stages of stand development, that is, succession (Table 2.2). The term 'climax' is fraught with difficulty as (following Clements[44]) it was formerly used for a hypothetical stable asymptote that plant communities might reach under a specific climate. This is not so in our contemporary concept of the mature phase, because individual tree mortality continues owing to competition and external catastrophe. Climax *species* as defined here differ greatly from one another in their response to light, but all differ from the pioneers in their requirements for establishment, and most participate in the mature phase of forest stands. Both climax successional and pioneer successional species therefore participate in stand development. ('Non-pioneer' might thus be a more precise term for what I mean by 'climax', but I cannot think of a term more economical than the latter to differentiate the pioneers from 'the rest'.)

Plate 2.6 The regeneration niche. **a,** A mosaic of fallen leaf mat and soil disturbed by wild boar provides diverse conditions for seedling establishment, Huai Kha Khaeng Wildlife Sanctuary, west Thailand; **b,** Wild rambutans establish in a gap between fallen leaves; **c,** Copper-coloured litter on yellow sandy soil, Lambir National Park, Sarawak. The dense soil surface leaf mat beneath of the MDF restricts establishment to those species with seed reserves sufficient for long radical extension through to the mineral soil; **d,** *Macrotermes* mounds offer a regeneration niche distinguished by proneness to drought, yet richer in nutrients than surrounding soil, Pasoh research forest, Negeri Sembilan; **e,** A landslip from the 1964 megastorm on the Belait sandstone, Lambir National Park, Sarawak, with regeneration still impeded by a blanket of resam (*Dicranopteris linearis*), 2010. The adjacent broken canopy bespeaks earlier landslips, whereas the denser more even canopy above, on gentler convex surfaces, reflects single tree mortality; **f,** Only shade tolerant seedlings can survive and emerge through trash left by a tree fall, Ulu Temburong, Brunei. (**a** H.H.; **b** E. Soepadmo; **c, e** C. Ziegler; **d** P.A.; **f** S.L. Sutton)

2.1. The regeneration niche: pioneer (and climax) species

Following germination, plant relative growth rate is negative while it depends on seed reserves more than on resources gathered from roots and photosynthesis. This period may last for months in shade-tolerant rainforest plants, the seeds of which are mostly hypogeal (in which the cotyledons remain within the seed at or below the soil surface or litter, and do not or only tardily emerge and photosynthesise) so that it takes longer for their leaves to emerge; and large seeds take longest. Seeds of rainforest trees vary in both size and dormancy. Both constrain the conditions under which seedlings can germinate and establish themselves. Large seeds, usually associated with climax tree species, are correlated with a higher proportion of mass in cotyledonary reserves for survival and growth.[45] Large seed size has often been regarded as an adaptation to enhance seedling survival in shade, awaiting increased light at gap formation. Carbohydrate requirements, however, are such that seed reserves alone would be unlikely to help a seedling survive without the ability to achieve net photosynthetic gain in low light, that is, without a low photosynthetic compensation point. Relative growth rates in unit dry mass of seedlings have, in fact, been found to be negatively correlated with dry mass of seed reserves, but positively with that of established leaf area.[46]

Species whose seeds germinate in response to canopy gaps have long been known as pioneers (Table 2.2).[47] Swaine and Whitmore concluded that the sole universal diagnostic characteristic of pioneers is the possession of dormant seed that germinate in response to sunlight,[48] but it is now considered that successful *establishment limited to gaps* is the primary criterion. All others then fall into the category of climax species. All pioneers require high light intensities to survive, and grow fast in response to them. Accumulating evidence supports the view that growth rates and survival among forest species increases along a continuum of overlapping photosynthetic responses to increasing light. This continuum extends from the slowest growing, most shade-tolerant hardwood climax species to the fastest growing pioneers. It is because the slowest growing species, which are the longest lived, inevitably eventually dominate stands in their mature phase that they are given the now archaic term 'climax'. As shall be seen, some climax species demand high light intensities to survive, and grow at rates comparable to the fastest growing pioneers in full or near-full sunlight. Therefore, although most pioneers grow in response to full sun faster than do climax species, it is solely their requirement for establishment that defines pioneers from climax species.

All pioneers examined nevertheless do possess dormant seeds (Table 2.2). Some have been found to germinate in response to the increase in the red/far-red ratio of the light spectrum that accompanies the change from diffuse canopy to direct sunlight;[49] others to increased daily temperature range in gaps where sunlight reaches the soil surface, and particularly the difference between the day and night temperatures they experience.[50] Their generally small seeds are mostly epigeal, that is their cotyledons emerge early, and photosynthesise, and their true leaves open rapidly so that their period of negative relative growth rate is short. A few pioneers in Asia – notably *Alstonia,* several of which eventually become emergent, and some vines – possess tiny seeds, wind-borne by tufts of long silky hairs. Pioneers possess various other adaptations to the gap habitat. Combinations of these in the trees as a whole, and changes over the course of their ontogenetic development, ensure their competitive success.[51] Importantly, pioneers start to reproduce when still small, and do so only in full sunlight in that respect like trees of the mature phase canopy. Dormant seeds of pioneer species – if not themselves eliminated by the gap-causing event – will germinate wherever sunlight penetrates to the forest floor. A few pioneer genera are large-seeded, though, including the candlenut (*Aleurites moluccana*), and climax *Artocarpus* species, several of which participate in early succession.[52]

It appears that small-seeded pioneers, which are the majority, only establish consistently within forest on bare mineral soil.[53] A few specialist pioneers, such as *Macaranga lamellata* and *M. kingii,* seem to be able to establish on raw humus, but most others, including *M. hosei* and *Trema cannabina,* are seen to grow only where the raw humus surface has burned and nutrients have been released. This is consistent with experimental evidence that soil nitrogen accelerates pioneer seedling growth.[54] In one test at Pasoh, seeds of early successional species remained variously dormant for less than five to more than 10 years; 77% remained dormant for at least a year.[55] Direct sunlight may be necessary to break their dormancy, but the small seeds are easily washed beneath litter and into shade. More perplexing, most subcanopy climax species' treelets (in Euphorbiaceae, Rubiaceae, Melastomaceae, Myrsinaceae and other families) have small seeds even though they pass their lives in at least partial shade. Indeed, the smallest-seeded plant known is a deeply shade-tolerant tropical forest herb, a *Sonerila* (Melastomaceae).[56] Grubb and his students[57] have shown that large seeds permit early investment in long radicles, which may enable seedlings to penetrate the dense litter of the understorey soil surface. Small-seeded subcanopy species establish only on surfaces free of litter, such as slopes, periodically flood-washed ground, and places where animals have disturbed the soil.[58]

Nevertheless, many cauliflorous monopodial subcanopy trees, including Myristicaceae, Annonaceae and *Garcinia* (Clusiaceae) do have larger seeds, implying that the costs of structural support of large fruit and seeds may constrain seed size when light is limiting[59]. Among dipterocarps, whose inflorescences are terminal or axillary on twigs, twig stoutness is positively correlated with nut volume.[60] Further, the majority of climax main canopy species are relatively large-seeded, and there is no inverse correlation among dipterocarps between seed size and either height at maturity or wood density (implying shade tolerance). For canopy species, the need to survive both in understorey shade as juveniles and in full sun as adults must exert great selective pressure on them to survive contrasting light conditions through high plasticity in this respect. I infer that the capability to extend a taproot early and also to survive at light intensities initially below the photosynthetic compensation point have both acted as selective influences

on seed size, but that the latter may constrain photosynthetic efficiency under full sun (photoinhibition, see below).

Those species whose seed do not germinate in response to light manifest a wide range in duration of seed dormancy (Fig. 2.9). Some climax species manifest delayed germination rather than true dormancy (Table 2.3). In one study, at Pasoh forest, 16% of species remained dormant for at least one year, though the proportion of pioneers is less than 10% (Table 7.6).[61] Notable in Asia are *Anisophyllea* (Anisphylleaceae) and certain large-seeded canopy legumes,[62] whose seeds are protected either by hard and fibrous pericarps or hard seed pods, and which germinate when these are broken or rot off. Some of their dull-coloured pods are consumed by forest ungulates, which later defecate the seeds, thereby dispersing them in welcome manure. These emergent legume species, like the dipterocarps, mast-fruit synchronously at intervals of several years. Their delayed germination leads to continuous but sparse production of seedlings in the years following supra-annual fruiting events. Leaves of leguminous trees have relatively high nitrogen concentrations and are therefore preferred by browsing animals over dipterocarp seedlings, for example, which are relatively low in nitrogen, resinous and rich in phenols, and suffer little browsing (Chapter 5).[63] For supra-annual mass flowerers possessing delayed germination, continuous release of a few seedlings thus appears to provide survivorship comparable to that achieved by dipterocarps lacking seed dormancy. In sum, trees in different families, possessing different phenological, nutritional and other attributes, have evolved different means to reduce the risk of predation.[64]

There appears not to have been any critical comparison of dispersal differences between pioneer and climax rainforest tree species.

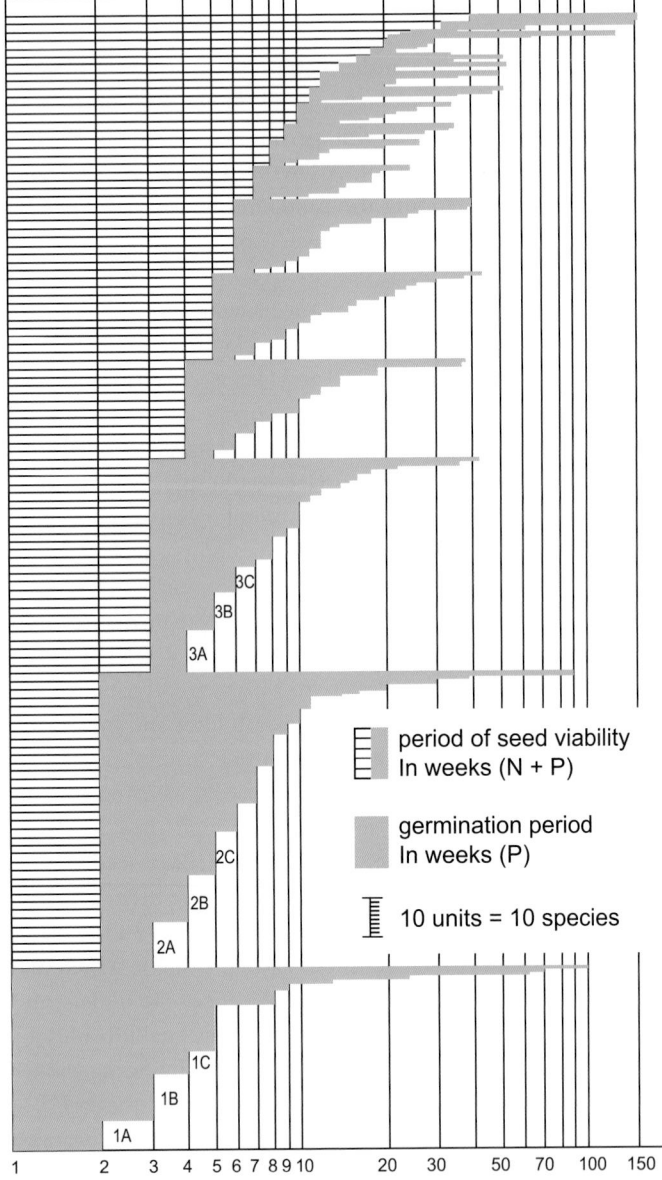

Fig. 2.9 The frequency of Peninsular Malaysian tree species according to their seed viability (potential dormancy: N + P) and range in times to germinate (P). (From Ng 1980)

Leguminosae		
Adenanthera pavonina		
Crudia curtisii		
Dialium platysepalum		
Intsia palembanica		
Koompassia malaccensis		
Callerya (Millettia) atropurpurea		
Pterocarpus indicus		
Sindora spp.		
Anisophylleaceae		
Anisophyllea corneri		

Table 2.3 Known western Malesian forest tree species with delayed seed germination. (After Whitmore 1975, Wyatt-Smith 1963, Ng 1980, 1991)

Plate 2.7 The gap phase. Plants of moist habitats have sufficient constant soil water available to build a dense upper layer of leaves (high LAD), thereby restricting survivorship of shorter competitors. **a,** *Merremia* climber blanket in a forest gap on moist clay loam, Sepilok, Sabah; **b,** *Macaranga siamensis* shoot with leaves overlapping like slates on a roof; **c,** Beneath the crown of a young *Macaranga winkleri;* **d,** Diffuse crowns of maturing *Macaranga hypoleuca,* species of edaphically drier sites, Danum, Sabah; **e,** The diffuse crown and plagiotropic branches of the light demanding successional climax species *Shorea trapezifolia* sapling, which those of *Dryobalanops* species also resemble. (**a, c** P.A.; **b** S. Davies; **d** C.-E. Lee; **e** M.A.)

2.2. *The gap phase and succession*

On entering the Pasoh research forest in northern Negeri Sembilan for the first time, I was struck by the amount of canopy damage. There are extensive patches of mature *red meranti-keruing* MDF on the low hills and better-drained sandy alluvium; but in the floodplain, large and small patches of uprooted giants are already being engulfed by tangles of sapling regeneration and vines. The roots of these large trees are surprisingly shallow, raised to the vertical

within their soil as discs when trees blow down (Plate 2.8). Closer examination reveals patches of small pioneer trees, apparently young but already flowering, with large thin leaves: *Macaranga* (Plate 2.7b,c) and others with orthotropic branches soon opening a dense single-layered large-leafed umbrella-like crown, *Glochidion* with spreading twigs carrying a dense carapace of tightly overlapping smaller leaves, and an occasional *Alstonia* sending up its vertical leader from which a leafy platform of horizontal branches periodically radiates (Plate 2.3a).[65] A recent tree-fall reveals that these species establish wherever there is bare mineral soil, in the tip-up craters and even under the root-disc, and on ground where the leaf litter has been swept aside by rain-wash or by pigs searching for larvae and truffles. In other places within the gap, cohorts of young dipterocarps are regenerating, each group a single species, the tallest almost as tall as the *Alstonia*. Wherever the soil surface has remained undisturbed, abundant seedlings of dipterocarps and other climax canopy species have already established and survived since the last fruiting. From the beginning, then, there is a rich mixture of pioneer and climax species.

Plate 2.8 The plate-like root-ball of a tree tipped up by wind, reveals its shallow roots on the periodically waterlogged alluvium of Pasoh forest. (P.A.)

The gap phase in zonal forests has been comprehensively described by Tim Whitmore, and I cannot improve on his masterly accounts.[66] The following description is based on static observation in many forests, including two transect censuses in a 50 × 25 m artificial gap at Pasoh, earlier created by Japanese botanists during a 1971 project of the International Biological Programme.[67] All living vegetation had been removed, and transects were surveyed four and 12 years afterwards. *This illustrates a recurring theme in our story: the details of forest function will not be learned until dedicated and persistent scientists track tagged trees over many decades, even passing on the baton to succeeding generations.*

2.2.1. *Gap origins* (Plates 2.8, 2.9)

Plate 2.9 Forms of canopy gap. **a,** A fresh landslip provides seed free subsoil near its top, debris and disturbed topsoil below, Lambir National Park, Sarawak; **b,** Linear gap caused by single tree fall, with minimal soil disturbance but much trash, Bukit Mersing, Sarawak; **c,** Single dead emergent left standing on a Lambir ridge, leaving a small canopy gap. Branches and trunk will collapse gradually, with minimum impact on the understorey; **d,** Looking up into the small gap left by the absence of main canopy crowns beneath that of a leafless dead emergent tree; hill dipterocarp forest, Temenggor Forest Reserve, Upper Perak; **e,** Mid succession in a large gap on clay loam soils; emergent *Koompassia excelsa* and *Shorea* at rear, Danum, Sabah; **f,** Several *Macaranga* species, spindly and with narrow diffuse crowns, co-exist but hardly compete on the droughty low nutrient soils of a Lambir ridge. S. Davies as scale. (**a, e, f** P.A.; **b** P. Hall; **c** C. Ziegler; **d** R. Harrison)

Natural gaps have several possible causes:

1. The smallest gaps are those caused by the death of a single standing tree. This may result from lightning strike, drought or – apparently only occasionally – disease. Such mortality causes minimal disturbance at the forest floor. Juveniles already established beneath may suffer damage as branches and eventually parts of the trunk fall, but these can resprout by shoot reiteration, so that their canopy and the litter layer beneath remain unbroken, depriving seeds of pioneer species of access to the direct sunlight and litter removal necessary for germination.

2. Larger gaps may be formed, in order of magnitude, by (i) local wind-throw, (ii) logging, (iii) landslips and floodwater, (iv) downdrafts and cyclonic storms and (v) volcanic lava and ash. In these cases, the soil surface becomes patchily or wholly scarified and established regeneration is damaged or removed. Unlike gaps caused by the death of standing trees, wind-throw and landslide-caused gaps can result in adjacent canopy instability, resulting in further falls and increasing gap size over time.[68]

Always, the tallest adjacent survivors exert a commanding influence on the light climate below them and consequently on the fate of shorter individuals, depending on their crown density. The size, nature and origin of a gap thus determine, to a considerable extent, the course of tree succession within it.

2.2.2. Early gap phase dynamics

When a gap is formed in the canopy by the death of a canopy tree, a patch of sunlight penetrates to the subcanopy for part of the day (Figs 2.6, 2.7; Plate 2.9b). The extent to which sunlight penetrates mediates the effectiveness of the gap in spurring regeneration, and it is determined as much by the stature of the forest in which it is set as by the diameter of the gap.[69] Every gap includes a diversity of light regimes, and the surviving adjacent canopy is also affected.[70]

Tropical rainforest dynamics differ fundamentally from those of other forests in that the great majority of species are climax and lack seed dormancy (Chapter 5) – though most do not germinate immediately, but rather over ensuing weeks and in some cases months (Fig. 2.9).[71] Only removal of established regeneration at gap formation, as in landslips, gives the advantage to those few species whose dormant seed remain, or that fruit frequently – namely the pioneers. In this case, the *gap phase* of a patch of forest is said to have been initiated.

If a gap is big enough in relation to the height of the adjacent canopy, the increase in sunlight at the forest floor may stimulate dormant seeds to germinate. Pioneer species – more fecund, fruiting more frequently, and in many cases with more effective seed dispersal mechanisms – at first have competitive advantage over climax species, because the climax species' seeds lack dormancy. But eventually they reinvade and their more shade-tolerant seedlings ascend through the pioneer canopy.

The leaves of pioneer species grow quickly to maximum size and change little ontogenetically. Thus, they resemble their mature crown leaves almost from the seedling stage. In most, they are presented in a single densely overlapping layer.[72] They develop secondary branching earlier, enabling greater foliage plasticity.[73]

The atmosphere in rainforest canopy gaps is generally still. The leaves of pioneer crowns, tessellated like slates on a roof, are immersed in a thick layer of humid air (Fig. 2.10). This could

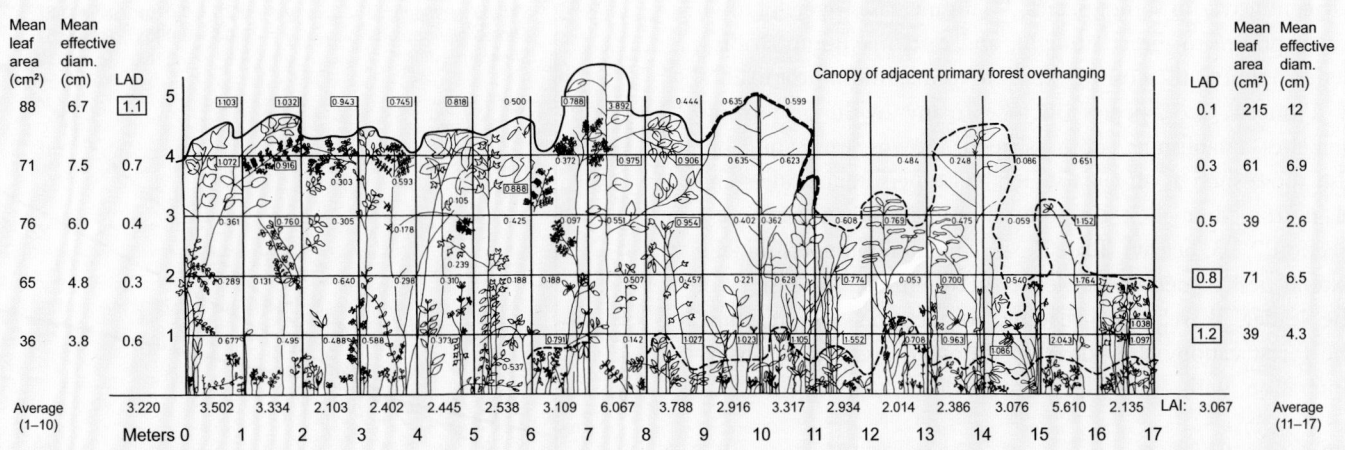

Fig. 2.10 Leaf area density (LAD) and leaf area index (LAI m^{-2}) of 4 year old pioneer regeneration, established from seed in a 20 x 100 m artificial completely cleared (canopy to ground) gap within MDF, Pasoh forest, Peninsular Malaysia. LAD increases again near the ground, owing to establishment of shade-tolerant seedlings beneath the pioneer canopy. (After P.S. Ashton 1978)

Plate 2.10 Wilting among pioneers: *Clerodendron*. **a,** In normal times, Temengong Caves, Sabah; **b,** Beneath the heat of the mid-day sun, Mulu National Park, Sarawak. (P.A.)

easily lead to overheating of the leaf in direct sunlight, and can only be overcome by high rates of transpirational cooling or by wilting, when the leaves hang and the competitive advantage of out-shading is lost. Meinzer and colleagues[74] found that control of transpiration by pioneer leaf stomata is low, and is governed as much by the thickness (and therefore conductance) of the boundary layer as by intrinsic physiological controls. They found a relatively weak correlation in pioneers between stomatal conductance and midday leaf water potential. Midday wilting is common (Plate 2.10), as are gnastic movements bringing leaves into positions that reduce incident energy. Continued expansion of the length of the developing petiole over a period as long as three months, and an increase in the angle of petiole incline from the vertical, reduces shading of adjacent leaves in some species.[75] This increases overall crown leaf exposure to direct sunlight, and casts maximum shading beneath, but at the cost of exposure to maximum heat load.

Pioneer tree species include many with orthotropic branches establishing dome-shaped crowns early on, but also those such as *Alstonia* which, with persistent leaders and plagiotropic branches, ascend into, or even above, the canopy (section 2.2.3. below) (Plate 2.3a). These tall pioneers initially extend tiers of densely-set leaves between lofty extensions of the leader, thereby out-shading competition. On reaching the forest canopy they develop a domed crown, usually by reiteration, with smaller leaves more diffusely arranged.

2.2.2.1. *Trade-offs between rapid growth, competitive exclusion, and adaptation to drought.* Pioneers grow most rapidly in gaps because their leaves attain high maximum rates of photosynthesis in response to direct light (Fig. 2.4). In areas of relatively high (but still diffuse) light intensities below the

canopy, their daily gross photosynthesis often falls beneath gross respiration, and pioneers cannot survive. Climax species, including those of late succession, have lower maximum rates of photosynthesis in full sun, but higher rates in shade than pioneers. In shade, slow growing shade-tolerant heavy hardwoods can sustain net photosynthesis in most shade light intensities (Fig. 2.11).

Fig. 2.11 Relative growth rates (grams per total mass in grams) in saplings of three *Shorea* species in relation to estimated sunfleck light availability: a 'light demander' (*S. parvifolia*), a 'shade tolerant' species (*S. glaucescens*) and a species of intermediate response (*S. xanthophylla*). (From Moad 1992)

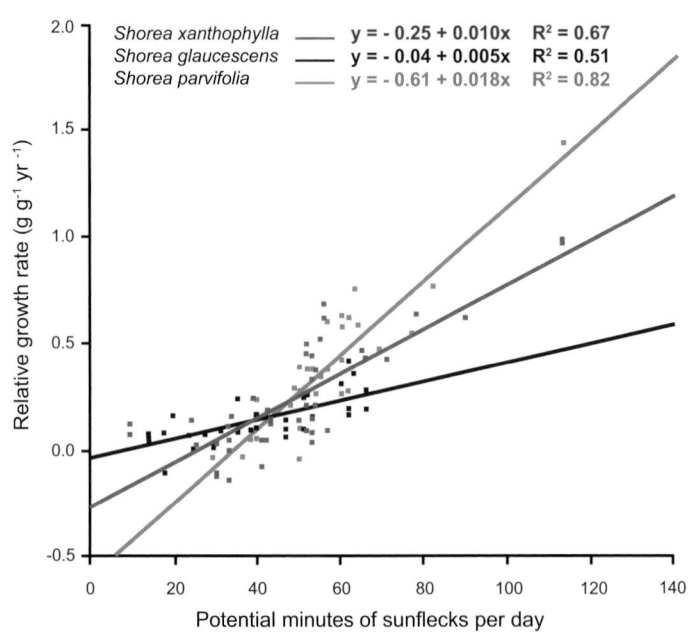

Pioneers and vines invest relatively more than do climax species in shoots than roots, and in wide diameter vessels that permit rapid movement of water up the stem, enabling rapid growth. These, and their dense shallow crowns, allow them to out-shade individuals of lower stature, but expose them to high risk of vessel cavitation (embolism) during drought. (Wide vessels are also sensitive to frost-induced embolism, thereby explaining the absence of such species in temperate forests.) Vessel embolism can recover under increased water pressure from the roots, but root pressures are not high and decline rapidly with height above the soil.[76] The risks are high of drought-induced mortality, predation and damage on account of their weak low-density wood.[77] The pioneer's high-risk life history strategy is mitigated by high fecundity and seed dormancy.

2.2.2.2. Early succession. The new seedlings, together with those established juveniles able to respond to the new conditions by increasing their growth rates, form the regenerating stand. *The regeneration-paths of climax species with differing light responses can at present only be inferred, pending long-term observation of individual trees and populations. The CTFS plots provide unique opportunities for such observation.*

Many canopy openings, however, do not lead to disturbance or mortality near the ground. These initiate a truncated stand succession in which climax species' juveniles survive and pioneers may play little or no part. Even among leaves opening in direct sunlight, full sun over long periods is inimical to the growth and survival of shade-tolerant climax species' seedlings, which may experience inhibition of photosynthesis (*Shorea glaucescens* (Fig. 2.4)). Established seedlings of such climax species will nevertheless be competitive in the mostly diffuse light environment towards the gap edge and in the adjacent subcanopy where it also experiences increased light.[78] The optimal gap size for late-successional light hardwood emergent *Shorea* in logged MDF, with perhaps 40 m tall canopy, has been found to be c.500 m²,[79] while that for slow-growing heavy hardwood *Shorea* and others is smaller.[80] (But it is the height of the adjacent undamaged canopy as much as the area of the gap that determine intensity and duration of sunlight at the forest floor.) The advantage always rests with those juveniles that survive the gap event and are tall enough to create a shadier light environment which shorter adjacent competitors and germinants must endure.[81] During the first five years the pioneers, where present, thus gain and retain ascendancy.[82] Early on, little vegetation survives beneath the thin but dense layer of leaves across their crowns, a few scattered seedlings of shade-tolerant climax subcanopy and canopy species excepted. But, as they grow, their crowns become more diffuse as the increasing weight of leaves bows their branches out (Plate 2.11). It is then that climax saplings arising from subsequent

Plate 2.11 The opening orthotropic branches of a maturing *Macaranga* (left), beside a young *Shorea argentifolia* crown with branches beginning to ascend from plagiotropic to orthotropic, Kuala Belalong, Brunei. (P.A.)

seedfall start to grow in height, eventually to emerge between the pioneers. *The proportion of these climax saplings already established when gaps open, versus the proportion germinating from more recent seed-fall, varies between forest types. This must be documented when designing silvicultural prescriptions, particularly as it determines the need for retaining seed trees.*

Gaps resulting in little direct light provide opportunities for the growth of species which have high survivorship in low understorey light regimes, for instance *balau* or *selangan batu* (*Shorea* sect. *Shorea*). These may have already reached pole size in an earlier gap phase. It is the calm and equitable micro-environments of the lowland tropical rainforest subcanopy (unlike in climates where fire, freezing rain or snow damage the leading shoot) which results in the tall columnar boles and symmetrical domed crowns paradigmatic of these forests.

Primary forests are therefore comprised of a mosaic of stands whose origins are diverse, and which are initiated at different times. In the Asian tropics, most gaps are caused by catastrophic climate-related events variously taking place at the local or landscape scale, or higher. At that larger scale, therefore, they tend to be synchronised and show a predictable long-term periodicity. The clumping so characteristic of emergent crowns, especially dipterocarps, may thus reflect shared beginnings as a stand in a gap. Differences in their size and abundance in the landscape reflect differences in the prevailing canopy disturbance regime.

2.2.2.3. *Deflected succession.* All vines in evergreen rainforests belong to one of two ecological categories: (1) shade vines, which in Asia are generally small, often attaching by adventitious roots to tree trunks; and (2) sun vines, which, being large and in the majority, may play a major part in succession. Sun vines, like pioneers, show high growth rates in response to light, further enhanced by reliance on other plants for support structures, and by vessels of exceptional diameter and length. These characters enable certain species, notably of genera *Merremia* (Convolvulaceae) (Plate 2.7a) and the exotic weed *Mikania* (Asteraceae) in Asia, to form blanketing canopies over competing trees. The gap or early successional phases may thereby persist for decades, eliminating regeneration of all but a few of the more light-demanding climax species beneath. Some vines can persist and grow large in the successional phase, but they remain more abundant in the mature phase main canopy than among emergents.[83] *Dicranopteris linearis,* a climbing and creeping rhizomatous fern widespread on degraded or infertile yellow soils in the wet lowlands, commonly invades landslips and degraded soils following cultivation. The fern rapidly forms a continuous canopy. Its litter, acid in reaction, accumulates a surface horizon of root-matted raw humus, increasing leaching in the mineral soil beneath.[84] Although it is low in biomass, *Dicranopteris* can account for half the uptake of nitrogen and phosphorus, releasing them only slowly through decomposition. Tree regeneration is inhibited and, once invaded, a gap can only be gradually reduced by shading from the edge. One landslip initiated at Lambir National Park after exceptional rain in 1964 still supported a cover of *Dicranopteris* in 2006, although spindly saplings of *Mastixia* aff. *eugenioides* and *Timonius flavescens* were emerging above it.

2.2.3. *Later succession*
At first the canopy of the regenerating stand increases in height as the stand grows, and the crowns within it do not conform to the stratification of the mature phase stands nearby. Until a stand approaches maturity, and a small group of individuals begins to dominate, the tree crowns within the stand continue to lack any vertical stratification other than an increasingly dense upper layer of foliage in their canopy. Their stand canopy structure therefore continues to obscure the overall stratification of the forest profile.

Gap succession in MDF, particularly when it starts from bare soil with pioneer species, early on establishes a shallow canopy of high leaf area density (LAD) but low leaf area index (LAI),[85] with all crown branches immersed beneath it (Fig. 2.10). The LAI builds up as the LAD of pioneers' outermost canopies declines. That is the opposite pattern to the one described by Horn in a temperate broad-leaved forest, where species of increasing upper crown leaf density, and decreasing LAI, replace one another in forest succession.[86] As succession approaches the height of the mature phase main canopy, emergent crowns become apparent, their leaves diffusely arranged or dense, according to habitat and species (see section 3.1 below).

Oldeman[87] described a morphological inversion surface, where crown branches arise at a uniform height within stands of trees of similar stature in early succession, but in which the bole height at which crown branches are initiated, and crown stature, become increasingly variable as the stand matures. He recognized that 'the canopy is the master regulator of the forest', that the number and distinctiveness of these inversion surfaces becomes stabilised for long periods among mature stands, and that they consequently determine the degree of vertical stratification that a particular forest attains. Their distinctiveness therefore reflects a forest's regime of canopy disturbance. Though he did not recognize that main and emergent species differ phylogenetically, or that canopy and subcanopy species reproduce under significantly different light conditions, Oldeman's understanding of the causes of forest stratification is consistent with that caused by leaf stratification presented here (Section 1.4).

2.2.3.1. *Climax species of mid-succession.* MDF differs in mid-succession from many other rainforest types in that, in MDF, the subcanopy and understorey are dominated by juveniles of canopy trees, mostly emergent dipterocarps. Within the decade following moderate-size gap formation therefore – and in the absence of vine cover – the new canopy becomes dominated by species that do not flower until their crowns have emerged at least partially into full sun.[88] These species are mostly faster-growing climax species whose seeds germinate either at the time of, or shortly after fruit fall, or at variable times following a dormant period independent of light or shade[89] (Chapter 5).

These light-demanding late successional climax canopy species maintain their dominance through a single leader until full height is approached. They achieve their height by expanding internodes over a long growing period, and bear many relatively short-lived leaves.[90] LAD and crown depth therefore both increase in juvenile and pole-sized successional dipterocarps,[91] but crown depth declines once the full domed emergent crown is formed. Many late successional climax individuals have orthotropic branches, or these develop as their crowns emerge above the canopy (Plate 2.11). Their leaves are of variable size and they generally hang, thus casting less shadow as they mature. This allows them to achieve a lower light extinction rate within the crown, and a higher LAI.[92] Within a given size class, maximum growth rates are proportional to light interception and to wood density.[93] Many slower-growing climax species nevertheless experience photoinhibition and a lowered rate of photosynthesis in full sunlight (Figs. 2.4, 2.16) (see section 2.2.5 below).

These faster-growing climax species that first overtop the pioneers are themselves relatively shade intolerant. They may achieve net photosynthesis rates comparable to those of pioneers, with growth rates of up to 15 mm in diameter and 2 m in height per year. Since their growth rate is inversely correlated with wood density, these species include the light hardwood mahoganies and merantis that comprised most internationally traded wood in the period of 1970–1995 (Chapter 8). The lowland rainforests of the perhumid Far East, including the Philippines, carried far higher volumes of these light hardwoods than forests elsewhere in the tropics; most were dipterocarps in the red and yellow meranti timber groups of *Shorea*. Up to 70% of the MDF canopy may be composed of these timber species on mesic sites.[94] They flower en masse at around five-year intervals (see Chapter 5); afterwards, the ground is dense with their seedlings. Although they suffer high mortality in the shade, enough seedlings generally survive the five years between fruiting seasons (Fig. 5.11).

It is the relatively high survival rate of these late successional climax dipterocarp juveniles in the perhumid tropics within which MDF is confined that distinguishes them in composition and dynamics from the rainforests of the seasonally dry climates of other tropical continents.

A suite of tall pioneer species may persist within stands into late succession. These notably include *Alstonia* (Plate 2.3a), but also *Dyera* (Plate 2.3b), *Pterocymbium*, *Octomeles sumatrana* and some *Terminalia*. Their wind-borne seeds germinate in response to light and may possess some dormancy.[95] Some early successional climax species, too, are persistent. These include a group with large seeds, either bird-dispersed (*Sterculia foetida*) or water-

dispersed (*Terminalia copelandii*, *T. phellocarpa* and others, many east of Wallace's Line). These species are 'pagoda trees',[96] conforming to several different architectural models but in each case gaining height rapidly by periodic massive extensions of a vertical leader while plagiotropic branches extend horizontally in false whorls, their dense foliage layers out-shading competitors. These species can gain height faster than

Plate 2.12 a, Mature stand of the pioneer *Octomeles sumatrana* on an alluvial bank, Ulu Segama, Sabah; **b,** A population of the heavy hardwood deep-crowned main canopy *Petersianthus quadrialatus* remains in forest cut for swidden, its wood too hard for the peasant axe, eastern Mindanao, 1975. (P.A.)

adjacent climax individuals. Unlike the predominant shorter pioneers though, they are found as scattered individuals or small groups in the forest; many gaps lack them. Furthermore, like pioneers, they are most abundant and achieve greatest growth rates on well-watered soils, including floodplains. They commonly persist in groves along riverbanks where their columnar juvenile and maturing crowns continue to receive lateral sunlight (Plate 2.12a). A mature *Alstonia* crown may eventually exclude all other canopy species beneath it.

Why, then, do climax species, such as the light hardwood *Shorea* species, during early succession develop deep multi-layered crowns of hanging small leaves on branches which become increasingly orthotropic as they reach the canopy, whereas many tall pioneers retain leaves arranged in dense monolayers on widely spaced plagiotropic branches? Probably because whereas the pioneers experience prolonged full sun from their seedling stages, the successional climax light hardwoods experience a daily light environment of great temporal variability, in which the competitive advantage of light exclusion gained by dense leaf layers is less than that gained by the increased photosynthetic area and reduced transpiration gained by a high LAD of small hanging leaves in many layers.

2.2.3.2. *Climax species of late succession and the mature phase*

What then of the slow-growing, heavy hardwood canopy species? In Sundaland these include not only the emergent *balau* and *cengai* (*Shorea* sect. *Shorea* and *Neobalanocarpus*), but also heavy red *merantis* like *S. curtisii* (Dipterocarpaceae), *sepetir*, *merbau* and *tapang* (*Sindora*, *Intsia* and *Koompassia* of Leguminosae). In the main canopy are ironwood (*Eusideroxylon*: Lauraceae), *petaling* (*Ochanostachys*: Olacaceae), *Petersianthus quadrialatus* (Lecythidaceae) (Plate 2.12b) and *ketiau* (*Ganua*: Sapotaceae). In Sri Lankan main canopies we find *Shorea dyeri* and *S. lissophylla* (*Shorea* sect. *Shorea*), *S. worthingtonii* (endemic sect. *Doona*), *Mesua ferrea* (Guttiferae), the famed calamander ebony (*Diospyros quaesita*) and *Palaquium petiolare* (Sapotaceae). Seedling studies indicate that these species can survive many years beneath the mature phase canopy.[97] *It has not been documented, though, whether they grow into the canopy more or less continuously over the 'lifetime' of a single canopy gap, or whether they grow in height intermittently, following a series of small canopy gaps and surviving between them in the subcanopy shade. This is a prime project for the long term tree demography plots of CTFS.* Inferential evidence for the latter might be found in: (1) the common occurrence of less-columnar, crooked boles amongst them, (2) population structure with a higher proportion of larger subcanopy individuals than in known successional species, and (3) low maximum growth rates. There is little evidence that the first two characteristics apply to heavy hardwood dipterocarps, but both are found in other heavy hardwood stands including

Eusideroxylon, Petersianthus, and *Mesua ferrea,* an abundant tall canopy tree on ridges in Sri Lankan MDF. In these species, lack of seed dormancy and shade tolerance permits saplings to survive, and gradually gain height in light shade. The already established height of larger surviving juveniles gives them at least initial competitive dominance over faster-growing but initially shorter competitors in new gaps, which they can out-shade. Surprisingly, though, both *cengal* and ironwood are able to grow by as much as 1cm diameter a year under exceptional conditions, comparable to a light hardwood red meranti. Saplings may therefore sometimes reach canopy height following a single canopy opening.

2.2.4. *The mature phase*

Stand maturity is said to be reached when the upper canopy species have reached full height, though they continue to grow in diameter. Their crowns then broaden to their full dimensions, while their leaves become concentrated in one shallow layer at the domed crown surface (Plate 2.13). In calm climates, this phase may occupy more than 80% of the canopy[98] but, as we shall see, the proportion varies both over time and with topography and soil. Without data on height growth its area can only roughly be inferred, particularly because the mature phase at landscape scale includes many stands with diverse histories. The main canopy intervening between the emergents may itself be mature and relatively stable; or it may represent emergent species in late succession. The relative area occupied by main canopy and emergents differs between forest types (section 3.3 below and Chapter 7) (Fig. 2.3).

2.2.5. *Plasticity in plants*

Plasticity is the capacity to adapt to contrasting conditions, either as an individual at one stage of its life history or as a species. Plasticity may refer to adaptive attributes such as the range of photosynthetic response to different light intensities and spectral ranges, or the parallel responses in leaf morphology and anatomy. It may refer to patterns of ontogenetic change from seedling to mature plant within a species, or to the range of soil types on which the plant or species has competitive advantage.

Climax canopy species have low light compensation points and saturation rates of photosynthesis as seedlings. They may retain these in large measure at maturity.[99] Differences in height at maturity are associated with different canopy light regimes, and plasticity is required for light adaptation during growth.[100] Shade-tolerant climax canopy species therefore possess the greatest plasticity during their ontogeny, from shade-tolerant seedling to fully exposed crown (Table 2.2; Plate 2.13). The leaves of most change little in shape, though they may change in size. Saplings bear generally larger, thinner leaves than mature individuals, with a shallow palisade layer and thin mesophyll spreading horizontally. They bear sparser indumenta than those of mature individuals, if present. Leaves of these saplings, like those of mature crowns, often experience photosynthetic inhibition when growing in full sunlight, a manifestation of the limits of their plasticity[101] (Figs 2.4, 2.16). In a study at Pasoh, mean photosynthetic rate was found to peak and then decline in relation to increasing *photosynthetically active radiation* (PAR) in the crown leaves of four canopy species.[102] However, Ninomiya[103] found that the light saturation point of some dipterocarps increased as the trees entered conditions of continuous sun. Crowns in the canopy of a late successional stand in Panama were found to experience reduction in carbon dioxide uptake, accompanied by reduction in stomatal conductance.[104] Leaf surface temperatures there had exceeded 35°C and vapour pressure differences between leaf and air were more than 40 m Pa Pa^{-1} (season and wind speeds were not noted). It was inferred that daily net photosynthesis might achieve higher levels beneath overcast skies. *Dryobalanops lanceolata* in Borneo, and to a lesser extent *Shorea megistophylla* in Sri Lanka, are remarkable in their photosynthetic plasticity. Their saplings are highly shade-tolerant, surviving in the understorey where the leading shoot hardly grows but the horizontal plagiotropic

branches extend diffuse layers of leaves. In this growth habit they resemble several subcanopy *Diospyros* (Ebenaceae) and *Polyalthia* (Annonaceae). Released into a gap, they achieve height and diameter growth rates comparable to the fastest pioneers.[105] *Saplings of Dryobalanops aromatica have similar growth characteristics, but await ecophysiological study. These features, if common to all Dryobalanops, may be major contributors to their success as gregarious emergents (Chapter 7).*

Subcanopy species share low light compensation points but, ill-adapted to full sunlight, they have less plasticity. Pioneers, being highly intolerant of shade, also have relatively low plasticity in these respects.

Plate 2.13 Emergent dipterocarp crowns. **a,** Shallow, near monolayer, crown approaching maturity of the heavy hardwood *Shorea superba*, shade tolerant as a sapling. Friable clay loam, Danum, Sabah; **b,** *S. parvifolia* emerging above the main canopy, with branches becoming orthotropic, Danum, Sabah; **c,** Mature *S. curtisii* with multilayered microphyll leafage concentrated on the periphery of the domed crown. Yellow sandy humult soil, Labi road, Brunei, 1959; **d,** The multilayered microphyll leafage of a mature *Dryobalanops aromatica* crown, close-up from above. Yellow sandy humult soil on a Lambir ridge. (**a–c** P.A.; **d** C. Ziegler)

2.3. *The ever-changing canopy*

2.3.1. *Mixed dipterocarp forest canopy structure*
MDF present a unique canopy structure. They include the tallest forests in the tropics, with the sole exception of stands of the conifer *Araucaria* in New Guinea. (There, as so often in forests worldwide, a single monopodial conifer species maintains apical dominance over angiosperm competitors from germination, finally outstripping them in height.) In MDF, the density of emergents is unparalleled owing to the abundance of dipterocarp numbers and species. And, except in narrow valleys humid with the spray of whitewater rivers, MDF canopies are only sparsely adorned with epiphytes. Asian forests lack tank epiphytes, while the giant *Platycerium* ferns and giant orchids such as *Grammatophyllum speciosum* are rare, other than at the forest edge (Plate 3.3). This contrasts with the upper Amazon forests, where emergents are scattered but the broken main canopy is dense with tank epiphytes.

The leaves, or leaflets, of rainforest canopy trees are variable in size (Table 2.1), though they differ somewhat in predominant size according to habitat (section 3.1 below). Overwhelmingly, though, they are elliptic or lanceolate, leathery, hairless, dark, and often shiny above but dull and sometimes hairy or waxy beneath. Petioles are short, stout, and do not articulate, holding the blades in rigid positions on the twig. The leaves do possess stomatal control at least following first expansion, but transpire relatively freely;[106] water loss is restricted by leaf shape, orientation, and arrangement within the crown (section 1.4.4 above).

2.3.2. *How does the two-stratum canopy develop?*
How emergent and main canopy strata become differentiated and sustained is one of the questions we cannot answer precisely until enough examples have been carefully monitored over time. Canopy structure must be interpreted from the interplay between the gradual process of tree growth and the usually sudden death-event. The following can be observed. Late-successional stands growing in large wind-throws and other gaps of similar size form a canopy which includes immature dipterocarps along with immature main canopy species, plus the occasional tall-growing pioneer. All will have been relatively fast growing to reach the canopy within the high light intensity of a gap. The emergent climax species, being genetically capable of achieving greater maximum heights, do eventually outgrow the main canopy species, which then survive between their crowns.

MDF experiences a range of frequency and intensity of catastrophic canopy opening. In addition, all MDF suffer regular mortality of individual standing emergents, but this does little to increase the sunlight penetration to the subcanopy. The crowns of adjacent main-canopy individuals then quickly expand to fill these smaller gaps, perpetuating subcanopy light conditions and inhibiting further height growth of tree crowns beneath, including that of immature dipterocarps. These suffer crown die-back and eventually mortality. Thus, over time, a stable main canopy stratum is formed between a scattered emergent stratum.

Droughts may be the nemesis of emergent pioneers in the wet tropics, and were seen to disproportionately kill the light-demanding *Shorea leprosula* on ridges in East Kalimantan during the 1982 ENSO-induced episode. But the final stature and dominance of emergent dipterocarps and legumes may be explained by their greater access to soil water during dry periods, thanks to their extensive ectotrophic mycorrhizal mycelia. Few climax species have the capacity for mature crown repair by epicormic branching. Trees with such reduced crowns do not recover maximum growth rates.[107] Competition among crowns thereby accentuates the stratification of the leafy layers.

It seems likely that the paradigmatic emergent canopy structure of lowland evergreen tropical rainforest, composed of more or less isolated individual and clumped giants, must reflect the vestiges of stands which arose from gaps of different median diameters. After reaching the canopy stratum, these individuals experience continuing individual background mortality through drought and/or disease, which affect species differentially, or from lightning, which does not. This background pattern is overlain by periodic catastrophic events: major droughts, landslips or volcanic eruptions. The pattern and density of the emergent canopy, and therefore the extent of the main canopy, largely reflect the history of canopy mortality, and specifically the median scale and frequency with which canopy gaps are formed. They are also specific to a habitat. *Only careful continuing observation, at the landscape scale, can test this proposition. An opportunity is now presented by the worldwide network of CTFS plots.*

2.4. *Mortality*

2.4.1. *'Trees of the past, present and future'*
All tree communities consist of individuals whose numbers decline as size-classes increase. Even among the fastest-growing individuals of a particular shade tolerance class, only a very few will ever reach reproductive maturity. This is because competition for light and space results in the early death of the vast majority. As a result, the great majority of individuals in a stand are growing at less than their maximum rates[108] – indeed many are not growing, but are trapped in a slow decline toward death. The leading factor correlating with decline in individual performance of juveniles is damage or poor development of the crown through out-shading.[109]

Capacity to survive low PAR intensities is as crucial to survival as the capacity to respond to increased light with rapid

Plate 2.14 Drought-induced mortality. **a,** *Dryobalanaops aromatica* seedlings on a Lambir ridge; **b,** Massive emergent mortality in primary unlogged MDF along low ridges, one year after the 1982 ENSO-related drought, near Wanariset, East Kalimantan. (**a** C. Ziegler; **b** W. Smits)

growth: there is a trade-off between the two. Pamela Hall and Lisa Delissio[110] monitored 8,500 seedlings of a range of mostly climax species over 10 years, in MDF on low-nutrient humult sandy soils at Lambir and Bako National Parks (Sarawak). Remarkably, half the individuals in those sites survived, and population samples of 13 species included a majority over 10 years old. But few grew substantially in height. Similarly, Gong Wooi Khoon found high survivorship among seedlings of the main canopy dipterocarp *Hopea pedicellata* on drought-prone granite-derived sandy soils in the coastal hill dipterocarp forest on Penang.[111] Nevertheless, the example of *Dryobalanops* (section 2.2.5 above) demonstrates that there is much latitude in the relative costs and benefits of adaptations favouring growth and those favouring survivorship. It is this latitude, rather than the place on a linear growth-survivorship gradient, that provides the great range of options for niche specialisation in relation to light – and these are further expanded by changes in response within species during ontogeny.[112]

Mean growth rates of a natural stand on a particular site give little indication therefore of the potential growth rates (or productivity) that can be achieved in a plantation, where competition is minimised by wide spacing of planted trees; nor can species' growth rates be so evaluated. But *maximum* growth rates achieved in natural stands do provide more reliable measures.

Oldeman introduced the concept of 'trees of the past, present and future'.[113] He considered these stages identifiable through static observation. He defined them respectively as those individuals that are in decline through competition or disease; those that have reached full height and reproductive maturity; and those in active height growth. Decline through senescence *per se* appears relatively rare in Asian rainforests, since it is generally pre-empted by natural catastrophe. This is consistent with vertical stratification according to canopy branching and onset of flowering, characteristic of his 'trees of the present', and with the ready identification of canopy dominants in successional stands. Identifying the 'trees of the future' among juveniles of shade-tolerant species is more challenging, and requires prior knowledge of their photosynthetic light response curves; but it is essential for silvicultural management in selection systems (Chapter 8). Percentage mortality at first remains constant as tree size increases, but eventually increases with it.[114] This may be owing to the increased requirement for light and leaf area as the respiratory cost of maintenance increases with tree size, as well as to increasing vulnerability to catastrophic events such as drought (Plate 2.14).

2.4.2. For how long can a rainforest tree live?

Species may achieve dominance through longevity following attainment of uncontested space at maturity. Such species, usually slow growing hardwoods, include the *balau* (see chapter 7), *Eusideroxylon,* and *Petersianthus* (Plate 2.12b). I sampled the heartwood of a ironwood (*Eusideroxylon zwageri*) 327 cm in girth, felled for roofing shingles in an East

Kalimantan swidden, but it was less than 250 years old and thus too young for carbon dating. But another assay using the same method yielded 877 (–1,000) years, with mean radial growth of 0.058 mm y^{-1}.[115] The novel, if complex, methods based on the signature left when wood tissue is laid down in years of high atmospheric radioactivity have not yet been applied to trees in the Asian tropics. Ng has argued that the tannins and other chemicals that protect hardwood from rotting have a species-specific longevity of effectiveness. If this is known (from the durability of house piling, for instance), then the age of surviving heartwood can be calculated from the width of its band, and the age of the tree from that plus the radius of the rotten core.[116] From his observations and on this basis, he estimated that few Malaysian lowland forest trees exceeded 400 years in age. By such methods, the largest *Shorea superba* in east Sabah might be over 600 years old; yet one individual of 43.9 cm radius aged by C$_{14}$ dating was estimated to have started growing 1,660–1,685 BP![117] At the other end of the scale, I have witnessed a cohort of *Trema* die after six years in *kerangas* in Brunei, and *Neolamarckia cadamba* after c.12 years on sandy alluvium in southern Cambodia. Canopy light hardwoods seem unlikely to live more than 300 years.

2.5. *Primary productivity*

Because mortality rates vary greatly over time, especially in those forests subject to occasional major catastrophe, net production rates also vary and are generally positive, even over decades, because most stands are recovering from some previous catastrophe (Fig. 2.12, Plate 2.14b). This variability greatly complicates estimates of long-term productivity and interpretation of site-related differences. It is currently of particular interest in the context of assessment of the influence of tropical forests on the global carbon cycle and climate change, so will be discussed in Chapters 8 and 9.

Tim Whitmore[118] reviewed productivity estimates in Far Eastern and other tropical forests, and I raise the issue here only to discuss their significance in the present context. Tatsuo Kira and his colleagues[119] estimated the net primary productivity of MDF at Pasoh in the most rigorous study of its kind. Three independent methods of estimation were used, after standing above-ground biomass had been laboriously calculated by destructive sampling of 0.2 ha samples of the mature phase. Emergent trees there reached 55 m in height. In two separate samples the above-ground standing biomass was 475 and 664 t ha^{-1}, substantially higher than rainforests elsewhere in the tropics and far higher than estimates from documented temperate broadleaved forests. These estimates are as high as North American *Tsuga* (hemlock) and *Pseudotsuga* (Douglas Fir) stands, but individuals not as high as the tallest individuals there. This biomass is held disproportionately in the largest-diameter trees of the emergent stratum (bearing in mind that biomass is related to the square of the bole diameter). MDF is now almost gone, but it was a major repository of globally circulating carbon stocks.

Nevertheless, Kira estimated that annual *net* biomass production was an unremarkable 27 t ha^{-1} at Pasoh, when losses via litter and dependent organisms were included.[120] This implies an annual rate of standing biomass turnover of only 1.0–1.5% between catastrophic events, since confirmed by 20 years of data on growth and mortality of the 350,000 individuals more than 1 cm diameter under periodic recensus in the 50 ha CTFS plot established in Pasoh since 1985–1986.[121] However, *gross* annual production, which additionally takes account of respiration loss, was estimated by Kira at a massive 80 t ha^{-1} y^{-1}. Thus, a stock of about 500 t biomass (equal to about 250 t carbon) recycles about 35 t carbon (about 125 t carbon dioxide) by photosynthesis and respiration annually. This represents about five times the rate of respiration recorded in temperate broad-leaved forests, and about four times their gross production rate.

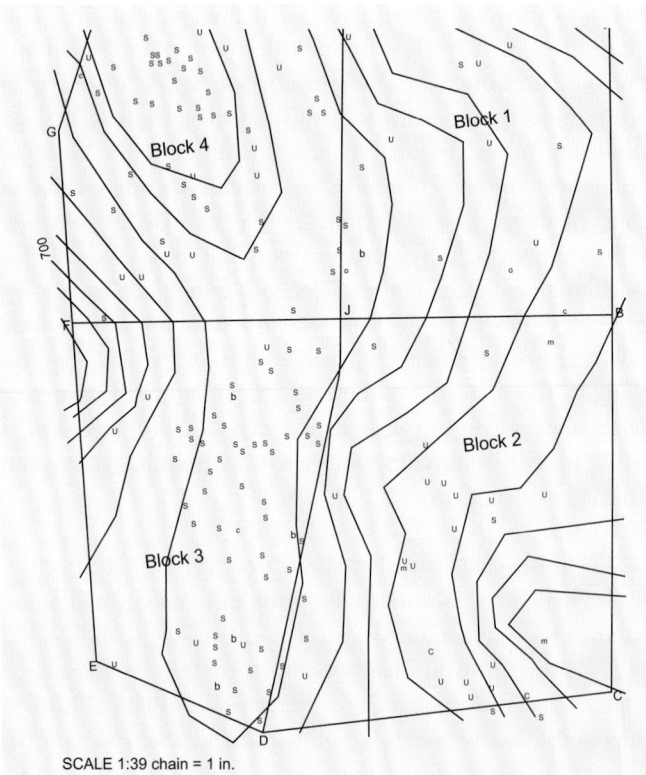

SCALE 1:39 chain = 1 in.

Fig. 2.12 Map of canopy tree mortality following a drought, along a ridge in hill MDF with canopy maturing 40 years after a shelterwood cut, Senaling Inas Forest Reserve, Peninsular Malaysia. S, *Shorea curtisii*; B, heavy hardwood *Shorea* sect. *Shorea*; C: *Neobalanocarpus heimii* (a heavy hardwood); M, meranti; O, other timber species; U, non-timber species. (From Tang & Chong 1979)

Maximum growth rates of climax species are similar in tropical and temperate broadleaved trees (but see the next two sections below). The rates of diameter growth and of biomass increment are similar, notwithstanding a growing season at least twice as long in MDF. Surprisingly, LAI was c.7–8 in Pasoh MDF compared, for instance, with 5.5–6 in a 20 m regenerating red oak stand in New England. The quite low LAI in MDF relative to its great stature likely reflects the stringent between-leaf competition for light in the calm climate of its canopy. Yoda,[122] in a now celebrated diagram, showed that the single leaf layer in the crown of an emergent (*Koompassia malaccensis*) reduced incident light by three quarters (Fig. 2.13). Relative illuminance gradually declined downward from the main canopy therefore, linearly when expressed on a logarithmic scale, to 0.2% at the forest floor. Nevertheless, mean leaf area density (LAD) was < 0.2 m^{-2} m^{-2}. LAD peaked at c.0.25 m^{-2} m^{-2} in the main canopy, then declined to c.0.1 before increasing again to 0.4 below 1.3 m, where canopy juveniles comprised the majority. Vines and stemless palms contributed c.40% of LAI. High respiration rates must reflect the cost of maintaining such tall forest and high biomass, and may in part reflect the cost of defence against predators and pathogens in a warm perhumid climate; but it will also be due to the impact of constant high temperatures on metabolic rates.

Bearing in mind that the range of *net* photosynthetic rates in tropical forests differs little from those in cool temperate forests if herbs are included, the massive difference in *gross* rates of photosynthesis might be due to: (1) a growing season nearly three times as long with favourable soil water conditions; (2) a modest contribution from a slightly higher LAI; and (3) a greater contribution from the extraordinarily large surface area, per unit area of ground below, receiving photon flux values in excess of maximum photosynthesis requirements, due to the domed crowns and two-layered canopy.

3. Structural and physiognomic variation within perhumid landscapes and regions

Climb from the trunk road onto the ridge that leads through and beyond the Lambir CTFS 52 ha plot in the Malaysian Borneo state of Sarawak. Your first impression will be of a dark understorey with distinct sunny patches caused by sharp breaks in the canopy, but once the main ridge is reached the scene becomes uniformly lighter. The understorey is dense with saplings of canopy dipterocarps and other species, and the extraordinary tunjang pipit palm (Eugeissona tristis subsp. gracilis) (Plate 2.16a), the living parts of its stems lifted on tall straight aerial roots like stilt walkers.

The ecological influence of this region's calm perhumid climates is uniform because the range of variation in their mean annual rainfall lies entirely above the threshold of the physiological requirements of trees. Western Malesia, including the southernmost Philippines, also constitutes the only extensive *windless* tropical region in Asia. In the perhaps unique presence of these climatic factors together, the influence of other factors on forest structure, dynamics and composition can be evaluated free of the influence of regional variation in climate, which is ubiquitous elsewhere.

In the perhumid tropical lowlands of the Far East, rainforests vary across the landscape in stature, vertical structure, physiognomy and dynamics more than has been observed anywhere else within a uniform climate. This cannot be attributable to variation in prevailing wind speeds and consequent leaf pruning, or to variation in local rainfall and therefore drought. In fact, the variation is arguably as great as between other evergreen broadleaved forests of *differing* climates! Within an effectively uniform rainfall regime, this variation is particularly marked on the steep ridges and siliceous rocks so widespread in sedimentary Borneo; but it is also marked on granitic Peninsular Malaysia and, albeit less so, on the ancient metamorphic mountains of south-west Sri Lanka where forests experience the south-west monsoon winds (Chapter 1). In many Neotropical forests studied, and also the northern and

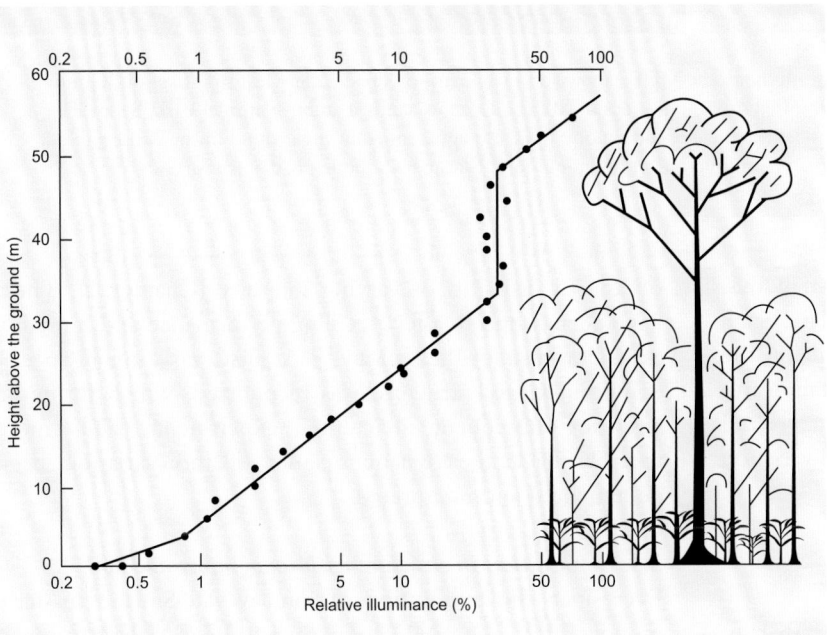

Fig. 2.13 Light extinction curve (percent total sunlight, logarithmic scale) beneath an emergent *Koompassia malaccensis*; MDF, Pasoh, Peninsular Malaysia. Note the high proportion of light absorbed (or reflected) by the emergent crown (bearing in mind the logarithmic scale of relative illumination). (After Yoda 1974)

Plate 2.15 The MDF subcanopy. **a**. The break in foliage beneath the dense MDF canopy on clay loam soil creates clear leaf stratification. Gunung Palung National Park, West Kalimantan; **b,** The less clearly stratified subcanopy trees of MDF on yellow sands: Lambir National Park, Sarawak*;* **c,** Leaf diversity beneath the Lambir canopy, on sandstone; **d,** The diffuseness of the canopy foliage is apparent from the ground in the dense multi-layered MDF on humult yellow sand over the Lambir sandstone: hemispherical photograph. (**a** T. Laman; **b** C. Ziegler; **c, d** S. Russo)

southern Far Eastern 'rings of fire', the hills bear base-rich soils, often over volcanic rocks or influenced by volcanicity. Such soils dampen this variability. Variability is enhanced, on the other hand, by the emergent canopy dominance of dipterocarps in most forests in the Asian tropical lowlands. Their stature and density in particular varies in a pattern consistent with variation in the physical landscape.

3.1. *The lowland mixed dipterocarp forests*

The geographical range of MDF coincides with that of perhumid equatorial climates, in the Far East and south-west Sri Lanka (see maps within the covers). These are defined by *mean* monthly rainfall in excess of estimated mean evapotranspiration rates, that is, 100 mm per month. Nevertheless, coastal regions and those exposed to the effects of the ENSO experience

occasional droughts of varying intensity (Chapter 1). These have been found to affect seedling performance and mortality as well as that of the canopy.[123] At first, variation in species composition and structure was recognized, which formed a basis for subsequent silvicultural experiments, Only more recently have the dynamic causes of this variation been documented and interpreted.

3.1.1. *Variation in structure with topography and soil*
Ever since the publication of John Wyatt-Smith's masterly monograph,[124] zonal MDF of perhumid West Malesia growing on yellow-red soils of moderate clay content has been known as *red meranti-keruing forest*. The upper stratum is dominated by light hardwood *Shorea* of the light red and yellow meranti field groups and species of *Dipterocarpus*, with scattered white meranti, all of which are light-demanders as juveniles. The

red and yellow merantis are almost confined to perhumid Sundaland and the Philippines. Only one species, *S. selanica,* occurs east of Wallace's Line (where it formerly dominated the emergent canopy in islands of the Moluccas), and none in seasonal Indo-Burma (Chapter 6).

Following Symington,[125] Wyatt-Smith recognized that the coastal hills of Peninsular Malaysia bore a floristically distinct forest (Chapter 3) differing silviculturally from inland forests in their dominance by slower-growing dipterocarp species of higher wood specific gravity. Most of this *coastal hill dipterocarp forest* was dominated by *balau* (*Shorea* sect. *Shorea*), but in some areas was dominated by slow-growing dark red meranti (*S. curtisii* sect. *Mutica*). Both authors also recognized that the forest on inland ridges above c.300 m was floristically and

structurally similar to this coastal forest, whereas the adjacent steep slopes supported forest floristically similar to the lowlands, but with patchier stands of mature emergent light hardwood dipterocarps (Chapters 3, 4).

Wyatt-Smith also recognized that some inland MDF also carry dense populations of single emergent heavy hardwood species, including *Neobalanocarpus* and *Shorea* sect. *Shorea* (*balau, selangan batu*), which there dominate over the light hardwood dipterocarps. Light hardwood species, in the absence of dipterocarps, are also subordinate to heavier hardwoods in the mature phase of seasonal evergreen dipterocarp forest, as in rainforests of seasonal climates elsewhere in the tropics.

In Brunei, the proportion of heavier hardwood species was found to increase from hillsides to ridges and from clay to sandy soils[126] (Table 2.4), implying that these species and their forests enjoy a competitive advantage on sites prone to soil water stress. Understorey density was also higher on ridges and sandy sites, implying higher subcanopy light conditions favourable to survival of shade-tolerant heavy hardwood juveniles but insufficient for species with higher compensation points (Plates 2.15, 2.16). The structure and dynamics of three MDF were documented over 20 years on a coastal site on leached humult yellow sands in Sarawak,[127] on inland sites

Table 2.4 Density and basal area (b.a.) per hectare of light and dark red meranti trees >9.7 cm dbh compared, in groups of five, 0.4 ha plots, in MDF along the primary ordination axes at Kuala Belalong and Andulau, Brunei Darussalam. The more shade tolerant dark red merantis have competitive advantage where diffuse canopies with small gaps dominate, by favouring these slower growing species through their higher survivorship at moderate understorey light levels. Lower nutrient levels and more frequent water stress also correlate. (From P.S. Ashton 1964a)

Kuala Belalong (udult clay loam soil, moderate nutrients)

Mean axis position		2.2	2.7	3.0	3.3	3.4	4.0	4.9	5.2	5.5	8.4
Topography				Slopes					Ridges		
		150 m				450 m	450 m				650 m
Density	Light	2.88	3,36	5.35	5.28	12.0	3.84	7.92	6.96	6.48	2.88
	Dark	0.25	-	-	-	-	-	0.24	0.72	0.72	-
b.a. (m²)	Light	1.23	2.14	5.31	5.27	5.89	3.72	8.61	7.99	7.26	2.32
	Dark	0.40	-	-	-	-	-	0.20	0.46	1.18	-
b.a. ratio: light/heavy		-	-	-	-	-	-	43.05	17.37	6.15	-

Andulau (humult yellow sandy clay soil, low nutrient)

Mean axis position		2.5	3.5	4.4	5.4	6.0	7.0	7.3	7.6	8.2	8.6
Topography						(.................................Ridges.....................150 m)					
					(.................................Slopes....................................)						
		(50 m ....Alluvium.....................)									
Density	Light	2.88	3.36	5.28	6.72	4.80	4.32	2.40	1.44	0.48	0.48
	Dark	-	1.92	2.88	3.36	9.60	5.28	6.24	10.08	9.60	12.96
b.a. (m²)	Light	2.15	2.0	2.42	2.71	2.37	0.84	0.81	0.28	0.35	0.10
	Dark	-	0.81	1.58	1.93	4.49	3.94	3.54	4.55	8.58	5.36
b.a. ratio: light/heavy		-	2.47	1.53	1.40	0.53	0.21	0.22	.06	0.04	0.02

Plate 2.16 The MDF understorey. **a,** The dense Lambir understorey in diffuse subcanopy light on sandstone: MDF within the CTFS plot at Lambir National Park, north-east Sarawak. Tunjang pipit (*Eugeissona tristris* subsp. *gracilis*), the predominant understorey palm on deep humult yellow sandy soils in MDF and deep podsols in *kerangas* in north-west Borneo; **b,** Diffuse MDF understorey beneath giant yellow (*Shorea gibbosa*) and red merantis in the dark subcanopy of a clay loam alluvial bench, Gunung Palung National Park, West Kalimantan. (**a** S. Davies; **b** T. Laman)

over deeper sandy soils, and on friable clay loam over Tertiary basalt (Fig. 2.14; Tables 2.5, 2.6). Mineral soil total nutrient concentrations were lowest at the coastal site and highest on the basalt-derived clay (Chapter 3). The range of structural variation within MDF in Sarawak was recorded by diagrams of the vertical profile of forest within 64 × 8 m strips (Figs 2.1, 2.12; Plate 2.17). The tallest forests were found on the deepest soils, on broad ridges and undulating land, but the densest emergent canopy stratum was confined to deep friable clay loams with good crumb structure and no mottling. Leaf sizes were larger on clay than on sandy soils and on slopes than on ridges (Fig. 2.18).

In perhumid south-west Sri Lanka, where MDF also dominates the lowlands, the forest is shorter, mostly 30–40 m tall, with a few emergent species or individuals reaching 50 m only in sheltered coves. Their relatively smooth canopy may be attributed to the strong breezes experienced throughout the

Plate 2.17 Canopy variation with topography at Lambir. **a,** The broken emergent canopy on slopes reflects windthrow and landslip frequencies; **b,** Ridges support a more continuous emergent canopy, broken by smaller gaps created by trees dying standing. (C. Ziegler)

year and gales during the wet south-west monsoon. However, the relationship between forest structure and the landscape of south-west Sri Lanka closely recalls that of inland Borneo.

The travelling naturalist soon discovers not only that forest structure constantly varies with habitat within a landscape, but that this variation is repeated in similar habitats, on similar land forms, and in similar climates regionally. This predictability of structure, we infer, can only be explained by the constraints that habitats variously impose, through soil nutrient limitations and proneness to water deficits on plant function, on species' life cycles and on the dynamics of the forest.

3.1.2. *Variation in forest and stand growth*

Even after 20 years, mean growth rates scarcely differed among the three Sarawak forests studied, but the growth rates of the fastest growing individuals, namely pioneer and successional climax species (Tables 2.5, 2.6, 2.7b), were markedly higher on fertile clay loams.[128] Within this group, competitive advantage is determined by response to nutrients as well as to light. Growth of pioneer species, in comparison to climax successional species, responds strongly to nutrient addition in pot experiments.[129] But individuals of late successional and climax species at the Sarawak MDF sites, although hardly

Table 2.5 Proportional growth rates in relation to soils nutrient levels over 20 years in MDF in four, 0.6 ha, plots at three sites, Sarawak, Malaysia. Plot data combined. Trees >9.7 cm dbh. Residents are trees that were measured from first census, recruits those that appeared in later censuses. Diameter categories refer to the trees at their onset of measurement. Growth rates in cm, computed for the total length of time indicated, except for recruits for which means were estimated for the time they were present. For all trees >9.7 cm dbh, 9.7–30 cm dbh, and >30 cm dbh; *: $P<0.05$, **: $P<0.01$, ***: $P:<0.001$. HCl-extractable nutrients from mineral horizons at 70–80 cm. (From P.S. Ashton & Hall 1992)

differing in maximum growth rates between sites, differed substantially in their height/diameter ratio (Fig. 2.14). This implies that growth rates in height and biomass productivity *did* vary, in relation either to nutrient levels or water stress. On the lower-nutrient, drought-prone sites where forest stature is relatively short, canopy trees are stouter for a given height, while the most slender trees are in the tallest stands, which are on both fertile loams and at the margin of peat swamps. This implies that it is their water economy, rather than nutrients, which mediates mature stand stature. Since

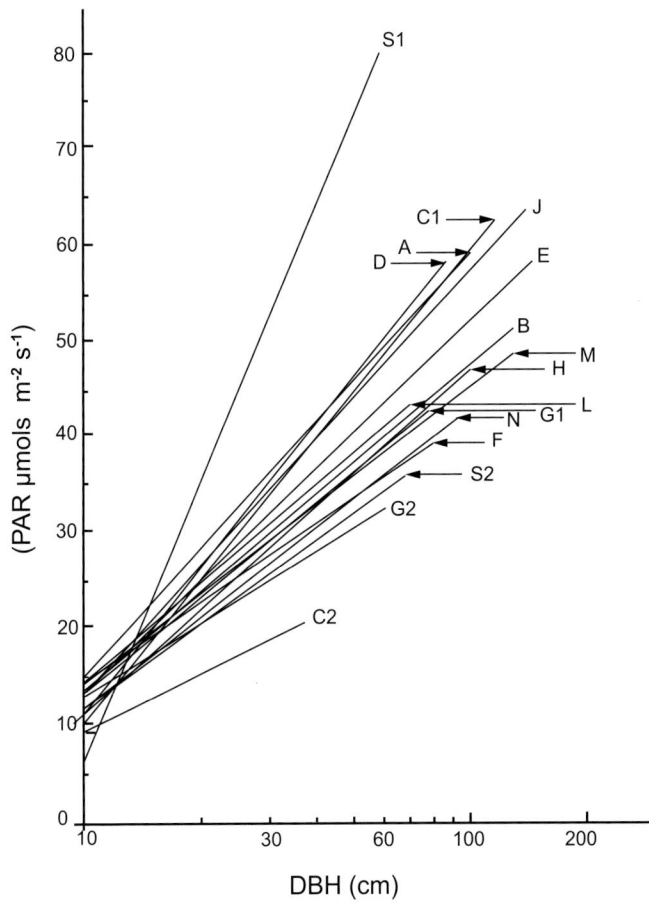

Fig. 2.14 Change in the overall slope of the height/diameter relationship, of trees >5 m tall, on different soils, Sarawak. Data derived from 64 x 4 m profile diagrams. Lines drawn using coefficients from a major axis regression of $\log_e$ transformed height and dbh, positioned from minimum to maximum recorded dbh. Site locations as in Fig. 3.9. Soils: shallow humult yellow sand on slope (B, F), deep yellow humult sand (A, L, M, N); sandy clay loam on shale and sandstone (H), on rhyolite slope (G1) and ridge (G2), and on dacite (D); deep clay loam on basalt (C1) shallow loam on basalt ridge, 900 m (C2), on shale low hills (E, J; outer *Shorea albida* peat swamp phasic community 2 (S1) and inner phasic community 4 (S2). The data were insufficient to test for departure from linearity, and it is possible that the trajectory of the lines is determined more by the height of the canopy, after which trees fail to gain height but do gain diameter, than by differences in the height/diameter relationship during height growth (From Ashton & Hall 1992)

Tree category	Phosphorus	Potassium	Magnesium
All trees (dbh >9.7 cm)			
1965–1985	0.44	0.58*	0.48
1965–1970	-0.49	-0.35	-0.42
1980–1985	0.63*	0.58*	0.69**
Survivors 1965–	0.32	0.51	0.40
Recruits	0.66**	0.65**	0.67**
dbh 9.7–30 cm			
1965–1985	0.48	0.61*	0.52
1965–1970	-0.37	-0.26	-0.31
1980–1985	0.69**	0.69**	0.74**
dbh >30 cm			
1965–1985	0.09	0.01	0.09
1965–1970	-0.54	-0.35	-0.49
1980–1985	0.19	-0.02	0.23

Site & data category	Whole stand	dbh class (cm)				
		10–20	21–30	31–40	41–60	>60
Bako						
Standing volume (m³ ha⁻¹) 1965 1985	402.1 464.8	55.7 45.9	81.0 70.2	76.2 84.5	117.7 157.8	67.4 103.2
Net change	+63.6	-9.8	-10.8	+8.1	+40.1	+35.8
% change	+15.9	-17.6	-13.3	+10.7	+34.0	+53.1
Recruitment	10.1					
Mortality	116.0					
Increment over 20 years	169.5					
Lambir						
Standing volume (m³ ha⁻¹) 1965 1985	957.4 968.8	59.7 54.8	81.8 78.3	94.0 94.0	198.3 184.3	523.5 557.3
Net change	+11.4	-4.9	-3.6	+0.1	-14.1	+33.8
% change	+1.2	-8.1	-4.3	+0.1	-7.1	+6.5
Recruitment	15.8					
Mortality	225.9					
Increment over 20 years	221.5					
Bukit Mersing						
Standing volume (m³ ha⁻¹) 1965 1985	1073.1 1173.6	27.5 28.6	46.5 50.7	61.7 62.9	157.5 147.3	780.1 884.1
Net change	+10.5	+1.1	+4.2	+1.3	-10.1	+14.0
% change	+9.4	+4.1	+9.1	+2.1	-6.4	+13.3
Recruitment	16.0					
Mortality	301.4					
Increment over 20 years	385.9					

Table 2.6 Changes in estimated volume over 20 years in MDF in four, 0.6 ha, plots at three sites in Sarawak, Malaysia. Trees >9.7 cm dbh. All figures represent volumes m³ ha⁻¹. Overall volumes have increased in all sites owing to absence of a wind- or drought-induced mortality event during the period of measurement, but most especially at Bako. (From P.S. Ashton & Hall 1992)

a(i)

species	dbh1	years measured	growth (mm y⁻¹)
Shorea macrophylla	181	5.5	99
Randia sp. 1	154	4.8	78
Dipterocarpus acutangulus	800	5.3	76
Shorea laxa	433	4.6	66
Dryobalanops aromatica	850	5.3	50
Gluta macrocarpa	346	4.7	45
Lithocarpus bancanus	223	4.7	44
Unidentified	29	5.8	43
Dipterocarpus confertus	601	5.7	43
Xylopia sp. ined.	28	4.7	42
Syzygium sp.	36	5.4	42
Dipterocarpus globosus	666	5.4	42
Dipterocarpus globosus	800	4.6	40
Dryobalanops aromatica	970	4.6	38
Vitex gamosepala	124	5.4	36
Macaranga gigantea	112	5.5	36
Dryobalanops aromatica	1115	5.4	36
Melicope glabra	162	5.5	34
Campnosperma auriculatum	164	5.3	34
Neonauclea paracyrtopoda	502	5.3	33
Shorea curtisii	1250	4.7	33
Melicope glabra	219	4.5	31
Santiria tomentosa	390	4.6	30
Macaranga gigantea	61	5.5	30
Macaranga gigantea	220	5.7	29
Parashorea parvifolia	205	5.2	28
Dipterocarus globosus	1200	5.4	27
Adinandra dumosa	140	4.6	27
Santiria laevigata	214	5.2	27
Calophyllum castaneum	250	5.5	27

a (ii)

species	dbh1	years measured	growth (mm y⁻¹)
Ficus sp. 1	2391	5.3	103
Ficus sp. 2	104.6	5.7	95
Tetrameles nudiflora	1000	5.7	95
Ficus geniculata	1178	6.6	87
Ficus altissima	1004.1	6.8	85
Ficus sp. 3	1940.8	5.2	58
Pterospermum grandiflorum	588.1	6.7	56
Ficus geniculata	210	5.5	53
Saccopetalum lineatum	373.1	6.7	50
Ficus geniculata	1267	5.2	44
Alphonsea ventricosa	227.1	5.2	41
Saccopetalum lineatum	631.5	5.7	37
Garcinia celebica	357.2	5.2	36
Sterculia hypochroa	250.3	6.8	36
Gluta obovata	440	5.4	34
Tetrameles nudiflora	316.8	6.8	34
Diospyros variegata	91.8	5.3	33
Saccopetalum lineatum	232.8	6.7	32
Champereia manillana	42.1	5.3	30
Macaranga siamensis	31.3	5.2	30
Protium serratum	172.5	5.3	29
Acer oblongum	308.6	5.3	28
Ficus altissima	340.2	5.7	27
Polyalthia viridus	54.8	6.8	27
Macaranga siamensis	44.8	5.2	27
Mallotus nudiflorus(Trewia)	242.3	6.8	27
Persea sp.	345.2	5.2	27
Persea sp.	307.2	5.3	27
Cinnamomum porrectum	210.2	5.3	26
Trema orientalis			

b

Species	Tree number 1965	Recruits	Mortality	Tree number 1985	Diameter increment mean, mm y⁻¹ (range)	
					5 years	20 years
Pioneers						
Octomeles sumatrana (Datiscaceae)	1	1	0	2	19–23	21
Pterospermum stapfianum (Sterculiaceae)	3	15	0	18	0–19	8.9±3.4 (0–19)
Macaranga hosei (Euphorbiaceae)	2	0	2	0	0–3	1.0 (0–2)
Late successional dipterocarps						
Dryobalanops lanceolata	100	13	19	94	5–26	3.8±2.6 (-1–14)
Parashorea macrophylla	10	1	4	7	2–20	9.5±7.5 (0–33)

Table 2.7 a, The 30 fastest growing trees >1 cm dbh, in CTFS 50 ha plots. Trees >1 cm dbh; dbh1: diameter at first measurement. (i) In MDF, Lambir National Park, Sarawak. Climax species, notably dipterocarps predominate and, once large, compete with pioneers (*Macaranga, Randia*) in diameter growth. Surprisingly, all but a few of the fastest dipterocarps (*Parashorea, Shorea macrophylla*) are low nutrient humult sand specialists, but perhaps their height growth (unavailable) does not compare with their diameter growth. (ii) In semi-evergreen forest, Huai Kha Khaeng Wildlife Sanctuary, Thailand. Remarkably, comparable rates to Lambir are achieved in a climate with five dry months. Figs (*Ficus*), both free-standing and hemiparasitic, dominate; few (*Tetrameles, Saccopetalum, Mallotus (Trewia), Vitex*) are deciduous and none of the deciduous leguminous trees that yield the valuable rosewoods, nor the abundant late successional *Lagerstroemia* or *Terminalia* species, are represented. Here no dipterocarps are among the fastest. **b,** Growth and mortality of leading pioneer trees and late successional dipterocarps, at Mersing forest, Sarawak, on basalt. (From P.S. Ashton & Hall 1992)

comparative diameter growth rates from peat swamp forest are lacking, it cannot yet be concluded that soil fertility mediates diameter growth, but the relative paucity of pioneer and early successional species on low-nutrient peat (as well as on podsols) implies that it does.

Mortality tends to be sudden and spasmodic, and is therefore difficult to interpret from only 20 years' observation. The distribution of dead canopy trees, and the form of mortality – whether the trees are standing, snapped or uprooted – are correlated with habitat and size of canopy gap.[130] In these MDF, dead standing trees are relatively more abundant, uprooted

trees less so, than in documented forests of the Neotropics.[131] Dead standing trees were relatively most abundant on sandy soils and on ridges (Table 2.8).

Remarkably, forest at one of the three Sarawak MDF long-term plot sites, coastal Bako, experienced a 20% increase in basal area over 20 years of census. This was concentrated among canopy individuals, implying that there had been major canopy mortality prior to the first census, likely due to a severe drought (Table 2.8, see also Table 2.7b). Tang and Chong[132] documented the development of the characteristic discontinuous emergent canopy of dipterocarp forest on freely

Table 2.8 Mortality one year following the ENSO-related 1982 drought in unburned MDF at Kutei National Park, East Kalimantan 0.6 ha plots. Number of individuals, percentages in parentheses. (After Leighton & Wirawan 1986)

Habitat	Woody climbers	Figs		Trees (dbh class)			
		Stranglers	Climbers	4–10	11–30	31–60	>60
Upland	23	86 (86)	100 (22)	20	24	43	71
Moist valley	6	n.a.	n.a.	16	14	0	11

draining soils, in a stand of dark red meranti (*Shorea curtisii*) thirty years after logging and silvicultural treatment by a shelterwood system. The forest had been clear-felled, following subcanopy thinning to ensure adequate regeneration. This treatment had increased the density of the meranti, but as competition led to variability in height and crown diameter, a drought killed many of the tallest trees (Fig. 2.12).

On freely draining but deep yellow-red sandy and sandy clay soils, the discontinuous emergent MDF canopy may still occasionally reach 70 m. The main canopy here reaches its greatest state of development and the understorey is also quite dense, implying that the light climate beneath is variable and overall brighter than on the clay loams. Canopy death on ridges and gentle slopes is mostly due to drought or lightning strike, affecting scattered single large trees that die standing, as in *kerangas* (Table 2.8). It is on these soils that the paradigmatic three-stratum structure of MDF is most developed. *Comparison of light profiles between differing forests including kerangas using Yoda's method is much needed.*

3.1.3. *Variation in species growth*

Forests, and the stands within them, vary within the landscape because they differ in species composition (Chapter 3), and species with different patterns of growth and mortality prevail in each. Climax species are the principal determinants of stand dynamics, so will be the focus here. Variables such as physiology, maximum growth potential, leaf longevity, number and turnover rate, crown structure and relative allocation of biomass to shoot and root *together* determine ecological distribution, including successional role.[133] The size and shape of gaps and the stature of the surrounding canopy provide a diversity of light conditions and soil moisture levels in which regenerating seedlings compete, setting the stage for niche differentiation[134] (Chapter 7). Stuart Davies found that, even among 11 species of the pioneer genus *Macaranga* sharing a single gap at Lambir (Sarawak), each had distinct light responses, minimum stature at first flowering, phenology and fecundity[135] (Fig. 7.4).

There are two sources of light near the ground in the mature phase understorey: gaps between crowns and branches, especially on hillsides, from which it receives periodic transitory patches of direct sunlight; and transmitted through the leafy canopy, by which the light has been depleted of PAR. These become at least partially mixed as the distance beneath the canopy increases, to form a patchwork of varying intensity and spectral composition, mostly of diffuse light. In forests where the canopy is itself diffuse, this moving patchwork is sufficient to support juveniles of shade-tolerant species with a diversity of light compensation points, often for decades. Moad[136] compared light-demanding to shade-tolerant dipterocarp seedlings in the subcanopy at Sepilok

Forest Reserve (Sabah). Relative growth rates of each species were correlated with minutes of sun-flecks experienced per day, as estimated from hemispherical photographs (Fig. 2.11). Growth rates were correlated with light compensation points, but there was a trade-off between higher growth rates and higher mortality rates with shorter leaf life span. Montgomery and Chazdon[137] found a change-over in the rank order of seedling performance as rated by growth and mortality, among unrelated seedlings in a Costa Rican forest, under different light intensities over 14 months. They concluded that the influence of light compensation point on mortality in different light regimes may be critical in differentiating species' performances. Physiological ecological research on individual climax species has overwhelmingly focused on the seedlings of emergent dipterocarp species. Stuart Davies' seminal work on the comparative ecology of co-occurring species of pioneer *Macaranga* is discussed in Chapter 7.[136]

Zipperlen and his students at Danum[138] also compared the photosynthetic rates of *Shorea leprosula* and *Dryobalanops lanceolata* at different understorey light intensities, and of photosynthetic induction from shade into light. This has implications for sun-fleck utilisation efficiency, that is, the amount of PAR utilised relative to that received on the leaf surface. They found that *S. leprosula* had a lower photosynthetic rate and higher mortality at continuous low photon flux densities (PFD), but slightly higher rates of photosynthesis and growth in high PFD. *D. lanceolata* had a higher sun-fleck utilisation efficiency potential and a marginally higher steady state photosynthetic rate in diffuse light, explaining its lower mortality rate in shade. The observed differences between these species in their light responses alone were insufficient to competitively discriminate between them. But *D. lanceolata*, in shade, shifted allocation of production to plagiotropic side branches, thereby maximising the capture of occasional sun-flecks, whereas *S. leprosula* continued to maximise height growth, a strategy that favoured it under high light. It is these differences in allocation that likely determine competitive advantage.

Brown, Press and Bebber[139] pointed out that, thanks to survivorship differentials in shade, juveniles of shade-tolerant species gain competitive advantage by creating, with their crowns, the light environment within which later-arriving seedlings must persist. Broadly speaking (though *Dryobalanops lanceolata* is a notable exception), growth response to light is inversely correlated with survivorship in shade, implying trade-offs between the energetic costs of the two adaptations. Changes in the rank order of competitiveness among species as they grow would substantially add to the number and diversity of niches within a community. Earlier investigators failed to find evidence of such changes. But Sack and Grubb[140] emphasised that such changes might be expected to occur with longer-term ontogenetic change, and concluded that

Fig. 2.15 Population distribution and structure in relation to altitude (therefore topography: lower slope to ridge) of two *Shorea* sect. *Doona* species in the 25 ha CTFS plot, Sinharaja forest, 430–565 m. alt., south-west Sri Lanka: **a,** the light demanding *S. trapezifolia*; **b,** the shade tolerant *S. worthingtonii*. (From Gunatilleke *et al.* 2004a)

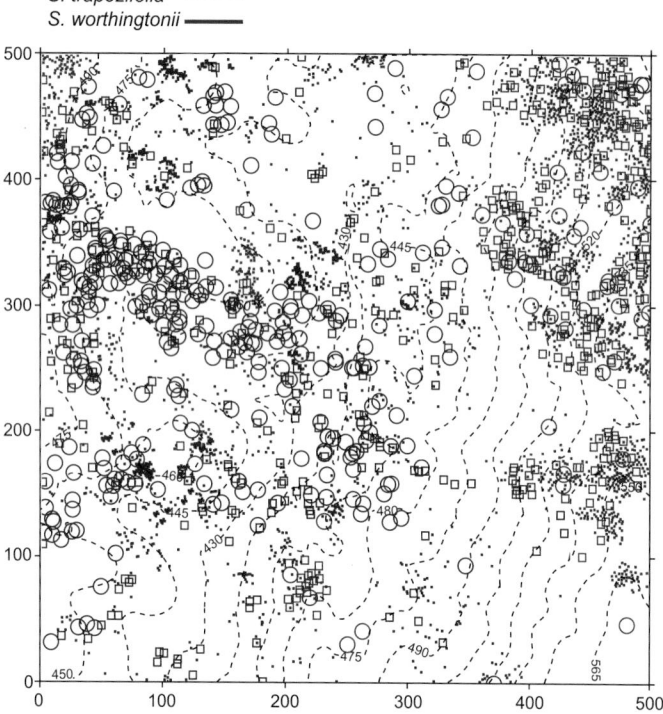

S. trapezifolia ━━━━
S. worthingtonii ━━━━

Spatial distribution of individuals (<10 cm = •, 10-30 cm = □, > 30 cm dbh = ○) in relation to elevation (m) in the 25 ha plot.

depth. Convex surfaces of upper slopes and ridges experience more frequent soil water deficits than do concave and valley slopes.[143] Raw humus is patchy and thin on ridges, absent elsewhere. Differences in nutrient status with topography are inconsistent and slight. There is evidence of past landslips along the sheer lines of the rock. Nevertheless, wherever the soil surface has survived canopy opening, established climax juveniles with developed roots and shade-imparting crowns gain competitive advantage.[144] Here again, canopy gaps along ridges are small, predominantly due to individual standing emergent tree mortality. Valley gaps, in contrast, are larger and more variable in area, mostly caused by single or group wind-throws or occasional slips. These larger gaps experience direct sunlight and high PFD for up to two hours longer each day than do small gaps (Fig. 2.6), and the difference is even greater in the diffuse light of cloudy skies than in direct sunlight. Large gaps tend to retain more soil water toward their centres, owing to lower densities of surviving roots. The canopy is most diffuse on ridges and upper slopes, and more lateral light reaches the subcanopy. Lateral light has been found to be an important factor in canopy regeneration on ridges.[145] This influx of relatively uniform subcanopy light is reflected in a dense understorey there of both canopy juveniles and treelets, as at Lambir.[146] Juveniles of shade-tolerant species have elswewhere been found to have deeper crowns and greater leaf areas in the understorey than their more light-demanding congeners.[147] Because the topographic gradient upon which these Sri Lankan data are based is general in the hills of the

Plate 2.18 Experiments into dipterocarp responses to topography, hence understorey light and soil water. **a,** Women from the village, with expertise in transplanting rice, successfully transplant dipterocarp seedlings to experimental plots in the Sinharaja forest, Sri Lanka; **b,** Mature foliage of the four *Shorea* sect. *Doona* whose seedlings were subjected to experiment: *S. disticha* (top left), *S. worthingtonii* (top right), *S. megistophylla* (bottom left), *S. trapezifolia* (bottom right). (**a** Mary Ashton; **b** Mark Ashton)

experiments need to be continued for long periods if valid tests of physiological differentiation are to be attempted. Philipson has now confirmed that prediction, by showing that allocations to growth by different species follow non-linear trajectories, with the consequence that their rank order in competitiveness in a specific environment may change over time.[141]

Research by Mark Ashton, B.M.P. Singhakumara, Nimal and Savitri Gunatilleke and their students, in the Sinharaja UNESCO World Heritage Forest (the site of a 25 ha CTFS plot in south-west Sri Lanka), has extended these observations to explain differences in species distributions in relation to topography[142] (Figs 2.15–2.17; Plate 2.18). The landscape is of steep-sided hills up to 500 m high, with narrow ridges supported by harder more siliceous Archaean metamorphic rocks, and narrow valleys bearing colluvium but little alluvium. Soils are well-drained clay loams with little mottling, yet increased soil moisture with

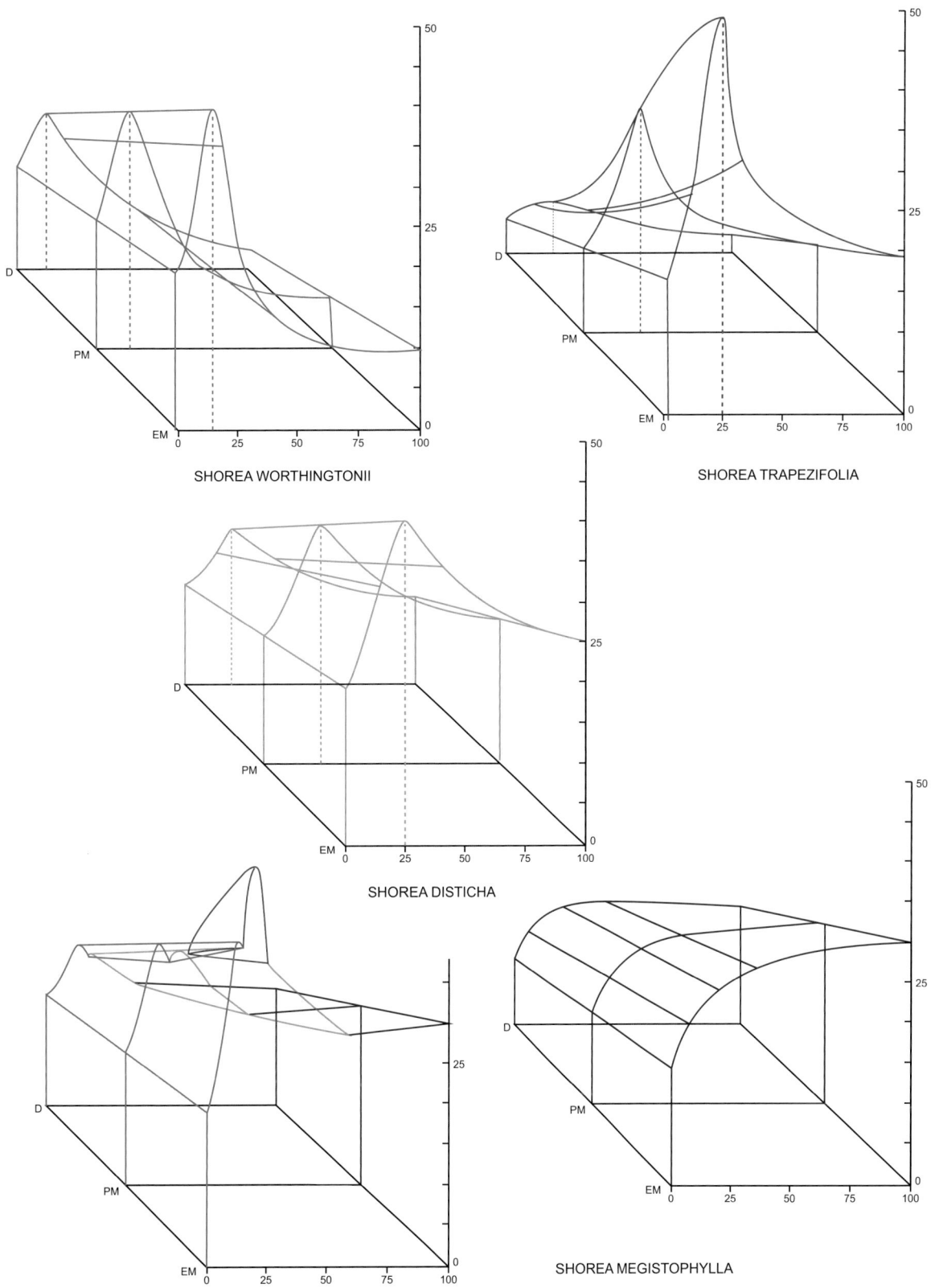

SHOREA WORTHINGTONII

SHOREA TRAPEZIFOLIA

SHOREA DISTICHA

SHOREA MEGISTOPHYLLA

Fig. 2.16 Growth of four *Shorea* sect. *Doona* species in response to gradients of light and water: schematic diagram of height growth (vertical measure) after 800 days. Soil moisture: D droughty, PM periodically moist, EM continuously wet; Light: daily PAR as percentage of full sunlight (horizontal measure); lower left diagram: extrapolated 'surface' of maximum height growth and physical conditions for the four species superimposed to reveal the conditions under which each prevails. (From P.M.S. Ashton 1990b)

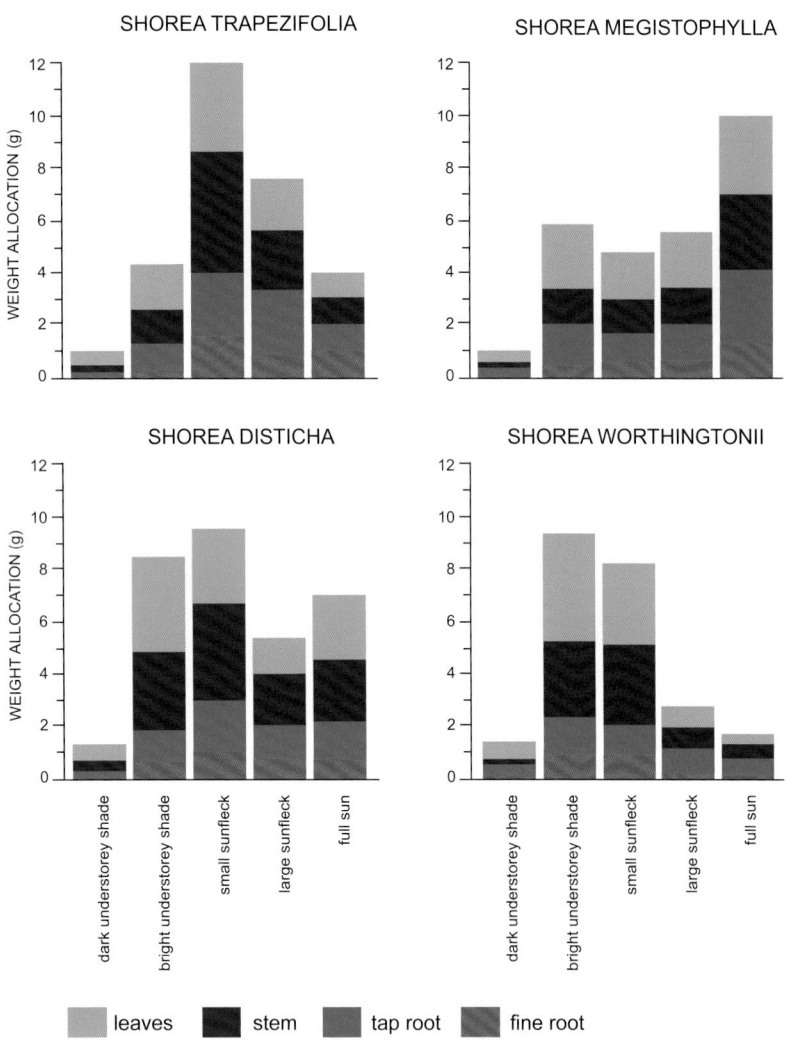

SHOREA TRAPEZIFOLIA
SHOREA MEGISTOPHYLLA
SHOREA DISTICHA
SHOREA WORTHINGTONII

leaves   stem   tap root   fine root

Fig. 2.17 Dry mass allocation among the four *Shorea* sect. *Doona* species under different light conditions. Note the high allocation to stem growth by the small gap light demander *S. trapezifolia,* and to leaves by the shade tolerant *S. worthingtonii* which comes to develop a diffuse but multilayer crown. (From P.M.S. Ashton 1995)

Far East, the results are generally applicable there as well. They therefore lay a foundation for understanding the major gradient in canopy disturbance in MDF, which in turn is a key to understanding both local and regional patterns of tree species diversity (Chapter 7). They further provide the foundation for a general practice of silvicultural management (Chapters 8, 9).

Sri Lankan MDF has a great advantage for comparative ecophysiological studies: it has only one-quarter as many tree species as Far Eastern MDF. Forest understorey density on ridges is higher than recorded in MDF in the Far East. There are nine emergent *Shorea* and two *Dipterocarpus* species in the Sinharaja forest studied. Four members of *Shorea* (sect. *Doona*) were studied[148] (Figs 2.16), with overlapping ranges

from ridge (a heavy hardwood, *S. worthingtonii*) to valley (the medium hardwood *S. megistophylla* and light hardwood *Shorea trapezifolia*) (Plate 3.2a). Field observations of planted seedlings across gaps on ridge, on slope and in valley were complemented with observations of potted seedlings in shade houses whose light regimes simulated those of the three gaps (Fig. 2.16). The mean area of individual leaves of each species consistently declined according to habitat, from valley to ridge. Photosynthetic compensation points were at a PFD of 300 for *S. megistophylla,* c.100 for *S. worthingtonii* and the other two species. Nevertheless, seedlings of the ridge species *S. worthingtonii* were still surviving in the understorey 10 years after planting, whilst 99% of those of *S. trapezifolia,* a light-demander of large gaps, had died within two years. The topographic distributions were explained by differences in their combined responses to light and soil moisture (Figs 2.16, 2.17). Only the moist valley species *S. megistophylla* escaped photoinhibition, as a decline in net photosynthetic rate above 800 PFD; the rate of the ridge species, *S. worthingtonii,* declined above 350 PFD.

Whereas *Shorea megistophylla* has a large canopy crown leaf with a thick spongy mesophyll and thick epidermis, the comparatively small leaf of the other valley species, *S. trapezifolia,* is thin, less adapted to the transpirational demands of full sunlight. The two valley species differ notably in survivorship in shade, *S. megistophylla* surviving remarkably well compared with *S. trapezifolia*. Both had higher leaf nitrogen concentrations than the other species, implying greater maximum growth rates. Although stomatal conductance increased from shade to full sun in both, transpiration rates in shade were higher in *S. megistophylla,* and transpiration peaked in both species at the light intensity at which their maximum photosynthetic rates were achieved.[149]

Survivorship in Sinharaja was higher in the shade-tolerant ridge species *Shorea worthingtonii* than in the lower slope species *S. megistophylla*. Light intensity measured at ground level in the understorey was greatest in the valley and lowest on the ridge, probably due to shade cast by the denser ridge understorey, which is dominated by saplings of canopy species. *S. megistophylla* has the highest leaf plasticity, being shade tolerant yet responding most in height growth in full sunlight. The heavy hardwood *S. worthingtonii* has a lower leaf plasticity, it survives long periods in understorey shade, but attains low maximum growth rates in sun, its leaves suffering photoinhibition. It is best adapted to the periodic soil water deficits on the ridges, while attaining maximum photosynthetic rates in the moderate light of single tree gaps. Although broadly negatively correlated with net assimilation rates, survivorship in shade thus provides a further axis along which spatial niche differentiation can occur.[150]

Further studies are being advanced using other series of congeneric species at Sinharaja, with comparable results[152] (Chapter 7). In the Western Ghats, highest growth rates of canopy trees have been found on steep slopes where the stand structure implied frequent disturbance.[151] Such seedling studies have yet to be made in Far Eastern MDF, but there appear to be many parallels in patterns of stand structure and forest dynamics. Dipterocarp distributions in relation to topography in Sri Lanka mirror those of Peninsular Malaysia.[153] Just as on clay rich soils in the Far East, for instance, MDF on lower slopes in Sri Lanka are prone to water-logging, with mottling prominent in the profile often within one metre of the surface. Rooting is shallow on the lower slopes in both, but the canopy experiences more frequent medium- to large-scale opening in comparison with slope forests on similar soils in the Far East. There are two major differences: (1) The variability of soils is greater in the Far East, where freely draining, leached and humult sands are widespread, whereas Sri Lankan perhumid zone soils are overwhelmingly clay or sandy clay loams. Comparative forest soil data from Sri Lanka remain unavailable, but the widespread cultivation of wet rice paddies on lower slopes, a technology almost confined to volcanic soils in the perhumid Far East, implies moderate to relatively high fertility; (2) The occurrence of prevailing winds, the south-west monsoon from late May until late July, is associated with episodic massive rainstorms, flooding, wind-throws and landslips on a scale unknown in the perhumid Far East outside the southern Philippines and occasionally in north-east Borneo (Plate 1.9a).

3.1.4. *Heavier hardwood species*

Heavy hardwoods are distinctive in two respects: shade tolerance (partly discussed in section 2.2.3.2 above), and drought tolerance. High wood specific gravity reflects a high density of lignified cell walls resulting from small diameter conducting cells. Small diameter vessels are less prone to cavitation, or vacuum pockets due to water tension. Cavitation reduces the height to which the tree can lift water, and therefore the height to which a tree can grow, since the effects of root pressure are limited above ground level. As might be predicted, heavier hardwood species are thus relatively more abundant in Sundaland and also in south-west Sri Lanka on narrow ridges, on freely draining sandy soils, and in the drier climates of coastal forests (Chapter 1) (Table 2.4), and in *kerangas* (see section 3.2 below) – all habitats prone to periodic severe soil water stress.

Most of the emergent heavy hardwood species of these habitats bear leaves with silvery, waxy or thinly felted undersides that reflect incident energy, increasing the boundary layer thickness and thereby reducing water loss (Plate 4.8a). The moderate light conditions beneath result in not only high mortality of light hardwood meranti and other species with higher light compensation points, but also survival of those species such as *balau,* whose leading shoots eventually extend through gaps beneath which light is sufficient for growth. The *balau* species *Shorea laevis* and *S. maxwelliana* are abundant along narrow shale ridges where the clay soil is shallow, the latter also on the freely draining sandy soils where, in Borneo, the majority of this section occur. *S. materialis* is the only *Shorea* confined to inland and coastal *kerangas; S. falcifera* and *S. glauca* to coastal hills. Each *balau* species successful in drought-prone habitats therefore has a specific ecological range determined by other factors.

Two other dense-wooded dipterocarps become dominant on freely draining soils in Peninsular Malaysia. *Neobalanocarpus heimii* (*cengai*), a heavy hardwood with large wingless seeds, becomes gregarious on freely draining soils especially, but not exclusively, on ridges. *Shorea curtisii* (*seraya*), a dark red meranti, is the dominant dipterocarp along ridges, especially on granite 300–800 m, and also on the coastal hills. Inland the limits of its range are set by environments of greater humidity, its lower limit coinciding with the ceiling of nocturnal mist, its upper limit with the cloud base and lower boundary of lower montane dipterocarp forest ('upper dipterocarp forest', see Chapter 4). In north-west Borneo, *S. curtisii* is locally common instead (though never dominant) in MDF on deep, yellow sandy soils below 400 m (Chapter 4).

The dominance of slower-growing, dense-wooded, shade-tolerant emergent species on ridges and sandy soils appears to be brought about in part by these species resistance to mortality caused by soil water stress. In part also, predominantly small gaps and diffuse canopy leafage imbue the subcanopy with a diversity of light intensities generally only sufficient for survival and growth of species with low light compensation points.

Individual heavier-timbered dipterocarp species can achieve high relative abundance in other habitats. Among the heavy hardwood *balau* in western Malesia, some are confined to mesic clay soils, including *Shorea falciferoides, S. inappendiculata* and *S. superba. S. superba,* a species of mesic clay loams, attains as tall a stature as any rainforest tree. These species appear to be at competitive advantage in large gaps where patchy soil scarification leads to development of an early blanket monolayer of large-leaved vines. Where a vine canopy becomes established early, having germinated on scarified soil, succession is slower, favouring shade-tolerant species including *Dryobalanops lanceolata* and heavy hardwoods such as *Hopea andersonii, Shorea* sect. *Shorea* species (including *S.*

superba and *S. inappendiculata)* and the ironwood *Eusideroxylon zwageri.* Among these, ironwood is a special case, as it alone has high capacity for epicormic branching within the crown, thereby recovering from killing droughts with apparently greater frequency than most other species. This may explain the formerly widespread occurrence of gregarious stands of this slow-growing heavy hardwood in southern Sumatra and East Kalimantan, the regions of West Malesia most prone to ENSO-induced severe droughts.

The Sinharaja research in Sri Lanka has defined the ecophysiological characteristics of the canopy species prevalent in two major classes of MDF, by structure, dynamics and floristic composition. These are the classes discussed in Chapter 3, occurring respectively on ridges and slopes, alternatively on freely draining humult yellow sandy and udult clay loam soils. Their management requires fundamentally different silvicultural treatment (Chapters 8, 9). Differences in the light response of co-occurring species sharing a common habitat and community are further considered in Chapter 3.

3.1.5. *The mature phase understorey*

Light in the understorey is diffuse overall, but of varying intensity due to variability in leaf density within the canopy above. It is variably interspersed with sun-flecks. Leaves of the juveniles of climax canopy species are slow to regain photosynthetic activity after prolonged shading, but once induced will remain responsive in varying light as sun-flecks move with the sun. Robin Chazdon[154] concluded that, despite the overall importance of sun-flecks for understorey denizens, small microsite variations in diffuse light also strongly affect carbon gain and thereby survival. Species that reproduce in the understorey must do so in predominant shade. Their leaves, and therefore they themselves, cannot survive without periodic sun-flecks, or at least with diffuse light intensities high enough often enough to permit daily net carbon gain. On mesic sites, where mature phase canopy density is relatively high, this is generally achieved through sun-flecks and canopy gaps. On more xeric ridges and on sandy soils, diffuse light beneath the canopy appears to be intense enough at sapling and pole level to support a diversity of subcanopy species among the canopy species regeneration.

3.2. *Kerangas and peat swamp forest*

The term *kerangas* originated in Borneo for land unsuitable for the planting of hill rice; its swamp analogue is *kerapa.* Whereas *kerapa* is always peat swamp, *kerangas* appears to be defined by the appearance of the primary forest cover: short, dense in all size classes giving the appearance of a forest of poles, relatively well lit in the understorey with a multitude of small sun-flecks and high diffuse light levels, and small-leaved with notophylls

or microphylls dominant (Fig. 2.18). Such forests are easy to identify by remote sensing imagery owing to the small uniform crowns making up their canopy (Plate 2.19a). They occur on the humic lowland podsols of raised Pleistocene beaches and on the shallow podsols of sandstone plateaux. But they are also ubiquitous on shallow soils on the narrow ridges of Borneo, irrespective either of substrate (which may include karst limestone[155]) or of the soil's clay content and fertility, except that the soil is always shallow (see Chapter 3).

Eberhard Brünig has described the structure of *kerangas* on freely draining white sand humic podsols in lowland Sarawak.[156] There is no clear canopy stratification in *kerangas.* The light requirements for flowering among *kerangas* species have yet to be documented. The canopy is usually continuous but with diffuse leafage, comprised of main canopy MDF genera, and generally 20–30 m in height, sometimes less (Plate 2.19b). The crowns of some short-stature dipterocarps, and *Koompassia malaccensis,* in scattered clumps, may emerge no more than 5 m above the main canopy, the local monopodial conifer *Agathis* excepted. Most dead trees remain standing, gradually rotting and collapsing branch by branch. The subcanopy is dense and small-leaved.

Kerangas structure and physiognomy (but not flora) is therefore similar to that of lower montane forest on shallow soils and at higher altitudes. The latter differs in its denser canopy with frequent presence of crown shyness (Plates 2.19a,c *vs* Plates 4.3d, 4.5a) and less dense understorey.

Kerangas tree roots appear confined to the surface acid organic humus, though a few branching crooked coarse roots descend to the B horizon which is impervious to both roots and water (Plate 1.12a). Roots are mostly fine, dense and intertwined. Few tree falls occur and most mortality is by death of single canopy trees through lightning and periodic drought. The leaves are smaller than in MDF with notophylls or microphylls dominant, rather thick, often dull, with short petioles and lacking drip-tips (Plate 2.19d). (Sister species also occur in MDF which differ in these characters, such as *Shorea parviflora* in MDF and *S. retusa* in *kerangas.*) Occasionally, distinct genotypes may be found, as on Berawan Hill in Tutong (Brunei, 600 m), where I saw a nanophyll individual of the normally microphyll *S. multiflora* with leathery rotundate leaves, lacking drip-tips.

Leaves in the *kerangas* canopy are held at a steep inclination to incident radiation, mostly ascending. Although they cluster near the twig endings, owing to their small size they impart an even, comparatively high diffuse light intensity beneath (Plate 2.19a,d). Leaves on the orthotropic canopy twigs are borne more densely and with larger numbers per twig than in MDF, owing both to probably greater leaf longevity and shorter internodes. Canopy leafage is therefore relatively deep, with high LAD.

Plate 2.19 *Kerangas*. **a,** The Lambir summits, 500 m: *kerangas* (small-crowned) on the dip slope (left), MDF below the scarp cliff (frontispiece); **b,** The *kerangas* subcanopy, Bako National Park, Sarawak; **c,** The *kerangas* canopy from beneath: *Gymnostoma nobile* (Casuarinaceae) (left), *Dacrydium elatum* (right), Bako National Park; **d,** Foliage in a *kerangas* gap: *Syzygium bankense* (nanophyll), *Whiteodendron moultonianum* canopy juvenile (mesophyll) (both Myrtaceae), Bako National Park.
(**a** H.H.; **b-d** P.A.)

Kerangas and peat swamp forest can be rich in epiphytes, and adorned with thin festoons of moss in hyper-humid environments, especially at the base of mountains and by whitewater rivers. Vines are few, and large vines absent. Overall, the structural character and range of variation of *kerangas* is remarkably similar to that of the *caatinga* forests on tropical podsols in the Rio Negro northern tributary of the Amazon, including its extreme form, *bana,* although there is little floristic similarity.

The *kerangas* and peat swamp soil surface remains undisturbed and dense with saplings in most canopy gaps, leaving little opportunity for pioneers. Pioneer trees with large thin leaves and dense monolayer crowns are uncommon

in *kerangas*. They are confined to severely disturbed areas, including those subject to fire. Most species are small and short-lived. Vines play little part, though *Uncaria* and *Smilax* species commonly form sparse canopies in early succession. Vines and large-leaved pioneers are confined to peat swamp margins where, unlike in *kerangas,* there are tall canopy pioneer tree species. These include *Dyera polyphylla* and the emergent *Alstonia pneumatophora,* both endemic to that habitat. Several successional species (including *Ploiarium alternifolium, Gymnostoma nobile, Tristaniopsis* spp., *Syzygium* spp., *Gaertnera vaginans, Timonius flavescens* and some other Rubiaceae) instead dominate early *kerangas* succession, while *Lithocarpus andersonii* may be abundant in forest succession in the outer

Plate 2.20 The wax-coated freshly emerged leaves of *Cotylelobium burckii,* on the periodically arid sandstone tableland of Bako National Park, Sarawak. Note the small size and inclined leaves of other species. *Dacrydium elatum* (Podocarpaceae) in the foreground. (P.A.)

experimental cutting of a sample plot at Gunung Pasir (Samboja, East Kalimantan).[159]

The transition from MDF to *kerangas* occurs abruptly when it is correlated with a sudden change in the substrate. It is more gradual where associated with change in soil depth on a single substrate along a ridge crest. In such situations the floristically defined MDF itself varies in structure, and may assume many *kerangas* characteristics (Chapter 3). On the geomorphologically young, mostly sedimentary topography of inland Borneo, many narrow ridges carry shorter, small-crowned, even-canopied *kerangas*-like MDF. Structural *kerangas* crowns the summit of Cenozoic basalt Mersing Hill (Central Sarawak, 950 m), and siliceous metamorphic Sinhagala (850 m) in Sinharaja World Heritage Forest (Sri Lanka). On each of these peaks, the soil is shallow over bedrock, and there is a thin layer of acid raw humus on the soil surface. The flora there is distinctive, yet more similar to MDF than to *kerangas* (Chapters 3 and 4).

There has been controversy whether the distinctive structure and physiognomy of *kerangas* is the result of adaptation to water stress or to nutrient deficiency.[160] Small leaf size and narrow shape, high volume/surface area ratio, frequent shiny or pale reflective surfaces (Plate 2.20), high inclination, and tight bunching on twigs yet diffuse distribution in crown and canopy all suggest adaptations for long-lived leaves to reduce water stress in a windless climate. Yet the small leaves, short internodes and consequent dense branching of these trees might also be explained by poor growth in low nutrient soil. It has also been suggested that this sclerophylly (that is, small thick leaves rich in sclerenchymatous tissue) may be an adaptation against herbivory.[161] This will be considered in respect of species composition in Chapter 3.

It is a curious fact that *kerangas* on humic podsols and peat swamp forest share many species in common (Chapter 3). Nevertheless, whereas *kerangas* rarely attains more than 40 m in stature, the outermost *Shorea albida* peat swamp community achieved 70 m (Plate 2.21). Nishimura and Suzuki[162] made an interesting comparison between tree species in peat swamp forest and *kerangas,* including two species common to both. Both sites shared similar acid organic soils with similar nutrient status, but the swamp had a permanently high water table whereas the *kerangas* soil was freely draining and drought-prone. They found that the *kerangas* trees invested more in roots relative to shoots, rooting depth, and in leaf biomass; the peat swamp trees in greater stem height and crown and leaf area, extending only a shallow root system. Experiments in which transpiration rates have been monitored in severed leaves, and in trees during droughts, have indicated that stomata of *kerangas* trees have little stomatal response to water stress.[163] In this, they resemble leaves in the mature phase canopy of MDF.

Becker[164] found that sap flow was linearly related to tree

phasic communities of peat swamp (Chapter 3). These are predominantly mesophylls or notophylls, and differ little from climax species in crown architecture or leaf orientation. Their germination and establishment deserves careful investigation. Where burned by humans, *Macaranga* invades, as does *Trema cannabina* – a stand I observed in Berakas forest (Brunei) established, grew, reproduced and died within six years!

Many trees of peat swamp and *kerangas* have a capacity to reiterate by coppice if snapped.[157] I have seen wind-thrown canopy individuals of the peat swamp dipterocarps *Shorea albida* and *Dryobalanops rappa* of up to 50 cm dbh send up rows of vertical epicormic shoots along their supine trunks.[158] Soedarsono Riswan compared regeneration in MDF and *kerangas* in East Kalimantan. He found that a higher proportion of *kerangas* species could regenerate from stump sprouts. He found this to be the main form of regeneration following

Plate 2.21 The diffuse emergent canopy, continuous but for crown shyness, and complete lack of a main canopy; in the *S. albida* phasic community 3, *alan bunga*, of the north-west Borneo peat swamp forests, Ulu Mendalam, Belait, Brunei. (A. Cobb)

crown area, but that there was no consistent difference in rates of transpiration, expressed in the ratio of rates between dry and wet periods, between MDF at Andulau forest and *kerangas* at Badas in Brunei. Though he observed that rate declined in the tallest *kerangas* trees, he did not take account of the taller canopy of MDF than *kerangas*. Matsumoto and colleagues[165] found that osmotic potential of *kerangas* tree leaves was remarkably high compared, for instance, with that of temperate forest canopy leaves. They concluded that the comparatively low photosynthetic rates observed are due to low *maximum* rates of transpiration (30–77, mean 200 mmol H_2O m^{-2} s^{-1}).[166] They concluded that this is not due to

stomatal resistance, but to whole-leaf and foliage adaptations, including small effective leaf width, reflecting surfaces, high inclination to incident energy reducing heat load, and dense canopy foliage excluding convectional air movement, all of which reduce water loss. Such adaptations are most effective in a generally windless climate.[167] Brünig[168] pointed out that the characteristically even canopy surface of *kerangas* will reduce convectional air turbulence in and above it, thereby reducing transpiration rates by minimising canopy surface area and reducing air movement, thus thickening the boundary layer. But this will come at the cost of increased leaf heat load, offset by the form and orientation of the leaves and twigs in the *kerangas* canopy, which both enhance reflection of and reduce radiation received. One gains the impression that diffuse light intensity is greater in *kerangas* understorey on podsols and in the inner peat swamp forest communities than it is in the outer, mixed peat swamp forests and elsewhere, which would be consistent with observations in analogous Neotropical *caatinga*.[169] The differences between the canopy architecture

111

and leaf characteristics of MDF and *kerangas* are thus seemingly explained, but not the floristic similarities between *kerangas* on podsols and peat swamp forest (Chapter 3), and therefore their leaf characteristics in part. That must surely be related to their similar soil nutrient economy.

Like other primary tropical forest canopies, *kerangas* canopy varies with site and (apparently) with soil water economy. However, the *cause* of the noticeable evenness of the *kerangas* canopy could be a relatively uniform genetically determined maximum stature achievable by the majority of canopy species rather than a proximal ecological influence. These species share high wood density, narrow water conducting elements and similar hydraulic architecture. In addition – as occurred in our Bako MDF plots – individuals which raise crowns above the canopy may be the first to die in droughts. The generally small leaf sizes of the *kerangas* on podsols and peat swamp forest understorey, where the diffuse light and high atmospheric humidity will exert less influence on leaf form, implies that nutrients are their principal mediator.[170] I conclude that, although intense periodic water stress appears to influence canopy structure and leaf size and arrangement, nutrient deficiency may also constrain leaf expansion rates and ultimate dimensions, and the range and identity of species that can endure the nutrient deficiencies. Both therefore likely influence canopy leaf density and consequently subcanopy stand density. As in MDF on the most mesic sites and fertile soils, floristic variation *within* both peat swamps forests and *kerangas* on podsols appears to be correlated with occasional water stress rather than with variation in a nutrient subsidy from the substrate, which in any case is likely non-existent[171] (Chapter 3).

Currently, the only dynamic data from *kerangas* or peat swamp forest appear to be that of Nishimura,[172] who found the latter to be the less resilient to water stress, but whose data were insufficient to yield comparisons with other sites. It may reasonably be supposed, based on the prevalence of species with wood of high specific gravity in *kerangas*, and the lower maximum growth rates of MDF on freely draining humult yellow sandy soils, that growth rates in *kerangas* are low. *Astonishingly, there has never been a permanent plot set out in kerangas.*

3.3. *Variation in canopy disturbance*

The density, distribution and stature of the emergents above the main canopy are the principle mediators of canopy structure, and therefore the light climate beneath.

Why do the emergent species not more often form a continuous canopy except for gap openings, thereby more or less excluding the main canopy? It could be argued that, when emergents form a continuous canopy, they are no longer emergent. I therefore reserve the term for species of the emergent stratum

discussed in Section 1, irrespective of the relative loftiness of individual trees. In this section I consider emergents as a group of species that as juveniles must compete on equal terms for light with the juveniles – and in due course the reproductively mature individuals – of one or more strata beneath them. At full height (unlike members of the strata beneath) they come to occupy space in which they are in direct competition for light, often minimally, with other mature individuals of their stratum only.

3.3.1. *Variation in emergent canopy density with topography and soil*

Steep slopes universally experience soil movement, landslips and frequent tree fall. As a result, fewer large emergents are seen in MDF, and they are often grouped on more stable surfaces. In contrast, on the clay loam soils of mesic undulating land a continuous emergent canopy, broken only by lightning strikes and narrow slices caused by domino-effects in local wind-throws, may prevail. Gaps are often large and elongate, created by group wind-throws (Plate 2.9e), as documented for *Shorea megistophylla* in Sinharaja There, the understorey light regime is highly contrasting between deep shade cast by the distant canopy above and direct sunlight in the sharp-edged gaps. When a severe squall or downblast occurs, a continuous emergent stratum is particularly prone to massive blow-down (Plate 3.9i).

The upland MDF with continuous emergent canopy was found on soils well-drained and oxygenated yet with high capacity for water retention, rarely, sometimes perhaps never, experiencing soil water deficits. Their emergent canopies were dominated by mixed late-successional *light* hardwood dipterocarps (merantis). Emergent trees and canopies on such soils can be exceptionally tall. On the most fertile soils a single species, *paji* (*Dryobalanops lanceolata*), dominated. In northern and eastern Borneo *paji* can reach 75 m, the associated heavy hardwood *Koompassia excelsa* (Caesalpiniaceae, *tapang, tualang, menggeris*), 84 m.[173] The tallest recorded broadleaved forests in the Asian tropics are MDF on the fertile, friable clay soils of north-east Borneo. Roman Dial, tree physiologist, ascended individuals by tree bicycle on the black volcanic loams north of Tawau and measured by hand the tallest.[174] This was a light hardwood yellow meranti, *Shorea gibbosa* (sect. *Richetioides*), growing on the upper slope of a low ridge and measuring 88.32 m. Situated nearby were two heavy hardwoods, *S. superba* and *S. maxwelliana* (sect. *Shorea*), respectively 83.49 m and 82.82 m, and another *S. gibbosa* at 81.11 m. *Koompassia excelsa,* also a heavy hardwood species confined to mesic clay sites, was recorded in Sarawak at 83.82 m, and only a little less at 80.72 m in Peninsular Malaysia.[175] *Dryobalanops lanceolata* commonly reaches 70 m, and the one around which a tree tower is built at Lambir is 75 m. Just slightly shorter dipterocarps are found on other soils too: 70 m *Dryobalanops beccarii* on leached sandy clay in the Maliau Basin (Sabah), 60 m emergent canopy on deep yellow humult sands in Andulau Forest Reserve (Brunei),[176] 70 m *S. albida* stands at the coastal margin of the Brunei peat swamps. Forests on deep friable clay loams, as also *S. albida* stands in peat swamps, display tall and extraordinarily even emergent canopies (Plates 2.2a, 3.9a,j). These heights are only exceeded among broadleaf trees by some *Eucalyptus* in warm temperate Australia.

In their dominance by light hardwood species, and in their exceptionally high basal area, these Bornean forests closely resembled the dipterocarp forests of sheltered sites in the Philippines, now all but gone. These will be discussed below. In another case still surviving in several locations in south-west Sri Lanka, the late successional light hardwood *Shorea trapezifolia* (at 400–1,000 m altitude) and its sister species *S. congestiflora* (in the lowlands) form monospecific drifts on red clay loams on slopes. In such Borneo forests, the main canopy is sparse and patchy; in the *paji* forest the understorey is of low density; but in the Philippine forests, exceptionally, there was abundant regeneration in all size classes.

I hypothesise that the almost continuous emergent canopy structure and the variation in composition of these upland forests with dense emergent canopies reflect late succession following canopy catastrophe at differing scales. Patches of continuous emergent canopy composed of light hardwood *Shorea* were frequent on well-structured clay loams in Borneo. I infer that these are the result of large-scale wind-throws consequent to occasional storm downdrafts, a conclusion also reached by Ho Coy Choke and colleagues in Jengka Forest Reserve (Pahang) on Peninsular Malaysia.[177] Their exceptional stature and extent must reflect long intervals between catastrophes, relative to the life history of the component species. Drifts of *Shorea trapezifolia* in the south-west Sri Lankan MDF, too, appear to be late successional, following landslips down the dip-slopes of the bedded metamorphic substrate. In 1964, Sarawak Forest Department botanical staff and I established five 0.6 ha plots on the exceptionally fertile lower slopes of Mersing Hill, a Cenozoic basalt outcrop in Central Sarawak, and censused them every five years for 20 years. Here, the continuous canopy of *paji* was broken periodically by sharp-edged elongate gaps where one or a small group of these giants had been uprooted or broken off (Plate 2.9b). The dominance of *paji*, with its unusually high plasticity with exceptionally fast growth in high light combined with high shade tolerance, appears possible only in a high soil nutrient environment.

Shorea megistophylla resembles *paji* in these respects, but as a *Dryobalanops*, *paji* also reproduces more frequently than other dipterocarps (Chapter 5). The plots show that in diameter growth, the fastest-growing *paji* juveniles exceed adjacent *Macaranga hosei*, a species that can reach 30 m. They compare

with the fastest-growing tree in Bornean forests, *benuang* (*Octomeles sumatrana*, Datiscaceae), an emergent pioneer.

Why, then, are emergent slender-vesselled *balau* (heavy hardwood *Shorea* sect. *Shorea*) so much less common in these majestic Bornean forests? Why do the heavy hardwood canopy species *Eusideroxylon zwageri* and *Hopea andersonii*, which flourish in these *paji* forests here and in East Kalimantan, not eventually come to dominate?

I hypothesise that mixed light hardwoods form a dominant emergent canopy on friable clay loams following windthrows where a vine blanket does not early develop: soil scarification would have been confined to upturned root balls, so that established light hardwood seedlings would have survived.

On a more limited scale, dense and tall emergent stands may occur along ridges with freely draining soils of moderate nutrient status[178] (Table 2.4). The hill dipterocarp forests of

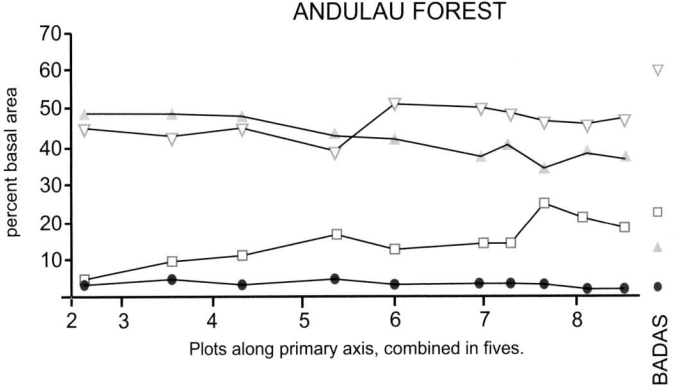

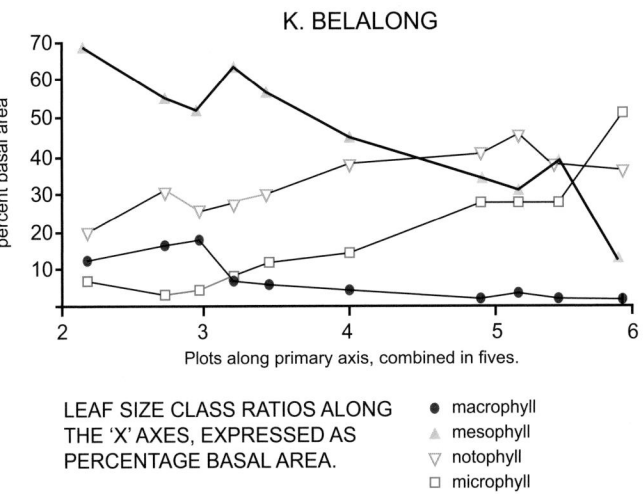

LEAF SIZE CLASS RATIOS ALONG THE 'X' AXES, EXPRESSED AS PERCENTAGE BASAL AREA.
- ● macrophyll
- ▲ mesophyll
- ▽ notophyll
- ☐ microphyll

Fig. 2.18 Change in leaf size representation along primary axes of Bray and Curtis ordinations of MDF in Brunei: on humic sandy soil (Andulau), clay loam (Belalong) and in *kerangas* (Badas). The axes correlate with altitude (50–100 and 150–650 m respectively), and with topography (slope to ridge). Each point represents 5, 0.4 ha plots aggregated, trees >9.7 cm dbh. By using basal area, canopy individuals dominate. (After P.S. Ashton 1964b)

Peninsular Malaysia, dominated by *Shorea curtisii* (Chapter 4) are a notable example. Here, though, the understorey is also dense, and the dominant canopy species are shade-tolerant medium or heavy hardwoods. Instead, therefore, lateral light penetrates beneath the canopy, which is itself diffuse and more microphyll and notophyll than mesophyll (Fig 2.18).

3.3.1.1. *Towards a physiological understanding.* It is claimed that the light-saturated maximum photosynthetic capacity of a leaf (A_{max}) increases linearly with leaf nitrogen content up to the maximum observed in nature.[179] This has been further confirmed when nitrogen, in the form of ammonium ions, was added to seedlings of six dipterocarp species.[180] At Pasoh though, Kenzo and colleagues[181] found that A_{max} varied between 7–18 µmol m^{-2} s^{-1} among five dipterocarp species. A_{max} was most strongly correlated here with chlorophyll content – specifically palisade structure – and only weakly with nitrogen content, leaf mass or thickness.

Studies by M.C. Press and his students[182] at Danum forest (Sabah) showed that seedlings of the light hardwood *Shorea johorensis* and *Dryobalanops lanceolata* responded to nitrogen addition in full sunlight by increased photosynthetic capacity per unit leaf area; yet only *D. lanceolata* responded at low nitrogen availability.[183] This implies that *D. lanceolata* has the competitive edge on low nutrient soils, in contrast with my field experience (see Chapter 3). Could their unexpected results be due to addition of nitrogen at higher levels in their experiment than are found in nature?

3.3.1.2. *Other cases.* Another exceptional case was the vast peat swamp forest that extended for thousands of square kilometres from Pontianak in north-west Borneo to Tutong District in Brunei. This forest, once dominated by the single emergent species *Shorea albida,* still survives as a magnificent intact peat swamp system in Belait District (Brunei) and a smaller area in Luagan Bunut National Park in the lower Baram valley (north-east Sarawak). (This forest will be discussed further in Chapters 3 and 6.) (Plates 3.9e–k). The canopy is continuous but for single-tree lightning gaps and the domino-like tree-fall strips caused by squalls. Beneath its continuous emergent canopy layer, it is the main canopy and its species guild which is sparse and clumped beneath the occasional gaps. The understorey is dense and small-leaved (Plate 3.9d,e). Unlike on clay loams, the emergent canopy crowns are diffuse-leaved. This implies that the density of the emergent crowns themselves, rather than foliage density, is the light-limiting factor for the main canopy, but that light beneath the diffuse-leaved crowns becomes sufficiently homogenised to support a dense subcanopy. The small-leaved subcanopy may instead be due to nutrient limitation. On deeper peats these forests recall *kerangas*. Pioneer species are few, and as in *kerangas,* they share

the leaf and crown characters of climax species. *Shorea albida* produces light hardwood and appears to be a light demander when juvenile, but toward the margins of the swamp domes individuals develop heavy timber after reaching the canopy. Early regeneration is patchy, indeed almost absent, in the tallest stands. The forest appears not to be in equilibrium, but clearly could not have grown following a single event. *Its recent history thus remains a mystery (see Chapter 3).*

3.3.2. *Major catastrophe regimes*

3.3.2.1. *Forests in recovery.* Even where past catastrophes are no longer discernible owing to attainment of full height by succession, they may remain manifest as continuing accumulation at landscape scale of biomass among the largest size classes, by diameter if not height growth. This is almost universal among the remaining unlogged rainforests. Pamela Hall and I were surprised to find that, after 20 years of monitoring, the basal area of four plots in MDF overlooking the coast at Bako National Park (west Sarawak) on infertile humult yellow podsolic soils had added as much as 20% to their original basal area (section 3.1.2 above).[184] The plots were spaced over more than a kilometre along the rocky slope, and all had gained. Nevertheless, similar plots established the same year at Bukit Mersing (central Sarawak) and Lambir National Park in the north-east, appeared to be approaching equilibrium in basal area and estimated volume within each cluster, although not within individual plots (Table 2.6) (though the Pasoh 50 ha plot has continued to slowly gain since census began in 1986). The gain at Bako was among the larger-diameter canopy trees; smaller trees were actually *declining* in number. Notable among the leading emergent dipterocarps were the medium hardwood yellow meranti *Shorea cuspidata* and the heavy hardwood *balau* species *S. falcifera*; no mature fully emergent individual of either species was present. It was apparent, then, that the forest was recovering from a past catastrophe. The most likely candidate was a major drought recorded 80 years earlier, when much of the *kerangas* on the sandstone plateau above is said to have burned. Severe regional droughts are rare in north-west Borneo but more frequent and more intense in eastern Borneo, facing the Pacific, where they are associated with unusually intense ENSO events[185] (Chapter 1). Mortality is highest in the understorey, especially among seedlings, and in the upper canopy where light hardwoods are said to be most susceptible[186] (Table 2.7b). Following a lesser drought at Lambir National Park (Sarawak) in Jan.–March 1998, juvenile mortality increased from 5% to 8% per year, and 30% of juveniles experienced a reduction in height through die-back.[187]

On a more local scale, sudden downdrafts can flatten forest areas in excess of 100 ha. I have observed the effects of such

Plate 2.22 Forests in typhoon country. **a,** Regenerating stand of MDF following typhoon blowdown several decades earlier, Bislig Bay, Mindanao, 1975; **b,** An old photograph of tall lowland MDF in Negros, Philippines, showing the typhoon-damaged open canopy and relatively abundant pole-sized juveniles; **c,** Cyclone blowdown: *Pinus kesiya* savanna in Nam Nao National Park, central Thailand; **d,** Typhoon-pruned even canopy of emergent *Shorea* in MDF at the Palanan CTFS plot, near the north-east Luzon coast; **e,** The shredded crown of an emergent *Shorea polysperma,* resprouting epicormically two years after a typhoon, 1995; **f,** Typhoon-blasted understorey near the CTFS Palanan plot site, rich in monocots, 1995. (**a,b, d–f** P.A.; **c** H.H.)

winds on the clay loam slopes of Ulu Temburong (Brunei). Another catastrophic downdraft destroyed our camp on the upper slopes of basalt Bukit Mersing (Sarawak) in 1963 when plots were being installed; our team survived by sheltering under a freshly fallen emergent trunk. Such downdrafts have been observed to occur sufficiently frequently to influence local forest composition on the clay loams of Sepilok Forest Reserve, and in the Danum forest in Sabah. They appear to be the cause of the stands of light hardwood meranti on similar soils in north-east Sarawak and eastern Mindanao (Plate 2.19a). Such a storm has been described in MDF, apparently on humic yellow sands, and in *kerangas* in the Barito Ulu (East Kalimantan).[188] In that case the MDF on ridges was most affected, while *kerangas,* also on a ridge but with roots densely matted and intertwined in the organic surface soil horizon, experienced less damage. Much early recovery from basal resprouting was noted in the *kerangas*.

3.3.2.2. *The Philippine typhoon belt*
Shipping water in a fishing boat in a stiff northeasterly across the Bay of Kompong Som (Cambodia) from the Chroieul Smach Peninsula to Sihanoukville (now again Kompong Som) in 1970, I was impressed by the wind-pruned shrublands over the bluffs of exposed promontories (Plate 2.23a). These are exposed to the strongest southwesterlies, and also to periodic typhoons (that is, cyclonic storms or hurricanes). Sun I-Fang has established a CTFS plot at the extreme south-east of Taiwan, at the very margin of the biological tropics yet in the face of the most intensive and frequent typhoons (Plates 2.23b,c).[189] He is documenting major differences in stature (always on the short side), dynamics and species composition between windward and leeward slopes.

To the north and especially east of Luzon's Sierra Madre, the forest structure is dramatically different from MDF in Borneo or Peninsular Malaysia.[190] Here, we see a continuous phalanx of emergent dipterocarps, representing a reduced range of species also found throughout the south. These trees form a smooth, dense, and continuous emergent canopy, shorter to windward, the twigs and branches reaching away to the west, recalling (in reverse) a grove of beech facing an English south-west coast (Plate 2.22d).[191]

In 1995 I first visited the eastern Sierra Madre, at Palanan. There had been a typhoon two years earlier, and another has struck since, in 2007 (Plate 2.22d). The canopy dipterocarps there grow on friable clay loam soil. None had fallen. Their trunks are of great diameter at the base, but taper strongly, the crown heights not exceeding 40 m. The foliage and many of the twigs had been shredded by the typhoon, but lushly reiterating new shoots clothed the canopy branches like insulation round a heating pipe, quite unlike the incapacity of Bornean dipterocarp crowns to reiterate following

drought (Plate 2.22e). The subcanopy light climate was one of extensive sunny patches, intercepted through the day by the leaf-clothed branches. Wild bananas, aroids, understorey palms and other large-leaved monocots formed a dense but heterogeneous canopy, recalling the forest at La Selva facing the Caribbean coast of Costa Rica but interspersed with abundant dipterocarp regeneration (Plate 2.22f). Only occasionally was there a break in the canopy, with *Macaranga* and other dicotyledonous pioneers. To the south, in the Sierra Madre at Real (Quezon), which I visited in 1975, I was struck by the deep, parsnip-like feeder roots extending at least 2 m down from the spreading buttresses of emergent dipterocarps, exposed by logging tracks on the slopes. The canopy trees hold in a typhoon like the masts of a windjammer; the light regime at ground level fluctuates wildly following each big storm, as the canopy resprouts.

Continuous emergent-canopied MDF appears to have been widespread on the predominant clay loam soils of the Philippines.[192] There, forests in some flat alluvial plains were dominated by one or a few light red meranti species (*Shorea negrosensis, S. polysperma, S. palosapis,* or *Dipterocarpus validus*) and extended in close-ranked emergent continuity over hundreds of square kilometres. They achieved estimated trunk volumes in sheltered sites exceeding 450 m³ ha⁻¹; these proved irresistible to the logger. The Philippine forests (except Mindanao) lie within the typhoon belt. This means that they receive storms increasing in frequency and intensity towards the north, and also persistent north-east trade winds Nov.– March. Each such storm can defoliate the canopy and, where relatively infrequent as towards the south and west, create massive wind-throw gaps. Each creates subcanopy light and soil surface conditions favourable to survival and regeneration of light hardwood seedlings and subcanopy juveniles. The frequent lesser winds cause breakage of expanding shoots: I was struck by the open canopies even in the tallest of these forests such as the former MDF of Negros, in the west, very different from the Borneo MDF canopy (Plate 2.22b). Cyclonic storms, variously called typhoons or hurricanes, affect the forests of seasonal regions most[193] (but see section 3.2 above), but also influence the perhumid eastern Philippines north of Samar as well as Taiwan. They occur there more frequently in the north than the south. Their visible effects are large-scale, covering broad swathes usually tens of kilometres wide. South and west of the eastern coastal Sierra Madre, occasional strong storms can leave extensive clear-felled forest on windward slopes and plains. Whitmore and Burslem[194] documented forest change and succession on Kolombangara Island in the Solomon Islands where typhoons were increasing in frequency over time – an exceptional observation in a forest apparently poorly adapted to resist them.

In combination with prevailing winds during the wettest

Plate 2.23 Forests in typhoon country. **a,** Wind-pruned woodland on the Chroueil Smach Peninsular, Cambodia; **b,** The canopy of the typhoon-stunted lowland forest of Nanjeng Shan, south-east Taiwan; **c,** Within the canopy at the CTFS plot, Nanjeng Shan; **d,** Monsoon wind-pruned stand of *Dipterocarpus indicus* at the Pilarkan sacred forest, Western Ghats; **e**. Stand of *Dipterocarpus costatus* with wind-pruned crowns in southern seasonal wet evergreen forest along an inlet at Riem Bay, Kompong Som, Cambodia; **f,** The river at Khao Ban Thad, Peninsular Thailand: Herbaceous vegetation marks the monsoon storm flood limit. (**a, e** P.A.; **b, c** S. Davies; **d** D. Willmore; **f** H.H.)

season, the frequency of typhoons are the likely cause of the unique and varying structure of MDF in the north and east of the Philippines, and the mainland to the adjacent west. These forests are often tall but retain an almost continuous, diffuse emergent canopy dominated by light red meranti. Below this are a sparse main canopy and then a subcanopy dense with meranti regeneration in the full range of sizes.[195] But on the windward slopes of the northern Sierra Madre the forest is shorter; in south-east Taiwan it is reduced to c.20 m. It appears, therefore, that a sequence of disturbance exists on clay loam soils up the Pacific coast from north-east Borneo and southern Philippines to southernmost Taiwan:

1. In the far south and on westward sheltered sites, occasional storms leave large patches of wind-throw.

2. Further north, in the path of the occasional violent typhoon, these patches coalesce into swathes of single-age stands, each consisting of the cohort that has regenerated since the last local catastrophe.

3. In north-east Luzon, more frequent typhoons create a forest of diminished stature that withstands hurricanes but is left after each storm with a variably shredded and defoliated canopy.

4. At the northern margin of the tropics, in south-east Taiwan, the forest is devoid of dipterocarps or indeed any emergent component, and retains a canopy dominated by main canopy species. This forest, though frequently defoliated, is highly resistant to wind-throw but achieves a stature no greater than 20 m.

Wyatt-Smith[196] described MDF in Kelantan (north-east Peninsular Malaysia), from whose short stature and the presence of many trees with deformed trunks he inferred the track of a typhoon perhaps a century earlier. These forests are but a short distance from Thai seasonal evergreen dipterocarp forests, to their north-west, which suffer more frequent typhoons (see 4.1. below, also Box 2.2).

Lowland rainforests in South Asia, both in aseasonal wet south-west Sri Lanka and the Western Ghats of southern India, rarely exceed 40 m in height, with canopy dipterocarps hardly emerging over main canopy species. In other respects their structure and physiognomy resembles lowland rainforests of the Far East, and varies in similar ways. A likely reason for their relatively short stature and rarity of true emergents is the influence of the three months of driving winds during the wet south-west monsoon (Plate 2.23d).

3.3.3. Implications for variation in forest dynamics therefore productivity

Pasoh MDF has near-average canopy disturbance and soil nutrient levels. I predict that productivity will be highest on mesic sites with fertile soils where canopy disturbance is on a large spatial scale, but at intervals long enough for fast-growing successional climax species to reach full height and reproduce. Such conditions reach their zenith, not in the perhumid lowlands and MDF, but in the moist valleys of the cyclone-prone wet seasonal tropics with their evergreen or semi-evergreen forests (section 5.7.3 below and Chapter 1); whereas species diversity is highest elsewhere (Chapter 7).

3.4. What, in summary, causes the diversity of forest canopy structure and stature within perhumid climates?

I argued that the broken emergent canopy of the paradigmatic MDF mature phase is caused by local emergent tree mortality, caused mostly by lightning or the selective impact of severe drought-induced water stress. Even-statured emergent canopies may indicate large stands regenerating from landslips and massive wind-throws (as at Bakong and Bukit Mersing in Sarawak), or wind pruning (as in the northern Philippines or south-west Sri Lanka). However, whether they are caused by wind pruning or succession in massive gaps, even-canopied forests can be found on any soil. Catastrophic dynamic events can neither explain the continuously even canopies of kerangas (Plate 2.19a) or peat swamp forests, nor of the emergent canopy over a whole landscape at Danum (Plate 2.2a): When viewed at a distance from a parallel ridge, the canopy of a mature MDF community appears remarkably uniform also. This is true although it is comprised of many emergent dipterocarp and other species, is fragmented by intervening stretches of the main canopy, and fluctuates with changes in the depth of soil beneath (Plate 2.2a,b). The taller a tree grows the longer it has lived, and the greater is the likelihood of its dying from one of a suite of causes. Nevertheless, many climax canopy trees persist, at times for centuries, at maximum achievable height while continuing to expand in diameter.

Kerangas saplings were found to experience reduced conductance during droughts, although those of the deep-rooted Dipterocarpus borneensis did not.[197] The saplings also then experienced reduced maximum rates of photosynthesis. Three out of seven observed species experienced pre-dawn leaf water potentials approaching the wilting point. (Comparable data from MDF were not collected.) Tyree and Becker[198] compared vessel cavitation rates of canopy trees in kerangas and in MDF on yellow sands in Brunei during a drought, and were surprised to find that they did not differ. They concluded that stomatal sensitivity did not differ among them. However, the emergent canopy of the MDF they investigated at Andulau Forest

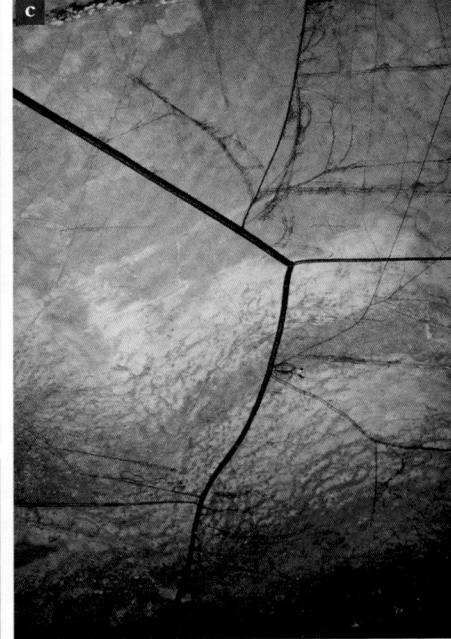

Plate 2.24 Some tap roots. **a,** Sinker roots descend from buttresses on friable clay loam. *Shorea inappendiculata* at Tanjung Redeb, East Kalimantan. I.G.M. Tantra and self as scale; **b,** Taproot of a young *Alstonia angustiloba,* a pioneer pagoda tree, Jerudong, Brunei; **c,** *Ficus* roots on a cave ceiling 20 m beneath the ground, Mulu National Park, Sarawak; **d,** A *Dipterocarpus tuberculatus* taproot penetrates sedimentary bedrock, deciduous dipterocarp forest, Khun Yuam, north-west Thailand. (**a–c** P.A.; **d** H.H.)

Reserve can reach 60 m, whereas the *kerangas* canopy studied at Badas is less than half that height. I infer that the height of these two canopies, and of others in their relatively windless region, is determined by the capacity of their species to lift water to the height at which severe soil water deficits threaten apical bud and twig mortality through vessel cavitation. This is consistent with the even reduction of canopy height from the edge to the centre of the *Shorea albida* peat swamp forests, from 70 to c.30 m, which I have observed to correlate with an increase in variability of the depth of the water table. The prevalence on drought-prone soils of species with wood of high specific gravity (therefore with vessels of small diameter) may indicate adaptations that reduce cavitation at a given negative pressure, thereby achieving maximum canopy height under a specific drought regime. The extraordinary stature achieved by both light and heavy hardwood *Shorea* on the black volcanic soils of Tawau, scaled by Roman Dial,[199] may therefore imply that these soils never experience water stress.

However, the classic broken emergent canopy and three-layered forest structure is seen most frequently on soils where either moderate frequency of drought or periodic waterlogging and shallow rooting, therefore proneness to windthrow, induces moderate mortality among emergents. The tendency to sudden canopy tree mortality appears due to the apparent inability of crowns of emergent species in the windless zone to regenerate through reiteration by epicormic branching. The classic triple-stratum MDF structure, with discontinuous emergent stratum, is widespread in western Malesia because the soil and climate conditions which result in its canopy disturbance regime are the most widespread there.

3.5. *Roots*

The depth and expanse of its rooting system determines the basic capacity of a tree to exploit the nutrients and water stored in the soil. The prolonged but intermittent wet conditions of the humid tropics can result in periodically anoxic water-saturated soil conditions at quite shallow depths. This may become visible in bleaching, where terracotta coloured iron and aluminium oxides are reduced to pale grey-blue hydroxides, except where oxygen penetrates along root channels (Chapter 1). Root systems then tend to concentrate here near the soil surface; indeed, this was long considered a distinguishing characteristic of rainforests. That this is by no means always the case will be shown below.

Earlier literature claimed that tropical rainforests invest more in above- than below-ground biomass, and that the root systems of tropical forest trees are shallow. The latter is true wherever soils become anoxic at shallow depths, on plains (especially periodically flooded alluvium) and on simple lattice clay soils such as the illitic soils, which are widespread throughout the humid tropics. There are notable exceptions, though, wherever soils are freely draining and deep. Roots in well-drained shallow soils may penetrate fissures in the rock substrate. Roots may be seen descending more than 15 m on the fresh face of limestone quarries, while those of *Ficus* emerge from cave roofs at even greater depths (Plate 2.24c). Sinker roots descend at least two metres vertically below the buttresses of tall emergent dipterocarps on friable clay loams (Plate 2.24a). These both increase their resistance to squalls – partially explaining the generally high density of the emergent canopy on these soils – and increase the volume of water available to them during droughts. Complex lattice clay soils with well-developed crumb structure have a high capacity to store water and make it available to roots. The short lattice clays and silty clays that are so widespread in inner Borneo lack this friability and become anoxid at depth, visible as mottlings of iron oxide in an increasingly bleached matrix. Roots become concentrated in upper mineral horizons and canopy trees become more prone to windthrow. Yellow and red sandy soils retain less water, although the iron sesquioxides present in these MDF-supporting soils (Chapter 3) do enhance water retention (Chapter 1). Well-drained siliceous sandy soils in the perhumid tropics are leached, bearing densely matted surface raw humus of varying thickness with fine tree feeder roots (Chapter 1). These, rather than sinker roots, bind to resist wind throw. The mineral soil beneath bears surprisingly few coarse roots, descending obliquely and much branched (Plate 1.11e). In the extreme case of the *kerangas* that characterises lowland humic podsols, fine roots are overwhelmingly restricted to the raw humus A_1 surface horizon. During droughts, soil water will be concentrated immediately above the impervious B horizon, where roots must descend to it; but this has rarely been observed.

Mycorrhizal mycelia extending from roots greatly increase their capacity in respect to water, phosphorus and perhaps some other nutrients.[200] These are thought to multiply water-absorbing surface area through their spreading mycelia, thereby reducing water tension in the trunk and enabling greater growth in height. Ectotrophic mycorrhizae on the root systems of dipterocarps and Caesalpiniaceae may underlie the emergent status of many, and the extraordinary height reached by some (Chapter 5).

Short-lived pioneers have shallow root systems and high above/below-ground biomass ratios. This is consistent with their 'high risk strategy' whereby they invest in rapid early growth and reproduction but remain susceptible as individuals to periodic drought, which their dormant seeds may enable their populations to overcome. But some tall-growing pioneer and successional light demanding climax canopy pagoda trees, including some *Alstonia* (pioneers) and *Terminalia* (climax species), are exceptional among rainforest trees in their development of massive tap roots (Plate 2.24b).

Examination of root systems requires extensive trenching, which has not yet been done in the Asian tropics (but see Chapter 5). More comparative observations can nevertheless be obtained from soil pits, along road cuts and in quarries.

3.6. *Biomass, productivity and carbon*

Substantial differences in these parameters must exist between primary forests within one climate; assessment remains a CTFS project for the future. Evaluation will be complex. Above-ground standing biomass and carbon will, I predict, be correlated with the basal area of the uppermost stratum, but will also be influenced by mean wood specific gravity (correlated with mean growth rates) and below-ground carbon (broadly increasing with basal area and stature), plus litter and soil carbon. Below-ground carbon and living biomass carbon will therefore be only partially correlated: living/dead ratios may be low on humult equatorial soils, and lower on peats than podsols, while both appear to reach high levels in some deciduous forests.

4. Conclusion

Forest structure reflects the life history of clumps of trees sharing a common history. Near the ground, seedlings and saplings either arise as a cohort from seeds on the ground at the moment when a gap opens, or as seedlings surviving in the understorey prior to gap formation. All respond together, and competition commences. Surviving individuals reach the canopy in approximately even-aged clumps. Openings are caused by canopy disturbance events, the 'motor' which drives the forest cycle and influences species composition at the local scale. The frequency and mean area of gap-forming disturbances vary predictably among habitats. Consequently, we see elements both of chance and of repeating, therefore predictable, patterns in disturbance. Both chance and pattern influence processes of differential survivorship among species, thereby influencing natural selection and ultimately evolution within the forest. Ascertaining the relative importance of chance and pattern in forests of differing composition and diversity is a fundamental goal in ecological research, and is a major focus of the first part of the next chapter.

Over the last 10,000 years, trees in the calm perhumid tropics have experienced stability of climate, but great diversity of topography and soils differing in nutrient availability and water deficit frequency. They face trade-off dilemmas between the comparative costs of exclusion of competitors by out-shading versus the costs of withstanding occasional extremes of water stress. Within the demands of each forest community and its habitat, species have evolved their own solutions to these dilemmas, reaching a given height and specific light environment optimal to its species; then, the additional cost of reproduction must be borne.

Conventional wisdom has long dictated that rainforests, at the scale of the landscape, are in dynamic equilibrium. By implication, their long-term dynamics are also equilibrial. Small plots, of course, never are because they are dominated by the history of one or two stands. No measurements have been attempted on landscape-scale changes in forest biomass over time, in part because of the difficulty in arriving even at crude estimates in the absence of individual tree height data. Diameter measurements can be converted to basal area (cross-sectional area at breast height) and, among habitats where forest height varies little, basal area can serve as a valid proxy for volume. Although our Bako (Sarawak) plots grew in basal area by 20% in as many years, our companion plots in the environmentally diverse Lambir hills varied substantially in their growth or loss, while our individual Mersing hill plots on clay loams approached an overall equilibrium over the 20-year period of observation. At the larger spatial scale though, none of the 50 CTFS tree demography plots that have been recensused have maintained a constant total basal area. Overall, they have gained, implying that the whole forest is in late succession, although patches within them have lost mass and others remain close to equilibrium.

Forests in which recruitment is exceeding mortality, or biomass is increasing, have been reported elsewhere in tropical Asia and Africa.[201] Overall gains have been widely reported from the Amazonian *hylaea* and Africa, and some have reasoned that this is a response to climate change, perhaps to an increase in atmospheric carbon.[202]

Again, we need to monitor forest change over decades. But careful analysis of change at the scale of the individual stand upwards, and deep studies such as those of Baker at Huai Kha Khaeng,[203] strongly suggest a more prosaic explanation, namely that few forests ever reach climax equilibrium. The rise of the main canopy during forest succession is accompanied by a change in the proportion of above-ground biomass from the smaller to larger trees; this can be ascertained in the change of tree abundances in different diameter classes.[204] Feeley showed that this is generally taking place in forests putatively

in equilibrium.[205] All forests are subject to natural catastrophes of varying periodicity and spatial scale, but generally frequent enough to set them back before equilibrium at the landscape scale is reached.

One impediment to understanding how rainforest stands mature is the absence of growth rings among evergreen rainforest trees, and therefore of any precise means to age putatively ancient individuals.[206] This is exacerbated by the rarity of large, dense-timbered, therefore by inference ancient, individuals whose trunks are solid to the core.[207]

Part II. Forests of the seasonal tropics
(see map inside back cover)

In November 1958 I took the Golden Blowpipe Express from Singapore to Bangkok, to attend a Pacific Science Congress. In those days, it was still possible to observe extensive unlogged forests clothing the hills much of the way. To be sure, the first part of the journey, 2° north of the Equator through Johor, was on low undulating land and we were hemmed in by monotonous ranks of rubber trees, but at Tampin the Main Range begins to rise to the east. I was struck by the majesty of the great trees following the ridges, and by the density of their huge domed grey-green crowns, presumably emergent but for the most part appearing contiguous. On the slopes, the forest canopy was different – broken, rich green, the emergent crowns scattered or in clumps on spurs and mostly of smaller diameter. Much further north we changed trains at Kuala Krai, not far from the Thai frontier. As the sun went down, we began passing through hills on whose ridges the grey-green crowns were absent, the emergents more scattered and less clearly raised above the rest of the canopy, while the slopes appeared more disturbed. Overall the forest appeared somewhat shorter, the crown diameters less, than in the south. Looking out of the train windows immediately on waking the following morning, I was startled to find myself in a different world. The humid heat and the cloudy afternoon Malayan sky had been replaced by a cool dry golden haze, the sun already reflecting off steep craggy hills. In the foreground were rice paddies in stubble, from which, a little further away, arose hills clothed in short sparse woodland which was beginning to change colour like a New England autumn: this forest was clearly deciduous. No more rubber trees, but scattered sugar– and occasional talipot (Corypha umbraculifera) palms. All was buff, gold, umber and terracotta, suffused with smoke from scattered brush and stubble fires. Only faintly, on the horizon, could higher ridges be made out, with patches of pale golden-green forest resembling that seen the previous evening albeit in a different light, between the ubiquitous hill rice swidden. A week later, I was again on a train, northward to Chiangmai (18° N) for a field excursion on Doi Suthep; the same landscape continued throughout that journey.

What I had seen that first evening was the change in canopy

structure that occurs from an aseasonal wet climate to one with a reliable dry season of less than four months. The following week I was in country with at least five dry months. There fire, for millennia overwhelmingly anthropogenic, invades the forest with increasing frequency the longer and more intense the dry season. Tall mixed evergreen forest is replaced with shorter, single-canopied deciduous woodland, the understorey variously dominated by short-lived shrubs, bamboo thickets or grass.

The forests of Asia's seasonal tropics are best known in India, where forest science was introduced 150 years ago, but that knowledge is heavily focused on silvicultural management, and increasingly on stands modified to increase timber yield. Only recently, with establishment of a regional network of large tree demography plots under the CTFS collaboration, are we beginning to understand how diversity of structure, phenology and species composition is shaped by dynamic forces and the environmental factors which mediate them.

H.G. Champion presented the first regional classification of forests types for former British India (that is, Pakistan, India, Bangladesh and Burma).[208] (Pakistan lacks tropical forests, although semi-arid thorn woodland penetrates the south-east.) This classification is still commonly adopted and works well especially for India. Although Champion introduced dynamic aspects, these were only broadly understood at that time. Between 1975 and 1993, under the inspiration of Henri Gaussen, researchers at the French Institute in Pondicherry have published detailed field accounts of the vegetation of peninsular India and Cambodia.[209] These were based on field observations correlated with climate using mean annual rainfall, length of dry season(s) and mean temperature of the coldest month. The forest classification adopted is a refinement and expansion of the floristic differentiation of Champion. Degraded forests are mapped, and seral stages described in detailed handbooks accompanying each map. This has allowed their authors to form hypotheses about the 'potential vegetation', or the inferred original natural vegetation, based on lists of surviving species combined with expectations of the prevailing climate at a site. Confirmation, through fuller understanding of successional processes and species' ecology, has been slow due to lack of permanent plot data. This is now being redressed in peninsular Indian lowland deciduous forest in Mudumalai Game Reserve (Tamil Nadu) by a team from the Indian Institute of Science under the leadership of Raman Sukumar, and in evergreen forest at a permanent plot in Uppangala Forest (Karnataka) by the Pondicherry Institute.[210] Blasco and colleagues have recently published a cartographically accurate but generalised map of residual forests for continental Asia.[211] Soils continue to have an important influence on forest structure and dynamics in the seasonal tropics, but have been little considered. I propose a classification of evergreen forests that takes soils into account.[212]

5. Seasonal evergreen dipterocarp forests

Champion[213] had no experience of MDF, which does not exist in India or Burma. He therefore failed to recognize that forests change fundamentally in flora, dynamics and phenology as soon as a regular annual dry season, during which evapotranspiration exceeds precipitation, intervenes. I therefore choose the name 'seasonal evergreen dipterocarp forest' over Champion's term 'tropical wet evergreen forest', partially so that the name applies solely to the forest, whose annual seasonal phenology distinguishes it, and not to an aspect of its assumed physical environment; partially also to distinguish it from MDF. MDF, and other mixed evergreen forests in full leaf during the monsoon as far as the tropical margins of South China, evapotranspire equally, at about 100 mm water vapour per month.[214]

It is the length of the season (or seasons), during which monthly rainfall fails to reach the average required to offset expected evapotranspiration, which most correlates with the structure, phenology, dynamics and composition of regional forest formations. Seasonal evergreen dipterocarp forests differ from MDF and associated forest types of perhumid climates in the general presence of deciduous canopy species, shorter average stature with their emergent stratum less prominently raised and, most consistently, in the replacement of mass flowering and mast fruiting by annual flowering and fruiting, albeit variable in intensity. Champion's classification of forest types in seasonal continental tropical Asia, including his montane forest types, does still hold. His nomenclature, though, assumes a strict correlation with mean annual rainfall, an assumption that ecologists now avoid and will be shown to be false. I accept his major forest classes, amplifying their phenological and floristic distinctions, but have changed their names except in cases where I have been unable to conjure an alternative.

5.1. Southern seasonal evergreen dipterocarp forests

Seasonal evergreen dipterocarp forest ('southern tropical wet evergreen forest' of Champion) extends in the south into climates with an annual dry season sufficiently short that fire does not penetrate primary stands: up to about four months. These forests were originally found wherever native wild durian species occurred and *Durio zibethinus* can be cultivated. They extend north from Kedah in Malaysia and the northern hills of Pattani province in south-east peninsular Thailand, up the Isthmus of Kra northwards in lowland peninsular Thailand to Chumphon, further north in Burma to the southern Dawna range (c.18°50' N), and spilling over along the frontier ranges into Thailand at Kaeng Krachan National Park in the frontier ranges west of Phet Buri as far as Sangka Buri

Plate 2.25 Southern seasonal evergreen dipterocarp forest. **a,** Khao Ban Thad, peninsular Thailand, hill slope: dipterocarps hardly emerge from the canopy, and are clumped between disturbed patches; **b,** Slope treefall gap, with abundant vines, Khao Ban Thad CTFS plot; **c,** Rich understorey of saplings other than dipterocarps, and Zingiberaceae, Khao Ban Thad National Park; **d,** Juvenile *Borassodendron machadonis,* one of several canopy palms in southern seasonal evergreen dipterocarp forest, Khao Ban Thad. (All: H.H.)

District (14°40' N)[215] (map inside back cover). They formerly continued up the coastal foothills of the southern Arakan Yoma, west of the Irrawaddy delta into the Chittagong hills,[216] close to 23° N. They are the dominant forest type in much of the Andaman Islands,[217] and exist in a modified form also in the Nicobars where, notably, dipterocarps are lacking (Chapter 6). To the east, there is a break across the Chao Phraya plains. Seasonal evergreen forests then resume on the lower slopes of granite Khao Luang in south-east Thailand, and on the coastal slopes of the southern sandstone range which continues into Cambodia as Phnom Kravanh and Phnom Kamchay (Cardamom and Elephant Chains), almost to the frontier at Kep; then across the lowlands to southernmost Vietnam; also on the offshore islands of Koh Chang and Phu Quoc. They then follow the eastern slopes of the Annamite range to north-central Binh Tri Thiên Province (c.17°30' N) where it seems they once changed to northern seasonal evergreen dipterocarp forest (described below). Little now remains at this latitude. A further belt of these forests occurs from the Tinnelveli hills inland of Cape Comorin, India's southernmost point, along the coastal foothills of the Western Ghats to the foothills north-east of Panaji, North Kanara (15°40' N).

Seasonal evergreen dipterocarp forests experience regular annual dry seasons of less than 4.5 months; but annual rainfall varies from c.2,000 mm to more than 5,000 mm where the summer monsoon confronts the coastal mountains along parts of the northern Western Ghats and Burma's Arakan coast. At Khao Ban Thad Wildlife Sanctuary (Khao Chong, near Trang) in southern peninsular Thailand, the seasonal evergreen dipterocarp forest experiences only two months in which evaporation and transpiration routinely exceed rainfall. The steep slopes of the granite hills face east; a 20 ha CTFS plot climbs from the lower slope to a ridge (Plate 2.25a,b). Large patches of climber tangle, the concentration of the larger emergent dipterocarps on rocky ridges and slopes, and the dramatic band of the tall grass *Saccharum spontaneum* to either side of the white-water river nearby testify to the onslaught of an occasional cyclone (typhoon) and attendant flooding (Plate 2.23f). Extensive patches on windward slopes have been flattened. Cyclones off the Bay of Bengal affect these forests on the western slopes of the peninsula with greater frequency. Cyclones occasionally stray into MDF to the south. Wyatt-Smith described MDF in Kelantan state, north-east Peninsular Malaysia, from whose stature and the presence of many trees with deformed trunks he inferred the track of a typhoon one century earlier.[218] In many areas tall bamboos establish early in wind-throws, retarding and perhaps deflecting succession. Bamboos invade forests following logging as far south as northern Perak (Malaysia), where I have seen forest logged c.20 years earlier dominated by bamboo and apparently devoid of dipterocarp regeneration. These Thai forests are richer in large vines than other forests in tropical Asia.

Coastal forests of south-east Indo-Burma and the Arakan coast recall MDF in the north-east Philippines, with a continuous even emergent canopy, often of dipterocarps. Forests on slopes of the Western Ghats exposed to the driving winds of the south-west monsoon present a similar profile (Plate 2.23d). In both cases, crowns are pruned to bend away from the prevailing wind. These forests rarely attain the stature of MDF, although scattered emergents may reach 50 m in sheltered coves. At 35–40 m, emergent canopy crowns are less clearly raised above the main canopy than in MDF, in this resembling MDF in perhumid Sri Lanka (see section 3.3.2 above). Both share breezy climates with storms and occasional cyclones during the south-west monsoon. The leaf size spectrum is within the range of MDF, with mesophylls predominant. Detailed study is awaited, but variation in canopy density on clay and sandy soils in Indo-Burma appears to mirror that described for MDF, as it does in the flora (Chapter 3). The peninsular Indian forest is solely on clay-rich soils – the same substrate as the Sri Lankan MDF.[219]

A notable difference between MDF and southern seasonal evergreen dipterocarp forest is the general presence in the latter of deciduous canopy species, albeit in low population density (Chapter 3). Pioneers include *Tetrameles nudiflora* and *Alstonia scholaris,* while *Lagerstroemia* spp. are present in later succession. The proportion of deciduous species in the canopy appears dependent on the frequency of canopy disturbance in which soils are scarified. They are less represented on freely draining acid, sandy soils of south-east Indo-Burma, and coastal and island forests, where litter breaks down more slowly.

Kadambi[220] described southern seasonal evergreen dipterocarp forests on the lower western slopes of the Western Ghats (at Agumbe, Shimoga, 13°27'–13°51' S). This is towards the northern limit of their range, with four dry months. He noted plentiful dew-drip Jan.–Feb. during the dry season. Kadambe observed that the seven months of wet monsoon winds 'decided the vertical distribution of the forest'. He distinguished five types, floristically defined albeit structurally different, which he correlated with topography and exposure to wind, leading either to greater canopy pruning and evenness, or to wind-throw and a higher proportion of deciduous species (Chapter 3).

5.1.1. *Dynamics*
The dynamics of seasonal evergreen forests remain poorly understood. Growth rates are not yet available for stand level comparisons, but there are few light hardwood species in mature phase stands, light hardwood yellow and red meranti species being absent (Chapter 3). This may imply lower mean species maximum growth rates (but see Table 2.7a,b). In many respects therefore these forests more closely resemble the evergreen rainforests of Africa and the Amazon Basin, which share a similarly seasonal climate, than they do the forests of perhumid Asia.

5.1.2. *Regeneration*
The understorey of seasonal evergreen forests is generally less dense in juveniles of canopy species. This results in lower overall forest densities, although not necessarily lower basal areas (Fig. 2.19). Notably, dense cohorts of juvenile dipterocarps are absent while subcanopy tree species' juveniles are proportionally more abundant.[221] Dipterocarps are often less abundant in the canopy, although here they flower more frequently, at least one species flowering each year in the more strongly seasonal regions (Chapter 5). An annual dry season is associated with more consistent annual fruit production and a higher density of browsing animals than are found in perhumid climates. Chengapa[222] found that *Dipterocarpus* regeneration responded to subcanopy clearing to 10 m. This implies that dipterocarps require moderate increases in light such as are provided by single tree gaps. Deciduous species, meanwhile, respond to larger gaps. The abundance of pioneer trees and vines then appear to pose a challenge to regeneration of the deciduous climax later successional species. Browsing also reduces sapling density, though the low densities at Khao Ban Thad, where browsing deer and other mammals have been extirpated by illegal hunting, cannot be thus explained.[223]

5.2. *Northern seasonal evergreen dipterocarp forests*

If you have known Hanoi in February you will not forget crossing the great bridge over the Red River in the cold penetrating driving rain of the northeasterlies, bright green cabbage vegetable gardens cramming the sloping banks down to the river, now at low water, against the gloom. The weather one expects in wintry London! These winds come off the China Sea, whose current sweeps up from the warm south. There is no frost in the air therefore, though snow occasionally falls on the lower montane forest above the old French hill station on Bavi (1,000 m), inland to the north-east.

During the cool dry season of inland Indo-Burma and North-East India, persistent dry north-east winds occur from Nov. until the end of March. Here, lowland north- and east-facing slopes are drier and cooler under the prevailing northeasterlies off the continent[224] (Chapter 1). The climate is calmer on

Fig. 2.19 a Mean basal area and stem density in CTFS sites in tropical Asia. Black part of density columns: trees >1–<10 cm dbh. LAM: lambir, Sarawak; PAS: Pasoh, Negeri Sembilan; BT: Bukit Timah, Singapore; SIN: Sinharaja, Sri Lanka; PAL: Palanan, Luzon; KC: Khao Ban That, peninsular Thailand; HKK: Huai Kha Khaeng, NW Thailand; XB: Xishuangbanna, southernmost Yunnan; MUD, Mudumalai, Tamil Nadu. (From Bunyavejchewin *et al.* 2011) **b** Stem density (above) and mean basal area (below) for plots in various Thai forests. DD: Deciduous dipterocarp forest; P: Deciduous dipterocarp-pine; MD: tall mixed deciduous; LM: lower montane; DSE: semi-evergreen; WSE: southern seasonal evergreen forest. (From Bunyavejchewin *et al.* 2011)

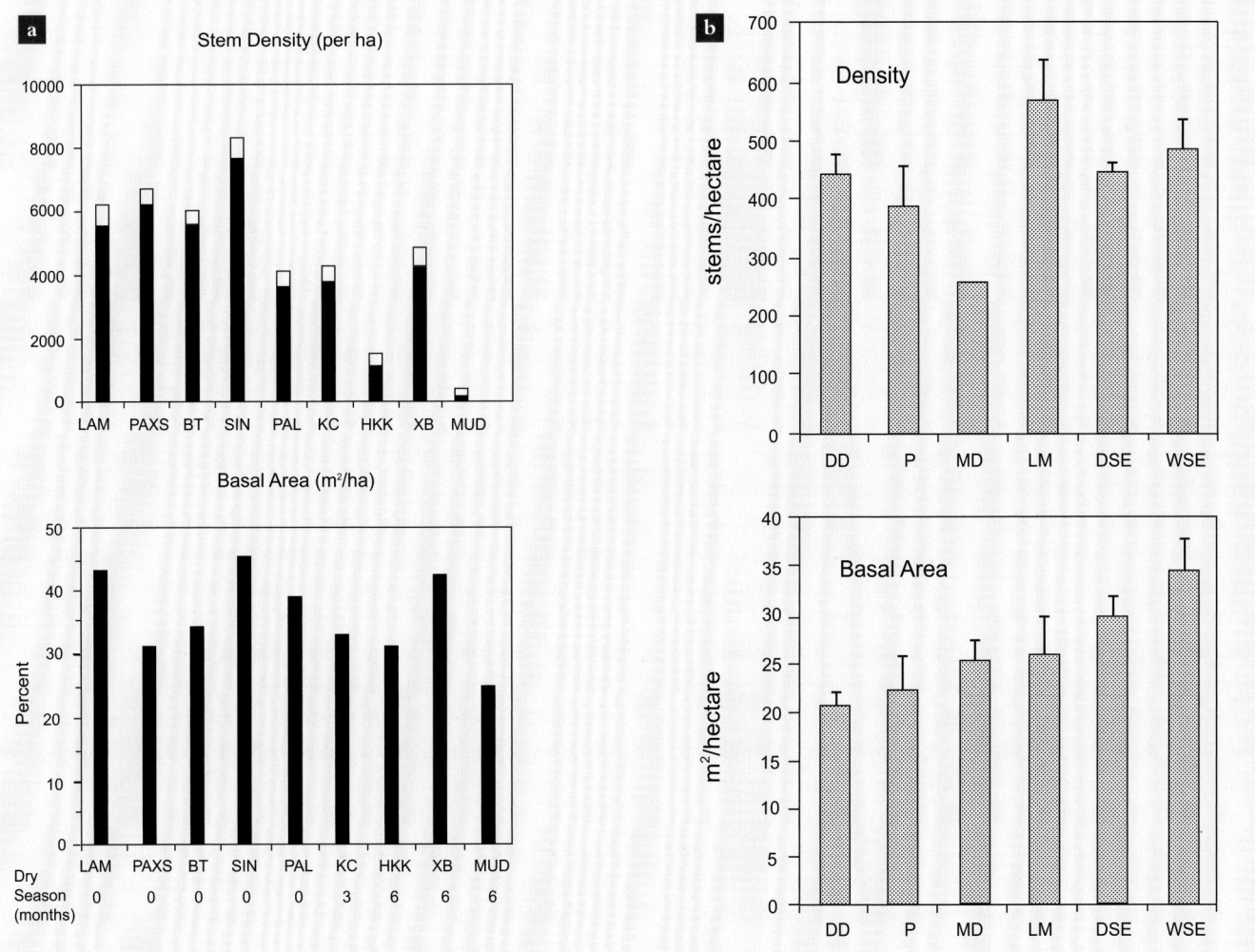

those facing south and west, and is moister due to nocturnal accumulations of fog above the forest canopy. Heavy nocturnal dewfall is experienced during the low rainfall cool season Dec.–March throughout the seasonal tropics north of c.10°, albeit abating during the hot dry season April–May. Dew extends longer as the tropical margin is approached (Chapter 1). In the higher tropical latitudes, the hot dry season also shortens. Inland, westwards to Arunachal in North-East India, dew bathes the forest canopy in the early hours from the end of the wet summer monsoon until the beginning of the hot dry season in April – and continues intermittently even then (Chapter 1). To the east, in Hainan and southernmost Taiwan, and south along the coastal lowlands of Vietnam, dewfall may be slight, but the north-east monsoon is wet and temperature is

moderated by the ocean. The south-facing slopes of southern Taiwan, Hainan, and coastal Guangxi at these latitudes, and the lowlands of southernmost Guangxi and northern Vietnam sheltered from the northeasterlies, are also kept intermittently moist even through the April hot dry season by orographic rainfall brought in by diurnal onshore breezes. Here overall, then, the south-facing slopes remain moist throughout the year. In contrast, further south (and in the mountain valleys of the inner Himalaya, Chapter 4), southern lower slopes are baked by the dry-season sun. They are prone to fire, and thus support deciduous forests.

Through these regions a distinct northern seasonal evergreen forest (the 'northern wet evergreen forest' of Champion) exists, in many respects analogous to the southern, in which emergent dipterocarps may also occur. The sole consistent difference from southern seasonal evergreen dipterocarp forest is floristic (described in Chapter 3, Part II section 3.1). Seasonal evergreen and semi-evergreen tropical forests occur as a separate northern belt from southern Taiwan westwards to Kumaon in the Central Himalaya of Uttar Pradesh; and from latitudes of 17–22° N on slopes facing the South China Sea to 29° N in Kumaon (see map inside back cover, and section 4.2.3 below). Along the eastern coast and on south-facing slopes the belt is typically narrow, except in lowland northern Vietnam and southern China (Chapter 3). Pruned in Taiwan and the Indo-Chinese coast by the frequent typhoons and driving winds of the north-east monsoon (see Chapter 3 under the tropical margin, also below), these evergreen forests achieve their greatest species richness in the east. Further west in the Central Himalaya, the dry season, and particularly the hot dry season increases in length; westwards, they decline gradually in richness also (Chapter 7). Seasonal evergreen forest has now been destroyed in Taiwan except on karst,[225] and fragmented elsewhere.

Floristic continuity is maintained throughout its range (Chapter 3), but the forests vary considerably in structure. In coastal areas subject to typhoons it may be no taller than 10 m with a dense even canopy, as at the CTFS plot at Nanjeng Shan (Taiwan) (see section 3.3.2.2 above). Coastal forests of northern Vietnam and Hainan rarely exceed 30 m in height, and emergent dipterocarps and other species hardly emerge above the main canopy. In this they resemble the characteristic structure of southern seasonal evergreen forests.

But on sheltered south-facing slopes at Mengla in Xishuangbanna Province (southernmost Yunnan), site of a CTFS plot,[226] and elsewhere east into Guangxi and south into northernmost Laos and north-west Vietnam, there are surviving patches of a magnificent triple-stratum evergreen forest in which the dipterocarp *Parashorea chinensis* reaches true emergent status at 50 m (Plate 3.12a). Structurally similar forests have been recorded from Arunachal and the north-east Burma valleys.[227] These forests are on sandy clay and clay loams, and even though they apparently experience more canopy disturbance than equatorial MDF on similar soils, they are equal in structural magnificence if not in species diversity (Chapter 7). Leaf size classes are similar, dominated by mesophylls. Although some deciduous pioneers and several successional species occur, the canopy is overwhelmingly evergreen. Fire does not penetrate primary stands.

6. Forests influenced by fire, drought and by storms

Flowering and death of bamboos ends in causing disastrous fires whose ravage is often indescribable; wholesale destruction or at least crippling of the forest tree stand takes place resulting in inestimable loss of timber.

K. Kadambi, 1949, *Indian Forester*.

6.1. *The nature and causes of forest fire*

Fire, consequently grass and ungulates including domestic cattle, are widespread in the semi-evergreen as well as deciduous lowland forests of continental Asia.

6.1.1. *Causes*

Fire is started in nature by lightning strikes when rain is not falling[228]. Additionally, our progenitors in Asia, as in Africa, have been burning forest, field and shrub layers wherever they are periodically combustible for almost two million years in order to raise game,[229] and for at least 4,000 years to encourage grass for cattle. In this respect, the history of fire as an ecological factor has been dramatically different from the Neotropics. There, only one hominid – ourselves – has been actively burning, and only for the last 20,000 years.

Fires can only start when there is dry combustible material available. They are rare in nature in perhumid regions. The few fire-resistent species there include *Melaleuca,* and *Dillenia suffruticosa* and *Ploiarium alternifolium* which sprout from coppice, of open grasslands, and *Combretocarpus rotundatus* which is abundant in the 'padang teruntum' savanna woodland on the summit of some peat swamps (Chapter 3). Fires are rare at the beginning of the dry season, but litter and ground vegetation becomes increasingly fire-prone as the season progresses. Rainfall seasonality influences fire incidence and intensity as well as influencing forest structure, phenology and growth. Importantly, increasing seasonality is accompanied by increasing variability, not only of the length of the wet monsoon and the amount of precipitation, but also in its date of onset and termination (Chapter 1). As regional aridity further increases, the production of annual biomass declines, and thereby the frequency and intensity of fire falls. From Thailand eastwards, there is increasing frequency of showers during the hot dry season starting in April, when trees begin to flush before the onset of the monsoon proper, and natural fire frequency also declines. As the tropical margin is approached, the cool dry season and dewfall extends closer to the beginning of the monsoon, and forests again become resistant to burning. (Fire-maintained *Imperata* grasslands are, of course, ubiquitous in wet regions, originating from degraded swidden).

6.1.2. *Types of fire*

Fires are of two kinds, differing in intensity and impact. *Ground fires* are of low intensity, occurring where combustible material is sparse and comprised mostly of leaf and twig litter or short grass. Ground fires only affect the field layer (Plate

Plate 2.26 Fire. **a,** Ground fire in tall deciduous forest, Mae Hon Son, north-west Thailand; **b,** Hot crown fire in short deciduous forest, exacerbated by the understorey shrub *Lantana camara*, Bandipur National Park, Karnataka; **c,** Low temperature ground fires can destroy canopy trees: burned out trunk of a mature *Hopea odorata* in semi-evergreen dipterocarp forest, Huai Kha Khaeng CTFS plot, west Thailand; **d,** Teak with epicormic shoots two years after a hot fire, 2010, Bandipur National Park. (**a, c, d** H.H.; **b** R. Matthias)

recall their tribulations (Plate 2.26d). Fire exclusion leads to accumulation of fuel, and increases the risk of crown fires, as early colonial foresters discovered to their cost (Chapter 8).

The fuel necessary for hot fires is of two kinds. One is combustible vegetation in the understorey. Bamboos – especially those many genera that reproduce and die synchronously over whole landscapes and even regions – are associated with devastating fires. Although dense bamboo germination and establishment ensues once the rains arrive, the opening of the often dense bamboo canopy after flowering provides the sole opportunity for tree regeneration. Tree species with seed dormancy and fire resistance, and seedlings with low photosynthetic compensation points have the competitive edge and survive thereafter, albeit patchily.[231] The Neotropical shrub *Lantana camara* has, over 150 years, invaded the deciduous forests of dryer parts of South Asia to become the dominant understorey shrub. As it matures, it becomes inflammable. But its canopy is neither so dense nor so tall, at less than 2 m, as to exclude all other species. However,

2.26a), occasionally entering the hollow trunks of canopy trees, especially when already wounded by crown fires. Temperatures may reach 400°C at 50 cm above ground. But coarse litter of dead tall grasses, fallen branches and other material, or combustible subcanopy vegetation feed a *hot fire*. Such fires, consuming combustible fuel including shrubs and bamboo, can ascend to the canopy, reaching 700–800°C and often killing mature fire-sensitive tree species[230] (Plate 2.26b). Fire-resistant species, including teak and sal, recover by epicormic branching, but their crooked trunks and branches

the fires which feed on it can reach the canopy of these short forests (Plates 2.26b, 2.29b).

Tall grasses in dense swards (notably *Imperata,* which can grow to 1 m) can cause fires of intermediate intensity, killing evergreen understorey but not reaching the canopy. The greater fire frequency associated with such grasses destroys all seedlings.

6.1.3. *Fire frequency*

Infrequent fires may provide conditions favoring invasion of bamboo. Vast regions of continental South-East Asia are nowadays dominated by forests with a bamboo understorey, and bamboos also invade landslips and heavily logged forests. Frequent fire, on the other hand, leads to the death of bamboo and its replacement by grassland. Frequent fire is associated with massive changes in tree species composition, phenology, and forest structure. These changes can transform multi-layered evergreen forest into deciduous forest with a single-layered canopy and an understorey of grass, bamboo, or in mesic sites, woody or herbaceous dicots.

Fires have increased in frequency, particularly over the last century, as human populations have grown and pressure on forest resources have mounted. In Indo-Burma, ground fires from burning rice stubble spread many tens of kilometres into the parks and sanctuaries. Many forests now experience almost annual ground fires, in India set to encourage grasses for the ubiquitous cattle, which also severely reduce regeneration by their browsing. There, the interval between fires in deciduous forests is estimated to have decreased over the last century from about 7–8 to 3.3 years. There were 6–10 fires in 15 recent years in the taller deciduous forests at Mudumalai, 2–8 in the shorter.[232] This has in part been due to the invasion and current dominance of flammable shrubs (Chapter 8).

6.1.4. *Interactions between fire and forest*

The impact of fires of differing severity and frequency on forests of differing stature, age, and species composition, although of central importance to forest management in the seasonal tropics, still awaits careful systematic observation and experiment. Other than *Melaleuca cajuputi* (Myrtaceae) – a widespread tree of fire-mediated grasslands on acid sulphate soils behind the mangrove barrier of seasonal tropical East Asia – there appear to be no evergreen tree species in continental Asia that can survive persistent fire as juveniles, although *Syzygium cumini, S. nervosum* (syn. *operculatum*) and *Mangifera indica* are more resistant than most and occur in forests that are otherwise deciduous.[233] Seedlings and young saplings of evergreen species are particularly sensitive. Larger individuals of many species retain thin bark. Persistent ground fires can wound the butt bark, leading to penetration of fire into the wood (Plate 2.26c). Fire may then smoulder and penetrate up hollow or resinous trunks, which smoke like giant cheroots and die, or become

invaded by wood-eating termites. Nevertheless, many species recover by coppice shoots if fires are infrequent or not too intense.[234] Widespread semi-evergreen forest pioneers such as the evergreens *Neolamarckia* and *Macaranga,* but also deciduous such as *Tetrameles* and briefly deciduous *Alstonia scholaris,* are particularly sensitive, and become confined to moist valley sites.

Rootstocks of some, perhaps most, seedlings of deciduous forest species can become resistant to ground fires by the end of the wet season in which they germinate. Tree seedlings are nevertheless often absent in deciduous, especially short deciduous, forests. The relative importance of burning, browsing or drought in inducing mortality differs with general forest condition. Deciduous forests are therefore comprised of a guild of fire-tolerant deciduous species which, as first-year seedlings, may be killed by fire but shortly thereafter can survive as rootstocks, resprouting from the ground each year until there is either a run of years without fire, or the rootstock has grown sufficiently to support a shoot that can grow tall enough to escape a ground fire in one season. Fire and browsing kill above-ground parts of young trees, but the root may continue to grow slowly, forming a large woody 'lignotuber'. Each year, the growing lignotuber has more resources for a reiterating leader to grow taller in the following dry season; eventually, following several years without fire, the sapling develops bark sufficiently thick to insulate living tissues from fire. Species with such bark are overwhelmingly deciduous. Fire resistance is therefore the primary motor behind the distinctive composition and phenology of tropical deciduous forest.

Because the density and nature of the understorey affects the amount of combustible material, hence the extent and intensity of fires, the most intense crown fires are in tall deciduous and the taller margins of short deciduous forests. Fires rarely invade thorn woodland owing to lack of fuel materials. Nevertheless, stand density at Mudumalai (Tamil Nadu) was found to correlate with rainfall, not with fire.[235] Increased fire frequency is currently leading to a decline in fuel accumulation and therefore in crown fires. This can be expected to lead eventually to changes in stand composition in favour of species of the shorter communities.

Ground fires have the benefit that they burn the leaf litter accumulated over the dry season, which can become matted during the rains thereby deterring successful penetration by the radicles of germinating seedlings. Unburned teak litter, falling at c.1,300 g m^{-2} y^{-1}, may take eight months to decompose during and after the rains.[236]

6.2. *Champion's classification requires renaming*

European foresters in the colonial period, trained to protect their temperate forests against all fire, were faced early on with a paradox. Overwhelmingly, the quality commercial hardwoods in India are deciduous species rare or absent in semi-evergreen and evergreen forests (see a fuller discussion in Chapter 8). In moister seasonal areas, deciduous species failed to re-establish following fire exclusion. If they did not fail to germinate, they were early out-shaded by faster-growing evergreen species, which spread up slopes from moist valleys and river banks where, free of fire, they survived and reproduced. The great forester Dietrich Brandis, posted to the Burmese teak forests as superintendent in 1856, noticed teak trees surviving in disturbed evergreen forests near villages. It was another 30 years before a forester, courageous enough to experiment with fire as a management tool, finally demonstrated that both germination and survival of teak depended on the exclusion of evergreens by fire. It is said that he was himself fired for such a seemingly outrageous claim (Chapter 8).

Champion[237] distinguished three forest categories in India and Burma by their degree of deciduousness. These, again, require renaming to avoid any implication that climate alone determines their nature and range.

1. 'Tropical semi-evergreen forests', often known in Thailand as dry evergreen forests, which I am terming *semi-evergreen dipterocarp and other tropical semi-evergreen forests*. These possess an evergreen mature phase, which may include fire-sensitive dipterocarps. Some pioneers and most successional climax tree species are deciduous, and include many which reach and persist in the mature canopy which is nevertheless mainly evergreen. Although Champion does not mention this, many of these forests do withstand periodic ground fires. At the regional scale, there is a continuum of forests with the characteristics of seasonal evergreen and semi-evergreen forests, broadly correlated with increasing length of the dry season if fire does not intervene; but on a local scale each type may be distinct, and occupy a different topographic element, separated by a narrow ecotone.

2. 'Moist deciduous forests', tall (generally 20–30 m) whose canopy is exclusively deciduous, but in which an evergreen understorey sensitive to hot fire endures a nomadic existence. I will call these forests *tropical tall deciduous*, to avoid a presumed correlation with climate or drainage.

3. 'Dry deciduous forests', entirely lacking evergreen species, and generally 10–20 m tall. I will call these forests *tropical short deciduous*.

This overall classification has endured because, although dependent on a dynamic interrelationship, it appears to provide a means of easy diagnosis from static observation. As we shall see though, increased fire frequency through human activity can lead to local extinction of evergreen species where seed sources are lost. Semi-evergreen and tall deciduous forests co-occur in climates with 4.5–6.5 dry months, short deciduous forests on uplands with up to 8.5 dry months.

Champion's classification does not bear close comparison with forest classifications of other continents. Putatively primary deciduous forests are restricted in area in the Neotropics, except in Central America where they have been the subject of regional classification. In southern Africa, the high biomass of grazing and browsing mammals, and the consequently more widespread open canopy savanna woodlands with short grass field layers, appears to have reduced the importance of fire as a determining factor, while evergreen species are a rarity there in the savanna woodland understorey.[238]

6.3. Semi-evergreen forests

Semi-evergreen and tall deciduous forests mostly occupy similar rainfall regimes, though the tall deciduous extends into somewhat drier regions. They differ in their proportion of evergreen trees.[239]

6.3.1. *Diversity and distribution*

Semi-evergreen forests extend north into regions with six dry months, from Chumphon (peninsular Thailand) northwards, and throughout the lowlands of Indo-Burma (map inside front cover). From Kumaon (north-east Uttar Pradesh) eastwards (map inside back cover) a patchy but widespread and floristically distinct semi-evergreen forest occurs on slopes below c.800 m, and along riparian fringes in the *terai*, in which dipterocarps are often absent and deciduous species are well represented in the canopy (see Chapter 3). To what extent dewfall supplements rain here during the cool dry season has yet to be recorded. Where this forest occurs on steep terrain, it appears to suffer more canopy damage than does the northern seasonal evergreen forest. It sharply abuts the tall deciduous sal forest, which dominates the adjacent *terai* terraces and plains. There, riparian semi-evergreen forests differ floristically from the sal forest. In both North-East India and Indo-Burma, the most extensive patches are on freely draining grey sandy soils, as at Huai Kha Khaeng and over the sandstones south-east of Bangkok, south of the Khao Yai mountain range which extends east to the Cambodian frontier and beyond. Small patches also exist north of it, as at Sakhaerat Research Station on the sandstone Khorat plateau. On clay soils, semi-evergreen forests have been greatly reduced by swidden. They now survive in North-East India and probably over the Burmese border in the Patkai hills. Elsewhere such forests can now be found only in patches on the lower slopes and alluvium of the deep northern valleys, where dominant stands of *Dipterocarpus turbinatus* betray their former presence. In the foothills on both slopes of the Western Ghats, semi-evergreen forests have been reduced to patches and modified by more than a century of logging, as at Mudumalai. Semi-evergreen forest also still survives between the seasonal evergreen and deciduous forests west of the Western Ghats between 11–15° N, along the eastern base of the Western Ghats, and in some moist valleys of the Eastern Ghats, particularly in the former hunting domains of the maharajas. Scattered across the Karnataka Deccan exist forest patches, often sacred groves, called *kan,* where a taller forest occurs including species otherwise found in less seasonal climates.[240] It is unclear whether these are relics of the forest type that prevailed before cattle and fire increased, or whether they occupy favorable mesic sites. In all such cases, the subcanopy is now generally degraded and largely replaced by deciduous species, including exotic weeds.

6.3.2. Ecology of semi-evergreen forests

Our understanding of semi-evergreen forest dynamics still substantially rests on the careful and exhaustive experiments of the silviculturist B.S. Chengappa, in the 1930s in the Andaman Islands[241] (Chapter 8). Chengappa, whose field background and training were in India, was vexed that the climate, towards the wet end of this forest's range with a 3–4 month dry season, was one where closed forest fires were unknown. Indeed, ground fires could not be ignited even after subcanopy vegetation had been cleared, so that brush had to be heaped before burning. Nevertheless the forests of the coastal plain, then still largely intact, were semi-evergreen. They differed from the seasonal evergreen forests of the steep inland hills in their generally lower stocking of dipterocarps (mostly *Dipterocarpus alatus*), as well as relative abundance of the successional deciduous genera characteristic of it on the continent (Chapters 3, 6), which are preferred for their timber. He was struck by the prevalence, in both semi-evergreen and seasonal evergreen dipterocarp forests, of huge 'overmature' canopy individuals, yet poor and often absent sapling and pole-sized regeneration. By means of numerous silvicultural experiments, he successfully devised a means to establish regeneration of the preferred deciduous species, by cutting and removing subcanopy trees less than 20 m tall. The forest was then logged, after which abundant deciduous species' regeneration occurred. This required early cleaning of the mass of simultaneous pioneer and vine regeneration that otherwise out-shaded the preferred species. Saplings grew well enough thereafter to outdo the competition.

Chengappa thereby showed that deciduous successional species, common also to tall deciduous forests, need not depend on fire for their regeneration, but do require freedom from (evergreen) competition, therefore an open understorey, in at least their first year. These semi-evergreen forests of the Andaman plains had been catastrophically affected by a cyclone in 1891,[242] from which they were still recovering in Chengappa's time. Occasional catastrophic wind-throws on landslip scale ensure abundance of deciduous climax successional species in the regenerated canopy. Whether subsequent fire is necessary in disturbances of this scale to suppress fire-sensitive vines and trees, seems not to have been documented. Deciduous successional species also seem to establish in nature in moderately small gaps caused by lightning and small wind-throws, providing too little light for reliable germination of pioneer dormant seed but enough for survival and growth of the deciduous late successional quality hardwood species which do not require light for germination and early establishment.

Wind-throws at the landscape scale are recorded by foresters from peninsular Burma as well as the Andaman Islands.[243] A massive wind-throw occurred in the 1990s at Nam Nao National Park (Central Thailand), in a *Pinus merkusii*/lower montane oak-laurel forest mosaic, where it was followed by an intense fire (Plate 2.22c).[244] Pine and deciduous successional species regenerated in the following year, before grass cover became continuous (Chapter 4).

It seems likely that semi-evergreen forests have survived better on freely draining sandy soils in part because the dense tall bamboo brakes – highly combustible after mass fruiting and mortality – are confined to moist sites among them, or to clay, thereby restricting crown fires to those sites. The semi-evergreen forest patches (*kan*), described by Pascal[245] on the Deccan Plateau are likewise restricted to freely draining, coarse yet well-watered sites. But the hilly regions where semi-evergreen forests survive, or are thought to have formerly occurred, on clay loams have been and still are the preferred lands of swiddening minorities. In the plains they have also

Plate 2.27 (right) Semi-evergreen dipterocarp forest. **a,** The broken canopy of a semi-evergreen dipterocarp tall deciduous forest mosaic, in the dry season, the air hazy with the smoke of distant stubble burning, Huai Kha Khaeng Wildlife Sanctuary; **b,** A similar mosaic on a hillside in south-east Thailand. The mauve suffusion is fruiting *Lagerstroemia calyculata;* **c,** Mature evergreen stand of *Hopea odorata* (large blackish boles), with sparse understorey of tree regeneration, Huai Kha Khaeng CTFS plot; **d,** Remnant patch of *Dipterocarpus turbinatus*-dominated semi-evergreen dipterocarp forest on clay soil; the *Dendrocalamus strictus* bamboo understorey suggests that fire periodically invades. Foothills of Doi Inthanon, north Thailand; **e,** Semi-evergreen dipterocarp forest on the sandstone Khorat Plateau, at Sakhaerat: hardly emergent canopy with *Hopea ferrea, Shorea henryana* (pale crowns); **f,** Deep coarse rooting in semi-evergreen dipterocarp forest, Huai Kha Khaeng Wildlife Sanctuary; **g,** Patrick Baker cored this ancient *Afzelia xylocarpa* in Huai Kha Khaeng Wildlife Sanctuary, and estimated it to have germinated near the beginning of the 18th Century. Its trunk, spreading in several branches from the base, betrays the open conditions in which it regenerated. (**a, c, f ,g** H.H.; **b, d** P.A.; **e** S. Bunyavejchewin)

been centres of civilisation for over 15 centuries (Chapter 8). At Angkor and its water source on the Phnom Kulen plateau, the present forests presumably regenerated following the collapse of the Khmer civilisation 700 BP.

With increasing seasonality, there is an increase in the representation of deciduous pioneer and successional species. Some, for example *Afzelia xylocarpa,* establish following fire and are long-lived (see Box 2.2; Plate 2.27g). A similar increase seems to be associated with increasingly frequent and widespread natural catastrophe, notably extensive cyclonic storm damage and increasing frequency of ground fires, which destroy early regeneration.

Semi-evergreen and tall deciduous forests subject to occasional ground fires may have an understorey rich in broad-leaved monocotyledons, bamboo and ginger (Zingiberaceae) species, as well as a well-developed understorey of deciduous shrubs and canopy saplings of both evergreen and deciduous species that, once attaining a certain diameter, can withstand ground fires (Plate 2.28a). In tall deciduous forest, by contrast, the evergreen component is more or less sparse, but tall bamboos can dominate the understorey while gingers remain abundant in moister sites (see section 5.4.1 below).

In summary, while semi-evergreen forest may persist in the absence of fire where storm damage is sufficiently drastic and frequent to ensure sufficient opportunity for regeneration of their deciduous component, tall deciduous forests in Asia are mostly dependent on fire for their existence. This is in apparent contrast with Africa, where fire is not considered essential. The higher density and diversity of browsing animals there may ensure that evergreen species are browsed to extinction during the dry season.

6.3.3. *Growth rates*

Whereas only c.6% of MDF species are pioneers, pioneers and early successional species comprise 25% of species in the semi-evergreen CTFS plot at Huai Kha Khaeng. Maximum growth rates achieved, at least of evergreen species, are similar to those achieved in MDF (Table 2.7a, b). The much higher soil nutrient levels of semi-evergreen forest soils should be borne in mind (Fig. 1.10), though also their growing season which may be no more than half as long. Among the fastest-growing species are many figs, both free-standing and hemi-parasitic (Table 2.7a). Although mean comparative growth rates are as yet unavailable, the abundance of pioneer and successional species at Huai Kha Khaeng implies that mean stand growth rates are higher there than at Pasoh. The Asian seasonal dipterocarp and semi-evergreen forests share a similar climate (including wind storms), and also an abundance of figs (and the associated arboreal vertebrates), with the CTFS plot at Barro Colorado Island (BCI) in Panama. These forests may approach similar mean rates of growth and turnover.

Box 2.2

The conundrum of the semi-evergreen – deciduous forest mosaic: Huai Kha Khaeng UNESCO World Heritage Forest and Wildlife Sanctuary

The Huai Kha Khaeng and Thung Yai Naresuan Wildlife Sanctuaries (15°0'–15°47' N), at 18,700 km² comprise the largest continuous area of protected forest in South-East Asia. It is recognized as a UNESCO World Heritage Forest. With an altitude 500–1,500 m, the area retains the full complement of vertebrates to be expected in this regional climate of 5–6 dry months. Huai Kha Khaeng consists of the headwater catchment and forests of the river (Huai) Kha Khaeng. The forests are rimmed on their west by low mountains bearing lower montane oak-laurel forest and blocks of semi-evergreen forest, and on the east and north by low uplands (Fig. 2.20).

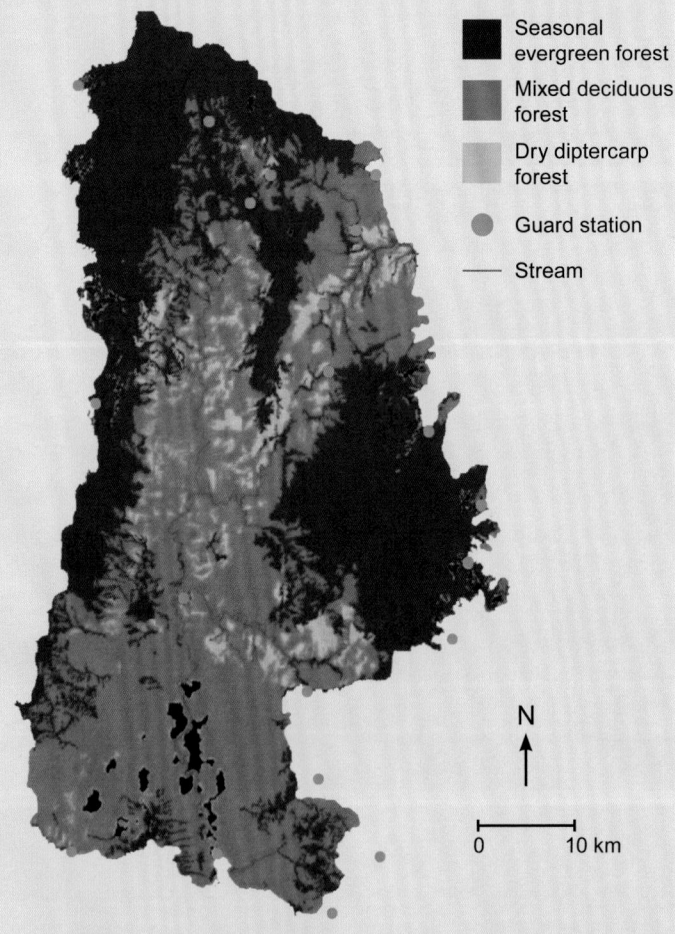

Fig. 2.20 Forest types within the Huai Kha Khaeng Wildlife Sanctuary, north-west Thailand. Note the penetration of semi-evergreen forest along water courses within the tall deciduous forests, and the concentration of deciduous dipterocarp forest within the latter. (From Bunyavejchewin *et al.* 2009)

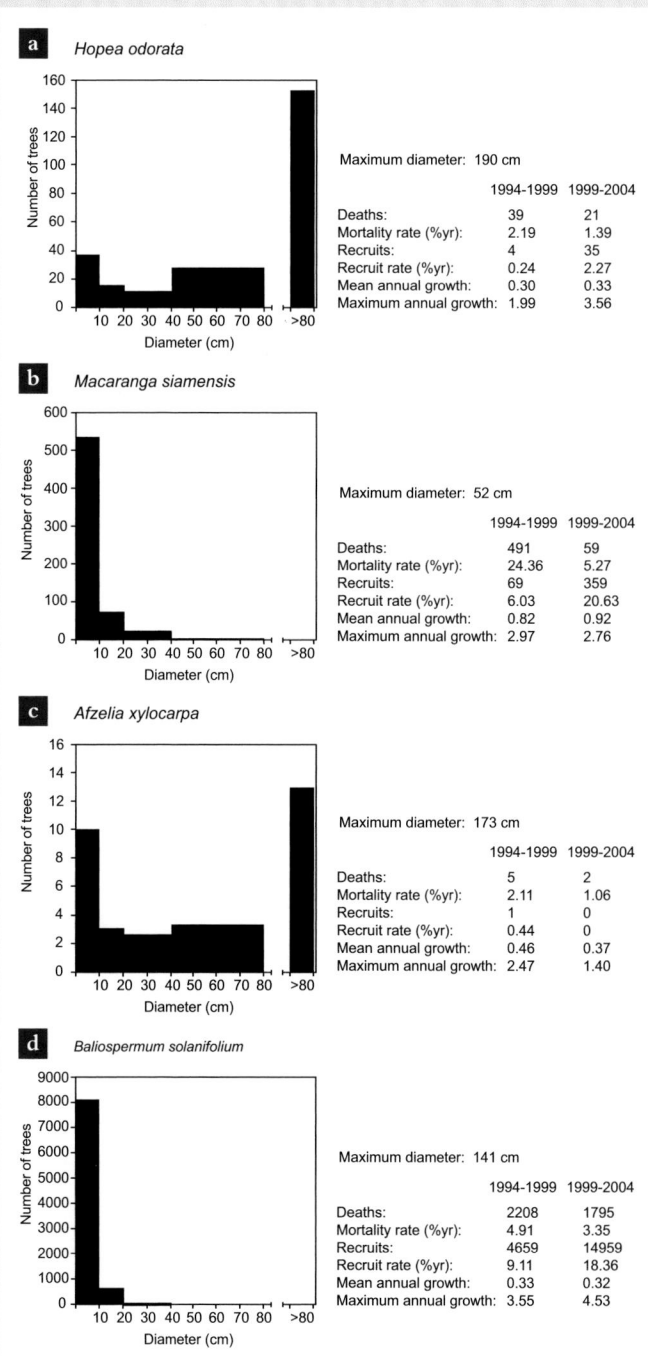

Fig. 2.21 Population structure of four tree species within the 50 ha CTFS plot at Huai Kha Khaeng Wildlife Sanctuary (HKK), Thailand: **a,** *Hopea odorata,* an evergreen canopy tree; **b,** *Macaranga siamensis,* an evergreen pioneer; **c,** *Afzelia xylocarpa,* a deciduous canopy species; **d,** *Baliospermum solanifolium,* an evergreen understorey treelet. Diameter (cm), diameter growth (mm). The two HKK climax canopy trees, *Hopea* and *Afzelia,* show the 'reverse J' structure that implies that conditions for successful regeneration are rare but widespread when they occur. This effect is heightened when it is realized that trees exceeding 40 cm dbh have reached the canopy so that canopy tree density in these species. *Macaranga,* and *Baliospermum* which regenerates after ground fire, have normal population structure at the scale of a 50 ha sample. (After Bunyavejchewin *et al.* 2009, Co *et al.* 2004)

Unexpectedly, the central alluvial lowlands and the low hills around the stream-heads are covered by tall deciduous forest, except where semi-evergreen forest penetrates along riparian fringes. Their soils are predominantly clay loams. Curiously too, the semi-evergreen forest is confined to freely draining greyish sandy loams which, although of moderately high nutrient status,[246] are drought prone. The semi-evergreen forests seem to have survived because their soils were shunned by farmers.

Bamboo brakes occur in patches through the tall deciduous forests alone. It appears that they may hold the key to the forest differentiation. Two genera predominate: *Dendrocalamus* and *Bambusa* (Plate 5.3b–e). Both are clump formers. *D. strictus* attains a moderate 10 m stature, and is scattered, forming brakes on drier sites. Although sometimes reproducing gregariously, clumps fruit and die individually in most years. Brakes thus do not become dense, and fuel within them rarely builds sufficiently to generate catastrophic fire. *D. strictus* occasionally occurs scattered in the semi-evergreen forest, which is otherwise almost devoid of bamboo. It also occurs in tall deciduous forest on more freely draining upland soils. *B. bambos,* and the less common *B. tulda,* in contrast, concentrate on moist sites and clay soils. They form dense thickets reaching a lofty 40 m at maturity; *B. tulda* is somewhat the shorter. Once established, a horrid battery of spiked horizontal axillary shoots defends *B.bambos* against browsers. Both species reproduce gregariously over whole landscapes, *B. bambos* at intervals estimated at 30–50 or 60 years.[247] Periodic death throughout the catchment leads to periodic infernoes, which would seem to explain the distribution of the deciduous forest. In contrast, the semi-evergreen forest has experienced only ground fires, though every 3–10 years in some areas in recent decades. These last penetrated widely in 1994.[248]

The site for a 50 ha plot at Huai Kha Khaeng, within the CTFS collaboration, was chosen in 1991 by the late doyen of Thai forest science, Tem Smitinand, and has been established and maintained by the Royal Thai Forest Department under the leadership of silviculturist Sarayudh Bunyavejchewin. Studying the semi-evergreen forest in and around the plot Sarayudh, with then-PhD student Patrick Baker,[249] they found that evergreen species dominate the canopy except in patches where deciduous species common to moist deciduous forest predominate. Unlike in CTFS plots in evergreen dipterocarp forests, some 20 abundant canopy species manifested a distinct population structure at scales from patch to landscape: larger diameter trees outnumbered smaller in many. As Chengappa also found in the Andaman Islands, these are species which would have profited from catastrophic canopy opening, but were part of a cohort whose canopy repressed further regeneration of their own species (Fig. 2.21). They include both evergreens (such as the light hardwood dipterocarp *Hopea*

Fig. 2.22 (see right) Leafing phenology of 85 tree species over 4 years in a Thai semi-evergreen forest on sandy soils: Huai Kha Khaeng Wildlife Sanctuary, north-west Thailand. Error bars represent standard deviation of interannual variation in the mean flushing date. (The period of leaf flush for *Dipterocarpus obtusifolius* (top) is truncated). The dry season begins here in November and ends in late April. Leaf fall is gradual in the forest, and late for many species. (From Williams *et al.* 2008)

odorata) and deciduous species (such as *Afzelia xylocarpa*). In contrast to MDF and seasonal evergreen forests, deciduous species dominate succession, though they vary much in the duration of their leaflessness and are eventually substantially overtaken by evergreen species in mature stands (Fig. 2.22). Many canopy trees have misshapen boles, evidence of recovery from past wind damage.

Baker and Sarayudh estimated the dates of periodic catastrophic events, leading to periodic forest recovery, over three centuries. They were able to use deciduous pioneer and successional species that do possess annual rings, notably the rosewood *Afzelia xylocarpa*. They also inferred the age of evergreen species, notably *Hopea odorata*, from growth rates and branching patterns of individual trees (Fig 2.23), and from the relative stature of species' cohorts in individual stands.

Baker concluded that this forest consists of a patchwork of stands, varying from small patches to hundreds of hectares, arising from catastrophes at

Fig. 2.23 Stem irregularities, used for inferring age from the pattern of branching once the timing of canopy catastrophes had been inferred from other evidence, in *Hopea odorata*. Vertical scale in metres. The individuals represented are specific examples from the 50 ha plot at Huai Kha Khaeng. The straight-stemmed large tree is typical of canopy individuals. (From Baker *et al.* 2005)

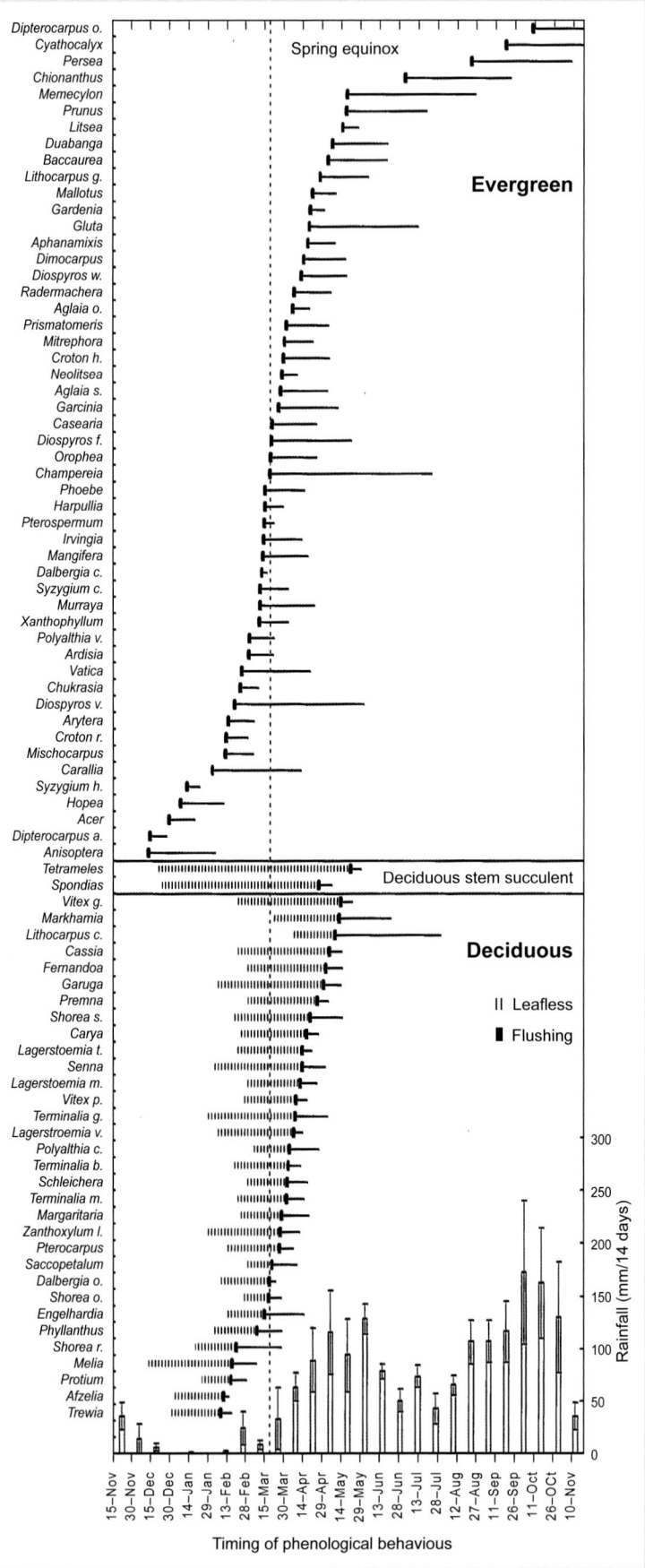

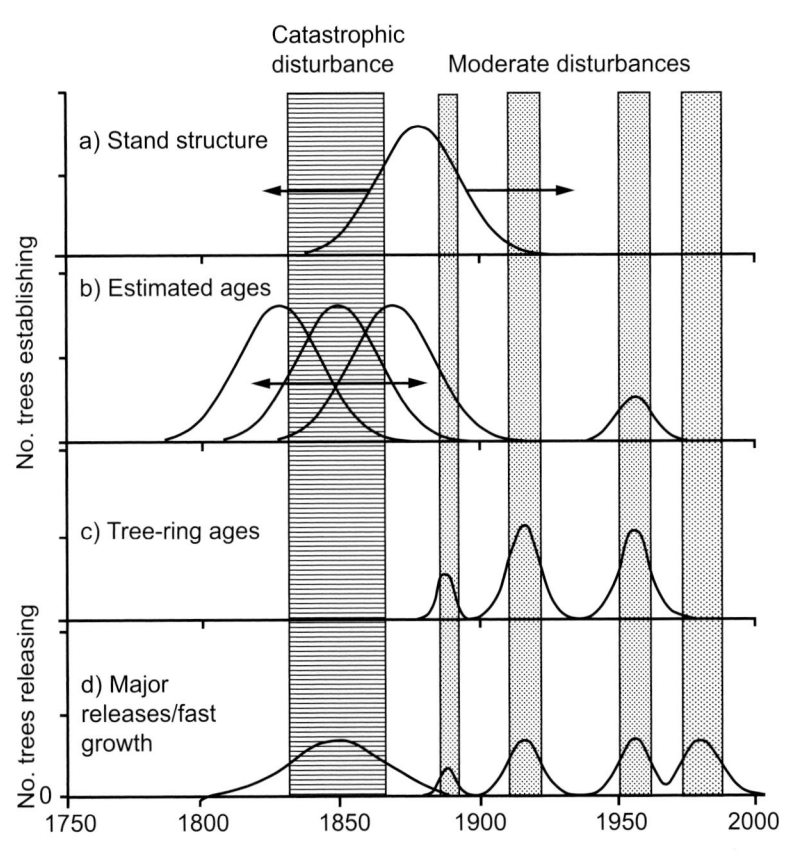

Fig. 2.24 Inferred history of semi-evergreen forest in and around the 50 ha CTFS plot at Huai Kha Khaeng Wildlife Sanctuary, north-west Thailand. **a**, Stand structure suggests a single cohort of canopy trees, but cannot indicate the timing of the canopy catastrophe from which they arose (thus the question mark); **b**, Estimates of ages of several species suggested that one cohort of *Hopea odorata* established in the 19th Century. Tree growth rings indicated peaks in **c**, recruitment, and **d**, growth in the second and third, and again in the fifth decade of the 20th Century, also a lesser episode in the late 19th Century and increased growth rates in the eighth decade of the 20th Century. Among older *Afzelia xylocarpa,* exceptionally high growth rates occurred mid-19th Century. In synthesis, a catastrophic stand-replacing canopy opening appears to have occurred mid-19th century, which has been followed by several lesser, probably more local, disturbances. (Estimates from each method for inferring age did differ slightly). (Baker *et al.* 2005)

different times and scales (Fig. 2.24). About 150 years ago a landscape-scale wind-throw led to conditions suitable for the establishment and growth of evergreen forest species, notably the dipterocarp *Hopea odorata*. They therefore inferred that it was at first followed by years without ground fires. This might have been caused by the rare penetration of a cyclone off the Bay of Bengal across the mountains to the west, followed by several years with shorter-than-average dry seasons. Thereafter, four subsequent wind-throw events resulted in smaller canopy openings in which deciduous species, but not evergreen *H. odorata* and its associates, successfully regenerated. Baker and Sarayudh concluded that these must have experienced fire, which would account for the absence of evergreen competition necessary for the success of deciduous successional species. This interpretation of events is consistent with Chengappa's findings, albeit different in detail.

At some time, several exceptionally dry years must have occurred which would account for the survival, within this semi-evergreen forest, of a group of species of deciduous dipterocarp forest (see section 5.5.3 below), including *Shorea siamensis* and *Dipterocarpus obtusifolius*. These would probably only have survived following several years of fire leading to the open conditions and the lack of competition they apparently need to establish in the canopy.

Sarayudh established two 16 ha plots in Huai Kha Khaeng, in which deciduous species dominate the canopy. In one, four *Lagerstroemia* species, as well as other deciduous species dominate, while *Miliusa lineata* (Annonaceae) and other evergreen species are patchily present in the understorey. It therefore conforms to tall (that is 'moist') deciduous forest. The population structure of these deciduous species is bell-shaped, implying that they are regenerating as single cohorts following catastrophic major opening of the canopy, perhaps by wind-throw followed by fire. Stand density at first census was relatively low, at less than 1,250 ha^{-1} individuals more than 1cm dbh, but more than 2,500 ha^{-1} recruits had appeared in the first recensus after five years. Two thirds of these recruits belonged to one species, the understorey tree *Baliospermum solanifolium,* which characteristically establishes after fire. Much bamboo is also appearing. In the second plot, the dominance of deciduous species in both canopy and the understorey, where *Dendrocalamus strictus* was also patchily abundant, led him to regard it as a 'dry' mixed deciduous forest, but its stature is tall, and equal to the first. This site is edaphically dry, but seemingly not more so than adjacent semi-evergreen forest. It may have experienced an intense fire in the past. The enigma of co-occurring 'dry' and 'moist' deciduous forests in Indo-Burma will be further discussed in section 5.5.1.

6.4. *Deciduous forests in India*

In India, the evergreen and semi-evergreen forests of the Western Ghats form an ecological island in a sea of deciduous forests and woodlands that otherwise occupy the lowlands of the entire peninsula.[250] Deciduous tropical forests extend uninterrupted north-west along the eastern and northern margin of the Desert of Thar, whose dry climate experiences winter frost, to the Himalayan foothill Hoshiarpur Siwaliks in Himachal Pradesh (at 32° N); also northeast up the Brahmaputra valley into Subansiri District of Arunachal (at 27°30' N), as well as along the southern flanks of the Khasi Hills in Meghalaya (map inside back cover). To the south, they occupy almost all of the lowlands of the Indian peninsula south of the Ganges Plain, the coastal hills of the south-east partially excepted (see section 7 below). Champion divided the deciduous tropical forests of the former British India into two principal categories. But, as we shall see here and in Chapter 3, they form a continuum; and indeed, anthropogenic fire and cattle browsing can degrade the former into a woodland more resembling the latter in structure. The two are nevertheless broadly distinguishable and serve as the generally accepted, indeed most appropriate, means to divide this most widespread of Indian forests for ease of description.

Plate 2.28 Tall deciduous forests. **a,** Tall deciduous teak, with *Zingiber*-rich field layer lacking saplings, implying frequent ground fires, Mae Ping National Park, north-west Thailand; **b,** Tall deciduous sal (*Shorea robusta*) on the Buxa *terai*, West Bengal. **c,** Sal bark in tall, and **d,** short, deciduous forest: the tall deciduous forest only experiences ground fires, the short deciduous forest periodic hot fires. (All: H.H.)

6.4.1. *Tall deciduous forests ('moist deciduous forests' of Champion)*

Tall deciduous forests differ from semi-evergreen forests primarily in the dominance of deciduous species in the canopy, and the frequency and intensity of fire. In tall deciduous forest, fire has eliminated evergreen species from the canopy, though juveniles and some subcanopy species capable of resprouting may persist in the understorey. The interrelationship between the two forest types is therefore dynamic and shifting. Tall deciduous forests may only exist in nature where hot fires are of sufficient frequency to suppress or eliminate evergreen tree regeneration. Ever since our ancestors learned to avail themselves of fire in Asia almost two million years ago,[251] these conditions have been maintained, and much later intensified by introduction of domestic cattle and the annual burning needed to encourage the young grasses on which they depend. But periodic hot fires create conditions favourable to bamboo, which in turn leads to further hot fires. The natural conditions under which deciduous trees can succeed in competition with bamboo remain imperfectly understood, but they appear quite demanding. Where deciduous species manage to close the canopy, bamboo density will decline. Thereafter, ground fires must be sufficiently frequent to eliminate early evergreen invasion, but they must periodically abate for long enough to allow deciduous seedlings to grow roots and shoots to fire-resistant size. This can only occur where browsing and trampling are light. Such conditions have become rare.

The deciduous species that regenerate in gaps caused by major wind-throws and occasional ground fires in semi-evergreen forests are light-demanders.[252] That is the case for all canopy tree species in deciduous forests. In Indian tall deciduous forests, nevertheless, the often dominant teak and sal are only deciduous for 1–2 months, Feb.–April. In effect, the gradient in light response, which dominates ecophysiological differences among species of evergreen forest, is replaced by gradients primarily related to drought tolerance, resistance to fire and resistance to browsing.[253]

6.4.2. *Short deciduous forests ('dry deciduous forests' of Champion)*

The short deciduous forests (Plate 2.29) as defined here differ from tall deciduous forests dynamically in that they do not revert to semi-evergreen forest if fire is excluded. Short deciduous forests lack the juvenile evergreens that persist in the tall deciduous understorey by stump-sprouting or spreading from refuges in nearby moist, fire-free sites. Tall deciduous forests regenerate periodically, predominantly by seed in large scale cohorts, while short deciduous forests regenerate more patchily, by coppice, and less from seed (at least, at present). Short deciduous forests other than where invaded by *Lantana* (see below), and thorn woodlands bear a sparse grassy

understorey, fueling less intense and less frequent fires, though increase in anthropogenic fire frequency in tall deciduous forests obscures that difference. Canopy species bear thin, sometimes smooth bark.

In India, semi-evergreen and tall deciduous forest has generally dominated regions with up to seven dry months, short deciduous forest from 6–9 dry months, and thorn woodland c.9–10 dry months. The ever-increasing number of domestic cattle is reducing a once-productive field layer of short deciduous forest to short sparse grassland and preventing tree regeneration by coppice, thereby impoverishing the tree flora. This promotes the expansion of forest types of drier into moister climates. Thus, increasing fire has led to widespread elimination of evergreen species, without whose seed tall deciduous forest, too, remains deciduous.

Lantana camara is now pervasive in the transitional gradient between tall and short deciduous forests of peninsular India. Although it is nibbled by ungulates and its shoots are eaten by elephants,[254] it achieves a dense continuous canopy which appears to impede tree regeneration, including stump sprouting. Its effect is comparable to that of bamboo, but some tree species, including *Kydia* and *Celtis,* regenerate well under *Lantana.* Short deciduous forest otherwise has a more sparse understorey, with a field layer dominated by short grasses, while in thorn woodlands the understorey and field layer is reduced to sparse short grasses. Saplings of taller trees may be protected from browsing within the crowns of low thorny bushes.

All species of short deciduous forest possess seed dormancy. Seed and seedling survival is much influenced by the intensity of drought and by the variability in duration of the dry season. Large seeds have been found to be more resistant to drought than small, although seed sizes in dry forests are small overall.[255] Seedling survival in short deciduous forest is very low owing to drought, fire, and especially to browsing – at present overwhelmingly by domestic cattle and goats. On the other hand, there is evidence that frequent fire, by accelerating nutrient release from litter, can also increase growth rates for the surviving trees in short deciduous sal forest. The distinction between light-demanding and shade-tolerant juveniles does not apply to these often open-canopied woodlands.

Regions of rugged topography dominated by tall deciduous forest, such as the Eastern Ghats and the Siwalik foothills, bear shorter forests devoid of evergreen species on convex and upper slopes. They are edaphically dry (see section 5.5 below). Such forests may also occur on gentle topography where they result from a history of human intervention. But their deciduous tree flora, albeit impoverished, is that of tall deciduous forest except where the moist-dry climatic transition is approached, while some evergreen species, notably *Syzygium cumini* and *S. nervosum,* are generally present where fire is restricted.

Plate 2.29 (right) Short deciduous forests. **a,** Climbing the eastern foothills of the Nilgiries, Tamil Nadu, Nov. 1990; **b,** Short deciduous forest following a hot fire. Bandipur National Park, Karnataka; **c,** The same forest type, three years after a hot fire, with *Lantana* returning: trees shooting epicormically – teak with dark bark, *Anogeissus latifolia* with pale bark; **d,** *Lantana camara*, habit; **e,** *L. camara* flowering inflorescence; **f,** Short facies of tall sal forest on the upper slopes of Meghasena, c.700 m, in Similipal National Park, Mayurbhanj, Odisha, experiencing leaf change in late March. (**a, c, f** H.H.; **b** R. Sukumar; **d, e** S. Davies)

A deciduous forested landscape: Mudumalai Wildlife Sanctuary
At Mudumalai Wildlife Sanctuary and Tiger Reserve (Tamil Nadu), a 50 ha plot within the CTFS network has been established in teak forest by the Ecology Division of the Indian Institute of Science under the leadership of H.S. Suresh and H.S. Dattaraja, with oversight by R. Sukumar.[256] The forest

within this plot is considered transitional between tall and short deciduous forest. It is dominated by teak (Plate 2.30). The rampant exotic shrub *Lantana camara* arrived in India about 1880, and the rarity of younger teak implies that it has suppressed regeneration. At the first plot census in 1987, following fire in the previous year, the field layer was dominated by the grasses *Imperata cylindrica, Themeda cymbaria* and *Cymbopogon contortus*.[257] Fire has subsequently been controlled. By 1993 the grass was being overtaken by *Lantana. Bambusa bambos* in swales had flowered and died, and was replaced by the exotic perennial weed *Chromolaena odorata*. The year 2000 was the last year of fire, although it was followed by several dry years. During that time, *Cromelina* dried up and *Lantana* invaded. Now, *Lantana* is in decline and the native shrubby understorey trees *Helicteres isora* and *Kydia calycina*, though favoured by elephants, are increasing, as is the spiny shrub *Xeramphis spinosa*.

These species are typical of short deciduous forests. Nevertheless, rainfall regimes at the two weather stations at Mudumulai are very similar in volume and seasonality to that in semi-evergreen forest at Huai Kha Khaeng, though the latter does experience more pre-monsoon hot dry season showers. Evergreen *Mangifera indica* occurs in the Mudumalai swales, *Syzygium cumini* on the uplands, while *Lagerstroemia* (Fig. 3.20) and *Terminalia,* paradigmatic genera of tall deciduous forests, are abundantly represented in the substantial part of Mudumalai that receives higher rainfall. In 2008, six years after the last fire, abundant seedling regeneration could be seen establishing around mature *Syzygium cumini*.

The forest has a long history of occupation by graziers, and a century of logging and management for timber – especially teak – which culminated in the 1970s. Stand density, estimated in plots throughout the sanctuary, correlates only weakly with fire frequency and rainfall. Variation in density of understorey shrubs, particularly the overwhelmingly dominant *Helicteres isora,* is the principal determinant of variation in overall density. *Helicteres* is browsed by elephants at

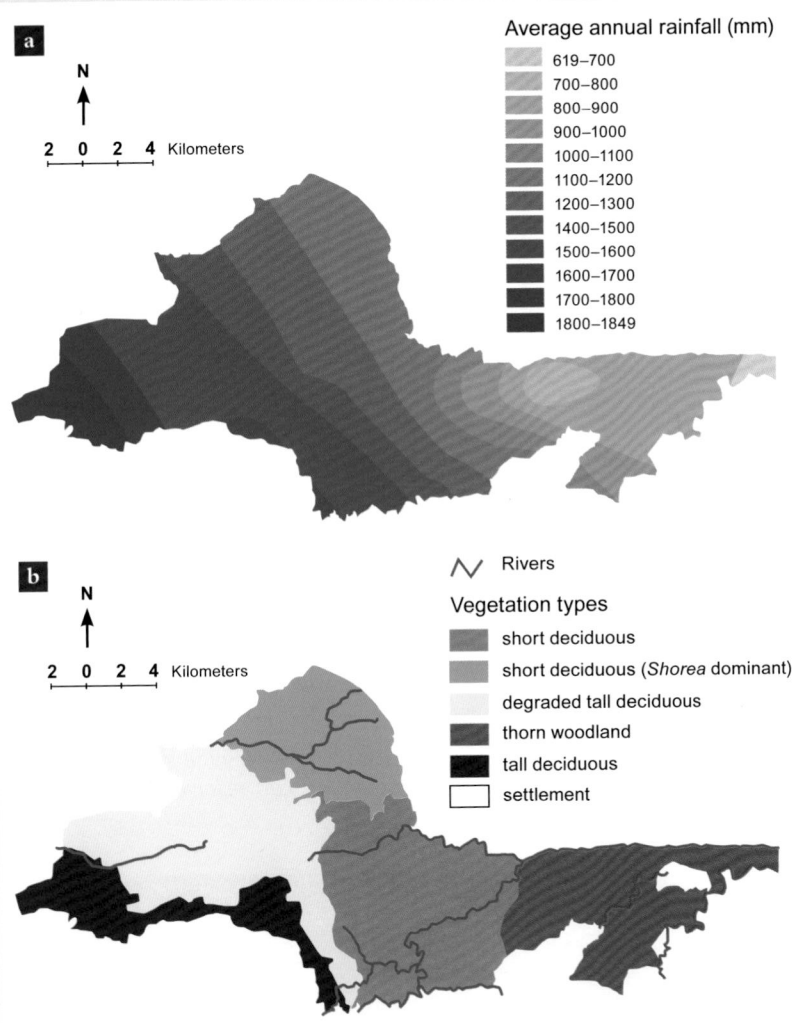

Fig. 2.25 Mudumalai Wildlife Sanctuary. **a,** Extrapolated distribution of mean annual rainfall, showing the steep decline at increasing distance east of the Ghats (which are to the immediate south-west of the Sanctuary). **b,** Vegetation map. *(from Suresh et al. 2011)*

Site	Nanophyll	Microphyll	Notophyll	Mesophyll	Macrophyll	Deciduous	Pinnate
Mudumalai	1	3	7	2	2	24	6
Giritale	2	4	11	7	1	2	6

Table 2.9 Number of species possessing different leaf characteristics, out of a total of 25 leading species in: Mudumalai Wildlife Sanctuary, Karnataka, India; and in semi-evergreen notophyll forest, Giritale, eastern Sri Lanka. (Derived from Suresh *et al.* 1996, Peeris 1975)

Plate 2.30 Short deciduous teak forest in March towards the end of the dry season (left) and early Nov. near the beginning of the dry season (right). Mudumalai Wildlife Sanctuary, Tamil Nadu. (H.H.)

varying intensities over a several-year cycle (Box 5.8). 90% of tree regeneration is from stump sprouts. *Lantana* thickets resist access by deer, thereby reducing the debarking of young trees caused by deer rubbing their horns against trunks.[258] Tree regeneration by seed and stump sprouting increase in the five years between a burn and establishment of *Lantana* cover. The oldest teak was aged by growth rings at 270 years in the 50 ha plot, but most are 150–190 years old, predating the arrival of *Lantana* in the 1880s. Juveniles of three of the most abundant species 1–10 cm dbh (small sample sizes precluded others) showed negative density-dependent recruitment and mortality with canopy individuals, those more than 10 cm dbh positive.[259] John and Sukumar have suggested that continuing negatively density-dependent mortality in these size classes implies that pathogens or predators are the cause.[260] Unlike evergreen forest, neither recruitment nor mortality related consistently with conspecific population density.

On the north-west margin of the Mudumalai reserve, in a rainfall regime of approximately 1,600 mm y^{-1} and five dry months, a semi-evergreen forest patch survives, sharply separated from the deciduous forest by a perennial stream. The canopy includes few deciduous species (though *Dalbergia* has been logged here in the past), and the understorey apparently none. There are few thorny plants. Leaf sizes are remarkably uniform, with mesophylls slightly dominating notophylls. Browsing animals are said to rarely enter there.[261] Large vines are patchy, but so also are Zingiberaceae. The canopy palm *Caryota urens* is present. Mudumalai Tiger Reserve is at 860–910 m; there is a band of semi-evergreen forest along its western margin somewhat higher up and against the Nilgiri slope, which includes lower montane forest elements. *Only long-term monitoring of this forest mosaic and fire exclusion experimentation can resolve the status of its deciduous forests.*

Plate 2.31 Tall deciduous forest with stand of *Lagerstroemia lanceolata,* April (left) and early Nov. (right); with *Lantana* and the grasses *Cymbopogon* and *Themeda*, Mudumalai Wildlife Sanctuary CTFS plot, Tamil Nadu. (H.H.)

6.5. Indo-Burmese deciduous forests pose puzzling questions

6.5.1. Is the Indo-Burmese 'dry deciduous forest' regulated by fire?

In Indo-Burma, regions with more than seven dry months (where short deciduous forests might be expected) are found only in the northern plain of the Irrawaddy up to Mandalay, and in a lone pocket in the central plain of the Chao Phya River in Thailand where little original forest survives. The vast region of Indo-Burmese deciduous and semi-evergreen forests thus shares a rainfall regime with 4.5–6.5 dry months, while those of more northerly latitudes, sometimes with seven or more dry months, receive some dew during the cool dry seasons.

In both India and Thailand, ecologists argue that the steep uplands between the valleys in this climate are too dry to support invasion of evergreen species if fire is excluded from them. They therefore recognize a putative topographically differentiated 'dry' deciduous-'moist' deciduous forest mosaic there.[262]

Problems of distinguishing between 'moist' and 'dry' deciduous forests here have arisen in part because millennia of anthropogenic fire, and more recently grazing by domestic cattle, have virtually eliminated the evergreen element on clay soils, although its usual persistence near streams betrays its likely wider past distribution. In Thailand and elsewhere in Indo-Burma, it seems likely that much of the deciduous forest, which there has been included in 'dry deciduous forest' by foresters, is actually tall deciduous forest that has lost its evergreen element through increased fire. This suspicion is supported by its tall stature, usually 30 m but sometimes reaching 35 m (Plate 2.28a). The understorey of *Dendrocalamus strictus* and *Bambusa tulda* bamboos or tree regeneration, patchiness and frequent absence of grass, absence of *Lantana* and local abundance of broad-leaved monocotyledons including gingers all place it there. These features of Indian tall deciduous forests abate toward the upper slopes of ridges, where the forest may decline to 15–20 m tall, bamboos become sparse and monocotyledonous herbs rare. However, even there, a grassy field layer is generally patchy and thin or absent, while the tree flora remains an impoverished subset of tall deciduous forest or deciduous dipterocarp forest (see Section 5.5.3 below). As in the deciduous upland forests of the northern Eastern Ghats, also short in stature, where evergreen *Syzygium cumini* and *S. nervosum* persist, so do scattered evergreens in some Indo-Burmese deciduous dipterocarp forests (see section 5.4.3 below, Chapter 3 sections 2.3.3 and 2.3.4.1.1).

Unlike in peninsular India, where one major forest type extends over vast landscapes of relatively uniform topography and geology, Indo-Burmese landscapes support a mosaic of forests with the structural and physiognomic attributes of semi-evergreen, tall and short deciduous forest. This mosaic is due to the heterogeneity of north central Indo-Burmese topographic landforms and sedimentary geology, quite unlike the vast extents of siliceous Archaean metamorphic rocks or the basalt Trap of peninsular India (Chapter 1). Deciduous forests described as 'dry' can be found amidst forests whose rich deciduous flora and stature qualifies them as 'moist'. This is apparently due to the local extinction of an evergreen understorey. But a truly short deciduous forest – lacking an evergreen element and prevailing along ridges with clay soils – appears always to have existed. It is caused here by water deficits of edaphic rather than climatic origin. It is also present in Indian teak and sal forests, though it was not identified as such by Champion, probably because of its relative rarity and continuity with taller forest in India's mature and rolling topography. These edaphically correlated short deciduous forests differ floristically from short deciduous forests that are climatically correlated, for they lack a distinct short deciduous forest flora, instead retaining an impoverished element of adjacent tall deciduous forest (Chapter 3).

A further factor that complicates the relationship between deciduousness and the physical environment is the higher proportion of deciduous species on clay loams than siliceous soils. This is apparent throughout Asia but results in most complex patterns in Indo-Burma. Within a climate of particular seasonality teak, which dominates clay loams, is leafless for a longer period than sal, which dominates siliceous soils (Chapter 3). Teak is deciduous for several weeks in the driest parts of its range, while outside of the very driest forests sal is only marginally deciduous. Indeed, in relatively moist stands, sal leaves may be retained even as the new flush emerges in April. Teak-bearing deciduous and semi-evergreen forests include more deciduous canopy species, or with longer annual leafless periods, than deciduous dipterocarp (5.5.3 below) or *Hopea odorata* semi-evergreen forests. Semi-evergreen forests occur on basalt in southern Vietnam, and clay loam in South Sumatra, in climates where seasonal evergreen forests otherwise prevail (Chapter 3 section 4.1); while even in north-west Bornean MDF, the only (shortly) deciduous species, *Chukrasia tabularis*, *Pterocymbium tubulatum*, *Firmiana malayana,* are restricted to limestone talus, basalt and, very rarely, turbiditic clay loam.

6.4.2. Problems of nomenclature

Champion had extended his classification of forests in seasonal India to Burma, and Thai foresters have attempted to adopt it, but inconsistencies have emerged. The French, in Indo-China, evolved a similar yet not identical classification. Several others, all modifications of the same basic foundation, have been used, including my own.[263] Excepting the Indo-Chinese, many have been usefully compared in a table by

Maxwell,[264] who also proposed his own. Persistent inconsistencies in the interpretation and classification of deciduous forests, especially between India and Thailand (and indeed all of Indo-Burma), are substantially due to insufficient attention to edaphic influences on seasonal water stress.

A second reason for confusion about forest nomenclature arises from a distinction being drawn in India, between the deciduous forests on siliceous soils and those on clay. Three classes of both tall and short deciduous forest have been recognized there: those dominated by teak, those dominated by sal, and those not dominated by any single species (though local dominance by other species in special habitats does occur) (map inside back cover). Teak forests occur, mostly in the west and the northern Deccan, especially, it appears, where soil calcium content is relatively high. Sal forests are confined to the north and east (map inside back cover), while the others are found in a variety of areas and soils, particularly in the south-east. Dominant species aside, these tall and short forests appear quite similar in their overall dynamics and phenology, though growth rates are lower and deciduousness more prolonged in the drier areas. The distinction between forests on clay loams and on siliceous soils has not been made in the same way in Indo-Burma, where a separate formation, deciduous dipterocarp forest, is recognized. This is the deciduous forest of siliceous soils – but it also occurs on others.

6.4.3. Deciduous dipterocarp forests

Throughout Indo-Burma occur the deciduous dipterocarp forests, called *indaing* in Burma,[265] in part analogous to the Indian sal forests (Fig. 2.26; Plates 2.32). They are dominated by one or more macrophyll or mesophyll dipterocarp species in the genera *Dipterocarpus* or *Shorea*. These forests are most widespread among the tall deciduous forests, but they extend to the margin of the Burmese thorn woodlands, while some of their species occur on karst in climates with less than three months south to north-west Peninsular Malaysia. Dipterocarps of deciduous dipterocarp forest are notable for their thick fissured bark, more so even than species of short and tall deciduous forest (Plates 2.28c,d, 2.33d). Deciduous dipterocarp forest varies greatly in stature, from 40 m (formerly recorded for stands of *D. tuberculatus* on ferricrete in north-west Burma) to less than 10 m on some sandstone ridges and plateaux in Thailand and Cambodia (where they are often dominated by *D. obtusifolius*, see Chapter 3). In Thailand, *Shorea obtusa* has been recorded at 32 m in height, *S. siamensis* at 25 m.[266] The taller deciduous dipterocarp forest occurs on freely draining, albeit sometimes shallow clay loam soils on upper slopes and ridges, also on plains and over ferricrete (Plate 2.34a). The latter are straight-boled, form a closed canopy, and lack grass in their woody understorey, though sometimes with scattered herbaceous monocots including gingers and palms. The deep

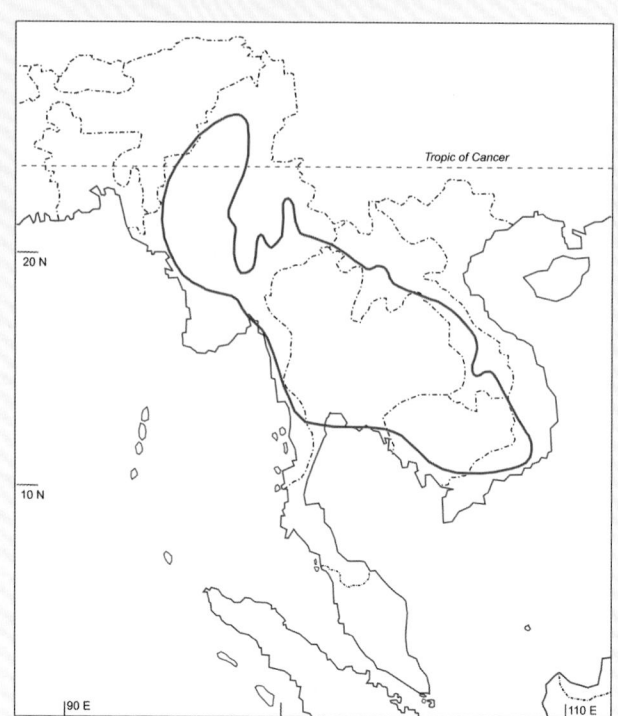

Fig. 2.26 The geographical range of deciduous dipterocarp forest. Two of its characteristic dipterocarps extend to north-west Peninsular Malaysia on karst limestone, including the off-shore Langkawi Islands, and Tarutao National Park islands, Thailand, adjacent to the north. These sites may be refuges from a more seasonal past climate. (The South Asian analogue of deciduous dipterocarp forest is deciduous sal forest; the distribution of sal is indicated on the forest type map within the back cover).

litter accumulation endures frequent ground fires. Short-statured deciduous dipterocarp forest usually carries an open canopy. It occurs variously on karst, where the understorey is sparse with patchy *Cycas*, palms and sparse grass; or on freely draining sandy soil where tall grasses including *Imperata* form a dense field layer. The thicket-forming bamboo *Vietnamosasa* is locally abundant in eastern and southeastern areas of open-canopied deciduous dipterocarp forest. Surprisingly, some of the shortest bear evergreen subcanopy species with fire-resistant rootstocks (Chapter 3).

The stature of deciduous dipterocarp forest therefore ranges from that of semi-evergreen to that of short deciduous forest, or even of thorn forest, in India. Overall therefore, the characteristics of Indo-Burmese deciduous forests turn Champion's classification on its head. They again imply that it is seasonal edaphic water stress, mediated by topography and substrate, rather than rainfall seasonality, which is the principal determinant of forest structural and physiognomic characteristics in Indo-Burma. The thick, fissured bark of deciduous dipterocarp forest dipterocarps has been interpreted

Plate 2.32 Deciduous dipterocarp forest **a,** Tall *Dipterocarpus tuberculatus* forest on freely draining alluvium, Mae Ping, National Park, north-west Thailand: dry season in late Feb., still with thick litter layer and semi-deciduous canopy (above), after the monsoon, Nov. (below); **b,** On sand, Mae Surin National Park, *Dipterocarpus obtusifolius* (left) with evergreen shrubs *Vaccinium sprengelii* and *Craibiodendron stellatum* (foreground). In the dry season, March (above) and the end of the wet season, Oct. (below). (H.H.)

as an adaptation against fire. Saplings are few, owing to frequent fire and browsing. The crooked boles and gnarled open crowns in short deciduous dipterocarp forest implies either that crown fires, or drought-induced bud abortion, periodically occur. Crown fires appear to have become rare in shorter deciduous dipterocarp forests, which nowadays lack sufficient understorey fuel owing to more frequent burning. However, a fire-exclusion experiment at Sakhaerat Forest Research Station (south-east Thailand) has resulted in a thicket of the bamboo *Vietnamosasa pusilla* tall and dense enough to create a crown fire. *Vietnamosasa* reproduces annually, after which culms die forming combustible fuel. It regenerates by means of fire-enduring underground rhizomes from which new culms shoot rapidly at the beginning of the monsoon.

At Huai Kha Khaeng (north-west Thailand), a similar short deciduous dipterocarp forest on sandy soils, free of fire for several years, experienced regeneration of a dense mostly deciduous subcanopy dominated by *Cratoxylon formosum*. In this case, too, re-entry of fire would likely create an inferno reaching the low canopy. Whereas other types of deciduous dipterocarp forest lack evidence of invasion by semi-evergreen forest tree species' juveniles, fire exclusion does lead to their appearance. Sarayudh[267] deduced that the predominant semi-evergreen dipterocarp forest on sandy soils at Huai Kha Khaeng eventually succeeds deciduous dipterocarp forest where fire

Plate 2.33 (right) Deciduous dipterocarp forest. **a,** Tall *Dipterocarpus tuberculatus* forest over red-brown sandy clay alluvium, with Zingiberaceous field layer; experiencing only low intensity ground fires. Mae Ping National Park, north-west Thailand; **b,** *Shorea siamensis* changing colour, Dec. 1990, Mae Sariang-Mae Hon Son road, north-west Thailand; **c,** Scorched but not killed: *Dipterocarpus obtusifolius* seedlings after a ground fire in pine-dipterocarp savanna, 800 m, Nam Nao National Park, central Thailand; **d,** The thick V-section fissured bark of *D. obtusifolius*; **e,** Stand of the dwarf bamboo, *Vietnamosasa pusilla* in degraded deciduous forest, Sakhaerat, Khorat Plateau, north-east Thailand; dead above-ground and combustable in the dry season. Many tree crowns arise from within the bamboo; **f,** Rhizome of *Vietnamosasa pusilla* showing adventitious shoots (5 cm scale). (**a, c, d** H.H.; **b** P.A.; **e, f** S. Dransfield)

144

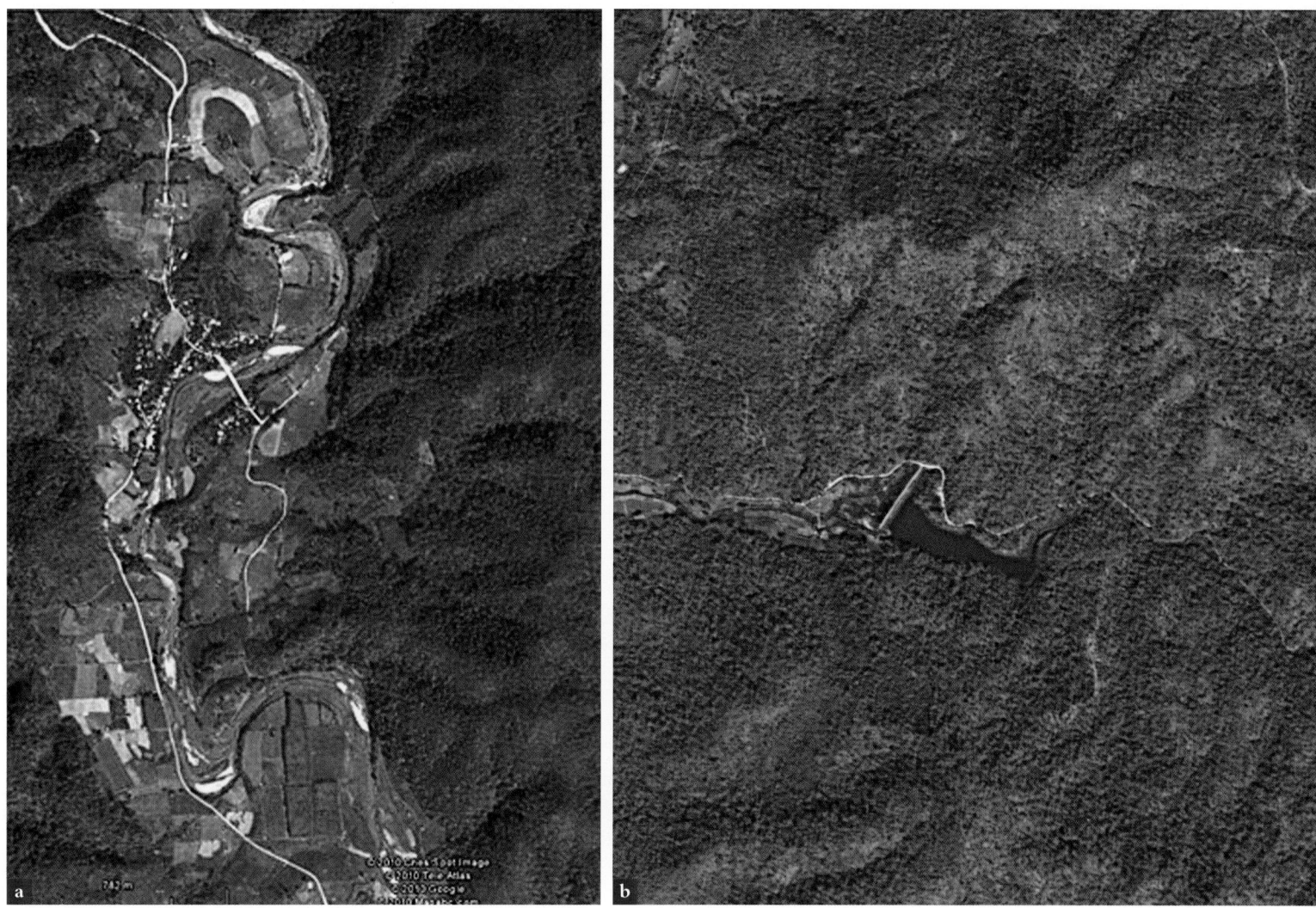

Plate 2.34 Deciduous forests on the north-west Thailand turbidite. Tall deciduous forests on north-facing slopes and clay loam soils, the crowns smaller towards the ridges as soils become shallower; deciduous dipterocarp forests on south facing slopes and sandy soils. **a,** The Mae Sariang valley; **b,** Salawin National Park. Scale: 1/5000 (Google Earth).

does not intervene, provided there is a seed source nearby. Patches of deciduous dipterocarp species occur in the midst of mature semi-evergreen dipterocarp forest in Huai Kha Khaeng and elsewhere[268] (Plate 2.27e; Fig. 3.24). In such patches they are straight boled and may reach 40 m in height. Patrick Baker[269] found no associated soils differences. I infer that, whereas Indo-Burmese semi-evergreen forest on clay loam is converted to tall deciduous forest following intensive burning, on freely draining sandy soils (especially those on convex surfaces) it becomes deciduous dipterocarp forest.

Deciduous dipterocarp forest canopies as a whole are seldom completely leafless. Whereas teak in Thailand is synchronously deciduous for 1–3 months, the deciduous dipterocarp forest canopy dipterocarp *Shorea siamensis* is synchronously deciduous for only 6 weeks, *Dipterocarpus tuberculatus* for 2–3 weeks. *S. obtusa* (a sister species of *S. robusta*), is slow to lose its leaves, while *D. obtusifolius* does so irregularly and sometimes hardly at all.[270] I observed much flushing of new leaves as early as 8 March in deciduous dipterocarp forest along the Pai-Doi Chiangdao road, two months before the onset of the wet monsoon. The complex floristics of deciduous dipterocarp forest will be discussed in Chapter 3.

Growth rates have not been documented, but the predominance of heavy hardwood *Shorea obtusa* and *S. siamensis* implies that they are slow.

6.6. *Deciduous woodlands of East Java and the Lesser Sunda Islands*

Climates appropriate for deciduous forests dominate the lowlands of this region. However, it seems that only a few degraded patches remain, although intensively managed teak forests survive in East Java. Their floristic affinities will be discussed in Chapters 3 and 6.

6.7. *Some general characteristics of Asian semi-evergreen and deciduous forests*

6.7.1. *Forest stature and structure*

The stature of primary seasonal forests declines with increasing length of dry season, and also along a soils gradient from deep clay mesic loams to shallow drought-prone soils irrespective of mineralogy or fertility. The tallest trees in semi-evergreen forest can reach 50 m, although the canopy more generally stands at about 40 m. Tall deciduous forest can reach 35 m but is more generally 20–30 m, while short deciduous forest is 10–20 m, but can be as low as 5 m. In semi-arid regions, thorn woodlands rarely exceed 10 m.

All Asian tropical deciduous forests recall temperate deciduous broadleaved forests in the shortness of the trunks of canopy individuals relative to the depth of their crowns; and the drier the habitat, the more extreme is this relation. Columnar trunks become rare in shorter forests; misshapen trunks may be a consequence of crown fires, of browsing by large mammals, or of early flowering of the leading shoot among individuals of species with terminal inflorescences, including teak.

6.7.2. *Leaf shape*

A diversity of leaf shapes exists in the semi-evergreen forest understorey, too, especially among canopy saplings. However, the leaves of pioneer and canopy trees of both evergreen and deciduous forests in Asia differ conspicuously from those of temperate deciduous broadleaved forests in the rarity of lobing – as seen among temperate oaks, maples and planes (*Platanus*) – and of toothed margins. (The simple entire margins of leaves of tropical oaks, of Asian tropical maples (*Acer laurinum* and *A. oblongum*) and of the rare tropical plane (*Platanus kerrii*) of north-east Indo-Burma, contrast with the great majority of their temperate congeners.)

An extraordinary feature of the tall and short deciduous forests, also deciduous dipterocarp forest, is that canopy dominants (teak, sal or *Dipterocarpus* species), albeit often only shortly deciduous, are macrophylls or large mesophylls with leathery leaves, dispersed rather sparsely in the crowns. Other tree species of deciduous dipterocarp forest also have large leaves, including *Aporosa villosa, Melanorrhoea usitata* and *Strychnos nux-blanda. The competitive advantages apparently lent by these large leaves among suckering species of fire-mediated deciduous forests remains a mystery.* The proportion of pinnate leaves with microphyll or smaller leaflets does increase in short deciduous forest and thorn woodlands of drier climates, reflecting the success there, as in other parts of the tropics, of leguminous and rutaceous tree species.

The functional significance of the diversity of leaf sizes in the canopy of tall deciduous and semi-evergreen forests in Asia and some other regions appears not to have been systematically addressed (Table 2.8). Nearly all leaf and leaflet blades are entire. Petioles are not disproportionately long, and leaves are held at various angles and dispersed within their crowns. Most species flush as atmospheric humidity increases in the hot dry season, beginning in early April, and are at their physiological peak during the south-west monsoon when skies are overcast for days and the atmosphere is saturated. The climate in these respects resembles that of the evergreen forest understorey. Higher diffuse light intensity during the wet monsoon, which is windy, may account for high net rates of photosynthesis. The wide spacing of canopy trees in shorter deciduous forests indicates relatively low above-ground competition, lessening ecophysiological constraints on phylogenetic traits, but implies intense root competition for water. The ensuing dry season is cool, with lower evaporative power.

6.7.3. *Growth rates and productivity* (see section 5.3.3 above)

The presence of a dry season restricts the length of the growing season. Evergreen species must adapt to withstand annual soil water deficits; species with narrow water conducting elements (and therefore heavy timber and slower growth rates) are favoured.

An unexpected but explicable finding of the CTFS network of large tree demography plots in Asia is that the maximum growth rates achieved by canopy tree species in the semi-evergreen forests at Huai Kha Khaeng, with 1,500 mm of rain restricted to seven months, is almost identical to that at Lambir in perhumid Borneo (Table 2.7). At Huai Kha Khaeng these fast growing trees number several *Ficus,* free standing as well as hemiparasitic. The fast-growing tree *Tetrameles nudiflora* (Datiscaceae) is a deciduous pioneer attaining canopy height. It is related to the Malesian *Octomeles sumatrana,* among the fastest-growing trees in its perhumid region and sometimes also briefly deciduous. *Octomeles* is confined in nature to flood plains, well-watered fertile loams on basic volcanic rocks, and the base of limestone hills. *Tetrameles* has a similar but wider ecological distribution. The widespread evergreen pioneer *Neolamarckia cadamba* achieved a height growth rate of 3 m y^{-1} over the first 8 years in the semi-evergreen forests of the Buxa Duars (West Bengal), and 53 cm diameter at 1.8 m after 16 years, that is, 3.3cm y^{-1}.[271] This growth rate slowed after 20 years.

The soils in these seasonal regions, even on unpromising substrates, are vastly more fertile than in perhumid regions (Fig. 1.10). Annual number of days of full sun also appears to provide an explanation for the formidable growth rates achieved by some evergreen tree species in the seasonal tropics. A contributory factor may be the reduced rate of carbon dioxide uptake and growth experienced by trees during the longer periods of cloudy weather in the perhumid tropics.[272] Similar observations have been made in West Africa,[273] and may in part explain the vastly higher rates of growth, recruitment and turnover at the BCI (Panama) CTFS plot, under a three month dry season, as compared to Pasoh.[274] Nutrient turnover rates have been estimated to be as high as 57–60% per annum in *Dalbergia sissoo* stands at the foot of the Himalaya, with five dry months.[275]

6.7.4. *Species of later succession and the mature phase*

Climax canopy species of the mature phase cannot be expected to compete with the extraordinary growth rates among pioneer species (and canopy figs). Most of the deciduous species, including sal and teak and most legumes, although adopting a successional role in semi-evergreen forests, are heavy hardwoods. Nevertheless, teak in managed tall deciduous forest may reach 50 cm dbh in c. 80 years, representing a mean diameter growth rate of 6.3 mm y^{-1}. Sal reaches 50 cm dbh in 70 years, or 7 mm y^{-1}.[276] These rates compare favourably with growth rates of MDF light hardwood dipterocarps in plantation or managed regenerating stands. Surprisingly again, sal forests are distributed on siliceous substrates, and their grey-brown soils are less rich in nutrients than the calcium-rich clay loams to which teak forests are restricted in nature. Water availability rather than nutrient concentration likely restricts maximum growth rates on the higher-nutrient soils of the strongly seasonal tropics.

The growing season of deciduous forests is extended beyond onset of the wet monsoon by the precocious leafing of most species during the hot dry season, which starts in March at lower latitudes. These early shoots are boosted by the previous reabsorption of nutrients from senescing leaves of the previous season: 20–100% of phosphorus may be retained.[277] Biomass has been calculated for a mature managed tall deciduous sal forest in the *Bhabar Terai*, with a remarkable estimated total biomass of 710 t ha^{-1}. This is higher than Kira's estimate for 50 m tall MDF at Pasoh. Net primary production, without the inclusion of litter, was 18.6 t ha^{-1} y^{-1}, of which the dominant sal comprised 12.8 t ha^{-1} y^{-1}. That represents a lower turnover rate than at Pasoh.

6.7.5. *Below-ground biomass*

Trees root deeply in the friable soils of the seasonal tropics wherever seasonal drainage and soil depth will permit. Roots remain stout and deep as the forest's above-ground biomass declines with increasing seasonal water stress (Plates 1.15a–c, 2.24d). Root/shoot biomass ratios therefore increase. Carbon content in mineral soil is much higher in seasonal semi-evergreen and deciduous than in evergreen forests (Chapter 1). Below-ground biomass in roots was estimated at 180–1,172 g m^{-2} in an Indian short deciduous forest, contributing half the total soil respiration. This is similar to that of temperate deciduous broadleaved woodland, and double that recorded from evergreen rainforest.[278]

6.7.6. *The regeneration niche*

Seed germination and seedling establishment occurs in seasonal evergreen and deciduous forests during the wet monsoon.[279] As the wet monsoon gets shorter, it becomes crucial for seedlings to germinate early in it, so that their leading root can extend deeply in their first wet season. Variability in the annual rainfall pattern lends selective advantage to seed dormancy and to responsiveness in germination to increased humidity.[280] The same is true for stump sprouting, which allows reserves to be built up in the roots so that a high root/shoot biomass ratio is achieved early, permitting maximum access to water and growth of a leading shoot that can survive in a relatively fire-free year. Foresters have found that, although fire kills first year seedlings of most species, the dormant seeds of many species are fire resistant. In the case of teak, fire actually increases the germination rate.[281] Furthermore, fire reduces the thick layer of large fallen leaves of teak and sal, which in the wet season adhere to topsoil as an impervious cover inimical to root penetration.[282] Fire exposes a friable surface rich in ash and favourable to seedling establishment.[283] Nevertheless, fire prevention is necessary for a few years initially if rootstocks, and coppice shoots in later years, are to survive.[284] Currently, though, fire frequency throughout fire-prone forests in the Asian tropics has increased so much that seedling establishment has become rare even in many semi-evergreen forests. Regeneration is thus increasingly dependent on stump sprouts, which themselves are increasingly repressed.

Whereas a regular dry season, its length and variability are the dominant influence on all aspects of the forests of the seasonal tropics, this influence sharply contrasts with the influence of the occasional supra-annual several-month droughts associated with anomalies in the ENSO on primary forests. There, although significant mortality occurs in both upper canopy and lower understorey, in the absence of fire in nature no permanent influence is apparent in primary forest other than a possible reduction in the abundance of certain successional species on the more drought-prone sites.

Bamboo brakes are patchy but can be extensive in some regions, especially North-East India and Indo-Burma. Brakes occur mostly on clay soils or in moist valleys. Bamboos are light demanders and do not survive under established canopies. Bamboos possess seed dormancy, germinating in response to light. Their seeds are dispersed by rodents and granivorous birds, and they are highly invasive. Bamboo brakes, where they occur in seasonal evergreen and tall deciduous forests, are a major impediment to forest regeneration. Regeneration is most successful during years immediately following mass bamboo fruiting, death and fire which opens the canopy. Seedlings of more shade-tolerant tree species survive beneath the bamboo canopy, hardly growing in height. Bamboo thereby favours survival of shade-tolerant slow-growing heavy hardwood tree species.

6.7.7. *Vines*

Vines are more abundant in the semi-evergreen and tall deciduous forests than elsewhere in the Asian tropics, but await comparative documentation. Large vines are particularly abundant in windfall gaps on clay and sandy clay soils and readily form the dense monolayer leaf blanket also seen in gaps on mesic sites in the perhumid tropics. They dominate the dense cover of large windfall gaps that are frequent in semi-evergreen forests in the absence of fire, as in the Andaman Islands[285] and in seasonally waterlogged semi-evergreen and tall deciduous forests on level ground in the Buxa *terai* (Duars). Many vines are deciduous. They become less abundant, and also shorter and more slender, in short deciduous forest, and are absent in thorn forest.

6.7.8. *Browsing and grazing*

Biomass of browsing and grazing vertebrates was formerly vastly greater in the seasonal than the aseasonal tropics, by virtue of the high seasonal production of accessible biomass in the herbaceous field layer, and as coppice.[286] Although the species diversity of ungulates in Asia is not great overall (but see Chapter 5), the impact of elephants especially, and also gaur (*Bos gaurus*), is immense. Grassland, whether induced by flooding, frost, or fire, attracts grazing and browsing herds. Bamboo attracts elephants (Box 5.8). Bamboo rhizomes resist fire, but the shoots and seedlings are also browsed by cattle and deer,[287] though some are poisonous. Browsing intensity is greatest where food and shelter are most abundant, in the semi-evergreen–tall deciduous forest mosaic. Browsing mammals have a variety of food preferences. Many vines, especially *Merremia* and other fast-growing species, are favoured by elephants, and these are attracted to areas of forest disturbance also by the fleshy growth of successional species. Their depredations may, however, provide conditions favouring deciduous successional species. Elephants are fond of *Artocarpus* in evergreen forests,[288] understorey *Kydia calycina*, *Helicteres isora* and young bamboo shoots in deciduous forest[289] (Box 5.8). Deer and langur monkeys favour sal regeneration.

Browsing intensity fluctuates in the deciduous forests with the seasons, and from year to year (Fig. 5.37). Although the best-managed parks in seasonal Asia still support close to natural populations of wildlife (although opportunities for migration have been greatly reduced), regional patterns of forest vegetation are no longer influenced by wildlife, but instead by the ubiquitous presence of cattle. *The relative impact of browsing versus fire in restricting juvenile survival and growth has not been assessed.*

6.7.9. *Drought mortality*

Increasing length of the dry season is accompanied by its increasing variability and the occasional occurrence of extreme droughts. Other than the anecdotal, little has been published on the impacts of such droughts on the deciduous forests. R.S. Troup, for instance, described and photographically documented the effects of the 1907 and 1908 monsoon failures on the *Shorea robusta* (sal) forests of the foothill *terai* and plains of the central Himalaya (Uttar Pradesh).[290] Thousands of trees, especially among the larger sizes, died (the smaller were presumably coppices with root systems deep and extensive relative to their above-ground biomass). Mortality was greater on the hills than on the alluvial plains where the water table remained within reach of the roots. Different species varied greatly in mortality rate, implying that these occasional events play a crucial role in determining species composition. Such catastrophes, in effect, turn back the forest to early successional stages.

7. Semi-arid vegetation

Short deciduous teak forest extends into central Gujarat and southern Rajasthan, in a climate with 950 mm y^{-1} of rain and nine dry months; short sal occurs in climates with 8.5 dry months in Madhya Pradesh. In the semi-arid climates with more than eight dry months, forest in India is increasingly replaced by thorn woodland ('tropical thorn forests' of Champion) nowadays severely overgrazed, which falls outside the scope of this book. This occurs in two distinct regions: in the lowlands of northwest India into north-east Pakistan, and

Plate 2.35 (right) Thorn woodland. **a**, Panorama at the Moyar Gorge, Karnataka, late March (above) and Oct.; pale crowns on far left slope are *Boswellia serrata*; *Euphorbia royleana* in foreground (cf. Plate 1.1a); **b**, Profile of dry thorn woodland with *Euphorbia*, sparse grasses, Tamil Nadu, December 1982; **c**, Neem (*Azadirachta indica*) sapling embedded in spiny *Catunaregam spinosa* and thereby protected from foraging goats, Tamil Nadu, 1982. (**a** H.H.; **b, c** P.A.)

149

in the peninsula east of the Western Ghats but inland from the eastern (Coromandel) coast. Most of the northwestern semi-arid lands bordering the Desert of Thar are subject to frost, and therefore temperate by the definition adopted here (see Chapter 3 section 9.2). The vegetation is too sparse for fires to spread; regeneration dynamics are dominated by grazing and browsing, nowadays by cattle (Plate 2.35b). Saplings may escape the browsing when they establish within the pervasive thorny shrubs (Plate 2.35c).

In the semi-arid northern plain of the Irrawaddy (Burma) where the dry season lasts 7–8 months (map inside front cover), as also on karst limestone in Bataan Province (western Luzon) with a 6-month dry season, *Tectona grandis* is replaced by the shorter species *T. hamiltoniana* (Chapter 3). This forest has been little studied, but in Burma appears to be analogous to short deciduous forest in India.

8. Semi-evergreen notophyll forests ('tropical dry evergreen forests' of Champion)

The distribution of this forest formation correlates with a rainfall regime that is rare in Asia but widespread in Africa and the Neotropics (Chapter 1). Owing to the distortion of prevailing wind patterns created by the monsoon in seasonal Asia, few areas experience two dry and two wet seasons. They occur up the southeastern Coromandel Coast of India (to c.18° N)[291] and down to northern and eastern Sri Lanka, together with south-east Vietnam. These regions receive cool season rain from the north-east monsoon, respectively, off the Bay of Bengal and the China Sea. That distribution corresponds to an unique climate experiencing rain from both south-west and north-east monsoons, yet varying greatly in mean annual rainfall: from 900–1,900 mm.[292] The hot season is always relieved by some precipitation from the south-west monsoon, but so is the cool season from the north-east monsooon.

In north and east Sri Lanka the predominantly grey sandy soils there are dominated by a closed forest resembling short deciduous forest in stature – it is usually only 10–20 m tall, of a single even-canopied stratum – but it is almost entirely evergreen and lacks a grassy field layer.[293] The deciduous species that do occur are characteristic of short deciduous forest and thorn woodland in India (Chapter 3). Their relative abundance is greater in India, where the woodlands are under greater browsing pressure. The dominant leaf size class is notophyll, as in *kerangas* (Table 2.8), but unlike *kerangas* the leaves are not thickly leathery. Most are acuminate and not predominantly erect around the twigs. The understorey beneath the single diffuse but deep forest canopy is sparse and includes mostly shrub species, some of which are spiny, rather than canopy juveniles. The arrangement, shape and thinness of the evergreen leaves may seem surprising in an ostensibly dry climate, and in comparison with *kerangas*. *These characters may at least in part reflect greater nutrient availability, and would seem to promise rewarding experimental study.*

No dynamic studies of semi-evergreen notophyll forest appear to have been undertaken. These forests merit comparison with the similarly closed canopy dry evergreen cerradón and coastal restinga forests of Brazil, which flourish in a similar environment.

Immediately east of the Sri Lankan mountain massif, increased rainfall and reduction to one dry season is correlated with taller forest in which additional deciduous canopy species occur, including *Alstonia scholaris, Chukrasia tabularis, Filicium decipiens* and *Tetrameles nudiflora* along with some semi-evergreen rainforest elements including *Dimocarpus longan* and *Walsura piscidia*.[294] Although retaining overall semi-evergreen notophyll forest affinity, this association has commonly been termed semi-deciduous.

9. The savanna question

A savanna is a woodland with a grassy field layer, in which the trees are isolated or in scattered clumps, like an apple orchard. Such forests are now widespread in continental Asia. With the exception of the pine forests (Chapter 4) and some deciduous dipterocarp forests (see section 5.4.3 above), tropical Asian savannas appear to be caused by human activity: overgrazing by domestic cattle, cutting especially for fuel, and over-frequent burning to aid grass regeneration.[295] The term *savanna woodland* is used for woodland in which the canopy is continuous, albeit diffuse and broken, yet with a grass-dominated field layer. Asian short deciduous forest meets these criteria, also some deciduous dipterocarp forests (Chapter 3).

In addition, natural open grasslands exist in the Asian tropics, where periodic flooding, probably through diurnal heating and deoxygenation of still floodwater, prevents tree regeneration. The most extensive examples are where major rivers, particularly the Brahmaputra, Tarsa, Teesta and Gandak surge out from the Himalaya into the plains (Plate 2.36e), as at Kaziranga National Park. There, remnants of the

Plate 2.36 (right) Semi-evergreen notophyll evergreen forest and savanna grassland. **a,** Semi-evergreen notophyll evergreen profile; Giritale, Sri Lanka. Savitri Gunatilleke as scale; **b,** Semi-evergreen notophyll evergreen forest subcanopy, with shrubs, treelets and litter, no field layer; Giritale, Niwal Gunatilleke as scale; **c,** A scattered emergent, *Manilkara hexandra,* in semi-evergreen notophyll evergreen forest; **d,** *Wyall* savanna in Mudumalai Wildlife Sanctuary, with Cyperaceae, *Imperata*; **e,** Savanna grassland in Duars, on seasonally flooded alluvium, where the Tarsa river issues from the Siwalik Himalayan front ranges; **f,** Sedge moor on periodically back-up flooded plain of a Sungei Belait tributary, Loagan Lalak, Brunei Aug. 1958. (**a–e** H.H.; **f** P.A.)

mid-Pleistocene large terrestrial mammal fauna survive; relics of the fauna found in the Trinil fossil beds in Java along with Java Man. Their survival here implies that grassy floodplains may have followed the great rivers that flowed in the vast valley north-east of the Sumatran-Javan 'ring of fire' at times of low ocean levels, when large grazing mammals and their predators occurred in Java (discussed in Chapters 1 and 6).

Small grassy floodplains, providing important grazing for mammals, exist in southern Indian deciduous forests (Plate 2.36d), where they are called *wyalls* (thus, Wyanaad is 'the land of the *wyalls*'). There, *Imperata cylindrica* and *Themeda* grasses mix with *Fimbristylis* and other sedges. Grasslands are currently rare in the perhumid tropics, and dominated by *Fimbristylis* and other sedges. One example is Luagan Lalak (Brunei), where the Belait River periodically backs up into a tributary (Plate 2.36f). Trees are confined to the banks of the meandering stream.

10. An epilogue

Dan Janzen was right in concluding that the forests of the perhumid lowlands of Asia are, by comparison with those of the Neotropics generally and Central America in particular, low-nutrient systems with low maximum tree growth rates and low vertebrate carrying capacity.[296] Even the fastest-growing pioneer on basalt loams at Bukit Mersing, *Octomeles sumatrana*, cannot vie with the fastest Central American pioneers. But this is a 'comparison of apples and oranges'. *If he had compared the growth rates or carrying capacity, at Pasoh or Lambir with those of Yasuni, on similarly low-nutrient soils of the perhumid Ecuadorian Amazon, would they be so different? These comparisons have yet to be made.* And similarly, how would these measures compare if the wet seasonal climates and late successional forests of *Falcataria moluccana* in basic volcanic Maluku (Moluccas) were compared with those in the Costa Rican Cordillera Central with similar rainfall seasonality and history?

In addition to the constraints imposed by nutrient limitation, even the mightiest of the contemporary Asian rainforests likely remain below their biomass potential. Most tropical forests are seemingly limited by their capacity to lift water from the soil in which they grow at the time of once-in-a-century drought. But the tallest of all Asian stands, in eastern Sabah, give indications that they might grow taller yet were it not for the probability of prior death by lightning, wind-blast or, yes, drought. In all old-growth forests, patches of canopy at apparent full height have different histories, and are studied in plots that include gaps born of more recent catastrophe. Each of these stands is engaged in a Sysiphean pattern of steady growth set back, in varying ways, by occasional episodes of destruction – from which they will ascend again.

References

[1]Whitmore 1984a; Richards 1958, 1996. [2]Whitmore 1984a, b. [3]Farjon 2008. [4]Hallé & Oldeman 1970, 1975. [5]Hallé & Oldeman 1970, 1975. [6]Hallé & Ng 1981; P.S. Ashton 1978. [7]Hallé *et al.* 1978. [8]Hallé & Oldeman 1970, 1975. [9]Metcalfe *et al.* 1998. [10]Thomas 1996a, b. [11]King 1990; Kohyama *et al.* 2003. [12]King 1995, 1998a, b, 1999; King & Maindonald 1999. [13]Kohyama & Hotta 1990; Kohyama 1993. [14]Ackerly & Bazzaz 1995a, b; Ackerly & Raich 1999; Ackerly *et al.* 2000. [15]Zotz 2004. [16]Penman 1948; Li *et al.* 2010; Kume *et al.* 2011. [17]Kenzo *et al.* 2007. [18]Medina *et al.* 1978. [19]Givnish & Vermeij 1976. [20]Horn 1971. [21]Myers *et al.* 1987. [22]Bloor & Grubb 2003. [23]Horn 1971; Terborgh 1985. [24]Givnish & Vermeij 1976; Givnish 1987, 1988. [25]P.S. Ashton 1982. [26]Ishida *et al.* 1996; Furukawa *et al.* 2001. [27]Furukawa *et al.* 2001. [28]Hallé *et al.* 1978. [29]Sasaki & Mori 1980. [30]Billings 1966. [31]D.W. Lee 2007. [32]Shukla & Ramakrishnan 1981. [33]Panditharathna *et al.* 2008. [34]Osada *et al.* 2002. [35]Thomas & Ickes 1995. [36]Tomlinson 1979, 2006. [37]Tomlinson 1979. [38]LaFrankie & Saw 2005. [39]Dransfield *et al.* 2008. [40]Wyatt-Smith 1963; Burgess 1971. [41]Dransfield *et al.* 2008. [42]Turner 2001a. [43]Whitmore 1984a, 1998a. [44]Clements 1936. [45]Green & Juniper 2004. [46]Bloor & Grubb 2003. [47]Whitmore 1984a. [48]Swaine & Whitmore 1988; Singhakumara *et al.* 2003. [49]Vasques-Yanes & Smith 1982. [50]Pearson *et al.* 2002. [51]Whitmore 1990. [52]Kiew 1992. [53]Bloor & Grubb 2003. [54]Jankowska-Blaszczuk & Grubb 2006. [55]Kanzaki *et al.* 1997. [56]Metcalfe & Grubb 1995. [57]Grubb 1998. [58]Metcalfe & Grubb 1995; Metcalfe *et al.* 1998. [59]Corner 1949. [60]Grubb *et al.* 2005. [61]Kanzaki *et al.* 1997. [62]Ng 1978, 1980. [63]Felker 1981. [64]Jankowska-Blascczuk & Grubb 2006. [65]P.S. Ashton 1978. [66]Bazzaz & Pickett 1980; Whitmore 1998a . [67]P.S. Ashton 1978. [68]Vasques-Yanes & Smith 1982. [69]P.M.S. Ashton 1992; Ediriweera *et al.* 2008. [70]Lieberman & Lieberman 1989; Raich 1989. [71]Ng 1978. [72]P.S. Ashton 1978; Ackerly 1996. [73]Sterck & Bongers 2001. [74]Meinzer *et al.* 1995. [75]Yaneda *et al.* 2000. [76]Tyree & Ewers 1996. [77]King *et al.* 2006a, b. [78]Raich & Gong 1990. [79]Tuomela *et al.* 1996. [80]P.M.S. Ashton 1995; P.M.S. Ashton *et al.* 1995. [81]Brown & Whitmore 1992. [82]Schwarzwäller *et al.* 1999. [83]Campbell & Newbery 1993; Muthuramkumar & Parthasarathy 2000. [84]Cohen *et al.* 1996. [85]P.S. Ashton 1978. [86]Canham & Burbank 1994; Horn 1971. [87]Hallé *et al.* 1978. [88]Appanah 1993; Kochummen 1966. [89]Ng 1980. [90]Boojh & Ramakrishnan 1982a, b. [91]P.S. Ashton 1978. [92]Kitajima *et al.* 2005. [93]King *et al.* 2005. [94]P.S. Ashton 1982, 1997; Wyatt-Smith 1954b. [95]Ng 1978, 1980. [96]Corner 1988. [97]P.M.S. Ashton 1995; P.M.S. Ashton *et al.* 2006. [98]Whitmore 1984a. [99]Ninomiya 1995. [100]Kitajima & Poorter 2008. [101]Langenheim *et al.* 1984; Chazdon *et al.* 1996; Lawrence 2001; Vincent 2001. [102]Furukawa *et al.* 2001. [103]Ninomiya 1995. [104]Zotz & Winter 1996. [105]P.S. Ashton & Hall 1992. [106]Tyree *et al.* 1998; Furukawa *et al.* 2001. [107]Dawkins & Philip 1998; Vincent *et al.* 2002. [108]Primack *et al.* 1985. [109]Bryan 1981; Chai 1988; Dawkins & Philip 1998. [110]Delissio *et al.* 2002. [111]Gong 1981. [112]Grubb *et al.* 1996. [113]Hallé *et al.* 1978. [114]King *et al.* 2006a. [115]Kurokawa *et al.* 2003. [116]Ng 2013. [117]Robertson *et al.* 2004. [118]Whitmore 1984a, 1989a. [119]Kato *et al.* 1974; Kira 1978. [120]Kira 1978; Luyssaert *et al.* 2007. [121]Condit *et al.* 1999. [122]Kira 1978; Kira & Yoda 1989. [123]Leighton & Wirawan 1986; Burslem *et al.* 1996; Delissio & Primack 2003; Potts *et al.* 2002. [124]Wyatt-Smith 1963. [125]Symington 1943. [126]P.S. Ashton 1978. [127]P.S. Ashton & Hall 1992; Gale 1997, 2000; Gale & Hall 2001; Delissio *et al.* 2002. [128]P.S. Ashton & Hall 1992. [129]Turner 2001a. [130]Gale & Hall 2001. [131]Gale 1997; Gale & Hall 2001. [132]Tang & Chong 1979. [133]Chazdon *et al.* 1996. [134]Chazdon & Fetcher 1984; Brokaw 1987; P.M.S. Ashton 1992; Brown 1993;

Barker *et al.* 1997. [135]Davies *et al.* 1998, Davies & Ashton 1999. [136]Moad 1992. [137]Montgomery & Chazdon 2002. [138]Zipperlin & Press 1996, 1997. [139]Brown *et al.* 1999. [140]Sack & Grubb 2001. [141]Philipson 2009. [142]P.M.S. Ashton & de Zoysa 1990; P.M.S. Ashton & Berlyn 1992; P.M.S. Ashton 1995; P.M.S. Ashton *et al.* 1995, 2006; Gunatilleke *et al.* 1996, 2004a–c. [143]Becker *et al.* 1988. [144]Brown & Whitmore 1992; Aiba & Nakashizuka 2007. [145]Bebber *et al.* 2002b; Robert & Moravie 2003. [146]Yamakura *et al.* 1996. [147]Sterck *et al.* 2001. [148]P.M.S. Ashton 1990b, 1995. [149]P.M.S. Ashton & Berlyn 1992. [150]Brown *et al.* 1999. [151]Robert & Moravie 2003. [152]Gamage *et al.* 2003; Singhakumara *et al.* 2003; P.M.S. Ashton *et al.* 2006. [153]Burgess 1969–1975. [154]Chazdon & Fetcher 1984; Chazdon *et al.* 1996. [155]Crowther 1982. [156]Brünig 1970; Whitmore 1984a. [157]Riswan 1982; Riswan & Kartawinata 1991. [158]Yamada & Suzuki 2004. [159]Riswan & Kartawinata 1991. [160]Seddon 1974; Turner 1994a; Turner *et al.* 2000. [161]Turner 1994b. [162]Nishimura & Suzuki 2001. [163]Peace & Macdonald 1981. [164]Becker 1996. [165]Matsumoto *et al.* 2000. [166]Matsumoto *et al.* 2000. [167]Mulkey *et al.* 1996a, b. [168]Brünig 1970, 1974. [169]Coomes & Grubb 1998. [170]Givnish 1988, Becker *et al.* 1999. [171]Newbery *et al.* 1986. [172]Nishimura *et al.* 2007. [173]Whitmore 1984a. [174]Roman Dial, pers. comm. 2008. [175]Whitmore 1984a. [176]P.S. Ashton 1964b. [177]Newbery *et al.* 1986. [178]Wyatt-Smith 1960. [179]Chazdon & Field 1987; Vincent 2001. [180]Norisada & Kojima 2005. [181]Kenzo *et al.* 2007. [182]Field 1988. [183]Bungard *et al.* 2000, 2002. [184]P.S. Ashton & Hall 1992. [185]Leighton & Wirawan 1986; Potts *et al.* 2002. [186]Woods 1989; Leighton & Wirawan 1986; Newbery *et al.* 1996; Becker 1996; Nakagawa *et al.* 2000; Bebber *et al.* 2003. [187]Delissio *et al.* 2002, Delissio & Primack 2003; Kume *et al.* 2007. [188]Proctor *et al.* 2001. [189]Sun *et al.* 1998; Sun & Hsieh 2004. [190]P.S. Ashton 1993; Newbery *et al.* 1999. [191]Everett & Whitford 1914. [192]Whitford 1909. [193]Boose *et al.* 1994; Foster & Boose 1995. [194]Burslem *et al.* 2000. [195]Wyatt-Smith 1954b; Whitmore 1984a. [196]Wyatt-Smith 1950. [197]Cao 2000. [198]Tyree *et al.* 1998. [199]Roman Dial, pers. comm. 2008. [200]Goltapeh *et al.* 2008. [201]Phillips *et al.* 1998; Lewis *et al.* 2004a. [202]Baker *et al.* 2005. [203]Baker *et al.* 2005. [204]Baker *et al.* 2005. [205]Feeley *et al.* 2007a, b. [206]P.S. Ashton 1981; Baas & Vetter 1989. [207]Ng 2013. [208]Champion 1936. [209]Gaussen 1978. [210]Pascal *et al.* 1998. [211]Blasco *et al.* 2000. [212]P.S. Ashton 1991, 1995a. [213]Champion 1936. [214]Li *et al.* 2010; Kume *et al.* 2011. [215]Maxwell 1995. [216]Biswas & Misbahuzzaman 2006. [217]Tripathi *et al.* 2006a. [218]Wyatt-Smith 1954a. [219]Rai & Proctor 1986. [220]Kadambi 1941, 1942. [221]Bunyavejchewin *et al.* 2001. [222]Chengappa 1934, 1937, 1944. [223]Kadambi 1941, 1942; Chengappa 1934, 1937, 1944. [224]Corlett 2009a. [225]S. Davies, pers. comm. 2009. [226]M. Cao *et al.* 2008. [227]De 1923; Kingdon-Ward 1937. [228]Anon. 1888, 1892. [229]V. Ling to R. Rangham, pers. comm. July 2011. [230]Stott 1986. [231]R.S. Singh 1992. [232]Suresh *et al.* in press. [233]Sukumar *et al.* 1998. [234]'Troup 1921. [235]Suresh *et al.* 2011. [236]Kumar & Deepu 1992. [237]Champion 1936. [238]White 1965. [239]Champion 1936. [240]Pascal *et al.* 1988. [241]Chengappa 1944. [242]Anon. (E.G.C.) 1892. [243]Anon. (E.G.C.) 1892; Chengappa 1944. [244]S. Bunyavejchewin, pers. comm. Feb. 2008. [245]Pascal *et al.* 1998. [246]Bunyavejchewin *et al.*, unpubl. ms.. [247]Gamble 1896; Seethalakshmi *et al.* 1998. [248]Baker *et al.* 2009. [249]Baker 1987; Baker & Bunyavejchewin 2006a, 2009; Baker *et al.* 2005. [250]Champion 1936; Chandrasekharan 1962a–c. [251]V. Ling to R. Rangam, pers. comm. July 2011. [252]Chengappa 1944; Baker *et al.* 2005. [253]John *et al.* 2002. [254]H. Suresh, pers. comm. 2008. [255]Troup 1921. [256]Suresh *et al.* 2011. [257]H. Dattaraja, pers. comm. 2008. [258]H. Dattaraja, pers. comm. 2008. [259]John *et al.* 2002. [260]John & Sukumar 2004. [261]Dattaraja, pers. comm. April 2008. [262]S. Bunyavejchewin, D. Swaine, pers. comm. 2008. [263]P.S. Ashton 1991, 1995a. [264]Maxwell 2004. [265]Stamp 1924. [266]Troup 1921. [267]Bunyavejchewin, pers. comm. Feb. 2008. [268]Bunyavejchewin 1986. [269]Baker & Bunyavejchewin 2006b; Bunyavejchewin *et al.* 2011. [270]S. Bunyavejchewin, pers. comm. Feb. 2008. [271]Troup 1921. [272]Graham *et al.* 2003. [273]Baker *et al.* 2003. [274]Condit *et al.* 1999. [275]Lodhiyal & Lodhiyal 2003. [276]Troup 1921. [277]Lal *et al.* 2001. [278]Behera *et al.* 1990. [279]Troup 1921; Chandrashekara & Ramakrishnan 1993. [280]Troup 1921; Ng 1991. [281]Troup 1921. [282]Smythies 1919. [283]Champion 1920. [284]Rautiainen & Suoheimo 1997. [285]Chengappa 1944. [286]Dinerstein 1987. [287]Champion & Seth 1968a. [288]Champion & Seth 1968a. [289]Suresh *et al.* in press. [290]Troup 1921. [291]Champion 1936; Champion & Seth 1968a; Partharathy & Karthikeyan 1997. [292]Mueller-Dombois 1968; Meher-Homji 1977. [293]Perera 1975; Gunatilleke 1975; Gunatilleke & Ashton 1987. [294]de Rosayro 1950; Koelmeyer 1957. [295]Yadava 1990. [296] Janzen 1974.

Mixed dipterocarp forest on the rolling hills
of Sinharaja forest, SW Sri Lanka (M. Ashton)

Chapter 3

PATTERNS OF SPECIES COMPOSITION IN ASIAN TROPICAL LOWLAND FORESTS

> The Bornean Flora assumes multiform aspects, and its components are of a varied nature according to the localities, the elevation, and the physical conditions of the soil. Thus distinct areas of varied extent can be recognized, some much restricted, on which a special vegetation grows, different from that of adjoining lands.
>
> O. Beccari, 1904, *Wanderings in the Great Forests of Borneo*, (in which he recounts, from field notes, his two years in Sarawak four decades earlier).

In this chapter I describe the variation in tree species composition across diverse landscapes of lowland tropical Asia. I identify patterns of variation at both local and regional scales, and discuss possible causes of these patterns. Broader floristic comparisons, up to the continental scale, will be considered in Chapter 6.

Some regional-scale variations in forest structure and phenology are correlated with patterns of rainfall seasonality. These are discussed in the previous chapter. Similarly, climate is associated directly or indirectly with many of the major floristic variations at regional and continental scales. Therefore I divide this chapter into two sections; Part I devoted to the mixed dipterocarp forests (MDF) and other Asian forest types of the perhumid regions, Part II addressing the influence of seasonality on forest formations. However, several other physical factors also interact powerfully with climate to produce the current patterns of floristic distribution. Finally, history – both of the forests themselves and of human activity within them – also imposes its influences and constraints, and these are examined more deeply in Chapters 6 and 8.

In temperate forests, soils play a major role in determining forest structure, dynamics and species composition at the scale of the local landscape.[1] The influence of soils on species composition is often particularly clear at the local scale, within a given climatic zone. It is therefore surprising that so little attention was paid by early foresters to the relationships between forests and soils in much of tropical Asia,

particularly in the lately forested terrains of the perhumid Far East. Due to an excess of mean monthly rainfall throughout the year, the climate there is effectively uniform over vast areas, from the ecological perspective (Chapters 1, 2). Within this perhumid zone the soils are seldom dry enough for nutrients in solution to remain unattached or unleached, let alone to be drawn upward and concentrated by capillary movements of solutes, as may happen during annual dry seasons (Chapter 1). Instead, relentless leaching enhances the differences between mineral soil types by altering the concentration of positive and negative charges which can hold nutrient ions.

Nevertheless, at time scales comparable to the lifespan of a tree, nutrients may become available to its roots. The forests may therefore be exceptionally variable (Table 3.1). Here we inquire how forests and soil variation are related and even interdependent.

Forest type	Number of tree species	Borneo endemism (%)
Sea shore	25	0
Mangrove	33	0
Back-mangrove, brackish river banks	50	5
Plains rivers, alluvial banks	200	15
Rocky, whitewater river banks	200	35
Floodplains (seasonally swamped)	600	25
Peat swamp forest	200	25
Kerangas	300	50
MDF, sandy, sandy clay soils	1,200	50
MDF, clay soils	900	35
Upper dipterocarp forest	600	40
Lower montane forest	500	45
Upper montane forest	100	50

Table 3.1 Approximate number of species in different vegetation types in a north-west Borneo region: Brunei. The figures in total exceed the number of species known from Brunei because many occur in more than one forest type. (After P.S. Ashton 2003b)

Part I. Forests of the aseasonal regions

1. The mixed dipterocarp forests

I went to Brunei, on the north-west coast of Borneo, as Forest Botanist almost one century after the great botanist Odoardo Beccari, but having first familiarised myself with the more recent work of Paul Richards.[2] From the work of the earlier explorer-naturalists and from foresters in the 1930s when he did his fieldwork, Richards knew that the apparently chaotic jumble of species in biodiverse rainforests was not homogeneous across the landscape. Richards placed single plots in two or three comparable forest types in Guyana, Nigeria and Sarawak. He found that forests in similar habitats worldwide shared structural features, but that there were differences as well. Richards was thus the pioneer who established the principles so greatly extended by the Center for Tropical Forest Science (CTFS). He documented differences in species among nearby samples on different soils, but could not confirm their constancy since that would have required multiple plots throughout each region. Botanical explorers, including Richards himself, continued to view the floristic variation within zonal forests, such as mixed dipterocarp forest in perhumid Asia, as determined solely by the vagaries of seed dispersal and the vicissitudes of germination and establishment.

It took me some six months to reliably identify the perhaps 200 commoner species in Brunei, from fallen leaf and characters of bark and slash as taught to me by my Iban Dayak team-mates. My main task was to prepare a manual of the Dipterocarpaceae, the dominant tree family in perhumid Asia west of Wallace's Line.[3] On well-drained yellow/red loam lowland soils within this zone, most of the dipterocarp genera are emergent trees, and they comprise 60–90% of all the trees in that stratum (Fig 2.1). At the time, Robb Anderson·was completing studies of the peat swamp forest,[4] Eberhard Brünig of the kerangas.[5] I decided to take on the hyperdiverse lowland mixed dipterocarp forest, representing the other − and the principal − lowland forest type of the region. (Table 3.2)

Mixed dipterocarp forest forms the core of the lowland rainforest throughout perhumid western Malesia including the Philippines, where it occasionally spreads westward into regions with less than six dry months. I have termed these forests mixed dipterocarp forests (MDF) because their canopy layer is dominated not by any single species, but by many, overwhelmingly in the family Dipterocarpaceae (and, as we shall see, there are also forests dominated by a single dipterocarp species).[6] MDF covers all freely draining yellow and red soils in perhumid western Malesia below c.700 m, and is replaced by a similar forest in New Guinea, as described below. Excellent general accounts of these forests have already been compiled by Tim Whitmore,[7] and a more detailed silvicultural account in the classic monograph of John Wyatt-Smith.[8] But as we shall see, a more comprehensive local and also regional understanding has developed over the decades since their

work, not least owing to the enormous new opportunities provided by the global and regional network of large forest tree demography plots of CTFS, and the work of the scientists who direct them (see the Prologue).

As soon as I could, I began to keep notes of the species, especially the dipterocarps, which my team and I encountered along the forest tracks as we collected specimens, for formal identification for an eventual forest flora. The hunters' paths followed undulating ridges in the forest, so did not equally represent all components of the landscape. I was nevertheless soon struck by the very different tree flora of the MDF in the Belait coastal hills, when compared with the steep inland ridges draining into the whitewater Temburong River. The Belait hills were soft young Mio-Pliocene sandstones, the Temburong older, harder and Lower Miocene shales. The soils were distinctly different, deep and sandy versus shallow and clay. Further exploration indicated that these early perceived floristic and landscape differences were in fact substantial, consistent, and precisely followed the substrate boundaries on the geological map (Fig. 6.17a,b). At Bukit Teraja and in the Ulu Belait, there are hills on whose sandy ridges and dip slopes I found recurring lists of sandstone species, while on some clay slopes the shale species of Temburong reappeared.[9] In a few localities there seemed to be a mosaic of alternating species and soil associations over the landscape. At the local level, too, more subtle distinctions appeared between ridge and slope. But always, and later when I came to explore other regions in Asia, I saw that the sandy habitats form islands within a regional matrix of clay soils; sandstone subordinate to shale, ridges surrounded by slopes. Even where extensive sandstone formations, and plateaux, occur, they were isolated from one another by a sea of clay loam-forming rocks. The forest department was interested to know more, because these observations implied major differences in timber stocking, and perhaps in silvicultural options. I therefore set out 50 single-acre (0.4 ha) plots at two locations, Andulau Forest Reserve in the sandy coastal hills of Belait District, and Kuala Belalong in steep terrain at the base of the Ulu Temburong mountains.[10] These were arranged systematically so that each plot was restricted to only one topographical unit: ridge, slope, or river bank or floodplain. Restricting myself to trees larger than 1 ft in girth (9.7 cm diameter), I recorded some 30,000 individuals from 760 species.

My aim, then, was to examine floristic variation within the two landscapes sampled, and to search for similarities between them. I returned to university uncertain, though, in the days before personal computers, how I would go about analysing such a massive data set. I was spectacularly lucky, because two plant ecologists in Wisconsin, J.R. Bray and J.T. Curtis, faced with a similar challenge, had devised an ingenious method of expressing the similarity between sample plots quantitatively and visually in two dimensions.[11] Samples are aligned along gradients of descending importance based on a quantitative measure of their similarity, and the results can be expressed in graphics depicting their positions along up to three gradients. Habitat measures, such as slope or soil nutrient concentrations, can then be superimposed

Table 3.2 a, Floristic similarity of major Sunda forest types: approximate percentages of shared species, in Brunei as an example. **b,** The ecological distribution of the dominant yellow–red zonal soils of hill and coastal dipterocarp forests in perhumid Sunda. (**a,** from Hasan & Ashton 1963, P.S. Ashton 1964b)

a

	1	2	3	4	5	6	7	8	9	10	11	12	13
1													
2	<10												
3	<10	20											
4	0	0	20										
5	0	0	0	<20									
6	0	0	0	<10	<20								
7	0	0	0	0	0	0							
8	0	0	0	0	0	0	<80						
9	0	0	0	<10	0	<10	<30	<60					
10	0	0	0	80	<20	90	0	<10	<35				
11	0	0	0	<10	<10	<40	<30	<30	<20	<60			
12	0	0	0	<5	<5	<5	<10	<10	<10	<20	<30		
13	0	0	0	0	0	0	0	<10	0	0	<10	<20	

b

	Clay soils, overlying shale, basic igneous rocks etc	'Humult'sandy soils, overlying sandstone, quartzite and on granite ridges
Continuously moist soils: inland, slopes etc.	'Udult' friable clay (long-lattice) to periodically waterlogged clay (short lattice) loams	Sandy clay loams, accumulating variably deep raw humus; over sandstone and quartzite only
Intermittently dry soils: coastal hills, narrow or high inland ridges below c.700 m	Clay loams, often shallow, accumulating shallow raw humus ('humult') at higher altitudes and on coasts	Leached yellow sandy soils, accumulating shallow (granite) to deep raw humus

in a search for correlations. This approach, which they called ordination, has now been used countless times to resolve flora-habitat relationships in tropical forests and elsewhere. After three weeks of 14-hour days on a hand calculator, coefficients of similarity between plots, for each site-based group of plots separately, were ready for ordination. The excitement was intense. As the plots were added to the figures, a pattern gradually emerged – rather the way an old photographic print used to appear under the hypo. It was amazing how well the patterns reflected the intuitions I had gained from tramping the forests (Fig. 3.1).

In ordination, my Brunei plots did indeed fall into groups according to their places in the topography. This pattern was strong on the higher, sharper 700 m ridges at Belalong, weaker on the rounded 100 m hills at Andulau. On Andulau's undulating topography there was much overlap between ridge and slope, though the bottomland plots were almost isolated. Only later did soil data become available. Soil nutrients were primarily correlated with geology, but partially also with topography. It was impossible at that time to distinguish between the influence of nutrients and water variability – both

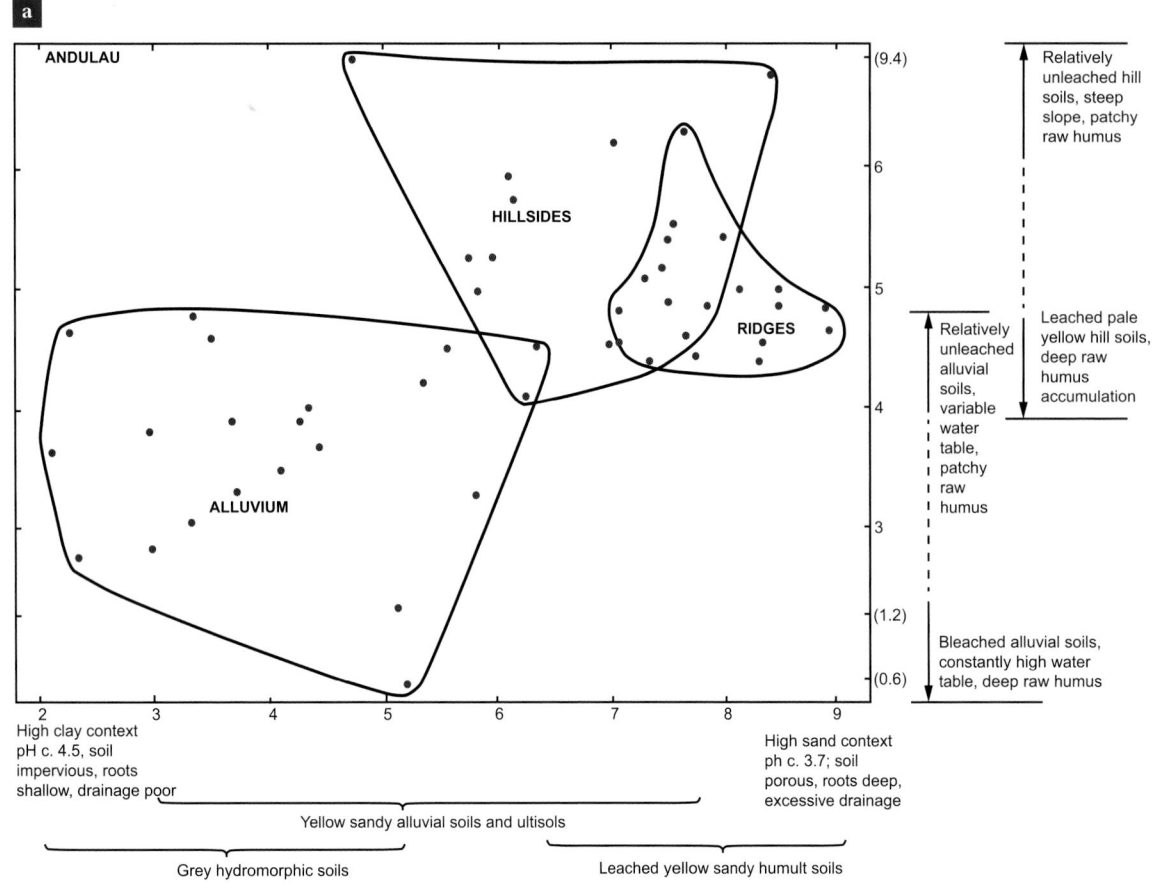

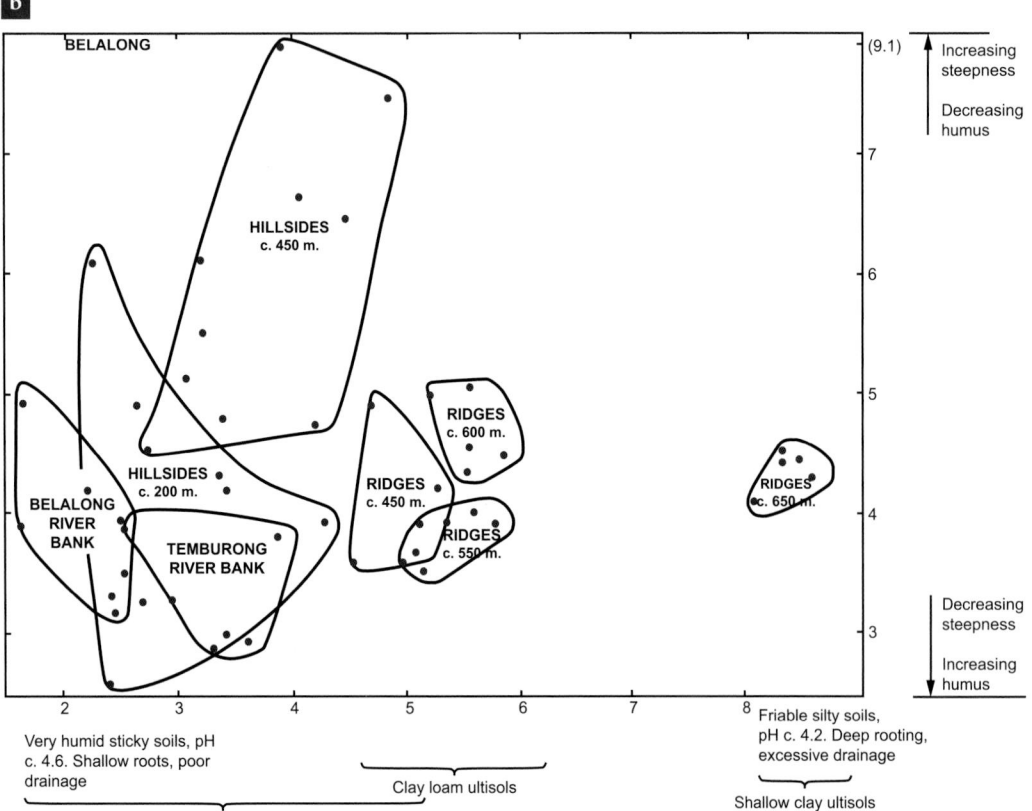

Fig. 3.1 Bray and Curtis ordination of 50, 0.4 ha, plots each, in MDF in two Brunei forests; **a,** on humult sandy soils and shallow peat at Andulau Forest Reserve; **b,** on clay loams at Kuala Belalong. (Ordination is a method whereby plots (dots) are arranged in mathematical space using data on their species, or vice versa, so that their distance apart reflects their similarity, and gradients of variation can be recognized along axes (the horizontal, 'x', and vertical, 'y' axes of the figure) which are ordered by their strength; tests for correlation with physical characteristics may then be attempted). (After P.S. Ashton 1964b)

of which would surely vary with topography – on species composition.

I could not have observed these patterns without reasonably reliable and accurate species identification. I was most certain of the dipterocarps, because I was preparing a manual on them and they were the leading family both ecologically and for timber. I found some patterns characteristic of genera, such as the concentration of *Dipterocarpus* species on ridge tops, where seven species each are focused at Andulau and Belalong. There were overall patterns within natural species groups or guilds: for instance, the heavy-timbered red merantis in *Shorea* (sects

Brachyptera, Mutica and *Rubella*) were better represented on sandy soils and ridges, whereas light red merantis clustered more densely on clay soils and slopes (Table 2.4). Striking, too, were the distinct but generally overlapping distributions of some related species. The most dramatic of these were the four species of *kapur* (*Dryobalanops*), two at each site (Fig. 3.2a), each of which had an ecological distribution so distinct that their ranges hardly overlapped. Yet others, such as many of the red merantis (Fig. 3.2b), overlapped a good deal.

The species list for the two sites enabled overall comparison of MDF with the *kerangas* and peat swamp lists prepared by

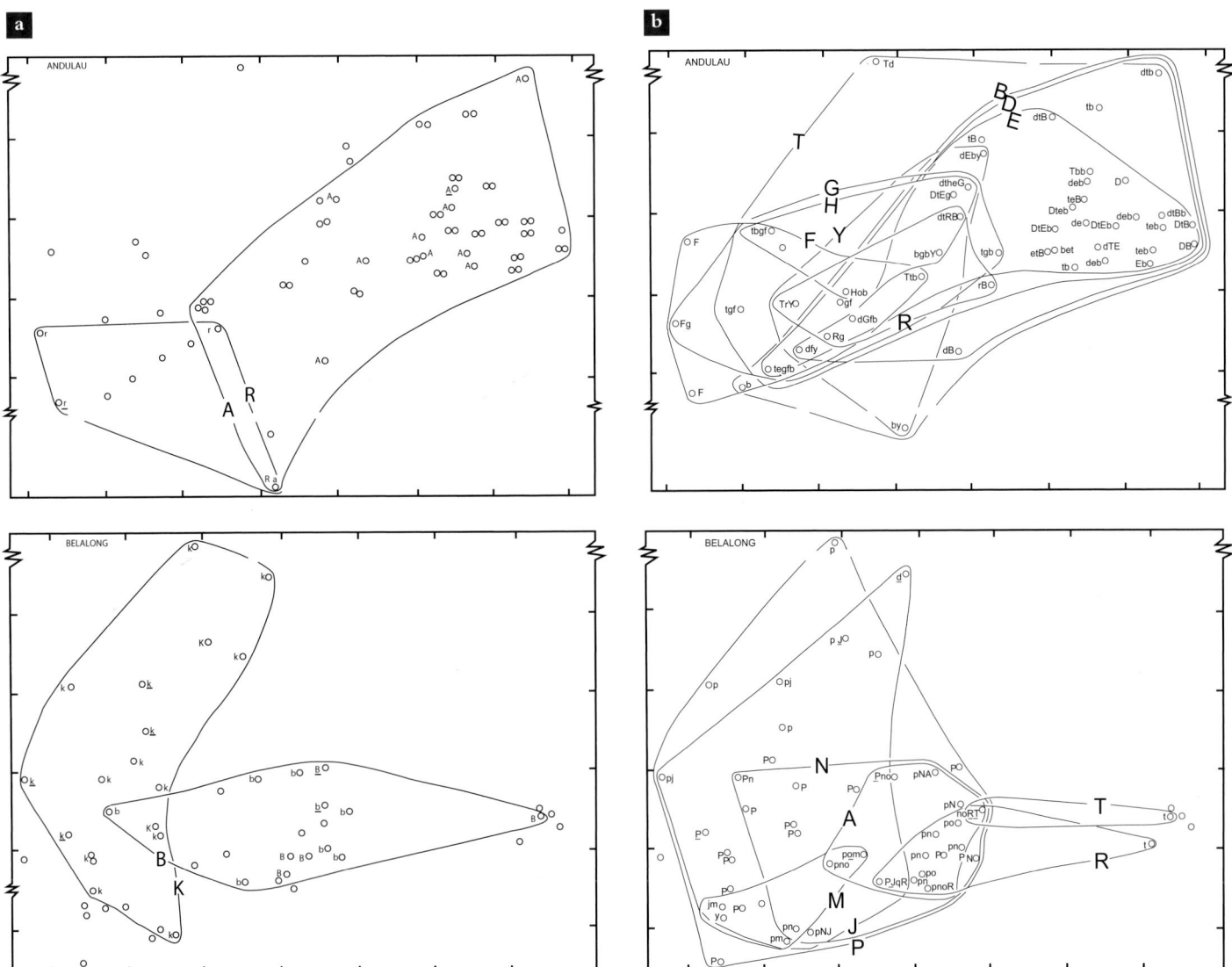

Fig. 3.2 Distribution of selected species in the Brunei ordinations: Letters indicate presences, in quartiles of basal area, from the highest: capital underlined, capital, lower casing underlined, lower casing. (After Ashton 1964): **a,** *Dryobalanops*, a small genus in which each species has a distinct, hardly overlapping, ecological distribution: A, *D. aromatica;* R, *D. rappa;* B, *D. beccarii;* K, *D. lanceolata;* **b,** Red meranti *Shorea* sects *Mutica* and *Ovalis,* in which up to 9 species occur at a site, each of which has a distinct distribution, but overlapping with others to differing degrees; clusters of species may share a habitat, with most species co-occurring at ecotones; but no plot includes more than 5: A, *S. argentifolia;* B, *S. acuta;* D, *S. curtisii;* E, *S. ovata;* F, *S. teysmanniana;* G, *S. scabrida;* H, *S. ovalis;* J, *S. leprosula;* M, *S. myrionerva;* N, *S. macroptera* ssp. *macropterifolia;* P, *S. parvifolia;* R, *S. rubra;* T, *S. quadrinervis;* Y, *S. slootenii.*

Anderson and Brünig. The MDF flora on clay shared less than 5% of species with either of these forest types, but the sandstone shared c.20% with *kerangas* growing on the same substrate. The total recorded complement of tree species for *kerangas* in Brunei as a whole is now estimated at c.300, compared with c.900 for MDF on clay and c.1,200 on sandstone.[12] There are c.200 tree species in the Brunei peat swamps, out of the c.4,000 species recorded from Sarawak and Brunei overall. About 80% of these also occur in *kerangas,* though many are confined to poorly drained *kerangas* (*kerapa*). Only 10% are found in MDF. Some of these species are confined to ecotones, that is communities and soils of intermediate character between those of the principal floristic associations that currently dominate the Brunei landscape (described in sections 1.2.1 and 2.1 below). In such forests, ecotones are usually sharp, often occurring over less than five metres at geological boundaries, but may be much more extensive, for instance between *kerangas* and peat swamp behind beaches. Some sandstone formations, such as the Balingian in coastal Central Sarawak, carry extensive partially podsolised yellow soils bearing a forest type intermediate between *kerangas* and MDF, and including a few specialists (Chapter 6).

Savitri Gunatilleke examined local and island-wide patterns of variation in Sri Lankan forests.[13] At two sites in the perhumid south-west (Kottawe in the southernmost lowlands, and Kanneliya 30 km to the north and at 250 m altitiude), ordination again indicated floristic variation correlated with topography, and also with soil phosphorus. In a more recent analysis of species spatial patterns within the 25 ha CTFS plot in the Sinharaja World Heritage forest, she and Nimal, her husband, showed that in their plot, 82% of the 125 species with more than 100 individuals greater than 1 cm dbh are significantly more abundant on one or two of eight topographic habitat categories than in the others (Fig. 3.3); canopy and understorey species were similarly correlated.[14] Nutrient levels are generally low on these clay soils, but are still loosely correlated with topography.[15]

Ecologists on other continents, seeking generalisations about Asian tropical forests for comparison, have focused on the forests of western Malesia and particularly the Sunda Shelf: the MDF. They have emphasised their great stature and the unique role of the emergent species of dipterocarps, the abundance of their juveniles in the understorey crowding out other species (Chapter 2), and their supra-annual mass flowering and mast fruiting (Chapter 5). Other distinctive characteristics of these forests include an extraordinary diversity of palms, especially the climbing rattans, yet rarity of canopy palm tree species and the comparative sparseness of epiphytes and absence of tank epiphytes (Chapter 2). As our chapters unfold, we shall see that many of these features are in fact peculiar to MDF, not to all the forests of tropical Asia.

Fig. 3.3 Some co-occurring sister taxa possess sharply distinct spatial distributions. Ecological distribution of two sister species: *Mesua* (Clusiaceae) in the 25 ha CTFS plot at Sinharaja forest, Sri Lanka. *M. nagassarium* (red dots), a co-dominant species in the main canopy, on the spurs; and *M. ferrea,* a subcanopy species, in intervening gullies. (After Gunatilleke *et al.* 2004a)

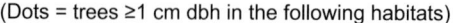

(Dots = trees ≥1 cm dbh in the following habitats)

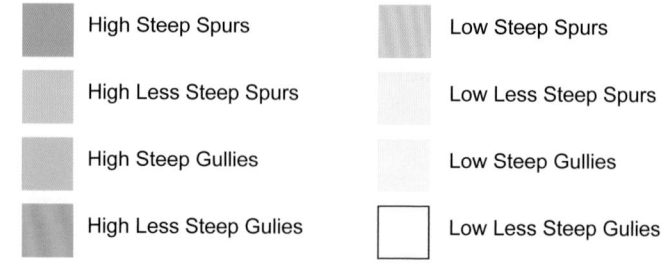

High Steep Spurs   Low Steep Spurs

High Less Steep Spurs   Low Less Steep Spurs

High Steep Gullies   Low Steep Gullies

High Less Steep Gulies   Low Less Steep Gulies

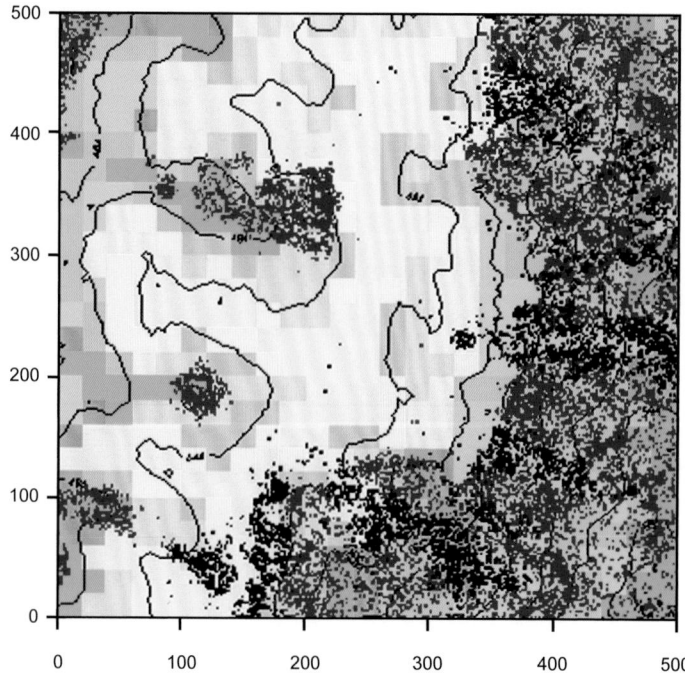

Ecologists have focused research on the most species-rich communities, of which the richest in Asia by far is the MDF of the Sunda lands. More is now known concerning floristic variation within MDF, and how species coexist within these hyperdiverse communities, than about others. Much of this understanding can nevertheless be employed in discussing differences between them and other, less diverse, forest types of more limiting soils and climates.

Quantitative analyses such as ordination have confirmed, for those places where it has been used, that rainforest tree species communities are habitat-specific; and that many species – most species in the case of northern Borneo and south-west

Plate 3.2 Floristic variation in mixed dipterocarp forest. **a,** A Sinharaja hillslope: *Shorea worthingtonii* crowns silhouetted on the ridge, a 'drift' of pale-crowned *S. trapezifolia* to the right on the upper slope, and the dark crowns of *S. megistophylla* in the swale at lower right; **b,** The massive v-section fissured trunk of *Shorea curtisii*, Lambir National Park; **c,** Bertam, *Eugeissona tristis*, dominating the understorey within the diffuse light cast by *Shorea curtisii* crowns along the granite ridges and coastal hills of Peninsular Malaysia, Semangkok Forest reserve, Selangor (cf. Plate 2.3f); **d,** Coastal hill dipterocarp forest with dominant *Shorea curtisii* on upper slopes and crests, Mukah Head, Penang, Peninsular Malaysia. (a, M.A.; b, S. Davies; c, R. Harrison; d, J.E. Ong)

Sri Lanka – are habitat specialists.[16] The pattern of tree species variation on the land is echoed in other plants. Notably, the forests on fertile clay loams where litter decomposes rapidly or is washed away are rich in ground flora (see section 1.2.1 below). These patterns are a major cause of species diversity at the landscape scale; the patterns of tree community species *richness* and their likely causes is the subject of Chapter 7.

Questions remain, though, concerning patterns of floristic variation.

1. In Chapter 2, variation in lowland tropical evergreen forest structure was attributed principally to spatial and temporal variation in soil water stress. What are the relative influences of water stress and nutrient availability on species distributions?

2. What is the nature of the competitive interactions that ultimately determine the observed ecological range of tree species within Asian rainforests?

3. To what extent does history of climate, of mountain building, of river capture and sea level change modify community composition at the regional scale by influencing regional extinction and opportunities for species migration? This will be the topic of Chapter 6.

Ordination can establish spatial patterns, and correlations between floristic and habitat patterns, but not causes. Ordination and other comparative analyses are invaluable for formulation of explanatory hypotheses, but these can only be tested by experiment.

when Stuart Davies and his associates ordinated subplots within the edaphically heterogeneous CTFS 52 ha plot established under Sylvester Tan's leadership at Lambir National Park, in nearby Sarawak.[18] The plot includes, along its 1.04 km length, the ridge of a mid-Miocene cuesta with the scarp and most of the dip slopes separately along each side. The sandstone strata overlie early Miocene shale. The shale is exposed to the ridge at its lower southern end; on the scarp; in dip slope valleys along stream gullies; and in one broad swale where sand is mixed with the clay in the soils[19] (Fig. 3.4). Akira Itoh[20] had shown

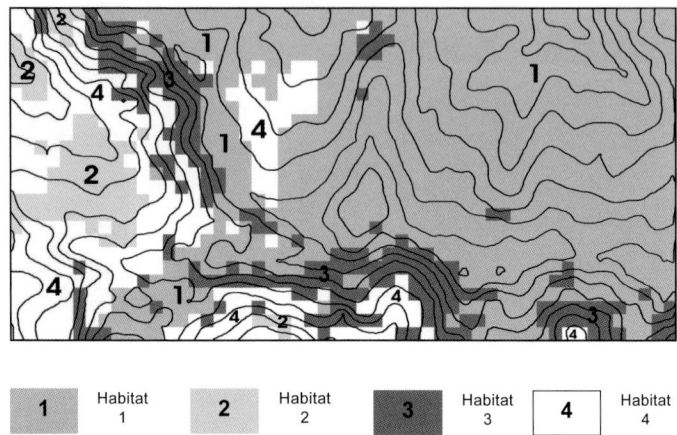

| 1 | Habitat 1 | 2 | Habitat 2 | 3 | Habitat 3 | 4 | Habitat 4 |

Fig. 3.4 Floristically correlated main habitat types within the Lambir, 52 ha CTFS plot, as used by Russo 2008). 1: Humic sandy clay loams on dip slope; 2: Clay loam ('clay loam'); 3: Yellow humult sandy soil along ridges ('sandy loam'); 4: Silty clay loam ('fine loam'), mainly on dip slopes. (From Russo *et al.* 2008)

1.1. *The nature of the habitat influences on species composition: intersecting influences of nutrients versus water stress*

The Brunei ordinations showed conclusively that topography and forest floristic variation is correlated, and they have since been supported by observations of broadly similar patterns in the Neotropics.[17] I had concluded, on account of correlations observed repeatedly throughout Brunei, that the substantial floristic differences between my two sets of plots must be due to their different substrates and soils. Less than one third of species were shared. However, on the plot data alone it could equally have been argued that the floristic difference between the two sites – which are 70 km and three river valleys apart – is due to restricted seed dispersal of the species. These alternatives could only be rigorously resolved

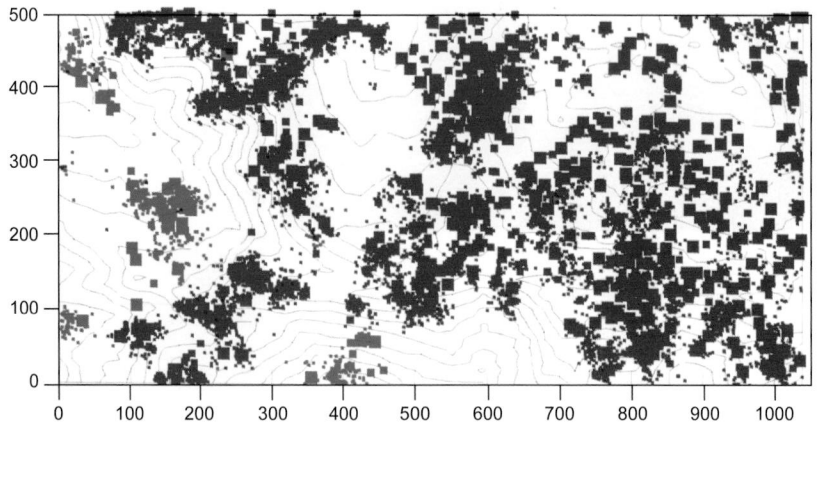

Fig. 3.5 Distribution of two *Dryobalanops* in the 52 ha CTFS plot at Lambir National Park, Sarawak: *D. lanceolata* (blue), *D. aromatica* (red). Symbols indicate diameter classes. (From H.S. Lee *et al.* 2002a)

■ D. aromatica

■ D. lanceolata

that the distinct and separate distributions of the two dominant *Dryobalanops* there are correlated with differences in both topography and soil (Fig. 3.5). Davies' ordinations produced a horseshoe pattern of plots, in effect two axes of floristic variation diagonal to the two primary axes of the ordination itself[21] (Fig. 3.6a). To the left, plots containing species of sandy soils clustered on upper slopes and ridges were aligned along a gradient of increasing elevation. To the right, plots scattered on clay soils along concave slopes of scarps and in shallow valleys, sorted along a nutrient gradient. An ordination of species in the plots divided them similarly (Fig. 3.6b).

Plots with sandy humult soils and the species recorded in the Andulau plots were at one end of the combined gradients of Davies' ordination, udult clay loam soils and species found at Kuala Belalong at the other (Fig. 3.6a). Species showed significant association with one or more categories of soil, mapped by Ian Baillie[22] from his survey (Fig. 3.4) and from systematic sampling and analysis. Remarkably, only 13% of the 764 commoner species (out of 1,192) used in the Lambir ordinations proved to be generalists whose distributions were

not associated with the soil categories. Recently, Rahayu Hj. Sukri[23] has examined 37 commoner dipterocarp species at the Kuala Belalong and Andulau sites. She found the ecological ranges of 19 to be significantly soil nutrient-associated. At Kuala Belalong, dipterocarp composition was associated with total and exchangeable calcium and magnesium, total nitrogen, phosphorus, potassium, and pH, whereas at the topographically more uniform Andulau forest only magnesium was correlated. Stuart Davies,[24] in separate research, found that even the 11 species of the mainly pioneer genus *Macaranga* in the CTFS 52 ha Lambir plot occupied niches in part partitioned by soil nutrient status (Fig. 3.7). Garry Paoli,[25] at Gunung Palung (West Kalimantan), similarly found that 18 of 22 abundant species were significantly associated with a specific geological substrate and soil type; available magnesium and calcium correlated here too. An overwhelmingly stronger influence of habitat variation over dispersal constraints at landscape scale, and geographical separation was thus implied. Again at Lambir, total phosphorus and available magnesium and calcium were the most highly correlated nutrient ions.

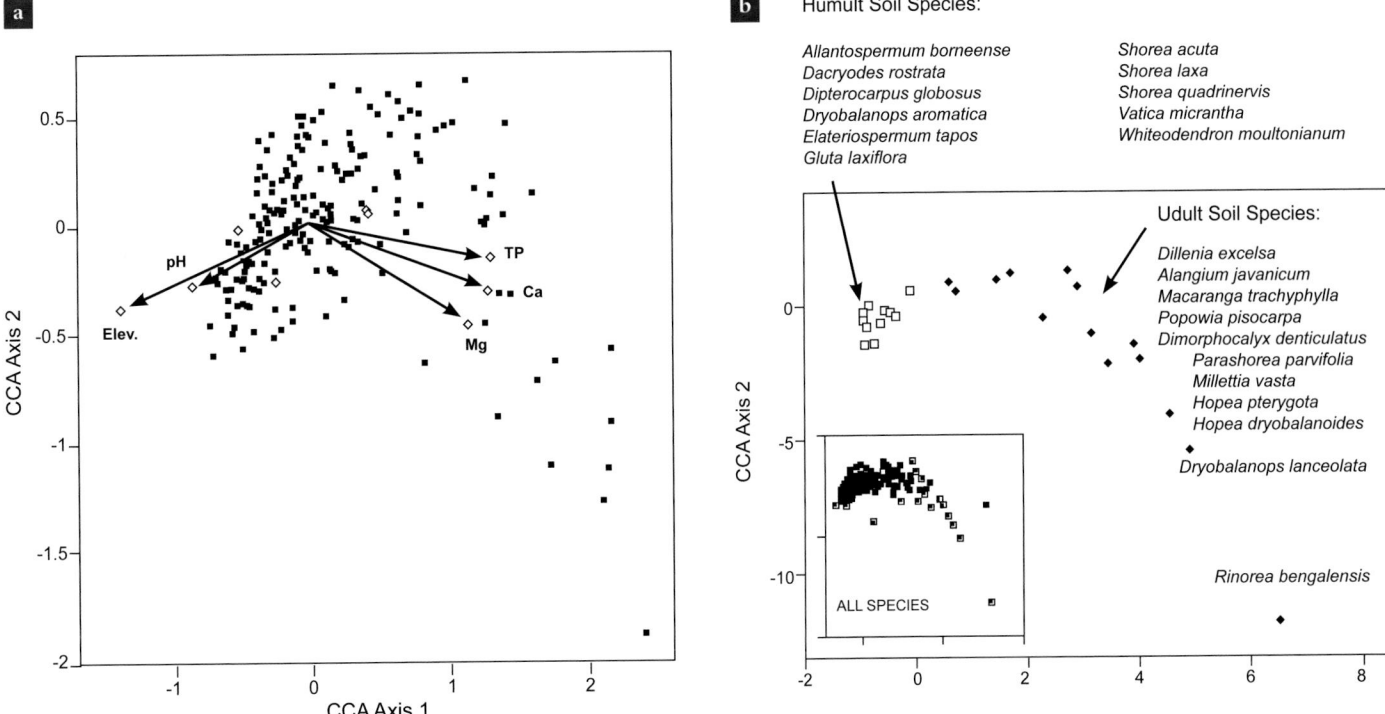

Fig. 3.6 Ordinations (canonical correspondence analysis (CCA)) of 200, 0.25 ha plots within the Lambir CTFS plot, based on their 776 species with at least 50 individuals >1 cm dbh. **a,** trends in physical habitat variables superimposed (soil nutrients represent total concentrations; TP: total phosphorus). Note that the plot points follow two trends: a dense cluster trending with pH and elevation, and a diffuse cloud trending with soils nutrients. Such two-armed distributions imply that two semi-independent gradients in variation have emerged, representing two floristic associations. **b,** ordination of their species based on the plots in which they occur. Selected species and (inset) all species. The horse-shoe pattern emerges again, here to distinguish species (and their floristic associations) on the two main soils. (From Davies *et al.* 2005)

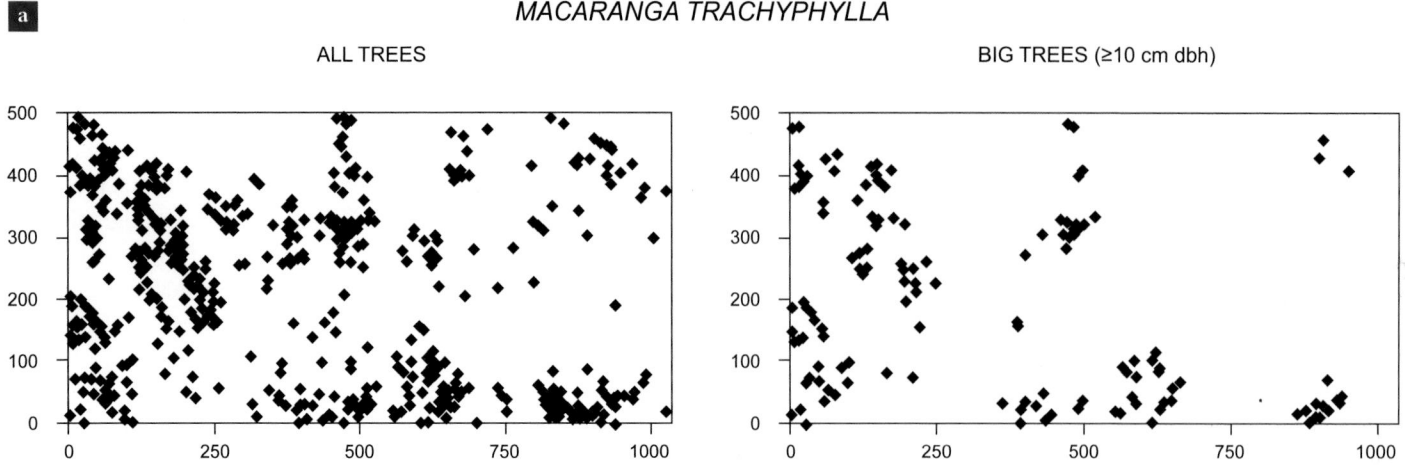

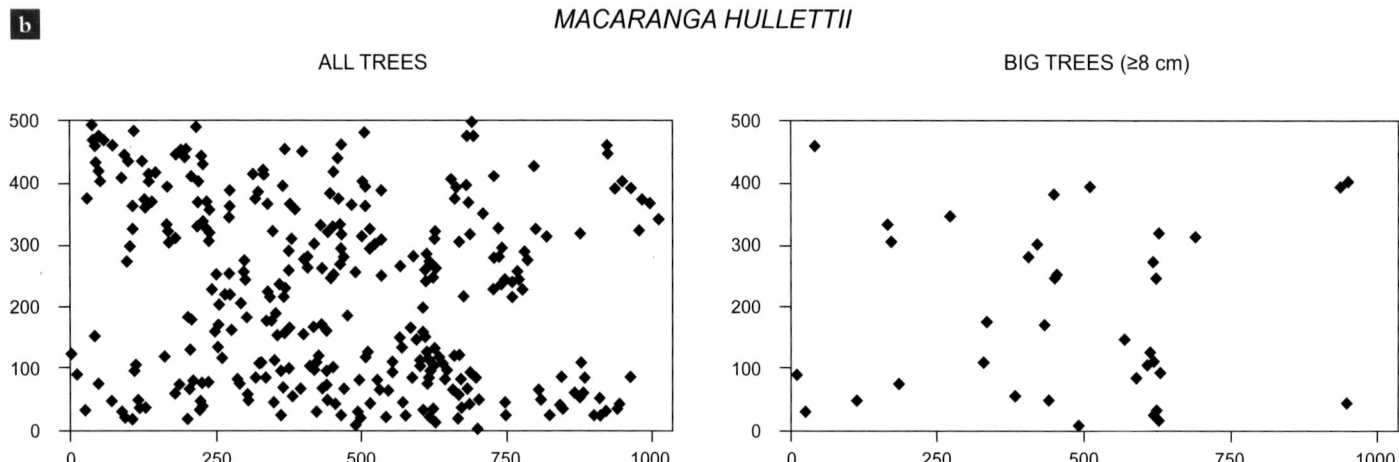

Fig. 3.7 Contrasting mature tree distributions among pioneer *Macaranga* in the Lambir CTFS plot. Smaller, juvenile, trees have broader soil ranges than larger, mature trees which are more restricted to specific soils implying selective mortality. **a,** *M. trachyphylla,* a species of clay loams; **b,** *M. hullettii,* a species of humult yellow sandy soils. (After Davies *et al.* 1998)

1.1.1. *Technical challenges in measuring soil nutrients*

The evaluation of soil–plant interdependencies has been controversial. This is because direct measurements of nutrient uptake, in which radioactively labelled nutrients are traced, are time-consuming and cannot be carried out on a routine basis over a landscape. Correlation of adequate samples of nutrient-rich tissues, such as fresh leaves, with soil nutrient assays has rarely been done. Most importantly, plant ecophysiologists customarily focus on 'available' nutrients, that is, nutrients that are immediately available to roots in solution in the soil.[26] Analyses of soil samples are snapshots, though, not measures of

fluxes (Chapter 1). The nutrients in humus are derived from the litter dropped by the trees themselves, only secondarily from the physical habitat. To establish whether forest species composition varies with soil nutrients, the nutrients derived from the rock substrate and conserved in the clay lattice and sesquioxides in mineral soil must be evaluated because it is they that provide the spatially variable subsidy to nutrients circulating within the ecosystem. The total nutrients in a soil can be estimated by drastic extraction with hydrofluoric acid, but this is expensive. Sarawak soil chemist John Bailey[27] used the cheaper concentrated hydrochloric acid. This method extracts the crucial element phosphorus less completely (Chapter 1), but nevertheless provides a reasonable estimate of the nutrients (including phosphorus) available from mineral soil over the lifetime of a tree. This measure, called *reserve* nutrients, had become standard for tropical tree crops in Brunei and Sarawak, and was used in all our floristic studies.

Importantly, these reserve soil nutrient levels are estimates of the mineral soil totals, and therefore only rough approximations of the subsidy from the substrate. Nutrient *flows* are difficult to measure directly. Samples analysed for exchangeable nutrients, always low in soils of the perhumid tropics, would likely give similar results, but require more intensive sampling to achieve comparable accuracy (Chapter 1).

1.1.2. *Topography mediates water stress*

Topography, by influencing slope and position on a hillside, and soil depth is the universal mediator of local variation in the soil's capacity to retain water. Both desiccation of litter and surface raw humus, and waterlogging by excess rainfall or run-off, impede decomposition and therefore nutrient release. Soil hydrological and nutrient characteristics therefore generally covary at the local scale, as exemplified by the spatial distribution of key species of Peninsular Malaysian hill dipterocarp forest (Fig. 3.8; see Chapter 4). The relative importance of soil water relations and nutrients in floristic diversification across the landscape, therefore, is not easy to tease apart. But the soil nutrient store is ultimately mediated by the presence or absence of rock decomposition and nutrient release from the geological substrate beneath. The mineral composition of this substrate is, of course, independent of the hydrological regime.

But mineral soil nutrient concentrations covary with topography. Level sites are less liable to surface erosion and soil truncation, and consequently are more liable to nutrient leaching. Soils, if freely draining, may accumulate more deeply there though, providing a larger volume for roots to occupy. Narrow ridges are supported by harder, up-tipped strata, usually sandstone in sedimentary formations, often set in a landscape dominated by shale and clay. Such ridges dominate the geomorphologically young terrain of inland Borneo and contrast with the broad and ancient ridges of Peninsular Malaysia. They bear shallow and, in the case of sandstone, low-nutrient humult soils. Their flora is distinctive but merges with *kerangas* where the sands are leached white. Broad, geomorphologically mature ridges may support the deepest soils on a hill.

Symington had recognized floristically distinct MDF on the coastal hills of Peninsular Malaysia, irrespective of geology, and on the ridges of the inland granite mountains between 300–800 m (Chapter 4). The coastal hill MDF association covers the larger Peninsular offshore islands, such as Tioman in the South China Sea. On the evidence of dipterocarp distributions, coastal hill MDF once also dominated the forests of the Anambas group and the Riau and Lingga archipelagos in the straits of Malacca.

These two Peninsular associations occur in habitats with similarly drought-prone soils but differing topographic profiles. They share characteristic species that are absent from

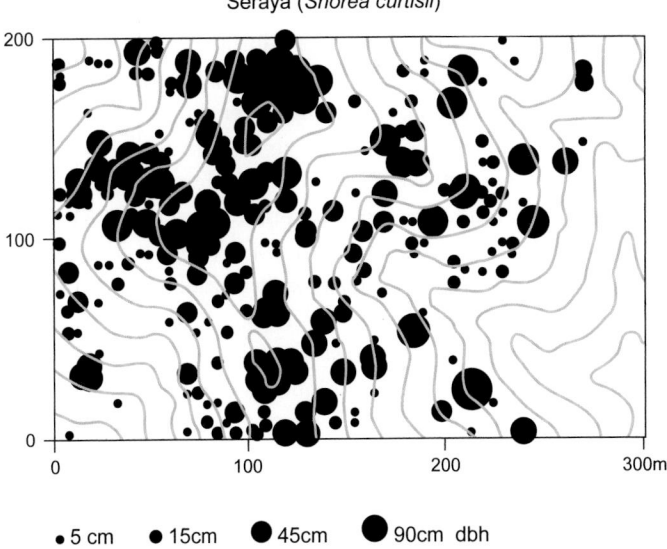

Seraya (*Shorea curtisii*)

• 5 cm   ● 15cm   ● 45cm   ● 90cm dbh

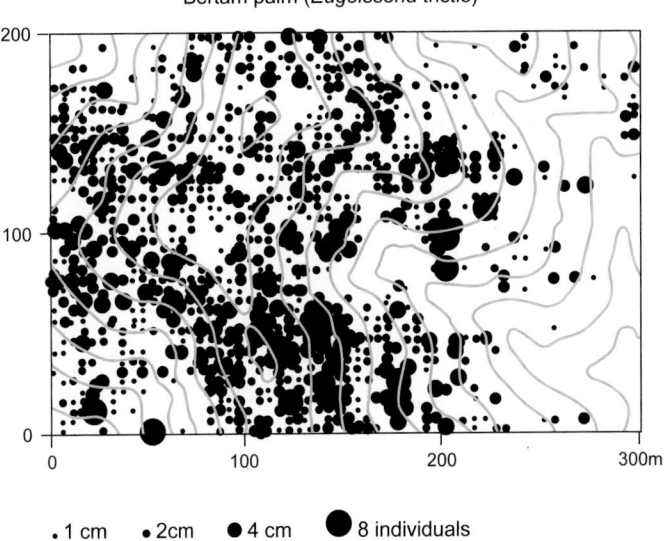

Bertam palm (*Eugeissona tristis*)

. 1 cm   • 2cm   ● 4 cm   ● 8 individuals

Fig. 3.8 Distribution of seraya, *Shorea curtisii* (above), and the stemless bertam palm *Eugeissona tristis* (below) on an edaphically dry granite ridge and spurs in hill MDF at Semengkok Forest Reserve, Selangor, in contiguous 5 x 5 m subplots; circles indicate diameter size. Both are concentrated on the same convex ridge and spur surfaces. Bertam creates a *moderately* shady understorey in which seraya seedlings have relatively high survival therefore competitive advantage over light-demanding meranti *Shorea,* but bertam increases in density following felling, impeding young seraya regeneration. (From Niiyama *et al.* 1999a)

other MDF associations.[28] Surface carbon/nitrogen ratios in each of these soils are high because they bear thin patchy horizons of acid raw humus, in which white fungal mycelia bind the fallen leaves before they become comminuted (Plate 1.11d). Coastal hills and islands are more drought-prone than inland (Chapter 1), while the upland ridges lack nocturnal mist and are also prone to drought during the breezy northeast monsoon. Indeed, soil surface desiccation may be at least partially responsible for retarding litter breakdown rates[29] (Chapter 1).

However, several of the species characteristic of MDF on such coastal hills and inland granite ridges of Peninsular Malaysia are also found in north-west Borneo (Table 4.2). There they are concentrated on deep yellow sandy humult soils. Localities include Andulau Forest Reserve in Brunei,[30] while a few occur in isolated inland localities in *kerangas* and MDF on sandy soils in southern Kalimantan (Chapter 6). Sarawak forestry technicians and I set out 105 plots of 1.5 acres (0.6 ha) each through the MDF of Sarawak from Miri to Kuching (Fig. 3.9). Twelve of these plots at three sites, which were made permanent and periodically re-censused, have been referred to in Chapter 2. Among the 13 plot sites, B and F on shallow yellow sandy humult soils were in more coastal locations, and G, K, L, M and N on mostly humult soils further inland. The latter are

physically close to early Holocene coastlines, as are the sites in southern Kalimantan – now often behind peat swamps – and all are below 300 m altitude. Families better represented in these forests than elsewhere in MDF, and reflected in the plots, include Myrtaceae, Guttiferae, Burseraceae and Ebenaceae. *Shorea curtisii*, the dominant species of the granite ridges of Peninsular Malaysia,[31] occurs in these lowland localities, but is very rare on Bornean ridges, so far being known there only from Brunei. It never dominates in Bornean forests. Many Bornean ridges share similar soils, but none experience the Peninsula's

Fig. 3.9 Site locations of 105, 0.6 ha, plots in Sarawak MDF. Number of plots at each site in parentheses. Sites: A1: Carapa Pila; A2: Ulu Mujong, base of Hose Mountains, Balleh; B: Gunung Santubong; C: Bukit Mersing, Anap; D: Ulu Dapoi, Tinjar; E: Bok Tisam forest, lower Tinjar; F: Bako National Park, G: Bukit Iju, Arip, Balingian; H: Bukit Raya, Kapit, Rejang; J: Ulu Bakong, Tinjar; K: Lambir National Park, north slope; L: Lambir National Park, south slope; M: Nyabau forest, Bintulu; N: Segan forest, Bintulu. (After Potts *et al.* 2002)

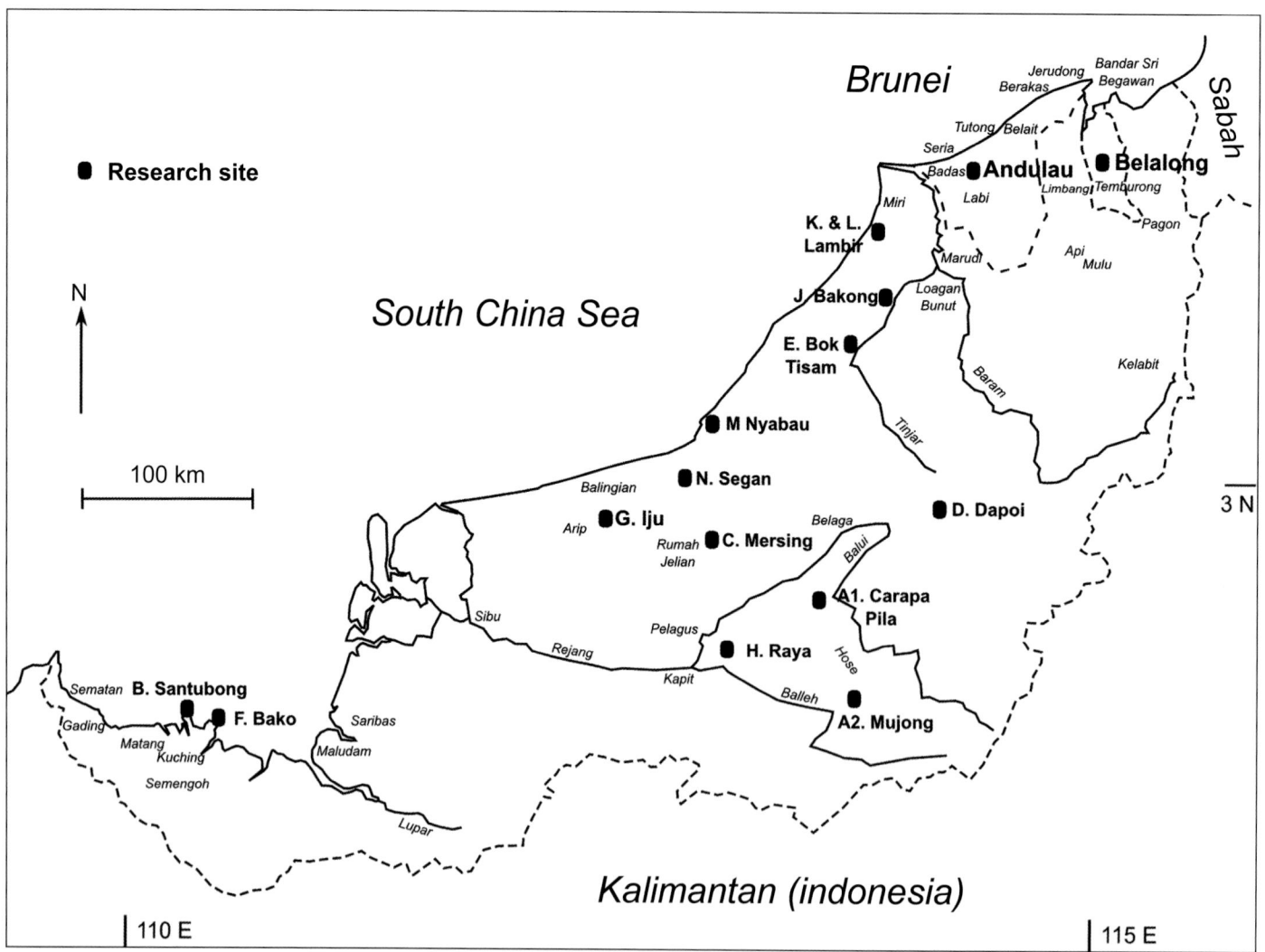

Plate 3.3 The giant epiphytic orchid of lowland forests, *Grammatophyllum speciosum*. (T. Laman)

dry northeasterly winds off Indo-China. The presence of a raw humus surface horizon in all these soils, albeit shallow on the granite ridges, will influence both the physical and chemical environment for seedling establishment and may discriminate between potential mycorrhizal symbionts (Chapter 5).

1.1.3. *Influences of nutrients and of topography remain difficult to distinguish*

It is impossible, from site comparisons alone, to conclude whether nutrients or soil water economy is the principal mediator of species distributions. Frequency of water stress and the topography across which it varies, by influencing canopy height and structure, influences subcanopy light (which is also influenced by lateral exposure along ridges), thereby influencing species composition. Mark Ashton, his students and colleagues demonstrated this in the case of dipterocarp composition on the steep topography at Sinharaja, where soil nutrient variation

is minimal.[32] In Chapter 2, I suggested that forest canopy height is governed by intensity and frequency of soil water stress in windless climates, and described how low-stature, even-canopied forests (in other words, forests with the structure of *kerangas*) can occur on a variety of shallow soils. Nevertheless, the *floristic composition* of such short-statured MDF on shallow yellow-red soils correlates more with their soil nutrient status than with their structure. *Kerangas* on humic podsols supports a distinct flora, which includes many species confined to that habitat[33] (Plate 3.5). Of the three floristic groupings that David Newbery obtained from Brünig's *kerangas* plots in north-west Borneo, one comprised MDF elements[34] (Fig. 3.10). Similarly, Kartawinata found that one of his samples, identified as *kerangas* on its structural characteristics, was floristically MDF.[35] Along the Brunei-Sarawak frontier range in the Belait headwaters, the narrow Miocene sandstone crests bear shallow yellow humult sandy soils. Their forest is of *kerangas* stature but, except where

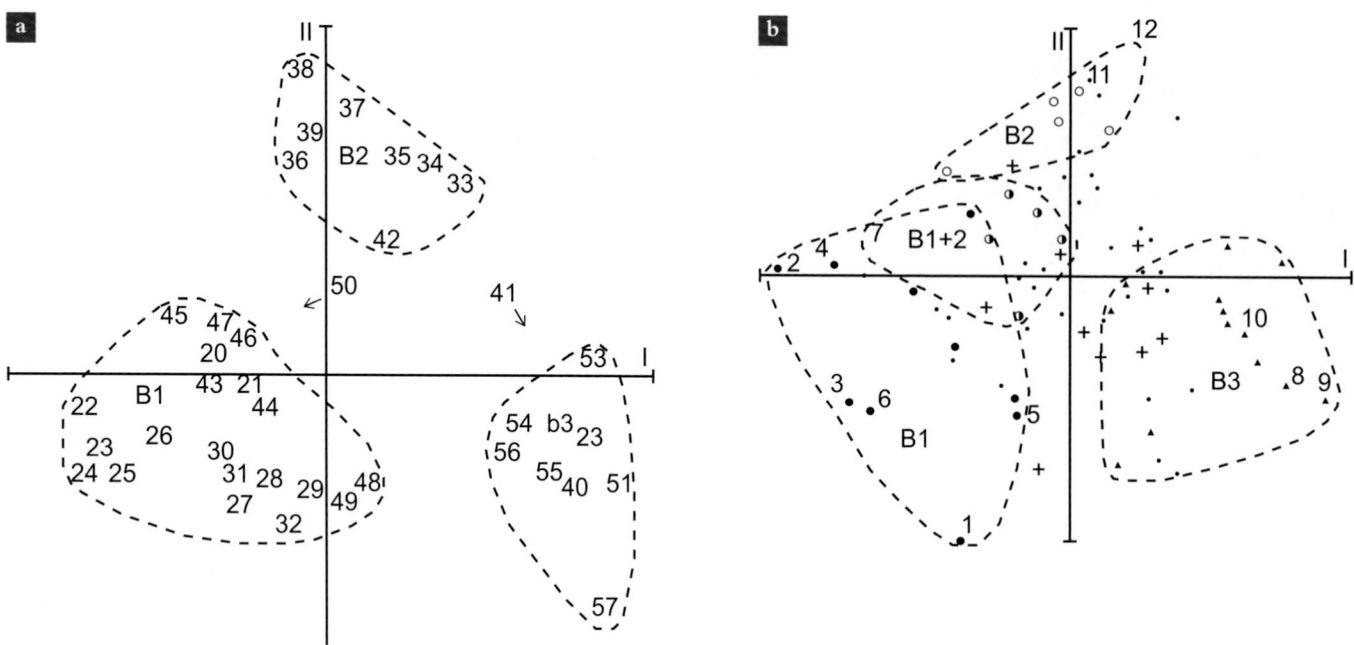

Fig. 3.10 Principal components ordinations of E.F. Brünig's plots in north-west Borneo *kerangas*, based on the 40 most abundant species in 38, ¼ ha plots. **a**, plot ordination: plots form three distinct clusters; **b**, species ordination, which reveals, from separate data on the ecological ranges of their species, that B1 clusters represent *kerangas* on deep podsols mostly on Pleistocene raised beaches, B3 cluster on shallow soils of variable drainage on sandstone plateaux (B3); while B2 represents ecotonal sites to MDF. (From Newbery 1991)

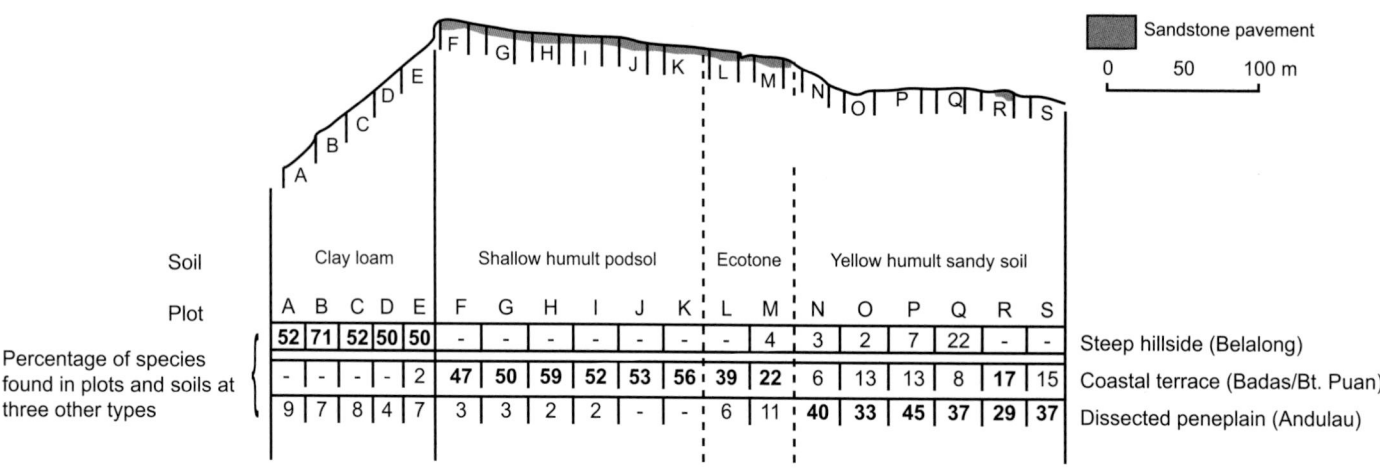

Soil	Clay loam					Shallow humult podsol						Ecotone		Yellow humult sandy soil						
Plot	A	B	C	D	E	F	G	H	I	J	K	L	M	N	O	P	Q	R	S	
Percentage of species found in plots and soils at three other types	52	71	52	50	50	-	-	-	-	-	-	-	4	3	2	7	22	-	-	Steep hillside (Belalong)
	-	-	-	-	2	47	50	59	52	53	56	39	22	6	13	13	8	17	15	Coastal terrace (Badas/Bt. Puan)
	9	7	8	4	7	3	3	2	2	-	-	6	11	40	33	45	37	29	37	Dissected peneplain (Andulau)

Fig. 3.11 Faithfulness of species to soils at Bukit Patoi, Brunei: transect across a sandstone plateau and shale scarp; with the percentage of species along the transect represented on similar soils at three other sites in each of which one soil type dominates. (D. Glick del. from P.S. Ashton 1964b)

the rock is exposed on knolls, is floristically most similar to that on deep humult sandy soil in Andulau forest to the north where the canopy reaches 70 m!

I have often observed that, where substrates and overlying soils change abruptly, the floristic transition from *kerangas* to MDF is immediate, whether or not there is any change in canopy stature[36] (Fig. 3.11). Eberhard Brünig,[37] a forester who apparently defined *kerangas* using structural criteria, censused a number of putative *kerangas* plots whose composition was at least partially that of MDF. Whether they were indeed MDF or were (as appears more likely) ecotonal, cannot in retrospect be ascertained.

Generally, the floristic ecotone from MDF to *kerangas* is

narrow and correlates precisely with the transition from yellow to white mineral soils. Yellow coloration is due to iron oxide coatings on the sand that, in these soils of low clay concentration, are critical in attracting phosphorus anions. These are tightly bound but available within the lifespan of a tree, and their release is facilitated by secretions from mycorrhizal hyphae. In contrast, the white mineral soil of tropical humic podsols is generally bereft of nutrients, leaving the vegetation entirely dependent on nutrients recycled from its litter. The ecosystem on these humic podsols is therefore fully oligotrophic, as in a peat swamp. Coomes and Grubb[38] argued that intensive below-ground competition for nutrients on these soils results in the high density of fine roots that are ubiquitous albeit confined to their deep surface raw humus horizon (Chapter 1), and the consequent allocation of relatively more biomass below than above ground (Chapter 2).

There is nevertheless floristic as well as structural variation within *kerangas*.[39] This has been somewhat obscured in the literature by the inclusion of MDF plots, sharing the short stature and structure of *kerangas*, in analyses. Excluding these, and examining only sites on humic podsols, Newbery and Brünig's ordinations showed a floristic correlation with soil water, rather than with nutrients[40] (Fig. 3.10). This is also the case in the floristic changes from the margins to the summits of peat swamp domes.[41] It is consistent with the way floristic composition has been observed to vary with microtopography and with raw humus depth in *kerangas* in Central Kalimantan.[42] But this is hardly surprising, considering the exceptionally low nutrient levels in *kerangas* mineral soil.

Floristic comparison of swamp and upland forest is also revealing. Periodically flooded alluvium at the edge of the Borneo swamps is generally rich in clay; soils are only oxygenated near the surface. However, the tree flora differs little from that on freely draining upland clays except that it has a higher proportion of early successional species. This is commensurate with the forest's shallow rooting habit and consequently greater dynamism through windthrow (Chapter 2). Beyond the reach

Fig. 3.12 Distribution of dipterocarp trees >9.7 cm dbh common to two sites, Kuala Belalong, Brunei and Lambir, Sarawak: **a,** within the 50 plots of the Belalong ordination (see Fig. 3.1), where all soils are clay loams but the species are concentrated on ridges; **b,** in the Lambir 52 ha plot, where the species are concentrated on slopes on clay loam and sandy clay loams, the ridges being sandy. At Kuala Belalong, the species avoid the less well drained slopes, whereas at Lambir they avoid the sandiest soils: they share the soils of the two sites, but not the topography. (Data from P.S. Ashton 1964b, and H.S. Lee *et al.* 2002)

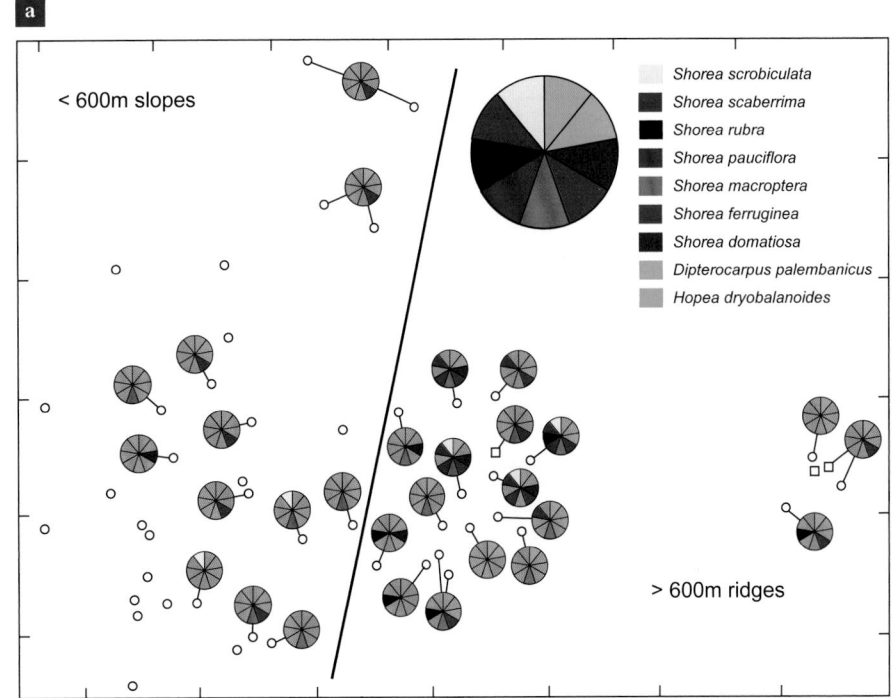

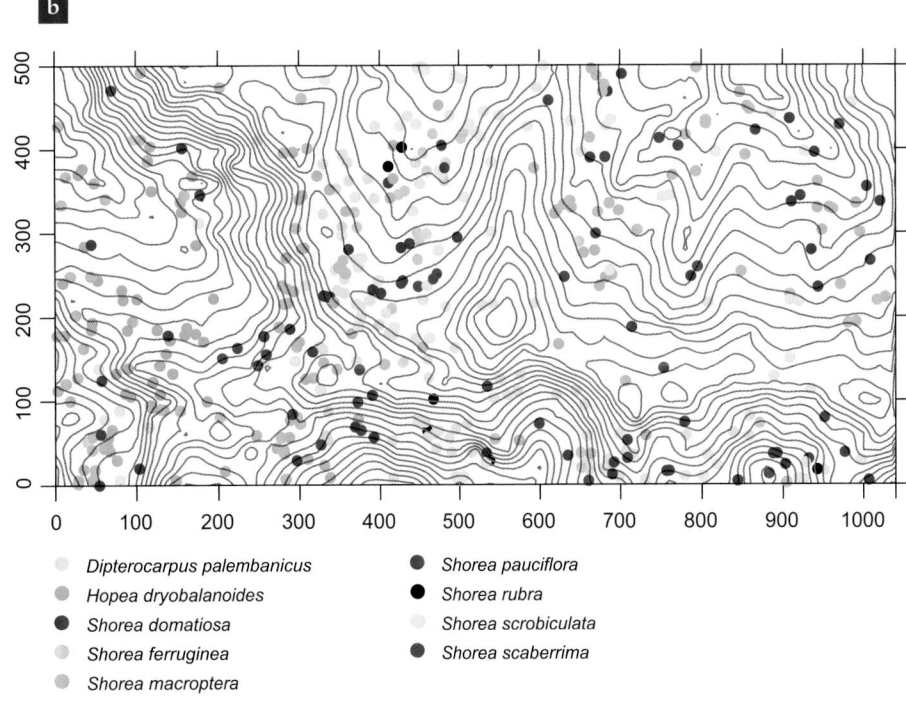

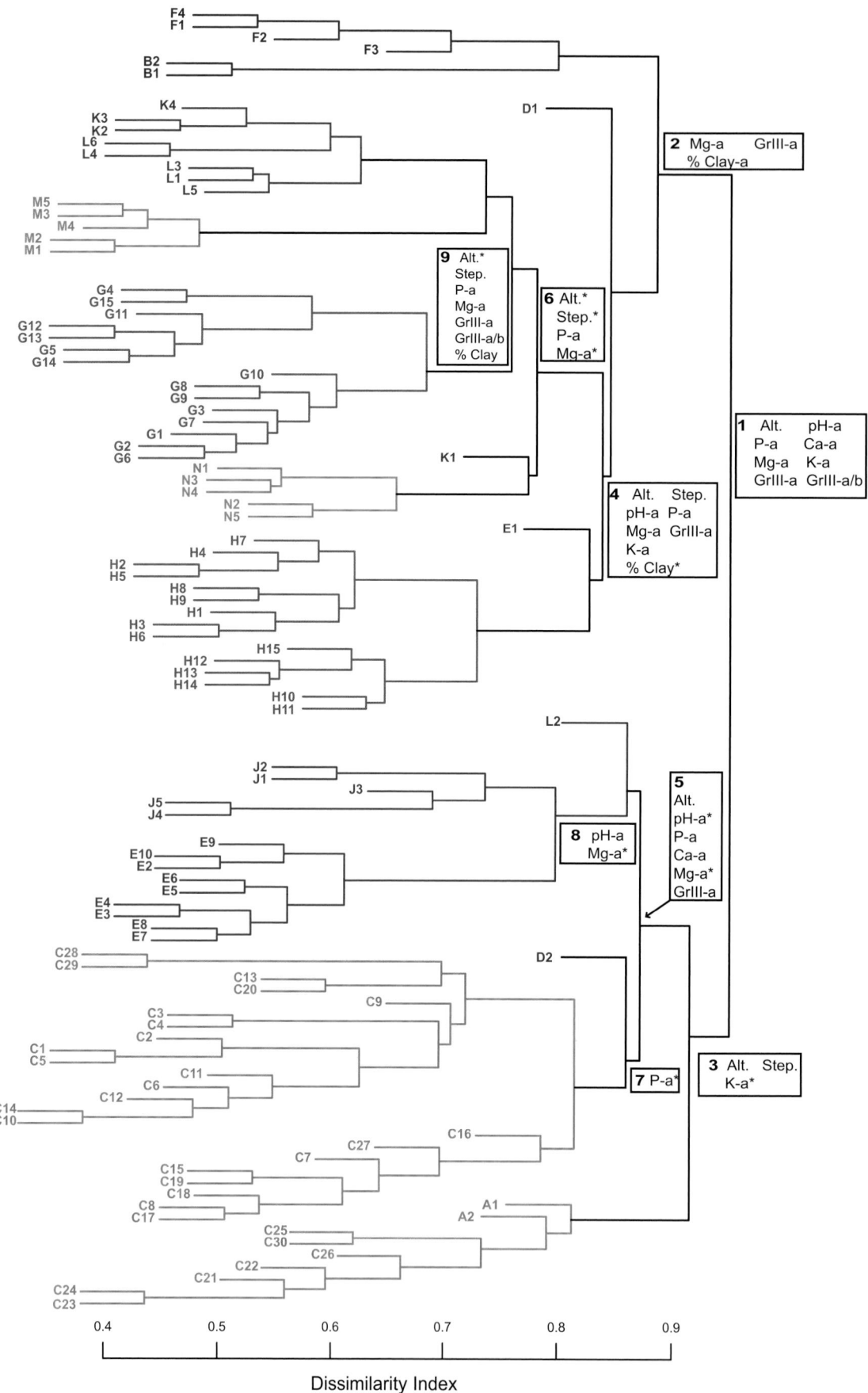

of floodwater, shallow peat forms with a permanently high water table, and the species change completely. Windthrow frequency declines due to binding of surface roots of stands within the surface raw humus; but, when squalls do break open this mat, domino-like lines of fallen trees can extend for hundreds of metres (Plate 3.9i). The floristic association near the peat swamp margin includes a suite of characteristic species (see section 2.4.3 below). Further in from the river again, a dense peat mat begins to form above the permanent water table. The tree flora continues to change[43] and *kerangas* elements appear. These increasingly dominate towards the peat dome. The dome itself experiences the greatest fluctuation in water table, even occasionally experiencing fire (see section 2.4.3 below). Under such conditions, water relations may again become influential.

Exceptions to the topography-mediated soil-water and nutrient correlations occur when ridges and slopes overlying different substrates, therefore soils, are compared. Sometimes strata with different mineralogies, and whose soils differ in clay/sand ratios, may be interbedded along the same ridge. In one example, several tree species found on ridges at Kuala Belalong are instead concentrated on slopes and in valleys 100 km to the west within the Lambir National Park 52 ha CTFS plot (Fig. 3.12). Similarly, *beraliya dun* (*Shorea worthingtonii*) is a ridge species in Sri Lankan MDF at Sinharaja[44], but occurs on yellow coarse sandy clay soils in association with species characteristic of such soils (including *Stemonoporus kanneliyensis*, *S. bullatus*, *Cotylelobium scabriusculum* and *Axinandra zeylanica*) on certain hillsides in Kanneliya Forest Reserve (south-west Sri Lanka). Adjacent slopes with red-brown clay loams bear the consistently slope-associated species *Shorea congestiflora* and *Dipterocarpus zeylanicus*; but also fertile clay loam specialists such as *Trichadenia zeylanica*. This suggests that the ecological range of these species is also primarily mediated by soil nutrients. *Mesua ferrea*, in contrast, is a dominant strictly ridge species in both of these forests and elsewhere in perhumid south-west Sri Lanka, implying that the soil water economy is the principal mediator of this species distribution.

Matthew Potts[45] found good correlations between Sarawak MDF plot groupings based on floristic similarity and soil reserve nutrient concentrations, including group III elements (iron,

Fig. 3.13 (left) Classification dendrogram of 105 Sarawak MDF plots. Correlated topography (step=steep), soil mean HCl-extractable Group III elements (Fe, Al) and nutrient ions at 20–30 cm indicated. The first division separates plots with humult sandy or sandy clay soils (above), from udult clay soils. Colours indicate geographical locations within Sarawak: blue, south-west; green, west central; yellow, east central; red, north-east (see Fig. 3.9). With the exception of plots from sites west of the Lupar river (B and F), the dendrogram classifies sites by habitat rather than geography, though plots tend to remain grouped within their sites. (After Potts *et al.* 2002).

aluminium) and especially phosphorus and magnesium which are components of the molecules involved in photosynthesis reactions (Fig. 3.13). Potts' analysis showed a primary division between samples on clay-rich soils and those on sandy and sandy clay soils. Topographic variation among plots within site clusters was poorly discriminated. More precisely, the division distinguished plots with at least a shallow surface layer of acid raw humus supporting a distinct mat of fine roots, from those generally lacking this mat and with litter directly overlying thin humus and mineral soil hardly discoloured by it. The dense mat of fine roots in the shallow surface raw humus on these yellow sandy and sandy clay soils represents the onset of a closed cycle of nutrients between the living forest and the organic soil surface horizon. In MDF, however, there are always additional nutrients to be taken up from the mineral soil, and therefore indirectly from the rock substrate.

Potts' dendrogram shows higher correlation of plot clusters with soil nutrients on the lower-nutrient humult than higher-nutrient udult soils where the 30 plots at Bukit Mersing (C), with exceptionally high but also variable nutrient levels, cluster together at a higher level of the dissimilarity index gradient than the humult plots.[46] Mantel tests were carried out to test the relative strength of correlations between plot scores along the primary axes of a Principal Components Analysis, representing their floristic relationships, and habitat (soil nutrients) and between-plot distance among all 105 Sarawak MDF plots. Habitat was seen to be the dominant influence over their distance apart. The same test was carried out on groups of 15 of these plots at a low (Iju), and high (Mersing) nutrient site (Table 3.3), where the relative strength among floristics, habitat and plot distance apart was assessed in relation to HCl-extractable phosphorus and magnesium, and topography. At Mersing, spatial distance was more highly correlated with distance apart (0.269) than either soil nutrients (0.190) or topography (0.030), implying a dominant influence of limited seed dispersal. But at Iju, where nutrient levels and variability were lowest, nutrients as a whole were equivalent to distance while reserve magnesium had almost double the correlation of distance; topography was least influential here also. Nevertheless Rahayu Hj. Sukri[47] found, at Kuala Belalong (Brunei), a topographically steep site with udult clay loam soils of intermediate nutrient levels over argyllic (shale) substrate, higher floristic correlation with nutrients than topography, although soil carbon, nitrogen and phosphorus covaried. The floristic influence of topography may be enhanced by the greater difference in levels of canopy disturbance on slopes from ridges in landscapes dominated by clay soils, where slopes are usually steeper and less stable.[48] Bukit Mersing, unlike Bukit Iju, bears soils at low altitudes lacking surface raw humus even on ridges, and relatively higher mineral soil reserve levels. On udult clay loams over argyllic sediments and gentle topography at Danum

(Sabah),[49] Gibbons and Newbery found correlations between the topographic distributions of the subcanopy Euphorbiaceous trees *Dimorphocalyx muricatus* and *Mallotus wrayi,* and their root characteristics and leaf osmotic potential, but not with soil nutrient levels. They interpreted these differences as adaptations to different frequencies of soil water stress.

Table 3.3 Mantel tests of the relative strength of correlation between floristic similarity and habitat (soils nutrients, slope), versus distance apart; at ten Sarawak MDF sites and two sites separately: Bukit Mersing (C, basalt) and Bukit Iju (G, rhyolite). All sites with >4, 0.6 ha plots, trees >9.7 cm dbh. *: $p<0.05$; **: $p<0.01$. (From Potts *et al.* 2002). Mantel statistics are above, partial Mantel below the diagonal. (Partial Mantel statistics were calculated with respect to floristics, habitat and distance.)

The Gunatillekes and their colleagues had obtained similar correlations of tree flora with topography at Sinharaja forest (south-west Sri Lanka), in spite of lower soil nutrient levels documented there than in similar Sarawak clay soils.[50] (However, wet rice is grown in nearby valleys there, implying higher fertility than in Borneo where this is not the case.) Topography and soil fertility are only loosely correlated there. They then compared the response of potted seedlings to added phosphorus and magnesium of eight of their *Shorea* (sect. *Doona*) species.[51] Both nutrients independently led to increase in dry mass yield. However, rate of yield increase declined when phosphorus was held constant and magnesium increased; while height and stem mass ratio increased when magnesium

Data	Floristics	Habitat	Distance	Floristics	Habitat	Distance
a. All 105 plots						
		Habitat PCA x axis			**Habitat PCA y axis**	
Floristics	-	0.651**	0.058	-	0.089**	0.058
Habitat	0.651**	-	0.168	0.083**	-	-0.127
Distance	-0.068**	0.172	-	0.0470	-0.123	-
b. Lowlands Bukit Mersing (site C, high nutrients)						
		Habitat PCA x axis			**Topography**	
Floristics	-	0.191**	0.269**	-	0.059**	0.269**
Habitat	0.190*	-	0.306	0.030**	-	0.113
Distance	0.269	-0.022	-	0.265**	0.101	-
		Residual Mg (ppm)			**Residual P (ppm)**	
Floristics	-	0.184*	0.269**	-	0.090*	0.269**
Habitat	0.200*	-	-0.031**	0.115*	-	-0.077**
Distance	0.280**	-0.085	-	0.278**	-0.106**	-
c. Bukit Iju (site G, relatively low nutrients)						
		Habitat PCA x axis			**Topography**	
Floristics	-	0.223*	0.271**	-	0.106**	0.271**
Habitat	0.221*	-	0.041	0.121**	-	-0.038
Distance	0.268**	-0.020	-	0.277**	-0.070	-
		Residual Mg (ppm)			**Residual P (ppm)**	
Floristics	-	0.420**	0.271**	-	0.130**	0.271**
Habitat	0.405**	-	0.121*	0.122**	-	0.049*
Distance	0.244*	0.009	-	0.267**	0.014	-

was held constant and phosphorus increased. Dry mass increase was greatest for the species found in nature on the moister lower slopes. No change in rank order of species response was observed in the experiments. Again, it was nevertheless concluded that soil water economy rather than nutrients determined the ecological distributions.[52]

Mantel tests showed that dispersal constraints were the dominant influence on species composition on the high nutrient basalt-derived soils at Mersing, whereas nutrients were the dominant influence on low nutrient rhyolite-derived soils at Iju (Table 3.3). There appears to be a threshold above which mineral soil nutrients no longer constrain competitive interactions between species, while evidence from Kuala Belalong, Danum and field observation suggest that soil water deficits may continue to do so. Above this threshold, which in Borneo appears to be at about 250 ppm reserve phosphorus and 800 ppm magnesium, floristic composition of forest varies less, and mainly in relation to topography. Below the threshold composition varies more, and mainly in relation to some soil nutrients. An alternative explanation, awaiting examination, may be that many lower nutrient, sandy soils of north-west Borneo are exceptional in their high magnesium/calcium ratio. It has been suggested that this may lead to competitive advantage of plants that have vacuolar carboxylates that may

enable differential accumulation of calcium against magnesium and other divalent ions.[53] Higher floristic differentiation and endemism on these soils may correlate with variation in that ratio.

Fertile clay soils, especially on slopes, are rich in large vines. This may be due in part to the relatively high nutrient levels and water availability, favourable to the rapid growth of the pioneer vines, in part to the relatively high level of canopy disturbance and large gaps on the unstable slopes.[54] Ground herbs also vary in species composition and abundance with the topography,[55] partly due to their shallow rooting on often poorly oxygenated clay soils downslope. In spite of deep shade beneath a dense canopy, these soils support the richest herbaceous flora in MDF, probably both on account of their areas of exposed mineral soil and on account of relatively high surface soil nitrogen content, endowing leaves with higher shade tolerance.[56]

1.1.4. *In summary: nutrients versus water*

At very low levels, nutrients appear to have little if any influence on floristic variation, which is instead mediated by soil water economy. As they increase in total availability, nutrients become correlated with increasing spatial diversity (that is, β-diversity) in tree species composition. Above a certain nutrient threshold, which may vary among regions, the correlation is once again lost. The floristic variation then becomes correlated with topography irrespective of soil nutrient levels, and thus, by inference, with soil water economy once again; while the influence of restricted seed didpersal becomes more apparent.

Except for those plots that were outliers in respect of soil nutrient levels, relative clay content, and presence or absence of surface raw humus, Sarawak plots were closely aggregated

Table 3.4 Correlations between topography and soil variables among 15, 0.6 ha plots at two MDF sites in Sarawak. Concentrated HCl-extractable nutrients from mineral soils at a, 20–30, b, 70–80 cm; Group III elements (Fe and Al): the ratio of their concentrations at 20-30cm (a) and 70-80 cm (b) is used as a measure of leaching: Mersing hill (C, basalt: relatively high nutrient levels) and Iju hill (G, rhyolite: relatively low nutrient levels). *: $p<0.05$; **: $p<0.01$. As altitudinal range is less than 300 m at Mersing and 200 m at Iju, altitude is probably a proxy for leaching. (After Potts *et al.* 2002)

	Residual P (ppm)	Residual Ca (ppm)	Residual Mg (ppm)	Residual K (ppm)	Residual group III elements (ppm)	Group IIIa/b (ppm)
Bukit Mersing, basalt						
Slope or ridge	NS	NS	NS	NS	NS	NS
Altitude	-0.352	-0.162	-0.219	0.105	0.257	0.010
Steepness	-0.114	0.019	0.019	0.057	0.019	0.152
Bukit Iju, rhyolite						
Slope or ridge	NS	NS	**	*	NS	NS
Altitude	-0.486*	0.371*	-0.343	-0.457*	-0.248	-0.095
Steepness	0.314	0.048	-0.010	-0.048	0.295	-0.162

by site in their floristic analysis. As the sites were separated by 20–105 km (mean: 40 km), these aggregations could reflect the influence of dispersal limitations at a more local scale. However, they might also reflect differences in nutrient concentrations between the different sites, since sites were mostly on different rock formations. A useful exception, though, were the Bakong(J) and Bok Tisam(E) plot groups, and plot L2 at Lambir, all of which overlay one formation, the Setap Shale, and are therefore likely to have shared the same soils. Although within 15 km of one another, and alone sharing a major class within the dendrogram, they still separated into distinct final clusters, thereby implying that dispersal limitations do place a significant, if generally secondary, part in determining local floristic differentiation at these spatial scales (Fig. 3.13).

A lower correlation between species composition and soil nutrients on higher nutrient and udult soils in part may explain why researchers in Peninsular Malaysia took so long to recognize that MDF does vary floristically, and that its individual species do have restricted edaphic ranges.[57] No landscape-scale or larger ordination has yet been carried out there, nor have quantitative soil-floristics correlations been attempted at that scale there or in other areas of the Sunda lands. Currently, conclusions can only be inferred from field descriptions in the literature, plus local plot-based research at one site, Pasoh Forest Reserve (Negeri Sembilan).[58] That forest lies on low rounded hills amid broad flat valleys with alluvium sediments

Fig. 3.14 Classification dendrogram, Crawford–Wishart method, of: **a**, MDF in a 100 x 80 m plot at Pasoh research forest, Negeri Sembilan, in which the 447 individual trees >9.7 cm dbh plus their 35 nearest neighbours are the 'plots'; **b**, as **a**, with habitats mapped within the plot; w, pig, we, pig or elephant, wallows; arrows indicate slope; **c**, as **a**, with major dendrogram classes mapped. (a, after P.S. Ashton 1976a)

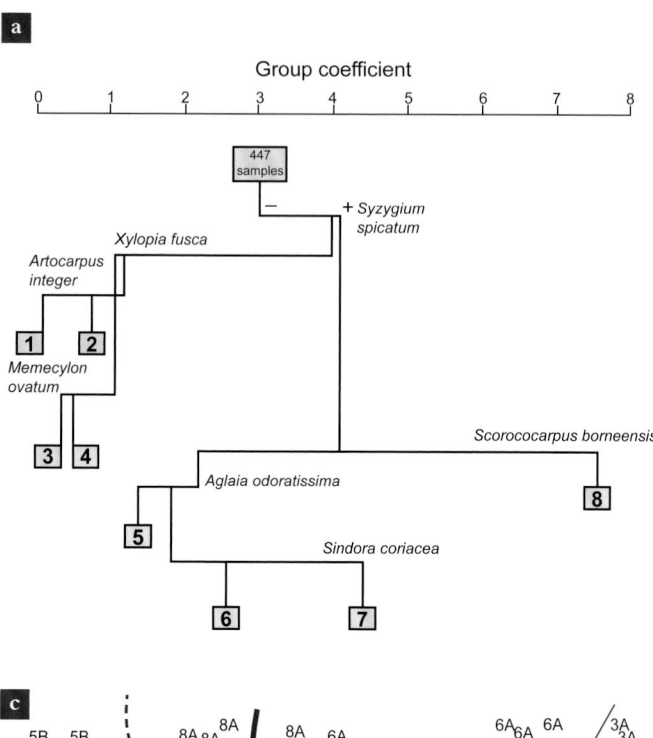

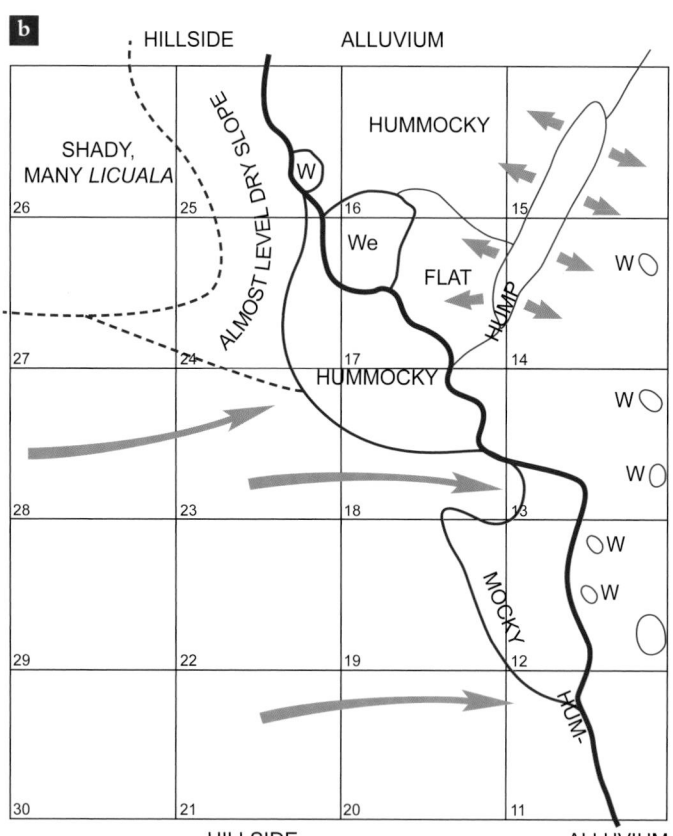

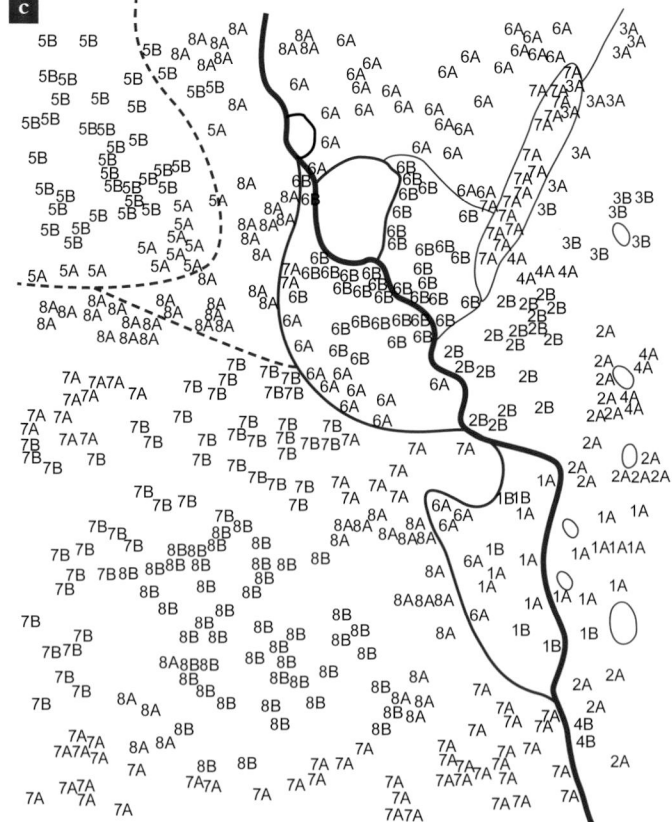

at two levels: (1) an upper level, well drained, rarely flooded and usually sandy but with yellow soils, deposited before the present pluvial period; and (2) the current alluvium, generally flooded at least once a year, bearing clay, and shallow peat in the lowest areas (Fig. 3.14a). Rooting is shallow on the young alluvium, and tree fall gaps are larger and apparently more frequent. Recent analyses have shown tree floristic variation between hills and alluvium, and between the old alluvium, the young alluvium along streams, and the peat.[59] There are sharp *overall* floristic changes at hill bases: surface disturbance by rooting pigs and former elephant wallows are concentrated there, and there are also floristic peculiarities (Fig. 3.14b,c). In the CTFS 50 ha plot at Pasoh, 129 of the 617 species with individuals of more than 1cm dbh are confined to the alluvium. However, all but five of these occur at densities below 1 ha⁻¹, their sample sizes therefore being too small to be confirmatory, while the stream bank species *Saraca declinata* alone exceeds 4 ha⁻¹. The MDF flora of Peninsular Malaysia and Sumatra shows the closest correlation with our H site at Bukit Raya (Ulu Rejang), on Paleocene shales and sandstone, and with the

udult group of sites generally.[60] The soils at the H site bear the thinnest and most patchy raw humus surface horizon among the plot clusters in the humult branch of Potts' dendrogram. Their tree flora most resembles that on the granite and Triassic sediments of Peninsular Malaysia, whereas the udult group flora more resembles that over volcanic substrates there and in Sumatra, and Sumatran soils influenced by volcanic ash. The tree flora of these Sarawak MDF sites has a higher proportion of widespread species and lower proportion of endemics than the other forests on humult soil (see Chapter 6, Table 6.16, also Fig. 6.17).

Thus, three widespread floristic associations can be recognized in Sunda MDF[61] (Fig. 3.15):

1. An association formerly continuous over vast areas of the lowlands, rich in species but with low endemicity, on sandy clay and clay loams of intermediate fertility on a variety of substrates. I have earlier termed this the *Shorea parvifolia* association; it approximates to the red meranti-keruing forest type of Wyatt-Smith.[62]

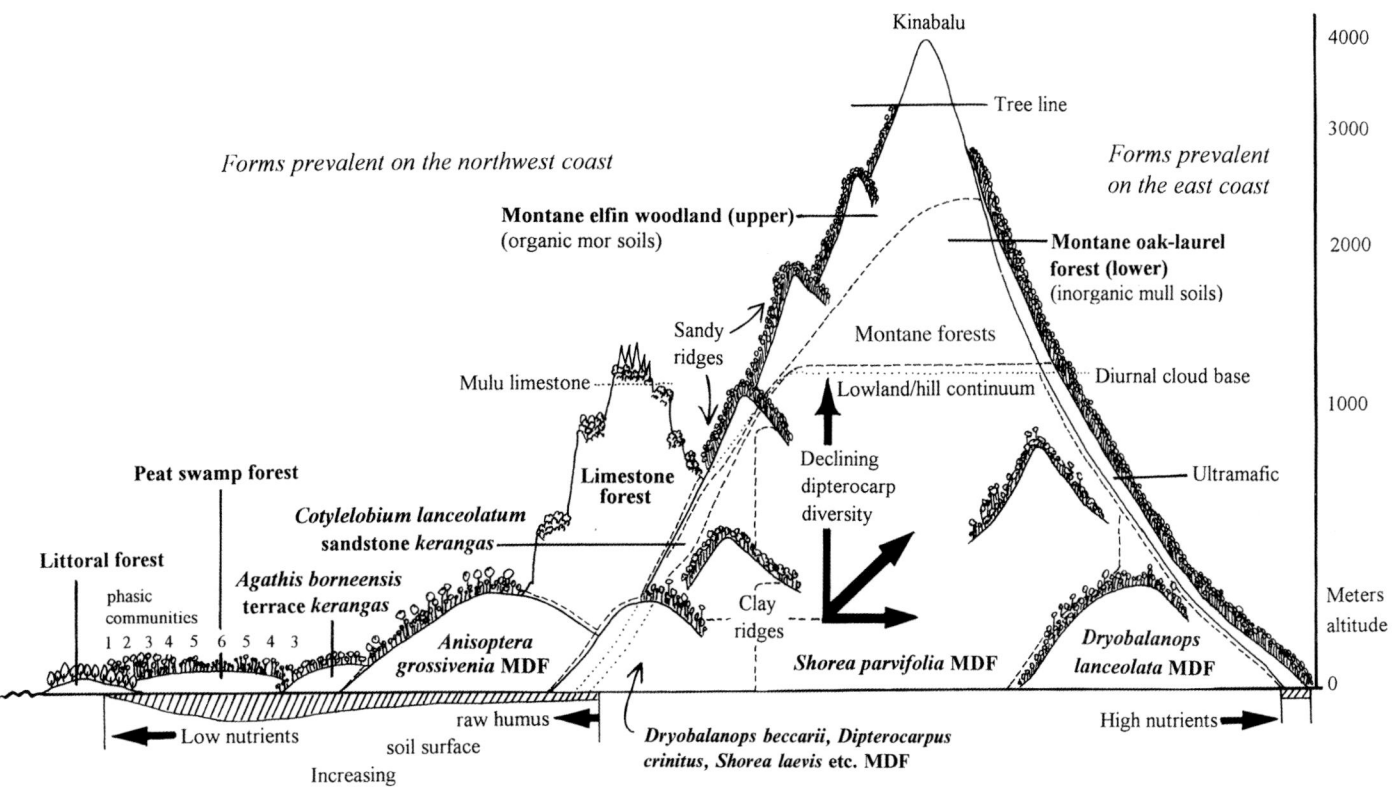

Fig. 3.15 Schematic diagram of the major floristic associations in the forests of northern Borneo. The *Anisoptera grossivenia* MDF is on humic yellow sandy loams, the *Shorea parvifolia* MDF on silty clay loam, the *Dryobalanops lanceolata* MDF on friable clay loam (all as at Lambir), while the *Dryobalanops beccarii* etc. MDF is on sandy clay loams as at Sarawak site H (Fig. 3.9). *S. parvifolia* and *D. beccarii* (as *Dipterocarpus crinitus*) MDF prevailed in the Peninsular Malaysian and Sumatran lowlands, and another analogous to the *D. lanceolata* MDF in Sumatra. The *A. grossivenia* MDF shares species with Peninsular Malaysian coastal MDF and hill *Shorea curtisii* MDF. (From P.S. Ashton 1995b)

2. A second, variably rich, association in which medium to heavy hardwood dipterocarps and the Riau Pocket flora are best represented and show high local to regional endemism. This is found on freely-draining leached siliceous thinly humult sandy clay loams over sandstone or siliceous igneous and volcanic rocks on drier sites. Within Borneo, I have called this the *Anisoptera grossivenia* association; in Peninsular Malaysia, where it is represented by the hill and coastal hill dipterocarp forests of Symington, and in Sumatra, it might appropriately be named after the sister species, *A. curtisii*.

3. An association dominated by light hardwood dipterocarps and relatively poor in species (Chapter 7) but rich in local to regional endemics, especially in Borneo. It is found on friable udult clay loam soils over argyllic and basic volcanic substrates. Within Borneo I have termed this the *Dryobalanops lanceolata* association.

1.1.5. *Differences in nutrient response among species are being tested by comparative observations of performance, and by experiment*

Diameter increment of smaller individuals (more than 9.7 cm diameter), in our groups of four 0.6 ha permanent plots at three sites in MDF on soils of contrasting nutrient status in Sarawak, were found to be correlated with soil nutrient status after five years[63] (Table 2.5). After 20 years, this correlation was becoming apparent in the whole forest stand, including late successional species as might be predicted by the higher representation of species with high wood density on low nutrient soils (Chapter 2). Nitrogen data were not available for the analysis of these, or the 93 other Sarawak plots. The first division of the 105 plots in Potts' analysis correlated with phosphorus, magnesium and calcium (at $P<0.01$) (Fig. 3.13). Among the first four divisions among plots on high nutrient udult clay soils, phosphorus and magnesium were twice correlated, calcium once; among the first four on low nutrient humult soils, magnesium was correlated with all and phosphorus with three, but calcium with none. Phosphorus appears to be a universally important mediator whereas magnesium may be more influential on infertile soils, calcium (therefore by implication nitrogen) on fertile soils.

Experimental attempts to relate performance of individual species to concentrations of exchangeable soil nutrients have inevitably been focused on seedlings and pot experiments. They at first gave inconsistent results.[64] Positive results were obtained in the case of pioneer species. David Burslem and associates, for instance, compared potted seedlings of the pioneers *Melastoma malabathricum* and *Dillenia suffruticosa,* grown in soil from primary coastal hill dipterocarp forest and adjacent degraded soils with *Adinandra dumosa*.[65] Although this study showed no effect of greater nutrient deficiency in the degraded soils, it did show that the effect of phosphorus on growth was far greater than that of other nutrient ions, consistent with the role that phosphorus played in differentiating the Sarawak MDF flora (Fig. 3.13). However, a later experiment[66] showed no such response of phosphorus among climax species. They had also earlier showed that nitrogen addition led to greater production of leaves and greater above-ground biomass relative to root biomass, whereas addition of magnesium had the reverse effect. Calcium is known to increase nitrogen availability through breakdown of litter and soil organic matter.[67] In another pot experiment in which seven pioneer species were grown separately and in mixture, and in which nitrogen, phosphorus, or nitrogen+phosphorus were added, the pioneer shrub *Melastoma malabathricum* out-competed all other species, co-opting most of the added nitrogen.[68]

David Burslem had found limited response of climax tree species' seedlings grown in low-nutrient soil to added phosphorus, yet he found that added magnesium enhanced growth.[69] Lack of clear early results were at least partially explained when Ian Alexander and colleagues, and Wim Smits working in Kalimantan,[70] found that the addition of mycorrhizae led to a greater growth response than addition of phosphorus. Gamage and his associates,[71] using vesicular arbuscular mycorrhizal *Syzygium* seedlings at Sinharaja, found that added phosphorus enhanced both growth and mycorrhizal infection rates in the low-nutrient clay soil there, whereas adding potassium or a combination of potassium, magnesium and phosphorus depressed both. Nevertheless, Peter Palmiotto[72] failed to show a growth response to added phosphorus among reciprocally transplanted seedlings of six climax species – five dipterocarps and a *Swintonia* (Anacardiaceae) – planted in MDF on humult yellow sands and udult clay at Lambir. Of these, two species were udult specialists, three humult specialists and one a generalist. Under gap light conditions, one humult specialist and the generalist performed similarly on both soils, but the others performed best on the soil to which they were specialised. Nevertheless, there was no change in rank order of performance among these during his three years of observation. Relative growth rate was strongly positively correlated with specific leaf area and leaf area ratio. The leaf/root ratios were higher on udult soils ($P<0.5$) for the generalist and the udult species, but only for *Dryobalanops aromatica* among the three humult species. The fine/coarse root ratio, though, was higher on humult soils among all but the vesicular arbuscular mycorrhizal humult *Swintonia* species. Interestingly, *D. aromatica,* which is dominant here on the humult soils, was the only humult specialist to achieve relative growth rates, leaf area ratios, specific leaf areas, and plasticity in allocation patterns comparable to the udult species (Chapter 7).

1.1.6. How stable are these floristic associations?

The immediate impact of different kinds of canopy mortality, and the size of canopy gaps, on forest dynamics and representation of successional species was discussed in Chapter 2; its eventual impact on species diversity will be examined in Chapter 7. The continuous leaching of clay and nutrients in stable soils within a hot perhumid climate implies that freely draining soils not experiencing surface truncation by rainfall run-off must undergo nutrient impoverishment over the span of centuries. Low-nutrient soils, protected from physical erosion by their root-matted surface raw humus, will be particularly prone, and may only be rejuvenated over prolonged periods by geomorphological processes of peneplanation: by the periodic collapse and retreat of scarps and the collapse of the banks of streams issuing from the hills.

Leaching may often be actively exacerbated by the tree flora itself.[73] Phenolic compounds in solution can reduce iron oxides in the soil, enhancing leaching of both iron and the nutrients occluded on the oxides.[74] They also precipitate proteins, thereby slowing litter decomposition by reducing bacterial activity in favour of slower-acting fungal activity.[75] Gong and Ong[76] found litter turnover rates in *Shorea curtisii*-dominated coastal hill dipterocarp forest, at 170% per annum, to be only half the 330% recorded by Ogawa at Pasoh.[77] Species with leaves rich in phenols are more abundant in low nutrient humult soils including coastal dipterocarp forest soils,[78] where leaf nutrient contents, notably calcium, are low (Table 3.5). Leaf nutrient element values for Borneo are notably lower than those recorded at the Barro Colorado CTFS plot (Panama), which partly explains the low vertebrate carrying capacity noted by Janzen.[79] Vines have been found to accumulate phosphorus in their leaves; this persists in the fallen litter, potentially providing patches richer in available phosphorus, which might influence competition between seedlings of different tree species.[80]

I have observed sharply defined sleeves of white sand beneath coarse surface roots of *Dryobalanops aromatica* in Andulau forest. Occasionally and at ecotones, soil profiles provide evidence of active podsolisation, where dead decayed tree roots are still visible beneath an impervious humic B

Table 3.5 a, Responses of fresh Borneo tree species' leaves to tests for protein precipitating compounds. Gelatine reaction explained in Ashton (1964). Strength of response indicated by +. (From P.S. Ashton 1964b) **b**, Nutrient ion concentrations in soil and leaves: two Sarawak MDF compared; leaves of both compared with Barro Colorado Island. Soil samples represent three bulked samples for 15 0.6 ha plots; cation estimates by concentrated HCl extraction (Bailey 1967). Leaf ion concentrations, Mg at Mersing excepted, greatly exceed those of the soil. Leaf ion concentrations are similar at the two Sarawak sites, Ca notably excepted, which is at much higher concentrations at the basalt than at the rhyolite sites. In leaves, the Ca values greatly exceed Mg values, but in the soils Mg values are higher than those of Ca. The BCI values exceed Sarawak's for NH_4 and K, are similar to Mersing for Ca, but are lower for Mg. There was some indication (not shown) that leaf differences were attributable more to different family abundances than different overall averages: high basal areas of families Dipterocarpaceae and Burseraceae (also Myrtaceae, Fagaceae) with low NH_4 and Ca leaf concentrations in MDF sites, and of Leguminosae and Meliaceae (also Moraceae) at BCI (see Table). These preliminary leaf analyses invite more extensive sampling. (P.S. Ashton, S. Riswan, J.B. Kenworthy, unpubl.)

a

Soil type	No. sampled	Response (percent)			
		+++	++	+	-
Podsol (*kerangas*)	19	58	21	16	5
Yellow humult sandy loam (MDF)	61	16	31	31	18
Yellow clay loam (MDF)	50	14	26	46	16

b

Site	Material	NH_4 (%)	K (ppm)	Mg (ppm)	Ca (ppm)	n
Bukit Mersing, basalt, C. Sarawak	Mineral soil	0.42	5,292	4,692	1,716	45 samples
	Leaves	2.29	9,560	3,080	12,330	21 species
Bukit Iju, rhyolite, C. Sarawak	Mineral soil	0.16	883	444	104	45 samples
	Leaves	2.25	8,920	2,910	4,070	20 species
Barro Colorado Island (BCI), base-rich sediments, Panama	Leaves	3.53	14,770	2,870	11,470	15 species

Table 3.6 Leading tree families in selected plots of the CTFS worldwide (leading families in each plot in bold): **a**, by percentage of individuals at least 1 cm dbh; **b**, by percentage basal area of individuals at least 1 cm dbh. Bold: leading family at each site. Dominant families by basal area, therefore canopy dominants, are by no means always dominants by population density, that is, by their juvenile densities. (From P.S. Ashton & CTFS Working Group 2004)

a

Family	Lambir	Pasoh	Bukit Timah	Palanan	Sinharaja	Huai Kha Khaeng	Mudumalai	Ituri, Congo	Korup, Cameroon	Barro Colorado Island	Yasuni, Ecuador
Annonaceae	4.3	7.1	3.7	5.1	1.3	21.5	0	2.8	3.3	7.3	3.1
Burseraceae	6.5	5.4	**14.4**	0.8	0	0.2	0.2	0.1	0.1	0.8	2.2
Combretaceae	0	0	0	0	0	0	**28.3**	0	0	0.1	0.2
Dipterocarpaceae	**15.4**	9.2	6.0	11.7	14.1	3.8	0.2	0	0	0	0
Euphorbiaceae	14.3	**13.1**	10.8	9.9	14.8	**21.6**	3.2	9.7	13.8	1.8	4.7
Clusiaceae	2.8	3.3	6.7	1.4	**15.5**	1.8	0	0.7	1.2	2.8	0.5
Leguminosae	2.1	3.3	0.9	1.7	11.0	3.1	7.7	15.6	5.9	7.5	**12.9**
Lythraceae	0	0	0	0	0	1.3	22.1	0	0	0	0
Meliaceae	1.9	2.7	0.9	**12.1**	0.6	2.2	0	0.7	0.5	7.3	5.3
Rubiaceae	4.5	5.8	2.9	3.5	7.9	5.3	4.4	2.8	7.0	**20.4**	4.5
Sterculiaceae	1.5	0.8	0.4	2.7	0	1.6	5.6	**44.3**	**22.0**	0.3	1.3

b

Family	Lambir	Pasoh	Bukit Timah	Palanan	Sinharaja	Huai Kha Khaeng	Mudumalai	Ituri, Congo	Korup, Cameroon	BCI, Panama	Yasuni, Ecuador
Annonaceae	1.6	3.3	1.4	0.8	1.5	19.5	0	1.7	2.6	2.3	1.6
Malvaceae/ Bombacoideae	0.9	0.6	0.1	0	8.8	0	0.9	0	0	**11.4**	4.9
Combretaceae	0.1	0.1	0	0.1	0	02.	28.2	0.2	0.9	0.6	0.6
Dipterocarpaceae	**41.0**	**28.2**	**38.4**	**52.8**	21.6	**21.2**	0	0	0	0	0
Euphorbiaceae	6.9	6.9	6.0	6.1	5.9	6.7	1.6	7.8	16.1	6.7	5.3
Clusiaceae	2.4	2.1	1.3	0.9	**26.7**	2.4	0	0.3	9	1.0	0.6
Leguminosae	2.1	8.6	3.5	2.3	0.9	2.5	2.4	**42.4**	6.6	9.9	**14.9**
Verbenaceae	0	0	0	0	0	0	**28.5**	0	0	0	0

horizon (Plate 1.12a). This process is exacerbated when sandy soils are cultivated and the nutrient store in the humus is lost, also when the raw humus is lost through burning or careless use of logging machinery. The case of the Singapore catchment forests, which were destroyed for vegetable cultivation by the starving population during the Second World War, is well documented.[81] Successional recovery there has been deflected and dominated by *Adinandra dumosa* and other sclerophyllous species associated with impoverished soils, including *kerangas*.[82] There is no evidence of reinvasion by the coastal dipterocarp flora, even where it is adjacent.[83]

In vivid contrast, the slumping of shale dip slopes due to water seepage through clay soil into fissures between

impervious strata, and the attendant natural cycle of soil and forest destruction and renewal, has been documented around Bukit Belalong.[84] Even on stable surfaces, absence of a root-matted organic horizon and the imperviousness of short-lattice clays to water combine to enhance surface run-off and soil truncation, therefore rejuvenation (Chapter 1).

Thus it appears that, thanks to surface truncation and landslips, the mosaic of soils where clay prevails is dynamic at local spatial scales, and may change within or not far beyond the life-time of a climax tree. In contrast, change on freely draining sandy soils takes place over millennia, and at larger spatial scales correlated with the substrate. Such long time periods suggest that rejuvenation may be mediated as much by orogeny and by climate change as by peneplenation. This is consistent with the larger, geographical, scale of floristic differentiation on these soils. In the near-total absence of surface erosion, such soils appear to be gradually changing – and usually degrading – at scales of up to several thousand square kilometres (see section 1.3.1 below, and Chapter 6).

Overall, this suggests that competition leads to dominance of slower-growing species with high root/shoot ratios on the less fertile humult soils, and faster-growing species with lower root/shoot ratios on more fertile udult soils. Much of the higher root biomass on humult soils may be concentrated in the fine root mat in their more humic surface horizons.

1.2. Competition influences species composition within habitats

1.2.1. Competition operates via differential survivorship

The visitor to the great botanical garden at Bogor (West Java) will be surprised to see a magnificent *Shorea teysmanniana,* a tree exclusively of peat swamps in nature, flourishing on the garden's freely draining fertile basic volcanic soil (Plate 3.4). I was similarly surprised when I found that foresters in Brunei had enriched logged forest on the

Plate 3.4 The peat swamp dipterocarp *Shorea teysmanniana* prospering, in the absence of competition, on clay loam over basaltic lava in the National Botanic Gardens, Indonesia at Bogor, Java. Rosniati Risna as scale. (P.A.)

low nutrient sandy soils of Andulau Forest reserve with the strictly clay loam species *Dryobalanops lanceolata,*[85] and that it was performing at least as well as the sandy soil specialist *Dryobalanops aromatica,* which had been planted in periodically cleaned adjacent lines. In natural forest, floristic composition within specific habitats is clearly differentiated by competition – in addition to the survival rates determined by availability of the minimum resources necessary to maintain a positive carbon balance.[86] Webb and Peart[87] found, at Gunung Palung National Park (West Kalimantan), that in 17 of 49 species, adult trees showed closer associations with particular habitats than did seedlings (5 of 22), implying that selective mortality was operating prior to achieving reproductive maturity. Yamada[88] observed that the five species of Sterculaceae co-occurring in the CTFS plot at Lambir have significantly different soil preferences and heights at maturity, and that soil conditions become more restrictive as the trees increase in size.

1.2.2. *Competition varies with genetic relatedness*
All in all, there is as yet insufficient evidence that the performance of closely related species can differ widely enough across any single physical gradient to provide a convincing basis for competition gradients among them *within* a uniform geological substrate or topography. There has, however, been some evidence for this when *less* closely related species were compared.[89] Thus, Press and his students[90] compared the response of seedlings of four red merantis – *Shorea leprosula* (sect. *Mutica*), *S. fallax* (as *S. oleosa*) and *S. johorensis* (both sect. Brachyptera), and *Dryobalanops lanceolata* – to added nitrogen. These potted seedlings were raised in udult clay soils with light level at maximum photosynthetic capacity, at Danum (Sabah).[91] All species responded to added nitrogen with increased growth and photosynthetic capacity per unit leaf area, that is, photosynthetic capacity per unit chlorophyll; and this increase was greater than the variance in capacity between them. Only *D. lanceolata* responded under low nitrogen. They found that *S. johorensis* showed greatest increase in light saturated rates of photosynthesis and chlorophyll content in the enhanced light levels of logged forest. They concluded that *S. johorensis* has the competitive advantage on higher nutrient soils and *D. lanceolata* on lower nutrient soils, and thus that nitrogen availability may affect seedling competition more than inherited differences in light response.

Other evidence also implies differentiation of overall competitive responses to physical resources among the four species studied by this group within the MDF landscape. *Shorea johorensis* and *S. fallax* are indeed most abundant on udult clay soils of moderate fertility derived from shale; but *S. fallax* is a dark red meranti of relatively high specific gravity, while *S. johorensis* has low specific gravity.[92] This implies

different growth rates and different light responses in nature, at least beyond the seedling stage. *S. leprosula* has wide edaphic distribution, occurring on relatively low udult sandy clay and leached illitic clay soils. It is the most widespread of the four species, being ubiquitous in the zonal MDF throughout Sundaland. *S. johorensis* is almost equally widespread, but is local and confined to higher nutrient soils. All four species occur on the highest-nutrient, basic volcanic soils, but in low densities. The exception is *Dryobalanops lanceolata,* which may become dominant surprisingly, on high-nutrient especially basic volcanic soils, achieving growth rates in height and diameter comparable to the fastest-growing canopy pioneers such as *Octomeles* and *Alstonia scholaris*.[93]

These basic volcanic soils are higher in calcium than those derived from sedimentary rocks, implying higher litter breakdown rates therefore available nitrogen levels. The understorey also appears more uniformly shady at these sites. Clearly, factors other than light are involved in mediating competitive advantage, at least in relation to such larger-scale soil diversity. Among these factors, the extraordinary plasticity of *Dryobalanops lanceolata* juveniles in relation to light may be crucial (see Chapters 2, 7). The possibility, too, of changes in rank order of response to light during ontogeny cannot be ruled out. This too can only be tested by long-term observation.[94]

The diversity of understorey light regimes across the topography was able to explain floristic differences among congeneric canopy species at Sinharaja by reference to the light-response of their seedlings[95] (Chapter 2). The same influence may also operate within the dynamic mosaic of a single forest community, permitting the coexistence of several species with similar nutrient responses and explaining the often wide overlap in their ecological distributions.[96] In this connection it should be borne in mind that competitive success is determined as much by survivorship as by performance. Either may be determined by a range of factors besides nutrient availability, one of which is response to light.

1.2.3. *Herbivory is also a factor*
Akira Itoh[97] found differential herbivory among germinating seeds of the two *Dryobalanops* species in a reciprocal transplant experiment across the ecotone from sandy humult to clay udult soils at Lambir. Differences in insect herbivory between rainforests on contrasting soils have been reported in the Neotropics.[98] Fine and colleagues[99] obtained similar results in a more ambitious reciprocal transplant experiment, involving seedlings of 35 species from three genera of the tribe Protieae, Burseraceae, including several congeneric pairs, in the upper Amazon. Here, 26 of the species were restricted in nature to one of three soils: white sand, yellow sand, or alluvial clay. They found that clay specialists grew faster than the others, but only when herbivores were excluded, even on clay soils.

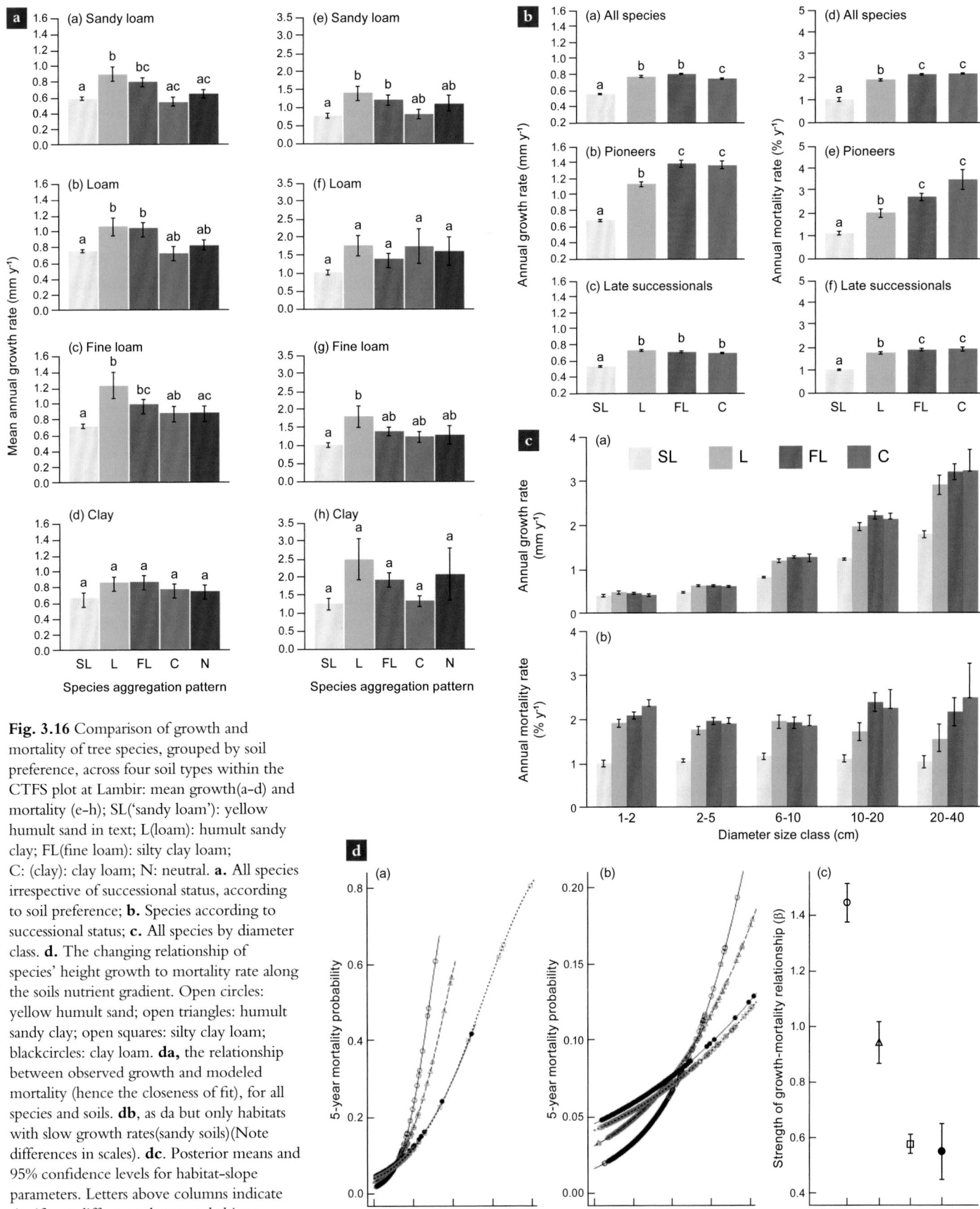

Fig. 3.16 Comparison of growth and mortality of tree species, grouped by soil preference, across four soil types within the CTFS plot at Lambir: mean growth(a–d) and mortality (e–h); SL('sandy loam'): yellow humult sand in text; L(loam): humult sandy clay; FL(fine loam): silty clay loam; C: (clay): clay loam; N: neutral. **a.** All species irrespective of successional status, according to soil preference; **b.** Species according to successional status; **c.** All species by diameter class. **d.** The changing relationship of species' height growth to mortality rate along the soils nutrient gradient. Open circles: yellow humult sand; open triangles: humult sandy clay; open squares: silty clay loam; blackcircles: clay loam. **da,** the relationship between observed growth and modeled mortality (hence the closeness of fit), for all species and soils. **db,** as da but only habitats with slow growth rates(sandy soils)(Note differences in scales). **dc.** Posterior means and 95% confidence levels for habitat-slope parameters. Letters above columns indicate significant differences between habitats. (From Russo *et al.* 2008)

1.2.4. *Competition leads to trade-offs*

Mark Ashton showed that different trade-offs between rapid growth and higher survivorship permit the co-existence of related species within MDF communities; and that differences in competitiveness can be mediated by differences in the light and soil water regimes of adjacent habitats.[100] He had the advantage of selecting four related common canopy species, of *Shorea* sect. *Doona,* which manifest differing abundances along the catena in species-poor MDF at Sinharaja forest. Here, soil nutrient levels are moderate, and change little in relation to topography; but soil depth does. He showed that, in this case, species performance was mediated by variation in the understorey light regime. Ridges and spurs experienced predominantly single-tree gaps and much lateral light, while the lower slopes and valleys had contrasting deep shade and bright light in the large gaps resulting from poor drainage and windthrows or landslips (Chapter 2 section 3.1.3).

Three quarters of the tree species in the CTFS plot at Lambir are significantly associated with one or more of four broad, contiguous soil fertility categories, from humult sandy loams to friable udult clay loams. These soil classes represent gradients both in fertility and in proneness to drought. Sabrina Russo[101] tested Palmiotto's and Davies' earlier experimental results in the forest setting, analysing growth and mortality among trees greater than 1 cm dbh within the plot, and comparing two censuses five years apart in a search for correlations of growth and mortality with soil nutrient levels. She found that growth rates of pioneers were lowest on the least fertile and most drought-prone soils, but growth rates of climax species did not differ significantly (Fig. 3.16). This outcome confirmed what Pamela Hall and I had earlier found at three MDF sites in Sarawak, although our study did find a weak growth response among climax species after 20 years.[102]

Russo found that mortality rates – which we had not examined – correlated with fertility among both pioneers and climax species grouped according to their soil preferences. Interestingly, *rank orders* of growth and mortality of species groups did not differ, whether the species was growing on the soil on which it was most abundant or on other soils among specialists of those soils. The species that were restricted to the least fertile soils had the lowest growth rate and the lowest mortality rate (Fig. 3.16a). There was thus little evidence of an overall performance advantage for a species where it was most abundant. Nevertheless, differential mortality (that is, relatively low mortality among the specialists to a soil type) occurred as tree diameter increased. This led to more frequent loss of generalists than specialists on all soil types, and consequently produced the floristic patterns observed in the landscape. Russo inferred that on less-fertile soils differential mortality was the main mediator of eventual

species composition, whereas on more-fertile soils differential growth rates also played an important part. This trade-off was steepest toward the lower end of the nutrient gradient (Fig. 3.16c,d). As tree diameter increased, sorting by soil fertility group strengthened. Some differentiation in response to light, however, did appear to correlate with the nutrient gradient. This implies that water stress is not the primary factor leading to differential selection along the gradient at the stage when mortality, and possibly selection, is greatest (although the gross leaf characters which distinguish canopy species become increasingly manifest as these approach maturity, and are consistent with adaptation to habitat). She emphasised that co-existence and differentiation in competitiveness along the habitat gradient must be governed by multiple inherited adaptations.

Russo's results have been subsequently further supported by pot experiments. Daisy Dent with David Burslem grew two specialists (of low and high nutrient soils, respectively) and one generalist at two light intensities and nutrient levels for one year.[103] All increased growth under higher light, with no change in rank order. But slower-growing low-nutrient species increased root allocation and reduced specific leaf area and leaf area ratio in response to reduced nutrient supply, thereby experiencing lower mortality than the faster-growing species. Dent found that differences in soil nutrients did lead to changes in the rank order of species performances. However, there remained the possibility that the same may occur with respect to light over a longer period of observation. This was confirmed by Philipson, who showed that growth of sapling parts is non-linear over time owing to changes in allocation, thereby facilitating changes in species competitiveness.[104]

The slopes of lines representing the trade-off between fast growth and mortality for poor versus fertile soil specialists cross over along the soil fertility gradient, the steeper slope being that of the poor soil specialists (Fig. 3.16d). Russo and collaborators concluded that adaptation for fast growth may entail a greater mortality risk if inherently fast-growing species fail to maintain a positive carbon balance when soil resources are scarce. Slow-growing species benefit from lower mortality when soil resources are limited, but are outcompeted when they are less so. Plant life-history strategies can therefore influence species distributions *across* resource gradients. Unexpectedly, they found no significant differences among juveniles of clay and sandy loam MDF species in their leaf structure, notably the stomates that regulate the balance between carbon gain and water loss, other than when correlated with incident radiation.[105] It would be interesting to at least compare anatomical differences between mature canopy leaves in these respects.

The diversity of life histories can also therefore potentially allow species to co-exist in floristic associations within which several resource gradients may be varying independently. Tilman[106] argued that species *within* otherwise uniform habitats may be sufficiently specialised to differing combinations of soil nutrient ion concentrations to partially account for the composition of communities. This would be exceedingly difficult to test in rainforest tree communities. In view of the vast number of species in each community, the varying seed-dispersal limitations, the extending radius of tree roots and the variety of soils they exploit as they grow, and the almost limitless range of potential nutrient combinations, such within-community selection appears improbable, at least beyond the sapling stage. Light appears to be the one physical factor capable of mediating within-community differential selection among individuals of different species, but it alone is clearly insufficient in hyperdiverse communities.

1.3. *Regional history also influences species composition*

1.3.1. *Diverse impacts at different spatial and temporal scales*

The role of historical accident in community composition can only be inferred by comparing communities in different regions that share the same habitat – that is, by keeping other variables constant. This is possible within regions sharing a common geological history and present climate, but geological variation at the scale of the major islands and continental regions makes a strict assessment of this kind impossible, because soil nutrient supply varies greatly with the mineralogy and rate of decay of the various substrates. Community composition at that spatial scale is obviously modified by regional patterns of endemism resulting from the sea barriers to dispersal between the land masses, a subject addressed in Chapter 6. But we saw in Chapter 2, and in the previous section, how similar topography and soils impose similar patterns of variation – albeit with more or less differing suites of species – on forests in different regions.

Of present interest is the extent to which local constraints on species movements (occasioned both by opportunity and by the effectiveness of their seed dispersal strategies) influences variation in community composition *within* a region of similar geological history. The longer the present climate of a region has persisted, the greater is the likely influence of limited seed dispersal (but also niche differentiation as a motor of speciation) on forest composition and local endemism. A suitable area for these investigations is, again, lowland north-west Borneo. This region of c.100,000 km² has experienced a unified geological history ever since the Sunda lowlands and their forests came into existence c. 22 Ma[107] (although the region is also fragmented by rivers of varying history). Furthermore, it seems likely that the regional climate has remained uniform

since the Late Miocene, perhaps 10 Ma (Chapter 6). It is a region of 'young' geomorphology and high topographic and soil diversity,[108] but also distinct geographic patterns of plant species correlating with the age of the major valleys that break the coastal hills.

In his floristic study of a set of forestry inventory plots spread over an area of c.150 km² in central Sarawak, Ian Baillie[109] showed a correlation between shale substrates and soil potassium derived from its clay. Reserve phosphorus (that is, nearly total phosphorus) was also correlated with shales. Sandstone-derived soils were correlated with elevated soil carbon, but low magnesium and especially calcium concentrations. There was some correlation between species composition and potassium, also with reserve magnesium in mineral soils, but not with phosphorus. There was also much random floristic variation, which could be due either to the influence of limitations on seed dispersal or to sample error due to small plot size. Baillie noted that exchangeable and reserve nutrient values were not closely correlated, but that reserve values correlated best with floristic variation.

Our 105 Sarawak MDF plots were laid out across the topography in the same manner as my Brunei plots, within 13 sites; but sites were spread over 500 km, and over an area of 80,000 km². At each, mineral soil was collected, dried and analysed for physical characteristics, pH, carbon content, and total concentrations of nitrogen, phosphorus, magnesium, calcium and sodium, as well as aluminium and iron (whose oxides are important long-term accumulators of other nutrient ions in soils low in clay), following the method of Bailey.[110] Nearly 2,000 species were recorded in these plots. In Potts' elegant analysis the plots became clustered by site, which were in turn correlated with soil values, notably percent clay and nitrogen, phosphorus, calcium on high nutrient clay sites and magnesium on sites low in nutrients and clay; and iron and aluminium sesquioxides which function as occlusion sites for nutrient ions (Chapter 1). Potts used a Mantel Test, which enables comparison of the contribution of separate factors in determining a pattern, to evaluate the relative role of these nutrients versus geographical separation on species composition of a plot subsample from each site. He found, that even within this single forest type, the influence of soils transcended the chance exigencies of seed dispersal (and therefore history) (Table 3.3a). Local species endemism, particularly west of the Lupar River (Fig. 6.16), is nevertheless strong enough to influence the path of the classificatory dendrogram (Fig. 3.13). I conclude that the predictions of the theory of island biogeography[111] are only fulfilled (in relation to tree species in floristically rich rainforests) over such very long time periods that climatic and geological events will always have intervened beforehand. This will be discussed in context of the central issue of Chapter 7: patterns of species diversity.

No similar study has yet been carried out in the region, in part because no comparable plot network exists, although Sarayudh is analysing a comparably extensive series throughout lowland forests in seasonal Thailand. In the MDF over the ancient continental rocks and geomorphology of lowland Peninsular Malaysia, where soils may reach 5 m in depth and are relatively uniform, the roles of history and habitat may have become reversed. Although correlation of floristic composition with the gentle topography and soil has been amply demonstrated at Pasoh, few if any species there are *confined* to a particular habitat, other than streamsides.[112]

1.3.2. *Seed dispersal patterns contribute to clumping of mature trees*

The impact of restricted seed dispersal is most clearly seen at the very local scale, in the clumping of juveniles around reproductively mature individuals and the clumping of mature trees within areas of apparently uniform or gradually changing habitat (for example, Figs 5.33, 5.34). However, habitat factors may also contribute to such clumping, as they appear to do in the clumping of plots within each site in Potts' Sarawak dendrogram. These might include patchiness of nutrient concentrations and sharing of host-specific symbiotic root fungi, the mycorrhizae (Chapter 5). Ectomycorrhizae in the lowland tropics are characteristically associated with dipterocarps and with Caesalpinioid legumes. These are the respective dominant emergent families of MDF. *The role of mycorrhizae in achieving abundance and emergent canopy dominance in these families merits further research.*

In the present context, the tree floras on different soils include differing proportions of the widespread Sundaland species. The most continuous lowland soils (clay and sandy clay udult) have the highest proportion, while the most fragmented (podsols and deep sandy yellow humults) have the least widespread floras and the highest proportion of endemicity, in part reflecting the influence of restricted dispersal opportunities but also, perhaps, long term edaphic change at landscape and larger scale through continued leaching of sandy humult soils (see also Chapter 6). Thus, only MDF on clay and sandy-clay soils shows anything like the pattern described in the similar perhumid climate of the upper Amazon, where groups of species form 'predictable oligarchies': regionally widespread associations, with high frequency of presence in plots and predictably high local abundance resulting in predictable rank orders of abundance.[113] In MDF, only the most abundant species manifest a predictable rank order (Chapter 7, section 2.3).

2. Inland forests of other habitats

Many tropical Asian landscapes are studded with patches of unusual habitat, recognized long ago by early travellers on account of their distinctive topography and flora.

2.1. *Kerangas* (Plates 3.5)

A definition of *kerangas,* with descriptions of its structure and physiognomy, is presented in Chapter 2 (Plates 2.21, 2.22). Here I again focus on *kerangas* in its narrow floristic definition. This formation is most widespread on the sandstone ridges and plateaux of inland Borneo, and on its raised beaches; but it also occurs on coarse granitic sands on Bangka and on raised beaches up the east coast of Peninsular Malaysia. It is said to occur along the foot of the central Maoke Range of Irian Jaya. It can further be found on raised beaches in the shortly seasonal climate between the back-mangrove and seasonal evergreen dipterocarp forest along the coastal foothills of the Cardamom mountains (south-west Cambodia) (Plate 3.5).

Kerangas occupies the same upland habitat as MDF, but where soil nutrient levels are too low to support MDF. These are tropical podsols with deep densely root-matted raw humus over a white sand mineral horizon lacking the capacity to hold nutrients, beneath which may be a more or less indurated horizon of sand infused with re-precipitated humic compounds, more or less impervious to water and roots. The *kerangas* ecosystem is oligotrophic, like the peat swamp (section 2.4.3 below). Floristically, though, it could be regarded as a low-nutrient MDF community, were it not for the absence of a ubiquitous emergent stratum of dipterocarps. Emergent dipterocarps occur, but in scattered clumps of few species; and they generally fail to emerge far above the comparatively even main canopy. *Kerangas* soil drainage varies from being free to impeded. Soils may be shallow over an indurated B horizon where this may impede drainage in flat topography, or lie directly over an exclusively sandstone substrate, usually on a ridge or raised plateau, where they will be drought-prone.

The *kerangas* tree flora is poorer than MDF. The same families are included, but some (notably Dipterocarpaceae, Meliaceae and Sapindaceae) are represented by few species, while others (notably Myrtaceae, Clusiaceae and Rubiaceae)

Plate 3.5 (right) *Kerangas*. **a**, *Kerangas* forest on humic podsol soil along the Veal Renh road, Cardamom coast, Cambodia, Nov. 1969. Canopy with *Dacrydium elatum* and *Dacrycarpus imbricatus*, *D. elatum* standing free by road; subcanopy with abundant *Tristaniopsis merguensis*; **b**, *Kerangas padang* caused by loss of soil and nutrients following ancient fires: *Ploiarium alternifolium* with *Dacrydium elatum*, *Gymnostoma nobile* and primary forest behind. Bako National Park, west Sarawak; **c**, Juvenile *Pholidocarpus borneensis*, canopy palm in the Bako National Park *kerangas*; **d**, The understorey palm *Johannesteijsmannia altifrons*, Bako National Park; **e**, *Kerangas* interior, raised Pleistocene beach, Badas Forest Reserve, Brunei; **f**, Stand of emergent *Agathis borneensis*; *kerangas* on raised beach, Bukit Puan-Labi road, Brunei, 1958. (All: P.A.)

are relatively species-rich. Conifers (*Agathis, Dacrydium, Podocarpus*) and Australian elements – including *Gymnostoma nobile* (Casuarinaceae), several Myrtaceae, and *Styphelia* (Epacridaceae) – are well represented (Chapter 6). *Kerangas* occurs in belts on raised Pleistocene sea beaches, often behind the peat swamps, or islands within MDF, or as lower montane forest, where it may include a distinct floristic element (Chapter 4). Floristically and structurally, *kerangas* is often ecotonal, enhancing local floristic variability. This effect is further enhanced by endemism, particularly within the Riau Pocket province of north-west Borneo (Chapter 6).

2.2. *Forests on limestone* (Plate 3.6)

Limestone in the wet tropics of Asia is generally of mid-Cenozoic age or older. The rock is hard, creating the dramatic karst topography seen in classical Chinese paintings. In perhumid northern Borneo, there is no mineral soil overlying the karst. In the seasonal dry tropics and at the tropical margins of northeastern Indo-Burma, udult red loam soils may form from limestone impurities and the topography beneath can become more rounded and gentle. Karst limestone hills include four distinct topographic elements, each with a distinct floristic association linked to different communities in other habitats:[114]

1. Around the base there is often calcium-rich alluvium, in many cases periodically inundated, supporting tall forest, which floristically is in part typical of fertile clay loams but also includes species most abundant on floodplains.

2. The karst towers are surrounded by prominent talus boulder-slopes, often steeper than 30°. In the humid tropics these generate little surface mineral soil, but the tree flora is nevertheless very similar to that on the friable udult red-brown loams of basalt hills. This tall forest includes few dipterocarps, but *Shorea guiso, Chukrasia tabularis, Firmiana* and *Pterocymbium* are almost confined to these habitats in Borneo, and *Toona sureni* in peninsular Malaysia, though they are widespread on the generally more fertile soils of seasonal Indo-Burma (see Chapter 6). There, the deciduous species (i.e., except for the evergreen *Shorea guiso*) occupy broader ecological ranges, including tall deciduous forest. Other species, such as *Hopea andersonii* and *H. plagata,* only occur on karst and on clay loams over basic volcanic rocks in Borneo. There is often also some local tree species endemism, the level of which varies among regions (Chapter 6). These endemics appear adapted both to drought and to high concentrations of calcium and other cations in solution.[115] It is astonishing that juveniles can survive on these periodically very dry surfaces until their roots penetrate deep into crevices, as can be seen on fresh quarry faces. The rocks and humus-filled fissures support a rich, often endemic herbaceous flora including Araceae, Gesneriaceae, Rubiaceae, Acanthaceae, Cannabinaceae and some Orchidaceae.

3. Cliffs rise above the talus, more or less vertical and sometimes exceeding 500 m in height. Largely bare, their fissures carry a rich herbaceous full-sun adapted flora including many local endemics, with a few trees and shrubs of the summit flora.

4. Karst summits provide more level summit surfaces, but the hard limestone is dissolved by rainwater into a honeycomb of sharp-edged vertical surfaces. In exceptionally wet areas the limestone may become eroded by solution into angular pinnacles up to 100 m in height, as on Gunung Api, Mulu (Sarawak) (Plate 3.6c) and the pinnacles area south of (warm temperate) Kunming (Yunnan). On exposed hills, limestone summit plateaux become very dry, posing in extreme form the question of how tree saplings become established. In regions of high humidity, especially where limestone is under the climatic influence of higher mountains, acid raw humus accumulates between and even over these rocks to depths of up to 50 cm. This supports a *kerangas* or lower montane *kerangas* flora (Chapter 4). Dipterocarps are represented as *kerangas* elements on peat-bearing summits, including *Cotylelobium lanceolatum, Shorea multiflora,* and the lower montane *S. carapae.* Organic soils over limestone occur, for example, throughout the Gunung Mulu limestone front-range in Sarawak, but also on the limestone upper ridge of Emei-shan (3,099 m) and elsewhere in warm temperate monsoonal East Asia. These drought-prone karst tops are especially prone to lightning-induced fires (Gunung Api – in Gunung Mulu National Park, Sarawak – translates as 'fire mountain'), which destroy the raw humus, leaving bare rock and even more drought-prone conditions. Human-induced fire has resulted in prevalence of such conditions elsewhere in Sundaland.

No floristically recognisable lower montane tree flora has been described on limestone. Upper montane forest may occur within the cloud base, sometimes with point endemics on exposed peaks but otherwise floristically typical for the region. For their latitude, Doi Chiang Dao in northern Thailand, Fan Si Pan in northern Vietnam and other high limestone hills in northern Indo-Burma carry an exceptional number of deciduous warm temperate tree species and herbs as well as gymnosperms (Chapter 4). Limestone is widespread in the snowy Sudirman Range (Irian) and elsewhere in upper montane and alpine New Guinea,[116] as well as the lowlands, but seems not to have been specifically investigated floristically.

Plate 3.6 Forests of karst limestone. **a**, The high karst of Mulu National Park, Sarawak, rising above 1,000 m. The three topographic elements of karst in the wettest climates is evident: the dry rocky summits with short vegetation on raw humus; the sparsely vegetated cliffs; and the tall forest covering boulders of talus slopes; **b**, The understorey at the base of a limestone talus. 200 m, Panay Island, Philippines; **c**, Solution of dolomitic limestone by rain alone: the 30 m tall pinnacles on a karst plateau at Mulu National Park; **d**, Limestone karst near Pai, north-west Thailand: deciduous forest on south facing slopes, semi-evergreen on less sunny surfaces; **e**, Deciduous forest understorey on a limestone scree, Xishuangbanna. (a, d H.H.; b, c T. Laman, e H. Zhu)

187

2.3. *Ultramafic (ophiolitic) substrates*

These ophiolitic rocks, called serpentine, are usually of Cenozoic origin in tropical Asia. They are rare outside the Far East, although they do occur along faults even in the ancient continental rocks of Indo-Burma and Sri Lanka[117] (Chapter 1). They are particularly well represented in north-east Borneo, the eastern Philippine mountains and parts of eastern Malesia, but are widespread throughout the Philippines (Figs 1.1, 1.4).

Ultramafic rocks vary widely both in base concentrations and in the physical characteristics of the soils derived from them. Nickel and chromium cation concentrations may be high, and some tree species accumulate them in their leaves. The unusually high magnesium/calicum ratios noted already in humult sandy soils within MDF (see section 1.1 above) can reach extreme levels in soils over ultramafic substrates. These soils are characterised by exceptional concentrations of bases, which may include magnesium, iron, cobalt, chromium and nickel – the last three being toxic to plants. Some species accumulate calcium differentially in leaves.[118] *The influence of such high cation concentrations on tropical 'ultrabasic' soils has been little investigated.* Species adapted to ultramafic soils synthesise chelating compounds, which inactivate unusual concentrations of heavy metal ions and accumulate them in mesophyll vacuoles.[119] John Proctor and his students found that trees had the ability to concentrate calcium relative to magnesium ions in their leaves, while one endemic dipterocarp, *Shorea tenuiramulosa*, at Gunung Silam (Sabah) accumulated an astonishing 17.7 mg g^{-1} of magnesium.[120]

Some ultramafic soils are reddish clay loams, bearing tall MDF floristically similar to basalt-derived soils[121] (Chapter 1) (Plate 3.7). Such soils and forests can be seen at the Ranau hot springs and formerly on Bukit Hempuan at the base of Kinabalu. More often, the soils are coarsely sandy and freely draining; black-water streams, rich in humic acids, emanate from them. Under these conditions the vegetation resembles *kerangas* in structure and physiognomy while its composition, in Borneo, includes many *kerangas* species. Elsewhere, it is characterised by high endemism of species manifesting leaf and branch physiognomy of *kerangas* forms. Mixed with widespread *kerangas* taxa are a high proportion of local or point endemic tree taxa, more than on any other substrate category[122] (Chapter 6). However, the forest is often even shorter than most *kerangas,* the short trunks branching low.

Plate 3.7 Forests on ophiolitic extrusions. **a**, *Kerangas* forests on ultramafic soil, with *Gymnostoma nobile*; **b**, the ultramafic main range of Palawan behind; **c**, Ultramafic soil profile, with *kerangas* above. Palawan, 1994. (P.A.)

2.4. *Riverine forests*

2.4.1. *Riparian forests*

These essays concern inland forests: coastal forests, notably the mangrove, have been well treated by others such as Tomlinson.[123] Beyond the influence of salinity, a sharp differentiation is invariably recognisable between forests – and flora – on the alluvial banks of the sluggish plains rivers, and those on the banks of usually swiftly flowing rivers hemmed in directly by their rocky substrate.

The habitat of plains river banks is unstable, ever-changing, as the river continuously erodes its outer banks at bends and deposits fresh sediment on the inner. Outer banks thereby tend to display exposed profiles of floodplains forests, while the inner banks support successional stands – often of a single pioneer or successional species (Plate 2.12a). The tree flora comprises both specialised riparian taxa, and pioneer and climax floodplains species (Plate 3.8). Composition is influenced, here too, by the chemistry and physical character of the sediment load of the river. Successional species dominate on clay soils especially, where strangling figs are often abundant (Plate 5.13a).

Rivers flowing directly over rocky substrates possess more stable banks, but rapids flow over continuously changing

Plate 3.8 Riparian vegetation. **a**, Riparian forest along an alluvium bank: the broken canopy of a highly dynamic forest. Lower Kinabatangan, Sabah; **b**, Mixed dipterocarp forest descending slopes to the river bank, east Sabah; **c**, Rheophytic *Nauclea* (Rubiaceae) along a rapid, north-east Borneo; **d**, *Khair (Acacia catechu)-Sissoo (Dalbergia sissoo)* deciduous woodland bordering floodplain grassland along a river issuing from the Himalayan front range, Buxa National Park, West Bengal; **e**, The front range of the Himalaya in Buxa National Park. Light demanding pioneer and early successional species along the river banks: *Tetrameles nudiflora,* foreground leaves upper right, emergent *Bombax ceiba* on the far bank. (a P.A.; b, d, e H.H.; c T. Laman)

shingle and sand bars. The characteristic tree flora of these habitats establishes in the frequently open but often rigorous habitat between the water's edge and the upper limit of more frequent flooding (Plate 3.8b). Above that limit, the forest hardly differs from that upslope.

Shrubs and juveniles of the riparian flora are recognized as rheophytes,[124] distinguished by their pliable twigs and branches, long narrow caudate leaves on short petioles, often large attractive flowers, and often precociously germinating seeds – some germinating even before they fall. Above the floodline, their crowns develop broader, albeit still generally long and narrow, leaves. Composition is influenced both by substrate and river chemistry; the banks of blackwater rivers include species of humult soils irrespective of the immediate substrate. River width, which has an influence analogous to canopy gap size, also influences species composition.[125]

The whitewater rivers of the Sunda uplands were glorious, in their seasons, with the orange or rich maroon-red flowers of Saraca: *each species in its distinct riparian habitat; with the orange-red young leaves of the many* Syzygium, *and the coppery-winged fruit of* Dipterocarpus oblongifolius, kesugoi, ensurai, neram, *brushing overhead as one toiled upstream; and in Borneo and the Philippines the tall pale-boled pale blue-green crowned* Swintonia acuta *high above. In Brunei and a few other parks elsewhere, you can still experience them, cooled by the daily downdraft off the mountains.*

2.4.2. Flood plain forests

Flood plain forests hardly survive in the seasonal tropics, having long ago been converted to rice cultivation. In perhumid Sundaland, particularly north-west Borneo, aseasonal flooding prevents irrigation (Chapter 8), while excessive rain leads to leaching and peat accumulation. In such areas, forests still occupy narrow belts between river banks and the vast peat swamps, becoming wider only where the rivers issue from the hills and flood most extensively. The great rivers to the east and south of Borneo, and elsewhere around the land masses of Malesia, flood seasonally as well as intermittently, and may have broader floodplains continuously enriched by fine sediments with their nutrients. Shallow rooting on these floodplains is associated with high frequencies of tip-ups and an irregular canopy with many large gaps. Tall pioneers including *Octomeles* (Datiscaceae), *Pterocymbium* and *Pterospermum javanicum* (Sterculiaceae), *Endospermum peltatum* (Euphorbiaceae) and *Alstonia scholaris* (Apocynaceae), and a diversity of large strangling figs are locally abundant, interspersed with scattered individuals of late successional species of upland clay loams. The presence of *Kleinhovia hospita* (Sterculiaceae) and deciduous *Lagerstroemia speciosa* (Lythraceae) along rivers as far north as the Kinabatangan reflect the onset of periodic rainfall seasonality. In more seasonal regions of India, the elegant layered crowns of *Terminalia arjuna* grace such sites.

2.4.3. Peat swamp forests

Plate 3.9 Peat swamp forests (see also Plate 2.21). **a**, The phasic community 2, *alan* forest as they were at Seria, Brunei, in 1959. *Shorea albida* at 70 m (Land Rover in background as scale); **b**, Phasic community 3, *alan bunga* profile, Sungei Liang-Labi road, Brunei; bracken fern, *Pteridium aquilinum,* establishing in a *ladang* cleared and burned for a crop; **c**, Vigorous *S. albida* sapling on the banks of a recent drainage channel, Seria, Brunei; **d**, Distinctly stratified mature peat swamp forest, phasic community 3, with continuous emergent canopy of monodominant *S. albida* and sparse main canopy merging with subcanopy. (cf. Plate 2.21); **e**, Profile within a *S. albida* stand (phasic community 4) *alan padang*, showing the complete absence of a main canopy species stratum; **f**, *S. albida* two-tiered root system: The main surface roots above normal water level, with descending roots below them finely branching into the waterlogged mire; Seria; **g**, Phasic community 5, *medang (Litsea palustris) padang keruntum* the trees hardly exceeding 30 m, forming lines of epicormic leaders when blown over. Seria swamp, Brunei; **h**, Phasic community 6: sedge savanna on the bog dome: *padang keruntum (Combretocarpus rotundatus),* with predominant tree *C. rotundatus* (Anisophylleaceae); Marudi swamp, Baram, Sarawak; **i**, Aerial photograph of *S. albida*-monodominant peat swamp with a gap caused by a massive downdraft, and tree-fall by 'domino-effect' scale 1/2,500; **j**, Panorama of the Belait-Baram peat swamp, with the Labi hills 15 km to the east, from an 80 m meteorological tower. Why is the canopy, with trees of varying age albeit all one species, so even in this calm climate? **k**, Beneath the shy crowns and diffuse shallow foliage of *S. albida* crowns in peat swamp, phasic community 3, Ulu Balai, Brunei; **l**, Satellite image of peat dome: pale brown area is *padang keruntum* savanna, the green ripples in it representing closed forest, seemingly caused by slumping of the dome summit surface. East swamp of Batang Baram, Sarawak (scale 1/30,000). (**a-c, g, h** P.A.; **d, e, j, k** A. Cobb; **f** J.A.R. Anderson; **i** Sarawak Survey Dept; **l** Google Earth)

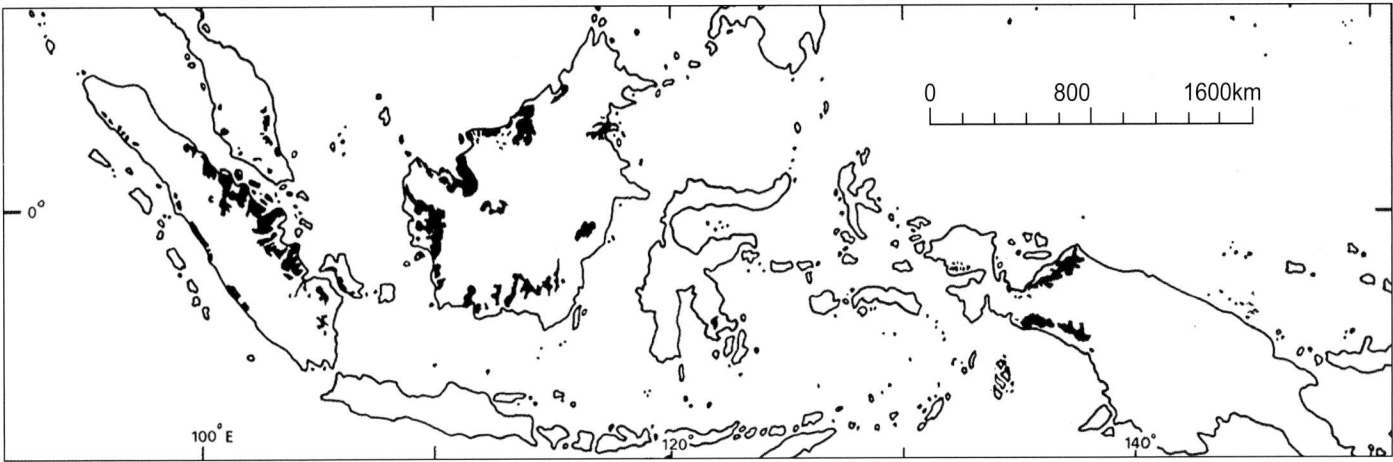

Fig. 3.17 Distribution of peat swamps in Malesia. (From Whitmore 1984b)

Peat forms only in tropical regions where evapotranspiration exceeds mean precipitation every month of the year.[126] These are the regions where, but only until recently, the technology – and low cost of energy - did not exist to convert plains forests to agriculture (Chapter 8). They are estimated to cover c.225,000 km² overall, down the coast of East Sumatra, on the southern, southeastern and northeastern coasts of Borneo, and on more limited areas on both coasts of Peninsular Malaysia and northwards to the limits of the perhumid climate in Narathiwat Province (Thailand)[127] (Fig. 3.17). They are also found on Sulawesi, Ceram, and on the coast of Irian Jaya,[128] while a small secondary peat swamp forest exists at Muthurajawella, between Colombo and Negumbo along the north-west coast of Sri Lanka's perhumid region, having regenerated following drainage by the Dutch in the 18ᵗʰ Century. Small areas occur on flat surfaces in the lower montane zone of Borneo. Peat swamps have been recorded nowhere else in the global tropics, though apparent mire habitats in the Peruvian Amazon may prove peaty. Peat swamp, being dependent on eustatic sea levels, is the most transitory and changeable tropical habitat over evolutionary time (Plate 3.9).

2.4.3.1. *How tropical peat swamps form.* The vast majority of the Far Eastern peat swamps were formed over sediments along the coasts of epicontinental seas, following slight declines in eustatic sea levels. The last major episode of peat accumulation started c.7,000 BP and resumed c.4,500 BP, as sea levels declined successively.[129] In Central Borneo, peat swamps called *kerapa* have also formed on gentle uplands above maximum Pleistocene sea levels. These are more ancient, and palynological evidence for their widespread presence here and in Java goes back to the beginning of the perhumid climate (Chapter 6). Their flora, in which *Dacrydium* and *Casuarina* (or *Gymnostoma*) are generally present, bears affinity with that of *kerangas*.

Coastal swamps form on the following pattern:

1. First, sand spits extend out from remaining coastal promontories or river mouths, where there is rapid sediment deposition following mixing of turbid fresh water with salty seawater. Spits follow the prevailing current and cut off shallow lagoons. Lagoons receiving major rivers rapidly silt up.

2. Soon, mangrove establishes, stabilising sediments and increasing deposition rates. Eventually, only spring and neap tides reach the inland edge. In aseasonal, rainy climates, salt leaches beneath the surface. In seasonal climates, salt instead builds to lethal concentrations, with sulphur in solution as sulphuric acid near the surface, toxic to most tree species but supporting a savanna or 'back-mangrove' (arrière mangrove) of sedges and grasses, some acidophilous herbs including the sundew *Drosera,* and the tree *Melaleuca cajuputi* (Myrtaceae). In perhumid climates, instead, mangrove litter eventually raises an organic surface above the highest tides, and peat begins to accumulate.

3. The peat accumulates into domes exactly as in a boreal quaking bog, where the only nutrient input is from dust and rainfall.[130] These domes may extend into massive bog systems up to 60 km in diameter. The peat may reach 20 m depth, and the mineral sediment base may lie below the current sea level. In north-east Sarawak and western Brunei, owing to slight distortions of the substrate, shallow lakes have formed along minor tributaries within the swamp. The top of the dome has the most variable water table; here, *kerangas* elements are most evident. In some, a sedge savanna forms, *alan padang,* in which two tree species, *Dactylocladus stenostachys* (Crypteroniaceae) and *Combretocarpus rotundatus* (Rhizophoraceae), predominate (Plate 3.9h). The often sharp edge of these savannas implies that they may be caused by occasional lightning-induced fire.

193

2.4.3.2. *Peat swamp flora and swamp structure.* The number of flowering plants and ferns recorded from peat bogs is small. Robb Anderson made a thorough study of the coastal peat swamps of Sarawak and Brunei, which comprise most of the north-west Borneo peat swamps. He recorded a mere 250 tree species, of which one fifth were confined to peat swamps and one third are endemic to that region.[131] Thus, these late Holocene communities may be the poorest in Sundaland – as might be expected from the relatively recent origin of the large coastal swamps, and the small size and isolation of the older upland peat areas.

Floristic change from the swamp margin inland correlates with flooding frequency, rather than with the depth of the peat.[132] Observations were made during the 1997 drought at Lahei, Palangka Raya (Central Kalimantan), when only 117 mm of rainfall was recorded for the five months June–Oct. (normal mean 247 mm per month). Tree uprooting rates decreased as expected, but mortality on hummocks doubled to 5% during the following year.[133]

The tree flora on the periphery of the peat lens and on the widespread shallow peat of the surrounding region is mixed, and as tall as MDF. Species composition varies with substrate and the neighbouring dry land forest. But where the swamp is isolated, and species restricted to it, between the river bank and deeper peat, there is a tall forest dominated by up to five dipterocarp species. The tree flora here is rich in species unique to it, and including some centres of endemicity although most species are widespread in western Malesia (Chapter 6). On deeper peat, single species dominance is common. *Shorea balangeran* dominates many swamps in Sumatra and southern Borneo. The similar *S. albida,* (alan) dominates in north-west Borneo to such an extent that, of the six tree 'phasic communities' recognized by Anderson[134] from the swamp edge inwards there, four are defined by its character and presence.

This remarkable tree formed monodominant stands 70 m tall – now all but gone – at the outer edge of the species ecological range in the swamps, yet less than 25 m tall at the inner margin (Plates 2.21, 3.9a-g). The astonishingly even and continuous canopy of all *Shorea albida* stands is punctuated by individual tree gaps, and more occasional wind-throw gashes, some in radiating clusters from a central massive collapse betraying a downblast (Plate 3.9i). The diffuse-leaved canopies of the *S. albida* consociations are unique in their complete absence of a main canopy stratum and flora (Chapter 2 section 3.2). Canopy individuals of *S. albida* in the taller stands are relatively stout for their height compared to those in the short stands – the opposite of what is seen when comparing the tall upland MDF and the short *kerangas* (Chapter 2). The explanation may lie in the fact that the tallest trees have dense timber but are all hollow. The shortest are solid, have light timber, but appear to experience frequent tip-ups and to regenerate by epicormic

leaders from the fallen trunks. Another peat swamp species, *Dryobalanops rappa*, is the only other dipterocarp observed to have this ability. It appears that the tall individuals of *S. albida* at first grew rapidly to full height, putting on soft, easily rotting timber. They then continued to grow in diameter, but more slowly, surviving long; whereas the short trees mostly failed to survive long after achieving full height. Trees of like size in the two stands should therefore be compared; but that is now impossible because the tallest stands at the coastal periphery are not regenerating (Chapter 2).

The peat swamp surface is dry except following storms, and consists of a platform of interwoven surface roots perched above the water table, on which raw humus accumulates, supported by coarse branching roots which descend into the mire beneath. The tallest alan trees reproduce, but few of their seedlings establish, and almost none survive; pole-sized trees hardly exist. Patches of abundant regeneration have only been seen on the exposed peat along banks of drainage channels. The reason for these extraordinary facts is unknown. Could they reflect increasingly prolonged periodic flooding, consequent to the currently increasing eustatic sea levels due to global warming (Chapter 8)?

2.4.4. Seasonally flooded swamp forests

Along the major rivers of slightly seasonal eastern and southern Borneo, also New Guinea, south-east Sumatra and south-east Johor[135] there are alluvial floodplains, usually narrow, where seasonal flooding prevents peat accumulation yet is insufficient to reduce anaerobic conditions in the soil profile. Here, and in contrast with the successional communities along the banks, species which develop intertwined surface roots and pneumatophores, yet not dense fine root mats, may flourish. Notable among them are *Lophopetalum multinervium,* recalling *L. wightianum* in similar albeit more restricted habitats in the Western Ghats, and *Terminalia phellocarpa.*

2.5. *Mixed evergreen forests east of the Sunda Shelf*

2.5.1. *Philippine forests*
Philippine lowland forests, now almost gone, represented a floristic and ecological extension of Sunda MDF. Predominantly on udult clay loam soils influenced by basic volcanicity, their canopies are dominated by light hardwood *Shorea* species in those sections with red wood, known in Malay-speaking regions as *meranti,* and in the Philippines as *lauan*. These forests are extraordinary in two respects:

1. Until quite recently, MDF extended north of Luzon to the Babuyan Islands at 19° N, in a climate with no more than two dry months; but also westwards into climates with

up to six dry months.[136] In each case these forests were distinguished by a small number of local endemics (mostly evergreen), but their core flora was an impoverished subset of the Sunda generic flora, yet with archipelago-wide species endemism (Chapter 6); that is, many of the same species that dominate the perhumid east and south-east, or their sister species, extend to climates with more than three dry months. Unlike in continental Asia, for example, the red merantis *Shorea negrosensis* and *S. polysperma* extend into strongly seasonal regions,[137] although other widespread Philippine dipterocarps, notably *Dipterocarpus grandiflorus* and *D. gracilis*, also extend to parts of Indo-Burma with up to 4.5 dry months, while *D. alatus*, rare in seasonal Luzon, is restricted to them there. Although *Pinus merkusii* occurs on the western island of Mindoro and *Tectona hamiltoniana* on karst in the western Luzon Bataan Province, the tree flora of the seasonal evergreen and tall deciduous forests of continental East Asia is hardly represented in seasonal regions of the Philippines[138] (Chapter 6).

2. In continental Asia, the line between perhumid and seasonal climates (and between MDF and seasonal evergreen dipterocarp forest) occurs only once, in the Isthmus of Kra, where it is quite distinct (Chapter 6). In the southern Philippines every major island has distinct rainfall seasonality regimes on its eastern and western slopes (map inside front cover). The Pacific sides support a narrow perhumid eastern slope with climates of less than two dry months, while a dry season of more than six months occurs in parts of western Luzon. This climatic variability must have changed cyclically in space and extent throughout the Pleistocene. *Whether the extension of MDF into seasonal climates is the result of reduced competition in the virtual absence of the continental deciduous tree flora, or to the enormous capacity of the prevailing deep base-rich volcanic clay loams to store water, is unknown. So little of this forest remains that opportunity to investigate it is running out.*

2.5.2. *Forests of eastern Malesia*

East of Wallace's Line, a complex pattern of seasonal and perhumid rainfall regimes extend into New Guinea. The southern islands tend to be drier and more seasonal owing to the dry south-east trade winds emanating from Australia. Beyond expedition field descriptions,[139] aerial surveys in Papua in which successional forest (judged by its composition) is distinguished from more local primary forest with dipterocarp canopy recognized by its larger crown diameters,[140] and a few single c.1 ha sample plots,[141] little ecological information on these forests exists. I will therefore be focusing rather on their historical biogeography, in Chapter 6. The northern islands, and slopes on some southern islands not subject to the dry southeasterly winds, mainly experience orographic rainfall and wet climates lacking a dry season. Volcanic soils, and locally karst, prevail (Fig. 1.4). The few plots available imply that floristic composition is also influenced by the lack of land bridges between many of the islands, thereby amplifying the exigencies of history (Chapter 6). There is a secondary evanescence of the West Malesian tree flora in lowland New Guinea. There, Sunda families and genera continue to dominate (Table 3.7), while some species also extend eastwards (Plate 3.10). *Ultramafic substrates and kerangas (which still awaits systematic documentation), are the sole lowland habitats where Australasian taxa are well represented.* Lowland mixed forest communities, although enriched by Australasian elements, are less species-diverse than those of the Sunda Shelf (Chapter 7). There are few dipterocarp species, and those that are present appear to be a random subset (Chapter 6). Surprisingly, dipterocarps rarely achieve canopy dominance, although in the Ceram lowlands the mighty *kayu bapa*, the dark red meranti *Shorea selanica*, was a notable exception – now almost logged out.

Plate 3.10 A widespread successional tree of moist clay soils, *Pometia pinnata*. **a**, Wanang, Papua New Guinea; **b**, Kuala Belalong, Brunei. The expanding shoots are brilliant with anthocyanin-red foliage. (**a**, S. Davies; **b**, P.A.)

Table 3.7 Composition of the tree flora from the CTFS and selected other plots in the Asian tropics, by family: the number before each colon is the number of genera, that after is the total number of species in those genera; the genus or genera with the largest number of species, and its number of species, is in parentheses; all for that plot. S SEDF: Southern seasonal evergreen dipterocarp forest.

Location	Pasoh	Lambir	Bt. Timah	Palanan	Sinharaja	Khao Ban Thad	Nam Cát Tiên	Huai Kha Khaeng
Area(ha.)	50	52	2	16	25	20	5	50
Min. diam.(cm)	1	1	1	1	1	1	10	1
Trees/ha.	1788	2664	2473	3379	5016	4110		1476
Basal area/ha(m²)	31	43	35	40	46			31
No. species	814	1192	347	336	205	596		291
Fisher's α	100	153		46	23	84		31
Forest type	MDF	MDF	MDF	MDF	MDF	S SEDF	Semi-evergreen	Semi-evergreen
Altitude (m)	100–125	200–325	100–150	100–150		150–300	150–200	549–638
Acanthaceae						01:01		
Acariaceae	08:13	09:20	01:01	01:03	02:02	05:09		04:06
Aceraceae								
Actinidiaceae		01:03		01:01				
Alangiaceae	01:04	01:03	01:03	01:01				01:02
Altingiaceae								
Anacardiaceae	11:32 (*Mangifera* 13)	13:33	07:09	04:08	4:8 (*Semecarpus* 5)	10:12	7:11 (*Mangifera* 4)	04:05
Anisophylleaceae	01:01	01:05	01:02					
Annonaceae	19:42	16:54	10:20 (*Polyalthia* 8)	10:12	03:03	12:17	06:09	8:11 (*Polyalthia* 4)
Apocynaceae	05:06	04:07	02:03	03:04		05:06	03:03	03:04
Aquifoliaceae	01:02	01:02				01:02	01:00	01:01
Araliaceae	02:02	02:02	01:01	01:01		01:01		01:01
Arecaceae	05:05	?	03:05	05:05		6:6??		
Araucariaceae								
Aristolochiaceae	01:01				01:01			
Asteraceae	01:01	01:01		01:01		01:01		
Berberidaceae								
Betulaceae								
Bignoniaceae	02:02		01:01	01:01		04:05	02:02	05:06
Bonnetiaceae	01:01	01:01						
Boraginaceae	02:02	01:01						02:03
Brassicaceae	02:24	01:01						01:01
Burseraceae	04:25	5:40 (*Santiria* 17)	03:10	02:05		04:10	01:01	03:05
Calycanthaceae								
Cannabinaceae	02:02	01:01	01:01	04:04		03:03		02:04
Capparaceae								
Caprifoliaceae								
Casuarinaceae								
Celastraceae	06:06	06:14	03:04	03:03	01:01	07:10	01:01	04:04
Chloranthaceae		01:01						
Chrysobalanaceae	04:07	05:08		01:01		01:01	01:01	
Combretaceae	01:04	01:03	01:01	01:03		1:3 (*Terminalia* 2)	2:4 (*Terminalia* 3)	2:4 (*Terminalia* 3)
Connaraceae		01:02		01:02				01:01
Convolvulaceae	01:01	01:03				01:01		
Crypteroniaceae	01:01	01:01	01:01	01:01	01:01		01:01	
Ctenolophonaceae	01:01	01:01	01:01					
Cunoniaceae								
Cyatheaceae								
Cycadaceae							01:01	

Location	Thai tall deciduous	Mudumalai	Halmahera	New Guinea	Xishuangbanna	Nanjeng Shan	Doi Inthanon	Gutianshan
Area(ha.)	16	50	0.5	1	20	3	15	24
Min. diam.(cm)	1	1	10	5	1	1	1	
Trees/ha		1250	742		4792	3582	4910	5827
Basal area/ha(m²)		25				36	40	
No. species		71	76		468	118	162	159
Fisher's α	10.5-23	9.2		65	42	16	20	18
Forest type	Tall deciduous	Tall/short deciduous	E.Mal.SEDF	E.Mal.SEDF	Northern SEDF	N. coast SEDF	Lower montane	Warm temperate evergreen
Altitude (m)	300–650	550–650	630	440-600	550–850	150–200		446–714
Acanthaceae					01:01			
Acariaceae		02:02	c.1:c.01	01:01	01:03			02:02
Aceraceae							01:01	01:03
Actinidiaceae					01:02	01:01		
Alangiaceae	01:01				01:01	01:01	01:01	01:01
Altingiaceae								
Anacardiaceae	02:02	04:04	03:03	5:7 (*Buchanania* 2)	05:06			03:03
Anisophylleaceae								
Annonaceae	2:3 (*Polyalthia* 2)		01:01	3:6 (*Xylopia* 3)	07:14			
Apocynaceae	02:02	02:02		02:02	03:04			
Aquifoliaceae				1:3 (*Ilex*)	1:2(Ilex)	1:6(Ilex)	1:2(Ilex)	1:10(Ilex)
Araliaceae				02:02	05:09	01:01	01:01	02:02
Arecaceae				02:02		01:01		
Araucariaceae			01:01	01:01				
Aristolochiaceae								
Asteraceae				01:01	01:03			
Berberidaceae								01:01
Betulaceae							01:01	01:01
Bignoniaceae	05:06	02:02			05:06	01:01		
Bonnetiaceae								
Boraginaceae	2:4 (*Cordia* 3)	1:2 (*Cordia*)			01:01	01:01	01:01	01:01
Brassicaceae	01:01			01:01				
Burseraceae	02:02	01:01	c.3:05 (*Canarium* 3)	3:7 (*Canarium* 4)	02:05		01:01	
Calycanthaceae								01:01
Cannabinaceae				01:01				
Capparaceae					01:02		1:2(*Capparis*)	
Caprifoliaceae							01:01	3:4(*Viburnum* 3)
Casuarinaceae				01:01				
Celastraceae		01:01	01:01	01:01	01:01	01:01	01:01	1:3(*Euonymus*)
Chloranthaceae								
Chrysobalanaceae			01:01					
Combretacerae	2:6 (*Terminalia* 5)	2:4 (*Terminalia* 3)		1:3 (*Terminalia*)	01:02		01:01	
Connaraceae								
Convolvulaceae								
Crypteroniaceae					01:01			
Ctenolophonaceae								
Cunoniacedae			01:01	c.1:15				
Cyatheaceae					01:01			
Cycadaceae								

Location	Pasoh	Lambir	Bt. Timah	Palanan	Sinharaja	Khao Ban Thad	Nam Cát Tiên	Huai Kha Khaeng
Daphniphyllaceae								
Dichapetalcaeae	01:01							
Dilleniaceae	1:03 (*Dillenia*)	01:05	01:02	01:03	02:02	02:06	01:01	01:02
Dipterocarpaceae	7:30 (*Shorea* 14)	8:88 (*Shorea* 56)	05:09	3:8 (*Shorea* 6)	3:11 (*Shorea* 8)	06:10	05:04	05:07
Ebenaceae	1:23 (*Diospyros*)	1:35 (*Diospyros*)	1:9 (*Diospyros*)	1:9 (*Diospyros*)	1:5 (*Diospyros*)	1:18 (*Diospyros*)	1:9(*Diospyros*)	1:6 (*Diospyros*)
Elaeocarpaceae	02:08	02:11	2:7 (*Elaeocarpus* 6)	01:02	02:02	01:05	01:02	01:02
Ericaceae								
Erythroxylaceae	01:01	01:01			01:01	01:01		
Escalloniaceae								
Euphorbiaceae	20:37	23:56 (*Macaranga*15)	???	11:19	7:11? (*Agrostistachys* 4)	14:24 (*Macaranga* 7)? 2:4		12:14
Fagaceae	03:15	3:21`	03:06			03:10		2:4 (*Lithocarpus* 3)
Gentianaceae	02:04	02:08	01:01			01:02		
Gesneriaceae			01:01					
Gnetaceae	01:01	01:01						
Gramineae						01:01		
Grossulariaceae								
Guttiferae	4:35 (*Garcinia* 17)	5:50 (*Garcinia* 23) *Calophyllum* 16)	3:16 (*Garcinia* 9, *Calophyllum* 6)	3:11 (*Garcinia* 6)	3:9 (*Garcinia* 4)	4:20 (*Garcinia* 15, *Calopyllum* 13)	4:9(*Garcinia* 6)	02:04
Hamamelidaceae								
Himantandraceae								
Hydrangeaceae								
Icacinaceae	05:09	05:09	01:01	01:01	03:03	02:03		02:02
Illiciaceae								
Irvingiaceae	01:01	01:01	01:01			01:01		01:01
Ixonanthaceae	01:02	01:01						
Juglandaceae	01:01	01:01						01:01
Lamiaceae	04:08	03:09	03:01	03:06				
Lauraceae	10:48 (*Litsea* 14)	12:78 (*Litsea* 29)	7:14 (*Litsea* 7)	6:17 (*Litsea* 8)	5:10 (*Cinnamomum* 4)	5:20 (*Litsea* 12)	8:15(*Beilschmiedia*, *Cryptocarya* 3)	7:11(*Beilschmiedia* 4)
Lecythidaceae	02:04	02:08	01:01	02:03		01:01		01:01
Leeaceae		01:01		01:07	01:01	01:02		
Leguminosae	15:28	11:24	07:11	08:09	02:02	13:16	10:12	11:16 (*Dalbergia* 4, *Albizia* 3)
Liliaceae	01:03	01:02	01:01			01:01		
Linaceae								
Lythraceae				01:01		02:02		1:6 (*Lagerstroemia*)
Magnoliaceae	02:02	01:06	01:01	01:01		01:01		01:01
Malvaceae	11:27	11:57	06:09	03:04	01:02	06:13	4:7(*Sterculia* 3)	7:15 (*Colona, Pterospermum, Stercula* 3)
Melastomaceae	02:03	03:08	01:01	03:05	02:03	02:02		
Memecylaceae	1:12 (*Memecylon*)	1:12 (*Memecylon*)	1:10 (*Memecylon*)	01:03	1:8 (*Memecylon*)	1:8 (*Memecylon*)		1:2 (*Memecylon*)
Meliaceae	8:43 (*Aglaia* 22)	10:55 (*Aglaia* 25)	5:10 (*Aglaia* 6)	5:9 (*Dysoxylum* 9)	03:05	11:38 (*Aglaia* 15)		8:11(*Aglaia* 3)
Monimiaceae	01:01	01:01	02:02	01:01		01:01		
Moraceae	05:23	5:45 (*Ficus* 28)	3:11 (*Artocarpus* 7)	2:18 (*Ficus* 16)	01:01	5:42 (*Ficus* 31, *Artocarpus* 8)	03:04	11:16 (*Ficus* 16, *Artocarpus* 3)
Musaceae								
Myricaceae								

Location	Thai tall deciduous	Mudumalai	Halmahera	New Guinea	Xishuangbanna	Nanjeng Shan	Doi Inthanon	Gutianshan
Daphniphyllaceae							01:01	01:01
Dichapetalcaeae					01:01			
Dilleniaceae			01:02	01:01				
Dipterocarpaceae	01:01	01:01	01:01		01:01			
Ebenaceae	1:4 (*Diospyros*)		01:01	1:2 (*Diospyros*)	1:4 (*Diospyros*)	01:01	1:2 (*Diospyros*)	1:2 (*Diospyros*)
Elaeocarpaceae			01:01	4 (*Elaeocarpus* 2, *Sloanea* 2)	2:18 (*Elaeocarpus* 17)	02:02	1:3 (*Elaeocarpus*)	2:4 (*Elaeocarpus* 3)
Ericaceae					01:01			04:10
Erythroxylaceae								
Escalloniaceae								
Euphorbiaceae	04:05	02:02	2:03 (*Macaranga* 2)	13:18	08:21	01:01	03:03	02:02
Fagaceae			02:02	1:4 (*Lithocarpus*)	2:14 (*Castanopsis* 9)	3:10 (*Lithocarpus, Quercus* 3)	3:8 (*Castanopsis, Lithocarpus* 3, Q. 2)	4:10 (*Castanopsis* 4, *Quercus* 3)
Gentianaceae	01:01					01:01		
Gesneriaceae								
Gnetaceae					01:01			
Gramineae		01:01						
Grossulariaceae								
Guttiferae	2:3 (*Cratoxylon* 2)		2:08 (*Calophyllum* 4, *Garcinia* 4)	2:18 (*Garcinia* 12, *Calophyllum* 6)	03:07	02:02	01:01	
Hamamelidaceae								04:04
Himantandraceae				01:01				
Hydrangeaceae								02:02
Icacinaceae				1:3 (*Stemonurus*)	04:06		04:04	
Illiciaceae								01:01
Irvingiaceae								
Ixonanthaceae				01:01				
Juglandaceae			01:01		01:03		01:01	
Lamiaceae		04:04		04:04	04:07	03:04	01:01	03:06
Lauraceae	01:01		c.4:07 (*Litsea* 2)	9:27 (*Cryptocarya* 11)	11:50 (*Litsea* 15, *Beilschmiedia* 9, *Phoebe* 7, *Cinnamomum*, *Cryptocarya* 6)	6:12 (*Machilus* 14)	8:18 (*Litsea* 7, *Beilschmiedia* 3, *Cinnamomum* 2, *Cryptocarya* 2)	6:14 (*Machilus* 4, *Litsea* 3, *Lindera* 3)
Lecythidaceae	02:02	01:01	01:01	01:01				
Leeaceae								
Leguminosae	10:16 (*Dalbergia* 6, *Albizia* 2)	11:13 (*Cassia* 2, *Dalbergia* 2)	02:02	3:5 (*Archidendron* 3)	08:18	02:02	01:01	03:03
Liliaceae				01:01				
Linaceae				01:01				
Lythraceae	1:6 (*Lagerstroemia*)	1:2 (*Lagerstroemia*)		1:1 (*Lagerstroemia*)	02:02	01:01		
Magnoliaceae					03:06	02:02	02:02	03:03
Malvaceae	9:12 (*Sterculia* 2)	5:6 (*Grewia* 2)		04:04	06:14	02:02	02:02	01:01
Melastomaceae				03:04	02:04	02:02	01:01	
Memecylaceae	1:1 (*Memecylon*)		1:c.01	1:2 (*Memecylon*)				
Meliaceae	02:02	01:01	1:2 (*Aphanomixis*)	5:21 (*Dysoxylum* 8)	9:24 (*Dysoxylum* 8)		02:02	
Monimiaceae								
Moraceae	3:6 (*Ficus* 4)	2:6 (*Ficus* 5)	01:01	4:16 (*Ficus* 10)	5:30 (*Ficus* 22)	1:7 (*Ficus*)	03:03	02:02
Musaceae								
Myricaceae							01:01	01:01

Location	Pasoh	Lambir	Bt. Timah	Palanan	Sinharaja	Khao Ban Thad	Nam Cát Tiên	Huai Kha Khaeng
Myristicaceae	5:31 (*Knema* 13)	4:40 (*Knema* 17)	4:14 (*Knema* 5)	04:08	01:01	3:8 (*Knema* 5)	01:01	02:02
Myrsinaceae	02:08	01:09	01:02	01:01	01:01	2:7 (*Ardisia* 5)		3:7 (*Ardisia* 5)
Myrtaceae	4:48 (*Syzygium* 45)	4:57 (*Syzygium* 54)	2:22 (*Syzygium* 21)	1:21 (*Syzygium*)	2:11 (*Syzygium* 9)	1:23 (*Syzygium*)	2:6 (*Syzygium* 5)	2:6 (*Syzygium* 5)
Nyctaginaceae								
Nyssaceae	01:02	01:06	01:01		01:01			
Ochnaceae	02:02	02:02	01:01		01:01			
Oleaceae	01:04	01:03		01:01	01:01	02:05	02:03	02:03
Opiliaceae	02:02					01:01		01:01
Oxalidaceae	01:02	01:02				01:01		
Palmaceae						6:6		
Pandaceae	01:02	01:01	02:04	01:01		01:01	01:01	1:3 (*Antidesma*)
Pandanaceae	01:02		01:01					
Phyllanthaceae	8:42 (*Aporusa* 13)	7:57 (*Aporusa* 22)	4:10 (*Aporusa* 8)	6:8 (*Baccaurea* 6)	02:04 (*Glochidion*, *Mallotus*)	18:28 (*Aporusa* 7)	3:5 (*Aporusa* 3)	8:13 (*Aporusa* 3)
Pinaceae								
Piperaceae		01:01				01:01		
Pittosporaceae								
Podocarpaceae	01:01	01:01	01:01			01:02		
Polygalaceae	1:10 (*Xanthophyllum*)	1:25 (*Xanthophyllum*)	1:6 (*Xanthophyllum*)	02:02		1:5 (*Xanthophyllum*)		01:01
Proteaceae	02:02	02:04		01:01		01:01		01:01
Putranjivaceae	01:08	01:12	01:05	01:03		01:02		01:02
Rhamnaceae		04:48	01:01	01:01		01:03		
Rhizophoraceae	02:02	03:05	03:03	01:01	02:02	03:03	01:01	01:01
Rosaceae	01:03	01:03	01:03	01:02		01:03	01:01	09:10
Rubiaceae	22:46	22:59	15:18	10:17 (*Psychotria* 5)	11:15 (*Psychotria* 4)	22:42 (*Ixora* 8)	12:15	14:15
Rutaceae	04:05	04:06	02:02	04:04	01:01	06:10	02:02	09:10
Sabiaceae		01:03						01:01
Santalaceae	05:06	07:05	04:04	01:01	01:01			01:01
Sapindaceae	10:20	09:19	04:04	01:03	04:04	6:13 (*Nephelium* 5)	03:03	10:12
Sapotaceae	06:14	5:33 (*Madhuca* 15)	05:13	02:10	3:7 (*Palaquium* 5)	04:08	03:03	01:01
Saxifragaceae	01:01	01:05	01:01					
Schizomeriaceae								
Scrophulariaceae								
Simarubiaceae	01:01	03:03	01:01			03:03		02:02
Solanaceae								01:01
Staphyliaceae	01:01	01:01		01:01		01:01	01:01	01:01
Styracaceae	01:01							01:01
Symplocaceae	01:05	01:07	01:02		01:03	01:02	01:01	01:01
Tetramelaceae			01:01			1:1 (*Tetrameles*)	1:1 (*Tetrameles*)	1:1 (*Tetrameles*)
Tetrameristicaceae		01:01						
Theaceae	02:02	04:11	02:02	02:02		01:01	01:01	01:01
Thymeliaceae	02:02	03:11	03:03		01:01	01:01		
Torricelliaceae						1:1 (*Aralidium*)		
Trigoniaceae	01:01	01:02						
Ulmaceae	01:03	01:03		01:01				03:03
Urticaceae				03:03				01:01
Violaceae	01:03	01:04				01:02		

Location	Thai tall deciduous	Mudumalai	Halmahera	New Guinea	Xishuangbanna	Nanjeng Shan	Doi Inthanon	Gutianshan
Myristicaceae			c.3:c.05 (*Gymnacranthera* 2)	4:17 (*Myristica* 9)	03:08			
Myrsinaceae				01:03	03:05	2:5 (*Ardisia* 4)	4:8 (*Ardisia* 4, *Maesa* 2)	
Myrtaceae	1:1 (*Syzygium*)	01:01	1:11 (*Syzygium*)	8:51 (*Syzygium* 41)	2:14 (*Syzygium* 13)	2:4 (*Syzygium* 3)	1:2 (*Syzygium*)	1:1 (*Syzygium*)
Nyctaginaceae				01:01				
Nyssaceae					01:01		01:01	01:01
Ochnaceae								
Oleaceae		01:01		01:02	02:04	02:02	03:03	02:02
Opiliaceae								
Oxalidaceae					01:01	01:01		
Palmaceae								
Pandaceae	1:4 (*Antidesma*)	01:01	01:01	01:02	01:02	01:01	01:01	1:1 (*Antidesma*)
Pandanaceae				01:01	01:01			
Phyllanthaceae	3:5 (*Aporusa*, *Bridelia* 3)	02:02		03:07	07:10	3:5 (*Glochidion* 3)	01:01	02:02
Pinaceae								02:02
Piperaceae					01:01			
Pittosporaceae				01:01			01:01	01:01
Podocarpaceae				02:03	01:02	01:02		
Polygalaceae				02:04	01:04			
Proteaceae					02:02	01:01		
Putranjivaceae	01:01				01:01	01:01	01:01	
Rhamnaceae	01:01	1:2 (*Zizyphus*)		01:01	01:02			02:02
Rhizophoraceae			01:01	02:03	01:02		01:01	
Rosaceae				01:07	03:05	02:02	2:4 (*Prunus* 3)	7:10 (*Photinia* 4)
Rubiaceae	06:06	05:05	3:04 (*Nauclea* 2)	08:13	20:27	6:10 (*Lasianthus* 5)	7:12 (*Lasianthus* 4, *Psychotra* 3)	07:07
Rutaceae	01:01			06:12	05:11	04:04	02:02	02:02
Sabiaceae								
Santalaceae		01:01			01:01			
Sapindaceae	04:05	02:02	02:02	08:11	09:09			
Sapotaceae			01:01	3:15 (*Pouteria* 12)	04:05		01:01	
Saxifragaceae					01:01		02:02	
Schizomeriaceae				01:03			02:02	
Scrophulariaceae								
Simarubiaceae					02:03			01:01
Solanaceae							01:01	
Staphyliaceae					02:03		01:01	01:01
Styracaceae					01:02	02:02		2:4 (*Styrax* 3)
Symplocaceae				01:01	01:02	1:3 (*Symplocos*)	01:03	1:5 (*Symplocos*)
Tetramelaceae	1:1 (*Tetrameles*)				1:1 (*Tetrameles*)			
Tetrameristicaceae								
Theaceae			03:03	03:03	04:05	07:08	4:5 (*Eurya* 2)	7:11 (*Camellia*, *Eurya* 3)
Thymeliaceae				02:02	01:01	01:01	01:01	01:01
Torricelliaceae								
Trigoniaceae								
Ulmaceae			01:01	01:02	04:07			01:01
Urticaceae							02:02	
Violaceae								

Part II. Seasonal forests: the influence of seasonality in rainfall and temperature overlays that of geology and soil

Worldwide, most lowland tropical forests occur in climates with rainfall that seasonally reduces to below the expected rate of evapotranspiration. This influences their phenology and floristic composition, strongly differentiating them from MDF and other tropical forests of perhumid climates[142] (Chapter 5). An excellent systematic survey of the indigenous forest types of Pakistan, India and Burma was published by Sir Harry Champion.[143] In Chapter 2, I described the major changes in forest structure, dynamics and phenology taking place at the continental ecotone between the mixed dipterocarp and other evergreen rainforest communities of perhumid Malesia and seasonally wet Indo-Burma, and also between the seasonal evergreen forests and deciduous forest types along the rainfall seasonality gradient as originally defined by Champion.[144] Here, I discuss the relationship between these gradients and floristic composition. The major floristically defined forest types, particularly those dominated by teak and sal (*Tectona grandis* and *Shorea robusta*) have subsequently been subject to continuing silvicultural research in modern India (summarised in Joshi[145]). Champion had no experience of perhumid Sri Lanka or Malesia and did not include MDF in his classification.[146]

3. Seasonal evergreen dipterocarp forests

3.1. *Southern seasonal evergreen dipterocarp and other forests* (refer to Chapter 2, section 4.1)

The major ecological disjunction across the Kangar-Pattani Line (6–7° N: see map inside back cover), by which that line is defined, is a drastic decline in the number of dipterocarp species in forest communities (Chapter 6), and the cessation of supra-annual canopy mass flowering and mast fruiting (Chapter 5). Whereas the CTFS plot at Pasoh (Negeri Sembilan) contains 30 dipterocarp species including eight red and two yellow light and medium hardwood merantis, the Khao Ban Thad plot in peninsular Thailand contains eight dipterocarp species and no meranti. An even narrower ecotone separates the remaining mixed dipterocarp forest at Oliyagankelle, 80°45' N in southern Sri Lanka, which includes five dipterocarp species, from the intermediate-zone forests lacking dipterocarps to its immediate east. Here the northern margin is more gradual: north of the Kalu Ganga in the western lowlands (6°25' N), the dominant *Shorea* (sect. *Doona*) occurs only in a belt at the base of the mountains, whereas *Dipterocarpus zeylanicus, Vateria copallifera* and two *Shorea* (sect. *Shorea*) persist in an increasingly species-poor

Pajanelia longifolia	Bignoniaceae
Radermachera glandulosa	Bignoniaceae
Anisoptera costata, A. scaphula	Dipterocarpaceae
Dipterocarpus costatus, D. dyeri, D. grandiflorus, D. hasseltii, D. kerrii	Dipterocarpaceae
Hopea helferi, H. pierrei	Dipterocarpaceae
Parashorea stellata	Dipterocarpaceae
Shorea guiso, S. gratissima, S. hypochra	Dipterocarpaceae
Lagerstroemia floribunda	Lythraceae
Syzygium siamense	Myrtaceae
Amesiodendron chinense	Sapindaceae
Scaphium scaphigerum	Achariaceae
Pterygota alata	Achariaceae

Table 3.8 Some more widespread tree species characteristic of southern seasonal evergreen dipterocarp forest.

forest northwards to c.7° N. Following Champion, I distinguish this seasonal evergreen forest formation, of both South and East Asia, as *southern* seasonal evergreen dipterocarp forest, as it differs floristically and in other respects from forests experiencing similar rainfall seasonality in the north (see section 3.2 below).

Seasonal evergreen forests are floristically quite distinct from the MDF growing in perhumid regions to the south of them (Table 3.8). Some areas of perhumid Malesia suffer occasional extreme droughts and perhaps locally even natural forest fires[147] (Chapter 1), but the floristic make-up of such areas shows little impact of these events. Southern seasonal evergreen dipterocarp forests have about three-quarters the number of tree species as does MDF (Table 3.7). But many of their Far Eastern species do also occur in Peninsular Malaysian MDF. A few canopy pioneer species are deciduous (e.g. *Tetrameles nudiflora, Lagerstroemia floribunda*). In some cases Dipterocarpaceae are less dominant, in others the forests are dominated by a single species, especially *Dipterocarpus indicus* in India and *D.costatus* and *D. grandiflorus* in the Far East.[148] (Their distinct phenology will be described in Chapter 5.) Notable is the absence of a dense understorey of dipterocarp juveniles (Chapter 2).[149]

3.1.1. *East Asian southern seasonal evergreen dipterocarp forests*

Although seasonal evergreen forests have been the subject of the path-breaking silvicultural researches of Kadambi in South India,[150] Chengappa in the Andamans,[151] and isolated ecological plot studies in the forests of north-east India, little has been published on local habitat-related floristic variation. The establishment of a 24 ha plot within the CTFS network at Khao Ban Thad sanctuary in Trang (peninsular Thailand), under Sarayudh's leadership, promises major increases in our understanding. Sarayudh has also established 42, 1 ha plots in primary or near-primary forest throughout Thailand, which include seven in this forest type, as yet awaiting detailed analysis. Most knowledge of this forest type's variance in relation to soils and topography therefore remains as yet limited to casual field observation.

Dipterocarpus costatus has an analogous ecological distribution to *Shorea curtisii* in MDF. It dominates the emergent canopy in some coastal forests of peninsular Thailand and southern Cambodia to southern Vietnam,[152] and maintains a scattered or locally dominant presence on ridges and spurs from c.200 m upwards into lower montane forest from northern Peninsular Malaysia to northern Thailand, and from the Andaman Islands across to Arakan Yoma in coastal Burma and through to Chittagong.[153] In the coastal hills it is confined to freely draining sandy soils, often with shallow fine-root-matted raw humus, while on ridges the soil physical character appears to be more variable, though raw humus is often present. In the lowlands, a distinct tree flora is associated which includes *Heritiera javanica, Swintonia schwenckii* (peninsular Thailand) and *S. pierrei* (south-east Thailand and Cambodia), and *Nageia* cf. *wallichianus*, with a few deciduous species in the canopy.

Forests on steep moist slopes, especially over granite, contrast in their greater degree of canopy disturbance and patchy abundance of large vines (Chapter 2). They also have a distinct tree flora, including the dipterocarps *Dipterocarpus gracilis* (*D. hasseltii* may here be synonymous), *Parashorea stellata* and *Hopea helferi*. A basic floristic division therefore occurs in these lowland forests in the Far East, which appears closely analogous to the primary division found in north-west Borneo and Peninsular Malaysia: that between communities on higher nutrient clay versus those on lower nutrient sandy soils. In this seasonally dry climate, though, neither soil bears continuous surface raw humus with its associated root mat. In this respect it differs from much upper dipterocarp *D. costatus* forest in perhumid Peninsular Malaysia (see Chapter 4).

Inland seasonal evergreen dipterocarp forests on undulating siliceous igneous and sedimentary substrates in the Andamans, in peninsular Thailand and in southern Vietnam appear to occupy an intermediate position, analogous to the Sunda *Dipterocarpus crinitus* MDF on sandy clay soils. Their flora includes species drawn from both sides of the above-described division, such as the ubiquitous *Irvingia malayana*. They also harbour distinct entities including *Shorea gratissima* in peninsular Thailand, as well as the more widespread *Dipterocarpus dyeri*, also *D. baudii* which extends southwards on sandy clay loams into Peninsular Malaysian and north Sumatran MDF (Chapter 6). Elements penetrate into the eastern Indian states north to Mizoram, but the relationship between this and semi-evergreen forest is complex here, and has not been studied. Seasonal evergreen forests with other species assemblages have been documented;[154] their interrelationships await research.

Outliers occur on the slopes of Khao Yai, south-west of the Khorat plateau in south-east Thailand; also in moister, less seasonal parts of Laos at least to Paksane (19° N), where elements of northern seasonal evergreen forests occur, including *Terminalia myriocarpa*[155] (see section 3.2. below) and the remarkable entire-leaved *Platanus kerrii*. Members of the flora, including *Hopea ferrea, Tetrameles nudiflora* and *Toona sureni* extend south to the peninsular Kangar-Pattani ecotone on coastal karst limestone.

3.1.2. *A unique seasonal kerangas*

At the coastal base of the Cardamom and Elephant mountains in Cambodia is an intermittent raised beach bearing humic podsol soils. Resembling *kerangas* in its even single-stratum canopy but hardly exceeding 20 m, the forest also bears floristic affinities with the Borneo *kerangas*. *Dacrydium elatum* dominates, along with the associated podocarp *Dacrycarpus imbricatus*, also *Syzygium zeylanicum, Tristaniopsis merguensis, Hopea pierrei*, and *Schima wallichii*, with *Alpinia* (Zingiberaceae) and *Nepenthes* in the field layer. This is the only *kerangas*-type forest known to me in the seasonal regions. It appears that, though peat cannot accumulate where there is even the briefest of annual dry seasons (see section 2.4.3.1 above), humic podsols with their associated raw humus can develop wherever dry seasons are of less than two months.

3.1.3. *South Asian seasonal evergreen dipterocarp forests*

Although patches still do survive, especially in sacred groves in the plains, evergreen forests west of the Western Ghats are now almost confined to the foothills. Their ecology has been investigated by S.N. Rai and Pascal and his colleagues, and Chandrasekharan, in the Western Ghats.[156] They decline in richness as the dry season lengthens from south to north. The decline is continuous, but Pascal[157] and others have recognized steps: at the southern end of the Palghat Gap in the Ghats, which appears to mark the northern limit of climates with less than four dry months; north of Mangalore, where *Diospyros* species increase in prominence and deciduous canopy species including *Lagerstroemia lanceolata, Terminalia bellirica* and *Chukrasia tabularis* become more frequent; and at 14° N, the northern limit of *Dipterocarpus indicus*, where *Hopea ponga* continues and deciduous species, notably *Vitex altissima, Spondias pinnata, Schleichera oleosa, Dillenia pentagyna, Tabernaemontana alternifolia* and *Sterculia guttata* become dominant in gap succession. The forest has thus become more strictly analogous to the semi-evergreen forests of East Asia. But here, predominantly evergreen forest nevertheless appears to extend into climates with up to six dry months. *The meteorological and ecological data here need careful, site-wise review.*

Pascal hypothesised that these western lowlands were originally covered in evergreen forest. Curiously, he does not explain how teak and its moist deciduous forest associates

Plate 3.11 Southern seasonal evergreen dipterocarp forests. **a,** Coastal forest, Tarutao, Bay of Bengal, Thailand. Barely emergent dipterocarp *Dipterocarpus costatus* (pale crowns); **b,** The widespread *Calophyllum polyanthum* (left), from South Asia to Sumatra, in southern seasonal evergreen dipterocarp forests over granite at Khao Ban Thad Wildlife Sanctuary, Trang, peninsular Thailand; **c,** Wind-pruned Southern seasonal evergreen dipterocarp forest dominated by *D. indicus,* on lower slopes of the Western Ghats at Pilarkan sacred grove, foreground; semi-evergreen forest below in the nearby foothills, mixed deciduous forest in the distance; **d,** Bank of a whitewater stream in the Katlekan Dark Forest, Western Ghats: southern seasonal evergreen forest with the endemic riparian *Myristica canarica* and *Pinanga dicksonii*. (**a,** G. Congdon; **b,** H.H.; **c, d** D. Willmore).

(including *Dalbergia latifolia, Xylia xylocarpa, Pterocarpus marsupium, Terminalia tomentosa, Lagerstroemia lanceolata, Schleichera oleosa, Anogeissus latifolia, Odina wodier, Gmelina arborea, Careya arborea, Buchanania latifolia, Stereospermum chelonoides* and *Erythrina indica*) could then coexist. It seems more likely that semi-evergreen and tall deciduous forest originally occupied most of the western coastal plains. The major dipterocarp genus of the Far East, *Shorea*, is absent. Two species of *Dipterocarpus* (Plate 3.11a,c), eight of *Hopea*, a *Vateria* and a *Vatica* partially make up for their absence: dipterocarps thus remain important canopy components but are overall less dominant than in Indo-Burmese analogues. These forests are rich in endemic genera and species, though the familial ranking of species richness differs little from the Far East. Some differentiation in relation to topography has been documented,[158] but none has been noted in relation to soil nutrients as it has in south-west Sri Lankan MDF. Soils are similar to those of south-west Sri Lanka too, with red-brown clay loams on slopes but, unlike in the Sri Lankan lowlands, widespread active laterite pans are present in the seasonal climate of the Kerala lowlands (Chapter 1).

The valleys running steeply down the Western Ghats, some opening into seasonal swamps, support a unique riparian flora, locally dominated by species of *Myristica*[159] (Plate 3.11d).

Saurauia roxburghiana	Actinidiaceae
Drimycarpus racemosus	Anacardiaceae
Choerospondias axillaris	Anacardiaceae
Polyalthia simiarum	Annonaceae
Markhamia stipulata	Bignoniaceae
Mayodendron igneum	Bignoniaceae
Dipterocarpus retusus ssp. *macrocarpus*	Dipterocarpaceae
Parashorea chinensis	Dipterocarpaceae
Shorea assamica ssp. *assamica*	Dipterocarpaceae
Sloanea tomentosa	Elaeocarpaceae
Glochidion ellipticum, G. lanceolarium	Euphorbiaceae
Castanopsis indica	Fagaceae
Gmelina arborea	Labiatae
Beilschmiedia roxburghiana	Lauraceae
Caryodaphnopsis spp.	Lauraceae
Chisocheton paniculatus	Meliaceae
Dysoxylum binectariferum	Meliaceae
Walsura robusta	Meliaceae
Artocarpus lakoocha	Moraceae
Ficus subincisa	Moraceae
Chionanthus macrocarpus	Oleaceae
Horsfieldia kingii	Myristicaceae
Saprosma ternatum	Rubiaceae
Sarcosperma kachinense	Sapotaceae
Xantolis burmanica	Sapotaceae
Leea compactiflora	Vitaceae

Table 3.9 Some more widespread tree species characteristic of northern seasonal evergreen dipterocarp forests.

Throughout their range these forests, when growing on moist clay loams, bear scattered emergent individuals of exceptional diameter and height in sheltered coves. These include species normally found in the main forest canopy: *Dysoxylum, Chukrasia* and other Meliaceae, *Calophyllum polyanthum*, some *Diospyros*.

3.2. Northern seasonal evergreen dipterocarp forests

A seasonal evergreen forest, dominated by one or a very few emergent dipterocarps, once ranged from Hainan and the adjacent coast into the upper valleys of northern Burma. It still exists in Arunachal west to Lakhimpur in the Brahmaputra valley of Assam, in North-East India. The overall range of these forests coincides with the northernmost and easternmost range of tropical lower montane forest, which it reaches at c.700–1,000 m altitude.[160] These forests exist thanks to copious dew deposition during the first three of the four dry winter months, Dec.–March (Chapter 1, Chapter 2 section 4.1.2).

Northern seasonal evergreen dipterocarp forests also survive still at their most majestic in small patches in Xishuangbanna District (Yunnan), where they have been most studied[161] but are rapidly being converted to rubber, and also in northern Laos and Vietnam. Here, the forest is dominated by the scattered emergent dipterocarp *Parashorea chinensis*, which can reach 50 m (Plate 3.12a). (This tree is now known only from southern Yunnan and Guangxi and northern Vietnam and Laos, although it may be the *Parashorea* whose fruits were noted by Kingdon-Ward on the road approaching Fort Hertz (Zemao) from the south in north-west Burma.[162]) In other such forests, *Dipterocarpus retusus* (whose northern range coincides with that of this northern dipterocarp forest), and *Shorea assamica* ssp. *assamica* (which is endemic to it) occur, but are shorter, albeit still emergent.[163]

A 20 ha plot in the CTFS network has been established by Min Cao and colleagues of the Xishuangbanna Tropical Botanical Garden, at Mengla (21°55' N, 101°15' E).[164] The climax tree flora is clearly divisible into three strata according to the presentation of their flowers, as in MDF – emergent, main canopy, and subcanopy –although the latter are difficult to recognize in the field owing to the general abundance of successional species (evidence of frequent disturbance). Unlike southern seasonal evergreen dipterocarp forests, dipterocarp regeneration is abundant in the vicinity of mature individuals here. Large vines are abundant where the canopy has been disturbed, much as in southern seasonal evergreen dipterocarp forests (Chapter 2).[165] Like MDF also, northern seasonal evergreen dipterocarp forest is rich in congeneric species series.[166] It possesses a rich endemic flora at species level, but few endemic genera, and these are overwhelmingly of Malesian affinity (Table 3.9). Rich local species endemism

is concentrated towards the east, in continental China and northern Indochina (Chapter 6). The Himalayan extension to the west, by contrast, is almost bereft of local endemics. There are exceptionally few temperate genera represented (though *Camellia* is widely represented and *Osmanthus, Fraxinus* and *Hydrangea* occur in the Taiwan CTFS plot at Nanjeng Shan), but warm temperate sister-species of tropical taxa may be present where these forests merge with warm temperate lowland Fagaceous 'lucidophyll' forests. One example is *Schima superba*, which appears at Nanjengshan. (The tropical *Schima wallichii* instead occurs at Mengla.)

Most notable, throughout these northern seasonal evergreen dipterocarp forests, is the presence of a significant lower montane component and lowland sister species to lower montane elements (Tables 3.7, 3.10). These include the widespread *Castanopsis indica* and the South China lowland endemic *C. boisii* (*megaphylla*) at Mengla. The lower montane element appears to prevail in forests at the base of slopes extending into lower montane altitudes. Their survival in the lowlands here in the north may be due to descent of fog during the south-west summer monsoon and to nocturnal dewfall during the northern cool dry season. Even in the wettest regions, though, there is a distinct element – most evident in the canopy – of widespread deciduous species of semi-evergreen and tall deciduous forests: *Acrocarpus fraxinifolius, Artocarpus lakoocha, Albizia chinensis, Alstonia scholaris, Neolamarckia cadamba, Chukrasia tabularis, Dalbergia* spp., *Erythrina stricta, Ficus variegata, Garuga pinnata, Lagerstroemia tomentosa, Pterospermum* spp., *Radermachera* spp*., Terminalia myriocarpa, T. bellirica, T. chebula, Tetrameles nudiflora, Toona ciliata, Turpinia pomifera.*

Within the CTFS Mengla plot, a remarkably narrow ecotone can be discerned between forests with canopies dominated by lowland and by lower montane species respectively. Subcanopy species also change abruptly along the same ecotone, and successional species suddenly decline in abundance. This line starts at 745 m on the main spur that enters from the north-east corner, but rises to 850 m at that corner (Fig. 4.3). The ecotone correlates with the change from steep slopes to more gentle convex ridges, and from udult yellow-brown clay loams free of surface humus to soils with a dark brown organic surface humic horizon.

North from Vinh Province, Vietnam (18°40' N), to south-west coastal Guangxi and southern Hainan island, is a floristically

distinct seasonal evergreen dipterocarp forest, rich in endemic species. It is distinguished by the frequent absence of *Parashorea chinensis* and presence of *Hopea hainanensis, Vatica glabrata* and *V. mangachapoi*. This forest coincides with the presence of winter rain and frequent typhoons off the China Sea.

These fascinating and sometimes magnificent northern seasonal evergreen dipterocarp forests are now reduced to patches (see Plate 8.13c). They are still declining in China, where Zhu Hua recognises two associations within northern seasonal evergreen dipterocarp forests and two within semi-evergreen forests.[167] Their presence, and that of the northernmost tropical lower montane forest above them (Chapter 4), indicates land suitable for *Hevea* rubber plantation.

Table 3.10 Species typical of tropical lower montane forest that also occur in the northern lowlands: **a,** at 600 m in the 50 ha plot Huai Kha Khaeng, north-west Thailand; **b,** below 200 m: in Cuc Phuong National Park, northern Vietnam and Namdampha National Park, Arunachal Pradesh, at the base of the India-Burma dividing range (a, from Bunyavejchewin *et al.* 2009, b, whole park record, from Biên & Thì 1992, 2.4 ha sample, after Nath *et al.* 2005)

a

Aesculus assamica (Sapindaceae)
Bischofia javanica (Phyllanthaceae)
Ulmus lanceifolia (Ulmaceae)
Schima wallichii (Theaceae)
Xantolis tomentosa (Sapotaceae)
Turpinia pomifera (Staphyleaceae)
Eriobotrya bengalensis (Rosaceae)
Castanopsis tribuloides (Fagaceae)
Engelhardia spicata (Juglandaceae)
Magnolia (*Michelia*) *baillonii* (Magnoliaceae)
Beilschmiedia gammieana (Lauraceae)

b

Species	Namdampha	Cuc Phuong
Acer oblongum		+
Aesculus assamica	+	
Altingia excelsa	+	
Bischofia javanica	+	+
Castanopsis indica	+	+
Cinnamomum bejolghota	+	+
Engelhardia spicata		+
Magnolia champaca	+	
M. griffithii	+	
M. hodgsonii	+	
M. hookeri	+	
M. mannii	+	
M. oblonga	+	
M. sp.		+
Sloanea sigun		+
Turpinia nepalensis		+

Plate 3.12 Northern seasonal evergreen dipterocarp and semi-evergreen forests. **a,** 'Drift' of emergent *Parashorea chinensis* (pale emergent crowns) in northern seasonal evergreen dipterocarp at Mengla, Xishuangbanna, Yunnan; **b,** Massive *Terminalia myriocarpa* at Mengla; **c,** Dense *Parashorea* and other sapling regeneration and treelets in the climax understorey; Mengla; **d,** *Terminalia myriocarpa* in flower; northern semi-evergreen forest above Phuntsholing, Bhutan. (a, b, H. Zhu; c S. Davies; d H.H.)

Plate 3.13 Semi-evergreen dipterocarp forests. **a,** Disturbed semi-evergreen forest on clay loam: *Dipterocarpus turbinatus* stand with deciduous canopy associates and largely evergreen, mainly secondary, understorey including pioneers; karst in background. Chiangdao, north-west Thailand; **b,** *Ficus stricta* in semi-evergreen forest over sandy loams, Huai Kha Khaeng Wildlife Sanctuary, west Thailand; **c,** *Bischofia javanica,* Huai Kha Khaeng Wildlife Sanctuary. A species confined to above c.700 m in perhumid climates and equatorial latitudes, descending into the lowlands on the most fertile soils and at higher latitudes; here at c.600 m; **d,** Abundant Zingiberaceae, sparse woody understorey, in semi-evergreen forest subject to periodic ground fires, Huai Kha Khaeng Wildlife Sanctuary; **e,** Massive evergreen *Mangifera indica* in semi-evergreen forest at 600 m, Similipal National Park, Odisha; **f,** Semi-evergreen forest beside a braided river bed issuing from the Himalaya, Buxa National Park. (All: H.H.)

4. Semi-evergreen forests

4.1. *Southern semi-evergreen forests: East Asia*

These are mixed forests with or without barely-emerging canopy dipterocarp species and an evergreen mature phase, but with predominantly deciduous pioneer and successional canopy gap-phase guilds.[168] They extend as far as northern Thailand, and from north-west Burma into North-East India along the Patkai range (27° N) (see Chapter 2). The composition of the dipterocarp forests of the Chittagong hill tracts of south-east coastal Bangladesh implies that they are semi-evergreen.[169] Patches of similar forest survive under climates with c.4.5–6 dry months in eastern Thailand and northeastern Cambodia, in the southern Laotian lowlands west of the Annamite range, and into Dong Nai Province (Vietnam).[170] At the southern boundary of this region, forests that are floristically similar but poorer in deciduous canopy species descend into regions of seasonal evergreen forest along the coasts and on offshore islands.[171] They also prevail in the northern and eastern lowlands of the Andaman Islands. Semi-evergreen forests lacking dipterocarps extend west in the *terai* and Siwalik front ranges of the Himalaya to Kumaon (Chapter 2 section 5.3)

In contrast with some seasonal evergreen forests (see below, also Chapter 7), there seems to be evidence of a decline in species richness at the community level in these Indo-Burmese forests along a gradient from west to east (Chapter 6). To cite one conspicuous example: there are six species of the dominant successional canopy genus *Lagerstroemia* in the CTFS plot at Huai Kha Khaeng in western Thailand. Compare this with the vast, though now fragmented, semi-evergreen forests stretching from the hills east of the Chao Phraya valley southeastwards into Cambodia and Dong Nai Province (Vietnam). Those forests are dominated by a single species, *L. calyculata,* sometimes growing to great size, albeit frequently with one to three subordinate *Lagerstroemia* species. The tree flora of these semi-evergreen forests also declines in richness with increasing rainfall seasonality (Chapter 7). There is a complex relationship with deciduous forests, and their relationship with ground fire has been addressed in Chapter 2.

Fire appears not to be the sole factor determining their

Fig. 3.18 The geographical distribution of *Dipterocarpus turbinatus*, a species characteristic of semi-evergreen forest on clay loam soils. (From Troup 1921, Smitinand 1969–1970, Smitinand *et al.*1990, herbarium collections and personal observation)

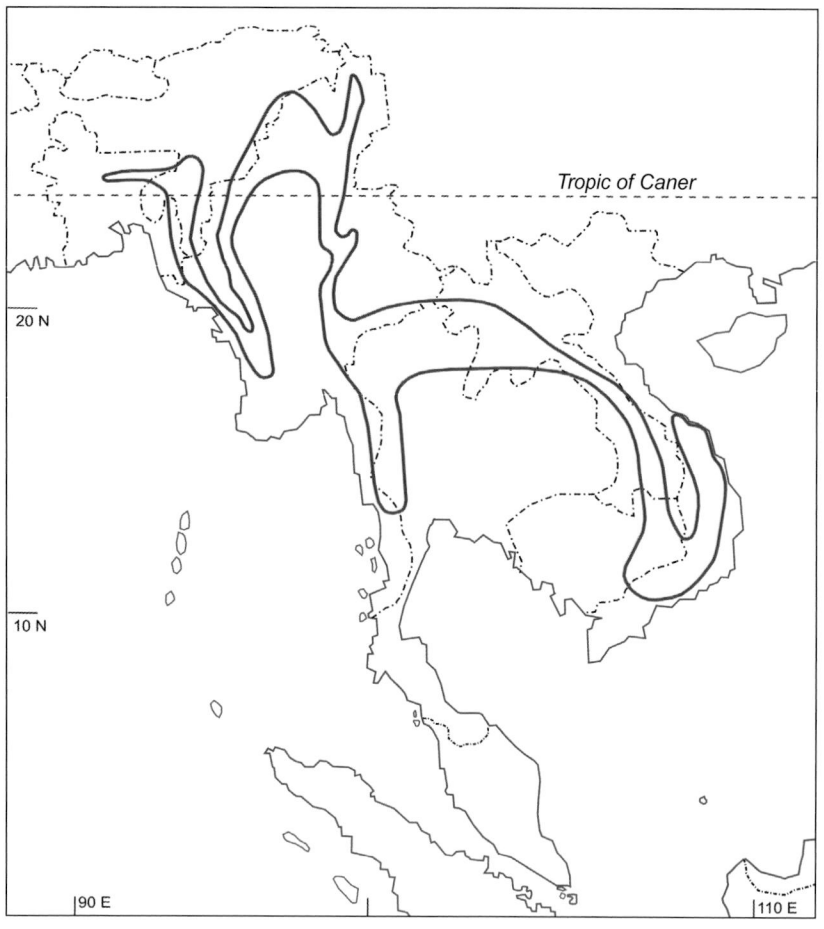

Tropic of Caner

20 N

10 N

90 E

110 E

☐ *Dipterocarpus turbinatus*

whose understorey is subject to much disturbance including burning, although sometimes dense with pioneers including evergreen species (Plate 3.13a). Upslope are mixed tall deciduous forests with *Xylia xylocarpa* and other species associated with turbiditic clay soils, sometimes including teak. Species distributions, notably that of *D. turbinatus* (Fig. 3.18), imply that this semi-evergreen forest association extends westwards to northern Burma and the southern Patkai range and the Chittagong Hill tracts at the northern, Bangladesh, end of the Arakan Yoma; also south to the northern Thai-Burmese peninsula. They persist in Indo-China, especially on the basalt of the Vietnam and north-east Cambodian lowlands and slopes of the Kontum Plateau, and elsewhere in the Annamite range.[174]

L. Blanc and colleagues found that four 1 ha plots on basalt in semi-evergreen forest, at Ma Da, Dongnai Dong Nai, southern Vietnam varied greatly in the proportion of evergreen canopy species, particularly in the abundance of the successional deciduous *Lagerstroemia calyculata*[175] (Plates 3.14Bb,Bc). Millet and colleagues compared the same forests using 25 circular plots, of less than 4 m radius, set at 10 m along the lines of a grid, with results from a 1943 inventory.[176] Stands on basalt were originally dominated by *L. calyculata* and *Xylia xylocarpa* ssp. *kerrii,* but the latter, valued for its timber, had disappeared, while the evergreen main canopy tree *Xerospermum noronhianum* was common. Floristic associations on siliceous substrates noted in 1943, with *Hopea odorata* and *Dipterocarpus alatus* on sandier soils and *Shorea roxburghii* and *Anisoptera* on sandy clay, had become degraded by the loss of commercial (canopy) dipterocarps, but with little increase in deciduous species. Although these authors mention a history of logging, but not of the lethal defoliation associated with war,[177] I infer that their plots had histories differing in the intensity of their canopy destruction.

These Indo-Chinese semi-evergreen forests on basalt derived clay loams, in their relatively high representation of deciduous species and Meliaceae, recall otherwise evergreen forests on base-rich soils recorded from slightly seasonal South Sumatra, which shares a similarly short dry season.[178] But Thai botanists not unreasonably term floristically similar forests in a more seasonal climate 'moist deciduous forests', along with the forests dominated by *Lagerstroemia calyculata*, *Afzelia xylocarpa* and often *Xylia xylocarpa* ssp. *kerrii* and *pterocarpus marsupium* in the canopy. French terminology In Indo-China varies

distribution relative to seasonal evergreen dipterocarp forests in the least seasonal climates in which they occur. In Cambodia and neighbouring southern Vietnam, forests occur on black montmorillonitic soils overlying basalt in which dipterocarps, notably *Dipterocarpus turbinatus* and *Anisoptera costata* occur. These exist in mixture with deciduous genera occurring in the tall deciduous forests on clay loam soils over sedimentary (turbiditic) substrate in north-west Thailand. That semi-evergreen forests might once have been widespread throughout western, northern and eastern Indo-Burma, wherever the dry season exceeds 4.5 months, is implied by the known range of *D. turbinatus,* an evergreen (or very shortly deciduous) emergent (discussed further in Chapter 2). Clear records of its occurrence in semi-evergreen forests exists for the Chittagong Hill tracts at the north-west, Bangladesh end of the Arakan Yoma,[172] northern Burma, Laos north to Luang Prabang, and south and central Vietnam.[173] In Thailand, *D. turbinatus* is now almost confined to gregarious stands within villages and agricultural landscapes between *forêt decidué, forêt semi-decidué,* and *forêt semi-humide mixte* according to rainfall regime and successional status.[179]

It is unclear whether semi-evergreen forest remains as extensive in Burma as the early literature implies it once was.[180] Extensive seasonal fresh water swamp forest dominated by *Lagerstroemia* inland from the Irrawaddy delta mangrove was included in maps; but it is not stated whether these forests included an evergreen canopy component. In north-west Thailand, the most extensive tract of semi-evergreen forest survives in the Thung Yai Naresuan–Huai Kha Khaeng wildlife sanctuaries, whose complex dynamics have been discussed in Chapter 2. That forest is distinguished, floristically, by its mixed dipterocarp association of *Dipterocarpus alatus, Anisoptera costata, Hopea odorata* and *Vatica odorata,* all of which can be locally abundant. At least six species of *Lagerstroemia* are present, including *L. calyculata* which, in the 50 ha CTFS plot established there by Dr Sarayudh Bunyavejchewin of the Royal Thai Forest Department, is scattered and concentrated in moist swales. But teak, *D. turbinatus* and *Xylia xylocarpa* are conspicuous by their absence. The sandy soils here are notable for the dark grey-brown humus discoloration of the mineral profile to depth, and for the local presence of earthworms, which are abundant on the soil surface during the late monsoon (Chapters 1, 2, Plates 1.15b–d). These are characteristics generally associated with lower montane forests at lower latitudes and with northern seasonal evergreen dipterocarp and semi-evergreen forests (see discussion in section 4.3 below) at higher latitudes (also see Table 3.10).

Sarayudh placed a 16 ha plot at Huai Kha Khaeng (western Thailand) in a similar largely deciduous stand on sandy soils, yet set in semi-evergreen forest with evergreen canopy species dominant. In this case, though, the stand was dominated by four *Lagerstroemia* spp., *Saccopetalum lineatum* and other deciduous species.[181] *Saccopetalum* and the *Lagerstroemia* species all lacked young juveniles and had bell-shaped population structures. The occasional presence in this case of *Hopea odorata,* the abundance of evergreen strangling figs and of rattans and juvenile evergreen species in the canopy as well as in the understorey, all indicated that the stand represents advanced succession of semi-evergreen forests following a major wind throw and possibly fire. This followed the pattern described by Patrick Baker[182] (Chapter 2). Frequent ground fires are associated here with an abundance of certain understorey shrubs and small trees, notably *Baliospermum solanifolium,* but their distribution is often patchy. The persistence of evergreen species in the canopy, including *Irvingia malayana, Mangifera* spp. and *Parinari anamensis,* would be atypical for tall ('moist') deciduous forest in India.

Indeed, southern Indo-Chinese semi-evergreen forests on basalt occur adjacent to seasonal evergreen forests on pale yellow sandy clay soils over schist, and in a climate which

elsewhere supports the latter rather than the former. Blanc's single plot on schist was situated on similar soils to those of the Huai Kha Khaeng CTFS plot in semi-evergreen forest in western Thailand, but had a dry season shorter by 1.5 months. Plots on the two substrates shared important evergreen canopy species, including the dominant dipterocarp *Dipterocarpus alatus,* also *Anisoptera costata* and *Hopea odorata.[183]*

Dense stands of *Dipterocarpus turbinatus* on turbiditic loams in northern Thailand recall Blanc's South Vietnam plot on basalt,[184] and appear to mark surviving fragments of semi-evergreen forest on relatively fertile clay loams over turbidite and metamorphosed igneous rocks. Floristic comparison between Burmese, Thai, and Indo-Chinese floras remains impeded by persisting differences in nomenclature of widespread species among taxonomists. Nevertheless, the floristic distinction between forests on sandy and clay loam soils does appear to be consistent.

Elsewhere in south-east Indo-Burma, forests relatively low in deciduous species occur on leached yellow sandy soils.[185] In this they resemble the semi-evergreen forest at Huai Kha Khaeng, but with distinct regional elements including the locally abundant dipterocarps *Shorea henryana* and *Hopea ferrea.* As at Huai Kha Khaeng and at Sakhaerat on the Khorat Plateau, these occur adjacent to dry deciduous dipterocarp forest dominated by *Dipterocarpus obtusifolius* and there again appears to be a dynamic interrelationship mediated by fire (see section 5.3 below and Chapter 2). Similar forest, on the evidence of species distributions, occurs in southern Laos.

Semi-evergreen forests share with tall deciduous forests the presence of several species of large crown vines, which persist in mature stands as relicts from their early succession in large tree-fall gaps. Both forest types also support localised abundance of gingers (Zingiberaceae) in the field layer, especially in moist sites and on clay-rich soils, and harbour a great quantity of *Macrotermes* termite nests. These conical mounds, rich in organic matter and nutrients but drought-prone, are said to differentially support seedlings of certain tree species[186] (Plate 3.15b).

At risk of over-simplification and to summarise, three widespread associations can be recognized in the mixed semi-evergreen and tall deciduous forests of Indo-Burma, each prevailing on soils of different substrates:

1. Associations in which *Lagerstroemia* species are many and abundant, and in which *Hopea odorata, Anisoptera costata* and *Dipterocarpus alatus* occur where evergreen stands survive. These occur in the Arakan Yoma, but otherwise are confined east of the Irrawaddy. They appear to be confined to sandy soils, over beds of sandstone and other siliceous substrates. They may occur in association with deciduous dipterocarp forests in which *Dipterocarpus obtusifolius* occurs, and which replace them on the most freely draining soils.

2. Associations distinguished by the presence and local abundance of *Xylia xylocarpa* ssp. *kerrii, Lagerstroemia calyculata* but few or no other *Lagerstroemia* species, and *Dipterocarpus alatus* where semi-evergreen elements survive. Confined to clay loam soils over turbiditic sediments, and other substrates including basic volcanic.

3. An association similar to 2, but with teak which is usually abundant, and an absence of *Dipterocarpus turbinatus*. Predominant wherever there are limestone beds or karst outcrops, therefore confined from the hills east of the Irrawaddy to the Golden Triangle area of north-central Laos, and north of 17° N.

Each of these soil-correlated associations vary greatly in their proportion of deciduous species in the canopy. The deciduous component increases both with increasing length of dry season and with disturbance; while it also increases with increasing clay and nutrient content of the soil. It is these triple influences on the proportion of deciduous elements in the canopy that most explain the complex patterns of so-called semi-evergreen and tall deciduous forests and stands, which once covered so much of Indo-Burma.

4.2. *Southern semi-evergreen forests: South Asia*

The Malabar coastal lowlands of western peninsular India have been intensely cultivated over such a long time that no distinct semi-evergreen forest belt survives. A floristic gradient occurs northwards along the coastal lowlands, where there is a decline in the number of evergreen tree species and a thinning in their abundance in the understorey as the dry season increases in length from 4 to 6 months, at the current limits of evergreen forest. North of c.15° N the canopy is increasingly dominated by deciduous elements. Pondicherry forest geographers recognize a semi-evergreen *Machilus-Holigarna-Diospyros* association. This is replaced further north of Goa by tall deciduous teak forest, called '*Tectona-Lagerstroemia-Terminalia* association' by them[187] (Table 3.11a) and 'South Indian tropical moist forest' by Champion and Seth.[188] This association occurs in a climate of 5–6 dry months. With its abundance of *Lagerstroemia lanceolata*, it is clearly analogous to semi-evergreen forest in the Far East, but the dynamic relationship between the two associations has not been documented.

Utkarsh and colleagues[189] ordinated trees more than 10 cm diameter in 96, 1 ha plots established by the Karnataka Forest Department on both sides of the Western Ghats (13°30'– 15°15' N) (Table 3.12). On the west, they distinguished

a

Champion and Seth's equivalent (approximate)	To the west and north (red-brown clay loams dominant)	To the east and south (grey-brown siliceous soils dominant)
Moist (tall) deciduous forests	*Tectona-Lagerstroemia-Terminalia* *Tectona-Adina-Anogeissus latifolia* *Anogeissus latifolia-Terminalia-Tectona* *Tectona-Terminalia* *(Hardwickia-A. latifolia)* *A. latifolia-Terminalia-Tectona* *A. latifolia-Terminalia*	*Shorea robusta-Dillenia-Pterospermum* *Shorea-Syzygium nervosum (operculatum)-Toona* *Shorea-Terminalia-Adina* *Shorea-Buchanania-Cleistanthus* *Shorea-Cleistanthus-Croton* *Shorea-Buchanania-Terminalia* South only: *Terminalia-A. latifolia-Cleistanthus*
Dry (short) deciduous forests	*A. latifolia-A. pendula* *Acacia-A. pendula*	*Hardwickia-Pterocarpus santalinus-Anogeissus pendula* *A.pendula-Chloroxylon*
Dry thorn woodlands	*Acacia-Capparis* *Salvadora-Prosopis* *Prosopis-Capparis-Zizyphus*	*Albizia amara-Acacia* *A.amara*

b

*Anogeissus pendula**	*Dalbergia. lanceolaria**
Bridelia retusa	*Dendrocalamus strictus*
Buchanania lanzan	*Diospyros melanoxylon**
Butea monosperma	*D. exsculpta**
*Chloroxylon swietenia**	*Pterocarpus santalinus**
*Cleistanthus collinus**	*Kavalama urens*
Cochlospermum religiosum	*Zizyphus nummularia**

Table 3.11 a, Primary deciduous forest associations in peninsular India recognized by the Pondicherry school along the rainfall seasonality gradient, listed in approximate sequence by increasing length of dry season and/or decreasing mean annual rainfall; **b,** species indicative of short deciduous forests. Aterisked species absent in Indo-Burma.

Table 3.12 Distribution of the 50 characteristic species within four recognized types of lowland evergreen and deciduous forests from sites either side of the Western Ghats, peninsular India. Relative densities (percent); 1: seasonal evergreen, 2, 3: semi-evergreen, 4: tall deciduous, 5: short deciduous. (After Utkarsh *et al*. 1988)

Species	Forest sites				
	1	**2**	**3**	**4**	**5**
Hopea erosa	1				
Baccaurea courtallensis	4	3			
Mesua nagassarium	3	3			
Holigarna nigra		2			
Canarium strictum	5	3	1		
Aglaia elaeagnoidea	5	6	1		
Knema attenuata	11	6	2		
Beilschmiedia dalzellii	6	3	2		
Harpullia arborea	3		1		
Atalantia racemosa				1	
Litsea stocksii	2		0.5		
Holigarna grahamii	4	1	3		
Dimocarpus longan	8	6	4		
Actinodaphne angustifolia	6				
Artocarpus lakoocha	2	5	2		
Tetrameles nudiflora	6	3	4		
Persea macrantha	7	3	3	1	
Hopea parviflora	4	5	3	1	
Cinnamomum spp.	5	7	4	1	
Mangifera indica	5	5	6	3	
Memecylon umbellatum	3	1	1	1	
Olea dioica	4	5	6	3	
Macaranga peltata	6	4	6	4	
Mallotus philippensis	2	4	5	3	
Xantolis tomentosa	1		1	2	
Syzygium cumini	3	1	2	3	
Vitex altissima	1	5	5	4	
Dillenia pentagyna		2	4	4	
Lagerstroemia lanceolata	2	5	5	7	4
Terminalia paniculata	2	2	5	5	2
Wendlandia thyrsoidea					
Xylia xylocarpa			2	4	
Schleichera oleosa		3	5	6	4
Terminalia bellirica	2	3	5	6	5
Careya arborea		1	2	4	4
Grewia tiliifolia		1	4	4	5
Wrightia tinctoria				1	
Bridelia retusa	1	1	0.5	6	7
Terminalia crenulata			2	5	7
Pterocarpus marsupium			2	7	9
Phyllanthus emblica	1	1		5	7
Tectona grandis			2	4	5
Wrightia (Holarrhena) antidysenterica					2
Morinda citrifolia				3	2
Madhuca longifolia var.*latifolia*				3	5
Butea monosperma				2	7
Anogeissus latifolia			0.5	1	11
Cassine glauca				1	7
Givotia moluccana (rottleriformis)					7

two associations with differing but substantial deciduous elements, and one that lacked them. But both semi-evergreen associations reached the northern limits sampled, with the more deciduous reaching south to 14°20' N, and at the lowest altitudes. A narrow band of semi-evergreen forest, containing genera largely shared with analogous forests in Indo-Burma, also survives in patches along the eastern base of the Western Ghats wherever moist deciduous teak forest meets the slope of the main dividing range; it is particularly evident along the northwestern foothills of the Nilgiris but has been little studied. There are also scattered patches of floristically similar forest on freely draining, but apparently well-watered, gravels set in the deciduous forest landscape of the Karnataka Deccan at middle altitudes (Chapter 2).[190] However, the highly restricted area of semi-evergreen forests in peninsular India likely reflects the many centuries of dense human occupation. This has resulted in their conversion to tall deciduous teak, sal, or even virtual elimination of evergreen trees, thereby degrading them in that respect to 'dry', that is short, deciduous forest, prevailing climate conditions notwithstanding.

4.3. *Northern semi-evergreen forests*

From southern Yunnan[191] to Central Nepal[192] there is also a distinct semi-evergreen forest, easily recognized by the prevalence of *Terminalia myriocarpa* (which also occurs, but scattered as a successional species, in northern seasonal evergreen dipterocarp forests) with its brilliant red inflorescences and copper-hued senescing leaves (Plate 3.12b,d).[193] That community is associated with drier conditions, but in the wettest localities seems to have replaced the dipterocarp-dominated seasonal evergreen forest following swidden. In such places it is associated with late successional tropical evergreen lowland species including *Pometia pinnata* f. *tomentosa* and *Duabanga grandiflora*. For these forests we have the plot documented by John Proctor[194] in Arunachal Pradesh towards the east. West of Lakhimpur, it is increasingly confined to moist valleys along slopes above the rivers that run out of the Siwalik front ranges, and on the freely draining *bhabar* and *terai* terraces bordering the Gangetic and Brahmaputra alluvial plain. At their westernmost extension, semi-evergreen forests are reduced to yet fewer species, most of which occur also in the tall deciduous sal forest that surround them. Indeed, sal may occur scattered within these *bhabar-terai* semi-evergreen stands, and there may be other consistent floristic differences from them on the steep adjacent foothill slopes. In the *bhabar* and *terai* terraces and slopes along the base of the Himalaya, soils are more often coarsely sandy. Though also lacking a clear humic horizon, the mineral soil is dark grey-brown. On the adjacent Siwalik slopes, clay loams again appear to prevail, though there are local limestone exposures. Further east, these

semi-evergreen forests appear confined to hill slopes, often steep, on yellow-brown clay loams, without a distinct humic surface horizon.

Northern semi-evergreen forests, and to a lesser extent northern seasonal evergreen forests, include several species which are confined to lower montane forests at lower latitudes (Table 3.10). *Terminalia myriocarpa,* remarkably, also occurs in lower montane forest in seasonal northernmost Sumatra, but elsewhere only the lowlands of northern Indo-Burma.

4.4. *Summary*

Floristic patterns within seasonal evergreen and semi-evergreen forests at the landscape (and larger) scales, and their relationship with topography, substrate and therefore soil, appear to parallel those in the perhumid tropics. The relative roles of soil water economy and nutrients awaits experimental investigation, although the influence of soil water stress might be expected to strengthen as the length and intensity of the dry season increases. This can also be inferred from work in analogous forests in Central America.[195]

5. Deciduous forests

5.1. *The floristic transition*

The floristic transition from evergreen to deciduous forests by way of semi-evergreen forests involves change in family dominance from Dipterocarpaceae to Lythraceae, Combretaceae, and Verbenaceae, while Fabaceae increase in importance (Table 3.7). The differentiation between evergreen and deciduous forest most often being determined by fire[196] (Chapter 2), a sharp boundary might be expected. However, this is only the case under certain circumstances: where fire-climax deciduous forest meets lower montane forest at the altitudinally constant cloud base; where topographic or geological-edaphic elements of the landscape carry forests which are differentially fire-prone; or where a moist valley or river act as a fire barrier. More often, the transition is from semi-evergreen to tall deciduous, and consists of a dynamic mosaic mediated by past changes in canopy damage and fire, as has been cleverly interpreted by Patrick Baker and Sarayudh Bunyavejchewin at Huai Kha Khaeng (Box 2.2). There, succession after large-scale canopy opening during cyclones can apparently follow contrasting paths. If the storm is followed by several wet years, evergreen canopy species regenerate predominantly, eventually forming patches of canopy abundance. These include *Hopea odorata*, the small seedling radicals of which cannot penetrate dense litter, implying that there may have been a ground fires prior to its final pre-storm fruitfall.[197] (Though Chengappa found fire unnecessary, see Chapter 2 section 5.3). But if the storm debris burns, generating a hot fire, juveniles of evergreen species are eliminated and species of the tall deciduous associations dominate succession to the canopy. Only later are these deciduous stands invaded, from the old gap margin, by shade-tolerant evergreen regeneration (Chapter 2).

5.2. *Tall and short deciduous forests of India ('moist' and 'dry' deciduous forests of Champion)*

Structural and phenological categories of deciduous forest in India have been described, and the need for their modification in Indo-Burma discussed, in Chapter 2. These are the tall and short deciduous forests and thorn woodlands. There is an increase in the representation of deciduous pioneer and successional species with increasing seasonality as well; many of these are deciduous. Some, e.g. *Afzelia xylocarpa,* establish following fire and are long-lived (Box 2.2). This increase seems to be associated with increasingly frequent and widespread natural catastrophe, such as extensive cyclonic storm damage and increasing frequency of fires which destroy early regeneration (Chapters 1, 2). The regions where these forests survive, and are thought to have formerly occurred, have also been centres of civilisation for over 15 centuries (Chapter 8).

Plate 3.14 A Indian deciduous forests. **a,** Tall deciduous sal (*Shorea robusta*) forest, Similipal National Park, Odisha. The large sal individual is 30 m tall, 1.4 m dbh; **b,** Tall deciduous sal (*Shorea robusta*) forest, with evergreen understorey; *Aglaia rohituka* on right. Buxa National Park, West Bengal; **c,** Short deciduous sal forest, with *Lantana camara* understorey. Jalpaiguri-Buxa road; **d,** *Terminalia crenulata,* Mudumalai Wildlife Sanctuary on the western Tamil Nadu Deccan; **e,** *T. bellirica,* Mudumalai Wildlife Sanctuary; **f,** *Pterocarpus marsupium; Schleichera oleosa* to left. Similipal National Park, Odisha. (All: H.H.)

Plate 3.14 B Indo-Burmese tall mixed deciduous forests. **a,** *Anogeissus acuminata* bole, Mae Ping National Park, north-west Thailand; **b,** *Lagerstroemia calyculata* in tall mixed deciduous forest on clay loams, Ta Ruang road, south-east Thailand; **c,** *L. calyculata,* semi-evergreen forest on clay soils, south-east Thailand; **d,** *Anogeissus acuminata* secondary mixed deciduous forest, lower western slopes of Doi Inthanon, west Thailand. (a, d H.H.; b, c P.A.)

5.2.1. *Floristic character and human influences*

Floristically, Indian deciduous forests are remarkably uniform. The canopy of Indian tall and short deciduous forests is variably dominated by six to ten widespread species[198] (Tables 3.11, 3.12). There are relatively few species characteristic of specific habitats, such as *Hardwickia binata*, which becomes gregarious on laterite caps and shallow clay soils, though it is also scattered in forests elsewhere (Fig. 3.19). Tree species distributions, teak and sal notably excepted, are rarely correlated with physical structure and fertility of soils, though there may be consistent variations in relative abundances. (This would be difficult to test today, when so many of these forests have been severely affected by increasing fire frequency and overgrazing.) Species of taller forests enter the shorter forest areas along river banks and swales.

Overall, the two principal floristic trends are correlated with increasing length of the dry season[199] and the associated frequency and intensity of fire[200] on the one hand (Fig. 3.20), and as a trend with latitude which may have historical biogeographic as well as ecological explanations. Northern tall deciduous forests alone are rich in evergreen elements of lower montane affinity, including *Castanopsis, Lithocarpus, Schima* and *Talauma* (Table 3.10). Only a few – such as *Bischofia javanica*,

Fig. 3.19 Range of two species of short deciduous forest in peninsular India with specific habitat preferences within it, *Albizia amara* and *Hardwickia binata*. (From Meyer-Homji 1970)

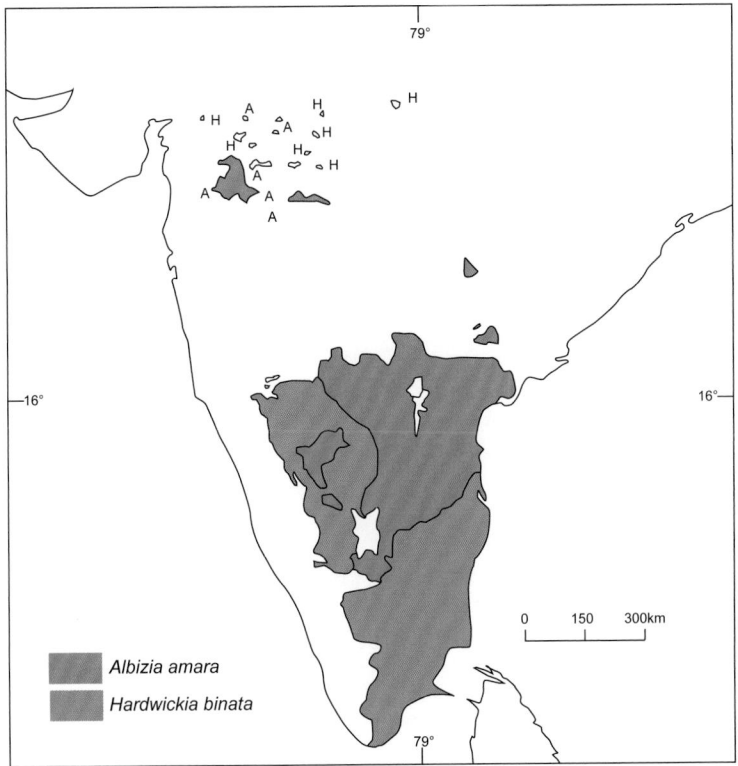

Fig. 3.20 Distribution (densities/ha) of *Lagerstroemia lanceolata* (*microcarpa*) (below) and *Anogeissus latifolia* (above), species of semi-evergreen, and tall and short deciduous forest respectively, in Mudumalai Wildlife Sanctuary, Tamil Nadu, India. (From Suresh *et al.* 2011)

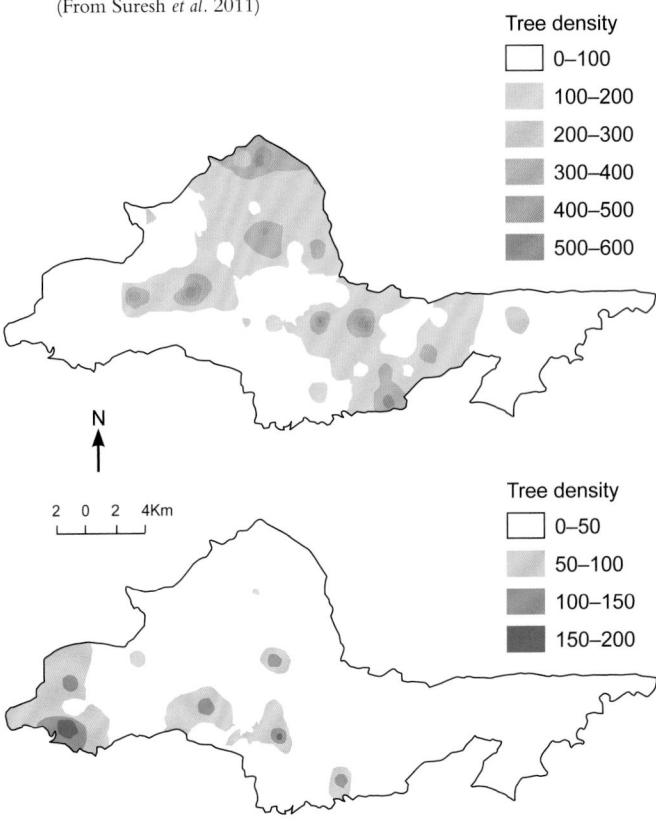

Litsea mysorensis, Persea macrantha, Actinodaphne angustifolia and *Michelia* – also occur in Indian peninsular forests. Some of these are present in Mudumalai at 850–950 m. Southern short mixed deciduous forests are distinguished by the dramatic umbrella crowns of *Acacia planifrons*.

The western coastal plain has been cultivated for so long that the original pattern of the forest is difficult to discern. The dominant clay loams now support tall deciduous teak forests south to Trivandrum and the southernmost ghats of the Tinnelveli hills, a region predominantly of only four dry months, which is wetter than the natural distribution of teak elsewhere. Unlike perhumid regions, where orographic rainfall predominates and the coastal regions are driest, the south-west monsoon dumps heavy rain when it crosses the coast (Chapter 1). There may then be some decline inland, before the second onslaught when the monsoon winds collide with the ghats. Currently, no natural vegetation exists along the coast here, but the plains are partially occupied by teak, which would have been outcompeted were evergreen forest formerly there. I infer that these wet seasonal plains were once clothed with a semi-evergreen and tall deciduous

Plate 3.15 Short deciduous forests in India. **a,** Short deciduous forest in Bandipur National Park, Karnataka; **b,** An old *Macrotermes* mound in Indian short deciduous forest; **c,** Short sal forest with *Syzygium cumini* (left centre), *Cochlospermum, Diospyros melanoxylon* (right centre), sal (left, with *Buchanania lanzan* to right). Similipal National Park, Odisha; **d,** Degraded short sal forest with *Cochlospermum religiosum* (taller tree) and *Cleistanthus collinus,* a common understorey tree. Similipal National Park; **e,** *Cochlospermum religiosum* in flower; **f,** *Dalbergia lanceolaria* in flower, March 2009. Mudumalai Wildlife Sanctuary, Tamil Nadu. (**a, c –f** H.H.; **b,** S. Yadav)

forest. As the ghats are approached, deciduous forest still gives way to a narrow band of semi-evergreen forest north of the Palghat gap, at least to Agumbe. Several useful accounts of Indian tall and short deciduous forests have been published.[201]

Whatever depredations have been wrought by human activity on the evergreen understorey in taller deciduous forests, a series of tree species replacements heralds the change from tall to short deciduous forest. Notable is the disappearance of evergreen elements in the understorey (Chapter 2). This can occur on drier upper slopes and ridges within tall deciduous forest regions, but the tree flora that remains in these less-rich communities is that of the tall deciduous communities below, and continues to include the occasional evergreen species such as *Syzygium nervosum (operculatum)* and *S. cumini* (Chapter 2).[202] These changes include the replacement of the conspicuous (though scattered) pale-barked canopy tree of tall deciduous forest, *Anogeissus latifolia,* by the abundant, often dominant *A. pendula; Lagerstroemia lanceolata* by *L. parviflora;* and *Dalbergia latifolia* by *D. lanceolaria.* Meanwhile, *Acacia leucophloea* and *Boswellia serrata* indicate the arrival of dry thorn woodland (Fig. 3.20; Table 3.11a; Plates 2.35a, 3.15f). The excellent vegetation maps compiled by the French Institute at Pondicherry and the Université Paul Sabatier (Toulouse) recognize a series of associations by the change in certain dominant and characteristic canopy species along the rainfall seasonality gradient[203] (Table 3.11a).

5.2.2. Teak *and* sal (map inside back cover)

Forests of the Indian lowlands have been the subject of silvicultural research for 150 years, more by far than any other tropical forest. Although they vary floristically, no rigorous research has yet addressed the mechanisms whereby even the dominant species replace one another in the landscape. That is testimony to the near-exclusive focus of forestry research on increasing sustainable timber production through empirical experiment. Nowadays, wherever teak or sal occurs in nature, they dominate overwhelmingly. This is a consequence of silvicultural manipulation – even though Troup[204] and others noted long ago that both species perform better and are less attacked by pests when growing in mixture with other species (Chapter 5). The two species are seldom found today other than as dominants, though there are forests in southern Odisha and Madhya Pradesh where one or the other occurs without dominating completely.[205] Both occur over a wide range of rainfall seasonality regimes.

Teak extends westwards into Gujarat, with eight dry months and 850 mm annual rainfall, and eastwards into Burma, Thailand and the Golden Triangle region of north-west Laos in climates with up to 1,800 mm annual rainfall and 5–6 dry months (maps inside cover). Mixed deciduous teak forest is now the subject of long-term research in a 50 ha CTFS plot

established in the Mudumalai National Park and Wildlife Sanctuary (Tamil Nadu) by R. Sukumar and his colleagues at the Indian Institute of Science[206] (Box 2.3; Plates 2.30, 2.31). It is placed within the tall-short deciduous forest interphase. Teak and several of its associates are gap pioneers in semi-evergreen dipterocarp forests at the wettest end of their ecological range in Burma.[207] The fruit of teak, a 15 mm diameter globe, has a spongy mesocarp and hard endocarp, and possesses dormancy. It germinates when the pericarp rots off or is abraded; fire increases the germination rate. Teak forests are associated with tall bamboo species, especially the widespread *Dendrocalamus strictus* and more upland *D. hamiltonii.*[208]

Tall sal forest (map inside back cover; Plate 3.14Aa,Ab) extends eastwards along the Himalayan *terai* to Darrang district (26°30′ N, 92° E); on the plains south of the Khasi hills eastwards to the Jaintia Hills (92°15′ E); south from the Santal Parganas on the southern banks of the Ganges, in Jarkhand state, to Ganjam, Jeypore and Pallenda in Vizagapatam (Tamil Nadu, 19° S); and west to Bilaspur, Mandla, Balaghat, Jubbulpur and Rajpur Districts in Madhya Pradesh. A 'preservation plot' in tall sal has also been recently established and censused by D. Swain in Similipal National Park (Odisha).[209] Sal is rare in evergreen forest, surviving as a relict from a previous period when drought and fire had permitted sal elements to invade. In the *bhabar* undulating terrain of the north-east Himalayan foothills, sal had survived as isolated individuals and spinneys in annually burned savanna before the introduction of forest management.[210] Extensive grasslands there, bordered by deciduous woodlands of khair (*Acacia catechu*) and sissoo (*Dalbergia sissoo*) still dominate the floodplains and provide sanctuary for large herbivores[211] (Chapter 6) (Plate 3.8d), as well as major sources of thatch and bedding for the local economy[212].

Sal fruit are winged. The pericarp is lignified and resinous, but the seeds lack dormancy and are killed by fire, as are the young seedlings[213] (Chapter 2). Establishment is impeded by litter, especially when waterlogged, therefore ground fire early in the monsoon and prior to seedfall promotes establishment.[214] Sal forest and some Indo-Burmese deciduous dipterocarp forests are found with short bamboo species (see section 5.3 below), short grasses and cycads. In the Siwalik foothill ranges of the north-west Himalayan tropics, *D. strictus* is also abundant in short sal and contributes c.45% of understorey cover there. The bamboo *Melocanna baccifera* occupies 1.8 million ha in north-east Indian tall deciduous sal forest, while *Phyllostachys bambusoides* covers 1.15 million ha in tall deciduous forest.[215]

Teak, and the mixed forests with which it is associated, follow the basaltic trap, Vindhyan and other rocks yielding red-brown fertile clay loams. *Sal* and associated mixed forests follow the siliceous granite, metamorphic schists and sandstones yielding sandy, freely-draining, low-nutrient soils.[216] Teak is tallest on deep loams with relatively high calcium content,[217] but sal also sometimes occurs on limestone outside the geographical range of teak.[218] No surface raw humus accumulates in these dry climates, and the nutrient levels of even the poorest soils are an order of magnitude higher than the most fertile upland soils in perhumid climates (Fig. 1.10).

The tree flora of teak- and sal-dominated forests differs quantitatively rather than qualitatively; no other species are unique to either. Though the ranges of these two species closely match the division drawn between major soil types of Sarawak MDF based on both physical and fertility criteria (see sections 1.1.3 and 1.1.4 above), here the floristic distinction still appears to be mediated by soil nutrients alone.

Floristically somewhat distinct mixed deciduous forests occur in floodplains associated with teak or sal, on adjacent hills and in certain special habitats throughout the ranges of these two dominant deciduous forest species.[219] Whether floristic variation is influenced by nutrients at the local or topographic scale has not yet been determined. E.A. Smythies[220] long ago claimed that it was so in the geologically diverse Kumaon Siwaliks.[221] This view was later upheld by others in that region and elsewhere, notably Mooney,[222] but was contrary to Champion's convictions (based on Schimper[223]) that climate and soil water are the overall determinants of vegetation.

Each forest type has been, in practice, somewhat arbitrarily divided into their two *facies,* one tall ('moist') and species-rich in climates with relatively long dry season, and one shorter ('dry'), less diverse, in relatively long dry seasons. Foresters recognize numerous other sub-types, some very local.[224] South of the range of sal, and of teak east of the Western Ghats, short mixed deciduous forests and thorn woodlands prevail except along the eastern coastal hills (see section 5.3.3 below). They are distinguished by the presence of the tall *Acacia amara* and *A. planifrons.*

5.3. *Deciduous forests of Indo–Burma*

Jules Vidal's 1956 monograph on Laotian vegetation remains the most comprehensive account of Indo-Burmese deciduous forests.[225]

5.3.1. *Tall deciduous forests*
Teak and its largely deciduous associates are set in the Indo-Burmese landscape (Plate 3.14B) where the mesic valleys and fertile clay loam soils – where the refuges of semi-evergreen forest must once have existed – have long been converted to irrigated rice paddies. Annual burning of forest litter to encourage grass for cattle is ubiquitous[226] and has been practiced for millennia. Indo-Burmese mixed deciduous forest composition is similar at the genus level to the tall deciduous teak forests of peninsular India, but there are significant differences at the species level within

many genera. Several of these genera are richer in species in Indo-Burma than in India[227] (Chapter 6). Species of *Lagerstroemia, Xylia, Dalbergia, Pterocarpus* and *Terminalia* are important in the canopies of both regions, though many of the species are different (Chapter 6; Table 6.13). Species associations appear to vary consistently along a complex soil continuum. Sarayudh[228] subjected 40, 0.1 ha plots scattered through deciduous forest in Thailand to an agglomerative classification and ordination. Leading species occupy increasingly distinct ranges in ordination space as they grow in size (Fig. 3.21a). He recognized four species associations with a primary division based on presence of either teak or *Lagerstroemia calyculata* (Fig. 3.21b). Teak appears to have competitive advantage over *Lagerstroemia* on calcium-rich soils (Fig. 3.21c), while *Xylia* and *Dalbergia* appear to regenerate more readily after windthrows. Today, a few successional evergreen species of the sub-canopy and even canopy strata persist only here and there in Indo-Burmese tall deciduous forests, mostly by streams and on the lower slopes – including *Duabanga grandiflora, Macaranga denticulata* and *Neolamarckia cadamba.* Along with *Tetrameles* and remaining stands of the evergreen or shortly deciduous *Dipterocarpus turbinatus,* these give some indication of what their original composition must have been.

Extensive climates with more than six dry months – that is, climates supporting short deciduous (including short teak) forests in South Asia – are confined in Indo-Burma to the northern Irrawaddy basin. Short deciduous teak forests are confined to that area, where a separate *Tectona, T. hamiltoniana,* also found on karst in dry western Luzon, occurs[229] (see section 5.3.3 below). *Cochlospermum religiosum,* an indicator species of short deciduous forests in India, is local in Thailand and concentrated in the northern plains of the Chao Phraya drainage where little natural forest remains. Teak itself is widespread in climates with 4.5–6 dry months in Burma east of the Irrawaddy, through northern Thailand to the great bend of the Mekong and north-west Laos. Although teak was mentioned by Brandis[230] as a component of the semi-evergreen forests of central Pegu, Indo-Burmese teak forests now generally lack climax evergreen species and their seed sources. As in India, teak appears to be associated with fertile clay loam soils where calcium content is relatively high, here over limestone beds within argyllic sediments. Elsewhere on clay loams where phosphorus concentrations are relatively high, teak is absent yet mixed moist deciduous forest with *Xylia xylocarpa* ssp. *kerrii* and many teak associates remain.

The use in Indo-Burma of the term 'dry deciduous forest' for tall deciduous forest lacking evergreen elements is inappropriate, as discussed in Chapter 2. These short forests differ in the absence of many species characteristic of short deciduous forests in South India. A few do occur in the region, but in deciduous dipterocarp forest (*Buchanania lanzan*), or degraded forests (*Butea monosperma*).

Fig. 3.21 Floristic categorisation of tall deciduous forests on clay loam soils in northern Thailand, in 20, 0.01 ha widely dispersed plots in each stand, trees >10 cm dbh: **a,** Agglomerative cluster analysis of 40 stands. **b,** Distribution of four leading species on the primary axes of a modified Bray and Curtis ordination of 40 stands, in which the range of Importance Values (cumulative relative density, relative frequency, and relative basal area) have been restricted thereby reducing the number trees of large size. Importance values (A) >100; (B) >50; (C) >25, as a measure of the decreasing size of the trees included: the species become increasingly ecologically distinct as they grow in size. (**a** after Bunyavejchewin 1983a; **b** after Bunyavejchewin 1985)

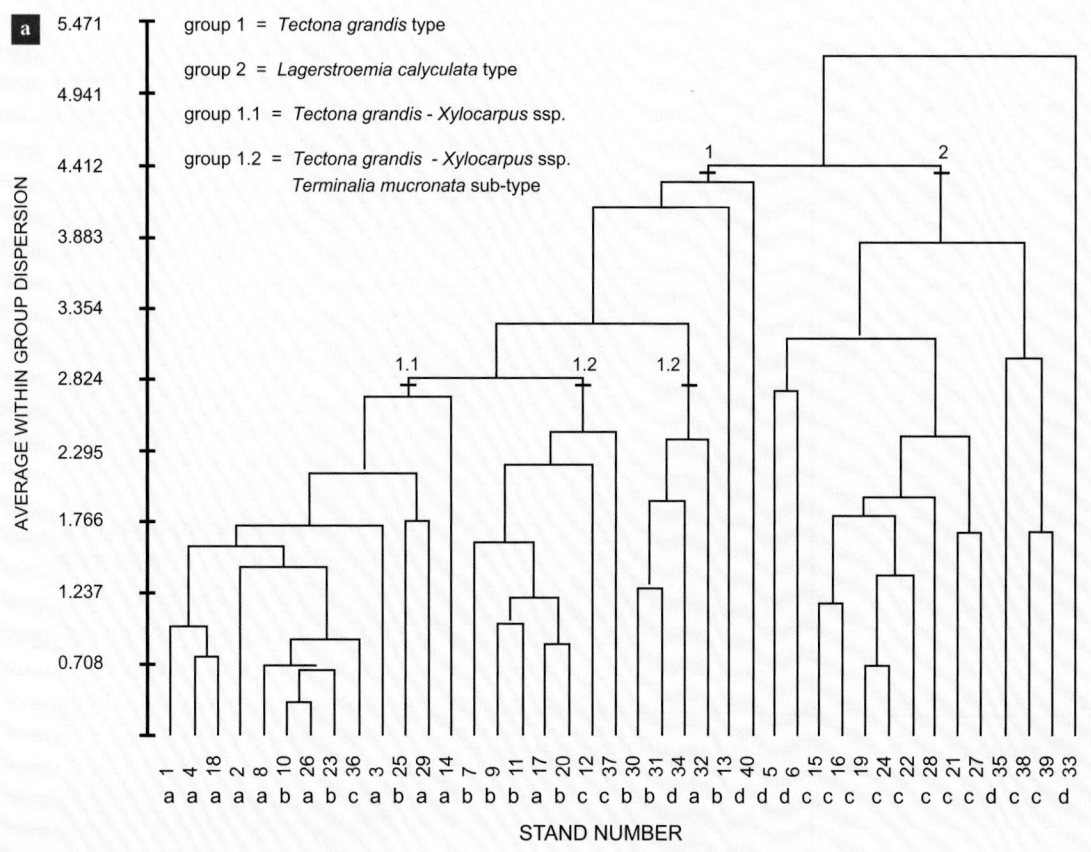

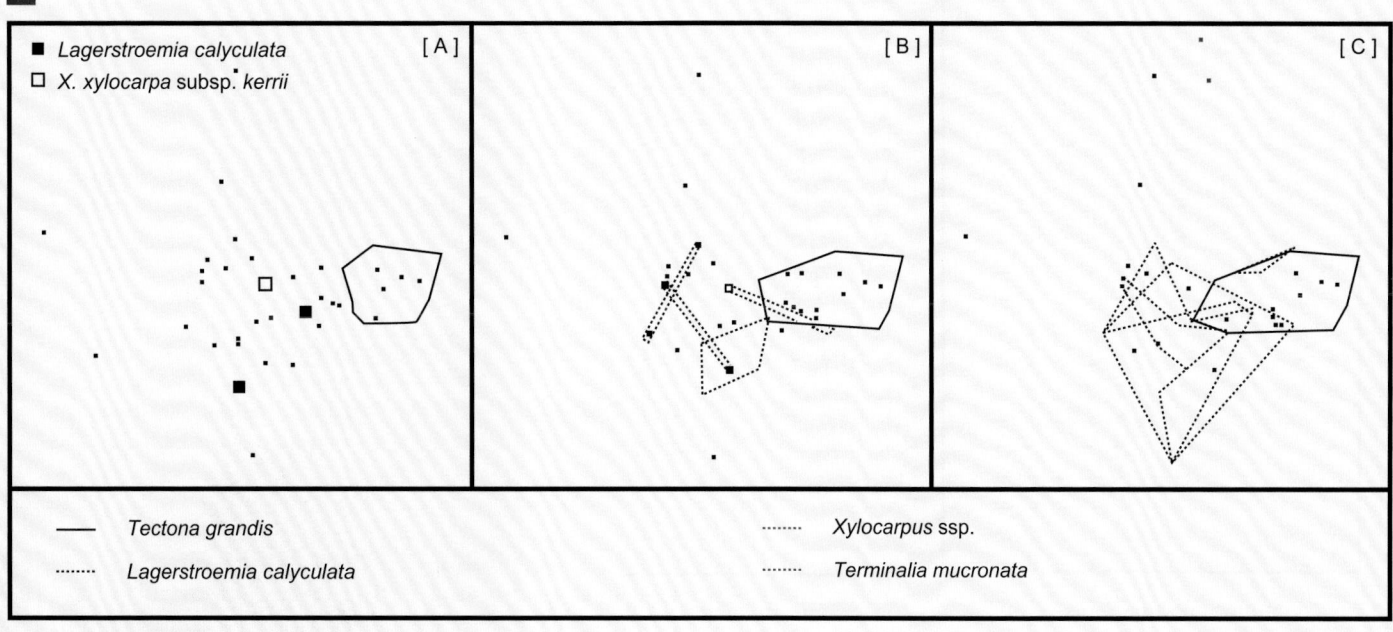

Fig. 3.21 c, Distribution of the four leading species along soil nutrient gradients: *Tectona grandis* tends to increase with soil calcium and phosphorus, whereas *Lagerstroemia calyculata* tends to decrease *Terminalia mucronata* and *Xylia xylocarpa* ssp. *kerrii* show no consistent trend. (**c** after Bunyavejchewin 1985)

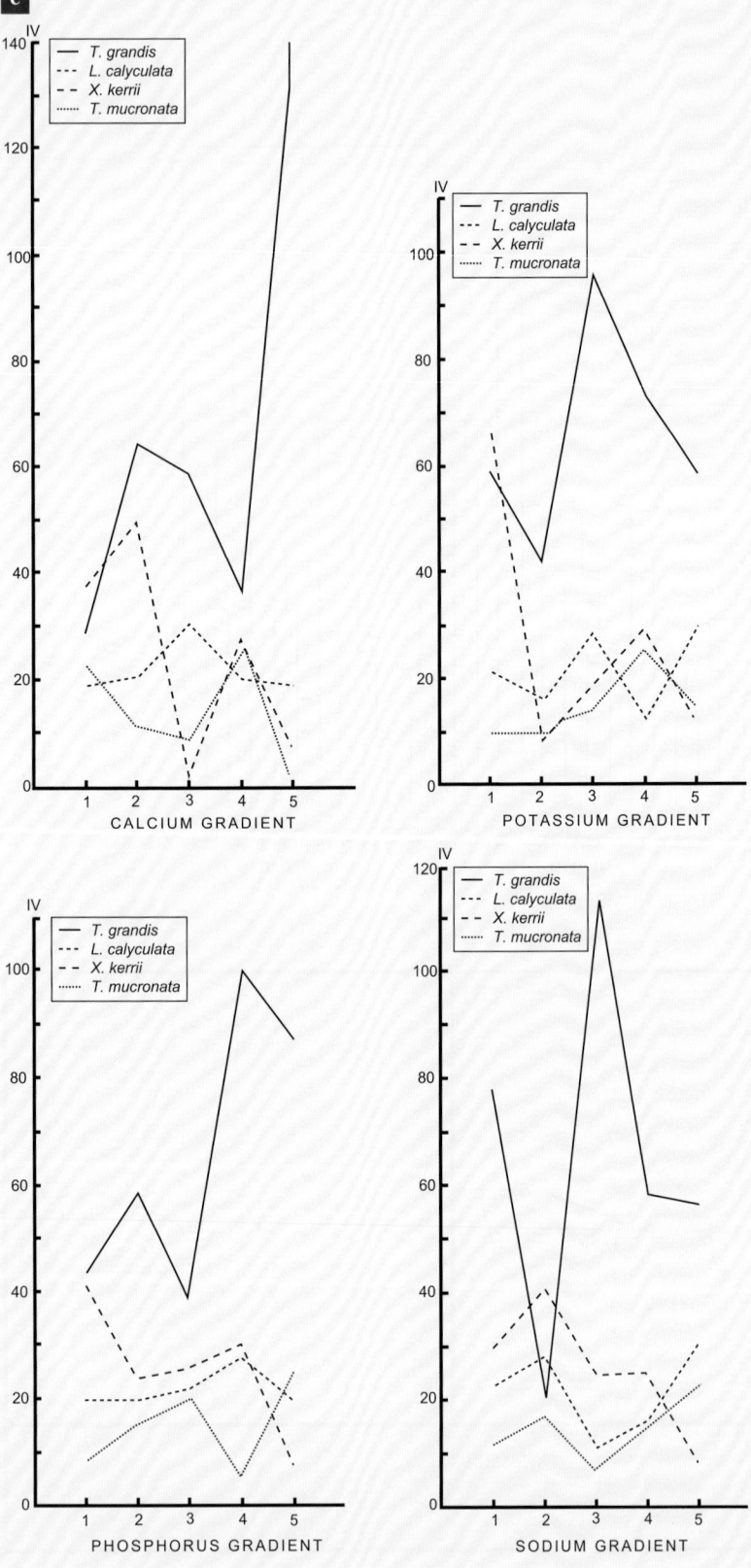

Many other Indo-Burmese deciduous forests do not include teak. In northern Thailand, the canopy may reach 30 m and include, among evergreen species *Dimocarpus longan, Polyalthia* spp., *Duabanga grandiflora, Syzygium* spp., *Elaeocarpus stipularis,* also deciduous *Terminalia, Albizia* spp., and *Tetrameles.* Eastern Thailand, central and southern Laos, Cambodia and parts of southern Vietnam bear a shorter deciduous forest, mainly lacking dipterocarps and with local and sub-canopy evergreen elements. They exist under climates with 4–6 dry months but low mean annual rainfall, sometimes less than 1,000 mm. These forests vary regionally in composition, but variation with soils has not been examined. Karst limestone bears soils on gentle surfaces throughout the seasonal climates of Indo-Burma. They are mostly shallow, and are generally bright rusty red and rich in clay. Their tree flora generally differs little from that of other shallow clay soils.

5.3.2. *Deciduous dipterocarp forests* (Plate 3.16) Deciduous dipterocarp forest has received several accounts based on the forests in individual Indo-Burmese nations.[231] Some of its characteristic species are found as far south, on drought prone substrates, as north-west Peninsular Malaysia[232] (Chapters 2, 6).

Plate 3.16 Deciduous dipterocarp forests. **a,** Forest gradient across the catena: degraded semi-evergreen riparian forest on sandy clay loam in foreground, fruiting *Lagerstroemia* sp. in canopy; background hillside of deciduous dipterocarp forest, *Shorea siamensis* dominant. Dec. 1997, Chiangmai-Mai Sariang road, north-west Thailand; **b,** Frequently burned *Dipterocarpus obtusifolius–Pinus merkusii* deciduous dipterocarp forests, Dec. 1997, Khun Yuam, north-west Thailand; **c,** *Cycas siamensis* in *D. obtusifolius-S. siamensis* dominated deciduous dipterocarp forest, Huai Kha Khaeng Wildlife Sanctuary, west Thailand; **d,** *Phoenix loureiroi* in *S. siamensis-S. obtusa* deciduous dipterocarp forest over karst limestone in Sarayudh's transect (Fig. 3.23), Mae Ping National Park, north-west Thailand; **e,** *Gluta velutina* in fruit, evergreen deciduous dipterocarp forest associate, Mae Surin National Park, north-west Thailand, Oct. 2009; **f,** Understorey of deciduous dipterocarp forest on sandy soil, with *D. obtusifolius* dominant: evergreen shrubs *Craibiodendron stellatum, Vaccinium sprengelii* (left foreground), March 2008 (left), Nov. 2008 (right).
(All: H.H.)

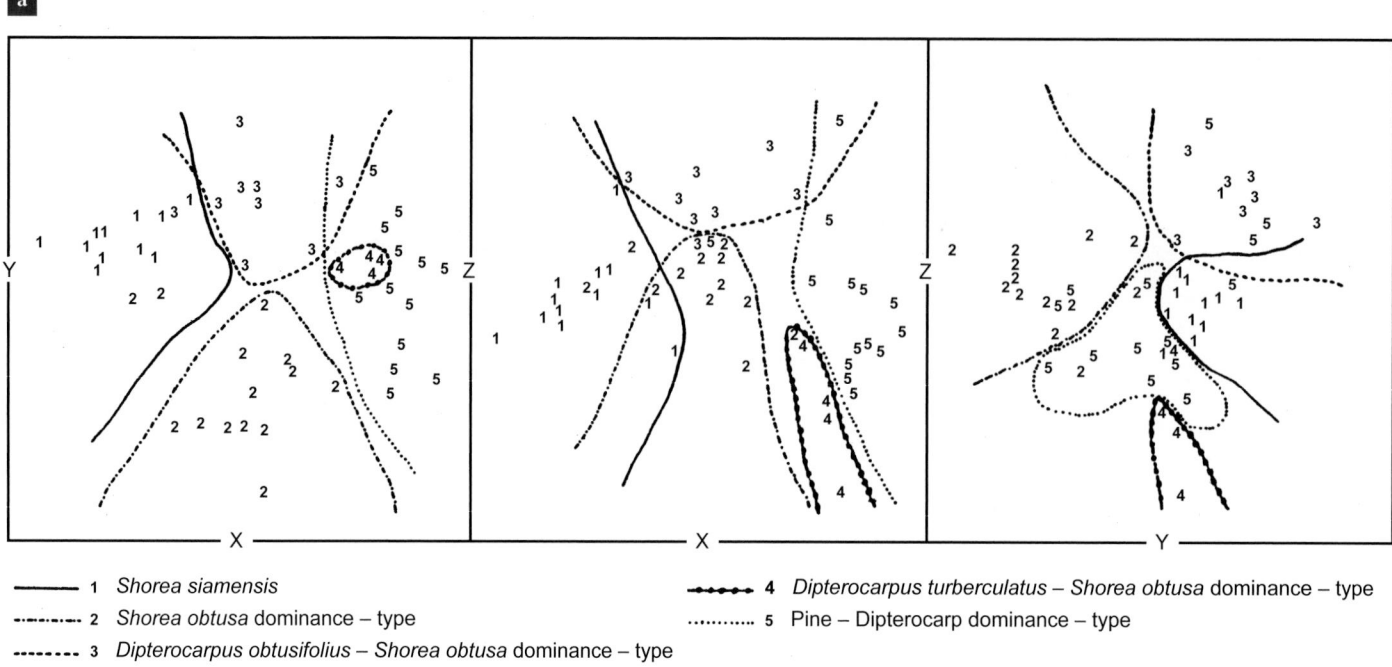

—— 1 *Shorea siamensis*

—··—··— 2 *Shorea obtusa* dominance – type

——————— 3 *Dipterocarpus obtusifolius – Shorea obtusa* dominance – type

•—•—•— 4 *Dipterocarpus turberculatus – Shorea obtusa* dominance – type

·············· 5 Pine – Dipterocarp dominance – type

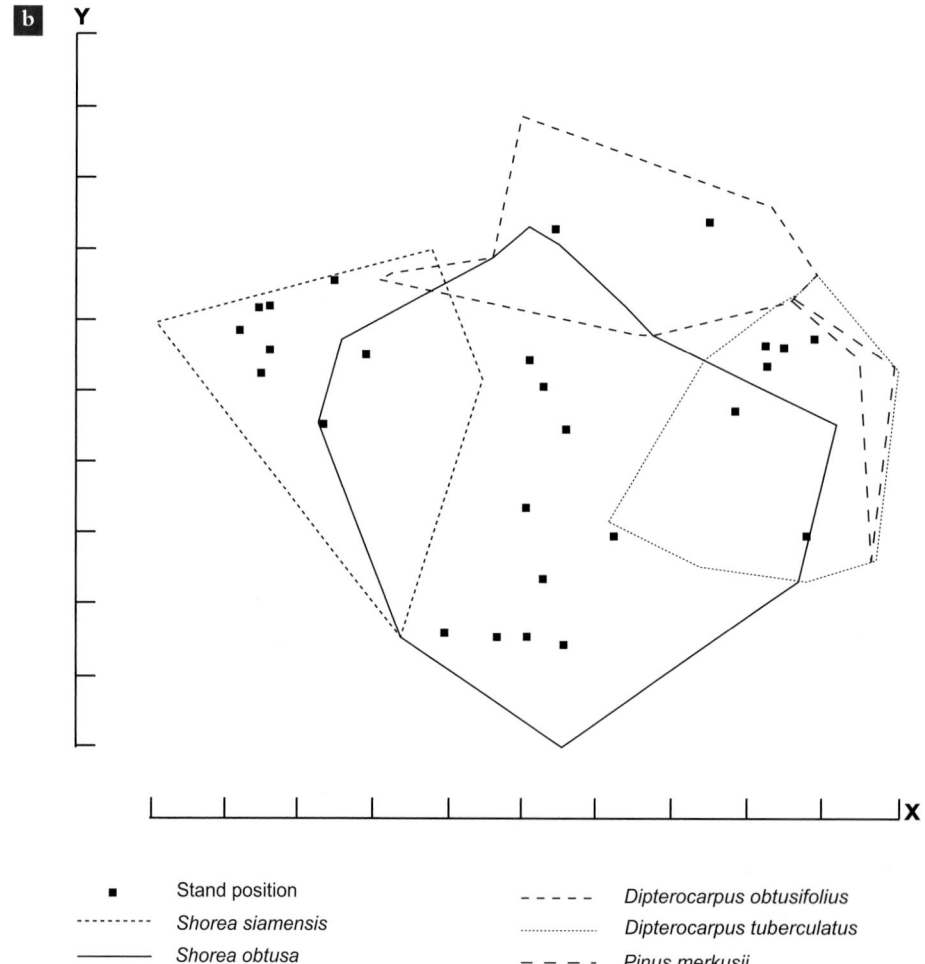

Fig. 3.22 Discrimination, in a Bray and Curtis ordination, of deciduous dipterocarp forest floristic associations on the basis of their five dominant dipterocarps in 50, 0.1 ha, plots; individuals >10 cm dbh: **a,** distributions of five floristic associations recognized on the three primary axes of the ordination; **b,** distribution, on the two primary axes of an ordination of 50, 0.1 ha plots, of the five leading species according to plots in which their importance values exceed 50; trees >10 cm dbh. (From Bunyavejchewin 1983b, technique modified after Bray & Curtis 1957)

■ Stand position

············ *Shorea siamensis*

—— *Shorea obtusa*

– – – – *Dipterocarpus obtusifolius*

················ *Dipterocarpus tuberculatus*

– – – – *Pinus merkusii*

Deciduous dipterocarp forest varies much in floristic composition as well as structure (Chapter 2) but all share a more or less deciduous canopy of one or more dipterocarp species. It contrasts with mixed tall deciduous teak forests, both in structure (Chapter 2) and in species composition, though they may share the same landscape. Deciduous dipterocarp forest usually occurs on edaphically drier sites than the mixed deciduous communities[233] and varies conspicuously in its dominant canopy dipterocarp. Five dominant species occur, either in mixtures with others or in pure monodominant stands. The mixtures occur on a perplexing array of soils. Sarayudh, on the basis of an ordination of 51 plots laid out in deciduous dipterocarp forest throughout Thailand, recognized five community types (Fig. 3.22a) based on these dominants. Although these have broad ranges within deciduous dipterocarp forest (Fig. 3.22a), each species dominates within a restricted range of habitats, and tends to occur on specific substrates on a local scale [234] (Figs 3.22b, 3.23):

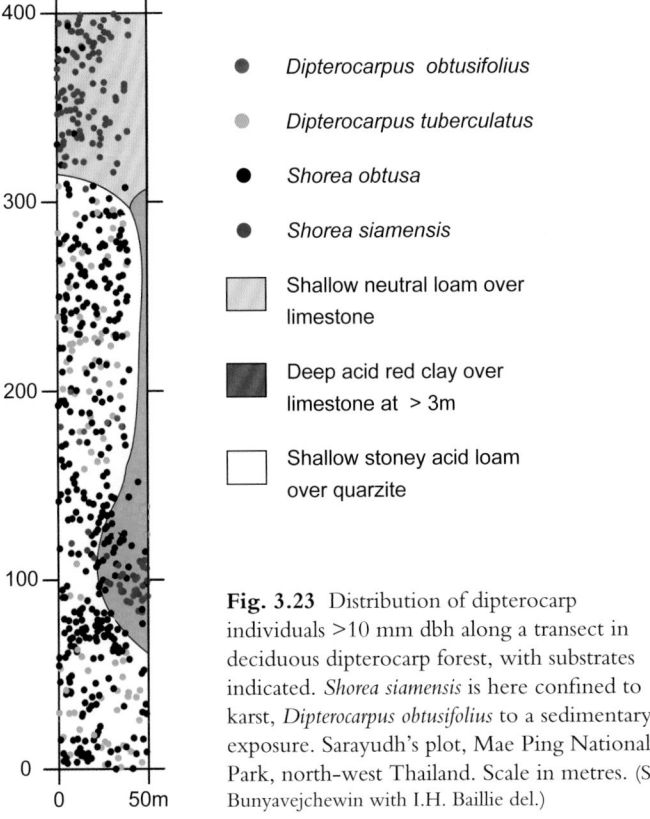

Fig. 3.23 Distribution of dipterocarp individuals >10 mm dbh along a transect in deciduous dipterocarp forest, with substrates indicated. *Shorea siamensis* is here confined to karst, *Dipterocarpus obtusifolius* to a sedimentary exposure. Sarayudh's plot, Mae Ping National Park, north-west Thailand. Scale in metres. (S. Bunyavejchewin with I.H. Baillie del.)

1. *Dipterocarpus tuberculatus* is the most widespread, occurring throughout Indo-Burma, though southwards only to the northern peninsula. It just crosses the Indian border where the upper Kabaw River, tributary of the Burmese Chindwin, enters. In north-west Burma, it forms monodominant stands on shallow clay soils over laterite pans, while in northern

Thailand and the Shan states it similarly forms stands on freely draining but seasonally moist clay loam soils over turbidites and intermediate to basic igneous rocks, where it can attain stature equal to the tallest deciduous forests. These stands are sometimes formed in association with *Shorea obtusa* or *S. siamensis,* often with *Dendrocalamus strictus,* both on plateaux and on ridges where mixed deciduous stands with or without teak occupy the slopes. *Dipterocarpus tuberculatus* occupies the greatest range of deciduous dipterocarp forest communities, forming tall straight-boled stands on sandy loam plains, but also in short open woodland, often with *D. obtusifolius* and *Pinus merkusii*, with rank grass or other understorey prone to hot fires.

2. The variably deciduous *Dipterocarpus obtusifolius* (Chapter 2) ranges eastwards of the Irrawaddy to southern Indo-China and south to north-west Peninsular Malaysia. It occurs to 1,300 m,[235] where it is usually in mixture with *Pinus kesiya*. It becomes monodominant on sandy soils that are dry, albeit sometimes waterlogged during the dry

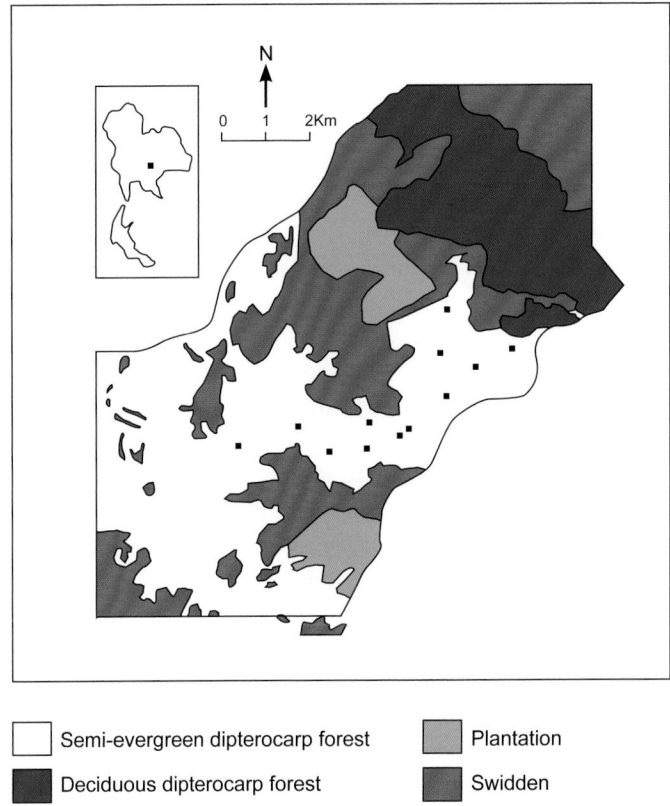

Fig. 3.24 A patch of deciduous dipterocarp forest is imbedded within a landscape of semi-evergreen dipterocarp forest on sandy soils, here dominated by *Shorea henryana* and *Hopea ferrea*, at Sakhaerat, Korat Plateau, north-east Thailand. (After Bunyavejchewin 1986)

season.[236] It is most often associated with *D. tuberculatus,* and often with *Pinus merkusii* (see Chapter 4). It forms open woodlands up to 20 m tall, with *Imperata* and other tall grasses, the dwarf bamboo *Vietnamosasa,* and deciduous shrubs prone to hot fires if infrequently burned (Chapter 2). Ecotones are therefore often sharp. It shares the same landscape as the semi-evergreen *Hopea odorata-Dipterocarpus alatus* forests at Huai Kha Khaeng, and the *Shorea henryana-Hopea ferrea*-dominated forest at Sakhaerat on the Khorat Plateau (Fig. 3.24).

3. *Shorea siamensis* and *S. obtusa* occur throughout Indo-Burma.[237] *S. siamensis* forms monodominant stands on skeletal limestone-derived clay soils, but it occurs across the full range of dry edaphic habitats, and is often locally common. Its range extends to north-west Peninsular Malaysia. Both species lose their leaves early, and are deciduous for at least two months.

4. *Shorea obtusa* rarely appears dominant, but has a wide edaphic range on clay soils in mixture with other species.

5. *Dipterocarpus intricatus* is confined to eastern Thailand, western Laos, Cambodia and southern Vietnam. It does not form monodominant stands, but occurs scattered with *D. obtusifolius* in deciduous dipterocarp forest, and within semi-evergreen forest on sandy yellow soils. These two deciduous dipterocarp forest species are the least deciduous, some leaves remaining at all times.[238]

Where sandy, soils are more compact than in adjacent semi-evergreen forest, but are of similar nutrient status.[239] Deciduous dipterocarp forests often include evergreen canopy species, especially on sandy soils – the geographically and ecologically widespread *Irvingia malayana* is among the most frequent, also *Mangifera caloneura* in the north.

The diversity of deciduous dipterocarp forest in structure and fire regimes has been discussed in Chapter 2. The tallest, straight-boled stands are of *Dipterocarpus tuberculatus,* on deep clay loam soils. The shortest, with most crooked boles and branches and open crowns, are of *Dipterocarpus obtusifolius,* often accompanied by *D. tuberculatus.*

The variable subcanopy composition of deciduous dipterocarp forest appears frequently correlated with the species dominating the canopy. *Cycas siamensis* appears ubiquitous. Stands dominated by *Dipterocarpus obtusifolius* are open and often densely grassy, whereas those dominated by other species may bear an exclusively woody understorey. The dwarf bamboo *Vietnamosasa pusilla* is ubiquitous east of the Chao Phya valley, *V. ciliata* above c.700 m in the north. Many northern stands are above 500 m altitude, often in mixture with *Pinus merkusii*

(which is locally common in the south at low altitude, also as far as Hua Hin in the northern Thai Peninsula) (Chapter 4). At these higher altitudes, evergreen shrubs and small trees are characteristic wherever *D. obtusifolius* dominates, including *Cassine* spp., *Ardisia crenata, Memecylon* spp., *Mammea siamensis, Syzygium cumini, Tristaniopsis burmanica, Craibiodendron stellatum, Quercus kerrii* and *Ternstroemia gymnanthera;* also the pioneer *Mallotus philippensis* and semi-deciduous *Aporosa villosa.* Many evergreens extend to lower altitudes in the north in *D. obtusifolius-D.tuberculatus-P.merkusii* deciduous dipterocarp forest, while many also extend into lower montane *Pinus kesiya* forest (Chapter 4). This echoes the prevalence of evergreen woody species on infertile soils in temperate regions, notably the shrub flora of North American pine forests. Vines are not generally abundant, but large vines are locally present, as they are in tall deciduous sal forests.

In drier parts of Indo-Burma, lowland deciduous forests on more siliceous soils are replaced above by *Pinus kesiya* savanna or, in the central and western Himalaya, by *P. roxburghii* savanna or dry warm temperate evergreen *Quercus*-dominated forest, especially those featuring ban oak (*Q. semecarpifolia*) and *Q. glauca* (Chapter 4).

In summary, whereas tall deciduous teak forests of India and Indo-Burma have much in common ecologically, and at the generic level floristically, in contrast, deciduous dipterocarp forests occupy a wider range of dry soils than tall sal forests, and a wider range of rainfall seasonality; their floristic similarity is weaker, and varies much more with the wider range of substrates that deciduous dipterocarp forest occupies.

5.3.3. *Burmese short deciduous forests and thorn woodlands*
Climates with 7–8 dry months are confined in Indo-Burma to the northern Irrawaddy plain, terminating just south of Mandalay. Short deciduous teak forest is widespread in northern Burma (though absent further east), and deciduous dipterocarp forest with *Shorea siamensis* and *Dipterocarpus tuberculatus* also enters these dry climates (Chapter 2). Stamp[240] noted that the distribution of 'dry' teak here is much influenced by topography, extending into wetter climates on steep upper slopes and ridges. In the driest regions of the northern Irrawaddy plain, a deciduous thorn forest is universally albeit variably clothed with the widespread *Acacia catechu* (*sha*), as well as *A. leucophloea, Dalbergia lanceolaria* ssp. *paniculata,* and *Bauhinia vahlii.* As yet little studied, and now much degraded by cattle browsing, these forests appear to differ greatly at species level from their Indian equivalents, being rich in endemics.[241] These include the locally abundant *Tectona hamiltoniana* (*dahat,* also found in one karst locality in Luzon), *Diospyros burmanica* (*tê*) and *Terminalia oliveri* (*than*). Here too, there is greater edaphic differentiation than observed in South Asia: the *than-dahat* association on stiff clays, *sha-dahat* on freely draining soils, and dense *sha* on alkaline soils.

6. Forests of seasonal climates east of Sunda

6.1. *The Philippines – an offshore enigma*

Early accounts[242] of the floristic pattern of the lowland Philippine forests suggest that the original forest was MDF throughout, in sharp contrast with continental Asia (see section 2.5.1 above). Species ranges do indicate some impoverishment from the perhumid towards increasingly seasonal climates, with only a few characteristically seasonal forest species (Table 3.13). The driest lowlands, in north-east Luzon, would support semi-desert *Acacia* woodlands were they on the continent, but few of the characteristic species have been recorded.[243] Widespread semi-evergreen forest dominated by *molave* (*Vitex parviflora*) in drier eastern areas may be ancient secondary. The explanation for these extraordinary differences from continental forests must be at least partly historical (Chapter 6).

6.2. *Seasonal forests of Java and east of Wallace's Line*

Long intensively managed, teak forests occur in East Java and south-west Sulawesi.[238] It is impossible now to determine their pristine composition. In Java they include several widespread continental species, but few endemics (Table 3.13, also see Chapter 6). The less disturbed forests of the Tanimbar Islands to the east, with a 4–6 month dry season, have recently been investigated by Yves Laumonier.[244] Surprisingly, he found few widespread continental Asian taxa in the prevailing semi-evergreen forest. Instead, the deciduous species appeared to be predominantly derived from evergreen genera of Sunda affinity, abundant in the seasonal evergreen forests of eastern Indonesia. Australasian seasonal forest elements are also few in the eastern Lesser Sunda Islands, albeit including *Eucalyptus alba*, *E. moluccana* and *E. urophylla*.

Table 3.13 Continental Asian deciduous forest tree species in the Philippines and Java.

Species in both Java and the Philippines

Species, family	Continental Range
Spondias pinnata, Anacardiaceae	Wide
Alstonia scholaris, Apocynaceae	Wide
Oroxylum indicum, Bignoniaceae	Wide
Cordia dichotoma, Boraginaceae	Indo-Burma
Diospyros montana, Ebenaceae	Wide
Mallotus philippensis, Euphorbiaceae	Wide
Trewia nudiflora, Euphorbiaceae	Wide
Flacourtia indica, Flacourtiaceae	Wide (cultivated)
Cratoxylum cochinchinense, C. formosum, Guttiferae	Indo-Burma
Vitex glabrata, Labiatae	Wide
Albizia lebbek, A. procera, Leguminosae	Wide
Cassia fistula, Leguminosae	Wide
Rhynchosia minima, Leguminosae	Indo-Burma (cultivated)
Syzygium cumini, Myrtaceae	Wide
Picrasma javanica, Simarubiaceae	Indo-Burma

Species present in Java but not the Philippines

Species, family	Range
Stereospermum chelonioides, Bignoniaceae	India only
Radermachera glandulosa, Bignoniaceae	Wide
Ehretia acuminata, E. laevis, Boraginaceae	Indo-Burma
Terminalia bellirica, Combretaceae	Wide
Dillenia pentagyna, Dilleniaceae	Wide
Diospyros vera (*ferrea*), Ebenaceae	Wide
Croton oblongus, Euphorbiaceae	India only
Margaritaria indica, Euphorbiaceae	Wide
Flacourtia indica, F. jangomas, Flacourtiaceae	Wide, cultivated
Callicarpa rubella, Labiatae	Wide
Premna tomentosa, Labiatae	Wide
Erythrina subumbrans, Labiatae	Indo-Burma, cultivated
Lagerstroemia indica, L. loudonii, L. floribunda, Lythraceae	India, Indo-Burma, Indo-Burma, all cultivated
Melia azadiracta, Meliaceae	Wide
Toona sureni, Meliaceae	Indo-Burma
Phyllanthus emblica, Phyllanthaceae	Wide
Psydrax diococco, Rubiaceae	Wide
Acronychia pedunculata, Rutaceae	Wide
Glycosmis pentaphylla, Rutaceae	Wide
Murraya koenigii, M. paniculata, Rutaceae	Wide, cultivated
Zanthoxylum limonella, Rutaceae	Wide
Helicteres isora, Sterculiaceae	Wide
Pterospermum diversifolium, Sterculiaceae	Wide
Sterculia foetida, Sterculiaceae	Wide
Tetrameles nudiflora, Tetramelaceae	Wide

Species present in the Philippines, but not in Java

Species, family	Range
Dipterocarpus alatus, Dipterocarpaceae	Indo-Burma
Tectona hamiltoniana, Labiatae	Indo-Burma
Pinus kesiya, P. merkusii, Pinaceae	Indo-Burma

Taxa missing from both:
Anogeissus (Combretaceae), *Afzelia, Xylia* (Leguminosae), *Chukrasia* (Meliaceae); Indo-Burmese species of (deciduous) *Dalbergia, Pterocarpus* (Leguminosae).

7. South Asian semi-evergreen notophyll forests

('dry evergreen forests' of Champion)

The best remaining examples of semi-evergreen notophyll forests are correlated with climate (Chapter 2; Plate 2.36a, b), as well as with freely draining grey-brown siliceous soils, in India frequently lying over gravel beds. It is unclear, however, whether this was always so in this densely cultivated landscape.

Table 3.14 The 25 leading tree species (Importance Value) >10 cm dbh in five, 0.5 ha plots within mature phase semi-evergreen notophyll forest, Giritale, Sri Lanka. (From Peeris 1975)

Nothopegia beddomei	Anacardiaceae
Polyalthia persicifolia	Annonaceae
Xylopia nigricans	Annonaceae
Ehretia microphylla	Boraginaceae
Connarus monocarpus	Connaraceae
Diospyros ebenum	Ebenaceae
D. oocarpa	Ebenaceae
Cleidion nitidum	Euphorbiaceae
Croton laccifer	Euphorbiaceae
Dimorphocalyx glabellus	Euphorbiaceae
Mallotus eriocarpus	Euphorbiaceae
Premna tomentosa	Labiatae
Vitex pinnata	Labiatae
Actinodaphne albifrons	Lauraceae
Alseodaphne semecarpifolia	Lauraceae
*Cassia fistula**	Leguminosae
Dialium ovoideum	Leguminosae
*Saraca indica**	Leguminosae
Strychnos nux-vomica	Loganiaceae
Memecylon petiolatum	Melastomaceae
M. sylvaticum	Melastomaceae
Ficus retusa	Moraceae
Syzygium cumini	Myrtaceae
Chionanthus zeylanicus	Oleaceae
Cleistanthus patulus	Phyllanthaceae

*Deciduous species

The great majority of evergreen species in these forests are common to both India and Sri Lanka, although so many deciduous dry thorn woodland species intrude in Indian stands, perhaps owing to their degraded state, that Meher-Homji[245] debated its separate status. The tree flora includes many genera and some species occurring elsewhere in semi-evergreen forest, but also some deciduous species shared with short deciduous forest[246] (Tables 3.13, 3.14). Dipterocarpaceae and Myristicaceae are notable in their absence, and the legumes such as *Acacia* and *Dalbergia*, which so dominate short deciduous forest in India, are scarce. On the other hand, *Drypetes* and *Manilkara* are now scarce in Indian deciduous forests, though they are abundant in Sri Lanka. These two genera, interestingly, are important components of African savanna woodland.[247] More intact stands remain in Sri Lanka – although they have regenerated there following the abandonment of the ancient dry zone irrigation systems some 500 years ago. Large deciduous trees, including *Terminalia* and *Careya,* are confined to riparian and other moist or open sites. Vines are abundant, and rich in species.[248] Little forest survives in the drier areas of these climates, since forest is generally replaced by open short grassland with widespread elements including species of South Asian short deciduous forest, and noxious weeds such as *Calotropis*. In degraded forests of moister areas in Sri Lanka, *Lantana* and *Chromolaena* prevail. In drier areas, the emergent evergreen canopy tree *Manilkara hexandra* becomes dominant. Unusually, it is semi-scandent when juvenile. *Syzygium cumini,* ubiquitous throughout Asia in semi-evergreen and moist deciduous forest, and occurring in the drier MDF of the western Philippines, becomes common in the more arid zones of north-west and south-east Sri Lanka.

The semi-evergreen notophyll forest tree flora is comprised of a core group of widespread species, enriched in regions with greater rainfall and in moist swales. Several proposed classifications have distinguished between mainly evergreen associations and others, particularly in the shallow valleys, where *Chloroxylon swietenia* and other short deciduous and thorn woodland species may grow to 20 m.[249] The following represents a consensus of what is recognized:

1. a dry climate facies, the *Manilkara hexandra* community;

2. a more widespread *Manilkara hexandra-Drypetes sepiaria* association;

3. an *Alseodaphne semecarpifolia-Berrya cordifolia-Diospyros* association, wherever streams or otherwise moister conditions prevail within the above; and

4. a mixed community in moist sites.[250]

8. Mechanisms underlying floristic change with increasing seasonality

The number of taxa within tree communities varies with soil type in the perhumid tropics, but the number also declines with the onset of rainfall seasonality and consistently declines as the dry season lengthens[251] (Chapter 7). This declining trend is clearest among species, least distinct among families except at the evergreen-deciduous forest interface. The decline is concentrated among the species belonging to a small group of genera, namely those in which many congeneric species co-occur within the same landscape or community. The species lost at each increase in seasonality tend to be climate specialists, while those surviving are generalists. Of the c.2,000 tree species in the CTFS Asian plots, only two are found across the entire climatic range, from the perhumid rainforests of Pasoh to the short deciduous forests of India. These are *Terminalia bellirica,* which enters the moister short deciduous forests, and the remarkable ambla tree, *Phyllanthus emblica,* whose range extends from Pasoh to the semi-arid thorn woodlands of peninsular India.

By transplanting seedlings reciprocally between plots, Jen Balzer[252] compared species present both in MDF at Pasoh and in seasonal evergreen dipterocarp forest at Khao Ban Thad, with a group of species occurring exclusively in one or the other. She found that species existing in both CTFS plots tolerated wider climatic variation by being conservative. Specifically, they were able to sustain living tissues at lower water potentials, that is they were more tolerant of drought; but they achieved this at the cost of lower maximum growth rates. Their conservatism thus reflects a loss of physiological plasticity. This finding mirrors those of Sabrina Russo who similarly compared specialist and generalist species across the soil nutrient gradient (see section 1.2.3 above). In both cases, one or more habitat factors come to limit opportunities for specialisation among the generalists. Balzer found no difference in mortality rates during the relatively short duration of her experiments, but long-term mortality rates likely differ substantially between her two sites due to the higher frequency of periodic severe droughts at Khao Ban Thad. Her results suggest that drought-induced mortality rates are likely to be highest among the Pasoh specialists. By implication, specialists on less favourable sites can make up losses from lower growth rates by experiencing higher survival rates.

9. At the margins of the tropics

Plate 3.17 Species at and beyond the tropical margin. **a,** Deciduous *Bombax ceiba* in karst country along the Ninming River, southernmost Guangxi, at the tropical margin, Feb. 1990; **b,** *Viola* sp., a temperate herb in pasture at the tropical margin, Ninming, Guangxi; **c,** *Mallotus philippensis* in fruit on Mount Victoria, Hong Kong; **d,** *Toona ciliata* in the frost zone at 1,900 m, Trongsa, bhutan; **e,** Betel nut, *Areca catechu,* plantations in the outskirts of Kao-Hsiung, Taiwan, at 23° N; **f,** Frost scorched cultivated bananas, 20 km north of Udaipur, Rajasthan, March 2013. (**a–c, f** P.A.; **d** H.H.; **e** S. Davies)

9.1. *Continental Asia*

32° N happens to be the latitude of both the northernmost outpost of tall deciduous sal *(Shorea robusta)* forest in north-west India, and therefore of the exclusively tropical Dipterocarpaceae, and *Kandelia obovata* in the bay of Kagoshima, southern Kyushu, representing the northern-most extension of the exclusively tropical Rhizophoraceae. West of central Nepal, lower montane forest is replaced by fire-climax frost-tolerant (therefore warm temperate) forest. Lowland tall deciduous sal forest therefore becomes liable to occasional frost scorching along its upper margins and where the canopy has been opened by logging. Some of the most detailed observations on the varying impact of established tropical trees of occasional frost and snowfall have been made at the New Forest of the Indian Forest Research Institute at Dehra Dun, 30°20' N.[253] Even at 32°04' N, in the Kangra natural sal forest, although frost may scorch foliage, it does not kill the established trees[254] (see also Plate 4.15f).

Only small pockets of forest remain in the lowlands of southernmost China and the coastal plains. They are difficult to interpret, but this effort will be greatly assisted by the recent establishment of large long-term forest monitoring plots within the CTFS network within the northern tropical margin and in southern warm temperate forest, Nanjeng Shan and Lienhuachih in Taiwan and Dinghushan, Gutianshan and Zhejiang Baishanzu in China.[255] From these, and from historical plant collections, the northern limits of many canopy trees with distributions in the tropics to the south may be inferred (see map inside back cover). Tropical forests appear to terminate over a narrow ecotone at the local scale, clearly manifest in the northern limits of the exclusively tropical family Dipterocarpaceae, as well as by an almost complete species turnover in the canopy family Fagaceae in the lower montane forests also, here to 1,900 m[256]. As canopy dominants, Fagaceae here descend from tropical lower montane to lowland warm temperate evergreen forest (Table 3.15). Myristicaceae, Lecythidaceae and Crypteroniaceae also disappear. The sudden descent of frost from c.2,000 m to near sea level is correlated with topography, the southern slopes being sheltered against the cold dry winter north-east monsoon. The Shiwan Dashan and coastal lowlands to its south and eastwards to Hainan, and the Yu Jiang tributary of the Pearl River from Nanning westwards, appear to be frost-free on the evidence of the riparian *Bombax ceiba* (though it was formerly cultivated for its kapok). The hills there once supported the dipterocarps *Hopea chinensis* and *Vatica mangachapoi*. But alluvial valleys of major rivers approaching the coast north of 22°30' N, now entirely and intensively cultivated, are frost pockets and unlikely to have supported tropical forest as was implied in the case of the Pearl River valley by Wang.[257] Rainforest pockets may exist

Key to columns

Ra.: Geographical ranges

> 1: occurring south of China, in one or more countries from Nepal eastwards
>
> 2: extending north at most to W. or S. Yunnan, S. Guangxi, or S. Guandong
>
> 3: north to Guizhou, Hunan, Jiangsu, or Fujian
>
> 4: north to Sichuan, Hubei, Anhui, or Zhejiang
>
> 5: north to Gansu, Shaanxi, Henan or Jiangsu
>
> 6: north of the Qin Ling Shan, including to Korea and Japan

Al.: altitudinal ranges

> 1: below c.800 m
>
> 2: c.800–1,700 m
>
> 3: c.1,700–2,600 m
>
> 4: above 2,700 m

Al. Bhutan: altitudinal range in the Flora of Bhutan and Sikkim

Ca: *Castanopsis*

Cy: (*Quercus* subgen.) *Cyclobalanopsis*

L: *Lithocarpus, Quercus* (subgen. *Quercus*)

Crosses (X) to the left or right of the centre of the columns indicate ranges towards the lower or upper part of a range.

Bold   print indicates species mentioned in Hu Shinshi's classification in subtropical forests.

Italics   indicate species in Zhu Hua's (2006) Appendix species list from lower montane forest at Mengsong, southernmost Yunnan.

Notes

- Additional species occurring in Zhou's plots in Yunnan tropical lower montane forest, all confined below 1,700 m and all in tropical southernmost China including southernmost Guizhou: *L. hypoglaucus, Cy. kerrii, Q. chrysocalyx, L. fordianus, Ca. tcheponensis, (L. gagnepainianus = pseudoreinwardtii). Cy. kerrii* is a curious addition as it is a species of dry forests in Thailand.

- The altitudinal categories appear to move down a category from the true tropics to zones to the north of them. The lowland category above the line running above *L. crassifolius*, especially, may be different from that below it, though not distinguished in the Flora. This explains the absence of any exclusively lowland species below that line.

- I conclude, in summary, that species listed above *L. crassifolius* are confined to or concentrated in warm temperate forests at a variety of altitudes including the lowlands; species below that line are confined to or concentrated in tropical lower montane forest; while *Castanopsis indicus* occurs only in the lowland-lower montane ecotone and below.

- [1, 2] The Flora states *Ca. hystrix* to occur between 1,500–2,500 m, whereas Schweinfurth states <1600 m.

Species name	Ra.1	Ra.2	Ra.3	Ra.4	Ra.5	Ra.6	Al.1	Al.2	Al.3	Al.4	Al.Bhutan
Q. aliena	X	X	X	X	X	X	X	X	X		
Q. acutissima	X	X	X	X	X	X	X	X	X		2,100–2,400
Cy. glauca	X	X	X	X	X	X	X	X	X		1,000–2,400
Cy. myrsinifolia	X	X	X	X	X	X	X	X	X		
Q. oxyodon	X	X	X	X	X		X	X	X		2,400–3,600
Ca. carlesii				X	X		X	X			
Ca. eyrei		X	X	X			X	X			
Ca. faberi	X		X	X			X	X	X	X	
Cy. gracilis		X	X	X			X	X	X		
L. hancei			X	X			X	X	X		
L. litseifolius v. litseifolius			X	X			X	X	X		
Cy. stewardiana		X	X	X				X			
Ca. ceratacantha	X			X				X			
Q. delavayi			X	X				X			
L. dealbatus	X			X				X			
L. craibianus	X			X				X			
L. variolosus	X			X							
Ca. orthacantha				X				X			
Q. stewardiana				X				X			
Cy. fargesii		X	X	X			X	X	X		
Ca. hystrix	X		X					(X)[1]	(X)[2]		1800–2400
Cy. helferiana	X		X				X	X	X		
Ca. chinensis	X		X				X	X			
Ca. fissa	X		X				X	X			
L. fenestratus	X	X					X	X			1,200–2,400
L. crassifolius	X	X								X	
Ca. echinocarpa	X	X						X	X		
L. thomsonii	X	X						X	X		
Cy. chapensis	X	X						X	X		
Ca. fleuryi	X	X				X		X	X		
Cy. lamellosa	X	X						X	X		
L. rhabdostachys	X	X						X	X		
Ca. tribuloides	X	X						X			
L.Xylocarpus	X	X						X			
L. pachyphyllus v. fruticosus	X	X						X			
L. truncatus	X	X					X	X	X		
Ca. ferox	X	X					X	X	X		
Ca. mekongensis	X	X					X	X	X		
Ca. calathiformis	X	X					X	X	X		
Ca. argyrophylla	X	X						X			
Ca. wattii	X	X						X			
Q. rex	X	X						X			
L.pseudoreinwardtii	X	X						X			
L. pachylepis	X	X						X			
L. echinophorus	X	X						X			
L. vestitus	X	X						X			
L. elegans	X	X					X	X			
L. fohaiensis		X					X	X			
L. echinolotus	X	X					X	X			
Ca. indica	X	X					X	X			300–1,800

Table 3.15 The geographical ranges of tropical and warm temperate Fagaceae in China (Nomenclature follows Huang *et al.* 1999)

there on south-facing slopes where frosted air descends above canopies protected by their diurnally accumulated latent heat, as implied by Len Webb's observations on the lower slopes of Dinghushan. The tropical margin bears interesting comparison at the other end of the Asian tropics, along the Yemeni coast of the Red Sea north to 15° N, where a tropical woodland dominated by *Anogeissus dhofarica,* albeit with Mediterranean elements including *Pistacia* and *Olea*, extend to above 1,000 m. This is confined to scarp slopes facing the sea: protected from cold winter northeasterlies off the desert, and where diurnal orographic rain falls in spring and late summer.[258]

Overall in the Far East, the lowland tropical margin is marked by turnover at the species level but much less at higher levels. In China the tropical-temperate frontier is mostly at c.22–23° N.[259] Exceptions are where tributaries of the Salween valley issue from Yunnan at 24° N, and in the Mekong valley rainshadow, c.23 –30° N, where tropical dry deciduous forest (with *Bombax ceiba, Anogeissus acuminata, Albizia procera, Bauhinia variegata, Phyllanthus emblica, Dalbergia fusca,* and *Eriolaena*) merges with Mediterranean woodland (containing *Olea* and *Pistacia*). Very few tropical canopy species extend north, one exception being *Engelhardia roxburghiana*, which descends to the lowlands. Species of paradigmatic tropical lower montane genera are replaced by congenerics whose ranges extend northwards, to varying extents, in the seasonally wet warm-temperate forests. These lower montane genera include *Bischofia, Nyssa, Schima*, the Hamamelidaceous genera *Altingia* and *Exbucklandia*, Fagaceous *Castanopsis, Lithocarpus* and *Quercus,* and innumerable *Ilex* and Lauraceae.

At this same latitude and location therefore, warm temperate species in those same genera of tropical Fagaceae and other families that dominate the canopy of tropical lower montane forest, come for the first time to dominate the *lowland* evergreen forests also, which Fagaceae continue to do northwards to the broad-leaved evergreen forest limits in southern Korea and Japan (Table 3.15). Mangrove, which is highly frost-sensitive, occurs naturally north-east up the China coast to southern Fujian, while it has been planted to southern Zhejiang. Tropical lowland inland forest taxa do also extend northeastwards along the coastal hills into southern Fujian (the coastal belt may be narrower than indicated on the map within the back cover), in several cases into locations in the inland provinces of Guizhou, Hunan, Jiangxi, and even to Sichuan and northwards.[260] These are generally confined below 550 m.[261] Dudgeon and Corlett[262] record a wide range of tropical taxa below 500 m in the traditional *Feng Shui* village woods of Hong Kong.

In the eastern Himalaya, the transition between frost-sensitive tropical lower montane and frost-hardy warm temperate evergreen forest takes place over a few hundred metres of altitude and slope: higher on ridges and upper slopes owing to nocturnal temperature inversions (Chapters 1, 4).

Above the transition, there is a continuum of floristic and phenological change, with increasing deciduousness to c.2,900 m where incidence of ground frosts appear to coincide with the appearance of dominant temperate conifer genera and the disappearance of evergreen canopy trees. This continuum is analogous to that in China, but there it extends in the lowlands from c.22–35° N, at least 1,500 km without a clearly discernible ecotone! There it is complicated by latitudinally changing altitudinal floristic and phenological zonation in which ecotones can be recognized, and others where local climate and soils can be influential. This explains why certain species may descend close to tropical lower montane forest limits in the eastern Himalaya but are confined to higher latitudes in Sichuan, Hubei, Jiangxi or Guizhou, and to higher altitudes in China. This is particularly true of deciduous species, including the archaic warm temperate *Tetracentron sinense* and *Euptelea pleiosperma*.

It is informative to compare intact remaining lowland tropical evergreen and warm temperate forests in inland South China: the CTFS 20 ha plot at Mengla (southern Yunnan, 21°37' N, 101°35' E) with the sacred forest at Dinghushan (23°08' N, 112°35' E).[263] Both include lowland predominantly evergreen forest and lower montane forest. The Mengla CTFS plot includes 12 canopy Fagaceae, though it lacks any *Quercus* – which are usually confined to drier sites in tropical Asia (Table 3.15). The Fagaceae in the Mengla plot are mostly restricted either to lowland or lower montane forest, but they are less abundant and do not achieve dominance in the lowland forest, where they are overtopped by *Parashorea chinensis*. No tropical evergreen canopy species has a range extending north of the narrow intermittent belt of northern seasonal evergreen dipterocarp forest in southernmost China.

A 0.5 ha plot established at 270–330 m elevation in Dinghushan temple forest experienced minimum night temperatures to -0.2°C during the decade 1975–1986.[264] This lowland forest sample is dominated by Lauraceae, Fagaceae and Theaceae, families which would only have dominance above 800 m in the tropics. Only three species of Fagaceae have been recorded there, although according to Hu Shin-Shi,[265] eight species are widespread in the evergreen broad-leaved forests to the south in Guangxi. The only Fagaceous species recorded both in Mengla and in Dinghushan (where it is doubtful) is *Lithocarpus fenestratus*. That species excepted, these Dinghushan Fagaceae extend north of the southern Chinese provinces into Guizhou, Hunan, Jiangsu or Fujian while four extend further, into more northern provinces. Only three of the eight local Fagaceae extend to the rainforest belt in the extreme south. The canopy at Dinghushan, lacking emergents, is dominated by *Castanopsis chinensis*, a species whose range extends north of Guandong into exclusively warm temperate regions at low altitude. *C. fissa* shares a similar geographical range and is also locally dominant in the reserve. But whereas the canopy Fagaceae are represented by warm temperate species, several tropical taxa, notably *Gironniera subaequalis,* are well represented beneath it. Lower, in a valley facing south over the Pearl River valley, Len Webb saw *Toona ciliata* and other tropical successional elements on a visit in 1986.[266]

The records of scattered meteorological stations suggest that the ecotone which separates sites like Mengla and Dinghushan coincides with the southern limits of periodic killing air frosts in the canopy – or more specifically, the occasional freezing of nocturnal winter dew to form hoar frost. *The distribution and impact of such frosts merit close investigation.*

The large-vesseled woody vine *Merremia boisiana* only reaches Hainan and southern Yunnan, Guangxi and Taiwan. While *Macaranga andamanica, M. denticulata, M. indica, M. henryi* do reach southernmost Guizhou, *M. sampsoni* reaches Fujian, and *M. tanarius* extends north-east to Okinawa, the four other pioneer *Macaranga* of China are confined to the far south. Southern coastal Fujian is the furthest north that commercial *Hevea* rubber is grown[267] and can therefore be accepted as the upland coastal tropical limit. But the successional tree species *Toona ciliata, T. sureni,* and the floodplain *Bombax ceiba* (which, admittedly, has been cultivated for its kapok), are all recorded from southernmost Sichuan and adjacent Guizhou. Here, where the Yangtse (Jiang) River emerges from a deep and steep mountain valley into the plains at Leibo, the size and flavour of the mangoes grown on the lower slopes have earned them an enviable reputation. It appears that pockets of frost-free or possibly drier climate acting as refuges for certain tropical canopy elements exist well north of the tropical evergreen forest margins.

Are these *tropical* trees, otherwise distributed in frost-free climates but restricted here to frost-free pockets in an otherwise frost-prone region – or are they mildly frost-hardy, and thereby defining an area that may be termed *subtropical*? The tropical East Asian woody species with more northerly distributions appear to be mostly either pioneer or successional and are frequently deciduous, or they are evergreen subcanopy species (Table 3.16). As with Fagaceae, they vary in their northern range limits. Most also occur in tropical lower montane forest further south. Taxa absent from or only just penetrating the lower ecotone of lower montane forest at higher latitudes within the tropics, including Chrysobalanaceae, Crypteroniaceae, Dipterocarpaceae, Lecythidaceae and Myristicaceae, fail to extend to these northern tropical pockets. At its southern limits in the lowlands, frost will follow a similar pattern to that on equatorial mountains: descending pre-dawn frosted air may not penetrate closed evergreen canopies warmed by a sunny previous day, but may do so on the shadier north-facing slopes and pond in plains. The botanical gardens at Guangzhou, for instance, just within the Tropic of Cancer and not far from

Gymnosperms	
Dacrycarpus imbricatus	NE Guangxi
Podocarpus neriifolius	Guizhou, Hunan, Jiangxi, Sichuan
Gnetum pendulum	SE Guizhou
Burseraceae	
Garuga pinnata	Sichuan (Leibo)
Meliaceae	
Toona ciliata	S. Sichuan
Munronia unifoliolata	Guizhou
M. pinnata	Chongqing
Malpighiaceae	
Aspidopterys esquirolii	SE Guizhou
A. cavalieri	Guizhou
Hiptage benghalensis	Guizhou
H. luodianensis	S. Guizhou
H. minor	Guizhou
Euphorbiaceae/ Phyllanthaceae	
Bischofia javanica	Anhui, Shaanxi
(B. polycarpa)	(Anhui, Shaanxi)
Phyllanthus emblica	Guizhou, Sichuan
Glochidion triandrum	Guizhou, Sichuan, Hunan
(G. pubescens)	(to Japan)
G. philippicum	Sichuan
G. wrightii	Guizhou
G. ellipticum	Guizhou
(G. daltonii)	(Anhui, Shandong)
G. eriocarpum	Guizhou, Hunan
Macaranga indica	SW Guizhou
M. henryi	S. Guizhou
Mallotus barbatus (esquirolii)	Guizhou
M. philippensis	Sichuan, Hubei
M. millietii	Hubei
(M. repandus)	(Anhui, Shaanxi)
M. dunnii	S. Hunan
M. barbatus	W. Hubei
(M. apelta)	(Anhui, Shaanxi)
M. oreophilus	Sichuan
(M. japonicus)	(to Japan, Korea)
M. lianus	S. Hunan

Table 3.16 Examples of the lowland tropical woody species that extend further north into China than exclusively tropical families. Presumed frost-hardy taxa in parentheses. (Data from Wu *et al.* 1994 –)

the Pearl River estuary, receive periodic winter hoar frost. But reliably frost-free pockets can be expected in sloping, south-facing coves on lowest slopes. It is also possible that the latent heat retained within the forest canopy can provide frost-free subcanopy micro-environments well north of the southern limits of canopy frost.

The plants most sensitive to frost include those with large vessels, which are especially vulnerable to cavitation when their contents freeze. Among these are large woody vines and many pioneers, also most palms owing to the absence of any dormancy or ability to recover by reiteration or secondary growth.[268] Corlett has described how certain species especially suffered in a rare ice storm.[269] The persistence of successional species is therefore enigmatic, though possession of frost-hardy dormant seed may explain the hardiness of some, winter deciduousness that of others. The survival of evergreen mid-successional species with large recalcitrant seeds, such as *Dracontomelum dao* at Dinghushan, is puzzling. The northernmost inland localities of several such late successional canopy species extend well north of the boundary of exclusively tropical families. *Bombax*

ceiba trees, whose massive trunks and spreading deciduous crowns form so characteristic a feature of the northernmost tropical floodplains, have been grown for their kapok as far north as Sichuan, but must surely occasionally succumb there to catastrophic frost.

Taiwan is also enigmatic:[270] Dipterocarpaceae are absent, probably because the deep Basho channel to the south is too wide for frequent dispersal from Luzon following Pleistocene extinctions. Nevertheless, Myristicaceae are present in the extreme south, and *Barringtonia asiatica* (Lecythidaceae) along the coast there. The remaining lowland forest, exemplified by the CTFS 3 ha plot at 300–500 m altitude at Nanjengshan (22°03' N), contains both tropical and warm temperate elements (Table 3.17). Cultivated tropical trees, such as the beetle palm, *Areca catechu*, are grown in cultivation north to Kao-hsiung (22°36' N) (Plate 3.17e). Perhaps difficulty of dispersal across the deep Basho channel north from Luzon has led to reduced competition in Taiwan south of 23° N, and permitted invasion of the well-represented warm temperate flora into a tropical lowland climate (Chapter 6).

Woody halophytes cannot survive interruption of salt conduction caused by freezing of their vessels. Mangroves extend up the China coast to the southern Zhejiang coast, coinciding with the northern limits of tropical upland woody taxa, and continuing intermittently northeastwards in the islands to Okinawa where upland tropical elements also persist (Table 3.18) and even southernmost Kyushu where the mangrove *Kandelia obovata* occurs in the Gulf of Kagoshima (31°34' N).[271] Some suspect it has been introduced; but introduced or not, its persistence over many decades indicates freedom from frost. *Avicennia marina* extends up the Red Sea only to the Tropic of Cancer.

Table 3.17 Species with tropical ranges to the south, and temperate ranges to the north, in the CTFS plot in lowland forest at Nanjen Shan, southern Taiwan. Asterisked taxa are more typically lower montane. (Data from Sun *et al.* 1998)

Tropical taxa	
Syzygium buxifolium	Ilex cochinchinensis
Syzygium euphlebium	Podocarpus rumphii
Syzygium kusukusuense	Garcinia multiflora
Syzygium densinervium	Lagerstroemia subcostata
Lithocarpus amygdalifolius*	Quercus longinux*
Castanopsis indica*	Ficus japonica
Ficus microcarpa	Glycosmis parviflora
Ficus virgata	Calophyllum inophyllum
Astronia ferruginea*	Melastoma candidum
Ternstroemia gymnanthera*	Mallotus paniculata
Bridelia balansae	Drypetes indica
Archidendron lucidum	Engelhardia roxburghiana*
Fagraea ceilanica	

Warm temperate taxa	
Castanopsis carlesii	Microtropis japonica
Schima superba	Machilus thunbergii
Machilus japonica	Viburnum odoratissimum
Eurya japonica	Cleyera japonica
Hydrangea chinensis	Itea parviflora
Osmanthus marginatus	Fraxinus insularis

Adinandra	Dodonaea	Macaranga
Adina[a]	Drypetes[a]	Melanolepis
Alchornea	Erythrina	Pongamia[a]
Bischofia	Erycibe[b,c]	Saurauia
Breynia	Excoecaria	Schefflera[b]
Bridelia	Garcinia	Syzygium[b]
Ceriops[b,c]	Glycosmis[b]	Tarenna
Crateva	Grewia	Wendlandia
Croton	Illigera	Wikstroemia
Dalbergia	Ixora[a,3]	

Table 3.18 Tropical tree genera which reach the Ryukyus.

[a] reaching Yakushima
[b] reaching Kyushiu
[c] possibly introduced

9.2. *The tropical margin in the semi-arid north-west*

Semi-arid thorn woodland species here appear to broadly transgress the frost line, contrasting with the pattern of frost sensitivity among the great majority of woody species of evergreen forests (see also Chapter 4). The pattern of frost incidence in the dry tropics is complex: Air temperatures, as happens elsewhere at the margin of the tropics, may drop to below 0°C but, because soil is dry and widely exposed between the trees, it heats during the day and retains latent heat during the night, warding off freezing at and above the ground in a manner comparable to the canopy of closed montane forest (Chapter 4). More densely vegetated areas, especially where moist during the early winter months, may therefore be more prone to occasional frost than more the woodlands of arid areas. Topography, where prominent, leads to much local variation in frost incidence. All these factors explain the wide incursion of the frost margin in north-west India indicated in the map within the inside cover. The northwestern winter is under the influence of northeasterly winds off the Himalaya, which can bring showers in hard winters and light snow in the north, cooling the soil. This may occasionally result in frosting in the thorn woodlands.

Although the few fleshy plants, such as cultivated bananas, suffer scorching of their leaves by frost (Plate 3.17f), no indigenous tree species appear to do so in the semi-arid regions east of the Desert of Thar. Even the leaves of many among the deciduous majority – notably *Acacia* species, which do not fall before the hot dry season when the winds return to south-west – are unaffected by these short freezing events.

9.3. *The southern tropical margin*

The tropics of Asia towards the south terminate in Australia. The margin there seems not to have received detailed examination. The mangrove *Avicennia marina* grows at almost 34° S in Sydney harbour, kept to c.1 m tall by air frosts above the warmer air over the water. Tropical rainforest taxa extend to 33° S in the coastal lands north of Newcastle, and at 29° S at 700 m in the Great Dividing Range at Lamington National Park on the Queensland-New South Wales border. West of the Range, tropical taxa appear restricted well north of the Tropic of Capricorn. Tropical tree taxa in Australia are overwhelmingly evergreen. Whereas these tropical taxa appear to have a defined southern limit, warm temperate taxa, also evergreen, extend well north into frost-free climates, and there appears to be no consistent northern limit.[272] The explanation may be in part historical: Much of the Australian flora had a warm temperate origin in the Cenozoic, while the weak Malesian floristic presence may result, in contrast to the situation in New Guinea, in relaxed competitive exclusion of temperate elements within the tropics (Chapter 6).

9.4. *Use of the term 'subtropical'*

We may delineate the northern margin of the Far Eastern tropics by the northern limits of exclusively tropical tree families, especially the canopy family Dipterocarpaceae and widespread, mainly subcanopy, family Myristicaceae; the descent of canopy dominance of Fagaceae to the lowlands, and an almost complete turnover of its species. They coincide with the limits both of northern lowland seasonal evergreen forest and tropical lower montane forest floristic associations. In that case, the boundary reaches 32° N at the base of the western Himalaya, trends south-east to 27–28° below the eastern Himalaya and in the northernmost lowlands of Burma, then sweeps south down the Sino-Burmese frontier to 24° in the upper Salween valley (northern seasonal evergreen dipterocarp forest with *Dipterocarpus retusus*) and the rain-shadow Mekong valley (short deciduous forest). It then fluctuates between 21–22° N from the southern foothills of Yunnan into Guangxi, there trending north in the coastal hills to Fujian and southernmost Zhejiang (28° N) (map inside front cover). It is strictly confined thereafter to the coast and offshore islands to c.32° N in southernmost Japan. The southern lowlands of Taiwan appear to be frost-free though little truly tropical forest remains.

Small apparently frost-free pockets serving as refuges for some tropical canopy tree species (but none in exclusively tropical families) remain well north of the main tropical-warm temperate boundary in South China. This contrasts with the absence of such pockets on the harsher side of every other climate-mediated ecological boundary in the Asian tropics. Dipterocarpaceae and tropical Fagaceae species, with limited seed dispersal and lacking dormancy neither of which occur in these pockets, would have been slow to advance and particularly prone to local extinction. These refuges have likely survived from the early Holocene, 14,000–4,500 BP, when the climate here was warmer and wetter.[273] I suspect that these otherwise tropical canopy woody species which occur north of the tropical-temperate boundary persist, or periodically reestablish, in localities or habitats which are persistently frost-free, rather than that they are mildly frost-hardy.

The term *subtropical* has been widely used by ecologists in China and India,[274] and more imprecisely by horticulturists in the West, to denote forests and species that occur in an intermediate zone between the tropics proper and warm temperate climates. Categories recognized may even include tropical lower montane forest among them, and as in 'subtropical monsoonal broad-leaved evergreen forest'.[275] Mean monthly minima, rather than the influence of occasional catastrophic freezing temperatures, have been relied on as the climatic differentiating correlate. But the tropical-temperate interface is quite narrow on the ground – although it is somewhat wider geographically owing to the persistence of tropical pockets in southern-most warm temperate regions – and few canopy species seem to transgress it. The southern ranges of most warm temperate lowland taxa (warm temperate canopy Fagaceae, for instance) do indeed coincide with this interface. But these have individualistic northern limits. There is a group of 'subtropical' sister entities of tropical lower montane species which occur to the southernmost lowland warm temperate evergreen forests but do not penetrate tropical lower montane associations. Again however, their differing northern range limits belie the presence of any clear northern boundary to what could be termed 'subtropical' (for example Table 3.15). There is therefore a continuous general changeover of species north of the limits of tropical seasonal evergreen dipterocarp forest and lower montane forest formations, and no discernable geographical or ecological floristic ecotone separating subtropical from warm temperate vegetation or floras. The term often seems synonymous with warm temperate evergreen forest in Chinese literature. I therefore conclude that the term 'subtropical' defies precise definition, and would be better dropped in formal ecology.

9.5. *A conclusion*

Whereas the penetration of tropical taxa into the temperate zone is restricted by lack of frost-hardiness, penetration of temperate taxa into the tropics is variably restricted by the effect of the energetic cost of frost-hardiness on their competitiveness.

Key to the main lowland forest types of Asia west of Wallace's Line

The types defined here are the most widespread on the prevailing yellow/red upland soils. Beware: This key does not include forests of special habitats, such as peat swamps or limestone. Forest most often varies continuously, and in real examples shares one or more attributes with those that share similar habitats. (This is not a proposed new forest classification; it extends and modifies an earlier attempt of mine.[276])

1.   Canopy entirely deciduous 2

1.   Canopy entirely or partially evergreen........................14

2.   Canopy dominated by dipterocarps, generally tardily deciduous 3

2.   Forest lacking dipterocarps, deciduous for at least three weeks 5

3.   Canopy dominated by sal, *Shorea robusta*: north and east India 4

3.   Sal absent; canopy dominated by other dipterocarps: Indo-Burma...................**deciduous dipterocarp forest**

4.   Forest stature generally at least 20 m; woody evergreen trees or shrubs in the subcanopy**tall** ('moist') **sal forest**

4.   Forest stature generally less than 20 m; understorey entirely deciduous......................**short** ('dry') **sal forest**

5.   Forest stature generally at least 20 m; woody evergreen trees or shrubs in the subcanopy 6

5.   Forest stature generally less than 20 m; woody plants entirely deciduous..10

6.   Canopy with teak, *Tectona grandis*, often dominant.....7

6.   Forests devoid of teak 9

7.   Teak forest with *Pterocarpus marsupium*......................
          .... **Indian tall deciduous teak forest**

7.   *Pterocarpus marsupium* absent................................ 8

8.   Teak forest with *Pterocarpus macrocarpus*.................. .
          ....**Indo-Burmese tall deciduous teak forest**

8.   Teak forest without *Pterocarpus macrocarpus*............. .
          ....**Javanese tall deciduous teak forest**

9.   Deciduous forest with *Lagerstroemia lanceolata*......... .
          ....**Indian tall deciduous forest**

9.   Deciduous forest with *Lagerstroemia calyculata* or *L. tomentosa* ....**Indo-Burmese tall deciduous forest**

10. Canopy closed, canopy trees rarely thorny 11

10. Canopy open, thorny trees abundant ....
....deciduous thorn woodlands 13

11. Forests with *Pterocarpus santalinus* or
Cochlospermum religiosum short deciduous forests 12

11. Forests without these species ....
short facies of Indo-Burmese tall deciduous forest

12. Forests with teak .... **short deciduous teak forest**

12. Forests without teak ....
.... **southern short deciduous forest (India)**

13. Woodlands with *Tectona hamiltoniana* and/or
Terminalia oliveri dominant ....
.... **Burmese deciduous thorn woodland**

13. Woodlands lacking the above; *Acacia* species and
other Leguminosae dominanat ....
.... **deciduous thorn *Acacia* woodland**

14. Canopy as a whole reproducing annually............... 15

14. Emergent or upper main canopy, especially
dipterocarps, reproducing mostly supra-annually 22

15. Notophyll forests with shrubby understorey................. .
.... **South Asian evergreen notophyll forest**

15. Mesophyll forests lacking a predominantly shrubby
understorey.. 16

16. Pioneer and successional species mostly deciduous....
....semi-evergreen forests 17

16. Pioneer and successional species mostly evergreen....
.... seasonal evergreen forests 20

17. Forests with abundant *Holigarna*, few dipterocarps ..... .
.... **South Indian semi-evergreen forest**

17. Forests with abundant dipterocarps, few or no
Holigarna 18

18. Forests with *Dipterocarpus indicus* present, often
dominant, *Holigarna* present.....................................
.... **South Indian semi-evergreen forest**

18. Forests with other semi-dominant *Dipterocarpus,
Holigarna* absent or rare..
.... **Burmese semi-evergreen forest 19**

19. *Dipterocarpus alatus, Hopea odorata* present, often
abundant; *Dipterocarpus turbinatus* absent..................
.... ***H. odorata* semi-evergreen forest**

19. *Dipterocarpus turbinatus* present, often dominant;
Hopea odorata absent ....
........................***D. turbinatus* semi-evergreen forest**

20. Forests with *Terminalia myriocarpa* ....
.. **northern seasonal evergreen dipterocarp forests**

20. Forests without *Terminalia myriocarpa* 21

21. Forests with *Dipterocarpus indicus*; *Shorea* absent ....
.............**South Indian southern seasonal evergreen
dipterocarp forest**

21. Forests with other *Dipterocarpus* species, *Shorea,*
present... .
.... **Indo-Burmese southern seasonal evergreen
dipterocarp forest**

22. Forests with single canopy stratum, emergents
scattered or absent..................................... ***Kerangas***

22. Forests with two canopy strata; emergent canopy
dominated by dipterocarps ....
.... **mixed dipterocarp forest (MDF)** 23

23. Forests with sect. *Doona* the leading emergent *Shorea*
group....... **Sri Lankan *dun* mixed dipterocarp forest**

23. Forest with red merantis the leading emergent *Shorea*
dipterocarp group.. 24

24. *Shorea* sect. *Brachyptera* the leading red meranti
group; yellow merantis (*Shorea* sect. *Richetioides*)
rare or absent ....
.... **Philippine *lauan* mixed dipterocarp forest**

24. *Shorea* sect. *Mutica* the leading red meranti group;
yellow merantis present..
... **Sunda mixed dipterocarp forest associations 25**

25. Dark red merantis, *selangan batu* (*Shorea* sect.
Shorea) abundant in canopy ....
**dark red meranti mixed dipterocarp forest, coastal
hill dipterocarp forest, peninsular hill dipterocarp
seraya (*S. curtisii*) forest, *kapur* (*Dryobalanops
aromatica*) forest**

25. Light red merantis dominating .... **(light) red
meranti-keruing mixed dipterocarp forest**

The primary phenological-physiognomic base of this classification largely follows the only other regional classification to have been published, that of Champion for continental Asia south of the Himalaya from Pakistan to Burma, Sri Lanka excepted. It differs in the following respect: Champion did not recognize MDF, as neither the forest nor its lowland perhumid climate exists in the region of his classification. His 'wet evergreen forests' are my seasonal evergreen dipterocarp forests. I recognize southern and northern seasonal evergreen dipterocarp, and semi-evergreen forests within Indo-Burma, distinct from those of peninsular India. I also define montane forests differently from Champion (see Chapter 4).

References

[1]Tansley 1939. [2]Richards 1958. [3]P.S. Ashton 1982; P.M.S. Ashton *et al.* 1995. [4]Anderson 1964 –1976; Anderson & Muller 1975. [5]Brünig 1974. [6]P.S. Ashton 1958a . [7]Whitmore 1984 –1998. [8]Wyatt-Smith 1963, 1966. [9]P.S. Ashton 1963, 1995b. [10]P.S. Ashton 1964b. [11]Bray & Curtis 1957. [12]P.S. Ashton 2003b. [13]Gunatilleke 1975; Gunatilleke *et al.* 2004a, b. [14]Gunatilleke *et al.* 2006. [15]Gunatilleke *et al.* 2004a–c; Baillie *et al.* 2006b. [16]P.S. Ashton 1964b; Austin *et al.* 1972; Gunatilleke *et al.* 2004a, b, 2006. [17]Duivenvoorden 1995; Tuomisto *et al.* 2003a, b; Fine *et al.* 2005; Duivenvoorden *et al.* 2006. [18]H.S. Lee *et al.* 2002a, b, 2004; P.S. Ashton 2005; Davies *et al.* 2005. [19]Baillie *et al.* 2006a. [20]Itoh *et al.* 2003b. [21]Davies *et al.* 2005. [22]Baillie *et al.* 2006a. [23]Sukri *et al.* 2012. [24]Davies *et al.* 1998. [25]Paoli *et al.* 2006. [26]Anderson *et al.* 1983. [27]Bailey 1967. [28]Turner 1989; Niiyama *et al.* 1999a; Lum *et al.* 2004; Symington 1943, Symington *et al.* 2004; LaFrankie *et al.* 2005. [29]Gong & Ong 1983. [30]P.S. Ashton 1964b. [31]Wyatt-Smith 1963, 1966; Whitmore 1984a; Niiyama *et al.* 1999a, b; Abdul Rahman *et al.* 2002; Symington 1943. [32]P.M.S. Ashton 1990b, 1995; P.M.S. Ashton *et al.* 1995. [33]Kartawinata 1980 . [34]Newbery 1991. [35]Kartawinata 1980. [36]P.S. Ashton 1964b. [37]Brünig 1974. [38]Coomes & Grubb 1998. [39]Riswan 1982. [40]Newbery *et al.* 1986; Newbery 1991. [41]Anderson 1964, 1976. [42]Mirmanto *et al.* 2003. [43]Anderson 1964, 1965; Morley 1981. [44]Gunatilleke *et al.* 2004a. [45]Potts *et al.* 2002. [46]Potts *et al.* 2002. [47]Sukri *et al.* 2012. [48]Dykes 1994a, b, 1996; Pélissier 1998. [49]Newbery *et al.* 1996. [50]Newbery *et al.* 1992, 1996; Gunatilleke *et al.* 2004a, b, 2006; Gunatilleke *et al.* unpubl. ms. [51]Gunatilleke *et al.* 1997. [52]Gunatilleke *et al.* 1997, 2004a, b, 2006. [53]Tibbets & Smith 1992. [54]Putz & Chai 1987. [55]Poulsen 1996. [56]Peace & Grubb 1982. [57]Wyatt-Smith 1963, 1966; P.S. Ashton 1976a; Wan Juliana 2001; Wan Juliana *et al.* 2009, and in prep. [58]Manokaran *et al.* 2004; Yaacob *et al.* 2007. [59]Wan Juliana 2001; Yaacob *et al.* 2007; Wan Juliana *et al.* 2009. [60]Wyatt-Smith 1966; Laumonier 1997; Potts *et al.* 2002; Manokaran *et al.* 2004; Symington *et al.* 2004. [61]P.S. Ashton 2004a .

[62]Wyatt-Smith 1963. [63]P.S. Ashton & Hall 1992. [64]Sundralingam 1983; Sollins 1998. [65]Burslem *et al.* 1994; Burslem 1996. [66]Burslem *et al.* 1996. [67]Watanabe & Osaki 2001, 2002; Davies & Semui 2006. [68]Watanabe & Osaki 2001, 2002; Davies & Semui 2006. [69]Burslem *et al.* 1994, 1996. [70]Alexander 1989; Alexander *et al.* 1992; Smits 1992. [71]Gamage *et al.* 2004. [72]Palmiotto *et al.* 2004a, b. [73]Askew 1964. [74]Bloomfield 1954; Inderjit & Malik 1997. [75]P.S. Ashton 1964b. [76]Gong 1982; Gong & Ong 1983. [77]Ogawa 1978. [78]Proctor *et al.* 1989. [79]Janzen 1974. [80]Cai & Bongers 2007. [81]Holttum 1954; Corlett 1990. [82]Sim *et al.* 1992; Grubb *et al.* 1994. [83]LaFrankie *et al.* 2005; Shono *et al.* 2006, 2007. [84]Dykes 1994a, b, 1996. [85]Hirai *et al.* 1997. [86]Itoh *et al.* 1995b, 1997, 1999, 2003. [87]Webb & Peart 2000. [88]Yamada *et al.* 1997, 2000a,b, 2006, 2007. [89]Chazdon *et al.* 1996. [90]Zipperlen & Press 1996, 1997; Bebber *et al.* 2002b. [91]Bungard *et al.* 2000, 2002. [92]Wood & Meijer 1964. [93]P.S. Ashton & Hall 1992. [94]Sack & Grubb 2001. [95]P.M.S. Ashton & de Zoysa 1990; P.M.S. Ashton & Berlyn 1992; P.M.S. Ashton 1995; P.M.S. Ashton *et al.* 1995. [96]Chazdon *et al.* 1996. [97]Itoh 1995; Itoh *et al.* 1995a. [98]Eichorn *et al.* 2008. [99]Fine *et al.* 2004. [100]P.M.S. Ashton 1995. [101]Russo *et al.* 2005. [102]P.S. Ashton & Hall 1992. [103]Dent & Burslem 2009. [104]Philipson 2009. [105]Russo *et al.* 2010. [106]Tilman 1982. [107]Morley 2012. [108]Liechti *et al.* 1960; Baillie & Ashton 1983. [109]Baillie *et al.* 1987. [110]Bailey 1967. [111]MacArthur & Wilson 1967. [112]Wong & Whitmore 1970; P.S. Ashton 1976a; Wan Juliana 2001. [113]Potts *et al.* 2002. [114]Henderson 1939; Anderson 1964; Chin 1977, 1979; Crowther 1982; Newbery & Proctor 1984. [115]Proctor *et al.* 1983a. [116]Blak unpubl. ms.. [117]Weerasinghe & Iqbal 2001; Rajakaruna & Baker 2004. [118]Tibbets & Smith 1992. [119]Tibbetts & Smith 1992. [120]Proctor *et al.* 1989; Baker *et al.* 1992. [121]Baillie *et al.* 2000. [122]Proctor *et al.* 1988, 1989. [123]Tomlinson 1986. [124]van Steenis 1981. [125]Corner 1954. [126]Gore 1983. [127]Niyomdham 1986, 1988; Phengkhlai 1989; Suzuki & Niyomdham 1992. [128]Hope *et al.* 1988. [129]Furukawa 1997. [130]Rundel *et al.* 2004. [131]Anderson 1980. [132]Mirmanto *et al.* 2003. [133]Nishimura *et al.* 2007. [134]Anderson 1964, 1976. [135]Corner 1978. [136]Whitford 1906, 1909; Merrill 1907, 1926;

Brown 1919; Anon. 1988; Luna *et al.* 1999; Co *et al.* 2004, 2006. [137]Whitford 1906, 1909. [138]Bedard 1958. [139]Paijmans 1976; Johns 1982. [140]Haantjens *et al.* 1964–1973. [141]Paijmans 1970, 1976. [142]Whitmore 1998a. [143]Champion 1936; Champion & Seth 1968a. [144]Champion 1936; Champion & Seth 1968a. [145]Joshi 1975–. [146]P.S. Ashton 1991. [147]Goldammer & Seibert 1989. [148]Congdon 1982. [149]Muthuramkumar & Parthasarathy 2000. [150]Kadambi 1941–1949; Rai & Proctor 1986. [151]Chenggapa 1934, 1937, 1944. [152]Dy Phon 1970; Rundel 1999. [153]Brandis 1907; Congdon 1982. [154]Promdej *et al.* undated. [155]Kerr 1933. [156]Rai & Proctor 1986; Chandrasekharan 1962a–c; Pascal & Pélissier 1996; Pélissier 1997, 1998. [157]Pascal 1986, 1988. [158]Inamati *et al.* 2007. [159]Varghese & Menon 1999. [160]Behera & Kushwaha 2007. [161]Cao *et al.* 1996, 2008; Cao & Zhang 1997; Zhu 1997, 2006. [162]Kingdon-Ward 1949. [163]Champion 1936. [164]Cao *et al.* 2008. [165]Cao *et al.* 1996; Zhu 1997, 2006. [166]Cao *et al.* 2008. [167]Zhu 1997, 2006. [168]Gardner *et al.* 2000. [169]Biswas & Misbahuzzaman 2006. [170]Bunyavejchewin 1983–1985; Santisuk 1988. [171]Kerr 1928. [172]Troup 1921. [173]Troup 1921; Smitinand *et al.* 1990. [174]Rundel 1999. [175]Blanc *et al.* 2000. [176]Millet *et al.* 2010. [177]P.S. Ashton 1986. [178]M. Jacobs, pers. comm. 1961. [179]Rundel 1999. [180]Brandis 1907; Stamp 1924. [181]Bunyavejchewin 1999; Bunyavejchewin *et al.* 2002–2009. [182]Sarayudh *et al.* 2004; Baker *et al.* 2005. [183]Bunyavejchewin *et al.* 2001, 2009. [184]Blanc *et al.* 2000. [185]Bunyavejchewin 1986. [186]Bunyavejchewin, pers. comm. Oct. 2008. [187]Gaussen *et al.* 1965a, b, 1966. [188]Champion & Seth 1968a. [189]Utkarsh *et al.* 1988. [190]Pascal *et al.* 1988. [191]Lü *et al.* 2010. [192]Stainton 1972; Cao *et al.* 1996. [193]Kadambe 1955. [194]Proctor *et al.* 1998. [195]Engelbrecht *et al.* 2007. [196]Champion & Seth 1968a. [197]P. Baker, pers. comm. 1993. [198]Kumar *et al.*1994; A. Singh *et al.* 1995; Kadavul & Parthasarathy 1999. [199]Dattaraja *et al.* unpubl. ms. a. [200]Dattaraja *et al.* unpubl. ms. b. [201]Chandrasekharan 1962a–c; Sukumar *et al.* 1992, 2004; Ilorkar & Khatri 2003; Khatri *et al.* 2004; Naithani *et al.* 2006. [202]Mooney 1938; Swain & Nanda 1997. [203]Gaussen *et al.* 1961–1992; Legris 1963; Meher-Homji 2001. [204]Troup 1921. [205]Lal 1942; Patra *et al.* 2007.

[206]Sukumar *et al.* 1992, 2004; Suresh *et al.* 1996. [207]Brandis 1903; Stamp 1924; Seth & Yadav 1959; Robbins & Kaosa-Ard 1981. [208]Troup 1921. [209]Swain & Nanda 1997. [210]Dinerstein 1979; Chauhan 1991. [211]Dinerstein 1992; Lehmkuhl 1994. [212]Bor 1942. [213]Rautiainen & Suoheimo 1997. [214]Troup 1921. [215]Smythies 1919; Singh 2002; Raina & Jha 2005. [216]Pandey & Shukla 2003. [217]Hamilton 1930. [218]Smythies 1919; Raina & Jha 2005. [219]Ilorkar & Khatri 2003. [220]Smythies 1919. [221]Champion 1920. [222]Mooney 1938. [223]Schimper 1903; Champion 1920. [224]Champion & Seth 1968a. [225]Vidal 1956. [226]Barrington 1931. [227]Maxwell *et al.* 1995. [228]Bunyavejchewin 1983a, 1985; Bunyavejchewin *et al.* 2011. [229]Stamp 1924. [230]Brandis 1903. [231]Stamp 1924; Champion 1936; Vidal 1956; Stott 1976. [232]Symington 1943. [233]Gardner *et al.* 2000. [234]Blanford 1915. [235]Rundel 1999. [236]Rundel 1999. [237]Kerr 1933; Smitinand *et al.* 1990. [238]Williams *et al.* 2008. [239]Sakurai *et al.* 1998; I.H. Baillie, unpubl. ms.. [240]Stamp 1924. [241]Stamp 1924; Champion 1936. [242]Merrill 1926. [243]Merrill 1926; van Steenis 1979. [244]Y. Laumonier, pers. comm. July 2010. [245]Meher-Homji 2001. [246]Gunatilleke 1975; Jayasuriya & Pemadasa 1983; Venkataswaran & Parthasarathy 2005. [247]M.D. Swaine, pers. comm. June 2012. [248]Reddy & Parthasarathy 2003. [249]de Rosayro 1950, 1956; Koelmeyer 1957. [250]de Rosayro 1950, 1955; Holmes 1956, 1957; Koelmeyer 1957–1960; Gunatilleke 1975; Gunatilleke & Ashton 1987. [251]Givnish 1999. [252]Balzer *et al.* 2008, 2009. [253]Griffith 1945; Seth & Dabral 1957; Qureshi *et al.* 1966. [254]Troup 1921; Hopkins 1924; Osmaston 1927; Nautiyal 1934. [255]Ma *et al.* 2009; Su *et al.* 2007. [256]Hou 1983; Hou *et al.* 1979. [257]Wang 1961. [258]S. Rhazanfar, pers. comm. June 2011. [259]Zhu 1997. [260]Wu *et al.* 1994–. [261]Zhuang & Corlett 1997. [262]Dudgeon & Corlett 2004. [263]Webb 1986; He *et al.* 1982; Cao *et al.* 2008. [264]Kong *et al.* 1997, 1998. [265]Hu 1979. [266]L.J. Webb, unpubl. ms.. [267]Wu 1961; Wu *et al.* 1994. [268]Tomlinson 2006. [269]Corlett 2009a. [270]Li & Keng 1950. [271]Oi 1953. [272]Webb *et al.* 1984. [273]Walker 1986; Wang *et al.* 1999. [274]Meher-Homji 1978. [275]Li & Walker 1986. [276]P.S. Ashton 1991.

Upper montane forest on Gunung
Mulu, Sarawak, NW Borneo (H. H)

Chapter 4

THE MOUNTAIN FORESTS: ABODE OF CLOUDS

At 6° N, Mount Kinabalu is the highest mountain in the Asian equatorial tropics. It is therefore the highest peak between the Himalaya, whose mighty peaks range beyond the tropics, and the snowy equatorial peaks of New Guinea. Kinabalu is a rough knuckle of granite fisting up through layered sediments, themselves up-buckled from a Cenozoic South China Sea (Plate 4.1; Chapter 1). Volcanic ultramafic rock extraordinarily rich in magnesium, iron and aluminium. Cenozoic lava inserted between shale and granite at 1,500–3,000 m testifies to Kinabalu's violent birth only two million years ago.

Plate 4.1 Mount Kinabalu, Sabah. Looking northwards from the subalpine zone, c.3,700 m. The road on the ridge is in lower montane forest, c.1,700 m. Subalphine woodland is brownish. (S.J. Davies)

Fifty years ago it was still possible to climb under continuous forest canopy from near sea level to the 4,094 m summit of Kinabalu. Today, swidden ('shifting cultivation': see Chapter 8) has reduced the lowland and hill forests (that once covered the surrounding shales of the Crocker Range up to 1,500 m) to isolated patches. In those days, ascending from the mighty dipterocarp forests of the coastal hills, one would have started quite suddenly up steep slopes on which the lush forest understorey gradually thinned. Along sharp spurs and ridges the emergents clustered more densely, though still somewhat shorter than the scattered giants on the lower slopes. By 800 m altitude, the number of familiar lowland species would have begun to wane and a few unfamiliar species, including

some from lowland families like the dipterocarps, would have appeared (patches of these still survive in some areas). As one approaches the mountain shoulder at 1,000 m, a great tableland, the Pinosuk Plateau, opens out. Today converted to a dairy farm and golf course, this fan of Pleistocene sediment had been the only extensive area in Borneo of *lower montane oak-laurel forest*, a forest of characteristic genera (Table 4.3), which extended at these altitudes throughout the East Asian tropics from Nepal and South China to New Guinea. Patches remain to the south of Kinabalu, down the Crocker Range.

Viewed from the forest floor, the appearance of these forests – crowned by a dense, even canopy rarely reaching 35 m, lacking in dipterocarps and with the few taller emergents clustered in sheltered valleys – is that of a warm temperate evergreen forest in eastern China, southern Japan or coastal Florida. The dominance of the oak family (Fagaceae), and the distinctive earthy, fermenting aroma, is common to all. A small but characteristic group of species and genera are present, rarely found at equatorial lower altitudes. These include *Bischofia javanica* (Bischofiaceae), *Nyssa javanica* (Nyssaceae), *Trigonobalanus verticillata* (Fagaceae), *Distyliopsis dunnii* (Hamamelidaceae), and higher up *Eriobotrya bengalensis* (Rosaceae) (Plates 4.4c,d, 4.6).

Above 1,500 m the real climb begins. Soon, the ridges and, especially, the ultramafic exposures are clothed in shorter woods with crooked boles, more densely twiggy crowns and small leathery leaves recalling those in lowland *kerangas*. The flora begins to change along these ridges: Myrtaceae and Podocarpaceae appear and become abundant, and Rubiaceae increase in the understorey. By c.2,200 m, few lower montane species remain. At around 1,200 m, mid-morning fog begins to enter the canopy, intermittent at first but pervasive by 1,800–2,000 m. It dissipates in the late afternoon and evening to lie in the surrounding valleys overnight (Chapter 1). By 2,000 m the way ahead is often obscured in sweeping mist, everything is drenched in dew; one's voice hardly carries through the muffled silence. From here upwards, the species become increasingly characteristic of this *upper montane forest*, while species numbers, small even from its lowest elevation, continue to decline. Tussocks and sleeves of moss engulf ground, trunks and branches. Even the epiphytic and ground orchids and ferns, enormously species-rich though not abundant in the lower montane forest, are here reduced to a few abundant species. By 3,000 m, no more than 100 tree species can be found, and the woods, now only exceeding 10 m tall in the remaining sheltered valleys, are dominated by one or a very few endemic species including *Leptospermum recurvum* (Plate 4.14b), *Lithocarpus havilandii* and *Schima brevifolia*. The wind picks up, occasional gusts increasing in strength as we continue to climb. The thick blanket of moss has become threadbare, signalling frequent dry conditions. By 3,400 m, areas of bare granite appear on convex slopes, and the woods are reduced to scattered lines of shrubs following fissures and rills. The summit area, extensive on Kinabalu, is bare but for occasional tussocks of narrow-leaved grasses of temperate genera.

Is this change with altitude in forest structure and composition stepped, or continuous? What physical characteristics may influence these changes? What role might competition, physiological constraints or catastrophe play in limiting the altitudinal distributions of tree species? Because mean annual temperature declines at a similar rate throughout the wet equatorial tropics, at c.0.6°C per 100 m, we might expect any forest ecotones to occur always at the same altitudes. Do they?

In this essay, I propose answers to these questions, and others, regarding structure and process within tropical forests as they range across altitudinal continua. My approach, as mentioned in the Prologue, remains primarily biogeographical and comparative. Mountain forests, as much as lowland, vary at many scales, from the purely local to the broadly regional; consistent patterns can likewise be detected at many scales. Several authors have used varying combinations of floristic data, structural or physical criteria, and altitude to name and describe forest types along altitudinal gradients in Asia and elsewhere in the tropics. Most of these categories are useful for some regions and less so in others. Can I, from broad Asian experience, find definitions for forest types based primarily on floristics, and agree on common labels that apply throughout the Asian tropics, and beyond? And if so, do the physical and structural characteristics of these forests correspond reasonably well to the floristic categories? I compare my own observations to those of other researchers, both historical and contemporary, and end by offering some general conclusions.

1. Forests altitudinally zoned

1.1. *Two mountains*

The forests of two Asian mountain massifs, Kinabalu and the Bhutan-Sikkim Himalaya – the former in the aseasonal wet equatorial zone and the latter at the very margin of the

tropics – are sufficiently well documented to permit detailed comparison of variation in floristic and structural zonation from the lowlands to the tree line on major mountains within one biogeographic region. For Kinabalu, John Beaman[1] has provided the most complete documentation of the flora of any single mountain in the tropics, while Kanehiro Kitayama[2] and his team have examined the ecology of forest zonation using the most comprehensive altitudinal sequence of plots to be laid out on a mountain in tropical Asia. In Bhutan, A.J.C. Grierson and H.J. Noltie and their colleagues at Edinburgh[3] have provided a less detailed albeit comprehensive account for a wider region, while Masahiko Ohsawa[4] with Rebecca Pradhan[5] have interpreted the ecology.

The Bhutan-Sikkim Himalaya lie at 27–28°, well north of the Tropic of Cancer. Their highest peaks reach well over 7,500 m, far exceeding the tree line (Plate 4.2). The Tibetan Plateau being the motor of the Indian monsoon (Chapter 1), hot moist summer winds drawn off the Bay of Bengal support tropical wet forests (that is, forests with a frost-sensitive flora) up to 1,800 m, even beyond on some slopes. Yet although the mountains protect from bitter dry north-east winter winds, nocturnal flows of heavy frosted air may descend the valleys

Plate 4.2 View north from 2,800 m, Dotula Pass, Bhutan. Inner valleys and frontier peaks. *Tsuga dumosa* cool temperate forest on ridge; below, the upper boundary of warm temperate broadleaf forest in foreground: *Acer, Populus, Persea elongata, Osmanthus suavis, Rhododendron falconeri, Lithocarpus thomsonii* etc. (H.H.)

to 1,500 m. As in the striking case on the west coast of New Zealand's South Island, where frost-sensitive *Podocarpus* forests extend along the slopes high *above* the Franz Josef Glacier (Chapter 1), temperature reversals may occur on slopes above these frost-prone valleys.[6]

Kinabalu, in contrast, supports gnarled, small shrubs in fissures and coves almost to the summit, but no temperate *woody* species. Beaman underestimates the number of species below 1,200 m because little primary vegetation remains. Here, I have extended the altitudinal ranges of species he includes down to their known lowest elevations in northern Borneo (Fig. 4.1a).

This compensates only partially for the deficit in lowland species in Beaman's account. At least an additional 1,000 tree species are likely to occur in the surrounding lowlands. A mere 52 ha of lowland forest in north-east Sarawak, for instance, contains nearly 1,200 tree species (Table 3.7).[7] Fig. 4.1a,b includes two histograms. In each of these, the data are standardised by use of percentages: (1) the percentage of all species occurring in each 300 m interval; and the percentage of all species recorded whose (2) lowest and (3) highest recorded altitudes occur in each interval, following the method adopted for the tropics by van Steenis.[8] Maxima in species present within altitudinal intervals may therefore indicate maxima in the lower and upper ranges of species, and therefore the location of ecotones between altitudinal vegetation formations; or they may indicate communities of exceptional richness. It is important to distinguish between these two.

1.1.1. *Structural and floristic simplification with altitude*

On Kinabalu (though not in the Himalayas, where temperate forests intervene above the frost line), the forests undergo a gradual but steady process of structural simplification with increasing altitude, and stature declines overall though it also varies with local habitat.[9] The emergent canopy becomes shorter, scattered or localised in sheltered valleys, finally disappearing. Eventually, a single canopy of light-demanding trees and shrubs forms a scrub below the tree line, nevertheless even here with scattered emergent *Leptospermum recurvum* (Plate 4.15e). The woody flora becomes gradually impoverished and the predominant leaf size declines (Fig. 4.2; Tables 4.5, 4.8). These trends are global on wet equatorial mountains.[10]

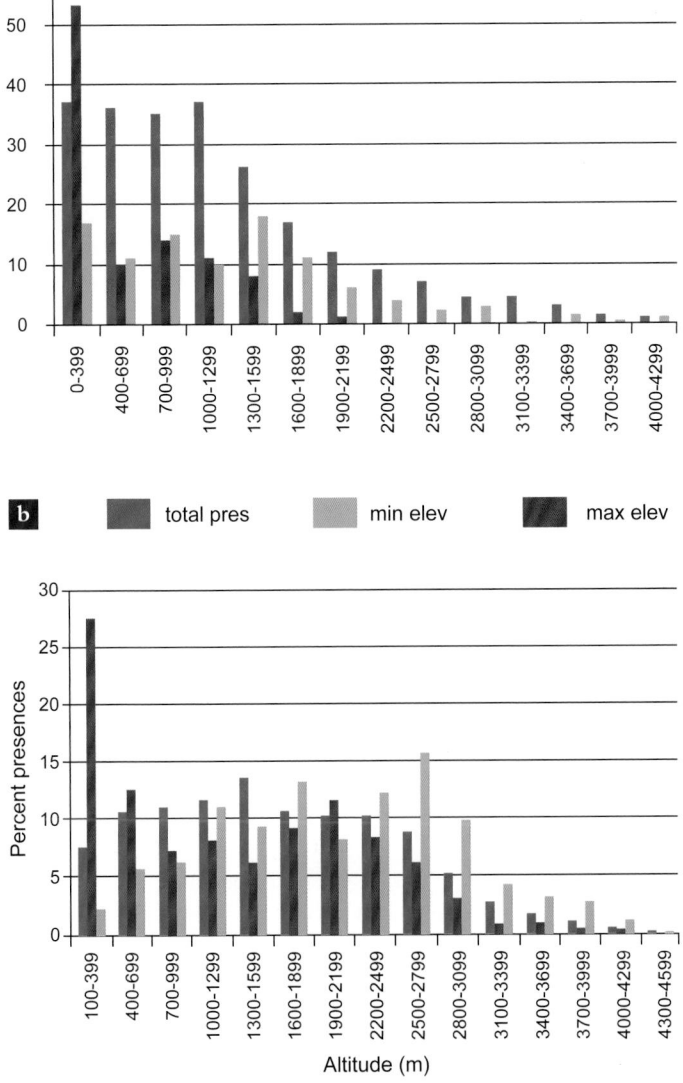

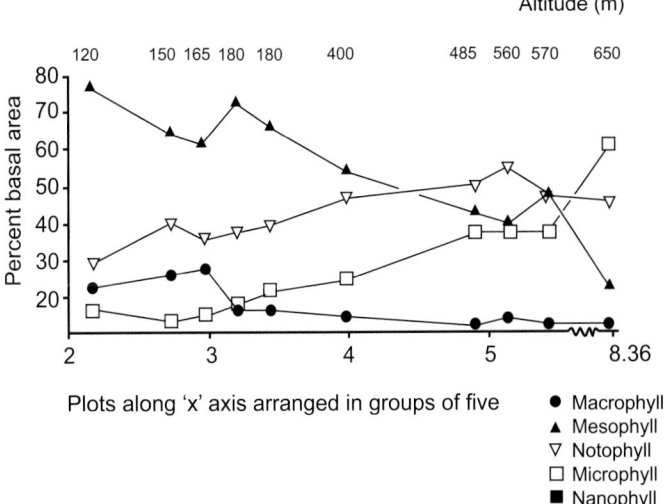

Fig. 4.1 Species zonation on two tropical Asian mountains. **a,** Gunung Kinabalu; **b,** the Bhutan and Sikkim Himalaya.(D. Glick del. modified after P.S. Ashton 2003a)

Fig. 4.2 Change in dominance of leaf size classes with altitude, Kuala Belalong, Brunei: percent basal area for all trees >10 cm dbh. (From P.S. Ashton 1964a)

1.1.2. *Floristic peaks and ecotones*

Within Kinabalu National Park, the number of species declines just slightly from the lowland estimate then increases slightly at 1,000–1,399 m. This pattern is almost certainly a consequence of under-recording because there is so little forest remaining in the park below 1,000 m. In Bhutan, species number increases to a maximum at 1,000–1,299 m (Fig. 4.1b), where the greatest number of high and low range limits – in other words, the highest rate of species turnover – is found. This pattern reflects the fact that the lowland flora is relatively impoverished in those lowland climates sufficiently seasonal for fire-climax deciduous and semi-evergreen forests to prevail. At Kinabalu, the greatest number of *upper* limits of species ranges clusters around 1,300–1,599 m, while fewer and fewer species *lower* range limits are encountered as one ascends. From 1,600–2,500 m we see a massive drop both in species present and in new 'arrivals' and 'departures'. In Bhutan, as elsewhere in the Eastern Himalaya,[11] there is a further peak in 'arrivals' and 'departures' (though not in total numbers present) at 1,600–2,500 m. This elevational band here marks the transition from mostly frost-free to entirely frost-prone sites (Chapter 1), and is therefore the transition from the tropical to a rich warm temperate Himalayan flora. Here the tropical elements survive on ridges and slopes sheltered from frost, whereas the temperate elements descend along the valleys. There are no temperate tree species on Kinabalu, even though there appears to be occasional frost above c.3,100 m (see section 6.2 below).

In summary, both these Asian mountains show maxima in tree species turnover between the lowland and lower montane floras (although this may be due to inadequate data at Kinabalu), while in Bhutan a second maximum marks the changeover from lower montane to temperate floras. Floristic change, and a decline in richness, has been found above 1,000 m up the slopes of Central American volcanoes[12] at 7° N, comparable to that on Kinabalu although seemingly continuous; but as in Bhutan, the maximum peak in richness is at 1,000 m, where lowland tree species are mixing with those of lower montane.

1.2. *Four forest zones*

No transects have ever been laid out in Asian *rainforests* extending all the way from the lowlands to the upper tree line. The longest are those of Isamu Yamada on the Pangrango volcano in West Java at 1,600–3,000 m,[13] and of Kitayama on Kinabalu at 700–3,100 m (discussed in sections 4 and 5 below), neither of which includes the species-rich lowland forests. Our understanding therefore rests substantially on relatively casual field observations and observed altitudinal species ranges. Using these methods, four forest zones have been recognized on the higher tropical mountains up to the tree line: *lowland, lower montane, upper montane* and *subalpine forests*[14] (Table 4.1). This zonation scheme, modified from the scheme proposed by Burtt-Davy[15] for the forests of New Guinea, has been recognized worldwide in equatorial and oceanic climates of similar structural-physiognomic characteristics[16] – though of course they differ in species composition.

Formation	Lowland forest	Lower montane forest	Upper montane forest	Subalpine woodland
Maximum canopy height	25–70 m	15–35 m	1.5–20 m	1.5–5 m
Emergent trees	Present to 70 m, or absent	Mostly absent, <50 m	Rare, <20 m	Rare, <5 m
Pinnate leaves	Frequent	Less frequent, a few species common	Uncommon	Absent
Dominant woody plant leaf size	Mesophyll or notophyll	Notophyll (microphyll)	Microphyll	Microphyll and smaller, cupped, shiny
Buttresses	Usually frequent, large	Uncommon, small	Absent	Absent
Cauliflory	Frequent	Uncommon	Absent	Absent
Big woody vines	Uncommon to abundant	Rare to common	Absent	Absent
Bole climbers	Locally common	Locally common	Rare	Absent
Vascular epiphytes	Usually uncommon	Abundant, species-rich	Abundant, less species- rich	Few
Non-vascular epiphytes	Local, species-poor	Frequent, increasing with altitude, species-rich	Abundant in equatorial regions and coastal, less so in seasonal forests, less species-rich, especially where seasonal	Sparse

Table 4.1 Characteristics that define the widely accepted zonation of tropical montane forests as found in Asia (expanded from Whitmore, 1984)

2. Lowland forests

Dipterocarpaceae, the dominant and one of the richest tree families in the perhumid lowlands of Asia, serve as the model for discussion throughout this book (Chapters 2, 3, 5, 7). They are one of the few tree families entirely restricted to tropical climates (Chapter 6). I recorded the altitudinal ranges of all 151 dipterocarp species in Brunei, 250 km south-west of Kinabalu (Fig. 4.6). Many lowland species there extend no higher than 400 m. Their distributions are correlated with the altitudinal limits of lowland broad-ridged topography, and still more with the deep yellow sandy humus-bearing soils of the younger Cenozoic coastal sediments[17] (see Chapter 1). Above them, only the steep shale slopes and narrow ridges of inner Borneo continue. This altitudinal species ecotone extends throughout Borneo, and is correlated with the limits of a topography and a soil type which are themselves correlated with geology and perhaps coastal climate, rather than with temperature *per se* (Chapter 3).

2.1. *Are lowland forests altitudinally zoned?*

In Peninsular Malaysia, Colin Symington[18] was the first to distinguish altitudinal forest zones in Sundaland, again based primarily on dipterocarp distributions. To a large extent, he recognized the same altitudinal zones and adopted a terminology that is now widely accepted.[19] However, he described lowland dipterocarp forest as terminating at c.300 m. He knew mostly the mountains of the Peninsular Main Range – granite with mature rounded topography – the same substrate, in fact, as the western lowlands. Above 300 m, he described a separate 'hill dipterocarp forest':

The main difference between the hill dipterocarp forests and the lowland dipterocarp forests is in species composition of the dominants of the upper strata of the vegetation. There is usually a slight diminution of size of the larger trees accompanied by a slight increase in number per unit area on ridge tops and towards the upper limits of the hill dipterocarp forests. Many of the lowland dipterocarps are represented in the hill forests, although they become scarce towards the upper limits of these forests, but many species appear in the hill forests that never, or very exceptionally, occur in the lowland forests. Predominant among these is Shorea curtisii, a member of the Red Meranti group that tends to be gregarious on ridges.

Shorea curtisii is a slow-growing microphyll tree of heavy red meranti timber, now known to enjoy competitive advantage as a sapling in single tree gaps and beneath the diffuse canopy of these ridge forests[20] (Plate 4.8a; Chapters 3, 8). Of the 48 dipterocarp species in these inland hill forests in Symington's chart (Table 4.2), 29 also occur in Borneo. All of these, including *S. curtisii,* occur mostly *below* 400 m in Borneo forests, and 23 are also in Peninsular lowland forests. Symington also stated:

…on isolated mountains, or on coastal ranges, this [lowland-hill forest ecotone] may be depressed to much lower altitudes. On Penang Island, for instance, hill dipterocarp forests develop on ridges almost at sea level.

Symington recognized a separate category of 'coastal hill dipterocarp forests' for these lower communities. It is inconceivable that mean annual temperature could explain this drop in species altitudinal zonation on coastal hills, since temperature would have to be depressed by as much as 2°C. Of the 29 dipterocarps listed on these coastal hills, 16 occur in Borneo (and another is represented by a sister species there), while nine extend up into Peninsular 'hill dipterocarp forests'. (Six of these nine are also known from lowland forests inland.) In Borneo, the ecological range of these species is defined more by soils than by elevation. The Peninsular coastal species there are, with two exceptions, confined to leached freely draining yellow sandy soils with densely root-matted surface raw humus. They occur on the late Miocene and Pliocene sediments of the coastal hills of Borneo's north-west. They are also found on older rocks in West Sarawak and West Kalimantan, where they extend far inland up the Kapuas Valley (see Chapter 6). One, *Cotylelobium lanceolatum,* in Borneo also extends up into uppermost dipterocarp forest on humult soils. The Bornean ranges of these 'hill dipterocarp species', both inland and coastal, therefore correlate primarily with freely draining low-nutrient soil rather than altitude, and irrespective of distance from the coast therefore perhaps with nutrient levels more than drought frequency.

The Peninsular Malaysian 'hill dipterocarp' species, again, are either confined to humult freely draining deep yellow sandy soils, which in Borneo are mostly confined below 400 m as are their species, or they are restricted to udult clay soils which occur at all elevations in Borneo. The species characterising Symington's 'hill dipterocarp forests' are concentrated on ridges and spurs in the Peninsula. The steep adjacent slopes on these hills bear a tree flora more typical of the lowlands, though lower in density of emergent individuals and lacking many species.[21] *Shorea curtisii* itself is rare on the inland ridges of Borneo, but is locally abundant on its coastal sandy hills (Table 4.2). *Dipterocarpus grandiflorus,* common in both Peninsular coastal and inland hill dipterocarp forest ridges, is confined to coastal forests in drought-prone southern and northeastern Borneo, though it is widespread on a diversity of soils in the more seasonal Philippines and Indo-Burma. The ridges in hill and coastal forests on which these species are concentrated in Peninsular Malaysia and Singapore also bear a thin, often discontinuous, fine root-matted layer of raw humus. This results from intense

	Peninsular Malaysia		Borneo					
	Coastal hill DF	Hill DF (300–800 m)	<400 m	200-800 m				>800 m
			Strictly coastal (dry)	Sandy humult soil	Clay soil	Humult clay soil	Clay soil (ridge)	Upper DF (humic soil)
Dipterocarpus crinitus		+						
D. eurynchus		+						
D. sublamellatus		+						
Hopea pedicellata		+						
H. semicuneata		+			+	+		
Shorea dasyphylla		+		+				
S. kunstleri		+		+				
Vatica mangachapoi		+		+				
V. nitens		+		+				
Anisoptera laevis		+		+			+	+
D. gracilis		+		+			+	
H. ferruginea		+		+	+		+	
S. balanocarpoides		+		+	+		+	
S. leprosula		+			+	+	+	
S. laevis		+			+		+	
S. lamellata		+			+			
S. longisperma		+			+		+	
S. parvifolia					+		+	
D. fagineus	+	+		+				
D. grandiflorus	+	+						
H. beccariana	+	+		+				+
S. curtisii	+	+	+	+		+		
S. multiflora	+	+		+		+		
S. ovata	+	+		+				+
Cotylelobium lanceolatum	+			+				
C. melanoxylon	+			+				
D. caudatus	+			+				
D. rigidus	+			+				
D. sarawakensis	+			+				+
H. griffithii	+			+				
H. nutans	+			+		+		
S. falcifera	+							
S. gratissima	+		+					
S. maxwelliana	+		+	+				

Table 4.2 Habitats in Borneo of dipterocarps of Peninsular Malaysian coastal hill, and hill dipterocarp forests.

leaching of the coarsely sandy granite ridge soils in a climate that is aseasonal yet experiences intermittent drought. Both the hill forest ridges, below the cloud base yet receiving dry trade-wind breezes, and the coastal hills where diurnal rain cloud is driven inland by onshore breezes, are prone to drying of soil surfaces (Chapter 3). *S. curtisii* and its associates rarely descend inland ridges below 200 m, and apparently never below 150 m.[22] That elevation correlates with the ceiling of nocturnal valley mist, which rarely forms on the coastal hills. In combination, the intense leaching (leading to surface acidification) and proneness to soil surface drying reduces the rate of litter decomposition and increases the relative role of fungal decomposition (Chapters 1, 3). This lowland *S. curtisii* association, best known on the hills of the north-west coast of the Malay Peninsula, extends down

its eastern coastal hills to Singapore also, particularly those of granite, and north to the slightly seasonal Pattani Circle in southeastern peninsular Thailand. This coastal hill dipterocarp forest is conserved in the remarkable Bukit Timah sanctuary within the City of Singapore, where there is a 2 ha plot of the CTFS collaboration. Its characteristic dipterocarp species seem mostly absent on the high inland Triassic siltstone and shale hills of the eastern Peninsula, which are dominated instead by *S. laevis*[23] and other species characteristic of the inland shale ridges of Borneo.

The stepped elevational ranges of Brunei's dipterocarps are therefore remarkably general in perhumid Sundaland. The altitudinal ranges of dipterocarps and other trees in perhumid Sri Lanka are similarly influenced by their edaphic ranges

(Chapter 3). I conclude that the floristic differences between 'lowland' and 'coastal hill' and 'hill' dipterocarp forests are mediated by soil, not by temperature, and that the ecotone at 200–300 m is attributable to a widespread change from undulating to steep topography, and sometimes to changes in geology. In general, I advocate the use of the more broadly applicable categories of *lowland forest* and *lower montane forest* within which floristic associations with distinct edaphic ranges occur, rather than the coastal hill and inland hill forest categories. Nevertheless, the MDF of perhumid Sundaland does not extend, in full species complement, beyond 400 m, somewhat higher in Sri Lanka. The series of co-occurring congenics of *Shorea* section *Mutica* and *Dipterocarpus*, in particular, decline in canopy dominance and local species array, sometimes being replaced by single dominant species.

2.2. *The lowland to montane forest transition*

Forest stature, albeit varying with topography, declines with altitude above 600 m, although trees up to 50 m tall may still occur in sheltered dells (Plates 4.3c, 4.4a). This decline is associated with a decline in canopy gap area, yet an overall *increase* in exchangeable soils nutrients.[24] How distinct is the floristic transition between these two forest types?[25] More than 100 species of the quintessentially lowland rainforest family Dipterocarpaceae are known from Kinabalu and adjacent hills. Only 52 are recorded from Kinabalu itself, certainly owing to deforestation below 1,000 m. Of these, 11 fail to ascend beyond 700 m, and a further 11 beyond 800 m; but 19 reach 1,300 m, ten reach 1,600 m and one reaches 1,700 m. A more comprehensive tally is available from Brunei (Fig. 4.6). Of the 1,442 known tree species in all families on Kinabalu, 979 are confined below 1,000 m. A majority of the well-represented lowland Borneo *genera* extend to 1,600 m, many to 2,200 m and a few to 2,400 m; at first, it is the *species* that change. Whereas the MDF of freely draining humult sandy soils is confined below c.400 m in Borneo by the altitudinal limits of its coastal Neogene hill habitat, the lowland MDF flora on the widespread clay loam soils begins to phase out at c.500 m notwithstanding the continuation of its soils to higher altitudes. Its decline accelerates beyond 700 m, yet the characteristic lower montane forest hardly appears below 750 m. There is therefore a zone of floristic impoverishment in the perhumid tropics at c.400–850 m (Fig. 4.1a).

In Bhutan and Sikkim, of the 503 known tree *species,* only 110 are confined to the lowlands below 1,300 m, reflecting both the predominance of species-poor deciduous forest in the lowlands and the rich tropical lower montane and warm temperate, mostly evergreen, tree flora above. Tree species richness is highest in the most continuously humid climates of Asia (Chapter 7); and, within regions with less than four

dry months, below 1,000 m and most especially below 400 m (Chapters 7, 9).[26] In the central and eastern Himalaya, the fire-climax deciduous forest can extend up to 1,200 m, and the ecotone to lower montane evergreen forest, which is toward the upper end of that in perhumid climates, is often sharp.[27] Of c.200 tree species in 62 genera widely represented in Far Eastern lowland rainforests, 85 occur to 1,600 m in Bhutan; among these, 62 species range up to 1,800 m, but none occur beyond 2,700 m. Within these genera, 28 other species occur *only* above 700 m. Thus, these represent montane forest specialists in Bhutan, though they also occur at lower altitudes in moister lowland climates further east.

Twenty-four of these 62 genera also occur on Kinabalu, and 15 of them contain species extending to 1,600 m, on one or both massifs. Although few species occur on both these Himalayan and Bornean mountain systems, there is overall parallel attrition of lowland species with altitude. In both cases, lowland mixed forest species fail to extend beyond the lower montane forest; but we will see that some *kerangas* species are exceptions in Borneo.

3. Lower montane forests

The lower montane forests were until recently the least studied of tropical forests, and there is still much that we do not understand about them. On both Kinabalu and in the eastern Himalaya, we find a substantial group of species (and a few genera) whose lower limit is 1,200 m. A few start from as low as 700 m (Fig. 4.1). On Kinabalu, 499 tree species are present at 1,000–1,900 m, and a further 307 above 700 m. Notable among the 499 are 35 of the 61 Fagaceae (Plate 4.3b) and 56 of the 106 Lauraceae known from the massif, whose abundance in both canopy and subcanopy at these altitudes has gained these forests the name *lower montane oak-laurel forest.*[28] Champion called them 'northern subtropical wet hill forests'. At the upper limit, only 90 among these lower montane species extend beyond 2,500 m. Remarkably – as Masahiko Ohsawa[29] first pointed out – the transition from lowland to lower montane forest, both structurally and floristically, generally occurs over a gradual ecotone at 750–1,300 m whether on the Equator or at the margin of the tropics, indeed on all mountains in the Asian tropics wherever fire does not intercede (Fig. 4.10).[30] Only occasionally, at the far northeastern tropical margin where lower montane forest exists only as a narrow band between lowland and warm temperate evergreen forest, does a narrow ecotone between lowland and lower montane evergreen forest occur, on steep topography with lower montane forest descending along the ridges (Fig. 4.3). Throughout the perhumid tropics, the altitude of the lowland-lower montane forest ecotone is consistently correlated with the predominant elevation of the diurnal cloud base. This is true, too, in seasonal tropical Asia during the wet summer monsoon. Thus the lowland-lower montane forest ecotone appears responsive to the influence of the climatic convergence zone, whose seasonal oscillation is responsible for most of the rain cloud in all regions (Chapter 1).

In those nations of seasonal tropical East Asia whose forests were once managed by British foresters, including the Himalaya, this tropical lower montane forest consisting largely of oak-laurel associations has been termed 'hill evergreen forest'.[31] The species belonging to this tree flora (but not the families or genera) are mostly different from those of the lower montane forest of perhumid Malesia. But a small number of widespread genera, monotypic or with few species in the Asian tropics and often in basal angiosperm families, are also paradigmatic of the lower montane oak-laurel forest of all tropical East Asia west of Wallace's Line, and in a few cases beyond (Table 4.3) (Chapter 6). In South Asia several important families, notably Fagaceae, are absent (Chapter 6). Following Burtt-Davy and later authors, we describe these forests throughout Asia as *lower montane forests*. Although Champion had designated these forests 'subtropical', they are frost-free and their tree families tropical.

Fig. 4.3 Distribution of **a,** a lowland species, *Parashorea chinensis*, and **b,** a tropical lower montane, *Castanopsis echidnocarpa*, in the 15 ha CTFS plot in northern tropical seasonal evergreen dipterocarp and lower montane forest at Mengla, Xishuangbanna, south Yunnan. (From Cao *et al.* 2008)

Plate 4.3 Tropical lower montane forest canopies. **a,** *Shorea gardneri* lower montane forest, Peak Sanctuary, Sri Lanka, 1,600 m. *Macaranga indica* to right; **b,** A *Castanopsis* population in flower along a ridge in lower montane oak-laurel forest, c.800 m, Mengla, Xishuangbanna, Yunnan; **c,** Lower montane forest, Doi Inthanon, Thailand, 1,700 m, canopy with emergent *Mastixia euonymoides*; **d,** Doi Inthanon: upper limits of lower montane oak-laurel forest, c.2,600 m, showing wind-pruned canopy and crown shyness. (**a, c, d** H.H.; **b** H. Zhu)

Plate 4.4 Lower montane forest profiles and trees. **a,** Massive boles of *Mastixia euonymoides* in Mamoru Kanzaki's 16 ha lower montane forest plot, Doi Inthanon, 1,700 m; **b,** Lower montane forest elements ascending a moist valley in tall deciduous upland sal, *Shorea robusta,* forest (in background), 800 m, Meghasena, Similipal National Park, Odisha, India; **c,** Upper limits of tropical lower montane oak-laurel forest, Zhemgang, Bhutan, 1,900 m. Profile with *Nyssa javanica* (pale crowns, right and back), *Castanopsis tribuloides* (centre, back), *Lithocarpus fenestratus* (front), a *Ficus* (left) and *Alnus nepalensis* (behind the *Ficus*); **d,** *Trigonobalanus verticillata* (Fagaceae), which is local in lower montane forest from Thailand to Sulawesi, here in Sarawak's Hose mountains. The mature trunk eventually collapses leaving a ring of coppices. (**a–c** H.H.; **d** P.A.)

Table 4.3 Ranges of lower montane species widespread west of Wallace's Line.

Species	Species range	Lowest elevation genus recorded in the tropics	Number of species in the genus, and principal distribution
Acer laurinum	Indo-Burma–Philippines, Sulawesi, Timor	30 m	Genus 111 spp., mostly temperate
Altingia excelsa	Bhutan, tropical S. China–W. Java, *not* Borneo	500 m	3 spp. warmest temperate S. China
Anneslea fragrans	Indo-Burma, top. S. China–Sumatra	Low	3 spp. temperate Taiwan, S. China
Bischofia javanica	Pen. India–Pacific	Low altitudes, moist soils at higher latitudes and humus rich substrates	1 sp.
Brassaiopsis glomerulata	Nepal, NE India.	400 m (China) S. China–Java, *not* Borneo	c.20 tropical Asia
Calophyllum polyanthum	S. & NE India, Indo-Burma–Sumatera, *not* Java, Borneo.	500 m, humus rich soil – 1,100 m China.	187 spp., mostly lowland trop. Asia
Castanopsis acuminatissima	Himalaya–New Britain	300 m, coastal mountains New Guinea	110 spp., mostly tropical, warm temperate Asia
Dacrycarpus imbricatus	Indo-Burma, S. China–New Guinea	Lowland podsols	9 spp., Indo-Burma–New Zealand, tropical, warm temperate
Dacrydium elatum	Hainan, Indo-Burma–Solomons	Low, acid raw humus: swamps, podsols	25 spp., as *Dacrycarpus*
Daphniphyllum glaucescens	Sri Lanka–Japan–Sulawesi	1,000 m (*D. griffithianum* (*laurinum*) on coastal hills)	2 tropical and 8 temperate S. China
Debregeasia longifolia	E. Himalaya, temperate S. China, N. Peninsula Malaysia	Low altitude at high latitude	5 spp. tropical, warm temperate Asia, Africa
Dipterocarpus retusus	NE India, S. China–Sumbawa *not* Borneo	Low altitude at high latitude	69 spp., lowland tropical Asia
Distyliopsis dunnii	Tropical lower montane forest, China–New Guinea, incl. Kinabalu	800 m	6 spp., 6 in tropical, warm temperate E. Asia
Engelhardia roxburghiana	E. Himalaya–Sumatra, Borneo	200 m, fertile loams	4 spp. tropical lowland, lower montane spp., S. China–New Guinea
Eriobotrya bengalensis	E. Himalaya, Indo-Burma, N. Peninsula Malaysia	(also rocky sea shores)	37 spp., tropical lower montane W. Malesia–Indo-Burma, warm temperate E. Asia
Eurya acuminata	NE India to S. China, Taiwan, W. Malesia	700 m	70 spp., montane warm temperate E. Asia, Pacific
E. nitida	Sri Lanka, E. Himalaya, trop. S. China–Pacific	1,000 m	
Exbucklandia populnea	E. Himalaya, tropical S. China–Java, *not* Borneo	500 m	2 in temperate S. China
Lithocarpus elegans	E. Himalaya–New Giuinea	Sea level	c.300, tropical E. Asia, plus some warmest temperate species in S. China, Japan
Lyonia ovalifolia	Kashmir–Japan–Peninsular Malaysia	1,200 m, frost pockets, high latitude	35 spp., warm temperate E. Asia, N. America
Maesa macrophylla	E. Himalaya, Indo-Burma– Peninsular Malaysia	1,000 m	100 spp., Old World tropical mostly lowlands
Meliosma pinnata	S. India–S. Korea, S. Japan–New Britain	Lowlands (separate ssp.),	28 other spp. in temperate East Asia

Species	Species range	Lowest elevation genus recorded in the tropics	Number of species in the genus, and principal distribution
M. simplicifolia	S. India–Japan–Peninsular Malaysia, Borneo	Lowlands (separate ssp.)	
Magnolia baillonii	E. Himalaya, S. China– Peninsular Malaysia, Sumatra(?), *not* Borneo	500 m	125 spp., northern temperate, tropical, esp. E. Asia
M. champaca	W. Ghats, E. Himalaya–Sumatra, Borneo	Low at high latitude	
Myrica esculenta	Indo-Burma–Philippines, Lesser Sundas	Coastal, sandy soils	
Nyssa javanica	E. Himalaya, S. China – Java, Borneo	1,000 m	5 spp., temperate E. Asia, N. America
Photinia davidiana	Nepal–S. China, Taiwan–Java, Borneo	900 m	40 spp. in warm temperate E. Asia
Pittosporum kerrii	Indo-Burma, S. China–Pen. Malaysia	700 m	c.200 spp., southern tropics (not S. America), warm temperate
Platea latifolia	Sikkim–Indo-Burma–Melanesia	Lowlands	5 spp., Malesian tropics
Podocarpus neriifolius	NE India–China, Japan–Fiji	Lowlands, sandy soils	94, mostly southern warm temperate
Nageia wallichiana	NE India, Indo-Burma–New Guinea	100 m, seasonal tropics	
Rhodoleia championii	Tropical S. China–Sumatra, *not* Borneo	100 m, tropical margin, quartzite	5 warm temperate China
Saurauia napaulensis	E. Himalaya, S. Chian– Peninsular Malaysia	400 m	c.300 spp., Asian, American tropics
S. roxburghii	NE India, S. China–Ryukyu–Pen. Mal.	To low altitudes	
Schima wallichii	E. Himalaya, trop. S. China–Bali, Borneo	Lowland equatorial coastal hills, northern seasonal evergreen and lower montane forest	2 other spp. warmest temperate China
Sloanea sigun	E. Himalaya, Indo-Burma, S. Yunnan–Java, *not* Borneo	800 m	c.100 spp., tropical lowland, lower montane, Neotropics (most), Madagascar,.Asia
Symplocos cochinchinensis	E. Himalaya, Taiwan, China–Japan–Fiji	Lowlands (Malaysia, S. China)	c.250 warm temperate and tropical species
S. glomerata	India, S. China– Peninsular Malaysia	1,200 m	
S. pendula	Sri Lanka–warm temperate China, Korea, Japan, Borneo, Java (not Sumatera)	500 m	
S. sumuntia	Nepal–warm temperate S. China, Taiwan–Japan– Peninsular Malaysia	100 m temperate regions	
Ternstroemia gymnanthera	NE India, tropical/temperate S. China, Taiwan–Peninsular Malaysia	<300 m high latitude	85 spp., tropics
Turpinia pomifera	E. Himalaya–Kinabalu, Philippines, Sulawesi	1,500 m.	10 spp., lower montane tropical to warm temperate E. Asia
Ulmus glabra	E. Himalaya–Java, Sulawesi (*not* Peninsular Malaysia, Sumatra, Borneo)	Lowlands, high latitude	18 spp., north temperate.

Defined floristically, lower montane forests vary in structure with altitude, and with topography and soils, as do lowland forests. The mature phase has been characterised as having two strata, a canopy lacking emergents and a subcanopy (Table 4.1), although scattered emergent species do occur at its lower altitudes, especially in sheltered valleys (Plate 4.3c). The main canopy is dense: the predominantly notophyll leaves are densely clustered and appear to manifest a higher leaf area density than in lowland forest (Plates 4.3d, 4.5a). The canopy varies in height from 20–40 m between different habitats, but it is relatively even in height within any one habitat. It becomes increasingly even with altitude. In the seasonal Far East, high canopy leaf density is associated with distinct crown-shyness at higher altitudes (Plate 4.3d). It is also correlated with a distinct gap in subcanopy foliage immediately beneath the main canopy, enhancing the distinction between canopy and understorey strata. Large-leaved pioneers are few. *Macaranga* does reach its greatest species richness in New Guinea, where 36 of 89 species occur on steep, unstable slopes at lower and a few at upper montane altitudes. Only six of 50 *Macaranga* species are exclusively montane in Borneo, and all of these are lower montane.[32] Three occur in tropical lower montane forest in the Bhutan Himalaya,[33] and one in Sri Lanka.

In the absence of buttresses, lower montane oak-laurel forest trees develop deep coarse roots, with a diffuse horizon of fine roots within 15 cm of the soil surface.[34] Fine roots are generally sparse beneath the litter, often owing to the presence of a clay surface horizon caused by erosion and trampling by boar and deer, abundant in places where hunting is absent (Chapter 1).

Plate 4.5 Lower montane forest canopy and foliage. **a,** Lower montane oak-laurel forest notophyll-microphyll canopy from below, Doi Inthanon, 1,700 m; **b,** Mesophyll *Castanopsis diversifolia* foliage, Doi Inthanon, 1,700 m; **c,** Lower montane forest canopy, Mount Kinabalu, 1,800 m; **d,** *Shorea gardneri* crown with bunched multilayered microphyll foliage, recalling *Dryobalanops aromatica* or *Shorea curtisii* of freely draining lowland soils but with crown shyness in its breezy climate; Peak Sanctuary, Sri Lanka, 1,700 m. (**a, b, d** H.H.; **c** P.A.)

3.1. *Lower montane forest types of Indo-Burma and western Malesia*

3.1.1. *Lower montane oak-laurel forest*

Best developed in the mountains of northern Indo-Burma, tropical lower montane oak-laurel forest once covered many thousands of square kilometres. In the north, it extended along a band from the Rapti valley in central Nepal east and across into northern Burma, where it spread over the high plateau of the Shan states, through northern Thailand and Laos into Vietnam, just crossing the Chinese frontier into southern Yunnan and southwestern Guangxi. Intermittently, it still clothes the higher peaks in the north-south-trending Indo-Burmese frontier and Arakan ranges, the Thai-Burmese frontier and the Annamite ranges. Everywhere, its soils have been favoured by swiddening communities, and latterly for commercial tea, whose large scale traditional cultivation exactly coincides with the distribution of the *warm temperate* oak-laurel forest to its north, which extend to north-east Honshu (Fig. 4.4). Tea is now also grown throughout the tropical lower montane zone of South Asia and in the Far East.

The main canopy of lower montane forest in East Asia and Malesia is dominated by Fagaceae. Single species often dominate in the relatively species-poor eastern Himalaya. *Castanopsis indica,* which occurs there and in the Duars immediately to the south in lowland semi-evergreen and tall deciduous forest, is often dominant between 900–1,600 m. *Castanopsis tribuloides* dominates at 1,500–2,100 m, and *C. hystrix* in the ecotone to warm temperate forest at 1,900–2,400 m.

These forests are taller at lower altitudes, particularly in sheltered valleys, the main canopy reaching 35 m on Doi Inthanon (Thailand) and in the New Guinea mountains. Scattered and occasionally clumped emergent individuals of a few species add a third, discontinuous and patchy stratum (Plates 4.3c, 4.4a). In the Crocker Range of Sabah, *Agathis endertii* once formed emergent stands in lower montane *kerangas,* before it was logged. Occasionally, emergents in Far Eastern forests may reach 50 m, and the crowns of these emergents are swept toward the north-east where there are prevailing wet summer southwesterlies.

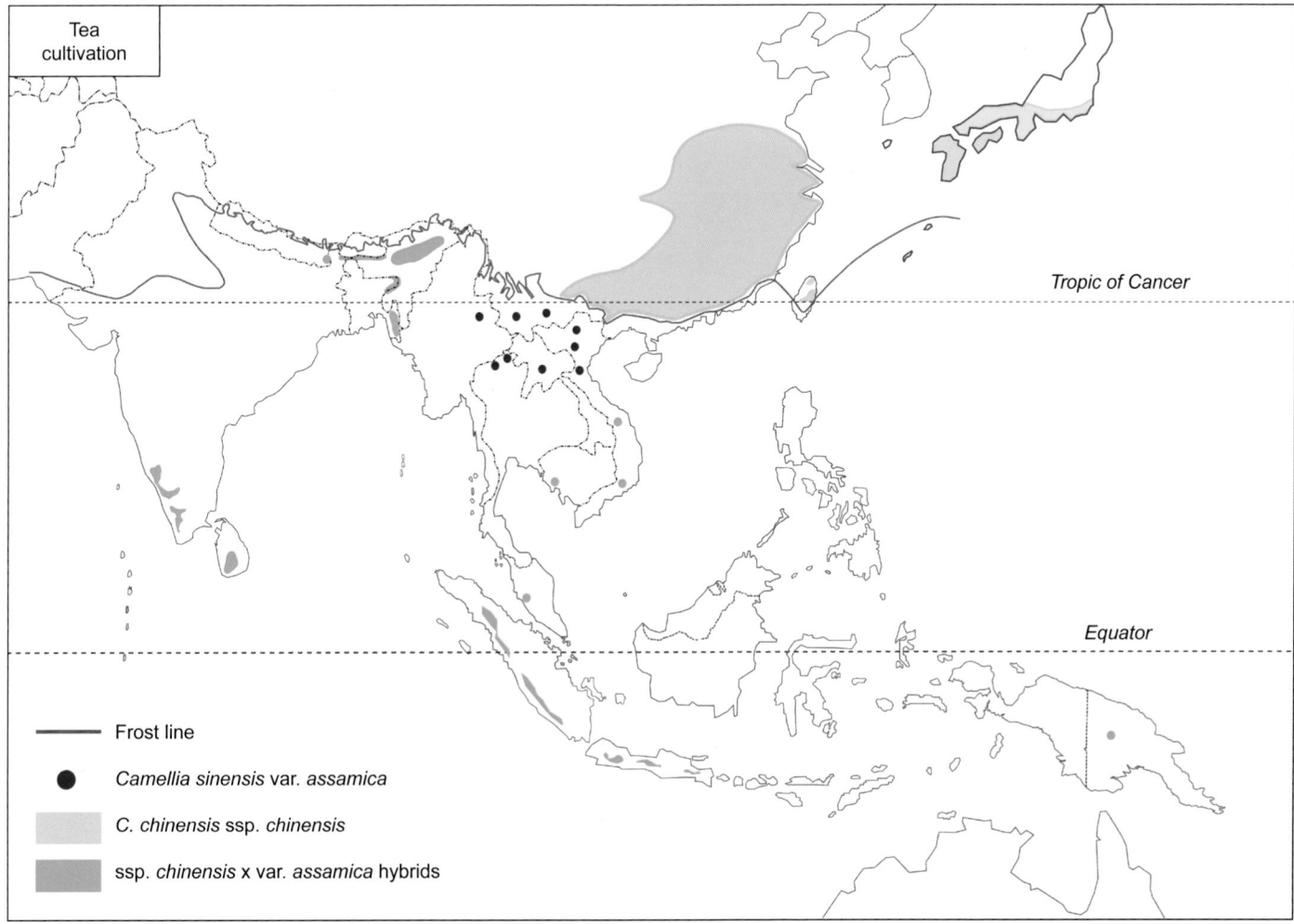

Fig. 4.4 The distribution of tea cultivation in Asia. (I.H. Baillie del.)

Box 4.1

An Indo-Burmese lower montane oak-laurel forest

Doi Inthanon (2,590 m, 18°35' N, 98°30' E) is the highest mountain in Thailand. Exposed in the south and west to the wet monsoon, it is clothed to its summit in evergreen forest. Fog, associated with rain, penetrates the canopy daily during the monsoon from late May to late Oct. During the dry season however, fog is intermittent, beginning by 11am and rarely descending below 2,000 m.[35] The wet monsoon is windy, and here emergent tree crowns trail to the north-east. The canopy becomes increasingly smooth, and crown shyness manifests with altitude. Northern and eastern slopes are drier.

The most detailed study of tropical lower montane oak-

laurel forest is progressing at Doi Inthanon under Mamoru Kanzaki and Sarayudh Bunyavejchewin,[36] focusing on a 15 ha plot at 1,700 m on a southwest-facing slope at a valley head over granite and gneiss, within the CTFS collaboration (Table 4.4). The soils are udult sandy clay loams with high carbon and nitrogen levels in mineral horizons relative to lowland zonal soils. The canopy rises to c.35 m, but scattered emergents, notably *Mastixia euonymoides* (Plate 4.3c, 4.4a) but also *Magnolia garrettii, Podocarpus neriifolius, Nyssa javanica* (Plate 4.6f) and some others, attain 50 m. Crowns are dense, and notophyll leaf sizes dominate in both canopy and understorey (Table 4.5; Plates 4.5a,b, 4.11a). Only 10 of the 126 species bear pinnate leaves. All trees, including *M. euonymoides, Magnolia garrettii* and *Calophyllum polyanthum*, which may exceed 1.75 m diameter, lack buttresses.

Rank	Species	Number of trees	% trees in plot	Species	Basal area (m²)	% b.a. in plot	% trees in plot
1	*Calophyllum polyanthum*	5995	8.1	*Mastixia euonymoides*	72.8	11.9	0.9
2	*Mallotus khasianus*	4939	6.7	*Quercus eumorpha*	51.5	8.4	1.6
3	*Castanopsis calathiformis*	3993	5.4	*Magnolia (Manglietia) garrettii*	35.7	5.9	0.9
4	*Psychotria symplocifolia*	2682	3.6	*Calophyllum polyanthum*	32.4	5.3	8.1
5	*Syzygium angkae* ssp. *angkae*	2647	3.6	*Quercus brevicalyx*	23.0	3.8	1.2
6	*Melicope pteleifolia*	2407	3.3	*Cryptocarya densiflora*	20.9	3.4	1.8
7	*Heynea trijuga*	1968	2.7	*Syzygium angkae* ssp. *angkae*	15.4	2.5	3.6
8	*Eurya nitida* var. *nitida*	1795	2.4	*Drypetes* sp.	15.1	2.5	1.4
9	*Symplocos macrophylla* ssp. *macrophylla* var. *sulcata*	1733	2.4	*Nyssa javanica*	14.5	2.4	0.4
10	*Lindera metcalfiana*	1588	2.3	*Mallotus khasianus*	13.9	2.3	6.7
	(Totals, these species)		(40.7)			(48.4)	

Table 4.4 The leading species in a 15 ha plot in lower montane forest, Doi Inthanon, Thailand: trees >1 cm dbh. (from Kanzaki & al. 2004)

Plate 4.6 Lower montane forest tree portraits. **a,** *Schima wallichii* (Theaceae), foreground left, Doi Inthanon, 2,500 m; **b,** *Alnus nepalensis* (Betulaceae), Trongsa, Bhutan, 2,000 m; **c,** Juvenile *Exbucklandia populnea* area (Hamamelidaceae), Chukha, Bhutan, 1,900 m; **d,** the winged fruit of *Engelhardia roxburghiana* (Juglandaceae), Shemgang, Bhutan, 1,600 m; **e,** *Beilschmiedia gammieana* (Lauraceae), Shemgang, Bhutan, 1,600 m; **f,** *Nyssa javanica* (Cornaceae-Nyssaceae), on the moist northern slope of a rainshadow valley, Shemgang, Bhutan, 1,900 m. (All: H.H.)

dbh (cm)	Leaf size class				
	n	Microphyll	Notophyll	Mesophyll	Macrophyll	Unknown
>10	2931	0	45	45	10	0
>40	626	1	53	40	0	7
>80	97	1	69	23	0	5

Table 4.5 Leaf size class representation of dominant tree species, Doi Inthanon lower montane forest (from Kanzaki & al. 2004)

The flora of the Doi Inthanon plot is overwhelmingly dominated by families and genera whose relative abundances remain consistent within tropical oak-laurel forest, from Bhutan to Wallace's Line (Table 4.3). Lauraceae are richest in species; Fagaceae have the greatest biomass though only eight species are present; while Lauraceae, Magnoliaceae and the sole Cornaceae are also high in biomass (basal area). Euphorbiaceae, as in the lowlands, remain important (Chapter 3). Among the core taxa of tropical oak-laurel forest, *Altingia, Exbucklandia* and *Bischofia* are all absent from the plot, though at least *Exbucklandia* occurs locally elsewhere on the mountain. *Calophyllum polyanthum,* widespread from the Western Ghats to West Java in lower elevations of lower montane forest, and also on fertile loams in the lowlands, is present. One third of the species extend to perhumid Malesia, two thirds are confined to Indo-Burma including the eastern Himalaya;[37] only one tenth are Thai endemics. Notophyll forest with Fagaceae and Lauraceae co-dominant extends to the summit of this mountain (2,590 m), where the tree flora is similar albeit less rich. The large lower montane vine *Melodinus cochinchinensis,* and smaller *Jasminum attenuatum* and *J. dispermum* also occur there.[38]

In Bhutan and Sikkim, 68 tree species occur at their lowest elevations at 1,300–1,800m. A further 18 species are added at 1,200–1,300m and 53 at 700–1,200m. Only 20 of these 139 range higher than 2,400m, which is the upper limit of frost-free sites. Nine of the 16 Fagaceae species occur at 700–1,800m, and none of these extends above 2,400 m. Seventeen of the 47 Lauraceae occur at 700–1,800 m, and only four extend above 2,400 m. Only a single species from among these two families is also found on Kinabalu: the polymorphic *Lithocarpus elegans,* found at 900–1,800 m on Kinabalu and at 1,300–2,400 m in Bhutan. In Bhutan there are 12 *Acer* species, all at least somewhat deciduous. Two of these occur in the tropical lower montane zone, at 1,000–2,200 m, while two further species occur from the lowlands up to 2,400 m; all others are temperate (among which only one extends beyond 3,350 m, to 4,880 m). The altitudinal ranges of the genera characteristic of lower montane forests here in the Himalaya are similar to those in the equatorial tropics. A separate group is confined to temperate forest at higher altitude, many of which occur also in Far Eastern warm temperate lowland evergreen forests.

Overwhelmingly though, genera of the woody flora of the Bhutanese lower montane forests extend from west of Wallace's Line to the warm temperate evergreen forests of East Asia, some even as far as Japan (Table 4.3). In contrast, the *species,* especially those in the larger genera, are regionally distributed. A few tropical species of eastern Himalaya and Indo-Burma range north into tropical southern China, but only a very few, such as *Quercus acuminatissima* and *Q. glauca,* extend yet further north into warm temperate east Asia and to Japan (Chapter 6). The margin of the tropics in wet East Asia is a less stringent barrier to tropical genera than it is to species.

In southernmost China, tropical lower montane forests near their northernmost limits support a rich tropical liane flora, distinct from the warmest temperate evergreen forest liane flora adjacent to their north, which at first is only slightly less rich.[39] Lower montane forests with a fully tropical tree flora containing tropical *Castanopsis,* Lauraceae and Theaceae still survive in a few localities, sometimes immediately adjacent to remnants of northern seasonal evergreen forest,[40] where earlier, in the Vegetation Map of China,[41] it had been termed 'mixed broadleaved evergreen forest'. It was included within 'subtropical evergreen broadleaved and sclerophyll forest' by Li and Walker.[42]

3.1.2. *A floristic boundary*
The boundary between the Indo-Burmese and Malesian lower montane oak-laurel forests occurs in the gap between the north-west Peninsular Malaysian Main Range, which penetrates southwestern peninsular Thailand, and Khao Luang ('Cloud Mountain', 1,786 m, 8°35' N, 99°40' E),

the southernmost mountain of the Thai-Burmese frontier range, lying 300 km to the north on the eastern side of the Peninsula. The tree species of a plot in lower montane forest on nearby Khao Phnom Bencha (8°15' N, 98°50' E) are typically Indo-Burmese. Lower montane oak-laurel forest is rare in Borneo and patchier in Peninsular Malaysia than implied by Symington.[43] The best known example is found in the broad valleys and summit ridges of Fraser's Hill (1,100 m, 3°40' N, 101°45' E). Here, as elsewhere in the perhumid Sunda lands, the oak canopy is interspersed with conifers including *Dacrycarpus imbricatus, Dacrydium elatum* (syn. *D. pierrei), Agathis borneensis,* elements that extend elsewhere into upper montane forest. But Fagaceae continue to dominate the lower montane forest canopy along the basic volcanic Barisan Range of Sumatra to west Java,[44] then burgeoning again in New Guinea where species of the southern beech, *Nothofagus* (now regarded as a separate family, Nothofagaceae, albeit with a somewhat similar ecology), come to dominate, especially at higher elevations.[45] Lower montane evergreen forests cling on as impoverished fragments, devoid of Fagaceae or Nothofagaceae, on the windward slopes of higher mountains along the Lesser Sundas east to Timor. There, *Engelhardia spicata,* widespread in lower montane New Guinea, also occurs.

3.1.3. *Lower montane kerangas*
Gunung Mulu (2,376 m) is a sandstone mountain in north-east Sarawak. Though isolated, it is representative of the mountains of Borneo's backbone.[46] The lower slopes bear MDF and, as one ascends the hiker's path to the summit, lowland species gradually drop out with altitude. The topography is steep, and the path climbs along a steep spur. Above 800 m a few new species appear, the dipterocarps become fewer, but the change is almost imperceptible until, at 1,200 m, the first true ridge is reached and the forest changes to forest resembling upper montane forest in its short stature and smaller leaf sizes. However, this forest lacks the characteristic species of upper montane forest, and is 15–20 m tall with dense mostly straight stems, in these respects more resembling *kerangas.* I therefore name it *lower montane kerangas* (Plate 4.7a–c). There are some species unknown in lowland forests. The mineral soil is shallow, a leached greyish sandy loam, and bears a variable but often thin horizon of raw humus dense with fine roots. Mosses and liverworts cover the trunks patchily but there are no tussocks until the high ridges are reached (above 1,900 m). On Kinabalu, the climb above the Pinosuk Plateau also passes through patches of forest on sandstone with the stature and structure of lowland *kerangas* 1,500–2,200 m, and a flora which also includes elements rare or local in the perhumid lowlands, such as *Dacrycarpus imbricatus, Dacrydium elatum,* and *Schima wallichii.*

Plate 4.7 Variants of lower montane forest. **a,** Lower montane *kerangas,* c.1,500 m; *Dacrydium elatum* with pale green crowns. Crocker Range, Sabah, Malaysia; **b,** Lower montane pole forest with *Agrostistachys borneensis, Stemonoporus angustifolius,* on friable humic yellow-red soil on a spur, 1,600 m, Peak Sanctuary, Sri Lanka. **c,** Lower montane pole forest interior, Peak Sanctuary. Note absence of festooning cryptogams; **d,** Summit forest subcanopy, Doi Inthanon, 2,590 m. Thinly mossy; Fagaceae, *Syzygium* etc. in canopy; *Maesa* and *Ardisia* in subcanopy; dense shrub layer of *Strobilanthes mucronato-producta, Ophiorrhiziphyllon macrobotryum* (both Acanthaceae); **e,** Seasonally dry montane forest, 1,500 m, Hakgala, Sri Lanka. (**a–d** H.H.; **e** I.A.U.N. Gunatilleke)

On ultramafic coastal Gunung Silam (840 m) in eastern Sabah, MDF with floristic similarity to MDF elsewhere on humult yellow sandy soils is replaced by a shorter forest with floristic affinity to *kerangas* but dominated by mesophyll leaf sizes to 770 m. Above this, notophylls dominate to the summit.[47] A point endemic subspecies of *Tristaniopsis kinabaluensis,* whose type subspecies is endemic on Kinabalu's ultramafics, is abundant in Silam's lower montane *kerangas* above 770 m. Along the Borneo mountain chain, the dipterocarps *Cotylelobium lanceolatum* and the low emergent *Shorea monticola* form gregarious stands in lower montane *kerangas* to 1,500 m, while *S. carapae* occurs where soils are poorly drained at 900–1,200 m. The isolated schist Gunung Jerai (Kedah, 1,140 m), bears lower montane *kerangas* above 800 m, including *Agathis, Dacrydium elatum* and *Tristaniopsis merguensis.*[48] Herbarium records suggest that c.1,000 m nevertheless marks the upper limit of many lowland *kerangas* forest species, as it does lowland mixed forest species.

In the seasonal Far East, the Cardamom and Elephant Mountains of southern Cambodia reach 1,563 m at Phnom Lumphor, but otherwise rarely exceed 1,000 m. Khao Yai (1,200 m) and the Dongrek Range in southeastern Thailand and northwestern Cambodia are other exceptions. All are hard Triassic sandstone, their summits often forming narrow plateaux with intermittently waterlogged podsolised sandy soils bearing acid raw humus (Chapter 1). These summit forests are not densely mossy. They have the structure of short *kerangas,* generally less than 20 m tall, with many straight boles, a dense understorey and an even canopy surface. On Kam Chay Mountain, in the coastal Elephant chain of southern Cambodia, the road climbs through upper slopes rich in Fagaceae, emerging suddenly onto a summit plateau at the ruined summit hill station of Popork Vul (1,000 m). The broad surrounding plateau supports gregarious stands of *Dacrycarpus imbricatus* and *Dacrydium elatum* on boggy organic soils – species also common in the lowland *kerangas* on freely draining white sand podsols near the coast below. On Khao Yai's upper slopes are found *Betula alnoides* and *Anneslea fragrans* (c.20 m in height). On its broad summit ridge we find *D. imbricatus, Nageia fleuryi* and *Podocarpus neriifolius,* over an understorey containing *Osbeckia* (Melastomaceae), and rich in Rubiaceae. All these taxa are typical components of Indo-Burmese lower montane forest elsewhere.[49] These summit associations comprise mosaics of lower montane oak-laurel and *kerangas* forests. Throughout the Far East, acid raw humus-bearing soils extend below the zone of persistent diurnal fog into the lower montane forests. These forests bear a canopy tree flora with several otherwise upper montane elements, notably gymnosperms. This, as will be seen, confounds a floristic definition of upper montane forest.

In no case does lower montane *kerangas* attain the height achieved by some lowland *kerangas* on the giant humult podsols of Pleistocene raised beaches (Chapter 2), but its stature is within the lower range found in the lowlands.

3.1.4. 'Upper dipterocarp forest'

Symington recognized the upper altitudinal limit of his 'hill dipterocarp forest' to be at 800 m. Immediately above this hill dipterocarp forest he distinguished 'upper dipterocarp forest'. He regarded upper dipterocarp forest as a category of lower montane forest, though he observed its upper limit to be as low as 1,300 m. Upper dipterocarp forest occurs as a distinct zone within the ecotone between lowland and lower montane oak-laurel forest, throughout perhumid Sundaland. It has the structure and dipterocarp dominance of short-stature MDF in which emergent dipterocarp species hardly exceed the main canopy, but includes some species common in lower montane forest such as *Schima wallichii.* It also includes a distinct flora of its own, notably the widespread Sunda dipterocarp *Shorea platyclados,* also *Dipterocarpus retusus* and *D. costatus.* These last two are also found in the northern seasonal lowlands.

Plate 4.8 Hill and upper dipterocarp forests. **a,** Peninsular Malaysian hill dipterocarp forest: pale crowns of the dominant, *Shorea curtisii,* 600m; **b,** Upper dipterocarp forest, *Dipterocarpus costatus* dominant, 900 m, Nam Nao National Park, central Thailand. *Beilschmiedia gammieana,* the dark crowned canopy tree at right, behind juvenile *Pinus merkusii.* (**a** P.A.; **b** H.H.)

On the mountains of Phang Nga National Park (Petchabun), on Doi Chiang Dao and elsewhere in both northern and peninsular Thailand, ridges at c.1,000 m bear local stands of massive, gregarious *Dipterocarpus costatus,* a species occurring in upper dipterocarp forest in Peninsular Malaysia[50] (Plate 4.8b). These may be mixed with lowland deciduous elements including deciduous dipterocarps (Chapter 3), and with lower montane oak-laurel elements. The large Indo-Burmese bamboo *Dendrocalamus hamiltonii* becomes abundant above 800 m in this forest type, and at lower elevations in oak-laurel forest, where it may form dense stands arresting succession in large gaps.

Upper dipterocarp forests at 800–1,300 m, and lower montane *kerangas* at 800–1,800 m, resemble facies of lowland MDF and lowland *kerangas* more than the lower montane oak-laurel forest.[51] In both, there is diminution of leaf size classes: from dominant mesophyll to notophyll (occasionally to microphyll) in upper dipterocarp forest, and notophyll to microphyll in upper *kerangas.[52]* These appear to correlate with slight soil changes, notably a patchy increase in depth of raw humus; but the soils continue to resemble analogous lowland soils more than the humic mulls of lower montane oak-laurel forest see section 3.6.2 below).

3.2. *Lower montane forests of eastern Malesia*

In New Guinea, the dipterocarp *Hopea forbesii* extends up to 1,000 m, but other dipterocarps have not been recorded above 800 m. Other lowland species, however, persist to at least 1,200 m.[53] Lowland forest gives way to a lower montane forest initially dominated by *Lithocarpus* spp. and *Castanopsis acuminatissima* (Fagaceae), which also occurs on Kinabalu. *Araucaria cunninghamii* (var. *papuana*) forms gregarious emergent stands on ridges and rocky defiles, while *A. hunsteinii* may dominate on former landslips down into the lowlands. Although that transition occurs at c.1,200 m, forests dominated by the very widespread *Castanopsis acuminatissima* with *Engelhardia spicata* occur exceptionally down to c.300 m on red soils over limestone on coastal hills near Lae,[54] but the lower montane flora otherwise does not accompany them. This is reminiscent of the forest along the Cenozoic basalt summit ridge of Bukit Mersing (Central Sarawak, 950 m), which is dominated by *Castanopsis paucispina*, other Fagaceae (Fig. 4.5) and *Engelhardia roxburghiana,* but also the dipterocarp *Vatica granulata. Castanopsis acuminatissima* is similarly associated with dipterocarps *Anisoptera thurifera* and *Hopea* spp., as well as with other lowland genera at these lower altitudes in New Guinea. Grubb and Stevens[55] found a high proportion of

species with bird-dispersed seeds in lower montane forest in New Guinea; but these are offset in Asian and Sunda forests by the abundance of Fagaceae and species with wind-borne fruit or seeds (for example *Acer, Engelhardia, Altingia, Exbucklandia* and *Ulmus*).

The diurnal cloud base on the major New Guinea massifs

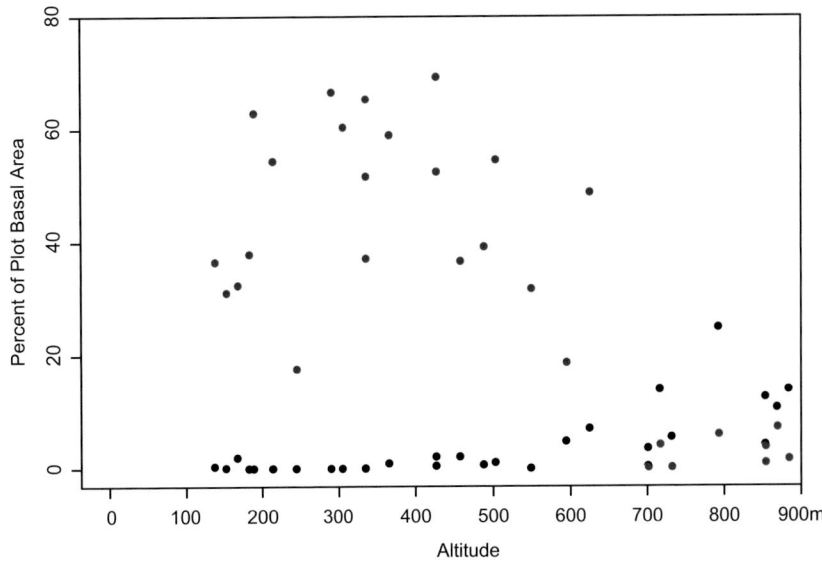

Fig. 4.5 Change in family relative dominance with altitude of Dipterocarpaceae on the basalt hill Bukit Mersing, Sarawak (orange: Dipterocarpaceae; black: Fagaceae). (P. Hall from P.S. Ashton unpubl.)

rises to c.2,800 m, and fog beneath is intermittent. Structurally and physiognomically defined mixed species interpreted as lower montane forest may extend to 2,850 m (Fig. 4.10),[56] becoming increasingly mossy with altitude and incorporating upper montane forest elements including *Papuacedrus papuana, Podocarpus brassii,* and Winteraceae. However, Robbins[57] considered lower montane forests on Mount Hagen to be restricted to below 2,400 m, above which 'beech' (*Nothofagus*) forests prevail. Southern beech, *Nothofagus,* species form dominant or codominant zones at various altitudes, but particularly above 2,600 m.[58] This southern element can be interpreted as constituting a uniquely tall form of upper montane forest at these higher altitudes, in part due to the presence of this canopy genus of southern warm temperate origin; but also because of the presence of some upper montane elements. However, its lack of dense mossiness, and its structure, more resembles lower montane forest. The uppermost such zone, dominated by *N. carrii* and the invasive shade-intolerant *N. pullei,[59]* occurs at 2,200–2,900 m. These species are microphyllous, but the underlying soils are humic mulls typical of lower montane forest elsewhere. Dominance appears to be sustained by group mortality of young canopy cohorts in stands of less than 2 ha, possibly occasioned by a soil pathogen, *Phytophthora.* Seedlings are unaffected, and saplings readily resprout. Ash[60] hypothesised that, as in northern American oak succession, mature *N. pullei* would be invaded and eventually succeeded by a mixed association of shade-tolerant species.

No systematic survey of any mountain between Wallace's Line and New Guinea exists. Sulawesi's mountains support several Sunda lower montane elements, including *Engelhardia roxburghiana*, *Lithocarpus*, *Castanopsis*, *Trigonobalanus verticillata*, *Acer niveum*, *Mastixia*, *Bischofia javanica* and *Ulmus lanceifolia*; but also elements which do not cross west of the Line, such as *Macadamia hildebrandii*. Lower montane *kerangas* appears to be widespread on ridges towards its limits at c.2,000 m, with the gymnosperms *Phyllocladus*, *Dacrycarpus*, *Dacrydium* and other Australasian elements becoming dominant (Chapter 6).

3.3. *Lower montane forests of South Asia*

3.3.1. *The Western Ghats*
On the wet west-facing slopes of the Western Ghats, which experience a 3–4 month dry season, the zonation resembles perhumid south-west Sri Lanka in altitudinal distribution (see next section), but the species are mostly different and there are fewer of them.[61] Lowland forest is replaced along a relatively narrow ecotone, at c.700 m in the south and at 850 m in the north, by a forest dominated by lowland genera though less species-rich. In contrast to Sri Lanka, this marks the upper limit of dipterocarps. These stands include characteristic upland elements such as *Palaquium ellipticum* (Sapotaceae), *Cullenia exarillata*, and *Mesua ferrea* (Clusiaceae), with abundant thickets of the understory bamboo *Ochlandra*.[62] North of 14° N, *C. exarillata* is replaced by *Poeciloneuron indicum* which becomes dominant on lateritic soils rich in sesquioxides of iron and aluminium. This zone in turn gives way, across a gradual ecotone (c.1,200–1,400 m) to mixed montane *sholas* rich in Lauraceae but lacking emergents, in which *Magnolia* (*Michelia*) (Magnoliaceae) and characteristic species of *Meliosma* (Sabiaceae) and others occur. The genera present in these forests are analogous to those of Far Eastern lower montane forests, being dominated (both in species numbers and in basal area) by Lauraceae. However, Fagaceae, Hamamelidaceae, Juglandaceae, *Nyssa*, *Acer* and all temperate deciduous tree genera, and conifers with the exception of *Nageia*, are absent (Chapter 6). *Garcinia* and *Calophyllum*, including the Sri Lankan upper montane emergent *C. walkeri*, Clusiaceae, a family well represented in the pre-collision Indian pollen record, are more prominent.[63]

In the Western Ghats, as elsewhere, there is a gradual decline in tree species richness with altitude. On western slopes, a smoother canopy and evidence of crown shyness on exposed surfaces reflect constant wind during the wet south-west monsoon. Forest stature declines on both slopes, but more markedly on the western. On the drier eastern slopes,

leeward of the south-west monsoon, forests are deciduous below c.1,200 m, where these change over to lower montane evergreen forests similar in structure to those of East Asia. The narrow, fire-mediated ecotone is today much obscured by farming and the extension of anthropogenic fire uphill.

Forest structure, similar in stature and profile to other lower montane forests at its lower limits, gradually changes with altitude to the short stature and twisted branches of upper montane forest on the Nilgiri and Palni summits at c.1,900–2,636 m. On these summits, mean annual rainfall declines below 2 m while there are up to five dry months, and winter frosts occur on the widespread grasslands. The *sholas* never gain a continuous mossy floor, instead often being richly vegetated with ground herbs and shrubs including Rubiaceae and *Strobilanthes* (Plates 4.7d, 4.11d). As on Doi Inthanon, the core generic element of the lower montane forest flora extends into the highest surviving forests. Mosses and other epiphytic cryptogams are patchy in the canopy, with the locally abundant lichen *Usnea*, indicative of periodic drought, hanging like Spanish moss. This will be discussed further in section 4.3.1 below.

3.3.2. *Sri Lanka*
In perhumid south-west Sri Lanka, south of the Peak Sanctuary and in the eastern Sinharaja, we find a distinct though narrow zone of tall forest between lowland MDF and lower montane forest (750–1,000 m). This forest is locally dominated by the single endemic dipterocarp *Shorea trapezifolia* and also includes several *Stemonoporus* species in the subcanopy. *Mesua ferrea* and *Durio exrillatus* are concentrated on ridges, occurring as low as 300 m. *Syzygium rubicundum* (syn. *S. spissum*) is conspicuous with its red young leaves in successional stands. The many lowland species and tall structure relate this forest to Far Eastern upper dipterocarp forest. In this case, though, *S. trapezifolia* is a fast-growing light demander of slopes, regenerating on landslips and windthrows and forming local monodominant 'drifts' (Chapter 5).

Some Sri Lankan wet zone ridges at 800–1,000 m bear a low forest cover resembling Far Eastern lower montane *kerangas* in structure and displaying its own distinct flora. This flora is rich in local endemics, including point endemic *Stemonoporus* species, which may form gregarious stands (Plate 4.7b,c). These ridges are narrow or exposed near the edges of the massifs. The soil is thinly covered with root-matted humus.

In a wide band at 950–1,800 m the Sri Lankan slopes were once covered by the great *rat dun* (*Shorea gardneri*) forests, also rich in other endemic dipterocarps such as *Stemonoporus oblongifolius*, *S. rigidus* and *S. revolutus*. Now confined to the slopes of the Peak Sanctuary and a few fragments elsewhere, this forest is dominated by the single endemic *Shorea gardneri* (Plates 4.3a, 4.5d). This massive tree, 30–40 m tall at maturity, is microphyllous. Its writhing branches are swept by the strong south-west monsoon winds here. Towards its upper limits, microphyll upper montane species are well represented in its understorey and gaps, which Yamada[64] also found in Java.

On the hyper-wet but south-west facing windy slopes of Sri Pada (Adam's Peak) itself, *Shorea gardneri* is replaced at 1,400–1,600 m on slopes and ridges by a lower-canopied forest intermediate in structure to upper montane forest and lacking dense mossiness. This is dominated by *Stemonoporus gardneri* with mid-canopy associates common to *Shorea gardneri* forest, including the endemics *Palaquium rubiginosum* and *Gordonia zeylanica*. Now threatened, *Stemonoporus affinis* once similarly dominated (likewise in association with other point endemics) on the northern isolated massif, the Knuckles range, where *Shorea gardneri* is absent. Both these *Stemonoporus* species are notophyllous trees, up to 20 m tall but showing the habit of upper montane trees, altitude notwithstanding. Throughout Sri Lanka's mountains, the drier northern slopes at lower montane altitudes bear forests rarely exceeding 20 m in stature, with an even but diffuse canopy and high microphyll representation. Here, *Calophyllum trapezifolium* is frequently dominant, in association with *Palaquium rubiginosum* and *Garcinia echinocarpa*.[65]

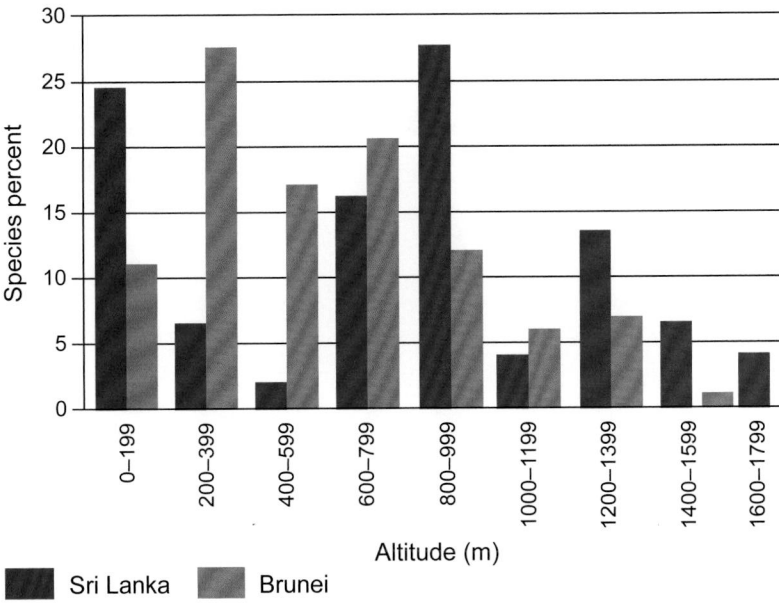

Fig. 4.6 Altitudinal range maxima of Dipterocarpaceae: Brunei and Sri Lanka compared. (D. Glick del. from P.S. Ashton unpubl.)

In summary, South Asian lower montane forests differ in the absence of major tree taxa, but are structurally broadly similar to those of East Asia. A lower zone exists in the Western Ghats, bearing some similarity to upper dipterocarp forest albeit lacking in Dipterocarpaceae, whereas narrow ridges at a similar altitude support forest analogous to Far Eastern lower montane

kerangas. Dipterocarps, which arrived in Asia by the Indian Noah's Ark in the Eocene (Chapter 6) replace the oaks, which likely arrived later in Far Eastern forests, as canopy dominants in perhumid lower montane Sri Lanka. In the absence of Fagaceae, it appears that Dipterocarpaceae evolved species that occupy the vacant niche[66] (Fig. 4.6).

3.4. Lower montane forests of the dry seasonal tropics

With increasing length of the dry season, lowland evergreen forests are replaced by floristically poorer deciduous forests (Chapter 2). These are fire-maintained, and continue up to the cloud base at 900–1,000 m on the northwestern slopes of the Western Ghats and to 1,300 m in Indo-Burma. In parts of East and South Asia where the dry season is at least five months, tall or short deciduous forest often climbs as high as 1,400 m, at least in part owing to human-induced fire. It then gives way to broadleaved evergreen lower montane forest at a sharply-defined fire limit. In South Asia and Indo-Burma, leeward slopes receive less monsoon rain than windward at higher elevations. Evergreen broadleaved forest in Indo-Burma and the northern Philippines is replaced by fire-climax monodominant pine savanna woodlands with a grassy field layer and, where severely disturbed, savannas (Plate 4.9) (see Box 4.1).

South of the Equator, in Java east of Gunung Lawu and the Lesser Sunda Islands, where the dry season exceeds four months, pines are replaced by a Casuarina (*Casuarina junghuhniana,* cemara).[67] This shares many pine characteristics, including requiring full sun and open ground for germination and establishment, leptocaul leaves reduced to scales on slender green shoots, and rough, flaky, persistent bark which attracts epiphytes in sheltered habitats. Fire, too, is a regeneration requirement; in this case, fire was originally often caused by volcanic activity. Cemara occurs above 1,300 m and can be found as high as the subalpine zone at 3,100 m. Repeated ground fires, partly dependent on rapid fine litter accumulation, sustain the regeneration of pure stands. Lower down we find a mixed narrow-leaved grassland, along with shrubs including *Dodonaea viscosa* (a pantropical species of both dry subalpine and sandy coastal vegetation), *Vernonia* spp., the tropical alpine herb *Anaphalis* and the ubiquitous fire-tolerant *Pteridium aquilinum* (bracken). Unlike the pines, *cemara* has an extraordinary ability to resprout from epicormic shoots deep in the bark, giving it resistance to crown fires which would be fatal to broadleaved montane forest species. Without fire, cemara forest is soon invaded by closed evergreen lower montane forest, beneath which, like tropical pines, it fails to regenerate.

261

Plate 4.9 Tropical Asian pines. **a,** *Pinus merkusii* in *Dipterocarpus obtusifolius*–dominated deciduous dipterocarp forest, 700 m, Mae Sariang, Thailand; **b,** *Pinus kesiya* forest, 1,200 m, Mountain Province, Luzon, Philippines; **c,** *Pinus roxburghii* stand along a ridge, 600 m, Sankhosh valley, Bhutan; **d,** Two, several year old, saplings of *P. merkusii* in the 'grass stage', left; single young *P. kesiya* sapling, right; 900 m, Ban Wat Chan road, north Thailand. (**a, b** P.A.; **c, d** H.H.)

Box 4.2

Tropical conifer ecology

Although several conifer genera of northern affinity have penetrated the Asian tropics, especially in Indo-China (see Chapter 6), elsewhere in the Far Eastern tropics there is a greater diversity of families, genera and species of conifers of southern origin (Chapter 6). Among the northern conifers, besides the two pines dependent on fire for their establishment, only *Taxus wallichiana* occurs in the Far Eastern tropics, extending to the upper elevations of mildly seasonal lower montane forests of Indo-Burma, the Philippines and northern Sumatra, often in association with the podocarp *Dacrycarpus imbricatus*.

Unlike the northern, the southern conifers prevail in lower montane forest (Fig. 4.7; Plate 4.10), where *Agathis* and *Dacrydium* reach their highest species diversity. Most occur on acid soils towards the upper elevations. *D. elatum* also becomes locally dominant in lowland *kerangas* on shallow peat, and is co-dominant with *D. imbricatus* in the seasonal *kerangas* of southern Cambodia (Plate 3.5a). *Nageia wallichiana* and *Podocarpus neriifolius* are exceptional in continental Asia, in occasionally also occurring on yellow-red lowland soils, most commonly in hill and upper dipterocarp forests. *Agathis* and *Araucaria* species, in particular, can form stands on yellow-red loams east of Wallace's Line where, presumably, angiosperm competition is lower. These two genera; like pines, have light, winged seeds, which contrast with the fleshy drupes of the podocarps and *Taxus*. These winged seeds disperse easily into fresh landslips, thereby gaining a march on their competitors.

Plate 4.10 Australasian lower montane gymnosperms on Kinabalu. **a,** Emergent *Agathis lenticula;* **b,** *Dacrydium xanthandrum.* (A. Farjon)

Fig. 4.7 Altitudinal distribution of conifers on Mount Kinabalu. (After Beaman & Beaman 1998)

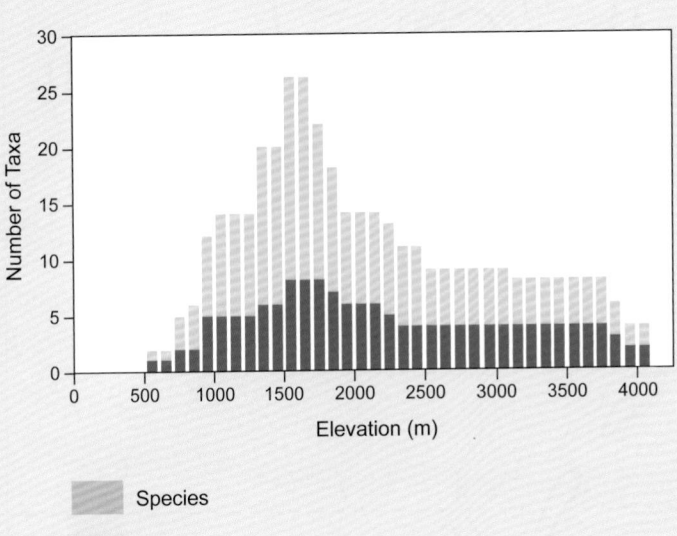

Deciduous forest conifers: the pines.

As dominants of much of the savanna woodlands in which they occur, pines play a special role in the drier forests of tropical East Asia (Fig. 4.8; Plate 4.9). *Pinus merkusii* (Plates 4.8, 4.9a,d), often named *P. latteri,* occurs widely in lowland Indo–Burma, also down the western slopes of Sumatra's Barisan Range and the western lowlands of Mindoro (Philippines). Seedlings at first develop long, supple grass-like needles, which increase their ability to photosynthesise in sward (Plate 4.9d).[68] *P. merkusii* forms an emergent overstorey in deciduous dipterocarp forest through northern Indo–Burma to the Cambodian coastal mountains and the northern Thai peninsula near Hua Hin (Chapter 3). This association reaches 1,200 m in the hills of northern Indo–Burma where lower montane oak–laurel forest species reinvade up valley heads.[69] These tropical pines are highly adapted to periodic fire. All species share thick bark even at seedling stage, and their seeds germinate following fire. In the absence of fire, regeneration of broadleaved species, many from fire-tolerant root stocks, engulf pine seedlings and reduce light at the forest floor below the compensation point of juvenile pines, and below the intensity required for their seed germination. Here, as in many other tropical pine forests, the original sources of fire were apparently end-of-dry-season lightning strikes. The presence of three pine species endemic to these forests, all of which require fire for their establishment, indicates that their

Fig. 4.8 The geographical range of four pines in Asia: *Pinus roxburghii* (blue), *P. kesiya* (green) and its warm temperate sister species *P. yunnanensis* (mauve), *P. merkusii* (red). (I.H. Baillie del.)

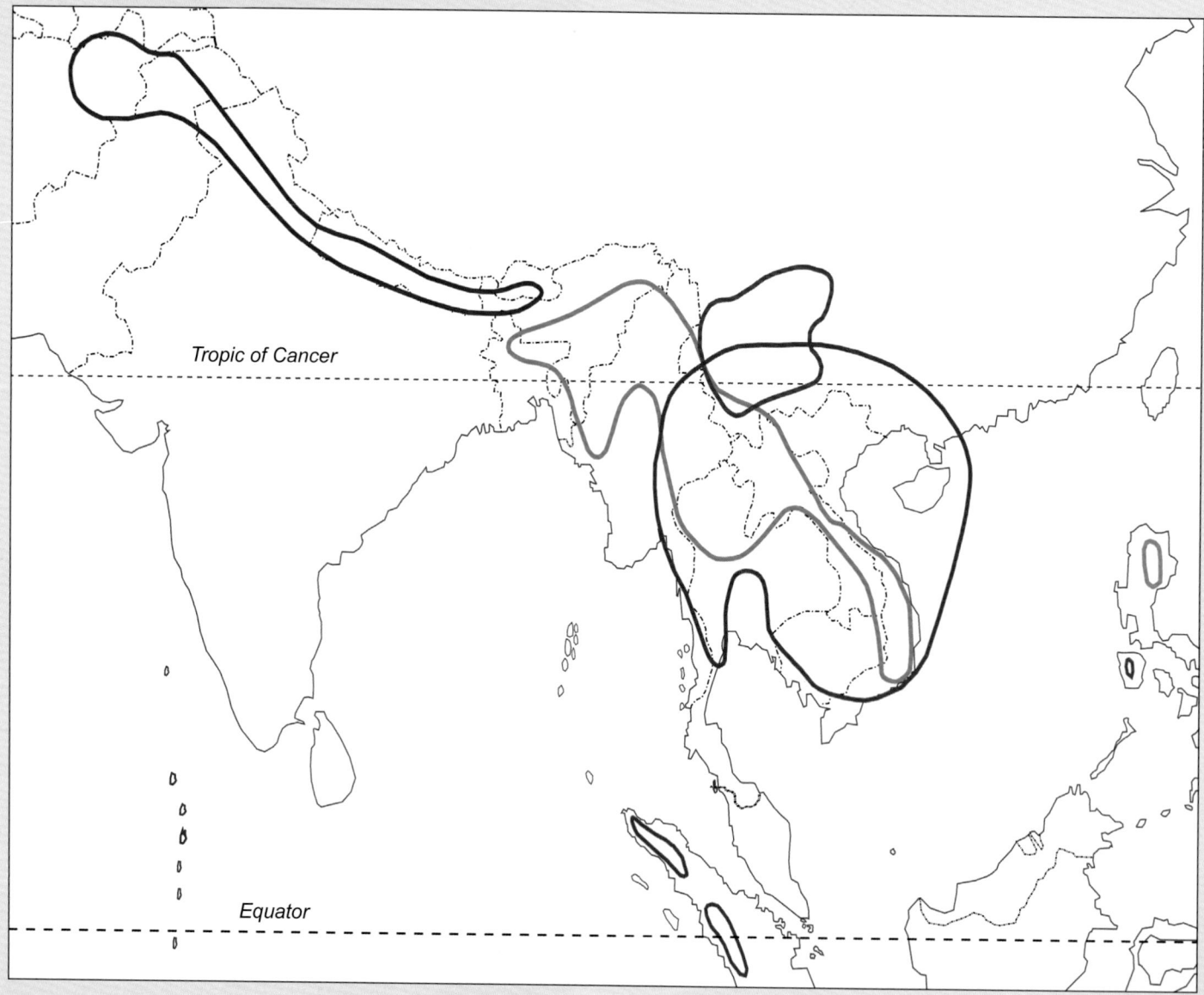

communities are indigenous and natural in origin, though their range has been greatly expanded by human activity.

Pinus kesiya savanna occurs throughout Indo-Burma from the Khasi hills eastwards, and in the central massif of Luzon, above the dry-season cloud base and therefore within the lower montane forest zone (Fig. 4.8; Plate 4.9b,d). It occurs especially on infertile, freely draining sandy soils and sunny slopes in the wet (south-west on the continent inland, north-east in Vietnam and Luzon) monsoon rain shadow, and on slopes prone to dry north-east trade winds.[70] Here, frequently burned *P. kesiya* savannas bear a characteristic grass flora including *Themeda triandra, Eulalia trispicata, E. quadrinervis, Miscanthus sinensis,* and also *Imperata cylindrica* and bracken fern (*Pteridium aquilinum*).[71] *P. kesiya* stands are occasionally subject to severe cyclonic storms which flatten whole landscapes (Plate 2.22c).

Pinus kesiya may descend to 1,000 m on sunny south-facing rain shadow slopes in northern Indo-Burma, becoming contiguous with the lowland deciduous forests and often mixing with *P. merkusii* in short deciduous dipterocarp forest at 850–1,200 m). On the Da Lat plateau of the southern Annamite range, *P. kesiya* occurs above 1,500 m in association with *Keteleeria davidiana.* It occurs at 1,400–2,200 m in the central mountains of Luzon, where its grasslands are being extended by human activity upwards into the ever-diminishing upper montane forests. The open grassland habitat of *P. kesiya* facilitates establishment of upper montane species otherwise rare at these altitudes, including *Anneslea fragrans, Ternstroemia gymnanthera, Eurya acuminata* and *Viburnum* spp. In China,

P. kesiya is confined to the extreme South and East, coinciding with the limits of tropical lower montane evergreen forest. To the north, in the mountainous southern provinces subject to occasional or more frequent frost, it is replaced by its sister species *P. yunnanensis.*

Pinus roxburghii (chir) occupies the habitat of *P. kesiya* from Subansiri District in Arunachal Pradesh (North-East India) to the western end of the Himalayan range of tropical lower montane forest. Being mildly frost-hardy, it extends into temperate winter-dry forests west to Afghanistan[72] (Fig. 4.9c). In the east, it is contiguous with the lowland deciduous sal (*Shorea robusta*) forests of the *terai*[73] (Chapter 3). Frosted nocturnal air occasionally descends from it into the sal, killing early regeneration where the canopy has been removed by logging (Chapter 3). At higher altitude, chir is replaced by *P. wallichiana* in warm temperate Himalaya, and at higher latitudes by *P. massoniana* in warm temperate ('subtropical'– see Chapter 3) South China. *Chir* is associated with several species of *Quercus,* which replace it in the absence of fire: *Q. lanata* and *Q. glauca,* in tropical environments; *Q. oblongata* (syn. *Q. leucotrichophora*) (ban oak) in temperate. Toward the upper limits of *chir* forest within the southern ranges of the Himalaya, rain shadow valleys and lower slopes often remain occupied by chir except in the moist steep gullies, while broadleaved forests rich in warm temperate genera and species clothe the slopes above. Toward the front ranges of the Himalaya, *chir* savanna may be sandwiched between lowland semi-evergreen forest and lower montane oak-laurel forest, as for instance above Tingtibi in Trongsa (Bhutan).

3.5. *Other plant life forms and habits*

3.5.1. *Shrubs*
The lower montane forests of Asia are home to one of the most spectacular of subcanopy shrubs: *Strobilanthes* (Acanthaceae) are found particularly in the seasonal tropics, especially in the upper elevations, and some species ascend into upper montane forest (Plate 4.11d). A few species occur in perhumid Sundaland, and some range into low altitudes. With over 200 species, most are monocarpic, like bamboos (Chapter 5). Each species comes into gregarious flower at different intervals of 5–15 years. At these times, whole hillsides may be set ablaze with white, blue or pink according to the species, recalling

mountain laurels in New England. Pollinated by honeybees and bumblebees – whose centre of species richness is in the Himalayas – their copious production of small dry seeds attracts birds and small mammals. As with the bamboos and North American mountain laurel, their dense continuous canopy impedes tree regeneration, confining opportunity for seedling establishment to the brief intervals when *Strobilanthes* populations die following seed-fall. Seedlings of shade-tolerant trees survive beneath the *Strobilanthes,* to emerge above it at the next mortality event. *Strobilanthes* are less prone to fire in nature than many bamboos. Consequently, many such lower montane forests bear a sparse and uniform understorey dominated by one or a few *Strobilanthes* species.

Plate 4.11 The lower montane forest subcanopy. **a,** Forest subcanopy, upper part of lower montane forest, c.1,900 m, Kinabalu; **b,** Lower montane forest reaches high altitudes in the seasonal tropics: D. Mohan Das at Tia Shola, Nilgiries, 2,100 m; **c,** Abundant ground herbs: *Elatostema macintyrei,* 1,700 m, 16 ha plot, Doi Inthanon; **d,** *Strobilanthes capitata,* 865 m, Gelephu, Bhutan; **e,** *Barleria* sp., 900 m, Similipal National Park, Odisha. (**a, b** P.A.; **c–e** H.H.)

3.5.2. *Vines*

There are many vine species in the lower montane forests of continental East Asia, in a range of sizes.[74] Vines decrease in size and abundance with altitude, peaking in richness at the lowland-lower montane forest ecotone in seasonal tropical Asia. A few species extend to the upper limits of forest, and where oak-laurel forests extend to high altitudes, as on Doi Inthanon, some are still large.[75] Vines may be abundant and large in disturbed or secondary lower montane forests at lower altitudes, particularly on mull soils as in the valley in the Doi Inthanon CTFS plot (1,700 m).[76] They are also well represented in continental southeastern Chinese lower montane forest.[77] Bole climbers and epiphytes are diverse, but are rarely abundant in Asian lower montane forest except in disturbed forest and in permanent openings, especially along riverbanks.

3.5.3. Tree ferns

Tree ferns (*Cyathea* and other less species-rich genera) are rare in the Asian lowland tropics but become common in lower montane forest. They continue to be common up into upper montane secondary grassland in New Guinea (3,800 m). Alone among trees, tree ferns absorb water directly through their trunks, reflecting the cool humid conditions of their habitats.

3.5.4. *Ground herbs, epiphytes*

Ground herbs and epiphytes greatly increase in abundance and diversity with altitude,[78] though less so in the more seasonal regions.[79] Epiphytic ferns and orchids, as well as terrestrial ferns and urticaceous herbs, reach maximal species richness above the cloud base,[80] in the lower elevations of lower montane forest.[81]

3.6. *The transition from lowland to lower montane forest: physical correlates*

3.6.1. *Climate*

It may seem extraordinary that the floristic and edaphic transition from lowland to lower montane forest[82] (Chapter 1) is found consistently at 800–1,200 m, whether on Bornean Mount Kinabalu, beyond the geographical tropics in Bhutan at 27° N, on the isolated Bornean coastal peak of Santubong, or on the slopes of the vast massifs of New Guinea. What is the cause, for only during the wet summer monsoon is the range of mean and diurnal temperatures the same at all these latitudes?

Any attempt to establish precise correlations between climatic and forest zonation is bedeviled by poor data, particularly for climate (see Chapter 1), and is further confused by range extensions of the fire-climax deciduous forest types

favoured by human activities. No long-term monitoring of climate at stations along an altitudinal transect exists in Asia. Ohsawa[83] claimed that there is a correlation between forest elevational change on tropical mountains in East Asia and Kira's index of warmth[84] (Fig. 4.9). He showed that this index correlates with altitudinal changes in species diversity, in tree height and in leaf size. Although he suggested that ecotones between forest formation zones coincide with certain measures of Kira's index, he did not suggest mechanisms by which distinct ecotones might occur between zones of more uniform composition. Kira's index declines more rapidly at higher than at lower latitudes in humid tropical climates. Within the tropics, summer temperatures do not decline with latitude, but winter night temperatures do. Nevertheless, the ecotone from lowland to lower montane forest remains remarkably constant at c.750–1,200m, as Ohsawa himself indicated[85] (Fig. 4.10). Thus, temperature cannot be the major determinant of this transition. The ecotone does not consistently coincide with the diurnal cloud base either. In isolated and coastal mountains the cloud base is frequently at this level or lower, but here the upper montane forest descends, truncating the lower (see section 5 below). On major massifs in perhumid equatorial regions the cloud base may be as much as 1,000 m higher, as it is on Kinabalu[86] and, during the dry season, on the South Asian (Himalayan) front ranges and the Indo-Burmese massifs; it is higher still on the larger New Guinea mountains.

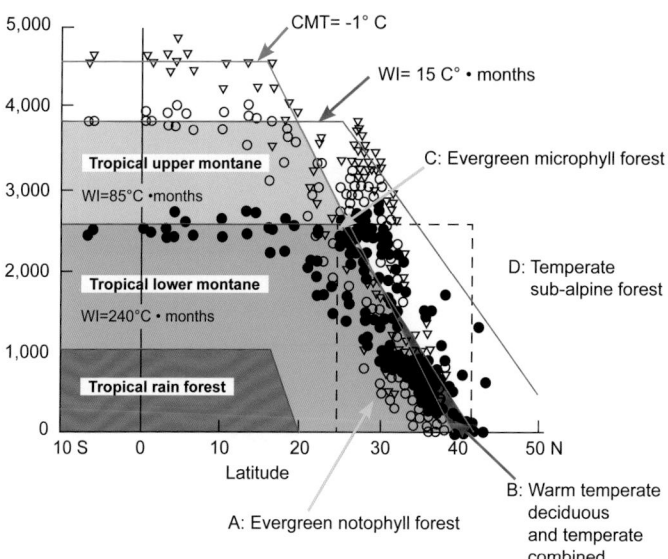

Fig. 4.9 Ohsawa's interpretation of altitudinal and latitudinal forest zonation on East Asian mountains in relation to Kira's index of growth. He combines temperate with tropical lower montane forest in S. China. CMT: coldest month mean temperature; WI: temperature sum (Kira's) index (see text). (From Ohsawa 2006a)

Fig. 4.10 Forest zonation on major mountains in the wet Asian tropics. Note: oblique ecotones indicate coexistence of two forest zones, only indicated on mountains observed by me. Horton Plains includes, for lower altitudes, the surviving sequence on the adjacent eastern Peak Sanctuary.

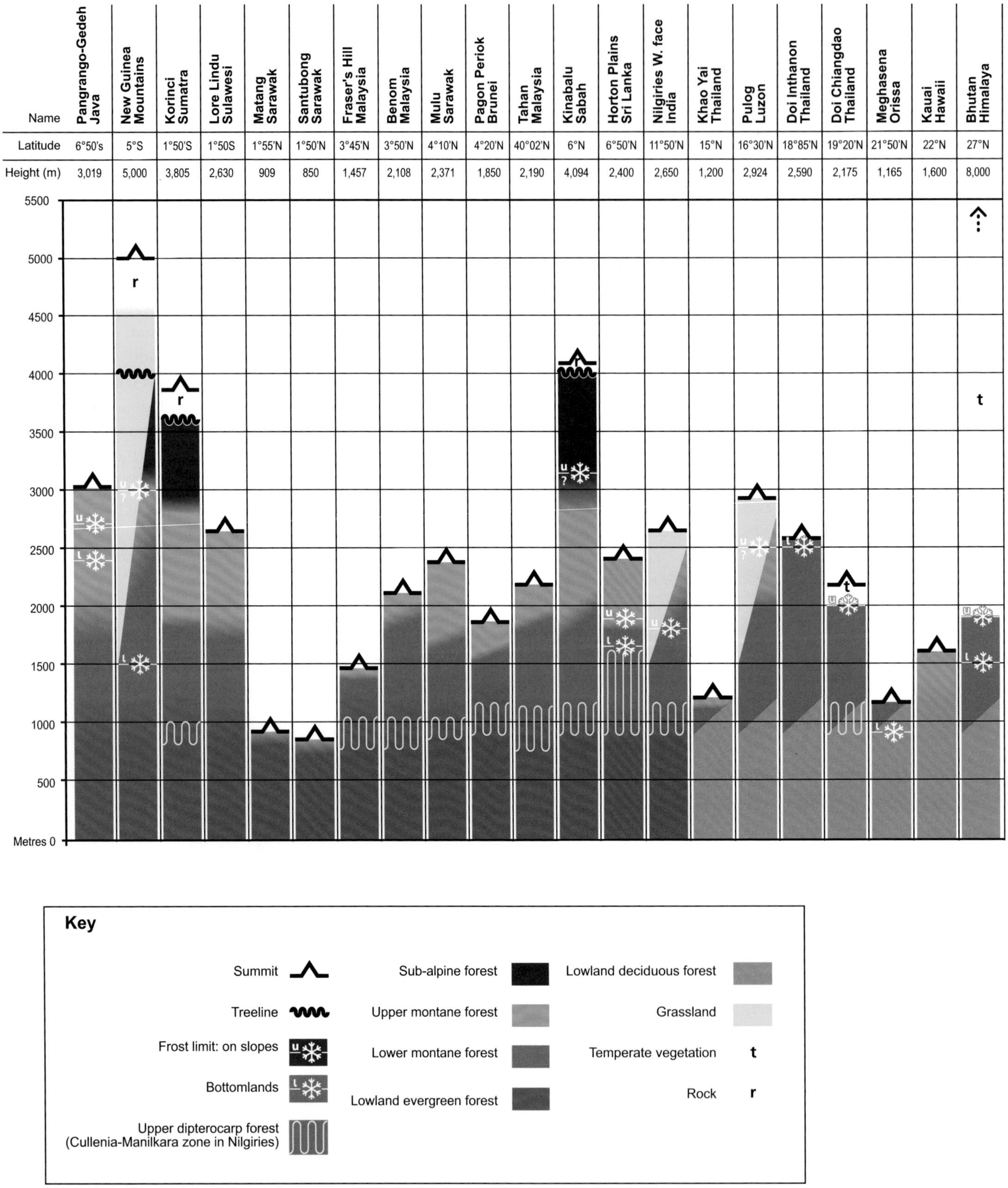

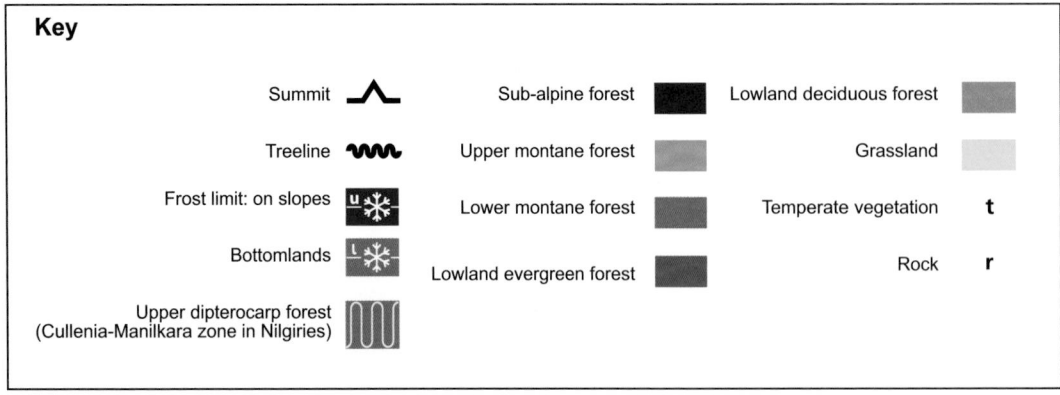

Name	Pangrango-Gedeh Java	New Guinea Mountains	Korinci Sumatra	Lore Lindu Sulawesi	Matang Sarawak	Santubong Sarawak	Fraser's Hill Malaysia	Benom Malaysia	Mulu Sarawak	Pagon Periok Brunei	Tahan Malaysia	Kinabalu Sabah	Horton Plains Sri Lanka	Nilgiries W. face India	Khao Yai Thailand	Pulog Luzon	Doi Inthanon Thailand	Doi Chiangdao Thailand	Meghasena Orissa	Kauai Hawaii	Bhutan Himalaya
Latitude	6°50's	5°S	1°50'S	1°50S	1°55'N	1°50'N	3°45'N	3°50'N	4°10'N	4°20'N	40°02'N	6°N	6°50'N	11°50'N	15°N	16°30'N	18°85'N	19°20'N	21°50'N	22°N	27°N
Height (m)	3,019	5,000	3,805	2,630	909	850	1,457	2,108	2,371	1,850	2,190	4,094	2,400	2,650	1,200	2,924	2,590	2,175	1,165	1,600	8,000

Key

Symbol	Label	Symbol	Label	Symbol	Label
⋀	Summit		Sub-alpine forest		Lowland deciduous forest
〰	Treeline		Upper montane forest		Grassland
u❄	Frost limit: on slopes		Lower montane forest	t	Temperate vegetation
L❄	Bottomlands		Lowland evergreen forest	r	Rock
〰	Upper dipterocarp forest (Cullenia-Manilkara zone in Nilgiries)				

Fog does not regularly penetrate the canopy of lower montane forest during the heat of the day. It becomes more frequent within it at higher altitudes, though, and there correlates with increasing mossiness on trunks and branches (less on the ground). The limit of fire-climax forest at 1,200 m on slopes exposed to the wet monsoon nevertheless implies that in such locations, humidity increases throughout the year at about this altitude. Cloud as fog is generally excluded from the forest interior by the diurnal upwelling of hot humid air following the canopy surface until rain commences (Chapter 1). This hot humid air does eventually cool and condense fog by 2,200 m, at somewhat lower altitudes on isolated mountains, and as low as 750 m on isolated coastal hills (Chapter 1). Massing around higher slopes and peaks, clouds do immerse forests above their c.1,000 m base in shadow, lowering diurnal surface temperatures.

Summer is the wet monsoon season of the seasonal tropics. Then, the cloud base remains at c.1,000 m, descending lower only during the incessant rains between June and October (Chapter 1). Fog (cloud) is driven by prevailing winds into the forest canopy during the monsoon, at night as well as by day. Winter is dry and sunny almost every day, with cloud restricted to high ridges (above at least 2,000 m) and to the afternoon. Winter rain in the mountains is generally lighter than the lowland orographic storm and monsoon rain. Nocturnal fog, condensing out of the humid air as cold dry air descends from higher altitudes, can accumulate in mountain valleys before dawn, bringing copious dew. Frequent cloud notwithstanding, lower montane forests of the seasonal tropics experience strong enough periodic droughts to cause regular juvenile mortality.[87] This may also in part account for the lack of emergents at higher elevations.

As anyone descending to an airport in the humid tropics will be aware, there is marked turbulence at altitudes immediately above the diurnal cloud base. On the mountain slopes above, diurnal winds are accelerated by air coursing convectionally upslope above the canopy. Lower montane forests in seasonal regions facing the Pacific, and warm temperate evergreen oak forests to the north, may experience defoliation and twig pruning during the frequent typhoons, but they appear resistant to wind throw. Elsewhere, lower montane forests receive diurnal upslope breezes increasing in intensity with altitude. They are also exposed to constant trade winds. By increasing the rate of evapotranspiration and elevating the risk of drought-induced mortality among the tallest individuals, these winds may also help explain the absence of emergent broadleaved angiospermous trees except in sheltered sites, and may thus be one cause of the increasingly smooth canopy structure seen in tropical montane forests with increasing elevation and exposure.

During the summer, mean monthly day and night temperatures at the margins of the tropics reach similar levels to those at the Equator. During winter nights, however, temperatures may occasionally descend close to freezing – much lower than equatorial temperatures, which rarely descend at night below 16° C and on average only to 20°C. At their highest latitudes in the Eastern Himalaya, tropical lower montane forests are concentrated on the most shady and moist slopes, often north-facing in rain-shadow valleys.

I conclude, therefore, that the ecotone from lowland into lower montane forest, being universally at the same elevation, is to be caused by an environmental change which is mediated by the one climatic factor common to all latitudes, and from coast inland: equatorial mean weather conditions and those of the wet summer monsoon.

3.6.2. *Soils*

Decline in mean ambient temperature within the shade of the understorey, at a lapse rate of 0.6°C, will lower temperature at 800 m, from that at sea level, by an unremarkable 4.8°C in what is a continuous gradient of decline with altitude. As mean temperature declines, so does the rate of chemical processes, including litter decomposition; but there is no evidence of a consistent overall change in the rate of litterfall. It is at around this altitude that rates of litterfall come into balance with rates of litter decomposition,[88] and udult soils come to resemble their temperate counterparts (Chapter 1 section 4.2.2). Soil change provides the most direct correlation with change in forest species at the lowland-lower montane forest ecotone. Nevertheless, there is also a continuous turnover of species with altitude within the zonal forest types (Fig. 4.1), suggesting that temperature may also have direct effects on species' competitive relationships.

I have often observed, along road cuts, tree roots including fine roots following the humus down through the mineral profile in the mull soils of lower montane oak-laurel forest, just as they do in temperate forests. This pattern contrasts with the concentration of finer roots seen near the surface in lowland tropical yellow/red soils, and in humult leached ultisols and podsols at all altitudes. The nature of changes in forest root systems with altitude awaits quantitative study. The edaphic correlates of such ecotones have not yet been studied in detail, but they cannot by themselves explain the loss of lowland species: they seem not to explain the turnover of species between lowland and upper dipterocarp and lowland and montane *kerangas* forests, where no major soils change is apparent; nor again the altitudinal ranges of *Pinus merkusii, P. kesiya* and their savanna woodland associates.

3.6.3. *Summary*

The lowland to lower montane ecotone is found uniformly at the same altitude, notwithstanding contrasting latitudes and consequent differences in mean annual temperatures and seasonal ranges (although the pattern has been obscured in many places by the upper limits of lowland agriculture, in Asia and throughout the tropics[89]). The transition marks a significant floristic and structural change, usually gradual but occurring consistently at 750–1,200 m. This ecotone represents the upper limit of the paradigmatic lowland mixed evergreen rainforest species assemblage of wet climates, although many lowland genera extend above 1,200 m. It is consistently correlated solely, and then only on loam-bearing substrates, with the change from lowland yellow-red soils, low in visible organic matter in mineral horizons, to humic mull soils in which the mineral horizons are pervasively darkened by humus. Lower montane *kerangas* differs little, other than in its stature, from its lowland analogue. It does include some species characteristic of these elevations and rarely found below 700 m, but no genera occur which are not also at least occasionally recorded in the lowlands. Little change is apparent in its humult podsolic soils, although they become more pervasive on the more stable convex surfaces of ridges with altitude. Floristically and structurally distinct lower montane oak-laurel forests may often interdigitate with upper dipterocarp forest or lower montane *kerangas* on the same slope. In seasonal climates, where deciduous fire-sustained forests dominate in the lowlands, the ecotone to lower montane forest is often sudden and dynamic, occurring towards or somewhat above the upper end of the lowland-lower montane evergreen forest ecotone. On freely draining and south-facing slopes in climates with six or more dry months, fire-climax lower montane *Pinus kesiya* or *Casuarina junghuhniana* forest, each over the same altitudinal range, replaces closed evergreen lower montane forest.

3.7. *Site-related floristic variation within lower montane communities*

Lower montane oak-laurel forest soils are erodible, varying greatly in depth with topography. Species composition has been shown to covary with topography both on Kinabalu,[90] and Doi Inthanon.[91] On Kinabalu, raw humus and associated upper montane forest species first appear on sedimentary rock ridges, sometimes as low as 1,500 m.[92] Takyu, Aiba and Kitayama[93] found that forest structure and composition varied most in relation to topography in oak-laurel forest on Pleistocene sediments, and least on ultramafic substrate where the densely rooted raw humus soil surface horizon resists surface erosion. This substrate supports a floristically distinct form of lower montane *kerangas*.

3.8. *Variation in tree heights and leaf sizes in lower montane forest*

Maximum heights of lower montane forest stands are generally uniform (and comparable to *kerangas*), with notophyll and microphyll leaf sizes predominating (Table 4.5; Plate 4.5). Why should this be so?

In Bhutan, deciduous as well as evergreen species of both tropical lower montane and warm temperate forests are predominantly notophyll. That implies that it is the wet summer climate, when rainfall is high and temperatures frost-free, that most influences leaf size. Universally, emergent crowns are pruned on their summer windward side. At that time, shoot expansion, which starts before the rains, is still continuing. Upwelling convectional air is humid, but its increasing velocity with altitude correlates with increasing canopy smoothness and crown shyness (Plate 4.3d). Although prevailing trade winds do occur at lower montane forest altitudes in the seasonal tropics, they only do so at the geographical margins of perhumid Sundaland. Increased soil organic matter in the mineral horizon of montane soils nevertheless increases soil water retention without reducing aeration. The desiccating effect of the wind, prevailing or orographic, may sometimes result in dominance of canopy species with microphyll leaf sizes. *The water relations in these forest canopies merit research.*

4. Upper montane forests

The transition from lower montane to upper montane forest differs from the transition from lowland evergreen to lower montane forest in two key respects: (1) It is more often distinct on the ground owing to a narrow lower ecotone, particularly in rugged topography in the perhumid tropics where it first appears along the crests; and (2) it varies greatly in altitude, both on different mountains and on different sites on the same mountain.

Upper montane forest appears on spurs and ridges first. Aiba, Takyu and Kitayama[94] found that its microphyll leaf sizes already dominated on ridges in their 1,700 m plot on Kinabalu, and leaf thickness (measured as leaf mass per unit area – which does not take account of air spaces) was also higher than on nearby slopes. The distinct upper montane tree flora was well represented. Upper montane forest, whether structural-physiognomically or floristically defined, becomes ubiquitous on Kinabalu by 2,200 m,[95] but some elements appear at lower altitudes on isolated and coastal mountains (Fig. 4.10). This is the tropical version of the 'mass-elevation effect', or Massenerhebungseffekt (see section 4.3.1 below).[96] The consequent interdigitation of lower and upper montane forests, in valleys and on ridges respectively, provides one explanation for the failure of documentation of species' altitudinal ranges alone to identify clear ecotones above c.1,500 m on Mount Kinabalu (Fig. 4.1).[97]

4.1. *General characteristics*

Upper montane forest at its greatest stage of development is recognized by its low stature and its single-canopy tree stratum, due to the decline and eventual disappearance of the woody subcanopy of trees or subordinate shrubs (Fig. 4.11; Plates 4.12, 4.13). Leaf-related characteristics of this forest type include dominance of microphyll and smaller leaf sizes (Tables 4.6, 4.8), leaves often scleromorphous (that is, fibrous and thick when fresh) and generally lacking drip tips, and an abundance of canopy species with brilliant red young leaves. Among the woody flora, the absence of clear boles and presence of unbuttressed boles contorted from the base, the twigs with short internodes and the dense twiggy sympodially branching crowns bearing a dense layer of small leaves – in part occasioned by frequent bud and shoot abortion – are typical structural characters. Pinnate leaves are generally uncommon in upper montane forests. However, the mostly pinnate Cunoniaceae are locally abundant, though rare at lower altitudes west of Wallace's Line. Pinnate genera of Rutaceae and *Turpinia* may also be well represented, especially in South Asia.

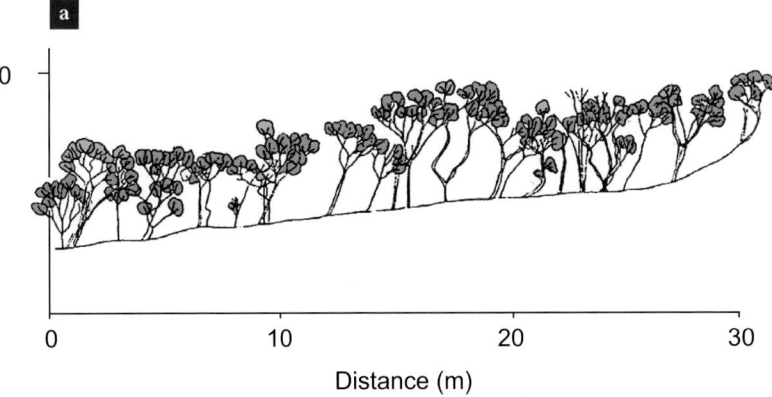

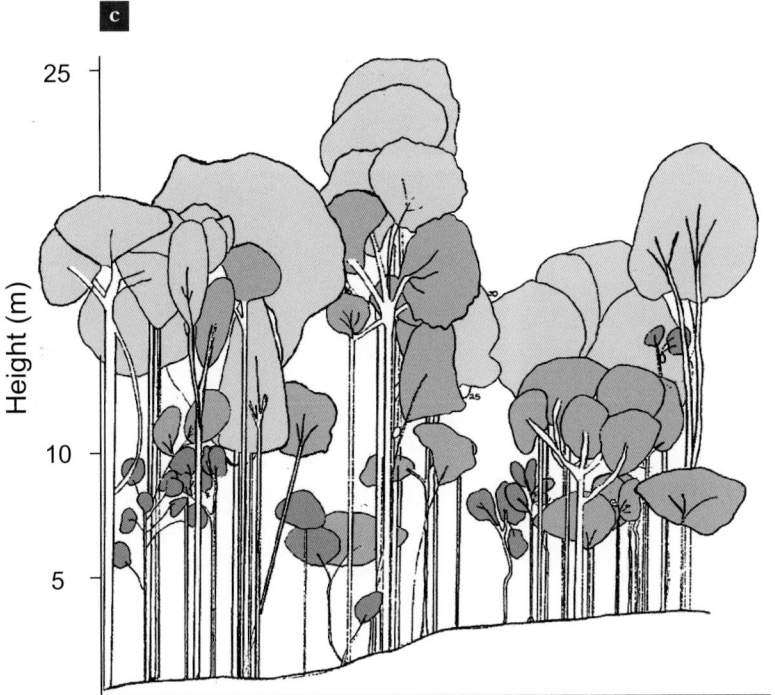

Fig. 4.11 Profiles of montane forest types on Kinabalu: **a**, lower montane forest; **b**, upper montane forest; **c**, subalpine woodland. (From K. Abu Salim, unpubl.)

Plate 4.12 Upper montane forest canopies. **a,** Gunung Mulu, Sarawak, c.2,200 m; **b,** Horton Plains, Sri Lanka, 2,100 m. *Elaeocarpus glandulifer* with flat crowns; **c,** Nilgiries, Western Ghats of Tamil Nadu, India, c.1,900 m; leaves in red flush; **d,** Upper montane sholas, on west-facing, cloudy slopes, Bangtipal, Nilgiries, 2,100 m. Grazier-induced fire now restricts most to steep moist gullies, the grassland experiencing periodic ground frost; **e,** *Usnea* festoons proclaim periodic drought, 2,100 m, Lover's Leap, Horton Plains; **f,** The ferns *Matonia pectinata* and *Dipteris conjugata* (lower left and right) in a gap, 1,800 m, Bukit Retak, Brunei, April 1958. (**a, b, e** H.H.; **c** K. Bawa; **d, f** P.A.)

Table 4.6 Leaf size distributions on some tropical Asian mountains: **a**, Kinabalu (from Kitasyama & Aiba 2002); **b**, other Asian mountains; in percentages (basal area percentages at Doi Inthanon in parentheses).

a. Kinabalu (source: Kitayama & Aiba 2002) 6.2 °N

	n	Compound	Lepto	Nano	Micro	Noto	Macro Meso	Macro
3,100 m:								
Other	24	0	8.3	12.5	54.2	16.7	8.3	0
Ultramafic	9	0	33.3	11.1	44.4	11.1	0	0
2,700 m								
Other	23	0	0	8.7	60.9	26.1	4.3	0
Ultramafic	26	3.8	11.5	11.5	65.4	7.7	3.8	0
1,700 m								
Other	121	6.7	1.7	1.7	36.4	36.4	23.1	12.1
Ultramafic	43	4.8	4.7	11.6	48.8	20.9	11.6	2.3
700 m								
Other	163	18	0.6	0	14.7	33.7	45.4	3.7
Ultramafic	161	14	0	1.9	23.6	41.6	28.6	2.5

b. Continental Asia (personal data, Bhutan excepted)

	n	Compound	Lepto	Nano	Micro	Noto	Macro Meso	Macro
Horton Plains, c.2,100 m, 6.8 °N	51	2.0	2.0	7.8	64.7	19.6	5.9	0
Nilgiris, 2,200 m, 11.5 °N:								
Bungtipal, facing W.	28	3.6	0	10.7	53.6	21.4	10.7	0
Tia shola, facing E.	37	0	0	2.7	43.2	43.2	10.8	0
Doi Inthanon, 2,200-2,320 m, 18.6 °N	34 (b.a.157m²)	(0)	0 (0)	0 (0)	21 (b.a.19)	48 (b.a.50)	31 (b.a.31)	0 (0)
Bhutan (source, the Flora), 27 °N								
Min. elev. 2,200-2,400 m	77	11.7	1.29	0	26.0	36.4	24.7	0
Max. elev. 1,600-1,800 m	120	16.7	0.08	1.7	14.2	25.8	35.8	5

Upper montane forest is further defined: by a general absence of woody vines, though herbaceous bole climbers may persist; by an abundance, but relative floristic poverty, of vascular epiphytes; and by a diverse and rich cryptogamic flora, especially of mosses, filmy ferns and liverworts. But forests with the structure and physiognomy of upper montane forests differ much in their density of epiphytes, both vascular and otherwise. Upper montane forest is also characterised by a distinct flora, albeit impoverished.[98] (Each of these characteristics will be discussed in greater detail in the sections below.)

4.1.1. *The vascular flora*

The vascular flora of upper montane forest is relatively impoverished, even on great mountains. Many lowland tree families are generally absent (Table 4.7a). It is characterised by a restricted number of those families and genera of Sunda lowland affinity, generally with bird-dispersed seeds.[99] These taxa are almost universally present, many of them represented by local or regional endemics but few of them species-rich (Table 4.7b). On Kinabalu, The Beamans and Anderson[100] recorded only 232 woody dicotyledonous species above the limit (at 2,200 m) of lower montane forest. This number declines with increasing altitude, as does the number of new species appearing (Fig. 4.1). At 3,100 m, they recorded only 73 tree species. Aiba and Kitayama[101] found that decline of species richness correlated with decline in forest stature. The most abundant tree and shrub families in upper montane forest on Kinabalu include Aquifoliaceae (4 species), Elaeocarpaceae (3 species), Ericaceae (c.20 species), Melastomaceae (c.10 species), Myrtaceae (c.15 species) and Podocarpaceae (c.10 species). There are only seven Fagaceae, all but one of which is also found in lower montane forest, and nine Lauraceae, all of which also occur at lower elevations. The same families prevail in New Guinean upper montane forests mountains,

a

Genus	Family	Borneo	Sunda	Indo-Burma	S. China	Temp. China	S. Asia
Ilex	Aquifoliaceae	+	+	+	+	+	+
Dendropanax	Araliaceae		+	+	m		
Schefflera	Araliaceae	+	+	+	+	+	+
Vernonia	Asteraceae		+	+	+	+	+
Viburnum	Caprifoliaceae		+	+	+	+	+
Euonymus	Celastraceae		+	+	+	+	+
Microtropis	Celastraceae		+	+	+	+	+
*Ascarina**	Chloranthaceae	+					
Clethra	Clethraceae		+		+		
*Weinmannia**	Cunoniaceae	+	+				
Daphniphyllum	Daphnyphyllaceae	+	+	+	+	+	+
Elaeocarpus	Elaeocarpaceae	+	+	+	+	+	+
*Styphelia**	Epacridaceae	+	+				
*Trochocarpa**	Ericaceae	+					
Diplicosia	Ericaceae	+	+				
Gaultheria	Ericaceae	+	+		+	+	+
Rhododendron (Rhododendron)	Ericaceae			+	+	+	+
Rhododendron (Schistanthe)	Ericaceae	+	+	+	m		
Vaccinium	Ericaceae	+	+	+	+	+	**+**
Polyosma	Escalloniaceae	+	+	+			
*Scaevola**	Goodeniaceae	+					
Platea	Icacinaceae	+	+	+			
Actinodaphne	Lauraceae	+	+	+	+	+	+
Lindera	Lauraceae	+	+	+	+		
Neolitsea	Lauraceae	+	+	+	+	+	+
Myrica	Myricaceae	+	+	+	+		
Memecylon	Myrsinaceae	+	+	+		+	+
Ardisia	Primulaceae	+	+	+	+	+	+
Embelia	Primulaceae	+	+	+		+	+
Maesa	Primulaceae	+	+	+		+	+
Myrsine	Primulaceae		+	+		+	
Rapanea	Primulaceae	+	+	+	+	+	+
Decaspermum	Myrtaceae	+	+	+	m		
Syzygium	Myrtaceae	+	+	+	+	+	+
*Leptospermum**	Myrtaceae	+	+				
*Xanthomyrtus**	Myrtaceae	+					
Helicia	Proteaceae		+	+	m	+	+
Photinia	Rosaceae	+	+	+	+	+	+
Prunus	Rosaceae	+	+	+	+	+	+
Hedyotis	Rubiaceae	+	+	+	m	+	+

a

Genus	Family	Borneo	Sunda	Indo-Burma	S. China	Temp. China	S. Asia
Lasianthus	Rubiaceae		+	+	m	+	+
Psychotria	Rubiaceae	+	+	+	m	+	+
Melicope	Rutaceae	+	+	+		+	+
Brucea	Simarubiaceae		+	+	m		
Turpinia	Staphyliaceae		+	+	+	+	+
Symplocos	Symplocaceae	+	+	+	+	+	+
Adinandra	Theaceae	+	+	+	+	+	
Eurya	Pentaphylacaceae	+	+	+	+	+	+
Schima	Theaceae	+		+	+		
Ternstroemia	Pentaphylacaceae	+	+	+	+	+	+
*Kelleria**	Thymeliaceae	+					
*Tasmannia**	Winteraceae	+					

Table 4.7 a, Geographical ranges of the major genera in upper montane forests in tropical Asia. m: present only in extreme southern China, possibly in frost-free pockets. Asterisked taxa of Australasian affinity. **b,** Angiospermous tropical tree families generally absent from Asian tropical upper montane forests.

b

Anacardiaceae	Dipterocarpaceae
Annonaceae	Ebenaceae
Apocynaceae	Euphorbiaceae
Bignoniaceae	Hamamelidaceae
Bombacaceae	Meliaceae
Burseraceae	Myristicaceae
Chrysobalanaceae	Polygalaceae
Clusiaceae	Sapindaceae
Cornaceae	Sapotaceae*
Dilleniaceae	Sterculiaceae

except Fagaceae which are absent and Melastomaceae which are poorly represented.[102] Some mountain masses in slightly seasonal regions may support genera and families not otherwise represented in upper montane forests. These include *Calophyllum* (Clusiaceae) on the Horton Plains, Sri Lanka and *Xantolis* (Sapotaceae) in the Western Ghats.

Taxa of southern, Australasian Gondwanan, origin are particularly well represented in Far Eastern, but almost absent from South Asian upper montane forests (Chapter 6). These include Podocarpaceae, Winteraceae, Cunoniaceae, *Scaevola* (Goodeniaceae), *Styphelia* and *Trochocarpa* (Epacridaceae), *Ascarina* (Chloranthaceae), and *Xanthomyrtus, Leptospermum, Seorsia,* and an exceptional number of *Syzygium* (Myrtaceae) which extend west of Wallace's Line (*Syzygium* to South Asia).

Only one angiosperm genus appears to be entirely restricted to upper montane forest on Asian equatorial mountains (though even it occasionally descends into taller forest immediately below on acid humult soils). That is *Tasmannia,* although *Rhododendron* (subgenus *Schistanthe*: 'Vireya') is particularly diverse and abundant there. Both are absent from South Asia. *Tasmannia* is a sister genus to the Neotropical *Drimys* and its allies in the vessel-less southern family Winteraceae, all of which are canopy trees and shrubs. The stomata of these trees are obstructed by plugs which, rather than reducing transpiration, function to block entry of water condensing from fog.[103] Several podocarp genera may also be thus restricted. Further ecophysiological studies of families thus restricted, such as the Winteraceae, are merited. (Two winteraceous subcanopy shrubs in New Guinea, *Bubbia* and *Belliolum,* do descend to low altitude in East Malesia, mostly apparently on humic podsols and ultramafic-humult soils in very wet foothill valleys.[104])

Steep slopes, especially in earthquake-prone New Guinea, occasion extensive landslips where soils are removed or truncated and organic horizons thin or removed. Early succession on landslips is dominated by a characteristic yet little-studied succession and flora, including many herbaceous and shrubby genera, in e.g. Rubiaceae and Acanthaceae. On Kinabalu, such disturbance gives rise to persistent thickets of scandent bamboos, especially *Racemobambos gibbsiae,* recalling *Chusquea* in the high Andes. The ferns *Gleichenia dicarpa* and *Dipteris conjugata* (which also occurs on moist cliffs in the lowlands) replace the *Dicranopteris* thickets at lower altitudes. A distinct *Gleichenia-Dicranopteris* zone at 2,400–2,900 m occurs on Gunung Kerinci in Sumatra's Barisan Range. The frequency of landslips in the precipitous and earthquake-prone landscape of New Guinea appears to explain the diversity of pioneer montane *Macaranga* there which, though mostly mesophyllous, bear smaller than average leaves for the genus.[105]

4.1.2. *Epiphytes*

Epiphytes may be more abundant in upper montane forest than in lower montane forests, but Wood, Parris and their colleagues[106] showed that species diversity of both epiphytic orchids and pteridophytes reach their peak at c.1,600 m in the lower to middle zones of lower montane forest on Kinabalu. Fernando and Ranasinghe,[107] found that of the 172 species of orchids (named and unnamed) in Sri Lanka, 75 species occur in lowland MDF, 126 in lower montane *Shorea gardneri* forest, and only 62 in upper montane forest, though orchid total abundance does not decline.

Cryptogamic epiphytes, particularly mosses and liverworts, become sparser at higher altitudes, and on mountains experiencing a dry season. There, lacking moss tussocks, they become confined within the canopy and eventually to the ground

4.2. *The problem of defining upper montane* **forest**

The defining attributes of upper montane forest are not universally correlated. This confronts us with a problem of definition, but also affords us the opportunity to establish separate environmental correlates with some attributes.

4.2.1. *Structure and physiognomy*

Upper montane forest is often popularly called 'elfin woodland'. On Kinabalu and other large massifs, forest up to 18 m tall and with the physiognomic characteristics and absence of dense epiphytic cryptogams associated with lower montane forest, but with more crooked boles and a mixture of lower and upper montane genera and species, occurs on shallow acid organic soils over bedrock to 3,000 m. In New Guinea,

similar forests up to 30 m tall are widespread below c.2,800 m altitude, and are often dominated by *Nothofagus* species (see discussion under section 3.2 above). Upper montane forests vary in stature as much as other tropical forests, therefore. On account of their frequently tall stature, upper montane forest is often regarded as starting at higher altitudes in New Guinea than elsewhere, even as high as 2,800–3,000 m.[108] However, subalpine forest may also descend to that altitude[109] and frosts frequently reach below 2,500 m and pond in valleys where only grassland and a few shrub species survive it. Frost during ENSO events may apparently occasion mortality among taller trees on the larger massifs (see sections 5.3.1 and 6.2.3 below).[110]

Johns[111] treated this upper *Nothofagus* forest as a distinct mid-montane zone, and observed that its lower limit coincides with the upper limits of cultivation at about 1,800 m on the warmer northern, 1,500 m on the cooler southern slopes. Grubb and Stevens[112] documented the presence, in this 'mid-montane' zone, of genera typical of lower montane forest in New Guinea, as well as a group of genera mostly restricted to upper montane forests elsewhere. (These include Winteraceae and some Podocarpaceae.) Tree stature is somewhat shorter than in lower montane forest at lower elevations. Lianas of all kinds are fewer and epiphytes more abundant. Structurally and physiognomically, however, the mid-montane forest appears to be a facies of lower montane forest, although microphyll species are better represented than at lower altitudes.[113] Whereas Grubb and Stevens imply that this upper facies of lower montane forest is beneath the diurnal cloud base, Johns[114] observed increasingly dense mossiness with altitude, in part owing to intermittent fog. On the Sumatran volcano Gunung Kerinci, Jacobs and Ohsawa[115] describe additional less clear zones between 'typical' lower and upper montane forest, while Laumonier[116] considers only one ecotone, at 1,800–1,900 m, to indicate a dramatic structural and floristic change. But that one sharp ecotone is exceptional for a volcano, and may coincide with a change in substrate or history of volcanic activity. The lack of clear ecotones carefully documented by Isamu Yamada on Gunung Pangrango (West Java)[117] reflects the continuity of slope and topography characteristic of young volcanoes, consistent with observations in Central America.[118] At the higher altitudes of upper montane forest on Kinabalu and other Borneo mountains, forest in moist swales reaches 15 m in height, whereas on nearby ridges it hardly exceeds 5 m, and may often be reduced to a c.2 m dense scrub.[119] Here, muck soils in the swales replace the coarse sandy organic soils of drier sites (Chapter 1).

In both South and East Asia, the overall smooth upper montane and subalpine forest canopy is broken by one or more taller-boled species which bear their mature crowns, albeit similar in leaf size and structure to other canopy species, distinctly above them. These include *Agathis*, Podocarpaceae, *Dacrydium pectinatum*, and *Leptospermum recurvum* and *L. javanicum* on Kinabalu, and *Calophyllum walkeri* (Clusiaceae) on the Horton Plains (Sri Lanka) (Fig. 4.11b; Plates 4.13d, 4.15e).

Montane forest on low isolated, especially coastal summits, though it may be gnarled and relatively short in stature, is structurally, physiognomically and floristically intermediate between upper montane forest and *kerangas*. It may occur directly above lowland forest, thereby truncating zonation, as at 850 m on Santubong. Mount Mariveles on Luzon's west coast, for example, experiences particularly strong monsoon winds. Here, stunted forest, including the upper montane elements *Eriobotrya, Leptospermum, Ligustrum* and *Symplocos,* is found down to 900 m.[120] Similarly, on the southern flanks of Adam's Peak, at the western end of the Peak Range in Sri Lanka's perhumid south-west, and other mountains directly exposed to the south-west monsoon there, and again on the isolated Knuckles range to the north, a short microphyll woodland descends to 1,500 m (section 3.3.2 above). But it differs floristically from upper montane forest there in the presence of characteristic species of genera occurring in lower montane forest, notably *Stemonoporus gardneri* and *Palaquium rubiginosum,* and *Stemonoporus affinis* on Knuckles. These woodlands are not notably mossy or rich in epiphytes.

The canopy of upper montane forest is diffuse, bearing mostly microphyll and smaller leaves (Tables 4.6, 4.8), inclined upwards or sometimes hanging. Catastrophic canopy opening such as landslips are infrequent, probably owing to the densely intertwined roots, and single tree mortality seems to prevail.

4.2.2. Upper montane forest and kerangas: similarities and differences

Whitmore[121] described how lowland *kerangas* shares several attributes of upper montane forest, including smooth canopy, sclerophylly, small leaf size (although in *kerangas* dominated by notophylls, thus differing from lowland mixed forest in which mesophyll leaf sizes are generally dominant) and the absence of drip tips. Buttresses are also small or absent in *kerangas*, large vines are rare, and epiphytes more abundant but fewer in species than in mixed lowland forests.[122] Some species are also shared.

However, trees in *kerangas* are straight-boled, internodes

Leaf size class	Upper montane 2,500–2,799 m (n=63)	Subalpine 3,400–3,699 m (n=66)
Mesophyll	9.5	1.5
Notophyll	22	27
Microphyll	57	35
Nanophyll	8	18
Leptophyll (including gymnosperms)	3.2	17
Cupped leaves (excluding leptophyll)	45	76
Shiny leaves (excluding leptophyll)	32	57

(percentages)

Table 4.8 Percentages of leaf characteristics of upper montane and subalpine forest species compared: Kinabalu (species from Beaman et al., 1998a, 2001, 2004)

are longer, bud and shoot abortion infrequent, and mosses, though locally abundant in 'premontane' habitats, do not form tall hummocks or sleeves along branches.

The similarities between *kerangas* and upper montane forests are boldly evident, however, only in sandstone regions with extensive lower montane *kerangas* forests, such as on Borneo and in the coastal mountains of Cambodia. Particularly on Borneo, *kerangas* occurs at all altitudes and is frequently continuous along ridges where it extends upwards as floristically distinct lower montane communities. There, it gradually takes on many of the features attributed to upper montane forest as altitude increases, as for instance on Gunung Dulit.[123] Kitayama's team[124] found that the structural-physiognomic ecotone between lower and upper montane forest had already been crossed by 1,700 m in a plot on ultramafic substrates where soils are deeply organic and humult, but not in another on sediments where the soils were not. Kitayama's plots were all sited on gently sloping land that was therefore free of landslips. They found the upper montane species *Lithocarpus rigidus, Quercus lowii, Ilex oppositifolia, Weinmannia fraxinea, Agathis kinabaluensis* and *Podocarpus gibbsii* at this altitude.

The upper montane forest species *Dacrydium pectinatum* also occurred in their other plot at 1,700 m in lower montane oak–laurel forest on sandstone, where it is most abundant in valleys.[125] The upper montane species *Phyllocladus hypophylla* and *Myrica javanica* were found in both ultramafic and sandstone plots at 1,700 m. Seven of the 15 most abundant species of the two plots at 1,700 m were shared between them, whereas only two of the ultramafic plot species, and five in the sandstone plot, occurred in their plots in upper montane at 2,700 m and subalpine forest and 3,100 m. The genera represented on the ultramafic site at 1,700 m occur elsewhere in both lower montane *kerangas* and upper montane forest. Thus, the lower ecotone of upper montane forest can be obscured by continuity of acid organic soils between lower and upper montane forests on ultramafic and sandstone substrates, and the gradual accumulation of upper montane attributes at increasing altitude within lower montane *kerangas*.

Short-statured forest with the twisted boles typical of upper montane forest, though with less twiggy crowns and few epiphytes vascular or otherwise, is found as low as 150 m on sandstone and limestone bluffs throughout perhumid regions.

4.2.3. Mossiness

Upper montane forest is often regarded as synonymous with montane *mossy* forest, and is also often called 'cloud forest', especially in the Neotropics.[126] Pervasive mossiness, particularly continuous bryophyte cover on soil surfaces free of litter, does appear to correlate with the typical upper montane tree flora of the equatorial tropics, floristically defined. Also correlated is the presence of abundant vascular epiphytes and bryophyte sleeves growing over a layer of fibrous, usually soggy, dead, fine organic matter on branches. On the mountains of Borneo, extreme mossiness is found on sheltered slopes and lesser ridges, while the short scrub at the summits is less mossy. In such continuously foggy sites, mosses form characteristic mounds around the bases of trees and extend along the main branches as sleeves resembling pipe insulation lagging. In extreme cases, these associations may extend from the soil, where they are broken only by wild boar tracks (Plate 4.13b), growing up trunks and into the branches so densely that the visitor cannot easily determine whether he is standing on the ground or in the canopy! This 'mossiness', characteristically, is rich in liverworts and filmy ferns. Examples of this include the Pagon range,[127] Ulu Limbang (Sarawak, c.1,600 m), and some ultramafic exposures on Kinabalu. The moss mounds themselves provide a unique substrate in mossy upper montane forest. On the upper slopes of the wettest Bornean mountains, as on Gunung Pagon Periok (Brunei, 1,950 m), the epiphytic and terrestrial niches thereby become indistinguishable; rhododendrons can establish on tree branches, epiphytic orchids on the ground.[128] On Pagon Periok,[129] on many New Guinea mountains and even on Kauai in Hawaii, a suite of larger tree species have boles that arise from the soil surface, but many smaller woody species initially establish on the moss mounds that surround their bases.

Dense tussock moss occurs at 2,200–2,700 m on the high equatorial mountains of New Guinea and Kinabalu, typically at 1,400–2,300 m on the lower mountains predominant in Borneo and Peninsular Malaysia, and lowest on isolated coastal peaks including 900 m on Kauai (Hawaiian Islands,

22° N). Mossy 'elfin' forests also occur in oceanic temperate climates such as the western flanks of New Zealand's South Island (42–45° S) and at 2,600–3,500 m from the eastern Himalaya to Yunnan, where this 'mid-mountain wet evergreen forest' with successional temperate deciduous species of the temperate broadleaved genera *Betula, Alnus* and *Acer* experience more than a month of frost annually, but less than a week of snow. In mossy forests at temperate latitudes, diurnal cloud and fog is pervasive during the drier seasons. In the eastern Himalaya, during the dry winter season diurnal cloud concentrates in the temperate forests, above 2,000 m; but during the summer monsoon the diurnal cloud base is at c.1,000 m prior to rain, during and after which it drops further as always (see section 3.6.1 above, and Chapter 1).

Dense moss is not restricted to upper montane forests in the tropics. Thick blankets and tussocks of moss can also be found in narrow valleys with whitewater rivers and falls emerging from higher mountains, as in New Guinea and Borneo. There, the air is daily saturated with water droplets, and cloud mists settle at night. Tussocks also occur in some woods along the eastern coast of Luzon, exposed to orographic rain and the incoming typhoons. Though the tree flora of these lowland mossy forests differs little from adjacent forests at the same altitude, their herbaceous ground flora is often dense and rich, especially in New Guinea.[130] Extreme bryophyte cover occurs only where there is good sunlight penetration beneath the canopy.

Mossiness and elfin structure are not consistently correlated, however. On coastal mountains whose summits are frequently shrouded in fog, bryophyte density is reduced and patchy on both ground and vegetation. Borneo examples include Santubong in Sarawak (850 m), Gunung Palung in Kalimantan Barat (1,100 m), and Gunung Silam on Sabah (840 m). On Santubong, where diurnal onshore breezes may blow cloud inland from the peak (Plate 1.8), the floristically impoverished summit forest includes the upper montane fern *Matonia pectinata,* and *Rhododendron jasminiflorum* (in the mostly East Malesian subsect. *Solenovireya),* but the mosses are dominated by drought-resistant *Leucobryum* and, around an intermittent summit pool, *Sphagnum.*

Sri Lanka's Horton Plains (7°30' N) carry forest fulfilling all the classical upper montane characteristics except that it retains a shrubby understorey (Plates 4.12b,e, 4.13c). At 10°–11°30' N on the high tops of the Western Ghats, however, mossiness has declined to a patchy cover. Structural-physiognomic upper montane forests in the seasonal climates of the Western Ghats[131] and Mount Pulog (16°58' N),[132] both of which experience a frequently cloudless annual dry season, are only sparsely mossy. The soils may lack raw humus and resemble those of lower montane forest. This is discussed further under sections 7.3.1 and 7.2 below.

On Kinabalu, mossiness declines gradually above c.2,900 m. Above 3,000 m, trailing strands of the lichen *Usnea* become increasingly evident, indicating periodic incidence of drought. The West Javan volcanoes become drier eastwards as they become increasingly sheltered from the wet south-west monsoon by other volcanoes to the west (Fig. 4.10). Under the influence of Australia to the south and east, *Usnea* likewise becomes abundant on the drier Javanese summits and is more abundant high in the crowns in moister climates there.[133] *Usnea* festoons the upper montane canopy on the Horton Plains (Sri Lanka) (Plate 4.12e), especially along its eastern scarp, which faces rainfall seasonality in the lowlands below. It becomes ubiquitous on the summits of the Western Ghats, and the lower montane forest at the summit of Doi Inthanon is similarly adorned and is also thinly mossy.

Mossiness on mountains does seem to be associated with the accumulation of surface raw humus and associated surface fine roots on clay-rich soils, but moss does not always cover a raw humus horizon.

There does appear to be some correlation between tree species composition and mossiness. Nevertheless, the same core group of tree species occurs at a given altitude on Kinabalu and on the West Javan mountains, whether moss cover is dense or sparse. *The possibility that the mossy soil surface may act as a 'regeneration niche' favouring establishment of these tree species merits research.*

4.2.4. *Species composition*

Forest with the composition and other characteristics of upper montane forest (including leaf size spectrum) may reach 18 m in stature, as already noted. At its lower elevations on Kinabalu for instance, the dense single-stratum canopy is mostly raised well above a distinct understorey of treelets and

Plate 4.13 Upper montane forests. **a,** Gunung Kerinci, Sumatra, 3,805 m, from tea estates at 1,700 m. The sudden change from lower montane to upper montane canopy may be associated with substrate change; the 'Gleichenia zone' is recognisable as pale patches immediately below it; **b,** Boar run in densely mossy elfin upper montane forest, Gunung Mulu, c.2,300 m. (**a** Google; **b** H.H.)

Plate 4.13 Upper montane forests. **c,** Upper montane forest subcanopy on lea slope, Horton Plains: mosses almost confined to sleeves on lower trunks; abundant *Ochlandra* bamboo; **d,** Emergent *Calophyllum walkeri,* 2,100 m, Horton Plains; **e,** Tall upper montane forest, 2,500 m, Mount Albert Edward, New Guinea: abundant moss festoons; **f,** Upper montane secondary grassland with fire-tolerant *Cyathea* fern. Mount Albert Edward, New Guinea, c.2,700 m. (**c–d** H.H.; **e** P.A.)

shrubs, including many Rubiaceae (*Hedyotis, Psychotria* etc.), Chloranthaceae (*Chloranthus*). In more seasonal South Asia, monodominant stands of the hapaxanthic shrub *Strobilanthes* may similarly occur.[134] Many of these species can descend across the ecotone into the upper elevations of lower montane forest, especially in the seasonal tropics, where they are another of several factors that obscure the distinction.

For example, the Horton Plains, at 2,000–2,400 m on the eastern scarp of Sri Lanka's central massif (6°50' N), experience cloudlessness during the south-west monsoon; yet they sustain upper montane forest according to all major criteria. Moss cover is patchy, however, and dense moss sleeves on branches are uncommon. South Asian upper montane forest lacks the characteristic Australasian element of the Sunda peaks, also *Myrica,* and is therefore less clearly defined floristically. Further north, elfin woodland structure persists on the summits and west-facing upper slopes of the Western Ghats, but bryophyte cover becomes sparse, and we find a distinct upper montane woody flora impoverished while lower montane taxa including *Meliosma, Isonandra* and *Turpinia* extend to the highest elevations.[135] The understorey, rich in Rubiaceae, *Strobilanthes* (Acanthaceae) and other shrubs, is compositionally hardly distinguishable from that of lower montane forest.[136] The highest peaks along the Annamite Range, Vietnam–Laos, support mossy upper montane forest, rich in Ericaceae, on their windward east-facing uppermost slopes and crests, here sometimes as low as 1,400 m,[137] but lower montane oak-laurel forest on west-facing slopes. At 2,650 m in the Western Ghats and at 2,590 m on Doi Inthanon, summits rise more than 500 m above the upper limits of lower montane forest on Kinabalu or the New Guinea mountains – but bear lower mountain forest entirely or on leeward slopes whether we define these limits floristically, structurally or by the absence of a dense continuous moss cover (Table 4.6).[138]

Change in forest structure and physiognomy is most sharply apparent only on steep topography and small mountains, where there is equally sharp floristic change. On larger mountains, change in floristic composition may not be clearly correlated with change in forest stature or stratification. The floristic ecotone, however, obviously is correlated with change in leaf size class and pinnate leaf representation (though microphyll representation is surprisingly high in some lower montane forests).

4.3. *Physical correlates of structurally defined upper montane forests*

4.3.1. *Climate and structure*
Of particular interest has been the observation that the

ecotone between lower and upper montane forest tends to be higher on larger and more continental mountains. This effect is also said to be more marked on increasingly high mountains: the mass elevation effect.[139] If it were due to temperature differences alone, this variation in ecotone elevation would require a doubling of the lapse rate in some cases. Since there is no evidence for this, causes of variation must be sought among other factors.

Dense tussock moss is restricted to those mountains that are bathed in diurnal fog in all seasons, with only occasional and unpredictable diurnal cloudlessness. These are confined in the tropics to oceanic climates, perhumid at least in the upper montane zone. They occur at the centre of the diurnal cloud cover, directly above the prevailing cloud base. Condensation – that is, cloud formation – occurs when the saturation deficit descends to zero as temperature declines with increasing altitude (Chapter 1). In the perhumid equatorial convergence zone, diurnal humidity saturation is reached as low as at c.800 m over the sea and along coasts (Plate 1.8), but at c.1,000 m or even c.2,000 m on slopes of inland mountains and major massifs, and higher in intermontane rainshadow valleys (Chapter 1). In the perhumid tropics, saturation can sometimes be reached at 1,000 m or lower, both over the plains and in the mountains irrespective of their size or isolation.

The occurrence of extensive moss cover coincides with consistent penetration of fog within the canopy during the heat of the day.[140] Fog penetrates the forest canopy when the subcanopy retains insufficient latent heat to resist condensation of water from saturated air. Diurnal penetration of cloud as fog into the forest (Chapter 1) may also coincide with the appearance of many upper montane forest tree species and their associated canopy structure and leaf physiognomy, but not consistently with marked reduction of stature. Van Steenis[141] noted that clouds and mossy upper montane forest are higher on the higher Javan volcanoes than on the lower. He observed the ecotone to upper montane forest as high as 2,500 m on the taller West Javan volcanoes, where updrafts ascend along unbroken and increasingly steep slopes on all sides. This is consistent with Yamada's detailed plot census at nine elevations on Gunung Pangrango, from 1,600 m to 3,000 m near the summit.[142] He observed lower montane elements, including *Schima wallichii* and *Acronychia punctata,* up to 2,400 m at which altitude the forest became increasingly mossy, culminating at 3,000 m in dense mossy sleeves.

On the mountainous peninsula of Santubong (Sarawak), the ecotone to upper montane forest coincides with intermittent diurnal cloud fog borne in by onshore breezes as low as 800 m, thereby truncating lower montane forest. On the large equatorial massifs of Kinabalu and New Guinea, on the other hand, mossy upper montane forest starts at 1,700 m on steep spurs and ridges, becoming prevalent by 2,200 m, again coinciding with fog penetration of the canopy. This mossy forest type coincides with the summer monsoon and winter orographic cloud base in the oceanic monsoon tropics such as the central mountains and eastern coastal Sierra Madre of northern Luzon (17° N), where it occurs at 1,400 m, and as low as 900 m on the Hawaiian island of Kauai (22° N).

Towards the latitudinal tropical margin, the incidence of winter frost increasingly truncates the upper elevations of inland tropical forest zonation, initially by eliminating the tropical upper montane forest habitat. Seasonal variation in diurnal temperature and the altitude of the cloud base further diminishes mossiness. Nevertheless, elements of the upper montane tree generic flora remain even above the tropical limits on the eastern Himalaya (Table 4.7a).

At first, increase in rainfall seasonality and in montane fog hardly affects the decline in forest stature and the increased crown twigginess with altitude, although bryophyte cover declines and epiphytic *Usnea* curtains increase. On the southeastern scarp of the Sri Lankan central massif at Horton Plains (2,100–2,400 m at 7°30' N), fog becomes intermittent Jan.–March. The Plains carry an upper montane forest displaying all classical characteristics, except that mossiness is comparatively reduced, *Usnea* festoons the forest canopy along the most drought-prone eastern scarp (section 4.2.3 above; Plate 4.12e).

On mountains that receive more rain and fog from the seasonal windy wet monsoons than from the relatively calm conditions of orographic rainfall, fog can enter the canopy as often at lower as upper montane altitudes, while intermittent dry periods become increasingly frequent. The northeastern valleys within the Sri Lankan massif, and its steep seasonal northeastern face, are in rain shadow during the south-west monsoon. However, they receive intermittent rain and fog during the north-east monsoon, and are therefore intermittently wet and foggy throughout the year, with wind-driven monsoon fog penetrating their forests to low altitudes. Soils are yellow-red, lacking raw humus accumulation except at the highest altitudes. They are clothed with a short (less than 15 m, taller in valleys) forest physiognomically resembling upper montane forest, with dense gnarled notophyll-microphyll crowns yet with thin moss cover and lacking the liverworts, filmy ferns and even the *Usnea* which so characterise equatorial upper montane forest. In a few places, as on slopes by the Hunasgiriya Pass, they may descend as far as c.1,200 m. Some upper montane elements such as *Eugenia rotundata* occur throughout the altitudinal range, while several species such as the endemic *Syzygium spathulatum* appear only at lower altitudes. The botanist Leslie Wijesinghe has appropriately named this 'dry evergreen montane forest'(Plate 4.7e).[143]

On the high tops of the Western Ghats (10°–11°30' N), mossiness declines to a patchy cover. A rich woody understorey persists, but exclusively upper montane tree species are fewer, and they are often subordinate to lower montane taxa ascending to higher elevations. Only 33 of the 92 tree species recorded there can be considered exclusively montane. In the Palni Hills (10°15' N), 62 of the 79 montane tree species recorded are confined to elevations above 1,600 m, but only 30 above 1,900 m. Fyson[144] records only 45 tree species from the Nilgiri summits (11°5–15' N), few of which appear to be strictly upper montane species. There, as throughout Indo-Burma, emergent and other taller tree crowns of southwest-facing montane forests are bent following the direction of the summer wet monsoon winds, when speeds reach their highest velocities during periodic cyclones (typhoons). Occasional droughts during the summer monsoon, when shoots are flushing and extending, can be devastating. Occasional winter frost in seasonal climates may also exacerbate bud mortality and the consequent twigginess (see section 5.3 below).

In the Nilgiris, where upper montane structure and leaf size are correlated, I compared two *shola* forests at 2,100 m: Tiashola on an eastern ridge and slope, and Bangtipal facing west. Bangtipal receives more fog during the dry season and strong winds during the wet monsoon. Whereas at Tiashola the canopy reaches c.20 m and the trees have well developed straight boles, at Bangtipal the gnarled trees with dense twiggy crowns hardly exceed 8 m. The two localities share the same tree flora (albeit in differing relative abundances). Microphyll and notophyll leaf types are about equally represented at the drier, less windy Tiashola, while microphylls dominate at the more continuously foggy Bangtipal, on slopes exposed to the wet and windy south-west monsoon (Table 4.6; Plate 4.12d). The tree flora in both forests is overwhelmingly lower montane, yet the leaf sizes of individual species are reduced at Bangtipal as compared to Tiashola. In the cases of *Litsea floribunda* and *Syzygium* cf. *grande,* for example, this change is from meso- to notophyll, and in two unidentified *Cinnamomum* spp. and *Turpinia cochinchinensis* (pinnate) from noto- to microphyll. (Leaves become similarly smaller and thicker with increasing drought and wind in the montane forests of Venezuela.[145]) Bangtipal is the mossier forest, though moss is not continuous or dense there. The soils are lower montane mulls in both forests, darker with humus discoloration at Bangtipal but both lacking raw humus.

The high rain shadow valleys of Papua New Guinea (at 2,200–2,800 m) experience frequent overcast skies but little fog penetration. The forests there are sparsely mossy. Structurally and floristically lower montane, they also include floristic elements which in Borneo are typically upper montane, including *Tasmannia* and *Ascarina* (Chloranthaceae).[146]

4.3.2. *Does climate determine the floristic boundaries of upper montane forest?*

I infer from the above that fogginess, and possibly wind, influence both leaf size and soil in upper montane forests, but not necessarily floristics. The appearance of much of the upper montane forest tree flora appears instead to be correlated with accumulation of raw humus (most apparent in stable soils; see section 4.3.3 below), but it is not consistently correlated with change in forest stature. But raw humus appearance at or near upper montane forest altitudes is often accompanied by an initially patchy moss cover, I infer from this that raw humus accumulation correlates with increase of frequency of diurnal fog canopy penetration, and may be caused by the increased nutrient leaching and acidification that would be induced by canopy drip precipitation. I observed the same floristic correlation with soil change and the appearance of *Weinmannia* and *Drimys* in the Venezuelan Andes. I conclude that fogginess, therefore climate and raw humus accumulation partially co-vary. But there is no exact correlation between the floristic and the structural-physiognomic transitions from lower to upper montane flora.

Upper montane forest is well represented in the central mountains of Luzon above 2,300 m,[147] and somewhat lower on the Sierra Madre (below 1,100 m) fronting the Pacific to the east. Both of these areas receive intermittent cloud throughout the year. Kerr[148] recorded short-stature forest rich in epiphytes and Ericaceae on the summit of Phu Bia (2,820 m, 19° N) in Laos, but mountains further north experience frost at this altitude – and as low as 2,000 m in southernmost China. The higher peaks of the Annamite range support a similar vegetation, north to Phu Bia.[149]

Separate lowland and montane or subalpine sister taxa, particularly those in species-rich genera, occur and may coexist in the same region. Winteraceae, though occurring in the very wet foothills of New Guinea (Chapter 6), appear otherwise to be the only exclusively montane family elsewhere.

The characteristic upper montane herbaceous elements, including many epiphytes, are generally absent below 1,000 m; but tree species have more frequently been recorded at low altitudes. *Myrica* occurs locally both in MDF on sandy soils and heath forest on skeletal soils. *Eriobotrya bengalensis* occurs in lower montane forest from Bhutan to Kinabalu, but also on off-shore islands of Peninsular Malaysia. So also do many *Rhododendron* (subgen. *Schistanthe*), such as *R. quadrasianum*, which is a free-standing shrub in upper montane forest but an epiphyte in lower montane forest.

Among Myrtaceae of Australasian affinity, *Xanthomyrtus flavida* occurs on the rocky coastal sandstone summit of Santubong and Gunung Lambir (Sarawak); *Leptospermum javanicum* on rocky headlands in Sabah; and *Baeckea frutescens* more widely in similar habitats in Borneo and Peninsular Malaysia. *Weinmannia fraxinea* (Cunoniaceae) has been found in lowland peat swamp in South Kalimantan, and occurs in lower montane *kerangas* as well as in upper montane forest. *Dacrycarpus imbricatus* is locally common in lower montane *kerangas* as on Kinabalu, occurs rarely in lowland *kerangas,* but is quite common in *kerangas* in coastal southern Cambodia. *Dacrydium elatum* is locally abundant there and in poorly drained *kerangas* in Borneo. The lowland habitats – peat swamp excepted – are all sandy or rocky (or epiphytic), and therefore acid with raw humus, and intermittently dry. Many species occur more abundantly, and in a wider range of habitats, at low altitudes east of Wallace's Line. Overall, the confinement of those upper montane taxa that occur at lower altitudes to restricted, demanding, habitats is consistent with the view that competition, rather than climate or other physical factors, constrains their altitudinal range.

Leaf size classes of Far Eastern upper montane species include leptophylls (*Baeckea*) and nanophylls, together with leptophyll, nanophyll or microphyll lowland *kerangas* species such as *Gymnostoma nobile, Calophyllum nodosum, Syzygium buxifolium* ssp. *perparvifolium* and *S. phryganodes* (Table 4.8). Nanophyll *Leptospermum recurvum, Xanthomyrtus flavida, Syzygium kinabaluense* and *Styphelia suaveolens* (sister to *L. javanicum, Syzygium buxifolium* and *Styphelia malayana* respectively), ascend Kinabalu to the highest altitudes. Here they are found among subalpine vegetation where there is also evidence for periodic intense drought. I infer that their nanophyll leaves are in part an adaptation to drought (Chapter 2 and section 5.3.3 below).

4.3.3. *Soil*

Soil surface horizons of mossy upper montane forests (floristically defined) in perhumid climates (except following landslips) are acid, with deep raw humus, though often also high total nitrogen and extractable phosphorus irrespective of substrate.[150] This is said not to be the case exceptionally on some young base-rich volcanic and calcareous substrates[151] (Chapter 1). There is a close correlation between the appearance of upper montane woody taxa and the first appearance of raw humus on the soil surface along Mount Kinabalu and high Bornean sandstone ridges. At about 1,600 m, thin raw humus begins to accumulate under the litter along ridges. I therefore infer that acid surface raw humus had begun to form in Kitayama's sandstone plot at 1,700 m, whereas it is deep and ubiquitous in his ultramafic plot. Further, the ultramafic substrate on Kinabalu is habitat for the majority of its upper montane woody endemics.[152]

The occurrence of upper montane woody species within the upper reaches of otherwise typical lower montane forest on Kinabalu seems to be at least partially explained therefore by the influence of the soil substrate. In contrast with the microtherm herbs, these low-altitude occurrences are confined to freely draining, often shallow, ridge soils. Kitayama and colleagues showed, by using plots placed along transects on ultramafic, sedimentary and granite substrates on Kinabalu, that structural and floristic zonation above 1,000 m is depressed on the ultramafic substrate.[153] Their plot on ultramafic substrate at 1,700 m had the same low proportion of compound-leaved species as that on sandstone, but whereas the sandstone plot had equal percentages of species with notophyll and microphyll representation, that on ultramafic substrate bore twice as many microphyll species. The number of plots was too limited to identify the presence and altitude of ecotones, though Kitayama implies on the basis of stand density and floristics that the lower–upper montane forest ecotone is at c.1,600 m on ultramafic, and at c.2,000-2,300 m on sedimentary and igneous rock. In both cases, the ecotone here appears gradual.

It is clear, both from Kitayama's plot and from the collections-based data of Beaman *et al.*[154] that there is mixing of lower and upper montane forest floras between c.1,400–2,200 m within individual stands. This is particularly evident on the ultramafic substrate on Kinabalu, but it is also widespread on siliceous substrates in the mountains of Borneo, and it cannot be explained by interdigitation of forest types along convex and concave slopes alone. In Borneo, the lower montane soil organic matter is commonly similar to temperate moder rather than the mull of oak-laurel forest, with fine-rooted organic raw humus mixing with the pale yellow sandy clay to c.20 cm, mineral soil often visible at the surface, and gleying frequent beneath (Chapter 1). Whether the mixing is due to soil faunal activity or lateral creep has not been observed.

Kochummen recorded a similar paucity of lower montane forest flora, but an abundance of heath and upper montane forest families including conifers, Myrtaceae, Clusiaceae, Theaceae and Sapotaceae, on the upper slopes of the siliceous and isolated coastal Gunung Jerai (Kedah, 1,200 m).[155] Kochummen did not observe the soils, other than the presence at this altitude of a thin covering of bryophytes and populations of ferns and herbaceous dicotyledons.

As in *kerangas,* the leaves of upper montane tree species contain high concentrations of the protein-precipitating compounds that inhibit bacterial decomposition. These depress rates of litter humification and contribute to the accumulation of raw humus. In upper valleys and swales, though, lateral water movement near the surface is associated with organic muck soils, often rich in earthworms. *Floristic variation between upper montane forest on freely draining slopes and its taller facies on valley muck soils has yet to be studied.* Below the altitude of persistent fog, in rain shadows on the major massifs of New Guinea and in the seasonal tropics, soils may differ little from those of lower montane forests,[156]

though they are often shallower and rockier. Upper montane forest soils are frequently waterlogged, but it seems likely that all short stature upper montane ridge forests experience occasional drought as well.[157] As in *kerangas* – but in contrast with other lower montane forests – rooting in upper montane elfin woodland is mostly shallow and confined to organic horizons. The almost ubiquitous blanket of densely rooted raw humus nevertheless endows mossy upper montane forests and soil with a uniformity lacking in lower montane forests (Plate 1.14c). Here, it is the scars of landslips and their associated flora which add edaphic heterogeneity.

4.4. *Summary*

All upper montane forest types are distinguishable from classic lower montane forest of the oak-laurel type, but vary continuously floristically and in some structural and physiognomic attributes with *kerangas* forests.

1. Paradigmatic upper montane forest has trees with crooked trunks, gnarled branches, and dense crowns in which micro- and nanophylls prevail. It varies in stature, though it is short in exposed sites. It includes a small diagnostic tree and angiospermous epiphyte flora, some species of which are abundant. It is densely mossy, with filmy ferns and liverworts abundant, rich overall in cryptogams.

2. It occurs where diurnal fog persistently penetrates the canopy throughout the year, and is generally associated with leached humult soils that may develop to true podsols, or mires in valleys, even on basic volcanic substrates. I hypothesise that fog is the primary cause of such upper montane forest habitat, and the motor of soil conditions.

3. Upper montane forest becomes universal above 2,000–2,200 m in perhumid climates. But where diurnal breezes blow fog into the canopy, notably on coastal mountains, upper montane forest – albeit depauperate in tree species and cryptogasms, including diagnostic species – may descend to near the cloud base at c.800 m, thereby truncating lower montane forest.

4. The more *intermittent* the diurnal fog penetration of the canopy in perhumid climates, the less mossy the forest and the less adorned with epiphytes. Physiognomically and structurally similar forest may occur where fog is occasional, perhaps even absent, on the skeletal soils of rocky summits down to low altitudes. Some taxa otherwise diagnostic of upper montane forest, notably in Myrtaceae, may descend to 500 m but here most of its tree species, and most or all of its epiphytes and cryptogams, are absent.

5. The more *seasonal* the fog penetration, the straighter the trunks and the less gnarled the branches; microphylls and notophylls come to dominate; mosses and epiphytes decline; species of lower montane forest enter and increase. Upper montane forest disappears where a fogless season exceeds about four months, and is replaced by lower montane or, if there is frost, warm temperate forest. Acid humult soils give way to humic loams, except where soils are sandy.

6. The ecotone between short-stature mossy upper montane forest and all other forest types, including other montane forests, is generally clear and narrow in the equatorial and oceanic tropics. Where slopes are constant, as on volcanoes, the ecotone is gradual, but it is more often narrow and distinct. It often coincides with a break in topography from steep upper slope to spur or ridge top. It is apparently correlated with reliable and altitudinally constant diurnal fog. Nevertheless, upper montane forest types lacking one or more of its characteristics form part of a continuum of variation in structure, physiognomy, life form and species composition with other forest types in sheltered valleys, and where fog is intermittent or seasonal.

7. Wherever the length of the dry season increases at higher latitudes in Asia, upper montane forest becomes confined to islands and coastal mountains, and it disappears entirely by 12° N inland, and by 20° N in easternmost coastal Asia, as winter frost increasingly penetrates its altitudinal range.

8. It remains unclear whether most upper montane tree taxa are restricted by diurnal fog frequency, or by presence of soil surface raw humus, because these two factors in part covary; though raw humus extends below the altitudinal zone of fog penetration on low nutrient soils and ridges, and some upper montane taxa with it. Few taxa appear to be restricted to habitats experiencing year-round almost continuous fog within the canopy, *Tasmannia* being the most likely case.

9. It is possible that most upper montane taxa are not restricted to high altitudes by temperature constraints; but rather that, though few tolerate extreme soil water stress, most are excluded by competition from suitable habitats in the Sunda lowlands. But the failure of some upper montane species to survive below 1,400 m could also be due to the influence of warmer temperatures on life history traits, rather than to the influence of soils. *At present, only montane herbaceous species have been shown to be microtherms. Why are there not more montane tree species growing in lowland botanic gardens?*

5. Tropical subalpine forests and the upper limits of equatorial forests

5.1. *Definition and distribution*

Subalpine forests are confined to those equatorial mountains high enough to support them. In New Guinea subalpine forests, floristically defined, appear at 3,650 m,[158] above 2,900 m on Kinabalu[159] but lower near the summit of nearby Trus Madi (2,649 m) in Sabah, and on the tallest volcanoes of West Java and Sumatra above 2,200–2,700 m. On exposed narrow ridges below those altitudes – as on the summits of Pagon Periok (1,850 m) in Brunei and Mulu (2,371 m) in Sarawak – upper montane forest becomes short, its canopy exceptionally dense and manifestly wind-pruned; but their flora remains upper montane. Leaf sizes show increasing nanophyll representation. Above 3,200 m in New Guinea at c. 6° S lat., above 2,900 m on Kinabalu at 6° N, and above 2,100 m on the Horton Plains Horton Plains at 7° 15' N, many species have cupped leaves with the upper surface convex and shiny, the midrib frequently sunken above, the margin revolute or the whole leaf cupped with the outer suface uppermost, apices obtuse or retuse, and internodes shorter (including greatly shortened rachises). In these characters they resemble the leaves of canopy trees in warm temperate evergreen forests (see below), but differ from them in their predominantly microphyll, nanophyll or leptophyll sizes (Table 4.8). Their presence at such low altitudes on the South Asian high hilltops heralds the truncation by frost of the upper limit of tropical montane forest with increasing latitude.

Mossiness becomes less dense on both ground and branches (Plate 4.14a,b). A distinct indicator is the presence of temperate herbaceous genera, which may include *Viola, Ranunculus, Potentilla* (Plate 4.14e), *Primula,* Apiaceae and agrostoid grasses. Temperate herbaceous elements often first appear in open sites in the upper reaches of upper montane forest, as they do on the Pangrango summit in West Java, on Kinabalu,[160] and in New Guinea. Continuous moss on litter-free soils and bryophyte sleeves on branches are not found above c.2,900 m, although scattered clumps of moss mixed with lichens continue to the highest altitudes.

Subalpine forest is distinguished on Kinabalu, as well as on the lesser summits of Trus Madi (Sabah, 2,649 m) and Kerinci (Sumatra, 3,805 m), by a distinct shrub flora including specialist *Rhododendron* sect. *Schistanthe* and other Ericaceae, and Epacridaceae; as well as on New Guinea where other southern elements, notably *Olearia* (Asteraceae) are present. Between 2,700–3,200 m new tree taxa, recognized as sister species or subspecies of taxa at lower altitudes, appear. Such pairs are found within large genera including *Syzygium* (Myrtaceae), *Memecylon* (Melastomaceae), *Rhododendron* and *Vaccinium*,

Elaeocarpus (Elaeocarpaceae) and *Ilex* (Aquifoliaceae), also *Leptospermum javanicum* with its subalpine and ultramafic sister, *L. recurvum,* both abundant on Kinabalu.[161] Some are endemic sisters to widespread upper montane species that may co-occur in the ecotone. *Molecular study combined with hybridisation experiments would be useful, for instance, to establish the relationship between the lowland Syzygium bankense and its several allies in the Kinabalu upper montane forest, or the two putative subspecies of Syzygium cordifolium in Sri Lanka.*

Aiba and Kitayama[162] documented the most abundant woody species in paired 0.06–0.25 ha plots, on sandstone and granite and on ultramafic substrates, at 2,700 m and 3,100 m on Kinabalu. By 3,100 m, leaf sizes in the canopy were substantially leptophyll and nanophyll, with some microphylls in a shrubby patchy understorey. Net assimilation rates did not decline with altitude on typical upper montane forest soils, consistent with observed increases in direct sunlight, intermittent with contrasting fog, at higher altitudes, and implying that changing leaf characters partially serve as adaptations to optimise the rate of carbon dioxide uptake relative to water loss.[163]

Plate 4.14 Subalpine vegetation, Mount Kinabalu. **a,** The soil surface, with *Diplycosia* sp.; **b,** *Leptospermum recurvum* (left and upper right) and other shrubs with nanophyll revolute shiny leaves; **c,** *Leptospermum recurvum*; **d,** *Ternstroemia lowii*; **e,** A temperate genus: *Potentilla polyphylla* var. *kinabaluensis*; **f,** The limits of woody vegetation, c.3,800 m. (**b, c, d** E. Soepadmo; **e,** S.J. Davies; **a, f** P.A.)

The transition from upper montane to subalpine forest is gradual and continuous.[164] This correlates with the highly variable disposition of the diurnal cloud caps in contrast with the relatively even cloud base.[165] Upper elevations are prone both to squalls and to continuous high wind speeds, occasionally associated with extreme droughts in ENSO years, as in 1995–1996 when there was extensive canopy mortality (see section 5.3.1 below). Soils on Kinabalu become increasingly skeletal above c.2,700 m, though deep humic muck soils occur in swales up to 3,100 m. Earthworms remain abundant almost to the summit.[166] On drier slopes, the organic surface horizon becomes mixed with mineral soil and gravel, implying increased down-slope creep.

Ultraviolet light levels are high during the frequent sunny intervals at highest elevations. The small leaves, highly inclined against incident radiation and with shiny upper surfaces, reduce penetration of ultraviolet light, and minimise rates of somatic mutation in expanding leaves.[167] Many upper montane and subalpine species bear vivid anthocyanin-crimson flushes, which enhance this effect.[168] David Lee[169] noted that anthocyanins absorb light of medium spectral wavelengths, thereby reducing photosynthetically active radiation and thus photoinhibition in full sun while also warming the expanding young leaves.

5.2. *The altitudinal limits of forest in equatorial Asia*

The tree line on tropical mountains is often sharp, but varies in elevation with topography. Kinabalu (4,094 m) is exceptional in that its summit is devoid of soil except in seepages and fissures, and in that there has never been human modification (Plates 4.1, 4.14). The tree limit there is unclear; a few shrubby trees extend almost to the summit along soil-filled fissures (Chapter 6). On summits with soil cover, forest is replaced by alpine dwarf shrublands, or in Asia more frequently with grasslands. Grasslands attract ungulates and their predators, while people have brought cattle for centuries and have lowered the altitude of the tree line by fire. In continental Asia, the ecotone is fringed by dwarf bamboos, *Hedyotis* and other Rubiaceae, and other woody pioneer shrubs and treelets.

5.2.1. *Alpine grasslands*

Mountain grasslands, dominated by narrow-leaved temperate grass genera and more or less rich in herbaceous communities belonging to temperate genera and families, are widespread on tropical Asian mountains. Grassy flats occur as low as 900 m in Javanese calderas at 7° S latitude, at 2,000 m in valley bottoms on Sri Lanka's Horton Plains (8° N) and lower in the Palni and Nilgiri mountains of the Western Ghats (to 12° N). In the Similipal hills of Odisha (22° N), grassy valley bottoms descend to 850 m. The lowest natural alpine grassland communities

Plate 4.15 Death and destruction near the altitudinal limits of tropical trees. **a,** Canopy frosting following the winter of 2007–2008: crowns recovering by epicormic shoots; Pema Wangda's plot at the tropical lower montane-warm temperate forest ecotone, 2,100 m, Gedu, Bhutan; **b,** Roadside pioneer *Melastoma normale,* bronzed by light frosting, Gedu, Bhutan, 2,100 m; **c,** Within a dead stripe of upper montane forest, *Rhododendron zeylanicum* with epicormic shoot, Kirigalpota, 2,200 m, Horton Plains, Sri Lanka. Here the branches suggest wind updrafts, which might have caused penetration of cold or dry air through the canopy; **d,** A dense even upper montane canopy proves air frost-resistent, in spite of being in a valley bottom: a shola in the Nilgiries, c.2,000 m, unaffected while the stand of the introduced Australian tea shade tree *Acacia mearnsii* on adjacent slopes,

with diffuse feathery crown, suffered almost complete mortality from 2007–2008 frosts; **e,** Dead emergent *Leptospermum* on Kinabalu, *Schima brevifolia* flowering beneath, c.3,800 m; **f,** Bottomland frost ponding, 900 m, Meghasena, Odisha: peripheral sal (*Shorea robusta*) scorched while *Salix monosperma* following the brook is unaffected; sal on the distant slopes also unaffected, the descending cold nocturnal air being repelled by the latent subcanopy heat; **g,** Stagnant nocturnal ponding of bottomland frost, 900 m, Meghasena, Odisha: Sal (*Shorea robusta*) coppice shoots recovering from deep stout lignotubers within a sward which includes *Corydalis* and other temperate herb genera; scorched tall deciduous hill sal forest fringe in background. (**a, b, f, g** H.H.; **e** P.A.; **c** W. Werner; **d** D. Mohan Das)

are confined to valley bottoms, and harbour temperate herbs. Their margins are sharply defined, their woody vegetation periodically scorched by frost (Plate 4.15e,f). They may at times provide habitat for temperate woody species. For example, a stand of *Lyonia ovalifolia* and *Rhododendron delavayi* (var. *delavayi*) surrounds the little grassland in the valley immediately below the Doi Inthanon summit at c.2,500 m, where several temperate herbaceous taxa also occur.

5.2.2. *Physical determinants of the tree line*

Körner[170] has argued that the tree line marks the altitude at which the production gains from photosynthesis are matched by the maintenance costs of respiration, rather than solely the limitation on production imposed by restriction of photosynthetic rates. This may provide a general explanation for the altitude of the tree line on wet equatorial mountains. It is consistent with the inference of Takyu *et al.*[171] that root/shoot ratios of trees on Kinabalu increase with altitude (also see sections 6 and 7.7 below). Leaf mass per unit area also increased on Kinabalu, but leaf area index decreased. Persistent diurnal fog and surface moisture, by inhibiting gas exchange, may reduce rates of photosynthesis in trees whose inherent photosynthetic capacities are not otherwise low.[172] However, fog becomes intermittent near the tree line[173] and diurnal surface temperatures are frequently high. Fog cannot therefore alone explain why physiological costs exceed gains there. Instead, increasing respiratory costs incurred by increased root/shoot ratio may eventually produce the debit.

5.3. *Frost and drought on equatorial mountains*

The natural distribution of montane grasslands correlates with frost-ponding in climates with a marked dry season

(Chapter 1), but has been greatly extended by human-induced fire for pasture. This extension continues down to lowland deciduous forest on the dry peninsular Indian mountains east of the Western Ghats.[174] Analogous cases are the grassy plateaux of the Plateau des Bolovens (15°30' N) and the Plaine des Jarres (19°50' N) on the leeward western slopes of the Annamite range in Laos (1,000–2,200 m). On Mount Pulog (Luzon) the grasslands have been much extended by fire and are now continuous from the *Pinus kesiya* savanna to the summit.[175]

Frost occurs in the subalpine zone of New Guinea above 3,600 m, and is a serious albeit occasional hazard to agriculture in valleys there down to 2,000 m[176] and occasionally even to 1,500 m.[177] Van Steenis[178] observed early morning hoar-frosts on the grasslands of the East Javan Diĕng Plateau (c.2,000 m), an area presumed once forested; yet the frost-sensitive trees in adjacent forest appeared unaffected. In South Asia the tree rhododendrons *R. nilagirica* and *R. zeylanicum* (often regarded as subspecies to the Himalayan *R. arboreum,* hardy in oceanic temperate gardens) resist both fire and frost, and extend into the grassland as scattered individuals. Their buds are apparently frost-resistant, while their leaves droop thermonastically at freezing temperatures.[179]

In an experiment in grassland adjacent to snow gum (*Eucalyptus pauciflora*) forest in Australia, seedlings in sward, when subjected to lower temperatures than others on bare ground, experienced photoinhibition with delayed recovery and scorching of growing stems and leaves.[180] The absence of tree juveniles beyond the forest canopy in tropical subalpine grasslands[181] – even on bare surfaces and when fire is excluded – implies that there is frost and that they are frost-sensitive.[182] The forest-grassland ecotone is clearly dynamic. Montane successional shrubs, especially *Hedyotis* spp., other Rubiaceae and short bamboo species, form marginal thickets around South Asian *sholas* and are periodically scorched by frost, although human-induced fire is often the principal cause of their mortality. Gillison,[183] from observations made at 2,700 m in Papua New Guinea, inferred that pioneers advance at the forest fringe between frost events, providing cover from the annual light frosts, and thereby allowing closed-forest species to invade a pioneer canopy within which latent heat maintains a frost-free climate.

Whereas grassland with frequent incidence of ground-frost starts on the New Guinea mountain slopes at c.3,500 m, Van Steenis[184] regarded winter frost as an annual event above 2,000 m in Java. Frost was at least once recorded as low as 900 m in seasonal East Java (7–8° S), while in the west it occasionally descends to 1,500 m only in valleys. Annual frost may truncate upper montane forest in East Java, but Gunung Halimun (1,929 m, 6°45' S, 106°20' E), and Pangrango (3,020 m, 30 km to its south-east) in West Java support upper montane forest to their summits. Gunung Salak (2,211 m,

6°45' S, 106°30' E) has a distinct albeit intermittent dry season; it bears a narrow band – just c.100 m – of truly mossy upper montane forest, giving way above to subalpine scrub.

Frosted grasslands are found at 2,000 m on the Horton Plains (Sri Lanka, 8° N), and at c.1,900 m in the Palni and Nilgiri Ghats (below 12° N). The upper valley of Meghasena in northern Odisha (22° N) endured January frosts as far down as c.600 m in the hard winter of 2007–2008. These frosts scorched the bordering juvenile sal (*Shorea robusta*) stands in the adjacent short facies of tall deciduous forest, which were regenerating after the previous frost, and killed back sal coppices in the grassy bottomland (900 m) to the ground, where they survive as rootstocks (Plate 4.15g). Such seedlings may recruit for several years between killing frosts, permitting their deepening rootstocks to endure and eventually become stocky and extensive (as they do in response to fire and browsing; see Chapter 2). It is unexpected to see the lowland tropical sal annually reshooting in grassland dominated by temperate grasses and herbs including *Corydalis* and *Viola*. Scattered seedlings of *Bombax ceiba* were similarly affected.

Several temperate grass taxa occur in and above the tree line even on Kinabalu, where *Anthoxanthum klossii* shelters in fissures toward the summit otherwise bereft of soil.[185] Van Steenis[186] introduced the concept of microtherm species for those temperate genera which occur in and above the tree line on equatorial mountains, but which may also be found in open places on loam soils in subalpine valleys. All genera confirmed to be microtherm by transplantation experiment are herbaceous. Their distribution is apparently associated with flows of cold and even frosted nocturnal air down the mountain valleys. Microtherm herbs, such as *Primula prolifera,* appear on Gunung Gede-Pangrango and become increasingly well-represented eastwards as the dry season lengthens. Many species descend well below the frost line in open places in the forest, particularly near streams. They descend lower on large mountains than small – the opposite pattern to upper montane forest – which may simply be due to the greater volume of cold night air flow down valleys of higher and more extensive massifs.

The South Asian montane grasslands betray their valley origins: forest is absent in upper valley alluvial hollows but persists as fragments on slopes, ridges, and especially in steep moist gullies.[187] Cold valley downdrafts create a nocturnal temperature inversion between lower and upper slopes.[188] Similarly in certain New Guinea mountain valleys, subalpine forest reaches its highest altitude as a fragmented belt mid-slope, with grassland in the valley below. The summit forests on Sri Lanka's Horton Plains (7°50' N), like those of the seasonally wet Western Ghats (10°–12° N) where lower montane *shola* forests reach the winter frost zone, have been largely replaced by burning with grasslands.[189] It was formerly assumed that the South Indian Ghat highlands had once been entirely forested, and that all their grassland is anthropogenic, but Sukumar *et al.*[190] showed that temperate festucoid grasses and other taxa occurred in a Nilgiri upper valley from the Pleistocene c.45,000 BP. Although his evidence is restricted to valley bottom habitats, the presence of an endemic grazing and browsing ungulate, the Nilgiri Tahr, *Nilgiritragus hylocrius,* along with west Himalayan floristic elements such as *Rhododendron, Berberis* and *Gaultheria* implies that grassland was more widespread.

Although we lack good meteorological data at present, the transition to subalpine forest appears to correlate with the onset of occasional winter frost. Even on the highest Far Eastern equatorial mountains, this may occur when skies remain clear for prolonged periods in ENSO years. Closed forest extends on slopes well above the lower limits of occasional killing frost in valley flats. In the Nilgiris, the escaped exotic shade tree *Acacia mearnsii* forms diffuse-canopied stands with open feathery crowns, its shoots continuing to expand throughout the cool dry season. While these stands suffered massive mortality during the unusually frosty 1995–1996 dry season, pockets of upper montane forest embedded in the *Acacia* were unaffected (Plate 4.15d). The typically dense closed canopy of upper montane forest may retain enough latent heat to exclude frosted air, as well as a warmer air layer immediately above the canopy, analogous to the fog-free air layer above lower montane forest within the cloud base.[191] Buds in the canopy remain dormant during the cold dry season in mildly seasonal climates. Unlike cultivated tea, which in South Asia shoots continuously, indigenous canopy species of closed upper montane forest flush and flower after the frosts, as the wet monsoon approaches. This pattern is consistent with adaptations both to escape nocturnal freezing and to delay diurnal warming, thereby ameliorating the potentially synergistic destructive impact of drought and frost on vessels and mesophyll.

The tree line appears uncorrelated, then, with frost incidence as such – which must be widespread over the canopy at these high altitudes – and frost does not therefore represent a potential altitudinal limit of forest on these continental mountains. Only on the high mountains of New Guinea, and probably on Kinabalu, does forest reach its altitudinal limit for equatorial mountains; and only on Kinabalu is the peak free of man-induced fire. It remains unclear whether occasional catastrophic frost or fire most defines the interface between alpine grassland (or dwarf shrubland) and forest. Further, it is possible that the tree line on Kinabalu, where grassland is absent, is also defined by occasional catastrophic frost, although the absence of scorching or mortality at tree limits there implies that it does not. *The physiological and environmental determinants to the altitudinal limits of subalpine trees on tropical mountains merit further research.*

At higher latitudes, mountain fog becomes increasingly intermittent and appears later in the day during the dry season (Chapter 1), while it may be almost absent in drier climates.[192] From the eastern Himalaya to Fan Si Pan in North Vietnam there are several months of continuous rainless sunny weather in winter. During the dry season clouds may still accumulate over moist orographic updrafts on southern slopes during the afternoon; but the dry north-east trades flowing off the continents truncate these clouds and lead to periodic droughts and proneness to fire. Frost is restricted to the highest grassy valley bottom on Doi Inthanon at c.50 m beneath the summit, and above c.2,000 m on Fan Si Pan (3,143 m, 22°20' N). Nevertheless, temperate herbs in grassland, and by inference frost, appear on the karst limestone upper slopes at 1,500 m on Doi Chiang Dao (2,200 m, 19°20' N). Tropical upper montane forest zonation thus becomes truncated on continental mountains with increasing latitude. The northernmost contemporary upper montane forest in East Asia is at 18° N in the Sierra Madre and central mountains of Luzon.

5.3.1. *Canopy death near the equatorial tree line*

Incidence of frost varies annually. The Kinabalu summit, usually frost-free, is said to have experienced frost and light snow in 1995. In 1858, the traveller Spencer St. John recorded a night temperature of 2.5° C and observed hoar frost at Paka Cave (3,100 m).[193] That is the altitude at which subalpine species and leaf characteristics appear, not the tree line. Kitayama recorded two days of air fost at 3,780 m, with ground temperature descending to -1.2° due to reradiation; while Gaku Kudo recorded temperature drops to below -0.5°C in three out of the four years 1998–2001, June–Sept., at the Donkey's Ears (3,940 m; Kinabalu is shielded by the Borneo uplands from the south-west monsoon).[194] The granite summit supports a sparse but surprisingly diverse population of microtherm herbs.[195] In Sri Lanka, air frosts occasionally descend as low as 1,800 m, scorching the growing shoots of high country tea. Furthermore, there is evidence of catastrophic canopy death near the upper limits of forest in many parts of the tropics.[196] Toward the summit of Doi Inthanon (northern Thailand), patches of larger trees in the upper facies of lower montane forest experienced crown die-back following clearance for road-building. On the drier eastern fringe of Sri Lanka's central massif (2,100–2,260 m), at Lover's Leap and on the eastern flanks of the Totapola summit on the Horton Plains near Ohiya, upper montane forests have suffered strips of total canopy mortality since c.1980, while patches of subcanopy *Strobilanthes* were also killed.[197] The strips ascend the vertical scarp from c.200 m below the rim and are more than 15 m wide (Plate 4.15c). Apparently no trees survived along the strips, and succession is currently dominated again by *Hedyotis* and *Strobilanthes,* other shrubs and dwarf bamboo. Few young trees are yet emerging.

In every case where such canopy death has been observed, air frost is also known or suspected to occur. The crowns of many trees on the Horton Plains are remarkable for high bud mortality inducing extreme twigginess. *Calophyllum gardneri,* which forms emergent clusters above the otherwise smooth closed canopy of upper montane forest there, has experienced extensive mortality in recent years. The tallest individuals seem most affected. In the Nilgiris, individuals of nanophyll *Syzygium calophyllifolium,* a semi-emergent tree in the high elevation forests at 2,200 m on both east and west slopes, experienced mortality on the drier eastern slopes in the hard winter of 2007–2008.

Kinabalu experiences increasing diurnal oscillations in atmospheric humidity in and above the subalpine forest.[198] Larger individuals of subalpine *Leptospermum* have experienced catastrophic mortality following an intense drought there in 1995–1996, and others earlier (Plate 4.15e).[199] This mortality could be caused by drought, possibly in combination with catastrophic freezing accompanying a decline in atmospheric humidity. In Sri Lanka and Thailand, the semi-evergreen lowland forests below have been converted to grasslands with lower evaporative power during dry seasons, thereby increasing the altitudinal lapse rate. *Careful long-term observation of winter temperatures, leaf flush phenology, and the causes of bud mortality could prove informative.*

By themselves, the physiological limits to growth of woody plants can hardly explain the continuing existence since Pleistocene times of grasslands with a temperate flora in the summit valleys of mountains in the seasonal wet tropics of Asia, which currently descend to 2,000 m and lower. Nor can it explain the observed occurrence of periodic catastrophic mortality of taller species of tropical affinity in the closed forest canopy near the tree line. Scorching of foliage and sudden mortality of woody species at the edge of montane alluvial flats where heavy nocturnal air can pond, is clearly due to frost. *What remains unclear, though, is the relative role of occasional catastrophic frost or drought versus chronic physiological limitations to growth on the slopes and ridges of the highest equatorial mountains.*

5.4. The upper limits of tropical montane forest, and transition to temperate evergreen montane forest at higher latitudes

In continental East Asia, tropical lower montane forests differ little in stature or in physiognomy from adjacent warm

Bhutan Himalaya	Lowland	Montane below frost line	Montane above frost line
	0–299 m (n=136)	1,300–1,500 m (n=181)	2,200–2,400 m (n=177)
Deciduous	35	23	28
Toothed	23	31	46
Toothed deciduous	38 (n=47)	35 (n=42)	68 (n=50)
South China	**Tropical** (n=101)	**Warm temperate** (n=79)	
'Crenate, dentate, serrate'	23	42	

Table 4.9 Percentages of leaf characteristics of lower montane and warm temperate forest species compared, Bhutan and China. Percentage toothed deciduous as percent of deciduous species.

temperate evergreen forest. The tree flora is at first similar at the generic level, except that temperate deciduous taxa appear; but there is a major turnover in species relative to the small number still remaining (Fig. 4.1). *Leaf sizes decline slightly, though the consistency of that trend requires more study* (Table 4.9).

Tropical forests generally extend further north in the eastern Himalaya and northern Burma, and some tropical tree species to the continental islands south of Japan, than elsewhere. This is owing to the warm ocean current which keeps to the coast as far north as Sendai (north-east Honshu), as well as to the protection from the cold continental winter northeasterlies afforded by the mountains to the north (Chapter 1). The transition from tropical to warm temperate forest in the lowlands is discussed in Chapter 3. In what follows, I address the tropical frontier at higher latitudes, notably the Himalaya and South China mountains.

5.4.1. *South and East Asia*

On the mountains of seasonal central and eastern peninsular India, where the annual dry season exceeds six months, lowland forest reaches summits at 1,600 m in the Shervarayan Hills (11°45–55' N, 78°11–20' E). There, semi-evergreen forest including the lower montane evergreen elements *Pittosporum napaulense, Ilex wightiana, Meliosma pinnata, M. simplicifolia, Schefflera stellata,* and *Wendlandia thyrsoidea* replaces short deciduous forest at c.1,200 m. Meghasena ('Abode of Clouds', 1,160 m, 21°36' N, 86°26' E), at the northern end of the Eastern Ghats, has a five-month dry season. Lowland tall deciduous forest, with sal, *Lannea coromandelica, Schleichera oleosa, Terminalia bellirica, Kydia calycina* and the vine *Bauhinia vahlii,* extends on slopes to the summits. *Magnolia (Michelia) nilagirica, Meliosma simplicifolia* and a few other lower montane forest species mix with *Toona ciliata, Alstonia scholaris* and other tall deciduous forest species to form a semi-evergreen forest in the valley at Chahala (530–950 m). Flat valley bottoms above 850 m support grassland with agrostoid grasses, *Viola betonicifolia, Corydalis, Thalictrum* and *Anaphalis.* On Parasnath (1,350 m, 23°56' N, 86°07' E) in nearby Bihar, tall deciduous forest extends to the summit.[200] Here and to the west on the Satpura mountains (21–22° N, 74–78° E),[201] grassy patches above 900 m support warm temperate microtherm shrubs, vines and herbs, variously including *Berberis asiatica, Buddleja asiatica, Clematis, Cardamine, Potentilla kleiniana, Rubus rugosus* and *Thalictrum.* To the west and at the desert fringe, Mount Abu (1,732 m, 24°41' N, 72°50' E) still bears temperate shrubland with *B. asiatica* and *Rosa involucrata* on its much-visited summit. However, no temperate genus or species of tree is indigenous to the South Asian hilltops.

Tropical lower montane forest extends westwards to the Rapti valley in central Nepal (25° N, 84° E).[202] There, lower montane elements including *Castanopsis indica* and *Schima wallichii* reach their western limit. The lower ecotone is with tall deciduous *Shorea robusta* forest; lowland semi-evergreen forest is by then restricted to coves. *Castanopsis tribuloides*, paradigmatic species of the northern tropical oak-laurel forests east to China and Vietnam, reaches west only to 89° E.

In Thailand, Doi Chiangdao (19°20' N) is lower at 2,200 m than Doi Inthanon to its west. It is embedded in mountains descending south from Yunnan, and is limestone towards its summit and in places down to the foothills at 350 m. On its moist western slopes, tall deciduous forest gives way to upper dipterocarp forest with drifts of *Dipterocarpus costatus.* This forest is comprised of both lower montane and lowland species at c.800 m, and lower montane oak-laurel forest from c.1,000–1,800 m in steep narrow valleys. On drier slopes, *Pinus kesiya* savanna woodland extends to 1,850 m. Above 1,500 m we find grassland on skeletal karst soils, bearing a flora rich in temperate herbaceous genera but, again, no temperate trees. Nowhere is there dense mossiness.

The northernmost tropical seasonal evergreen lowland forests in North-East India, and in north-east Burma south of Putao, are at 27° N.[203] Along the southern hills of Yunnan and Guangxi, at 22–23° N near the Tropic of Cancer, the northern margin of lower montane tropical forest follows the northern margin of MDF (see Chapter 3).[204] Here, an ecotone to montane temperate broadleaved forest appears, but is locally complex. This is possibly on account of the cold winter northeasterlies, which here absorb humidity off the China Sea and Pacific, bringing sleet to lowland northern Vietnam and snow to summits as low as Mount Bavi (1591 m, 21°20' N) (Chapter 3). From the Rapti in Nepal to the China Sea, cool humid winter nights experience drenching dewfall during the cold winter dry season, until April when hotter weather and periodic orographic storms start. This dew ameliorates the apparent drought recorded in open rainfall stations (Chapters 1, 3).

The crests of Vietnam's Annamite Range descend from the warm temperate mountains of Guangxi south to Da Lat (12° N). Lower montane forest occurs there, as elsewhere at this latitude, to 2,250 m.[205] Driving cold wet winter northeasterlies off the China Sea are associated with cool temperate species southwards at least to Fan Si Pan (3,143 m, 22°30' N), where stands of the Sino-Himalayan cool temperate genera *Abies delavayi* and *Tsuga dumosa* exist above 2,400 m. Beneath is a zone where broadleaved evergreen forest in which warmest temperate evergreen elements, including *Burretiodendron hsienmu, Cardiodaphnopsis tonkinensis* and *Exbucklandia tonkinensis,* occur.[206] The region is geologically diverse and includes much karst. *Whether there is an altitude at which air frost regularly commences has yet to be ascertained.* The mountainous island of Taiwan, the southern third of which is south of the Tropic of Cancer, must once have carried lowland seasonal tropical evergreen forest, but only fragments remain (Chapter 3) and the ecotone to lower montane forest is obscure.[207] An ecotone to warm temperate forest seems to occur at c.1,200 m.[208]

5.4.2. Bhutan as an example

The forests of the northern tropical frontier remain most intact in the Bhutan Himalaya, where they have received recent study especially by M. Kanehiro Ohsawa and Rebecca Pradhan.[209] In these mountains, winter hoar frost descends in valley stubble down to c.1,500 m.[210] The temperate genera *Alnus, Betula, Carpinus* and the winter-deciduous mildly frost-hardy *Quercus griffithii* appear at this altitude, starting on slopes exposed to the dry northeasterly winter winds. Tropical lower montane elements decline on slopes above 1,800 m and disappear (except in a few sheltered sites) by c. 2,400 m – the altitude at which annual air frost, and ground frost on open land, becomes pervasive (Fig. 4.1).

The altitude of greatest tree species richness in Bhutan (1,900–2,200 m) coincides with the limits of frost-free habitat, except for a few sheltered slopes where some tropical species extend higher yet (Fig. 4.1; Plate 4.15a,b). This high species richness is due to the appearance in the valleys of temperate species that do not survive in the tropics below, yet the survival of tropical elements on adjacent slopes above. Lapse rates are higher on dry rain shadow slopes than on moist ones, while nocturnal valley down-drafts result in temperature inversions. On the ground, the change often takes place over a narrower ecotone; the high number of species at this altitude is therefore due to β-, not α-diversity. In Bhutan, the frost boundary appears to coincide with the appearance of *Castanopsis hystrix,*[211] which curiously occurs *below* the frost line down to 700 m in the lower montane tropical forests of southern Yunnan hills, and in southernmost Taiwan.[212] Indeed, the distinction between tropical lower montane and warm temperate evergreen forest has not been consistently recognized by Chinese botanists.[213]

Ohsawa and Nitta[214] observed a peak of *Symplocos* species and abundance in Bhutan, with *Eurya, Ternstroemia, Rapanea* and *Ardisia*, at these altitudes. Each of these genera includes several species associated with tropical upper montane forest, but here they are in warm temperate communities.. Ohsawa found them in moist sites on mountains at c. 2,000–2,800 m, ranging up to and beyond the upper limit of tropical woody species. He interpreted these forests as the northernmost extension of tropical upper montane forest. But these genera occur to low altitudes in warm temperate South China. *Symplocos* is toxic to cattle in the Bhutan Himalaya where cattle are nearly ubiquitous, and regenerates abundantly where they are present. The forest does become thinly mossy, especially above 2,500 m but as low as 2,000 m in the southern front ranges.

Notably missing, too, are certain prominent genera of tropical upper montane forests, including *Syzygium, Rhododendron* (sect. *Vireya*), and *Dacrydium*, along with other upper montane genera of Australasian affinity. Physiognomically, the ecotone between 1,800–2,200 m in Bhutan is accompanied by an increase in the proportion of deciduous species belonging to temperate genera. Temperate winter-deciduous species of *Acer, Betula, Prunus, Quercus, Cornus, Acacia, Corylopsis, Fraxinus, Sorbus, Populus, Salix, Styrax* and *Celtis* confirm the strong temperate affinity of Eastern Himalayan forests above 2,200 m. There is an increase in species with toothed leaf margins – especially among deciduous species, where their proportion is almost doubled (from 35% to 68%, Table 4.9). Unlike the classic tropical upper montane forests, these forests are not significantly shorter than the tropical lower montane forest below 1,800 m. There are fewer species of mesophyll leaf size class, but most of the incoming temperate species are notophyll, not the microphyll considered diagnostic of tropical upper montane forest (Table 4.6). Notophyll and mesophyll Fagaceae and Magnoliaceae remain dominant in the canopy. Both floristically and structurally therefore, this is a warm temperate forest, still predominantly evergreen. A similar transition seems to occur at c.2,000 m at the northern tropical margin in Central America.[215]

Tagawa and other Japanese botanists[216] describe both tropical lower montane and warm temperate evergreen Fagaceae-dominated forests as 'lucidophyll' (Latin *lucidus*: shiny) in reference to the shiny upper leaf surfaces evident, especially in canopy species and subcanopy evergreen species in moist temperate forests with a deciduous canopy. Furthermore, these shiny leaves are more coriaceous than their lower montane counterparts, frequently glabrous and tilted upward and cupped or revolute, attributes also found in many equatorial subalpine trees and shrubs. In the warm temperate evergreen species, however, these characters are not associated with extreme reduction in size.

Hadfield[217] compared the two ecotypes of domestic tea, *Camellia sinensis*: (1) the Chinese var. *sinensis*, a short canopy tree (to 6 m) of warm temperate forest, in which leaves are small (5–9 × 2–3cm), upward-tilted and shiny (Plate 4.16a,b); and (2) the Assam-Upper Burma form, f. *assamica* (actually a geographical and ecological subspecies), which is a slender subcanopy tree (to c.15 m) of tropical lower montane forest, in which the leaves are larger (8–14 [–22] × 3.5–5.5 [–7.5] cm), hardly tilted, and often somewhat more dull, appearing velvety (though Hadfield observed his to be more shiny – as indeed all *Camellia* leaves are to some extent) and spreading (Plate 4.16c,d). He found that the Chinese form supported a leaf area index of 5–7. This he inferred was because the smaller up-tilted leaves cast less shadow and also reflect photosynthetically active radiation down from their shiny surfaces. The tropical Assam form carried a leaf area index of 3–4. The Chinese form experienced a reduction of net assimilation rate when in shade, because its lower leaves received insufficient light, whereas the Assam form did so in full sun when its leaves became heated beyond the optimum for photosynthesis. Hadfield's concern was the impact of shoot form on leaf production, but productivity may not establish competitive advantage, for a mature canopy tree in nature, over the gain achieved by outshading with a thin canopy of spreading foliage (Chapter 2). Eden[218] comments that temperatures below freezing 'are inimical to tea', particularly when nocturnal frosts are followed – as they often are in monsoon Asia – by a rapid rise in temperature accompanying a cloudless morning, which can scorch expanding shoots. As Michelle Hollbrook commented to me, shininess reflects incident energy early on sunny winter mornings, thereby reducing the potentially lethal effect of a rapid increase in leaf temperature before transpiration can be restored in frozen vessels.[219] Further, being set at a greater angle, they will exchange more energy with the nocturnally warmer subcanopy.

Mamoru Kanzaki[220] observed to me that the canopy species at the Doi Inthanon summit (2,590 m) include *Quercus eumorpha* as subdominant. It has shiny leaves that are notably thicker than those of canopy species in his lower montane forest plot below. A densely leafy evergreen canopy provides a formidable protection against the penetration of frosted air beneath, and explains how an appropriately adapted upper

Plate 4.16 *Camellia sinensis,* tea, which has both lower montane tropical ('Assam', f. *assamica*) and warm temperate ('China', ssp. *sinensis*) varieties. **a,** The China variety in Cibodas mountain garden, Java, introduced to the garden from Japan in 1908; **b,** Ascending foliage of the China variety: Tregothnan tea estate, Cornwall, England; **c,** The Assam variety: survivors of the original provenances transferred from Assam as seedlings, c.1860, in Hakgala Botanic Garden, Sri Lanka; **d,** Foliage of the Hakgala tea trees. (**a** P.A.; **b** J. Jones; **c** P.A.; **d** N. Gunatilleke)

montane forest canopy can remain unaffected while existing beside frost-affected grasslands with microtherm herbaceous perennial herbs.

Shiny and cupped leaves, though here generally microphyll or smaller and coriaceous, also predominate in moist equatorial subalpine forests, and are less evident in tropical lower and upper montane forests there. But further west in the Himalaya, canopy species of the warm temperate forests dominated by *Quercus oblongata* (syn. *Q. leucotrichophora*) and exposed to a longer dry winter season, bear dull, often tomentose leaves. This is also commonly the case among *Ilex* and other evergreen subcanopy genera occurring in both seasonally dry and moist warm temperate climates.

Ohsawa and Nitta[221] note a convergence in growth phenology between tropical lower montane and warm temperate forests. They documented changes in bud and shoot architecture in *Cinnamomum, Symplocos, Maesa* and *Quercus,* from tropical to temperate evergreen forests and their climates. In lower montane forest, primary shoot extension is protracted and buds are naked or protected by fewer scales (cataphylls). This habit may continue in warm temperate understorey species, but the buds of canopy species are protected by abundant scales. Continuous growth and unprotected buds variously give way in frost-prone climates to interrupted growth and buds protected by scales or cataphylls (Fig. 4.12). The microphyll and nanophyll leaves of many temperate *Rhododendron* species – which grow in full sun and

are exposed to nocturnal re-radiation – are shiny and rigid, whereas the mostly mesophyll leaves of subcanopy species are dull and exhibit nastic flaccidity during prolonged periods of low temperature.[222] Shoot growth among the former may be continuous, intermittent or manifold, while the latter show intermittent growth or a single spring flush with bud break accompanied by abscission of the previous year's leaves.

Broadleaved forests above 2,000 m in the Eastern Himalaya are within the summer rain clouds, but above the frost line in the dry winter season. During the exceptionally cold winter of 2007–2008, following hard frosts I observed at 2,050 m at Gedu (Bhutan), many taller trees were scorched (Plate 4.15a). They recovered by epicormic shooting. These warmest temperate evergreen forest Himalayan communities are best developed in the central valleys, which are in rain shadow and below the cloud base. Soils are typical temperate loams, and cryptogamic cover is low. Therefore, neither mossy tropical nor warm temperate 'elfin' upper montane forests appear in the Bhutan Himalaya.

Ground frost and probably occasional rather than regular air frost within the forest canopy marks the climatic margin of the tropics. It marks an ecotone between largely evergreen notophyll tropical lower montane forest (including dry-season-deciduous tropical species) and temperate moist evergreen forest, also predominantly notophyll but frost-hardy and partly winter-deciduous. This forest hardly merits the designation 'upper montane' because it gives way at higher altitude to temperate conifer, then cool temperate deciduous forest at the tree line. The further rise in species maximum elevations at 2,500–2,800 m in Bhutan coincides with the upper limits of this warm temperate broadleaved forest and the lower limits of canopy dominance by the cool temperate conifers *Tsuga, Abies* and sometimes *Picea*. It also coincides with the winter diurnal cloud-fog base and the accumulation of deep raw humus beneath the conifer stands. The few deciduous temperate broadleaved species (especially *Betula* and *Salix*) which continue to the tree line (at 4,500–5,000 m) are pioneer and successional species of steep unstable valley slopes, where soil surface raw humus becomes dislodged.

In the Bhutan Himalayan foothills (26.5° N), dry in winter and shielded by the mountains from cold northeasterlies, hoar frost descends annually to 1,500 m in dry valley fields. Ground frost on open land therefore descends almost 1,000 m lower than the upper altitudinal limits of tropical woody taxa, and c.300 m below the lower limits of temperate woody taxa there. Evergreen stands in the warm temperate forest also apparently minimise frost from the subcanopy and soil. The upper tropical forest limit must therefore be attributed to incidence of air frost, although incidence of ground frost in adjacent open land would seem to correlate with the lower limit of temperate canopy tree species in the valleys where they first establish.

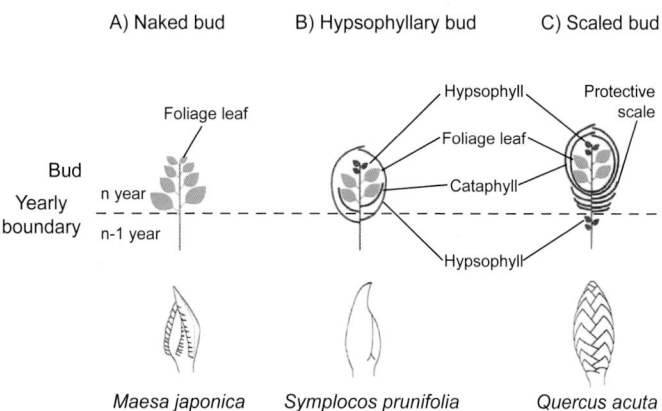

Fig. 4.12 Changes in bud protection with change in minimum winter temperature. There are two categories of foliar organs: foliage leaves and protective organs (bud scale, cataphyll, hypsopyll). Schematic illustration of three bud types of evergreen broad-leaved trees. Internal structure (upper) and external appearance (lower) of each bud type: **a**: naked bud (*Maesa japonica*); **b**: hypsophyllary bud (*Symplocos prunifolia*); **c**: scaled bud (*Quercus acuta*). (From Ohsawa & Nitta 2002)

5.4.2. *A horticultural implication*

This may come as a surprise to horticulturists – so many tree species whose native habitats are *above* the frost line in East Asia seem to be sensitive to frost in moist climates at higher latitudes. There appear to be two possible explanations. First, the daily range of leaf temperatures will be greater at lower latitudes, where nocturnal canopy frost thaws daily, fog rarely accumulates before noon and the mornings are warm and sunny. Second – especially relevant for the deciduous species – summer temperatures in oceanic climates at higher temperate latitudes are relatively cool. Lack of hardiness of species like *Acer campbellii* may thus be due to lack of summer heat. (A cognate case is the group of North American species, including *Cornus florida* and *Cercis canadensis*, which are barely hardy in oceanic North-West Europe.) The protection afforded by early snow in the Himalaya cannot be a further reason, because in the Himalaya frost starts relatively early in the dry season, in November, whereas snow rarely falls at warm temperate altitudes there before late December. Snow is then intermittent, and does not settle long below 3,000 m.

5.5. *Departures from previous montane forest classifications*

The terminology of my classification of montane forests differs from that generally adopted in the continental tropics of Asia, as follows:

1. Champion's 'subtropical broadleaved hill forest', often called 'hill evergreen forest' in Thailand, is termed by me *(tropical) lower montane forest*, following Whitmore. At the northern end of the Western Ghats on the little plateau of Mahableshwar, 18° N, Champion's 'western tropical hill forest', occurring above c.1,000 m to c.1,500 m, includes several elements of lowland tall deciduous or semi-evergreen forest. Some of these, such as *Terminalia chebula*, are deciduous. Here, as on the hills of more seasonal parts of Indo-Burma, analogous fire-mediated forests reach high into the lowland-lower montane ecotone. This forest differs from the Indo-Burmese in its short and gnarled stature, in this resembling upper montane forest; but it is only sparsely mossy and, though endowed with species of *Syzygium, Memecylon* and other genera, these occur in both upper montane and lowland forests. It may be regarded as an intermediate type, probably associated with its isolated hilltop location and relatively long dry season, notwithstanding its exceptionally high mean annual rainfall.

2. Champion's 'subtropical dry broadleaved evergreen forest', of the Central to West-Central Himalaya, is frost-prone and by my criteria *warm temperate*.

3. Champion's 'southern montane wet temperate forest', along the spine of the Western Ghats above c.1,800 m, is regarded here as *tropical upper montane forest*, containing several taxa unknown in temperate forests, and entirely evergreen.

4. Champion's northern, that is, 'Himalayan montane wet temperate forest' (also just present in the Khasi hills), in which most widespread Laurasian temperate deciduous tree genera are present and increasingly common with altitude, I regard as *warm temperate evergreen forest*.

6. Dynamics of montane forests

6.1. *Productivity*

Mergen *et al.*[223] found, by a growth chamber experiment, that tree growth can be increased by increasing the diurnal in relation to nocturnal leaf temperatures. Whereas diurnal leaf temperatures are achieved in nature by incident radiation ameliorated by transpiration and may therefore rise higher than diurnal ambient temperatures, nocturnal temperatures will not be so influenced and will follow the ambient temperature lapse rate with altitude. They predicted therefore that there is a more gradual altitudinal decline in diurnal than nocturnal leaf temperatures, and that gains that would thereby be achieved by photosynthesis over lowered nocturnal respiration rates would lead to increased net production rates at higher elevations. The impressive performance and stature achieved above-ground by introduced *Eucalyptus* in plantation below equatorial upper montane elevations might thus be explained, though it may also be explained if there is no commensurate or greater increase in below-ground biomass. There have been few comparative studies of lowland and montane forest growth and dynamics since Brown's classic study in the Philippines.[224] He recorded *declining* growth rates with altitude. Species compositions imply the same throughout the region. Persistent cloud cover and fog, together with a consequent decline in diurnal leaf temperatures with altitude, may be expected to reduce rates of photosynthesis.[225]

On ultramafic Gunung Silam (Sabah), John Proctor[226] found median diameter growth rates declining from 1.56 mm y^{-1} at 280 m to 0.77 mm y^{-1} at 870 m in the notophyll lower montane *kerangas* summit pole forest. Magnesium exceeded calcium concentrations here, as it can on humult yellow sandy and sandy clay soils in lowland northern Borneo (Chapters 1, 3). This ratio is, however, reversed in the leaf-litter.

Kitayama and his colleagues Takyu and Aiba, in the most comprehensive study of altitudinal vegetational change on any equatorial mountain, estimated above-ground net primary productivity (ANPP) at four altitudes at 700–3,000 m on

Kinabalu.[227] They found that ANPP declined with declining forest stature and LAI, while estimated biomass declined with increasing altitude (Table 4.10). Within lower montane forest at 1,700 m they had six sites, on slopes and ridges on three substrates (not presented in Table 4.10). They found that ANPP was correlated with LAI at each site. Leaf mass per unit area increased from slope to ridge on all substrates, as mean individual leaf areas decreased. This implies that the outer canopy becomes more diffuse, allowing deeper crown foliage and denser subcanopy (see Chapter 2). Differences were greatest between the alluvial and ultramafic substrates. The pool size of soluble phosphorus and inorganic nitrogen in topsoils decreased from alluvial through sedimentary to ultramafic substrates at 1,700 m, but surprisingly did not do so consistently with increasing altitude. Mean leaf phosphorus and nitrogen contents, when expressed by unit leaf area, increased with altitude on the sedimentary substrate (Körner gained the same results for nitrogen, but not phosphorus[228]),

but differed among substrates only in the case of phosphorus. Stand-level nutrient use efficiency increased with altitude only on the ultramafic substrate.

From these observations, Kitayama's team inferred that net assimilation rates may be maintained with increasing altitude, except on ultramafic substrate owing to its phosphorus deficiency. They therefore suggested that the general decline in ANPP and above-ground biomass with altitude on tropical mountains may be attributed to increasing relative investment in below-ground biomass. These results differ from those of Vitousek and colleagues,[229] while Tanner found in Jamaica that nitrogen and phosphorus concentrations decline above 1,500 m (that is, in upper montane forest). These workers conjectured that upper montane forests are more limited by nitrogen than phosphorus, whereas lowland forests are more limited by phosphorus. On acid raw humus-bearing (humult) soils, where leaching is increased by the phenolic compounds released from the litter,[230] gradual long-term nutrient impoverishment is likely (Chapters 1, 3).

Table 4.10 Changes in forest productivity and soils with altitude: Kinabalu. (after Takyu *et al.* 2003)

Property	Site							
Altitude (m)	700		1,700		2,700		3,100	
	Sed.	Ultra.	Sed.	Ultra.	Sed.	Ultra.	Sed.	Ultra.
Soils								
pH (H$_2$0) (Standard deviation)	4.1 (0.1)	4.5 (0.2)	4.0 (0.2)	5.4 (0.1)	3.4 (0.2)	5.1 (0.1)	4.9 (0.1)	5.3 (0.1)
C/N ratio (S.d.)	13.8	11.4	13.7	12.1	19.2	9.9	11.4	13.3
Ex. Ca (S.d.)	17 (10)	29 (12)	83 (13)	630 (132)	61 (18)	299 (147)	734 (336)	375 (147)
Vegetation								
Total AGB (kg m^{-2})	43.7	55.4	29.4	23.8	30.8	12.2	21.5	3.7
LAI (m^2 m^{-2})	4.8	5.6	3.8	3.6	3.5	3.8	2.4	1.6
ANPP (g m^{-2} y^{-1})	1913	1715	1222	813	780	725	816	199
NAR (g m^{-2} y^{-1})	797	613	643	542	446	382	680	248
Ln (PUE) (g gm^{-1})	8.47	8.42	8.83	9.73	8.53	9.62	8.60	10.23

Sed: sedimentary substrate; Ultra: ultramafic substrate; Ex.: exchangeable; AGB: above ground biomass; LAI: leaf area index; ANPP: above ground net primary productivity; NAR: net assimilation rate; Ln (PUE): phosphorus use efficiency.

6.2. Determinants of species' altitudinal ranges

A dramatic example that could imply that competitive exclusion may set upper limits for many taxa is the domination of lower montane forest in Sri Lanka by Dipterocarpaceae of the canopy genus *Shorea* (one species), and several species of the subcanopy *Stemonoporus* (Fig. 4.6), where Fagaceae are absent for historical reasons. I have inferred that this is due to competitive advantage elsewhere of Fagaceae.[231] Another such example, in this case for setting lower limits, is the appearance of Fagaceae-dominated forest low on some coastal mountains, 300 m in New Guinea and 600 m on Mount Maquiling, Luzon, where the lowland mixed forest flora – notably Dipterocarpaceae – is relatively impoverished. In both these cases, it is noteworthy that the invading family is represented by a single dominant species in the canopy.

Savitri Gunatilleke and colleagues[232] tested the very same monodominant Sri Lankan lower montane *Shorea*, *S. gardneri*, by growing its seedlings with five other species in its section, *Doona*, in uniform soil, in shade houses at 125 m, 580 m and 1,060 m. Species performance rank order with respect to height and dry mass changed with elevation (Fig. 4.13). Notably, height growth declined with altitude in all species but for the lower montane dominant rata dun (*S. gardneri*), indicating that physiological constraints imposed by temperature decline had a limiting influence on the majority. Height growth of *S. gardneri* was slow throughout, and only once matched another species, *Shorea disticha*, though at 1,050 m. The results implied that rata dun's performance at lower elevations reduced its competitiveness. I infer that the slow-growing *S. gardneri* is the arboreal tortoise that – subject to the demanding cloudy conditions within the diurnal cloud base – eventually wins over the hares! These particular results are consistent with Peter Grubb's conclusion[233] that physiological constraints on growth or reproduction, mediated especially by temperature decline, set the upper altitude of most species, and impose an underlying continuum in the upper limits of species' altitudinal ranges. Such physiological constraints might explain the individualistic upper limits of lowland species above c.400 m, some of which extend to the upper elevational limits of lower montane forest. And it has also been shown that tropical alpine herbs of 'microtherm' (that is temperate) genera fail to reproduce in the lowland tropics when grown ex situ. This is the case also with temperate fruit trees.

Many but not all species of lower and upper montane forest may also be found in scattered habitats at lower altitudes. This implies that competition, rather than physiological constraints, sets limits on their *lower* altitudinal abundances. Upper montane elements of Australasian affinity especially, including *Weinmannia* and *Xanthomyrtus*, occur occasionally in physically limiting habitats west of Wallace's Line, especially

Fig. 4.13 Performance of seven species of *Shorea* sect. *Doona* along the altitudinal gradient over three years. Changes in letters indicate significant differences in figures within a species. The lower montane *S. gardneri* succeeds by sustaining height growth at altitude: a tortoise among the hares! (From Gunatilleke *et al.* 1998)

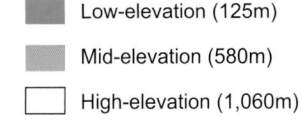

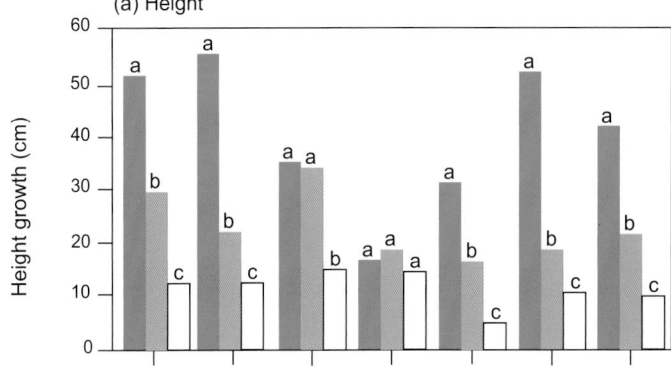

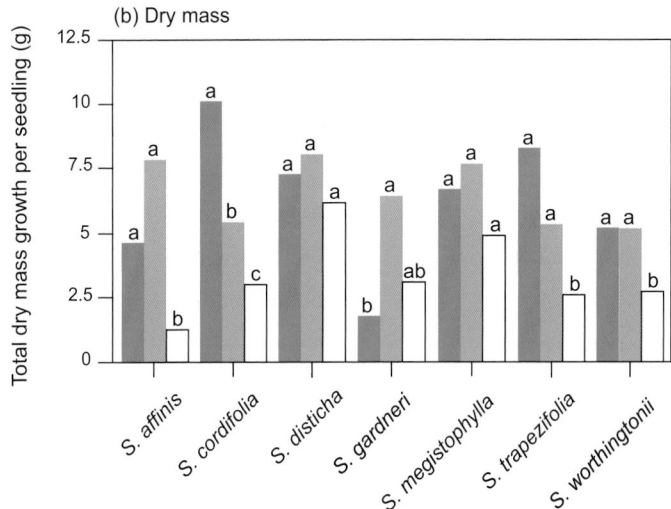

rocky headlands with thin humult soils; but they become more widespread in lowlands east of the Line. Clearly, there is no physiological constraint on the survival of these particular species in the lowlands, but competition seems to be relaxed. In addition, in northern Indo-Burma, South China and the eastern Himalayan margin of the tropics, several species characteristic of lower montane forest on mull soils descend regularly into the lowlands (Tables 3.8, 4.3). In this second case, the lowland soils are relatively rich in humus and earthworms are abundant. In the Duars, excellent tea is grown on a large scale at only 100 m elevation. These latter distributions appear to be explained by the greater similarity, at higher tropical latitudes, of lowland and lower montane environments (Chapter 3).

Wet zone lowland fruit tree species	Family	Highest altitude (m) at which trees yield fruit		Remarks
		Sri Lanka	Bhutan	
Manilkara zapota	Sapotaceae	100	none	
Cocos nucifera	Arecaceae	600	350	
Aegle marmelos	Rutaceae	600	250	
Annona reticulata	Annonaceae	600	250 in garden	Planted along riverbank at 1,050m in drier eastern Bhutan, where fruit smaller
Annona muricata	Annonaceae	600	none	
Annona squamosa	Annonaceae	600	250	
Averrhoa bilimbi	Oxalidaceae	600	none	
Averrhoa carambola	Oxalidaceae	600	none	
Citrus grandis	Rutaceae	600	1,500	
Cynometra cauliflora	Fabaceae	600	none	
Diospyros discolor	Ebenaceae	600	none	
Garcinia mangostana	Clusiaceae	600	none	
Garcinia quaesita	Clusiaceae	600	none	
Garcinia xanthochymus	Clusiaceae	600	1,000	
Hevea brasiliensis	Euphorbiaceae	600	none	
Nephelium lappaceum	Sapindaceae	600	None	But *N.hypoleucum* to 300 m
Punica granatum	Punicaceae	600	1,200	But grows to 2,400 m, with small hard fruit, in Bhutan
Spondias dulcis	Anacardiaceae	600	none	
Theobroma cacao	Sterculiaceae	600	none	
Durio zibethinus	Bambacaceae	700	none	
Syzygium jambos	Myrtaceae	700	600	
Areca catechu	Arecaceae	900	300	
Chrysophyllum caimito	Sapotaceae	1,000	none	
Musa x paradisiaca	Musaceae	1,200	1,200	
Annona cherimola	Annonaceae	1,200	none	
Persea americana	Lauraceae	1,200	1,800	
Carica papaya	Caricaceae	1,200	1,500	
Mangifera indica	Anacardiaceae	1,200	1,300	
Muntingia calabura	Tiliaceae	1,500	none	
Artocarpus heterophyllus	Moraceae	1,600	1,500	
Pyrus communis	Rosaceae	2,000	2,600	
Annona cherimola	Annonaceae	1,200	1,000	

Table 4.11 Maximum elevation at which cultivated (mostly exotic) lowland tropical fruit tree crops will yield, Sri Lanka, 6°N, and Bhutan, 27°N. Species restricted below 600 m in Sri Lanka are generally restricted lower in Bhutan, implying that minimum temperature may be the limiting factor; 600 m is the altitude above which many MDF species on the widespread clay soils begin to disappear. *Citrus grandis* and *Punica granatum* may be limited in Sri Lanka's wet zone by year-round cloudiness. (C.V.S. & I.A.U.N Gunatilleke, R. Pradhan)

Where many species' elevational limits coincide, they are clearly directly constrained by one or more physical factors, though they themselves are indirectly mediated by ambient temperature. Thus, many tree species are constrained by edaphic and topographic as well as climatic changes at lower elevations, particularly at c.400 m and 750–1,200 m, and by the persistent diurnal fog at higher, albeit varying, elevations. The generally strict confinement of Winteraceae, and also several isolated genera of the lower montane forest, to their zonal forest types in equatorial regions suggests that ecophysiological constraints may sometimes be as important as competition in determining both upper and lower altitudinal ranges.

The lower montane flora ascends to exceptionally high altitudes on the seasonally foggy summits of continental tropical Asia, and in the relatively fog-free upper valleys of New Guinea. This again implies that other climatic factors, in this case diurnal fog, rather than the direct influence of low temperature may often establish the physiological barrier to species' altitudinal range. Nevertheless, altitudinal floristic zonation, being correlated with soils ecotones, is likely to be mediated also by changes in species competitiveness (Chapter 3).

Constraints imposed by physical, especially climatic changes with altitude are also nicely demonstrated by the upper altitudinal limits of yield in tropical orchard-grown fruit tree species; but although each has distinct limits, these again aggregate at certain altitudes, the constancy of altitudinal lapse rates of temperature notwithstanding (Table 4.11). At low latitudes as in perhumid south-west Sri Lanka, the main aggregation is at 600 m, below the lowland-lower montane forest ecotone, but the altitude at which many lowland MDF species of clay loam soils begin to disappear despite the continuation of their soils to higher altitudes. In Bhutan, at the margin of the tropics, the aggregation of those few of these species that occur there is at lower altitude, implying that it is mediated by winter temperature. In Sri Lanka another aggregation occurs within the lowland – lower montane forest ecotone. This is less clear in Bhutan where two species especially, *Citrus grandis* and *Persea americana,* continue to higher altitudes where they are mildly frost hardy. Here I infer that the year-round diurnal cloudiness of this altitude in Sri Lanka, relieved in Bhutan by the sunny winter months, is the constraining factor. But, this is the altitude at which temperate vegetables and tea begin to flourish, due to the appearance of temperate humult loam soils.

6.3. *Succession*

Among the 162 tree species in the CTFS 15 ha plot at 1,700 m on Doi Inthanon, less than one tenth are pioneers although a further 15% can be considered successional.[234] That proportion is comparable to the lowland MDF plots at Pasoh and Lambir. In Bhutan, five species of *Macaranga* are found and only one

of these is confined below 1,000 m, while *M. denticulata* is restricted to montane forest and grows as high as 2,100 m. Among the few other montane pioneers in Bhutan are *Mallotus philippensis,* and at warm temperate elevations the deciduous *Hovenia* and *Tetradenia*. But pioneers rarely dominate in natural early succession in the lower montane forest. Instead, species with late successional characteristics, including the archaic genera *Daphniphyllum* and *Exbucklandia* (Plate 4.6c), also *Adinandra, Elaeocarpus, Symplocos* and *Turpinia,* may become locally dominant in early succession.[235] Some Lauraceae require canopy gap light conditions for germination.[236]

On the other hand, in New Guinea over half of the 83 *Macaranga* species extend up into lower montane forest.[237] Forty of these appear confined to montane forest but eight reach 2,500 m and five as high as 2,800 m, though whether in mossy upper montane forest is not recorded. These are young earthquake-prone mountain systems with steep, landslip-prone surfaces. On the ancient mature rounded topography of Peninsular Malaysia, only four *Macaranga* species occur in lower montane forest, three of which are restricted though to montane forest elevations.

Himalayan and Far Eastern lower montane forests seem, nevertheless, generally poorer in patches of gap succession dominated by a diversity of pioneer species than lowland evergreen forests.[238] Few species with the suite of pioneer attributes (Chapter 2) are found in most upper montane forests, as in *kerangas*. This is in contrast both to New Guinea upper montane forests, where the similarly steep landscape supports a rich *Macaranga* flora,[239] and to Himalayan cool temperate broadleaved forest, where pioneer and successional species of birch and willow are prevalent. In the Himalaya and the Indo-Burmese mountains, monotypic stands of *Alnus nepalensis* (in moist sites, Plate 4.6b) and *Betula alnoides* (on steep rocky slopes) do dominate early succession on landslips in lower montane forest. On drier lower montane slopes in New Guinea, *Araucaria cunninghamii* (var. *papuana*) establishes on moist landslips in lower montane forest which, with its light winged wind-borne seeds, it invades before most broadleaved pioneers.[240] *Araucaria* individuals can exceed 70 m; once they gain height above the angiosperms, they have no competitors.

In Bhutan at 1,700–2,000 m, lower montane forest succession 20 years after swidden was estimated to have recovered 60% of the basal area of primary forest, but the largest individuals less than 15 m height and only 15 cm maximum diameter. Maximum diameter growth was only 7 mm y^{-1}.[241]

6.4. *Patterns of mortality and regeneration*

Otherwise, Asian tropical lower montane forests mainly seem to experience single-tree canopy mortality.[242] Lower montane Fagaceae require gaps for successful recruitment

– though they are often initially outgrown by early successional species that, as in temperate deciduous forests, they eventually overtop. No Fagaceae are included among the truly pioneer and early successional species in wet lower montane forests. However, *Quercus lanata,* and in moister conditions *Q. glauca,* do establish early in *Pinus roxburghii* savanna. The tropical pines *P. kesiya, P. merkusii* and *P. roxburghii* are light-demanding pioneers, dependent on fire to eliminate or postpone the fire-sensitive broadleaved competition at establishment and early regeneration. In Indo-Burma, certain Fagaceae, including *Castanopsis argyrophylla, C. calathiformis, Q. kerrii* and *Q. kingiana,* similarly participate in broadleaved pioneer vegetation in *P. kesiya* savanna. Fagaceae species establishing in such savannas are sensitive to fire above-ground, but their rootstocks resist fire and resprout.[243] Then, if several years pass without fire, they come to engulf the maturing pines, which are unable to regenerate beneath their shade. Above and to the west in Bhutan, these oaks are replaced in frost-prone stands by the ban oak (*Q. oblongata* (syn. *leucotrichophora*)).

Upper and lower montane forests consist overwhelmingly of species capable of coppicing into maturity – Myrtaceae and particularly Fagaceae, whose coppicing ability has been traditionally exploited by the herding tribes of the Himalaya and Indo-Burma[244] (Chapter 8). In the eastern Himalaya and Indo-Burma, the evergreen Fagaceae and many of their associates are noted, like their temperate deciduous counterparts, for their stump sprouting, both in nature and following logging and browsing or *jhum* (swidden). In *Trigonobalanus verticillata* (Fagaceae), which is confined to lower montane oak-laurel forest forest from Thailand to Sulawesi, the mature trunk eventually collapses leaving a ring of coppices (Plate 4.4d).[245] *Unfortunately, coppicing has been little studied in the perhumid regions.*

In those few areas where hunting is controlled, lower montane oak-laurel forests seasonally support high densities of boar and deer, which come to consume the acorns. In Borneo, the bearded pig formerly migrated in hordes to lower montane oak forests during mast years[246] (Chapter 5). *Unlike dipterocarps in the lowlands though, Fagaceae seem not to mast synchronously at family scale, although systematic observations have not yet been made. Fagaceae seeds are dispersed over limited distances.*[247] For centuries, cattle grazing and browsing have been ubiquitous in the lower montane oak-laurel forests of seasonal tropical Asia. Nevertheless, where it is less intense, oak seedling regeneration is abundant near reproductives.[248] The coppicing ability of juvenile Fagaceae also gives them competitive advantage in forests with high biomass of indigenous browsing mammals, which may be one reason for their dominance on the fertile mull soils of lower montane forests.

7. Discussion and conclusions

7.1. *Which previously published zonal characteristics have been confirmed?*

1. *The substantial range of canopy height* within montane zones has long been recognized, although maximum heights in the lower parts of zones have been underestimated. Emergent species can reach 50 m in lower montane forest. (By way of comparison, lowland emergent dipterocarps have been measured to 88 m.) Upper montane forest species can reach 25 m in New Guinea and exceed 15 m on Kinabalu.

2. *The number of forest strata* can be rather precisely reported: there are three in sheltered lower montane forests where emergent species are present, but two there otherwise; two in the taller upper montane forests on big massifs, but essentially only a single (canopy tree) stratum in short-stature equatorial upper montane and subalpine forest – except, again, where a single shortly emergent species may commonly occur (Plate 4.13d).

3. *The dominant canopy leaf size class* is notophyll in lower montane oak-laurel forest but microphyll in, for instance, the lower montane forests of Sri Lanka. This recalls *Shorea curtisii,* dominant in MDF along coasts and granite ridges in Peninsular Malaysia, whose microphylly I attributed to periodic water stress (section 3.1 above); but that seems unlikely in the Sri Lankan mountains. Similar microphyll dominance occurs at the highest elevations of South Asian and Indo-Burmese forests, which appear on other structural and floristic criteria to be lower montane. Tree leaf size class representation is, of course, highly correlated with the flora representing it. Microphyll floras dominate upper montane forest, but also appear in the upper elevations of structurally lower montane *kerangas* and on rocky knolls in the lowlands. Nanophyll and smaller leaf sizes increase in subalpine forests.

4. *Pinnate leaves* are well represented among successional species at the lower elevations of lower montane forest, and are notably present in upper montane forest among Cunoniaceae, Rutaceae and Staphyleaceae, albeit not often abundant.

5. *Buttresses* are, as already recognized, almost universally short or absent in montane forests, irrespective of tree size.

6. *Rarity or absence of cauliflory* in montane forests remains valid.

7. *Distribution of woody and bole vines* is broadly valid. Bole and herbaceous vines are more abundant than large woody vines in lower montane forests except at the lowest altitudes: Woody climbers (vines), sometimes large, can be abundant in secondary and successional lower montane forest. Vines are generally small and sparse in upper montane forests, and absent in alpine forests.

8. *Vascular epiphytes* are more abundant in upper than in lower montane forest, but their highest floristic richness is reached at the lower elevations of lower montane forest.

9. *Ground herbs* are both more diverse and more abundant in lower than in upper montane forest.

7.2. If our primary definition of tropical montane forest zones is floristic, how closely are structural and physical attributes correlated with it?

Tree species composition is broadly correlated with zonal forest types recognized on structural and physiognomic criteria. Nevertheless, plasticity in key criteria (notably canopy height, bole height and leaf size class) results in partial independence of structure and floristics, especially at ecotones. For example, upper montane elements begin to appear in the flora of lower montane *kerangas* at relatively low altitudes, associated with humult soils below the altitude of penetrating fog. The structural-physiognomic ecotone between lower montane *kerangas* and upper montane (elfin) forest – floristically defined – does begin lower down than in other lower montane forests. This ecotone is more gradual than that between lower montane oak-laurel and mossy upper montane forest (floristically defined), even though upper montane forest elements can appear on scattered convex surfaces in lower montane oak-laurel forest hundreds of metres below the structural-physiognomic ecotone.

Myrtaceous upper montane elements descend along dry rocky crests to quite low altitudes. Several of these species also extend up into subalpine forest, where they experience periodic drought. Their short, gnarled but relatively cryptogam-free lowland forests recall the dry evergreen montane forests of the north-east massif in Sri Lanka, which are also intermittently drought-prone. Only a small group of upper montane angiospermous trees, notably *Tasmannia piperita,* are confined to upper montane forests.

Another upper montane forest characteristic is the occurrence of non-vascular epiphyte sleeves formed by the accumulation of dead organic matter between the living bryophyte layer and the tree bark substrate. Continuous cover of bare ground by bryophytes, many of which can also live as non-vascular epiphytes, is characteristic of equatorial upper montane forest. The mossiness of paradigmatic upper montane forest is broadly correlated with the persistence of diurnal fogginess.

But the relative role of fogginess and intermittent drought in determining upper montane forest structure, dynamics and physiognomy has yet to be determined.

7.3. What changes in physical habitat correlate with changes in the montane vegetation?

1. Floristic composition is primarily correlated with soil characteristics, specifically the nature of the organic matter content (see section 4.3.3 above). Lower montane forests occur on humic-clay and sandy-clay mull soils, as well as on acid raw humus-bearing humult soils over sandstone and ultramafic rock; they vary floristically and physiologically accordingly. The sole constant environmental variable in sites at these altitudes and latitudes is the climate of the summer monsoon. Upper montane species may appear in advance of major structural or physiognomic change, where soils on well-drained convex surfaces begin to accumulate surface raw humus or where wet organic muck appears on concave valley slopes. Several tree species are found, with relatively little variation in leaf characteristics, in upper montane (in the broad sense) forest, in lowland *kerangas* and at the centre of domed peat swamps. These environments differ greatly in water economy but all present raw humus-bearing soils, implying that soil nutrient constraints may be at least as influential as water stress in determining species distributions and leaf morphological characteristics. But raw humus accumulation appears to correlate with mossiness as well as substrate, and apparently frequency of diurnal penetration of fog within the canopy. I therefore conclude that climate *does* influence the floristic transitions – but that it does so in large part *indirectly*, through its influence on soil formation processes and the nature of the soil surface on which seeds germinate and establish. Nevertheless, upper montane species and forests have been observed on mull soils in New Guinea, and such soils do occur on well-drained base-rich substrates as well as on landslips. At all altitudes, the direct and indirect influences of topography and substrate on the positioning of ecotones are more important than is commonly acknowledged.

2. Upper montane fog is moved by updrafts or the prevailing winds of the seasonal tropics, when the vegetation, its surface area enhanced by epiphytes, rakes precipitation by physical entrapment of the water droplets.[249] This is dramatically evident in the Nilgiris, where forests at above 1,900 m on humic mull loam soils adopt the structure and physiognomy of upper montane forest on foggy windward

slopes, yet a lower montane structure and flora remains dominant on seasonally fog-free leeward slopes. The structural and physiognomic ecotone between lower and upper equatorial montane forests is correlated with the constancy of diurnal fog, irrespective of soil type, though fog itself creates the condition for rapid soil acidification, leaching, and raw humus accumulation. The zone of persistent diurnal fog in the perhumid equatorial and oceanic tropics varies in altitude on mountains of different sizes and degrees of isolation. It also correlates with the presence of highest cryptogam densities and the complete mossy cover of litter-free soil.

3. Lower montane *kerangas* undergoes a transition at its upper elevation from dominant noto- to microphylly. Since this happens below the daily fog zone, but well within the cloud base and in areas rarely if ever subject to drought, it implies that microphylly is induced here by nutrient deficiency; which is nevertheless increased by fog which restricts transpiration and translocation of metabolites. The presence of single-stratum stands of gnarled trees on narrow rocky ridges down to low altitudes implies that periodic water stress may also play an important part in determining these structural characteristics. Dominant microphylly is not unique to upper montane forest: it also occurs in some lowland *kerangas* displaying elfin structure, on drought-prone ridges and along coasts exposed to frequent monsoon gales (Chapter 3).

4. Tropical montane forests – especially equatorial upper montane and subalpine forests – bear exceptional abundances of species with young shoots rich in anthocyanins (section 5.1 above). I accept that these act as barriers to the high incident radiation at their altitudes, which would otherwise induce photo-inhibition in leaves which must also endure long intervening periods of fog and saturation.

5. Equatorial subalpine forests, like warm temperate notophyll evergreen forests, contain a group of species whose mostly glabrous leaves display convexly cupped, shiny upper surfaces. These characters will lead to the lowering of diurnal and raising of nocturnal average leaf temperatures. But the two forest formations differ greatly in predominant leaf size class. Subalpine forests are microphyll, nanophyll or leptophyll, while warm temperate forests are predominantly notophyll in both evergreen and deciduous species. They differ, too, in canopy structure: subalpine forest canopies are open, with leaves mostly inclined steeply upwards, while warm temperate evergreen forest canopies are mostly densely closed with spreading leaves (China tea notwithstanding). I infer that leaf surface, shape and inclination are adapted to

resist freezing nocturnal temperatures in both forest types, whereas the leaf size and canopy structure of subalpine forests are mainly adaptations to withstand occasional severe water stress in windy environments that are periodically above the cloud belt. Such drought conditions are rarely experienced in moist warm temperate forests, although I have seen some rocky ridges in China where these characters were manifest (and where China tea grows in nature).

6. Within the canopy, the upper montane to subalpine forest ecotone is signalled at first by the loss of moss sleeves, then by the loss of mossy ground-cover. The organic surface horizon of mostly shallow subalpine soils may become mixed with the mineral material below. This change in the soil surface must alter the regeneration niche.

7. Despite these qualifying observations, there can be no doubt that decreasing temperature with elevation, and variably diffuse and limited light conditions, must play a leading role in the energy-balance and physiology – and therefore in the structure and competitive relationships – that differentiate altitudinal forest zones on wet tropical mountains.[250] However, temperature decline is continuous, whereas floristic and structural-physiognomic change on mountains of tropical Asia is not. Zonation therefore cannot be explained by the direct influence of temperature, but instead indirectly, through its influence on the balance between litter accumulation and decomposition and other soil-forming processes, and on atmospheric water condensation.

These conclusions, and forest-grassland inversions observed in alpine valleys,[251] should warn palaeontologists against inferring Pleistocene mean temperatures and lapse rates in the tropics from tree pollen. The alpine herbaceous pollen record may provide reliable evidence, but very few tree taxa, among them Winteraceae (whose pollen is unfortunately thin-walled and fossilises poorly) and some Chloranthaceae and Hamamelidaceae, would yet appear to be reliable indicators of past altitudinal distributions (Chapter 6).

7.4. Are changes in forest structure and composition continuous, or is altitudinal zonation real?

The floristic and structural-physiognomic transition between the lowland forest and lower montane forest types is gradual, but even this gradual change is often punctuated by definable small steps. The floristic and structural onset of lower montane forest may occur as low as 700 m, and becomes more noticeable at 900–1,200 m. At 1,300–1,700 m, and as high as 2,900 m in New Guinea, paradigmatic lower montane forest (as distinct from either lowland or upper montane forest) is recognisable throughout the Asian tropics. This forest type is characterised by dominance of certain families, genera and some species, and a suite of characteristic taxa – though few of these can be found both in the Far East and in South Asia where lower montane forests (*sholas*) are recognized by different species. The ecotone between lowland and lower montane forest is distinct only where the lowland forest is deciduous and mediated by fire, and at high latitudes, as in southernmost inland and southeastern coastal China and Taiwan, where lower montane Fagaceae descend ridges locally to 725 m while lowland forest persists 100 m higher on adjacent slopes.

Defined structurally and physiognomically, lower montane and upper montane forests are clearly distinguishable in most equatorial and oceanic regions. On most mountains of equatorial Asia, the two forest types are generally separated over a narrow ecotone accompanying a change in topography. Upper montane forest is generally, and densely mossy upper montane forest is always, correlated with persistent fog. The floristic transition is less distinct, in part owing to the influence of substrate and topography on soil and therefore floristic composition (see section 8.3 above), in part because most of the limited group of genera and species characterising upper montane forest are well represented in other, more diverse, forest types. In addition, many upper montane floras are characterised by local endemic rather than regionally widespread species, in contrast to lower montane forests (see Chapter 6). Upper montane forests become both floristically and structurally increasingly indistinct at higher latitudes in continental regions, as fog becomes increasingly seasonal or intermittent.

The floristic ecotone is often correlated with different habitat factors from the structural-physiognomic ones. Volcanoes are exceptions to this: their slopes and ridges tend to rise with gradually increasing steepness, and an overall transition in both structure and flora may extend over more than 100 m. The vast high-altitude New Guinea massif is another exception, for here the cloud base often sits as high as 2,800 m, and fog is often more intermittent below this altitude than in other regions. At 2,200–2,850 m, New Guinea forests undergo a transition between lower and upper montane forest types. Lower montane forests display many of the floristic and physiognomic attributes of upper montane forest, such as epiphytic bryophyte densities increasing with altitude; but gentle slopes continue to support tall forests of upper montane flora, while epiphytes including bryophytes are sparse in high rain shadow valleys.

In general, there is no sharp ecotone between upper montane and subalpine forest. Tropical subalpine forest, which in Asia is confined to equatorial mountains, differs quantitatively from upper montane forest in tree stature, leaf size class representation and shape, and epiphyte abundance, and

qualitatively in both the presence of a distinct species element and the absence of many upper montane species. It is correlated with decline and qualitative changes in precipitation, increase in sunshine hours and periodic drought, and decrease in soil organic matter accumulation. It is also correlated with lower ambient temperatures and the possibility of occasional frost.

In summary, subtle and continuous transitional zones exist, not all primarily related to elevation. Nevertheless, floristic differences distinguish four major forest zones (lowland, lower and upper montane and subalpine) which correlate broadly (though not consistently) with structural and physiognomic criteria. As Janzen claimed,[252] these zones are indeed substantially more distinct than on temperate mountains. The forest type terminology that I have adopted here avoids presumed interdependencies with physical factors, for reasons that should be apparent.

7.5. *Do forest ecotones occur at consistent altitudes?*

The question of the elevational constancy of forest ecotones has been hampered by problems in the definition of the zonal forest types. Symington regarded upper dipterocarp forest as a type of lower montane forest.[253] If that is accepted, a consistent though gradual ecotone between lowland mixed forest and lower montane forest (as classically defined on structural and physiognomic grounds) occurs at 750–1,300 m. It accompanies a floristic ecotone at the species level and less consistently at the genus or higher taxonomic level. There is a continuous decline in species numbers, and a continuous compositional turnover, between these altitudes (Fig. 4.1). This ecotonal band marks the upper limit of almost all lowland species, and the lower limit of a smaller proportion of lower montane species. (Several of the latter are widespread, and a few of them occur in mesic humus-rich habitats in the moist high-latitude tropical lowlands (Table 4.3).) When that is accepted, the transition from lowland forest to lower montane forest appears remarkably consistent throughout the moist tropics of Asia, occurring gradually where upper dipterocarp forest occurs, yet often more sharply where it does not, and especially where fire and tall deciduous forests extend into the summer cloud base (Fig. 4.10).[254]

At the species level, the upper limit of the lowland mixed forest tree flora on zonal clay and sandy clay yellow-red soils appears to fall overwhelmingly at 700–900 m, although a few species do extend to the upper limits of lower montane forest. *This ecotone has yet to be examined in kerangas and on podsols, but relatively more lowland species appear to transgress it in these environments. Species of deep yellow humult sandy lowland soils disappear at 400 m (or as high as 800 m on granite). Claims that lowland forest extends to 1,500 m in the Neotropics and New Guinea have yet to be substantiated by floristic census. In* summary therefore, there is both stepped impoverishment, and more limited enrichment, of lowland tree flora with elevation, substantially mediated by soils distributions.

A mass elevation effect has been demonstrated on wet tropical mountains *only* in the case of the mossy facies of upper montane forest. The effect is due to its consistent immersion in moving fog within the diurnal cloud base. The altitude of the cloud base is determined not by ambient temperature alone, but by atmospheric humidity and the influence of the length and area of a slope on the volume of updrafted air; and consequently by the humidity and temperature of that air at the altitude at which the dew point is reached. This is why mossy elfin upper montane forest occurs at higher altitudes on higher and bigger massifs than on lower or more isolated ones. The presence of *persistent* diurnal fog correlates most clearly with the change to dense mossiness on ground and stems, and a distinct vascular and non-vascular flora.

Equatorial subalpine forests have altitudinal lower limits correlated with mean ambient temperature and apparent incident energy, but perhaps also with minimum night temperatures and periodic frost. The natural tree line (broad flat valley bottoms excepted) lies well above the main diurnal cloud concentration and is controlled by factors associated with available energy and possibly nocturnal temperatures Both are therefore, and in contrast with upper montane forest lower limits, less affected by the character of the mountain. Highest tree lines – that is, those which strictly mark the ecotone between subalpine forest and alpine herbaceous or dwarf shrub vegetation on surfaces free from the ponding of frosted nocturnal air – appear remarkably constant in the perhumid equatorial tropics at 3,800–4,000 m. Both upper and lower limits of subalpine forests appear to be correlated with a consistent temperature regime. The tree line in the wet seasonal tropics declines in altitude with increasing latitude, until it rises again with the advent of the temperate tree flora. Alpine grasslands maintained by nocturnally ponding frosted air are found at lower elevations, and they descend lower on large and continental than on small and isolated mountains.

The altitudes of forest zone ecotones, and their ranges of variation on Asian equatorial mountains, are consistent with those on Neotropical mountains in similar climates.[255]

7.6. What are the relative roles of competition and physiological constraints on the altitudinal distribution of tree species?

There is evidence that physiological constraints restrict both upper and lower limits to Asian tropical tree species' altitudinal ranges. This is most apparent in the broad ecotone between lowland and lower montane forest, where species

differ individually in their altitudinal limits and without any clear aggregation. In all these cases, competition can and probably does further restrict these limits; but I conclude that physiological constraints become paramount in constraining upper limits when sudden climatic changes, notably fog and frost, both influenced by temperature, intervene.

7.7. Conclusion: determinants of the altitudinal and latitudinal limits of tropical forest

I conclude that the altitudinal limits of tropical forests are indeed established in part by the limits to photosynthetic gain as rates of photosynthesis decline and respiratory costs increase, reflected in part by the increasing root/shoot ratio with altitude. These processes are brought about continuously in organelles within the cytoplasm. On tropical mountains – more than on temperate ones where seasonal change in temperature is marked – mean temperature varies little seasonally and consistently declines with altitude. This combines with a decline in diurnal cloudiness as the tree line is approached to produce an increase in daily temperature range. Acting in concert, these factors appear to be the primary mediators of altitudinal limits.

In addition, though, some slow-growing trees of tropical subalpine forests – and seedlings of nearly all when they establish in the open beyond the forest fringe – appear unable to overcome the effects of exceptional water stress acting synergistically with freezing temperatures. Vessel cavitation and bud abortion result, apparently causing the periodic mass mortality of taller individuals of certain species growing near but not necessarily at the upper limits of montane forests. Frost also scorches tree regeneration invading sward or dwarf shrub-land adjacent to subalpine and upper montane forest, irrespective of topography, thereby preventing closed forest species from invading grass- and shrubland. This creates a dynamic but sharp ecotone dominated by frost-hardy successional shrubs, and constitutes the effective tree line wherever these vegetations are prevalent – such as currently on the majority of high mountains in tropical Asia. The synergistic effects of drought and frost may also be major determinants of tree stature and canopy evenness, and of the dominance of nanophyll and smaller leaf sizes in equatorial forests approaching their altitudinal limits.

References

[1]Beaman & Beaman 1998; Beaman *et al.* 2001, 2004. [2]Kitayama 1991, 1992, 1995; Aiba & Kitayama 1999, 2002; Aiba *et al.* 2004; Kitayama *et al* 2000, 2004, 2012; Takyu *et al.* 2002, 2003. [3]Grierson & Long 1983–2001; Noltie 2000. [4]Ohsawa 1983–1996; Ohsawa & Nitta 1997, 2002. [5]Pradhan 2001 and pers. comm.. [6]Wangda & Ohsawa 2006a, b. [7]H.S. Lee *et al.* 2004. [8]van Steenis 1935. [9]Aiba & Kitayama 1999. [10]Lieberman *et al.* 1996; Vásquez & Givnish 1998. [11]Behera & Kushwaha 2007. [12]Lieberman *et al.* 1996; Vásquez & Givnish 1998. [13]Yamada 1975, 1976a, b, 1977. [14]Grubb *et al.* 1963. [15]Burtt Davy 1938. [16]Whitmore 1984a. [17]P.S. Ashton 1964b. [18]Symington 1943. [19]Burtt Davy 1938. [20]Burgess 1969. [21]Wyatt-Smith 1963, 1966. [22]Symington 1943; Wyatt-Smith 1966. [23]Burgess 1969; Symington 1943. [24]Ediriweera *et al.* 2008. [25]Beaman & Beaman 1989; P.S. Ashton 2003a. [26]Maxwell 1988; Maxwell *et al.* 1995, 1997; Davidar *et al.* 2005. [27]Smitinand 1969–1970; Xie 1997. [28]Symington 1943. [29]Ohsawa 1991a, 1993, 1995 b, 1996. [30]Brass 1941; Kingdon-Ward 1949; P.S. Ashton 1964b, 1988, 2003a, b; Robbins & Smitinand 1966; Smitinand 1966, 1968; Sawyer & Chermsirivathana 1969; van Steenis 1972a, b, 1984; Kochummen 1982; Li & Walker 1986; Stadtmüller 1987; Santisuk 1988; Maxwell 1988; Elliott *et al.* 1989; Hyndemann & Menzies 1990; Mangen 1993; Pendry & Proctor 1996; Maxwell *et al.* 1997; Hamann *et al.* 1999; Pradhan 2001. [31]Champion 1936. [32]Whitmore 2008. [33]Grierson & Long 1983; Behera & Kushwaha 2007. [34]Arunachalam *et al.* 1996. [35]Robbins & Smitinand 1966. [36]Kanzaki *et al.* 1999, 2004; Hara *et al.* 2002. [37]Kingdon-Ward 1930, 1937, 1949; Stainton 1972. [38]Robbins & Smitinand 1966; Khamyong *et al.* 2004. [39]Cai *et al.* 2009. [40]Zhu *et al.* 2005, 2006; Cao *et al.* 2008. [41]Hou *et al.* 1979. [42]Li & Walker 1986. [43]Symington 1943; D.R.Wells, pers. comm. March 2004. [44]Jacobs 1958; Meijer 1959a, b; Yamada 1975, 1976b, 1977; Laumonier 1994, 1997. [45]Symington 1943; D.R.Wells, pers. comm. March 2004. [46]Anderson *et al.* 1982. [47]Proctor *et al.* 1988. [48]Kochummen 1982. [49]Smitinand 1968. [50]Sawyer & Chermsirivathana 1969; Smitinand *et al.* 1979; Maxwell 1988. [51]P.S. Ashton 1964b. [52]P.S. Ashton 1964b; Aiba & Kitayama 1999. [53]Johns 1982. [54]R.J. Johns, pers. comm. 2007. [55]Grubb & Stevens 1985. [56]Grubb & Stevens 1985; Mangen 1993. [57]Robbins 1960. [58]Hynes 1970; Johns 1982; Ash 1988. [59]Johan *et al.* 1994. [60]Ash 1988. [61]Ash 1988. [62]Kadambi 1939; Blasco 1971; Nair *et al.* 2001. [63]Greller & Balasubramanyam 1993. [64]Yamada 1975, 1976a, 1977. [65]Greller *et al.* 1987; Greller & Balasubramanyam 1988; Werner 2001. [66]P.S. Ashton 1988. [67]van Steenis & Schippers-Lammertse 1965. [68]Goldammer & Peñafiel 1990. [69]Santisuk 1997. [70]McClure 1975; Armitage & Burley 1980; Goldammer & Peñafiel 1990. [71]Goldammer & Peñafiel 1990. [72]Troup 1921. [73]Kumar *et al.* 2004. [74]Brown 1919; Werner 1984; Beaman *et al.* 2001, 2004. [75]Khamyong *et al.* 2004. [76]Khamyong *et al.* 2004. [77]Zhu *et al.* 2005. [78]Poulsen & Pendry 1995. [79]Zhu *et al.* 2005. [80]Fernando & Ranasinghe 1997. [81]Wood *et al.* 1993. [82]Dames 1955; Whitmore & Burnham 1969. [83]Ohsawa 1995b. [84]Kira 1977. [85]Ohsawa 1995b; P.S. Ashton 2003a. [86]Beaman *et al.* 2001, 2004. [87]Kume *et al.* 2007. [88]Dames 1955. [89]Ohsawa 1990, 1991c, 1995b. [90]Takyu *et al.* 2002, 2003; Aiba *et al.* 2004. [91]Sri-ngernyang *et al.* 2003. [92]Beaman *et al.* 2001, 2004. [93]Takyu *et al.* 2002; Aiba *et al.* 2004. [94]Aiba *et al.* 2004; Takyu *et al.* 2002. [95]Beaman *et al.* 2001, 2004. [96]Richards 1958. [97]Beaman *et al.* 2001, 2004. [98]Whitmore 1984a, 1998; Grubb & Stevens 1985; Hamilton *et al.* 1995; Richards 1996; Aiba & Kitayama 1999; Bruijnzeel *et al.* 2011. [99]Grubb & Stevens 1985. [100]Beaman *et al.* 2001, 2004. [101]Aiba & Kitayama 1999. [102]P.G. Grubb, pers. comm. June 2012. [103]Field *et al.* 1998; Milliken & Proctor 1999. [104]Vink 1970. [105]Whitmore 2008. [106]Parris *et al.* 1992; Wood *et al.* 1993; Beaman & Edwards 2007. [107]Fernando & Ranasinghe 1997. [108]van Royen 1967; Grubb & Stevens 1985. [109]van Royen 1967. [110]Ash 1988. [111]Johns 1982. [112]Grubb & Stevens 1985. [113]Grubb & Stevens 1985. [114]R.J. Johns 1982, pers. comm. 2009. [115]Jacobs 1958; Ohsawa *et al.* 1985. [116]Laumonier 1994. [117]Yamada 1957–1977. [118]Lieberman *et al.* 1996;Vásquez & Givnish 1998. [119]Corner

1964a, b. [120]Whitford 1906. [121]Whitmore 1984a, 1998a. [122]P.S. Ashton 1964a. [123]Richards 1996. [124]Aiba & Kitayama 1999. [125]A. Farjon, pers. comm. 2011. [126]Werner 1984, 2001; Bruijnzeel & Veneklaas 1998; Bruijnzeel *et al.* 2011. [127]P.S. Ashton 1958b. [128]Lawton & Putz 1988. [129]P.S. Ashton 1958b. [130]R.J. Johns, pers. comm. 2007. [131]Aiyar 1932. [132]Jacobs 1972. [133]van Steenis 1972a, b. [134]Matthew 1999. [135]Fyson 1915. [136]Fyson 1915; Pascal 1988; Aiyar 1932; Davidar *et al.* 2007. [137]Kerr 1933; Rundel 1999. [138]Santisuk 1988. [139]Richards 1958. [140]Bruijnzeel *et al.* 1993; Bruijnzeel & Veneklaas 1998; Bruijnzeel *et al.* 2011. [141]van Steenis 1972a, b. [142]Yamada 1975–1977. [143]I.A.U.N. Gunatilleke, pers. comm. 2012. [144]Fyson 1915. [145]Sugden 1985. [146]Grubb & Stevens 1985. [147]Jacobs 1972. [148]Kerr 1933. [149]Averyanov 2003. [150]Askew 1964; Blaker 1980; Bruijnzeel & Veneklaas 1998; Roman *et al.* 2010. [151]Grubb & Stevens 1985; I.H. Baillie, pers. comm. March 2011. [152]Beaman 1998; Beaman *et al.* 2001. [153]Kitayama 1992; Aiba & Kitayama 1999. [154]Kitayama 1992; Beaman 1998; Aiba & Kitayama 1999; Beaman *et al.* 2001. [155]Kochummen 1982. [156]Grubb & Stevens 1985. [157]Askew 1964. [158]Wade & McVean 1969; Gillison 1970; Haantjens *et al.* 1970; Hope 1980; Mangen 1993. [159]Askew 1964; Wade & McVean 1969; Gillison 1970; Haantjens *et al.* 1970; Hope 1980; Mangen 1993. [160]Yamada 1977. [161]D.W. Lee & Lowry 1980b. [162]Aiba & Kitayama 1999. [163]Wade & McVean 1969; Lowry *et al.* 1973; Bruijnzeel & Veneklaas 1998. [164]Kitayama 1995. [165]Hope 1980. [166]Askew 1964; Sims & Easton 1972. [167]D.W. Lee & Lowry 1980a. [168]Flenley 1995. [169]D.W. Lee 2007. [170]Körner 1998. [171]Takyu *et al.* 2002, 2003. [172]Cavalier & Goldstein 1989; Bruijnzeel & Veneklaas 1998. [173]Brookfield 1964. [174]R.K. Srivastava *et al.* 1999; Srivastava 2002b. [175]Jacobs 1972. [176]Kalkman & Vink 1970. [177]Brown & Powell 1974; Barry 1980; Johns 1986. [178]van Steenis & Schippers Lammertse 1965; van Steenis 1968. [179]Nilsen 1992. [180]Ball *et al.* 1997. [181]Sukumar *et al.* 1993. [182]van Steenis 1968, 1984. [183]Gillison 1970. [184]van Steenis 1972a, b. [185]Askew 1964. [186]van Steenis 1935. [187]Matthew 1999. [188]Wangda & Ohsawa 2006a, b. [189]Balasubramanyan *et al.* 1993. [190]Sukumar *et al.* 1993. [191]Gates 1962; Geiger *et al.* 2003. [192]Brookfield 1964. [193]St. John 1863. [194]Gaku Kudo, unpubl., Kitayama 1996. [195]Smith 1970. [196]Mueller-Dombois 1988; Daniels & Veblen 2004. [197]Szechowycz 1959; Perera 1971–1972, 1978; Werner 1988. [198]Kitayama 1995. [199]Lowry *et al.* 1973; Kudo & Suzuki 2004. [200]Hooker 1848. [201]Khatri *et al.* 2004. [202]Stainton 1972. [203]Kingdon-Ward 1949; Proctor *et al.* 1998. [204]Zhu 2006. [205]Rundel 1999. [206]Rundel 1999. [207]Chang-Fu *et al.* 1998. [208]Hoang 1993; Editorial Committee 1994–2000; Chang-Fu *et al.* 1998. [209]Ohsawa 1991a, b; Wangda & Ohsawa 2006a, b; R. Pradhan, pers. comm. Nov. 2008. [210]R. Pradhan, pers. comm. Nov. 2008. [211]P. Wangda, pers. comm. Nov. 2008. [212]Su 1984; Zhu *et al.* 2005, 2006; Cao *et al.* 2008. [213]Wu 1961; Li & Walker 1986; Zhu & Cai 2004; Zhu *et al.* 2005. [214]Ohsawa 1995a, b; Ohsawa & Nitta 1997, 1997. [215]Hartshorn 1988. [216]Tagawa 1995, 1997; Kanzaki *et al.* 1999; M. Kanzaki, pers. comm. June 2011. [217]Hadfield 1968, 1974. [218]Eden 1976; Wilson & Clifford 1992. [219]M. Holbrook, pers. comm. Nov. 2009. [220]M. Kanzaki, pers. comm. July 2011. [221]Ohsawa & Nitta 1997, 2002. [222]Nilsen 1992. [223]Mergen *et al.* 1974. [224]Brown 1917, 1919. [225]Graham *et al.* 2003. [226]Proctor *et al.* 1988, 1989. [227]Takyu *et al.* 2002, 2003. [228]Körner 1989. [229]Bruijnzeel & Veneklaas 1998; Sollins 1998; Tanner *et al.* 1998; Vitousek 1998; Waide *et al.* 1998. [230]Bruijnzeel *et al.* 1993; Kitayama *et al.* 1997. [231]P.S. Ashton 1988. [232]Gunatilleke *et al.* 1998. [233]Grubb 1971, 1973, 1977b. [234]Hara *et al.* 2002. [235]Boojh & Ramakrishnan 1982a. [236]Kriangsak *et al.* 2003. [237]Whitmore 2008. [238]Chandrasekhara & Ramakrishnan 1993. [239]Whitmore 2008. [240]Hope *et al.* 1988. [241]Caldecott & Caldecott 1985. [242]Hara 1991a. [243]Boojh & Ramakrishnan 1982a, b. [244]Khan *et al.* 1986, 1987; Khan & Tripathi 1986, 1989; Tripathi & Khan 1992. [245]Soepadmo 1976. [246]Caldecott & Caldecott 1985. [247]Ash 1988; Read *et al.* 1990. [248]Tripathi & Khan 1992; Thadani & Ashton 1995. [249]Bruijnzeel *et al.* 2011. [250]Kira 1977. [251]Wade & McVean 1969. [252]Janzen 1967. [253]Symington 1943. [254]P.S. Ashton 2003a. [255]Frahm & Gradstein 1991.

Indian black-headed oriole servicing the flowers of the leguminous tree
Erythrina stricta, Khao Pandan Wildlife Sanctuary, NW Thailand (H. H.)

Chapter 5

TREES AND THEIR MOBILE LINKS:
THIRD PARTIES THAT MEDIATE NATURAL SELECTION

1. The indispensible mobile links

Among sessile organisms such as plants, the number of species in a community may easily exceed the maximum number of neighbours allowed to any given individual by space and geometry. In that case, opportunities for consistent direct physical contact and competition between species must inevitably be few. Interspecific competition, then, must generally operate under one of two possible conditions: either competition must be relatively relaxed, with different species functioning substantially as ecologically complementary; or competition must be mediated by third parties, or *mobile links* – mobile organisms that themselves exploit particular plant species, thereby positively or negatively influencing their competitiveness. In the case of plants, some mobile links *increase* fitness: dispersers of seeds or pollen, insects that defend them against herbivores, and the spores of symbiotic fungi; others *reduce* fitness, such as herbivores and predators, of seeds or of plant tissues (especially those of young plants), and pathogens.

In principle, continuously warm and equable humid climates present a relatively narrow range of physical challenges such as droughts and temperature extremes. Such conditions should therefore minimise the influence of physical parameters on plant fitness and competitiveness, while effectively enhancing the roles of the mobile links. If so, these will tend to exert strong – though indirect – influences on plant diversity. As primary producers, plants synthesise a vast array of complex molecules, exceeded in their diversity (and complexity) only by the products of fungi and of microbes themselves. These may act as specific attractants to dispersers, or as detractants against herbivores, predators and pathogens. In this way, animals and microorganisms – dependent on plants for the exploitation of solar energy in the conversion of carbon and water to biomass and food – may themselves be the major mediators of interspecific interactions among plants in diverse communities.

If, therefore, plant species within diverse communities are niche-specific, the competition between them that defines their niches could largely be mediated by mobile organisms. However, the species richness of many mobile link groups is much lower in Asian than Neotropical rainforest communities.[1] Within individual forest formations, the number of species of primates, birds, bats, bees and butterflies is substantially fewer again.[2] This is true even of lowland Asian MDF, where tree species richness is comparable to that of the Neotropics while stature and attendant structural complexity is generally greater.

On the other hand, certain other plant-dependent groups are particularly species-rich in tropical Asia, including the scale insects, fruit flies and the ectotrophic mycorrhizae. Toward the margins of the tropics, species richness of vertebrate browsers and grazers is greater in Asia than in the Neotropics, and comparable to that of parts of Africa.[3] Dispersing organisms in species-rich tree communities tend to service multiple plant species. Over time, tree species are likely to evolve characteristics that attract their undivided attention, such as interspecific phenological differences. Trees may sustain their genetic integrity by high interspecific pollen incompatibility or by low hybrid fitness. (Temperate congeneric and ecologically sympatric tree species are often partially pollen-compatible.)

The study of those mobile links that disperse by air – which are the majority – is demanding. Much of the action takes place in the forest canopy, where our own non-brachiating species is awkward and insecure (Plate 5.1).[4] Not all mature individuals in a population will flower in a given season. By the time incipient flowering has been detected, few days remain before close observation must begin. It may be necessary to access the outermost twigs of canopy species for flower observations, especially for cross-pollination experiments. For our canopy research at Pasoh, we gained the assistance of oilfield engineers in designing tree prosthetics: up to 20 m alloy booms, in portable parts assembled on site, which could be lifted into the canopy where they served as artificial branches with one end piercing the tree crown.[5] Other researchers have become dexterous with rapidly assembled climbing tackle.[6] As a consequence, the number of detailed studies is still limited to a few species, and it is too early to draw general conclusions regarding many issues. It is for that reason that I focus in this essay on individual cases, which I assign to boxes within the text.

Plate 5.1 How to brachiate in the rainforest canopy if you have no tail: ladders and tree towers give little access to flowering twigs; towers and canopy walkways are unmoveable structures; the tower crane gives access to many tree crowns, but few of one species; our three extensible booms did solve the problem for controlled crosses, but are inconvenient albeit feasible to move between trees. **a,** Savitri Gunatilleke and student cross-pollinate *Shorea megistophylla*; **b,** Hauling up the tree boom in Pasoh; Appanah supervises; **c,** The tree boom in action; **d,** Tamiji Inoue's canopy walkway at Lambir, and one of its towers; **e,** Tohru Nakashizuka's tower crane; **f,** A Kyoto University student studies with insect nets dipterocarp pollinators atop an emergent, Lambir. (**a** N. Gunatilleke; **b** P.A.; **c** M. Moffet; **d, e** C. Ziegler; **f** T. Inoue)

Observations of the dispersal of mobile links are critical, but they are easy only where pollen and seed numbers are greatest: around the source. The tail of declining abundance from that source is often impossible to document directly: The occasional pollen grain, seed, or fungal spore that disperses exceptional distances and germinates successfully may redefine the range of a tree species, or that of a gene sequence. But these abundances and distances must be inferred from indirect, including molecular genetic, evidence.

2. Challenges in achieving effective cross-pollination in hyperdiverse forests

2.1. *Floral ecology, pollinators, and pollination*

Charles Darwin recognized that, for evolution to occur through natural selection, there had to be heritable diversity among members of species and their breeding populations within which that selection is possible. We now know that genes are the currency in that diversity. For genetic diversity to be sustained at high levels in populations, genes must be continuously exchanged between genetically different individuals. The more the exchange is confined to related individuals, the less the genetic diversity of these local population segregates. The exchange takes place by the meeting and fusion of two gametes, each of which carries a copy of half the parent's genes, thereby making a new whole, a zygote. In higher plants, including trees, these gametes are the pollen grain and the ovule. The pollen grain is mobile and therefore regarded as male, while the ovule remains within the parent flower, housing the embryo as it develops inside the seed, and is therefore considered female.

Among most north temperate canopy tree species, pollen is borne on the wind, reaching fertile stigmas only at random. But this is an enormously wasteful process: only a tiny proportion ever reaches its destination. Flowers are among the most nutrient-rich parts of plants, and their production is energy-intensive. Many trees hardly grow during flowering and fruiting years.[7] One of the greatest mysteries of the species-rich but nutrient-poor rainforests has been how (and whether)

pollen is transferred from one distant individual to another, and how nutrient loss is minimised.

2.2. *The nature of the problem*

Pollen is rich in amino acids, therefore expensive in the nitrogen-starved environment of high temperatures and rainfall-leached soils. Wastage of pollen is inevitably immense wherever cross-pollination necessitates transit between individuals often distantly separated by the dense foliage of a host of other species. Here then, more than anywhere, parsimony in pollen expenditure is called for. In this competitive setting, attracting a pollinator requires expenditure of nutritious or sugar-rich nectar, or even sacrifice of some proportion of the pollen itself as food to lure the disperser. It may often entail energy-expensive synthesis of pollinator-specific aromatic compounds to attract particular species, including moths,[8] bats[9] and even individual beetle species.[10]

It was earlier assumed that rainforest trees were self-compatible and overwhelmingly self-pollinated.[11] Were this so, local populations would become increasingly genetically uniform and capable of generating new variation only through viable somatic mutation. That is because diversity declines as alleles increasingly become identical at each gene locus (homozygous) due to random extinction of alleles in small populations. This reduces the genetic variability on which natural selection must act, and therefore reduces the potential rate of change in gene frequencies and the rate of evolutionary change. It therefore increases the rate of competitive species extinction.

Supposing that the rainforest environment itself changes slowly or hardly at all, then natural selection may remain effectively neutral, its outcomes varying randomly about a mean set by that environment. This situation would then favour self-pollination, minimising waste of pollen and ovules without trading off adaptability.

The rare dispersal of fertilised seeds over long distances and severe dispersal barriers[12] has achieved major impacts on the population ecology and genetics of many self-compatible species. In the case of pollen, however, it is not the maximum distance that it can disperse that is critical, but rather the concentration achieved in a given area of receptive stigmas.[13] While pollen dispersed to the nearest conspecific individuals may result in increased gene fixation between half-sibs, only that which is dispersed in quantity to greater distances will maintain genetic variability at the highest levels.

2.3. *Tree breeding systems*

Breeding systems are the primary determinants of genetic variability. The predominant breeding systems in a plant community signal the prevailing levels of genetic diversity among its members.[14] In an early paper which laid the foundation for my later work in this area,[15] I argued that the presence of great genetic diversity in such an environment – where each seed successfully produced is costly in both energy requirements and nutrients – would provide the ultimate evidence for persistent stringent selection in response to competition between organisms. It would imply the existence of continuing and ever-changeable niche specificity among co-occurring species. Sex was assumed to be necessary for the maintenance of genetic diversity (although the overall evolutionary function of sex itself remains somewhat enigmatic[16]).

2.3.1. *Outbreeding predominates*

Most pioneer research on breeding systems in tropical trees had been carried out in the evergreen forests of Central America. There Kamal Bawa and his associates found an overall pattern of dominant outbreeding among canopy species.[17] Obligate outcrossing among bisexual tree species appears rare, although partial autogamy and geitonogamy, including a variety of monoecious systems, are widespread; but outbreeding dominates even where self-compatibility in stigma or style is prevalent. In Asia, tests for compatibility have been carried out on few species.[18] Appanah and Rogstad found extraordinary supporting evidence whereby sufficient outcrossing is achieved in the understorey wild rambutan *Xerospermum noronhianum* (Sapindaceae) and *Polyalthia* (Annonaceae) respectively (Boxes 5.2, 5.3).[19] Among dipterocarps, which have bisexual flowers, controlled pollination experiments with *Shorea* have shown that trees are partially self-compatible, but this varies among species[20] (Box 5.1). Chan found some evidence of high self-incompatibility in *S. leprosula* at Pasoh.[21] At Sinharaja in Sri Lanka, *S. congestiflora* trees were found to have higher but variable self-incompatibility compared with *S. trapezifolia* and *S. megistophylla.*[22] In *S. megistophylla,* seed abortion during development was higher among selfed than crossed flowers and there was additional isozyme evidence implying the same in that species and in *S. trapezifolia.* Murawski, Dayanandan and Bawa found high self-incompatibility in *S. megistophylla,* but only moderate in the successional species *S. trapezifolia.*[23]

Bee-pollinated *Shorea megistophylla* trees isolated by logging were found on isozyme evidence to have lower outcrossing rates (0.7 versus 0.87 in unlogged forest).[24] Isolated trees of thrip-pollinated *S. leprosula* were casually observed to produce less fruit, while trees in a clump yielded fruit roughly according to their prior flower number.[25] Selfed flowers of *S. siamensis* in Indo-Burmese deciduous dipterocarp forest were found to have reduced pollen tube growth; the proportion of selfing increased with the distance between nearest flowering trees.[26] The proportion of seed reaching maturity varies between species, but is generally low (Box 5.1). Storms can greatly reduce fecundity.[27]

Chan found that the variance of leaf characters among *Shorea leprosula* individuals increased to an asymptote at about 100 m between reproductively mature trees.[28] This was consistent with isozyme variation analysed by Gan Yik Yuen.[29] These findings imply that the effect of crossing among individuals related by recent common ancestry declines to a level close to panmixis at c.100 m in a thrip-pollinated winged-seeded species. But Kenta and colleagues[30] inferred from molecular evidence that pollen was dispersed a mean distance of c.200 m in moth and bee-pollinated *Dipterocarpus tempehes* in both mast and non-mast years. There was also some evidence of kin mating among *S. megistophylla* groups in unlogged forest.[31] S.L. Lee and colleagues[32] found that the coefficient of inbreeding was lower in mature trees than juveniles in *S. leprosula*, implying that progeny resulting from selfing were less fit and were selected against during regeneration. Harata[33] found evidence of greater inbreeding depression among the progeny of small- (mostly small insect pollinated) than large-fruited (mostly bee and Lepidoptera-pollinated) dipterocarps, and a negative relationship between inbreeding and the distance between trees.

The spatial pattern of cross-fertility can also yield key insight into inbreeding depression. Elizabeth Stacy[34] recorded survivorship of seedlings of *Syzygium rubicundum* and *Shorea cordifolia* in south-west Sri Lanka over one year, comparing those produced via selfing with those arising from cross-pollination. *S. rubicundum* is pollinated by a diversity of insects, and the seed is dispersed by birds, whereas *S. cordifolia* is bee-pollinated and dispersed by gyration and wind. Pollen was collected from trees at distances of 2, 12 and 35 km from the base population sample. There was very low seed set from selfed flowers in both species. Stacy found that mean survivorship peaked at 1–2 km between the mother tree and the pollen source for the former species, and 1–10 km for the latter. She concluded that inbreeding strongly reduced survivorship, and that partial reproductive isolation beyond 10 km again reduced fitness. In *S. cordifolia,* though, the rate of height growth of successful germinants continued to increase with distance, implying hybrid vigour. Populations of *Pterocarpus macrocarpus,* an Indo-Burmese species with bee pollination and mammal-dispersed seed, was found to have 82% polymorphism at isozyme loci, although heterozygosity was moderately low (0.22)[35] (although isozymes do underestimate levels of genetic diversity). There was some evidence of local inbreeding. Again, there was correlation, here highly significant, between genetic and geographic distance, with high differentiation between populations.

On evidence of isozyme diversity, overall levels of genetic diversity among dipterocarp populations are comparable to that of Neotropical and temperate zone tree species,[36] indicating high levels of outbreeding. Phenological and pollination studies in Asia and the Neotropics indicate processes of surprising precision, implying intense selection for cross-pollination and maintenance of genetic variability. The rainforest environment is climatically one of the most favourable for plant growth, and the most equable and unchanging over evolutionary time. Today, most tropical tree species give evidence of being sensitive and intolerant to even slight changes in rainfall seasonality (Chapter 6). This means that populations have survived and competed successfully within relatively narrow physiological limits. Nevertheless, while doing so they have evolved complex and resource-demanding reproductive systems whose primary advantage seems to be in furnishing populations with high levels of genetic variability. What powerful yet ever-changing force drives this great investment in variability?

2.3.2. Monoecy and dioecy

The emergent canopy of Asian MDF is comprised almost entirely of species that bear bisexual flowers. In the main canopy and beneath it – as throughout Central America, where a distinct and more or less continuous emergent stratum is unusual – various forms of monoecy are prevalent (Fig. 5.1). *Monoecy* is the production of separate male and female (or bisexual) flowers on each individual; *dioecy* is the separation of male and female sexual functions on the flowers of different individuals. For several reasons, dioecy has been considered an inefficient means of reproduction in sedentary organisms.[37] The sex of individuals remains constant (among species examined),[38] so the number of seed-bearing individuals is effectively halved within the mature population, while the next generation depends entirely on the success of the pollination process. Mack,[39] who observed 64 trees of the dioecious *Aglaia* cf. *flavida* in 260 ha of forest in New Guinea, found that female trees less than 200 m from a male yielded heavier crops than did those further away, regardless of their size. Population bottlenecks and Allee effects (the reduction of fecundity induced by constraints of cross-pollination) can significantly reduce the number of recruits in obligately dioecious populations, and when they do so, there is no turning back.

It is therefore all the more surprising that dioecy is so pervasive in evergreen rainforest. In Asia, dioecy is concentrated overwhelmingly beneath the canopy, and it is the rule in many of

Fig. 5.1 Representation of breeding systems by tree size (diameter classes) in highly structured MDF on sandy clay soils; Bukit Raya, central Sarawak: species with bisexual flowers, with unisexual or effectively male and bisexual flowers within individual trees, and dioecious. (From P.S. Ashton 1969)

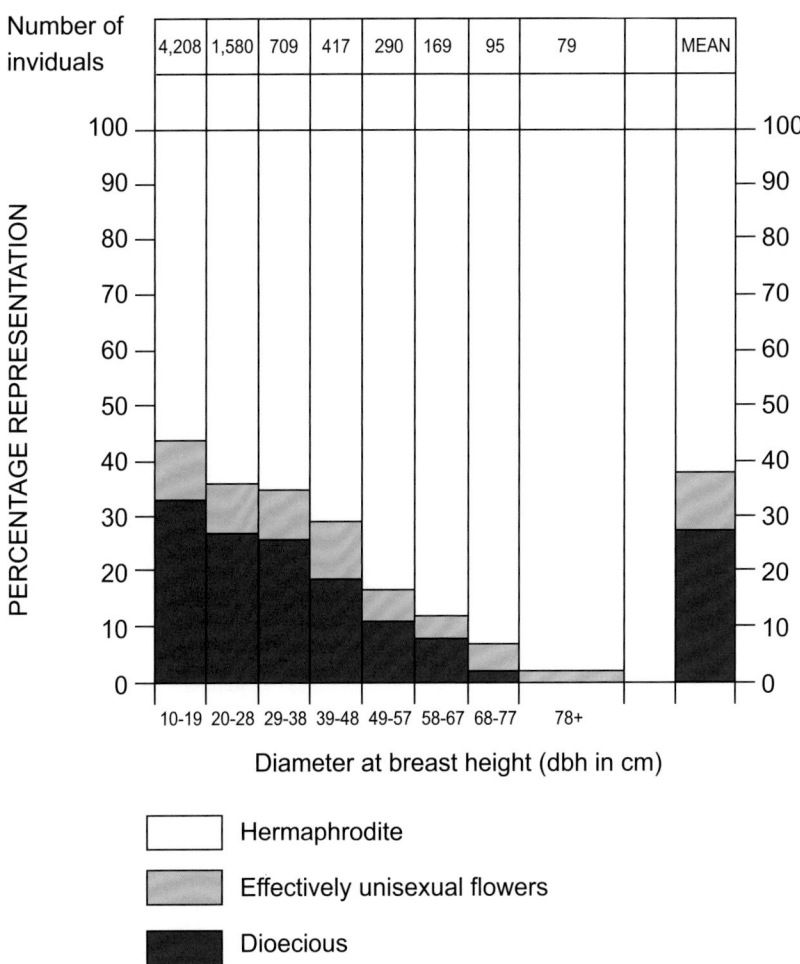

the largest genera that contribute series of co-occurring species. Surprisingly, there is no evidence from the 50 ha CTFS plots in hyperdiverse MDF at Lambir and Pasoh that the proportion of dioecious species declines with species' population densities. As many as a third of subcanopy species in MDF are dioecious (Fig. 5.1). These include the three large Myristicaceae genera, many Euphorbiaceae including *Aporosa, Baccaurea* and *Macaranga* (mostly pioneer), also *Diospyros, Garcinia, Aglaia,* many *Chisocheton* and *Dysoxylum,* and *Hydnocarpus.* Cryptic dioecy is also likely widespread.[40] The great height of the MDF canopy has permitted the evolution and co-existence of congeneric species-series reaching reproductive maturity at different heights within the canopy (Fig. 5.2; see Chapter 6). Many of these

Fig. 5.2 Flower production relative to tree size in three dioecious *Baccaurea* (Euphorbiaceae) species. (From Thomas 1993)

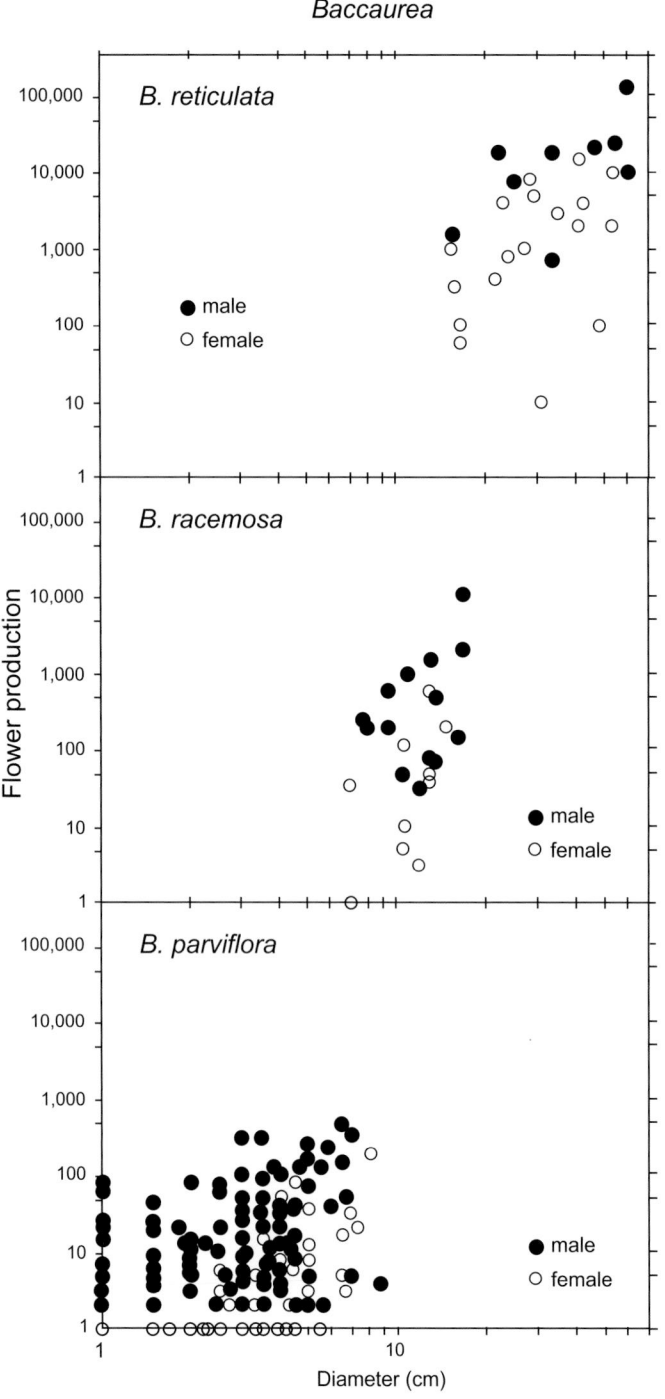

There may be a selective advantage in concentrating the colourful juicy fruit, dispersed by primates and birds, into a few individuals, which attract by making a tempting show in the sea of green. Kamal Bawa and Paul Opler, and Tom Givnish argued that dioecy, by concentrating fruit on at most half the reproductively mature individuals in a population, attracts seed dispersers more efficiently, and this has been supported by empirical observations.[41] Another hypothesis proposes that dioecy is advantageous where pollinators are unspecialised and have restricted foraging ranges; but Renner and Feil[42] found that dioecious species bear diverse flower morphologies and are visited by a diversity of pollinators, often quite specialised. It is striking, though, that these species are obligate outcrossers in the understorey – precisely the habitat within which visual attraction of pollinators is most problematic.

There may also be an advantage relating to the energetic costs of reproduction.[43] The cost, in energy and nutrients, of post-anthesis ovule or seed abortion can become vastly higher than that of producing sufficient pollen to ensure sufficient fit seed production to sustain the population. In Australian *Myristica insipida,* where open pollination was found to result in only 1% seed set, female trees were estimated to incur 420% higher energetic costs than males.[44] Under demanding ecological conditions, it would therefore be predicted that relatively more pollen per ovule would be produced. In bisexual flowers, the ratio is genetically limited by floral structure, and is therefore inflexible. In monoecious and dioecious species, the ratio can be modified in a particular season by the proportion of male flowers produced or the relative proportion of male trees that flower. That confers an additional selective advantage: dioecy allows for flowering of lower-cost male individuals in smaller individuals, and earlier in the season (Box 5.2; Fig 5.16). This assures a greater abundance of male gametes in species whose individual flower pollen/ovule ratio is immutable. Given these advantages, it is less surprising that dioecy and male-biased sex ratios at flowering (if not in populations as a whole) are prevalent in species restricted to the limited light of the understorey.[45]

Androdioecy is the production, on different individuals, of male and bisexual flowers (Box 5.2). It is rare in nature, but androdioecious *Xerospermum intermedium* (as are apparently most members of its tree family, Sapindaceae) has achieved greater abundance than any other tree species at Pasoh. As with monoecy, it has been interpreted as permitting additional allocation to the production of male gametes where successful outcrossing is fraught with difficulty and extra pollen therefore at a premium, in plants whose flowers have an inflexible pollen/ovule ratio. Androdioecy may be advantageous in species with low selfing rates (since male reproductive allocation thereby increases relative to the overall resource devoted to flowering), or where the relative cost of ovules per

congeneric species share pollinators. The prevalence of dioecy among most or all species in a genus, and whole families in the case of Myristicaceae and Ebenaceae, implies that the condition is generally ancient and stable (assuming it was derived from the bisexual state in the ancient past). What, then, are possible sources of the apparent strong selective advantage of dioecy?

flower is high and the chances of setting fruit from a pistillate flower are thereby increased.[46]

2.4. *A contrary enigma*

Bisexual-flowered genera attracting pollinators with visual clues do so within a range of visual environments. The size and abundance of flowers, and the pollen/ovule ratios produced, might each be expected to increase in more complex biotic environments. Forest structure increases in complexity from deciduous to evergreen canopies, and from the canopy to the understorey. However, the general pattern of floral complexity is the opposite. Among Dipterocarpaceae, the *Shorea* species presenting the largest flower with many stamens is the deciduous *Shorea robusta* of seasonal northern India. The *Dipterocarpus* with the largest, most conspicuous flowers are those of the short deciduous dipterocarp forest. In perhumid Malesia, *Shorea* (sect. *Richetioides*), most of whose species belong to the main rather than emergent canopy, bears the smallest anthers, and in some species anthers are reduced from 15 to 10.[47] The same is true in many *Hopea*. Among the main canopy and subcanopy *Stemonoporus* of Sri Lanka, three subcanopy species have their stamens reduced from between 10 and 15 to 5, and the number of ovules from 6 to 4.[48] Similar reduction in the size of flowers and number of stamens in many perhumid region subcanopy species is observed in the speciose genera *Syzygium* and *Calophyllum* with bisexual flowers, and the dioecious *Garcinia*. There is as yet no evidence for a shift in lure or pollinator in these examples, which appear inexplicably to run counter to prediction; yet their existence demonstrates that these species strategy succeeds under the circumstances in which it is found. *This would be a promising avenue for further research.*

2.5. *Hybridisation*

Experienced field botanists concur that equatorial rainforest tree species are extraordinarily constant in those characters – many of them seemingly trivial – by which related species can be differentiated. That is, mature individuals in nature very rarely display a blend of the physical characters of two co-occurring species. This implies that either successful hybridisation events are rare,[49] or hybrid individuals are rarely fit enough to survive to maturity. Morphological constancy is particularly noticeable within populations, but can even be seen over vast distances (but see section 5.2 below). Morphological intermediates are markedly more common in the seasonal tropics, however, having been recorded among *Dipterocarpus* species in the seasonal evergreen and semi-evergreen forests of Burma and Thailand. These appear most frequently between semi-evergreen forest species (especially *D. turbinatus*) and deciduous dipterocarp forest species.[50] They are usually found in groups, on sites where species of both forest types occur which are ecotonal and dynamic, and where fire is involved (Chapter 2).

Morphological intermediates in hyperdiverse MDF appear rare indeed. Mature trees intermediate between the lowland *Shorea leprosula* and the hill *S. curtisii* are known from three localities in Peninsular Malaysia. Fruit examined by Yik-yuen Gan and me from a site at Kepong (Selangor) proved empty of seed, implying that the hybrid cannot establish a reproducing population. I have twice seen seedlings morphologically intermediate between *S. parvifolia* and *S. acuminata*. In Sri Lanka, Nimal Gunatilleke found hybrid juveniles between the lower-slope *S. megistophylla* and the upper-slope *S. disticha*.[51]

Despite the apparent rarity of interspecific hybridisation in contemporary forests, Koichi Kamiya has provided molecular evidence for frequent interspecific hybridisations in the evolution of four species in *Shorea* (sect. *Brachyptera*) at Lambir (north-east Sarawak).[52] In recent research on four species in *Shorea* (sect. *Mutica*) co-occurring in the surviving coastal MDF fragment at Bukit Timah (Singapore), Kamiya[53] also found frequent hybridisation among seedlings, and some among reproductive individuals: 13 of 64 *S. curtisii*, 12 of 15 *S. leprosula*, 1 of 4 *S. parvifolia*, but all hybrids were of the F_1 generation. Absence of subsequent hybrid generations could imply either lack of competitive fitness – which appears likely – or perhaps some genetic compatibility barrier acting at that stage. Whichever is the case, this arrest of potential introgressive hybridisation is fundamental to the survival of co-existing species series, most or all of which appear to flower at least partly concurrently.

Among dipterocarps in the lowland and lower montane forests of Sri Lanka, *Stemonoporus* contains more than 20 morphological species, most of which are point endemics of restricted co-occurrence. This is a striking contrast to *Shorea* (sect. *Doona*) there, all species of which are widespread in the 10,000 km² area of perhumid climate to which their MDF are confined. *Stemonoporus* species are of the main canopy or more often subcanopy, and are often abundant within their restricted ranges. The fruits are wingless, apparently lacking dispersers, and relatively few of the generally large seeds are borne by each tree. I observed flowers to be *Trigona* bee-pollinated, gene flow therefore taking place largely through pollen dispersal. Murawski, Hamrick and Bawa[54] found a very high (84%) multilocus outcrossing rate on isozyme evidence in *Stemonoporus oblongifolius*, though facultative apomixis was inferred in one individual. Genetic differentiation was detected in populations 10 km apart. They conjectured that the high outcrossing rate might be due to high population density, possible introgressive hybridisation among sympatric 'species', and retention of ancient allozyme polymorphisms. *Shorea* (sect. *Pachycarpa*) is the sole section endemic to Borneo and has 10 recognized species. This section is exceptional in that although populations and taxa are morphologically remarkably uniform, entities of intermediate (albeit also uniform) morphology occur over extensive regions. *This exception invites investigation.*

The persistence of series of partially interfertile sympatric sister species of dipterocarps, and probably other families, must rest on the competitive elimination of hybrid progeny. It recalls Palmer's[55] observations on hybrid oaks in North America which, like dipterocarps, produce fertile seed but whose progeny are reduced by competition except along narrow ecotones, and in secondary forest. The two cases serve equally as persuasive evidence for natural selection among closely related taxa.

2.6. *Apomixis*

When a pollination test indicates self-compatibility, there is also the possibility that the embryo is being formed asexually, from maternal tissue either of an unfertilised ovule, or of tissues surrounding the ovule. This adventive embryony is a form of asexual reproduction or *apomixis,* and is well known among temperate perennial herbaceous weeds and pioneer trees, notably North American *Crataegus* (Rosaceae). Stebbins

argued that these habitats favour rapid production of genetically uniform and therefore uniformly fit propagules, because competition is solely intra-specific, for occupation of an open and unchanging physical habitat.[56] Surprisingly, adventive embryony also appears quite common among rainforest tree species; precisely how common is as yet unknown because, to distinguish it from self-compatibility, careful microscopic embryological study is necessary. *But thanks to new microscopic technology, exciting opportunities now exist to explore this important phenomenon further.* It is therefore generally only inferred, on the evidence of multiple apparently adventives embryos. It has been noted in Asian fruit tree genera such as *Mangifera, Lansium,* and *Citrus,*[57] and among dipterocarps[58] (Fig. 5.3; Table 5.1) and legumes.[59] Sean Thomas[60] documented a population of the dioecious *Garcinia scortechinii* that was entirely female, implying obligate apomixis, already well known in the cultivated *G. mangostana.*[61] Multiple embryos are also recorded among members of co-occurring congeneric species series including *Syzygium,*[62] *Garcinia, Calophyllum,*[63] *Diospyros, Memecylon, Shorea* and *Hopea,*[64] but these rainforest taxa manifest neither the extraordinary proliferation of locally distinct metapopulations nor (therefore) the plethora of quasi-species so characteristic of temperate apomicts.

 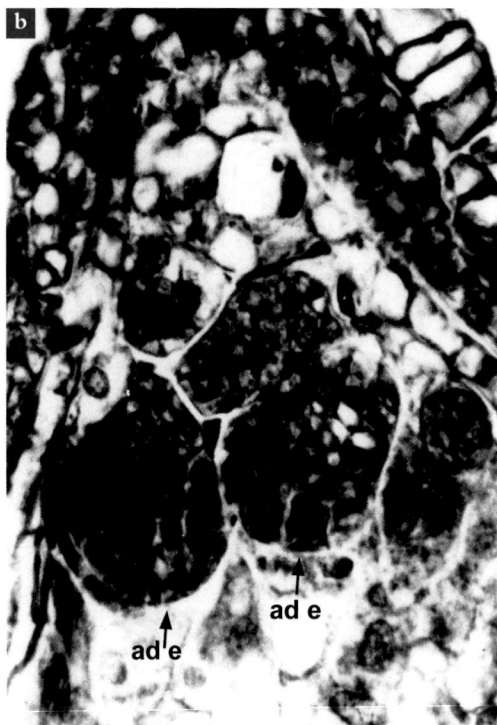 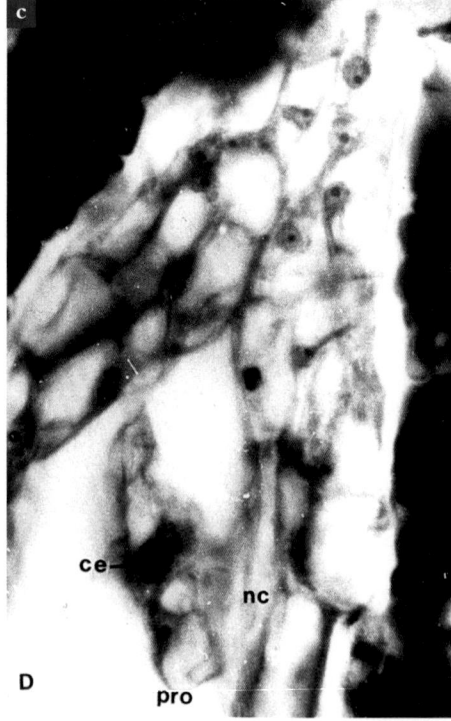

Fig. 5.3 Apomixis through adventive embryony among dipterocarps. *Shorea ovalis,* a tetraploid:
a, Longitudinal section (l.s.) of the ovule showing degeneration of the embryo-sac (arrows), x450; **b**, l.s. of the micropylar area of the ovule showing adventives embryos (ad, e) arising from nucellar cells surrounding the embryo sac cavity, x175. *Shorea agami,* a diploid: **c**, l.s. seed showing adventive pro-embryo arising from one third down the side of the embryo-sac (ce: free endosperm nuclei, nc: nucellar cells, pro: pro-embryo), x1,500. (From Dalbir Singh 1977, Kaur *et al.* 1986)

Plate 5.2 Dipterocarp flowers. **a,** *Shorea siamensis* (sect. *Pentacme*)(scale: 5/1); **b,** *Trigona* bee approaching *S. siamensis* flowers (scale: 2/3); **c,** Beetles on *Shorea* sect. *Mutica* flowers (scale: 3/1); **d,** Thrips on *Hopea nutans* flower (scale: 1/15). (**a, b** H.H.; **c** S. Sakai; **d** M. Moffett)

Table 5.1 Incidence of multiple seedlings and triploidy among Peninsular Malaysian dipterocarps. Individuals can vary, as in the case of *Shorea macroptera*. (From Jong in P.S. Ashton 1982)

Species	Observed multiple seedlings (%)
Diploid (2=14)	
Shorea argentifolia	2
S. lepidota	0
S. leprosula	0
S. macrophylla	0
S. multiflora	0
S. parvifolia	0
S. pauciflora	6
S. siamensis	0
S. stenoptera	0
S. macroptera	
tree 18	0
tree MM 4	<30
S. agamii	18–40
Triploid (3*x*=21)	
Hopea odorata	50
H. subalata	21
S. resinosa	c.100
Tetraploid (4*x*=28)	
S. ovalis	15–47

Among dipterocarps, the first species of *Shorea* (sect. *Mutica*) to flower at Pasoh is *S. macroptera*. As many as 17 seedlings occurred in fruit with at most six ovules! This is only possible by adventive polyembryony (apomixis), which results in genetically identical seed. Dipterocarp flowers generally bear six ovules, only one of which generally survives to yield a single seed in the fruit. Fruit yielding more than one seedling therefore indicate the possibility of apomictic duplication through adventive embryony, *but this requires confirmation via anatomical examination* (Table 5.1). If common, apomixis can cause significant reductions in population genetic variability (see below). Of the 68 dipterocarps examined 62 were diploid,[65] but a tetraploid, *Shorea ovalis* (subsp. *sericea*), was confirmed to be apomictic. Triploidy, which has also been found in three *Hopea* species and *Shorea resinosa*, makes apomixis obligatory.

How can apomictic species populations persist in an environment where all other evidence suggests that variability has high survival value? Apomixis should erase genetic variability in a population, but the slim available evidence suggests that this is not necessarily the case. Wickneswari and colleagues[66] found that, on isozyme evidence, asexual reproduction indeed appeared to prevail in four *Hopea odorata* populations sampled, yet 40% of isozymes were nevertheless polymorphic and evidence overall indicated that heterozygotes

were well represented, perhaps implying greater fitness (Chapter 6). Awtar Kaur,[67] in her study of apomixis among Malaysian dipterocarps, found indirect evidence of genetic variability within the Pasoh population of *Shorea macroptera* ssp. *macroptera*: Nearby individuals differed by as much as 70% in the proportion of seeds with multiple seedlings. *S. macroptera* and *S. parvifolia*, in which facultative apomixis has been inferred, are among the most regionally diversified members of their section,[68] contrasting with the regionally morphologically uniform *S. leprosula* in which it has not.

Adventive embryony is an imprecise process, and usually generates a series of embryos associated with a single ovule, in descending size and all smaller than a sexual embryo. These might be expected to be of lesser fitness, but multiple embryos may also reduce loss of a genotype when fruit are partially attacked by seed predators. *Shorea macroptera* is the first species to flower during mass flowering, before the thrip pollinators have built up full abundance. In this case and also that of *Hopea subalata*, selection may favour apomixis where pollinator shortage reduces fecundity, as when a species has freshly invaded or is on the point of extinction and is in exceptionally low numbers. An isolated individual of *S. megistophylla*, planted outside its range in Peradeniya Botanic Garden (Sri Lanka), was found to produce abundant fruit, some of which contained multiple embryos.[69]

Facultative apomixis may increase the rate of fit genotypes in founder populations, yet maintain heterosis as a population continues to expand. Abundant species like *Shorea macroptera*, largely reproducing apomictically, may be in recovery from a recent bottleneck; but it is hard to explain how they could remain both apomictic and competitive over evolutionary time, since persistence of large numbers of identical homozygous genotypes would seem to expose them to catastrophic pathogenic infection (see section 6.4 below). Apomixis may therefore suggest the occurrence of 'firework displays', that is, genetic diversification of populations into niches which are temporary in evolutionary time, while the core outbreeding metapopulation – often a widespread sister species – persists. *Whether those congeneric species series so characteristic of forests within the perhumid tropics of Asia are, at least in part, thus evolved remains a tantalising field for future investigation.*

2.7. *How do rainforest tree species successfully cross-pollinate despite the evolutionary conservatism of flower morphology?*

The extraordinary success and diversity of seed plants is ascribed in part to the co-evolution of their elaborate and versatile flowers with an ever more diverse suite of pollinators. It may seem curious, then, that the basic ground plans of flower structure are generally consistent within rainforest tree taxa:

species within speciose genera and even some speciose families share the same details of flower and fruit morphology, and the same families of pollen and seed dispersers. This implies that these characteristics are both ancient and conservative. Insect diversification accelerated during the Cretaceous in tandem with the diversification among flowering plants, and it is clear that insect diversity evolved in intimate communion with the proliferation of chemical attractants – and defences – among flowering plants.[70] Co-evolutionary relationships are nevertheless often extraordinarily conservative. Many basal flowering plant families remain associated with primitive pollinators, such as Annonaceae and Myristicaceae with beetles, and Dilleniaceae with flies. Among congeneric series of tropical trees, most share the same pollinator higher taxa, and frequently species. In larger genera with single dominant pollinators, including *Shorea* and *Macaranga*,[71] subgeneric taxa rather than species attract different dominant pollinators. Figs are a major exception (Box 5.7). How then did these enormous species-series, which contribute so greatly to overall tree species diversity (Chapter 7), arise, and how are they sustained? How do these series of often ecologically sympatric tree species avoid pollen wastage and pollution of their often tiny stigmatic surfaces with foreign pollen? Put another way, how do related species avoid wasteful introgressive hybridisation?

This enigma is greatest among archaic tree families, such as the Magnoliaceae, wild nutmegs (Myristicaceae) and soursops (Annonaceae), and the cycads of course, whose species are mostly pollinated by that ancient insect group, the beetles. (There are exceptions, such as the understorey Gymnospermous treelet *Gnetum gnemon* (var. *tenerum*), which is pollinated by a pyralid moth.[72]) But Steven Rogstad[73] has shown, astonishingly, that co-occurring sister species can attract *different* beetle pollinators (Box 5.3). Chemically, therefore, these ancient trees turn out to be highly sophisticated! According to van der Pijl,[74] taxa pollinated by flies are also among the most highly pollinator-specific.

In contrast, species observed within large genera of more advanced families, such as co-occurring species of *Dipterocarpus, Shorea* (sect. *Mutica*) and *Syzygium,* share pollinators. The occurrence of sequential yet overlapping flowering times within such co-occurring species series presents the possibility that competition for pollinators could lead, through differentiation of flowering times, into *sustaining* tree species diversity through the mutual advantage gained in maintaining pollinator numbers (Box 5.1). In other respects, these co-occurring tree species could either be ecologically complementary – as could yet be the case in some *Shorea* (sect. *Mutica*) (but see Chapters 2, 3) – or they could differ additionally in, for instance, light response and height at reproductive maturity, as in *Syzygium*.[75] Sequential flowering has been observed in *Shorea* (sect. *Mutica*) and *Parkia* (Figs 5.1, 5.4, 5.23), but there is little evidence so far that overlapping sequential flowering occurs among

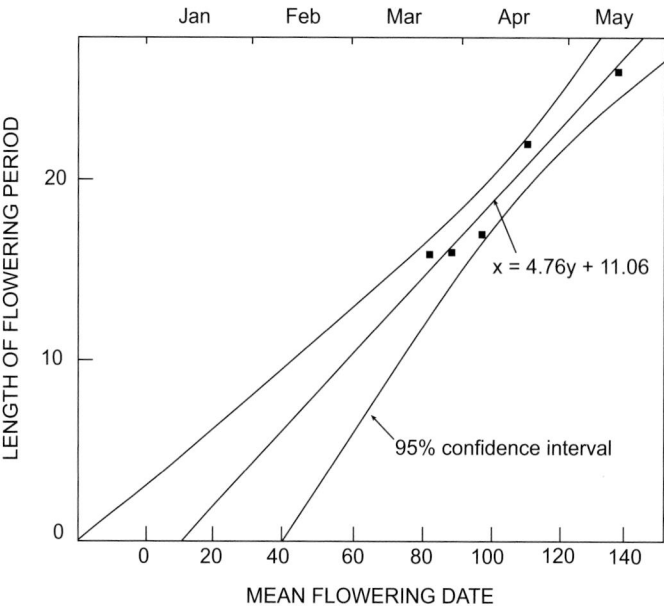

Fig. 5.4 The method of predicting the date when flowering was induced in *Shorea* sect. *Mutica* species at Pasoh: extrapolation back of the regression line relating species' mean flowering dates to the length of their flowering periods; the date of the induction is when the flowering periods coincide at zero. (After P.S. Ashton *et al.* 1988)

the many other series of co-occurring congeneric species sharing floral colours and morphologies. In this respect, Asian experience appears to agree with Wheelwright's observations at Monteverde (Costa Rica).[76]

2.8. *Flowering phenology*

Although differences in the time of fruit fall may be caused by heritable differences in the rate of fruit development, and can also be modified in individual seasons by weather, the triggers to the process of reproduction overall must occur prior to flowering. Once set in motion, the sequence of events cannot be stopped except when weather or predation causes catastrophic abortion. Thus, the search for a climatic trigger of any temporal variation in the stages of reproduction from anthesis to fruitfall must be made during the time before flowers appear.

2.8.1. *Triggers to flowering*

Nevertheless, the climatic triggers to flowering remain obscure.[77] For dipterocarps, it is a trigger to cell division to begin forming an inflorescence primordium. In other families, such as in some Meliaceae and Sapindaceae, the trigger leads to expansion of an already formed dormant inflorescence for which there may or may not have been a prior climatic trigger. Water stress itself has been suggested as a principal trigger to initiation of the reproductive primordium,[78] but this seems unlikely in the perhumid tropics. Flowering times there are uncorrelated with physical characteristics of soils or with topography. Mass flowerings have been observed in mixed peat swamp forests with permanently saturated soils in the same years as upland forest mass flowerings. (Flowering in the peat dome interior appears less frequent, *but has not yet been systematically observed.*)

An old villager in Tanjung Malim (northern Selangor) once informed me that dipterocarp flowering followed the occasional arrival of a dry north wind: dry north-east trade winds, after dropping their rain on the Annamite range and crossing onto the Khorat Plateau, are sometimes diverted southwards by a climatic anomaly associated with ENSO.[79] The resulting clear skies and drier atmosphere are associated with lowered night temperatures. A likely principal trigger to canopy flowering in the seasonal tropics, as well as to mass flowering events in the perhumid Sunda lands, remains a depression of night temperature to below 20°C for several days,[80] in spite of some lack of consistency in meteorological records;[81] but this cannot be the sole trigger (Box 5.1).

2.8.2. *Mass flowering*

The perhumid regions of the Sunda Shelf are celebrated for the periodic mass flowering of the dipterocarps (Box 5.1). Most observations have been made in Malaysia, and at Gunung Palung National Park (West Kalimantan).[82] Thanks particularly to Tamiji Inoue and his former students, Shoko Sakai and Kuniyasu Momose, the event has been studied at the level of the whole community over a span of several years at Lambir National Park, Sarawak.[83] More recently, Francis Brearley has documented reproductive phenology over a decade in MDF in central Borneo,[84] while David Burslem and his students and associates have completed the most detailed phenological and pollination observations of dipterocarps to date, at Sepilok forest, Sabah.[85] Mass flowering periodicity is correlated with ENSO drought years and therefore have a regular periodicity of about five years. Some species also flower in the year preceding or following a mass flowering.[86] Minor mass flowerings may occur which are uncorrelated with ENSO events, and which may vary in their timing even within small regions. These provide relief for those starving vertebrates that can forage over long distances.

Community mass flowering in the Sunda lands is concentrated in the forest canopy, and to a lesser extent in other habitats exposed to direct sunlight. It is by no means a highly standardised event. Most canopy species flower with exceptional intensity during these events, but even dipterocarp species (Box 5.1) rarely flower *exclusively* in those years – and even those species do not come into flower during *every* mass flowering year[87] (Fig. 5.5). Canopy species, including

dipterocarps, flower more often between mass flowering events when their crowns become exposed after logging events, at forest edges, or when planted as specimen trees.[88] Subcanopy, and particularly understorey shrubs and trees, flower more or less annually but may even flower somewhat *less* during mass flowerings.[89]

The geographical extent of mass flowering in the Far East remains unknown, as no consistent records have yet been made in New Guinea, nor long enough term in the Philippines – the latter of particular interest as it is the only region where exclusively mass flowering *Shorea* (sect. *Brachyptera*) extend into the seasonal tropics. Interestingly though, Hamann[90] observed flowering phenology over four years at the lowland-lower montane ecotone in northern Negros (10°41' N). The area experiences a dry season of 2–4 months or even less, yet the mass-flowering red merantis *Shorea polysperma* and *S. almon* are present, as is *Parashorea malaanonan*. All three species flowered supra-annually. Both *Shoreas* also flowered in two successive years, together and apparently synchronously each year, and more heavily the second year.[91] *P. malaanonan* similarly flowered in two years, but one year later. No exceptional decline in minimum temperature was recorded prior to flowering, rather the annual decline correlated with the beginning of the dry season.

From records at Khao Ban Thad National Park and elsewhere in peninsular Thailand, where seasonality is similar to the Negros example, and from observations in more seasonal climates (see Box 5.1) it appears that Dipterocarpaceae flower annually in the seasonal climates of continental Asia, but different species flower in different years. However, some years witness heavier general canopy flowering than others.[92] *Here again, long term monitoring of marked trees would prove invaluable.*

Mysteriously, the carbon cost of mass flowering – estimated at 3% of net primary production[93] – has been estimated as greater than that of wind pollination. It has been observed that the cost of mass reproduction events is correlated with a following drop in growth rates.[94] Onset of reproductive maturity in tropical trees may result in permanent reduction of leaf size, which might imply a permanent change in carbon allocation from growth to reproduction,[95] besides adaptation to full sun in canopy species. Semelparous (that is, once in a lifetime) gregarious reproduction is a widespread phenomenon in the Asian tropics, prevalent in the omnipresent bamboos (Box 5.4) and *Strobilanthes* (Acanthaceae) (Plate 5.3f). Like the dipterocarps, these genera produce abundant undefended seed which, by their occasional heavy production, saturate the appetite of the seed predators in their habitat.[96] Different species of *Strobilanthes* have life cycles of 4–14 years, but like some dipterocarps and bamboos, a few may flower in the year prior to mass flowering. In this case, the sparse seed is eliminated by predators.

Fig. 5.5 Phenology of different plant life forms within the Lambir MDF over a five-year period, during which one year, 1996, experienced mass flowering and mast fruiting. (After Momose *et al.* 1998)

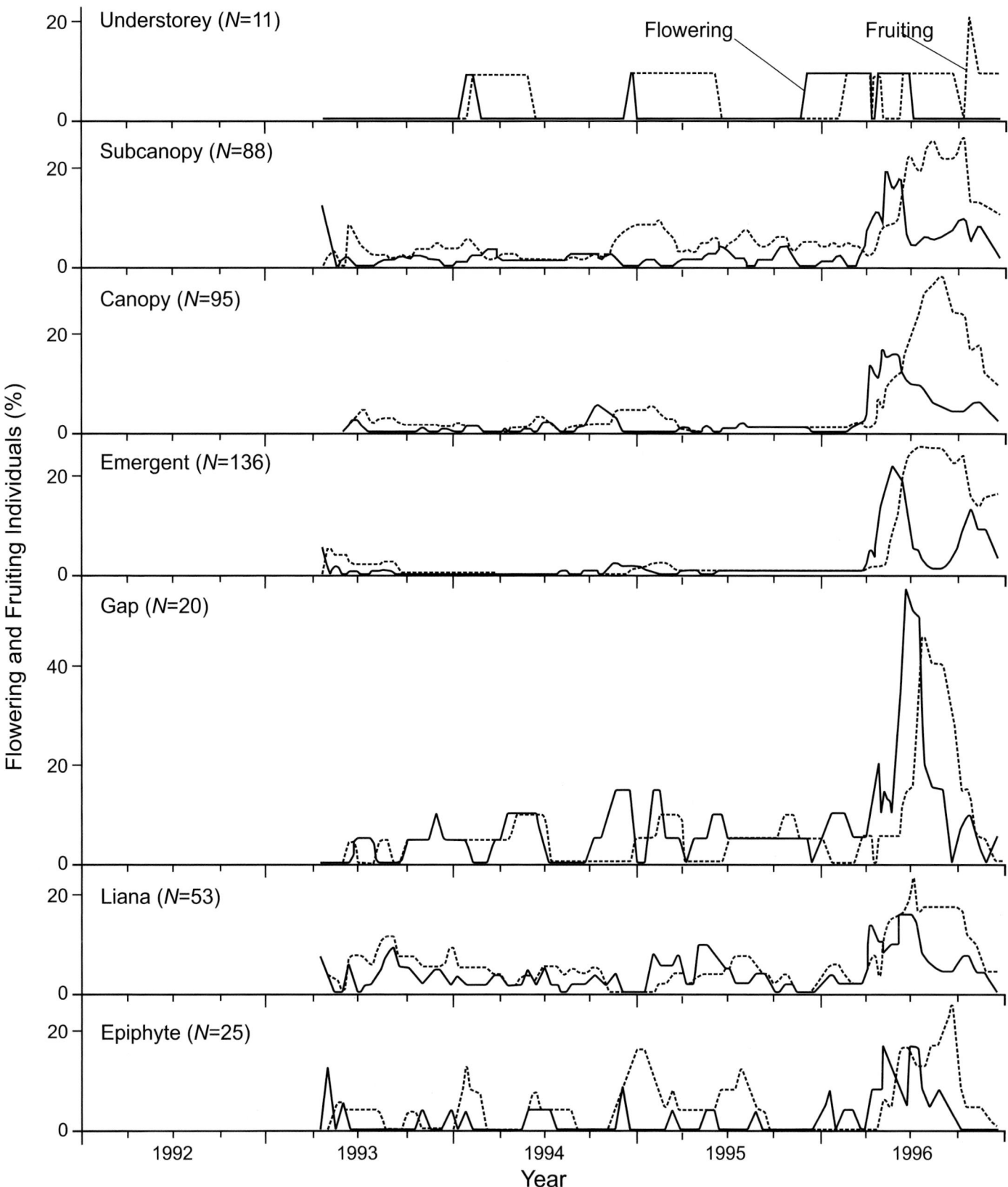

Plate 5.3 Mostly monocotyledons. **a,** The largest inflorescences among flowering plants: avenue of hapaxanthic *Corypha umbraculifera* still flowering synchronously, 1977, the third generation from seed planted for Queen Victoria's 1897 diamond jubilee; **b,** *Dendrocalamus strictus* browsed by elephants, Mudumalai Wildlife Sanctuary, Tamil Nadu; **c,** *Bambusa bambos* thicket, Huai Kha Khaeng Wildlife Sanctuary, west Thailand, with sharp axillary shoots; **d,** *D. strictus* grove, Huai Kha Khaeng; **e,** Dead seeding *D. strictus* culms, Huai Kha Khaeng; **f,** Bee-pollinated synchronously flowering hapaxanthic *Strobilanthes* in the Nilgiries. (**a** P.A.; **b,d,e** H.H.; **c** Mary Ashton; **f** K. Bawa)

322

Box 5.1

An Asian particular: the dipterocarps

It is the dry north wind that brings the flowering of
the dipterocarps.

A village elder, 1973, Tanjung Malim, Selangor.

Dipterocarpaceae are a pantropical tree family which likely originated in western Gondwana (Chapter 6); but the subfamily have diversified in perhumid Asia alone into 13 genera and more than 450 species. As rich mixtures of genera and species, they have come to dominate the canopy of the lowland MDF.[97] Several genera include mostly subcanopy species, while in some forests, in North-East India for instance, single species may dominate (Chapters 3, 7). The capacity of dipterocarp species, and the family as a whole, to dominate is not therefore consistently linked with their species diversification. Because of the economic importance of the dipterocarps as the leading timber trees of perhumid Asia, foresters in several countries, notably Malaysia, have long been trained to identify species for inventory. Thus their geographical and ecological distributions and silvicultural ecology are uniquely well documented (Chapters 2, 3, 6).

Subfamily dominance of the forest canopy is shared with Fagaceae, which dominate tropical lower montane and warm temperate forest in East Asia (Chapter 4). There, dipterocarps may be subordinate or absent in perhumid regions, even more than Fagaceae are subordinate in lowland dipterocarp forests. In Sri Lanka, where Fagaceae are absent, dipterocarps replace Fagaceae in montane forest, implying a competitive relationship[98] (but see Chapter 4). Both families have limited seed dispersal mechanisms, dipterocarp fruits being dispersed by gyration, wind gusts, or through scatter-hoarding by rats,[99] although in a few by flotation. Along with Caesalpinioid leguminous trees, both families are also exceptional in the humid tropics for their association with ectotrophic mycorrhizae (see section 6.5). Dipterocarp Basidiomycete (ectotrophic) mycorrhizae are dominated by the family Thelophoraceae, while Russulaceae are also well represented in MDF and appear to be associated.[100] In the Neotropics and Africa, Caesalpinioids show a similar relative abundance and high survivorship of juveniles compared with other tree families. But, there they also display the tendency to canopy including emergent canopy dominance, but it is usually monospecific. All three families are most dominant on soils with moderately low nutrient levels, where dipterocarps are also richest in species[101] (Chapter 7). These characteristics are consistent with the probable competitive advantages bestowed by ectotrophic mycorrhizae most in these habitats. In hyperdiverse rainforest, spores may not disperse far in the windless understorey and rapidly lose viability, but limited seed dispersal, and especially contact with parent roots,[102] increases the likelihood of infection by the assemblage of mycorrhizal species apparently characteristic of each dipterocarp species.

The reproductive cycle

In the seasonal tropics, dipterocarp species flower annually with varying intensity, or supra-annually but independently and without familial or generic synchronicity. The best known dipterocarp, the gregarious sal (*Shorea robusta*) of Indian deciduous forests, flowers annually from Feb. well into April, usually over c.4–5 weeks, but with extra intensity every few years. Troup recorded flowering years between 1891 and 1915 at many localities throughout its range.[103] Widespread flowering years were at c.5 year intervals, and heaviest flowering every decade. In seasonal climates, species flower at varying times, for one month or more within the cool dry season. Each year, clumps of certain species flower.[104] Sarayudh[105] has recorded dipterocarp phenology over several years in semi-evergreen dipterocarp forest at Huai Kha Khaeng World Heritage Forest and Wildlife Sanctuary (north-west Thailand). Every species flowered at least once during this time, and some occasionally flowered twice in a season.

In perhumid climates of Asia, most canopy dipterocarps flower en masse as a family,[106] whereas subcanopy species of the archaic Sri Lankan genus *Stemonoporus* may flower sporadically in all seasons. These mass flowerings of dipterocarps occur at varying intervals of about five years, broadly coinciding with the drought years occasioned by the ENSO (Chapter 1) (Fig. 5.6). Their timing varies regionally in relation to the season in which drought is experienced.[107] At least since systematic observations began at Pasoh in 1958, heavy flowering years have tended to alternate with weaker years in the five-yearly cycle. Other canopy families similarly flower heavily in these years, but also more frequently in the years intervening. The climatic trigger, which activates the flowering gene, leading to development of reproductive primordia in the bud, is imperfectly understood.[108] It is broadly correlated with periods of drought, but not directly with soil water deficit as flowering does not vary with local topography. Canopy dipterocarps only flower once their crowns are exposed; they flower at smaller size, and more frequently, in arboreta and logged forest.[109] There appears to be some correlation with a depression of minimum night temperatures below 20°C for several days,[110] but this correlation too is inconsistent, and it appears likely that trees must first build sufficient sugar reserves.

Where several sister species, such as several *Shorea* of one section, co-occur they flower in overlapping sequence, each flowering alone for a period (Figs 5.7–5.10). Simmathiri Appanah and Chan Hung Tuck found at Pasoh that populations within sect. *Mutica* flower with high synchronicity, the first for

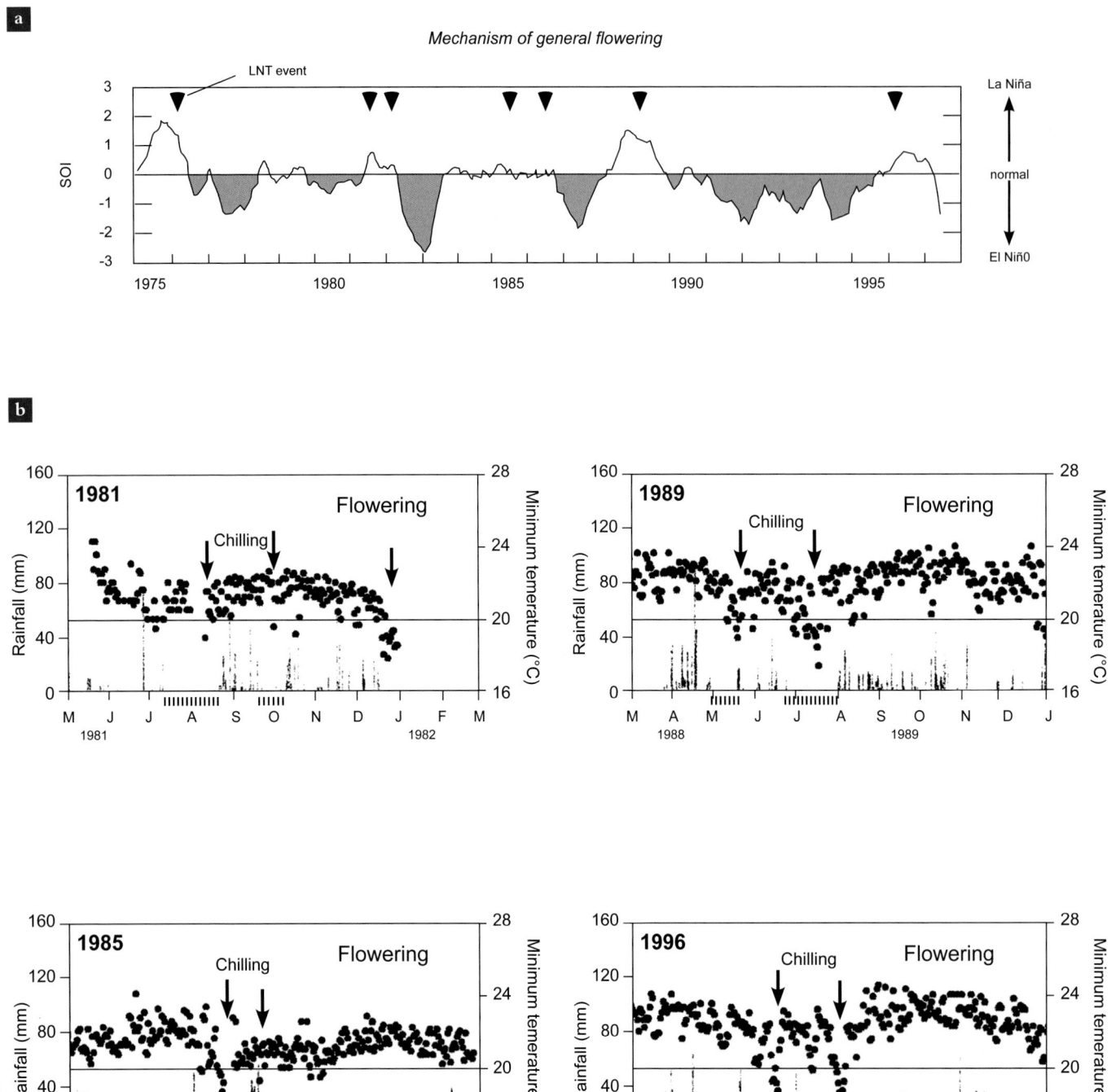

Fig. 5.6 Mass flowering and the El Niño Southern Oscillation (ENSO): **a,** correlation of ENSO events and low night temperatures over 27 years in Peninsular Malaysia. LNT: low night temperatures; SOI: southern oscillation index; **b,** the relationship between ENSO events and LNT occurrences at Pasoh forest, Negeri Sembilan: minimum night temperatures (black, exceptionally low arrowed dots), daily rainfall (vertical bars), rainless periods (broken bars below), and flowering in four mass flowering years (black horizontal bars). (**a** after Yasuda *et al.* 1999; **b** after Numata *et al.* 2004)

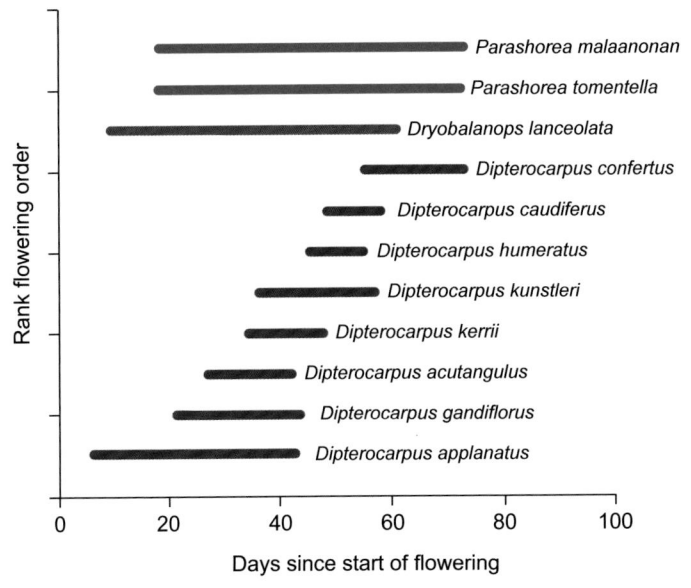

Fig. 5.7 Flowering phenology of the large-flowered dipterocarps *Parashorea*, *Dryobalanops* and *Dipterocarpus* (two species of which were observed to flower nocturnally) at Sepilok. (C. Maycock unpubl.)

about two weeks, then successive populations over an increasing period to the sixth and last, which flowered for 3.5 weeks (Fig. 5.4). The differentials in flowering times imply that each species has a different rate of floral development, following simultaneous firing of their flowering gene. The dates provided the means, by extrapolating backwards, to identify the date at which all these developmental trajectories originated – the date when the flowering gene was triggered – and therefore to search for any climatic event correlated with it. The flowering sequence has subsequently been confirmed in other seasons and sites, where species co-occurring in different regions within Sundaland flower in the same sequence.[111] It is inferred, therefore, that different species are genetically set to develop at different rates until anthesis. In the most species-diverse forests, such as Lambir and Semengoh in Sarawak (where there are 15 and 13 *Mutica* species, respectively, co-occurring in research plots) there are nevertheless at most six co-occurring in any one habitat defined by soil type, as at Pasoh. (Therefore at least half their diversity is explained by differences between the habitats, that is, ß–diversity.)

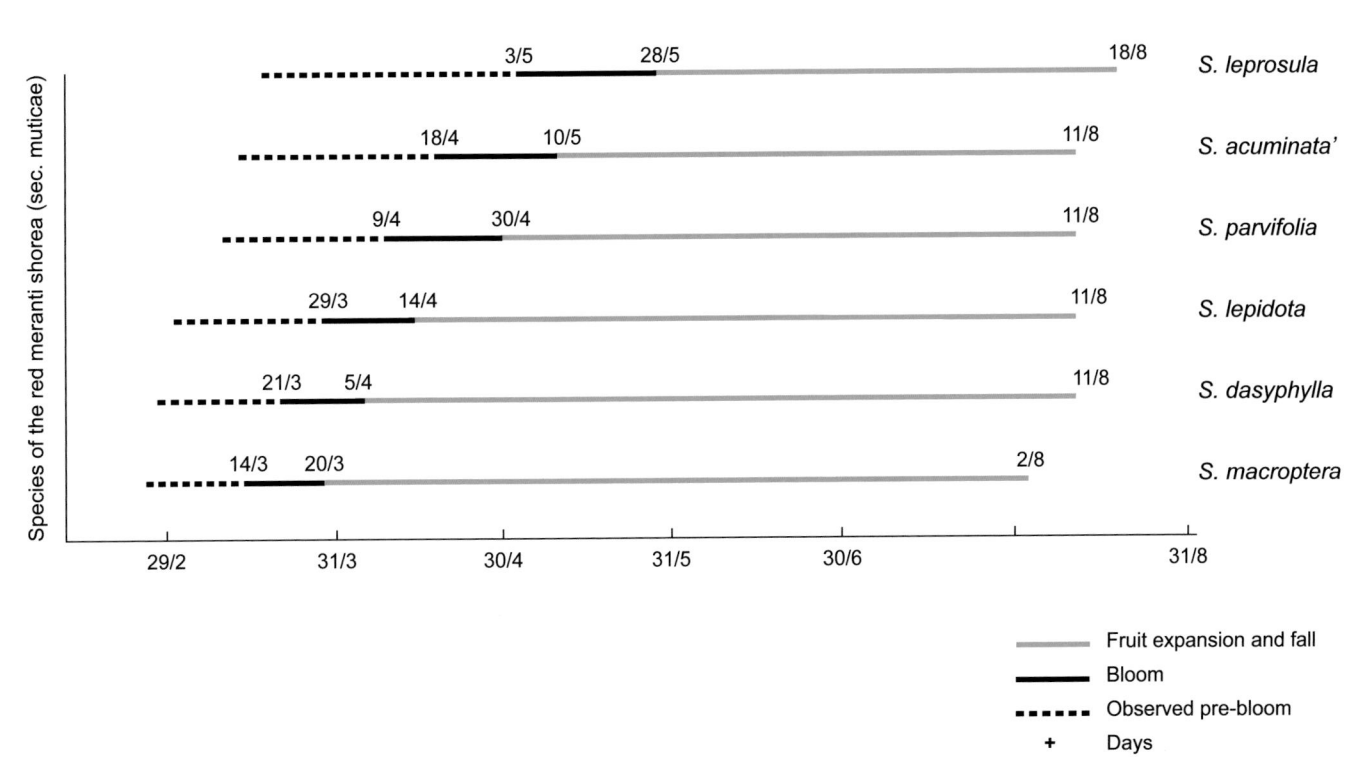

Fig. 5.8 Reproductive phenology of six co-occurring *Shorea* sect. *Mutica* species populations at Pasoh forest, 1976, from first observation of buds to start of ripe fruit fall (dates above: dotted lines, bud expansion; black continuous line: anthesis and corolla fall; blue continuous line: fruit expansion and fall of unripe fruit. (After Chan 1977, Chan & Appanah 1980)

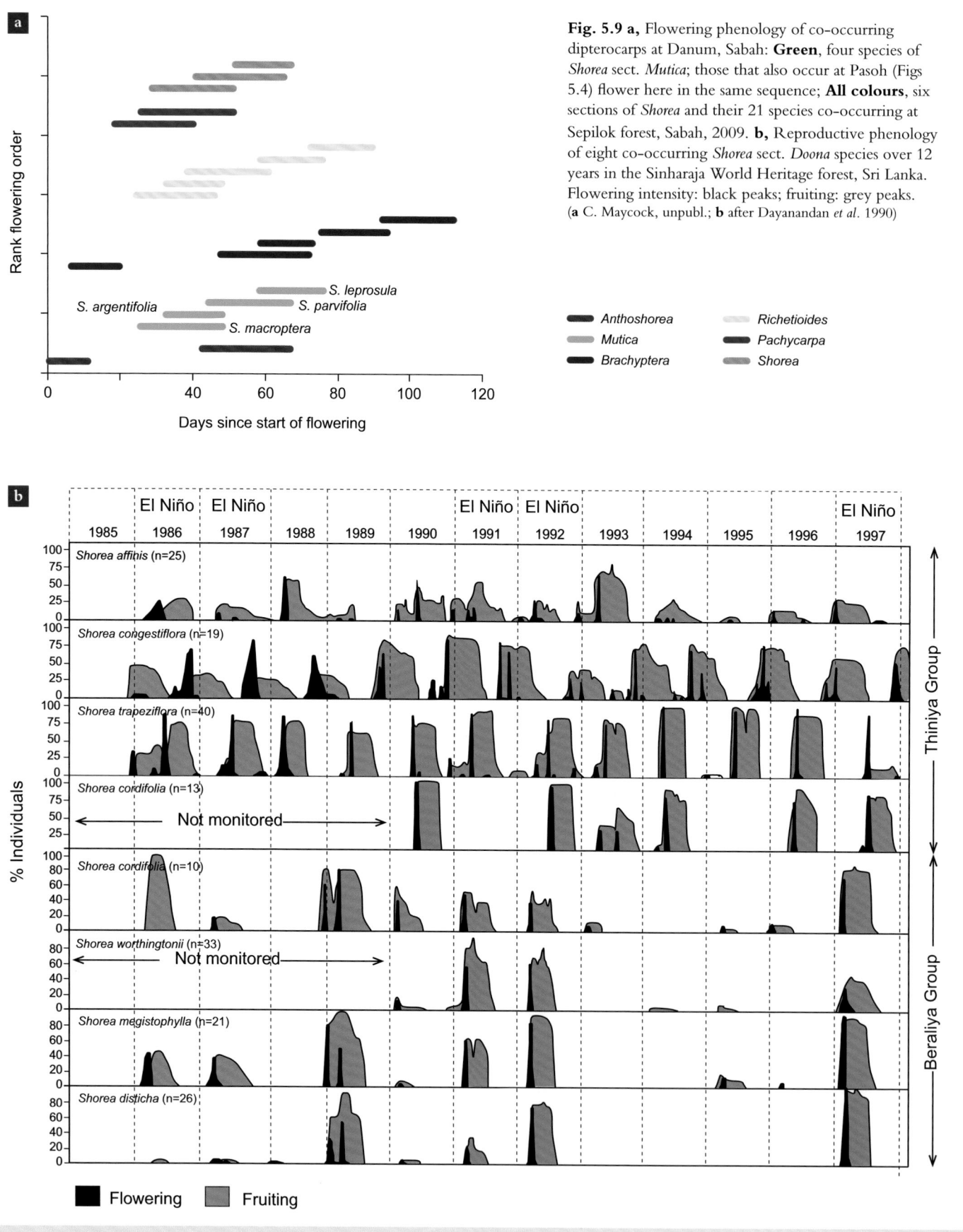

Fig. 5.9 a, Flowering phenology of co-occurring dipterocarps at Danum, Sabah: **Green**, four species of *Shorea* sect. *Mutica*; those that also occur at Pasoh (Figs 5.4) flower here in the same sequence; **All colours**, six sections of *Shorea* and their 21 species co-occurring at Sepilok forest, Sabah, 2009. **b,** Reproductive phenology of eight co-occurring *Shorea* sect. *Doona* species over 12 years in the Sinharaja World Heritage forest, Sri Lanka. Flowering intensity: black peaks; fruiting: grey peaks. (**a** C. Maycock, unpubl.; **b** after Dayanandan *et al.* 1990)

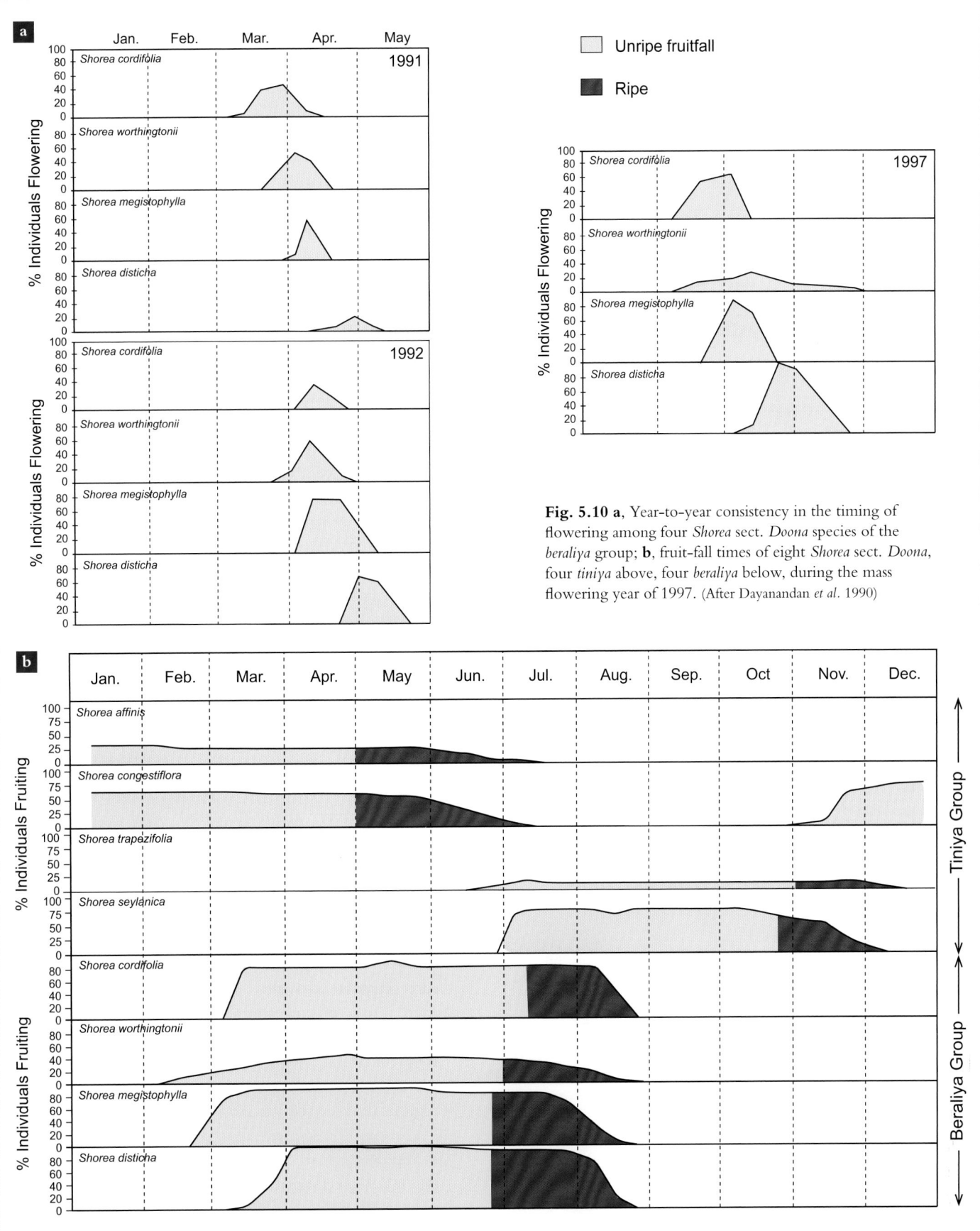

Fig. 5.10 a, Year-to-year consistency in the timing of flowering among four *Shorea* sect. *Doona* species of the *beraliya* group; **b**, fruit-fall times of eight *Shorea* sect. *Doona*, four *tiniya* above, four *beraliya* below, during the mass flowering year of 1997. (After Dayanandan *et al.* 1990)

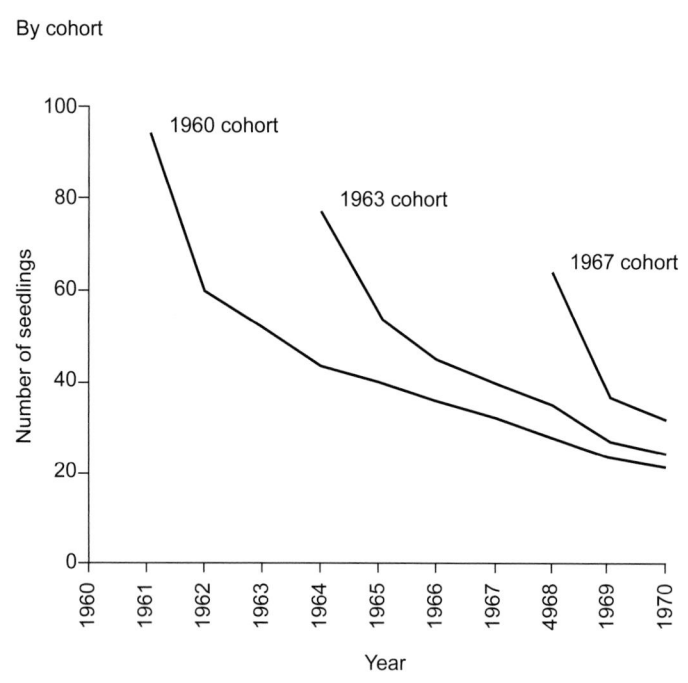

By cohort

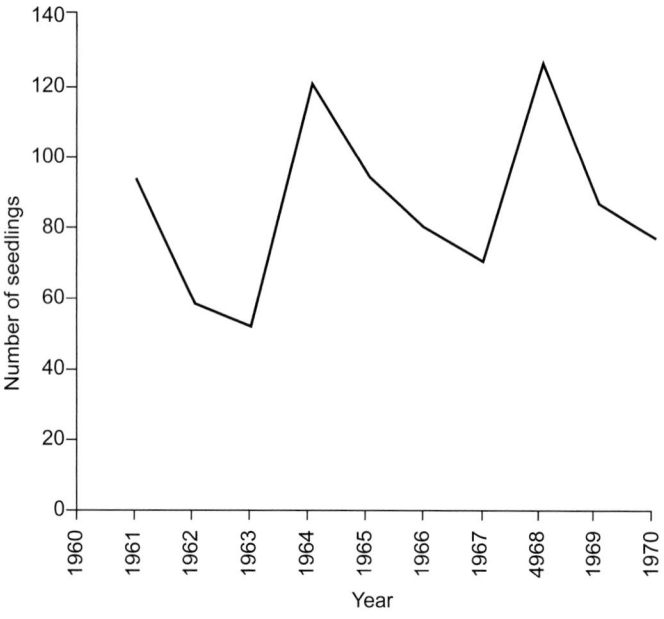

Whole population

Fig. 5.11 The pattern of seedling mortality in *Parashorea tomentella,* following supra-annual fruiting events: **a,** by individual cohorts following mass flowerings; **b,** whole population. (From Fox 1972 in Whitmore 1984a)

Later, Savitri and Nimal Gunatilleke, their students and the local school teacher carried out similar studies of the eight wild honeybee pollinated *Shorea* (sect. *Doona*) in MDF of the Sinharaja World Heritage Forest (south-west Sri Lanka).[112] They found that the eight species fell into two equal groups: one which also mass-flowers at intervals of about five years, but another in which the species flower more frequently – indeed, some individuals of two species sometimes flowered and set fruit twice annually[113] (Fig 5.9b. Both East and South Asian supra-annually flowering *Shorea* species series flower in overlapping sequence, but with much greater overlap in sect. *Doona* than in sect. *Mutica.* Individual *Doona* species populations had no periods during which they alone were in flower. Clearly, different climate triggers must be involved in the two *Doona* groups. Individuals of some taxa in the Far East, notably *Dryobalanops* species and *Dipterocarpus crinitus,* may flower more frequently than others in the intervening years between mass flowerings.

Colin Maycock, in David Burslem's team, found at Sepilok forest (Sabah) that several congeneric series of small-flowered dipterocarps flowered in overlapping sequence[114] (Fig. 5.9a). This was also true of the large-flowered *Dipterocarpus* species, at least some of which flowered nocturnally. Species of *Shorea* (sect. *Mutica*) occurring at both Pasoh or Lambir and Sepilok flowered in the same order, and this occurs at Lambir too. As with comparatively large-flowered *Shorea* (sect. *Doona*) at Sinharaja (Sri Lanka), the flowering times of large-flowered bee-pollinated species of *Parashorea* and *Dryobalanops* at Sepilok overlapped more thoroughly (Fig. 5.17). Although several Sri Lankan *Shorea* (sect. *Doona*) and Sunda *Dipterocarpus* species do occur within a community, large-flowered and large-anthered dipterocarp taxa with strongly overlapping flowering phenology fail to accumulate the numbers of co-occurring species populations achieved by *Shorea* sections *Mutica* and *Richetioides.*

Bearing in mind that dipterocarps are predominantly outbreeders,[115] the integrity of species in the midst of such extreme diversity can only be sustained where there is little or no introgressive hybridisation. Although, on morphological evidence, apparent hybrids are quite frequent among *Dipterocarpus* in the seasonal tropics,[116] there are few cases of maturing hybrids in perhumid regions (but see section 2.5).

Among the sequentially flowering species and genera of dipterocarps in the perhumid tropics, mass flowering extends over four months or less, but is followed by mast fruiting over a period of no more than four weeks. During this period, those species series that had flowered in sequence fruit synchronously (Fig. 5.8). Those species which develop from expanding bud to anthesis in the shortest time take the longest time to ripen their fruit, and *vice versa*: an astonishing manifestation of the reconciliation of antagonistic selective pressures through competition!

In many dipterocarp species, a developing inflorescence takes the place of a leafy shoot, so that reproduction occurs at a time of reduced photosynthetic leafy surface. A similar reduction in the number of leaf primordia consequent to mass flowering has also been observed in Himalayan Fagaceae, where it is counteracted by increasing the longevity of leaves already opened.[117] The overall cost of dipterocarp reproduction in energy and nutrients is formidable, and its consequences include both reduced production of leaf shoots and reduced stem diameter growth. Nevertheless, among nutrients only phosphorus, for whose uptake the ectotrophic mycorrhizae are critical, has been found to accumulate in the stem prior to flowering, when it is released.[118] The wing-like fruit sepals characteristic of most dipterocarps, especially the emergent species, are densely furnished with stomata, and act as substitutes for leaves.[119] Carbon requirements are met by concurrent photosynthesis – in which the photosynthesising wings of the developing fruit appear critical. Many flowers develop into photosynthesising winged fruit but prematurely abort.

Dipterocarps – their abundant seeds rich in starch and fats yet chemically undefended other than by the resin in their leathery pericarps – are classic examples of masting trees.[120] Good mast years are times of plenty for seed predators, and these are the years when many, such as the orangutan, reproduce.[121] Lisa Curran monitored 2,367 individuals of 54 dipterocarp species over almost six years, 1985–1993, in Gunung Palung National Park (West Kalimantan).[122] Masts occurred in 1986, when it was preceded the previous year by a minor fruiting in which there was no seed survival; and in 1990. Phenological observations were supplemented by planting the large seed of *Shorea stenoptera,* an illipe (see Chapter 8), at varying distances from mature individuals, in logged and unlogged forest. Her astonishing care and industry revealed that, once seedlings became established, predation declined. The mast indeed saturated resident vertebrate seed predators but the dominant predator, the bearded pig, is migratory and became saturated at an undefined wider, provincial scale while, at the local scale observed, pigs arrived after germination and seedling establishment and did little damage. Seed survival was therefore by escape, and the amount of seed *locally* produced was irrelevant to survivorship. Of the germinants originally established in the earlier mast year, 65% were surviving four years later at the second. Logged forest, in which the proportion of dipterocarps in the canopy had become much reduced, suffered higher predation rates. The impact on vertebrate seed predation patterns of the reduction of lowland

MDF to isolated parks and other forests awaits documentation.

In off-years, dipterocarps do flower sporadically, particularly where their crowns are exposed in arboreta or logged forest;[123] but persistent seed predation generally precludes successful establishment at such times.[124] The entire seed-crop of trees such as *Dipterocarpus crinitus,* which sometimes flowers out of mass flowering years, is often destroyed by predation. During these off-years, figs serve as an important famine food (Box 5.7). Nevertheless, frugivore and granivore numbers are reduced by starvation and predation.[125]

Major seed predation is also attributable to Curculionid beetles, and Staphylinid weevils of the genera *Alcidodes* and *Nanophyes,* notably *N. shoreae.*[126] Eggs are laid early in seed development, and early seed abortion results. Scolytids also predate seed following fall. Seed and seedling predation does vary among species. Akira Itoh,[127] who compared two *Dipterocarpus* with two *Dryobalanops* species, found that the first pair experienced higher predation (c.90% versus 60–70%) between dispersal and the shedding of cotyledons. Thereafter, mortality declined in all species, and other causes of mortality came to outweigh predation. The Gunatillekes and I observed that seeds of mass-flowering *Doona* species, recognized by villagers as *beraliya*, were predated upon in seed trays, whereas seeds of the annually flowering species, recognized by the name *tiniya*, were not. The seeds of *beraliya* were once collected and cooked to form gruel in famine years, but *tiniya* seeds were rejected as too resinous. Frequently flowering *Shorea* (sect. *Doona*) and *Dryobalanops* may be protected by their more resinous pericarps. The clumped distribution of mature individuals so characteristic of Asian dipterocarps is enhanced by heavier fruiting and juvenile survivorship within clumps.

Once established, dipterocarp juveniles experience comparatively little herbivory, but their numbers continue to decline between mast fruiting years (Fig. 5.11). Light demanding species decline more rapidly than shade tolerant ones which, in the case of case of forest transplant experiments with *Shorea worthingtonii* in Sri Lanka, can persist for more than a decade in the understorey (Chapters 2, 3).[128] *The relative roles of herbivory and the physical environment in determining survivorship have not been examined.* Dipterocarp leaves and stipules, especially those of juveniles, bear secretory glands which function as the organs expand, and attract ants which defend them against herbivores (Plate 5.20a). Many also bear domatia in the axils of leaf veins. These house mites which are predators both of arthropods and of epiphyllic fungi.[129] Dipterocarp juveniles can experience positive (as well as negative) density-dependent mortality (see section 6.4), and mature individuals also tend to be grouped in clusters in MDF. Both of these characteristics are consistent with nutrient sharing through species-specific mycorrhizae.

But that is contrary to the density-related mortality found in so many other families and generally attributed to host-specific pathogens. *The potentially conflicting impacts of pathogenic and symbiotic mycorrhizal fungi on dipterocarp demography and co-evolution remain an 'abominable mystery' demanding the attention of ingenious minds!*

The co-ordinated flowering and fruiting phenology of many congeneric dipterocarp species-series, whose species share pollinators and whose synchronised fruit-fall lends them high seed survivorship through predator satiation,[130] provides perhaps the most convincing evidence for stringent interspecific competition in hyperdiverse rainforest being mediated by mobile links.

It implies that the extraordinary diversity of dipterocarp species in species-rich rainforest communities cannot be sustained by chance alone. Their fruit-set, seed dispersal and juvenile establishment are *least* certain at times when few other species are engaged in reproduction, when opportunities for their establishment would therefore seemingly be optimal.[131] The multitude of dipterocarp species in hyperdiverse forests, such as Lambir in Sarawak, therefore appear to be sustained: (1) at landscape scale by their soils specificity (Chapter 3); (2), within communities in uniform habitats by their growth responses to light combined with their survivorship in shade; and (3) by their place at reproductive maturity in the vertical structure of the forest. In some cases, though – including several *Shorea* (sect. *Mutica*) species at Pasoh and elsewhere in Peninsular Malaysian MDF – species appear to be at least partly ecologically complementary in these respects. Such populations of species in co-occurring congeneric series may then be differentiated in niche occupancy by a fourth factor: sustained competition for a shared pollinator. Reduced pollination success leads to lower fecundity during periods when interspecific flowering overlaps, and thus to selection for sequential flowering; while interdependence on high pollinator numbers leads to shared fecundity rates and population densities. This might imply that these species, at least, could be physiologically complementary, but their numbers checked by competition for a shared pollinator. In this respect the spatial predictions of neutral theory may prevail among their populations. Yet they are interdependent in that they also rely on each other to build sufficient pollinator numbers! (see also section 6.4).

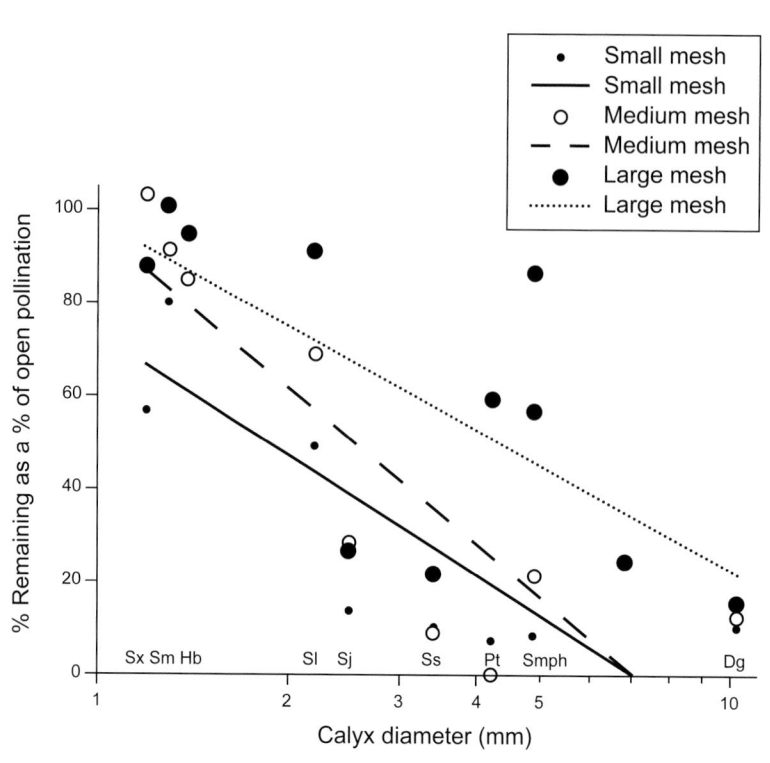

Fig. 5.12 The small flowered dipterocarps *Shorea* sects *Richetioides* (Sx: *S. xanthophylla*) and *Mutica* (Sm: *S. macroptera*, Sl: *S. leprosula*) achieve higher pollination success in small mesh bags than is the case in larger flowered dipterocarp taxa, implying that their pollination is more successfully achieved by small insects such as thrips. Hb: *Hopea beccariana*, Sj: *S. johorensis* (*Brachyptera*), Ss: *S. superba* (*Shorea*), Pt: *Parashorea tomentella*, Smph: *S. agamii* (*Anthoshorea*), Dg: *Dipterocarpus grandiflorus*. (C.J. Kettle 2012, unpubl.)

Pollination biology

Indeed, species within each dipterocarp genus (and within each section of the large genus *Shorea*) do appear to share the same principal pollinator. The flowers of most dipterocarps open early in the morning, and corollas start to fall in large numbers around noon. The isolated and basal genus *Dipterocarpus* is pollinated by Lepidoptera;[132] bees also visit, but likely as pollen robbers.[133] Burslems's group carried out pollination-exclusion experiments enabling them to confirm that, though all dipterocarp taxa are visited by diverse insects, flower size and morphology (especially of the androecium) act as filters, selecting different predominant effective pollinators for each genus and infrageneric species series (Fig. 5.12).

The anthers of Asian dipterocarps are found in two forms: narrowly oblong and yellow, containing many pollen grains; or ellipsoid and cream coloured, smaller and with fewer grains.

The larger oblong yellow anthers are found in the following taxa:

- The widespread *Dipterocarpus*.

- *Vateriopsis,* endemic to the Seychelles, and also the basal genera *Vateria* and *Stemonoporus,* endemic to South Asia.

- *Upuna,* endemic to Borneo.

- The more widespread *Cotylelobium*.

- The Sundaland endemic *Dryobalanops,* basal in the Tribe *Shoreae*.

- *Neobalanocarpus,* endemic to Peninsular Malaysia and possibly of hybrid origin between *Hopea* and another taxon.[134]

- *Parashorea,* widespread in the Far East and basal to the red meranti sections that are endemic to perhumid Malesia.

- *Shorea* sect. *Doona,* endemic to Sri Lanka.

- *Shorea* sect. *Pentacme* of Indo–Burma and the Philippines.

- The red meranti sect. *Rubella* of Borneo and the Philippines.

- The flowers of all dipterocarp appear to attract Chrysomelid beetles. *Dipterocarpus* flowers are visited by butterflies, and the nocturnally flowering species *D. tempehes* by moths. Those others which have been observed (*Vateria, Dryobalanops, Neobalanocarpus* and *Shorea* sections *Doona* and *Pentacme,* all of which are canopy trees) are pollinated by honeybees, although many other insects visit in small numbers including Chrysomelid and Scarabid beetles, and *Trigona* bees in *S. siamensis* (*Pentacme*), and other taxa when honey bees are absent.[135] Dayanandan found that *Shorea (Doona) megistophylla* flowers have many visitors, although bees prevail (Table 5.2).[136] The primarily subcanopy *Stemonoporus* is pollinated by Meliponid *Trigona* sweat bees (pers. obs.).

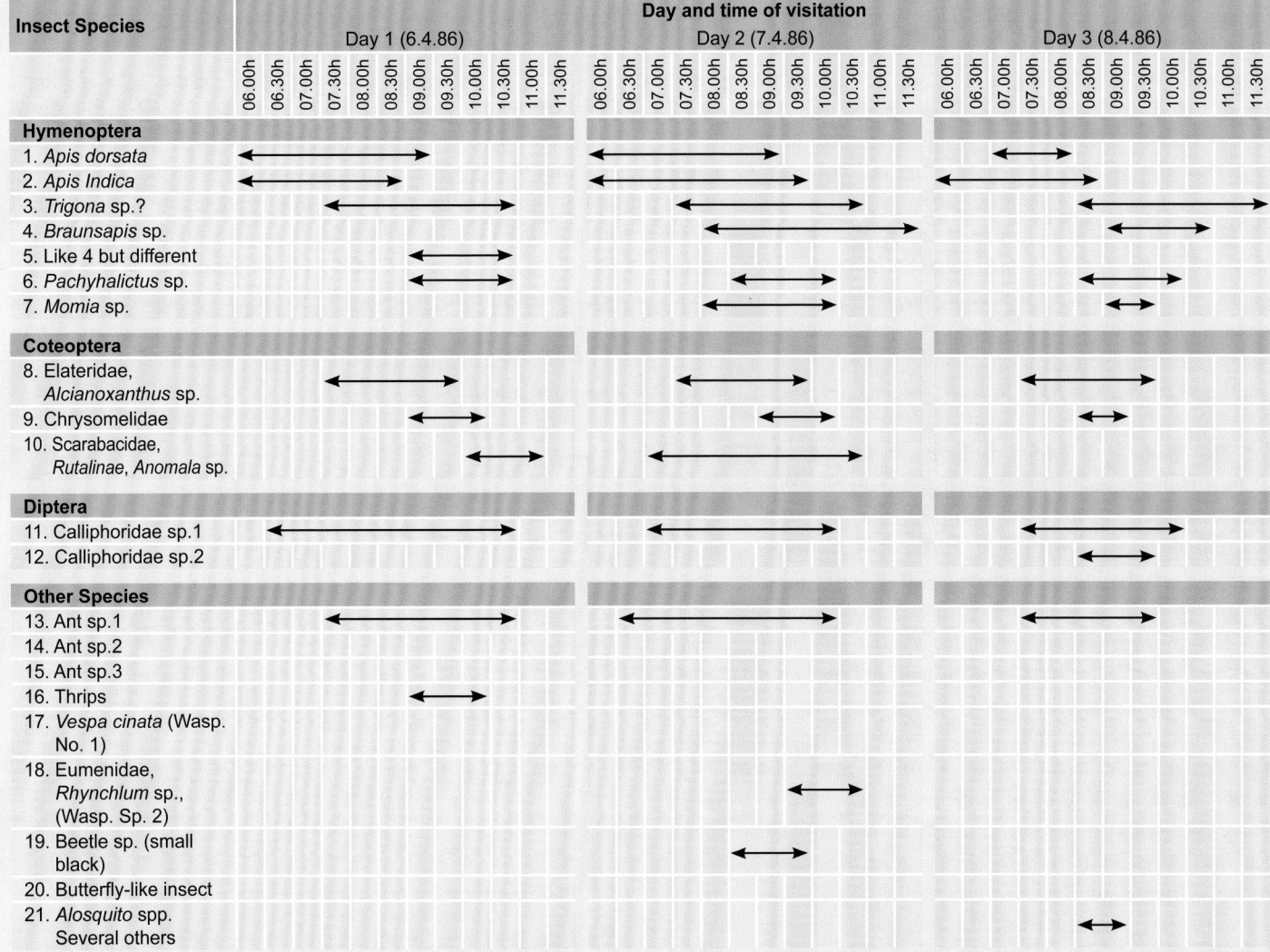

Table 5.2 Floral visitors to *Shorea megistophylla* (sect. *Doona*) over three consecutive days during full anthesis. (From Dayanandan *et al.* 1990)

The Asian dipterocarps possessing the smaller white anthers include all subcanopy dipterocarp taxa in hyperdiverse Far Eastern rainforests, as well as several canopy clades:

- All *Hopea*, *Anisoptera* and *Vatica* species.

- *Shorea* sections *Shorea, Anthoshorea, Brachypterae, Pachycarpae, Mutica, Richetioides* and *Ovales*.

The flowers of *Hopea*, *Shorea* sections *Mutica*, *Brachyptera*, and *Richetioides* are abundantly visited by thrips, and also Chrysomelid beetles.

Simmathiri Appanah and Chan Hung Tuck revealed the marvellous story of pollination among Shorea *(sect.* Mutica*), thanks to their careful observations at Pasoh Forest (Negeri Sembilan).[137] Both students were awaiting a mass flowering in order to complete their doctoral dissertations. In Feb. 1976 they were alerted by the expanded racemes and elongating buds of a* S. hemsleyana *tree (from the same section) in the arboretum of the Malaysian Forest Research Institute at Kepong. Taking some buds back to their laboratory at the university, they were surprised the following morning to find the still-closed petals punctured by minute, neat round holes, and tiny insects trapped within the bag: thrips! Could these prove to be the long-sought pollinators? Thrips have been observed as pollinators of the pioneer* Macaranga hullettii.[138]

When the populations of the six Shorea *species in the Pasoh MDF successively came into flower, no other floral visitor was found. Sticky traps suspended in the forest canopy had at first recorded a steady population of thrips, which started to build up as the first species to flower came into bud. Appanah, growing them in culture, found that their life-cycle was completed in eight days. It was later calculated that this permitted three generations of thrips to propagate on petals in bud before the first flowers of the tree species opened. (In* Macaranga hullettii *the thrip life-cycle was 17 days.[139])*

The flowers open during the evening and spiral away (thanks to their contorted corolla) the following morning, carpeting the forest floor and permeating the shade with their heady scent. On board are thrips in all developmental stages. A haze of mature individuals rises erratically into the canopy each morning, their bristle-like wings cleave the air – viscid at the scale of their tiny bodies – like oars (Fig. 5.13, 5.14). Larvae continue to feed on the fallen but still loosely adhering petals for several days, till they too mature and rise into the canopy. The overlapping petal bases of the Mutica *corolla at anthesis form a kind of urn, gated at its apical opening by the connectival appendages of the anthers closed within it (Fig. 5.15). This gate traps the thrips until the corolla falls. Crawling within the corolla, they receive pollen grains on the backs of their thoraces. Like us with a mosquito on one's back on a hot night, the thrips struggle to reach these grains – to them as big as oxygen cylinders – in their vain attempts to clean. Free of the corolla, they fly, as it were, like scuba divers, blundering into the sticky traps where they are found. Caught by a gust, thrips have little control over their movement. Their pollination services are nevertheless* *marginally more targeted than the wind since, once close to a flower, they can land with their burden directionally.[140]*

Thrips and Chrysomelid beetles were pervasive in dipterocarp flowers at Sepilok. Thrips have been found in the flowers of *Hopea* and of the yellow meranti *S. multiflora* (sect. *Richetioides*), both of which have *Mutica*-like corolla and anther structure. Chrysomelid beetles were found abundantly on flowers of a *S. parvifolia* at Lambir, and were seen as potentially effective pollinators[141] (Plate 5.2c). Their populations are sustained in years between masts as they feed on other flowers and young leaves.[142] But the *Mutica* flowers seem less well adapted to them. Indeed, a striking though circumstantial piece of evidence points to thrips as the principal pollinators of *Shorea* sections *Mutica*, *Richetioides*, and perhaps *Brachyptera* and the genus *Hopea* (Fig. 5.14). Although their perianth morphology otherwise differs, the androecium is always arranged so that the stamens with slender filaments bearing tiny anthers and bristle-like appendages, which in *Mutica* species especially curve back at anthesis entrapping the newly emerging thrips, remarkably resemble the flowers of heather, *Erica*, which the Hagerups[143] so vividly confirmed to be thrip-pollinated (Fig. 5.14).

Chan[144] found that one emergent, *Shorea leprosula* (sect. *Mutica*), which has small, white flowers, with a c.8 mm diameter open corolla and small white anthers, produced up to 654,000 flowers in a single day at peak flowering, culminating in up to four million flowers being presented by an individual tree over 3.5 weeks. He found that individual *S. leprosula* trees yielded an estimated 4,000–114,000 fruit with germinable seed. Therefore as many as 35 flowers are yielded to produce a single fruit and seed.[145] (But bear in mind that a single successful progeny over a lifetime, on average, is sufficient to sustain the population!)

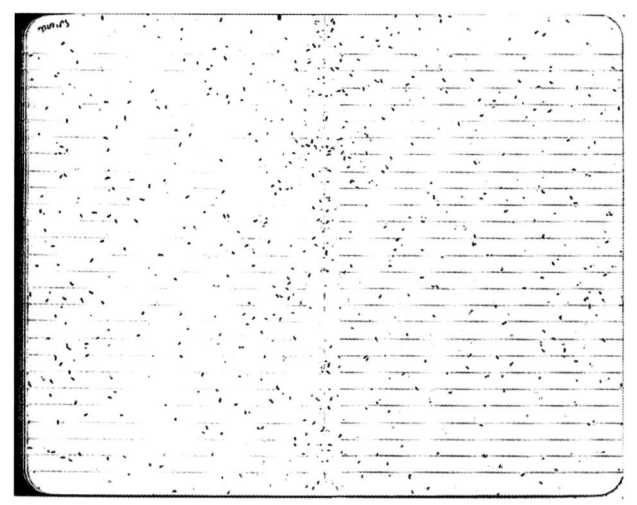

Fig. 5.13 Thrips on the pages of Tim Laman's notebook, left open beneath a flowering *Shorea parvistipulata*, Gunung Palung National Park, West Kalimantan. (Laman unpubl.)

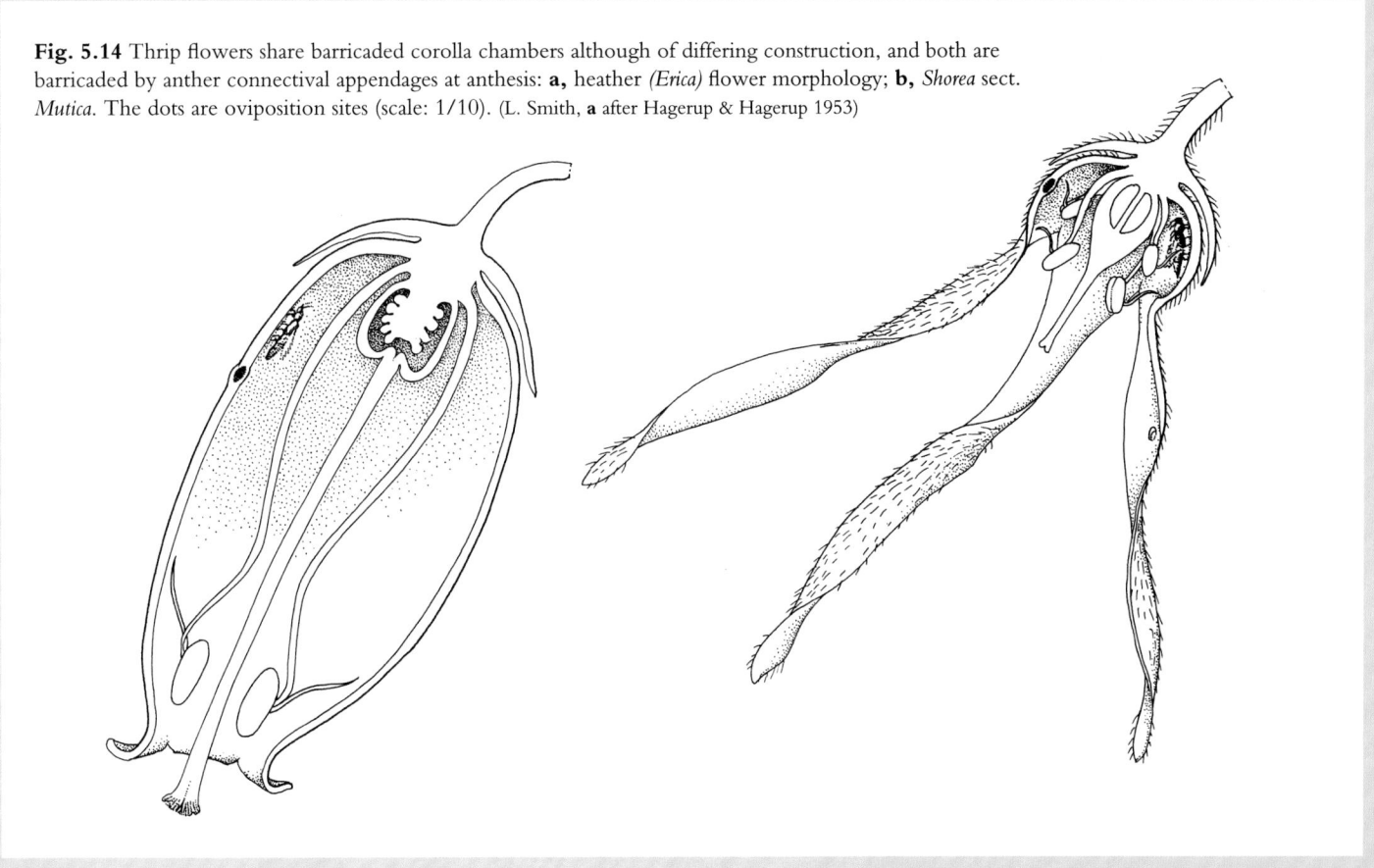

Fig. 5.14 Thrip flowers share barricaded corolla chambers although of differing construction, and both are barricaded by anther connectival appendages at anthesis: **a,** heather *(Erica)* flower morphology; **b,** *Shorea* sect. *Mutica.* The dots are oviposition sites (scale: 1/10). (L. Smith, **a** after Hagerup & Hagerup 1953)

Plate 5.4 Dipterocarp flowers. **a,** *Dryobalanops aromatica* in full flower, Lambir Hills, Sarawak, 1996; **b,** Crowns of emergent *Koompassia excelsa* (coppery) and dipterocarps in very early fruit, east Sabah, March 2010; **c,** *Dipterocarpus obtusifolius* (scale: 2/5); **d,** *Shorea* sect. *Doona* species with *Trigona* visitor; **e,** *Shorea* sect. *Doona* species with the giant honey bee, *Apis dorsata;* **f,** *Dryobalanops aromatica* with *Apis dorsata* (d-f: scales c. 1/1). (**a, f** T. Inoue; **b** H.H.; **c** J. Ghazoul.; **d, e** N. Gunatilleke)

Bee-pollinated *Shorea* (sect. *Doona*) bear larger flowers than *S. leprosula* with a c.12 mm diameter open corolla and large yellow anthers. Only one fifth as many (larger) flowers are presented, while less than 20% of hand cross-pollinated flowers yielded fertile seed.[146] Appanah[147] estimated that *Neobalanocarpus heimii*, with flowers similar in size and morphology to *Shorea* (sect. *Doona*), presented fewer than 104,000 flowers per day, and less than 637,000 flowers in a season. Ichie and colleagues[148] found that the flower/fruit ratio in three *Dipterocarpus tempehes* individuals (with large flowers pollinated by Lepidoptera and giant honeybees and producing larger fruits than *Shorea*) varies widely, between 6:1 and 65:1. The large emergent *Dryobalanops* trees, with similar large bee-pollinated flowers to *Neobalanocarpus* but presented in vastly greater numbers (albeit not carefully documented), provide the exception; their high fecundity may in part explain their local dominance (see Chapter 7).

Selfed flowers were associated with a lower flower/fruit ratio, but yielded fewer fruits overall. The apomictic tetraploid *Shorea ovalis* presented a mere 2,300 flowers (of intermediate size) per day,[149] but, though not precisely documented, I have observed that it achieves high seed production.

The flowers/fertile fruit ratio appears to be similar in *Shorea leprosula*, which bears vastly more flowers with smaller anthers producing fewer grains, and *Dryobalanops*. The small-flowered canopy dipterocarps of *Shorea* (sect. *Mutica*) dominate late succession in the MDF forest cycle. Their abundance constituted the backbone of the Far Eastern timber industry,

Plate 5.5 Dipterocarp fruit and seedlings. **a,** Fruiting crown of *Shorea amplexicaulis*, Gunung Palung National Park, West Kalimantan; **b,** Diversity of size and wing/seed weight ratio among Lambir dipterocarps (scale: 1/4); **c,** First year *Shorea* seedlings; **d,** Multiple apomictic seedlings emerge from a *Shorea resinosa* fruit (scale: 2/3). (**a** T. Laman; **b** S. Davies; **c** C. Ziegler; **d** A. Kaur)

before oil palm replaced their forests; *Dryobalanops* achieves this only on a local scale. I infer that, by decking their broad hemispheric emergent crowns with millions of pale flowers like a flag, sect. *Mutica* taxa successfully attract a ubiquitous pollinator whose short life-cycle enables it to build up sufficient numbers to achieve relatively high fecundity through cross-pollination. The fecundity of bee-pollinated species, though, is limited by a pollinator life cycle too long for it to respond to a mass flowering by build-up of numbers. Instead, swarms are drawn in from other habitats (Box 5.5) to trees decked with fewer larger flowers, but such opportunities are more limited. Here, it seems, the greater specificity of the pollinator gains a similar fecundity in terms of fruit/flower ratio, but at the cost of lower seed yield per tree. *Dryobalanops* has achieved its *local* canopy dominance in part by attracting bee swarms from afar with the snowy summits of their large flowering clumps, but presentation of their many large flowers must incur a high energetic cost which must be met by some other selective advantage. The fact that many large-flowered taxa which present fewer flowers per individual (notably *Dipterocarpus* and *Neobalanocarpus*) achieve local high population densities (Chapter 7) must be explained by other aspects of their life history.

These observations have been expanded by David Burslem, Colin Maycock and Chris Kettle at Sepilok Forest (Sabah). They found that the higher the number of nearby conspecific trees in flower, the lower the number of flowers aborting following anthesis and the higher the fruit set.[150] This confirmed similar observations in *Dryobalanops aromatica* at Lambir.[151] They also found that species of *Shorea* (sects *Mutica* and *Richetioides*) produced more flowers but achieved lower pollination success and lower pollen dispersal distances than taxa with larger flowers and pollinators, including *Dipterocarpus* and *Parashorea* species. Kettle concluded that these trade-offs nevertheless resulted in similar fecundity, delaying or negating competitive exclusion among these taxa during this part of their life cycle.

Plate 5.6 *Xerospermum noronhianum*. **a,** *Euploea* visitor on male flower; **b,** *Idea* butterfly on male flowers; **c,** Hermaphrodite flowers with paired ovaries (young fruit) and decomposing anthers (scale: 2/1); **d,** Fruit opened to reveal seed with fleshy sweet sarcotesta (scale: 2/3). (All S. Appanah)

The dipterocarp ovary is three-celled, each containing two ovules. Surprisingly for outbreeding species, the number of pollen grains produced per ovule appears overall to have *reduced* over evolutionary time and as the family has diversified into hyperdiverse communities. Especially beneath the canopy, where pollen transfer must surely be most hazardous, this seems counter-intuitive (see the main text for other examples).

Box 5.2

An Asian particular: *Xerospermum noronhianum*, Sapindaceae, a subcanopy tree and the commonest tree in the Pasoh forest

Xerospermum noronhianum, known in Malaysia as Rambutan pacat ('leeches rambutan'), is characteristic of the climax subcanopy and the shadier parts of the main canopy of MDF (but attains the canopy of semi-evergreen forest, see Chapter 3) (Plate 5.6). In the 50 ha plot at Pasoh (Negeri

Sembilan, Malaysia), in 1986 there were 8,962 individuals at least 1 cm dbh. (The second most common species, *Shorea maxwelliana*, had 5,682). This constituted 2.7% of total tree individuals, and about 12 *X. noronhianum* >10 cm dbh ha⁻¹. *X. noronhianum* was studied by Simmathiri Appanah and Yap Son Kheong.[152] The flowers are small and inconspicuous, cream-white, borne on short axillary spike-like racemes among the leaves. Appanah found that, like some Acer species in temperate forests,[153] the trees are androdioecious. (This appears to be general among Sapindaceous trees in the Far East[154]). They have either male flowers up to 5 × 2.6 mm in diameter, with four anthers but lacking ovaries; or bisexual flowers up to 3.6 × 3.8 mm, with two-lobed ovaries and anthers. The anthers of the bisexual flowers fail to dehisce, lacking a suture. If the stigma receives no foreign pollen, however, the anthers bend onto it and their walls decompose to release the pollen. Flower-bagging experiments established that this process can

achieve self-pollination, thereby sustaining some fecundity when cross-pollination fails. Androdioecy is said to be rare among seed plants,[155] and appears invariably to involve the flowers of ovary-bearing individuals retaining their functional anthers.[156] (See section 2.3.2)

Rambutan pacat populations flower annually, but for varying periods of 2.5–7 months and in varying intensities, beginning July–Dec. and ending Dec.–March. Inflorescence buds develop early and independently of flower development, resting dormant for varying periods until a flowering trigger is released and the inflorescence starts to expand. Populations flower with low synchronicity for 2–13 weeks; individuals flower for a shorter period of c.5–6 weeks during intense flowerings. The male trees flower more frequently; they start earlier and flower longer, thereby skewing the population's sex ratio (Fig. 5.15). During an intense flowering episode, individual male trees presented c.500,000 flowers overall or up

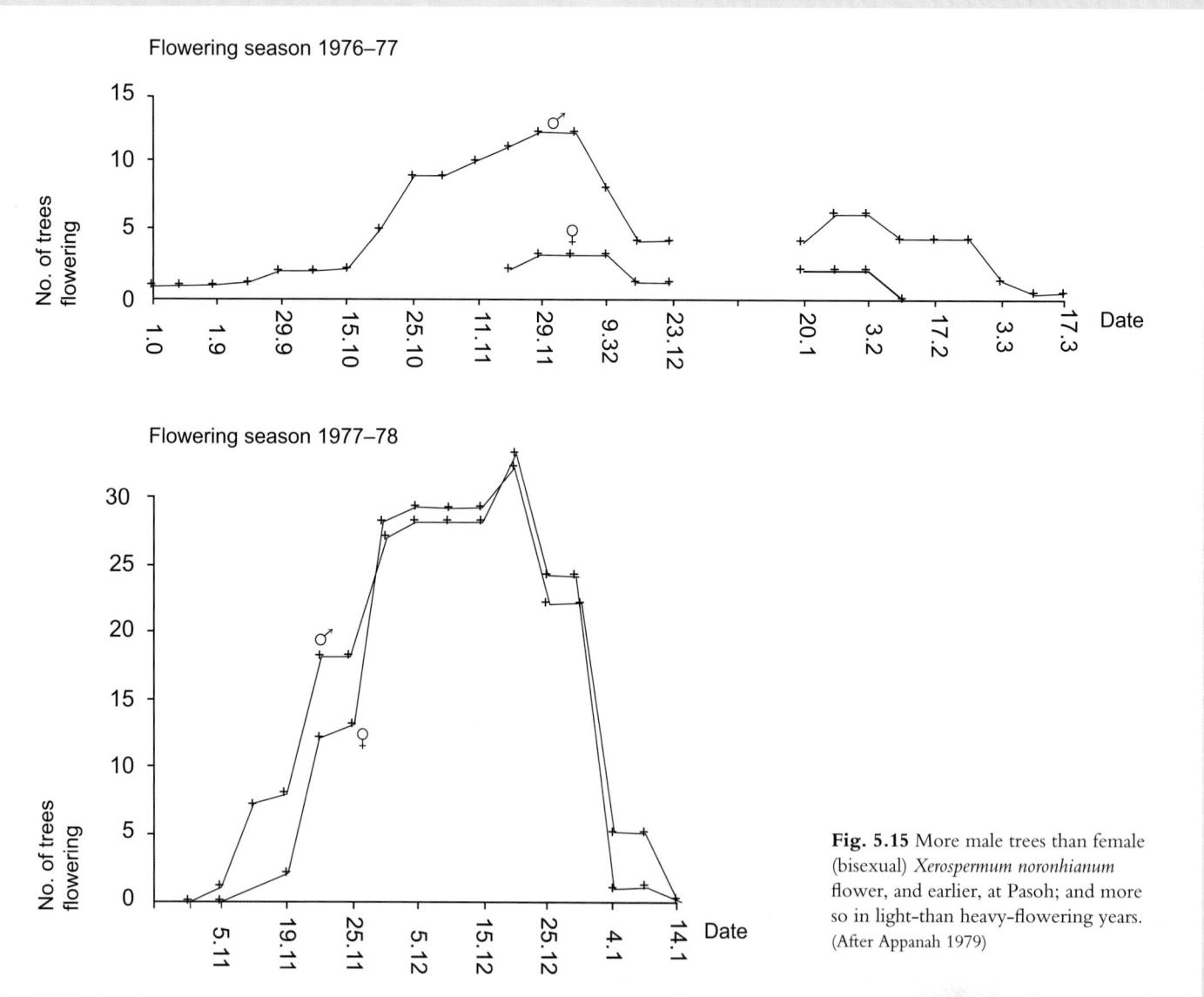

Fig. 5.15 More male trees than female (bisexual) *Xerospermum noronhianum* flower, and earlier, at Pasoh; and more so in light-than heavy-flowering years. (After Appanah 1979)

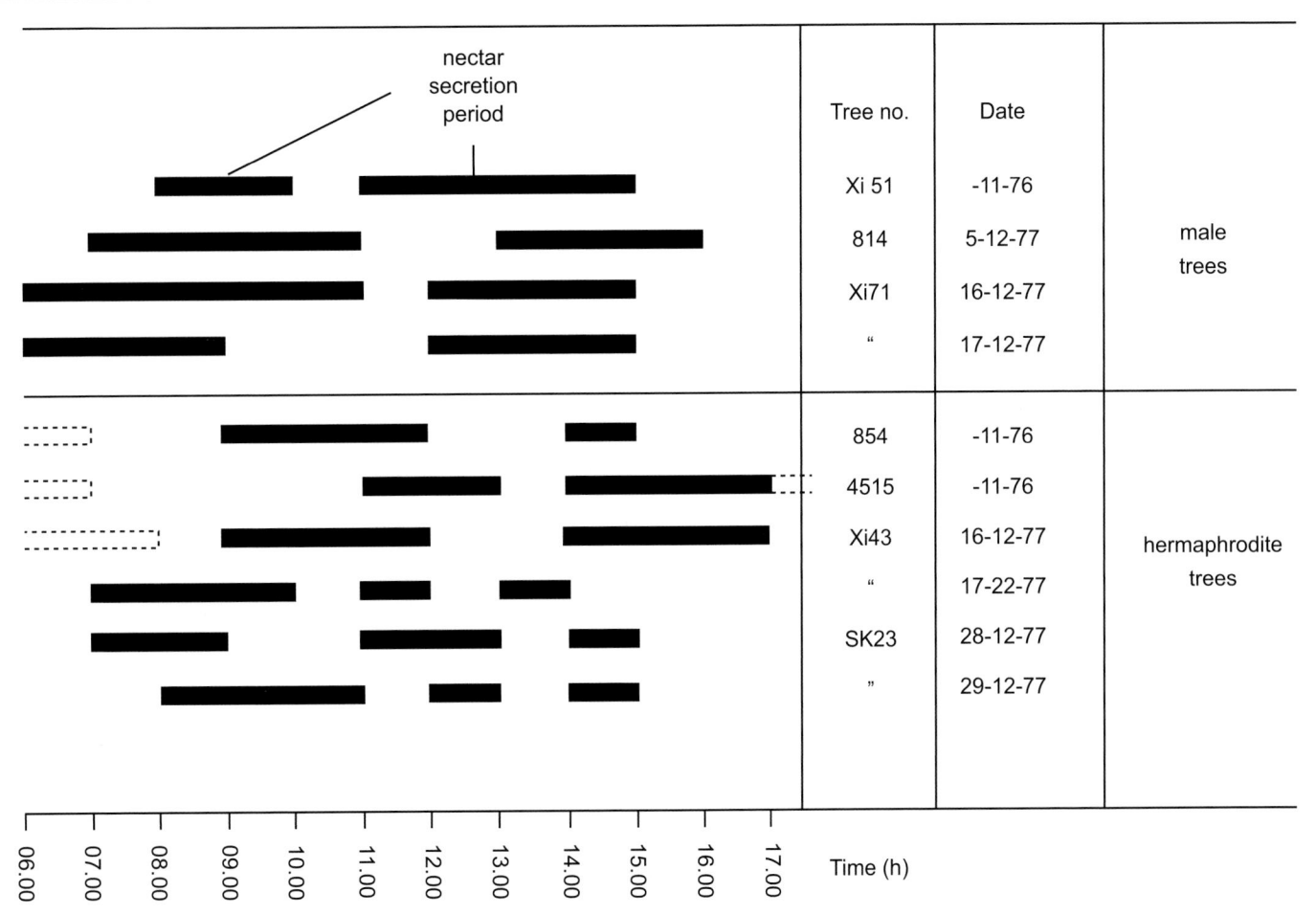

	Tree no.	Date	
	Xi 51	-11-76	
	814	5-12-77	male trees
	Xi71	16-12-77	
	"	17-12-77	
	854	-11-76	
	4515	-11-76	
	Xi43	16-12-77	hermaphrodite trees
	"	17-22-77	
	SK23	28-12-77	
	"	29-12-77	

Fig. 5.16 Phenology of nectar secretion among male and female (bisexual) *Xerospermum noronhianum* trees, Pasoh. Males secrete first, and in two episodes; bisexuals secrete later, but often in three episodes. (After Appanah 1979)

to 5,000 a day, whereas bisexuals presented c.80,000 flowers, and somewhat fewer per day.

Anthesis in both sexes begins after dawn, and continues through the morning when opposite pairs of anthers successively dehisce. The stigma remains receptive for up to three days. The flowers are faintly sweet-scented, and were visited by a wide range of diurnal insects including 13 species of *Trigona* sweat bees, three species of *Apis,* one *Xylocopa,* one *Nomia* bee; also two species of wasp on male trees only, presumably taking pollen. Butterflies were next best represented, especially small mixed flocks of the Danaids *Euploea* (seven species) and less often *Idea* (three) and two *Danaus,* five Nymphalidae and four others; also some beetles, thrips, Orthoptera and Hemiptera. *Trigona* were found on both sexes and appeared to be the principal pollinators. Only seven of 21 butterfly species, mostly Nymphalids, were seen on trees of both sexes.

Both flowers offer nectar; the concentration of amino acids was higher in bisexual than male flowers. Astonishingly,

the male trees secreted nectar in two diurnal peaks, and the bisexual trees often in three: one preceding, one between and one following the male peaks (Fig. 5.16). Insect visitors were attracted to, and later abandoned, trees of each sex at time periods correlated with their nectar flows (Fig. 5.17), the pattern thus eliciting optimal opportunities for cross-pollination. Notwithstanding these adaptations, there were vastly more visitors to male than to bisexual trees, especially during heavy flowering years, whereas the number visiting bisexuals remained nearly the same each year.

From this activity, an estimated 11,500 fruits were yielded from a bisexual tree; therefore about one of eight bisexual flowers successfully developed a fruit. Either one or both ovary lobes may develop. The fruits are ellipsoid, up to 2.5 × 2 cm, with a leathery knobbled pericarp and sugary but thin sarcotesta containing a single ellipsoid seed.

It took Yap, who carried outcrossing experiments, two seasons to validate his own species epithet, sapiens *(wise), over the observational*

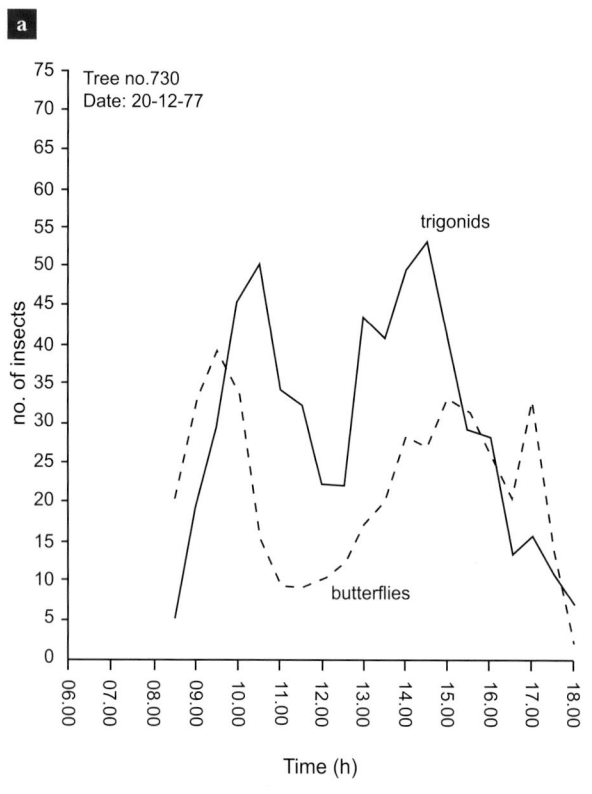

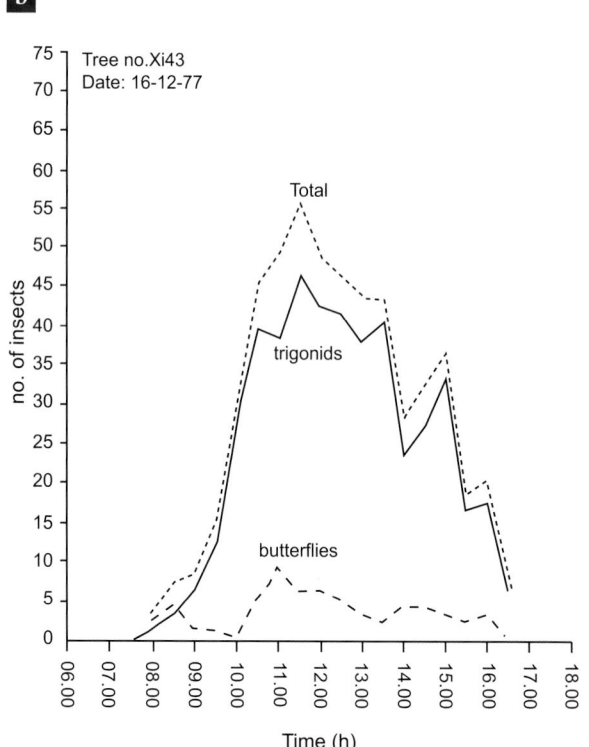

Fig. 5.17 Diurnal flower visitation of two potential pollinators to a: **a,** male; **b,** female (bisexual) *Xerospermum noronhianum* tree. (From Appanah 1979)

acuity and ingenuity of the species primary seed disperser at Pasoh, the long-tailed macaque, Macaca fascicularis – *although the success of a doctoral dissertation rested on it! To the relief of us all, he eventually succeeded. The young green fruits ripen yellow, apparently at night because the macaque had moved in while the student was still at breakfast. But eventually, the student discovered that the seed becomes germinable before the fruit turns yellow, when a drop of sugary liquid is secreted where it joins the stalk.*

Macaques feed in small groups, stuffing their cheek pouches with shelled seeds and then congregating for a feast, often on the parent tree. Seeds are not swallowed, so little dispersal is effected. How unlike the solitary gibbon, a species no longer present in Pasoh, who brachiates the canopy of his territory (up to 1 km in diameter), impatiently swallowing a few seeds whole from each tree before moving on, eventually to deposit them with a nourishing squid of dung (see section 5.3).

Box 5.3

An Asian particular: reproductive ecology of an archaic subcanopy clade: the *Polyalthia hypoleuca* group

Annonaceae flowers are generally protogynous. They are mostly pollinated by beetles, but some by flies and possibly thrips. One group of eight tree species in the Old World genus *Polyalthia,* known as the *Polyalthia hypoleuca* complex, now separated as a genus, *Maasia,* all have leaves characterised by pale glaucous, densely papillose undersurfaces. Their leaves differ somewhat in size and other respects, but they are nevertheless so similar that the species cannot be reliably identified on vegetative characters alone. Steven Rogstad made an exhaustive study of the three species occurring in the Pasoh forest, Malaysia, and more limited observations of these and others in Borneo and New Guinea.[157] He was able to show that although they overlapped in each group of distinguishing characters when taken alone, the species can nevertheless be clearly distinguished by a combination of vegetative and flower or fruit characters. Co-occurring species, for example, can be recognized by discontinuities in the relative numbers of their stamens and carpels.

The eight species occur, widespread or locally, from Indo-Burma to New Guinea. At Pasoh, the co-occurring *Polyalthia hypoleuca, P. sumatrana* and *P. glauca* differ in height at maturity, water relations, tolerance of drought and flooding[158] and in flower biology. The flowers are cauliflorous. Fruit are indehiscent, globose or ellipsoid, ripening to red or black, and dispersed by birds and rodents. Flowering is highly synchronised within populations, but there appears to be no correlation, such as consistent sequential flowering, between

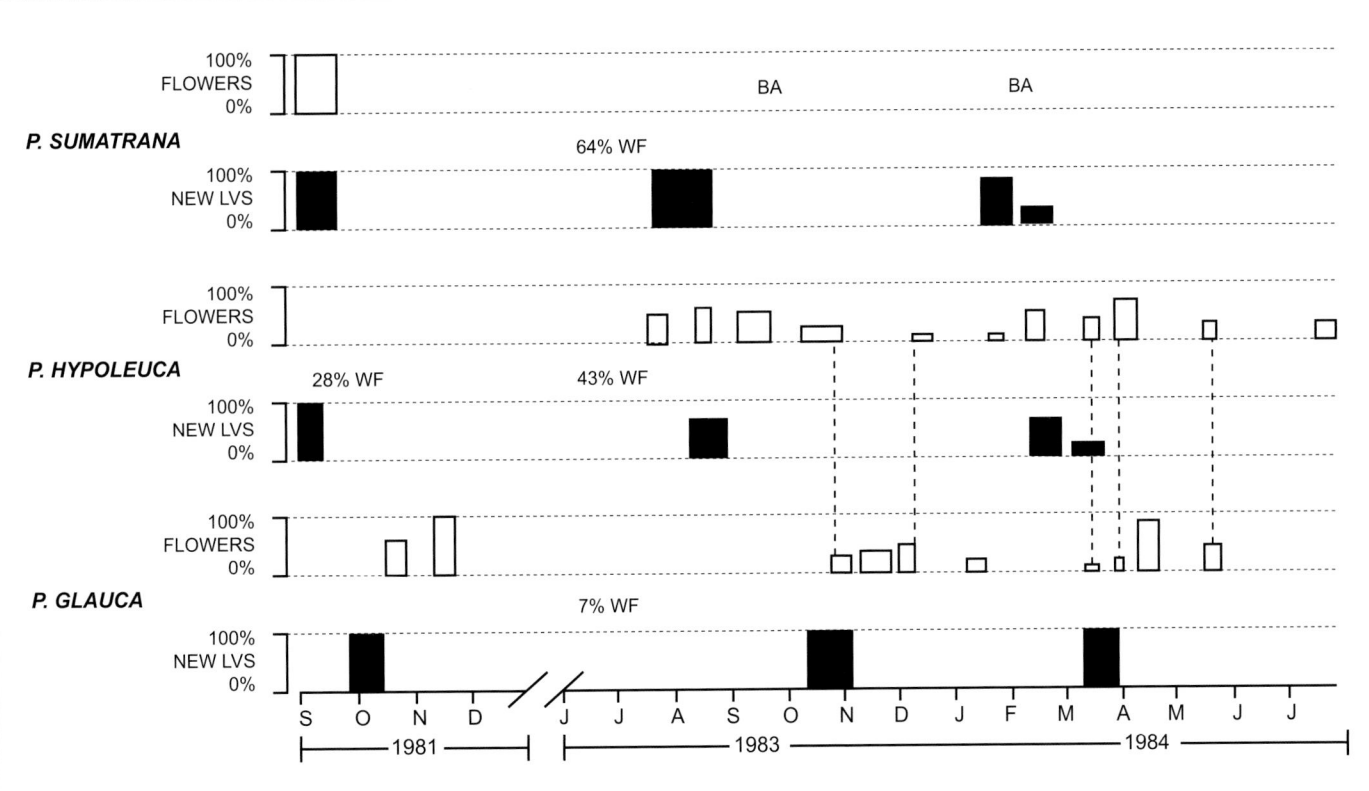

Fig. 5.18 Reproductive phenology of populations of three *Polyalthia* species in Pasoh forest, Negeri Sembilan. There is overlap in anthesis times between *P. hypoleuca* and *P. glauca*. BA: Buds abort; WF: with fruit. (From Rogstad 1986)

the flowering times of co-occurring species even though there is a season when flowering is heaviest. Overlap of flowering between species occurred during three seasons of observation at Pasoh, but only when few flowers were at a stage when crossing was possible. (Two species observed in New Guinea, however, flowered together.) Yet each species at Pasoh had a detectably unique floral odour and a unique beetle pollinator; a finding that was replicated for the three other species examined.

Polyalthia phenology, floral development and anthesis proved to be immensely complex, adding up to virtually insuperable barriers to self-pollination (Figs 5.18–5.20). Like *Lansium* (Meliaceae) and many other subcanopy genera, flower bud initiation and anthesis are temporally decoupled, the buds accumulating and remaining dormant on the trees for weeks or months before coming into flower together in a wave. Each species at Pasoh flowers with a different frequency, over two shorter or two longer seasons, or supra-annually. Anthesis occurs synchronously within populations, in up to ten waves per year. Individuals may participate in one, some or all of these waves. Anthesis, observed on *P. hypoleuca* and *P. glauca* at Pasoh, proceeds on each individual over 16 days, the flowers progressing from the dormant bud reserve in successive cohorts

such that pistillate and staminate stages never coincide. After 14 days the tepals, which have remained green while gradually expanding, turn yellow though they remain hanging, the flowers become distinctively fragrant and the pistil receptive. The pistil remains receptive for 38–40 hours, ceasing at noon on day 16. Two and a half hours afterwards, the tepals fold back and the stamens open! They abscise and fall after only two hours, when the petals turn black. The next flower cohort then proceeds.

The other species flower in overlapping flower cohorts. The individual trees are in effect geitonogamous with separate periods when stamens are released and stigmas are receptive, but individuals within a population are out of synchrony between male and female phases. They are also self-incompatible.

The pollen is sticky, coated in thin mucilage that binds it in clumps. The pollinating beetles, which enter the flower while the stigma is receptive and leave as the petals shrivel, can thus depart for another tree whose flowers are at the pistillate stage, their bodies covered in pollen clumps. Thus these tree species have a precisely synchronised changeover from female to male condition and back every other day; some species even switch in a single day.

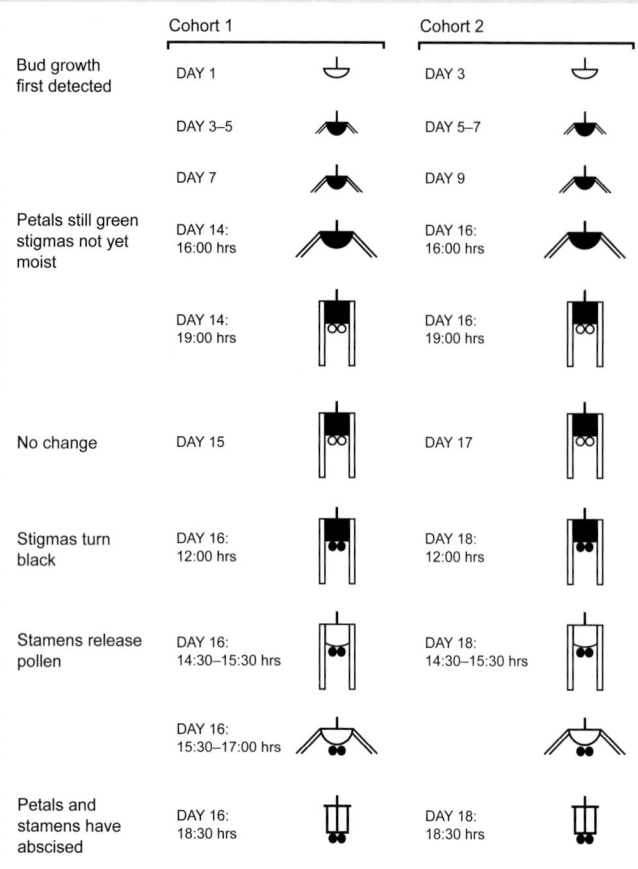

Fig. 5.19 Timing of simultaneous development of two successive flower cohorts within an individual *Polyalthia hypoleuca* tree. (From Rogstad 1994)

It is difficult to imagine a more elegant example of apparent competitive differentiation among co-occurring sister species in the quest to achieve obligate outbreeding for each, yet also prevent introgressive hybridisation. The Gaussian hypothesis is here magnificently demonstrated, and among species of an ancient, basal line of flowering plants.

Box 5.4

An Asian particular: the bamboos (Plates 5.3b-d)

Bamboos are perennial woody grasses. All but a few of the 45 genera are Asian; the centre of bamboo species diversity lies in the moist seasonal forests of tropical lowland and lower montane Indo-Burma and the warm temperate forests of South China. There are 138 species in India alone. In Haryana State, India, three species contribute 80% of bamboo cover.[159] Of these, *Dendrocalamus strictus* alone contributes 45%, occupying 4 million ha; *Melocanna bambusoides* occupies 1.78 million ha; and *Bambusa bambos* 1.15 million ha. They contribute raw materials for a vast array of traditional artefacts (Chapter 8). The culms (hollow stems) of some may reach 35 m, whereas *Vietnamosasa* of the short deciduous dipterocarp forest (Chapter 3) and *Yushania* near the eastern Himalayan tree line forms thickets hardly exceeding 2 m. Many bamboos flower annually and some gregarious taxa are semelparous, flowering communally over whole landscapes at intervals of many years (up to 55

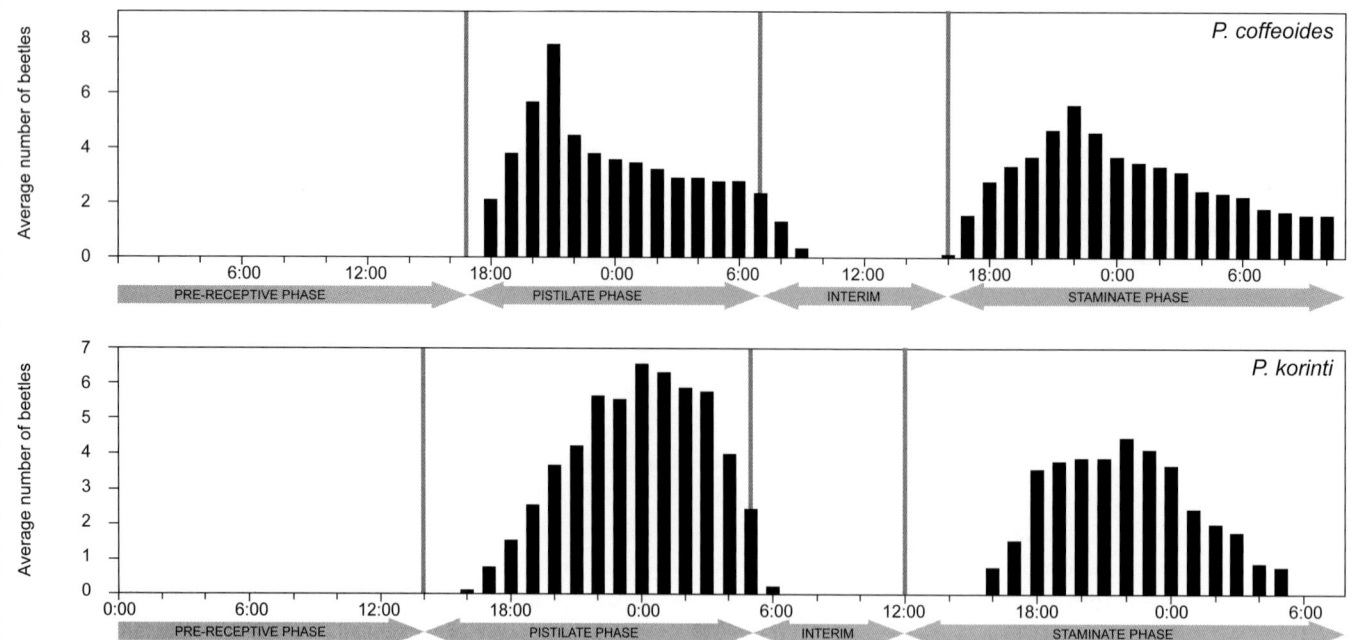

Fig. 5.20 Visitation rates of *Endaeus*, a Staphylinid beetle, and presentation times of female (pistillate) and male (staminate) phases of flowers in two *Polyalthia* species, *P. coffeoides* and *P. korinti*, of the Western Ghats. (From Ratnayake *et al.* 2006)

years in *Bambusa bambos*) or patchily in clumps. In this they resemble the lower montane shrub *Strobilanthes* (Acanthaceae) (Chapter 4). Indian forestry literature is replete with reports of the occasional mass flowering events;[160] among which those of Kurz[161] and Brandis[162] remain outstanding. Dan Janzen provides a more recent and exhaustive review of records of bamboo mass flowerings.[163] Like masting trees such as oaks and dipterocarps, bamboos yield vast quantities of nutritious comestible seed. A 40 m² clump of *D. strictus* was observed to yield 150 kg of seed. This feast attracts small vertebrates, especially rodents and birds including pheasants; large animals such as elephants are said to be uninterested. In spite of this predation, vast numbers of seedlings, whose foliage is often poisonous, manage to establish. Famines among human populations once followed the flowering of bamboo over large areas when rats, having multiplied in the year of fruiting and abundance, moved into villages and grain stores.

Bamboos are highly combustible, especially after death. The death of abundant tall bamboo such as *Bambusa bambos* can result in natural or human-induced crown fires in deciduous forests where trunks of trees are short and crowns low; also following blow-downs in tall forest. If such fires occur at shorter intervals than the life-cycle of fire-sensitive evergreen trees, they establish the conditions for tall deciduous to replace semi-evergreen forest (Chapter 2). Thickets of *Vietnamosasa*

can likewise create crown fires in short deciduous dipterocarp forest. These fires kill seedlings and above-ground parts of tree regeneration. Only those species that can regenerate from stump sprouts therefore survive. While the occasional mass seeding of gregarious semelparous bamboo species constitutes a classic case of predator satiation,[164] it is also clear that it imposes phases when survivorship of the dense groves of adults is low, due to both mass mortality and fire. The death of a bamboo thus provides a narrow window of opportunity for canopy tree regeneration, whether from seed or shade-tolerant seedling[165] (Chapters 2, 3). Bamboos possess dormant seed. Following canopy gap formation, their seedlings soon form dense stands wherever their seed is present in seasonal evergreen, semi-evergreen and tall deciduous forest, increasing in pervasiveness in that order. The abundance of bamboo seed, and their rapid regeneration, results in extensive 'bamboo brakes' where trees have been completely excluded in semi-evergreen and tall deciduous forests. *Vietnamosasa* itself survives by means of rhizomes from which new culms shoot annually (see Plate 2.33d). The challenges of prolonged survival under the dense bamboo canopy followed by burning is likely a major reason why light-demanding, fast-growing pioneer and *early* successional species are rare in seasonal forests compared with MDF (Chapter 3), and become entirely absent in deciduous forests.

3. The pollinators and their flowers

3.1. *Foraging patterns*

The foraging behaviour of pollinators influences the effectiveness of cross-pollination and the distance that pollen is dispersed, thereby influencing the genetic structure of plant populations.[166] In closed-canopy rainforests, tree pollination is accomplished almost entirely by animals, mostly insects. Unlike orchids, most tropical trees are visited by a diversity of potential pollinators (Plates 5.7, 5.8c,e, 5.9f).[167] For instance, bat flowers may be visited by birds, others by various butterflies (Plate 5.9 e,f), or yet by a single group such as swallowtail butterflies (Plate 5.8f). Most flowers with spreading petals are visited by a variety of bees. In the perhumid tropics, relatively few tree species are pollinated by birds or bats (Box 5.6), but there are exclusively bird-pollinated canopy species in semi-evergreen and deciduous forests (though the flowers are also visited by rodents), where flowering takes place when the trees are leafless (Plates 5.8, 5.9).[168] Other trees are principally pollinated by insects but are also visited by birds.[169]

Plate 5.7 The diversity of birds that visit the flowers of an *Erythrina stricta,* Khao Pandan Wildlife Sanctuary, north-west Thailand. Other visitors included a black crested bulbul and a minivet, but no flowerpeckers or sunbirds were seen. **a,** Hair-crested drongo; **b,** Lineated barbet; **c,** Indian black headed oriole; **d,** Grey headed parakeet, a likely robber. (All H.H.)

Plate 5.8 Pollinators and other flower visitors. **a,** Fly on *Hortonia floribunda* (Monimiaceae); **b,** Flies on *Eriolaena candollei* (Sterculiaceae); **c,** *Pinanga* understorey palm with beetles and other insects; **d,** Flower of *Polyalthia coffeoides* in interim phase (stage VIII), with proximal inner petal removed, showing curculionid beetle (scale: 2/1); **e,** *Xylocopa* bumblebee, also *Trigona* and small flies, on *Dillenia obovata;* **f,** Swallowtail butterfly (*Papilio memnon*) on *Ixora*. (**a, b, e** H.H.; **c, f** Saw, L.G.; **d** R.M.C.S. Ratnayake)

A promiscuous pattern is particularly seen among species that present excess pollen as a lure; an expensive strategy most often found on the fertile soils of the seasonal tropics and on river banks. But most – perhaps all – tree species have a 'lead pollinator' species distinguished by its faithfulness during foraging excursions. A few tree families contain species that attract a single dominant visitor almost exclusively: these include the archaic Myristicaceae and *Annonaceae* (Box 5.3).

It is not easy, however, to determine which floral visitors are the most effective pollinators. They must be found to carry pollen between trees, and between flowers in such a way that it will reach a stigma. Among effectively dioecious species, therefore, they must frequently visit the female trees. They must demonstrate high consistency to a single host species during each flight. Among pollinators that respond to visual cues, bees are exceptional in this respect, although they are also pollen feeders and therefore robbers. Other such species that may be effective pollinators include bats (Box 5.6) and Danaid butterflies which 'trapline'.[170] This foraging pattern increases the likelihood of visiting subsequent trees of the same species as the animal retraces the same or similar circuit over the course of a day. Many beetles are pollen robbers whose clumsy activities within the flower and random foraging patterns may reduce cross-pollination. Some species of beetle appear to be attracted by specific odours, often concentrating their foraging on a single plant species and thereby effecting pollination.

3.2. *Pollinators in Asia and the Neotropics*

Much has been made of the lower diversity of pollinators in the Asian tropics, especially compared with the Neotropics where there is greater diversity among vertebrates and the promiscuously foraging Euglossine bees. About 2,000 species of bird are said to regularly visit flowers in the Neotropics, of which two thirds are flower specialists. These are dominated by the hummingbirds (Trochilidae), a staggering 97 species of which occur in Venezuela alone.[171] Vastly fewer visit in the

Old World. Besides sunbirds, spiderhunters and flowerpeckers, the main groups likely to pollinate effectively are bulbuls and minivets, though parakeets and others visit as flower predators (Plate 5.7). The total number in tropical Asia is in the hundreds. Much of this difference may be due to the far fewer epiphytes, in both abundance and species, and absence of Bromeliad and Cactaceae epiphytes in Asian lowland rainforests (Chapter 2), so many of which are bird pollinated in the Neotropics; in part also to the paucity of flowers between mass flowering years in perhumid regions. Supra-annual flowering has likely restricted

Plate 5.9 Pollinators and other flower visitors. **a,** Ceylon White-eye (*Zosterops ceylonensis*) on *Rhododendron zeylanicum*, Horton Plains; **b,** Oriental white-eye (*Zosterops palpebrosus*) on *Canthium dicoccum*, Mudumalai Wildlife Sanctuary; **c,** Squirrel on *Careya arborea*, Mudumalai; **d,** Little spider-hunter (*Arachnothera longirostra*) on *Careya arborea*; **e,** Birdwing butterfly (*Troides* sp.) on *Mitragyna rotundifolia*; **f,** *Euploea* butterfly on *Mitragyna rotundifolia*, Mae Ping National Park; other butterflies, bees, wasps also visited. (All H.H.)

the diversification of specialised local pollen dispersers. Tropical Asia is particularly rich in pollinators from archaic insect taxa, notably beetles but also flies[172] (Fig. 5.21; Table 5.2). In the seasonal tropics of Asia, wide-ranging pollinators including honeybees, Xylocopid and *Amegilla* bees and a diversity of pollinating birds that are nevertheless generalised feeders are relatively more abundant, albeit not rich in genera or species.

Perhaps as a consequence of the relative scarcity of epiphytes in Asian lowland tropical forests – and because Asian nectarivorous bats forage at distance by eyesight, not by scent as do the Neotropical nectarivorous Microchiroptera – most vertebrate pollination is concentrated in the canopy, and in open sites including riverbanks and canopy gaps. None regularly enters beneath the canopy (Box 5.6).

Pollinators that forage over long distances or migrate may sustain population numbers in perhumid regions sufficient to service their flowers in mass flowering years. There are more

tree species in Asian rainforests pollinated by each of these groups than there are by birds. Pollinators of more restricted range must either have alternate food sources, as have leaf beetles – which visit flowers,[173] or high fecundity and short life cycles, as have thrips (Box 5.1; Fig. 5.14).

The late Kuniyasu Momose and his associates observed the pollinators of all plants in the mixed dipterocarp forest at Lambir National Park (north-east Sarawak) over 53 months, including a mass flowering, in the most comprehensive study of its kind.[174] He has elaborated his results in detail, which will therefore only be summarised here[175] (Figs 5.5, 5.21). He showed that, compared with a Central American site,[176] Lambir was less endowed with species of far-ranging solitary bees (*Amegilla* substituting for Euglossines), birds and bats, and Lepidoptera; but with more frequent visitation by other bees, beetles and by a diversity of other insect visitors (Fig. 5.21). Using their comprehensive list of floral visitors and tree taxa, I was struck by the high proportion of tree species exclusively visited by Curculionid, and especially Chrysomelid leaf beetles. These included all but one of the 19 species of *Hopea* and *Shorea*. These dipterocarps of Shoreae tribe represented two genera and eight sections within them, each consistently distinguished by a distinct stamen and ovary structure. Unless these beetle species are lured by specific odours – which seems unlikely – their constancy of visitation, even to the species in different genera (and therefore their efficiency as pollinators), is questionable.

Vertical stratification of mature forest stands by species guilds distinguishes them in part according to where they present their flowers and fruits (Chapter 2). In the open landscape and full sun of the canopy, plants attract with visual cues. Most bright-coloured flowers and many such fruits in Asian forests are found on canopy, gap, or riverside tree, vine and epiphyte species. Compared, in particular, with the Neotropical hummingbirds, the nectarivorous Asian sunbirds and flowerpeckers are few in species, poorly diversified and less host specific. Although they are the exclusive pollinators of mistletoes in the canopy and some terrestrial *Zingiberaceae* of gaps and secondary forests, they visit few tree species.[177] One such exception is trees of the small genus *Saraca,* whose brilliant orange flowers bedeck the forest canopy along the banks of upland streams. Another is *Durio kutejensis* which, with its large brilliant red flowers, is visited by spider-hunters as well as the giant honeybee.[178] Flowerpeckers and spider-hunters become important pollinators of *Rhododendron* and other shrubs and trees in upper montane and, particularly, subalpine forests.[179] The Australasian lorikeets are represented in Sunda MDF by only one species, the Blue-Crowned Hanging Parrot (*Loriculus galgulus*), although more are found east of Wallace's Line. In seasonal regions, several tree species including *Butea, Erythrina, Firmiana* and *Pterocymbium* – all

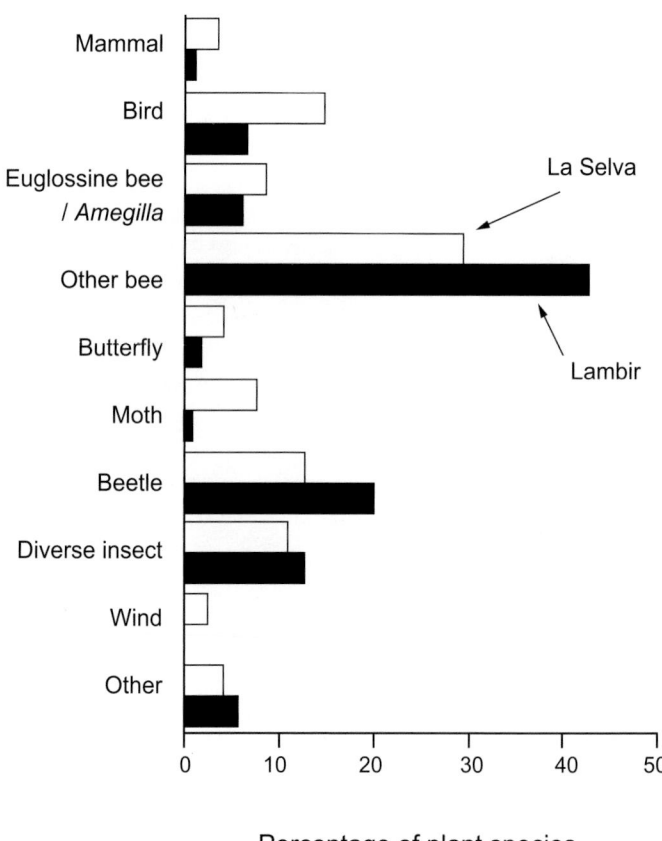

Fig. 5.21 Comparison of frequency (percent of plants visited, all life forms) of main pollinators, Lambir and La Selva, Costa Rica. (Evidence for wind pollination was not sought at Lambir.) (After Momose & Hamid Karim 2005)

deciduous and with brilliant coral red, and the last with orange, flowers – attract a wide diversity of feeding generalist passerine birds, including mynahs, orioles and bulbuls (Plate 5.9) as well as sunbirds and spider-hunters. These trees, however, are rare or absent in perhumid climates, only occurring occasionally on nutrient-rich soils. The extreme dominance of the dipterocarps in the MDF emergent canopy, none of which are pollinated or their seeds dispersed by arboreal vertebrates, likely partially explains the comparative poverty there of both vertebrate species and numbers in it.[180] But the ENSO-induced cycle of multi-year famine between single years of excess must also play a major part.

Flowers of many deciduous tree species are large, providing visual attraction during the dry season when trees are leafless (for example Plates 5.8e, 5.9c,d). Several, such as the mahwa, *Madhuca longifolia* (Sapotaceae), present flowers rich in nectar, and in this case fleshy united corollas which ripen at the same time as the epipetalous stamens. They attract various vertebrates, and also honeybees which here may act mainly as pollen robbers[181] (Box 5.5). Some tree genera with species whose mature crowns differ widely in height such as *Syzygium*[182] and *Canarium* and other Burseraceae[183] have flowers that attract a wide range of visitors, mostly insects; or are pollinated by generalised pollinators, such as flies and beetles.

A major ecological difference between the Asian tropics and the Neotropics, then, is that vertebrate pollinators in Asia, including the macroglossine bats and the honeybees, respond at distance to visual cues, and thus forage almost entirely in the canopy, in large gaps, and along forest edges especially riverbanks[184] (Box 5.6). Only the Meliponid sweat bees forage in the shady subcanopy, *Apis* and *Xylocopa* being primarily canopy, gap, or forest-edge feeders. Compared with Neotropical forests with their *Passiflora* and the bright-flowered Acanthaceae and Rubiaceae, the Asian subcanopy is a uniform green. There are exceptions: the bright nutmeg and other fruits and, especially under the diffuse canopy of *kerangas* and on ridges, the gleaming brick-red swallowtail-butterfly pollinated *Ixora* (Rubiaceae) – ecological analogue of the dwarf evergreen rhododendrons of warm temperate East Asia – and gingers of the genus *Globba*. The Danaid butterfly genera *Idea* (Plate 5.6b) and *Euploea* (the latter taxon rich in ecologically sympatric species (Plates 5.9f) are important flower visitors and presumed pollinators in Asian forests, though they are no comparison in diversity with the Neotropical Ithomiidae and Heliconiidae. Both genera are canopy foragers, but *Euploea* frequently penetrates beneath as well. Meliponid bees and butterflies may be directional foragers (Box 5.2). The subcanopy of Asian rainforest is therefore the realm of pollinators that respond to olfactory, chemical cues (Fig. 5.21).

Though hawk (sphinx) moths are known to pollinate epiphytic orchids and rhododendrons, Momose[185] recorded only one sphinx moth visitor, to the understorey tree *Barringtonia sarcostachys* whose white, nocturnal brush-like flowers with many stamens on long filaments – nectar-rich and acrid-smelling, presented on hanging racemes – rather appear to be adapted to attracting bats (but see Box 5.6). Other moths, though, can be specialised. Moth-pollinated trees, including Sapotaceae and *Diospyros* (Ebenaceae), are abundant in species and diverse at least in the size of their flowers.

Parakeets and flying foxes (*Pteropus* bats, see Box 5.6), also most common in the seasonal tropics, feed on the flowers themselves as well as their pollen and nectar, though they may doubtless inadvertently achieve cross-pollination.

3.3. *Wind as a pollinator*

Wind pollination is rare in the Asian tropics, including in seasonal regions with prevailing winds. In perhumid Malesia it is confined to those habitats where wind is prevalent, including:

1. Diffuse-canopy coastal stands (*Casuarina equisetifolia, Podocarpus polystachyus*),

2. Near-coastal kerangas and peat swamps which experience daily on-shore breezes (*Agathis, Dacrydium beccarii, Podocarpus neriifolius, Quercus kerangasensis, Engelhardia serrata, Gymnostoma nobile*),

3. Riverbanks (*Octomeles sumatrana*), and

4. High ridges and karst summits in the lower montane forest beneath the cloud base (*Engelhardia roxburghiana, Quercus, Trigonobalanus verticillata* and *Gymnostoma rumphianum*).

4. Tree reproduction in the seasonal tropics

In view of the importance of tree reproductive biology for understanding the maintenance of species diversity, most research into the reproductive biology of tropical trees in Asia has been focused on species-diverse MDF, and I too have emphasised lowland evergreen rainforest in the foregoing. Trees of seasonal forests in tropical Asia differ notably in that they flower annually, although the intensity of reproductive events does vary interannually (Figure 5.22, 5.23).[186] As in other tropical regions, flowering in seasonal tropical Asia takes place overwhelmingly during the dry season – especially so in deciduous forests.[187] Flowering often starts shortly after the dry season commences and reaches a

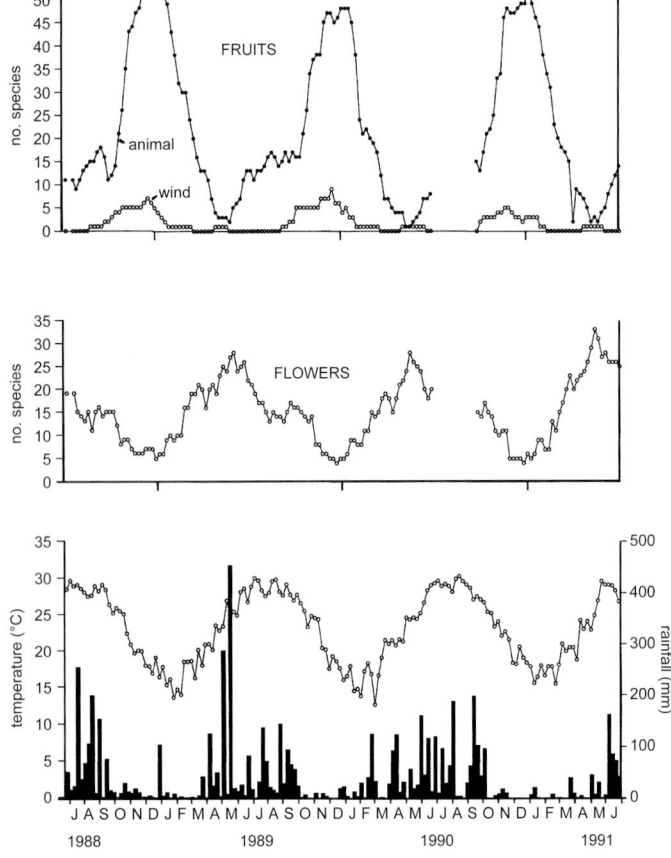

Fig. 5.22 The number of woody plant species flowering and fruiting over 42 months at the tropical margin (Hong Kong, 22° N). (After Corlett 2009b)

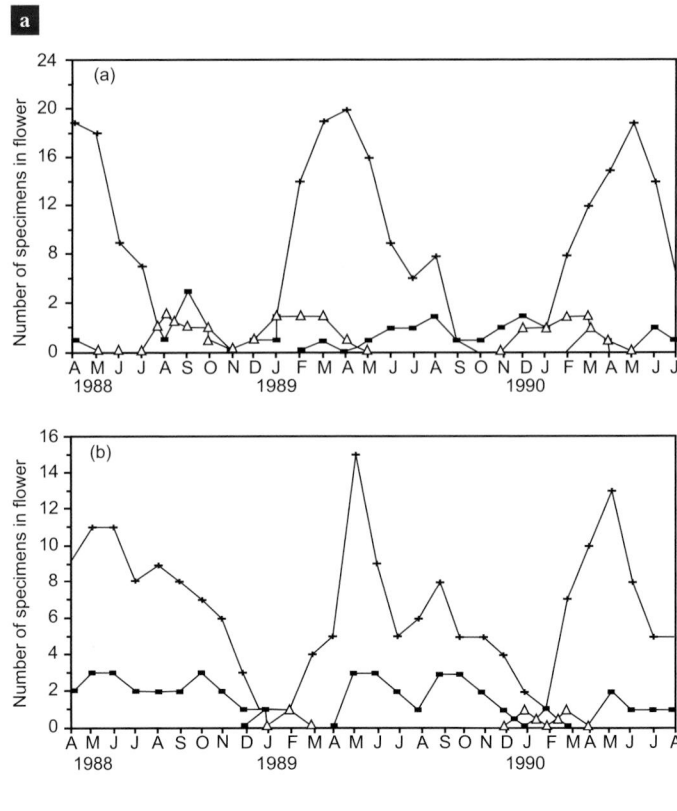

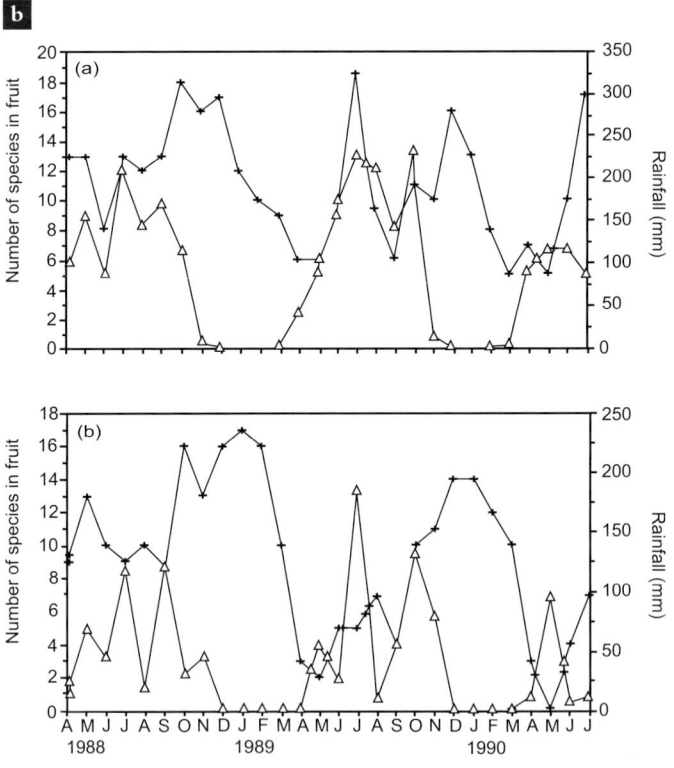

peak toward the end of the cool dry season, March–April.[188] Canopy flowering therefore peaks during the most leafless season and fruitfall towards the end of the dry season, but there is much intergeneric variation, especially beneath the canopy. Flowering times differ somewhat among groups of species visited by different pollinators.[189] Synchronicity is loose in many deciduous species, and trees may flower and fruit simultaneously, as Kingdon-Ward[190] observed in *Stereospermum*. Interestingly, fruiting peaks shortly after the flowering, at the height of the dry season implying that those of the majority possess seed dormancy; though there may be evidence of a secondary peak – perhaps of species with seeds of limited dormancy – as the onset of the rains approaches.[191] By that time, nutrients will have been reabsorbed from the senescing leaves of many deciduous species (Chapter 2).[192]

Studies in the Neotropics have been more extensive than those in Asia, and have yielded similar results.[193] However, there has been as yet little study of apomixis there, and none on the influence of ENSO on reproductive biology.

Fig. 5.23 Reproductive phenology of trees in tall (above) and short (below) deciduous forest, Mudumalai Wildlife Sanctuary, Tamil Nadu: **a,** flowering (pollinators +: insects, Δ: birds, squares: wind); **b,** fruiting (Δ: rainfall). (From Murali & Sukumar 1994)

Box 5.5

An Asian particular: the ecology of honeybee pollination

As with birds, the Old World lacks the Euglossine and other Neotropical bees whose diversity appears coevolved with important plant groups (Plate 5.10). The nearest equivalent in Asia is the small solitary bee genus *Amegilla,* species of which forage over long distances but do not appear to be exclusive pollinators of any tree. Our pollinating bees are overwhelmingly in three categories: Xylocopidae (pantropical bumblebees), Apidae (the honeybees, which are indigenous only to the Old World), and Meliponidae (the pantropical sweat bees). None of these are truly species-rich in Asia. The centre of true bumblebee (Bombicidae) species diversity is in the temperate Himalaya and East Asian high mountains. They do not descend below the limit of alpine herbaceous communities in Asia, athough they do on a modest scale in the Amazonian forests.[194]

Bees, like bats, feed on nectar but also rob pollen to feed their larvae. Bee flowers offer excess yellow pollen as an attractant. The honeybees (*Apis*) and the big Xylocopid bees forage in full sunlight.[195] They have eyesight of appropriate spectral sensitivity,[196] although some Xylocopids are said to be nocturnal.[197] *Xylocopa* are found visiting the flowers along riversides and roadsides including herbaceous weeds (*Crotalaria, Stachytarpheta*), in secondary forest and in large canopy gaps, and among the crowns of canopy species.[198]

Xylocopid bees are solitary foragers, and are said to range several kilometres on a flight. They appear to visit a narrower range of flowers than honeybees, visit at intervals and appear to trapline.[199] They are fond of large mauve, white and yellow-orange flowers yielding nectar liberally. During mass flowering events they desert the roadsides for the canopy.[200] There they have been observed on Fabaceae and *Xanthophyllum* (Polygalaceae), the keel of whose zygomorphic flowers serve to restrict access to nectar.[201]

Honeybees visit a wide range of canopy flowers of varying size, many of which offer less nectar than Xylocopid flowers. They occasionally visit bat- and bird-pollinated flowers, but they are unlikely to pollinate these effectively. They are ever-present above and around the forest,[202] but the giant honeybee, *Apis dorsata,* becomes particularly prevalent during mass flowering years and during the flowering season in the humid tropics, after which its nests are said to be most replete with pollen and honey.[203] Honeybees are opportunistic pollen and nectar foragers. As beekeepers know, it takes time

Plate 5.10 Bees. *Apis dorsata* colony in *Koompassia excelsa* crown: a characteristic sight in Borneo and southern Peninsular Malaysia; **b,** The nest of *Apis dorsata,* the giant honey bee (scale: 1/5); **c,** *A. dorsata* on *Phyllanthus emblica,* Mudumalai: a tree species which occurs from India's short deciduous forests to the MDF of Borneo (scale: 1/2); **d,** *Xylocopa* on *Stereospermum chelonioides* (Bignoniaceae), Mudumalai Wildlife Sanctuary (scale: 1/5); **e,** *Trigona* nests, Pasoh (scale 1/5); **f,** *Trigona* on *Nephelium* inflorescence (scale: 1/2). (**a, b** T. Inoue; **c, d** H.H.; **e, f** L.G. Saw)

for a hive to build up workers in response to increased floral resources – much longer than it takes a population of thrips to expand. Honeybees, like wild boar (Box 5.9), appear in unusual numbers when food becomes abundant, in this case during mass flowerings; but it is unlikely that their numbers can be sufficiently increased by reproduction alone within the duration of these events to qualify them as lead pollinators, and therefore a role particularly as pollinators among sequentially flowering congeneric species (Box 5.1).

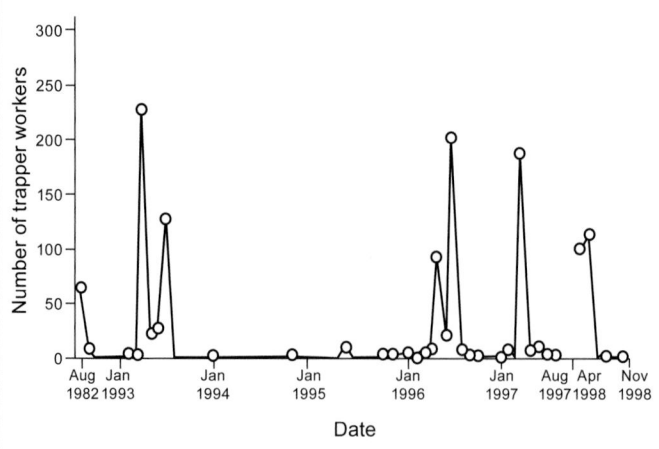

Fig. 5.25 Light catches of *Apis dorsata* in light traps monitored over seven years, 1992–1998, at Lambir National Park, Sarawak. (From Itioka *et al.* 2001)

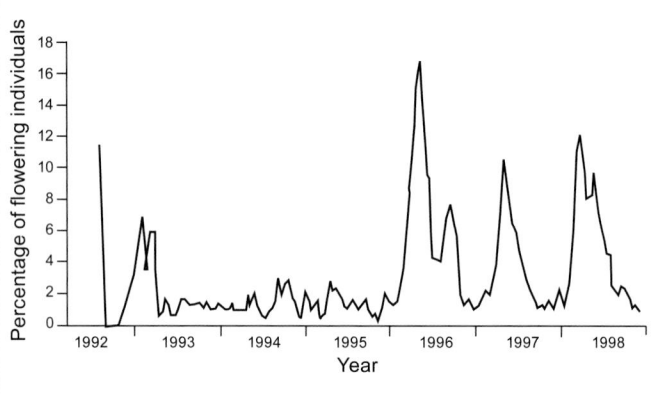

Fig. 5.24 Changes in flowering intensity over seven years 1992–1998, as percentage of 420 monitored individual trees, at Lambir National Park, Sarawak. (From Kato *et al.* in Itioka *et al.* 2001)

Takao Itioka[204] monitored the giant honeybee for six years at the Lambir Forest (Sarawak). He found few workers or nests in non-flowering years, but the numbers of both workers and nests in a locality increased simultaneously with the number of trees in flower (Figs 5.24, 5.25). He inferred that swarms may move long distances to major new food sources. In today's world, when mature forests are islands in a sea of secondary vegetation where many bee-pollinated plants flower annually or even continuously, this would seem easier than formerly when the whole regional landscape might experience simultaneous flowering. But congregate they do, responding quickly to new food sources in vast numbers – such as my shirt, salty with sweat, drying on a line in camp during a minor flowering in the upper Temburong (Brunei) in 1958! Itioka found that the smaller honeybee, *Apis koschevnikovi,* followed the same trend but with less intensity, retaining more workers during non-flowering years and with little evidence of swarm migration. The two *Apis* species were attracted to lures up to 640 m away.[205] Kenta and colleagues[206] found that the *Dipterocarpus tempehes* population achieved successful cross pollination at greater distances when it flowered in a non-mass flowering year (1998) and was pollinated by its customary moths, than in the 1996 mass flowering year when the giant honeybee was the principal visitor.

Appanah, in Malaysia, observed bee visits to *Shorea ovalis, Neobalanocarpus heimii* and *Dryobalanops aromatica,*[207] while Savitri and Nimal Gunatilleke observed them on *Shorea* (sect. *Doona*) species at Sinharaja Forest (Sri Lanka) (Box 5.1).[208] They all found that anthesis begins shortly before dawn, except in *S. ovalis* which flowered in the late afternoon. *Apis dorsata* was first to visit, shortly after sunrise. That species was followed quickly by successively smaller bees, which seemed to chase off their predecessors: on *Doona* by *A. indica* and later by *Braunsapis* sp., (Table 5.2). In both regions occasional *Nomia* and abundant *Trigona* bees were present. Appanah observed the sudden arrival of many *T. canifrons* on *N. heimii*. This aggressive species appeared to attack all other bees and drive them away.

The sweat bees, Meliponidae, are less diverse in Asian than in Neotropical forests, being represented by few genera of which *Trigona* is much the most important. Twenty-nine species of *Trigona* are recorded in Peninsular Malaysia and Sarawak, 24 of which were found in Lambir alone.[209] Appanah[210] observed 13 species visiting the flowers of *Xerospermum* (Box 5.2) at Pasoh. They are even more opportunistic feeders than honeybees. Although they are most numerous below the forest canopy, they also visit honeybee canopy flowers in smaller numbers, including those of emergent dipterocarps. Individual species appear to have some floral preferences. Although their photoreceptors are less effective than the larger bees, they are stronger in the bluish-green range, enabling them to forage more effectively in the understorey.[211]

Their nests, entered by a resin vestibule (Plate 5.10e), tube or bugle, characteristic in form to each species, are built up in layers, each layer studded with the pollen grains currently being collected. *Careful examination of this pollen might reveal the seasonal floral preferences of each bee species.* Sweat bees are generally weak flyers, and workers of each colony are thought to keep within a restricted area. Nevertheless, when Appanah examined pollen on individuals visiting lures close to the centre of the Pasoh research forest, he found some bearing oil palm pollen from outside the forest over 1 km away.

An Asian particular: the 'fruit' bats (Megachiroptera) and their flowers

Bat-pollinated plant species have independently evolved similar means of attraction, including scent (Plate 5.11).[212] All bat flowers – Neo- and Paleotropical – have several common characteristics. Neotropical bat flowers cultivated in Asia attract Asian Macroglossine bats, and *vice versa*. Seven of eight species were found to contain sulphur in their nectar, imparting the characteristic 'sour milk' odour. Three have mushroom-like smells imparted by a fatty acid derivative.[213] They open in the evening, and the corolla falls the following morning. They are large, and in Asia are borne so as to present a silhouette in the late evening light (Plate 5.12a). They are pale coloured, white or mauvish and often striped, enhancing their visibility at close range against dark foliage. Because all Asian nectarivorous bats land to feed, the flowers bear leathery corollas or other surface perches to allow access to the nectar. Some naturalists insist that bat flowers smell like the bats themselves, while others (including Tony Start, on whose studies much of this box is based) deny that there is any common odour and assert that some *Sonneratia* are entirely lacking in human-detectable scent. *Barringtonia* (Lecythidaceae), with its 'shaving-brush' flowers whose stench is evoked by its Malay name (*putat* = fart), has other bat-flower-like qualities, but appears to be visited by bats only occasionally, and has been observed being visited by a sphinx moth.

All bat flowers yield copious nectar during anthesis.[214] They are of two basic forms in Asia, named 'shaving brushes' or 'stamen balls', and 'campanulate'. Shaving-brush flowers present abundant pollen on stamens with long spreading filaments, which are offered as part of the lure (Plate 5.12c). The campanulate forms, typical of Bignoniaceae, bear few stamens, offering nectar alone: usually, only a few open each night on any tree, and their structure is finely adapted to avoid cross-pollination (Plate 5.12b). Of course, there are many exceptions. Bat-pollinated wild banana flowers may remain open during the day, attracting sunbirds. Durian flowers combine many stamens with a campanulate corolla shape (Plate 5.11a). Also, bats sometimes visit flowers lacking any of these characteristics, such as those of mango (*Mangifera*), jack (*Artocarpus*), *Rhizophora* mangrove, coconut and occasionally *Arenga* palms (Fig. 5.26; Table 5.3).

Plate 5.11 Bats that visit flowers and fruits. **a,** The bat *Eonycteris spelaea* feeds on *Durio zibethinus* (Malvaceae-Bombacoideae) flowers (scale 1/5); **b,** But *Durio kutejensis* is visited by honey bees: *Apis dorsata* (scale: 3/2); **c,** *Macroglossus minimus* feeds from a banana flower in Sarawak (scale: 3/2); **d,** A Kosrae flying fox (*Pteropus ualanus*) on *Pandanus tectorius* fruit, Kosrae Island, Micronesia (scale: 1/8); **e,** The Philippine endemic Greater Musky Fruit Bat (*Ptenochirus jagori*) consumes a fig, Sierra Madre National Park, Luzon (scale: 1/4); **f,** A fulvous fruit bat (*Rousettus leschenaultii*) eats durian flowers (scale: 1/3). (**a** A. Start; **b** T. Yumoto; **c** C.-E. Lee; **d-f** T. Laman)

Plate 5.12 (left) Bat flowers. **a,** Inflorescences of *Fernandoa adenophylla* (Bignoniaceae) silhouetted against the sky; **b,** Fallen flower of *F. adenophylla,* with four stamens and central style (scale: 2/3); **c,** The protruding brush flowers of *Duabanga grandiflora* (Sonneratiaceae (Lythraceae)) (scale: 2/3); **d,** *Parkia javanica* (Leguminosae-Mimusaceae) inflorescence: male flowers surround the peduncle apex, hermaphrodite cover the bulb; *Eonycteris* alights on the bulb to feed on the nectar flowing from the male flowers into the collar round the top of the bulb, sprinkling pollen from a previous visit onto the hermaphrodite flowers (scale: 2/3); **e,** Wild bananas in a forest gap, Salawin (Salween) National Park, north-west Thailand; note the pendent red inflorescence of a bat-pollinated species; **f,** Wild banana inflorescence with bat clawmarks on its outer spathe. (**a–e** H.H.; **f** A. Start)

Species	Bat observed	Pollen in guano
Musa malaccensis (Musaceae)	+	To genus
M. truncata	+	To genus
Musa sp. (cultivated varieties)	+	-
Cocos nucifera (Arecaceae)		To species
Arenga sp. (Arecaceae)		To genus
Mangifera spp. (Anacardiaceae)		To genus
Oroxylum indicum (Bignoniaceae)	+	To species
Pajanelia longifolia (Bignoniaceae)	[a]	-
Bombax valetonii (Bombacaceae)	+	To species
Ceiba pentandra (Neotropical Bombacaceae)	+	To species
Durio zibethinus (Neotropical Bombacaceae)	+	To genus
D. graveolens (Neotropical Bombacaceae)	+	To genus
Asteraceae sp.		To family
Barringtonia sp. (Lecythidaceae)		To genus
Parkia javanica (Leguminosae Mimusoideae)	+	To genus
P. speciosa	+	To genus
P. singularis		To genus
Artocarpus sp. (Moraceae)		To genus
Syzygium malaccense (Myrtaceae)	+	To genus
Sapotaceae sp.		To genus
Rhizophora sp. (Rhizophoraceae)		To family
Duabanga grandiflora (Lythraceae Sonneratoideae)	+	To species
Sonneratia alba (Lythraceae Sonneratoideae)	+	To species
S. caseolaris (Lythraceae Sonneratoideae)	+	To species
S. ovata (Lythraceae Sonneratoideae)	+	To species

Table 5.3 Tree species whose flowers are visited by the bat *Eonycteris spelaea* in Peninsular Malaysia. [a]Claw marks only: possibly *Macroglossus*. (From Start 1974)

In addition, species bearing bat flowers form two phenological groups, those that flower continuously and those that flower seasonally, once a year or more often. There are two principal genera in the first group: *Sonneratia* of the mangrove, and the wild bananas (*Musa*) that form small groves in forest gaps and secondary vegetation. *S. alba* and *S. ovata* flower in intermittent, partially overlapping flushes throughout the year, but there are periods when neither flowers. The estuarine *S. caseolaris* populations present some flowers at all times. Banana shoots develop, flower and die continuously within a grove. Bat-pollinated bananas bear hanging inflorescences, and the bats land on the inflorescence bud after a bract curls back in late afternoon, exposing the large oblong anthers at its base and allowing the copious nectar to flow down the bud. (In contrast, bird-pollinated bananas, chiefly *M. violascens* in Peninsular Malaysia, hold their inflorescences erect.) Individuals within a population lack synchrony, a few always being in flower.

Megachiropteran bats are confined to the Old World. They are best known for the large fruit bats (*Pteropus*), which forage and roost in flocks like parakeets. They act as pollen robbers, often damaging flowers by also eating their soft parts, although they have been documented as the primary pollinator of one *Freycinetia* (Pandanaceae) vine on Samoa,[215] and may accidently pollinate the flowers of other species that they visit. *Pteropus* appear to be important dispersers of the seed of many soft-fruited trees.[216] It is less well known that order Megachiroptera also includes several genera of small bats, including all of the nectar- and pollen-feeding bats of the Old World.

There are three exclusively nectarivorous genera of small Megachiropteran bats: *Eonycteris, Macroglossus* and *Syconycteris* (the last limited to East Malesia, northeastern Australia and Melanesia). They belong to the tribe Macroglossini. Like flower-visiting Asian birds, but unlike Neotropical Microchiropteran nectarivorous bats and hummingbirds, they do not hover, instead landing on flowers to feed.[217] Unlike these Microchiroptera, they feed exclusively on nectar and pollen, which they ingest by rapidly flicking a long narrow hairy tongue in and out. Small arthropods seem to pass through the gut undigested.[218] The contrasting foraging behaviour of the two Malaysian genera exemplifies that of other seed and pollen dispersers such as primates and bees: *Eonycteris* and *Macroglossus* have been the subject of an exhaustive and careful study by Tony Start, unfortunately mostly still unpublished.[219]

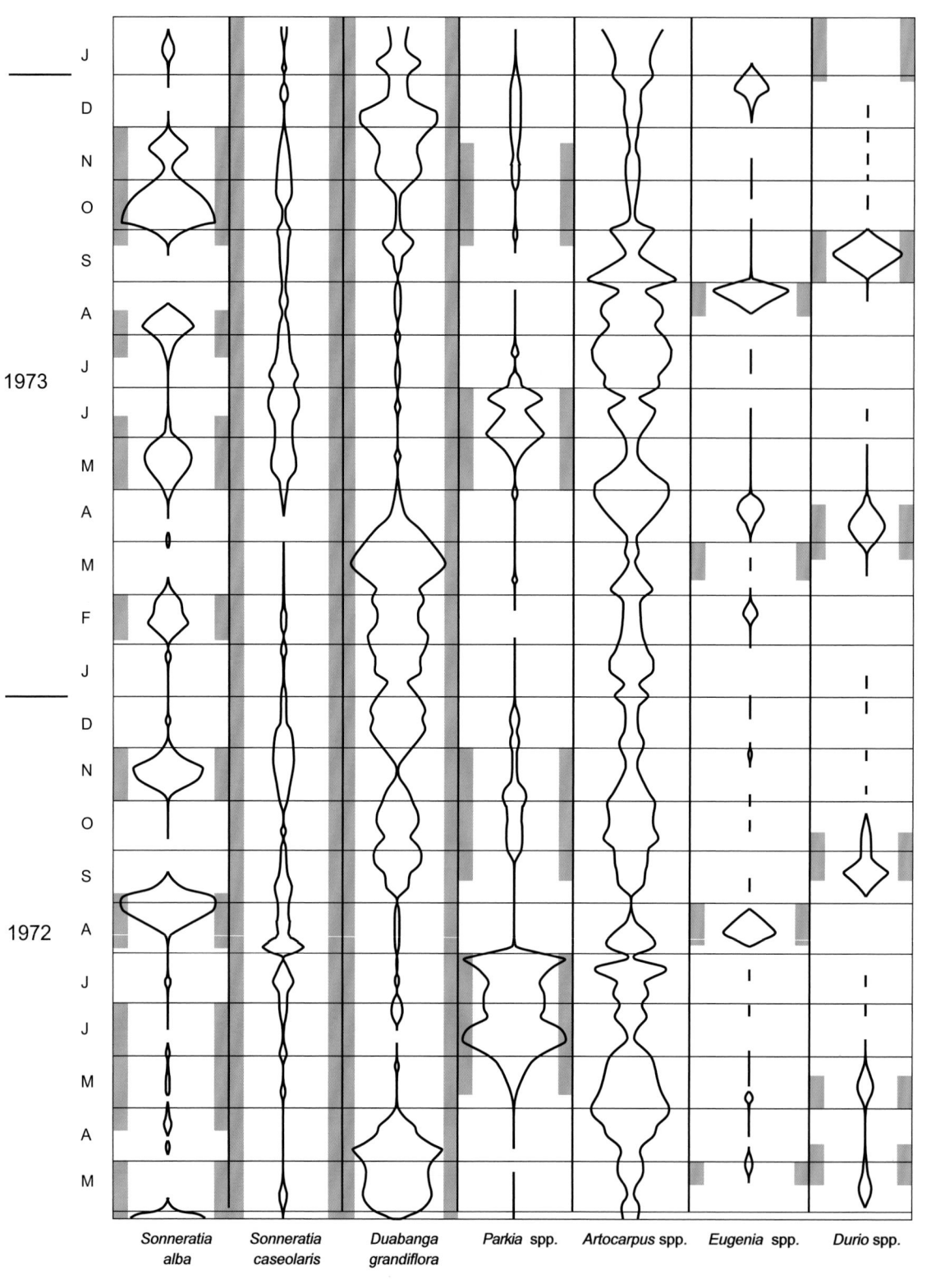

1973

1972

Sonneratia alba   *Sonneratia caseolaris*   *Duabanga grandiflora*   *Parkia* spp.   *Artocarpus* spp.   *Eugenia* spp.   *Durio* spp.

Fig. 5.26 Kite diagram of proportional frequency, over almost two years, of pollen of tree taxa collected weekly from guano in one 1 m² tray beneath an *Eonycteris spelaea* roost in Batu Caves, Selangor, Peninsular Malaysia. The green bands on either side of each column indicate observed flowering dates of each tree species. Note the synchronised changes in visitation of the tree species indicated in the diagrams; these might indicate either successional flowering of the tree hosts or – more likely because of their short-term fluctuations in relation to the host's flowering periods – the feeding preferences on particular flights. (After Start 1974)

Unlike Microchiroptera, Megachiroptera have good eyesight. They navigate to food, and to their roosts, by eye. They therefore feed in the evening and by moonlight and, unlike Microchiroptera, which impose the prevailing pattern in the Neotropics, mostly forage in the forest canopy, edge, and large gaps.[220] In tropical Asia, besides *Pteropus,* smaller fruit bats of several genera, especially *Rousettus* and the more solitary *Cynopterus,* are important seed dispersers.[221] *Eonycteris* occurs from South Asia to Malesia west of Wallace's Line. It includes several species but only one in Peninsular Malaysia, *E. spelaea.* Like the Malesian frugivore *Rousettus amplexicaudatus,* it roosts by the thousands in the roofs of caves. Tony Start collected fresh guano from individuals and from roosting groups in Batu Caves, near, Kuala Lumpur, ascertaining that different groups fed on much the same species nightly. Some interchange among groups may serve to communicate the availability of newly flowering trees by the pollen on their coats.

Eonycteris spelaea is medium sized (40–76 g), possessing the general form of a swallow: short narrow curved wings and short fur. It flies fast, but can only land on a flower by braking, and leave it by somersaulting. This species visits all flowers known to be foraged by Peninsular Malaysian bats (Table 5.3). It feeds primarily on the flowers of trees that flower seasonally. There is some evidence of food preferences, indicated by differences between feeding preferences and the food availability governed by trees' flowering phenology (Figs 5.26, 5.27). *Artocarpus,* for instance, produces copious pollen, though its cylindrical or globose capitate inflorescences hardly conform to the bat-flower model. (Indeed, it has been found to be pollinated by gall midges attracted to the male flowers

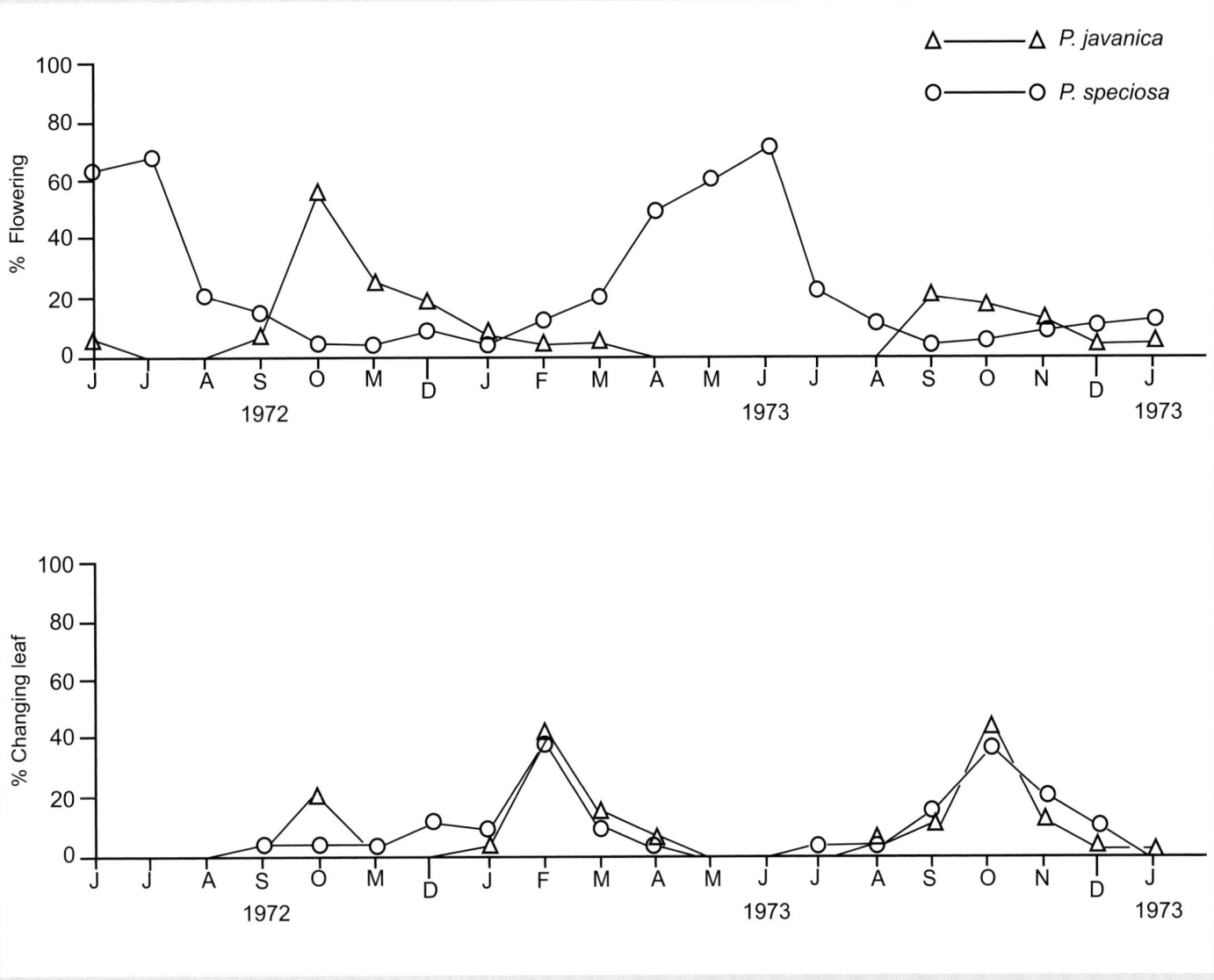

Fig. 5.27 Flowering and leafing phenology over 2.5 years of two co-occurring *Eonycteris*-pollinated *Parkia* species, Selangor, Malaysia: triangles, *P. javanica;* circles, *P. speciosa.* (After Start 1974)

by a commensal fungus.[222]) Nevertheless, this genus appears to be preferred over *Duabanga,* with its paradigmatic bat flowers, when both are available.

Most obviously, the bats only visited the continuously flowering *Sonneratia* species, in the coastal mangrove over 38 km away from Batu Caves, when closer seasonal forest sources were unavailable (Fig. 5.28). Bats in another cave, 60 km inland, never visited this major continuous food source, thereby indicating that they can survive without it, and also the limit to the distance over which they are likely to effect cross-pollination. Distant expeditions to the mangrove take at least

an hour, by which time night has fallen: the bats must then rely on moonlight, or reflection in the water, to locate their flowers visually. Bats do repeat these long-distance pathways, and may therefore in part be trapliners (see section 5.3). Remarkably, and in contrast to primates and bees which are heavily dependent on masting years, *Eonycteris spelaea* was found to breed in all months. The frugivorous bat *Rousettus amplexicaudatus* – like so many vertebrates relying on banyans for a continuous food supply – does the same. Despite some evidence of seasonal peaks, successively flowering tree populations appear to sustain relatively stable bat populations. *The relative nutrient value of*

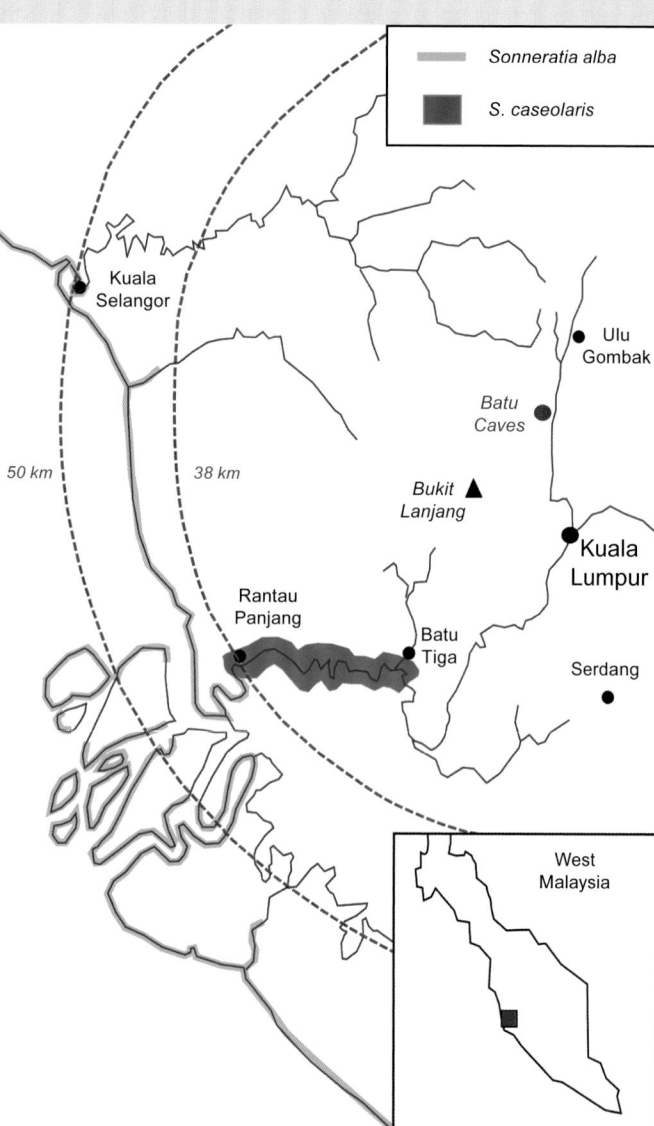

Fig. 5.28 Foraging distance of *Eonycteris spelaea* (inferred from the distance from its roost at Batu Caves) and populations of its mangrove food sources *Sonneratia ovata* and *S. caseolaris,* pollen of which appeared in guano traps beneath the roosts. (I.H. Baillie del. after Start 1974)

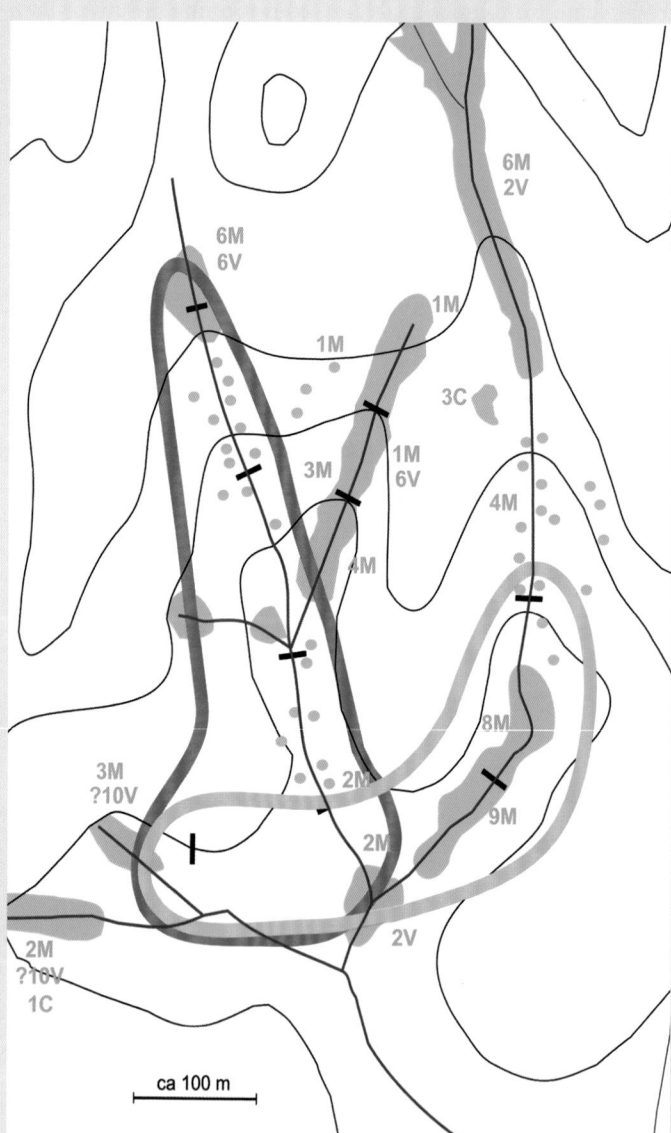

Fig. 5.29 Known ranges of two tagged *Macroglossus sobrinus* in glades along brooks near Kajang, Selangor, based on bat recapture records (net sites, across streams, indicated by lines); with banana clumps, *Musa malaccensis* (M: *Macroglossus*-pollinated) and *M. violascens* (V: bird-pollinated), also indicated. (C: cultivated bananas). (I.H. Baillie del. after Start 1974)

Asian bat flowers would merit investigation.

There are two species of *Macroglossus* in Peninsular Malaysia, similar to one another but distinguished by details of their mouth parts. *M. minimus* extends throughout Indo-Burma and perhumid Malesia to the Solomon Islands; while *M. sobrinus* also occurs in Indo-Burma, but in Malesia only from the peninsula south-east down the Sunda chain to Bali, not in Borneo or other islands to the north. These tiny bats (16–26 g) could hardly be more different from *Eonycteris*. They are solitary, roosting alone close to or within their food sources, which flower continuously, although they will occasionally visit seasonally flowering species when available nearby. They occupy overlapping territories of no more than 1 km in diameter (Fig. 5.29). Their characteristic habitats and foraging behaviour are reflected in flight pattern and morphology.[223] Their flight is slow and fluttering but agile, on relatively broad wings. They can hover, but usually choose to land on flowers to feed at ease.

On the peninsula, *Macroglossus minimus* feeds almost exclusively on the flowers of the three *Sonneratia* species of the coastal mangrove. This constrains its distribution, although it will occasionally visit cultivated banana (which yield only nectar), coconut and other village bat flowers. Along the Peninsular coast, *M. minimus* was found to breed throughout the year with no consistent seasonal peak. At Bako (west Sarawak), where only the periodically flowering *S. alba* occurs, and a few coconut palms are almost the sole inland food source here, there is evidence of a seasonal breeding pattern. *M. sobrinus* has an inland distribution and feeds almost exclusively on wild banana flowers in Peninsular Malaysia: *Musa acuminata* ssp. *malaccensis* in the lowlands and *M. acuminata* ssp. *truncata* in the hills. The two bananas hybridise only where they overlap, testimony to the limited foraging range of *Macroglossus sobrinus,* which is their only significant pollinator. In regions where *Macroglossus sobrinus* is absent, such as Borneo, *M. minimus* occupies its inland range and ecology, and is more catholic in its flower visitation (see Plate 5.11c), implying competitive niche occupancy.

Pteropus, the true fruit bats, are said to be effective pollinators of the brush-flowers visited by *Eonycteris,* including those of durians.[224] But whereas the smaller *Eonycteris* feed on pollen and nectar, *Pteropus* feed on a wider range of flowers, eating the stamens and fleshy bracts and perianths and often likely damaging the pistil.[225] Other mammals visit bat flowers and may inadvertently pollinate where flowers offer appropriate rewards, like the fleshy perianth in the South Indian durian relative, *Durio exarillatus*: Ganesh and Davidar observed this species to yield little nectar, but six mammal species and seven birds visited to consume the fleshy sepals. Though almost 40% of flowers were thereby destroyed, the rest were successfully pollinated,[226] implying that *Pteropus* does also achieve pollination.

Box 5.7

An Asian particular: figs

The genus *Ficus,* the figs, is with c.800 species among the largest genera of woody plants (Plate 5.13). The vast majority occur in tropical Asia to Australia and the Pacific islands. All figs secrete a sticky white latex and produce an inverted inflorescence resembling a fruit, hollow inside where the many flowers are attached to the inner face. The pollinators, specialised fig wasps of the family Agaonidae (Hymenoptera: Chalcidoidea) (Fig. 5.30), enter through a terminal pore, the osteole, which is armed with scale-like bracts to restrict entry and escape.

Ficus is perhaps the most diverse in form of all the woody plants, incorporating free-standing pioneer trees, successional and climax trees, stranglers (banyans), and vines, many with aerial roots; also understorey shrubby species including species which bear figs on specialised rhizomes as much as 10 cm below the soil surface; small climbers; and even six rheophytes including the Bornean *F. macrostyla*, which presents its figs below normal water level among rocks in Bornean whitewater riverbeds.

Figs are unique in their pollination biology, and remarkable among woody plants for the near-absolute specificity of their pollinators. Four subgenera have been recognized on morphological evidence, three of which are monoecious, bearing male and female flowers within each fig. These female flowers vary in the length of their styles. The female fig wasps have specially adapted needle-like ovipositors with which they penetrate the style. They prefer flowers with shorter styles, which thereafter become sterile gall figs, the ovule nevertheless swelling and thereby providing food for the wasp larvae, which bite their way out of the ovary at maturity. When a male wasp emerges within a gall fig, it loses its rudimentary wings, mates with one or more (subadult) females and dies. The mature female wasps escape through the scale-lined osteole, and their bodies pick up pollen from the male flowers that mature just as the females are preparing to leave. Because figs develop synchronously on each fig plant, and because the figs are maturing when the fig wasps emerge, populations of monoecious figs only survive because their populations reproduce continuously yet different individuals are simultaneously producing gall-figs or seed-figs: each individual reproduces several times a year[227] (Fig. 5.31).

Different monoecious fig species also bear styles of different mean length, and are pollinated by different fig wasps with correlated ovipositor lengths. One fig subgenus is morphologically gynodioecious but functionally dioecious, bearing either figs solely with female flowers or with male flowers and sterile female gall flowers. The gall flowers again

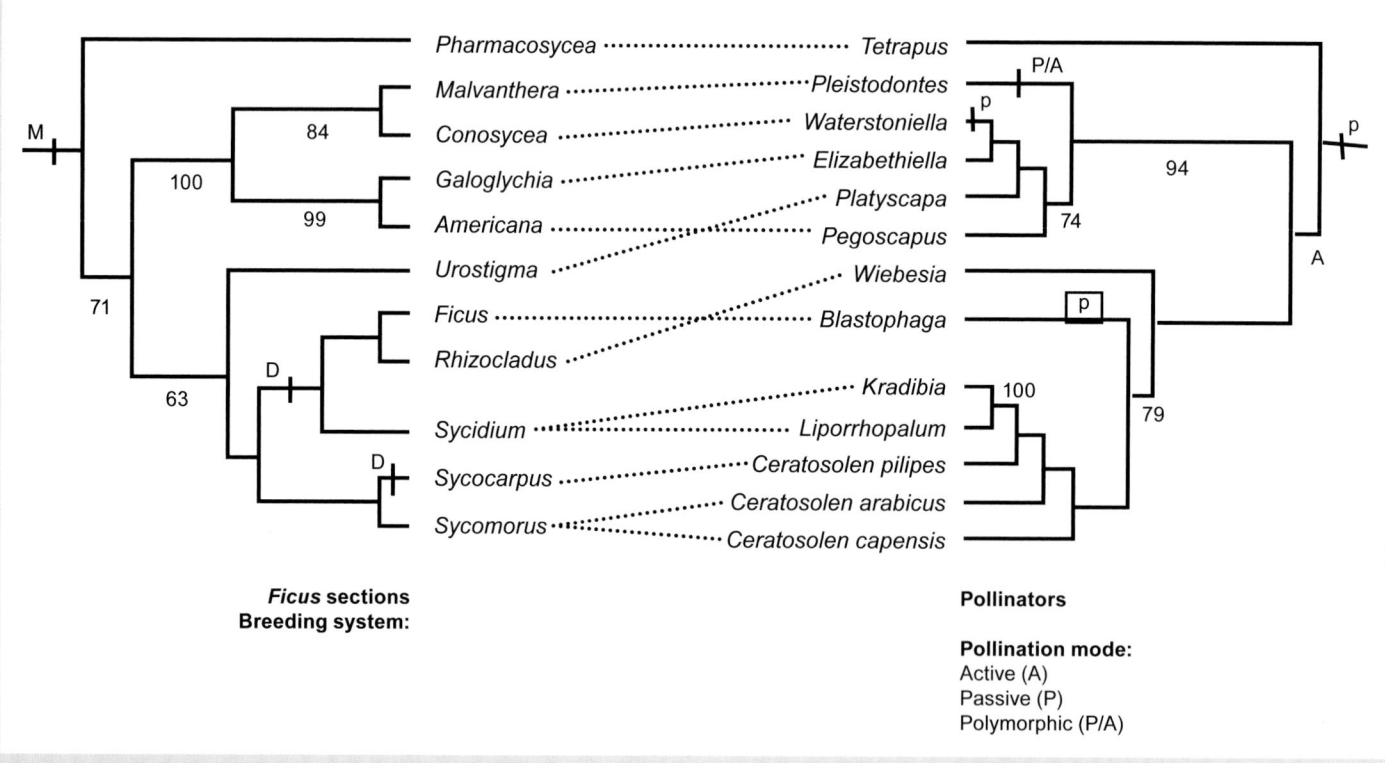

Plate 5.13 Figs. **a,** Free-standing fig tree in riparian forest along the Kinabatangan river; **b,** Strangling fig at Huai Kha Khaeng; **c,** Ramiflorous free-standing *Ficus variegata,* south Kinabatangan, Sabah; **d,** Ramiflorous free-standing *Ficus minahassae,* northern Sierra Madre National Park, Luzon; **e,** Female Agaonid wasp (*Waterstoniella masii*) emerging from the fig of the strangler *F. crassitamea* ssp. *stupenda,* by an opening previously made by a male wasp, Gunung Palung National Park, West Kalimantan (scale: 3/1). (**a, c** P.A.; **b** H.H.; **d, e** T. Laman)

Fig. 5.30 Co-phylogenetic history of the main sections of *Ficus* and their wasp pollinators. Numbers near branches are indicators of the strength of analytical evidence supporting them ('bootstrap values'). (After Herre *et al.* 2008)

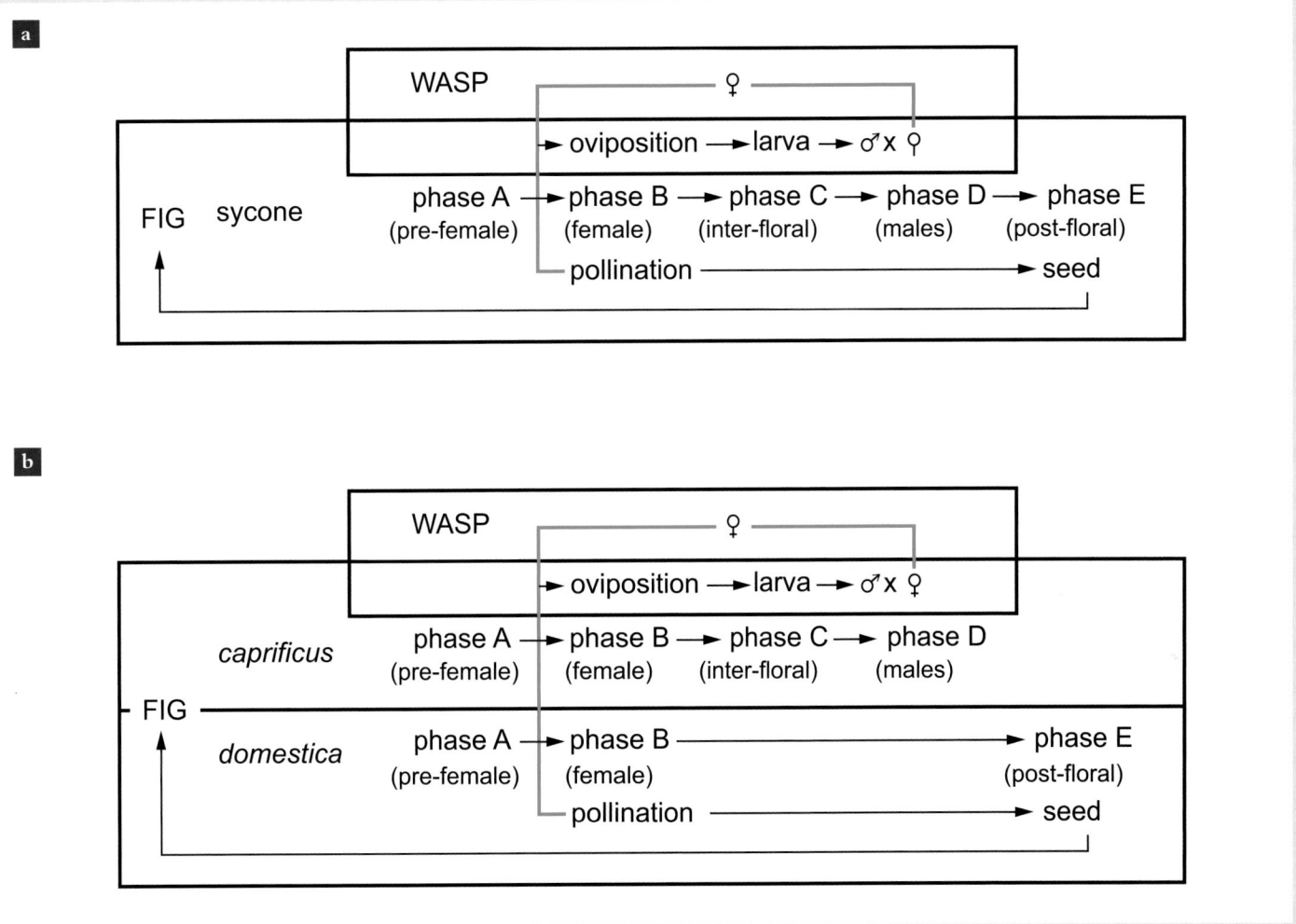

Fig. 5.31 Generalised life cycle of fig wasps (top of each diagram) in relation to the pollination and development of their host: the fruit-like inflorescence or *syconium,* which is a fig (below). Developmental phases of both indicated, those of the fig by letters A–E. The dispersal of pollen by the female wasp, following emergence (from a galled male fig in the dioecious case) is indicated by blue lines: **a,** monoecious figs; **b,** a dioecious fig, the domestic *Ficus carica,* in which the female (*domestica*) fig develops from A through pollination (B), direct to seed development (E), and fruit; whereas in the male (*caprificus*) phase B is a galled fruit within which larval wasps develop through to phase D at which the wasps emerge. (After Herre *et al.* 2008)

bear shorter styles than the fertile female flowers on the female trees. The female fig wasps visit both flower types equally, but their ovipositors are of the short style length, and they fail to successfully parasitise female flower ovules – but they do inadvertently pollinate them.

Figs originated at the end of the Cretaceous period, c.110 million years ago.[228] George Weiblen[229] found, in a molecular phylogeny of figs, that the accepted subgenera are not monophyletic; all crown groups in his phylogeny occur in Asia. But he found substantial congruence among molecular phylogenies of figs and their wasps (Fig. 5.32), whereas there was little correlation between the fig phylogeny and the less species-specific folivorous and sap-sucking fig specialists.

The wasps, with their vastly briefer life cycles, evolve more rapidly than their fig hosts whose comparative evolutionary conservatism constrains wasp diversification within the requirements set by the fig for their successful reproduction. Nevertheless, molecular evidence from the dioecious figs *Ficus variegata* and *F. racemosa,* widespread throughout tropical Asia, indicates allopatric divergence of these figs on either side of Wallace's Line, where they become associated with different *Ceratoxylon* wasp pollinators.[230] These facts, and the extraordinary way in which the dioecious figs have diversified into a multitude of species, yet the fig-wasps consistently fail to recognize female from bisexual figs despite gaining no reward from the former, led Weiblen and others to conclude

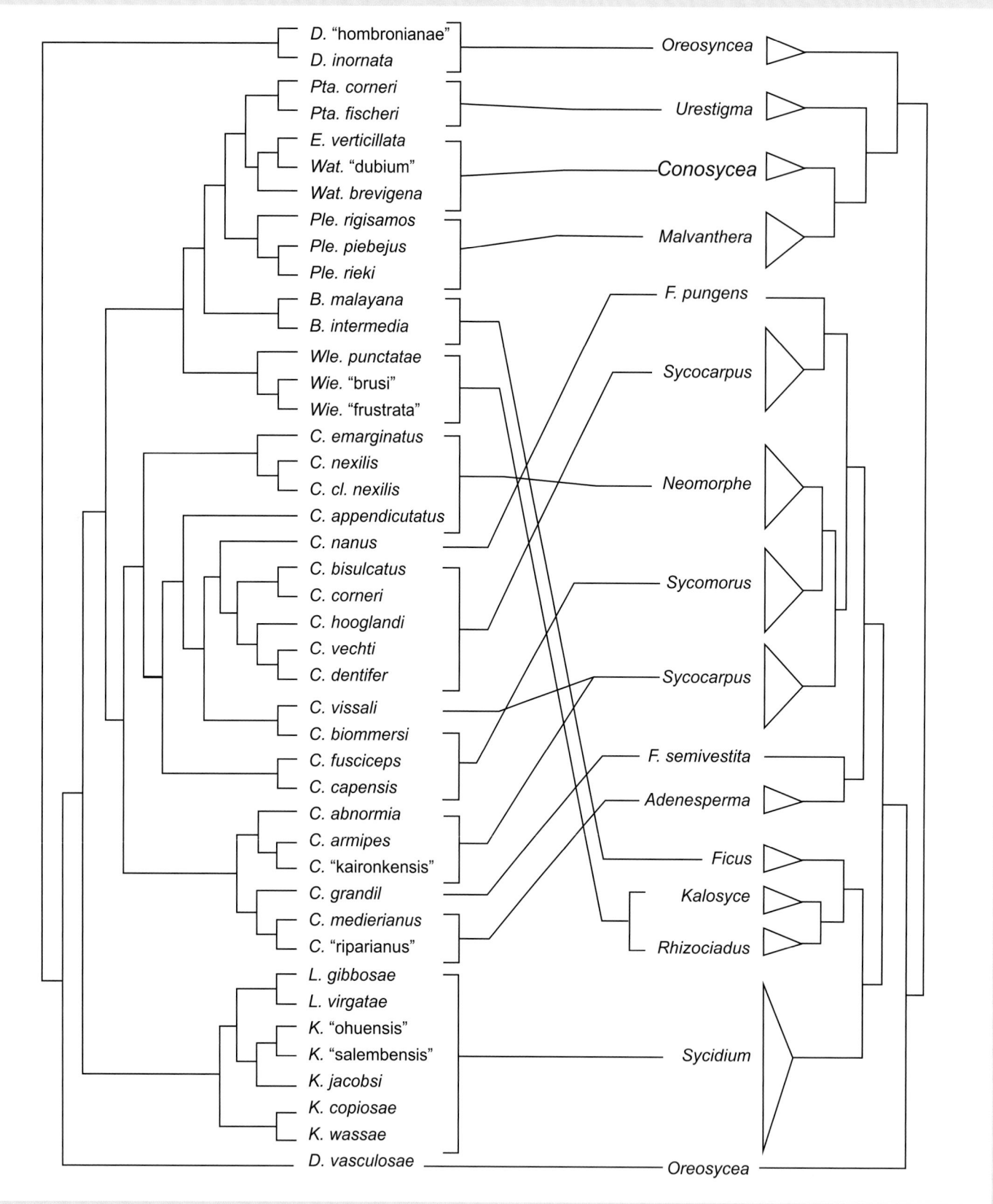

Fig. 5.32 Evolution of specific wasp pollinators (left) among Indo-Australian fig classes (right). Bracketed pollinator groups are uniquely associated with Corner's (1965) *Ficus* sections. Most pollinator phyletic groupings are associated with a single fig host section. (From Weiblen 2001)

that the association between fig and wasp presents a uniquely stringent demonstration of co-evolution.

Dioecious figs, which evolved more than once from monoecious ancestors[231] (Figs 5.31, 5.32), are predominantly short, broad-crowned pioneer trees of large canopy gaps. The wasps of dioecious figs are found mainly low in the canopy where their hosts' crowns occur. They do not generally disperse far but are locally abundant. Their host fig species tend to cluster in local populations and many are local endemics.[232] The wasps of the large monoecious strangling figs, in contrast, fly at full canopy height and occur in low densities. Caught in prevailing breezes they can be carried many tens of kilometers.[233] Some monoecious figs are among the most wide-ranging woody plants in tropical Asia; relatively few are local endemics. Monoecious figs were among the early immigrants and reproducing colonisers of Krakatau, equidistant and over 50 km from both Java and Sumatra, following its devastating eruption in 1884.[234] By contrast, Rhett Harrison found that fig wasps of dioecious species became locally extinct in the drought of 1996 at Lambir (north-east Sarawak).[235] Of the 58 dioecious fig tree species in Malaysian Borneo, only two – Ficus aurata and F. grossularioides – are common on the drought-prone sandy coastal soils.

The monoecious figs, particularly the banyans with their immense canopy crowns (Plate 5.13b), their populations reproducing continuously,[236] often fruit heavily in seasons when other canopy trees do not.[237] They thus provide famine food for the herbivorous vertebrates, which incidentally disperse their seeds. Vast numbers of tiny seeds are produced which, unlike those of most rainforest trees, may be dispersed over great distances. On the other hand, Tim Laman observed that fig populations are greatly constrained by the rarity of suitable moist establishment sites among the branches of the forest canopy (see below).[238]

Frugivorous birds of all sizes congregate on figs, notwithstanding the low lipid content of the fruit.[239] On the other hand, the fig being a fleshy inflorescence unlike the mesocarp of other fruits, it is supplied with vascular bundles, therefore to the nutrient ions that are transported within them. The abundant and frequent reproduction of banyans incurs enormous energetic and nutrient costs. Banyans are most abundant and diverse on the high nutrient soils of continental Asian semi-evergreen forests, where they sustain the rich avian fauna. There, free-standing canopy fig trees and banyans are among the tree species with the fastest growth rates (Table 2.7a). In the perhumid tropics they concentrate on riverine periodically flooded alluvium, and on fertile substrates. Rhett Harrison monitored all epiphytic banyans whose aerial roots reached the ground 1998–2005, in MDF in a plot in the relatively drought-prone coastal hills of Lambir (Sarawak).[240] This period immediately followed a major drought. He found

that banyan mortality rate (4.73% y^{-1}) was vastly greater than the overall tree mortality rate (1–2.5% y^{-1}) during the same period. A mere 1.33% banyan recruitment rate (not including recruits that may have died between censuses) was recorded. Harrison commented that this could be due either to drought or to hunting out of the seed-dispersing imperial pigeons, hornbills and primates.

Krau Game Reserve (Pahang, Peninsular Malaysia) overlies a diorite dyke and is a preferred research site for primatologists. This is owing to the number of banyans and to the more nutritious, less fibrous foliage found there, which are favoured by folivorous leaf monkeys.[241] Gunung Leuser National Park, another such site on basic volcanic substrate, is richer in vertebrates than comparable sites in Peninsular Malaysia. Orangutan populations are centred on such sites. In contrast, only Ficus calophylla and F. retusa are common banyans on sandy soils in hyperdiverse northern Borneo forests, with F. borneensis and the monoecious tree F. spathulifolia only occasional. Foresters in India plant fast-growing fig trees for fuelwood by establishing cuttings in chicken manure.[242] This suggests that the larger monoecious figs, by offering an abundance of food to vertebrates, may attract a rich nitrogen source in exchange.

Tim Laman documented the spatial distribution of semi-epiphytic strangler figs in lowland forest at Gunung Palung National Park (West Kalimantan), where 20 species co-occur.[243] There is significant and regular variation in the height, within the canopy, at which individual populations of the common species are found. Since potential establishment sites appear to outnumber figs, Laman concluded that this pattern cannot be due to interspecific competition for space. Good establishment sites – which are themselves uncommon – are crotches between tree branches, rotting pits where branches have snapped, and other places where there is humus collection and less stringent water stress. Laman showed that the rarity of seedlings even in suitable sites is likely due to seed predation by Pheidole ants – which may nevertheless inadvertently drop and thereby disperse seed.

Interspecific competition for the services of dispersing vertebrates may nevertheless occur. Figs differ in size and colour when ripe and, though all are consumed by a diversity of vertebrates, different species are predominantly dispersed by birds with a range of gape sizes, and bats and other animals, according to sensory differences and their locomotory physiology and foraging height (Plate 5.14).[244] Small-fruited species are consumed by virtually all frugivorous vertebrates,[245] but these animals may predominantly use different parts of the forest canopy. Seed is invariably ingested along with the figs; dispersers then differ in the distances to which they disperse, thereby influencing the breadth of effective dispersal barriers, and consequent endemism.

Plate 5.14 Fig fruit feasts. **a,** Long-tailed Macaque (*Macaca fascicularis*) feeding in a fruiting *Ficus stupenda* crown, Gunung Palung National Park, West Kalimantan; **b,** A Red (Maroon) Leaf Monkey (*Presbytis rubicunda*) with a fig of the strangler *F. dubia*, Gunung Palung; **c,** A Prévost's Squirrel (*Callosciurus prevostii*) feeds al fresco in a *F. stupenda* crown, Gunung Palung; **d,** A Thick-billed Green Pigeon (*Treron curvirostra*) sups on *F. crassiramea* ssp. *stupenda* figs, Gunung Palung; **e,** A young rhinoceros hornbill (*Buceros rhinoceros*) tosses a large fig (*F. dubia*) into its gullet, Gunung Palung; **f,** A coppersmith barbet (*Megalaima haemacephala*) purloins several small figs, Mudumalai, Tamil Nadu. (**a–e** T. Laman; **f** R. Matthias)

5. Seed dispersal

The dispersal of a single seed and subsequent successful recruitment of one individual to reproductive maturity is potentially equivalent to the dispersal of millions of pollen grains a similar distance.[246] It is commonly believed that seeds of tropical forest trees are dispersed over short distances compared with those of temperate forests. Limited seed dispersal leads to clumping of juveniles around mother trees. Some of these juveniles survive and become clumps of reproductively mature individuals. Among species which have higher fecundity than the average, this can also lead to local aggregations of a species through 'reproductive pressure', independent of interspecific competitive advantage in relation to the physical habitat (*Dryobalanops* for example, see Chapter 7). *To date, few comparative data for seed production among species have been gathered from tropical forests, though some conclusions can be drawn from the relation between the relative abundance of species and their fruiting phenology and effectiveness of seed dispersal.*

5.1. *Tropical versus temperate tree seed dispersal*

Hubbell's Unified Neutral Theory of Biodiversity and Biogeography[247] (see Chapter 7) arose partly in response to the contrasting predominance of light wind-borne seeds among temperate canopy tree genera, especially conifers, whose seeds are widely dispersed. The tropical conifers *Agathis* and *Araucaria* are similarly wind-dispersed, but they are common only locally. If only broadleaved species are compared, the differences in seed dispersal mechanisms between temperate and tropical trees are less emphatic. Canopy dipterocarps and maple fruits are similarly dispersed by gyration, though maple fruits are generally lighter and, in the windier temperate climates, their dispersal is more generally effective. Post-fruit-fall, dispersal of dipterocarps by rodents and squirrels is limited compared with the extensive dispersal of acorns by squirrels and frugivorous birds such as jays.[248] Temperate pioneer trees with light wind-dispersed seeds, such as birch and willow, are matched by *Alstonia*, *Octomeles* and *Tetrameles* in tropical Asia. These tropical trees are tall canopy or emergent pioneer trees of moderate longevity, and are scattered sparsely in primary forest at maturity (Chapter 2). The short-lived tropical pioneers are small-seeded but overwhelmingly animal-dispersed, but so are many early temperate pioneers such as *Berberis*, *Prunus*, and *Rubus*. Tatsuhiro Yamada and Elei Suzuki[249] observed that pioneer and climax species of Sterculiaceae differ in fruit allometry: the fruits of the pioneers (*Pterospermum*, also

successional *Pterocymbium*) are winged, with light seeds, whereas climax *Heritiera* has greater seed mass per wing, and is therefore less effectively dispersed. In nature, tropical pioneer species occur in habitat patches of varying areas and distances apart.

A major difference between temperate forests and tropical rainforests is the richness of the tropical subcanopy, including understorey species. Tropical subcanopy and main canopy species mostly produce animal-dispersed seeds. In Asian forests, animal-dispersed fruits and seeds occur predominantly among species of the main canopy and subcanopy, although several in tropical Asia have small dry explosively dispersed seeds, a rare mechanism among temperate trees. Many subcanopy tree genera bear fleshy fruit of moderate size, and are dispersed by birds and mammals,[250] but a surprising number are barely fleshy or dry, including the several Euphorbiaceae dispersed by their explosive carpels (*Cleistanthus, Elateriospermum, Mallotus, Koilodepas*) (Fig. 5.33). In the rainforest understorey, small fruit with small seeds predominate. The fruit of emergent dipterocarps are generally dispersed by gyration, sometimes enhanced by squalls. Some emergent durian species bear fruits beneath their crowns, on the trunks as well as branches. Emergent leguminous trees bear dull coloured pods and seeds; some are thought to be dispersed by ungulates, but *Koompassia* fruits are thinly flanged and gyrate.

Fig. 5.33 Population distribution of *Koilodepas longifolium* (Euphorbiaceae), an understorey species with small explosively dispersed seeds, in the 50 ha CTFS plot at Pasoh forest, Negeri Sembilan over 20 years; trees >1 cm dbh. The distribution of the smallest size class implies that clumps may be expanding, but that occasional dispersal beyond established clumps also occurs. (Little change in demography occurred among 16 such species over the 20 years of census data currently available). (From Manokaran *et al.* 1992)

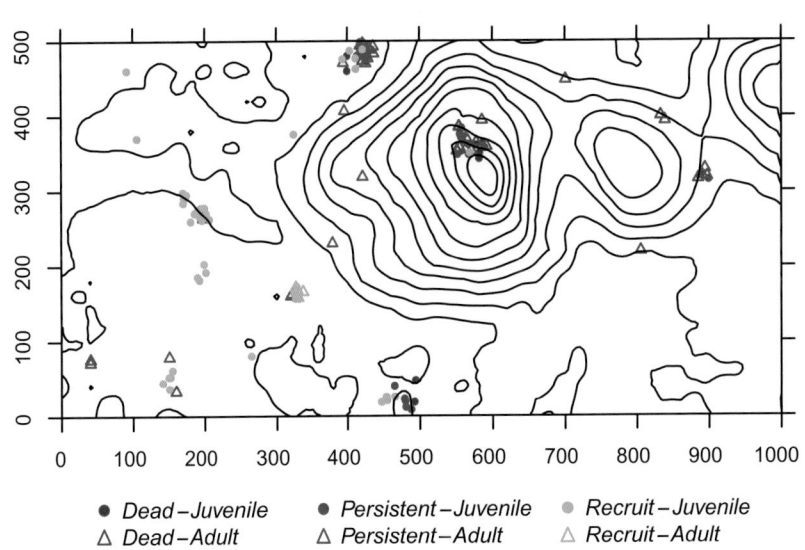

● Dead–Juvenile   ● Persistent–Juvenile   ● Recruit–Juvenile
△ Dead–Adult   △ Persistent–Adult   △ Recruit–Adult

The supra-annual mast fruiting of the canopy species guilds of MDF are synchronised to the same season in all participating families, unlike the oaks in North American forests: though stimulated to flower by the same trigger, white oaks take a year longer than red oaks to mature their fruit. Mast fruiting is weak in the MDF understorey where fruit production, though varying from year to year, does occur annually and species are less synchronised (Fig. 5.5). The seeds of most of those related subcanopy species being dispersed by vertebrates, it might be expected that competition among them for dispersers would result in their fruiting in overlapping sequence, as is the case in the flowering of many mass flowering dipterocarp taxa that share a common pollinator. But there seems to be little evidence for this, even among those species in perhumid Asia whose time of fruit ripening is unlikely to influence establishment success, and where food resources are limited compared with many tropical regions elsewhere.

Moreover, tree species diversity in Far Eastern MDF relative to the richest Neotropical forests is greatest among canopy species and least beneath the canopy (Chapter 7). Species diversity would be enhanced when species release their seed at different seasons, thereby eliminating early interspecific competition; but in Asia's most species-diverse vertical species guilds, those of the canopy ripen their seed in synchrony, contrary to expectation.

5.2. *The seed dispersers*

The diversity of dispersing animals is, of course, far greater throughout the continental tropics than in temperate climes. In tropical Asia the predominant frugivores and seed dispersers include among mammals elephants, tapirs, ungulates, rhinoceros, bats, macaques and gibbons, and civets, and among birds the barbets, broadbills, hornbills, fruit pigeons, corvids, muscicapids, bulbuls, white-eyes, laughing thrushes and flower peckers.[251] Less rich in frugivorous birds than the Neotropics, the Asian tropics are better endowed, and comparable to Africa at local scales, with browsing and seed-dispersing mammals. Fish, ants and even tortoises and monitor lizards have been implicated. Parakeets and (following fruit-fall) pigs, both seed predators, also accidentally disperse seeds (Box 5.9). Specialised frugivores consume the pulp surrounding seeds but not the generally toxic seeds themselves.

Birds, fish and primates can sense colour, and seeds dispersed by them are often vivid red, blue or black. The small bird-dispersed seeds and fruits of *Eonymus* and *Sambucus* are familiar in temperate forests, but the diversity of size and form in tropical forests is extraordinary (Plate 5.15). These coloured fruits and seeds stand out against the prevailing green within the canopy, thereby presumably attracting the disperser. Many subcanopy legumes, Euphorbiaceae and others yield seeds with contrasting

Plate 5.15 Seed dispersal. **a,** Fruit diversity in a Lambir mast fruiting year: *Nephelium,* primate dispersed (top left), *Artocarpus* cf. *odoratissimus,* primates (centre left), *Dipterocarpus sublamellatus,* gyration (top right), *Baccaurea parviflora,* birds and primates (centre to lower left, and opened pericarps), and unidentified others (scale: 1/15); **b,** An orang utan enjoys the fruit of *Polyalthia hypoleuca*; **c,** Giant Malabar squirrel (*Ratufa affinis*) and *Grewia tiliifolia* in tall deciduous forest, Mudumalai; **d,** Colours which attract birds: *Sterculia* sp., *buah ayan antu sebayan* (Iban): 'the fruit that was gazed on by the spirits of the dead' (scale: 1/4); **e,** Koel *Eudynamys scolopacea* and *Diospyros montana*; **f,** A black hornbill (*Anthracoceros malayanus*) and fruiting *Aglaia*. (**a** S. Davies; **b, f** T. Laman; **c, e** H.H.; **d** E. Soepadmo)

black and red coats or arils which remain attached, dangling on long funicles, after the fruit has split open. The brilliant seeds of some *Sterculia* fruit, released from five dehiscing satiny vermillion follicles, themselves dangling and spread like fingers, is named by Iban Dayaks *buah ayan antu sebayan*: 'fruit (that are) stared at by the spirits of the dead' (Plate 5.15d).

Asian rainforests in both perhumid and seasonal climates support lower biomass of vertebrate-dispersed fruits than in the Neotropics. The years between the mast fruiting that follows mass flowering events are often years of famine for frugivores.[252] One third of sunbears died during such a famine period in Sabah,[253] while orangutan experience muscle deterioration and higher mortality.[254] Ganesh and Davidar[255] found, even in Western Ghats forests with a four-month dry season and no mast fruiting episodes, that only half the 650 kg ha^{-1} y^{-1} of fruit produced was edible to vertebrates. The Sunda MDF mostly covers moderate to low nutrient yellow soils, while the extensive domed peat swamp forests and *kerangas* support communities in which fruit size is small.[256]

Deposition in warm dung may trigger germination. Foresters in 19th Century India, perplexed by regeneration problems in teak, found that germination was triggered when the seed was immersed in cow dung.[257] The dung also enhances seedling performance.[258] *Many of the larger hard-surfaced dull-coloured fruits currently suspected to have no disperser may be found to be consumed and dispersed by ungulates and other larger terrestrial vertebrates when sequential camera trapping is attempted.*[259]

5.3. *Seed dispersal distances*

Temperate and tropical closed forests may not differ as greatly as often supposed in respect of the dispersal distances of broadleaved tree seeds, or in the spatial pattern of seed rain around trees. Much more is currently known about dispersal around tropical than temperate trees, particularly because the enormously greater species diversity of lowland tropical rainforest enables easier monitoring of seed rain from isolated trees of individual species.

The average dispersal distance of a species' seed will be correlated with the foraging distances of the dispersing agents.[260] Ridley[261] estimated that the winged fruit of most dipterocarps would not be dispersed more than 100 m, and this estimate is supported by genetic analysis (see section 2.3.1 above). But few will disperse even half that distance in the generally windless climate and dense canopy of MDF. The wingless seed of many dipterocarps, and other species which seem to have no consistent disperser, will mostly fall within the area of the maternal crown (Fig. 5.34; see Chapter 6). The several genera of Euphorbiaceae, mostly of the subcanopy, with hard shiny seeds bereft of fleshy tissue and dispersed by the explosive opening of the pericarp, are generally only dispersed over a few metres, as reflected in their highly clumped distributions (Fig. 5.33). Seeds furnished with elaiosomes, such as those of many *Macaranga,* are dispersed by ants, and this too will be limited in distance. Different species of ants tend to disperse seeds of different

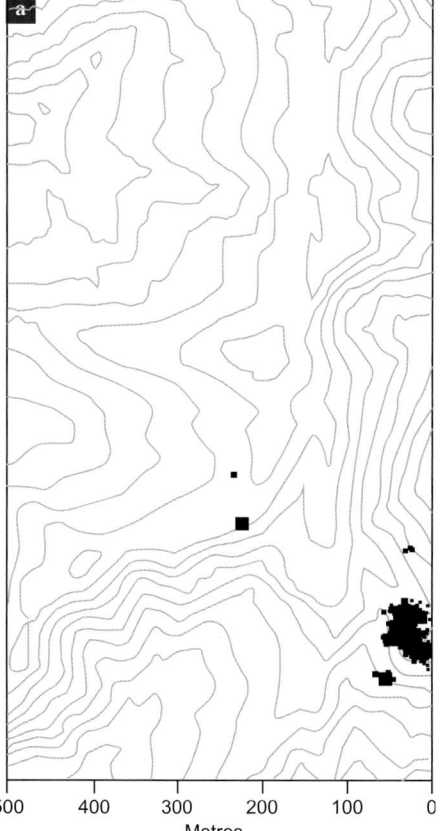
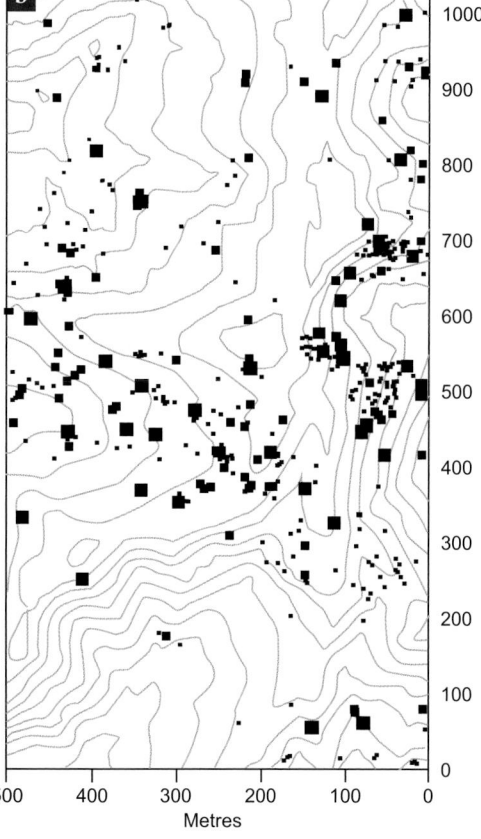

Fig. 5.34 Population distributions of two *Shorea* sect. *Shorea* within the CTFS 52 ha plot at Lambir National Park, Sarawak: **a,** *S. geniculata,* an emergent with large wingless fruit; **b,** *S. scrobiculata,* a main canopy species with small winged fruit. Dots: trees 1–10 cm; small squares: 10–30 cm; large squares: >30 cm dbh. (From H.S. Lee *et al.* 2002)

plant species, according to the size of the seed relative to that of the ant.[262] In contrast, bats and hornbills may easily carry fruit a distance exceeding 1 km to roosting sites, where they eat them and defecate,[263] but these germinating seedlings may often be even more densely clumped than if they had fallen directly from the tree crown. Sabrina Russo,[264] studying South American spider monkeys, found that they scatter seed that pass through their gut when foraging, yet deposit seed in clumps while roosting. The resulting clumping of seedlings survives to the sapling stage, notwithstanding higher mortality rates in dense seedling cohorts. Some seedlings survive to maturity, explaining the commonly observed clumping of mature trees of the same species in species-rich forests. Macaques likewise sit in tree crowns once their cheek pouches are full of fruits, the flesh of which they remove at leisure without leaving the tree (Box 5.2).[265]

David Chivers documented the social and feeding ecology of the siamang gibbon (*Hylobates lar*) at the Krau Game Reserve, Kuala Lompat, Pahang and one other site in Peninsular Malaysia.[266] Siamang consume fruit, leaves, and occasionally small animals. Figs comprised 13% of trees visited, but yielded the dominant food. One ape would explore c.8 ha of forest each day, with a linear range of 320–2,860 m within a maximum radius of 700 m. Individuals rarely returned to the same tree within a day. Young leaves were consumed, and also the more fibrous mature leaves during periods of dearth. Gibbons and orangutan swallow fruit and may travel up to 1 km before defecating, although it has been estimated that most seed passes through the gibbon gut in less than one hour.[267] In Central Borneo, McConkey[268] found that 88% of seeds consumed by gibbons were predated, though all but 0.45% of surviving seed successfully germinated. Seeds swallowed by sambur deer pass through their gut, whereas they are regurgitated by chital.[269]

Short-distance dispersal is relatively easy to observe and document.[270] It has therefore been emphasised in mast-fruiting dipterocarps and others; but it is the seed that disperses over a longer distance and whose progeny occasionally survive, that expands a species range and holds breeding populations together in rich forests and heterogeneous terrain.[271] Elephants, which feed on a variety of fruits, have a vast home range, estimated at 170 km² in Peninsular Malaysian MDF where food is particularly limiting and widely dispersed; it is c.60 km² in more productive secondary forest.[272] Rhinoceros ingest the large dull fruit of *Trewia nudiflora* in tall deciduous forest, potentially dispersing it considerable distances.[273] Occasional successful long-distance dispersal, critical to speciation in continuous landscapes and negating an assumption of the neutral theory (see Chapter 7), may be more frequent than supposed.[274] Such occasional dispersal must be reconciled with the occurrence of sister species in different adjacent habitats (Chapter 6). Even

transatlantic seed dispersal has, it seems, played an important role in the origins of the Neotropical flora[275] (Chapter 6). At a smaller scale, periodic recensus of explosively dispersed species at Pasoh has revealed the sudden establishment of new clumps that could only be due to an unknown disperser (Fig. 5.33). Scatter-hoarding of dipterocarp seed by rodents probably plays a larger part in their dispersal than expected, the seed sometimes being carried tens of metres from the tree.[276] Many dipterocarp fruit can be surprisingly widely dispersed by gyration induced by their twisted wing-like sepal lobes, and the large surface area of wing relative to the mass of the seed.[277] Jan Muller[278] recounted how, during the 1958 mast fruiting, the yard drains of his residence in the coastal oilfield town of Seria (Brunei) received numerous winged seeds of the *Shorea albida* dominating the peat swamp – which was by then more than 1 km to the south. A squall was once observed to waft abundant dipterocarp fruit up and over a ridge in the Ampang reservoir catchment forest outside Kuala Lumpur.[279]

At the same time, the case of Krakatau, a volcanic archipelago whose islands, some 50 km from adjacent Sumatra and Java, were variously immersed in ash in the eruption of 1883, serves to caution against any facile presumption of frequent long-distance dispersal among rainforest trees.[280] Bird species numbers reached an asymptote about 1970, but plants still remain limited to wind-dispersed taxa and a few widespread bird-dispersed species including figs. Clearly also, most species require specialised establishment conditions which have not developed there as yet.

Since chloroplast DNA is solely transmitted by seed, but nuclear markers like microsatellites are transmitted by pollen as well, the relative dispersal distances of seed and pollen can now be evaluated. Harata and his group examined greater fine-scale spatial genetic structure (FSGS), using satellite DNA, among 10 emergent dipterocarp species.[281] They were thereby able to infer dispersal distances of both pollen and seed. They concluded that bee dispersed pollen increased the radius of male inherited FSGS from individual trees relative to that dispersed by thrips, and that species with small winged seed did likewise for female-inherited FSGS relative to species with heavy seed (Fig. 5.34). The emergent bee-pollinated, large wingless-seeded, climax heavy hardwood dipterocarp, *Neobalanocarpus heimii,* has been the subject of a particularly apposite examination.[282] Pollen flow of more than 400 m was observed directly, and the average mating distance between trees was estimated at 524 m. Absence of shared seed microsatellite haplotypes among neighbouring trees implied that an unknown seed disperser was also active. There was significant, but weak, negative correlation between genetic relatedness among trees and their distance apart up to 1 km, implying continuous gene flow and seed migration over at least that distance.

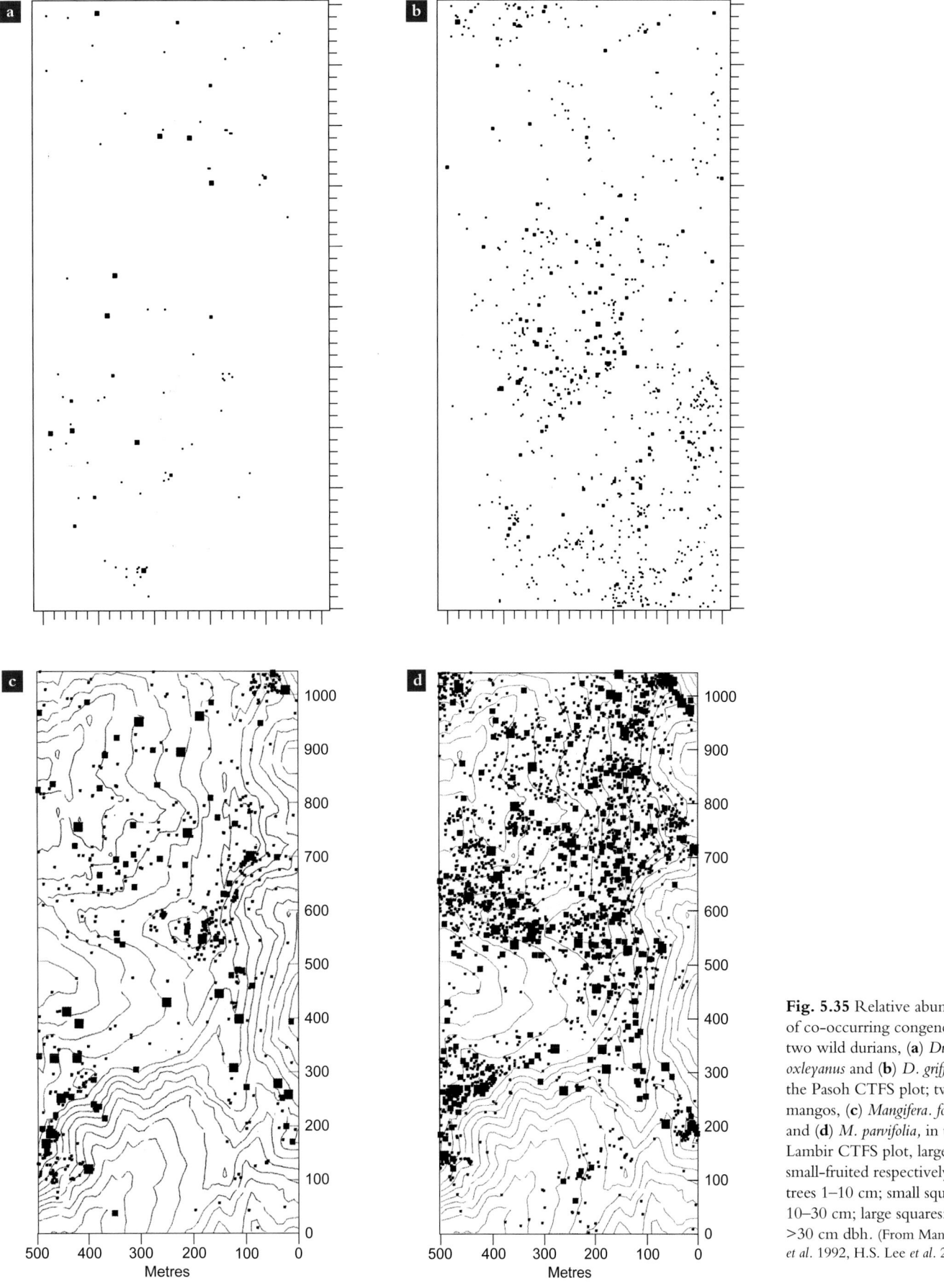

Fig. 5.35 Relative abundance of co-occurring congenerics: two wild durians, (**a**) *Durio oxleyanus* and (**b**) *D. griffithii,* in the Pasoh CTFS plot; two wild mangos, (**c**) *Mangifera. foetida* and (**d**) *M. parvifolia,* in the Lambir CTFS plot, large- and small-fruited respectively. Dots: trees 1–10 cm; small squares: 10–30 cm; large squares: >30 cm dbh. (From Manokaran *et al.* 1992, H.S. Lee *et al.* 2002)

Certain forms of fruit and seed dispersal mechanisms are consistently correlated with rarity, providing some predictability to the rank order of species abundances in rainforest communities and contrary to expectations of neutral theory. Species with wind-dispersed seed, for example, being pioneers, occupy temporary island habitats of low relative area, and are therefore consistently in low density in primary forest. But in windy lower montane sites, wind-dispersed *Engelhardia* may become abundant. Sugary-fruited tree genera of the continuous clay and sandy-clay soils of the perhumid Sunda lands are mostly widespread, with low local endemism[283] (Chapter 7). Species bearing large fruit with vertebrate-dispersed seeds, including *Mangifera, Artocarpus,* many *Durioneae* and the palm *Pholidocarpus,* produce few fruit and seeds, and are consistently at low density at landscape scale in primary MDF, albeit mostly widespread and with low local endemism.[284] Small-fruited species in these same genera, such as *Mangifera parvifolia* and species of *Durio* (subgen. *Boschia),* are equally widespread but are generally locally abundant (Fig. 5.35). Grubb suggested that large seeds experience higher predation, which offsets advantage gained by greater resources available for their establishment (Chapter 2).[285] Nevertheless, evidence for an overall relationship between abundance and seed size among climax species is not strong (Fig. 5.36). The most abundant species within the Pasoh 50 ha CTFS plot, the

subcanopy tree *Xerospermum noronhianum* with more than 10 individuals >10 cm dbh ha^{-1}, bears fruit with a single medium sized pulp-coated primate-dispersed seed in fruit; while the second-most abundant species, with more than 3 individuals >30 cm dbh ha^{-1}, is the emergent *Shorea maxwelliana,* whose smaller winged fruit with dry seed are mostly dispersed by gyration and squalls.

5.4. *Fecundity and competitive success*

Many widespread or locally abundant species, and also quite a few local or rare species, remain at a broadly consistent rank order of abundance in their chosen habitats (Chapter 7);[286] and there appears to be no more than a loose correlation between means of seed dispersal and population density. High fecundity though, as measured by fruit production, appears to maintain some species at consistently high densities relative to others of similar ecology (see Chapter 7 section 2.3.1), while species with low fecundity remain at low densities. Another group of species appears to fluctuate in abundance over time and space, and the possibility exists that some of such fluctuation is random, as neutral theory predicts.

A long-standing debate among biogeographers concerns whether speciation, which is assumed to occur when a barrier divides metapopulations, is most often driven by the occasional dispersal of founder populations to hitherto unoccupied yet suitable habitats, or alternatively whether it occurs through vicariance – that is, the division of the habitat itself as by tectonic drift, sea level rise or river capture. Here, we have emphasised factors important in determining the first driver, but the second is equally probable when circumstances have favoured it, as will be discussed in Chapter 6.

If species within a community are ecologically complementary and enjoy equal reproductive advantage locally, they could spread over time simply by persisting within communities by chance and continually generating new patches of local numerical dominance. Such a pattern of restricted ranges of seed dispersal could facilitate species co-existence, and thus be a leading determinant of floristic structure and species richness. This hypothesis forms the basis of Stephen Hubbell's Unified Neutral Theory of Biodiversity and Biogeography.[287] Hubbell has demonstrated that this is both theoretically feasible and consistent with the structure and dynamics of the species-diverse rainforest community samples as large as 50 ha censused in the CTFS plot network. According to the theory, seed dispersal limitations should also influence speciation rates and diversification

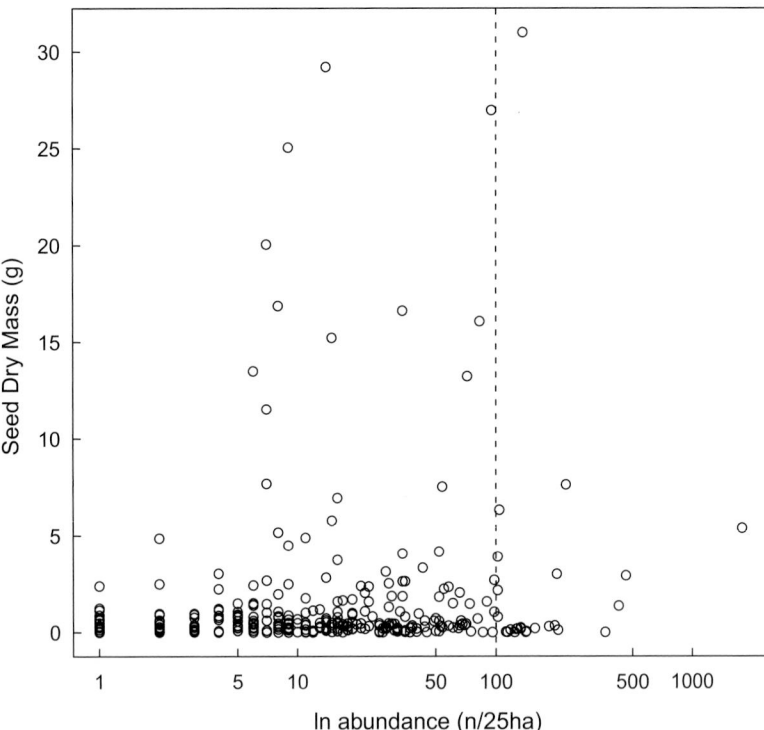

Fig. 5.36 Few Large-seeded tree species occur in high densities in the Yasuni 25 ha CTFS plot, Ecuadorian Amazon. (Data of N. Garwood plotted by S. Queenborough, unpubl. 2010)

within continuous landscapes (Chapter 6). Local fortuitous extinction would exceed competitive extinction in frequency, leading to local populations potentially becoming isolated long enough to become genetically distinct, even to speciate. Such random patterns of species distribution could obscure the coinciding ranges of species limited by common ecological barriers to dispersal. Neutral theory thus predicts that rank order of abundance of species in a community should vary unpredictably. What evidence we have suggests the contrary among some of the more abundant, as well as some that persist in low population densities (but see Chapter 7).

5.5. *Dispersal and genetic differentiation*

Sewell Wright[288] showed theoretically that, while very small populations experience inbreeding depression through increasing homozygosity, fairly small populations of freely outcrossing species may maintain high levels of heterozygosity. Isolated but expanding small populations will rapidly diversify, either at random or through differential selection pressure. Formerly, it was predicted that random effects would dominate unless selection was stringent, and that the likelihood of advantageous genes spreading in natural populations was less than 1%. Selection is now known to be much higher.

Trees are characterised by long life spans, high outcrossing rates, and moderately low numbers of highly stable chromosomes (although polyploid series do occur among taxa of some rainforest families). We have seen that levels of heterozygosity have been found to reach levels at least as high as in temperate trees.[289] Based on molecular data, Koichi Kamiya inferred that the populations of four species of *Shorea* (sect. *Brachyptera*) he examined had started small.[290] Periodic coalescence of isolated populations, such as would be expected among common species in continuous landscapes, should lead to sharing of accumulated genetic variation in shared alleles and increasing overall genetic uniformity of a metapopulation. Isolation for long enough to achieve speciation can be brought about by ecological barriers to pollen and seed dispersal, but neutral theory predicts that isolation through fortuitous decline in species abundance, leading to greater distances between surviving populations could lead to speciation through random drift within them. Such speciation might dominate among the predominant low densities of tree species in biodiverse tropical forests. It would, while in process, be manifested as clinal genetic and morphological variation. *Until the development of microsatellite DNA markers and other recent molecular techniques, such patterns of variation have been difficult to analyse, and no investigation has as yet been undertaken in the Asian tropics. They would provide an invaluable test of the neutral theory.* But then, the only likely cause of extinction of a species in perhumid climates, Allee effects, has yet to be observed (see Chapter 7).

I know of no convincing example of clinal morphological variation among tree species occupying continuous habitats, although putatively infraspecific discontinuous variation among species divided by ecological barriers, or occupying diverse habitats, is common (Chapter 6). A useful comparison of genetic diversity within and between isolated and continuous populations has been offered by the deciduous forest tree *Anogeissus dhofarica* of the desert fringe woodlands of southern coastal Arabia. Genetic fingerprinting (by AFLP) revealed no significant differences between the isolated and continuous populations.[291] This was attributed to the relatively short, c.6,000 year duration of fragment isolation since the last period of higher humidity, and the cloning capacity of the tree through root suckering, rather than continuing long-distance pollen dispersal. Such a time scale coincides with the age of many Far Eastern peat swamp forests, and is half the duration since the last Pleistocene low sea level event (Chapter 6). It does imply, therefore, that the time required for evolution of significant clinal genetic variation may be greater than the intervals of constant climate and land connectivity that whole regions have experienced in tropical Asia. However, such variation might still be revealed in putative Pleistocene refugia such as north-west Borneo (Chapter 6).

In contrast, there are several examples of morphologically recognized pairs of allopatric dipterocarp sister species, such as *Shorea seminis* and *S. quadrinervis* in Borneo and *S. sumatrana* and *S. acuminata* in Sumatra and Peninsular Malaysia, respectively. Koichi Kamiya[292] has found that *S. curtisii* occurs as separate haplotypes in Borneo and Peninsular Malaysia, while *S. leprosula* shows deep molecular divergence. These examples would appear to be evidence of vicariant speciation, following the separation of metapopulations of widespread species occupying continuous habitats during Pleistocene pluvial interludes (Chapter 6). *S. quadrinervis* and *S. acuminata* differ in soil preference, but not *S. sumatrana* and *S. seminis*. In both pairs, habitat isolation rather than chance local extinction appears to have separated them.

Shorea leprosula, a thrip-pollinated emergent with small winged fruit is confined to the most continuous habitat, of clay and sandy clay lowland soils (Chapters 3, 6), and is one of the commoner species on them throughout perhumid Sundaland. Study of its population variability has been restricted to the use of isozymes as proxy for evaluating genetic variation. The species is highly heterozygous.[293] We have seen (see section 2.3.1) that mean variation in both polymorphic isozymes and leaf morphology among mature *Shorea leprosula* trees increased with distance up to c.100 m, when variability reached the level of the overall population.[294] Gan Yik Yuen compared genetic variability between population samples of *S. leprosula* up to 100 km apart in Peninsular Malaysia, and found no greater variation between them than within her Pasoh sample.[295]

Lee and colleagues,[296] using RAPD and isozyme markers, then confirmed this conclusion in a wider sample over Peninsular Malaysia, and found little further differentiation in a sample from Lambir National Park (north-east Sarawak). However, isozyme assays have been shown by subsequent more refined techniques, such as microsatellite or chloroplast DNA markers, to underestimate levels of genetic differentiation. They do yield interpretable relative differences, and these results imply, for this once continuously distributed species, that the wind carries enough thrip pollinators and/or enough seed far enough to maintain ongoing genetic exchange over these long distances. That may explain why it is impossible to distinguish, on morphology alone, *S. leprosula* specimens from southern Thailand and Sabah, or Sarawak and northern Sumatra. There is contrasting microsatellite evidence of continuous drift in gene frequencies in mahogany populations in Central America,[297] but this was in a seasonal climate where small changes in rainfall could have resulted in selective change. *The continuous clay soils of lowland perhumid Sundaland - its aseasonal and excessive rainfall effectively assuring uniform climatic conditions (Chapter 1) - may be the only area where the necessary continuity of overall habitat has existed at the regional scale to test for genetic continuity in widespread putative metapopulations.*

Although some partially apomictic dipterocarp species show allopatric differentiation (see section 2.6 above and Chapter 6), the genetic differentiation revealed by isozyme analysis of four wild populations of the partially triploid riverbank species *Hopea odorata* was unrelated to their distance apart.[298] A similar examination of 15 seedling families, each from a single tree, was made for the diploid riparian *Shorea macrophylla,* whose large short-winged fruits float in water.[299] Trees were sampled from five localities in West and Central Sarawak. The proportion of polymorphic loci identified was similar to *H. odorata* (41% and 43% respectively). In this case, though, genetic differentiation was greatest across the basins of the Lupar and Rajang Rivers, consistent with overall tree species' biogeographic patterns (Chapter 6).

6. Other third parties

6.1. *Herbivory*

Herbivores, from insect larvae to elephants, may influence plant competitiveness by reducing their photosynthetic capacity, but this will only affect species competitiveness if the herbivore is relatively specific to it, unless the plant is already rare. Elephants are well known to prefer some browse over others, such that they can influence the demography of the browsed species (Box 5.8). Harvesting of tree juveniles by pigs for nest building is a major cause of mortality in some forests (Box 5.9), but appears fairly non-specific. Most specific

herbivores are insects. The diversification of the major insect groups from the Cretaceous onwards is broadly correlated with the diversification of flowering plants and of their chemical defences,[300] as well as their attractants.

Many examples of variably host-specific insect herbivores exist in temperate forests, such as the woolly adelgid (*Adelges tsugae*) on the eastern hemlock (*Tsuga canadensis*), or the less specific gypsy moth (*Lymantria dispar*) larva on oaks. No definite example of herbivore-induced species-specific reduced survivorship or mortality has yet been recorded from species-diverse Asian rainforests, though they may be expected to exist. During the 1960s, sharp-edged canopy gaps, sometimes more than 1 km in diameter, appeared in monodominant *Shorea albida* stands in Sarawak and were attributed by villagers to a hairy caterpillar (Plate 5.16c). These gaps might have been expected to expand and eventually eliminate these unique and anomalous monodominant stands, but they did not continue, perhaps because the herbivore itself was attacked. They again underline the extraordinary unpredictability of plant-herbivore and plant-pathogen interactions.

Longicorn beetles (Cerambycidae) were a particular collecting interest of Alfred Russel Wallace, and are of extraordinary diversity in tropical Asia. Host specificity varies among species, though narrow specificity is unlikely[301] – as is currently exemplified by the introduced Asian longhorn beetle, *Anoplophora glabripennis,* in North America. The widespread generalist *Hoplocerambyx spinicornis* becomes abundant in the pure stands of sal created by foresters, killing larger trees and reducing their density.[302] Overall, though, recent observations indicate that host specificity of herbivores is no higher in tropical rainforest than in temperate forests, and that the high species diversity of insects in tropical forests is due to the high diversity of their plant hosts alone.[303] Lepš and colleagues[304] found that the relatedness of host plants explained 56% of variation in herbivorous insect diversity in New Guinea, whereas habitat (including succession) explained only 4%. This implies, again, that host specificity is at a higher systematic level than species.

Insect herbivory is most intense at the height of leaf flush immediately before and during the monsoon in deciduous forest, increasing during the season as insect numbers build.[305] Here, insect herbivory has been found to be most intense on the most abundant host species.

It is well known that the abundance of browsing and grazing mammals and their predators is relatively low in equatorial rainforests, and increases with increasing seasonality. It reaches its maximum in deciduous forests, where periodic fire supports a partially herbaceous field layer and abundant young sucker shoots, while *Saccharum spontaneum* and other dense tall grasses occupy floodplains unfavourable to tree establishment.[306] There, elephant, rhinoceros and ungulates including gaur, wild cattle and several species of deer are (or

Plate 5.16 Predator and pathogen mortality. **a,** Dead mature *Mesua ferrea* in the Sinharaja CTFS 25 ha plot; **b,** Dead *Cinnamomum ovalifolium*, leaves dead but still attached, Horton Plains; **c,** Satellite image of regenerating peat swamp forest in a gap attributed to crown predation of the monodominant emergent *Shorea albida* by a hairy caterpillar (see Anderson 1961); canopy crown shyness also viewed from the air. (**a, b,** H.H.; **c** Google Earth)

have been) major browsers of tree juveniles (Plate 5.17). Tree juveniles are anyway reduced by fire, but are well adapted here to coppice (Chapter 2). Browsing stimulates further above-ground growth, but reduces long-term root extension below ground.[307] Here, any density-dependent seedling mortality due to host specific herbivores and predators is obscured by generalised causes.[308]

Mammalian herbivory has also been limited in Asia, in comparison with Africa, by the smaller area of natural savanna and grassland (Chapter 2).[309] Today, human-induced fire has created widespread species-poor savanna. Overgrazing by cattle is well-known to lead to the increase of unpalatable plant species, such as *Lantana, Chromolaena* and the native weedy shrub *Calotropis procera* (Asclepiadaceae) in tropical dry zone Asian grasslands and open woodlands. Mammalian browsing may therefore reduce rather than increase tree species diversity overall. No evidence so far exists for any increase in numbers of unpalatable tree species in heavily browsed Asian closed-canopy forests.

Almost ubiquitous over-hunting has further reduced indigenous mammal biomass in Indo-Burmese forests, and may be leading to changes in forest succession.[310] Insect herbivory does greatly increase, though, during the exceptionally heavy leaf flushing that follows the end of ENSO droughts.[311]

Plate 5.17 Some other browsers. **a,** Gaur browsing *Lantana,* Mudumalai; **b,** Sambur browsing *Lantana,* Mudumalai; **c,** *Ardisia* leading shoot browsed by deer, Huai Kha Khaeng; **d,** What orang utan eat between mast years: bark, bamboo foliage, palm hearts, available fruits. Gunung Palung National Park, West Kalimantan; **e,** Two young orang utans stripping bark; **f,** Orang utan predating *Dipterocarpus sublamellatus* fruit. (**a–c** H.H.; **d–f** T. Laman)

Box 5.8

An Asian particular: the Asian elephant, *Elephas maximus*

The elephant, by far the largest of Asian browsing and grazing mammals, is justly celebrated for its intelligence and its calm dignity – maintained most of the time! In the Hindu-Buddhist tradition the elephant deity, Ganesh, is the spiritual manifestation of wisdom with humility (the lofty condition to which we researchers aspire). The ecology of the Asian elephant has been the subject of ongoing research by Raman Sukumar, author of a magisterial monograph on both African and Asian species.[312] I have derived much of the following summary from his texts.

Elephants, as large-bodied herbivores, are adapted to eat foods that are available in abundance. They must therefore be generalists, eating in bulk to gain the ingredients needed to sustain health. Elephants occur from the desert fringes to the equatorial rainforests, usually below 1,000 m; the current surviving species once occurred in warm temperate and Mediterranean climates. They range widely according to food availability. Their most provident habitat is tall deciduous forest. In Peninsular Malaysian rainforests, they have been found to consume as many as 400 plant species, 350 of which were palms and rattans.[313] The generally low nutrient content of this foliage, however, means that elephant population densities are usually much higher in secondary forests and riverine rainforest habitats: There they find many pioneer and successional trees whose foliage is comparatively rich in nitrogen and whose fruits are fleshy and nutritious. Today, the

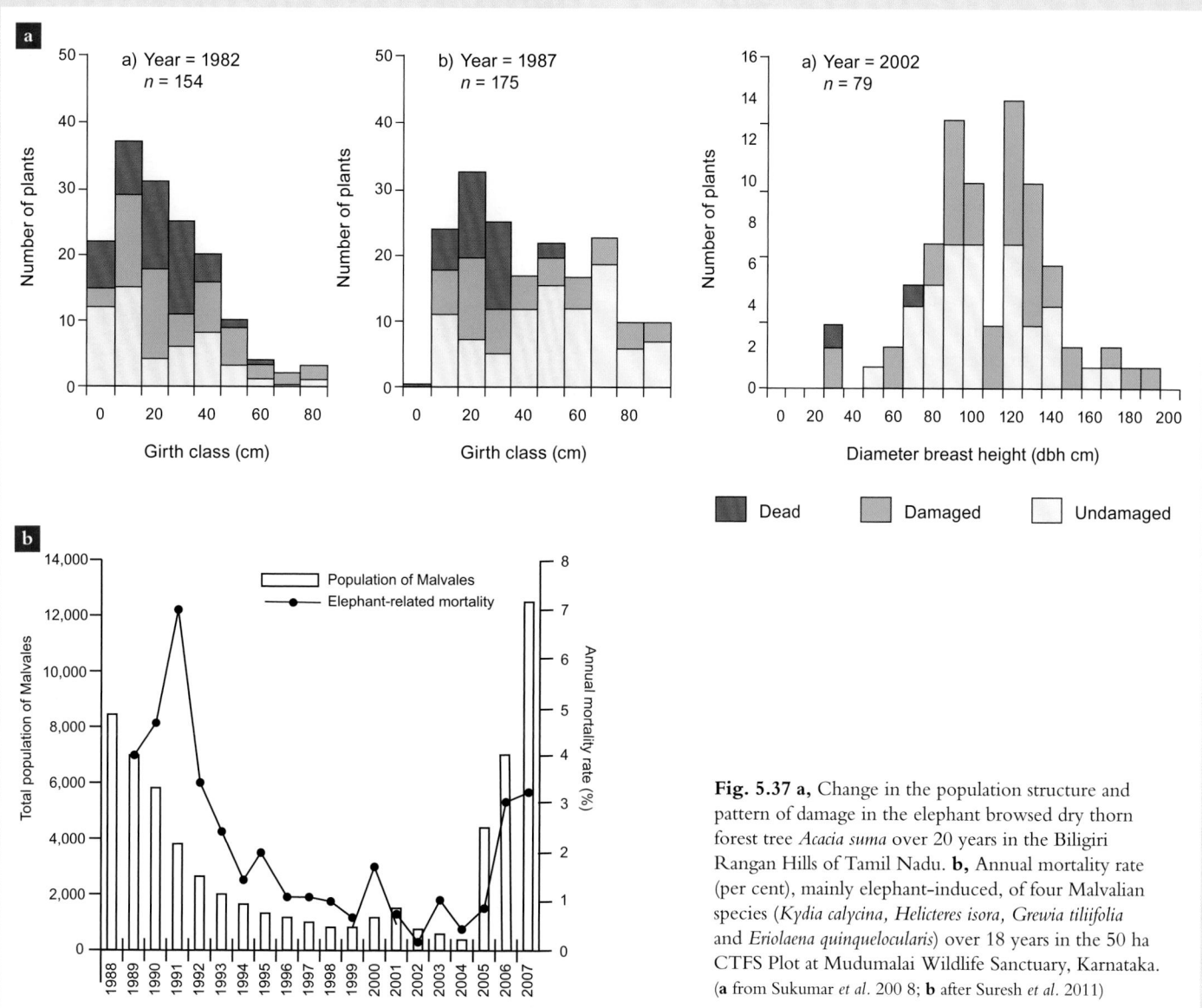

Fig. 5.37 a, Change in the population structure and pattern of damage in the elephant browsed dry thorn forest tree *Acacia suma* over 20 years in the Biligiri Rangan Hills of Tamil Nadu. **b,** Annual mortality rate (per cent), mainly elephant-induced, of four Malvalian species (*Kydia calycina, Helicteres isora, Grewia tiliifolia* and *Eriolaena quinquelocularis*) over 18 years in the 50 ha CTFS Plot at Mudumalai Wildlife Sanctuary, Karnataka. (**a** from Sukumar *et al.* 200 8; **b** after Suresh *et al.* 2011)

Bornean population is confined to the fertile lowland soils of the north-east, yet they were formerly widespread in Sumatra, perhaps because there were wider areas of fertile soil where forest forage was of good quality.[314] The former Bombay Burma Timber Company attempted to log MDF in Central Sarawak with their customary elephant gangs, transferred from Burma in the 1950s, but was forced to abandon the venture owing to a lack of forage in this region of leached sediment-derived soils.

Elephants must consume a staggering 1.5–2% of their own body weight in forage daily – or 60 kg fresh weight (34–44 kg dry weight) – over half of which is usually browse. A herd of 200 individuals has been estimated to consume 22,800 t fresh weight of vegetation from 20 km² per annum. In the dry zone forests of South Asia, elephants live mostly on basal grass shoots during the monsoon, while in tall deciduous forest, to which they retreat during the dry season, they concentrate on tree bark and forage. They must migrate between the forest types, and their home ranges may thus reach 3,900 km².

Browse is essential, for it supplies more than 70% of the collagen required as well as essential amino acids. Bark is high in many nutrients, though grass provides more calcium. In South Asian deciduous forests, elephants may consume up to 112 plant species, but concentrate on about 25. These include Leguminosae, and subcanopy trees and shrubs especially in the Malvales, notably *Kydia calycina* and *Helicteres isora* (Tiliaceae), also *Eriolaena quinquelocularis* and *Grewia tiliifolia*. Other preferred foods include palms, bamboos and grasses, and sedges. They have been found to decimate, in succession, populations of *Kydia* and *Helicteres* and, in dry thorn forest, *Acacia suma* (Fig. 5.37) (see Chapter 2). In one season in the Biligiri Rangan Hills, north-west Tamil Nadu, *Kydia* and *Grewia* suffered 80% damage, or as much as 15% mortality among juveniles under 10 cm in diameter. But there is evidence that elephants move from one preferred forage to another over several years, resulting in a pattern of cyclical recovery recalling the Lotka-Volterra predator-prey cycle but apparently more erratic: what has been called the 'Caughley cycle' after its proposer. This leads to greatly varying inter-annual elephant densities, which are also influenced by fluctuations in local climate and fire frequency. In drier regions, elephants have less choice and increasingly depend on grass, and on the nitrogen-rich foliage of woody legumes including *Acacia* and *Albizia,* and *Ficus*.

Major sources of elephant forage are the larger bamboos, including *Dendrocalamus* and *Bambusa,* most abundant in tall deciduous and semi-evergreen forests, and palm leaves. Bamboo seedlings and inflorescences are poisonous to them: in the Tamil epic, Sangam, female elephants that fed on bamboo seedlings aborted their calves. Instead, elephants forage young foliage growing on culms and clumps, and this may comprise one fifth of their diet. *Bambusa bambos* early hardens

Plate 5.18 Elephants. **a,** Browsing the Mudumalai understorey; **b,** *Kydia calycina* with bark stripped by elephant (Mary Ashton as scale); **c,** *Helicteres isora,* a favourite browse of elephants, regenerating in 2008 (H.S. Dattaraja as scale); **d,** Elephant damage in lower montane woodland, Nam Nao National Park, Thailand. The fallen tree at front is *Lithocarpus lindleyanus*. Hans Hazebroek photographing. (**a** P.A.; **b** P.A; **c, d** Cynthia Lobato)

axillary shoots into hard sharp spears, at elephant height, on its tall culms.

The potentially critical ecological role that elephant browsing might have played before the herds were reduced or eliminated in seasonal evergreen forests has been too little considered. Bamboo in seasonal evergreen and tall deciduous forests, and the stemless palm bertam (*Eugeissona tristis*) in Peninsular Malaysian seraya (*Shorea curtisii*) MDF can severely impede regeneration after logging. Doubtless, elephant browsing would have formerly limited their competitiveness.

Elephants eat the fruits of many species, and are important dispersers of large-seeded trees. Asian elephants do little harm to canopy trees and only occasionally push over smaller trees, often to strip their bark (Plate 5.18b,d). In evergreen forests especially, their apparently indiscriminate depredations, often pushing over trees for no obvious purpose, can lead to semi-permanent patches of secondary forest or even grassland, – which they happen to favour.

Box 5.9

An Asian particular: wild boar

Pigs of the genus *Sus* are restricted in the wild to the Old World, being replaced in the Neotropics by peccaries. Pigs are omnivorous, and are the major predators both of seeds, which they must occasionally also inadvertently disperse, and saplings (Plate 5.19).

Sus scrofa, the domestic pig of temperate Eurasia, extends as an indigenous species throughout the continental Asian tropics, but in the Malesian archipelago is found apparently only in domestication.[315] The native Peninsular Malaysian *Sus cristatus* is sedentary, roaming only locally. Where not hunted, they may be abundant. At Pasoh, where a forest remnant of 1,100 ha is surrounded on three sides by oil palm plantations, the pigs now possess a ready subsidy during inter-mast famine years. This has resulted in high densities in the forest where the sows give birth, and much seedling and sapling predation.[316] Kalan Ickes, in an enclosure experiment, found that the density of saplings more than 1 m tall was three times greater inside than out after three years, and taller on average. Smaller juveniles were unaffected. He found that saplings less than 3m tall would be snapped to make crèches, which sows prepare twice annually (Plate 5.19a). There are two crèches per hectare, amounting to over 170,000 saplings damaged per km² y⁻¹. Ickes found that most did resprout but grew slowly in the shade of the understorey; yet one third died within three years compared with 9% of an undamaged control sample. Canopy vine and tree species, especially dipterocarps and Euphorbiaceae, had

Plate 5.19 Wild boar (*Sus scrofa*) and another seedling browser. **a,** Boar crèche, Huai Kha Khaeng; **b,** Soil rootling by boar for worms and truffles; **c,** Boar wallow; **d,** Langurs feasting on sal (*Shorea robusta*) seedlings, Similipal National Park, Odisha. (**a, b, d** H.H.; **c** E. Soepadmo)

lowest survival rates. Soil grubbing and seed predation were also important (Plate 5.19b; see also Fig. 3.14). The overall effect is likely to eventually change the dynamics and stand structure of the Pasoh forest to one more similar to seasonal evergreen forest.[317] (Chapter 2).

The bearded pig of Borneo, *Sus barbatus*, was celebrated for its migrations[318] (Fig. 5.38). *To observe hundreds of these stocky creatures, the boars with fearsome incisors, hurtling down a steep slope and across the rocky bed of a narrow inland torrent — like a crowd trying to catch the last bus — was one of the most thrilling and signal sights of inland Borneo. But now they are in pathetic decline, as the forests have been converted or trashed by loggers and* overhunted. Whereas the lowland forests experience mast fruiting years, the lower montane forests are less affected and said to reproduce twice annually.[319] From March–May the acorns are falling abundantly in the hills. Hordes of pigs migrate tens or hundreds of kilometres up from the lowlands, fording the cataracts in their hundreds in the quest for a feast. *Sengelang*[320] *told me that this is the time to shoot the sows, who are waxing fat and nubile as their scrawny lovers give them the best Fagaceous treats. But by June the sows are pregnant, less lovely and themselves becoming scrawny, as their former admirers lose interest ('just like us'): it is their turn to go partying but only, it seems, if there is a mast.*

Fig. 5.38 One former mass migration of the bearded pig, *Sus barbatus*, during the fruiting of oaks in the mountains of north-east Sarawak. Green dots: lower montane forest rich in Fagaceae; black dots: longhouses. (I.H. Baillie del. after Caldecott & Caldecott 1985)

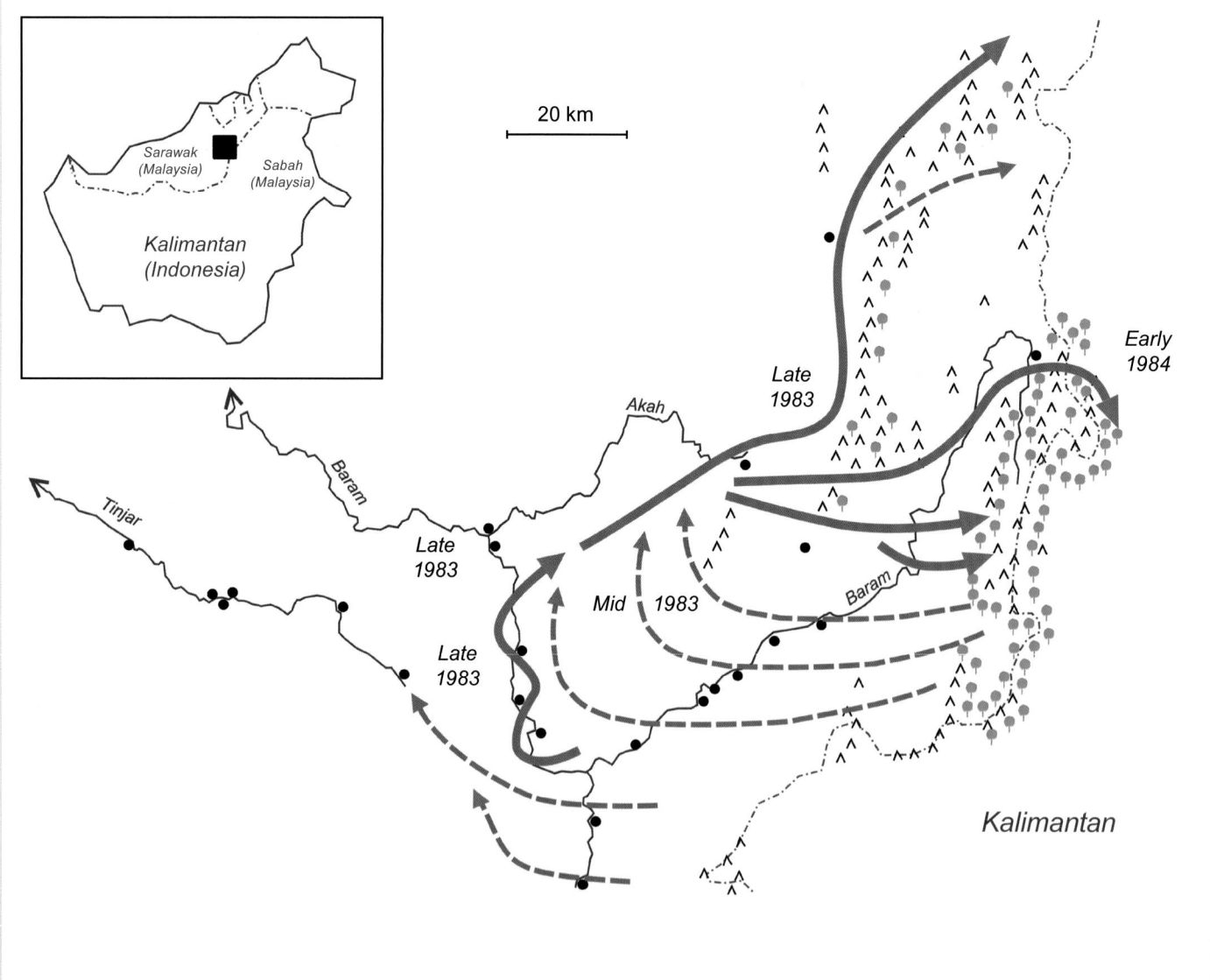

6.2. *Insects as defenders*

Many tree species possess multicellular secretory glands – extrafloral nectaries – on leaves and young shoots, or other attractants to insects that are predators on insect herbivores.[321] Notable among these 'defender' predators are ants. Dipterocarp seedlings especially, possess extrafloral nectaries which, secreting a sugary liquid, attracts ants that defend the young leaves (Plate 5.20a). The nectaries dry up as the leaves mature. In *Shorea acuminata* and perhaps other species, the stipules, which are cupped and do not fall immediately, similarly secrete and thereby attract aphids, which the ants 'milk'.

Among these symbiotic relationships, those between ants of the *Decacrema* clade of the genus *Crematogaster* and pioneer species of *Macaranga* are the best known in Asia, thanks to the work of Fiala, Itino and Quek Swee Peck and their colleagues.[322] All *Macaranga* possess extrafloral nectaries along their leaf margins, but these only secrete sugars in the non-myrmecophytic species that lack strict mutualism and attract a diversity of ants. The mutualistic myrmecophytic *Macaranga* are populated by single *Crematogaster* ant taxa, which obtain their sugars indirectly by milking aphid colonies housed within the host trunk.

Some tree species, including *Macaranga,* attract ants by producing energetically costly multicellular oily food bodies[323] (Plate 5.20b,c). However, the mutualistic gain is linked with a reduction in other defences, chemical and physical (such as leaf toughness), compared with other species.[324] *Crematogaster* ants harvest the food bodies to feed their aphids. Itioka and

co-workers[325] at Lambir found declining ratios of ant/plant biomass in *Macaranga winkleri, M. trachyphylla* and *M. beccariana* respectively. Of the three species, *M. winkleri* occupies the most fertile clay loams and presents the most abundant food bodies. *M. beccariana,* producing the least, occupies the least fertile yellow-sandy soils and occurs at the margin of *kerangas,* while *M. trachyphylla* occupies an intermediate position. The ant entities – 'biotypes' – are highly host-specific obligate symbionts, identifiable only on molecular evidence, each dependent on local populations of a *Macaranga* species.

Most of the myrmecophytic *Macaranga* species are found in two sections, *Pachystemon* and *Pruinosae*. They differ in two respects important to the symbiotic relationship. First, they either have food bodies hidden within stipules on their abaxial surfaces and develop hollow stems with domatia which attract the ants (*Pachystemon*), or they have food bodies exposed on stipular adaxial surfaces while the ants burrow into solid stems, which stimulates domatia (*Pruinosae*). Second, either their stems are waxy and difficult to climb, or they are non-waxy.

The ants, notwithstanding their exceptionally high host specificity, were only found to have co-diversified at a broad level. Basal *Decacrema* ant taxa are associated with basal *Pachystemon* species, which are the *Macaranga* with waxy stems. *Decacrema* later invaded the solid-stemmed *Pruinosae*. Molecular phylogenies showed that co-speciation was subsequently low within these groups, the ant biotypes apparently repeatedly invading the *Macaranga* populations opportunistically, at a level varying locally among host species.

Plate 5.20 Commensals and symbionts. **a,** *Dipterocarpus tuberculatus* seedling leaf with sugar glands and a defending ant, Mae Ping National Park, north-west Thailand (scale: 2/1); **b,** Lipid bodies within the stipule of a *Macaranga* sect. *Pachystemon,* Lambir National Park, Sarawak (scale: 5/1); **c,** Hollow stem of *Macaranga* section *Pruinosae* with commensal *Crematogaster* ants storing lipid bodies (scale: 4/1); **d,** Rooted cuttings of *Shorea* with ectomycorrhiza-infected roots (scale: 1/5); **e,** *Scleroderma* ectomycorrhiza-infected rootlets of *Shorea leprosula* (scale: 5/1); **f,** *Russula* ectomycorrhiza-infected rootlets of *Shorea leprosula* (scale: 5/1). (**a** H.H.; **b, c** C. Ziegler; **d** W. Smits; **e, f** S.S. Lee)

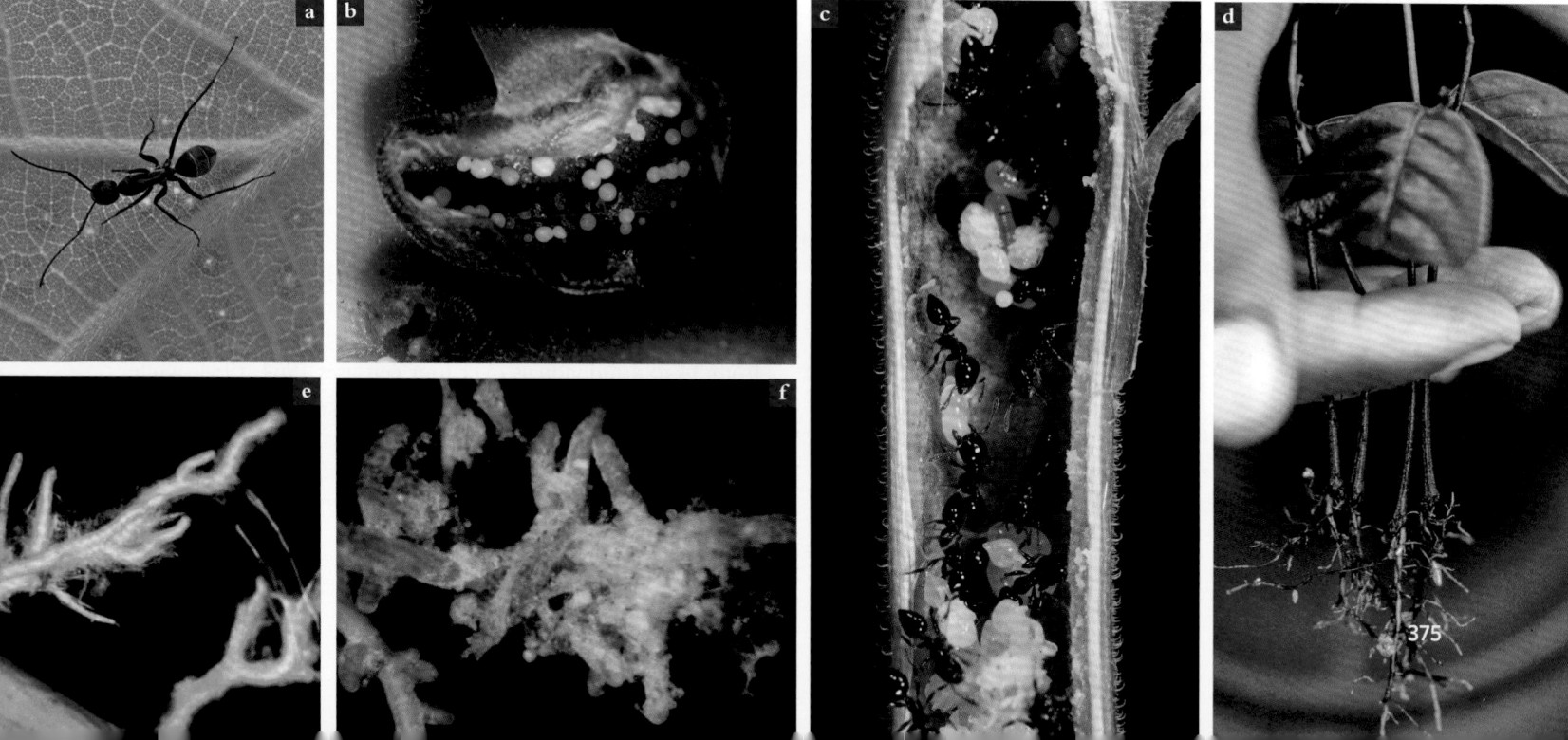

6.3. *Insect predation of seeds and seedlings*

The entomologist Dan Janzen and marine ecologist Joe Connell independently and simultaneously hypothesised that seed and seedling predation may be a major cause of tree species diversity in tropical forests[326] (Chapter 7). They argued that host-specific predators would preferentially consume seeds and seedlings where they are easiest to locate, that is where they are found in high density. This would then reduce density, opening new vacant space for species that they do not attack.

An overall pattern of species' juvenile mortality positively correlated with their spatial density has been demonstrated in both Neotropical and Asian forests.[327] Few cases of *consistent* density-dependent juvenile mortality have been identified in Asian tropical trees, although it has been inferred in *Swintonia schwenkii*,[328] and density-dependent predation inferred in some dipterocarps.[329] Few if any vertebrate seed predators or seedling browsers are host-specific. Janzen was studying Bruchid beetles, which are host-specific predators of leguminous trees.[330] So far though, few other host-specific insect seed or seedling predators have been examined. Although they may be expected to exist for tree taxa with chemically defended seeds, their influence on tree population fecundity and demography remains unknown. In species-diverse MDF, the lean periods between mast fruiting years set strict limits on predator population numbers and may severely limit host-specific seed predation among tree species that mainly or only fruit at these times. Dipterocarp seeds are predated by several species of weevil, none of which are host-specific (Box 5.1).[331]

Seed predation by weevils may nevertheless be a major cause of reduced fecundity of *Dipterocarpus* and other species that participate in the annual flowering of dipterocarps in the seasonal tropics[332] (Box 5.1). It may also lead to failure of out-of-mast fruiting in the perhumid tropics, including among species such as *D. crinitus* that frequently flower out of season. Insect predation levels fluctuate from year to year, and they do increase during mast years.[333] But the *proportion* of seed destroyed by beetles, like that of vertebrates, is reduced in mast fruiting years as the beetles clearly have no means of increasing their numbers prior to the event. *Comparative studies between forests are needed, for which the CTFS plots are well suited.* I conclude that predation of seeds and seedlings may cause local extinctions, but is unlikely to instigate more widespread species extinctions in Asian forests, nor to be a primary cause of density-dependent, species specific, seed mortality.

6.4. *Pathogens*

The demographic impact of individual species-specific pathogens remains unexplored in Asian forests. Current empirical and experimental research in the Neotropics implies that, by

preventing tree species' population densities from exceeding low thresholds – as stipulated by the Janzen-Connell hypothesis – it may be tree species-specific pathogens rather than predators that provide the recurrent unoccupied habitat spaces into which great numbers of species can fit.[334] Conversely, this evidence warns that attempts to grow native tree species in densities greater than the greatest observed in nature may eventually be fraught with difficulties (Chapter 9). In contrast, phenolic chemical defences may be positively associated with leaf toughness and fibre, which reduce *general* species rates of herbivory.[335]

Pathogen-induced density-dependent mortality appears to be overwhelmingly concentrated among juveniles. Cam Webb and David Peart[336] found density-dependent juvenile mortality in MDF at Gunung Palung National Park (West Kalimantan). Halton Peters[337] and Wills and colleagues[338] showed that 80% of species in the Pasoh 50 ha plot experience juvenile density-dependent mortality (that is, of individuals 1–2 cm dbh), which is the same proportion as in the plot in less-rich semi-evergreen forest at Barro Colorado Island (Panama).

Lisa Comita[339] has now shown that the strength of density-dependent mortality increases the lower the tree species' population density, implying that pathogens may be the leading determinant of consistent rank orders of abundance observed in dipterocarp forests.[340] The implication of this for neutral versus niche-specific explanations for tree species co-existence in hyperdiverse is discussed in Chapter 7.

Mass mortality among mature trees of a single species, such as occurred among temperate elms and the American chestnut, has seldom been observed in tropical forests. One striking exception occurred in c.2001, in and around the CTFS tree demography plot at Sinharaja (Sri Lanka). There, the most abundant species is the canopy tree *Mesua nagassarium* (Plate 5.16a), a species strongly concentrated on high ridges where it is locally dominant. In that year over one hundred mature individuals died in and near the 25 ha plot, but no more juveniles than would normally be expected. A sister species (*Mesua ferrea*, only slightly less abundant and concentrated in adjacent upper gullies) appeared unaffected (Fig. 3.3). The cause of mortality is unknown, but a species-specific pathogen is suggested. A similar wave of mortality among mature trees seemed to be occurring among the upper montane *Cinnamomum ovalifolium* on the Horton Plains (Sri Lanka) in 2007 (Plate 5.16b). In four 0.6 ha MDF plots spread over 1 km in Bako National Park (Sarawak), the number of individuals over 1.7 cm dbh of the canopy tree *Pimelodendron griffithianum* (Euphorbiaceae) declined from 34 to 4 individuals over four censuses 1964–1984, while the subcanopy *Cleistanthus pseudopodocarpus* (Euphorbiaceae) declined from 15 to 0.[341] Both have explosively dispersed seed, and individuals are more or less tightly clumped (as in Fig. 5.33). Such clumps, like flocks of birds, may reduce periodic mortality by increasing distance between clumps compared with individuals of a randomly dispersed population of the same density (provided co-occurring species suffer different causes of mortality), but 25 years of observation at Pasoh has not as yet revealed an example.

These observations point to a major research priority, until now hardly addressed in Asian forests. The paucity of studies of pathogens in species-rich rainforest is hardly surprising. It is known from temperate examples that their numbers fluctuate and their appearance is unpredictable. This is because the pathogens are born by vectors, usually insects, which have their own predators and parasite populations. The latter increase as the local host population increases, but with a time lag, so that in some cases every host individual is eventually infected, and its local population is consumed to local extinction: the 'Lotka-Volterra' effect.[342] Pathogens, like pollen or seeds, are dispersed mostly within the local host population, yet occasionally over longer distances to other populations. In this way, mutual extinction is generally avoided. In the largely windless climate of species-rich forests, neither freely dispersing spores nor vectors will often be blown off-course. This could lead to a peak in the mean between-tree distances of the species, irrespective of the species richness of the forest (Fig. 7.8). The trade winds and cyclones of the seasonal tropics will lead to increasing dispersal distances both of fungal spores and their insect vectors.

6.5. *A crucial symbiont*

Mycorrhizae are fungi that form close cellular associations with tree roots, allowing translocation of some minerals (notably phosphorus) in solution and increasing water uptake.[343] In exchange, they obtain carbon in the form of sugars synthesised by their photosynthesising hosts, most of which benefit from the full sun of the canopy at maturity. The fungal hyphae vastly increase the effective absorbing surface area available to roots, and thus the volume of exploitable soil. Mycorrhizal hyphae can effect translocation of nutrients between roots.

Ectotrophic mycorrhizae (that is, Basidiomycete root symbionts whose hyphae do not penetrate living root cells) are considered more host-specific than the vesicular arbuscular mycorrhizae (V.A.) almost universal among tropical trees.[344] Although still unclear owing to the technical difficulties posed by their study, differences in mycorrhizal species, or species associations, may lead to interspecific differences in translocation efficiency, and may thus restrict translocation to conspecific individuals, in particular to parent trees and juveniles. Ectotrophic mycorrhizae are found on all conifers and many temperate tree species, but among tropical families they are predominant only in Caesalpinioid legumes, Dipterocarpaceae, Fagaceae and Gymnosperms (including *Gnetum*).[345] These angiosperm families all have mast fruiting patterns in perhumid

Asia, and generally limited seed dispersal. This may serve to facilitate fungal infection of the seedling roots, including direct infection from mother tree mycorrhizal mycelia.[346]

Ectotrophic mycorrhizae are also considered more efficacious than vesicular arbuscular mycorrhizae in water and nutrient uptake. Higher uptake therefore energetic efficiency, combined with greater host specificity, could explain the monodominance of many Caesalpinioid trees in Neotropical and African forests, where they usually occur on low-nutrient soils.[347] Newbery has hypothesised that their ectomycorrhizal fungal associations have permitted Caesalpinioids to accumulate the photosynthate and nutrient surplus required for mast fruiting. This has in turn given them the exceptional juvenile survivorship that leads to dominance. Indeed, in the tropics worldwide, ectotrophic mycorrhizal tree taxa tend toward single-species, or single family in the case of Dipterocarpaceae in MDF, canopy dominance in seasonal and low-nutrient habitats (Box 5.1). The possible role of mycorrhizae in the exceptional species diversity of dipterocarps is discussed in chapter 7 section 2.5.4.

It has been suggested that mycorrhizal infection may also convey resistance to pathogenic fungal infection.[348] This could be achieved indirectly, by co-evolution with the tree of pathogen-specific chemical defences.

Dipterocarps are almost ubiquitous in the Asian wet tropical lowland evergreen and semi-evergreen forests, but they reach their apex of species richness on leached humult yellow-red soils in perhumid Sundaland (Chapters 3, 7). This richness is correlated with an extraordinary diversity of ectotrophic mycorrhizal mushroom species there (Plate 5.20 d–f). Indeed, the species richness of the basidiomycetes in the perhumid Asian lowlands as a whole goes far beyond their diversity anywhere else in the tropics. Kabir Peay and colleagues,[349] using molecular methods of hyphal identification, analysed ectotrophic mycorrhizal communities across the ecotone from clay to sandy soils within the 52 ha CTFS plot at Lambir National Park (north-east Sarawak). They found that their generic representation and taxonomic diversity resembled that in temperate forests. Russulales were most diverse with 30 species, then Boletales, Agaricales, Thelephorales, and Cantharellales with 12 species, in order of diversity. The less-fertile sandy soil contained 50% more ectotrophic fungal taxa, though the clay soil was more intensively sampled. Between-soil differences explained 25% of the fungal diversity between samples. There is therefore a clear shift in mycorrhizal representation across the ecotone. Whether that is due to fungal response to soil conditions, or to a shift in the soils preferences of host trees and thus their specific mycorrhizal associates, cannot yet be ascertained. That crucial question, too, will be difficult to answer by experiment.

Mycorrhizae may thus be a primary cause both of the

canopy dominance of the Dipterocarpaceae and Fagaceae in tropical montane and temperate broadleaved forests (Chapters 3, 4), and also their exceptional species richness (Box 5.1). This proposition is, unfortunately, exceptionally difficult to test. While fungal pathogens are air-borne above the canopy or dispersed by flying arthropods, mycorrhizae, living in soil and bearing carpophores that arise from it in the windless understorey, have limited dispersal opportunities in closed rainforests. McGuire showed experimentally that seedling infection by mycorrhizae was positively density-dependent, and occurs through sharing of mycelia when seedlings are close enough.[350] Molecular evidence indicates that individual species of ectomycorrhizae co-occur on phylogenetically related hosts,[351] but that the array of fungal species that comprise mycorrhizal associations as a whole may be host-specific. Not only is their diversity enormous, but variance in host species susceptibility to ectomycorrhizal infection is great. Hence, opportunities for diversification of co-evolutionary relationships, especially where soils vary, are nearly endless.

Ectotrophic mycorrhizal symbiosis may thus act in opposition to host-specific pathogenicity, by enhancing positive rather than negative density-dependent survivorship. Density-dependent juvenile mortality has not always been evident among dipterocarp seedlings and, where present, may be due to other factors such as shading by the parent crown,[352] while enhanced performance observed among adjacent conspecifics has been attributed to sharing of host-specific mycorrhizae.[353] *Greater understanding of this enigma would well repay ingenious investigation.*

7. Conclusions

Rainforest tree species maintain high genetic diversity within continuous habitats in their populations, even when restricted to ecological ranges of extraordinary narrowness and precision (Chapters 2, 3). This diversity – widespread apomixis apparently notwithstanding – argues that natural selection has played a dominant role in the dynamics and structure of the forest and in the evolution of its species. That is not to say, though, that chance does not also play a role, as is obvious in the strikingly limited dispersal of most species' seeds and the seedlings resulting from them. The relative roles of chance and the determinism wrought by natural selection will be further addressed in Chapter 7.

Although we may expect that species diversity in tropical rainforests would be the result of highly specific interdependencies between plants (here mostly trees) and other organisms, that expectation has so far only occasionally been supported by rigorous investigation. Among the groups of organisms that directly influence tree species diversity, the most notable are: (1) those that disperse pollen and seed, thereby enabling high genetic diversity; (2) the pathogenic micro-organisms and fungi which, it is becoming clear, are major determinants of community diversity; and (3) the ectotrophic mycorrhizal fungi which appear to have enabled dipterocarps in the lowlands (and Fagaceae in the mountains) to achieve the high levels of canopy dominance so noticeable in Asian tropical forests (therefore predictable high ranking in family, and often species order of abundance), as well as contributing to the evolution of their species richness. Opportunities to rigorously investigate these interrelationships are fast declining, as the continuous lowland forests are fragmented and converted. *I believe them to be among the most promising of all future research questions, as they would provide the ultimate test of the applicability of neutral theory in plant ecology and biogeography.*

References

[1]Corlett & Primack 2011. [2]Janzen 1974. [3]Corlett 2009a; Corlett & Primack 2011. [4]Moffett 1993. [5]P.S. Ashton *et al.* 1995. [6]Moffett 1993. [7]Corner 194. [8]Knudsen & Tollsten 1993. [9]Knudsen & Tollsten 1995. [10]Rogstad 1986. [11]Corner 1954. [12]Dick *et al.* 2007. [13]Bateman 1947. [14]Stebbins 1977. [15]P.S. Ashton 1969. [16]Charlesworth 2007. [17]Bawa 1974; Bawa *et al.* 1982. [18]Chan 1981; Hamrick & Murawski 1990; Kress & Beach 1994; Murawski 1995; C.T. Lee *et al.* 2011. [19]Appanah 1979, 1982; Rogstad 1994. [20]Murawski *et al.* 1994a. [21]Chan 1977. [22]Murawski *et al.* 1994a, b; I.A.U.N. Gunatilleke 1999. [23]Murawski *et al.* 1994a, b. [24]Murawski *et al.* 1994b. [25]Chan 1977. [26]Ghazoul *et al.* 1998. [27]Chan 1977. [28]Chan 1977. [29]Yap 1976. [30]Kenta *et al.* 2004. [31]Gunatilleke *et al.* 2004a. [32]S.L. Lee *et al.* 2000a, b, 2001. [33]Harata *et al.* 2012. [34]Stacy 2001a, b; Stacy *et al.* 2001. [35]Liengsiri *et al.* 1995. [36]Gan *et al.* 1977; Hamrick & Murawski 1990. [37]Grant 1981. [38]Thomas & LaFrankie 1993. [39]Mack 1997. [40]Kevan & Lack 1985; Lack & Kevan 1987. [41]Bawa & Opler 1975; Bawa 1990; Givnish 1980, 1982. [42]Renner & Ricklefs 1995; Renner & Schaefer 2010. [43]Cruden 1977. [44]Armstrong & Irvine 1982. [45]Thomas & LaFrankie 1993. [46]Charlesworth 1984. [47]P.S. Ashton 1982. [48]P.S. Ashton 1980b. [49]Symington 1943. [50]Parkinson 1932; Pooma 2002. [51]I.A.U.N. Gunatilleke, pers. comm. 1995. [52]Kamiya *et al.*, unpubl. ms. [53]K. Kamiya, pers. comm. 2007. [54]Murawski & Hamrick 1991; Bawa 1994. [55]Palmer 1948. [56]Stebbins 1957. [57]Treub 1911; Ha *et al.* 1988a, b; Richards 1990a–c; Richards *et al.* 1996. [58]P.S. Ashton 1982. [59]Treub 1911; Richards 1990a–c; Richards *et al.* 1996. [60]Thomas 1997. [61]Treub 1911; Richards 1990a–c. [62]Tiwary 1926; van der Pijl 1934. [63]Stevens 1980. [64]Grant 1981; Kaur *et al.* 1978, 1986. [65]Kaur *et al.* 1986. [66]Wickneswari *et al.* 1994. [67]Kaur 1977; Kaur *et al.* 1978, 1986. [68]P.S. Ashton 1982. [69]Murawski *et al.* 1994a. [70]Farrell & Mitter 1998; Farrell 1998. [71]Fiala *et al.* 2011. [72]Kato & Inoue 1994. [73]Rogstad 1986, 1994. [74]van der Pijl 1972. [75]Singhakumara *et al.* 2003. [76]Wheelwright 1985b. [77]Chapman *et al.* 2005. [78]Reich & Borchert 1982; Sakai *et al.* 1999. [79]Yasuda *et al.* 1999. [80]Yasuda *et al.* 1999; Numata *et al.* 2003. [81]P.S. Ashton *et al.* 1988; Yasuda *et al.* 1999. [82]Medway 1972; P.S. Ashton *et al.* 1988; Appanah 1993. [83]Inoue & Abdul Hamid 1994, 1997; Sakai *et al.* 1999. [84]Brearley *et al.* 2007. [85]Burslem 2010. [86]P.S. Ashton 1989. [87]Yap & Chan 1990. [88]Ng 1981; Yap & Chan 1990. [89]Sakai *et al.* 2005. [90]Hamann 2004. [91]Heideman 1989. [92]S. Bunyavejchewin, pers. comm. 1999. [93]Roubik 1993. [94]Primack *et al.* 1987a. [95]Thomas & Ickes

1995. [96]Robinson 1935; Matthew 1971. [97]P.S. Ashton 1982. [98]P.S. Ashton 1989. [99]P.S. Ashton 1982; Becker *et al.* 1985; van der Meer *et al.* 2008. [100]S.S. Lee 1998. [101]Richards 1958, 1996. [102]S.S. Lee & Alexander 1994. [103]Troup 1921. [104]Marod *et al.* 2002. [105]S. Bunyavejchewin, pers. comm. 1999. [106]Wood 1956; Cockburn 1974; P.S. Ashton *et al.* 1988; P.S. Ashton 1989. [107]Wood 1956; Cockburn 1974; P.S. Ashton *et al.* 1988; Numata *et al.* 2003. [108]Chapman *et al.* 2005. [109]Appanah 1993. [110]P.S. Ashton *et al.* 1988; Sakai *et al.* 1999; Numata *et al.* 2003. [111]Yasuda *et al.* 1999. [112]Gunatilleke *et al.*, unpubl. ms. [113]Dayanandan *et al.* 1990; Gunatilleke *et al.* 2004a. [114]Maycock *et al.* 2005; C. Maycock, unpubl. ms. [115]Gan 1976; Chan 1977; Gan *et al.* 1977; Chan & Appanah 1980; Murawski & Hamrick 1991; Murawski *et al.* 1994a, b; Kenta *et al.* 2002. [116]Palmer 1948; Pooma 2002. [117]Singh *et al.* 1990. [118]Ichie *et al.* 2005. [119]Kenta *et al.* 2003. [120]Janzen 1974. [121]Knott 1998. [122]Curran 1994; Curran & Leighton 2000. [123]Toye *et al.* 1992; Curran 1994; Curran & Leighton 2000; Curran & Webb 2000; Maycock *et al.* 2005. [124]Curran 1994; Curran & Leighton 2000; Curran & Webb 2000; Blundell & Peart 2004; Maycock *et al.* 2005. [125]Knott 1998; van Schaik *et al.* 1993. [126]Browne 1966, 1968; Singh, Daljeet 1974; Momose *et al.* 1996. [127]Itoh *et al.* 1995a. [128]P.M.S. Ashton, pers. comm. Nov. 2010. [129]Rozaria 1995. [130]P.S. Ashton 1989; Appanah 1990; Curran 1994; Curran & Leighton 2000. [131]Chesson & Warner 1981. [132]Appanah 1979, 1990. [133]Kettle 2012. [134]Jong in P.S. Ashton 1982; Kaur *et al.* 1986. [135]Ghazoul *et al.* 1998; Momose *et al.* 1996. [136]Dayanandan *et al.* 1990. [137]Chan 1977; Chan & Appanah 1980; Appanah & Chan 1981. [138]Moog *et al.* 2002. [139]Moog *et al.* 2002. [140]Kirk 1988. [141]Sakai *et al.* 1999. [142]Hagerup 1950; Kishimoto-Yamada & Itioka 2008. [143]Hagerup 1950; Hagerup & Hagerup 1953. [144]Chan 1977, 1981. [145]Chan 1981. [146]Dayanandan *et al.* 1990. [147]Appanah 1979. [148]Silvertown 1980; Ichie *et al.* 2004. [149]Appanah 1979. [150]Maycock *et al.* 2005. [151]Itoh *et al.* 2003a. [152]Yap 1976; Appanah 1979, 1982, 1990. [153]Kunth 1906. [154]Yap 1989. [155]Kunth 1906. [156]Charlesworth 1984. [157]Rogstad 1986, 1994. [158]Rogstad 1986, 1990. [159]Seethalakshmi *et al.* 1998; Singh 2002. [160]Kurz 1876; Brandis 1897; Kwe-Tu-Wet-Tu 1903; Bradley 1914; Nicholson 1922; Blatter 1929, 1930, Win 1951; Yadav 1963; Janzen 1976a; Simmonds 1980; Gadgil & Prasad 1984; Stott 1986, 1988; Stott *et al.* 1990. [161]Kurz 1876. [162]Brandis 1897. [163]Janzen 1974. [164]Janzen 1976a; Silvertown 1980. [165]Takeda 1995; Keeley & Bond 1999. [166]Bawa *et al.* 1990. [167]Binoy & Matthew 2001; Soehartono & Newton 2001; Singh & Singh 2003. [168]Rangaiah *et al.* 2004; Solomon Raju *et al.* 2004. [169]Nicodemus *et al.* 2005. [170]Janzen 1971. [171]Meyer de Schauensee & Phelps 1978. [172]van der Pijl 1972; Faegri & van der Pijl 1979; Momose *et al.* 1998. [173]Toye *et al.* 1992. [174]Momose *et al.* 1998. [175]Momose *et al.* 1998. [176]Kress & Beach 1994. [177]Kato *et al.* 1993; Yumoto *et al.* 1997; Momose *et al.* 1998; Sakai 2000. [178]Yumoto 2000; Itioka *et al.* 2001. [179]Stevens 1976. [180]Janzen 1974. [181]Kuruvilla 1989. [182]P.S. Ashton 2011. [183]Momose *et al.* 1998. [184]Francis 1994. [185]Momose & Hamid Karim 2005. [186]Troup 1921. [187]Koelmeyer 1960; Sivaraj & Krishnamurthy 1988. [188]Troup 1921; Selwyn & Parthasarathy 2006. [189]Selwyn & Parthasarathy 2006. [190]Kingdon-Ward 1949. [191]Murali & Sukumar 1994; Selwyn & Parthasarathy 2006. [192]Lal *et al.* 2001. [193]Bawa 1974; Bawa & Opler 1975; Bawa & Hadley 1990; Bawa *et al.* 1982. [194]Chavarria Villaseñor 1996. [195]Roubik *et al.* 1995. [196]Peitsch *et al.* 1992. [197]D. Swain, pers. comm. Feb. 2005. [198]van der Pijl 1954. [199]Janzen 1971. [200]S. Appanah, pers. comm. 1975. [201]Appanah 1979, 1990. [202]Eltz 2004. [203]Asah ak Unyong, pers comm. 1958. [204]Itioka *et al.* 2001. [205]Roubik *et al.* 1995. [206]Kenta *et al.* 2002. [207]Appanah 1979, 1993. [208]Gunatilleke & Gunatilleke, unpubl. ms. [209]Inoue & Abdul Hamid 1997. [210]Appanah 1979, 1982. [211]Menzel *et al.* 1989; Peitsch *et al.* 1992. [212]Knudsen & Tollsten 1995. [213]Knudsen & Tollsten 1993. [214]Start 1974;

Gould 1978. [215]Cox 1984. [216]Gould 1977, 1978. [217]Hopkins 1983. [218]Start 1974. [219]Marshall 1983. [220]Francis 1994. [221]Tan *et al.* 1998. [222]Sakai *et al.* 2000. [223]Hodgkison *et al.* 2004. [224]Gould 1978. [225]van der Pijl 1956. [226]Ganesh & Davidar 1997. [227]Medway & McClure 1966. [228]Herre *et al.* 2008. [229]Weiblen 1999, 2001, 2004; Ronsted *et al.* 2008. [230]Harrison 2001. [231]Weiblen 1999, 2004; Herre *et al.* 2007. [232]Harrison & Rasplus 2006; Kochummen & Go 2000. [233]Harrison & Rasplus 2006. [234]Whittaker 1989. [235]Harrison 2001. [236]Janzen 1979; Terborgh 1986. [237]van Schaik 1986; van Schaik *et al.* 1993; Patel 1996; Shanahan & Compton 2000. [238]Laman 1994. [239]Lambert & Marshall 1991. [240]Harrison 2001. [241]Waterman *et al.* 1988. [242]S.N. Rai, pers. comm. Nov. 1989. [243]Laman 1994, 1996a, b; Shanahan & Compton 2001. [244]Shanahan & Compton 2001. [245]Wheelwright 1985a. [246]Grant 1981. [247]Hubbell 2001. [248]Curran & Leighton 2000. [249]Yamada & Suzuki 1999. [250]Grubb 1998; Givnish 1999. [251]Corlett & Primack 2011; Corlett 2002, 2009a, b. [252]P.S. Ashton 1969. [253]Wright *et al.* 1999. [254]Wong *et al.* 2005. [255]Ganesh & Davidar 1997. [256]Janzen 1970. [257]Troup 1921. [258]Nchaji & Plumptre 2003. [259]Prasad *et al.* 2010. [260]Corlett 2002, 2009a, b. [261]Ridley 1930. [262]Ness *et al.* 2004; Pfeiffer *et al.* 2006. [263]Kitamura *et al.* 2004b. [264]Russo 2003; Russo & Augspurger 2004. [265]Yap 1976; Yumoto *et al.* 1998. [266]Chivers 1971, 1972, 1974a, b. [267]Touranont 2000. [268]McConkey 2005. [269]Suresh, pers. comm. Oct. 2008. [270]Muller-Landau *et al.* 2008; Corlett 2009b. [271]Corlett 2009b. [272]Olivier 1978. [273]Dinerstein & Wemmer 1988. [274]Clark 1998. [275]Richardson *et al.* 2001. [276]Becker *et al.* 1985; van der Meer *et al.* 2008. [277]Osada *et al.* 2001a; Suzuki & Ashton 1996. [278]J. Muller, pers. comm. 1960. [279]Webber 1934. [280]Whittaker 1989; Thornton 1996. [281]Harata *et al.* 2012. [282]Konuma *et al.* 2000. [283]P.S. Ashton 1969. [284]P.S. Ashton 1998a. [285]Grubb 1998. [286]Potts *et al.* 2002. [287]Hubbell 2001. [288]Wright 1965. [289]Hamrick & Loveless 1986; Murawski 1995. [290]Kamiya *et al.*, unpubl. ms. [291]Oberprieler *et al.* 2009. [292]Kamiya *et al.* 2012. [293]Gan *et al.* 1977; S.L. Lee *et al.* 2000a, b. [294]Yap 1976; Chan 1977. [295]Yap 1976. [296]S.L. Lee *et al.* 2000a, b, 2001b. [297]Novick *et al.* 2003. [298]Wickneswari *et al.* 1994. [299]Kanzaki *et al.* 1996. [300]Farrell 1998; Farrell & Mitter 1993. [301]Beeson 1941. [302]Tewari & Kala 2007. [303]Fiedler 1998; Heckroth *et al.* 1998; Novotny *et al.* 2002; Novotny & Basset 2005. [304]Lepš *et al.* 2001. [305]Murali & Sukumar 1993, 1994. [306]Dinerstein 1987, 1992. [307]Pandey & Singh 1992. [308]John *et al.* 2002. [309]Janzen 1974. [310]Srikosamatara 1993. [311]Itioka & Yamauti 2004. [312]Sukumar 2011. [313]Olivier 1978. [314]Payne *et al.* 1985 [315]Menon 2009. [316]Baker 1985; Ickes *et al.* 2003. [317]Ickes *et al.* 2003. [318]Pfeffer 1939; Hislop 1949; Anon. 1953; Caldecott & Caldecott 1985. [319]Kimura *et al.* 2001. [320]Sengelanganak Nantah, pers. comm. April 1958. [321]Gadrinab & Belin 1981; Fiala & Linsenmair 1995. [322]Fiala & Maschwitz 1991, 1992; Fiala *et al.* 1999; Quek *et al.* 2007. [323]Heil *et al.* 1997. [324]Nomura *et al.* 2000. [325]Itioka *et al.* 2000. [326]Connell *et al.* 1979; Connell & Lowman 1989. [327]Webb & Peart 2000; Peters 2003. [328]Suzuki & Kohyama 1991. [329]Takeuchi *et al.* 2005; Takeuchi & Nakashizuka 2007. [330]Vir & Jindal 1994. [331]Browne 1966, 1968; Momose *et al.* 1996. [332]Senthilkumar *et al.* 2009. [333]Nakagawa *et al.* 2003, 2005. [334]Augspurger 1983a, b, 1984; Augspurger & Wilkinson 2007. [335]Coley 1983; Coley *et al.* 1985. [336]Webb & Peart 2000. [337]Peters 2003. [338]Wills *et al.* 2004. [339]Comita *et al.* 2010. [340]P.S. Ashton 1998a; Potts *et al.* 2002. [341]Hall 1991. [342]Lotka 1925; Volterra 1926. [343]Goltapeh *et al.* 2008; Brearley 2012. [344]Smits 1992, 1994; McGuire 2007; Horton & Bruns 2001; Singhakumara *et al.* 2003. [345]Onguene & Kuyper 2001. [346]Smits 1992, 1994; McGuire 2007. [347]Richards 1958; Connell & Lowman 1989; Hart 1990; Newbery *et al.* 1998b; D.M. Newbery 2005. [348]Herre *et al.* 2007. [349]Peay *et al.* 2009. [350]McGuire 2007. [351]Horton & Bruns 2001; Yamashita *et al.* 2007, 2012; Brearley 2012. [352]Itoh *et al.* 1995a, 1997; Miyamoto 1998. [353]Stoll & Newbery 2005.

Vateriopsis seychellarum in flower in the
greenhouse of Cruickshank Botanic Garden (P.A.)

Chapter 6

THE PALIMPSEST OF HISTORY, AS REFLECTED IN GEOGRAPHICAL DISTRIBUTIONS

Plants differ from most animals in ways that powerfully affect their geographical distributions. Some of these differences are both basic and obvious: for instance, plants are sessile and have limited means of dispersal. Consequences flowing from these facts were addressed in Chapter 5: plants have evolved a great variety of reproductive systems and strategies, and these are much more diverse among flowering plant genera and families than they are in most animals. Other differences are slightly less obvious but nevertheless have extensive implications. For instance, most tree species are potentially longer-lived than most animals, although vanishingly few individuals ever reach maturity. Those that do, often survive for a century while a very few may exceed a millennium in age (Chapter 2). These characteristics (and others) are associated with high ecological dependence on particular physical habitats (Chapters 2, 3) combined with limited opportunities for direct interspecific competition. As we saw in Chapter 5, one consequence is that natural selection through interspecific competition exerts its influence on plant populations via a significantly more *indirect* set of forces.

Together, these factors result in distinctly different – and often more *local* – patterns of geographical dispersion in plants than in animals. Seeds and pollen are overwhelmingly dispersed locally, though they do occasionally achieve greater distances. The length of time over which the required local habitat conditions of a species persist is therefore vitally important. Historical contingencies leave enduring imprints on plant distributions; and that is the subject of this chapter.

1. Interpreting the past

1.1. *Sources of evidence*

Fossils provide confirmation of the existence of a plant species in a particular place and time. The further back in the geological past one searches, the more the fossils differ from extant plants, and the more difficult it is therefore to grasp their relationship to modern taxa. Fortunately, pollen grains and spores are among the organic remains most resistant to decomposition, particularly if deposited under anaerobic or acid-reacting conditions. Fossil history has been enormously amplified by the analysis of pollen in sedimentary rocks and macrofossils in volcanic deposits. When pollen is deposited in swamps or lakes, for instance, it is generally derived from adjacent sites and can illuminate the floristic composition of neighbouring communities. If pollen yield of the species can be estimated, even community structure can be approximated; but that is only possible in younger deposits, where species are similar to or the same as those now extant.

The fossil record is limited in important respects. Fossils can confirm the minimum age of a species, but rarely a maximum. The record can never be comprehensive, because the micro-conditions favouring fossilisation are extremely rare – other, perhaps, than in the case of volcanic eruptions, whose effluvia may 'freeze' an entire community in time. The ecological conditions for fossilisation are generally limited to habitats where decomposition has been at least partly arrested. In the case of plants, these are found most commonly in the anaerobic conditions of peat bogs, swamps, lakes and marine muds. Certain flowering plant families – unfortunately including several archaic groups – possess thin-walled grains, which decompose rapidly and are therefore missing from the record, or produce pollen indistinguishable from those of gymnosperms.

Plant species origins and the history of migration routes can be inferred by analysis of current distributions in relation to their phylogeny, derived from molecular analysis of their genetic similarities combined with information from the fossil record – the developing science of molecular phylogeography. These data can provide valuable information at all levels of an evolutionary tree, but the proportion of taxa missing from the record through extinction increases the deeper one peers into the past.

It is an often-overlooked fact that migration routes of organisms with restricted capacities for dispersal, like plants, are constrained by more than sea barriers. On continents, equally stringent barriers exist as belts of climate unfavourable for a taxon's survival. Mountain chains my act as barriers, or alternatively as corridors linking temperate and tropical regions. The Andes do this, as does the eastern Himalaya as it trends southeastward into the Malesian 'ring of fire' (Chapter 1). In Asia at present, we can trace extensive geological formations whose dominant soil-cover may similarly act either as barrier or as corridor for plant migration, depending on how well-adapted a particular species is to the soil type. On the other hand, molecular evidence is showing that many plant species – up to 20% of the flora of Africa and the Neotropics – may have arrived there from across the Atlantic, by rare long-distance dispersal events *subsequent* to the drifting-apart of those continents.[1] Yet more remarkable, the Australian baobab (*Adansonia gregorii*) originated from African cousins during the Cenozoic period and spread by long-distance migration of its winged seeds *against* the prevailing easterly trade winds.[2] It is difficult to directly document long-distance dispersal events of pollen or of seeds which usually disperse over short distances. Such dispersal is often grossly underestimated, both in historical biogeography and in evaluating speciation processes – a subject to which I will return below.

Sometimes forgotten, too, is the fact that early arrivals provide the vegetational structure within which late-comers must compete, survive and diversify. Each of these structures is in some measure unique to a particular region, due to the accidents of its particular biogeographic history.

The *youngest* morphologically identifiable entities in the taxonomic hierarchy we derive from the fossil record are the lower-level groupings: species and sub-species. Being relatively young, these taxa were influenced in their patterns of distribution by the relatively recent events that either provided opportunities for migration or caused extinctions. A few individual taxa, of course, are truly ancient: the *Nypa* palm of brackish waterways, a single species now confined to Asian coasts, is recorded from all tropical continents as far back as the Cretaceous (c.70 Ma).[3] Multiple similar species may then have co-existed – and one of them may even have been the species surviving today. But past events have influenced *all* species, and all taxa at least up to family rank, for all taxa are in some measure constrained by ecological conditions. Patterns of migration and diversification driven by recent geological and climatic events are the most clear and most easily interpretable. Events that occurred in the deep past also leave traces in plant geographic patterns, but they are increasingly less distinct. The more ancient the event, the older must be the taxon that retains its imprint; and therefore the higher its position in the organizational hierarchy of life. The most primordial traces of all will be detectable only in sub-families and families.

1.2. *Origins*

There appears to have been a major episode of diversification among the angiosperms during the Maastrichtian age,

toward the end of the Cretaceous (70–65 Ma).[4] The prodigious diversity of the Asian tropical flora – documented earlier than that of the Americas – was once thought to point to Asia as the seat of angiosperm origins.[5] Both the fossil record and contemporary distributions, however, show this to be unlikely, and the Asian tropical angiosperm flora has subsequently been found to have been less diverse than the American. Flowering plants are now known to have originated in the Late Jurassic (c.150 Ma), in a region that remains uncertain, although the claimed earliest reproductive parts recognized as angiospermous from China are Cretaceous (Aptian).[6] Remains of the Magnoliids, which on molecular evidence are basal to all flowering plants, have been found widely in eastern North America, North West Europe and West Africa. But the Eudicots, basal to the dicotyledons and therefore the broadleaved trees,

Family	Lowland tropical	Lower montane tropical	Upper montane tropical	Warm Temperate
Austrobaileyales				
Schizandraceae (incl. Illiciaceae)		+		+
Chloranthaceae	+	+	+	+
Core eudicots				
Winteraceae*			+	+
Aristolochiaceae	+			+
Piperaceae	+	+		+
Calycanthaceae				+
Hernandiaceae	+			
Lauraceae	+	+	+	+
Monimiaceae*	+	+	+	+
Annonaceae	+	+		
Eupomatiaceae				+
Magnoliaceae	+	+	+	+
Myristicaceae	+	+		
Eudicots				
Berberidaceae				+
Circaeasteraceae				+
Eupteliaceae				+
Lardizabalaceae				+
Menispermaceae	+	+		+
Sabiaceae	+	+	+	+
Platanaceae		+		+
Proteaceae*	+	+		+
Trochodendraceae				+
Buxaceae	+			+
	13	12	6	20

Table 6.1 The habitats and ranges of basal flowering plant families in tropical and temperate Asia and Australasia. (Groups from Angiosperm Phylogeny Group 2009)

originated and early diversified during the Cretaceous[7] and were almost certainly concentrated in the tropical (possibly seasonal tropical) climates of western Gondwana – that is, South America and Africa before they divided (c.100 Ma). By mid-Cretaceous times, the angiospermous tropical rainforest may already have been well developed.[8]

East Asia is currently unique in that there are more 'basal', that is archaic, flowering plant families concentrated in its moist warm temperate than its tropical forests (Table 6.1). Bob Morley[9] has written a marvellous text on the history of plants of the tropical rainforest, which I will only summarise here, and only for our region. The global fossil record, combined with evidence from molecular phylogenies, indicates a peak of diversification during the Maastrichtian into families that are still extant today. The Indian subcontinental fragment of Gondwana had become isolated at temperate latitudes in the southern part of the Tethys Sea during the early Cretaceous (Aptian), when global climates were warmer and its climate likely more tropical (Chapter 1). Until the Indian Plate collided with the Laurasian in the Eocene c.50 Ma, the forests of the tropical Far East contained few flowering plant families. These families had presumably arrived earlier by means of a northern migration, from North America by way of western Eurasia, when tropical climates extended to high latitudes and sea barriers were narrower. The tropical Asian angiosperm flora has therefore overwhelmingly arisen from immigrants.

Many pan-tropical families are archaic and could already have existed before the final break-up of Gondwana; others are relatively advanced. Until the Oligocene period (25 Ma), there were still intermittent connections between the Americas and between North America, Europe, and East Asia (Fig. 6.1). These bridges were tropical as far north as c.50° N. But Africa was still far to the south, and Europe was separated from Asia by a sea running south from the Arctic Ocean, the Turgai Strait.

2. A series of immigrations into tropical Asia

Although obscured by subsequent events, and by the overarching influences of current climates and other physical forces, patterns which may in part reflect the ancient tracks through which flowering plants originally arrived into tropical Asia can still be detected. The fossil record indicates that immigrations occurred in waves at different periods.

Attempts to create a 'phylogeography' by quantitative correlation of molecular with biogeographic and other evidence is difficult, not least because the earliest fossils rarely coincide with a taxon's origin. Here, I merely attempt to interpret phylogenies in the direct context of current geographical ranges and known fossils.

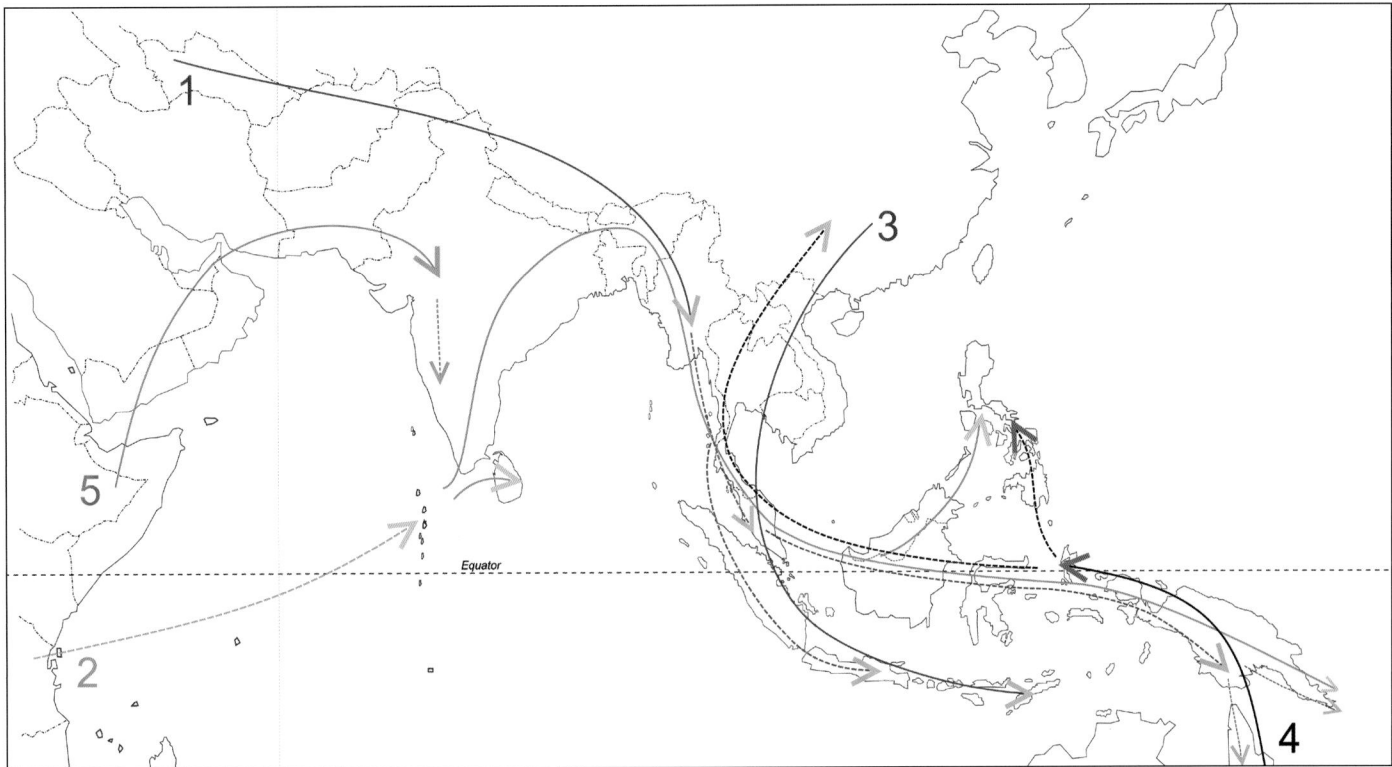

Fig. 6.1 Immigration tracks of flowering plants into tropical Asia, mapped onto current continental configurations. Numbers indicate their sequence over geological time.

2.1. *Five distinct geographical patterns among higher taxa*

2.1.1. *Pan-tropical distributions*

Three main opportunities for the spread of West Gondwanan land plants eastward into the Asian tropics seem to have arisen. Each of these was probably intermittent, owing to changes in climate and breaks in land connections. The first was via North America into Europe and southwestern Laurasian Asia, especially when they were connected during the mid-Cretaceous to early Paleocene periods (90–c.60 Ma). The Cretaceous invasion of tropical plants into Asia has left no discernable pattern, such as plant distributions coinciding along a geographical frontier. Better evidence comes from the fossil record and from the new science of molecular phylogeography. This discipline, sifting through molecular genetic evidence, has already done much to clarify relationships between the phylogenies of current genera and their geographic distributions. Late Cretaceous sediments from Borneo indicate a flora dominated by ferns, gymnosperms (including *Ephedrites* and *Araucariacites* of worldwide distribution), and *Clavatipollenites* (a mysterious fossil pollen of archaic construction which may represent a flowering plant related to the surviving worldwide tropical and warm-temperate archaic family Chloranthaceae)[10] (Table 6.2). In the Paleocene, pollen forms probably related to extant taxa appear, including the advanced families Sapotaceae, Myrtaceae and Olacaceae and the estuarine palm *Nypa*, and the *Brownlowia* type of Malvaceae. However, many of the most characteristic elements of the present flora were still apparently missing, including dipterocarps, durians and their allies, Fagaceae, Euphorbiaceae, Rhizophoraceae, Gonystyloideae, Icacinaceae (now divided between Aquifoliales and Garryales, but treated

as formerly here), Cardiopteridaceae, Stemonuraceae, and Lythraceae (Sonneratiaceae). Warm temperate (or lower montane) elements had appeared in the fossil flora of China, notably Altingiaceae in the Late Cretaceous (Turonian 88 Ma);[11] even the advanced taxon *Ilex* (Aquifoliaceae) had arrived.

In modern tropical Asia, there are about 125 recognized angiosperm families that include trees.[12] That is about equal to 127 families in Neotropical West Gondwana. Of the modern Asian families, c.94 are pantropical, and all but a very few also include temperate species. Few angiosperm families occur in the Cretaceous or Paleocene fossil floras of tropical East Asia (Table 6.2), yet these include advanced families like the Asterid Sapotaceae and the Aquifoliaceae, reflecting early diversification,[13] though current *Ilex* (the only Aquifoliaceae genus) species, and probably others, appear to have resulted from Miocene diversification following earlier extinction events.[14] Climate appears to have been periodically dry, and the presence of northern gymnosperms and Laurales has been interpreted as indicating the presence of high mountains. This first wave of immigrants was only possible owing to close land connections still surviving at that time between North America and Asia by way of Europe, and the penetration of tropical climates to their high latitudes during that period.

During the early and middle Eocene also, when there appears to have been a continuation of this northern and western wave, or a resurgence of it, the global climate was exceptionally warm. By the Eocene the north Atlantic had opened sufficiently for a nascent Gulf Stream to sustain temperatures far north along western European shores. There was land continuity from the Malay Peninsula at least as far as western Borneo and eastern Java – at that time the wettest region in tropical Asia.[15] Although a connection between North America and Europe still existed then, a wide sea channel, the Turgai Strait, separated Europe from Asia until the Indian collision. Further though, the Hadley climate cells (see Chapter 1) have always remained at the same latitudes, so that the southern lowlands of Laurasia would have been subject to dry northeasterlies off the continent for much of the year, although onshore breezes doubtless caused orographic storms around uplands.

The very rich Eocene flora preserved in North West European sediments – the 'London Clay' – includes many tree families with pantropical as well as warm temperate distributions, which could therefore have earlier reached tropical eastern Laurasia by the Atlantic track. Charles Davis[16] has inferred such a route

Cretaceous
Clavatipollenites (cf. Chloranthaceae)
Cycadaceae
Araucariaceae
Ephedraceae

Palaeocene
Ilex (Aquifoliaceae)
Myrtaceae
Sapotaceae
Anacolosa (Olacaceae)
Ulmaceae
Nypa (Palmaceae)

Table 6.2 Fossil taxa recorded from East Asia prior to the collision of Indian Gondwana (those putatively related to modern taxa only).

Fig. 6.2. (right) Phylogeny of Annonaceae, a pantropical lowland archaic family. Pantropical ranges of some basal taxa, such as *Anaxagorea*, could imply early spread when pantropical migration corridors existed, but later long-distance sweepstakes dispersal cannot be ruled out; however the majority appear to have arrived more recently after a gap, possibly from Africa via India. (D. Glick del. from Couvreur *et al.* 2011)

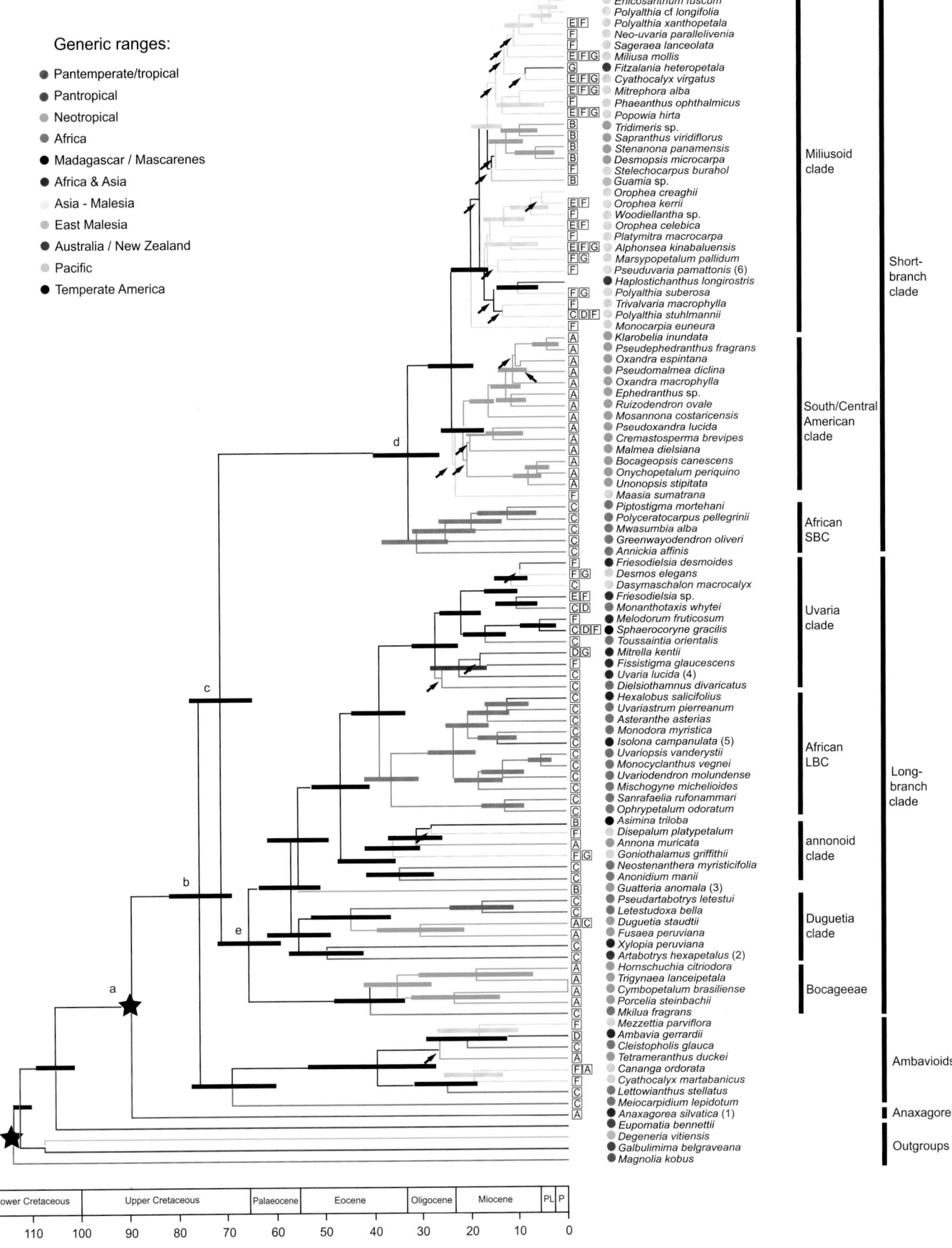

Generic ranges:

- Pantemperate/tropical
- Pantropical
- Neotropical
- Africa
- Madagascar / Mascarenes
- Africa & Asia
- Asia - Malesia
- East Malesia
- Australia / New Zealand
- Pacific
- Temperate America

for Malpighiaceae, a family richest in the Neotropics but also including vines in the Asian tropics and a few warm temperate representatives. The basal genus of Annonaceae, *Anaxagorea*, is pantropical but molecular evidence suggests that it originated in the Neotropics (Fig. 6.2).[17] Abundant *Polyalthia*-like seeds in the London clay could be interpreted as evidence for Laurasian dispersal of Annonaceae. Lauraceae, well represented in Far Eastern warm temperate forests, as well as lowland rainforests throughout Asia, could have followed a similar track.[18] Among the currently exclusively tropical families occurring in Asia, the European fossil record for Myristicaceae rests on a single fruit.

[19] Dipterocarpaceae are represented by a twig with anatomy resembling *Anisoptera* but of a size far smaller than extant species, while pollen of Lecythidaceae has only been recorded from the late Eocene/early Oligocene of Hungary.[20] The other exclusively tropical tree families Chrysobalanaceae and Crypteroniaceae have not as yet been detected, and neither has Lythraceae (Sonneratiaceae) – though its archaic fossil form, *Florscheutsia*, has been recorded from the Aptian age of West Africa. This leaves some possibility that the inland climate was frost-prone, therefore warmest temperate, and that tropical elements (notably *Nypa*) were confined to coastal uplands as they are today in South China north of the Tropic of Cancer.

2.1.2. *The Indian Noah's Ark*

A second wave of immigration of rainforest taxa occurred by way of the South Asian/Gondwanan subcontinent. This began at the time of its first contact with the southern Laurasian coast during the Eocene, when the meeting was near the Equator and likely perhumid. It terminated at least with the rise of the Himalaya, probably in the late Oligocene (c.25 Ma) when the global superhot period ended. The continuing movement of the Indian plate northwards, at 7 cm y^{-1}, would have increasingly led the migration route eastwards, north of the Bay of Bengal, out of the wet tropics; following this, the start of the monsoon further restricted opportunities for wet evergreen forest taxa by initiating a long dry season. The absence of so many Sunda genera in perhumid South Asia suggests that this occurred before Sunda became widely perhumid, perhaps 23 Ma.

As I described in some detail in Chapter 1, South Asia broke free from the warm temperate Gondwanan supercontinent in the mid-Cretaceous (Aptian) – that is, at the earliest stages of Eudicot evolution or before; its flora resembled the Australasian of that time. The South Asian 'Noah's Ark' drifted north into the equatorial regions, along the Madagascan and East African coast – at that time adjacent to it. During that slow transit, direct migration was initially possible to South Asia from Madagascar, but the tropical African flora would have immigrated later, when the Indian plate was more distant but had entered wet equatorial climates. The South Asian land connection was likely made with the southern Laurasian

coast c.45 Ma, at equatorial latitudes. The collision initiated the Himalayan bow-wave, but high mountains do not appear to have risen before the Oligocene and the Indian monsoon may have originated as late as 12 Ma. But South Asia continued its northeasterly transit, away from the Equator (see section 3.1.4 below). Therefore, this initial period may well have witnessed the first and last corridor of equatorial perhumid climate overland from India to the Far East. This would have provided the only opportunity for the migration of taxa confined to that climate yet still surviving in perhumid south-west Sri Lanka and Malesia, including *Axinandra* (Crypteroniaceae), *Trichadenia* (Achariaceae), the palm genus *Eugeissona* formerly in Africa but now confined to the Far East, and possibly the dipterocarp *Cotylelobium* and Malvaceae (tribe Durioneae). Climate may not have become widely wet again at higher northern tropical latitudes in Asia until the upward push of the high Tibetan Plateau c.12 Ma and the likely start of the monsoon. Perhumid climates in Malesia, and seasonally wet in eastern Indo-Burma, had already spread 10 million years earlier, when the eastern Indonesian ocean throughflow became obstructed. But the later wet migration corridor from South Asia was almost certainly seasonal and monsoonal rather than perhumid, and it did not last (section 3.1.4 below).

Bob Morley[21] has marshalled direct and inferential fossil evidence arguing that the Asian lowland rainforest flora arrived principally via the Indian route (Table 6.3). The palynological record indicates a peak in arrivals of flowering plant families from the middle Eocene to early Oligocene, following the collision of the Indian continental plate with the Laurasian coast.[22] The presence of many angiosperm families on the Indian plate is currently being confirmed by Vandana Prasad[23] and colleagues at the Birbal Sahni Institute, who have access to rich pollen floras in upper Paleocene and lower Eocene coals, formed in swamps that existed along the northern Indian continental shelf before the collision (Table 6.3). A remarkable array of pollen types of taxa still extant in the rainforests of South Asia are found. These include the distinctive *Durio*-type of the Malvacean tribe Durioneae (now endemic to tropical Asia but formerly also in Africa) with one genus, *Cullenia*, endemic to South Asia and four in the Far East. Arecoid palms appear, including genera endemic to Madagascar and the Seychelles (both Gondwanan fragments) but also well represented today in South Asia.[24] Other finds include the genera *Garcinia* and *Mesua* of Clusiaceae and Calophyllaceae respectively, certain Meliaceae and Euphorbiaceae, and the genera *Dipterocarpus* and *Shorea* of Dipterocarpaceae.

Dipterocarp resin, as amber, has been confirmed.[25] In light of current Asian dominance in the flora it is remarkable that we find Podocarpaceae (*Nageia*, *Dacrydium*) and a Proteaceae in the South Asian fossil record, currently richer in Australasia. These elements therefore apparently arrived in Asia in separate

Fossil taxon	Similar modern genus
Podocarpidites	Podocarpus (Podocarpaceae)
Arengapollenites echinatus, Neocuperipollis echinatus	Arenga (Arecaceae)
Dicolpopollis fragilis	Calamus (Arecaceae)
Longapertites, Quillonipollinites	Pinanga (Arecaceae)
Albertipollenites, Foveotricolpites alveolatus, Dipterocarpuspollenites retipilatus	Dipterocarpus (Dipterocarpaceae)
Tricolpites reticulatus	Hopea (Dipterocarpaceae)
Intrareticulites brevis	Shorea (Dipterocarpaceae)
Dipterocarpuspollenites retipilatus	Vateria (Dipterocarpaceae)
Sastripollenites trilobatus	Calophyllum (Calophyllaceae)
Clavaperiporites heteroclavatus, Polybrevicolporites cephalus, P. indistinctus	Garcinia (Clusiaceae)
Cupanieidites	Syzygium (Myrtaceae)
Incrotonipollis burdwanensis	Croton (Euphorbiaceae)
Margocolporites sahnii	Agrostistachys (Euphorbiaceae)
Lakiapollis ovatus	Cullenia (Durioneae)
Meliapolis naveleii	Toona/Walsura (Meliaceae)
Striacolporites	Bhesa (Celastraceae)
Paripollis broachensis	Lophopetalum (Celastraceae)
Pellicieropollis langenheimii Polygalacidites	Alangium (Alangiaceae) Xanthophyllum (Polygalaceae)
Pseudonyssapollenites kutchensis	Gomphandra (Icacinaceae)
Retimocolpites thanikaimonii	Myristica (Myristicaceae)
Rhoipites	Nothopegia (Anacardiaceae)
Tetracolporopollenites brevis	Palaquium (Sapotaceae)

Pollen from East Asian sediments
Podocarpaceae, Eugeissona (Arecaceae, Alangium, Beauprea (Proteaceae), Sapindaceae, Ctenolophon, Durio, Gonystylus, Ixonanthe, Polygalaceae.

Table 6.3 Evidence for the Eocene flora in tropical Asia. Pollen from East Asian sediments and Gondwanan Indian coal, putatively associated with modern genera. (from Morley 1999; Prasad *et al.* 2009)

phases. Some 20 lowland rainforest families are now recorded from the Paleocene of the Indian plate, but more may be expected as research continues.

Dipterocarpaceae occur as three subfamilies, sometimes regarded as families: (1) the monotypic Pakaraimoideae on the ancient Guyana Shield sandstones of northern South America; (2) the Monotoideae, consisting of one monotypic neotropical genus and two African genera with 3 and c.35 species respectively; and (3) Dipterocarpoideae (Fig. 6.3), with 13 genera and c.475 species, now only in Asia and the Seychelles[26] (though the genus *Dipterocarpus* also occurred in north-east Africa during the Miocene[27]). Molecular phylogenies indicate their affinity with Sarcolaenaceae, now endemic to Madagascar but formerly in Africa; also, unexpectedly, with Cistaceae.[28] All three families have associations with ectotrophic mycorrhizae,[29] otherwise confined among tropical families to Caesalpinioid legumes and Fagaceae (Chapter 5). All dipterocarpoid genera (but for the monotypic *Upuna* of Borneo) are either still in South Asia or represented by Cenozoic fossils there. Two genera, *Vateria* and *Stemonoporus*, are endemic to South Asia and another, *Vateriopsis* (related to *Stemonoporus*), survives also in the Seychelles. All three are confined to wet climates, *Stemonoporus* to perhumid. *Cotylelobium*, with winged fruit (except one South Asian species) and the same climate constraints, occurs as two species in perhumid Sri Lanka, and as three species in perhumid Sundaland and in slightly seasonal southern peninsular Thailand. Comparative morphology and anatomy[30] is consistent with molecular phylogenies in implying that basal taxa are concentrated in South Asia, and that the Neotropical and African taxa are basal to the Asian.[31] The main genera of the tribe *Shoreae*, as currently recognized with *Hopea* imbedded between sections of *Shorea*, seem to have originated early, in South Asia before or soon after the collision. One clade, *Shorea* (sect. *Doona*) with nine species, remains endemic to perhumid Sri Lanka.[32]

Dipterocarp seeds lack dormancy, and most are killed after prolonged immersion in salt water. Although most are winged, they are heavy and disperse mainly by gyration (Box 5.1). Vertebrates predate but hardly disperse them. However, the fruit of Mascarene *Vateriopsis* and the South Asian genera *Vateria* and *Stemonoporus*, together forming a clade endemic to Indian Gondwana, lack wings. None of these characters argue for efficient dispersal across open ocean waters, and the latter two are the only dipterocarp genera to have failed to migrate to the Far East. Madagascar is the current home to Sarcolaenaceae, sister to the dipterocarps. The absence of basal Dipterocarpoids there implies that dipterocarps migrated to the Indian plate subsequent to its separation from Madagascar c.80 Ma. Nevertheless, one species of the African winged-fruited dipterocarp genus *Monotes* is endemic to Madagascar, currently lying 400 km off the African coast. This implies that occasional

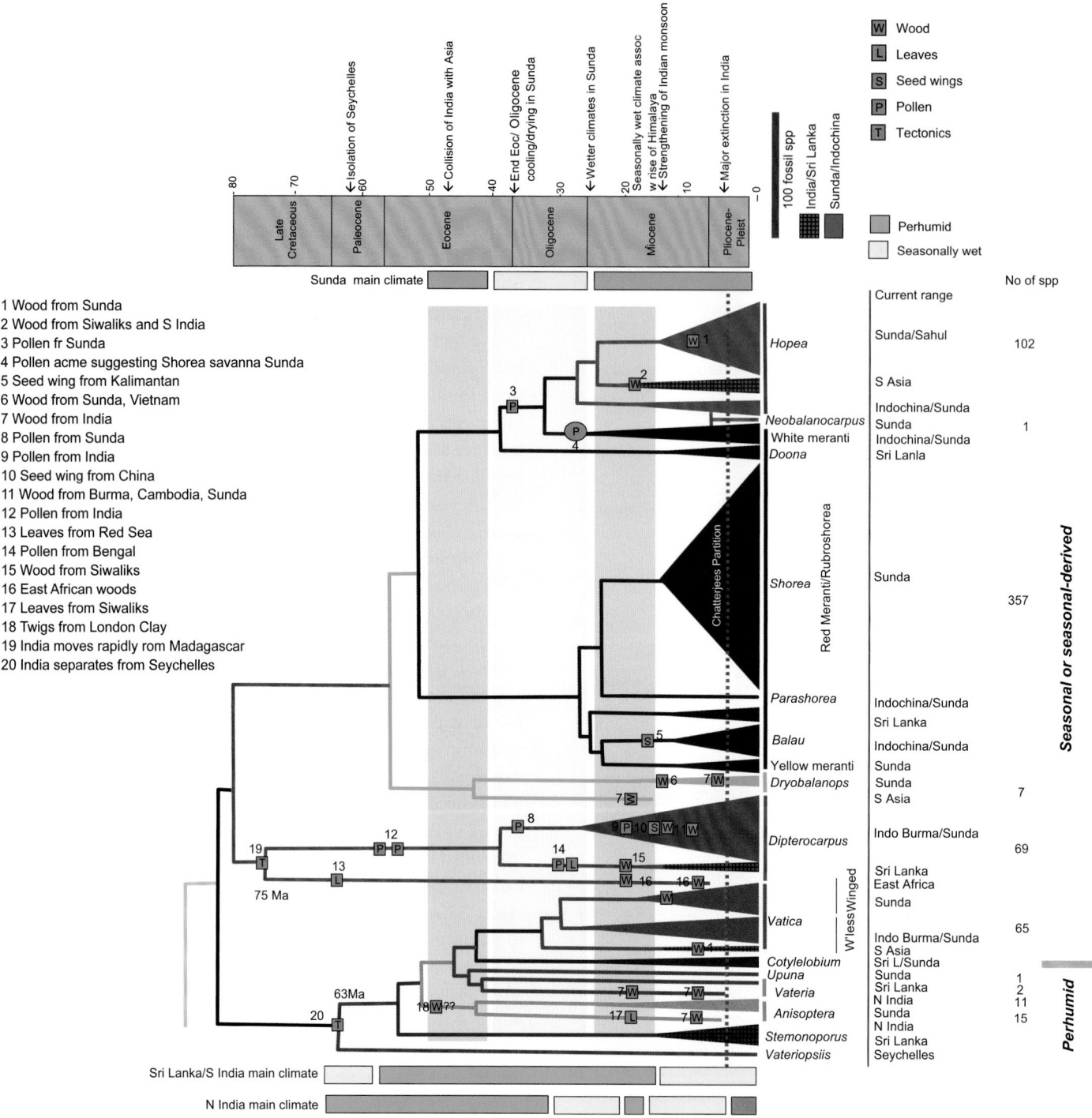

1 Wood from Sunda
2 Wood from Siwaliks and S India
3 Pollen fr Sunda
4 Pollen acme suggesting Shorea savanna Sunda
5 Seed wing from Kalimantan
6 Wood from Sunda, Vietnam
7 Wood from India
8 Pollen from Sunda
9 Pollen from India
10 Seed wing from China
11 Wood from Burma, Cambodia, Sunda
12 Pollen from India
13 Leaves from Red Sea
14 Pollen from Bengal
15 Wood from Siwaliks
16 East African woods
17 Leaves from Siwaliks
18 Twigs from London Clay
19 India moves rapidly rom Madagascar
20 India separates from Seychelles

Fig. 6.3 Schematic history of Dipterocarpaceae subfamily Dipterocarpoideae. Among its two large clades, all genera of the lower clade occur in South Asia or the Indian Ocean with the exception of monotypic Bornean *Upuna*, or were present in its fossil record (*Anisoptera*); *Vateriopsis* is confined to the Seychelles, *Stemonoporus* and *Vateria* to South Asia, and all three are confined to wet seasonal or perhumid climates. Among the top clade, *Dipterocarpus* fossils occur in East Africa; *Dipterocarpus*, *Hopea* sect. *Hopea*, and *Shorea* sects *Shorea* and *Anthoshorea* occur in South Asia, while they, *Parashorea* and *Shorea* sect. *Pentacme*, both confined to the Far East, occur in seasonal climates; whereas *Hopea* sect. *Dryobalanoides*, *Shorea* sect. *Anthoshorea*, and the red meranti sections, are confined to perhumid or only weakly seasonal climates (with one Philippines exception). Three species excepted, no *Dipterocarpus*, *Shorea*, *Parashorea*, or *Hopea* sect. *Dryobalanoides* crosses Wallace's Line eastwards. Although the eastward migration route from Indian Gondwana by North India is depicted as having a long late Cretaceous to early Oligocene perhumid period, and a short early Miocene perhumid period, the latter is unlikely because the continued trajectory of the Indian plate northwards would have carried the northerly migration route beyond perhumid equatorial into seasonal climates; the second period was therefore almost certainly entirely seasonal, and preceded by a distinctly dry period. (R.J. Morley & P.S. Ashton, phylogeny following Gamage *et al.* 2006)

Plate 6.1 Gondwanan elements in South Asia and the Mascarenes, and a Neotropical dipterocarp. **a,** *Hortonia angustifolia* (Monimiaceae-Hortonioideae) in its lowland riverside habitat, Sinharaja forest, south-west Sri Lanka; **b,** *Hortonia angustifolia* in flower and fruit; **c,** *Pakaraimaea dipterocarpacea*, Dipterocarpaceae, Santa Elena, Venzuelan Guyana; **d,** *Vateriopsis seychellarum*, endemic Seychelles dipterocarp in flower in the greenhouse of Cruickshank Botanic Garden, University of Aberdeen, 1977; **e,** *Vateria copallifera*, fruiting crown; Pitakelle, Sinharaja, Sri Lanka; **f,** *Schumacheria castaneifolia*, Dilleniaceae, Sri Lanka endemic genus, Sinharaja forest; **g,** Endemic Agamid lizard *Ceratophora stoddartii*, Horton Plains, 2,000 m, Sri Lanka. (**a, b, e** N. Gunatilleke; **c, d** P.A.; **f, g** H.H.)

389

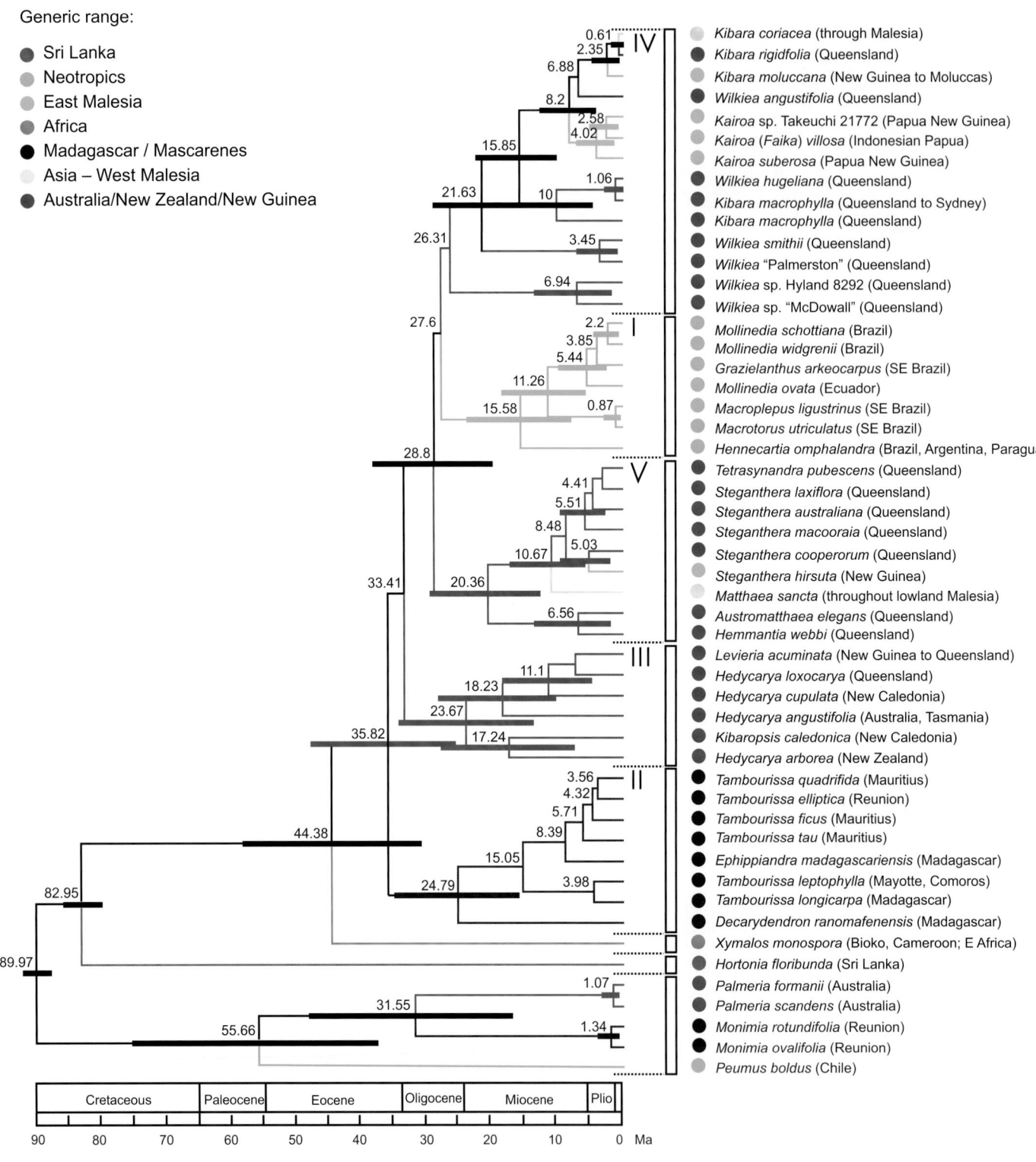

Generic range:

- ● Sri Lanka
- ● Neotropics
- ● East Malesia
- ● Africa
- ● Madagascar / Mascarenes
- ● Asia – West Malesia
- ● Australia/New Zealand/New Guinea

Fig. 6.4 Phylogeny of Monimiaceae. Although representation of many taxa in north–east Australia, and from thence into Malesia, implies that easternmost Gondwana is the primary source in tropical Asia, the extraordinary age of subfamily Hortonioideae, endemic to Sri Lanka, is consistent with its origin in Madagascar by way of the Indian Noah's Ark. The riparian lowland habitat of one of its two species might explain how this taxon of wet mountain affinities survived the dry latitudes on its way north. (D. Glick del. after Renner *et al.* 2010)

chance dispersal (or 'sweepstakes' dispersal, in the terminology of Simpson[33]) of dipterocarp seeds must have occurred over that distance. It seems likely that as India moved north into the wet tropics, island chains persisted, linking India and Madagascar with continental Africa and facilitating the dispersal of some dipterocarps.[34]

Dipterocarpoideae provide particularly informative inferences on their migration and diversification histories: they are endemic to Asia but for *Dipterocarpus;* they are unambiguously of Indian Gondwana origin; they are diverse; and there is well-correlated molecular and morphological evidence for their diversification. Whereas taxa of wet and perhumid climates confined to South Asia are basal in the subfamily, the six red and the yellow (*Richetioides*) meranti sections of *Shorea*, and *Hopea* (sect. *Dryobalanoides* and sect. *Hopea* subsect. *Pierrea*), are all derived and confined to the Far East. Opportunity for migration of taxa confined to perhumid climates likely ceased by the early Oligocene. With very few exceptions these taxa, though reaching the Philippines, have failed to cross Wallace's Line into Wallacea. But a number of species in perhumid south-west Sri Lanka have sister species in the wet seasonal Far East, implying more recent migration by a wet, but seasonal, route (Table 6.11). Opportunities would have peaked during the mid-Miocene warming phase that preceded the Pliocene cooling. Fossil *Dipterocarpus* and *Anisoptera*, taxa currently present in seasonal evergreen forests in the Far East, occur in the Miocene sediments of the Siwaliks in north-west India.

Tropical Asia is exceptional in lacking isolated endemic angiosperm families, but this is consistent with its relatively young flora. The morphologically distinct and apparently anciently diverged Scyphostegiaceae, with a single species, are endemic to Borneo, and are currently included in Salicaceae.[35] *Sonneratia*, one of the two genera in former Sonneratiaceae (now included in Lythraceae), occurs along the coasts of East Africa and Madagascar and east to the Pacific. But the other Sonneratiaceae, *Duabanga*, which is not closely related, consists of two inland species and is restricted to Indo-Burma and Malesia.[36]

Among other higher taxa endemic to tropical Asia are Monimiaceae (subfamily Hortonioideae). *Hortonia* is an isolated taxon in a Gondwanan family with two, perhaps three, large shrub species found in the upper montane forests and beside lowland clear-water rivers in perhumid south-west Sri Lanka (Plates 6.1a,b). *Hortonia* apparently represents the only endemic plant taxon with truly ancient origins on Gondwanan India, its Madagascan and Mascarene confamilials being distantly separated (Fig. 6.4). Its partially riverine ecology could explain how it might have persisted through the lowland desert climate that must have prevailed while the Indian plate rafted through the Horse Latitudes. Madagascar and the Cenozoic volcanic Mascarene Islands are still rich in Monimiaceae, and are its presumed provenance.[37] Other Monimiaceae genera occur in South America, and in Australasia from whence they have immigrated into Malesia from Cenozoic times onward (see sections 2.1.4 and 3.1.1 below). Crypteroniaceae, also pantropical, contains three genera (*Crypteronia*, *Dactylocladus* and *Axinandra*) that are endemic to tropical Asia.[38] Of these, *Axinandra* alone is found in Sri Lanka. The Borassoid palms, notably subtribe Lataniinae, manifest phylogenetic links from the rainforests of Africa through Madagascar and the Mascarenes to tropical Asia.[39] The Arecoid palm *Rhopaloblaste*, of the Nicobars, Peninsular Malaysia, and New Guinea to the Solomons,[40] represents an isolated lineage allied to tribe Areceae which includes *Loxococcus*, endemic to Sri Lanka. It appears related to genera outside Areceae and sister to *Dictyosperma* of the Mascarenes, implying ancient roots. *Orania*, comprising the tribe Oranieae, occurs throughout Malesia including the Philippines, but also on Madagascar. African and Asian *Terminalia* species occupy separate clades within the genus except, inexplicably, the most widespread inland Asian species, *T. bellirica*, which shares a clade with a Neotropical species (Fig. 6.8).[41] These species are more likely to have immigrated via the Indian than the Laurasian track that opened from Africa in the Miocene, which was likely semi-arid.

The Gondwanan family Myrtaceae and its tribe Syzygieae now occur from Africa to Australasia, have their centre of species diversity in Malesia and eastwards to tropical Australasia, and may have originated as late as the Eocene. The presence of a relatively advanced *Syzygium* clade in Africa, as well as their presence in South Asia, indicates long-distance dispersal *westwards*, likely by birds, from South Asia back to Africa (Fig. 6.7).

2.1.3. *An eastern Laurasian connection*

By the Oligocene, sediments from the southern South China Sea start to yield pollen of the temperate conifers *Abies* and *Picea*, also *Pinus* and *Alnus*. Their presence continues up to the mid-Miocene. Initially, they may have originated from Indo-Chinese mountains extending south from temperate China, being rotated south by pressure from the Indian subduction far to their west. Towards the north, Indo-China would likely have initially been warm temperate, but would have warmed as it was pushed south. Later, these fossils were probably derived from uplands associated with horst and graben topography, which characterised the northern Sunda Shelf at that time and were most extensive at times of seasonal climate (Chapter 1). The opening up of the South China Sea by rifting down the East Asian coast from this time would have increased the mild maritime humidity of the climate in that region. But as the *global* climate started to cool, four climatic phases, interpreted from four distinct pollen assemblages, are recognisable from the same South China Sea basins. They suggest intermittent freshwater swamp and dry or moist seasonal forest climates; periodic increases in temperate elements imply cooler periods.

The phytogeographic connection between temperate eastern North America and North-East Asia has been well documented since Asa Gray drew attention to it over a century ago.[50] The East Asian temperate flora is currently far richer than the American. This is in part due to greater land continuity within humid warm temperate East Asia, which has allowed north–south migration during Pleistocene climate changes. It is also in part on account of the warm current which, in contrast to the Atlantic Gulf Stream, hugs the East Asian coast north to northern Honshu, thereby ameliorating winter temperatures; and to the strength of the wet warm summer-rain monsoon. These together support a rich warm-temperate evergreen forest as far north as Ibaraki prefecture in Honshu. The current extent of this climate originates with the closure of the Indonesian ocean throughflow and the initiation of perhumid conditions in Sundaland, 23 Ma. Much of it apparently persisted through the Pleistocene.[51] Bearing in mind that Hadley cells and their prevailing wind directions have remained as constant as the westward spin of the Earth, the climatic influence of the Pacific Ocean has also long remained constant in the Asian tropics. The richness of the East Asian warm temperate flora, especially in archaic families, is therefore due in part to the ancient and uninterrupted persistence of its associated climate system. The flora extended as far as Europe in the Cenozoic, and became isolated from it, and the American after the Bering Straits region cooled, in the early Miocene.[52] It includes many families endemic to the Far East, several of which are Magnoliids or Eudicots, basal in the angiosperm phylogeny (Table 6.1). Many of the Eudicots are confined to temperate forests in Asia, but about half penetrate the tropics. There they are predominant in lower

Plate 6.2 Northern elements in the East Asian lower montane flora. **a,** *Maingaya malayana*, Hamamelidaceae, north-west Peninsular Malaysian endemic, in flower at the Forest Research Institute; **b,** *Eriobotrya bengalensis*, Rosaceae, lower montane forest, 1,700 m, Thimpu-Trongsa road, Bhutan; **c,** *Berberis aristata*, north-west Himalayan microtherm element in the South Asian upper montane grassland flora, 2,000 m, Horton Plains, Sri Lanka; **d,** *Magnolia* lanuginosa, Magnoliaceae; lower montane forest, Thimphu-Trongsa road, Bhutan; **e,** *Schima wallichii*, Theaceae, lower montane forest, 1,700 m, Kintamani, Bali; **f,** *Schima wallichii* flower. (**a, e, f** P.A.; **b–d** H.H.)

montane forests (Plate 6.2), but in several cases are also well represented in the wet lowlands. East Asia is unique in that basal families predominate in warm temperate, not tropical, climates. This may again reflect the strength of early northern warm temperate rather than tropical connections (although it must equally reflect the longer persistence of their climates in that region than elsewhere).

Fagaceae occur in both Asian and American tropics, but not in Africa;[53] but they are only abundantly represented in lowland rainforests, in both species and numbers, in perhumid Malesia. The evergreen genera, all tropical or warm temperate, are the older.[54] The pollen of Fagaceae can rarely be definitively distinguished from that of several other rainforest families, so it is rarely quoted; but it is mysterious that none of their ecologically associated basal lower montane families appear in the fossil pollen record, even when temperate conifers are well represented. Molecular evidence indicates a complex history for *Ilex*, whose pollen is also absent from the African record,[55] while it has been recorded in the Turonian of Australia.[56] Although African extinction cannot be excluded, there remains the possibility of mid-Cenozoic invasion of Asia by *Ilex*, directly westward by long-distance dispersal from North America via Europe. Both Fagaceae and Juglandaceae include evergreen genera. These are presently found in northern warm temperate forests, but are the only form represented and occur widely in Far Eastern lowland rainforests. Among Asian tropical evergreen Fagaceae, *Trigonobalanus* and *Castanopsis* are represented by single species in the New World. *Castanopsis* is represented by temperate species in both regions, while American leaf impressions, if correctly identified, imply a former temperate presence of *Trigonobalanus* there.[57] The wind-pollinated *Quercus* subgen. *Cyclobalanopsis* is endemic to East Asian tropical and warm temperate evergreen forests, where it is best represented on drier sites and more seasonal climates (Fig. 6.5). Ectotrophic mycorrhizal Fagaceae overall become richest in Asia in numbers (though not species) in the East Asian lower montane tropical and warm temperate evergreen forests. They dominate these forests, thus bearing interesting comparison with Dipterocarpaceae in the lowlands (Chapter 4).[58]

Other than the Magnoliaceae, predominantly temperate Laurasian families are notably absent from tropical upper montane forests. Cornaceae[59] are rich in Far Eastern warm-temperate with a few in tropical lower montane elements, yet *also* well-represented in lowland rainforests (Table 3.17). But Magnoliaceae appear to have diverged originally in the tropics in both the New World and Asia in the Eocene, with temperate taxa diverging only later.[60]

Among families predominantly lower montane in the tropics, Hamamelidaceae is represented in both Laurasia and Gondwana. It includes five tribes, two of which occur

in temperate North America but are otherwise confined to East Asia (Fig. 6.6).[61] *Rhodoleia* and *Exbucklandia*, each alone in their tribe though in the same subfamily, are endemic to tropical lower montane forests and to the warmest of warm temperate forests; *Rhodoleia* occasionally occurs on siliceous soils in the lowlands. They continue down the lower montane oak-laurel forest zone through peninsular Thailand and Malaysia, and on to Sumatra to West Java; also to the

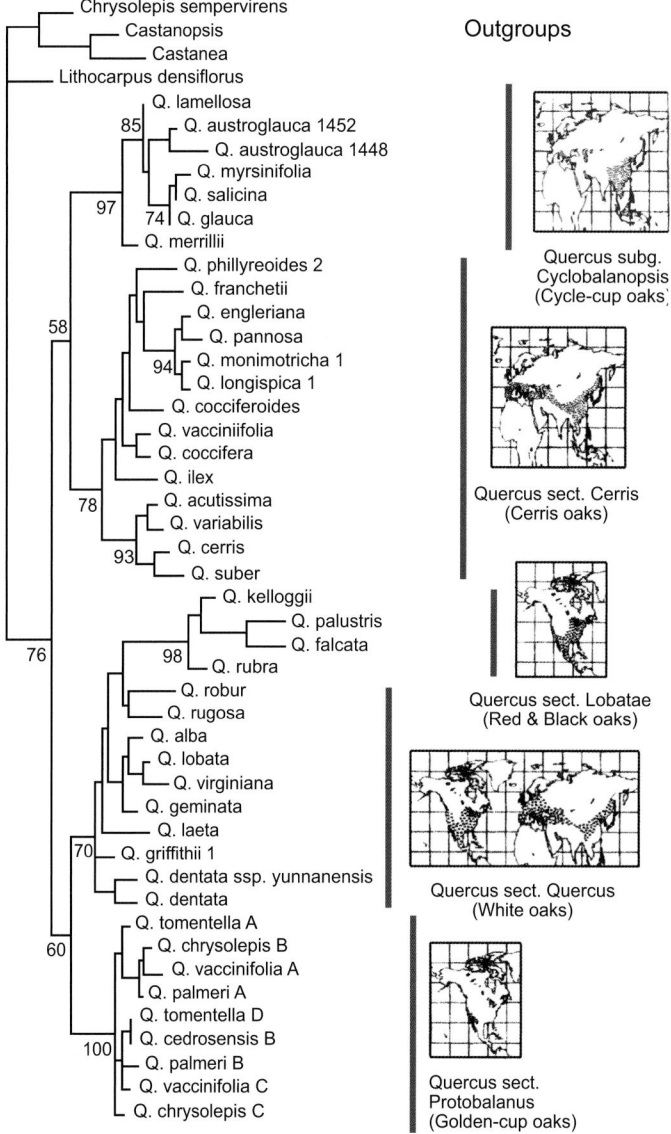

Fig. 6.5 Phylogeny of *Quercus*, Fagaceae, the oaks. All oaks of warm temperate and tropical East Asia are evergreen, and belong to subgenus *Cyclobalanopsis*, endemic to the region but absent from the Philippines and east of Wallace's Line. All *Quercus*, seemingly are wind pollinated, and are concentrated on ridges and other windier habitats in perhumid Sundaland. Although the only genus of clear East Asian warm temperate affinity that has richly speciated in the tropics, that diversification is concentrated in the Sunda climate, itself becoming widespread no more than 23 million years ago. (From Manos & Stanford 2001)

393

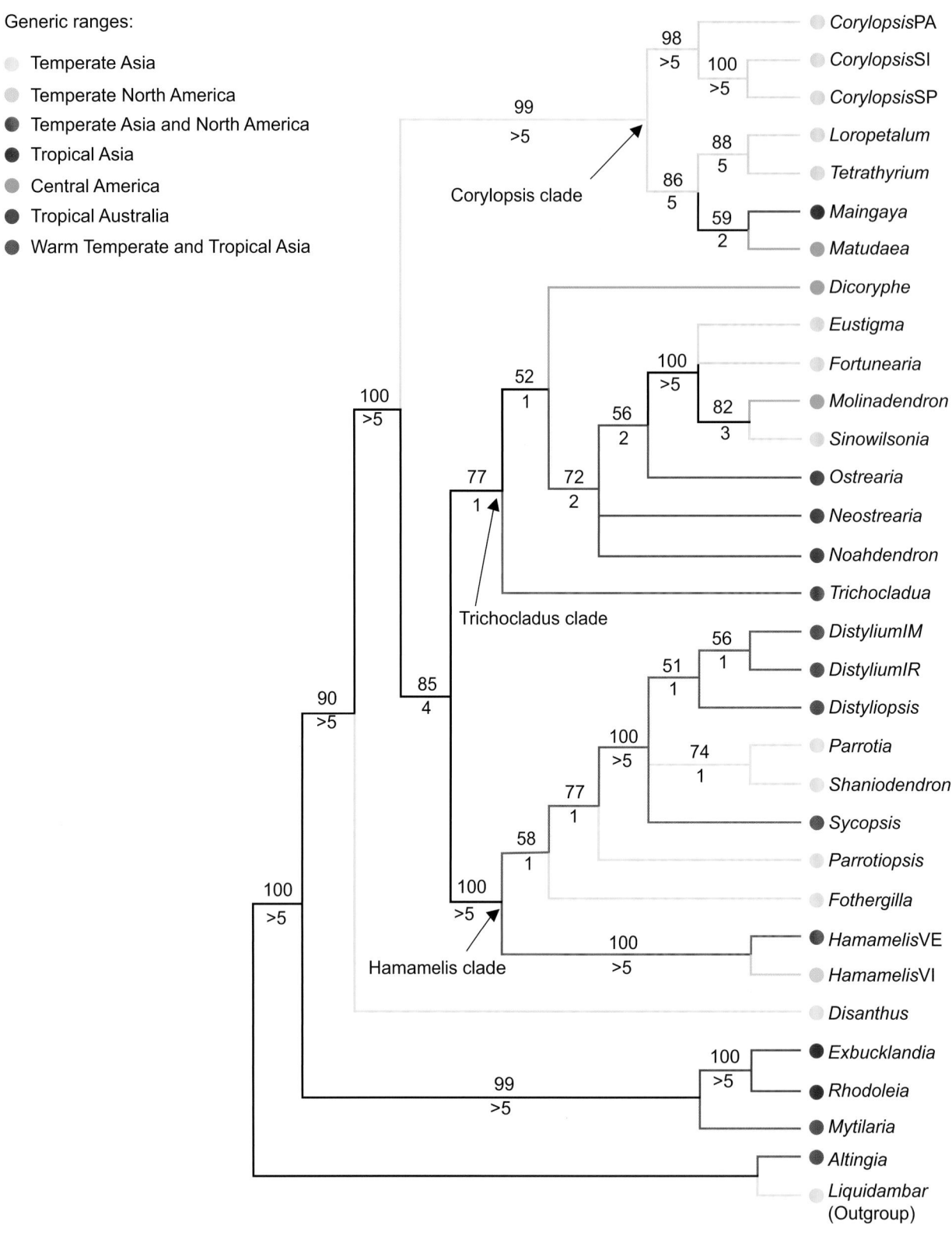

Fig. 6.6 Phylogeny of Hamamelidaceae, a predominantly temperate family. The predominantly tropical East Asian *Altingia, Exbucklandia, Rhodoleia and Sycopsis*, although apparently ancient, are predominantly lower montane and sisters to temperate genera which are the majority, implying temperate origins. They appear to have invaded the tropics intermittently, and over a prolonged period, although they have hardly speciated in the tropics. (D. Glick del. from Li *et al.* 1999b)

lower montane Kinabalu forests of Borneo. One endemic East Asian genus, *Embolanthera*, is found in Palawan but nowhere else in the Philippines: possibly rafted there as the South China seas opened. The phylogenetically isolated genera *Noahdendron*,[62] *Ostrearia* and *Neostrearia*, comprising the subfamily Hamamelidoideae, occur in the Queensland rainforest.

Wind-dispersed Ulmaceae – overwhelmingly temperate – are represented by *Ulmus lanceifolia*, extending through the lower montane forests of Indo-Burma and recurring in East Java and east of Wallace's Line in Sulawesi and Timor; while the wind-dispersed *Engelhardia* (Juglandaceae), reaches Timor and New Guinea.[63] Interesting as examples of ecological conservatism, all of these are among the few mainly temperate families also found in New World lower montane tropical forests, down the cordillera across the temperate forest limit at the Isthmus of Tehuantepec. Yet Fagaceae, Juglandaceae, Ulmaceae, Hamamelidaceae, Betulaceae, Nyssaceae and Altingiaceae, all warm temperate and also paradigmatic of tropical Far Eastern lower montane forests, are absent from the South Asian mountains, implying that the Gangetic plain has remained a powerful ecological barrier.

Box 6.1

Gymnosperm geography

Gymnosperms are widely represented in tropical Asia. As a more ancient taxon than the angiosperms they are of particular historical interest.

Cycads occur throughout the region, and in warm temperate climates north into Japan and south into Australia. *Cycas* species are confined to open habitats including coasts, deciduous forests with open understorey (where they are fire hardy), limestone summits, and more locally in *kerangas* on humult podsols and ultramafic substrates.

Ephedra does not enter the tropics, but its pollen occurs in early Miocene sediments off northern Borneo, implying that high mountains extended south at that time.[42] A great early Miocene mountain range, thrusting down from the north and surviving as the Annamite Range, had carried temperate moist boreal genera well into the tropics. They disappear before the appearance of podocarp and other Australasian pollen. In contrast, pan-tropical *Gnetum* occurs today (mostly as vines) throughout the lowland evergreen and the lower elevations of lower montane forests from South Asia to New Guinea, reaching north into the warm temperate forest of southernmost China (Chapter 3). There, all species are endemic.

The pines, *Pinus*, are confined to northern latitudes, except in Sumatra where they extend in a finger of seasonal climate just south of the Equator. There are six species in the Asian tropics (Chapter 4). Three are endemic to the Vietnam-Laos mountains. Only *P. merkusii* occurs in the lowlands. Like the lower montane *P. kesiya* and *P. roxburghii*, *P. merkusii* depends on fire to regenerate (Box 4.2). It is restricted to the savanna woodlands of seasonal East Asia and the Philippines.[43] Most *Pinus* species thrive in climates sufficiently seasonal for fire in nature, but occur also occasionally on limestone karst in wetter regions.

The mountains of warm temperate and tropical East Asia, fringing the South China Sea from South China and Taiwan into Vietnam, have in part acted as a corridor of warm temperate climate south into the tropics. They host an extraordinary richness of conifers:[44]

1. There are two endemic pines in the lower montane forests at the southern end of the Vietnamese Annamite range: *Pinus dalatensis*, similar in ecology to *P. kesiya*, and the extraordinary *P. krempfii* with its flat and narrow leaves, which does not appear to require fire for establishment and which is unusually shade tolerant[45] (Chapter 4).

2. A fir (*Abies delavayi*) and a hemlock (*Tsuga dumosa*) are found on granite peaks in northern Vietnam to 22°30' N under cool but equable temperatures with winter frost.

3. *Keteleeria davidiana*, one of the few tree species to have been successful in both cool tropical and warm temperate climates, extends throughout this region from Zhejiang and Hubei south to the southernmost Vietnamese mountains at Dalat, and into the mountains of north-east Thailand.

4. The rare *Fokienia hodginsii* is confined to open woodland on karst limestone, apparently only within the finger of warmest temperate climate that penetrates the northern Vietnamese mountains.

5. The single small population of *Taiwania cryptomerioides* in northern Vietnam occurs in dense forest on metamorphic substrate at c.2,000 m; whether subject to frost or not is unclear, but is likely.

6. The recent discovery of a new conifer (*Xanthocyparis vietnamensis* (Plate 6.3d)) on limestone karst summits in northernmost Vietnam – in a genus previously represented by a single species in the boreal forests of North-West America – provides compelling evidence that the forests of this region, now almost gone, are a refuge for ancient taxa of global importance.

7. But the only northern conifer to extend widely into tropical evergreen forest in Asia is *Taxus*, which reaches northern Sumatran and Luzon seasonal lower montane forests.

Plate 6.3 Indo-chinese conifers. **a,** *Keteleeria evelyniana*, Dalat, Vietnam; **b, c,** *Pinus krempfii*, Dalat and Congtroi, Vietnam, with laminar leaves. The trees, with diffuse crowns on the ridge, are imbedded in lower montane forest, and it is unclear how they regenerate; **d,** *Xanthocyparis vietnamensis*, only other member of a genus otherwise in boreal North America, in the temperate-tropical interphase of the lower montane forests of northern Vietnam. (**a** P.A.; **b, c** P. Thomas; **d** A. Farjon)

Araucarioid conifers, now of exclusively southern affinity, occurred worldwide before the angiosperms evolved, and are well represented in the fossil record in Borneo until the Late Cretaceous period. Gymnosperms are currently poorly represented in lowland mixed evergreen forest on zonal soils, except in New Guinea where *Araucaria hunsteinii* regenerates in dense stands on landslips. In the Far East, *Podocarpus neriifolius* and *Nageia motleyi* occur locally as scattered individuals in MDF. Nevertheless, Kitayama and his group have found that podocarps on Kinabalu are light demanders that compete with angiospermous pioneers by secreting acid phosphatases. These mobilise organic phosphorus, and attract saprophytic fungi that may facilitate its uptake as symbionts. Although slow growing, they are long-lived thereby increasing the chance of successful seed dispersal and establishment.[46]

The families Araucariaceae and Podocarpaceae have therefore penetrated contemporary Asia to varying extents: most since the arrival of Australasia and the islands of Wallacea during the early Miocene. Among the Araucarioids with winged seeds, *Araucaria* occurs north only to New Guinea, while *Agathis* extends to Peninsular Malaysia and has diversified throughout its broad Malesian range, especially into regional lower montane and lowland closed forests on humult podsols and ultramafic soils.

Podocarpaceae are fleshy-fruited, but migrated north-east into Laurasia to varying extents (Table 6.4). *Podocarpus* pollen is also present on pre-collision Indian Gondwana.[47] One segregate genus, *Nageia wallichiana*, occurs from the Sunda Hills to the Western Ghats. *Dacrydium* arrived in East Asia late in the Oligocene, probably from South Asia where there is an early pollen record, and expanded during the favourable warm climates of the early to mid-Miocene, reaching Japan at that time.[48] *Dacrydium* pollen became abundant in the Sunda lands, associated with *Casuarina* (whose pollen cannot be distinguished from *Gymnostoma*) in upland peats, as *D. elatum* (*beccarii*) is at present with *G. nobile*. *Dacrydium elatum* currently extends from Borneo to the lower montane forests of Hainan, Guangxi, northern Vietnam and north-east Thailand,

Locality	Number of genera	Number of endemic species	Total number of species
Philippines	5	2	13
Borneo	7	13	30
Java	3		5
Sumatra	4		13
Malay Peninsula	5	1	16
Burma, Thailand	3		3
Indo-China	3	1	5
Taiwan	1	2	4
Southernmost China	4		7
Southern China	3		4
Japan	2		3

Table 6.4 The range of Podocarpaceae west of Wallace's Line.

and occurred further north into South China during the last northern glaciations.[49] There are three *Dacrydium* species in the Philippines.

Among other podocarps, *Dacrycarpus* and *Phyllocladus* arrived early on the rising New Guinea land mass, c.20 Ma in the Miocene, both spreading to Borneo by the Pliocene, whereas *Dacrycarpus* spread more widely in the Far East during the Pleistocene. *D. imbricatus* occurs to northern Thailand and Vietnam in lower montane forest, and extends into southern warm temperate China (Chapter 3). With *D. elatum*, it locally dominates the *kerangas* on humult podsols bordering the Gulf of Thailand in the briefly seasonal wet climate south of the Cardamom Mountains. *Dacrydium* especially has diversified on the young mountain Kinabalu, as have *Podocarpus* and *Dacrycarpus*, while *Phyllocladus hypophylla* has reached the Philippines and other Borneo mountains. *Falcatifolium* reaches Peninsular Malaysia (Table 6.4).

I infer that Wallace's Line has not proved a permanent barrier to gymnosperm migration, except where *ecological* barriers have persisted.

2.1.4. *Australasian arrivals*

The Australasian flora of the perhumid tropics first crossed Wallace's Line west after the perhumid corridor between East and South Asia was severed. *Syzygium* excepted, it is unrepresented in South Asia. Late Oligocene sediments from the Java Sea indicate a transition from very wet to more seasonal climates. Remarkably, again, Australasian elements had already begun to appear while Australia remained far away, likely still in a warm temperate

climate: southern *Podocarpus* (*polystachyus* type) originally appears in the West Malesian Eocene, *Austrobuxus* (Picrodendraceae), *Dacrydium* (Podocarpaceae) and *Casuarina* (Casuarinaceae) in the early Miocene. Morley[64] associates these, by analogy with modern vegetation, with coal yielding shallow peat swamp pollen therefore perhumid climate. *Dacrydium*, however, may also have occurred on coastal *kerangas* in a moderately seasonal climate, as it currently does in Cambodia (see Box 6.1 above and Chapter 3).

Fig. 6.7 (left) Phylogeny of *Syzygium*, Myrtaceae, which occurs from East Africa eastwards, and genera currently united with it. *Eugenia* is mostly in the Neotropics & likely arrived by way of Indian Gondwana; whereas all others, including all montane genera, appear to be of Australasian origin. The African clade of *Syzygium*, imbedded in otherwise Asian/Australasian clades and clearly younger than the arrival of India in Asia, shows that long-distance dispersal does occasionally occur, in this case back from India to East Africa. (D. Glick del. from Biffin *et al.* 2010)

Ranges:

- Pacific
- Australia
- New Guinea – Australia
- East Malesia
- Sunda
- East and South East Asia
- Solomons
- Madagascar – Malesia
- Continental Asia – Australia
- Africa
- South East Asia and Sunda
- New Caledonia
- Lord Howe Island
- New Zealand

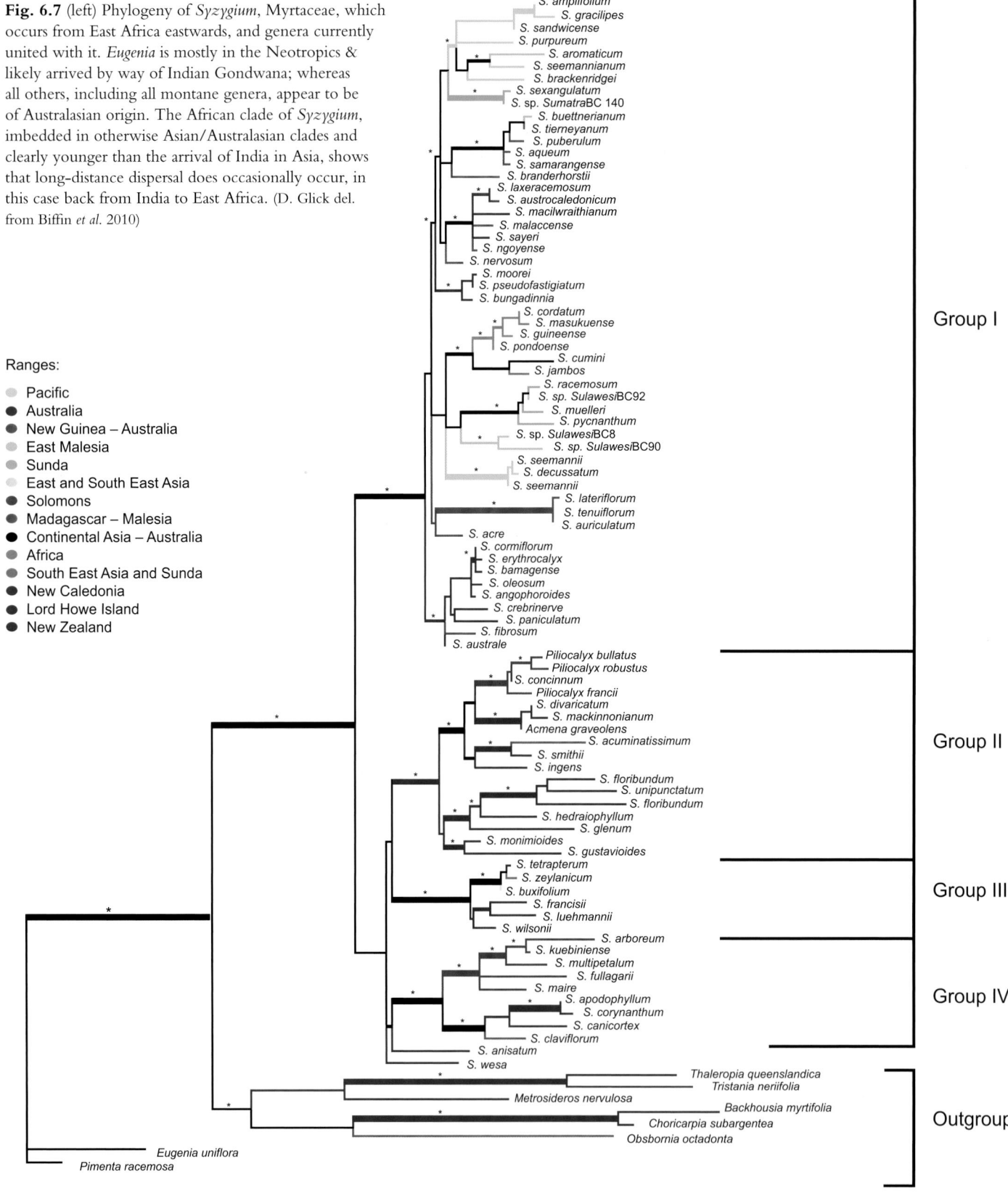

5 changes

The collision of advance fragments of the Australasian plate, which includes most lands west of Borneo, with the Sunda continental shelf occurred in the early Miocene, c.20 Ma. It reduced free ocean flow from the Pacific to the Indian Ocean, with resultant accumulation of warm seas in the south-west Pacific and the initiation of perhumid climates in archipelagos east of mainland tropical Asia (Chapter 1). One early Miocene site in southern Vietnam revealed abundant *Lagerstroemia* pollen, likely indicative of seasonal riverine swamp: seasonality in this region of the South China Sea again appears to have become more intense over time. But fossil pollen of the southern gymnospermous tree family Podocarpaceae (*Dacrycarpus*) appears in the South China Sea records in the Pliocene (c.3.5 Ma)[65] after the disappearance of northern conifer pollen, and the erosion of their mountains.[66] Perhumid taxa such as the lower montane podocarp *Phyllocladus* appear first only in the Pliocene and Pleistocene. Java Sea deposits meanwhile indicate peat swamp there in the early Miocene, with *Austrobuxus*, *Blumeodendron*, *Neoscortechinia* and *Melanorrhoea*. Swamps became extensive again in the late Miocene, at the start of the further cooling that culminated in the northern ice ages. Cyperaceae (suggesting seasonal inundation) alternated with the fern *Stenochlaena* in the Far East (implying continuous waterlogging, peat and perhumid climate).

Given that Australasia arrived only relatively recently at a proximity close enough for substantial plant dispersal, its plant geographic imprint should be relatively more recognisable – and it is. Sea barriers correlate with breaks in plant distributions between the Asian Sunda and the north-east Australasian Sahul continental shelves. The most dramatic of these breaks is the line running immediately east of the Philippines, down between Borneo and Sulawesi and (according to its discoverer, Alfred Russel Wallace) between the Sunda islands of Bali and Lombok[67] (map inside back cover). Terrestrial Gondwanan vertebrates such as marsupials and rattite birds are confined to the New Guinea/Australia Sahul continental shelf, east of a zoogeographical boundary recognized as Weber's Line. Primates, other than some adventurous macaques and ourselves, are confined west of the Line.

Modern north-east Australian lowland rainforests remain dominated by indigenous families and genera, where Proteaceae (24 genera, 47 species of tree), Monimiaceae (9 genera, 16 species of tree), and Myrtaceae (32 genera if *Syzygium* segregates are recognized, 129 species of which 67 are *Syzygium* in the wide sense) (Fig. 6.7).[68] But in the lowland forests of New Guinea and Wallacea these taxa, *Syzygium* excepted, are ecologically subordinate to Asian taxa and are mostly confined to marginal habitats, particularly humult soils. This is yet more so in the lowlands west of Wallace's Line. Everywhere in the tropical Far East, Australasian elements in the tree flora are best represented in montane forests (Chapter 4), where Cunoniaceae, Podocarpaceae, *Agathis* (Araucariaceae) and Myrtaceae of Australasian affinity frequently dominate the canopy, and where the Australasian monotypic Himantandraceae (*Galbulimima*) and Trimeniaceae are confined. Cunoniaceae and the southern gymnosperms originally occurred worldwide.[69] These taxa (*Galbulimima* excepted) are now widespread east of Wallace's Line at low altitudes on ultramafic exposures and siliceous soils (Table 6.5), where the Australasian Casuarinaceae are also best represented. West of Wallace's Line, Monimiaceae of Australasian affinity, *Matthaea* and *Kibara*, are scattered in the understorey (Table 6.6). The podocarp genera *Podocarpus* and *Dacrydium* also *Agathis* are widespread (Table 6.4), sometimes locally dominant in lowland *kerangas*. *Weinmannia* (Cunoniaceae), *Leptospermum*, *Baeckia*, and *Xanthomyrtus* (Myrtaceae) are occasionally found at low altitudes there (Chapter 4), whereas *Tasmannia* (Winteraceae) is confined to upper montane forests, and *Ascarina* (Chloranthaceae) above 1,400 m (Table 6.7, chapter 4). I infer that the current ecological distribution of Australasian and Indian-Laurasian elements in the Far Eastern and Malesian flora reflects the climate and soils of their lands of origin, as well as competitive success.

When their increasingly seasonal and dry climates are taken into account, the Sunda Islands east of Java show no distinct phytogeographic boundary between Bali and Lombok (contrary to Wallace). The Australasian element is surprisingly weak in a region also weak in continental Asian deciduous forest elements (see 3.1.3. below), although the rainforest species *Eucalyptus deglupta* and others reach the

Podocarpaceae	*Podocarpus neriifolius*
Casuarinaceae	*Gymnostoma papuana*
Cunoniaceae	*Ceratopetalum succirubrum*, *Weinmannia fraxinea*
Goodeniaceae	*Scaevola oppositifolia*, *S. sericea*
Myrtaceae	*Decaspermum*, *Corymbia papuana*, *Lophostemon suaveolens*, *Myrtella beccarii*, 2 *Tristaniopsis* spp., *Octamyrtus insignis*, 1 *Rhodamnia* sp., 1 *Rhodomyrtus* sp., 22 *Syzygium* spp., *Xanthostemon petiolatus*
Monimiaceae	*Kairoa suberosa*, *Kibara oblongata*, *Levieria nitens*, *Steganthera hirsuta*
Proteaceae	*Bleasdalia papuana*, *Finschia chloroxantha*, *Grevillea papuana*, *Helicia obtusata*, *Stenocarpus moorei*
Winteraceae	*Zygogynum* sp.

Table 6.5 Australasian taxa in a lowland forest on ultramafic substrate, Papua New Guinea. (After Takeuchi 2003)

Name	Geographical range	Altitudinal range (m)
Trimeniaceae (1 genus)		
Trimenia (3 spp. in Malesia)	W. to Sulawesi	1,000–3,000
Atherospermataceae		
Dryadodaphne (2 spp.)	New Guinea	500–2,800
Monimiaceae	W. to Nicobars (Hortonioideae in Sri Lanka)	
Palmeria (11 spp.)	W. to Sulawesi	(100–) 3,200-4,000
Levieria (7 spp.)	W. to Sulawesi	(500–) 1,200–3,000
Wilkiea (1 spp.)	New Guinea (Papua)	2,400–2,600
Faika (1 spp.)	W. New Guinea	100–1,250
Parakibara (1 spp.)	Halmahera (1 sp.)	?
Kibara (39 spp.)	W. to Nicobars, 2 spp. W. of Wallace's Line	0–2,800
Kairoa (1 spp.)	New Guinea (Papua)	
Steganthera (17 spp.)	1 sp. W of New Guinea to Sulawesi	0–1,200
Matthaea (6 spp.)	1 sp. through Sundaland, 6 spp. in Philippines/Talaud	100–2,280 (–3,250) Lowlands
Himantandraceae (1 genus)	W. to Sulawesi	
Galbulimima (2 spp.)		1,200–2,700
Winteraceae (c.2 spp.)	New Guinea	
Zygogynum (c.20 spp.)	W. to NE Borneo	
Tasmannia (c.1 sp., variable)		100–3,000 1,200–3,000

Table 6.6 Basal Australasian taxa in East Malesia.

Name	Habitat
Philippines only	
Eucalyptus (Myrtaceae)	Lowlands
Aphananthe (Cannabinaceae)	Lowlands
Acacia subgen.Pyllodinae	Lowlands
Clianthus (Leguminosae)	Lowlands
Diploglottis (Sapindaceae)	Lowlands
Metrosideros (Myrtaceae)	Lowlands
Papualthia (Annonaceae)	Lowlands
Pleiogynium (Anacardiaceae)	Lowlands (ultramafic)
Quintinia (Grossulariaceae)	Lower montane
Sararanga (Pandanaceae)	Lowlands (coastal)
Schuurmansia (Ochnaceae)	Lowlands
Stackhousia (Celastraceae)	Lowlands
Pimelia (Thymeliaceae)	Lowlands
Xanthostemon (Myrtaceae)	Lowlands (ultramafic)
Borneo	
Matthaea (Monimaceae)	Lowlands
Kibara (Monimaceae)	Lowlands
Xanthomyrtus, Seorsus, Leptospermum (Myrtaceae)	(Mainly) upper montane
Baeckia, Uromyrtus, Tristaniopsis, Melaleuca (Myrtaceae)	(Mainly) lowlands
Ascarina (Chloranthaceae)	Lower montane
Weinmannia (Cunoniaceae)	Lower, upper montane
Helicia (Protaceae)	Lowlands, lower montane
Styphelia, Trochocarpa (Epacridaceae – Ericaceae)	(Mainly) upper montane, subalpine
Mackinlaya, Osmoxylon (Araliaceae)	Lowlands
Aralidium (Torricelliaceae)	Lowlands, lower montane
Perrotetia (Celastraceae)	Lowlands, lower montane
Scaevola (Goodeniaceae)	Coastal, upper montane
Vavaea (Meliaceae)	Lowlands
Pittosporum (Pittosporaceae)	Lowlands, lower montane
Alphitonia (Rhamnaceae)	Lowlands
Rhysostoechia (Sapindaceae)	Lowlands
Casuarina, Gymnostoma (Casuarinaceae)	Lowlands, lower montane

Table 6.7 Australasian genera occurring west of Wallace's Line.

Ranges:

● Neotropics

● Africa and Neotropics

● Africa

○ Asia

● Australia

○ Pacific

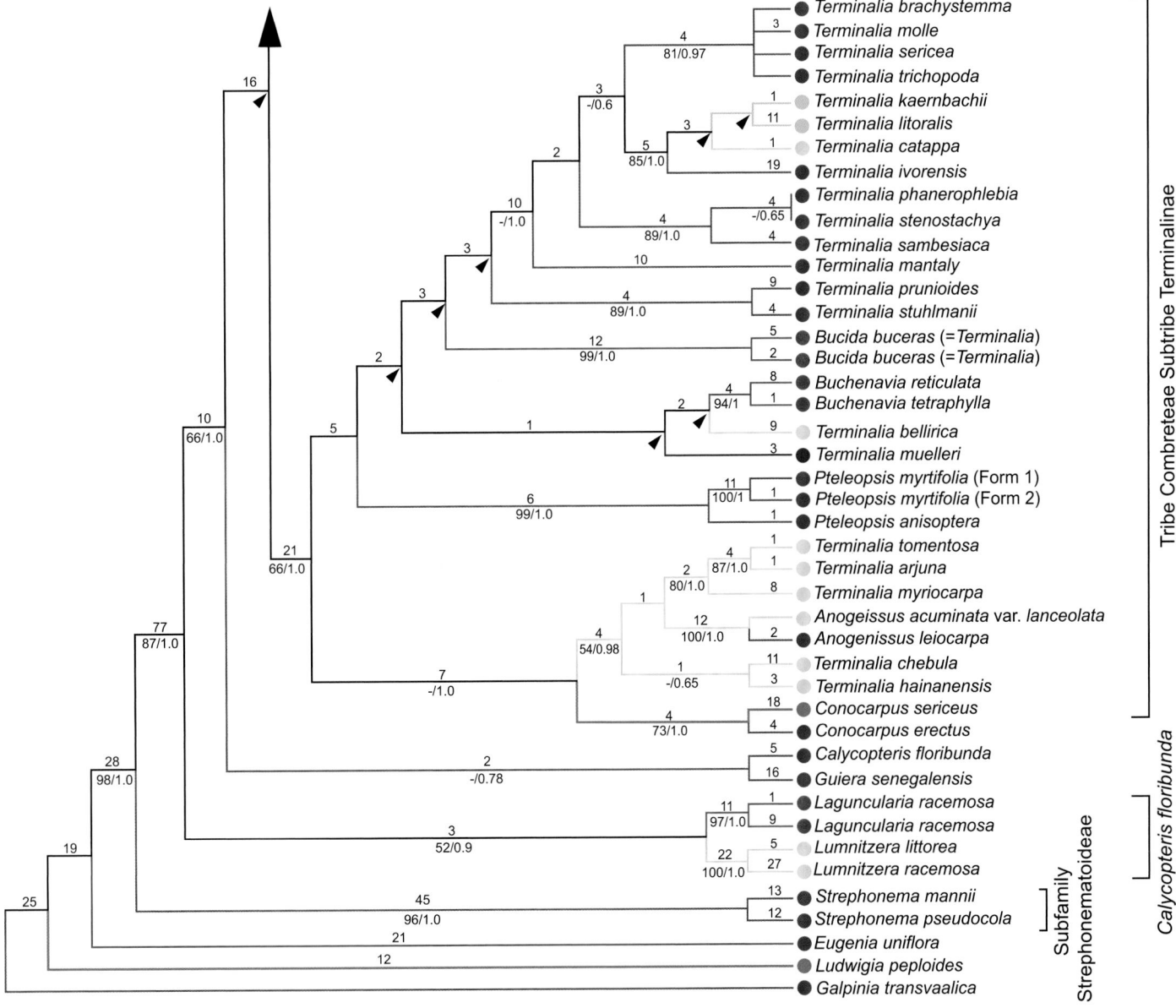

Fig. 6.8 Part of a phylogeny of Combretaceae. Numbers above the branches are measures of distinctness of the divisions, those below of confidence. Most Asian *Terminalia*, within which *Anogeissus* is imbedded, are separated from the African and Neotropical species. The exceptions are the two most widespread: the littoral *T. catappa*, and *T. bellirica*. Migration from Africa to Asia was likely in part by island hopping onto India as it moved north, but in part also during more humid intervals following the collision of Africa with Laurasia. (D. Glick del. after Maurin *et al.* 2010)

Philippines along a perhumid northern track. The savanna species *E. alba* occurs in the eastern Lesser Sundas but not further west than Wetar.

2.1.5. *The Saharo-Sindian element*

Many tree families occur in the wet Asian and New World tropics but not in Africa. That reflects the drying of Africa in the Miocene, and its consequent well-known floristic impoverishment before land routes, albeit dry, were established from Africa across the Middle East to the Asian tropics. In addition, those land routes soon became temperate. Only Pandaceae and the monogeneric Irvingiaceae occur exclusively in the rainforests of Africa and Asia, arriving presumably via India. Trees of Salvadoraceae and the monogeneric Moringaceae are distributed in sub-Saharan Africa and in arid and semi-arid South Asia. Many genera and some species in several families are shared, comprising what has long been recognized as the Saharo-Sindian element in the South Asian flora.[70] Particularly well represented are *Acacia* and other Fabaceae, *Anogeissus* (Combretaceae) (Fig. 6.8), Rhamnaceae, Capparaceae, also Zygophyllaceae and Hernandiaceae – widespread taxa with some representation also in lowland tropical rainforests.

2.2. *Subsequent diversification in tropical Asia*

In all five immigration patterns, very few species remain common to their sources as well as colonised regions. Those that are, such as the widespread Australasian *Tasmannia piperita* of Malesian upper montane forest and *Acacia arabica* of North-East African to South Asian semi-arid woodlands, have effectively dispersed seeds and there are few large gaps between currently existing habitats. Our current Far Eastern rainforest tree species appear to have originated from the lower Miocene (23 Ma) when perhumid climates began to become widespread, and their major diversification to have continued at least into the early Pleistocene, as will be discussed (section 3.3 below).

2.3. *In Asia, where are the ancient flowering plant families concentrated?*

Only eight pan-tropical archaic tree families are widespread and diverse in the lowland tropics of Asia. It is striking that a mere 14 of the c.31 archaic families of tropical and warm temperate Asia and Australasia (excluding New Caledonia) have penetrated into the tropics. In East and North-East Asia, only five of the 16 basal orders, including Saxifragales (Table 6.1), have penetrated. Of those 16, only Lauraceae are well represented in the tropical lowlands. It would appear that early temperate migration routes into Asia have been more

favourable than early tropical, north or otherwise.

Whereas temperate deciduous taxa, such as *Ulmus*, deciduous *Prunus*, *Fraxinus*, *Rhus*, *Alnus* and *Betula* reach similarly low latitudes in both Old and New Worlds, tropical lower montane evergreen Hamamelidaceae and *Nyssa* do not occur in the Neotropics, while evergreen *Quercus* only reach northern Colombia. The distribution of these families implies a northern warm temperate rather than tropical origin.

Nooteboom[71] summarised evidence for a Late Cretaceous North American origin for the genus *Magnolia*, at low altitudes and 45–60° N latitude, which would then in part have been warm temperate. (Although this view has subsequently been countered (see section 2.1.2).) They would then have spread eastwards in similar lowland conditions under the warm conditions of the Eocene, with subsequent Cenozoic diversification in temperate regions and into the tropical lowland and lower montane forests following climate cooling in the Oligocene and Neogene.

It is not necessary to posit a humid tropical track for the immigration of most families of Laurasian origin to the Asian tropics. Phylogenies of Magnoliaceae, Aquifoliaceae (*Ilex*), and Hamamelidaceae (Fig. 6.6), for example, indicate complex histories, the current taxa having diverse origins implying multiple invasions and massive extinctions. Most species appear to be Upper Cenozoic in origin though their genera are of varying age.[72]

Indeed, a similar trend can be seen among Australasian immigrants: among the five basal orders, only Proteaceae and Monimiaceae currently contribute tropical lowland representatives in Asia (though there is a rich tropical African fossil flora). It is apparent that the diversification of flowering plants that migrated to tropical and temperate Asian climates happened very early, and that capacity to reverse that specialisation has been limited. The distinct clustering of families in the larger orders into single distributions and a broadly defined habitat emerge clearly in Table 6.1. Early-evolving familial traits – many of which may appear to have lost adaptive significance (Gould and Lewontin's 'spandrels of San Marco'[73]) – thus continue to impart a fundamental ecological conservatism. Limits to ecological diversification were set early, and may continue to accumulate over evolutionary time. Their constraints on successful migration have been insufficiently appreciated.

2.4. *Summary*

Pan-tropical families had already arisen by the end of the Mesozoic era. The evidence of current distributions can be usefully interpreted in the light of our knowledge of fossil history, albeit still incomplete, and by molecular phylogenies, particularly those that have been calibrated against a time scale by dated fossils. These distinct strands of evidence imply that accumulation of the current tree flora occurred by waves of immigration which occurred principally by five routes:

1. A first wave occurred during the Late Cretaceous (Santonian), when both tropical and warm temerate connections to Asia by way of North America and Laurasia were at their height. The families that arrived are mostly still pan-tropical but not exclusively tropical. A steady but modest trickle of such families may have continued via North America and the southern coast of Laurasia up to the end of the Eocene, when the cooling climate, combined with the increasing breadth of the Atlantic Ocean, halted further arrivals of tropical elements.

2. Temperate families have continued to arrive in warm temperate East Asia since the Cretaceous, at first eastwards from North America but waning as the Atlantic expanded; and latterly from the Eurasian north-east and finally east. A minority of these families diversified within the tropics. Migration westward across the Bering Strait was still possible for them also, until after the mid-Miocene global temperature maximum (c.15 Ma).

3. A major influx of new tropical families came via Indian Gondwana, arriving on the southern Laurasian coast in the Middle Eocene (c.45 Ma). Some of these had invaded India via Madagascar probably mostly during the Late Cretaceous; but most likely arrived onto the Indian Noah's Ark by island-hopping from Africa after it had crossed the southern dry tropics.

4. There was a lesser immigration from Australasian Gondwana, peaking c.16 Ma following collision with Sundaland, of warm temperate families with tropical elements, with some immigrants during the Oligocene (such as *Dacrydium*) and others as late as the Plio-Pleistocene (such as *Dacrycarpus*, *Phyllocladus*). Elements adapted to infertile humult soils prevailed.

5. Immigration of elements from the more seasonal drier and arid regions of northern Africa by way of the Middle East followed the collision of Africa with Laurasia (c.16 Ma). It was facilitated ecologically by the closure of the Indonesian ocean throughflow through Indonesia at that time, and the rise of the high Himalaya, which together originated the south-west monsoon and led to further drying of north-west South Asia (as well as the spread of the perhumid climate of the Sunda lowlands).

3. Our present era: diversification through speciation

3.1. *Barriers to dispersal as drivers of speciation*

Introductory texts on evolutionary biology take it as axiomatic that changes in forest structure or composition, or both, may occur across barriers to dispersal such as ocean straits or high mountain ranges. We can add that sometimes major river valleys and their floodplains play the same role. Such barriers may also arrive *after* the spread of a species, thereby leading to separate diversification of the metapopulations isolated on each side. On the other hand, diversification may have taken place owing to variations in gene frequencies among small founder populations, which originated by chance dispersal of seed across a barrier sufficient to otherwise confine a species to one side. Changes in forest structure are generally correlated with changes in the physical environment (Chapter 2), and can also facilitate diversification. A clear example is the stature and well-developed emergent canopy of forests on zonal yellow-red soils in the perhumid Far East, which has facilitated the dominance and diversification of both canopy dipterocarps and the species beneath them.

And it is also true – and too often ignored by those with limited field experience – that diversification may be driven by variations in soil and topography among regions within a common climate regime. Especially in continental contexts, where a large species pool mixes relatively fluidly, such range boundaries are never absolute. Nor have they generally lasted long enough over evolutionary time (except under exceptional tectonic circumstances) to produce clearly defined, coincident range limits of higher taxa. Although certain genera are delimited in this way where they have become associated with changes in vegetational structure or phenology, such boundaries are mostly apparent at the *species* level.

Several major floristic regional boundaries of this type can be recognized in tropical Asia. Among these, Wallace's Line is the most celebrated, owing to its association with the genesis of the theory of evolution through natural selection – but partly also because it, alone among those to be discussed, is defined by a strait of open ocean alone. But others, less recognized, may occur on land and correlate with changes in the physical environment.

3.1.1. *The Australasian frontier: Wallace's Line*
The biogeographic frontier, recognized by Alfred Russel Wallace 150 years ago, is still recognized by zoogeographers as Wallace's Line (map inside back cover). Indeed, surprisingly little of the indigenous Australasian flora has spread into the Far Eastern lowland tropics west of its Australasian continental, Sahul, shelf. Nonetheless, Australasian Gondwana has re-entered the

403

perhumid tropical zone for 20 million years, shortly after the expansion of perhumid climate through the expanding Sunda lowlands (albeit both had extensively wet tropical climates during the Eocene global temperature maximum). Most of the lowland wet tropical Asian plant families have spread across the Line, though fewer extend into the Bismarck Archipelago immediately to the east, or into Australia. Of the more than 20,000 flowering plant species in families centred in the perhumid Far East, about half are located east of Wallace's Line.[74]

The central track of the line, between Borneo and Sulawesi, marks the eastern edge of the Sunda continental shelf. But the south-east arm of Sulawesi originated as part of south-east Sunda Borneo (Chapter 1), and these land masses were united overland in Eocene times, but appear to have been separated by 47 Ma. That bridge may have served as an initial toe-hold for the eastward spread of the Laurasian tropical flora once Australasia had approached. But the major quantitative and qualitative differences that exist between the floras of Sulawesi and Borneo imply that there has been much subsequent extinction in Sulawesi, and little more than later sporadic sweepstakes dispersal to it of all but the most effectively dispersed taxa.

3.1.1.1. *The Sunda lowland flora moved east.* At the level of the genus, the composition of lowland mixed forests in New Guinea is remarkably similar to that west of Wallace's Line – even with regard to the genera that have independently become most speciose on either side of the Line (Tables 6.6, 3.18). It is at the species level that Wallace's Line has proven effective as a partial barrier to dispersal and as a constraint on hybridisation, thereby facilitating speciation on either side.

East of Wallace's Line, lowland and lower montane forest is currently defined, first, by the dominance of Sunda families and genera; second, by the absence of many Sunda genera combined with burgeoning speciation and endemism of others within the large island of New Guinea (Table 6.8), and to a lesser extent in the intervening northern Sulawesi and Moluccas; and thirdly, by differences at all taxonomic levels among the smaller islands between the Sunda and Sahel shelves:

1. Certain genera of Sunda origin have undergone exceptional diversification in and around New Guinea (Table 6.8). Most of these are also diverse west of the Line. The largely pioneer genus, *Macaranga*, is richer in lower montane taxa in New Guinea than anywhere else in its African, Madagascan and Asian range, apparently in response to the steep slopes and frequent earthquake-induced landslips of that mountainous island.[75]

2. Other genera have diversified in Sundaland and have reached New Guinea, but have not diversified there. These include *Grewia, Knema, Memecylon, Xanthophyllum, Aporosa, Dacryodes, Polyalthia, Mangifera* and lowland Fagaceae.

3. Many widely-dispersing pioneer and successional species, such as *Alstonia scholaris, Octomeles sumatrana, Tetrameles nudiflora* and *Pometia pinnata*, occur on both sides of the line. The lattermost species has evolved an East Malesian subspecies.[76] (Fig. 6.13; Table 6.18; Plate 3.10).

4. The two tree families in tropical Asia with the most poorly-dispersed seeds are Dipterocarpaceae and Fagaceae. Their ability to migrate is constrained not only by their limited means of seed dispersal and lack of dormancy, but likely also by their dependence on host-specific ectomycorrhizas, which must migrate with them. There are eight Fagaceae in Sulawesi (all endemic) and 141 in Borneo; *Quercus* is absent from the Philippines and east of the Line. The seven Sulawesi dipterocarps (in three genera) hardly compare with the 269 species in Borneo. Several have close sister species in the Philippines and/or Borneo, but only two species are found in both places.[77] Both Fagaceae and Dipterocarpaceae have had difficulty in dispersing to Sulawesi, which split from the Sunda Shelf 47 Ma, long before the first appearance of these two plant families in the Far East during the Oligocene (Chapter 1).

5. Among Fagaceae, *Quercus* has failed to spread to the Philippines (with the exception of Palawan, the single island of the archipelago on the Sunda continental shelf).[78] Neither does *Quercus* extend along the Sunda island chain east of the wet West Javan rainforests, even though lower montane forests are present on Sumbawa.[79] *Engelhardia spicata* and *Ulmus lancefolia* successfully reached Timor, possibly because once established they are tolerant of fire, but possibly also because both have wind-born seed.[80]

6. Five of Sundaland's nine dipterocarp genera presently occur in the West Javan southern seasonal evergreen forest, and one *Dipterocarpus* species does cross east to Sumbawa.[81] Dipterocarp fossils are abundant and widespread in the Javan Pliocene, and a tree of *Hopea sangal* still survives a moist pocket within the lowlands of strongly seasonal Pasuruan Regency, East Java. The presence of *Dipterocarpus* in Sumbawa east of Wallace's Line implies that the confinement of Fagaceae west of that Line is due to an ecological barrier as much as dispersal limitation.

7. Deciduous forest with *Tectona* is present in East Java and in south-west Sulawesi, where it has long been cultivated and may have been introduced, but *Tectona* does not occur east of Java along the Sunda island chain, even though suitable habitat is abundant (and see section 3.2.5 below).

8. The Lesser Sunda Islands have been subject to dry southeasterly trades ever since continental Australia entered the tropics, whereas the north-east trades have always come in off the Pacific, except in the lee of the Philippine islands. It is therefore likely that the eastward spread of the Sunda lowland rainforest flora —already more than five million years in development and therefore potentially highly competitive with its newer, less rich Australasian analogue — took place across the north through Sulawesi, the Moluccas, and latterly the Philippines. This reflects current distributions.

Calophyllum (Calophyllaceae)	50
Canarium (Burseraceae)	29
Cryptocarya*, Endiandra*, Litsea (Lauraceae)	87,52,55
Dysoxylum, Aglaia (Meliaceae)	
Elaeocarpus* (Elaeocarpaceae)	85
Fagraea* (Loganiaceae)	20
Ficus (Moraceae)	156
Garcinia (Clusiaceae)	60
Harpullia*(Sapindaceae)	15
Helicia* (Proteaceae)	81
Hopea section, subsect. *Hopea* (Dipterocarpaceae)	13
Horsfieldia, Myristica* (Myristicaceae)	100
Macaranga* (Euphorbiaceae)	87
Medinilla, Memecylon (Melastomaceae)	85,50
Melicope *(Rutaceae)	59
Palmeria* (11), *Kibara** (36), *Steganthera**(16) (Monimiaceae)	
Pandanus (Pandanaceae)	60
Prunus (Rosaceae)	14
Psychotria, Timonius* (Rubiaceae)	179, 97
Rhododendron sect. Schistanthe* (Ericaceae)	143
Saurauia (Actinidiaceae)	100
Syzygium (Myrtaceae)	208
Schefflera (Araliaceae)	60
Terminalia (Combretaceae)	50

Table 6.8 Species-rich tree genera in New Guinea, with the number of species in each genus. *Nothofagus*, *Helicia* and the Monimiaceae are of southern affinity, while the Laurasian taxa represent a selection, but do not include all, those which are also rich in the Sunda lands, while *Harpullia* is less rich in Sunda. Asterisked genera are exceptionally rich compared with those in West Malesia. (Partly supplied by L. Jennings)

Kerangas
Agathis borneensis (Araucariaceae)
Dacrydium elatum (Podocarpaceae)
Gymnostoma nobile (Casuarinaceae)
Styphelia (Epacridaceae)
Baeckia, Uromyrtus, Tristaniopsis (Myrtaceae)
Scaevola (Goodeniaceae) (1 coastal)

Upper montane forest
Agathis spp., (Araucariaceae)
Dacrydium, Podocarpus, Dacrycarpus, Falciforme, Phyllocladus (Podocarpaceae)
Gymnostoma (Casuarinaceae)
Ascarina (Chloranthaceae)
Weinmannia (Cunoniaceae)
Scaevola (Goodeniaceae)
Leptospermum, Soersus, Xanthomyrtus (Myrtaceae)
Drimys piperita (Winteraceae)

Table 6.9 Australasian taxa in the Sundaland *kerangas* and upper montane forest.

3.1.1.2. *The spread of the Australasian flora westward proved more challenging* In the Asian lowlands today, most Australasian taxa are relegated to marginal habitats in rainforests both east and west of Wallace's Line. They are only common in habitats where they receive less direct competition from other species even in New Guinea, such as in *kerangas* along the coastal foothills of the Lorentz mountains, on limestone karst on the Vogelkop Peninsula, and on the widespread ultramafic exposures,[82] and in the upper montane and subalpine forests (for example Tables 6.8, 6.9).

New Guinea and Australia ride the same Sahul continental shelf, and were connected by land as recently as 12,000 years ago.

The rainforests of Queensland – well within the tropics to the north – are about as seasonal as south-east New Guinea, yet the number of tree genera shared with West Malesia are only 220 out of a modest total of about 600 (depending on one's definition of a tree).[83] Parts of eastern Australia had been wet until the Pliocene. The relative poverty of the Queensland rainforest tree flora (encompassing only 600 tree species), in spite of the long history of wet tropical climate, must be attributed to a Pleistocene history of periodic aridity and massive extinction of indigenous elements and retreat, recalling the Western Ghats. Malesian elements are meagerly represented probably in part because the lowlands, formerly connecting Queensland to New Guinea, have been seasonally dry since the time when Australasia rafted north into the tropics.

Today, dry tropical savanna dominates in the Asian Far East only on the central-southern New Guinea peninsula between the Digul and Fly Rivers, but similar vegetation extends in the Lesser Sunda Islands in the lea of Australia. The contribution of Australasian elements to the surviving flora of the Lesser Sunda Islands and southern Wallacea is nevertheless meagre, and human influences have further obscured their presence. The Autralasian *Eucalyptus* occurs in savannas from southern New Guinea west to Timor and Wetar. Van Steenis[84] lists few Australasian elements that reached Java. Among the woody plants, he mentions only *Coprosma* (Rubiaceae) whose only recent geological record implies distant dispersal; and *Styphelia* (Epacridaceae) and *Weinmannia* (Cunoniaceae) – all taxa of wet forests that might have arrived via a northern track.

There is a southern concentration of archaic evergreen flowering plant families in the tropical and warm temperate rainforests running down the eastern Australian coast. Austrobaileyaceae, and Australian Hamamelidaceae are endemic to north-east Australian rainforests. Himantandraceae, Trimeniaceae and *Nothofagus* (Nothofagaceae: southern ecological analogue of northern *Quercus* subgen. *Cyclobalanopsis*) have reached New Guinea and the Australasian continental shelf (Weber's Line), while *Trimenia*, *Quintinia* and *Sphenostemon* have reached Sulawesi. Winteraceae, many *Podocarpus*, *Agathis* (Araucariaceae), *Ascarina* (Chloranthaceae), and Monimiaceae all represent archaic southern groups that have crossed Wallace's Line into Sundaland.

The advanced southern families Casuarinaceae, Cunoniaceae and Goodeniaceae, and the dry fruited Australasian Myrtaceous taxa *Babingtonia*, *Baeckia*, *Leptospermum*, *Uromyrtus* and *Xanthomyrtus*, also successfully invaded westwards. The most successful taxon of probable Australasian origin is the vertebrate dispersed *Syzygium* (Myrtaceae), largest genus of trees in Malesia, with many species in South Asian evergreen forests at all altitudes also, and a presence in East Africa where it must have arrived from the east by long distance dispersal as South Asia moved north into Laurasia. East of Wallace's

Plate 6.4 The Australasian element. **a,** *Gymnostoma nobile* along the crest of the sandstone Lambir summit, 465 m, Sarawak; **b,** *Helicia* sp. (Proteaceae), Daintree lowland forest, Queensland. (**a** P. Hall; **b** P.A.)

Line, Himantandraceae and Trimeniaceae rarely occur below 500 m, while many other Australasian elements are restricted in the lowlands to open or limiting environments, including *kerangas*, ultramafics, limestone karst, and rocky coastlines.

West of Wallace's Line, elements of southern warm temperate families in tropical lowland MDF include, besides *Syzygium*: (1) *Matthaea* and *Kibara*, the only Monimiaceae in Sundaland, both of which are rare understorey treelets except in the Philippines where *Matthaea* has diversified; and (2) the two tropical Asian Proteaceae *Helicia* (Plate 6.4) and *Heliciopsis*. The former of these may have entered via India, along with a sister species in Madagascar (*Malagasia*).[85] By contrast, the northern tropical-warm temperate archaic *Exbucklandia* (Hamamelidaceae), *Nyssa*, and *Altingia* (Altingiaceae), and the largely temperate *Ulmus*, invaded the wet tropics through lower montane forest (see section 2.1.3 above). Australasian flowering plant genera are again best represented in the species-poor communities of upper montane forest and of extreme habitats, especially rocky defiles and seashores (*Casuarina equisetifolia*, *Scaevola suaveolens*) in the lowlands; here, *Baeckia* is widespread and locally common, *Leptospermum* and *Xanthomyrtus* both rare, and *Weinmannia* (Cunoniaceae) only once recorded in the

lowlands in a Central Kalimantan peat swamp forest. Consistent with prevailing low nutrient soils in their land of origin, several Australian gymnosperms have become abundant in relatively species- and nutrient-poor habitats in the lowlands, although the canopy trees *Podocarpus neriifolius* and *Nageia motleyi* are locally well represented even in MDF (also in Indo-Burmese lower montane forest, the latter possibly of Indian origin). *Agathis borneensis* is locally dominant in *kerangas*, *Dacrydium elatum* (*beccarii*) (Podocarpaceae) in coastal mixed peat swamp in Borneo, while *Dacrycarpus imbricatus* and *Dacrydium elatum* (*pierrei*) are still locally abundant in the remaining Pleistocene coastal beach *kerangas* in southern Cambodia.

I infer that Wallace's Line has been less of a constraint on the dispersal of plants than it has of terrestrial vertebrates, and that the constraint has been substantially ecological. Whereas the tropical Asian flora has invaded the lowlands of eastern Malesia, the Australasian flora has mostly spread westwards along the chain of upper montane forests leading into continental South-East Asia (see section 2.1.4 above). The current plant geography of eastern Malesia should be interpreted to be at least as much determined by climate and soil as by history: compare the map inside the front cover with Fig. 6.9.

3.1.2. *The Philippines as a stepping stone*

The Philippines appear to have been on or close to the main eastward route for the Malesian lowland rainforest flora since their arrival as an oceanic volcanic island arc from the north-east. The biogeographer E.D. Merrill in 1926 estimated that, of 356 Sunda genera unknown east of Wallace's Line, 218 (61%) do occur in the Philippines including Palawan.[86] But there is also a strong Australasian wet forest presence at all altitudes there, including *Eucalyptus deglupta* in the Mindanao lowlands: 25% of the c.225 genera otherwise unknown west of Wallace's Line are recorded from the Philippines. Merrill at first argued for the major plant geographic dispersal barrier to be west of the Philippines, except for Palawan which is on the Sunda shelf.[87] But the Philippine lowland forests (Palawan probably excepted) were until recently clothed with MDF almost identical at the generic level to the Bornean, and with similar generic relative abundances. Some Asian genera, including evergreen *Quercus* (*Cyclobalanopsis*), are missing. Their forests were less species-rich – although no census has yet been taken in the richest forests of the south-east. The MDF in the CTFS 16 ha Palanan plot on the eastern coast of northern Luzon is about half as rich as a Bornean equivalent (Table 3.7, Chapter 7). *Ecologically* considered, the Philippine MDF are firmly West Malesian – extensions of the Sundaland forests in spite of the infusion of Australasian elements.

There are few endemic *genera* in the Philippines: Merrill recorded only 47 (4%) endemics of 1,308 total genera.[88]

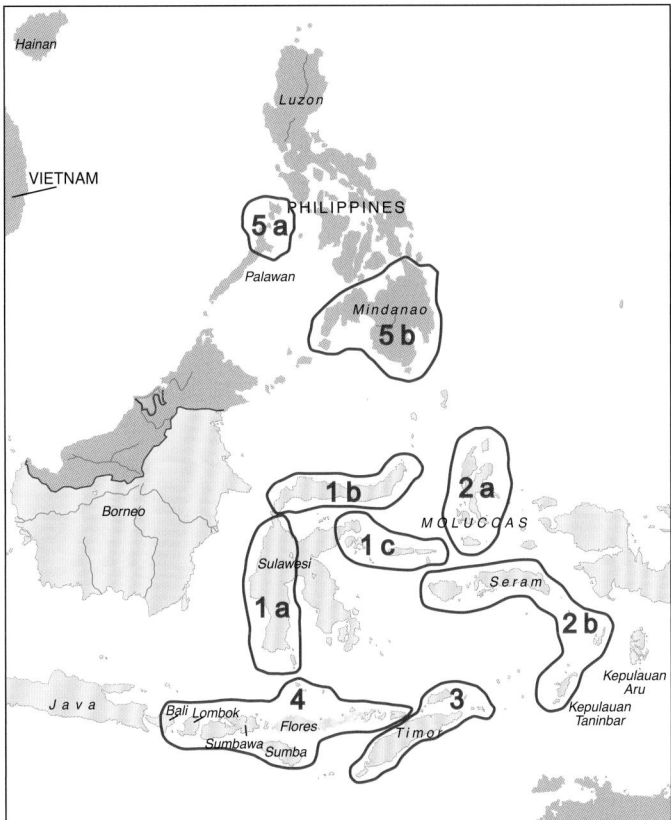

Fig. 6.9 Plant geographical regions in Wallacea according to Michaux (2010). Some, at least, are more correlated with differences in rainfall seasonality (see map within front cover) than tectonics, and may therefore have an ecological rather than historical explanation. (I.H. Baillie del.)

However, he estimated *species* endemism at 75% of a total of 7,620 angio- and gymnosperms. Tree species endemism in those families already revised by *Flora Malesiana* still exceeds 70%. These percentages are comparable with those for perhumid south-west Sri Lanka (see section 3.2.9.1 below). Unlike plants with shorter life cycles, tree species endemism is low *within* islands of the Philippine archipelago, and species ranges tend to be mediated instead by rainfall seasonality (Chapter 3). (In this, as so often, Palawan is the exception – see section 3.1.2.2 below.) Many Philippine endemics, however, are closely related to sister species in Borneo that differ only in minor characters (Table 6.10).

Philippines	Borneo
Anisoptera aurea	*Anisoptera grossivenia*
Dipterocarpus orbicularis	*Dipterocarpus confertus*
Parashorea malaanonan (also in NE Borneo)	*Parashorea lucida*
Hopea malibato	*Hopea dryobalanoides*
H. brachyptera	*H. bracteata*
H. acuminata	*H. sangal*
H. basilanica	*H. andersonii*
H. samarensis	*H. megacarpa*
H. philippinensis	*H. tenuinervula*
H. mindanensis	*H. enicosanthoides*
Shorea malibato	*Shorea hypoleuca*
S. astylosa	*S. domatiosa*
S. polita	*S. gratissima*
S. negrosensis	*S. dispar*
S. almon (also in NE Borneo)	*S. parvistipulata*
S. polysperma	*S. johorensis*
S. palosapis	*S. fallax*

Table 6.10 Sister dipterocarp species in the Philippines and Borneo.

In contrast to the case of the vertebrates, it is hazardous to infer original plant migration tracks across the two island chains connecting Borneo and the Philippines,[89] owing to the present ecological barriers of ultramafic and limestone substrates in Palawan in the north, and basalt on Basilan in the south. Remarkably, the distance between Banggi Island, with its (impoverished) Bornean flora, and Balabac, the nearest island with a true Philippine flora, hardly exceeds 50 km of open water. This nevertheless implies frequent and recently prolonged archipelago-wide separation

from Borneo, at least since the Pliocene c.3 Ma (see section 3.1.2.2 below). Given what this implies about the ability of many tree species to disperse across ocean barriers, it is perhaps unsurprising that dipterocarps and many other tropical taxa are entirely missing from the southernmost tropical lowlands of Taiwan north of Luzon (22° N), even though the Babuyan islands 300 km to the south, across the Basho channel but just north of Luzon, formerly included populations of *Anisoptera*, *Hopea*, *Shorea* and *Vatica*. The poverty of the Taiwan seasonal rainforest tree flora must be attributed to its long isolation on the northeastern side of the expanding South China Sea.

3.1.2.1. *A seasonal climate in the west and north, but with a poor seasonal forest flora*. The poor representation in the Philippines of Indo-Burmese deciduous forest species is well known, notwithstanding widespread areas (including several islands) of climates with five and six dry months (Table 3.11).[90] The presence of the seasonal evergreen and semi-evergreen forest species *Dipterocarpus alatus*, *Berrya cordifolia*, *Lagerstroemia speciosa*, *Phoenix loureiroi*, *Pinus kesiya* and *P. merkusii*, *Pistacia chinensis*, *Protium connarifolium* (endemic), *Radermachera pinnata*, *Schleichera revoluta* (endemic), *Tectona hamiltoniana*, *Terminalia calamansanay* and *Toona sureni*, either Indo-Burmese or sister to Indo-Burmese taxa, and bearing fruit with diverse dispersal mechanisms, must be attributable to sweepstakes dispersal over a sea barrier. As must the apparently random assemblage of northern and montane elements including *Anneslea*, *Buxus*, *Taxus*, *Fraxinus*, *Gleditsia*, *Photinia* and many montane grassland herbaceous genera.

3.1.2.2. *Palawan: odd island out*. The Philippine island of Palawan lies on the Sunda continental shelf, but its nearest neighbour to the north (Mindoro). Connections to its south, from north-east Sabah across to Basilan and south-west Mindanao via the Sulu islands, all lie across deep water channels. Past migration among them may always have been over narrow sea barriers, and for most rainforest taxa all connection may have been curtailed since at least the Miocene. Palawan has enjoyed periodic contact with Borneo during periods of low sea level since it reached its present position 10 Ma, and is currently perhumid south of its dividing range, yet it lacks – remarkably – the largest dipterocarp genus, *Shorea*. This genus has no less than 135 species in Borneo and 17 in the Philippines, although relatively few in seasonal Mindoro and north-west Luzon. There is only a single, tentative record of *Hopea* from Palawan also (a sterile specimen in the Philippines National Herbarium), but 12 from elsewhere in the archipelago.

Much of Palawan is ultramafic or limestone, with each substrate bearing the distinctive vegetation and flora associated with it, but there are lowland plains – now deforested –

which once bore MDF dominated by *Dipterocarpus hasseltii* and *D. grandiflorus*. Palawan also has a Hamamelid genus, *Embolanthera*, which also occurs in Indo-China but not elsewhere. *Embolanthera* may have originated before the South China Sea opened up in the Oligocene, when Palawan would have been attached to the Asian mainland close to northern Vietnam. Interestingly, it grows along brooks, as does *Protium connarifolium*, sole member of this genus of continental Asian deciduous forests. Palawan's ultramafic vegetation has strong Borneo connections, including *Gymnostoma nobile* (Casuarinaceae) and *Vatica mangachapoi* (Dipterocarpaceae).

Its impoverished MDF flora, compared with other Philippine islands, might be explained by Palawan's relatively recent arrival at its present position c.10 Ma, before which the archipelago otherwise was closer to Borneo. But it might be more parsimoniously explained by arid conditions on the island during the long Pleistocene low sea level periods (evidence for such conditions has been found in fossil bat guano[91]). Even at present, most of Palawan has a 2–5 month dry season. Owing to Palawan's past climate, as well as its unfavourable substrates, migration of the lowland Sunda MDF flora from Borneo into the Philippines most likely took the perhumid southeastern route through the Tawi-Tawi archipelago into Basilan and Mindanao.

3.1.3. *Still-enigmatic Wallacea*

Wallacea's plant geography is complex, and understanding it has not been helped by the poor state of knowledge of some islands. A plethora of islands of varying size, geology and tectonic history are isolated between the Sunda and Sahul continental shelves. It is unsurprising that they are floristically relatively poor, yet they differ greatly from one another both in overall flora and in dominant taxa. Many islands support areas of seasonal climate, increasing toward the south as they come under the climatic influence of northern Australia. Michaux[92] has recently related their floristic geography to geology, but current climate seasonality (judged by the few meteorological stations in the region) also correlates, and may provide – at least in part – a more parsimonious explanation.

3.1.4. *The isolated rainforests of South Asia*

Upon completing ten years of fieldwork in Borneo, in 1970 I visited Sri Lanka's southwestern perhumid zone for the first time. Forest ecologists in Malaysia had until then worked in isolation, so nothing had prepared me for the great similarity of their lowland rainforests. Notwithstanding big differences in the size of their tree floras, the forests proved to be so similar as to represent the same formation: mixed dipterocarp forest (MDF). The families and genera were overwhelmingly the same, and ranked in similar order of abundance; only the species were different. Sri Lankan and Sunda MDF are

separated by 2,000 km and an ocean, and they are situated on land with divergent tectonic histories. Yet, if one delves deep enough, they must share a remote history.

Thirty years later, I first visited the 50 ha plot established by the Indian Institute of Sciences, under R. Sukumar, H.S. Suresh, H.S. Dattaraja and their colleagues at Mudumalai Wildlife Sanctuary (Tamil Nadu). Set in the ecotone between tall and short ('moist and dry') deciduous forest not far from the western base of the high Nilgiris, and with a rich fauna, Mudumalai is paradigmatic of the taller South Asian deciduous forests. Teak, once dominant but currently recovering from a harvest in the 1970s, recalls analogous forests – and similar soils – in northern and western Indo-Burma. So do many other of its tree species, but the abundant species Anogeissus latifolia, Terminalia crenulata, Ougeinia oojeinensis, Dalbergia latifolia and Pterocarpus marsupium were all unfamiliar to me. None of these occurs in Indo-Burma, where they are replaced by similarly abundant congeners.

The South Asian rainforest flora bears the imprint of isolation.[93] Compared with the Far East, it is depauperate in tree families, genera and species. Remarkably, apart from *Magnolia* (*Michelia*) (Magnoliaceae) and *Ilex* (potentially of Gondwanan origin), none of the northern lower montane families, archaic or otherwise, currently appear in the Ghats rainforests. Most dramatically, Fagaceae, which dominate the canopy of both tropical lower montane and warm temperate Himalayan forests, are absent. The temperate deciduous genera *Alnus* and *Betula* occur south to 18°50' N in Indo-Burma, yet fail to reach Meghasena (21°40' N) in the Eastern Ghats. The genera present are broadly the same as in Far Eastern forests. But in the lowland rainforests there are absences too, notably the canopy and emergent leguminous genera *Koompassia*, *Sindora* (though with one African species) and *Parkia*. With the absence of *Shorea* (the two South Indian species, *S. roxburghii* and *S. tumbuggaia*, are confined there to short deciduous forest) the forest is consequently often devoid of true emergents, which give them a distinctive facies even in wind-protected valleys; though *Dipterocarpus* is present.

One unequivocal Gondwanan element remains yet in the Sri Lankan mountains. The endemic and isolated subfamily Hortonioideae of the archaic Gondwanan Monimiaceae contains a single genus, *Hortonia*, with a montane and a whitewater river bank species (see section 2.1.2 above). This taxon is ancient (Fig. 6.4; Plates 6.1.a, b).

For its size (65,610 km²), Sri Lanka is the richest area of South Asia, with 1,052 genera and 3,210 species of flowering plants. Ninety percent of these species are confined to the c.10,000 km² of the perhumid south-west[94], where an extraordinary 95% are endemic to it or other South Asian rain forests.[95] Endemism in MDF tree communities exceeds 70%.[96] Sri Lanka possesses a full complement of dipterocarp genera, but few endemic genera overall – and all of them restricted to its southwestern perhumid zone. Individual tree communities follow the same pattern of species richness and abundance as that of the Far East (Table 3.7) but, at c.200 species, individual communities are only one quarter as diverse (Chapter 7). Endemic species are few in the seasonal three-quarters of the island.

The tall deciduous and evergreen forests of the Western Ghats, by comparison, harbour 58 endemic genera.[97] Forty-nine of these are monotypic, though *Nilgirianthus* (Acanthaceae) has 20 species. Of the c.5,000 flowering plant species there, only 34% are endemic; yet of c.500 tree species exceeding 10 cm gbh, 352 (70%) are endemic.[98] Individual lowland seasonal evergreen dipterocarp forest communities in the Western Ghats, under a short dry season, may have 50% tree species endemism.[99] They may include c.150 species, three-quarters as many as in perhumid south-west Sri Lanka; each, again, has only one quarter the species richness of their Far Eastern analogues.

Regional and community tree species diversity in South Asia's rainforests is impoverished by global standards, either on account of their relatively small current area or past catastrophe. When the monsoon abated during temperate ice ages (Chapter 1), rainforest would have retreated to refuges where onshore winds brought reliable orographic rainfall. Fossilised lion remains from Pleistocene Ratnapura, in the centre of south-west Sri Lanka's perhumid lowlands,[100] and widespread laterite pans, clearly inactive and often buckled, in the soils of its southwestern lowlands, imply that evergreen rainforest retreated to upland valley refuges. It may have survived only in diurnal cloud shadow areas or within the cloud base on mountain slopes. The pattern of tree species richness in Sri Lankan rainforest landscapes may also be suggestive of Pleistocene extinctions: unlike anywhere else in Asia, diversity is highest at c.500 m, not in the lowlands.[101]

Plate 6.5 Shared genera between MDF in Sri Lanka and western Malesia. **a,** *Agrostistachys borneensis*, Euphorbiaceae, Sinharaja forest, Sri Lanka; **b,** *Diplodiscus verrucosus*, Tiliaceae, in semi-evergreen notophyll forest at the Giritale tank, Sri Lanka. (H.H.)

South Asia		South-East Asia	
Name	**Habitat**	**Name**	**Habitat**
Dipterocarpus hispidus	MDF	*D. baudii*	MDF, Southern seasonal EDF
D. zeylanicus	MDF	*D. turbinatus*	Semi-EDF
D. indicus	Southern seasonal EDF	*D. turbinatus*	Semi-EDF
D. glandulosus	MDF	*D. costatus*	Upper dipterocarp forest, southern seasonal EDF
Hopea brevipetiolaris	Semi-EF	*Hopea reticulata*	Northern seasonal EDF
H. ponga	Semi-EF	*H. oblongifolia*	Southern seasonal EDF
H. parviflora	Southern seasonal EDF	*H. odorata*	MDF, Southern seasonal EDF-Semi-EDF
Shorea dyeri	MDF	*Shorea guiso*	MDF, Southern seasonal EDF
S. stipularis	MDF	*S. hypochra*	MDF, Southern seasonal EDF

MDF: mixed dipterocarp forest; EDF: evergreen dipterocarp forest; EF: evergreen forest

Table 6.11 Sister dipterocarp species in South and South-East Asia, and their forest formations.

Habitat	Endemism	Peninsular & NW India	NE India/E. Himalaya	Burma[a]
Predominantly evergreen forest	Regional	19	13	34
		1		27
	Widespread	8		
Predominantly deciduous forest	Regional	25	6	34
		4		10
	Widespread	21		

[a] Endemic to Burma, or from Burma eastwards

Table 6.12 Tree species ranges in continental seasonal Asia, shown as percentage of species that are regionally endemic through to widespread by forest type; N= 4302. (Percentages estimated from Brandis 1907, Kress *et al.* 2003)

The rainforest flora of South Asia shares c.95% of its tree genera with Far Eastern rainforests but less than a quarter of its species.[102] This is true despite the fact that the Western Ghats and the westernmost Far Eastern forests in North-East India, Bangladesh and Burma share similar rainfall regimes. Few genera *restricted* to perhumid climates are shared (Plate 6.5). However, morphologically close sister species in the two regions are common in the larger genera. Interestingly, species in perhumid Sri Lanka are, as a rule, sisters to Far Eastern species of the seasonal evergreen and dediduous forests, rather than of the perhumid dipterocarp forests (Tables 6.11, 6.12).

There are relatively few and scattered Cenozoic sediments in South Asia, except in the Siwalik foothills of the Himalaya. The Siwaliks yield fossil evidence of a greater representation there of Far Eastern tree genera up to the mid-Miocene thermal maximum (c.12 Ma: the dating of earlier reports has been revised by Morley[103]). Morley[104] suggests that the return of an at least seasonally wet climate was associated with the rise of the Himalaya, which reached heights approaching the present only at the mid-Miocene thermal maximum, when the Indian monsoon is thought to have originated. The southern Laurasian coast, near the Equator at the time of the collision and then surely in a perhumid climate, has gradually shifted north and beyond the tropics to its current place in the Himalaya. In the present era, the Gondwanan continental plate continues its northward creep beneath the mountains, carrying moist premontane lowlands at the foot of their southern flanks. This creep has therefore carried the well-travelled east–west biotic corridor into increasingly cooler and more seasonal climates. The continuing northward movement of Gondwanan India, the rise of the high Himalaya from the early Miocene to heights comparable to the present day, and the consequent inception of the monsoon, resulted in rarer wet tropical corridors between South and East Asia, and exacerbated rainfall seasonality. On the evidence of current species distribution patterns (Table 6.11, Fig. 6.3), a corridor with less than four dry months ceased to exist before the origin of the extensive perhumid Sunda lowlands (c.23 Ma). The monsoon has been erratic and even absent over long periods during the last two million years, and the current wet period of high sea levels is likely to be almost as moist as at any time since the Pliocene and earlier. Fossils nevertheless indicate a rich generic dipterocarp flora in Central Java, even now with too long a dry season for evergreen dipterocarp forest.

3.1.6. *A neglected biogeographical frontier: Chatterjee's Partition*

The pockets of moist semi-evergreen forest that exist in valleys of the higher hills of the Eastern Ghats north to Meghasena (Odisha) contain a flora with greater affinity to that of the Western Ghats (1,200 km to the west-southwest) than to that of the Khasi hills of Meghalaya (half that distance to the north-east). None of the Himalayan tree families that are absent in the Western Ghats are present in the Eastern Ghats, though most are present in the Khasi Hills, isolated from the Himalaya by the Brahmaputra valley, including Hamamelidaceae. The plant geographic boundary between the evergreen forests of India and Indo-Burma is therefore a partition made by the deciduous forests of the Gangetic Plain and lower valley.[105] But whereas the rainforests of South Asia and the Far East are currently separated by over 1,500 km, the tall deciduous forests (which include the dry dipterocarp and sal forests of the two regions) are separated by a gap of only a few tens of kilometres: that is, the evergreen forest barrier between the dry valleys that

run west into India and east into Burma along their political frontier. (Although only a single collection of sal, from Imphal (at Kew), is known to me from the western Indian frontier regions, it is claimed by foresters to be more widespread there.) Seasonal evergreen dipterocarp forests that catch the monsoon rains separate these valleys. They run down the high ridges, which curve south from the Himalaya, through Manipur and Mizoram down the frontier into the Chittagong Hill Tracts of Bangladesh, down the Burma coast as the Arakan Yoma, and in the Bay of Bengal where they occur again as the Andaman Islands (Plate 6.6a). Their ridges are the bow-wave of the still-advancing Indian Plate (Chapter 1).

The greatest opportunity for exchange for rainforest elements eastwards must have existed immediately following the South Asian collision, when the collision zone was still in the wet equatorial tropics and the Himalaya were as yet low hills. Opportunities for these elements must then have gradually diminished as the Tethys receded, and almost entirely ceased as India continued to move north into the seasonal tropics and the Gangetic plain became increasingly dry. The development of the south-west monsoon 12 Ma meant that the Indus-Ganges lowlands and plains turned more strongly seasonal. But the first rising of the Himalaya in the late Oligocene, and of the India-Burma frontier hills associated with the advancing Indian subduction front perhaps 12 Ma, must have acted as a major brake on lowland deciduous forest plant migration between South and East Asia. But for the seasonal evergreen forests these ridges were by contrast a highway for migration south and eastwards.

The lowland and lower montane tropical evergreen forests of the central and eastern Himalaya do not comprise

Plate 6.6 The parallel turbidite ridges, with clay loam on slopes, that likely acted as migration routes for the tree flora of Far Eastern rainforests. **a,** The northern Arakan Yoma, Burma; **b,** Ulu Rejang, central Sarawak, Malaysian Borneo. Scales: 1/20,000. (**a** Second World War photography; **b** Sarawak Lands & Survey Dept.)

a floristically distinct province from that of northern Indo-Burma, as had been thought by earlier plant geographers of India.[106] The present boundary, as I define it here, was first recognized by D. Chatterjee in 1940,[107] although he did not attempt to map its precise position. Nor did he recognize that the transition from Indian to Indo-Burmese floras does not occur along a single line, but on either side of *two* ecological barriers: (1) A north Indian belt of deciduous forest, including the Gangetic Plain, separates South Asian Gondwanan from Laurasian evergreen forest elements. (2) A north–south belt of evergreen forest along the Indian-Burmese frontier and adjacent ranges separates deciduous floras which became isolated as the Indian subduction threw up these hills from the Oligocene onwards (see map inside back cover). The earlier plant geographers J.D. Hooker and C.B. Clarke recognized, as did Chatterjee, that botanical and political boundaries coincide at the Indian-Burmese frontier. It was Chatterjee who recognized the relationship between eastern Himalayan and northern Burmese evergreen floras.

All three authors described separate plant geographic provinces for Burma, the Khasi Hills and the Himalaya. These provinces were seen as separate because of their distinct herbaceous floras. But in respect of their tree flora – that is, the taxa with longer life cycles that evolve more slowly – they form a unity. Like previous authors, Chatterjee did therefore fail to recognize that the Indo-Burmese lowland semi-evergreen rainforest and its flora extends westward along the Himalayan foothills through Nepal and just into Kumaon (Chapter 3). With their associated lower montane forests, these northern forests are truly part of the Indo-Burmese floristic province. Although somewhat distinct from those of Meghalaya (for example, the Khasi hills) and the Indo-Burmese frontier ranges to their south, they form an almost seamless continuum with the evergreen tropical forests of northern Indo-Burma through to South China (Chapter 3). The overriding change, as one travels eastward toward the China Sea coast, is an ongoing accumulation of species numbers.

Remarkably, the tall deciduous forests, separated by the sometimes narrow India-Burma frontier boundary of evergreen forest, remain floristically distinct from each other (Table 6.13). This distinction is at the species rather than higher levels, though some genera – notably *Afzelia* and *Sindora*, *Dipterocarpus* and *Irvingia* – are absent from the Indian deciduous forests, while *Hardwickia* is absent and *Holigarna* rare in Indo-Burma. Indeed, several of the genera (including *Diospyros*, *Pterocarpus* and *Terminalia*) are pan-tropical, whereas others (such as

Table 6.13 (right) Chatterjee's Partition: the natural ranges of tree species in the two large genera of semi-evergreen and tall deciduous forest, *Dalbergia* and *Lagerstroemia*. Note: North-East Indian semi-evergreen forests are on the Chatterjee Partition. (From LeComte *et al.* 1907-1951, Brandis 1907, Aubréville & LeRoy, 1960, Kress *et al.* 2003, Smitinand 2001.)

	India	Indo-Burma
Dalbergia		
sissoo	NE	
latifolia	S., N.	-
rimosa	NE	-
ovata	-	NW Burma
cultrata	-	Wide
lanceolaria	Wide	Wide
assamica	NE	Wide
kurzii	-	North
cana	-	DDiF, wide
sericea	NE	Wide
oliveri	-	-
frazeri	-	Wide
wattii	Manipur	NW Burma
reniformis	NE	-
burmanica	-	Burma
yunnanensis	-	Burma
henryana	-	Burma
kingiana	-	NW Burma
lacei	-	NW Burma
obtusifolia	-	Burma. Wide
peguensis	-	Burma, wide
pinnata	-	Wide
prainii	-	Burma, wide
cochinchinensis	-	Burma, wide
darlacensis	-	SE
malabarica	Peninsular India	SE
parviflora	-	N. Thailand
junghuhnii	-	Peninsular
suthepensis	-	Peninsular
		N. Thailand
Lagerstroemia		
flos-reginae	Wide	
lanceolata	Peninsular India	Wide
thomsonii	Peninsular India	E. Thailand
rottleri	Peninsular India	
parviflora	Peninsular India	
hypoleuca	-	N. Burma Andamans
calyculata	-	Wide
macrocarpa	-	Wide
speciosa	-	Wide
venusta	-	Wide
villosa	-	Wide
tomentosa	-	Wide
floribunda	-	Wide, Sunda
collinsae	-	Thailand
siamica	-	Thailand
spireana	-	Thailand
undulata	-	Thailand
cuspidata	-	E Thailand
duperreana	-	E Thailand
indica	-	E Thailand
loudonii	-	E Thailand
balansae	-	E Thailand
ovalifolia	-	E Thailand
angustifolia	-	Vietnam
thorelii	-	Vietnam

Anogeissus) are shared with the woodland savannas of Africa. More clearly than at other plant geographic boundaries in the continental Asian tropics, the floristic differences between Indian and Burmese deciduous forest communities appear to be attributable to historical rather than ecological factors: there is no significant difference in climate between the two deciduous associations. Yet on the Burmese side there is a greater number of species belonging to the characteristic tall deciduous forest genera (found also in semi-evergreen forest), notably *Dalbergia*[108] and *Lagerstroemia*[109] (Table 6.13). Although many species are found on both sides, others are represented by sister taxa on either side. Examples from among the important timber trees are *Pterocarpus marsupium* and *Xylia xylocarpa* in South Asia, sister species respectively of *P. macrocarpus* and *X. xylocarpa* ssp. *kerrii* (known as pyinkado) in Indo-Burma.

Forest type	The main continuous evergreen forest blocks (ranges and associated lowlands)		
	NE India–W. Burma frontier ranges	E. Burma–W. Thailand frontier ranges south to Kangar-Pattani Line	Laos–Vietnam frontier ranges
Southern wet seasonal evergreen	7	29	30
Northern wet evergreen	3	4	9
Southern semi-evergreen	3	3	4
Deciduous dipterocarp forest	2	5	7
Total	16	41	50
Number of spp. common to all	12	12	12
Number of endemic spp.	1	6 (all in south)	15 (8 in north)

Table 6.14 Dipterocarp distributions down the three main north-south mountain chains of Indo-Burma, shown as dipterocarp species number in each forest type. (From Brandis 1907, Smitinand 1969–1970, Smitinand *et al.* 1970)

A second north–south trending range of mountains separates the Burmese from the Thai lowlands, but the only detectable floristic change eastwards is a continuing increase in species richness.

Overall, tall deciduous communities of Indo-Burma are richer in species than those of South Asia (Table 3.7). Local endemicity and floristic diversity is fairly low in the deciduous forests of both floristic provinces, but it increases toward the east in Indo-China (although it may receive an additional boost there by separate traditions of plant taxonomy). This trend is most apparent in the seasonal evergreen dipterocarp forests (for example Table 6.14).

The *Tectona hamiltoniana* thorn-dominated woodlands of the northern reaches of the Irrawaddy plain in Burma in part experience a similar rainfall regime to Indian short deciduous forest; yet they have little in common with it floristically, other than a similar abundance of *Diospyros* species. *T. hamiltoniana* is unknown in India, but is now considered conspecific with *T. philippinensis* of the dry karst of Bataan, in the Philippines!

Likely causes of the comparative floristic poverty on the Indian side are periodic extinction events in South Asia, caused by weakening of the south-west monsoon during north temperate glaciations (Chapter 1). Such climatic events were likely also the cause of the attrition of species numbers and community diversity, in the northern seasonal evergreen and lower montane forest, as they approach their western limits in the northeastern to central Indian foothills and plains immediately south of the Himalaya; also, the successively *increasing* richness of southern lowland seasonal evergreen forests following successive north–south trending mountain systems eastwards, from the Indo-Burmese frontier and Burmese-Thai frontier ranges to the Lao-Vietnamese Annamite frontier mountains.

The differences in floristic richness across Chatterjee's Partition do also have ecological causes. The greater diversity of geomorphology, geology and prevailing soils on the Indo-Burmese side (Chapter 1) have likely driven higher rates of speciation and diversification. The competitive relaxation that would have resulted from Pleistocene extinction events in South Asia must also have contributed to the remarkably low floristic differentiation there among deciduous forests in relation to their soils. The Indo-Burmese deciduous dipterocarp forests are floristically diverse at least partly due to substrate, and occupy an overall habitat unrepresented west of the partition.

In summary, Chatterjee's Partition is as comparably distinct, in relation to the woody and especially the deciduous flora, as Wallace's Line, but has not received comparable attention. It appears to have acted since its dual origins both as a continuing *barrier* to deciduous flora migration, and as a periodic migration *track* of evergreen forest flora from and to India via the northern Gangetic Plain and down the Indo-Burmese frontier ranges and the Isthmus of Kra, alternatively the Andamans and Sumatra, to Sundaland. *This lack of research attention may be partly because the Partition runs overland, and because it consequently fails to display a distinct zoographical signal - though more study of insects might be rewarding in this regard.* When the absence of so many Himalayan lower montane tree families in the hills of South Asia, and of Monimiaceae in the Himalayas, is taken into account, the overall, dual, boundary is yet more significant.

It may seem at first truly extraordinary that so many tree families apparently failed to cross the Gangetic Plain. It is less surprising, though, if it is true that the plain had been acting as a climatic and edaphic boundary since before the main rise of the Himalaya. The additional possibility that some families were able to cross, but were subsequently eliminated during drying periods, cannot be excluded.

The India-Burma frontier ranges continue south as the Arakan Yoma Burmese coastal range, to submerge beneath the Bay of Bengal, then arise as the Andaman Islands before descending and arising again as Sumatra's Barisan Range. The Andaman tree flora is more comparable to that at their latitudes in the upper peninsula than the Arakan to the north, but depauperate in species among the big deciduous genera. Sea depths are currently too great for any Pleistocene connection, and this tree flora is consistent with immigration ceasing after the hotter moister seasonal Miocene. The Nicobars to their south bear a more generalized, oceanic, flora.

3.1.7. *The tropical margin as a plant geographic frontier*

3.1.7.1. *In the wet tropics.* In the eastern Himalaya, winter hoar frost may settle in valley fields down to 1,500 m (Chapter 4). The floristic ecotone from tropical lower montane to warm temperate forest on slopes and ridges lies higher, occurring over a c.300 m-wide vertical band at around 2,000 m (Chapters 1, 4). Crossing this ecotone, there is only a small change in generic composition, but a major one in species – becoming a massive one if a greater elevational range is included (such as Table 3.13). The floristic change is greatest in the canopy. All the species within Fagaceae change, conspicuously altering the forest's visual aspect (Chapter 4). (This floristic frontier, incidentally, is not at all matched among the mammals: many tropical mammal species range to the northern continental limits of warm temperate evergreen forests on the southern flanks of the east–west-trending Qin Ling Shan in China.[110])

In South China, there is also a quite narrow northern margin to forests of almost exclusively tropical species composition, but the warm temperate forests to their north change along a gradual continuum over hundreds of kilometres as temperatures cool (Chapter 3). Nevertheless, many *genera* are shared between tropical and warmest temperate evergreen forests. Three types

of tropical forest survive at the tropical margin on south-facing slopes and mild coastal lands (see Chapter 3 for a fuller discussion): (1) lowland northern seasonal evergreen dipterocarp forest, (2) semi-evergreen dipterocarp forest, and (3) lower montane forest rising above it to as high as c.2,000 m (Chapters 3, 4). Of the genera, 90% are of Indo-Burmese or Malesian affinity, while 75% of species are of Indo-Burmese. Zhou[111] has argued for a late Cenozoic origin for this rainforest flora. On the other hand, the apparently persistent wet tropical climate over geological time, and the survival of endemic tropical lower montane gymnosperms such as *Pinus krempfii*, implies that its origins might have been earlier, perhaps coincident with the widening of the South China Sea during the Oligocene–early Miocene (32–15 Ma).[112] At that time, a periodically dry climate prevailed inland and in the mountains, while moist tropical conditions likely developed along the coastal lowlands.

A large number of genera on either side of the tropical margin contain species endemic to either tropical seasonal evergreen or to warm temperate forests. It may seem counterintuitive that the temperate tree and shrub genera tend to be richer in species than the tropical ones (and this does appear to be a real pattern, not a consequence of different traditions in species concepts!). The pattern forms a sharp contrast with Orchidaceae, for instance, in which the number of species in Thailand exceeds those of all China.[113] Tree species, being soils-specific, may have diversified prolifically in South China owing to its exceptional topographic and geological diversity, including 'archipelagos' of karst limestone in which their dispersal limitations may also have enhanced diversification. Regional variation in frost frequency and intensity may have provided further impetus.

Currently, warm temperate evergreen forest replaces tropical over short distances as low as at sea level, in areas seemingly exposed to cold air from the north or descending nocturnal winter frost (Chapter 3). A few tropical families (notably Dipterocarpaceae, Myristicaceae and Lecythidaceae) and many genera disappear abruptly at the tropical forest frontier, but several tropical genera extend northwards along the coastline, and some are found in pockets north to Fujian and coastal Zhejiang, and inland as far as south-west Sichuan. I hypothesised (Chapter 3) that their occurrence is due to frost-free pockets persisting there during times of northern temperate glaciations (for example, from 16–70 ka and also earlier climatic oscillations).

There is evidence that, in contrast to South Asia, East Asia experienced a mild climate and a northerly extension of the wet tropics during the Pleistocene glaciations.[114] Liu and colleagues[115] found pollen of *Dacrydium*, a frost-sensitive Asian species, in peat dated at 36,000–20,000 BP from Menghai (southern Yunnan, 22° N, 1,250 m). The locality currently experiences frosts Nov.–March, implying that its climate may have been milder, possibly wetter, when other regions were drier and colder. The tropical margin would have extended further north. But the restricted range of species in the current apparent northern tropical refugia may have other explanations (see Chapter 3). *More palynological evidence is required. More extensive documentation of the current local occurrence of rare killing frosts in this region would be informative for agriculture and forestry, as well as of interest to biogeographers.* In addition though, archaic temperate families including Hamamelidaceae and Magnoliaceae contributed to the tropical flora, while certain tropical families such as Theaceae, Rutaceae and Sapindaceae have made successful transitions into temperate climates in East Asia. This must be due to the uninterrupted continuity – unique for the northern hemisphere – of moist tropical and temperate climates up the western rim of the North Pacific through much of the Cenozoic era, at least since China's arid period in the Oligocene and through much of the Neogene.[116]

3.1.7.2. *In the dry tropics.*
The dry tropics experience the greatest seasonal temperature extremes (Chapter 1), yet the floristic margin is less clear than in the wet tropics. Short deciduous sal forests, extending north-west to 32° N (Chapter 3), survive periodic air frosts analogous to South Asian upper montane forest. Short deciduous forests and thorn woodlands of semi-arid north-west India extend into air-frost prone regions of Rajasthan along the southeastern fringe of the Great Indian Desert; but north-west of the desert the tropical tree taxa are replaced by Mediterranean elements, *Olea* and *Pistacia*.

3.1.8. *A conclusion: a challenge to the historical interpretation of fossil evidence*
Palaeontology appears to have taken little account of the overall similarity in composition at the genus level between tropical lower montane and warm-temperate lowland evergreen forest in East Asia. The following biogeographic observations need to be integrated into taxonomic and historical scenarios currently being developed from the North-West European Eocene London Clay fossil flora:

1. The *current* warm temperate flora of East Asia is richer in basal flowering plant families than is its tropical flora. Many of these families arrived early, and their presence almost certainly reflects the long persistence of its moist temperate climate (and therefore its ability to serve as a refuge when climates changed elsewhere). It seems likely that these families spread eastward to Asia from western North America via Europe, rather than westward across the Bering Strait – which only provided a landbridge by the Oligocene.[117]

2. The tropical families missing from the East Asian lowland warm temperate forest are also the ones missing from the London Clay flora. These include Chrysobalanaceae, Combretaceae, Lecythidaceae, Moraceae (*Ficus* possibly excepted), and Rhizophoraceae, while Dipterocarpaceae and Myristicaceae are based on single records.

3. Apart from the conspicuous absence of Fagaceae and Aquifoliaceae in the London Clay, the presence of *Sloanea*, *Eurya* and other Theaceae, Lauraceae, Magnoliaceae, Nyssaceae, Cornaceae, Symplocaceae, Primulaceae-Myrsinoideae and even Cyclanthaceae (now confined to the Neotropics) imply prevalence either of lowland warm temperate or lower montane tropical forest.

4. The presence in the clay of some upland tropical taxa (the palms *Caryota*, *Livistona*, *Sabal* (though some identifications may be questionable); *Dracontomelon* (Anacardiaceae), Protioideae (Burseraceae), *Lagerstroemia* (Lythraceae), Icacinaceae (in the former sense) and Sapotaceae does imply a source nearby of seasonally dry tropical semi-evergreen forest taxa.

5. Taxa of brackish coastal habitats, notably *Nypa* but also *Ochrosia* (Apocynaceae), suggest onshore or tropical estuarine conditions of deposition.

6. The fruits appearing abundantly in these depositions (identified as the exclusively tropical lowland *Polyalthia* (Annonaceae), never common in contemporary forests) may represent a now-extinct coastal genus.

7. It must be recalled that the quintessentially tropical Rhizophoraceae mangroves, along with upland tropical genera such as *Macaranga* and *Garcinia*, presently occur north to Okinawa (26°30' N), while there is a stand of *Ceriops* (Rhizophoraceae) in southernmost Kyushu (31°30' N) (though there is some doubt whether it is native).

Overall, I suspect that these fossils were deposited in or near coastlines and estuaries under conditions similar to those currently prevalent in South China. The Hadley cells remained at the same latitude over geological time (Chapter 1), therefore Western Europe would have remained under the influence of south-westerly winds. Predominant southwesterly winds off the Atlantic Ocean – then warmer – would likely then have engendered a climate similar to that of modern South China, on the western coast of the Pacific and subject to the East Asian monsoon. In the absence of a Tibetan plateau in the Eocene, there would have been no Asian monsoonal climate and summers would likely have become drier as the Tethys retreated eastwards and the Indian subcontinent rafted north. The plains would have increasingly supported semi-evergreen or deciduous tropical lowland forest and associated estuarine vegetation near the coast, but warm temperate evergreen forest to the north in inland Laurasia. This interpretation poses further questions as to what opportunities might have remained for an exclusively tropical West Gondwanan flora to migrate along a tropical northern coastal Tethys Laurasian route, especially subsequent to the Paleocene.

3.2. *The regional rainforests and their origins*

3.2.1. *Origin and diversification of the Sundaland mixed dipterocarp forest*

The lands of West Malesia – that is, the lands of the Sunda continental shelf (Java and southernmost Sumatra excepted) – are unique in that they currently experience a single ecologically uniform climate, hardly varying seasonally or diurnally in temperature while receiving an excess of rain throughout the year (Chapter 1). The margins of this climate lie overland through peninsular Thailand and Malaysia, and patchily at the eastern margin of West Java's mainly seasonal wet lowlands (map inside front cover). The perhumid climate extends as far as the eastern slopes of the Philippine islands, where its margin is complex and sometimes more gradual. This climate coincides with the range of MDF, peat swamp and other associated forests.

3.2.1.1. *Varying climate regimes*. Although the Sunda lowlands have been in existence from the early Miocene (23 Ma), the extent of area under perhumid climate has varied since that time. Cycles of seasonality and low sea levels, associated with changes in Earth's eccentricity, started during the late Pliocene and Miocene and continue to the present. They led to the Pleistocene northern ice ages and hence to variation in sea levels. Initially, and since 900 ka, these lasted each about 100 ka, but were 40 ka in the intervening period. Morley presents evidence from East Kalimantan and Java that, notwithstanding prolonged periods of moderate seasonality affecting the Sunda lands as a whole, there remained a perhumid equatorial core, at least in Borneo.[118] It is during this 23 million years, and in this oscillating and variably fragmented Sunda region of perhumid and more seasonal climates, that the current rainforest flora of Sunda both originated and diversified. Its current status, owing to high sea levels, is therefore geographically, if not climatically, refugial.[119]

The most extensive hyper-diverse lowland forests in the Old World are circumscribed by the Sundaland perhumid climate and by Wallace's Line. All told, there are 117 endemic genera

and an impressive c.15,000 endemic species in Sundaland and the Philippines, or 60% of the total c.25,000 endemic species of Malesia (these estimates have been increased by some 20% in recent decades, by both new discoveries and a narrowed species concept). It was formerly claimed that these forests are ancient, and even harboured the origins of flowering plants.[120] But there are no endemic tree families; although in the large Sunda island of Borneo grows *Scyphostegia borneensis*, forming an isolated clade in Salicaceae, a shrubby tree that grows on shingle and disturbed vegetation along fertile whitewater river banks in upland Borneo. Instead, many families are vastly richer in species within the limits of the perhumid climate, generating the highest community diversity in Asia (Chapter 7). Notable are the Dipterocarpaceae, which have 45 species in seasonal Indo-Burma; yet perhumid Peninsular Malaysia, contiguous to the south, boasts 160 species. This change in family richness coincides precisely with a change from annual flowering (of fluctuating intensity) to supra-annual flowering and mast fruiting at about five-year intervals in most species (Chapter 5).

Bob Morley[121] has argued, though, that the Far Eastern tropics must have been seasonal from the late Eocene to the end of the Oligocene, that is, *after* the families, which rafted to Laurasia on Gondwana India, had landed. But the opportunity must have existed after initial contact for eastward migration of taxa confined to perhumid climates, while northern India was still near the Equator; and they must have survived in perhumid refugia in the Far East thereafter because the regions between South and East tropical Asia soon became seasonally dry, and a perhumid corridor likely never returned (see section 3.2.1 above and this section, below). The Sunda region was also mountainous into the mid-Miocene. Extensive perhumid lowlands would have been a prerequisite for the diversification of the current tree flora – comparably rich to that of the eastern Andean foothills of the Amazon hylaea, which shares the same perhumid climate[122] (Chapter 7). This modern flora of the mast-fruiting evergreen forests of Sundaland can therefore be no more than 23 million years old. Certainly these communities are old in comparison with those of cool temperate and boreal forests, which experienced massive changes in distribution during the ice ages (2 million–10,000 BP), as well as catastrophic extinction where opportunities for escape through dispersal did not exist. Nevertheless, the tropical communities too would have diversified and changed much with changes in the extent of the perhumid climate and geological and topographic changes, and as their species evolved over more than 20 million years.

3.2.1.2. *Migratory linkages*. The survival of genera known exclusively from MDF and perhumid climates in both Sri Lanka and Malesia implies that there must once have been a perhumid corridor between them. Some of these genera, such as *Trichadenia* (Achariaceae), are isolated within their families.

417

Others, too, are clearly ancient.[123] Two examples are (1) the dipterocarp *Cotylelobium*, in a basal group of the Asian subfamily with two species in Sri Lanka and three in the Far East, and (2) *Axinandra* (Crypteroniaceae) with one species in Sri Lanka and three in the Far East. Both of these genera are concentrated on siliceous soils of moderate to low fertility, yet *Trichadenia* is confined to fertile loams (although an undescribed Borneo endemic which has been ascribed both to this species and to the closely allied genus *Eleutherandra*, is confined to humult yellow sands). It seems that the corridor for dispersal eastwards may have been edaphically heterogeneous.

It is as yet not possible to distinguish the seasonal from the perhumid fossil evergreen dipterocarp forest floras, but it appears likely that the corridor for migration of exclusively perhumid regional elements was closed well before the rise of the high Himalaya in the early Miocene (see section 3.2.1 above). It may have already closed by the end of the Eocene–Oligocene megathermal period, because the relentless northeastward rafting of South Asia would have carried the migratory route far north of perhumid equatorial climates. The gap in the lowland seasonal evergreen rainforest corridor is now

c.2,000 km wide, though late Oligocene coals with a Malesian generic flora occur in upper Burma; while dipterocarp fossils, of genera currently occurring on seasonal evergreen dipterocarp forest as well as MDF occur northwestwards to the Siwaliks during the Miocene warm period. The last time the seasonal evergreen dipterocarp forest corridor was bridged is unknown, but bearing in mind that we are currently in a warm period it is likely to have been pre-Pleistocene, or more than 2 Ma. *It is therefore of interest that many South Asian dipterocarps appear to have sister species of Far Eastern seasonal evergreen forest, though this awaits molecular genetic confirmation* (Table 6.11).

3.2.2. *The Kangar–Pattani Line: a historical or purely current ecological frontier?*

The Isthmus of Kra has long been known to mark a major turnover in plant and animal species. A clear plant geographic boundary at the southern end of the Isthmus of Kra was recognized a century ago.[124] Symington[125] registered its ecological importance as marking the northern limits of the emergent light hardwood red and yellow meranti *Shorea* species that dominate the MDF canopy in western Malesia

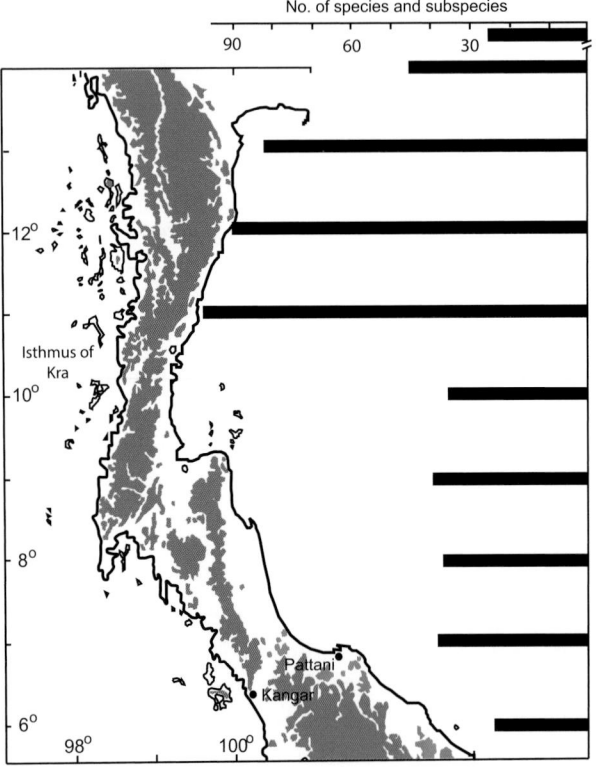

Fig. 6.10 The mountain ranges over 500 m down the southern Isthmus of Kra. The bars on the right indicate range limits of resident birds, north or south. Whereas the tree species changeover is across a line between Kangar and Pattani coinciding with incidence of rainfall seasonality to its north, the bird changeover just north of the Isthmus of Kra coincides with the arrival to its north of semi-evergreen and tall deciduous forest. (I.H. Baillie del. after Woodruff 2003)

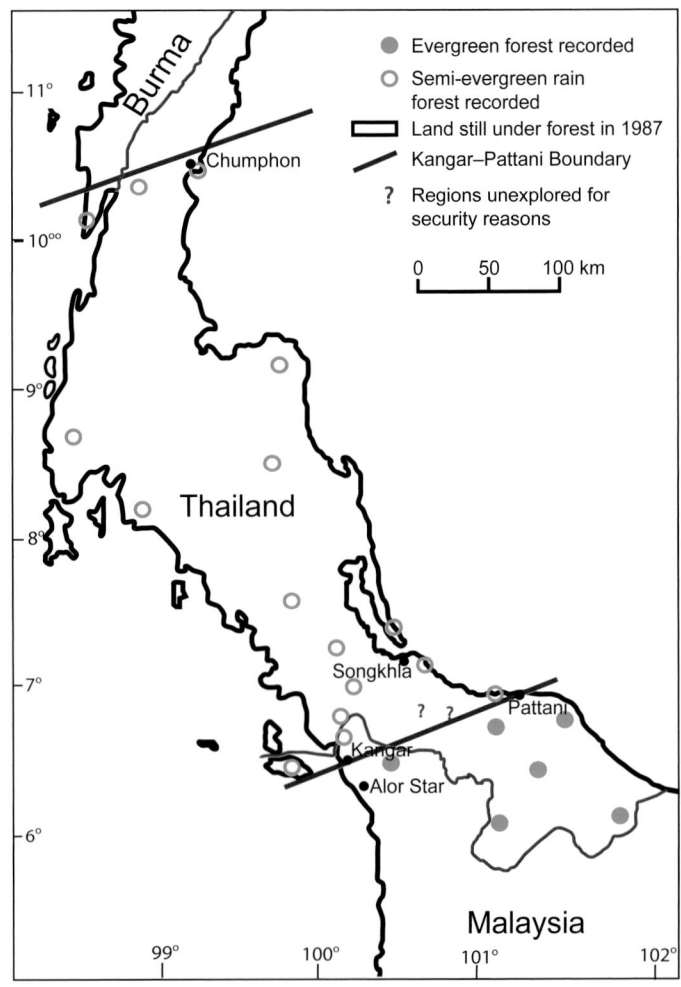

Fig. 6.11 Recorded ranges of Dipterocarpaceae down the southern Isthmus of Kra: **a,** *Shorea* sect. *Anthoshorea*: 1, *S. assamica* ssp. *globifera*; 2, *S. bracteolata*; 3, *S. farinosa*; 4, *S. gratissima*; 5, *S. henryana*; 6, *S. hypochra*; 7, *S. roxburghii*. **b,** the red meranti sections of *Shorea*; 1, *S. curtisii*; 2, *S. dasyphylla*; 3, *S. lepidota*; 4, *S. leprosula*; 5, *S. macroptera* ssp. *macroptera*; 6, *S. ovalis*; 7, *S. ovata*; 8, *S. palembanica*; 9, *S. parvifolia*; 10, *S. singkawang*; 11, *S. pauciflora*; 12, *S. kunstleri*. **c,** *Dipterocarpus*: 1, *D. baudii*; 2, *D. caudatus* ssp. *penangianus*; 3, *D. chartaceus*; 4, *D. concavus*; 5, *D. cornutus*; 6, *D. costatus*; 7, *D. costulatus*; 8, *D. crinitus*; 9, *D. dyeri*; 10, *D. gracilis*; 11, *D. grandiflorus*; 12, *D. hasseltii*; 13, *D. kerrii*; 14, *D. kunstleri*; 15, *D. obtusifolius*; 16, *D. retusus*; 17, *D. elongatus*; 18, *D. eurynchus*; 19, *D. fagineus*; 20, *D. lowii*. (I.H. Baillie del. from Symington *et al.* 2004)

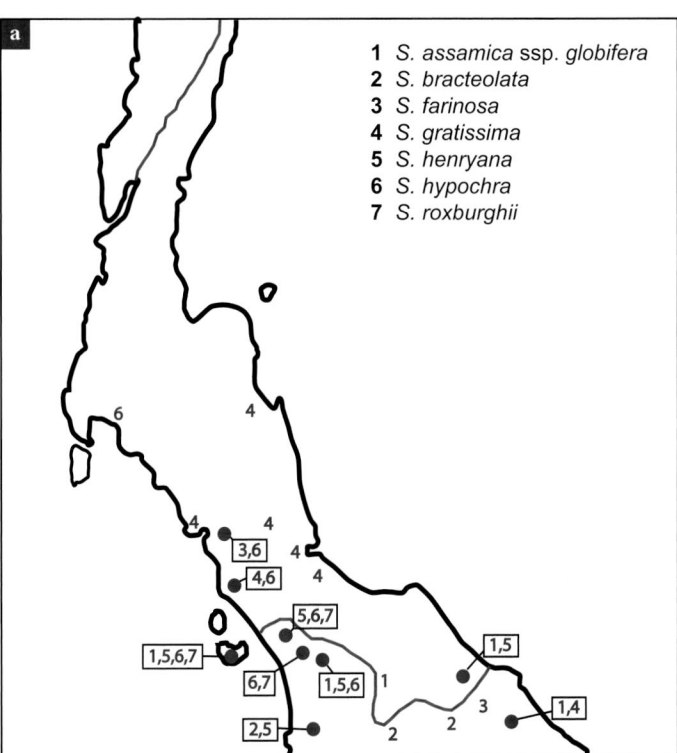

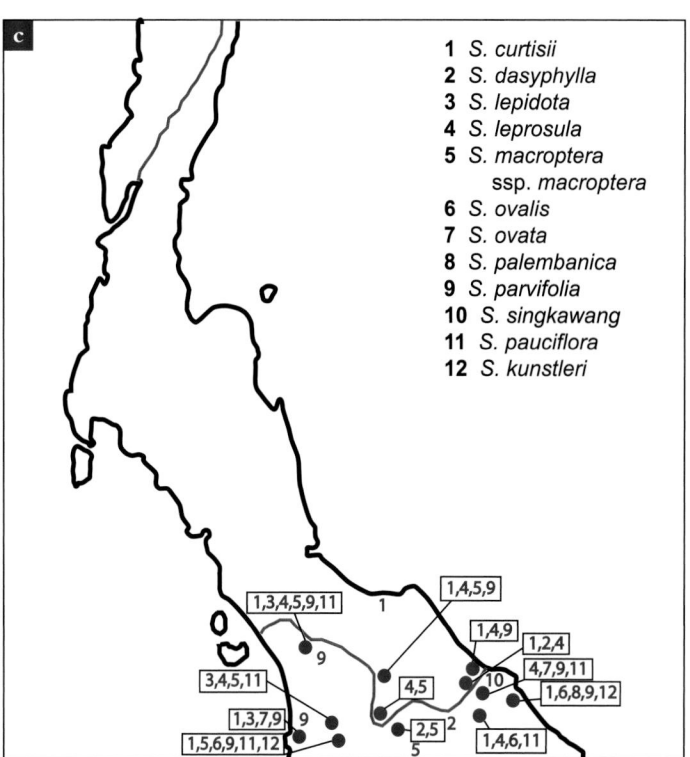

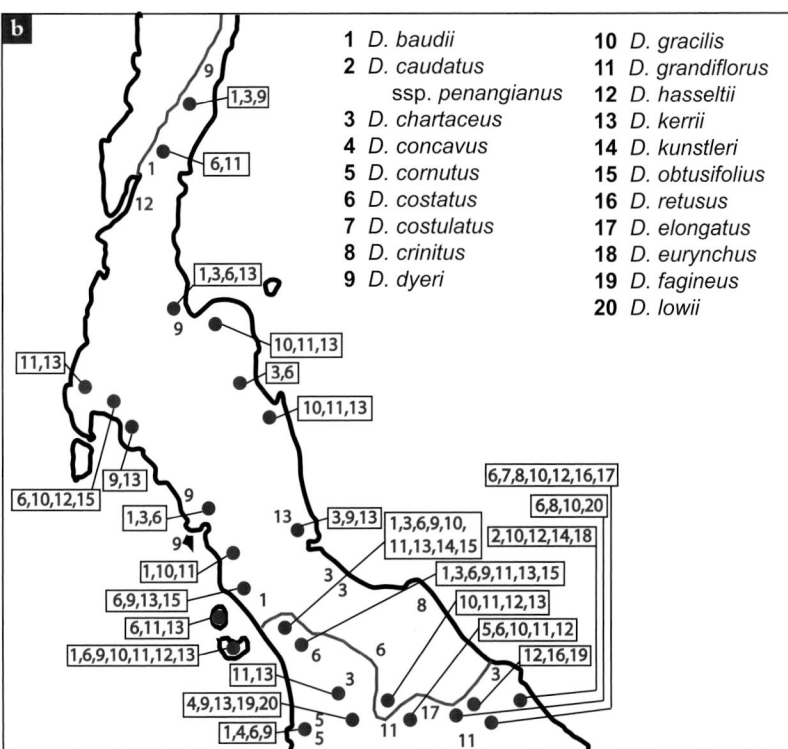

and the Philippines (and also dominated the timber trade during the decades of their exploitation and destruction). The line runs south-west to north-east, from Kangar (6°15' N) in the Malaysian state of Perlis to Pattani (6°50' N), capital of the Thai Province of that name (Figs 6.10, 6.11). It was Tim Whitmore who named it after these towns.[126] Remarkably, even lower montane species change at the line: of 64 Fagaceae in Peninsular Malaysia, only four species are found on both sides.[127] Sarayudh's plot in lower montane forest at Khao Phnom Bencha (8°12' N) contains a typically Thai tree flora.[128] The intrusion of Indo-Burmese seasonal evergreen dipterocarp forest elements into northern Peninsular MDF, such as *Amesiodendron chinense*, *Bombax anceps*, *Chukrasia tabularis*, *Excoecaria oppositifolia*, *Syzygium siamense* and *Celtis rigescens* is restricted to the lowlands, below the cloud base.

Significantly, the Kangar–Pattani Line is not recognized by zoologists, although entomologists have yet to investigate it in detail. Ornithologists recognize a line in the north of the isthmus, near Chumphon, distinguishing an Indo-Burmese fauna in which species of two-stratum deciduous forests are prevalent, from a Malesian fauna in which rainforest species dominate. Woodruff[129] has identified changes in small mammal fauna across the low country between the northern limits of the Malaysian Main Range and the start of the Thai peninsular mountains. He proposed an intermittently sea-filled strait during the Pliocene and Pleistocene, but sea levels are not thought to have risen sufficiently high during this period (see section 3.2.3 below).

The Kangar–Pattani Line coincides with the northern limit of the perhumid climate and the onset of a reliable annual dry season of more than two months (map inside front cover). It therefore coincides with the northern limit of forest-wide mast fruiting (Chapter 5). A shorter and less consistent dry season is experienced along the western coastal lowlands of Peninsular Malaysia south to Penang Island. This change in climate follows the topography: the last western outposts of MDF clothe the foothills of Gunung Jerai, continuing northwards along the northwestern slopes of the main range into Thailand, where they veer north-east. Unlike the tropical margin in South China, there appear to be no significant local pockets of the characteristic MDF red and yellow merantis to the north, but several Indo-Burmese elements extend further south. Some (including *Hopea odorata*, *Shorea hypochra* and *S. guiso*) are widespread in the northern and central Malaysian lowlands both east and west, but most are concentrated in the west, where they extend south to Selangor and, in several cases, to Malacca. This distribution does not appear to correlate with any present climatic zone, but does coincide with the distribution of eroding ferricrete crusts. These are undatable (Chapter 1), but their topographic position – bearing in mind that they must have been formed on flat surfaces subject to seasonal waterlogging – suggests that they are at least early Pleistocene. Lowland pine savanna existed south at least to Selangor during the last low sea level and rainfall period.[130]

What is left of the forest flora of coastal lowland Perlis in far north-west Malaysia (with abundant *Schima wallichii*, also *Anisoptera costata*, *Shorea henryana*, and *Vatica cinerea*) is characteristic of southern seasonal MDF. The short deciduous forest species *Dipterocarpus obtusifolius* has been recorded in the past. Other species present (including *S. roxburghii*, *Lagerstroemia floribunda*, *Careya arborea*, *Dillenia obovata*, *Fernandoa adenophylla*, *Terminalia calamansanay*, *Cratoxylon cochinchinensis* and the

bamboo *Gigantochloa nigrociliata* (*Oxytenanthera nigrociliata*)) are usually indicative of successional semi-evergreen forest following fire, and generally associated with a dry season exceeding four months. *S. siamensis* occurs with *Hopea ferrea* on the karst of the Langkawi Islands off the coast, and occasionally inland. Again, this flora suggests survival from a previous drier period.

3.2.2.1. Conclusion: a true Indo-Burmese floristic province.
Chatterjee's Partition, the tropical margin, the continental coast of the South China Sea and the Kangar–Pattani Line together define a distinct Indo-Burmese floristic province. Its forests range from semi-arid deciduous thorn to seasonal evergreen lowland forests and lower montane forests, yet share a unique floristic imprint. Altogether, these habitats are estimated to contain c.13,500 vascular plant species of which, remarkably, about half are endemic to this province.[131]

3.2.3. Other current frontiers of mixed dipterocarp forest
The little that remains of the lowland forest of West Java indicates seasonal evergreen forest rather than MDF. The species present are characteristic of the fertile clay loams that prevail there. The presence of endemics in special habitats (such as *Vatica bantamensis*) implies that the present climate probably survived in refuges throughout the Pleistocene; while the survival of *Hopea sangal* in Central and East Java implies extension of seasonal evergreen forest eastwards during humid Pleistocene or early Holocene periods. South Sumatra bears MDF rich in red meranti, albeit with a somewhat impoverished dipterocarp flora. Some southern lowland areas are rich in Meliaceae and Sapindaceae,[132] recalling forests on basalt in Sarawak, southern Vietnam and north-east Cambodia (Chapter 3), and possibly on soils influenced by basic volcanic ash.

The northern and eastern margins of MDF in Sri Lanka are dissimilar. (1) In the south-east, the margin appears to coincide exactly with the onset of two distinct dry seasons: Oliyagankelle Forest Reserve, on a low range surrounded by rice paddies, still includes two *Shorea*, *Dipterocarpus zeylanicus*, and as yet unidentified mendora (*Vateria* or *Stemonoporus*). The next surviving forests to the east are typical notophyll evergreen; but *D. zeylanicus* survives in pockets in Matara and Moneragala Districts in the biseasonal east, also extending further north than other dipterocarps. Although apparently tolerant of a shortly seasonal climate, these range extensions again imply past less seasonal periods. (2) The northern fringe of Sri Lankan MDF is bordered by a narrow band with *Filicium decipiens*, *Mesua ferrea* and other species associated with semi-evergreen forests in the Western Ghats. However, the endemic dipterocarps *Stemonoporus* and *Shorea* (sect. *Doona*), paradigmatic species of the Sri Lankan MDF canopy, extend in the western lowlands no further north than the Kalu Valley

(6°30' N). (One of these species, *S. disticha*, was collected over a century ago on Lanegal; Kande in the now-deforested foothills inland at 7°25' N.) *Hopea brevipetiolaris*, thought to be endemic to this northwestern semi-evergreen fringe (though with a sister species in Vietnam!), has recently been discovered on coastal hills far south in Kalutara District in the perhumid zone, recalling distributions in northwestern Peninsular Malaysia. The Knuckles Mountains, separated by the Mahaweli Valley from the main massif, still shelter *D. zeylanicus*, and formerly *Shorea dyeri* and *Vateria copallifera* in their foothills, while the upper dipterocarp forest there harbours *Stemonoporus laevifolius* and the lower montane forest *S. affinis*.

3.2.4. *Land continuity across the Sunda Shelf*

Current sea levels are probably within 5 m of the highest they have been since the Miocene, but occasionally they have fallen to as much as 120 m below the present level (Fig. 6.12). The level was c.2 m higher in the early Holocene and declined successively about 6,000–7,000 BP, 4,500 BP and 1,500 BP, generating the extensive coastal peat swamps of eastern Sumatra, western Peninsular Malaysia and Borneo.[133] Thus, coastal swamp forests are the youngest natural rainforests of the region and among the youngest on Earth, though the upland peat swamps (*kerapa*) of south-central Borneo, dominated by *kerangas* elements *Gymnostoma nobile* and *Dacrydium elatum* (*beccarii*) but with much of the dome flora of the coastal peat swamps, have existed since the Oligocene and were then more widespread (Chapter 3).

Counterintuitively, unique swamp species including provincial endemics occur in the riverine mixed peat swamps. These soils are among the shallowest and most recent alluvia and peats, dating from less than 1,500 BP.[134] Otherwise, most of the swamp flora is widespread. I infer that, whereas the forests of the peat domes must have fluctuated greatly in extent and location during the Pleistocene, those swamp forests immediately adjacent to the major rivers must have been more persistent.

3.2.5. *Fossil evidence for a drier Pleistocene past on Sunda*

The current perhumid climate and island landmasses of South-East Asia are exceptional for our epoch.[135] During the northern ice ages, when eustatic sea levels fell, Sumatra, Peninsular Malaysia, Borneo and the Philippine island of Palawan became a single landmass almost as extensive as modern

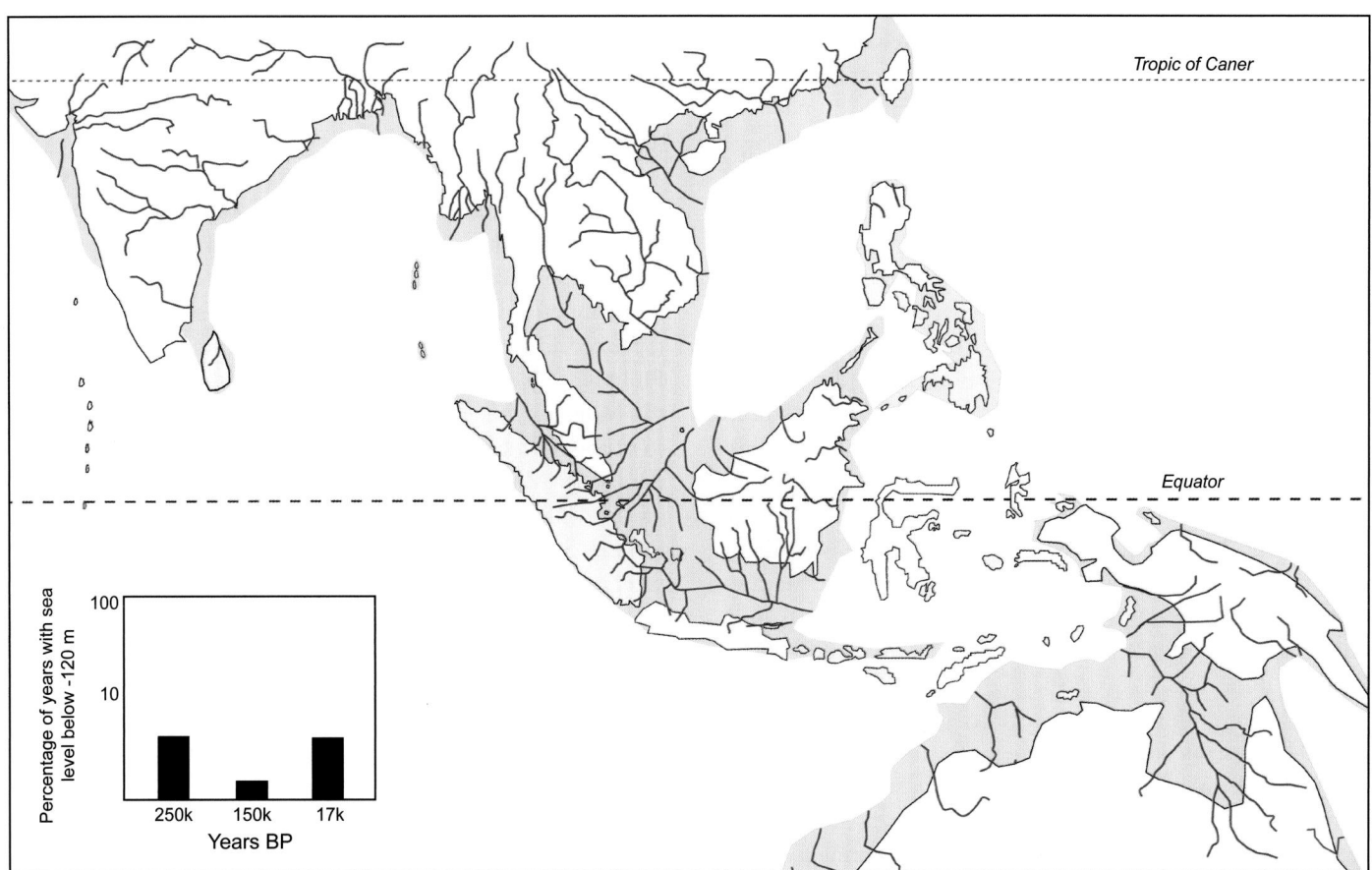

Fig. 6.12 The furthest extent of Pleistocene land linkages in tropical Asia: river beds of most recent time schematically indicated. The bar graph indicates the estimated percentage of time, over the past time period indicated, that sea levels were at or lower than illustrated, that is 120 m below present. (I.H. Baillie del. after Voris 2000)

northern South America (Chapter 1). This was drained to the north-east by a mighty river, the Proto-Mekong (Fig. 6.12; see also Chapter 1). Sea levels dropped to their lowest – 120 m below the present level – at the climax of the last northern ice age, though they rarely fell by more than 30–40 m at other times.[136] The tropics experienced a string of drier periods especially when sea levels were low; in South Asia these were probably exceptionally dry owing to failure of the monsoon. This would have affected South-East Asia less than South Asia, owing to the ever-present moist trade winds off the Pacific and orographic rainfall in the coastal lowlands. Nevertheless, the weakening of the south-west monsoon, which in the Far East may also have veered more to the south-east, combined with the further restriction of southward flow of waters from and consequent warming of the South China Sea, could have had profound impacts on climate there, possibly increasing rainfall inland in South China and Vietnam (see section 3.1.7 above). Despite this, good fossil evidence exists of lowland pine savanna as far south as Kuala Lumpur.[137] However, this is only 300 km south of the current southern limit of the dry deciduous dipterocarp forest associate of *Pinus merkusii*, the now apparently extinct *Dipterocarpus obtusifolius*.

Based on fossil evidence from the mid-Pleistocene beds at Trinil and elsewhere in Java,[138] there appear to have been open, probably at least semi-evergreen, forests with an abundance of savanna ungulates, browsers including hippos, and large predators including 'Java man', *Homo erectus*. Several of the large mammals recorded from the Pleistocene of Java now survive in the floodplains and terraces along the eastern Himalayan *terai*, especially where the Brahmaputra descends out of the Himalaya at Kaziranga National Park. Does their past presence indicate a periodic Pleistocene corridor of semi-evergreen forest, or even savanna, across the Equator from Indo-Burma to Java?[139]

The onset of consistent rainfall seasonality, at the Kangar-Pattani line and elsewhere, is currently correlated with a drop in tree community diversity of about one third, particularly in the canopy and among those series of co-occurring congeneric species that characterise MDF (Table 3.7). *Shorea* sect. *Mutica* and *Richetioides* species series, which reach their northern limit there and which dominate the MDF canopy, are also dependent on ENSO as trigger for their supra-annual mass flowering. The *Mutica* series is ubiquitous in Sunda MDF, and the three species participating in all three major land masses flower in the same sequence (chapter 5), implying that ENSO has sustained them since their probably pre-Pleistocene origin. Interestingly also, a separate *Mutica* series also occurs in north-west Borneo on humult yellow sandy soils, where the flora could migrate onto the extensive continental shelf to their north during low sea levels. In contrast to the strict lack of effective rainfall seasonality in MDF regions, their mean annual rainfall can vary from 5 m (Sinharaja) through

more typical 2,500 mm (2,700 mm at Lambir), to 1,800 mm at Pasoh, which shares Lambir's community species diversity (chapter 7): MDF could have suffered substantial falls in mean rainfall provided aseasonality, and ENSO, persisted.

3.2.5.1. *Plant geographic evidence for a drier Sunda past.* This fossil mammal evidence demands systematic explanation, for no other direct paleontological evidence is available to more precisely document the penetration of seasonal climates into West Malesia during the Pleistocene epoch. The current impact of severe ENSO-associated droughts and consequent human-initiated fires in south-east Borneo has led some to infer that fire has long persisted in these areas.[140] Yet there are no fire-adapted, usually deciduous, forest tree species in the primary inland forest flora there today. The rocky coasts, coastal islands and isolated islets of Malesia carry widespread species with ranges that extend far into the seasonal tropics, such as *Shorea gratissima*, *Serianthes grandiflora*, *Manilkara kauki*, and *Mallotus philippensis*, which also occurs on dry karst summits where others, such as *Celtis rigescens* and *Chukrasia tabularis*, persist.[141] It is difficult to explain the presence of a very few species of a seasonal rainfall climate in isolated localities in the slightly drier climates (Chapter 1) around the coasts of Peninsular Malaysia and Borneo and on isolated karst hills, other than by greater past continuity of a seasonal – or simply drier – climate. This applies especially in the case of the poorly dispersing dipterocarps.

Shorea gratissima, widespread in southern seasonal evergreen dipterocarp forest in peninsular Thailand and scattered in Kelantan (north-east Peninsular Malaysia), recurs at Jugra Hill on the Selangor coast, Singapore, Gunung Palung (West Kalimantan) and at several sites on the east coast of Sabah separated from it by more than 1,000 km. It is sister to the Philippine species *S. polita*, and *S. montigena* in Sulawesi and the Moluccas. *S. guiso* is widespread from peninsular Thailand to Cambodia, and again in the Philippines, also in Sumatra and Peninsular Malaysia, but is known in Borneo only at the base of karst limestone in Sarawak, and in several localities along the east Sabah and north-east Kalimantan coasts. *Dipterocarpus kerrii*, widespread from the Andamans and Arakan Yoma in southern seasonal evergreen dipterocarp forest to mostly coastal sites in Peninsular Malaysia, recurs in coastal north-east Borneo. Mysteriously, a few majestic trees of *Pterygota horsfieldii* (Sterculiaceae) dominate a narrow diorite dyke on the Santubong Peninsula, jutting out into the South China Sea in West Sarawak (Plate 6.7a). This species is widespread in seasonal evergreen and semi-evergreen forests of China and Java, common from the Philippines to New Guinea and present in south-east Borneo, but is entirely absent from Peninsular Malaysia and Sumatra. It is represented by a sister species in south-west Sri Lanka, *P. thwaitesii*. Likewise, several

Plate 6.7 (left) The north-west Borneo refugium and the Riau Pocket. **a,** Santubong, peninsula on the north-west Borneo coast (Bako National Park is the low hills on the horizon to its right): home of a relict population of *Pterygota horsfieldii*, isolated from other populations by hundreds of kilometers; **b,** Segari Melentang Forest Reserve, coastal Perak hills; **c,** The north-west Borneo endemic *Shorea geniculata*: along the former Labi road, Brunei, 1958. (**a** T. Laman, **b, c** P.A.)

species common, even locally abundant, in southern seasonal evergreen dipterocarp forest, including the dipterocarps *Anisoptera costata*, *Hopea sangal*, and *Dipterocarpus gracilis* survive as isolated individuals in Borneo MDF, even in the perhumid north-west Borneo refugium, implying a more seasonal or drier past.

Note that these species, which are all of clay loam soils, follow a markedly different geographical distribution from the Riau Pocket flora within perhumid Sunda. No member of the Riau Pocket flora currently extends into seasonal climates (see sections 3.2.6 and 3.3.3.2 below).

Van Welzen and colleagues ordinated the ranges of some 7,000 Malesian flowering plant species.[142] This resulted in geographical divisions that correlated with climate seasonality and forest types associated with it. Wallacea was placed in a predominantly seasonal cluster equal in rank to Sunda and Sahul Shelf clusters. But Java, mostly seasonal, was included in the Wallacea cluster. Java's placement in a cluster otherwise east of Wallace's Line can hardly be justified on grounds of the east Javan lowland woody plant flora, as van Steenis had earlier indicated.[143]

3.2.5.2. *A corridor of savanna across the Equator?* The large grazing and browsing fossil mammal fauna, dependent on seasonal grass swards, was recorded from Java from 2.4 Ma. Its composition confirms that it must have immigrated across the Equator from seasonal Indo-Burma. Subsequent long periods of isolation led to allopatric differentiation of a distinct fauna there.[144]

The perhumid Sunda region currently excludes some widespread pioneers which are common even in the most slightly seasonal climates (Fig. 6.13). But curiously, although the West Javan seasonal evergreen dipterocarp forest is rich with Indo-Burmese species unknown in South Asia, including regional endemic genera such as *Reevesia* (Malvaceae subfamily Sterculioideae), the deciduous forest to its east possesses none. Instead, this tree flora, now much diminished by a millennium of dense human settlement, is overwhelmingly South Asian or widespread in continental Asia (Tables 6.15). Many species, such as *Terminalia arjuna* (otherwise endemic to southern India and Sri Lanka), teak, and the ornamental *Cassia fistula* (widespread in tropical Asia), may have been introduced in ancient times as were several comestible fruit trees (Chapter 8): *T. arjuna* has been identified in the bas-reliefs at Borobudur

Table 6.15a Geographical affinities of deciduous species in Thai deciduous and semi-evergreen forests, including fire resistant evergreen species of deciduous dipterocarp forest.

Genera and species in Thailand	Those of which also occur in:			
	South India	**NE India**	**Java**	**The Philippines**
122 genera, 189 species	49 genera, 57 species	88 genera, 101 species	43 genera, 51 species (of which 25 are in cultivation there)	21 genera, 35 species (of which 8 are in cultivation there)

Notes
- c.20 Indo-Burmese species are DDiF specialists, among which 4 also occur in North-East Indian semi-evergreen and tall deciduous sal forest.
- c.8 Indo-Burmese endemics are confined to the Irrawaddy arid zone, c.10 to northern semi-evergreen and wet seasonal evergreen forests, i.e. 38 to forest types unique to the region.
- c.18 endemic genera in S. India.

b Lowland species endemic to both Indo-Burma and North-East India. DDIF: Deciduous dipterocarp forest

Ternstroemia gymnanthera (DDiF-sal)	*Rhus javanica* subsp. *chinensis*
Heteropanax fragrans	*Vitex peduncularis, V. glabrata, V. vestita, V. canescens* (out of 8)
Annesleia fragrans (DDiF)	*Xylia xylocarpa* ssp. *kerrii*
Viburnum colebrookianum, V. foetidum, V. odoratissimum (out of 5)	*Litsea khasiana*
Reevesia pubescens	*Dalbergia assamica, D. sericea, D. stipulacea* (out of 21 Indo-Burmese spp.)
Olea dioica	*Ocotea lancifolia*
Colona flavida	*Terminalia chebula, T. myriocarpa* (northern) (out of 12)
Heterophragma (DDiF)	*Aporosa octandra* (DDiF)
Hiptage bengalensis	*Anogeissus acuminata*
Callicarpa arborea, C. heterophylla (out of 3)	*Duabanga sonneratoides*
Artocarpus lakoocha (northern)	*Cycas pectinata*
Lagerstroemia parviflora (out of 9)	
Castanopsis indica (northern)	

Summary
Chatterjee's Partition follows the southern and western margins of northern semi-evergreen and seasonal evergreen forests

Plate 6.8 Species shared between East Java and continental Asia. **a,** the yellow-flowered *Cassia fistula*, Mudumalai Wildlife Sanctuary, Tamil Nadu; **b,** *Parkia timoriana*, Khao Chong National Park, peninsular Thailand; **c,** *Terminalia arjuna*, here indigenous, alongside the 14th Century Elahera irrigation canal, Sri Lanka. (All H.H.)

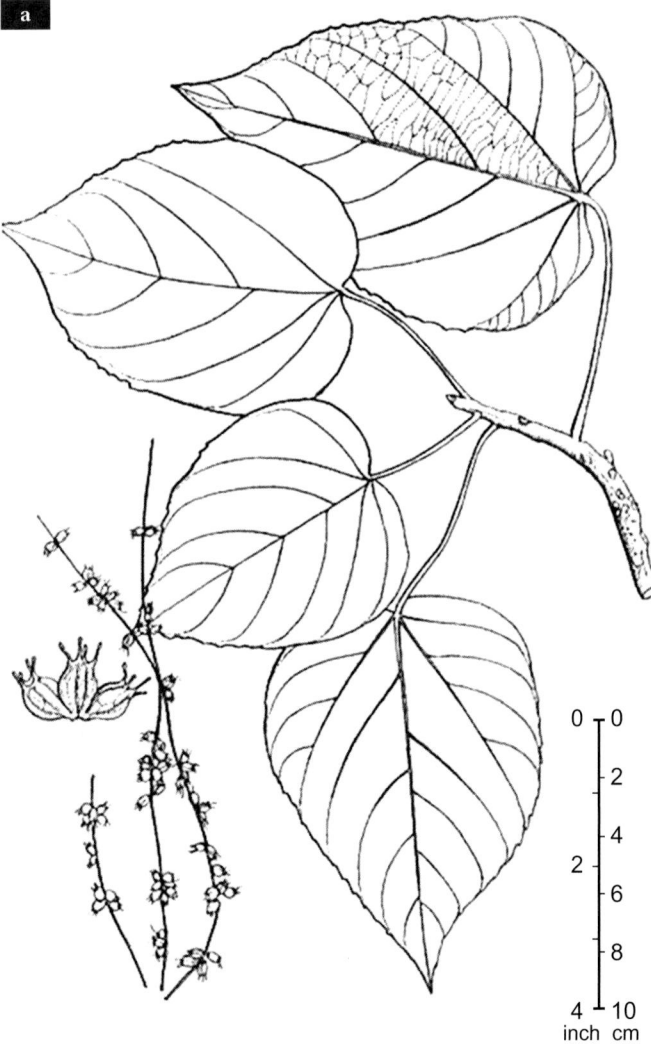

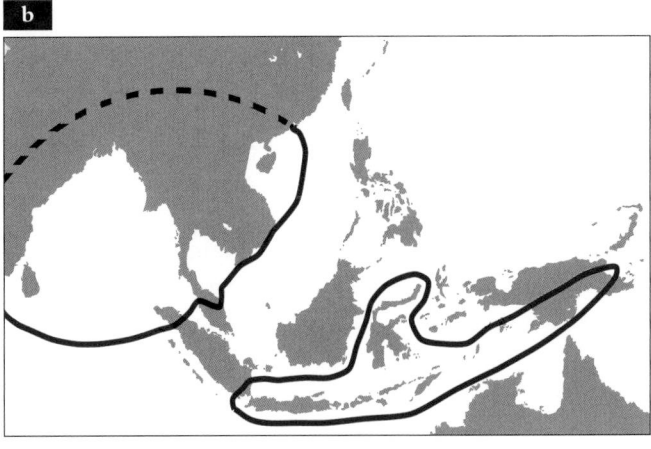

Fig. 6.13 The emergent pioneer *Tetrameles nudiflora*, among the very few deciduous trees, all pioneers, which occur from Sri Lanka to New Guinea. (From Whitmore 1973)

in Java (Plate 6.8c). *Molecular genetic comparison of South Asian and Javan populations could resolve their origins.* No Australasian trees occur in the West Javan seasonal evergreen forests either, though *Weinmannia fraxinea* (which may have spread via Borneo and Sumatra), *Casuarina junghuhniana* (which occurs east of Wallace's Line along the Sunda Islands), and several southern shrubs and herbs[145] are found in Javan montane forests. Striking, overall, is the low number of widespread continental deciduous tree taxa that co-occur either in east Java or the dry western Philippines (Table 3.11); although east Java does have more continental species not occurring in the Philippines than vice versa. Bearing in mind that there never was a Pleistocene land connection from continental Asia to the Philippines, I infer that most East Javan deciduous forest tree species also arrived by long distance dispersal (including by human effort), though a few might have arrived overland. The absence of Indo-Burmese deciduous forest endemics, notwithstanding continuity of suitable soils through Sumatra, implies that this mammalian invasion spread occurred well before the Pleistocene, or followed the likely only seasonally flooded plains of the major tributaries of the Pleistocene Sunda river system.

Whether or not there was a Far Eastern seasonally dry corridor across the Equator during dry periods of low sea level in the Pleistocene continues to be debated.[146] Savanna, that is, tropical grassland exists in Asia either as a fire-mediated grassy field layer in deciduous woodland or as floodplain grassland (chapter 2). True savannas, that is, open grassland with scattered trees, likely never existed – before mankind – in the Asian tropics. But it should also be borne in mind that association of man with fire in continental Asia is known to date back 1.8 million years (see Chapters 2, 8). Our ancestors could have induced a grassland corridor as they penetrated south-east through the plains to the east and north of the Barisan and Javan 'ring of fire' during drier continental periods. They could have been accompanied by mammals of the savanna, but these corridors could have persisted for too brief a time for any but a few common tree species to have migrated.

The hypothesis that a deciduous forest or savanna grassland corridor may have stretched down the Straits of Malacca as far as Java at certain times during the Pleistocene (100,000–12,000 BP) is consistent with the following points:

1. Fossil ferricrete crusts (Chapter 1) exist in the lowland rainforests of Kedah, and from Selangor to Malacca in Peninsular Malaysia, Palembang and Jambi in south-east Sumatra, Singapore and Batam Island south of Singapore. Their presence implies periodic soil inundation and desiccation, and forest canopies sufficiently open during the dry season to permit the necessary accumulation and oxidation of iron and aluminium sesquioxides. Their absence in Bangka and Belitung, whose current climate approaches seasonality, is likely due to granite and other

425

substrates there being insufficiently endowed with iron and aluminium sesquioxides. Ferricrete is itself undatable (Chapter 1), but occurs in part on the Peninsular Malaysian 'old alluvium', a late Pleistocene regional feature.[147]

2. A number of species of Indo-Burmese seasonal evergreen dipterocarp forest penetrate the perhumid zone down the Malaysian Peninsula, following the ferricrete distribution to Selangor and sometimes Malacca (section 3.2.2 above, Chapter 7). Whether these represent survivors from a more seasonal past or merely slight differences in current conditions cannot be ascertained.

3. Open meadowlands with trees confined to the banks of watercourses are found in the perhumid climate of Brunei (Plate 2.36f). There, shallow water heats diurnally and expands to produce periodic but irregular shallow flooding. This inhibits tree establishment and produces a sedge-dominated sward. They occur there where minor tributaries receive and retain shallow floodwater backed up from headwater flooding of their major river downstream. Their soils are low in sesquioxides, but it appears that conditions would be favourable for ferricrete accumulation, were the substrate rich in them. The Brunei meadowlands therefore indicate that floodplain grassland does not require rainfall seasonality, provided other conditions are suitable.

4. The palynological evidence for pine savanna on the old alluvium, dating from the height of the last ice age, near Kuala Lumpur.[148] However, these currently exist as ecological islands on freely draining sandy soil (Chapters 2 and 4).

5. The floristic poverty of the Bangka *kerangas* (*padang*), which lacks *kerangas* dipterocarps and indeed any members of the Riau Pocket flora (see section 3.2.5 above).

6. The striking distribution of the whitewater riparian *Dipterocarpus oblongifolius*, rheophytic when juvenile, whose current restricted distribution – in Borneo and Peninsular Malaysia but absent in Sumatra – suggests a slow reinvasion of a formerly dry Malacca Straits region from perhumid late Pleistocene refuges facing the South China Sea. Widespread in river systems to the east of the Peninsular Main Range, its sole locality to the west is along streams issuing from Gunung Bubu (Perak).

And yet, other observations argue against this scenario:

1. The currently submerged continental Sunda river system meanders, with its incised courses implying the presence of

trees to hold river banks. Strong seasonality and flooding would result instead in braided river systems.

2. The low floristic diversity of East Javan deciduous forests: lacking Indo-Burmese endemics and with little floristic similarity to their analogues in Indo-Burma; prevalence of species with continental ranges extending to South Asia; little more floristic similarity in the tree flora to continental forests than forests of the north-west Philippines with similar rainfall seasonality which have *never* had direct overland contact with Indo-Burma (Plate 6.8).

3. The lowland MDF clay soil flora of east Sumatra strongly resembles that of inland Peninsular Malaysia (for example, only 8% of Sumatra's 104 lowland dipterocarps are endemic, while 86% also occur in Peninsular Malaysia), serves as evidence of Pleistocene continuity rather than separation even by a corridor of seasonal evergreen, let alone an extensive or continuous deciduous forest or savanna.

4. Floristic elements of the Riau Pocket and peat swamps persist in the Riau Archipelago itself, and in the Perak coastal hills and lower Perak swamps (Fig.6.14, Plate 6.7b), respectively in the centre and adjacent to a putative savanna corridor in the Straits of Malacca. None of these species extend into the seasonal climate to their north. These include the local endemics *Shorea hemsleyana* (subsp. *hemsleyana*) and an endemic form of *S. macrantha* in the swamps, and *S. lumutensis*, *Dipterocarpus perakensis*, a local form of *D. lowii* and *Vatica pallida* in the coastal hills. *V. perakensis* and *Parashorea globosa* are confined there and in northern Sumatra[149] (see section 3.2.6 above). Peat swamp also does not form in seasonal climates, presently extending no further north than Narathiwat in southeastern peninsular Thailand, nor do Riau Pocket species currently exist in them.

5. Similarly, the substantial point endemic woody flora on the Klang Gates quartzite ridge north of Kuala Lumpur[150] can hardly have originated in the Holocene, and is therefore evidence of a continuing presence of perhumid climate there (although plains immediately to the west could have been seasonal). This flora includes the dipterocarp *Hopea subalata* and the sole locality west of the Main Range of *Dryobalanops aromatica*, as well as the sole lowland site of *Rhodoleia championii* (Hamamelidaceae).

6. Lowland pine savanna – that is, deciduous dipterocarp forest dominated by *Dipterocarpus* species and *Pinus merkusii* – is currently confined to freely draining sandy soils. These lowland soils are found in discontinuous ecological islands

and are unlikely to offer a continuous savanna corridor from the northern to southern seasonal tropics. Floristically defined short deciduous forest, which may also have an open canopy and a grassy field layer (especially after human intervention) and could once have formed continuous vegetation on the predominantly clay floor of the Straits of Malacca, currently exists on gentle topography in nature only where the dry season exceeds seven months. That is hardly compatible with the survival of the straits Riau Pocket flora. Short deciduous forest is widespread in South Asia but is currently confined in Indo-Burma to the semi-arid upper plains of the Irrawaddy and perhaps to north-east Thailand prior to human intervention.

In overall conclusion, it remains difficult to visualise in detail how the weakening of the Indian south-west monsoon during periods of low sea level would have affected Far Eastern climates. Plant geographical and geomorphological evidence suggest that perhumid (albeit perhaps overall drier) conditions must have continued in the foothills on either side of the Malacca Straits – and especially in the north and south between the Malaysian Peninsula and Bangka – even though seasonal conditions may have prevailed in the northern Malacca straits and from Bangka southwards. I infer that there was more likely a reduction of average rainfall, rather than any distinct increase in seasonality. Thus the intermontane valley of Negeri Sembilan, with rainfall below 2,000 mm annually and one month in which the mean just falls short of 100 mm, supports MDF of typical composition but with the addition of a few coastal species such as *Shorea glauca*, *Dipterocarpus kerrii*, *Myristica guatteriaefolia* and *Pouteria obovata*, which may be indicators of frequent drought.

It nevertheless remains difficult to escape the overall conclusion, on current species distributions, that there were times during the Pleistocene when MDF and its perhumid lowland climate retreated to patches, some of them large such as Borneo north of the Equator, but most smaller such as Peninsular Malaysia east of the main range and Perak in the north-west, and coastal eastern and northern Sumatra. The penetration of several species of seasonal evergreen dipterocarp forest down the inland western lowlands south to Selangor implies that lack of rainfall seasonality, albeit likely frequent aseasonal drought, was confined to near-coastal sites of that time.

From current tree species distributions I infer that there was a corridor east and north of the Sunda 'ring of fire' mountain chain (now the northern Malacca Strait) where, in the plains, a climate with at most four dry months penetrated. However, climates in the surrounding foothills and in the catchment region between rivers running north-west and south-east must have remained perhumid (see section 3.2.6.1 below).

The presence of hippos (dependent on open freshwater) in the Javan fossil record implies that there was a major riverine connection, whereas the absence of lions (though possibly recorded from last interpluvial deposits in Sri Lanka's perhumid south-west[151]) implies the absence of a short deciduous forest corridor. Periodic prolonged shallow flooding in the plains of the Proto-Mekong upper tributaries could have led to extensive seasonal grasslands. The browsing mammals of the Trinil megafauna, and protohuman use of fire, would have extended these grasslands and savannas into the adjacent uplands as they foraged during the wet months. It is possible to revisit some hint of that scene by visiting Kaziranga National Park in North-East India (see Chapter 2). There the mighty Brahmaputra (Zangbo), emerging out of its Himalayan Gorge, seasonally swamps a vast grassland savanna in the eastern *terai* and *duars* lands.[152] The largest remaining population of Asian one-horned rhinoceros co-exists here with elephant, tiger, and numerous ungulates.

3.2.6. *The Riau Pocket*

Assuming drier climates prevailed periodically, how might the rich endemic MDF flora, now strictly confined to perhumid climates, have survived? Apparent support for the conclusions arrived at in the previous section comes from the many species which share broadly coincident distributions and

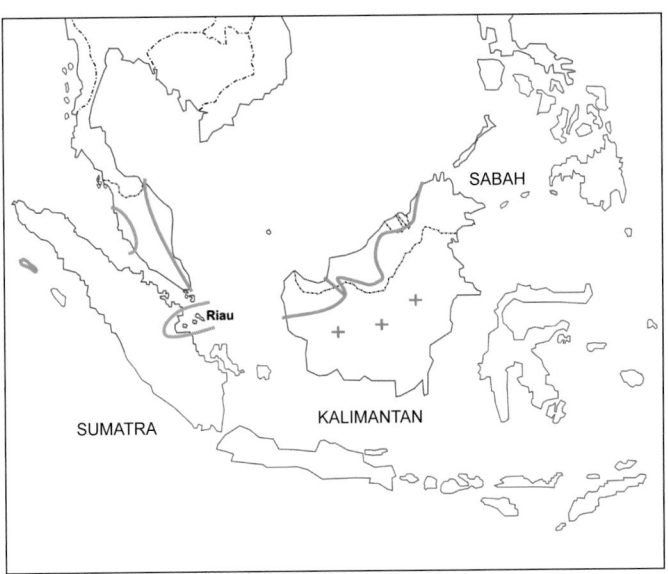

Fig. 6.14 The expanded Riau Pocket plant geographical province. Three localities beyond which Riau Pocket dipterocarps occur, in the southern Borneo foothills, indicated. The records from most are from drought-prone sites, but also include freely draining humult sandy soils inland, But the north-west Borneo subprovince includes many also from udult clay soils, implying that this area, especially, was a more general Pleistocene MDF refuge. (I.H. Baillie del. modified from Morley 1999)

Humic podzols

Humult yellow sands

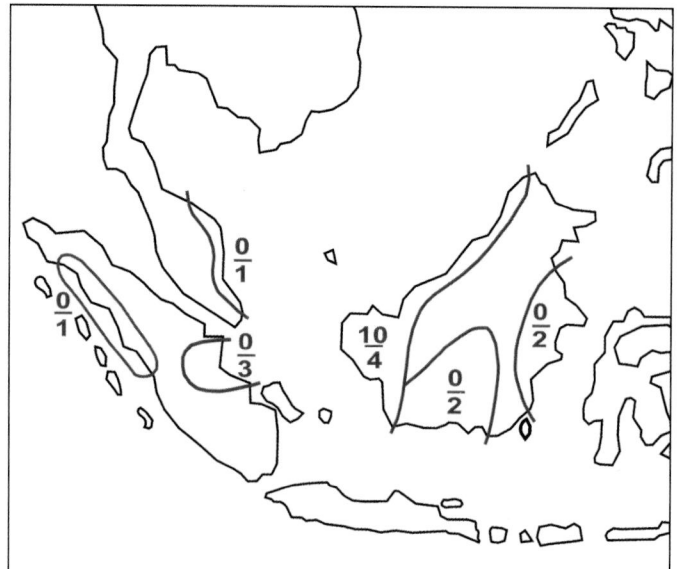
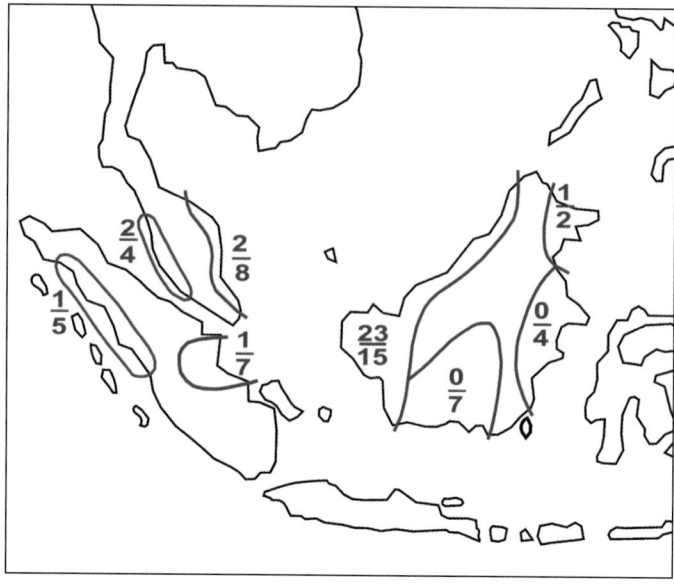

Fig. 6.15 The number of dipterocarp species endemic (above) and widespread (below) in each of the lowland regions of Sundaland in which podsols (left) and humult yellow sandy soils (right) prevail. (I.H. Baillie del. from P.S. Ashton 1984)

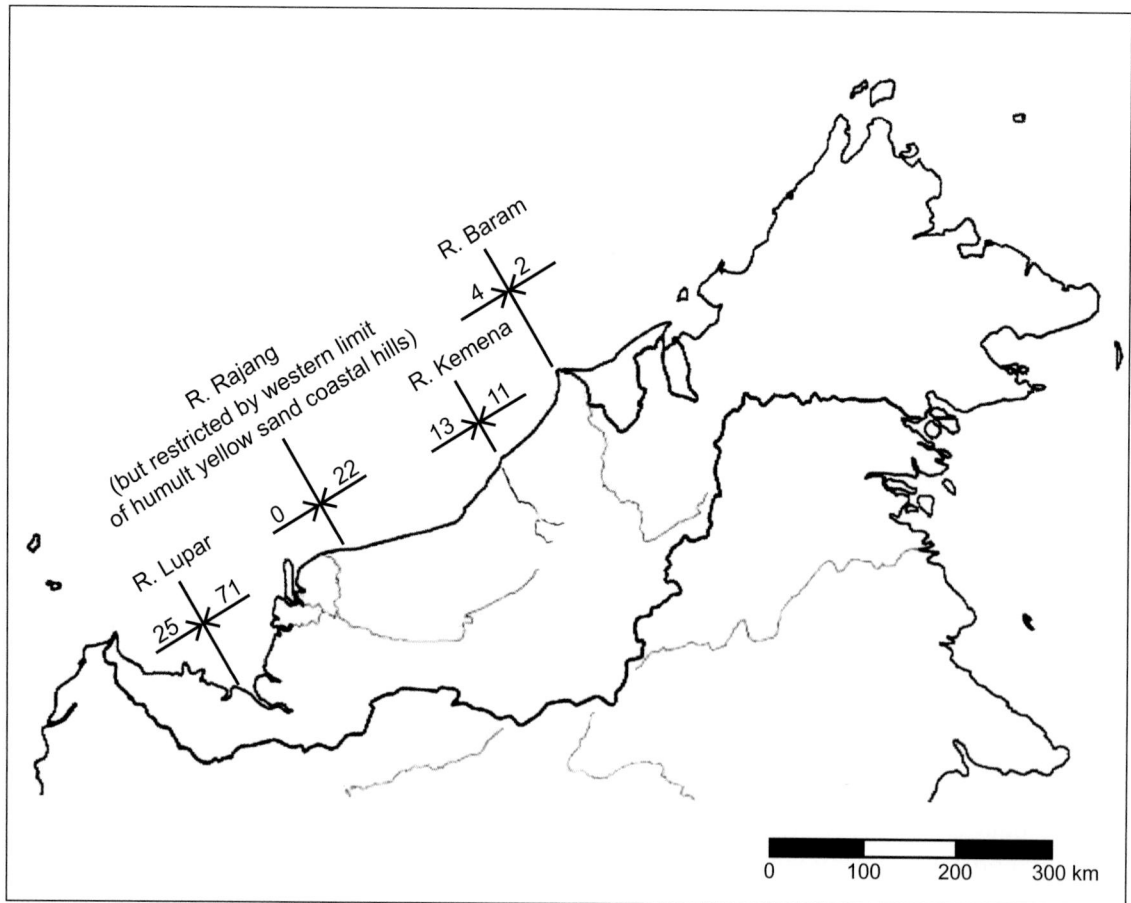

Fig. 6.16 River valley barriers and the distribution of 201 dipterocarps in north-west Borneo. Of 299 taxa in Malaysian Borneo, 116 are widespread, 148 restricted by one of the major river systems or the geographical limits of their geology or soil; 24 are restricted to the north-east; only 11 are narrow endemics. (I.H. Baillie del. modified from P.S. Ashton 1972)

define a floristic province which includes north-west Borneo, eastern coastal Peninsular Malaysia and its west coast from the mouth of the Perak River north to the seasonal zone, and the Lingga and Riau archipelagos (Fig. 6.14). Yet the regions within this overall area are also exceptionally rich in endemic species. North-west Borneo, the largest, is particularly rich both as a whole and in distinct provinces within it, separated by rivers and the associated valley habitats (Figs 6.15, 6.16). It is also the only one in which many endemics specific to a wide variety of soils exist.

The northwestern Peninsular Malaysian MDF flora, centred on Perak state, is distinguished by several dipterocarp geographical subspecies, one of which occurs on Singkep in the Lingga Archipelago, and several in north-west Borneo. There is a strong floristic connection between the overall region defined here, and the plant geographic region as originally recognized and named the 'Riau Pocket' by John Corner.[153] Some species with this distribution extend as far as the easternmost Sumatran lowlands near Pekan Baharu, and I add Simueluë Island off the north-west Sumatran coast as an outlier.

Evidence from sediments on the bed of the South China Sea,[154] and from the Niah Great Cave (north-east Sarawak) where human occupation extends before 40,000 BP, beyond the last sea level minimum[155] and which yielded middens with vertebrate bones, indicated persistence of the current forest types. At Niah, I identified pod fragments of the evergreen forest

fruit tree *Pangium edule*, still prospering on nearby talus slopes. South-east Johor, rich in the sole Peninsular Malaysian localities of Bornean species and those sister to Bornean species, including some specialists of clay loam soils as well as humult yellow soils and shallow peats, is remarkable for its small area. All the Riau Pocket regions would have been ecologically separated by the wide alluvial habitats of the plains of the Proto-Mekong and its tributaries, either subject to flooding or covered in peat as at present. They would have remained more isolated than the intermittently continuous riparian and hinterland uplands with their predominantly clay soils and associated flora (Plate 6.7), much of it common to seasonal swamp (which always overlies fine sediments). Indeed, the MDF on clay and sandy clay soils is less rich in endemic dipterocarps, and more so in widespread species than is MDF on humult yellow sandy soils, and *kerangas* (Fig. 6.17a,b; Table 6.16). On the other hand, the MDF flora of ecological islands of sandy soil within the Riau Pocket appear to manifest more past extinction events than the clay flora, as might be expected. The absence of the widespread Bornean MDF sandy soil specialist *Upuna borneensis* from the Lambir hills may be interpreted as a random extinction from a small ecological island (nevertheless with a hyperdiverse flora); but the conspicuous absence of both the widespread *Dryobalanops aromatica* and *Shorea curtisii* from Sarawak west of the Lupar river and West Kalimantan, though they are locally abundant and even dominant elsewhere in their ranges, might imply a past extinction induced by a climate change.

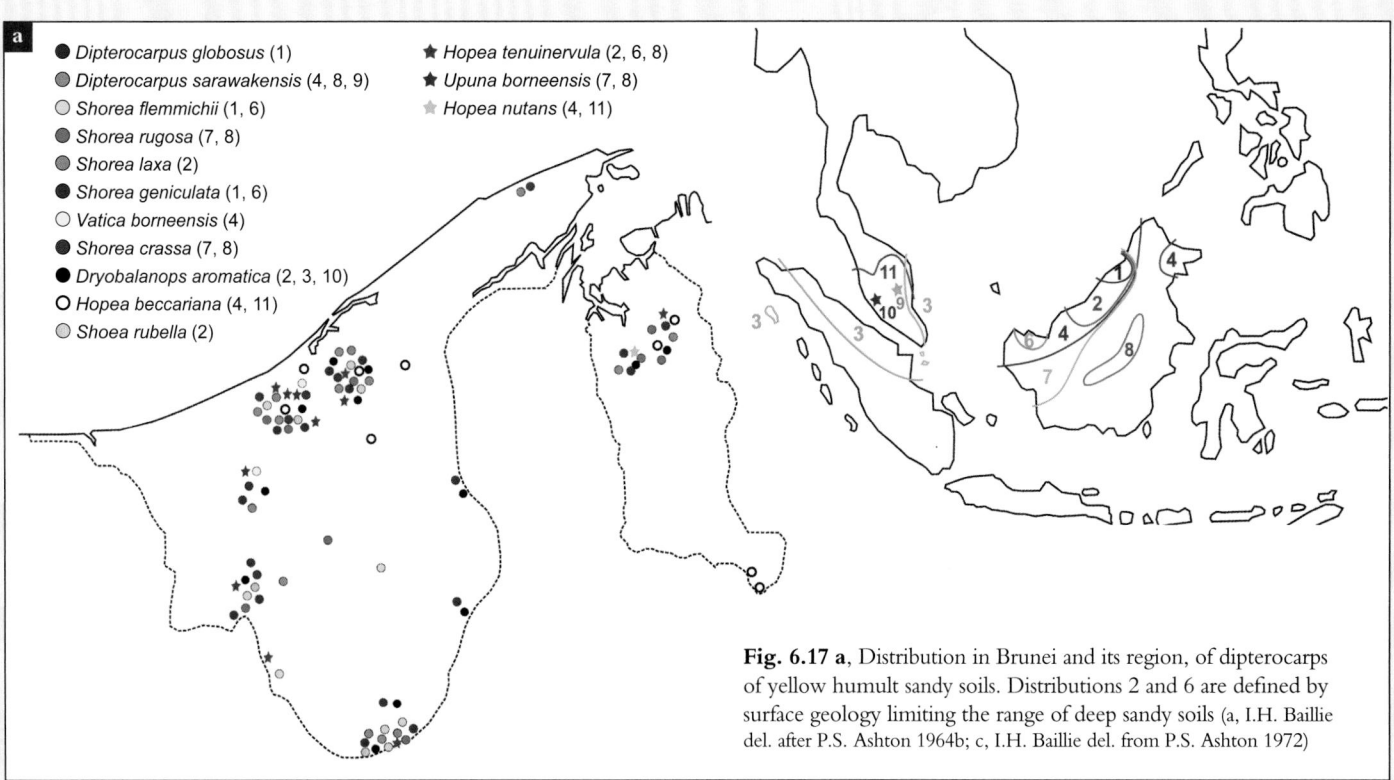

Fig. 6.17 a, Distribution in Brunei and its region, of dipterocarps of yellow humult sandy soils. Distributions 2 and 6 are defined by surface geology limiting the range of deep sandy soils (a, I.H. Baillie del. after P.S. Ashton 1964b; c, I.H. Baillie del. from P.S. Ashton 1972)

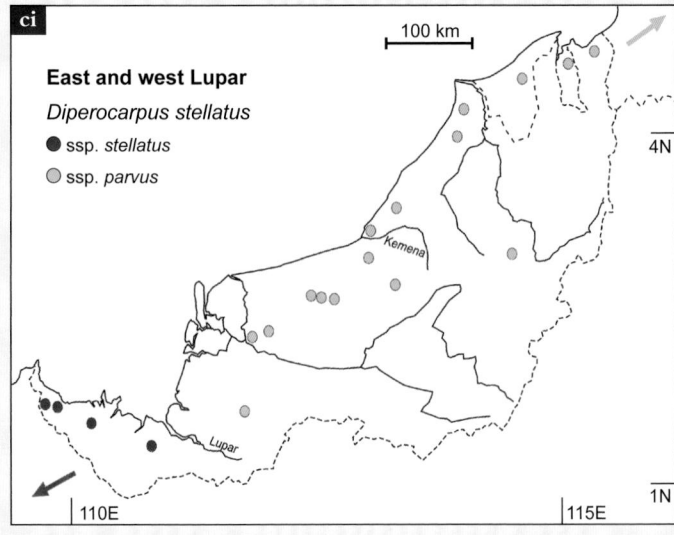

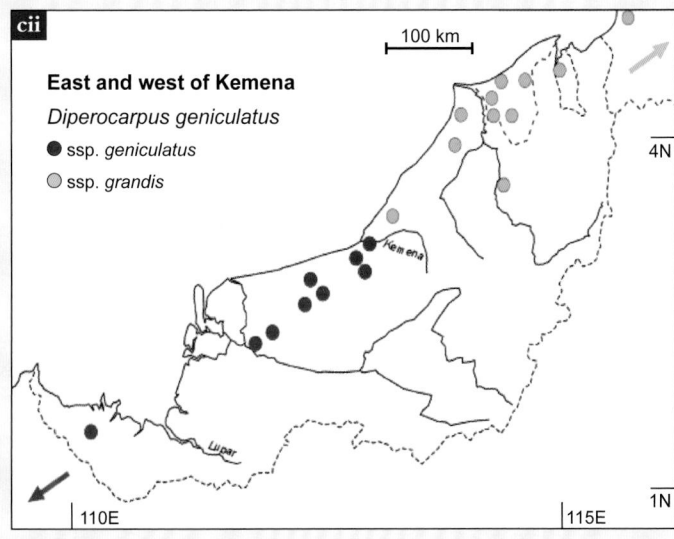

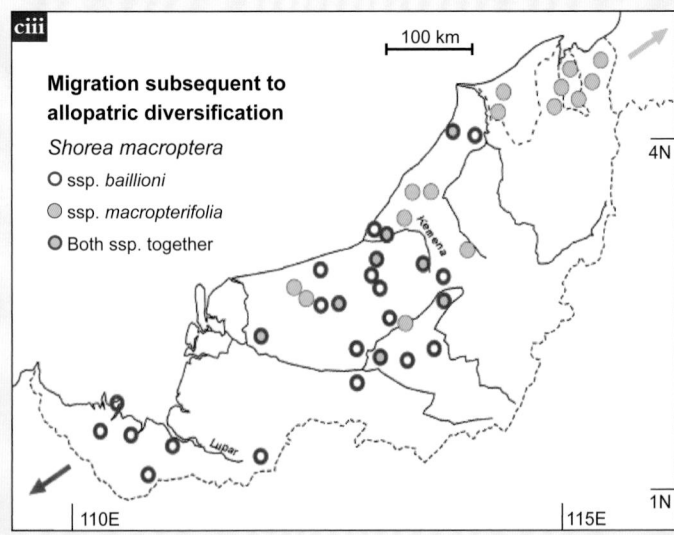

Fig. 6.17 b, Distribution, in Brunei and its region, of dipterocarps of clay loam soils. **c**, Examples of infraspecific allopatric differentiation in north-west Borneo: **i**, *Dipterocarpus stellatus*; **ii**, *D. geniculatus*; **iii**, *Shorea macroptera*. (b, I.H. Baillie del. after P.S. Ashton 1964b; c, I.H. Baillie del. from P.S. Ashton 1972)

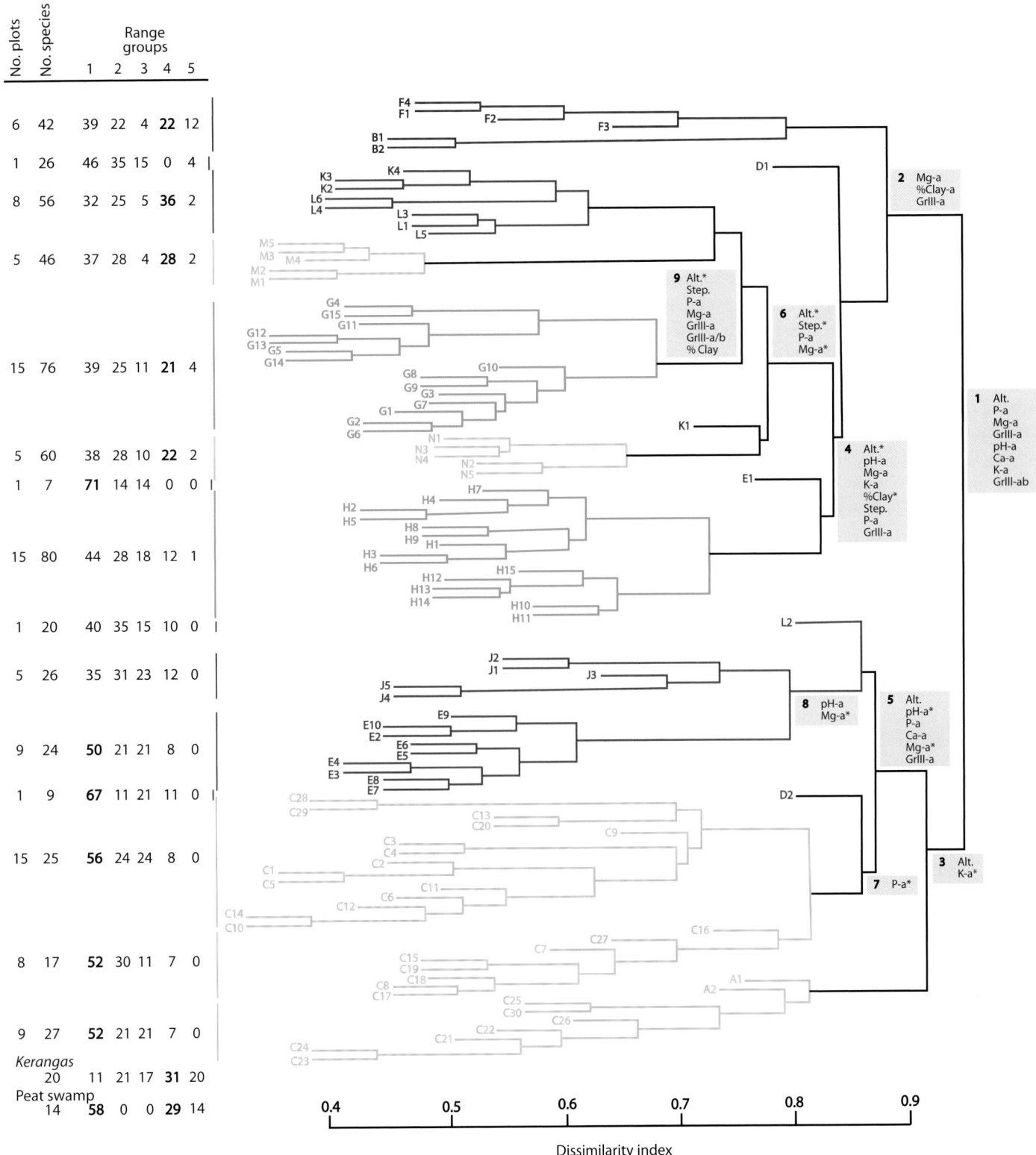

No. plots	No. species	Range groups				
		1	2	3	4	5
6	42	39	22	4	**22**	12
1	26	46	35	15	0	4
8	56	32	25	5	**36**	2
5	46	37	28	4	**28**	2
15	76	39	25	11	**21**	4
5	60	38	28	10	**22**	2
1	7	**71**	14	14	0	0
15	80	44	28	18	12	1
1	20	40	35	15	10	0
5	26	35	31	23	12	0
9	24	**50**	21	21	8	0
1	9	**67**	11	21	11	0
15	25	**56**	24	24	8	0
8	17	**52**	30	11	7	0
9	27	**52**	21	21	7	0
Kerangas 20	11	21	17		**31**	20
Peat swamp 14	**58**	0	0		**29**	14

Dissimilarity index

Table 6.16 Categorisation of the average geographical ranges of dipterocarp species within the Sarawak 105 MDF plot clusters, shown as the percentage of dipterocarp species within each range group. Range groups (numbered columns): 1, widespread in Sundaland; 2, Borneo endemics; 3: endemic to Borneo north of the Equator; 4: Riau Pocket endemics; 5: endemic to north-west Borneo. *Kerangas* and *peat swamp* denote total numbers of species known from these forest types in north-west Borneo. The bold figures draw attention to the highest concentrations of widespread species on predominantly clay loam soil sites (plots and groups L2-C, beneath) even when an individual plot is on humult sandy soil, also on peat swamps but not *kerangas*; and the concentration of species of the Riau Pocket in plots on humult sands (plots and groups F-G above, with H somewhat intermediate) including *kerangas* podsols (see also Chapter 3 section 1.1.3). The clay loam sites do include some species of the NW Province of the Riau Pocket. (Dendrogram classification of Potts *et al.* 2002)

Kamiya and colleagues[156] examined haplotype differentiation between Peninsular and Bornean metapopulations of the sandy-soil Riau Pocket, and Peninsular granite ridge species *Shorea curtisii*. They found divergences so great as to be on the scale of morphologically cryptic speciation, in contrast to the more homogeneous populations found in the clay-soil species *S. leprosula* and *S. parvifolia*.[157] They inferred that there had long been no long distance dispersal or introgression between the two *S. curtisii* metapopulations subsequent to their original separation.[158] I infer that the great plains exposed in Sundaland by low sea levels would likely have included periodically flooded clay alluvium. This could have carried much of the MDF clay tree flora, whereas freely draining humult sandy soils would have been more local, isolated and transitory, on the banks of faster flowing rivers emerging from the uplands. It is remarkable that *S. curtisii*, also several Riau Pocket species of sandy soils, occurs on Singkep and the Riau Archipelago itself, in the Malacca Strait, and has recently been discovered where the Barito River (Central Kalimantan) emerges from the uplands; but it remains unknown from Sumatra – though appropriate soils occur on the Sibumasu basement rocks towards the north. The only morphological distinction between Peninsular and Bornean *S. curtisii* is in flower colour; cream and dark red respectively. Whereas molecular evidence from Peninsular populations of *S. curtisii* indicated past expansion, compatible with its current widespread presence there along the inland granite ridges, north-west Bornean populations appear to have experienced bottlenecks, perhaps a consequence of their fragmented habitat. Population samples of *S. lepsosula* or *S. parvifolia* indicated neither expansion nor restriction.

It is difficult to imagine either purely historical or purely ecological explanations for these complex distribution patterns. North-west Borneo is easily the richest part of the Riau Pocket[159] (in the extended usage we employ here) including in local endemic vertebrate subspecies[160] – but then, it is also the *largest* area. The whole Sunda region is unified by its perhumid climate, but some lowlands also experience relatively frequent droughts owing to their coastal position, or exposure to ENSO-induced droughts (Chapter 1). But both the southern, inland part of the north-west Bornean province, and the ENSO-influenced Barito and Mahakam hinterland Riau Pocket refuges, serve to show that shared soils are more highly correlated than shared occasional droughts. Indeed, proneness to occasional aseasonal drought itself enhances the raw humus accumulation, which characterises the habitat of so many Riau Pocket endemics (Chapter 1). The coastal climate is associated with thin but distinct soil surface raw humus accumulations (Chapters 1, 3). This coastal tree flora, particularly in Borneo, has strong associations with the flora of similar humult yellow-red soils on siliceous substrates further inland (Figs 6.15–6.17a), but it occurs on humult coastal soils

on a variety of substrates and in varying sand/clay ratios (see Chapter 1). Slik[161] found that widespread Sunda genera were more associated with 'coarse sand' than clay soils, which is the opposite of that occurring in species associations. Generic ranges are likely to pre-date the Pleistocene, and may still be correlated in part with their earlier geomorphological history (see section 3.3 below).

3.2.6.1. *Refuges from seasonality?* Nevertheless, there is lesser but substantial provincial endemism in the Riau Pocket flora on other, non-humic soil types[162] (Table 6.16). This implies that some floristic provinces of the Riau Pocket must have acted as a refuge for a perhumid Pleistocene flora in general, by being the area that experienced the least climate change.[163] This idea receives support from the much poorer representation and endemism of dipterocarps confined to humult yellow sand and *kerangas* south and east of the Bornean mountains (Fig. 6.15) (though this could also be due to their more humid inland current habitat). The Riau Pocket flora occurs mostly in coastal regions prone to lower rainfall and more frequent drought but does not extend to seasonal climates. Yet there are also inland stations, as in the upper Kapuas valley and the upper Barito and Mahakam. Quek Swee-Peck,[164] in an important molecular phylogeographic analysis of *Crematogaster* ants symbiotic with pioneer *Macaranga* species throughout Sundaland, found a primary centre of origin and survival in northern Borneo at the north-east end of the north-west Bornean floristic province of the Riau Pocket. She also found evidence that in Peninsular Malaysia – with the exception of one east coast sample – *Crematogaster* appears to have survived only in the mountains during periods of low sea level. These sources of evidence are consistent with the persistence of a widespread perhumid climate through the Pleistocene owing to embayment and warming of the South China Sea during low sea level periods.

The Sunda lowland *species* flora is likely to be mostly of late Miocene to Pliocene origin and, judging from their present range, few species could have survived a seasonal climate. Yet the Riau and Lingga archipelagos are in the centre of the Malacca Strait, which is the putative seasonal migration corridor for the Trinil fauna. There must also have been floodplains near the lower Perak River no more seasonal than Narathiwat in Pattani Changwat, on peninsular Thailand, where mean monthly rainfall exceeds 100 mm throughout the year. Otherwise the local endemic Perak subspecies of the mixed peat swamp *Shorea macrantha* and *S. hemsleyana*, and the exceptionally rare *Dipterocarpus semivestitus* there, would not have survived. *Comprehensive mapping of the islands of Riau Pocket endemicity should prove informative.*

The occurrence of several species of the Riau Pocket lowland humult yellow soil flora on the granite ridges of

Peninsular Malaysia to 800 m (Chapters 3, 4), where soils are also freely draining, sandy and humic, and where dry southwesterly breezes, having dropped their rain on Sumatra's Barisan Range, prevail June–Aug., may explain how much of the Riau Archipelago MDF flora survived the Pleistocene interpluvials: in a climate of reduced rainfall but hardly increased seasonality. Interestingly, there is evidence that Peninsular populations of the hill dipterocarp forest dominant, *Shorea curtisii*, have been through a genetic bottleneck but have subsequently expanded, implying a severely restricted range prior to a likely Holocene spread.[165] Peninsular population samples of the two clay soil species, *S. leprosula* and *S. parvifolia*, do not appear to have experienced similar fluctuations. But unlike *S. curtisii*, their Peninsular haplotypes are also present in Borneo, where endemic haplotypes are less allopatrically differentiated. The two species, in contrast to *S. curtisii*, are members of the core clay loam Sunda MDF flora whose regional continuity is so much greater than that of the Riau Pocket flora.[166]

3.2.6.2. *Why is the large mammal fauna of Borneo so poor?*

Borneo seems to have harboured few or no tigers (but for a prehistoric tooth from human residues in the Niah Cave (north-east Sarawak) and a record from neighbouring Palawan, into which it must have migrated from Borneo[167]); no leopard other than the arboreal clouded; no siamang; and only a diminutive elephant (albeit a distinct subspecies) and the Sumatran rhinoceros, now endangered but never common. Tapir, now extinct, may also have been rare.[168] Yet Borneo is rich in tree species. Sixty years ago the old Bombay-Burma Timber Company decided to try their hand at commercially logging MDF in the Rejang Valley (central Sarawak), bringing in a team of elephants from Burma. They failed because they could not maintain a supply of forage for the elephants. Sambar deer are uncommon now, and the *tambadau* (banteng) rare. It is no accident that the range of elephant and banteng largely coincide on the fertile volcanic and sedimentary loams of east Sabah and north-east Kalimantan (Kutei). Elsewhere, the leached soils under incessant rains and over the serially peneplained and uplifted sediments that dominate Borneo provide an arena for successional tree communities of exceptional species diversity but low productivity. They are short commons for terrestrial browsing animals, although upland riparian forests would have provided some sustenance.

3.2.6.3. *Evidence for a less seasonal past climate?*

The borderlands of perhumid Malesia and south-west Sri Lanka provide little evidence for the survival of MDF species in refugia surrounded by seasonal evergreen dipterocarp forest. The persistence of *Dipterocarpus zeylanicus* in isolated patches at the base of the Knuckles range to the north and in the 'intermediate' moist seasonal climate zone east of the central Sri Lankan massif is one. A more convincing example is the survival of southern seasonal evergreen dipterocarp forest species in the eastern Indian states of Tripura and adjacent southern Assam, a region where semi-evergreen dipterocarp forest prevails with *Dipterocarpus turbinatus*, a species of clay loams not known to occur in seasonal evergreen dipterocarp forest elsewhere (Chapter 3). Dipterocarps recorded there, not otherwise found associated with *D. turbinatus*, are *Dipterocarpus costatus*, *D. gracilis*, *D. kerrii*, and *Shorea assamica* which also occurs as a subspecies in northern seasonal evergreen dipterocarp forest from Arunachal to Thailand. The latter excepted, these are northwestward extensions of species ranges from the southern seasonal evergreen dipterocarp forest of the Burmese coastal Arakan Yoma, variously 150–300 km to the south. Their arrival there could be explained by a lengthening of the south-west monsoon rainy season during the early Holocene. *Thorough examination of the relationships and extents of forest formations in these localities is needed while opportunities last.*

3.2.7. *Sunda centres of tree species richness and community diversity*

In Borneo, Ferry Slik[169] identified the south-east as the region richest in tree genera, although his sampling was weak for the south and south-west. Southern Borneo would have been most directly connected by upland habitats to Java and Sumatra, and thence to Peninsular Malaysia. Its generic richness would reflect its connection to the central core of the Sunda flora. Additionally it might reflect survival, in the south, of a higher proportion of seasonal forest elements than elsewhere in Borneo. Even now, a few species otherwise absent or rare in Borneo but penetrating northern Peninsular Malaysia (including *Kleinhovia hospita* and *Lagerstroemia speciosa*) occur along the eastern rivers from the Kinabatangan southwards.

On the basis of herbarium collections, Niels Raes[170] showed that the centre of *species* diversity in Borneo is the north-west lowlands, encompassing the richest region of the Riau Pocket (Figs 6.15, 6.16). The 50 ha CTFS plot at Pasoh (Peninsular Malaysia) has almost exactly the same number of families and genera as the 52 ha plot at Lambir (north-west Borneo) – 82 and 83 families and 288 and 287 genera, respectively[171] – but while the Pasoh plot includes 817 species, Lambir has 1,192. The sandstone and shale-derived soils at Lambir share only about one third of their tree species, whereas the soils at Pasoh are relatively uniform. Therefore, the difference in species richness can be explained entirely by the plot's differences in habitat diversity, without resorting to a historical explanation. This observation is only strengthened when it is borne in mind that the Pasoh hills are capped with laterite, implying Pleistocene climate change. With its current mean annual rainfall below 2,000 mm, Pasoh almost certainly experienced a drier, more

Locality	Number of species	Widespread Sunda lands	Widespread Borneo (Lambir) or Peninsular Malaysia (Pasoh)	Local: Riau Pocket (Lambir), southern Peninsular Malaysia (Pasoh)
Lambir Hills, Sarawak, Borneo	780	530 (68%)	138 (18%)	112 (14%)
Pasoh Forest, Negeri Sembilan, Peninsular Malaysia	463	379 (82%)	60 (13%)	24 (5%)

Table 6.17 Tree species endemicity within the Pasoh 50 ha and Lambir 52 ha CTFS plots. Percentage of total within plot in parentheses. (Families revised to date in Soepadmo *et al.* 1995–)

seasonal climate through much of the Pleistocene. Its flora must therefore represent a later reinvasion from a refuge to the east. Raes' Borneo-wide analysis fails to pick up the very different geographical distributions of species on different soils, which vary within as much as between landscapes (Chapter 3), because the areal scale of his sample units is too big. He consequently overemphasised climatic and historical over local and ecological influences on plant geography.

As a general principle, the proportion of endemic species gives rough evidence of the length of time over which a locality or local region has experienced constant climatic conditions. Here, Pasoh with 5% of species confined within the south in Peninsular Malaysia differs from Lambir, with 14% confined to north-west Borneo – similar to that at Pasoh compared to the Peninsula as a whole (Table 6.17). The widespread lowland clay- and sandy-clay loam soils, which prevail at Pasoh, are likely to have been continuous throughout the Sunda Shelf for long periods during the Pleistocene. Individual communities sampled on similar soils appear to indicate similar species diversity throughout Borneo, although endemism is higher in the north-west and the north-east as far south as Tidung.[172] MDF communities on humult yellow sandy soils and *kerangas* communities appear to be richest in dipterocarp species, including endemics, from the Kapuas valley northward in the west, and in the lowlands north-west of the central Bornean mountains as far as extreme southwestern Sabah (Sipitang, Beaufort District) (Fig. 6.15). (However, collecting intensities have also been greatest in the Malaysian states and Brunei.) This is consistent with the distribution of the most intense ENSO-related droughts since records began: these have not affected Sarawak west of Bintulu, and have affected other parts of lowland Sarawak, Brunei, southwestern Sabah and north-west Kalimantan less severely than the rest of the island's

lowlands. In contrast, dipterocarp communities are depauperate on sandy soils south and east of the Bornean mountains where ENSO droughts may hardly penetrate; patches currently exist in the mountain foothills of the Ulu Barito and Mahakam (Long Bleh). This is consistent with north-east and north-west Borneo acting as refuges during periods of climate change.

Sumatra's perhumid eastern lowland MDF shows strong connections with Peninsular Malaysia. Nevertheless, before its almost complete conversion to commodity plantation, lowland Sumatra harboured a relatively poor tree flora and lower endemicity. To take the comprehensively collected Dipterocarpaceae as an example,[173] there are 163 species in Peninsular Malaysia (132,600 km²) of which 28 (17 %) are endemic, while 11 (10 %) of the 108 species known from the much larger Sumatra (524,100 km²) are endemic. These compare with 157 (58 %) endemics among the 270 species in Borneo (754,000 km²). Among Fagaceae, 8% of 50 species in Sumatra, 28% of 46 in Peninsular Malaysia, 33% of 142 species in Borneo are endemic.[174] This again implies frequent and recently prolonged archipelago-wide separation by sea of other Sunda landmasses from Borneo, at least since the Pliocene c.3 Ma. As in Indo-Burma, species richness and endemicity both increase toward the Pacific and the influence of monsoon rains from both hemispheres. Both also decline with the presumptive increasing frequency, extent and intensity of Pleistocene changes in rainfall, though Sumatra's floristic poverty may also be partly a consequence of the massive Toba eruption 75Ka.

3.2.8. *Evidence for cooler Pleistocene periods*

3.2.8.1. *In the mountains.* Many MDF tree species are restricted to below 400 m owing to geological and edaphic constraints on their ecological range (Chapter 4). If a lapse rate of ambient mean annual temperature is conservatively estimated, in a perhumid climate, at 0.6°C/100 m ascent, and a sea level drop of 120 m is accepted during the cool interpluvials, this implies that these species would have been extinguished had mean Pleistocene temperatures dropped more than 3°C in the lowlands.[175] This is consistent with estimates of 2–2.5°C based on changes in the composition of marine pelagic fauna documented from bottom sediments.[176] Documentation of minimum Pleistocene elevations of terminal glacial moraines in New Guinea,[177] and maximum descent of the treeline and alpine vegetation there derived from the pollen record,[178] imply a drop of the latter from c.4,000 m in the early Holocene to c.2,300 m at the height of the last glaciations. This suggests a temperature drop of more than 10°C. Pollen cores from

West Javan and Sumatran mountain lakes at 900–1,300 m have been interpreted as indicating a drop, even at that rather low altitude, of c.900 m (though the best evidence is from Danau di atas and Toba lakes at 1,500 m).[179] Assuming that species current altitudinal limits are substantially set by minimum temperature limits, this too would have led to the extinction of most of the lowland tree flora. That there has been a drop in temperature, and that this drop has increased with altitude, cannot be doubted, but these estimates may be exaggerated. In the Sumatran cores, the evidence is derived from changes in the abundance of *Dacrycarpus*, *Altingia* and *Engelhardia*, which, judged by their current distributions, could equally have indicated a more seasonal rather than a colder climate. In the Javan case, which predicted a drop of 1.5–5.2° C, the evidence is from *Dacrydium* and *Dacrycarpus* pollen that was 800 m lower than at present; but this could instead indicate wetter and hence more acid soil conditions as much as temperature decline, as seen in their current lower altitudinal range and climate on Bornean mountains (Chapter 4). It must also be recalled that valleys conduct rivers of frosted air hundreds of metres to 1,000 m below the natural frost-sensitive tree line on adjacent slopes, and frost may pond on flat alluvium where the sampled lakes exist. Also important is the correlation of the lower limit of lower montane forest with the altitude of the cloud base, which would have remained constant then, as now, from the Equator to the tropical margin, and from perhumid to seasonal climates, during whatever rainy period persisted in the season when the convergence zone was overhead. The steep decline in MDF richness above 400 m, especially from 600–900 m, especially the decline in richness and dominance of *Shorea* sects *Mutica* and *Richetioides* is to be expected as this zone would have contracted by the amount that the cloud base may have descended during these cooler periods (Chapter 4). Currently, Fagaceae-Lauraceae dominated forests descend below 750 m only in warm temperate, frost-prone, South China.

Whatever the actual degree of tree line descent, cooler temperatures and the associated forests must have been vastly more extensive in many montane areas than they are today. Montane forest corridors between currently isolated peaks along ranges such as the Western Ghats and Sumatra's Barisan must have been more continuous. John Flenley has suggested that upper montane forests may have completely disappeared during drier Pleistocene periods.[180] Current upper montane species distributions imply that, if they did disappear, it would have been during times of increased rainfall *seasonality* (Chapter 4). Given current levels of local endemicity in upper montane forest, it seems probable that diurnal orographic cloud would have persisted throughout the year in much of currently perhumid Malesia.

The survival of an endemic goat, the Nilgiri tahr (*Nilgiritragus hylocrius*), on the Western Ghat summits confirms the long persistence of shrubby and grassy rocky summits, and probably more widespread grasslands in frost pockets there. The pollen record confirms the natural origin of the Nilgiri montane grasslands, and indicates that drier conditions did prevail there 16(-22) kya during the last interpluvial maximum, after which the monsoon returned.[181] This would satisfactorily explain the dispersal along the South Asian mountains of *Rhododendron arboreum* subspecies, and of a small but distinct north-west Himalayan (that is, warm temperate and seasonally dry) element in the montane grassland flora.

3.2.9. *Lowland floristic hotspots*
Lowland hotspots of exceptional floristic richness for their regions are:

1. South-west Sri Lanka.
2. Tropical lowland north-east Indo-Burma including southern lowland Yunnan.
3. Montane Guangxi and adjacent northern Laos, and Vietnam south Montane down the Annamite chain to Dalat.
4. The Cardamom and Kamchay mountains along the southeastern Cambodian coast.
5. South-east Peninsular Malaysia.
6. North-west and north-east Borneo.
7. The southeastern Philippine islands of eastern Mindanao, Basilan, and Samar.
8. The Fly-Digul region of southern New Guinea.

The Sri Lankan, Philippine and New Guinean hotspots correspond with the most constantly wet – though not necessarily the highest overall – rainfall in their regions. The Peninsular Malaysian and Bornean hotspots at the centre of the Riau Pocket are areas of exceptional habitat diversity (notably in soils), as well as likely Pleistocene refugia, and have been discussed. The north-east Indo-Sino-Burmese regional *tropical* hotspot correlates with a seasonal rainfall climate made continuously moist by winter dew; and by low year-to-year and evolutionary time-scale climate variability owing to its proximity to the Pacific Ocean, and distance from the vagaries of the Indian monsoon. It is also a region of exceptional habitat diversity owing to rugged topography and abundant karst islands, topographic variation in rainfall seasonality in rain shadows under the north-east trade winds, and temperature and soil variation at the margin of the ecological tropics. Abutting it to the north is the richest tree flora in the temperate world, sharing more than half its genera with its richer tropical neighbour. Both floras have provided gene pools from which novel species have arisen through migration across the climatic boundary (for example, see Table 4.3).

3.2.9.1. *Centres of lowland endemism*. Although plant species endemism is negatively correlated with their length of life cycle, it is not clearly correlated with species successional role. Some pioneer species are exceptionally widespread on account of their tolerance of a range of rainfall seasonality (Table 6.18), while certain pioneer genera, such as *Macaranga*, *Glochidion*, and dioecious *Ficus* are rich in local endemic species. On the other hand, many climax species, including dipterocarps, are as widespread as their chosen habitats.

Hotspots of species richness are often, but not always, centres of endemism (Table 6.19). The ancient and geologically uniform landscape of perhumid south-west Sri Lanka is a major repository of tree species endemism (estimated at 70%), although endemic genera are few. Lowland point endemism there is difficult to document owing to the area's long history of cultivation. In a flora of species mostly wide-ranging within the zone, the endemic dipterocarp *Stemonoporus*, with over 20 species, is the exception.[182] Confined to lowland and lower montane forest, most *Stemonoporus* species are also confined to the most humid region immediately south and west of the Peak Sanctuary range. Thus their ranges are restricted both altitudinally and horizontally – in many cases to less than 100 km². Only a few species are relatively widespread, such as the montane *S. gardneri* (occurring throughout the main massif) and the lowland riparian *S. wightii*. With *Mesua thwaitesii*, *S. moonii* is known only from one of the very few remaining freshwater swamps, near Bulatsinhala just south of the lower

Kalu Ganga. *S. bullatus* and *S. kanneliyensis* occur together with the more widespread but local *Axinandra zeylanica* on humult coarse sandy soils in the lowlands of Kanneliya forest. They are sister- or subspecies of two occurring just to the north but in the higher-elevation Sinharaja forest. Such co-occurrences are a common pattern among rare species throughout tropical Asia, as a result of specialisation within a rare habitat. It has important implications in conservation planning (Chapter 9). *In addition, it appears that Stemonoporus – by reason of some characteristic of their breeding system or population genetics yet to be ascertained – possesses an exceptional rate of diversification.*

Point endemism in the East Himalayan to South Chinese tropical margins is also difficult to evaluate, for the same reason that has reduced indigenous forests to fragments. In South China in particular, there appear to be multiple records of tropical taxa at the same inland localities north of the limits of the northern seasonal evergreen dipterocarp forest (see Chapter 3). I interpreted these as frost-free refuges, isolated during the retreat of tropical forest here at the end of the last northern glaciations or early Holocene (Chapter 3). The topographic, edaphic and climatic diversity must also contribute.

In the perhumid Far East, remarkably, there is comparatively less tree endemism at island level, even among the Philippines where endemism among plant life forms with shorter life cycles is high: Palawan is the major exception, partly owing to its extensive ultramafic exposure but also due to its original

Name	Range	Ecology	Seed dispersal
Alstonia scholaris, Apocynaceae	Sri Lanka–Solomons	Pioneer, clay	Wind
Anisoptera costata, Dipterocarpaceae	Chittagong–Wallace's Line	Climax emergent, clay	Gyration, wind
Shorea roxburghii, Dipterocarpaceae	S.India–Pen. Malaysia	Deciduous forest	Gyration, wind
Macaranga tanarius, Euphorbiaceae	E. Himalaya–Ryukyu–SW Australia–Vanuatu	Pioneer, clay	Birds, ants
Croton caudatus, Euphorbiaceae	Sri Lanka–Wallace's Line	MDF understorey, clay	Explosive fruit
Drypetes longifolia, Euphorbiaceae	Malesia, wide	MDF understorey, clay	Gravity, (deer?)
Castanopsis acuminatissima, Fagaceae	E. Himalaya–New Britain	Lower montane forest, clay	Birds
Lithocarpus elegans, Fagaceae	E. Himalaya–Sulawesi	Lower montane forest, clay	Gravity, vertebrates
Engelhardia spicata, Juglandaceae	E. Himalaya–Timor	Lower montane forest, clay	Wind, gyration
Syzygium zeylanicum, Myrtaceae	Sri Lanka–New Guinea	Coastal, secondary forest	Birds
Neolamarckia cadamba, Rubiaceae	Sri Lanka–New Guinea	Pioneer, clay	Wind
Pometia pinnata, Sapindaceae	Sri Lanka–New Guinea	Successional, clay	Vertebrates
Tetrameles nudiflora, Tetramelaceae	Pioneer, MDF	Pioneer, clay	Wind

Table 6.18 Examples of widespread species in tropical Asia. Although many are wind dispersed pioneers, there are also climax species albeit more confined by climate; widespread lowland evergreen forest species are all found on the only continuous soil corridor: udult clay loams.

Region	Approximate number of tree species	Percentage endemism
SW Sri Lankan lowlands	850	70
Montane Sri Lanka	500	70
Western Ghats	800	60
Lower montane northern Indo-Burma, Meghalaya	1,000	65
Lowland northern Indo-Burma, South China	1,500	60
Indo-Burmese southern seasonal evergreen forest	2,500	60
Andaman Islands	1,000	30
Lowland Peninsular Malaysia	3,000	30
NW Borneo	4,000	35 (to Borneo)
NE Borneo	4,000	35 (to Borneo)
Borneo: Kinabalu	2,300	25 (to the mountain)
Sulawesi	600	40
Lowland New Guinea	3,000	65
Montane New Guinea	3,000	75

Table 6.19 Hotspots of tree species richness and endemism in tropical Asia. Figures are approximations. (Partly after van Dijk *et al.* 2004)

placement against Vietnam, perhaps evidenced by the presence of *Embolanthera* (Hamamelidaceae). Volcanic Basilan – first in a string of perhumid stepping stone islands leading south-west to Borneo – is richer in endemics than most.

Endemism in the large island of Borneo is concentrated in the northern half, consistent with palynological evidence that it has acted as a refuge of perhumid climate since the Miocene.[183] All but two of the 269 dipterocarps in Borneo occur in the Malaysian Borneo states, Brunei, and West Borneo south to the Kapuas drainage. Ninety-three (58%) of Bornean endemic dipterocarps are confined to Brunei and the Malaysian states of Borneo – that is, the putative northern refuge. This can also be seen by comparing endemicity in the plots at Pasoh (Peninsular Malaysia) and Lambir (north Borneo). At Pasoh, only 10% of the 30 dipterocarps are Peninsular endemics, and whole-flora endemicity stands at 20%, whereas at Lambir 54 (61%) of the 88 dipterocarp taxa are endemic to Borneo, and over 55% of the flora as a whole (Table 6.17). Habitat heterogeneity

accounts for a large part of dipterocarp endemicity at Lambir, with the humult sandy soils particularly rich in endemics, but the whole tree flora is richer in endemics in the Bornean than the Peninsular Malaysian plot. Twenty-five dipterocarp species are in Borneo confined to the north-western region defined by the Lupar river fault and Kapuas valley, of which six are endemic to it; 18 also occur in Peninsular Malaysia. Two locally abundant species in Peninsular Malaysia and elsewhere in north-west Borneo are absent: *Dryobalanops aromatica* and *Shorea curtisii*.

Local and more regional endemism of the Riau Pocket flora, in Borneo's north-west, is replaced by a second centre of endemism in Sabah excepting the extreme south-west; this extends in the east south as far as Sankulirang, beyond which it declines without a clear boundary.[184] This north-east Bornean floristic province includes the exceptional physical environments of Kinabalu (see section 3.2.10 below) and also the several islands, of differing area, of ophiolitic extrusion and ultramafic soils. These support endemic sister taxa of widespread species (see section 3.2.9.2 below) and several distinct taxa of Australasian affinity (*Scaevola chanii*, several taxa on Kinabalu, perhaps *Borneodendron* in the lowlands).[185] Overall, however, the province is defined by the transition in north-east Sarawak and Brunei from the Rejang series sedimentaries to lower Miocene shales folded into steep hills and low mountains, yielding clay loam soils of higher base status (represented by the MDF plots at Kuala Belalong, also E, J and L$_2$ on the Setap Shale Formation of the southern Lambir Hills: see Chapter 3) than either the sandy soils of the north-west coastal hills or yet the sandy clays and clays of the Rejang series. This region, therefore, has likely served as a refuge as much as the north-west Borneo province, but its soils link the well-represented widespread element of its flora to the general Sunda MDF flora on clay. In contrast, the Riau Pocket flora predominates, in north-west Borneo as elsewhere (albeit not exclusively) on the more fragmented terrain of humult, mostly sandy soils.

3.2.9.2. What accounts for patterns of endemism? Limited seed dispersal and ecological specialisation may each influence tropical tree community composition within and between local habitats. *The relative strength of these influences is a focus of contemporary research, and remains incompletely resolved.* We have currently no universal physical, or biotic, explanation for spatial patterns which may serve as evidence for the influence of a chance factor such as limited seed dispersal; but this may also simply reflect insufficient knowledge. Fragmented habitats, for instance, clearly set constraints that influence the geographical ranges of individual species. Henry Ridley[186] suggested that canopy dipterocarps generally disperse their seeds no more than 100 m (Chapter 5). At that rate, assuming that these trees reach reproductive maturity after 50 years on average, populations would spread on average no more than 2 m y^{-1}. But dipterocarps apparently returned from perhumid refuges in mountain valleys over distances of 800 km to southernmost Borneo since the end of the last interpluvial, just 12,000 years ago. If dispersal had been as severely limited as Ridley observed, they would have needed 400,000 years! Clearly, occasional longer distance dispersal is crucial in determining species distributions and opportunities. Although local clumping of tree populations must surely be partially caused by dispersal constraints, the influence of these constraints on species distributions at a larger scale and community composition does not appear to be great (see section 3.2.9.3 below).

The major Sunda land masses support mammal faunas whose richness is well-correlated with their area.[187] Not so with tree species[188] (Table 7.1). The Sumatran lowlands are lowest in endemism among the major Sunda landmasses. The endemicity of its tree flora, which mostly is a subset of that of Peninsular Malaysia, stands at c.20%. Bornean endemicity, at 35%, may in part be due to its separation from Peninsular Malaysia and Sumatra by the Proto-Mekong valley; in part because the perhumid climate seems to have persisted in its north-west and north-east, which have thereby served together as a major refuge. Also, in part because Borneo includes the most extensive ecological islands of lowland podsols, and leached humult yellow sandy soils which, in the northwestern refuge arguably support the most species diverse forests in the Old World (Chapter 7). North-west Borneo appears to be the only region within the Sunda lands where some rivers act as barriers to the migration of tree species, such as dipterocarps, with poorly dispersed seeds (Figs 6.15, 6.16, 6.17c). Among them, the Lupar valley is exceptional, marking the boundary, either east or west, of the ranges of almost one third of the dipterocarp species in northern Borneo, and many others in other families.[189] These include many, though fewer, species on soils other than white and yellow sands. There is also higher local endemism in the low mountains to the west of the Lupar than in the often higher mountains of central and north-east Sarawak.

Point-endemism in the perhumid lowlands within the main Sunda landmasses is often correlated with exceptional habitats: limestone throughout the region; and ultramafics in north-east Borneo and south-east Philippines especially Sibuyan Island, to which the isolated monotypic genus *Borneodendron* (Euphorbiaceae),[190] and *Xanthostemon* (Myrtaceae) of Australasian affinity, are confined.

There is some evidence that west Sarawak, and West Kalimantan south to the Kapuas valley, differ floristically in part also owing to a past extinction event experienced there but not further to the east. *Dryobalanops aromatica* and *Shorea curtisii* are widespread species shared with Peninsular Malaysia and confined in Borneo to its northwestern Riau Pocket refuge

– except that they do not occur west of the Lupar even though suitable habitats are widespread. In Borneo, *D. aromatica* and *S. curtisii* are locally gregarious but never dominant on humult yellow sands. In contrast, they become dominant over vast areas in Peninsular Malaysia, respectively on siliceous soils in the eastern lowlands and on friable soils along inland ridges, where the tree flora characteristic of these habitats, notably the dipterocarps, is poorer than in Borneo. Conversely however, the Malayan dipterocarps *S. dealbata* and *S. resinosa* are *only* known to the west of the Lupar River.

But of the 58 species of dioecious *Ficus* trees in Malaysian Borneo, only three are confined to the island habitats of sandy soils.[191] Rhett Harrison[192] observed that, whereas monoecious fig pollinators are omnipresent and widely dispersed, those of dioecious species are more variable in time and space. He found that these dioecious species suffered local extinction during a severe drought at Lambir.

The strip of wet forest in the premontane lowlands north of strongly seasonal south-central New Guinea, at the source of the Fly and Digul rivers (currently undergoing logging and conversion) appears to be a major centre of endemism, but is poorly documented.

Throughout perhumid regions, limestone karst and sometimes basalt may act as refugia and stepping stones for the intrusion (or survival) of elements of drier climates into wetter regions, as well as for local speciation into specialised habitats. Notable are the extensive karsts north of the Red River in northern Vietnam, the Bau limestone in west Sarawak and the Mersing basalt of central Sarawak. In northern Borneo, *Shorea guiso*, *Glyptopetalum quadrangulare* (Celastraceae), deciduous *Firmiana malayana* and *Chukrasia tabularis*, and the pioneer *Pterospernum javanicum*, all locally occurring or common in seasonal East Asia, survive as rare small groups of individuals in these habitats. The role of basalt here implies that these areas in perhumid regions serve as ecological refugia for species of seasonal climates because their soils tend to be more fertile and therefore more comparable to those most general in seasonal climates.

3.2.9.3. *Endemism and seed dispersal*. It is well-established that the combination of limited seed dispersal and lack of seed dormancy leads to disproportionate clumping of juveniles around parent stems, although the intensity of this effect declines as the trees mature (Chapters 5, 7). One might expect the effectiveness of seed dispersal to influence levels of species endemism. Peninsular Malaysia, the most comprehensively documented of the Sunda tree floras,[193] lacks a large wind-dispersed family. Among bird and mammal-dispersed tree families in the Pasoh plot,[194] none of the 23 species in Moraceae (*Artocarpus*, *Ficus*) in the plot are endemic, but other such families have above-average Peninsular endemicity: Anacardiaceae (25% endemic of 32 species), Meliaceae (23%

of 43), Myristicaceae (23% of 31), Sapindaceae (30% of 20), Sapotaceae (21% of 14) and Ebenaceae, *Diospyros* (50% of 23). Meanwhile the two families with least effective seed dispersal, the ectomycorrhizal Dipterocarpaceae (30 species) and Fagaceae (15 species) remarkably show only 10% and 13% endemicity respectively. Other factors, then, are clearly involved.

Among the 242 Bornean dipterocarp species and subspecies occurring in upland habitats at low altitude, 127 occur principally on udult clay or sandy clay soils, 116 principally on humult sandy soils. Whereas 58% of the clay species are island-wide endemics, sandy soil endemics attain 71%. Fifty-one of the sandy soil species lack calyx lobes extending in fruit into wings to aid gyration and therefore dispersal (Chapter 5). In this case, all but five of these wingless species are endemic to the island. In each of the major taxa containing wingless upland species, the number of wingless taxa on the fragmented sandy habitats is greater than that on continuous clay or sandy clay (*Hopea* 5/3, *Shorea* (sect. *Shorea*) 4/0, *Shorea* (sect. *Richetioides*) 11/6, red meranti *Shorea* 6/2). An exception is the mainly subcanopy genus *Vatica*, where the proportion is equal. Comparing rates of endemicity between the two largest tree genera in Borneo as a whole – *Syzygium* with c.175 species and bird-dispersed seeds in fleshy fruit, and *Shorea* with 138 species with gyration-dispersed fruit – reveals almost identical island-wide endemicity, at 66% and 63% respectively. Wilson and colleagues[195] point out that it is the fleshy-fruited Myrtaceae that have diversified, not the species with dry fruits whose seeds are generally wind-dispersed. It appears that the dry-fruited Myrtaceous taxa are limited by being restricted to open, often isolated habitats; whereas the potential of the fleshy-fruited taxa to diversify into physical and biotic niches has not been so restricted: the ecological pressures to diversify have overcome the tendencies of efficient dispersal to dilute them (Chapters 3 and 5).

I draw two inferences from these comparisons. First, the ranges of rainforest species are only constrained by seed dispersal when means of dispersal are severely limited, which is unusual among the flora as a whole. The universal clumping of juveniles near the crowns of reproductives becomes dissipated at landscape scale and beyond, except where means of dispersal is severely limited. That is not to say that historical accident cannot produce anomalies. Examples of this would be the absence of a common species characteristic of certain habitats (such as *Upuna* in the Lambir forest), or their presence in unusually high numbers. Second, reduction of dispersal ability, as in dipterocarps (except *Vatica*) where short sepal lobes are a derived trait, may have selective advantages where habitats are fragmented or small. This would confer advantages on seeds that germinate close to their successfully established parent and share the parental soil and mycorrhizal array.

3.2.10. *Montane endemism*
Mountain floras occupy ecological islands of varying age and fluctuating degrees of connectivity. During the northern ice ages, Tropical high mountain forests of the seasonal Asian tropics must have been reduced by frost to refuges at lower altitudes and in valleys. This was especially true in the Himalaya, where the Tibetan ice-cap imposed aridity. Refuges within the Himalayan chain likely became smaller and more isolated toward the west, as the limits of tropical forest would have retreated eastwards.

Today, upper and lower montane forests present opposite patterns of tree endemism. Lower montane forests include several genera endemic to them (Chapter 4); but most of these genera have few species and low local endemism. Lower montane *kerangas* does support several locally endemic species, notably in Myrtaceae, in northern Borneo. Even in the Far East, the upper montane flora has few taxa endemic to this forest type other than Winteraceae (*Tasmannia*); but a number of genera are rich in local endemic species, including sister species of lowland taxa. Notable among these are *Rhododendron* (ect. *Schistanthe*), *Syzygium* and *Hedyotis*. Along with the Myrtaceous genera in the Far East, *Leptospermum*, *Seorsus* and *Xanthomyrtus*, few of these endemics extend down into the lower montane forest ecotone, although widespread Myrtaceous sister taxa – though like *Tasmannia* confined to perhumid climates – occur on freely draining rocky knolls in the coastal lowlands (Chapter 4).

Point endemism is highest in the least seasonal montane climates, reaching its zenith in the most extensive and isolated massifs: those of New Guinea. Here, endemism among the flowering plants increases with altitude, standing at an exceptional 46% overall.[196] Regional and local species endemism of montane tree floras differs among Sri Lanka, Peninsular Malaysia and Borneo. Relative to the size of the local flora overall, endemism is greatest in Sri Lanka. There are separate centres of endemism in the western Peak Wilderness, Knuckles, Horton Plains, and in the higher eastern Sinharaja mountains, with the Peak Sanctuary the greatest[197] (Table 6.20). The ancient mountains of Sri Lanka, built of metamorphosed Precambrian continental basement rocks, have been intermittently uplifted since the Indian collision with Laurasia 35 Ma. Together, these mountains shelter 51 endemic tree species, and a further 12 occur only here and in the Western Ghats.[198] The Knuckles and Sinharaja massifs, modest in height but rich in point endemics, are separated from the main massif by deep valleys. Though each no more than c.20 km long, they may be at least 10 million years old. Diversification is particularly high among those genera which are also diverse in the Malesia, and which occur as co-occurring species series in communities there, notably *Syzygium* and *Memecylon*.[199] Rates of endemism among trees are matched here by several animal taxa.[200] These include the

Place	Knuckles	NE & N massifs	Namunukula	E massif-Horton Pl.	Adam's Peak (Sri Pada)	Eastern Sinharaja
Number of species confined to a massif (% endemic)	28 (78)	35 (69)	0	7 (71)	32 (91)	1 (100)
Number shared (% endemic)		9 (56)				
		11 (98)				
			19 (63)			
			65 (45)			
				6 (16)		
			11 (91)			
			21 (62)			
				32 (63)		
				3 (33)		

Table 6.20 Sri Lankan centres of upper montane tree species endemism. Percentage endemism refers to global endemism. Knuckles and Namunukula are isolated, the rest united by intervening lesser ridges. Range patterns reflect the trend to more seasonal climate east- and northwards, where climates become more similar to the Indian Western Ghats to the north.

Agamid horned lizards (*Ceratophora*) (Plate 6.1g), possibly of Gondwanan origin, with one endemic species on the Knuckles, two on the eastern Sinharaja, and one on the central massif.[201] The pattern of tree species endemism in Sri Lanka markedly resembles that of frogs in speciose genera there.[202] Isolated Namunukula Mountain to the east and Ritigala to the north, each with summits exposed to seasonal drought, support few members of the upper montane flora and endemics (Table 6.20). Seasonality is greatest in the north and the east, less in the centre and the south-west, least in the Peak Sanctuary and Sinharaja.

The degree of local endemism in the upper montane forest flora of Sri Lanka, and to a lesser extent of the Western Ghats, is remarkable in comparison to what we find on Sunda mountains of similar stature. This is most likely due to their isolation as moist refuges during periods of severe drought, when the south-west monsoon abated in response to Pleistocene north temperate glacial periods. Even here, though, the uppermost zones appear to have been drier and likely colder.[203] It has been suggested, on account of the absence of subalpine species pollen in late Pleistocene peat in New Guinea, that the forest type may have suffered extinction during these periods.[204] The persistence of upper montane endemism in South Asia trenchantly argues against this, however. The pollen data could more likely imply a change from perhumidity to seasonal rainfall in the mountain investigated, namely Mount Wilhelm (5°48' S).

In the perhumid tropics, connectivity between subalpine and upper montane forests would have increased to the extent that forest zones declined in elevation during cooler periods; but lower temperatures would have led to a *narrowing* of the lower montane forest zone because its lower ecotone, broadly correlated with the cloud base as well as the temperature, is unlikely to have changed elevation above sea levels much (Chapter 4). The summit forests of Peninsular Malaysia occupy less than half the area of those of Borneo, but unlike the lowlands appear to be richer in endemic tree species – the exceptionally high Mount Kinabalu excepted. There is evidence from the *Crematogaster* ant phylogeny that the Peninsular lower montane forests acted as refuges for rainforest biota during the interpluvials,[205] implying that they must have remained perhumid. These granite and Triassic sediment mountains have been raised intermittently in the north since before the origin of flowering plants, though Cretaceous sediments overlie some southernmost hills. Major orogenic activity seems to have ceased before the Cenozoic (c.65 Ma). In contrast, the current Bornean mountain chains – also likely to have remained humid through the Pleistocene – have arisen over the last 2 million years.[206] This may explain their lower point endemicity (Kinabalu, which is twice as high, excepted), at the same time providing a conservative time scale for the rate of speciation among montane trees. Similarly, the montane tree floras of the Sumatran and Javan 'ring of fire' volcanoes are quite low in endemics. Sumatra's Barisan arose following the middle Miocene, the Javan volcanoes in the Pliocene. But since then, the continuous process of erosion and new volcanism has apparently provided insufficient stability on individual peaks for diversification of long-lived species. The high volcanoes are themselves younger than Kinabalu. Although Gunung Kerinci in Sumatra's Barisan range reaches 3,805 m, it bears few endemic tree species and lower endemicity overall, probably attributable to its youth as much as to its edaphic uniformity.

Quek Swee-Peck[207] found that Crematogaster ant species, which are specific to clades of their Macaranga hosts, are more diversified in the northern Borneo lowlands than elsewhere in Sundaland, where diversity is more associated with lower montane Macaranga species. This suggests widespread lowland extinction elsewhere, and is interesting because the Macaranga host species in question are confined either to lowland or montane forests.[208] She inferred that Peninsular montane forests acted as refuges for the ants during dry Pleistocene periods.

Patterns of tree endemism are less well-known in the mountains of New Guinea, whose alpine flora is also of late Miocene-Pliocene origin.[209] Island-wide tree species endemism is high, but rates of point endemism appear to be less than in Sri Lanka and local endemism comparable to that in Borneo.[210] *Patterns of local endemism have been best documented in Rhododendron* [211] *and* Tasmannia (Drimys),[212] *but the mountains remain insufficiently explored to define the full extent of ranges.*

Montane endemism is markedly higher in plants with shorter life cycles and in shrubs than trees. *Rhododendron*, in spite of its easily dispersed dust-like seed – but perhaps aided by its uncommon capacity for hybridisation – is particularly rich in endemic species. In the absence of competition, as on the newly-prepared airstrips in New Guinea, hybridisation initially becomes rampant.[213] The most widespread species are self-compatible. In Borneo, *Rhododendron* endemicity matches patterns among lowland trees.[214] Thus *R. jasminiflorum*, found in tiny patches of upper montane forest on several near-coastal mountains west of Batang Lupar, has a sister to its east, *R. chamaepitys*, on Gunung Lambir. Overall, point endemism in the Bornean mountains is low except on ultramafics (especially on Kinabalu) and on other specialised substrates. Similarly, only half of the 26 Philippine species of *Rhododendron* are endemic to single islands and most of these are on ultramafics such as Mount Guiting Guiting, Sibuyan, and Palawan.

Kinabalu is exceptional in its tree endemicity.[215] This can be attributed to its unique stature, at 4,094 m exceeding other mountains in Borneo by more than 1,000 m. It is undoubtedly also owing to its edaphic diversity, especially to its extensive ultramafic exposures at 900–3,200 m. These support the majority of the endemics – though many species also stray onto adjacent sandstone and granite substrates.[216] Kinabalu's substrates are also at least twice as old as existing volcanoes of the 'ring of fire'.

3.3. *The age and mode of speciation of rainforest tree species in Asia*

3.3.1. *Species of widespread habitats*

The ranges of widespread species are constrained by the continuity of their preferred climates and soils, apparently less so by their means of dispersal (Table 6.16). The clay and sandy clay loams of the perhumid Sunda lowlands comprise the only periodically continuous inland habitat on the main islands of the Far East. Interrupted only by the river systems, the upland clays predominate across the narrow upland valleys, but many clay soil species would also have occurred on the predominantly clay alluvial plains wherever free of peat. The tree flora of their MDF has the lowest level of endemicity of all the upland forest types. Sandy soils do follow along the turbidite and granite Sunda ridges in fragmented ribbons, but are elsewhere confined to more isolated patches set in the vast matrix of regional clay loam soils: in the plains where coarser textured soils originate near faster flowing water, and along the coastal hills fragmented by the estuaries of the rivers. This widespread lowland flora of clay loam soils may include the progenitors that fanned out into the emerging Sunda uplands and lowlands when they became widespread in the early–mid Miocene. By that time, the Oligo-Miocene thermal high was past its peak, and opportunities for westward interchange with South Asia of taxa confined to perhumid climates were already long past. In that sense, although these soils have continued to accumulate species as they diversified regionally, they carry the oldest rainforests in the Far East. It is these forests and their flora that are now closest to extinction owing to the suitability of their soils for commodity crops (Chapters 7, 8, 9).

Of the 108 dipterocarp species of Sumatra, 84% (89 species) occur mainly on clay loams. If species confined to the seasonal climate of the north-west are excluded, only 8% of Sumatran dipterocarps are endemic to the island and only 17% of Peninsular Malaysian, yet 59% (159 species) of Bornean, are endemic. This is substantially due to regional differences in habitat heterogeneity: clay loams dominate the Sumatran lowlands, sandy clay the Peninsular, while Borneo is exceptional in its edaphic diversity, therefore the fragmentation of its habitats.

Of the three large perhumid Sunda landmasses, Peninsular Malaysia is the best known, and the dipterocarps are by far the best documented tree family (Table 6.21). The continuity of the Peninsular Malaysian inland MDF on udult clay and sandy clay ultisol loams is extraordinary. The 50 ha plot at Pasoh is paradigmatic of this flora, containing half the tree species known from its state, and fully one quarter of the total tree flora of Peninsular Malaysia on its rather uniform udult sandy clay soils. If Peninsular habitat specialists are excluded, I estimate that this latter figure would exceed one half! In the case of the Peninsular Malaysian inland forests, the well-known continuously ascending species-area curve of tropical rainforest levels off much earlier than has generally been assumed. Even in Peninsular Malaysia though, half the dipterocarps (79 of 161) are confined to specific habitats: to the seasonal north-west, to montane forests, coastal hills, peat swamps, limestone or to sandy soils. I conclude that present or past environmental constraints dominate over randomness in dipterocarp distributions in this case.

(N=102)	
Widespread	51
Riau Pocket localities only (wide and local)	25
Seasonal evergreen dipterocarp forest, NW only	7
Towards N (6) or S (13) only	19
Local within widespread continuous habitat	1

Table 6.21 Ranges, within Peninsular Malaysia, of MDF dipterocarp species.

I know of no example of clinal variation in the morphology of dipterocarp species – or indeed of species in other families. Changes in gene frequencies greater than the variation observed within populations has been observed to be stepped, occurring on either side of physical barriers; but so have the samples themselves. *It would be useful to examine clinal gradients in the population genetics, particularly of adaptively neutral microsatellite DNA, of metapopulations over the continuous clay loam soils of major landmasses.* The ecologically uniform climate of excess rain in all months and the continuity of the lowland soils may permit a test of neutral community drift expectations. Theory would predict that the distributions of species not occupying the whole area of these soils would vary at random, or disperse to varying distances from some historical point of re-entry, whereas competitive selection would lead to significant repetition of species metapopulation ranges, coinciding with some subtle distinction in habitat.

Evidence from patterns of habitat specialisation and endemism imply that selection and speciation rates for habitat specialisation override any eventual balance between immigration and extinction rates in determining floristic diversity in a region. By this reasoning, any floristic province with an area exceeding, say, 10,000 km² would still accumulate tree species via unequal extinction and immigration rates, combined with some internal speciation rate. But that rate of accumulation through further speciation would decline over time in a series of episodes following fresh opportunities for competitive survival provided by a prior extinction event.[217] The rate of accumulation would be so slow that climate or other change would intercede long before an asymptote in regional species diversity is reached.

3.3.2. Phylogenies and community structure

We can expect to find correlations between phylogenetic relatedness and ecological similarity among forest communities. A measure of the overall evolutionary relatedness of a community of interacting organisms can be used to interpret the contemporary ecological processes that structure its composition. Campbell Webb,[218] using plots from MDF on differing substrates within Gunung Palung National Park (West Kalimantan), showed that species within one habitat manifested *greater* phylogenetic similarity than expected by chance. He later suggested that such investigations can add a community context to studies of trait evolution and biogeography. *This approach, which awaits further exploration, appears particularly promising for unravelling the history and current bases for coexistence among the many congeneric species series that characterise the forests of the perhumid Far East.* Many genera, too, are most abundant over specific habitat ranges, while related genera may either share them or differ. At higher phylogenetic levels, though, results may become increasingly difficult to interpret, as the coexistence of members of families with vastly different phylogenetic interrelationships, including archaic and derived taxa, may rest on one or a very few character traits which may be challenging to diagnose by this method.

3.3.3. How old are current lowland rainforest tree species in perhumid Asia?

3.3.3.1. Gondwanan origins. It seems likely that perhumid regions existed in the equatorial northern regions of the Indian Gondwana plate as it approached Laurasia. In these refuges, taxa confined to perhumid climates, such as *Axinandra* and *Trichadenia*, could have survived transit from East Africa. They could then have spread eastward relatively expeditiously following contact with the Laurasian coast. Only a few have survived the subsequent widespread climate drying in South Asia, especially severe between the late Eocene and late Oligocene to mid-Miocene global temperature maxima, and which again periodically increased following the start of the Indian monsoon c.12 Ma.[219] Mountainous landscapes prevailed in the Sunda Shelf until the mid-Miocene (c.15–20 Ma). The opening thereafter of extensive perhumid lowlands would have allowed the spread of species from earlier restricted distributions into new terrain, under conditions of reduced competition. Many genera, and probably most of these ancestral species, were initially adapted to seasonal evergreen forest, and became the ancestors in a major wave of speciation following the closing of the Pacific-Indian Ocean throughflow c.23 Ma, and subsequent spread of the perhumid Malesian climate. Many of our widespread regional species of clay loams would have originated then and thereafter, at a gradually slowing rate as community species diversity increased. At the present, these communities seem to have reached a diversity equilibrium whose maintenance will be discussed in Chapter 7. The current hyperdiversity, and especially the evolution of congeneric co-occurring species series, must have originated consequent to the origin of ENSO and are partially sustained by it.

The only Asian tropical tree taxon whose history of diversification is fully documented in the fossil record is the small mangrove genus *Sonneratia* of the Indian and Western Pacific ocean coasts.[220] Along with the upland *Duabanga*, ancestral *Sonneratia* (*Florschuetzia*) appears to have diversified from Lythraceae (with affinity to *Lagerstroemia*, also occurring from Asia to Australasia) during the late Eocene. At that time Indian Gondwana, approaching the Laurasian coast, had entered the tropical zone and had been receiving immigrants from tropical Africa. The two genera may have diverged during the early Miocene (c.22 Ma), their modern species during the course of the Miocene.

More generally, the rather low level of *local* tree species endemicity in the Sunda lowlands – especially within the continuous habitat provided by clay loam soils – implies that many tree species of continuous habitats within the main landmasses are at least as old as the last connection between these masses, c.15,000 BP; but that is hardly surprising. Among these are three of the six *Shorea* (sect. *Mutica*) which flower in the same sequence whether at Pasoh in Peninsular Malaysia, Lambir in Sarawak, or Sepilok in East Sabah; *S. macroptera*, *S. parvifolia* and *S. leprosula*. This consistency, notwithstanding the participation of other, more local, species in each local flowering sequence (Chapter 5), implies that the sequence evolved at some original point of their co-evolution, and the intriguing possibility that the ENSO-related trigger to their differentiated rates of floral development both came into existence by that time and has consistently persisted since then. That time is likely to be several million, rather than tens of thousands, of years. *Molecular genetic evidence for the age of their origins might yield evidence for the age of ENSO.* There is the possibility that many of these surviving widespread species (some with sister species in Indo-Burma and even South Asia), have been the progenitors of many of those currently existing in more isolated habitats, and with higher local endemism. *Molecular phylogenetic methods are currently advancing to a state where such exciting possibilities can be rigorously investigated.* Were this supposition correct, it implies that the forest communities on widespread, continuous substrates may indeed have been reaching an asymptote of richness: a climax to an evolutionary episode…now ended by the interventions of mankind.

3.3.3.2. *Asymmetric floristic affinities*. Why does the Peninsular Malaysian MDF flora show greater floristic similarity with Sumatran MDF than with Bornean? The high rate of endemism in lowland Peninsular Malaysia and especially Borneo, relative to Sumatra, must in part be due to the greater edaphic homogeneity of the Sumatran lowlands. Quite possibly this is also in part the result of asymmetric migration barriers. A lesser Proto-Mekong tributary flowed between Sumatra and the Peninsula, whereas the vast mainstem Proto-Mekong floodplain separated Borneo from Sumatra and the Peninsula since before the Pleistocene. But it probably also stems in part from more severe drying of the Sumatran lowlands during Pleistocene interpluvials. Sumatra lacked the extensive persistent perhumid refuges that Peninsular Malaysia would likely have had on its eastern flanks facing the South China Sea. Isolated areas of specialised habitat in lowland eastern Sumatra, such as yellow and white sand soils, would have suffered particularly acutely.

3.3.3.3. *An idiosyncratic dipterocarp subgroup*. Periodically but aseasonally dry, freely draining soils are widespread in the Riau Pocket as expanded here. A subgroup of species typical of these habitats also occurs outside the Pocket, in coastal hill dipterocarp forest and on the drought-prone ridges of the Peninsular granite hills. This subgroup includes the dipterocarps *Anisoptera curtisii*, *Hopea beccariana*, *Dryobalanops aromatica*, *Shorea curtisii*, and *S. kunstleri* (Chapter 3). This group of dipterocarps may have originated early in the history of the Sunda lowlands. But their major centre of species diversity is on the Neogene sediments of north-west Borneo (Fig. 6.15), rocks which, thanks to a history of repeated erosion and resedimentation in the sea to the north, are predominantly sandstones (Chapter 1). They were uplifted from the sea in a series of cycles starting at the time when the Sunda lowlands were expanding, c.23 Ma.[221]

Most podsol and humult yellow sandy ultisol specialists are widespread throughout the north-west Borneo Riau Pocket, notwithstanding the barriers posed by several major river valleys. The one valley large enough and ancient enough to limit migrations was the Lupar valley, probably dating at least from the north-east Borneo uplift which raised the uplands, around and including Mount Kinabalu, at least 3 Ma and possibly much earlier. Abutting the Neogene basin sediments immediately across the Lupar valley to its west is the Kayan Formation of Paleocene sandstone, which following uplift could have served as a cradle for some of this flora. Koichi Kamiya carried out a molecular phylogeny based on populations of four species of *Shorea* (sect. *Brachyptera*) in Lambir forest, 400 km north-east of the Lupar Fault. Two concentrate on sandy soils, two on clay loams over early Miocene shale.[222] He estimated all four local populations to have originated 3–2

Ma during the Pliocene. This thus coincides with the uplifting of coastal sediments laid down during the Miocene.[223] Three (*S. smithiana*, *S. fallax* and *S. bullata*) are north-east Borneo endemics and one (*S. kunstleri*) is widespread in Borneo and in the coastal forests of northern Sumatra and Peninsular Malaysia, and likely older. Populations of the local endemic *S. bullata* are genetically less variable than the other, more widespread species, implying a later origin than the other species. Sampling throughout the species ranges would have given greater age to their diversification. Remarkably, they diversified at a similar time to the elephants,[224] and later than the start of the extraordinary diversification of certain frog genera in Sri Lanka. (This appears to have started at a similar time to that of the dipterocarps in perhumid Sunda, following the initial spread of the perhumid lowlands there.[225]) I infer that the arrival of suitable climatic and other physical conditions, and their persistence, can explain the timing of origin, *and current survival* of these clades of related tree species, rather than the length of their life cycles, or rates of mutation.

3.3.4. *Summary and discussion*

I conclude that the increase in species richness through diversification of the Sunda MDF tree flora had been declining. Even without human intervention, we may currently have been approaching a state of equilibrium in the species composition and richness of the Sunda forest flora. This flora originated in surges of speciation following successive openings of extensive new habitats, starting c.23 Ma when the perhumid climate spread through the region. A core group, mainly of descendents from Indian Gondwanan ancestors, spread out into the new perhumid lowlands. Those widespread species, often rare on more fertile clay loams of perhumid Western Malesia but common in seasonal evergreen dipterocarp forest, such as *Anisoptera costata*, *Dipterocarpus gracilis*, *D. retusus*, *D. hasseltii*, *Shorea assamica* and *S. guiso*, may descend from this time. The exceptional *regional* richness of the northern Bornean tree flora, unmatched elsewhere in the Old World with c.4,500 species, is in part due to the diversity of lowland soils there, but partly also to persistence of its perhumid climate from the time of its original spread through the Pleistocene to the present. These two factors most distinguish northern Borneo from other areas, and is exemplified there by uniquely high diversification between certain river valleys.

New opportunities for further spread and speciation have arisen intermittently – notably eastward into the Philippines and East Malesia. Such opportunities must also have arisen during the Miocene (c.15 Ma), especially in Borneo which remained orogenically active, postdating the close approach of the Australasian lands. Rates of endemism among dipterocarp and Fagaceae floras are high in the Philippines, at 52% and 62% respectively.[226] Combined with the many examples of

sister-species relationships with Borneo, this suggests continuous separation since at least Pliocene times, perhaps following an earlier overland connection during the late Miocene. In general, tree species in the Philippines show low rates of island endemism – even among poorly dispersed taxa such as the dipterocarps and the ballistically dispersed subcanopy genus *Cleistanthus* (Euphorbiaceae).[227] This implies that islands were periodically connected through the Pleistocene. Higher local endemism on limestone karst and particularly on ultramafic substrates, especially in northern Borneo, may represent continuing speciation since that time. Such speciation could have been facilitated by high extinction rates within the small areas that they represent.

3.3.5. *How do rainforest tree species originate?*

3.3.5.1. *Founder event or diversifying metapopulation?*
In principle, rainforest tree species could have originated by dispersal of founders across a barrier, or by vicariant diversification of a metapopulation which was subsequently divided by a barrier to introgressive hybridisation. It is impossible to infer which it was from current species ranges alone. The absence of clinal morphological variation might seem to argue for the inertia of landmass-wide populations to genetic change. Given a constant climate and an established edaphic range, this suggests that vicariant diversification is unlikely. The separation of the subspecies of *Dipterocarpus geniculatus* and *D. stellatus* by the Batang Lupar in west Sarawak (in a valley clearly more ancient than these taxa) might seem to imply that they have originated on one side and dispersed across before diversifying. However, there is only sparse morphological evidence of sister species taxa across this or other river barriers in north-west Borneo (but see section 3.3.2 below). Most species separated by these barriers appear more distantly related. On the other hand, many dipterocarp sister species link Borneo and the Philippines (and some link Borneo with Peninsular Malaysia and Sumatra). In these cases, the lack of diversification of taxa throughout the Philippine archipelago that are nevertheless endemic to it suggests origins as small founder populations, followed by rapid expansion whenever opportunities arose.

Koichi Kamiya[228] presented evidence that all four of the *Shorea* (sect. *Brachyptera*) species whose molecular phylogeography he analysed arose from very small populations. This was consistent with a prediction of Egbert Leigh on theoretical grounds.[229] The time of the origin of the populations sampled, coinciding with the uplift of Neogene (Mio-Pliocene) sediments, yielded a range of novel soil substrates into which they successively diversified. This seems consistent also with processes inferred in a Neotropical case.[230] It could again imply speciation in response to strong ecological pressure even in the absence of a physical barrier to introgressive hybridisation.

Sympatric speciation remains controversial,[231] though some have argued that genetic differentiation could lead to sympatric speciation where ecological selection is exceptionally strong.[232] Kamiya found evidence of past hybridisation within the genomes of his morphological species,[233] which could have facilitated their adaption to new habitats by increasing genetic variability. There seem to be few current examples of metapopulations at the point of early, presumably local, morphologically recognisable diversification in tropical Asia, and there are few new physical habitats into which species could diversify (although our own species may be in the process of creating them, especially if we ourselves fail to persist). The Myrtaceous genus *Syzygium* is particularly rich in sister taxa which inhabit different habitats within a shared landscape; 25 examples occur in northern Borneo alone.[234] The most persuasive are on the Sabah ultramafics where vertebrate-dispersed *Syzygium georgeae* occurs on ultramafic extrusions adjacent to its sister, the widespread *S. castaneum*. Similar pairs are *S. ultramaficum* and *S. penibukanense*, and the wind-dispersed Myrtaceous *Tristaniopsis merguensis* (ssp. *merguensis* on clay, and ssp. *tavaiensis* on ultramafics).

Perhaps equally persuasive of sympatric speciation through strong ecological selection are *Leptospermum javanicum* and *L. recurvum*, also Myrtaceae, on the high Sabah mountains (*L. recurvum* is said to occur in Sulawesi also).[235] Whereas the former is widespread in upper montane forest, and occurs on coastal peaks at lower altitudes, the latter has a higher albeit overlapping altitudinal range and is commoner in subalpine woodland and on ultramafics in upper montane forest. Sister species in several other genera have similar overlapping elevational distributions on Kinabalu. These recall the case made for sympatric speciation of the two species of endemic palm *Howea* on Lord Howe Island between Australia and New Zealand.[236] Other Bornean examples include *Hopea vacciniifolia*, endemic to *kerangas* on a single fragmented Pleistocene raised beach in Brunei and adjacent Sarawak, yet sister to the adjacent and widespread *H. bracteata* of MDF. *Tristaniopsis whiteana* (ssp. *monostemon*) is endemic to humult sandy soils on the banks of blackwater rivers of north-west Borneo and sister to *T. whiteana* (ssp. *whiteana*), widespread on clay and sandy-clay ultisols on landslips throughout the Sunda Shelf.

Sean Thomas found evidence of cryptic species pairs – that is, taxa identical in the herbarium but differing in reproductive size and reproductive phenology – co-occurring in the Pasoh MDF.[237] The existence of such pairs implies that ecological differentiation can lead to spatial differentiation of populations within a common landscape. This in turn may lead to genetic differentiation of metapopulations, which may then lead to declining fitness of introgressive hybrids. Koichi Kamiya discovered that juvenile F_1 hybrids between related co-occurring *Shorea* species are common, but F_2 hybrids

are absent.[238] This explains how the species remain distinct and mature hybrid individuals rare (Chapter 5). It may thus provide an example of the final stage of such a speciation process.

Such speciation, initiated by genetic differentiation accompanying ecological differentiation within a meta-population inhabiting a heterogeneous landscape or even forest structure, could result in co-occurring sister species which continue to hybridise but whose F_1 progeny fail to successfully compete to reproductive maturity. That is consistent with a currently emerging consensus that this is a common process of plant and animal speciation.[239]

3.3.5.2. *Breeding systems and geography*.

Differences in breeding systems must account for some patterns of variation among families. Morphologically defined species are clearly delimited among dipterocarps, while there is wide geographic and edaphic variation within certain widespread species of *Syzygium*, such as *S. urceolatum* or *S. borneense*. These have bisexual flowers visited by various insects and birds. Among Sapindaceae, in which most Asian species are andromonoecious,[240] certain taxa show distinct geographical and ecological forms, often denoted as species or subspecies but variably distinct. These include *Pometia*[241] (occurring from Sri Lanka and China to Melanesia) and the possibly monotypic pantropical genus *Allophylus*.[242]

Rainforest tree species may also originate by polyploidy or other processes that prevent subsequent introgressive hybridisation in nascent populations. This likely happened in the tetraploid *Shorea ovalis* and in *Hopea subalata*, a triploid in which recombination is anyway impossible (Chapter 5). This mode of speciation appears uncommon. I found the morphospecies *Shorea rotundifolia* growing as several isolated populations along shale ridges in inner north-west Borneo. These could represent local subpopulations of the widespread Borneo endemic *S. amplexicaulis*, each of which originated through mutation of the same gene, which might arise then disappear after a few generations – or become more widespread, thereby creating a metapopulation recognisable as a true species. Their presence as small populations, as opposed to single individuals, suggests that they do not hybridise introgressively with the generally co-occurring putative parent species. *Shorea retusa* occurs as several isolated populations, some of them hundreds of kilometres apart, on humult siliceous soils on headlands along the northwestern and eastern coasts of Borneo, and along the early Holocene coastline, now inland, of the Ulu Barito foothills in the south. Could these populations be the consequence of several independent diversifications from a widespread core species, perhaps *S. scabrida*, under similar ecological selective pressures? *These relationships merit molecular examination.*

Apomixis, especially by means of adventive embryony, appears to be remarkably widespread among MDF tree taxa (Chapter 5).[243] It is particularly evident in north-west Borneo among those *Shorea* (sect. *Mutica*) species that are comprised of ecological or biogeographic subspecies, sometimes separated across river valleys (Fig. 6.18) (Chapter 5). Apomixis has been best known among temperate perennial pioneer species. It has been suggested to facilitate rapid production of fit seed for habitats in which physical selective factors may be limiting

but constant, while biotic selection is low.[244] This is clearly *not* the case in tropical rainforests, where outbreeding is the norm. In such habitats, apomixis might instead increase reproductive fitness by maintaining fecundity and preserving heterozygosity in populations whose densities are so low that cross-pollination is inhibited (Allee effects). It may alternatively facilitate the rapid spread of fit genotypes where competition is comparatively relaxed, for example when a founder crosses a barrier into forests where its species is absent.

4. Overall summary and conclusions

The tree flora of tropical Asia has its origin in late Cretaceous Gondwana, from which it has invaded indirectly, by five principal routes. Regional diversification has been accelerated differentially by arrival of migrants, in differing numbers and from different primary sources, to different regions. *In situ* diversification has taken place in regions separated by physical barriers of differing longevity and at differing periods, on land defined by climate, geology and soil, and topography, as well as by marine separation or incursion. Immigrant species have arrived in particular locations through accidents of history, but whether or not they have succeeded and dispersed has been determined by ecological opportunity. Although opportunities for fresh immigration into the Sunda land masses and MDF hardly exist, there is some evidence that we are witnessing the later stages of an evolutionary episode of speciation within the perhumid Sunda lands.

The survival of species, common in the seasonal tropics, in exceptionally low numbers in the most equable, perhumid climates, poses the question of how final extinction can occur there, if Allee effects can be overcome by self-compatibility or apomixis, an issue I will return to in the next chapter.

Tropical forest tree species in Asia are neither as ancient nor as young as has sometimes been suggested. Regardless, current human depredations are leading to extinctions that the slow labour of future evolution will require many times the age of mankind to repair.[245]

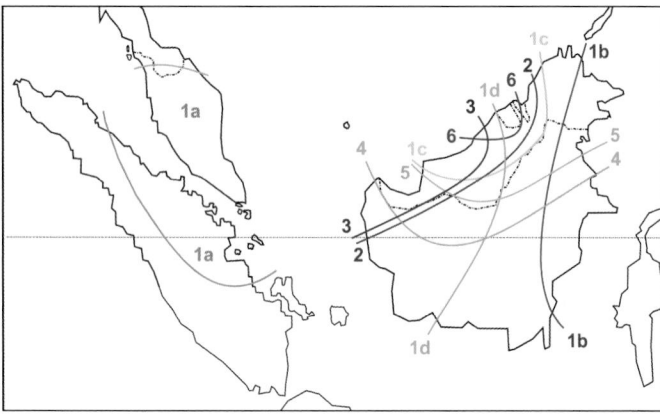

1. *S. macroptera* subsp.
1. a. *macroptera*
1. b. *sandakanensis*
1. c. *macropterifolia*
1. d. ssp. *baillonii*

2. *S. slootenii*
3. *S. sagittata*
4. *S. ferruginea*
5. *S. myrionerva*
6. *S. acuta*

Fig. 6.18 The geographical distribution of taxa in *Shorea* sect. *Mutica* subsect. *Auriculatae*, a group in which multiple seedlings have been observed and adventive embryony (apomixis) inferred. 1, *S. macroptera*: a, ssp. *macroptera*; b, ssp. *sandakanensis*; c, ssp. *macropterifolia*; d, ssp. *baillonii*; 2, *S. slootenii*; 3, *S. sagittifolia*; 4, *S. ferruginea*; 5, *S. myrionerva*; 6, *S. acuta*. (I.H. Baillie del. from P.S. Ashton unpubl.)

References

[1]Dick *et al.* 2003; Pennington & Dick 2004; Petit *et al.* 2008. [2]Alverson *et al.* 1998. [3]Morley 1999. [4]Raven & Axelrod 1974. [5]Corner 1954; Takhtajan 1969; Sun *et al.* 1998; Sun & Hsieh 2004. [6]Sun *et al.* 1998, 2002; R. Morley, pers. comm. April 2012. [7]Morley 1999. [8]Davis *et al.* 2004, 2005. [9]Morley 1999. [10]Muller 1968. [11]Zhou *et al.* 2001. [12]Angiosperm Phylogeny Group 2009. [13]Muller 1968; Morley 1999. [14]Manen *et al.* 2002, 2010. [15]Morley 2012. [16]Davis *et al.* 2002, 2004, 2005. [17]Doyle *et al.* 2004. [18]Chanderbali *et al.* 2001 [19]Wu *et al.* 1994; Qian & Ricklefs 1999, 2000, 2004; Doyle *et al.* 2008. [20]R. Morley, pers. comm. April 2012. [21]Morley 1999. [22]Muller 1968, 1972; Morley 1998. [23]Prasad *et al.* 2009. [24]Dijk in Mitteremeier 2004. [25]Rust *et al.* 2010. [26]Appanah & Turnbull 1998. [27]Bancroft 1933; Chiarugi 1933. [28]Angiosperm Phylogeny Group 2009. [29]Ducousso *et al.* 2004. [30]P.S. Ashton 1982. [31]Kajita *et al.* 1998; Dayanandan *et al.* 1999; Gunasekera *et al.* unpubl. ms. [32]Kajita *et al.* 1998; Dayanandan *et al.* 1999. [33]Simpson 1944. [34]Morley 1999. [35]Angiosperm Phylogeny Group 2009; Xi *et al.* 2012. [36]Morley 1999. [37]Philipson 1986; S. Renner, pers. comm. 2010. [38]Tobe & Raven 1987a, b; Conti *et al.* 2002; Angiosperm Phylogeny Group 2009. [39]Dransfield *et al.* 2008. [40]Norup *et al.* 2006. [41]Maurin *et al.* 2010. [42]Muller 1972; Morley 1999. [43]Santisuk 1997; Werner 1997; Lusk 2008. [44]Farjon, pers. comm. 2010. [45]Farjon *et al.* 2004; Brodribb & Field 2008. [46]Kitayama *et al.* 2012. [47]Morley 1999, 2012. [48]Morley 2012. [49]Lin *et al.* 1986; Liu *et al.* 1986; Sun *et al.* 1986; Walker 1986. [50]Gray 1846. [51]Wang *et al.* 1999. [52]Nooteboom 1993; Morley 1999. [53]Morley 1999, 2011. [54]Manos & Stanford 2001. [55]Morley 1999, 2011. [56]Martin 1977. [57]R.J. Morley, pers. comm. April 2012. [58]P.S. Ashton 1988. [59]Eyde & Xiang 1990. [60]Nie *et al.* 2008. [61]Li *et al.* 1999a, b. [62]Endress *et al.* 1985. [63]Jacobs 1960. [64]Morley 2012. [65]Morley 2011. [66]Morley 2012. [67]Whitmore 1982, 1984a, 1987. [68]Hyland *et al.* 1999. [69]Florin 1963; Schönenberger 2001. [70]Mani 1974. [71]Doyle *et al.* 2004; Nooteboom 1993. [72]Li *et al.* 1999a, b; Azuma *et al.* 2001; Nie *et al.* 2008; Manchester *et al.* 2009; Manen *et al.* 2002, 2010. [73]Gould & Lewontin 1979. [74]Whitmore 1982, 1987. [75]Whitmore 2008. [76]Jacobs 1962. [77]P.S. Ashton 1982. [78]Soepadmo 1976. [79]Soepadmo 1976. [80]Jacobs 1960. [81]P.S. Ashton 1982. [82]Takeuchi 2003. [83]Hyland *et al.* 1999. [84]van Steenis & Schippers-Lammertse 1965; van Steenis 1972a, b, 1979. [85]Johnson & Briggs 1975. [86]Merrill 1926. [87]Merrill 1926. [88]Merrill 1926. [89]Jones & Kennedy 2008. [90]Merrill 1926; van Steenis & Schippers-Lammertse 1965; van Steenis 1972a, b, 1979. [91]Bird *et al.* 2007. [92]Michaux 2010. [93]Meher-Homji & Misra 1973; Mani 1974; P.S. Ashton & Gunatilleke 1987; Hockings 1989. [94]Gunatilleke & Gunatilleke 1991. [95]Gunatilleke & Gunatilleke 1991. [96]Gunatilleke & Ashton 1987. [97]Mittermeier *et al.* 2004. [98]Ramesh *et al.* 1991; Ramesh & Pascal 1997. [99]Giriraj *et al.* 2008. [100]Deraniyagala 1958. [101]P.S. Ashton & Gunatilleke 1987. [102]Brandis 1907; Ahamedullah & Nayar 1986. [103]Morley 2012; Morley *et al.* 2013. [104]Morley 1999. [105]Clarke 1898. [106]Clarke 1898; Hooker 1907. [107]Chatterjee 1939. [108]Niyomdham 2002. [109]Smitinand 2001. [110]Mackinnon 1996. [111]Zhu 1997, 2006, 2008. [112]Morley 1999. [113]P. Cribb, pers. comm.

Aug. 2011. [114]Lin *et al.* 1986; Liu *et al.* 1986; Sun *et al.* 1986; Walker 1986. [115]Liu *et al.* 1986. [116]Qian & Ricklefs 1999; Qian 2002. [117]Biên & Thì 1992; Morley 1999. [118]Morley 2012. [119]Cannon *et al.* 2009. [120]Corner 1954; Takhtajan 1969. [121]Morley 1998, 1999. [122]ter Steege *et al.* 2003; Valencia *et al.* 2004. [123]Angiosperm Phylogeny Group 2009. [124]Boden Kloss 1920. [125]Symington 1943, Symington *et al.* 2004. [126]Whitmore 1984a. [127]Cockburn 1972; Soepadmo 1976. [128]Bunyavejchewin *et al.* 2009. [129]Woodruff 2003; Woodruff & Turner 2009. [130]Morley & Flenley 1987. [131]van Dijk *et al.* 2004. [132]M. Jacobs, pers. comm. 1958. [133]Haile 1971. [134]Geagh *et al.* 1979; Furukawa 1997. [135]Cannon *et al.* 2009. [136]Voris 2000; Cannon *et al.* 2009. [137]Morley & Flenley 1987; Morley 2012. [138]Whitten *et al.* 1996. [139]Bird *et al.* 2005. [140]Goldammer & Seibert 1990. [141]Turner *et al.* 1997b. [142]Welzen *et al.* 2011. [143]van Steenis 1972a, b. [144]Tougard 2001; Tougard & Montuire 2006. [145]van Steenis 1979. [146]Bird *et al.* 2005; Wurster 2010. [147]Teeuw *et al.* 1999. [148]Morley 2012; Morley & Flenley 1987. [149]P.S. Ashton 1982. [150]Wong 2011. [151]Deraniyagala 1958. [152]Brown 1997. [153]Corner 1954. [154]Wang *et al.* 1999. [155]Higham *et al.* 2009. [156]Kamiya *et al.* 2012. [157]Cao *et al.* 2006. [158]Kamiya *et al.* 2011. [159]Davis *et al.* 1995. [160]Hou 1983. [161]Slik *et al.* 2011. [162]P.S. Ashton 2010. [163]Morley 1999. [164]Quek *et al.* 2007. [165]Kamiya *et al.* 2012. [166]Ishigawa *et al.* 2008. [167]Piper *et al.* 2008. [168]Cranbrook & Piper 2009. [169]Slik *et al.* 2003. [170]Raes 2009. [171]Manokaran *et al.* 1992; H.S. Lee *et al.* 2002. [172]P.S. Ashton 1972, 1992, 1997, 2001; Wong 2011. [173]P.S. Ashton 1982, 2004a. [174]Soepadmo 1976; Soepadmo *et al.* 2000. [175]Kerr 1997. [176]CLIMAP 1976. [177]Löffler 1972. [178]Walker 1970, 1972; Flenley 1979. [179]Morley 1982; Newsome & Flenley 1988; Stuijts *et al.* 1988. [180]Flenley 1979. [181]Sukumar 1986; Vasanthy 1988; Sukumar *et al.* 1993. [182]P.S. Ashton 1980b; Kostermans 1992. [183]Morley 2012. [184]Wong 2011. [185]Wong 2011. [186]Ridley 1930. [187]Heaney 1984. [188]Roos *et al.* 2004. [189]P.S. Ashton 1972. [190]Airy-Shaw 1975. [191]Kochummen & Go 2000. [192]Harrison 2001. [193]Whitmore *et al.* 1972–1989. [194]Manokaran *et al.* 1992. [195]Wilson *et al.* 2005. [196]Roos *et al.* 2004. [197]Bremer 1979, 1983; P.S. Ashton & Gunatilleke 1987. [198]Werner 2001. [199]Bremer 1979, 1983. [200]Senanayake *et al.* 1977. [201]Cruz 1973. [202]M. Meegaskumbura, pers. comm. and in press. [203]Sukumar *et al.* 1993. [204]Flenley 1979. [205]Quek *et al.* 2007. [206]Liechti 1960. [207]Quek *et al.* 2007. [208]Davies 2001a. [209]Smith 1977. [210]R.J. Johns, pers. comm. 2008. [211]Sleumer 1966. [212]Vink 1970. [213]J.S. Womersley, pers. comm. 1971. [214]G. Argent, pers. comm. 2002. [215]Beaman & Beaman 1989. [216]Beaman & Beaman 1989, Beaman *et al.* 2001, 2004. [217]Simpson 1944. [218]Webb *et al.* 2006. [219]Quade *et al.* 1989. [220]Muller 1978; Morley 1998. [221]Liechti 1960; Wilford 1961. [222]Kamiya *et al.*, unpubl. ms. [223]Liechti 1960. [224]Sukumar 2011. [225]Meegaskumbura, unpubl. ms. [226]Soepadmo 1976; P.S. Ashton 2004a. [227]Dressler 1999. [228]Kamiya *et al.*, unpubl. ms. [229]Leigh *et al.* 2004. [230]Fine *et al.* 2005. [231]Nosil 2008. [232]Via 2002. [233]Kamiya *et al.* 2005, 2011. [234]P.S. Ashton 2011. [235]Lee & Lowry 1980b. [236]Jacobs 1962; Leenhouts 1967. [237]Thomas 1993. [238]Kamiya *et al.* 2011. [239]Lowry 2012. [240]Leenhouts 1986. [241]Jacobs 1962; Leenhouts 1986. [242]Leenhouts 1967. [243]Kaur *et al.* 1986. [244]Stebbins 1977. [245]Richardson *et al.* 2001; Kamiya *et al.* 2005.

Species diversity among juvenile trees in the understorey of mixed dipterocarp forest on humult yellow sand, Lambir National Park Sarawak. The large pinnatifid leaf is an *Artocarpus* (S. Russo)

Chapter 7

FOREST AND TREE TAXON DIVERSITY: WHY DOES IT VARY, AND HOW IS IT MAINTAINED?

The argument between niche-assembly and dispersal-assembly perspectives is long-standing. It has persisted so long precisely because each perspective has strong elements of truth and because reconciling them is non-trivial.

Stephen Hubbell, 2001. *The Unified Neutral Theory of Biodiversity and Biogeography.*

It is now widely recognized that tropical rainforests constitute the habitat for over half of the world's terrestrial biological diversity, or 'biodiversity'. Though the concept of biodiversity is complex, in this context it means simply the number of species of terrestrial plants, animals and microorganisms. Tropical forests probably also contain the majority of the world's genetic diversity, as expressed in the form of genotypes and gene sequences. Most of this genetic and species diversity is harboured among the insects and microorganisms, the great majority of which are as yet unnamed, *let* alone documented.[1] Plants, as the major producers of biomass through photosynthesis, are the fundamental sources of energy for terrestrial biodiversity. Most living plant biomass in forests is contained in trees. The richness and diversity of plant species appears to be broadly correlated with that of insect herbivores, of seed predators and especially of plant pathogens (see Chapter 5 and section 2.5.3 below).

In what follows, I will maintain a distinction between *species richness* and *species diversity*. Together they make up the species-level component of biodiversity. Species richness is a simple approximation of the number of species. Species diversity, on the other hand, takes into account the evenness, that is relative abundance, of the species represented. Richness, as understood here, is therefore but one aspect of diversity. Both can be further subdivided into sub-categories: α–diversity (or richness) refers to that with a single community or habitat; β–diversity represents species 'turnover', or the degree to which adjacent communities within a landscape mosaic differ from one another in species composition; while γ–diversity represents the total species-pool of a landscape or region.

The variable of ultimate interest to conservationists is often genetic diversity (see section 2.6 below), but this is impossible to quantify precisely in highly biodiverse

ecosystems. We therefore often resort to *species richness* as a reasonable approximation. That calculation is based on two assumptions: (1) that there are barriers to interspecific breeding among related species, so that each species represents a distinct and unique assemblage of shared gene frequencies; and (2) that there are mechanisms for the exchange of genes among individuals, and therefore the possibility both of generating new allelic combinations and of dispersing them within and among the metapopulations that define each species. In the context of hyper-diverse rainforest, we can also make the related assumption that, overall, populations of different tree species have similar levels of variability. Unlike many other types of organisms, most tree species have now been identified and formally named in the Asian tropics. Measures of tree species richness and diversity can therefore provide a simple and practical 'proxy' measure of overall biodiversity (Chapter 9). But first, it has been essential to assess whether trees are outbreeders, and whether their populations and species are comparable in genetic variability to those of other organisms. That has been addressed in Chapter 5.

To understand how biodiversity is maintained, one must know something about how it originated in the first place, and how it continues to be generated. A discussion of the origin and maintenance of tree species biodiversity requires consideration of all the subjects addressed in the previous chapters. It thus provides a convenient vehicle for a synthesis of our essays so far. In Chapters 5 and 6, I discussed what we can infer about processes of speciation among tropical trees, that is, how species diversity originates. Here, I describe regional and local variation in diversity – focusing particularly on tree *species diversity* and *genetic variability* – and suggest how it may be maintained over time.

1. Is species diversity maintained by ecological equivalence in harmonious environments, or by specialisation?

There can be no doubt that the exigencies of current and geological history influence population spatial patterns and species richness. This is shown in Asia by multiple lines of evidence, from the limited dispersal distances achieved by most seeds and the noticeable clumping of seedlings around parent trees of virtually all tree species (Chapter 5), to the substantial differences in species richness among communities occupying ostensibly similar physical habitats in hyper-diverse Sundaland as compared with Sulawesi or South Asia (Chapter 6) (Table 3.7). Whether and to what extent niche-breadth varies with species diversity is the question central to understanding how species diversity has arisen and is maintained.

Naturalists take it for granted that animal species, even insects, each play a unique role in ecosystems, such roles, or 'niches' imbuing competitive advantage. But it is different for plants, especially large plants like trees, because opportunities for dispersing their progeny as seed may be strongly constrained in both time and space, and this may severely limit occasions for ongoing physical competition between individuals of any one pair of species. Furthermore, the number of different ways in which such physical interactions between sedentary plants can be played out is limited. Above-ground competition among trees is generally resolved by one individual overtopping the other, excluding it from direct light, and – in the absence of subsequent catastrophic crown damage to the victor – killing it gradually. Direct competition therefore occurs only when individuals are at comparable heights. Simple geometry, however, dictates that an individual can be in direct crown competition with at most six others at a time. Over its life history from seedling stage until reproductive maturity, an individual which eventually becomes emergent is thus unlikely to compete directly with more than about 20 individuals, and almost certainly with a smaller number of species. Most seeds drop close to the parent crown, so most competition is at first intraspecific; but interspecific competition increases as the tree grows and occupies a larger area.

Opportunities for mating are similarly restricted by limits on pollen dispersal. These constraints are particularly stringent in the most species-rich communities such as MDF, where a single community on uniform terrain may include as many as 750 tree species. This observation has led some botanists to conjecture that most of these tree species must be ecologically equivalent, and that they survive in mixture because dispersal limitations restrict the distance that seeds from each individual are likely to travel.[2] I have elsewhere suggested that were this so, the great energetic and nutrient costs of ensuring genetic diversity through cross-pollination would be selected against, and self-pollination or apomixis would be the rule (Chapter 5).[3]

Stephen Hubbell, on the basis of elegant analyses of data from the Barro Colorado Island (Panama) and Pasoh (Malaysia) CTFS 50 ha plots, has shown that spatial distributions of tree populations within such samples of species-rich rainforests are indeed consistent with the assumption that tree species are ecologically equivalent, and that the constraints on their seed dispersal are sufficient to account for the observed spatial structure of these forests.[4] Hubbell's *Unified Neutral Theory of Biodiversity and Biogeography* is thus an elaboration of MacArthur and Wilson's theory of island biogeography,[5] extending it to communities, landscapes and regions *within* continental land masses. It assumes ecological complementarity and uniformity of distribution over the landscape at every scale. But how far does the relative role of ecological equivalence — 'dispersal assembly' over 'niche assembly' — extend beyond the 50 ha tested to larger areal scales? My essays have presented arguments, including current evidence from breeding systems and population genetics, as to why this assumption cannot be accepted as realistic, even though limited seed dispersal undoubtedly constrains the options for competition between individuals of different species in hyperdiverse communities on uniform terrain. The roles of natural selection and niche specificity on the one hand, and random factors including history and dispersal constraints on the other, are of paramount theoretical and also practical importance. How predictable, for example, is species composition within a specific habitat irrespective of its history? How much confidence can be invested in the outcome of management interventions based on such predictions? To what extent and how fast will reduction or fragmentation of a forest area lead to species extinctions within it? Hubbell's theory provides us with a rigorous null hypothesis against which to test conclusions reached from our empirical observations. The neutral model that he proposed, however, fits many hypotheses of niche differentiation as well as those arguing coexistence by fitness equivalence.[6]

Ultimately, these practical questions can only be resolved via the testing of hypotheses by rigorous experiment. Testable hypotheses, though, must first be formulated, and these must be based on careful and systematic natural history, including observations of individuals and populations in nature over time. Results from statistical analysis of quantitative data derived from defined plots, monitored over decades, and inevitably limited in number, must themselves be tested for consistency with the careful observations of explorer-naturalists, thereby setting them in their broader context.

This essay therefore places the conclusions of previous chapters in a dynamic and evolutionary context, using tree species richness as the central focus. The following broad questions are addressed:

1. What patterns can be discerned in the variability of tree species richness and community diversity among the forests of tropical Asia, from the scale of the local landscape to that of the whole region?

2. What historical, physical or biological environmental factors are correlated with these patterns, and what causal relationships can be hypothesised?

3. How do the hypotheses thereby generated conform to or differ from those already in circulation, which seek to explain how species diversity accumulates and is maintained?

Species diversity can be estimated by a number of measures. Some measures are highly sensitive to the size of the sample, and some assume random distributions of individuals that are in reality clumped within populations (but the larger the plot, the less serious is this problem). The measure most commonly used for plant communities is Fisher's α, which takes into account the first problem but not the second.[7] It may therefore underestimate species diversity where populations are highly clumped. This little affects its validity in comparisons among different tropical evergreen forests within a region however, since seed dispersal distances vary more between genera than between species, and regional forests share most of their genera.[8] We find fairly constant numbers of genera in the evergreen forests of tropical Asia (Table 3.7), while the numbers of species are similarly consistent in the deciduous forests. Evergreen forests tend to have a higher proportion of species with short-distance dispersal patterns than do deciduous forests, so clumping of individuals may lead to *relative* underestimates of richness in evergreen as compared with deciduous forests. Nevertheless, it is unlikely to distort comparisons *within* either forest formation.

2. Spatial patterns of species richness and community diversity within the tree floras of tropical Asia

2.1. *Patterns of richness and diversity at the regional scale*

2.1.1. *Accidents of history*

Species richness and community diversity may vary at the regional scale owing to the influence not only of differences in their area of uniform climate, but also to different histories of extinction and opportunity for immigration. In tropical Asia, the most recent period of catastrophic climate change, the Pleistocene, has been particularly influential (Chapter 6). These historical influences are plainly fortuitous, occurring at random within the context of natural selection and niche specialisation.

The theory of island biogeography[9] provides one of the few hypotheses in biogeography which is amenable to rigorous testing. It predicts that, for every order-of-magnitude increase in the area of an island, there should be a doubling of the number of species, provided immigration rates are equal. It is a 'neutral' theory in that it treats species as ecologically equivalent and

complementary, assuming therefore that they are equally likely to occupy any particular space. Neutral theory also predicts that restricted seed dispersal will lead to distance-correlated changes in the composition of plant communities occupying common habitats, and that these changes will be measurable at regional as well as local scales. Evidence is presented in section 2.1.3 below.

2.1.2. *Climatic gradients in species diversity may differ among regions*

It has been claimed that there is a globally consistent relationship in flowering plant floras between community α (that is, species diversity) and mean annual temperatures, when annual water deficits are taken into account.[10] This relationship might imply that diversity is somehow dependent on the favourability of conditions for growth and production. It may be insufficiently appreciated that great tropical biodiversity is in fact concentrated in a relatively small area of the wet tropics, in the lowlands below c.700 m and especially below 300 m. Growth rates in these wet lowlands do not differ consistently from those in seasonal climates (Chapter 2), where diversity is much lower in spite of several months annually with little or no rain. In this case, I inferred that high productivity there may be due to the higher fertility of soils in seasonal climates (Chapter 1). Species diversity also varies greatly between different tropical forest ecosystems, even within the most biodiverse equatorial regions. The short deciduous forest communities of Asia, for instance, are poorer in species than those of the indigenous temperate mixed deciduous forests of eastern North America or North-East Asia! Worldwide (and even at the local scale), centres of lowland tree species diversity in the tropics are correlated with rainfall aseasonality[11] (Table 3.7). This correlation itself implies specialisation, and is therefore contrary to neutral theory at the regional scale – though not necessarily at the local scale.

Evergreen forest communities are known to be generally more species-rich than deciduous ones. Comparison of Fisher's α among CTFS and other plots established in lowland forests indicates an overall decline in species diversity with increasing rainfall seasonality (Table 3.7). Within Asia, this decline occurs both in the Far East[12] and in South Asia, where lowland deciduous forests are only a quarter as diverse as those of the lowland evergreen rainforests in aseasonal regions.[13] This can broadly be attributed to increasing constraints directly or indirectly imposed by climate on species survival. Such constraints might be imposed by the direct influence of climate, or fire for instance, or through the influence of climate, for example, on the dispersability of pathogens or predators. But rainfall seasonality is also correlated with between-year variability in rainfall amount, duration, and date of onset (Chapter 1),[14] and therefore with the frequency of killing droughts (Chapter 1). The plunge in tree species richness from semi-evergreen to tall deciduous forest is due firstly to a decline in evergreen canopy species caused by more frequent burning (Chapter 2), then to a further decline in species numbers with increasing length of dry season. The latter is associated with more frequent catastrophic droughts, which exacerbate seed and seedling mortality (Chapter 5). These limiting factors act both by reducing survivorship frequently and severely, and by continuous constraints to survivorship and growth analogous to low soil nutrient levels.

Change along the seasonality gradient in the frequency and intensity of catastrophic cyclonic storms must also be considered. *The potential influence of each factor on species richness and community diversity needs to be teased apart.*

Remarkably, the maximum level of tree species diversity attained within a homogeneous climate and soil environment is similar on all three tropical continents, irrespective of the varying absolute amounts of forest area under a given rainfall regime in each. Nevertheless, overlying and distorting the rainfall seasonality/diversity gradient there is another pattern, apparently caused by historical geography, varying dispersal opportunity, and regional histories of catastrophes such as climate change. Furthermore, within-climate similarities in diversity between continents may have a range of causes. These observations merit further discussion below.

2.1.3. *Species richness and geological history*

Species richness may vary with the geological age of community habitats: *how much do limits to dispersal govern species diversification?* Gradients in floristic richness from tropical to boreal regions have been correlated with their history, especially their history of climate changes and the resultant extinction events.[15] In Asia, expectations from island biogeography theory appear to be contradicted when the number of tree species in the rainforests of south-west Sri Lanka (less than 1,000 species in c.10,000 km²) is compared with the number in the Western Ghats evergreen forests (c.800 species in c.40,000 km²). If we confine our comparison to their lowland evergreen forests, the total of each is reduced by a comparable proportion and the ratio hardly altered. Assuming island biogeography is correct, these areas must either have experienced different immigration rates or different extinction rates – perhaps in the form of historical extinction events from which they are still recovering and have not yet reached equilibrium.

From what we know of continental and plate tectonic movements, South Indian forests cannot have received many immigrant tree species from outside the region over the last 10 million years, while they must have suffered massive extinction during the northern temperate ice ages ending c.10,000 Ma (Chapters 1, 6). During these dry periods, both South Indian and Sri Lankan rainforests must have retreated to refugia in the steep inner valleys of the mountains. The two regions thus have important historical factors in common. Modern south-west Sri Lanka, however, has a perhumid climate, whereas the Western Ghats are variably seasonal.

In perhumid West Malesia, as in south-west Sri Lanka, mean rainfall exceeds evapotranspiration throughout the average year – its impact is therefore ecologically uniform. These regions serve as a rare instance in which climate and island biogeography – the two fundamental influences on species diversity – can be teased apart. The total tree flora of Sumatra (c.2,500 species in 524,100 km²) appears relatively poor (with the caveat that it has been less intensively collected) compared with that of much smaller Peninsular Malaysia (3,100 species in 131,565 km²). Neither is the tree flora of Borneo (c.6,000 species in 757,050 km²) richer than that of Peninsular Malaysia in the proportion predicted by island biogeographic theory, despite its greater soil diversity (Table 7.1). In another striking example, the estimated combined tree flora of lowland Sabah, Sarawak and Brunei in perhumid northern Borneo (c.5,500 species in c.75,000 km²) is as great as – perhaps greater than – the tree flora of the perhumid lowlands of upper equatorial Amazon (c.150,000 km²). These data are not consistent with predictions based on area according to the theory of island biogeography (and therefore neutral theory), which predict a doubling of species richness with each tenfold increase of land area. The greater number of species in Borneo than in Peninsular Malaysia might be explained by the greater diversity of its habitats and associated flora. On the other hand, Peninsular Malaysia shares with Sumatra a more similar diversity of climate and soil. But Malaysia is the more species-rich. This contravenes simple predictions from neutral theory, though it can be argued that Borneo is richer than expected because it comprises an archipelago of differing habitat islands. It seems likely though that Sumatra may have experienced greater climate change during the Pleistocene, from which its flora – since isolated by the Straits of Malacca – has had insufficient time to recover: That would be consistent with simple neutral theory expectations.

Land mass	Area (km²)	Number of tree species
Peninsular Malaysia	131,565	3,100
Sumatra	524,100	c.2,500
Borneo	757,050	c.6,000

Table 7.1 The area and estimated number of tree species of the three major Sunda land masses.

The area of Sundaland under perhumid climate has probably experienced substantial periodic shrinkage over geological time, but enough refuges have likely persisted on all the main land masses – probably with greater connectivity than at present – to sustain much of their flora (Chapter 6).[16] Most island habitats, though, bear floras far in excess of predictions based on their respective areas (Table 7.2), even taking into account their potential for enrichment of up to 30% by species from adjacent habitats. One notable exception is peat swamp forest, which is poorer than predicted by theory; but this can be explained by the relative youth of this habitat type, which expanded only 4,500 years ago though existing since the early Miocene, and its highly limiting low-nutrient waterlogged soils.

Locality	Number of species confined to the dominant habitat
Island habitats: coastal hills, deep humic sands	
Andulau Hills, Brunei, habitat c.300km²	41(out of 65 in all habitats)
Lambir Hills, Sarawak, habitat c.500 km²	38 (out of 82)
Widespread habitat: turbidite, clay loams	
Kuala Belalong, Brunei, habitat c.8,000 km² locally	49 (out of 58)

Table 7.2 The size of lowland dipterocarp floras in habitat islands and the widespread equatorial Sunda habitat compared.

Differences in relative size of the regional tree species pools of north-west Borneo and Peninsular Malaysia versus southern Kalimantan and Sumatra, and also between all of these and South Asia, are also most parsimoniously explained by what we know of climate change and extinctions occurring during Pleistocene periods of lower rainfall and possibly greater seasonality (Chapter 6). Significant opportunities for perhumid climate specialists to immigrate from elsewhere have rarely, if ever, existed since the period following the Indian collision in the early Cenozoic. Therefore, any increase in number of species within the region has had to depend on speciation. If we assume 50 years to represent an average generation among tree species, then the Holocene period has witnessed c.250 generations. That is probably not enough to achieve a new equilibrium through novel speciation of trees. And indeed, there is little biogeographic evidence for significant speciation having occurred during this period (Chapter 6).

In summary, then, the number of tree species known from each of these regions contravenes simple island biogeographic expectations, but do so for reasons consistent with another aspect of neutral theory: tropical tree species disperse and evolve slowly, delaying predicted equilibria.

A key question here is whether distribution limits are unique to each species, and vary at random throughout a region – which would imply different species ages or dispersal rates – or whether species ranges are limited by common major geographic barriers. The latter would imply that geographic barriers define the effective limits of 'habitat islands', and that dispersal rates have set the principal limitation on species ranges. If dispersal occurs at rates on the scale of evolutionary time, clinal variation should be detectable within species. *Much more work of this nature is needed while opportunities last.* Although spatially-patterned morphological variation occurs among many species, among the commoner and better-known species it consistently appears in a 'stepped' fashion, defined by geographic – current epicontinental seas over major Pleistocene valleys (Chapter 1); and especially, at the local scale, by current broad river valleys; or by ecological barriers. In north-west Borneo, these river valleys have been estimated to vary in age between c.500,000 to more than two million years[17] (Fig. 6.16). This is consistent with population genetic evidence (Chapter 6): but the ubiquitous Sunda endemic, *Shorea leprosula*, is a species of clay soils and continuous distribution within large land masses. Distance-related genetic variability irrespective of physical dispersal barriers has been found in this species, though among small population samples, and based on isozyme variation and RAPD markers,[18] (Chapter 5), even though there is no detectible morphological variation across populations. This, again, would be consistent with neutral theory.

Neutral theory would also predict that, owing to the effect of limited seed dispersal, there is continuous turnover of species ranges with distance from a given point, and that the range sizes of those species which have yet to exploit the full extent of their available habitat would vary geographically at random. Rick Condit[19] found that species turnover rates were too low to match neutral theory predictions at regional scales in the Neotropics, but were consistent at scales of less than 50 km². Similar results have been documented over uniform terrain at the regional scale in West Africa.[20] They suggest that original species migration rates must have been more rapid than is required by neutral theory.[21] Such uniform terrain in the Asian lowlands is now all under commodity crops, but evidence consistent with the impact at local level of dispersal constraints has been reported from Pasoh.[22]

The only data available from Asia to test this neutral prediction are the distributions of dipterocarps from herbarium records (more comprehensive than other families owing to their timber value), and personal observation. Sumatra remains insufficiently well known, but ranges in Peninsular Malaysia and Borneo do not vary at random, even for species of the continuous inland habitat provided by the matrix of yellow/red clay and sandy clay udult soils (but bear in mind that, even on these soils in their landscapes, at least the commoner species vary in relative abundance in a manner consistent with topography (as at Pasoh, Chapter 3). For example, 67% of the 55 dipterocarp species that occur on the continuous clay and yellow-red sandy clay habitats of lowland Peninsular Malaysia occur throughout that habitat's range and perhumid climate, while the remainder concentrate towards the perhumid south or somewhat more seasonal north, only one being local (Table 6.21). Rather, substantially coincident ranges result in areas of similar species richness within ostensibly similar climate and soils.

Such ranges may be constrained by some as-yet undocumented habitat factor such as the extent of very occasional extreme droughts, or they may reflect only partial recovery from the last major climate change, when the areas of range concentrations were refuges. If the latter, current dipterocarp ranges would be expected to correspond closely to the dispersal limitations imposed by the size of the nut in proportion to its winged calyx. However, we find such correspondence only among a set of species occupying ecological islands of MDF on yellow humult sandy soils (see Chapter 6). Range extents vary not randomly but again in a stepped fashion. In Borneo, these discontinuities coincide either with river valleys that have acted as barriers to migration, or with the extent of certain geological formations. Remarkably, each of the large lowland plots of the CTFS network captures, within 25 ha, approximately one half of its known *local* tree flora. Each also contains fully a quarter of its entire *regional* flora, as defined by a major barrier such as the sea or the climatic frontier at the Kangar–Pattani Line (Chapter 6).[23] This serves as testimony to the remarkable capacity of tropical trees to disperse over time, notwithstanding apparent short-term constraints imposed by their dispersal mechanisms. In short, Henry Ridley's observation that dipterocarp fruit disperse no further than 100 m is doubtless true for the vast majority of individual fruits in this poorly-dispersed family, but there are occasional opportunities for much longer-distance dispersal. These negate the influence of dispersal constraints on species ranges, except where habitats exist as 'islands' isolated by at least several kilometres of non–habitat. There has been enough time in the Holocene for ranges to re-occupy continuous habitats within large island masses such as Peninsular Malaysia and Borneo. Even then, truly rare events may have allowed dipterocarps to hop islands from Sunda to New Guinea – or even from Africa to the Indian 'Noah's Ark', and vice versa in the case of *Syzygium* (Chapter 6). On the very limited molecular genetic evidence so far available, it seems that clinal genetic variation may accumulate along continuous regional habitat gradients, but that sufficient seed dispersal ensures that continuity of populations is maintained and speciation is restrained or prevented.

Among those speciose rainforest tree genera that have been examined, dated phylogenies imply that most speciation has been quite recent. It seems to have taken place mostly between the late Miocene and Pliocene eras (c.10–3 Ma) and to have declined since that time[24] (Chapter 6). There is some support for the concept of 'evolutionary episodes', spurts of speciation starting when areas of new habitat become available, such as the Sunda lowlands following earlier peneplanation and upon initiation of the perhumid climate regime. Thereafter, rates of speciation may slow down as available niche space becomes saturated, even where the climate continues uninterrupted. But species accumulation may not cease, especially if causes of extinctions among species in very low populations are reduced (see section 2.3.2.1 below) There is also evidence that species have originated as very small populations, for instance by long distance dispersal, rather than by division of large metapopulations as a consequence of new physical barriers (Chapter 6).

If these scenarios are correct, species occupying the continuous loam habitat in southern Borneo must have spread out during the Holocene from refugia over distances of up to 500 km, therefore migrating at astonishing mean rates of perhaps 50 m y^{-1}. In the case of dipterocarps this greatly exceeds my estimate of one million years to disperse 500 km, based on Ridley's 100 m maximum seed dispersal distance and an estimated 50 years to reach reproductive maturity. It implies either that dispersal constraints are far lower than those assumed by neutral theory[25] or that late Pleistocene climate change was minimal and required shorter reoccupation distances. Crucially for the current context, these dispersal rates far exceed the limited dispersal rates assumed by neutral theory if dispersal constraints are to influence variation in species composition at regional scales in continuous habitats. Dispersal rates of tropical trees in fact resemble those found among North American trees following the last glaciations – which are again far faster than calculated from field observation of the dispersal distances of most seeds.[26]

Numbers of genera and families differ among regional species pools separated by dispersal barriers, as well as among those that differ in climate (Table 3.7). Using existing plot data, Ferry Slik[27] analysed the pattern of generic richness throughout Borneo. He found the highest generic richness in the south-east, the region most prone to catastrophic ENSO droughts. This region is also climatically diverse, with slightly seasonal and perhumid climates on either side of the Meratus Mountains (map inside front cover). In addition it is the closest hilly land to West Java, and therefore to Pleistocene upland land bridges northwards to continental Asia. At the landscape scale, it appears less rich in *species* than north-west or north-east Borneo[28] – due, at least in part, to the relative poverty of its ecologically insular *kerangas* and MDF floras on yellow sandy soil.

The Pasoh CTFS plot in Peninsular Malaysia includes 820 species in 290 genera and 80 families; in north-west Borneo, Lambir's nearly 1,200 species are in only 296 genera and 78 families. The main reason for the differences in species richness within the CTFS plots in tropical Asia is variation in the size of their congeneric species-series. Among plots sharing a current climate and soil type, time since the establishment of current climate apparently differs, so opportunities for continued speciation, immigration and survival have varied. One explanation for the differences in ratios of species to genera and families is that areas with a high proportion of species to higher-level taxa – such as northern Borneo – have had relatively uninterrupted opportunities for speciation since the later Miocene (c.8 Ma).[29] But much of this congeneric species richness is due to diversification across the highly diverse north-west Borneo soil mosaic (see section 2.2.4.1 below and Chapters 3, 6). The genera absent from such regions as Lambir might then be relicts from more seasonal climates, which either (1) lacked opportunities for immigration since the perhumid climate originated (that is, over the last 20 million years in the case of north-west Borneo), or (2) experienced unusually high rates of extinction during the Pleistocene. The latter, however, seems less likely. Exceptionally high species richness at the provincial scale in northern Borneo does therefore appear to be partially due to the region's unusual climatic stability over evolutionary time. This has both allowed higher taxa to persist and fostered ongoing speciation within them. High rates of local endemism may thus reflect high rates of survival, rather than high rates of speciation (Chapter 6). *More convincing evidence of active evolution through natural selection would be provided if unusually high levels of genetic variation were shown to be present in local populations.*

2.1.3.1. *Habitat islands.* Some habitats are naturally more fragmented within a region than others. In Borneo, white sand podsols are confined to plateaux, ridges, and raised Pleistocene beaches, forming 'islands' rarely greater in area than 50 km². Yellow humult sandy soils occur on exposed geological formations rarely exceeding 1,000 km². Udult clay loams, however, form the zonal matrix with continuity throughout the uplands. About 40% of the species of *kerangas* (floristically defined – see Chapter 3) are confined to its white sand podsols. Yellow sandy and clay loam soils both support MDF, and about 30% of MDF species occur on both soils. Most landscapes and larger areas support only one or a limited range of soils defined by surface geology. They thus constitute effective ecological islands, limiting immigration and dispersal to many of the species.

North-west Borneo provides further evidence for a part-

ially historical explanation for regional and community species richness. The most species-rich Sunda floras of habitat islands, *kerangas* and the MDF flora on yellow sand, occur here – but this richness is substantially due to the high *provincial* rather than island-specific endemism represented in them (Figs 6.14, 6.15; Tables 3.2, 7.2, 6.16, 6.17). No plots are available from the sandy soils of Borneo south of the central mountain massif, but species distributions imply that communities there are substantially poorer than in the north[30] (Fig. 6.15). Communities I examined on podsols at Semboja, Samarinda, and on yellow sands in Ulu Mahakam revealed the presence of species that in north-west Borneo occur solely on clay loams. I infer from this that their distribution in north-west Borneo is mediated by competition, and that competition is relaxed in regions where – possibly owing to high Pleistocene extinction rates – ecological islands there are depauperate in habitat specialists.

2.1.3.2. *The Riau Pocket.* North-west Borneo has the richest tree flora in the Old World. There are 3,100 tree species in Peninsular Malaysia, including the north-west where there is a seasonal climate and Indo-Burmese species increase the total number; whereas in northern Borneo alone there are now estimated to be c.4,500 species.[31] The richest individual communities there are on the yellow sandy humult soils of the north-west (Fig. 6.15). These soils, and humult soils on the coastal hills, harbour the majority of the species of the distinct plant geographic province, the Riau Pocket, throughout their range (Chapter 6). It is they that are principally (though not solely) responsible for the uniquely high species diversity. Yet these soils are discontinuous through the region, ecological islands isolated by river systems. The species richness of these islands appears unrelated to their size, beyond very small areas of a few hundred hectares. The Miocene sandstone of the Lambir Hills (north-east Sarawak), for instance – an ecological island of less than 500 km² – contains an estimated 2,500 tree species (including species on clay as well as sandy soils), or half the estimated total for all of northern Borneo. The species of the Dipterocarpaceae are habitat-specific and their seed dispersal is constrained by lack of animal dispersers and limited effectiveness of gyration in a generally windless climate. Nevertheless, four-fifths as many species are confined to the ecological islands of deep yellow humult sands of the Lambir and Andulau hills (each c.500 km² and separated by 60 km of peat swamp) as are restricted to the clay loams which continue unobstructed for c.50,000 km² across north-east Borneo. The extraordinary slow-growing emergent heavy hardwood, *Shorea geniculata* (Fig. 7.1; Plate 6.7c) – whose heavy wingless fruit the size of a golf ball has no known means of dispersal – occurs in some of these habitat islands from west Sarawak to Brunei, 500 km along the coast. Given Ridley's[32] estimate of seed dispersal

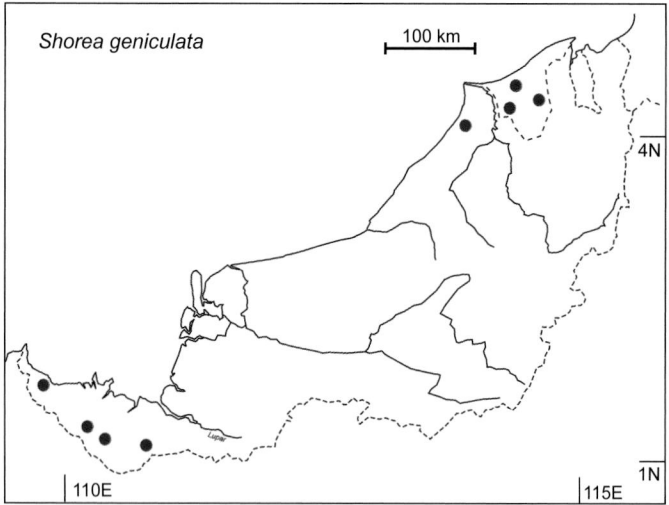

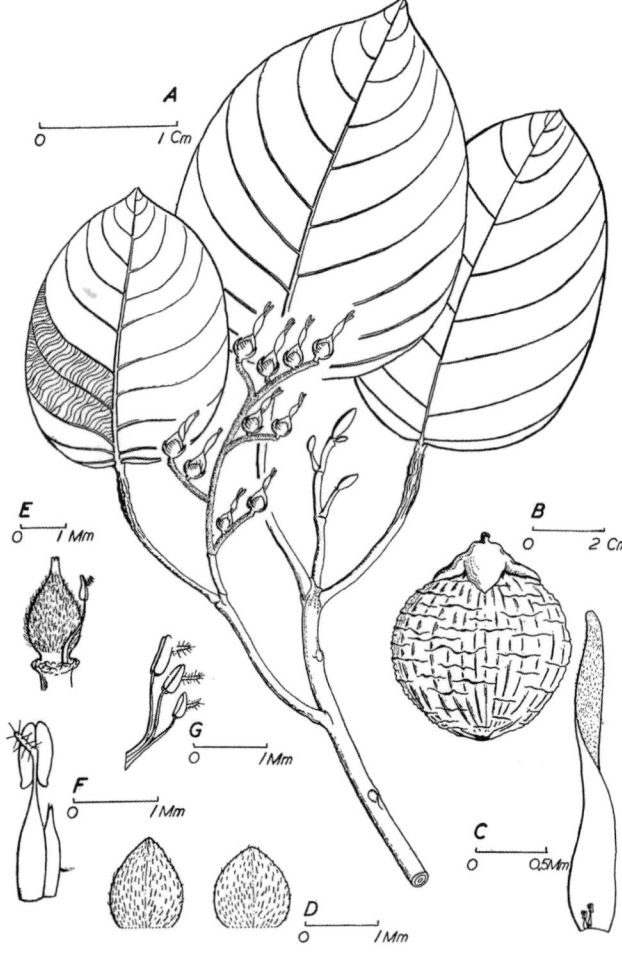

Fig. 7.1 The widely disjunct geographical distribution of the extraordinary *Shorea geniculata*, a species with wingless fruit the size of a golf ball, endemic to the ecological islands of near-coastal leached humic yellow soils of the north-west Borneo refugium.

Lowland	Montane
S. kudatense, S. multibracteolatum	S. pterophorum
S. napiforme	S. adenophyllum
S. castaneum, S. paludosum	S. castaneum ssp. altocastaneum
S. beccarii	S. ampullarium
S. bankense	S. bankense, S. nummularium
S. fusticuliferum	S. chaii
S. velutinum	S. dasyphylla
S. glanduligerum	S. erythranthum
S. punctilimbum	S. hypsipetes
S. subcrenatum	S. steenisii
S. napiforme	S. flagrimonte

At two altitudes on Kinabalu and the north-eastern Borneo mountains	
S. myrtillus and S. myrtilloides	
S. kinabaluense subspecies	
S. houttuynii and S. houttuyniifolia	

Rheophytic	Riparian or upland
S. medium	S. jambos
S. odoardoi	S. fastigiatum
S. pycnanthum var. angustifolium	S. pycnanthum

Lowland zonal	Ultramafic/limestone
S. claviflorum	S. cornuflorum, S. claviflorum ssp. tavaiense
S. penibukanense	S. silamense, S. ultramaficum, S. subisense
S. castaneum	S. georgeae
S. racemosum ssp. racemosum	S. racemosum ssp. calcimontanum
T. merguensis ssp. merguensis	T. merguensis ssp. tavaiensis (lowland ultramafic), ssp. kinabaluensis (lower montane ultramafic)

Clay soils	Sandy soils
S. hirtum	S. villiferum
S. rosulentum	S. praestantilimbum
S. urceolatum ssp. palembanicum	S. urceolatum ssp. kuchingense, ssp. urceolatum
S. nigricans ssp. nigricans	S. nigricans ssp. phaeophyllum
T. whiteana ssp. whiteana	T. whiteana ssp. monostemon (blackwater river banks)

Table 7.3 Allopatric species pairs of *Syzygium* and *Tristaniopsis* (Myrtaceae) in Sabah and Sarawak that share localities but differ in habitat. (From P.S. Ashton 2011)

Chapter 7

Forest and tree taxon diversity: why does it vary, and how is it maintained?

distance (see section 2.1.3 above), and an optimistic estimate that this slow-growing tree can reach reproductive maturity in 50 years, the species should take at least a million years to cover such a distance (in the absence of unpredictable events). This figure would undoubtedly be much extended by the rarity of opportunities to cross large rivers. Yet *S. geniculata* persists on ecological islands of less than 500 ha. Extinctions apparently caused by chance do occur though, as the absence of *Upuna borneensis* – endemic to Borneo but widespread on humult yellow sands – from the Lambir forest testifies.

The varying geomorphological age of the Sunda mountains is broadly correlated with their degree of local montane tree endemism (Chapter 6), thereby providing a rough estimate of both the minimum time required for speciation and the maximum time available for dispersal without morphological diversification. It appears to be about one million years.

2.1.3.3. *The case of Syzygium versus Shorea.* There is also some evidence of unusually high geographic and habitat-specific morphological variation within tree species in north-west Borneo, only partially explained by the greater intensity of botanical collecting in that region.[33] Patterns of endemism and local variation within the two most species-rich genera of trees in Borneo, *Shorea* (Dipterocarpaceae) and *Syzygium* (Myrtaceae), provide an instructive example of these patterns[34] (Tables 7.3, 7.4). The two genera differ dramatically in their means of dispersal. *Syzygium* species are bee-pollinated, and some species also attract a range of other pollinators including butterflies and birds. Their seeds are dispersed by birds, primates and small arboreal vertebrates such as squirrels. They are therefore as well-dispersed as any tree species of the forest's mature phase (Chapter 5). The molecular phylogeny of *Syzygium* indicates that the several species occurring in East Africa as a distinct subclade could not have arrived there before the Oligocene, long after India had moved northeastward to unite with Asia (Chapter 6, including Fig. 6.7). They could therefore only have arrived there by distant dispersal across the Arabian Sea. However, Biffen and colleagues recognized that

fleshy-fruited Myrtaceae include all the species-rich genera except *Eucalyptus*: they have diversified on a much larger scale than genera with small-winged seeds and more limited dispersal in windless climates[35] (Table 7.3).

By contrast, Bornean *Shorea* species in some sections appear to be pollinated mainly by thrips – no more capable of dispersing pollen over distance than the wind itself – and in other sections by other small and fecund insects (Chapter 5). Their fruits are either winged and gyrate to the ground, only occasionally being caught in squalls, or are wingless and dispersed by water or merely by gravity. Upland *Shorea* species are therefore among the *least* effectively dispersed species in the forest. I predicted, therefore, that the poorly dispersed *Shorea* species would be more habitat-specific, with more fragmented distributions and higher local endemism.

However, the surprise is that patterns of island-wide and local endemism among the two genera, and their degrees of local intraspecific morphological differentiation, are almost identical (Table 7.4). This might be explained by a much-underestimated role of 'fat-tailed' dispersal in *Shorea*, that is, occasional dispersal beyond the normal range.[36] But counter-intuitively also, the main difference between the two taxa is that *Syzygium* contains more local endemics than *Shorea* – even when montane species, in which *Syzygium* excels, are excluded. These endemics are habitat specialists, and sometimes, as in Kinabalu's upper montane forests and Sabah's ultramafic exposures, are sister species to others growing in adjacent habitats. I infer that the strength of natural selection in this example – and I believe it to be typical – overrides any observable effect of their contrasting dispersal efficiencies, and that limiting biological attributes of *Syzygium* override its apparent dispersal advantages over *Shorea*.

Table 7.4 Degree of endemism of Bornean *Syzygium* (Myrtaceae), with vertebrate dispersed fruit, and *Shorea* (Dipterocarpaceae), with fruit dispersed by gyration or not at all, compared. Numbers of species, with percentages of genus total in parentheses. (From P.S. Ashton 2004a, 2011)

Genus	Occurring beyond Borneo W. of Wallace's Line	Widespread in Borneo only	Borneo N. of the Equator	Local Bornean habitat island(s)
***Syzygium* (N=181)**				
Totals and percentages	72 (40)	31(17)	29 (16)	49 (27)
Montane alone	0	0	8 (4)	15 (8)
***Shorea* (N=296)**				
Totals and percentages	106 (36)	44 (15)	109 (37)	37 (13)
Montane alone	2 (1)	0	11 (4)	3 (1)

Forest and tree taxon diversity: why does it vary, and how is it maintained?

Chapter 7

Site (altitude, m)	Area (ha)	No. species	No. individuals	Spp/500 individuals	Fisher's α
Borneo					
MDF					
Andulau, Brunei (50)	0.96	256	621	235	163
Kuala Belalong, Brunei (250)	1.0	231	550	220	150
Wanariset, E. Kalimantan (50)	1.6	239	866	168	109
Lempake, E. Kalimantan (60)	1.6	209	712	171	100
Kerangas					
Mulu, Sarawak (20)	1.0	123	708	109	109
Bukit Sawat, Brunei (15)	0.96	121	734	114	114
Malay Peninsula MDF					
Sungei Menyala, Negeri Sembilan (30)	2.02	232	953	177	97
Pasoh, Negeri Sembilan (90)	50	683	26,450	201	125
Bukit Timah, Singapore (30)	2	182	847	123	71
Sumatran MDF					
Batang Ule (150)	3	502	1885	263	224
Ketambe (50)	1.6	116	861	97	36

Table 7.5 Species diversity (Fisher's α), recorded by various workers in Sunda MDF. (From Turner 2001b)

2.2. *Patterns of diversity at the community scale*

2.2.1. Expectations from theory and results in practice

If, following neutral theory expectations, communities consist of a random mix of species constrained only by variation among species in their evolutionary age and means of dispersal, we might predict that species diversity *within* communities would, in Asian forests, be correlated with the size of their regional flora – i.e., the species pool, which is the number of species available for immigration.

It turns out that this is only partly the case[37] (Table 3.7). The 25 ha CTFS plot in Sinharaja Forest (Sri Lanka) contains only 208 tree species, barely a quarter of the number at Pasoh though that site has a more uniform habitat of gentle hills. (The 25 ha plot at Pasoh captures 781 species.) Similarly, Fisher's α among Sarayudh's plots in Indo-Burmese tall deciduous forests has a range of 10.5–23, higher than that of tall deciduous forest at Mudumalai in South India (Fisher's α 9.2) (Table 3.7).

In these cases, differences in species diversity (as measured by Fisher's α) do reflect differences in the richness of their regional tree flora. But the decline in community species richness from perhumid Pasoh to Khao Chong in Peninsular Thailand with its short dry season – from communities with c.750 species and Fisher's α of 100, to communities of c.550 species and α of 84 – parallels the decline in community richness from c.200 species in perhumid Sri Lanka to c.150 species in

the somewhat seasonal southern Western Ghats (assessed from data in forestry working plans),[38] and is little related to the declines in the size of their respective overall tree floras.

Communities with higher species richness than those in north-west Borneo have been reported from South Kalimantan and from the Tigapuluh mountains in East Sumatra, and with similar richness on loams in East Kalimantan (Table 7.5). The Sumatran sample is particularly surprising because the known lowland tree flora of all Sumatra is considerably smaller than that of either east or north-west Borneo alone. Comparing plots enumerated by different workers is notoriously difficult, however. At Kuala Belalong, a plot enumerated by Axel Poulsen[39] on the site of four of my own ridge plots – though his census included slope habitat as well as ridge – revealed more than double the number of species I had recorded! This should not, though, invalidate comparisons between plots censused concurrently by the same worker or team. Periodic recensuses, such as those carried out in the CTFS plots, provide the best opportunity for accurate estimation of diversity. The north-west Borneo small plots imply that species richness accumulates to an asymptote, (that is, a maximum) specific to a habitat (Fig. 7.2). In the Sumatran case, in what is a narrow and fertile range of soils relative to Borneo, it appears that, for a given habitat, communities may accumulate from a smaller regional species pool up to a similar asymptote to that of Borneo.

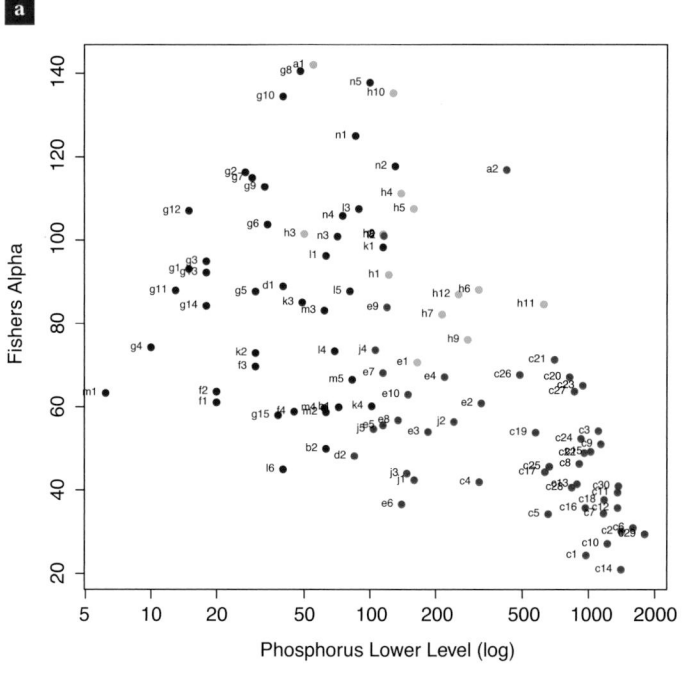

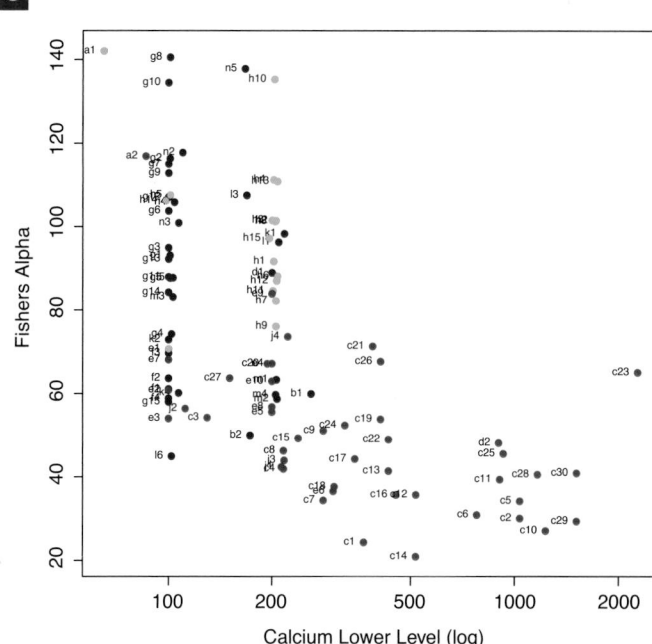

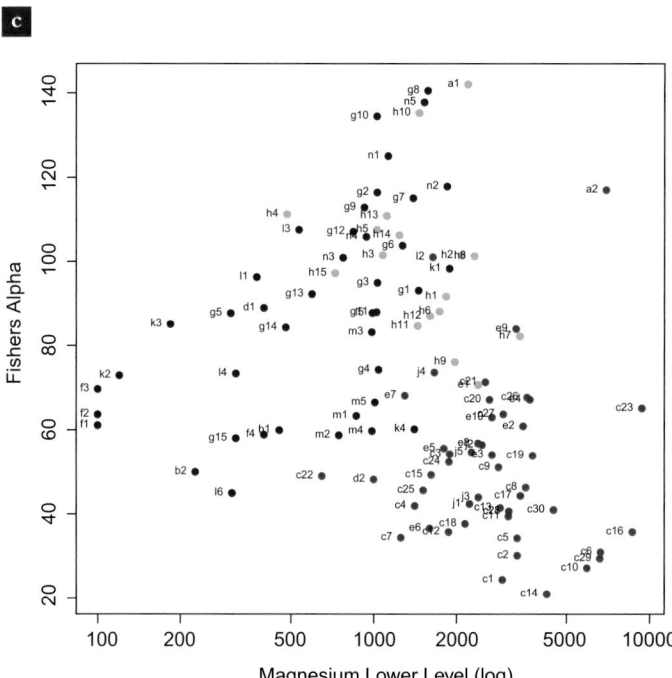

Soils

● humic

● intermediate

● nonhumic

Fig. 7.2 The relationship between species diversity (Fisher's α) in 105, 0.6 ha plots (trees >9.7 cm dbh) and total concentrated HCl extractable soil nutrients in the mineral soil at 70–80 cm: **a,** phosphorus; **b,** calcium; **c,** magnesium. The decline at high nutrient concentrations is correlated with increasing stand basal area and emergent canopy density (see text). Although maximum α values show a peaked distribution along the nutrient gradient, there is much variation beneath them, which is due to variation through gap succession within the communities. The lowest values correlate both with landslips, and with mature stands where emergents can be locally densely clumped in all communities. Plots A2 and C23 are transitional to lower montane forest. (P. Hall)

On the evidence of Brunei and CTFS Sunda plots, the maximum number of tree species in a community or uniform habitat is about 750 species of trees exceeding 1cm dbh. Fisher's α for individual 1 ha squares within the Lambir 52 ha plot does not vary consistently among soils (Fig. 7.3aa,ad). The most consistent correlation of high α-diversity is with edaphically-mediated floristic ecotones. (Estimates of α among the 105 plots in MDF throughout Sarawak also show the effects of variation in available nutrients in different soil types (Fig. 7.2). This will

be discussed later in this chapter.) Remarkably, the range of Fisher's α in 20 × 20 m (0.4 ha) subunits of the physically more uniform Pasoh CTFS plot (42–162 for all trees >1 cm dbh) falls *within* the range in the Lambir plot (14–191): these two forests, on currently separate land masses and 1,400 km apart, comprising floras nearly identical in families and genera yet with only one quarter of species in common (22% of Lambir species at Pasoh, 32% of Pasoh species at Lambir), manifest the same richness in shared habitats.

The only available estimate of Fisher's α for MDF in Sumatra was 224, a global record (Table 7.5).[40] But this estimate was certainly exaggerated by the inclusion of very high habitat diversity, as both α- and β-diversity was inflated as the sample was dispersed in a series of subplots. It may also have been biased by over-identification of sterile juveniles. Instead, it seems probable that Sumatran MDF communities are again similar in diversity to those elsewhere in the Sunda forests. By contrast, coastal MDF at Bukit Timah (Singapore), in a more drought-prone climate, does show lower species diversity. But it seems that the community species diversity of inland MDF, especially on the zonal yellow/red clay and sandy clay udult soils, is remarkably uniform – which is surprising in view of the major differences in their available floras. This implies that an asymptote exists in maximum available niche space in MDF on zonal udult soils. But judging also from the many small (0.2–1.0 ha) plots in MDF throughout Malaysia in which we can rely on the accuracy of species identification, there remains some species diversity of individual communities which correlates with their differing habitats. *But more evidence is required.*

If a ceiling in community diversity is reached in a particular habitat irrespective of area or age, does this imply that species can only immigrate by competitively excluding an existing member? Why does this apparent asymptote vary with soil and climate? *Long term monitoring of immigration and extinction rates at community scale is possible in the large tree demography plots of the CTFS network.* At Pasoh, there have been 11 losses and gains over 20 years; while only 4 of the 814 species represented there in 1985 have disappeared. At Sinharaja, there have been no losses or gains over 15 years. By comparison with plots in other climates, that is slow in proportion to the number of its species. At the Mudumalai 50 ha plot in deciduous forest, though, with its strongly seasonal and variable climate, only one immigrant and two losses have been recorded, but in a total count of only 65–71 species, in annual censuses over 12 years.[41]

Differences in community composition among sites lying wholly within a forest type (within lowland evergreen forest, deciduous forest, or montane forest respectively) but in different historic biogeographic regions are consistently seen at the species level, rather than at the generic or familial level. In particular, we find differences in the number of species within genera that accumulate series of multiple co-occurring species within individual plots (Table 3.7). Members of these congeneric series are the most ecologically similar taxa within communities. Thus, they must be either ecologically *complementary* or among the most severely *competitive* species. Evidence for competition as the explanation of species diversity within rainforest communities has accumulated in previous chapters, and will be summarised in relation to the evolutionary maintenance of community diversity in the sections that follow; while the specific case of congeneric series is addressed in section 2.4 below.

2.2.2. *Species diversity and variation in niche breadth*

Is increasing species diversity negatively correlated with niche breadth? Momoru Kanzaki and his colleagues[42] found that 70% of species distributions in lower montane forest at Doi Inthanon were correlated with topography, but 37% at the less-diverse warm temperate forest at Kasugayama (Nara, Japan).

The often prominent spatial clumping of individuals within rainforest tree populations shows that, at least at the local scale, restricted seed dispersal does influence species diversity. Joshua Plotkin and colleagues,[43] analysing the spatial distributions of five species at Pasoh, showed that the relative importance of (1) spatially expressed ecological specialisation (or 'niche breadth') and (2) dispersal constraints, varied among spatial scales and among species, according to their life histories and modes of dispersal. By itself, neither factor can fully explain how diversity is maintained.[44]

2.2.3. *Species richness and diversity and the temperature gradient*

Can we detect overall differences between communities that correspond to differing habitat influences? Mean winter temperatures decline with increasing latitude, but summer monsoon temperatures vary more with atmospheric humidity (Chapter 1). In Indo-Burma, the northern seasonal evergreen forests appear somewhat less rich than the southern, but there may be reasons for this other than mean temperature, such as history and degree of isolation. Overall, any temperature-derived pattern of richness among the lowland forests of tropical Asia is obscured by other factors. Although it is still poorly documented, we can detect a decline in the species richness of evergreen forests with altitude in perhumid regions.[45] Ediriweera, a student of Singhakumara, found that the size of gaps declined with elevation at 100–1,200 m in south-west Sri Lankan MDF; while declining canopy height was associated with greater radiation at the ground in the centre of gaps of similar diameter, but less radiation in the understorey.[46] Such changes overall could explain decline in species richness with altitude in lowland forests as much as any influence of declining mean temperature.

This decline with elevation is stepped in montane forests owing to altitudinal limits to species ranges set by limiting factors: (1) the prevailing altitudes of summit ridges, (2) the cloud base at c.1,000 m, (3) the diurnal penetration of fog on summits above c.800 m, (4) incidence of frost and drought and (5) edaphic zonation (Chapter 4). An individual species upper altitudinal limit is set by its capacity to maintain a positive carbon balance. This limit is defined along the elevational gradient by mean temperature, and the continuous decline in number of lowland species with altitude is not fully compensated by the appearance of montane species. Constraints on photosynthesis imposed first by cloud shadow and then by fog – affecting both light climate and diurnal ambient temperature – appear to account for two steps in the diversity decline within montane forests. The frequently documented *increase* in tree species diversity at 800–1200 m in seasonal climates[47] is occasioned by the reduced diversity in forests below this threshold there, which endure a more intense dry season than those above the cloud base. Increased precipitation caused by 'raking' of fog leads to increased leaching of soil nutrients, which may also limit species richness.

Takyu and colleagues[48] found that the supply of soil nitrogen and phosphorus declined with increasing altitude on all substrates in the montane forests of Mount Kinabalu, and was associated with a decline in both α-diversity and β-diversity among tree communities, as well as a general decline in forest stature. Givnish[49] argued that the decline in soil nutrients outweighs the effects of rainfall and cloud as a cause of decline in species richness with altitude. By itself this seems unlikely, since the levels of available nutrients are higher in the humus-rich, earthworm-prolific loams of temperate lower montane forests than they are in lowland rainforest soils (Chapter 1). The oligotrophic soils of upper montane forests are indeed low in available nutrients, but this is again reversed in the more eutrophic muck soils of subalpine forests, rich in worms like lower montane humic mulls.

The reduction of habitat area which occurs with altitude on mountains, and their isolation as 'islands' of more or less limited area, should influence both the size of their floras and the species richness of their forest communities. *No careful documentation exists to test this prediction*, but the differences appear slight except in very small areas. For example, the high floristic richness of Mount Kinabalu is overwhelmingly the result of endemism concentrated in two habitats regionally unique to it: upper montane and subalpine forests on ultramafic soils.[50]

2.2.4. Patterns of species diversity in mixed dipterocarp forest

2.2.4.1. Patterns of species diversity in three CTFS mixed dipterocarp forest plots: some specific examples. Variation in species diversity as Fisher's α, can be mapped as subplots as small as 20 × 20 m (400 m²) within CTFS MDF plots because they include, on average, about 300 individuals > 1 cm dbh (Fig. 7.3). That is more than are in the 105, 0.6 ha, Sarawak MDF plots, in which only trees > 9.7 cm dbh (12 inches

girth) are included. This has the advantage of approximating the area of the larger individual emergent crowns, and is therefore at a sufficiently small scale to capture variation due to stand successional stages.

Of the three plots, Pasoh, with gentle topography, relatively uniform sandy clay soils on alluvium or Triassic sediments, has the least variation in species diversity. The other two have steep topography and are more comparable. The Lambir plot's ridge consists of a sandstone slab, with freely draining lower nutrient soils sloping down to the west (i.e. towards the top in Fig. 7.3aa,ab,ac, see also Fig. 1.4), over softer shale bearing higher nutrient clay loams which are exposed to the south and on the scarp to the east. No consistent gradient in soil nutrients or physical composition has been found along the catena at Sinharaja. Soils on ridges are shallow at Sinharaja, sandy and deep and freely draining at Lambir, except to the south, and therefore also drought-prone though more stable. They become increasingly moist, gleyed and therefore anoxic below the surface, though more deep, and prone to landslips down steep slopes in both plots. Among emergent trees, mortality is mostly of single trees on ridges in both plots, leaving small gaps. Down-slope, gaps become variable and many are large, while the much-fragmented mature phase has a dense canopy.

Comparison of variation in species diversity over these three landscapes provides revealing similarities, and also differences. In both Lambir and Sinharaja, the lower slopes coincide with clay loam soils of moderate nutrient level and frequent ground movement and canopy disturbance. They are among the most species-diverse habitats. But on the gentle slope in the south of the Lambir plot (to the left in Fig. 7.3ab) – also of clay loams but here friable and stable – the density of emergents is high and disturbance low. Here late successional light hardwood dipterocarps (including *Dryobalanops lanceolata*) have built a dense continuous emergent canopy (at vertical by horizontal axis positions 80–180 × 140–260 m) (Fig. 7.3ab–ad). This is among the least species-diverse areas at Lambir. In the Pasoh plot, there is an area of similarly low richness on the flat 'upper alluvium' left of centre, with deep well-watered yet oxygenated sandy clay soils, where the heavy hardwood *Shorea maxwelliana* achieves canopy dominance. At Lambir, stable ridge-top sites with shallow or sandy soils also support less diverse communities than the unstable lower clay slopes. However, it is on the gently sloping sandstone Lambir dip-slope, with deep *stable* low-nutrient sandy soils and mostly small single-tree gaps, where the most species-diverse subplots are found. Perhaps importantly, these low-nutrient sandstone soils support mature phase canopies in which leaf sizes are comparatively small with notophylls dominating. Foliage here is diffuse, allowing penetration of abundant sun-flecks which vary in intensity according to the species and the

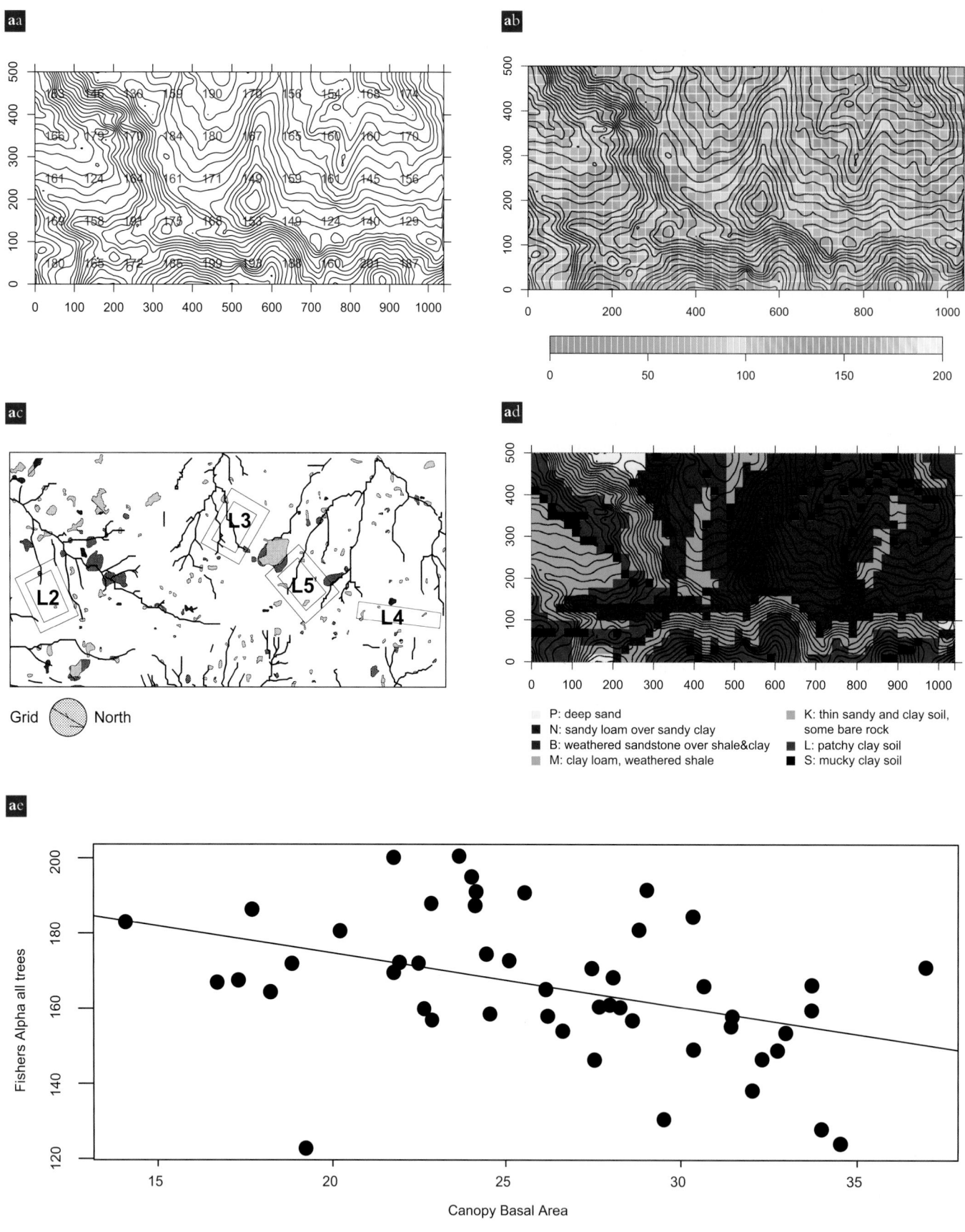

ae

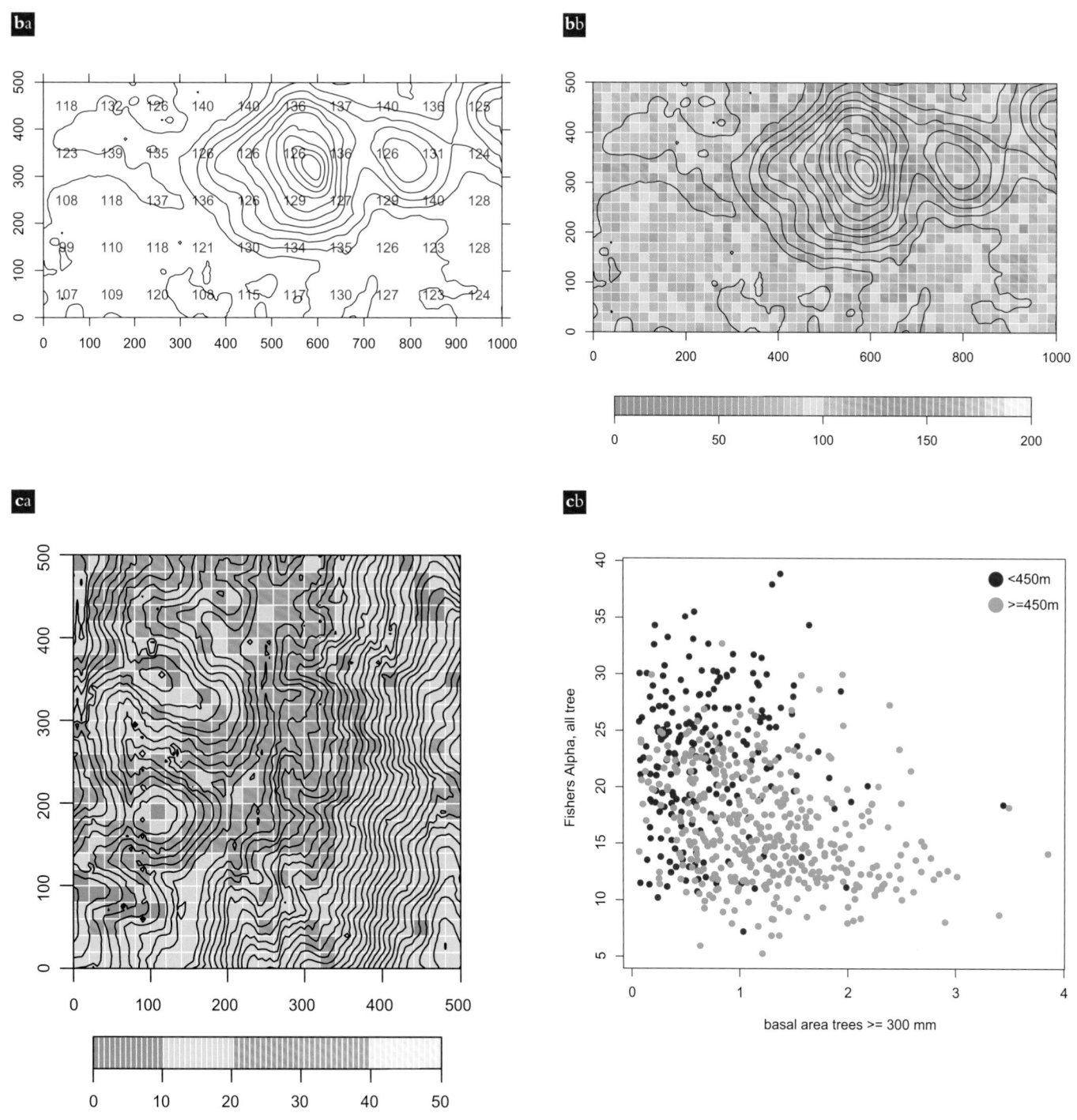

Fig. 7.3 Patterns of species diversity show similarities, but also differences between plots in the CTFS collaboration: Species diversity (Fisher's α) mapped onto topography at **a,** Lambir, **b,** Pasoh where the scale of α is as at Lambir, and **c,** Sinharaja; trees >1 cm dbh:
aa, **b**a. In 1 ha (100 x 100 m) subplots;
ab, **b**b, **c.** In 0.04 ha (20 x 20 m) subplots;
ac. Map of the Lambir 52 ha plot indicating recent or clear-fallen (cross hatched) and old (or standing dead) canopy gaps, and stormwater courses. Those 0.6 ha plots censused in 1964 (L series) also indicated.

ad. Map of seven major soils groups within the Lambir plot as recognized by I.H. Baillie.

Species richness is negatively correlated with the basal area of larger individual trees, >300 mm dbh:
ae. At Lambir (1 ha subplots).
cb. At Sinharaja (20 x 20 m subplots), where valley plots (red dots) include those with highest α but show a wide range of basal area, while upper slopes and ridges (green dots) include those with lowest α but highest basal area (R²=0.158). (**ac,** T. Ohkubo; all others P. Hall).

maturity of individuals (Chapter 2). Patches of low diversity at Lambir, on the sandstone dip-slope at 600 m along the long access and 260–300 m along the short, and on the scarp at 100-200 m along the long and 0–60 m along the short, coincide with relatively recent large gaps, while the larger area of low diversity in the south-west (top left in Fig. 7.3ab) is difficult to interpret, but is one again of dense understorey and few large trees. The forest here may have been damaged when the trunk road to the immediate south of the plot was being built.

Overall, it appears that long-term canopy disturbance regimes play a major part in mediating species diversity, which is low both in early and late gap succession, within habitats defined by geology and soil; but soil nutrients, perhaps by influencing canopy foliage density, also appear influential. The first inference is supported by the observation that basal area (as an approximation of biomass where canopy height does not vary greatly) correlates strongly (negatively) with species diversity in 0.04 ha subplots within the 25 ha CTFS plot at Sinharaja (Fig. 7.3cb), and loosely (negatively) with species diversity in 1 ha plots at Lambir. However, significant correlation with basal area of larger trees at Lambir is confined to the more abundant sandy soil plots ($R^2=0.158$) (Fig. 7.3ac). Subplots at Sinharaja show high variation in α, which declines and becomes less variable with increasing basal area of larger trees. This is less so at Lambir where a single subplot with both low basal area and α is an outlier at 1ha scale. This appears to represent a recent slip. Basal area increases in forest succession, as it becomes increasingly dominated by the largest canopy trees (Chapter 2). Basal area is highest (and diversity is relatively low for these soils) on Lambir ridges where soils are sandy, and their minimum mature phase diversity on sandy soils is reached here (see section 2.3.1. below).

Savitri Gunatilleke and her colleagues found that Fisher's α declines from ridges and spurs to valley at Sinharaja (Fig. 7.3c) in all species structural guilds.[51] The ridges show patchy dominance of the short emergent *Shorea worthingtonii* and main canopy *Mesua nagassarium*. Soils hardly vary in nutrient levels from valley to ridge here, but are shallow on ridges and upper slopes where rock is patchily exposed; while lower slopes are moist and prone to slips. Sinharaja's perhumid climate also experiences the south-west monsoon, and mean annual rainfall exceeds 5 m. Prevailing south-west monsoon winds are associated with a relatively uniform canopy height along ridges, and true emergents are confined to sheltered valleys (Plate 3.2a). Canopy gaps on ridges are small and canopy foliage is diffuse, as at Lambir. However, upper canopy structure here bears closer resemblance overall to that of *Shorea albida*-dominated peat swamp forest (Plate 3.9d,e); but the lateral light received along both Sinharaja and Lambir ridges, unlike in the peat swamp forest, is associated with the presence of a main canopy guild.

Reduced canopy heterogeneity may explain why, unlike in Far Eastern MDF, subcanopy species diversity at Sinharaja is higher than canopy species diversity. Lower slopes at Sinharaja do manifest a wide diversity of gap sizes, as well as a heterogeneous mature phase canopy including emergent dipterocarps. Thus, canopy structural and dynamic diversity appears to account for the greater overall tree species diversity of valley forests here.[52] At Lambir overall, species diversity is higher on the less-fertile sandy soils where single tree mortality predominates than on clay loams where larger gaps from wind-throw prevail. Richness can be low both in recent windfall gaps and on steep unstable slopes on either clay or sandy soils (as in Fig. 7.3ab, ac: at horizontal × vertical axis locations 100–210 × 0–60 m, 20–280 × 380–500 m). I observed that diversity is lowest on sandy soil at Lambir only where gaps are caused by landslips, other gaps remaining high, apparently owing to persistence of understorey species and canopy regeneration. The great variation in species diversity in subplots with low basal area appears to occur in canopy gaps. It seems to be substantially due to the cause of the gap: landslips lead to bare surfaces devoid of established regeneration, which subsequently takes years to build up species diversity; while succession following a canopy gap which hardly damages the understorey may start low in diversity if the majority of individuals are less than the 1 cm lower census threshold, but will rapidly assert very high diversity, which then declines as competition reduces the number of individuals, and as the canopy with its subcanopy shade on clay soils is re-established.

In all these cases, the influence of topographic diversity and/or soil nutrient levels on *mature phase* canopy structure is sufficient to override whatever impact there might be on richness from gap succession *within* communities. In contrast, Pasoh has gentler topography than Sinharaja and less-variable soils nutrient levels than Lambir. In Pasoh, variation in canopy basal area is mostly influenced by within-community successional phases and a characteristic mature phase canopy structure (Fig. 7.3ba, bb). Only on the raised alluvial bench, however, does the mature phase accumulate sufficient basal area among large trees to depress diversity.

2.2.4.2. *Species diversity and niche differentiation along the canopy disturbance, and consequent gap and subcanopy light gradient*

2.2.4.2.1. Vertical niche partitioning. Mean light intensity and frequency of sunflecks within forest canopies increase with height above ground, creating a roughly vertical resource gradient. Takashi Kohyama,[53] using data from a plot in MDF near Palangka Raya (Central Kalimantan), showed that the distribution of tree species maximum heights was consistent with his hypothesis that differentials in species inherent heights at maturity can maintain stable coexistence of species of both similar and different statures. This has been confirmed in a global context.[54]

Sean Thomas showed that subcanopy species within series of co-occurring congenerics do indeed differ in the height at which reproduction commences, and also in light response according to their height. Juveniles of taller species on average respond in diameter growth to higher light intensities than do shorter species.[55] Trees (like fish) continue to grow throughout life in both diameter and crown size, but reach an asymptotic height, as damage – such as the effects of periodic drought – increasingly constrain its further extension. Thomas found that variation in the asymptotic maximum height of selected subcanopy and main canopy species correlated positively with their growth rates, while wood density, and the relationship between height and diameter did so negatively. Remarkably, Stuart Davies also found similar reproductive differentiation among 11 pioneer *Macaranga* species at Lambir (Fig. 7.4; Plate 7.1).[56] But species-specific competition for light alone cannot determine the success of rainforest individuals: clearly, there are far too many understorey species – let alone juveniles of canopy species – for species-specific physical interactions to be consistent. Interestingly, Thomas found no significant partitioning of heights at maturity *among* the genera he studied, only *within* them.

Relative species diversity appears greater in tropical than in temperate forest understoreys. This could be due to greater amplitudes of height, or of variation in light along the vertical or horizontal dimensions.[57] In practice, though, the range of species heights at maturity is constrained by canopy stratification (Chapters 2, 5), and this also influences the range of available pollen and seed dispersers (Chapter 5). Climax species in Asian forests are specialised at maturity to one of up to three tree strata; among these, height diversity is greatest in the subcanopy stratum (Chapter 2). It is least in the emergent stratum, although variation in overall mature phase canopy height at maturity appears to be greatest in forests on soils of intermediate fertility, where species diversity also reaches its peak (Figs 2.3, 7.2, 7.3).

The vast majority of tree species, even in the most species-diverse forests, can be identified by leaf morphology alone. But interspecific differences among co-occurring congenerics are clearest at tree maturity and among the emergent species. Leaves of congeneric emergent dipterocarps differ in size, number of veins, hairiness, and waxiness and albedo of the undersurface; also angle of presentation and internode length.[58] Waxiness and high albedo are strongest in species with competitive advantage on xeric sites: ridges and sandy soils (Chapter 2). As a group, these variations represent contrasting adaptive pathways to surviving water stress. I

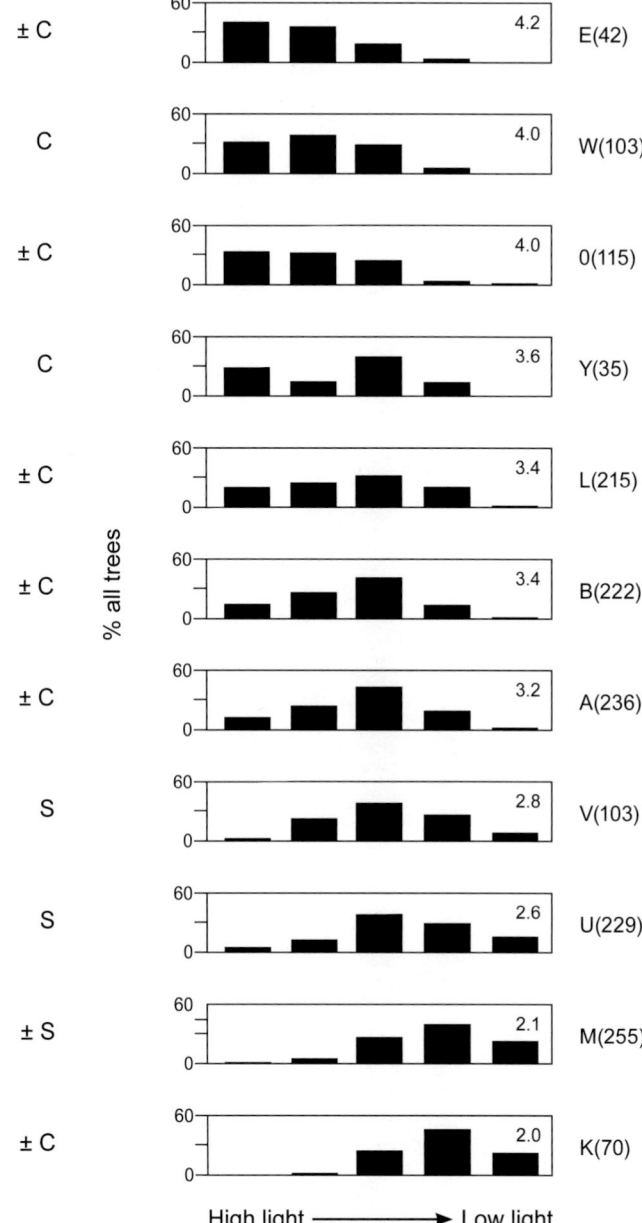

Fig. 7.4 Diversification among 11 pioneer *Macaranga* in the Lambir CTFS plot along the light gradient: the percentage of individuals occurring in each of five crown illumination classes, in order of a decreasing crown illumination index. Sample sizes in parentheses. Soil preferences on left: S: yellow humult sand. E: *M. gigantea*; W: *M. winkleri*; O: *M. hosei*; Y: *M. hypoleuca*; L: *M. triloba*; B: *M. beccariana*; A: *M. trachyphylla*; V: *M. havilandii*; U: *M. hullettii*; M: *M. lamellata*; K: *M. kingii*. (From Davies *et al.* 1998)

infer that among evergreen species in perhumid climates, the full rigour of water stress is only experienced by leaves in the outer canopy and in full sun. Values of mean specific leaf area tend to increase with an increase in available nutrients in the soil below.[59]

Plate 7.1 Leaf morphology is just one manifestation of the extraordinary morphological and ecophysiological diversity which distinguishes many species of pioneer *Macaranga*. Some of this diversity, when taken in context of their arrangement in the tree crown, appears to reflect adaptation which optimises the capture of light and overshading of competitors under their prevailing soil water economy and proneness to drought; yet some variation may be adaptively neutral inheritance – 'spandrels of San Marco'. (S. Davies) **a**. *M. grandifolia* (x¼), **b**. *M. siamensis* (x⅛), **c**. *M. depressa* (x⅓), **d**. *M. heynei* (x⅔), **e**. *M. havilandii* (x¼), **f**. *M. conifera* (x¾),

Fig. 7.5 a, Species diversity in relation to structural complexity and HCl–extractable soil Mg, at 70–80 cm depth, as a proxy for nutrient status. Profile diagrams (64 x 8 m, trees >5 m); Mg (mg g^{-1}) and Fisher's α from nearest 0.6 ha plot, from 10 sites in MDF, one (C27) in lower montane forest, Sarawak, Borneo. Sites as in Fig. 3.9. Dots indicate position of profiles on the axes. Blue dots: humult, red: udult, green: intermediate; brown: lower montane humic mull soils. Species diversity peaks at c.1,000–1,500 mg g^{-1} mg, where structural complexity is great, and main canopy well represented between clumped emergents, but there is much variation of α not obviously related to structure or stature. Diversity drops at high Mg levels, where soils are friable loams with high water retention, in tall forests where emergents are notable dense; but also in the short lower montane forest where understorey is dense. Sites B, and F where the forest is recovering from widespread canopy mortality (Chapter 2), share the same geology but the taller profile B, on deeper soils, has similar diversity to F in which emergent crowns are barely emerging from the main canopy. But B1, F2 and M1, all with low α, are also coastal forests, prone to drought. G12, H5 and C27 are on ridges; C27 at 900 m. Canopy leaf size increases with nutrient levels, especially on friable loams, and from ridges to lower slopes (Fig. 2.18), while subcanopy tree density declines and casual observation indicates that subcanopy shade increases (solid data yet to be collected). **b,** Species diversity is partly correlated with structural complexity in MDF. Species diversity in the three profiles asterisked in Fig. 7.5a: species diversity expressed as numbers of species >9.7 cm dbh per 1,000 individuals in adjacent plots; percentage of the total within a site in parentheses. (Compare with Fig. 2.3). (**a,** from P.S. Ashton 2008; **b,** P. Hall)

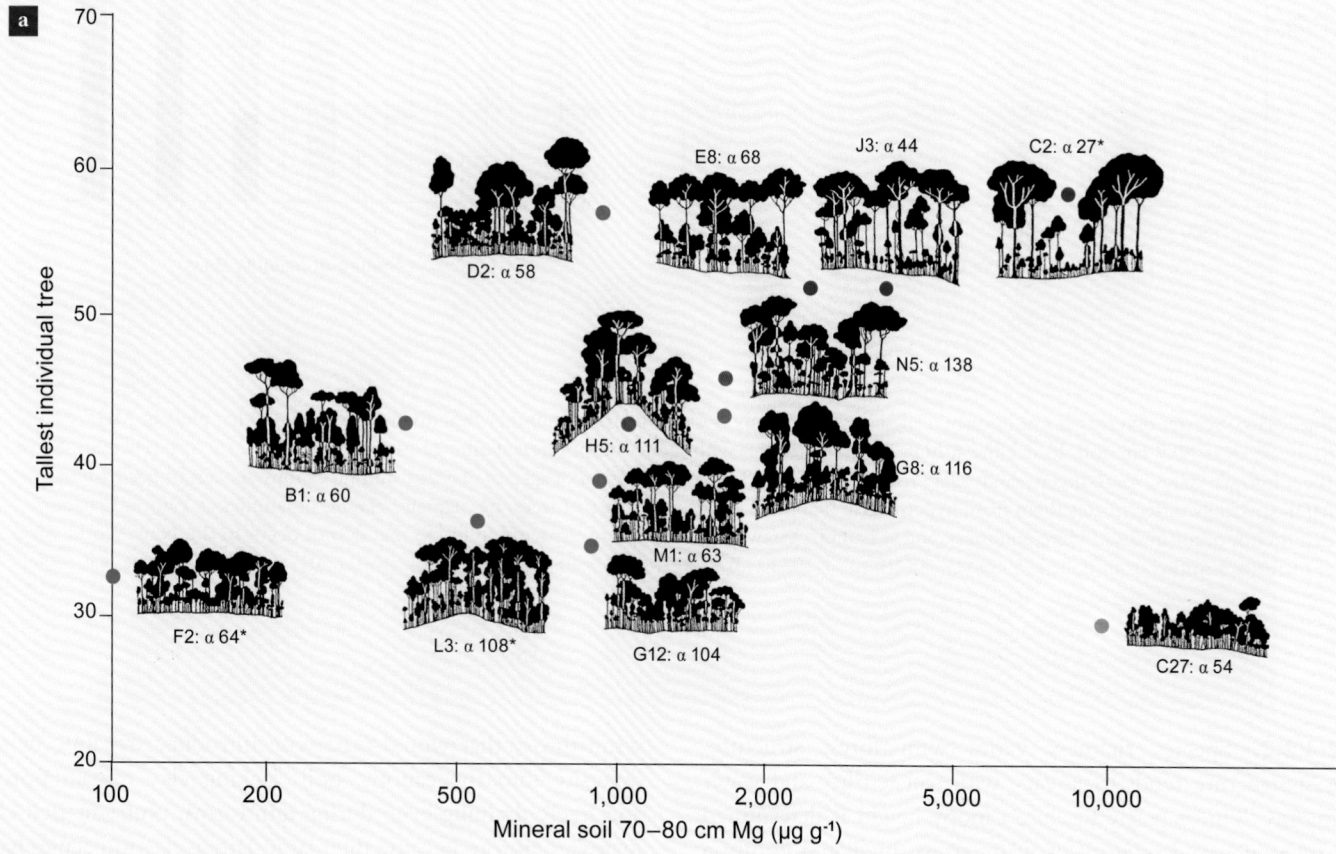

Locality	Bako N.P.	Lambir N.P.	Bukit Mersing, Anap
Site characteristics	Drought-prone coastal slope; infertile freely draining sandy soil	Moderately drought-prone moderately infertile sandy loam soil; ridge and gentle slope	Mesic lower slopes; deep fertile basaltic clay loam soil
Number of species	223	321	143
No. pioneer spp. (%)	18(8)	30(9)	15(10.5)
Emergent spp. (%)	10(4.5)	40(12.5)	8(5.5)
Main canopy spp. (%)	111(50)	147(46)	43(30)
Subcanopy spp. (%)	84(38)	104(32)	77(54)

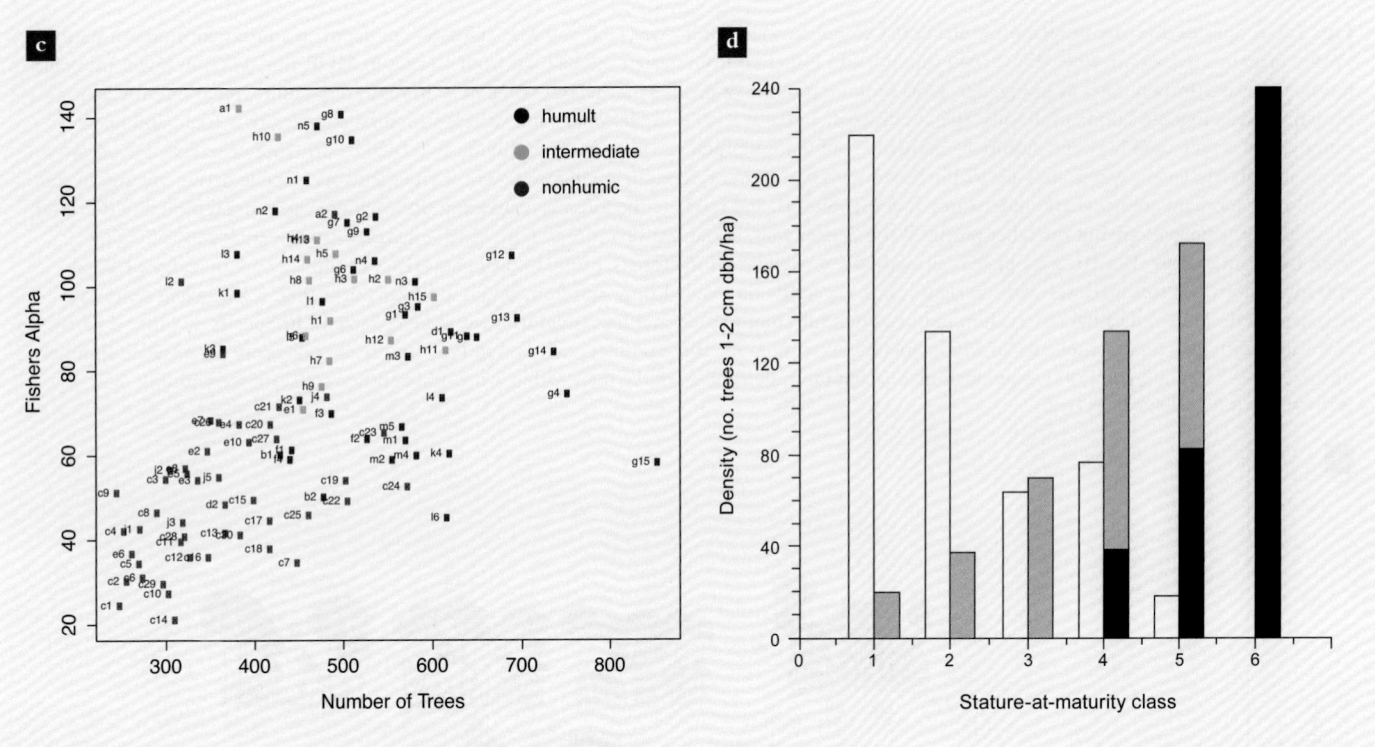

Fig. 7.5 c, Species diversity is not linearly related to stand density: Fisher's α against number of individuals in 105, 0.6 ha, plots in MDF, Sarawak; trees >9.7 cm dbh. Its decline at high density correlates with reduction in forest stature and simplification of structure (see text). Soils: Blue: humult yellow sand; green: sandy clay; red: clay loam. **d,** The concentration of species diversity among canopy species in Far Eastern MDF strikingly differentiates them from hyperdiverse Neotropical forests: density of trees, by diameter classes, of families at Yasuni, Ecuador that do not occur in Asia (white columns), and at Lambir, Sarawak (grey, Dipterocarpaceae in black) that do not occur in the New World. Size at maturity assumed from dbh: 1–5 cm (1), >5–10 (2), >10–20 (3), >20–40 (4), >40–80 (5), >80 (6). (P<0.0001). (**c**, from LaFrankie *et al.* 2006).

2.2.4.2.2. Species richness varies with forest structure. Forest structure tends to covary with soil fertility and water economy. This is partly because canopy height is determined by frequency and severity of water stress, mediated within a given climate regime by the soil's capacity to store and release water (Chapter 2). Mature forest stands on shallow freely draining drought-prone soils in the perhumid tropics lack tall emergent species. Their stature is short and their canopy even, restricting the range of heights at maturity into which canopy species can specialise (Fig. 2.3). Reduction to a single dominant species is nevertheless rare, implying that the several species are ecologically complementary in respect of mature height. Mature stands with closed, even-statured canopies also occur on the most fertile soils, but there the canopy is comprised entirely of emergent species, while main canopy species are few and scattered in low density (Chapter 2). Both mature forest stand types share secure rooting and resistance to the small-scale

wind-throw that would otherwise introduce dynamic height diversity into the canopy. Soils lacking clay remain porous and freely draining even during rainstorms. Nutrient-rich clay loams are porous because their large-lattice clay molecules have abundant charges which lead to agglomeration of the clay into crumbs, providing space for water and oxygen – and therefore roots – to freely penetrate to depth (Chapter 1). By contrast, medium-fertile short-lattice and silty clay soils, in which rooting is shallow, tend to build stands more prone to frequent wind-throw. These are the most widespread soils in the wet tropics; they account for the patchy emergent canopy characteristic of lowland rainforests worldwide.

The peak of species diversity coincides with the forests displaying their most complex structure (and their peak in heterogeneity of mature stand structure) on udult short lattice clay and udult and humult sandy clay soils. Our MDF data show that in all three canopy height guilds, species diversity is greatest on these types of soil (Fig. 7.5a,b). These are forests with intermediate stand density (Fig. 7.5c). These are also mineral soils with intermediate nutrient levels (section 2.2.4.3 below). From field observations, I infer that on these two soil types, humult and udult, which differ in species composition (Chapter 3), there are separate structural and physiognomic causes of high species diversity and subcanopy density (see also Chapter 2):

1. On deep sandy humult soils, continuity of the tall emergent canopy is broken by occasional drought-induced single tree mortality, creating gaps in which a diversity of mature main canopy species can survive. On both deep and shallow leached humult soils, relatively frequent water stress and/ or low nutrient availability are associated with a diffuse notophyll-dominated canopy. This provides a varied, but rarely deep, shade regime in which subcanopy species can diversify.

2. On shallowly-rooted leached silty clay soils with thin and patchy surface raw humus, canopy gaps are larger and more frequent, resulting in a higher proportion of late successional mesophyll components in the apparently mature stand, and persistence of patches of main canopy species.[60] Mature stand light conditions are yet more diverse, with dark shade better represented.

By comparison, on the most friable fertile udult clay loams, extensive patches with almost contiguous emergent canopy and sparse main canopy and subcanopy prevail. Canopy gaps are occasional and sharply defined, generally of moderate but variable and sometimes large size.

A number of inferences emerge from these comparisons. First, species richness is likely correlated with the *extent* and *diversity* of both gap and subcanopy light climates bright enough for net photosynthesis. Different species will respond to these at the ground level and at all heights up to the emergent canopy. In contrast with yellow humult sands, mature-phase MDF on friable clay loam soils has a generally dark understorey. This restricts the range of species (including canopy juveniles) that can survive there once their seed resources are exhausted. Median gap size is large, resulting in large areas of full sunlight with a contrasting but also rather narrow range of opportunities for specialisation. These opportunities are even further reduced by the tendency toward early dominance by rapid height growth on fertile, well-watered soils (Chapter 2).

Second, species diversity is enhanced by mature-phase structural diversity. This too is an indirect consequence of both the prevailing canopy gap regime and the diverse vertical light profile, which leads to three vertically distinct guilds of species at reproductive maturity.

Third, there can be no linear relationship between the productivity of primary forest and its species richness or diversity. Productivity is highest in the late successional phase on the most fertile − namely, well-watered and nutrient-rich − clay loams. But these habitats, in the Far East where the soil nutrient levels are sufficiently variable to compare, are not the most species-diverse.

Unfortunately, data from other regions and climates, which would permit such detailed comparisons of diversity and structure, are as yet unavailable. Decline in species richness with increasing rainfall seasonality is broadly correlated with a decline in overall stature. There is a transition from semi-evergreen canopied forest with a persistent three-stratum vertical structure (with increasing blurring of the distinction between the two canopy guilds), to tall deciduous canopy forest in which the emergent stratum has ceased to exist as a distinct guild. As seasonality increases, subcanopy tree species are replaced by deciduous shrubs, and the community shows a sharp decline in species richness (Table 3.7). Understorey density tends to be low in deciduous forests owing to fire and to browsing (Chapter 2). Seasonality is also accompanied by the presence, and rise and fall in intensity and frequency, of cyclonic storms, and local to regional massive canopy destruction.

Overall species diversity, clearly, is achieved in different communities by individual structural and dynamic guilds showing different proportions of species diversity. One striking example: species diversity at Lambir is similar but a little lower than that in the 25 ha CTFS plot at Yasuni (Ecuadorian Amazon), which has a similar rainfall distribution. Soils and topography at Yasuni are more uniform.[61] Lambir's richness at the whole-plot scale is partially attributable to its high β-diversity, correlated with its diversity of soil types (Chapter 6). But whereas tree species diversity is concentrated in the main canopy at Lambir, it is greatest in the *understorey* at Yasuni − where the emergent canopy is scattered and species-poor[62] (Fig. 7.5d). Neil Gale found that, among the 84% of tree deaths whose mode he had been able to ascertain in an Ecuadorian forest, 40% were uprooted, 41% snapped, and only 18% died standing; whereas in MDF at Kuala Belalong and Andulau (Brunei), 30–80% died standing (Chapter 2).[63] I was struck by the openness of the main canopy at Yasuni. Its trees have deep ellipsoid crowns, strikingly different from the diverse but overall convex dome-shaped crowns at Lambir. At Yasuni, the stubs of broken crown branches are evident, and frequent branch break appears to play an important part in causing canopy openness. This openness and irregularity creates an extremely variable light regime beneath, which may account for much of the environmental diversity underlying the subcanopy richness. Branch break may be a consequence of the occasional violent downdrafts during *friagem* (cold fronts off the Andes), perhaps in combination with the high biomass of Bromeliad tank

Forest and tree taxon diversity: why does it vary, and how is it maintained?

Chapter 7

epiphytes – a historically and biogeographically determined factor absent in Asian forests. It has also been claimed that the great abundance of dipterocarp juveniles so characteristic of Asian MDF prevents the accumulation of a rich understorey tree and shrub flora there (but see next section 2.2.4.2.3).

Curiously, and in contrast to Far Eastern MDF, species diversity at Sinharaja is about as great among subcanopy trees as it is in the main canopy. The subcanopy species – some of them light-demanding – total 93 (46.5%) by my estimate, while the main canopy species total 95 (47.5 %). This recalls the pattern at Yasuni. Interestingly, juvenile dipterocarps, measured as individuals 1–2 cm dbh, comprise a surprisingly low 8% of all individuals throughout the Sinharaja plot. At Pasoh they also comprise only 9%, but they comprise 15% on the shallow soils of the Sinharaja upper slopes and ridges. It seems unlikely that crowding by juvenile dipterocarps can account for relatively low species richness in the Far Eastern MDF understorey.

2.2.4.2.3. Species diversity varies with forest dynamics. Each community comprises a mosaic of stands of different ages, each with a partial suite of the species characteristic of its successional stage – always partial because stands are never large enough to include all species, and because at the scale of the stand, the stochastic influences of seed dispersal are likely to play a significant role in determining species composition. Interpretation of the relationship between species diversity and canopy gap characteristics therefore requires further careful observation.[64] Community species diversity and richness, if it does change during forest succession, will vary to differing degrees according to the shape and area of sample plots,[65] but will decline predictably to the mean as the sample dimension increases.

Every species differs to some degree in (1) its photosynthetic response to light, (2) its growth and allocation of biomass to shoots, reproduction and root systems, and (3) its capacity to survive in shade (Chapter 3). Species may also differ in seed dormancy and germination in response to light and warming. Pioneers thus play different roles from climax species in both spatial and temporal dimensions (Chapter 2).

The extent to which species populations survive in diverse rainforest communities through partitioning of the light gradient remains unclear, although it was a favoured hypothesis in early studies.[66] I questioned these conclusions in Chapter 2. *Resolution must await results from long-term demographic research, in combination with experiment.*

Forests of the calm perhumid tropics. Current evidence from plot analyses and pot and field transplant experiments[67] suggests that, although pioneer and climax species do show differing competitive success along the light gradient due to differences in both physiology and resource allocation, species sharing the

same dynamic guild often do *not* display such differences at a significant level – and this is particularly true of congeneric climax species.[68] There are vastly too many co-occurring genera in the dynamic guilds of rainforest trees for the light response gradient alone to partition more than a fraction of them.[69] Nevertheless, when changes in light response during ontogeny, and variation in the relationship between light responses and mortality rates are taken into account, the potential exists to sustain a substantial number of species by these factors alone (Chapter 2).

	Pasoh, 50 ha	Lambir, 52 ha
Number of pioneer species	46	63
Pioneers as percentage of total number	5.7	5.2

Table 7.6 The low representation of pioneer species in MDF. (From Manokaran *et al.* 1990; H.S. Lee *et al.* 2004)

Fresh gaps and pioneer vegetation generally occupy only a relatively small area in primary forest,[70] although they may *periodically* occupy larger areas in the cyclone belts of seasonal Asia. The proportion of putative pioneer species is low: 5.7% in Pasoh, 5.2% in Lambir, 6.3% in Sinharaja plots (Table 7.6). Instead, species diversity is determined principally by variation in the numbers of late successional and climax species.

Tree communities in various habitats do appear to manifest different median canopy gap sizes and different proportions of successional species – and consequently differ in the ranges of their maximum growth rates (Chapter 2). *These differences have yet to be carefully documented. Kerangas* on freely draining lowland podsols experiences mainly single-tree mortality and the lowest maximum individual growth rates, therefore the lowest range of maximum growth rates among its species.[71] Emergent shade-tolerant heavy hardwoods, notably *Shorea* (sect. *Shorea*) species (*balau, selangan batu*) are well represented in these communities, but also those with highest species diversity at Pasoh and on the humult sandy soils at Lambir. Vertical stratification of mature stands is most diversified where canopy structure or lateral light on ridges increases both mean and variance of subcanopy light, providing opportunities for high survivorship of juvenile shade-tolerant species as well as light demanding climax species.

The low species diversity of *kerangas* is in part due to the absence or poor representation of an emergent species guild; in part by the dominance of slow-growing shade tolerant heavy hardwoods, absence of pioneers, and sparse representation of light hardwood successional species. In MDF on yellow sandy

Chapter 7

Forest and tree taxon diversity: why does it vary, and how is it maintained?

or sandy clay soils of intermediate fertility, the volume of soil available to roots is generally greater, rooting is deeper and the soil has a higher capacity to retain water. Forests on these soils reach tall stature, but occasional intense droughts differentially kill emergents along with seedlings and saplings. Matthew Potts[72] found that rare species experience *higher* survival rates than common ones following such droughts, so that in this case, the droughts themselves contribute to the maintenance of high species richness. Pioneers are a significant component of the rare category. Also, periodic emergent mortality is a major cause of the incomplete and generally clumped emergent stratum, thereby clearing space for a diverse main canopy guild.

On well-watered friable clay loams, the dense, even, continuous emergent mature-phase stratum is dominated by a few species of late successional light hardwood giants, with occasional large gaps (more common in regions subject to landslips or cyclones). The main canopy and subcanopy are diffuse and relatively depauperate in species; overall richness is consequently depressed. I infer that such stands often represent late succession from massive wind-throws. Mixed among these emergents and contributing to α-diversity are varying numbers of heavy hardwoods, shade-tolerant as juveniles. I suspected, in Chapter 3, that these are not late intruders into the established light hardwood stand, but rather that as juveniles they survived the catastrophe that initiated it, and gained subsequent advantage by growing successfully beneath the dense blanket of vines that so often invade large gaps on well-watered fertile soils.

The CTFS plots can provide the opportunity to document the periodicity, process and duration of recovery of these massive wind-throws. The influence of variation in size and frequency of gaps, combined with the density of the mature canopy on species diversity, is most clearly seen across the topography in the CTFS Lambir and Sinharaja sites[73] (Fig. 7.3a,c). Habitat specificity among the species of perhumid south-west Sri Lanka appears as stringent as that of north-west Borneo, despite the fact that the available species pool and habitat-specific richness are each no more than a quarter of that in north-west Borneo. Neither site includes high-nutrient loams or loams on undulating land such as occur in the Mersing and Bakong plot sites in Sarawak (see Chapters 2, 3). In both, then, high species diversity is associated (though only in part at Lambir) with compact clay loams of intermediate nutrient levels which experience more frequent and intense canopy disturbance levels, including from landslips, than elsewhere in these plots.

Forests subject to cyclones. Tim Whitmore and David Burslem monitored the impact of the varying frequency of cyclonic storms on the composition of different forests on Kolombangara Island in the Solomon Islands.[74] Cyclones had recently increased in frequency. They found differences in space, and over time, among these forests. Baker and Sarayudh[75] showed how, at Huai Kha Khaeng, occasional catastrophic storms, with and without ensuing fire, enhanced α-diversity by producing a variety of successional pathways. They identified an increase in environmental variability along a gradient of canopy disturbance frequency and intensity: and the *most* disturbed stands here supported the highest species diversity. This is consistent with predictions derived from lottery competition models. According to this family of ecological models, communities are in a permanent state of post-catastrophe disequilibrium. Their α-diversity is enhanced by the persistence of less-competitive species that are in gradual decline until another catastrophe provides opportunity for reinvasion.[76] At Huai Kha Khaeng, the most subject of CTFS plots to major canopy destruction, the 5–8 month dry season nevertheless imposes a limiting influence on species richness and diversity which are lower than in more equable climates.

There is a diminishing trend in tree species richness within MDF following the perhumid east coast of the Philippines from Mindanao in the south into southern Taiwan. This appears correlated with increases in the frequency and severity of typhoons *(though this is still insufficiently documented as there are no sample plots from Samar or Mindanao at the southern end)*. For one family example, 14 dipterocarp species have been recorded from north-east Luzon, of which eight occur in the CTFS plot; whereas 28 species are known from eastern Mindanao. I observed that some forests in the south appear never to experience typhoons; these forests are tall, with a discontinuous emergent stratum, suggesting the influence of periodic killing drought as in Borneo. Other forests nearby on windward slopes, though, experience massive wind-throws over many hectares, and regenerate as stands of pioneers or successional light hardwoods initially of uniform height, casting deep shade in the understorey (Plate 2.22a).

At the CTFS Palanan plot, on the east coast of northern Luzon exposed to frequent typhoons, the MDF canopy is continuous and comprised of emergent dipterocarps (Table 3.7; Plates 2.22d,e). It is relatively low, at c.40 m. The trees are deep-rooting, with tapering trunks of great basal diameter; the main canopy is poorly represented. The canopy experiences defoliation during strong typhoons at about ten-year intervals, and the understorey is patchily dominated by light-demanding species, particularly monocots including bananas, aroids and palms (Plates 2.22f). Nevertheless, pioneer tree species richness is unremarkable at 24, or 7.5% of the total species count. Species richness overall, on the evidence of species geographical distributions, is almost certainly less than in the south in all strata, but here at Palanan it is the leaves and young twigs rather than the trees themselves which experience most mortality. The same is true in the extreme south of Taiwan, where Sun I-Fang has established a CTFS plot. The canopy here does not exceed 20 m, and is comprised of main canopy taxa, not emergents (Plate 2.23b). Canopy and understorey are indistinctly separated spatially, recalling upper montane forest (Plate 2.23c). Species richness is lower again than at Palanan, but the Bashi Channel between Taiwan and Luzon is deep and represents a major biogeographic boundary. The differences in species composition and richness, and perhaps the absence of emergent species, must therefore in this case be in part historically determined. But the latitudinal gradient in species richness and dynamics in the forests fronting the Pacific coast recalls the transect across coral reefs which inspired Joe Connell to propose his 'intermediate disturbance hypothesis'.[77] This gradient also extends from Bornean MDF on deep friable clay loams, which may never experiencing killing droughts, to the Bornean MDF on more drought-prone soils in which canopy gaps are commonly small, through the south-west Philippine MDF which experience occasional major wind-throw, to north-west Philippine MDF and south-west Taiwan forests which have adapted to resist frequent destructive typhoons. The Bornean MDF on loams appears analogous to the reef habitat within the calm of the Connell's lagoon, with species diversity depressed in both cases. The depression of diversity in the Taiwan forest, in contrast, recalls that of the reef edge receiving the full impact of the waves.

Overall, the peak in species richness – and diversity – in forest types of the perhumid tropics, is reached where canopy disturbance is modest to moderate, caused either by drought or squalls; that is, in MDF in Borneo on deep yellow humult sands or short lattice silty clay where gaps are predominantly small and frequent, and in the southeastern Philippines where typhoon damage is rare.

2.2.4.2.4. A conclusion. These canopy disturbance gradients, each correlated with gap frequency and intensity, appear to broadly agree with Connell's hypothesis[78] that community species richness and diversity peaks at intermediate levels of disturbance. Connell maintained that diversity is reduced under rigorous marine environments – such as the upper tidal limit which experiences daily drought and the upper edge of the reef where wave action is strongest – but also at low tide and in calmer reef waters, where one or a few species come to dominate the rest. Intermediate disturbance levels maximise the number of dynamic phases to which species can adapt, while at the same time preventing competitive exclusion by a few. In Asian forests, this predicted peak in species richness occurs in the perhumid tropics, where climatic effects are kept constant. Within them, the *least* diverse MDF in Sarawak (on the deep friable basalt-derived soils of Bukit Mersing, with Fisher's α at less than 55) is comparable in species diversity to that of the drought-prone coastal dipterocarp forest at Bukit

Timah near the other end of the canopy disturbance gradient within calm climates, consistent with Connell's hypothesis. However, in species richness Mersing surpasses the dipterocarp forests in seasonally dry Khao Chong and Huai Kha Khaeng, both of which experience occasional typhoons (Table 3.7). Therefore, where large-scale disturbances such as wind, drought or fire increase in frequency or intensity, often associated with increasing climatic seasonality along a geographic gradient, their influence as mediators of overall species richness come to outweigh the influence of the events that create intermediate disturbance levels at the local scale such as lightning, topography-correlated drought, landslips and wind-throw.

Data from the Philippines and Taiwan are insufficient at present to firmly conclude that intermediate levels of canopy disturbance mediate the change in α-diversity down their eastern coasts from north to south. *Particularly, we lack comparative data from leeward Philippine sites with similar rainfall regimes to those experiencing typhoons, but which have low rates of typhoon canopy catastrophe.* The Philippines-Taiwan example does imply that species diversity declines in parallel with the increasingly simplified canopy structure typical of extreme disturbance levels. That decline is not, however, due to restriction of species to the pioneer phase in an environment of frequent catastrophe. Rather, it seems due to structural simplification of forest canopies and tree crowns in response to gale-force winds through resistance rather than tree fall. That results in a fusion of the emergent with the main canopy, and a decline in species diversity of each, and a decrease in forest stature reducing the opportunity for niche differentiation by asymptotic height beneath the canopy (Chapter 5). Species diversity in MDF is therefore indeed greatest where canopy disturbance is at an intermediate level. This peak in diversity could be interpreted in two different ways:

1. On the one hand, intermediate canopy disturbance levels provide the greatest diversity of light climates in the mature phase subcanopy, within which climax species can adapt and partition light-response niches.

2. On the other hand, the early successional phases in forests on either side of the peak in diversity, whether it be along gradients of topography and soil within the humid tropics or the gradient of increased seasonality, take the form of an archipelago of 'islands' that are relatively small compared with the larger areas of late-successional and mature phases. Thus, the peak in diversity would be found in landscapes where the coverage area of successional phases approaches that of the mature phase, permitting greater accumulation of pioneer and especially successional species within them.

2.2.4.3. *Variation in species diversity with soil nutrient levels.*
Our evidence indicates that species distributions are largely determined at the regional scale by rainfall seasonality, but within a given rainfall regime by soil nutrient levels and – perhaps to a lesser extent in wet climates – by frequency and intensity of soil water stress (see Chapter 3). Similar results are now being reported from the Neotropics.[79] In Sarawak MDF, probably in Peninsular Malaysia and elsewhere in the wet aseasonal or seasonal tropics, the primary influence of soil nutrients appears to be to create a distinction between fertile non-humult loams lacking surface raw humus, and less-fertile humult soils on which a densely-rooted surface mat of raw humus is present (Chapter 3). The regeneration niche (that is, the physical and chemical environment in which seeds germinate and into which seedlings establish) is substantially different on these two classes of soils. Regions in which these classes co-occur consist therefore of 'mosaics' or 'archipelagos' of different habitat types, generally delineated by the underlying lithology and each with characteristic species assemblages (Chapter 6).

Unlike the large CTFS plots, the 105, 0.6 ha plots in lowland MDF in Sarawak were carefully placed in 13 uniform topographical sites, along ridges or on slopes. Some attempt was made to confine them within mature phase stands. As we saw in Chapter 3, nutrients are correlated with variation in species composition, particularly on the less fertile soils (Fig. 3.13). Species diversity in these stands also showed a distinct pattern in relation to soil nutrients, especially phosphorus, magnesium and calcium. In Fig. 7.2, Fisher's α is plotted against three mineral soil nutrient concentrations, on a logarithmic scale in which low values are accentuated relative to high. Maximum diversity values reach a peak toward the low end of the nutrient range in all cases.[80] (Similar peaks have been reported in Central America.[81]) The *maximum* values of Fisher's α peak along soil nutrient gradients at spatial scales both of 400 m² (the scale of 20 × 20 m subplots within the CTFS 52 ha Lambir plot) (Fig. 7.3ab, ad) and 6,000 m² (the scale of the 105 Sarawak plots), from which I infer that nutrient limitation *does* influence species diversity. However, the scatter, in the 105 plots, of diversity below these maxima (indicated in Fig. 7.2), and the decline of maximum richness at higher nutrient levels, may imply that emergent density – therefore forest structure and dynamics rather than direct interspecific competition for soil resources – is also driving richness. At Pasoh and Lambir overall, the values of α at the 400 m² scale, 14–191, vary in a manner that appears related both to soil nutrients, and to topography, and therefore, in these cases, to drought-induced canopy heterogeneity (Fig. 7.3a, b). Variation at this scale at Lambir is the more chaotic. In part it is difficult to interpret and possibly an artefact of the small sample size. But the maximum values still relate in

the same way as in the 0.6ha plots. Importantly, the Pasoh range of α at this scale, 42–162, is entirely within the Lambir range. This reflects the greater homogeneity of the Pasoh ecosystem, but also the fact that, although only 66% of species are shared, a similar level of species diversity is achieved in similar habitats. This is consistent with the hypothesised existence of a species community diversity asymptote.

There remains much variation in diversity within each soil type, as in the 105 plots of 0.6 ha. That is apparently due to variation in successional status (that is, the degree of past canopy disturbance) in the stands represented (see section 2.2.4.1 above). Correlates of soils and diversity also relate to the maximum, and therefore range, of growth rates achieved by species on soils of different nutrient availability. The *highest* growth rates occur among pioneer and early successional species and on the most fertile friable clay soils; among the 0.6 ha plots in Sarawak MDF, these are at Mersing (Table 2.7b). The *lowest* growth rates are on the low-nutrient humult sands of Bako. *Although data are as yet unavailable, the* narrowest range of growth rates are likely to be found in *kerangas*, that is on freely draining oligotrophic soils, because maximum rates are lowest there. Its mature phase lacks emergents, and has low stature, with high height/ diameter ratios (Chapter 2). Predominantly single-tree canopy mortality there provides unfavourable conditions for pioneer establishment, and pioneers are anyway rare or absent. Species diversity, again, is not where the range of growth rates is greatest, but where it is relatively, but not extremely limited, in MDF on the leached humult yellow sandy and udult silty clay soils (Chapter 2).

Existence of a distinct pattern of species richness correlated with variation in a factor of the physical habitat contravenes a fundamental requirement of neutral theory, namely that species distributions are limited solely by seed dispersal constraints. On the other hand, the peaked distribution of maximum species richness values in the forests' mature phase against soil nutrient concentrations was predicted in one of the leading hypotheses proposed to explain patterns of variation in species diversity. David Tilman reasoned that diversity will be low in any habitat in which resources (such as nutrients) are limited, because the range of options for differential resource exploitation is constrained, limiting the number of available ecological niches.[82] As nutrient availability grows, variations in availability among the many essential nutrients also grow, forming a spatial mosaic of nutrient availability within which there are increasing opportunities for niche specialisation. Tilman recognized that, as nutrient (particularly nitrogen) availability grows, and where other limiting factors (notably water stress) are low or constant, conditions will increasingly favour those species that are most competitive in capturing light. Tilman argued that increased maximum growth rates of a few canopy species will suppress species richness overall.[83] In that, Tilman and Connell agree. Tilman's hypothesis is consistent with Sabrina Russo's observations that species of low-nutrient soils compete through survivorship rates, whereas those on high-nutrient soils do so through growth rates (Chapter 3). In mature tree communities, the cost of height growth trades off with that of developing an overarching, light-excluding, leafy crown which outshades competitors. We have seen in Chapters 2 and 3 that such crowns gain advantage in rainforests, especially among the early pioneers, wherever soil water is reliable and eutrophic. Species with greater shade tolerance replace the pioneers during succession by investing more in height growth, and only later also by developing dense expansive crowns (Chapter 2). Eventually, this suite of species dominates all others in the quest for light, and overall species diversity declines. This decline with an increase in canopy dominance (see section 2.3.1.3 below) has been confirmed in the Asian tropics,[84] as nutrient concentrations increase to the maximum observed on the basalt substrate at Bukit Mersing.

Field data to help elucidate the precise relationship between tree performance and diversity in relation to nitrogen, the nutrient most closely correlated with growth, is notoriously difficult to collect. This is in part due to the great instability of nitrogen in soils, and to the rapidity of the chemical changes that ensue following sample collection over weeks and during transportation – formerly often downriver in open boats over many days – to the laboratory (Prologue Plate 2b). (That is the reason why I was unable to obtain reliable nitrogen estimates from my Sarawak plots). Field and shade-house experiments, however, have yielded results broadly consistent with Tilman's own experiments,[85] and this 'resource-ratio hypothesis' has become accepted as theory through testing in a variety of temperate sites.[86]

Returning to the Asian tropics, we had found evidence that nutrients are more tightly correlated with species distributions on the low and intermediate soil nutrient levels than on the more fertile loams, and that correlations between species composition and nutrient availability are lost on the most fertile soils over basic volcanic rock (Chapter 3, Table 3.4). Trenching experiments in low-nutrient Amazonian *caatinga* (*kerangas*) forest have yielded increased growth following root separation – evidence for stiff below-ground competition.[87] However, Tilman first tested and confirmed his hypothesis using unicellular aquatic algae in tanks to which varying concentrations of nutrients were added. It beggars belief that, for mixtures of over 500 species of large sessile organisms, dispersal limitations would not constrain opportunities for individual species to occupy (and specialise within) patches offering distinct nutrient levels![88]

2.2.4.4. Summary. The extraordinary number of tree species within a particular rainforest habitat therefore means that Tilman's resource ratio hypothesis cannot be the sole explanation for how such species richness is maintained. It is also obvious that forest stature and structure, and canopy disturbance regimes, are major influences, but variation in them appears to be itself influenced by soil nutrient levels as well as physical characteristics. Our data could nevertheless be consistent with Tilman's hypothesis, if spatial partitioning were taking place not among individual species but via more general competition among 'functional groups' of species whose attributes allow them to persist at particular nutrient levels. These functional groups are the 'species guilds' whose members occupy the different vertical strata at their maturity: structural diversity is greatest under conditions of intermediate disturbance. Species diversity is therefore greatest where intermediate levels of nutrient availability and disturbance coincide. In such a scenario, individual species within each functional group might show high (though not necessarily complete) ecological equivalence in respect of competition for soils resources. Soil nutrients can therefore enhance species β-diversity (that is, the composition and number of species turning over *among* nearby habitat patches), but hardly α-diversity (that is, variation in species diversity *within* communities and habitats uniform in all respects other than canopy disturbance level). Nevertheless, the number of interspecific encounters between individuals of any given pair of species must tend to increase over time, and their consequences vary with changes in light conditions occasioned by each disturbance regime. The possibility of such competition leading to changes in gene frequencies in the more abundant species cannot therefore be discounted.

As a perhaps oversimplified hypothesis, I suggest that the level of disturbance (that is, the size, nature and frequency of canopy gaps, and stand response to drought) determines a forest's structural complexity and the relative representation of its structural guilds; while nutrient levels – when below a certain threshold – influence both the species composition and diversity of those guilds. For example, in the Lambir CTFS plot the comparatively low-nutrient deep yellow sandy humult soils harbour higher species diversity, but diversity is reduced when samples include recent canopy gaps, especially when caused by landslips (Fig. 7.3ab). The correlation between species diversity and soil nutrients does not, therefore, exactly parallel the correlation between species diversity and degree of canopy disturbance. Both independently lead to peaks at intermediate levels, but those peaks only sometimes coincide within a single habitat. The influence of each factor may therefore be identified separately.

The influence of soil nutrient and disturbance levels on species diversity has yet to be specifically defined. From our MDF data overall, I infer:

1. Low nutrient levels reduce maximum growth rates, while the pioneer guild is increasingly reduced (Chapter 3). The soils are generally freely draining and liable to relatively frequent water stress, resulting in low forest stature, with emergent species absent or barely emergent at maturity. Structural guilds are thereby reduced. Dense interlocking surface roots in the raw humus, and deep albeit sparse feeder roots, result in predominantly single-tree canopy gaps caused by drought and lightning. These, combined with smaller canopy leaf sizes, lead to an understorey light climate that is generally sufficient for survival only of more shade-tolerant species.

2. Moderately low-nutrient humult sandy upland soils are freely draining, but variably deep and therefore variably liable to drought. Where the canopy is tall and heterogeneous, it forms a sufficiently variable mosaic of understorey light to differentially influence species survivorships and thereby enhance diversity.

3. Moderately low nutrient short lattice udult clay and silty clay soils are less freely draining, and their rooting mostly shallow. This is associated with similar or somewhat denser canopy foliage (Fig. 2.18), but more frequent and variable, generally somewhat larger canopy gaps achieved by more windthrow than on humult soils. Their species composition differs from that of humult soils. High structural and species diversity is achieved here by high diversity of subcanopy light climate.

4. Fertile clay loams yield exchangeable nutrient levels above the threshold below which they selectively limit species survivorship. They are friable and, where deep, may never experience water deficits. Here, species composition and richness is determined by the long-term regime of canopy disturbance, frequency and scale, in which occasional blow-downs, sometimes large, and landslips are the leading drivers. On the most fertile and deep loams, species with exceptional maximum growth rates outcompete others and thereby suppress richness, initially by more rapid height growth, and then by outshading with dense foliaged crowns.

5. Species richness is therefore greatest on soils of intermediate nutrient levels, and where levels of disturbance result in the greatest dynamic and structural complexity.

Both Connell and Tilman are right!

2.2.5. Enrichment of richness through subsidy

The two main soil classes in MDF, humult and udult, are each well represented within the Lambir CTFS 52 ha plot in Sarawak. Plenty of opportunities exist for the accidental establishment of individuals in soils where their species cannot support a self-reproducing population. This enriches communities on each soil class. This process of community enhancement along ecotones by continuous subsidy from source populations was first identified in Mediterranean plant communities by Schmida and Wilson,[89] and then defined theoretically by Pulliam.[90] This is known as the 'Pulliam effect',[91] and was seen to influence the number of co-occurring *Shorea* (sect. *Mutica*) species at Semengoh and Lambir (Box 5.1).[92] Another example has been documented among eight species in two genera of Sterculiaceae, whose populations inhabit the diverse topography and soils of the Lambir CTFS plot.[93] Sabrina Russo,[94] also at Lambir, was the first to demonstrate that the Pulliam effect, through partial competitive exclusion, is a major contributor to the stock of rare tropical tree species in a landscape where a mosaic of nutrient-mediated β-diversity is high. Russo's observations are also consistent with the lottery hypothesis of Chesson and Warner, which predict that richness is enhanced where opportunities for establishment differ among patches due to differing opportunities, in their example for seed to reach them.[95] Potts' demonstration that rare species at Lambir (many of which would be pioneers owing to the small area of gap there) increased following a killing drought is similarly consistent. It provides a further explanation for the higher maxima of Fisher's α at Lambir than in the more edaphically and topographically uniform Pasoh plot.

At the Lambir plot, a range of major soil classes encompasses each of two geological lithologies (and formations), shale and sandstone. A conservative approximation of the number of species able to maintain viable populations on each can be estimated by reference to two groups of one acre plots, 50 each at Andulau Hills and Kuala Belalong in adjacent Brunei (the two sites c.60 km apart). Each of the Brunei sites is situated on one or the other of the two geological formations within the Lambir 52 ha plot, and is surrounded by the associated soil type to a radius of at least 15 km. Therefore, species confined to one or the other soil type and also occurring at Lambir provide a conservative estimate of the soil-specificity of a subsample of Lambir species. From this, the number of species within a community on a specific habitat can be estimated.

It turns out that about a third of each of the Brunei site's species are shared, therefore found on *both* soil types. This is similar to the proportion shared in the same two soils at Lambir. Differences in species composition correlated with distance between sites, which could result from constraints on seed dispersal, are therefore minimal. This implies that the Pulliam effect is *not* strong where markedly differing soils abut.

Species name	Soil preference	No. plots occurring	Sites (no. plots)
Shorea macroptera ssp. *baillonii*	H	25	G(13), H(10), N(2)
Koilodepas longifolium	G	24	C(50), E(2), G(3), H(7), J(4), L(1), N(2)
Vatica micrantha	H	22	G(9), H(3), K(1), L(3), M(4), N(2)
Allantospermum borneense	H	22	G(8), K(3), L(3), M(5), N(3)
Shorea balanocarpoides	G	19	G(12), H(3), J(2), M(1), N(1)
Dryobalanops aromatica	H	17	G(9), K(3), L(4), M(1)
Dryobalanops lanceolata	U	17	C(9), E(8)
Callerya (Millettia) nieuwenhuisii	G	16	E(9), H(3), J(1)
Mangifera parvifolia	H	15	F(1), G(3), H(3), L(6), M(2)
Diospyros curranii	U	15	C(6), E(7), J(2)
Teijsmanniodendron sarawakanum	U	15	C(14), D(1)
Dipterocarpus caudiferus	U	13	C(11), J(2)
Dacryodes expansa	H	12	G(5), K(3), L(4)
Dryobalanops beccarii	H	12	B(2), D(1), F(3), H(4), N(2)
Elateriospermum tapos	G	12	E(7), G(1), K(2), L(1), N(1)
Mallotus wrayi	G	11	C(8), H(3)
Gonocaryum gracile	H	11	F(2), H(2), K(2), M(3), N(2)
Diospyros sumatrana	U	10	C(4), D(1), E(5)
Mesua myrtifolia	H	9	G(6), M(2), N(1)
Hopea andersonii ssp. *basalticola*	U	9	C(9)
Drypetes microphylla	U	9	C(9)
Syzygium valdevenosum	U	9	A(2), C(7)
Alangium javanicum	U	8	C(2), E(6)
Polyalthia cauliflora	U	8	C(8)
Lophopetalum glabrum	G	8	C(1), G(4), H(3)
Dipterocarpus globosus	H	8	K(2), L(5), M(1)
Pimelodendron griffithianum	G	8	E(1), F(2), G(3), H(2)
Gymnacranthera contracta	H	8	G(2), K(3), M(1), N(2)
Shorea macroptera ssp. *macropterifolia*	G	7	D(1), G(3), J(1), N(2)
Shorea pinanga	G	7	H(6), J(1)
Shorea faguetiana	H	7	G(1),H(5),N(1)
Hancea penangensis	G	7	A(1),E(1),H(2),J(2)L(1)
Hydnocarpus polypetala	G	7	C(4), E(2), H(1)
Hydnocarpus woodii	U	7	C(5), E(2)
Eusideroxylon zwageri	G	7	C(4), E(2), H(1)
Bouea oppositifolia	H	6	L(1), M(5)
Hopea dryobalanoides	U	6	C(4), G(2)
Shorea polyandra	U	6	E(5), J(1)
Hopea mesuoides	G	6	J(2), N(4)
Shorea faguetioides	G	6	G(2), H(2), J(2)
Shorea parvifolia	G	6	D(1), E(3), J(1), K(1)
Shorea pauciflora	H	6	G(2), H(3), K(1)
Ptychopyxis arborea	U	6	C(5), E(1)
Mallotus leucocalyx	U	6	C(6)
Madhuca barbata	H	6	B(1), F(1), G(4)
Teijsmanniodendron holophyllum	H	6	H(4), N(2)
Melanorrhoea wallichii	H	5	B(2), G(1), H(2)
Gluta laxiflora	H	5	L(1), M(4)
Orophea sp.	U	5	A(1), C(3), E(1)
Popowia pisocarpa	U	5	C(5)
Dipterocarpus rigidus	H	5	M(5)
Shorea pilosa	H	5	G(1), H(4)
Shorea rubra	H	5	H(5)
Shorea xanthophylla	U	5	C(4), L(1)
Vatica odorata	G	5	C(1), H(4)
Vatica vinosa	G	5	A(1), C(3), N(1)
Hancea (Mallotus) griffithiana	H	5	G(3), M(1), N(1)
Cryptocarya densiflora	U	5	C(5)
Nephelium mutabile	U	5	C(4), E(1)
Teijsmanniodendron sinclairii	H	5	G(5)

Table 7.7 Species rank-order of abundance at 14 sites (A: Andulau, B: Gunung Santubong; C: Bukit Mersing, Anap; D: Ulu Dapoi, Tinjar; E: Bok Tisam forest, lower Tinjar; F: Bako National Park, G: Bukit Iju, Arip, Balingian; H: Bukit Raya, Kapit, Rejang; J: Ulu Bakong, Tinjar; K: Lambir National Park, north slope; L: Lambir National Park, south slope; M: Nyabau forest, Bintulu; N: Segan forest, Bintulu.) in MDF throughout Sarawak. Species among the ten most abundant in more than five of 105, 0.6 ha plots; trees >9.6 cm dbh. H: humult soil specialist, U: clay loam (udult) specialist, G: generalist. (After Potts et al. 2002)

Chapter 7

Forest and tree taxon diversity: why does it vary, and how is it maintained?

This is most evident in north-west Borneo in the dramatic and sudden floristic change that accompanies the transition from MDF, on yellow soils, to *kerangas* on podsols (Chapter 3). A pronounced Pulliam effect would imply relaxed interspecific competition, as is assumed by neutral theory. But Russo and Yamada showed that soil specificity increased with tree size as a consequence of competition.[96]

2.3. *Rank order of abundance*

Neutral theory, because it assumes that co-occurring species are ecologically complementary, predicts that there should be no consistency in the rank order of abundance of species populations within a community other than at short distances where seed dispersal limitations produce a concentration of the leading species progeny. We have already shown that species composition is surprisingly consistent among communities sharing the same soil in Bornean MDF, irrespective of distance between sites (Figs 3.11, 3.13; Table 3.3).[97]

2.3.1. *Dominance*

Some species are common in nature, others are rare. Among the more common species we often find a considerable degree of consistency in their rank order of abundance, even in hyper-diverse MDF.[98] Forty-seven species in Table 7.7 occurred among the 10 most abundant in at least two sets of plots more than 100 km apart, as well as locally. This consistency can only be explained by parallel ranking in competitively significant aspects of their life history traits – which is again contrary to neutral expectations. Numerical dominance of one species over others in mixture necessarily reduces the available space for other species occupying the same spatial guild, thereby reducing overall species diversity. Dominance by a dense-crowned species may reduce species richness by outshading. A clear example is afforded by the exotic weedy shrub *Lantana camara*, whose distribution in the intermediate phase between tall and short deciduous forest at Mudumalai is associated with lower species diversity than that of short deciduous forest where it is absent.[99] A curious characteristic of upland tropical wet evergreen forest everywhere – and a partial cause of its extraordinary diversity in coexisting species – is the general absence of individual species achieving sole dominance in canopy or understorey. There are exceptions, though, and these demand explanation.

2.3.1.1. *The dominance of the dipterocarps.* The consistent canopy dominance of family Dipterocarpaceae in some Asian forests is unmatched by other families elsewhere in the tropics (Figs 2.1, 7.6). In Asian lowland forests, Caesalpinioids are numbered among the tallest trees, but nevertheless appear to be outcompeted by dipterocarps for numerical dominance

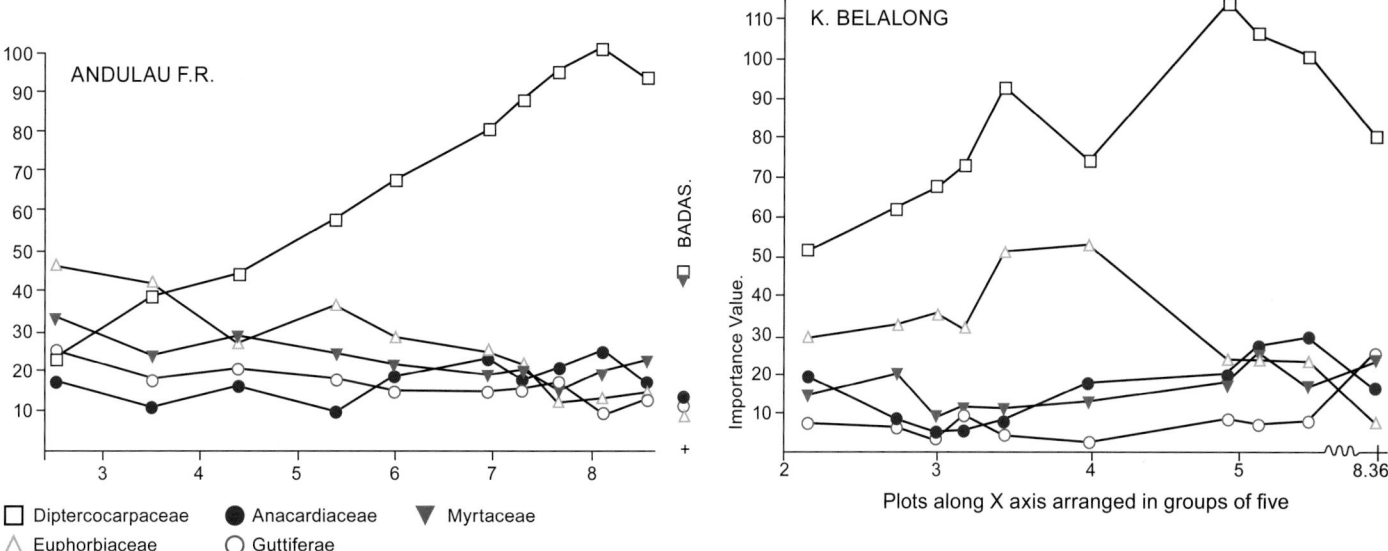

Fig. 7.6 Relative dominance (Importance Value) of Dipterocarpaceae and four other leading families, along the primary axes of Bray and Curtis ordinations, in MDF in two Brunei forests: Andulau (yellow sandy humult soils), and Kuala Belalong (clay loam soils); trees >9.7 cm dbh. The axes follow topography from valley (left) to ridge; dipterocarps increase to the right. (From P.S. Ashton 1964a)

(Plate 2.2a). The cause may be inferred by examining differences between forests where dipterocarps are dominant and those situated close by where they are not. Dipterocarp family dominance is confined to lowland yellow/red soils, and climates in which each month (on average) receives more precipitation than it loses in evapotranspiration. As a family, dipterocarps do not dominate the canopy of the distinctive *kerangas* forest, and their dominance is reduced (and often confined to a single dominant species) in wet lowland forests with a seasonal rainfall distribution. Individual dipterocarp species, usually drought-adapted, do achieve local dominance in both cases, while sal (*Shorea robusta*) is the regional dominant of the deciduous forests on the mostly siliceous freely draining soils, low in calcium and phosphorus, of North-East India (Chapters 2 and 3).

Asian dipterocarps, that is subfamily Dipterocarpoideae – in contrast with American and African species – differ in an important respect from the other tropical tree families with co-evolved mycorrhizal dependencies and tendency to canopy dominance. Excepting the locally dominant *Dryobalanops* in Borneo, their family dominance can be achieved *without* any significant loss of overall community species numbers (that is, α-diversity). Only in the severely limiting habitats of deciduous forest and peat swamp forest is complete dipterocarp monodominance seen (see below, also Chapter 3). For this reason, I distinguished the lowland dipterocarp forest on zonal upland soils as mixed dipterocarp forest (MDF).[100] No hypothesis has yet been presented to explain why Asian dipterocarps differ in this respect. I suspect, though, that it may be connected to two unique attributes among Dipterocarpoideae – but not other dipterocarp subfamilies – namely uniquely synchronised mass flowering and mast fruiting combined with their possession of resin canals and copious resins, whose diverse terpenes and other phenolic compounds offer such vast opportunities for the co-evolutionary diversification of chemical defences. These may also provide extensive opportunities for diversification of their ectotrophic mycorrhizal associations. Individual fungal taxa appear to be associated with more than one dipterocarp genus,[101] but W. Smits observed different carpophore (that is, mushroom) assemblages following the roots of co-occurring sister *Shorea* species in East Kalimantan.[102] This might reflect differential competition for nutrients. Conifers, by way of comparison, also have resin canals but lack the physiological capacity of dipterocarps and other angiosperms. This allows the latter to maintain higher growth rates under soil water deficits and to outshade most conifers (see Box 4.2). Vesicular-arbuscular mycorrhizae are ubiquitous in tropical forests,[103] but are thought to have low host-specificity.

Subfamily species richness is concentrated on appropriate soils in regions where dipterocarps have long existed (Chapter 6), such as Malesia west of Wallace's Line (notably the Riau Pocket, especially northern Borneo) and south-west Sri Lanka.

Species numbers dwindle to a handful, or even to a single dominant species, under certain conditions:

1. in perhumid climates on podsols and peat (Chapter 3);

2. in upper dipterocarp and lower montane forests on raw-humus bearing soils where they come into competition with Fagaceae (Chapter 4);

3. in deciduous forests; and

4. in eastern Malesia, where they arrived more recently.

The tendency toward dipterocarp dominance in the aseasonal lowlands is likely enhanced by their unique reproductive biology (further described in Chapter 5). Three specific aspects may be involved:

1. A capacity to saturate seed predators, generating vast drifts of seedlings which, being resinous, are unpalatable.

2. A generally very limited pattern of seed dispersal, facilitating reinvasion of complex ectomycorrhizal guilds from nearby parental roots.

3. Seedlings may on that account exhibit an unusual pattern of negative density-dependence, that is, higher survivorship at higher densities (see 2.5.3. below).[104]

4. A capacity for sustaining coexisting populations of closely related species, each likely exploiting their community's light environment in different ways, through sequential flowering and barriers to hybridisation.

In the seasonal tropics, dipterocarp canopy species do not fruit synchronously or flower supra-annually. The lack of that defensive 'strategy' against seed predation may largely explain why dipterocarp regeneration is so much less abundant in seasonal regions, and therefore why family dominance is not attained. *Careful long-term demographic research at the CTFS 25 ha plot at Khao Chong in peninsular Thailand, under a three-month dry season, should further illuminate dipterocarp demography.*

Several dipterocarp species do achieve monodominance in the emergent canopy (Plate 7.2):

1. Shade-tolerant heavy hardwood species of *Shorea* sect. *Shorea*, balau, achieve numerical dominance through drought tolerance combined with longevity: *Shorea robusta*, sal, in deciduous forests with more than four dry months in South Asia. Balau species may attain local dominance

Plate 7.2 Some dipterocarp species frequently occur in monodominant stands. **a,** *Dipterocarpus alatus* on floodplain alluvium in peninsular Thailand, where it is the sole dipterocarp represented; **b,** A selectively logged stand of *Dryobalanops rappa* in the Anduki Forest Reserve peat swamp, Brunei. Here too, this species becomes dominant as the back of sea beaches descend beneath the peat, a habitat lacking other dipterocarps. But all *Dryobalanops* tend towards local monodominance; their shade tolerance yet strong response to sunlight, and relatively low seed predation rates may together explain their success; **c,** *Dipterocarpus zeylanicus* in species-poor MDF at Kottawe Forest Reserve, south-west Sri Lanka; **d,** *Shorea trapezifolia* forms monodominant 'drifts', here in flower along the southern slopes of the Peak Sanctuary foothills, at c.800 m, south-west Sri Lanka; as does its lowland sister species, *S. congestiflora*. Seedlings of both species are light demanders whose seedlings experience high mortality in shade. But both also flower annually and experience low seed predation rates, enabling them to occupy landslips and other large canopy gaps. (All: P.A.)

in coastal forests and *kerangas: Shorea glauca, S. materialis, S. falcifera* (see Chapter 2). Likewise cengal, the emergent durable heavy hardwood-bearing *Neobalanocarpus heimii,* can become the dominant emergent on the freely draining sandy loams of the upper alluvium in Pasoh MDF, where species richness is also lower (see section 2.2.4.1 above).

Macrophyll *Dipterocarpus tuberculatus* (Plate 2.32a,b) and *D. obtusifolius* (Plate 3.16b) gain dominance in Indo-Burmese deciduous dipterocarp forest; *Dipterocarpus costatus, D. turbinatus* and *D. alatus* (Plate 7.2a) in seasonal evergreen forests of peninsular Thailand and the coastal forests of Cambodia; *D. zeylanicus* in Sri Lankan MDF (Plate 7.2c), and *D. indicus* in the peninsular Indian seasonal evergreen dipterocarp forest.

2. A dramatic case is the monodominance of alan (*Shorea albida*), which until recently covered thousands of square kilometres within the north-west Borneo peat swamps (Chapter 3, Plates 2.21, 3.9a-g). The related species *S. balangeran* similarly dominated many coastal peat swamps of south Borneo and east Sumatra. The swamps at the margins of the vast raised bogs (up to 60 km in diameter) support as

many as five co-occurring dipterocarps – all *Shorea* species – but alan and *S. balangeran* are not among them. *Each of these species appears to have evolved ecophysiological characteristics yet to be investigated* which permit it to occupy a habitat at maturity in which there are no competitors. I infer that alan and *S. balangeran* are competitively excluded from the surrounding mixed swamp forests, but dominate the bog interiors on account of unique, but as yet unknown, ecophysiological qualities.

3. Four analogous examples are:
   i. *Shorea gardneri* which, uniquely among dipterocarps, dominates the lower montane forests of perhumid south-west Sri Lanka (Chapter 4, Plate 4.3a), thereby occupying the role of the Laurasian family Fagaceae, which are absent in South Asia, elsewhere in tropical Asia;

   ii. *Shorea curtisii*, monodominant in the Peninsular Malaysian hill and coastal hill dipterocarp forests on the freely draining granite-derived soils, where the rich dipterocarp flora of similar soils in north-west Borneo is local or missing (Plates 3.2d, 4.8a);

iii. Dryobalanops aromatica, formerly occurring in extensive monodominant stands on sandy leached soils up the east coast plains of Peninsular Malaysia and Sumatra where, again unlike in north-west Borneo, other dipterocarps of similar ecology are absent; and

iv. Shorea selanica, one of only three *Shorea* east of Wallace's Line, which dominated the lowland forests of Ceram in the absence of congeneric competitors.

5. The seven species of kapur (*Dryobalanops*) are exceptional among Far Eastern dipterocarps; all are confined to perhumid Sundaland. All occur in north-west and north-east Borneo, where each occupies a distinct habitat with no overlap at maturity (Fig. 3.2a). In its own habitat, each forms gregarious 'drifts' in which it is dominant in the canopy, although only in the case of the peat swamp species *D. rappa* is it the sole canopy species, with relative densities comparable to some Caesalpinioids in African rainforests (Plate 7.2b).[105] *Dryobalanops* species are emergent but for the north-east Borneo endemic *D. keithii* and the Peninsular and Sumatran form of *D. oblongifolia* (subsp. *occidentalis*). The two best-known species, *D. lanceolata* and *D. aromatica*, both achieve local dominance in the CTFS Lambir plot MDF (Chapter 3, Fig. 3.5). Although statistical tests failed to show a significant negative correlation between population densities of either species and Fisher' α, both do occur in habitats where Fisher's α is lower than on similar soils nearby.

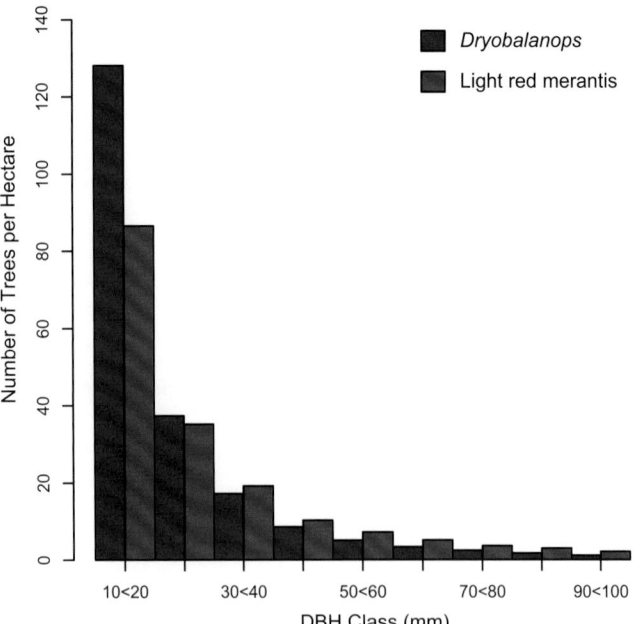

Fig. 7.7 Mean juvenile population structure of *Dryobalanops aromatica* and late successional light red meranti *Shorea* species compared: trees 10–100 cm dbh in Lambir National Park, 52 ha CTFS plot, Sarawak. Blue: *Dryobalanops*; red: light red meranti *Shorea*. (P. Hall)

Dryobalanops species are pollinated by honey bees, and flower more frequently than the ENSO-related mass flowering years (Chapter 5). In the case of *D. aromatica*, Tim Whitmore[106] suggested that dominance might be explained by relatively higher 'reproductive pressure'. This is now supported by evidence from population structure, especially the studies of Akira Itoh at the Lambir CTFS plot which show that they establish higher densities of juveniles than other light hardwood dipterocarps[107] (Fig. 7.7). The two species in the plot have almost exclusive habitats: *D. aromatica* on ridges and upper slopes on sandy soils, *D. lanceolata* on lower slopes on clay soils (Fig. 3.5). Itoh compared seed insect predation rates of the two Lambir *Dryobalanops* with two *Dipterocarpus* species. He found that *Dipterocarpus globosus* had 2–3 times higher pre-dispersal predation rates than the *Dryobalanops*. Between dispersal and the fall of cotyledons 90% of *Dipterocarpus* seeds, 60–70% of *Dryobalanops*, were predated. Thereafter there was little difference. Negative as well as positive density-dependent mortality has been found (see this section and 2.5.3 below).[108] These species are exceptional among light hardwoods for fast growth under high light intensities, yet also for their shade tolerance – in spite of relatively low leaf nitrogen concentrations.[109] While in shade they continue to grow by branching, rather than by extension of the leading shoot,[110] growing little in height but extending long false whorls of horizontally branching small-leaved twigs — in this respect resembling some understorey Annonaceae and *Diospyros*[111] (see Chapter 2). In gaps, growth switches to the leader, and both exhibit rapid growth in diameter and height to the canopy.[112] Whether they change in rank order of photosynthetic gain in growth overall, relative to other light hardwood dipterocarps, is unknown. *The full explanation for this extraordinary plasticity among Dryobalanops seedlings therefore remains a challenge for future researchers.*

In summary, individual dipterocarp species appear to achieve dominance through one of two strategies: either by high survivorship (via competitive advantage and/or high fecundity) *before* reaching maturity (*Dryobalanops* for example), and/or by long survivorship (allowing continued accumulation of individuals) *after* reaching maturity (for example, *Shorea* sect. *Shorea* and *Neobalanocarpus*).

Positive density-dependent mortality will, of course, inhibit achievement of high population densities. Akira Itoh,[113] however, found a negative relationship between mortality and density among juveniles in *Dryobalanops aromatica* at Lambir. The precise cause is unknown, but the pattern could be engendered if juveniles benefit from early inoculation of species-specific mycorrhizal symbionts from their parents. But if this were so, it would be puzzling why other dipterocarps do not gain in the same manner. *Dryobalanops* species, even when dominant, are conspicuously patchy within their habitats, and often absent in landscapes where they might occur. The reasons are as yet unidentified, but presumably involve a pattern of periodic population failure at the landscape scale. Potential examples of this dynamic are the patchy mortality of mature *Shorea albida*, attributed to a moth caterpillar (Plate 5.16b), also observed in the locally dominant main canopy *Mesua ferrea* (Calophyllaceae) at Sinharaja, perhaps attributed to a pathogen at whole-community scale (Chapter 5, Plate 5.18a). *These phenomena deserve systematic enquiry.*

2.3.1.2. Dominance among climax species in limiting environments. Single-species dominance can occur within individual stands all across the nutrient gradient within the perhumid tropics; but dominance at *community* scale is most frequent toward the two ends of the nutrient gradient. Single-species dominance is more common with increasing seasonality (Chapter 3). Such dominance is usually confined to one vertical species guild, and is often associated with depressed species richness within that guild.

On less fertile or more drought-prone soils, one species may be the sole occupant of the emergent stratum. Examples are (1) the conifer *Agathis borneensis* in *kerangas* on raised beaches in perhumid western Malesia, and (2) *Shorea albida, Shorea curtisii,* and *Shorea robusta*. In each of these cases, the dominant species appears to be the only one in its region adapted to sustain itself in the uppermost stratum under the particular constraints of its habitat. In north-west Borneo, *Shorea curtisii* occurs on similar soils to those in Peninsular Malaysia *without* achieving exceptional abundance. This suggests two developments that could occur over evolutionary time: that other species may arrive to compete and reduce dominance in those areas presently dominated by one or two species; or that dominant species may be reduced by disease (Chapter 5). The first of these alternatives implies that monodominant conditions are comparatively young, which is likely in the current extent of the coastal peat swamps, and can be inferred in Peninsular *S. curtisii*

2.3.1.3. Dominance in favourable habitats. Canopy dominance of *Dryobalanops lanceolata* or a small group of light hardwood *Shorea* species (Chapter 3) on well-watered fertile soils occurs in stands with depressed richness in all vertical guilds, as the contiguous crowns of the emergents characteristically form a dense canopy, casting deep shadow beneath (see Chapter 2, also section 2.2.4.2 above). *Shorea gardneri*, though, achieves such dominance in Sri Lanka's lower montane forests on fertile humic loams in the absence of the Fagaceae that is dominant elsewhere. The dominance of teak on fertile soils in mixed deciduous forest may be analogous. The exceptional abundance of the main canopy heavy hardwood known as Borneo ironwood (ulin or belian wi, *Eusideroxylon zwageri*), appears to be a special case. Its large fruit with no known means

of dispersal leads to local aggregation of the species. Achieving dominance in MDF on clay loams in those parts of south-east Borneo and south-east Sumatra most affected by catastrophic ENSO-initiated drought,[114] its success appears due at least in part to extraordinary powers of crown reiteration by epicormic shoots, combined with great longevity (but see Chapter 2).

2.3.1.4. Subcanopy dominance. Another kind of dominance occurs where a single understorey species may outcompete all others and, in the process, suppresses the regeneration of canopy species. In the Asian lowland tropics, the pre-eminent examples are the bamboos and bertam, the stemless palm *Eugeissonia tristis*. Bertam has been known to foresters as the major impediment to successful hill forest regeneration for a century[115] (Chapters 3, 8). It forms continuous populations along ridges in the Peninsular Malaysian granite Main Range and the coastal hills, and is associated with a dominant emergent dipterocarp, seraya (*Shorea curtisii*), whose shade- and drought-tolerant seedlings do survive beneath its fronds where not dense. *Bertam* produces large fruits in low quantities but at a continuous rate; these appear to be dispersed over limited distances. Populations increase following selective logging. *The dynamic relationship between bertam and seraya has yet to be fully deciphered.*

In seasonal Asia, dominance of genera or single species of bamboo in semi-evergreen and tall deciduous forests, and of the speciose shrub genus *Strobilanthes* (Acanthaceae) in the lower montane forests, is so widespread as to define the dynamics of those forests (Chapter 2). Several species of bamboo, in several genera, have similar effects in deciduous and evergreen seasonal forests as far south as the northern Peninsular Malaysian states[116] (Chapter 5). Their dominance is associated with reproductive characteristics in common: synchronous flowering, fruiting and ensuing death, often on landscape scale, at intervals of many years.

Exotic weeds may reduce tree species richness in forests subject to frequent natural or anthropogenic disturbance. The gradual decline in species richness with increasing length and variability of the dry season suffers a dip in the gradient between the moister tall deciduous forests and drier short deciduous forests of India, where the Neotropical shrub *Lantana camara* gains subcanopy dominance following frequent fire.[117]

2.3.1.5. Dominance in frequently disturbed habitats. The intermediate disturbance hypothesis predicts that, where disturbance is frequent or intense, there will be low species diversity, and therefore higher opportunities for single species dominance.

Araucaria hunsteinii achieves dominance in New Guinea on landslips where its light wind-borne seeds arrive early; *Octomeles sumatrana* and *Terminalia copelandii* on fresh fertile alluvium in the Malesian archipelago, and *Shorea congestiflora* and *S. trapezifolia* in mid-succession on landslips in Sri Lanka. They may therefore become dominant under special conditions, such as frequent canopy damage, where slow-growing shade-tolerant species rarely have the opportunity to outcompete them; or they may represent a temporary phase in succession.

2.3.1.6. Summary. Dominance is achieved through one or more competitively superior life history traits, differing among species. In perhaps the most striking example, high fecundity and juvenile survivorship appear to be attributes of the locally gregarious and dominant species of *Dryobalanops*, but *Eusideroxylon zwageri* achieves dominance while producing few large fruits, and probably achieves its dominance by higher survivorship at maturity. There must be a finite energetic cost to achievement of dominance. Why do so few species achieve it? And how can a group of closely related species all, apparently, have the ability? Why does a pathogen not reduce them? *These questions remain unanswered.*

2.3.2. Rare species in the mature phase

Rarity – the occurrence of species at low frequencies – occurs at a range of spatial scales.[118] A species may be regionally rare, or rare at the landscape scale, because its habitat is scattered or occupies only a small area; but such species may be common and even dominant within that rare habitat. The converse attribute and that which is of concern here – unique to speciose tropical lowland forest – is that *most* species persist in low population densities within their optimal communities (Fig. 7.8). Such populations may be disposed in clumps as a consequence of seed dispersal limitations; it is their mean density that is low. I considered the consequences of rarity in the context of population genetics in Chapter 5.

One cause of this kind of rarity has been alluded to in Chapter 2. In habitats experiencing occasional catastrophe, conditions immediately following a catastrophe will favour the establishment and regeneration of a suite of species. As the stand matures, conditions will change and opportunities for their establishment will cease until the next catastrophe. This can occur at the scale of the landscape, or locally as it does in most canopy gaps. Many pioneer species, especially those that grow to canopy height, exist in exceptionally low mean densities. At Pasoh, populations at low densities (whose dynamic roles within the community were not identified) were found to have higher mortality rates in the intervals between catastrophic events.[119] Occasional catastrophic canopy opening may lead to their resurgence, thereby increasing local species diversity. Matthew Potts[120] observed that the lowest-density populations at Lambir showed superior performance following the 1996–1997 drought. He did not identify the species, but

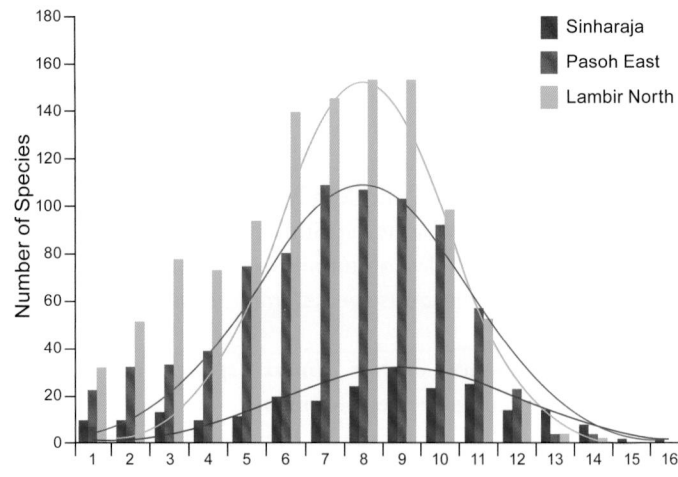

Canonical Distribution of Species Population Densities in 25 ha

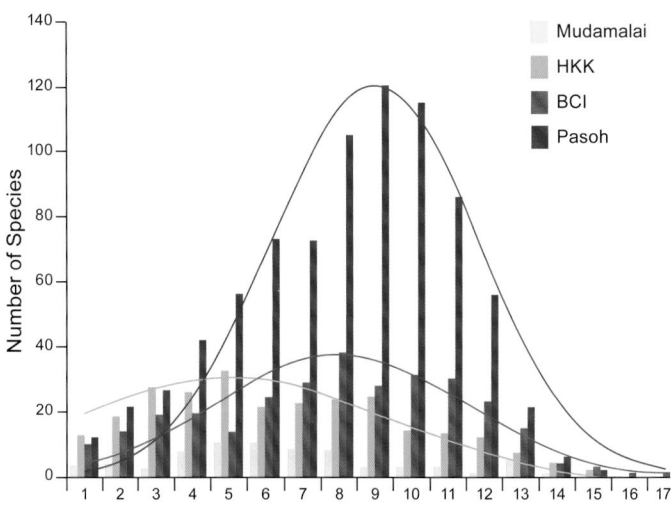

Canonical Distribution of Species Population Densities in 50 ha

Fig. 7.8 The number of species according to their population densities in CTFS plots, arranged by classes of doubling abundances (canonical distributions): **a,** MDF plots within perhumid climates (25 ha plots): blue: Sinharaja (8,200 trees >1 cm dbh ha⁻¹), red: Pasoh (6,700 trees >1 cm dbh ha⁻¹), green: Lambir (6,100 trees >1 cm dbh ha⁻¹). The peaks in species abundance in the three MDF plots differ only slightly, reflecting differences in stand densities less than their different species diversity would imply. **b,** in 50 ha plots in climates of increasing rainfall seasonality: yellow: Mudumalai (353 trees >1 cm dbh ha⁻¹), green: Huai Kha Khaeng (1,450 trees >1 cm dbh ha⁻¹), red: Barro Colorado Island, Panama (4,600 trees >1 cm dbh ha⁻¹); blue: Pasoh (6,700 trees >1 cm dbh ha⁻¹). The peaks of species numbers move towards greater rarity in closed canopy forests of increasing seasonality, correlating with declining stand density. The number of species in very small numbers ('rare species'), the deciduous forest at Mudumalai excepted, differ little among forests of varying seasonality: therefore, species-poor forests have proportionately more 'rare' species. (P. Hall)

it is likely that pioneer and successional species would have responded strongly to the opportunities presented by the increase and expansion of canopy gaps (Chapter 2).

Across the gradient of rainfall seasonality, large canopy gaps of catastrophic origin occur most extensively and frequently in seasonal evergreen and semi-evergreen forests. Patrick Baker and Sarayudh Bunyavejchewin,[121] by cleverly integrating several lines of direct observation and inference, have convincingly demonstrated how the semi-evergreen forest at Huai Kha Khaeng may be interpreted in terms of a history of occasional canopy catastrophe, patchily distributed at the landscape scale (Chapter 2). Massive wind-throws there appear sometimes to have been followed by years of high rainfall, when evergreen species (notably *Hopea odorata*) have regenerated. At other times, blow-downs have been followed by dry weather and fires, leading to strong regeneration of tall deciduous and deciduous dipterocarp forest species (including *Afzelia xylocarpa*, *Chukrasia tabularis*, *Toona ciliata*, *Dipterocarpus obtusifolius* and *Shorea siamensis*), all of which require large gaps for regeneration. Currently, these fire-climax dipterocarps are reduced in semi-evergreen dipterocarp forest at Huai Kha Khaeng to a few large individuals with little regeneration. *Hopea odorata*, whose juveniles are fire-sensitive, is more abundant but almost entirely devoid of juveniles (Fig. 2.21). A consequence of this disturbance regime is an exceptionally high β-diversity, and an exceptionally large proportion of species persisting at low population densities – even higher a proportion than at Pasoh or Lambir. This, however, is insufficient to offset the overall decline in species richness in this seasonally dry site due to limiting factors *other* than disturbance, notably annual drought.

Some rainforest species always occur in low population densities but are nevertheless widespread through Sundaland. All these appear, from my years of field experience as a botanical explorer, to be as habitat – notably soils – specific as more abundant species, and in that respect niche-specific. Wild mango species, and Malvaceae (tribe Durioneae) – including the genera *Neesia*, *Coelostegia*, and some but not all *Durio* (subgen. *Durio*) – are always rare. *Coelostegia neesiocarpa*, for instance, has distinctive large woody fruits that may take years to decompose after falling, yet I only saw that species once in ten years of fieldwork in Borneo; only two collections have ever been made in Sabah or Sarawak. All canopy Durioneae are likely chiefly pollinated by the wide-ranging bat *Eonycteris spelaea*, wild mangoes by the equally wide-ranging giant honey bee (and other bee species, see Chapter 5). These species flower heavily supra-annually, producing large seeds in enormous fruit in low numbers at intervals of several years. The larger-fruited wild mangoes are also always rare in hyperdiverse forests (Chapter 5). The most abundant among the six species in the Pasoh CTFS 50 ha plot has one individual

>1 cm dbh ha⁻¹.[122] Many are widespread also. Their seeds are dispersed by hornbills, primates and other large mammals.[123] Though these trees are apparently rare on account of their low fecundity, they appear able to sustain populations at extremely low densities yet with high outcrossing rates on account of their widely itinerant dispersal agents; but their rarity can only be thus explained if their mortality rates are comparable to their fecundity.

Although many consistently exceptionally rare individual species are found, and in most families, in the forests of perhumid Sundaland, the large-seeded *Cullenia exarillata* (Durioneae) and *Mangifera indica* are often common, even gregarious in seasonal evergreen forests, flowering annually.

2.3.2.1. Survivorship among rare species. A potential explanation for the coexistence in hyperdiverse rainforests of so many species in very low population densities may be that they suffer relatively high levels of density-dependent mortality (section 2.5.3 below; Chapter 5).[124] If widely confirmed, this may prove to be a key mediator of species richness in hyperdiverse communities. It could imply that this great majority of rare species is – in every respect *other* than pathogen-specificity – ecologically complementary *within their guilds*. This suggests the possibility that species that can maintain fecundity in very low population densities, by attracting pollinators and seed dispersers that forage over long distances, alternatively by self-compatibility and including apomixis, may survive in equable climates until the climate itself changes. Such species would be ecologically complementary and would conform with neutral rules.

One such possible category of consistently rare species is a group of species, widespread and often common in the wet seasonal tropics, that occur as scarce individuals, often of enormous size, on fertile clay loams in MDF. Examples among the Dipterocarpaceae include *Anisoptera costata*, *Dipterocarpus gracilis*, *D. hasseltii*, *Hopea plagata*, *H. sangal*, and *Shorea guiso* (see Chapter 6), also the emergent pioneer *Alstonia scholaris* (Apocynaceae) in Borneo. The coastal species *S. glauca* occurs inland near Pasoh Forest (Negeri Sembilan, Peninsular Malaysia). In the latter case, individuals of another coastal species, *Pouteria obovata*, have been recorded in the vicinity.[125] At Tasek Bera, a freshwater lake immediately to the south, peat was found to have initiated above mangrove in the early Holocene period. Water flows out of the lake both eastwards and westwards to the coast, and it appears to represent the remnant of what was a marine channel until c.6,000 BP.

Other such species (such as *Pterygota horsfieldii* (Malvaceae), *Shorea gratissima* and in Borneo *S. guiso*) survive as tiny populations on distant promontories, while still others (including *Chukrasia tabularis*, *Firmiana malayana* with its vermillion bird-pollinated flowers, and the magnificent strangler *Wightia borneensis* decked in mauve flowers) are confined to isolated and often minute populations on basic volcanic soils and basal screes of limestone karst. These species may represent survivors from a climatically more favourable epoch. If so, this implies that some part of α-diversity in communities of the perhumid tropics is attributable to its uniquely equable climate, where extinction of species in small populations is a very slow process in the absence of the killing catastrophes so influential in other habitats.[126] *Comparisons of population genetics throughout the climatic range of such species might prove informative.*

2.3.3. *Summary*

There is more consistency in the relative abundance of the *commoner* species within a specific habitat than neutral theory would predict, and the number of these may be greater in reality because of underestimation due to small sample size. This consistency is due in varying degrees to intrinsic competitive assets of the species themselves and to the prevailing disturbance regimes in the landscapes and communities they inhabit. But a group of species with exceptionally low population densities in the humid tropics, which include some common elsewhere, may inexplicably stave off extinction at very low population levels. Some may even be holdovers from a time when the climate was drier or more seasonal, and they were more abundant (Chapter 6). These species, where rare, may become ecologically complementary to more abundant species when in very low numbers, and conform to the predictions of neutral theory. It remains unclear what proportion of overall species diversity they represent. They may be gradually accumulating over time in equable climates, and a major cause of species richness in regional refugia.

2.4. *Co-occurring series of congeneric species*

We have seen that most of the variation in species diversity among similar forests of a region, for instance the evergreen dipterocarp forests of seasonal Indo-Burma and perhumid West Malesia (also south-west Sri Lanka), is due to variation in the number of co-occurring species within a small number of genera in several different families (Table 3.7). In each forest, the variably speciose genera come from the same small generic pool, and the number of species in congeneric series substantially determines overall species diversity of a community. Furthermore, with a few exceptions, the genera which vary in their number of co-occurring species even remain the same across major biogeographic boundaries such as Chatterjee's Partition and Wallace's Line, where overall community richness changes (Table 3.7). Therefore, this pattern can only be explained by something intrinsic to the genera themselves.

The pool is substantially different between evergreen and deciduous forests because deciduousness and evergreenness are mostly constant within genera, and the speciose deciduous genera are rare or absent from evergreen forests. Semi-evergreen and tall deciduous forests gain some species from each, compensating for their general poverty of species especially in the larger evergreen genera. Not surprisingly, these more closely related species are generally more similar morphologically and ecologically than co-occurring species in other genera and families. If niches are partitioned among them, they will therefore be subject to the most stringent competition. Such species often appear to differ, especially when examined in the herbarium alone, in characters such as flower size or colour. These attributes seem unlikely to influence competitive advantage when, as often, they share the same pollinators. Fedorov[127] suggested that these species series, so characteristic of diverse tropical rainforests, are ecologically equivalent and therefore adaptively neutral. Any evidence for competition leading to niche differentiation among species co-occurring within a community is thus a critical test of the role of neutral factors in structuring species α-diversity.

Certain genera appear neither to form co-occurring species series nor to be rich in endemics. Such genera include those whose seeds are dispersed by primates and other large mammals (such as *Durio*, *Mangifera*, *Nephelium*, among which up to four may co-occur); and genera whose flowers are bat-pollinated (among which *Durio* (subgen. *Boschia*) is again included). Givnish[128] suggested that the dense subcanopy of evergreen forest harbours more sedentary frugivorous birds, resulting in short seed dispersal distances and therefore higher local endemism among bird-dispersed genera. Many congeneric species series do indeed belong to predominantly subcanopy, bird-dispersed genera, including Myristicaceae and Ebenaceae, and the predominantly pioneer *Macaranga*. But greater local endemism among them is not apparent. Other congeneric species series with similar species ranges, however, such as those in Burseraceae and *Syzygium*, present fruit predominantly in the canopy itself, where bird movement is seemingly no more restricted than in less species-rich forests in leaf; and then there are the rich congeneric species series of mostly emergent *Shorea* and *Dipterocarpus*. In Chapter 2, I described field observations and experiments by Mark Ashton and B.M.P. Singhakumara and their students at Sinharaja[129] which showed that co-occurring species within several genera (*Shorea* (sect. *Doona*), *Dipterocarpus*, *Syzygium* and *Mesua*) do indeed partition the physical habitat in relation to gradients of subcanopy light and frequency of soil water stress. In these cases, niche differentiation resulted in differing centres of spatial distribution along the topographic gradient. Nevertheless, there was much spatial overlap. Co-occurrence here is apparently due to the highly variable mosaics of light and water conditions (the former perhaps more important) within the topographical elements of a landscape. However, it could also be due to the

Pulliam effect.

Sean Thomas' demonstration of different asymptotic heights among co-occurring subcanopy and canopy species in several genera at Pasoh was discussed earlier in this Chapter. At Lambir, Yamada and colleagues found that populations of eight species from two genera of Sterculiaceae are partitioned by topography and, when two share the same topography, they differ in reproductive height.[130] Five *Scaphium* species occurring in evergreen forests throughout Sundaland and southern Indo-Burma are partitioned by climate, topography or soil.[131] At Lambir again, juveniles of *Scaphium longipetiolatum* are found in mature MDF stands on fertile clay loam where subcanopy light intensities are low. They branch earlier and bear smaller leaves than juvenile *S. borneense* and *S. macropodum* – both confined to humult yellow sandy soils, where diffuse mature forest canopies prevail with more light in the understorey.[132]

Differences in seed dispersers and pollinators cannot exclude the possibility of otherwise ecological equivalence: they must also have distinctly different attributes that potentially diversify their interspecific competitiveness at some stage of their life cycle. Co-occurring figs (*Ficus* spp.) bear fruit of different crop sizes, fruit dimensions and colours, attracting different though overlapping guilds of frugivores. They are dispersed by birds differing in average size, although there is considerable overlap.[133] Such differentiation, however, appears rare among congeneric tree species. These figs also differ in site preferences for establishment.[134] Three subcanopy *Polyalthia* species at Pasoh are so similar in growth habit and leaf morphology that they can hardly be distinguished when sterile (Chapter 5). Most co-occurring congeners have similar floral morphologies and share pollinators; but Steve Rogstad[135] found that these three species of the archaic family Annonaceae had subtle quantitative differences in floral morphology and were pollinated by different beetles.[136] They also had distinctly different habitats and heights at maturity.

Debski[137] examined the relationship between the spatial distributions of 17 species of the subcanopy genus *Aporosa* (Euphorbiaceae) at Lambir and Pasoh. He concluded that a quarter of the populations were spatially differentiated. They also vary a lot in maximum size achieved. At Lambir, 15 species of *Macaranga* occur within the 52 ha CTFS plot (Plate 7.1). Remarkably, 13 of these are pioneers and all but one of them require mineral soil for seedling establishment. Although their seeds do possess dormancy, they establish preferentially in the pits resulting from wind-throws or on landslips where rain has washed the soil surface. Stuart Davies demonstrated, by field observation and shade house experiment, that each species had a distinct albeit overlapping range along gradients of (1) physical

Table 7.8 Population densities, given as numbers of individuals >20 cm dbh ha⁻¹, of *Shorea* sect. *Mutica* species in their chosen habitats in four Malaysian MDF: Pasoh Research Forest, Negeri Sembilan, Peninsular Malaysia, udult sandy clay loam soil; Semengoh Nature Reserve, west Sarawak, Malaysia, weakly humult silty clay loam; Lambir National Park and two nearby sites in north-east Sarawak, udult clay loam; and Lambir National Park, humult sandy loam; (After P.S. Ashton 1984, H.S. Lee *et al.* 2004, Manokaran *et al.* 1992)

Shorea species	Pasoh, Peninsular Malaysia	Semengoh, W. Sarawak	Lambir, NE Sarawak, udult	Lambir, NE Sarawak, humult
acuminata	2.5	-	-	-
acuta	-	-	-	7.7
argentifolia	-	-	0.2	-
curtisii	-	-	-	2
dasyphylla	0.2	3.5	-	-
ferruginea	-	-	9.6	-
hemsleyana	-	0.05	-	-
lepidota	2.7	-	-	-
leprosula	4.1	0.2	0.1	-
macroptera ssp. baillonii	-	4.6	-	1.7
macroptera ssp. macroptera	1.8	-	-	-
macroptera ssp. macropterifolia	-	-	2.3	-
myrionerva	-	-	0.05	-
ovata	-	1	-	1.5
parvifolia	2.9	0.9	3.9	-
quadrinervis	-	4.7	-	2
rubra	-	0.1	-	0.4
rugosa	-	0.1	-	-
scabrida	-	4.7	-	0.2
slootenii	-	1.1	-	0.4

and chemical soil characteristics, (2) light, and (3) reproductive height and phenology. Davies' results, and later Sabrina Russo's, show that niche differentiation becomes apparent only when the totality of life history characteristics of a species series are taken into account.[138] They are consistent with experience in species-diverse temperate forests.[139]

But there are 52 putatively distinct taxa of *Syzygium* within the CTFS Lambir plot, and 45 within the edaphically and topographically more homogeneous Pasoh plot. It beggars the imagination to accept that each species might have a significantly distinct set of life history characteristics! Six *Shorea* (sect. *Mutica*) species occur at Pasoh, 14 at Lambir, and 11 in the Semengoh forest (west Sarawak). However, there are only four species with population densities of reproductives at more than 4 ha⁻¹ at Pasoh, and at most four also on each predominant soil type in the other forests (Table 7.8; Box 5.1). The rest are intruders from adjacent soils in which they maintain viable populations (owing to edaphic diversity within the sample rather than the Pulliam Effect).[140] Most species existing in low numbers in one habitat are more abundant in another. *What proportion there are of those species which can sustain low numbers in the absence of a more favourable habitat, and which may therefore survive through chance by sustaining fecundity (see section 2.3.2.1 above), merits further study*; though confirmation of that will prove elusive.

2.4.1. *Conclusion*

Evidence from the most closely observed co-occurring congeneric species suggests that they comprise a core group that are sufficiently distinct in light response (Chapter 3), in reproductive height or in phenology (Chapter 5) to sustain balanced population densities through competition within a given physical habitat. Others – often the majority – may only survive through persistence in parts of a habitat mosaic, or perhaps continuing subsidy from adjacent different physical habitats in which they are competitively successful. These congeneric series should therefore be largest in heterogeneous landscapes – which, as at Lambir, indeed they are. But survival of some co-occurring congeneric species in very low population densities may be due also to maintenance of fecundity alone; in the absence of any force for extinction they may survive even though ecologically complementary.

The co-existence of so many species-rich subcanopy and canopy *genera* also remains unexplained. One possibility is that the genera differ in their pathogens, such that the number of individuals of species populations within each whole genus (clade) is also limited[141]. This would have the consequence that species within each clade survive with species in other co-occurring clades under reduced competition, in effect with greater ecological complementarity in respects other than their pathogens.

2.5. *Competition may not be detectable in spatial patterns*

Mutation, aided by the manner of speciation inherent in dense animal-pollinated forest, must have rigorously selected in this tumultuous process of trees seeding upon trees, and has resulted in the unbelievable variety of present-day tropical forest.

E.J. H. Corner, 1954, *The Evolution of Tropical Forest*

In 1964, UNESCO sponsored a conference at Kuching (Sarawak) to discuss progress in ecological research towards understanding how forests in the Far Eastern tropics function. Some of the world's leading specialists attended. There was a field excursion to the Santubong peninsula, where Robb Anderson, then directing forestry research in Sarawak, was guiding us. Robb tested our ability to recognize tree species in the field. He brought one to us that stumped us: spirally arranged palmately pinnate leaves, white sap, what could it be? None of us guessed, but it was a Hevea rubber tree – a species that had been introduced 80 years before from the forests of the Amazon. By the 1960s, it had replaced almost half the lowland MDF, but never before or since have we seen it invade indigenous forest!

In all climates worldwide, exotic plant species consistently fail to invade closed-canopy indigenous forest undisturbed by human activity. David Tilman[142] found, from grassland experiments, that increased species diversity is associated with reduced likelihood of invasion by others, thereby increasing community stability. The great botanical garden at Bogor is situated on a volcanic slope. The magnificent loamy soils are nutrient-rich and freely draining; yet specimen trees of peat swamp specialists such as *Shorea teysmanniana* and others from a diversity of habitats, free of competition, grow to maximum stature here (Plate 3.4). On the coastal hills and humult yellow sandy soils of Andulau Forest Reserve (Brunei), foresters have planted dipterocarp saplings in forest degraded by logging, without consideration for their ecological distribution in nature. Surprisingly, planted and tended saplings of the clay loam specialist *Dryobalanops lanceolata* are growing just as well as *D. aromatica*, which is naturally abundant in this forest. The implication is that successful establishment from seed and early seedling survival, constituting the regeneration niche,[143] presents the most stringent requirements for survival in this case.

Competitive species replacement during succession in gaps was discussed in Chapter 2. Mark Ashton, Singhakumara and their colleagues[144] and Stuart Davies[145] have shown, by controlled experiments under varied light conditions combined with experimental planting in gaps, that co-existing congeneric species differ in their responses to light, to canopy gap size and to their position within gaps. Demonstrating competitive differentials on different soils via nutrient manipulation has proven tricky; results that seemed inconsistent at first are now becoming clearer (Chapter 3).

At Pasoh and Lambir, individual soils and topography-defined habitats include up to 750 species. Not surprisingly, the prevailing opinion had at first been that the composition of hyper-diverse rainforest may vary in relation to the spatial heterogeneity of soils and subcanopy light, but that within a community and in a single dynamic phase, there must still remain a substantial residue of diversity that can only be explained by historical accident operating through limitations on seed dispersal.[146]

Nevertheless, the possibility remains that competition for space between individuals of the commoner species may be sufficiently frequent to result in changes in gene frequencies within their populations, leading to differentiation, and likely narrowing, of their habitat specificities over several generations (see Chapter 6). Detection of such changes, and proof of their cause, is central to defining the relative roles of chance and determinism in mediating the structure of hyper-diverse plant communities – but it is fiendishly difficult to achieve! Competition *for space* cannot in any case take place between the majority of species, which survive in low population densities. General competition among individuals for spatially variable resources (light, water, nutrients) is more likely to form complex local mosaics *within* communities. Sabrina Russo again has shown how differentials in survivorship and performance *together* can mediate the varying outcome of multispecies competition in a mosaic of soils of differing fertility, thereby leading to changes in the rank order of abundance at least of commoner species occurring on different soils.[147] The same principle has also been demonstrated in seedling experiments with temperate trees.[148]

Rainforest communities in equable equatorial climates are subject to minimal physical environmental constraints (nutrient availability excepted) to which a species may become adapted in the context of a particular habitat and community; but opportunities for co-evolution among species, and for competitive diversification among related species and their mobile links such as dispersers and pathogens, are limited only by the dispersal distances of the mobile links (Chapter 5). Therefore, while consistent interspecific competition for space between tree species cannot exist in these hyper-diverse communities, we can expect to find an exceptional array of indirect competitive interactions via mobile links, for which co-occurring sister species may differentially compete. Although differences in pollen- and seed-dispersal agents among related genera are common, species within congeneric series generally share such mobile links. But differences

in reproductive phenology among species that rely on a common disperser could permit coexistence among species that are otherwise ecologically complementary. These forms of indirect interspecific competition for space will eventually lead to changes in gene frequencies among the competitors, and therefore to increasing niche-specificity; but will not produce interspecific differences in spatial patterning on the ground.

Limitations on seed dispersal lead to aggregation of seeds and seedlings around parent trees. This spatial pattern will have little influence on gene frequencies except in respect to intraspecific competition, and is therefore neutral in relation to other species. The *relative* strength of neutral versus deterministic influences on competition in hyper-diverse forests is central to our understanding of how species diversity is maintained, and to what level of predictability management protocols can be designed. The enigma remains: absence of habitat-correlated spatial patterning must imply random (and therefore ecologically neutral) interspecific competition, until other evidence is found to the contrary.

2.5.1. *An example of niche specificity not manifested in space: differentiation of flowering phenology*

During mass-flowering years, four to six co-occurring species of dipterocarps each in Shorea sections *Mutica* and *Doona*, in Sunda and Sri Lankan MDF respectively, flower in a sequential yet overlapping pattern[149] (described in Chapter 5). This implies that a given species must experience reduced pollination success when more than one species is flowering (Chapter 5); and that this enhances and sustains separation of flowering times over evolutionary time. Such patterning could only arise where there is a barrier to introgressive hybridisation, or that hybrid progeny are less fit in competition.[150] Almost all the morphological evidence for survival of hybrids to maturity among dipterocarp species has in fact been reported from *Dipterocarpus* species in the species-poor associations of seasonal Indo-Burma.[151]

Co-occurring species in *Shorea* (sect. *Mutica*) of MDF have been found to hybridise, but few progeny survive beyond the F$_1$ generation (Chapter 5). These species depend on one another for maintenance of their thrip pollinator populations, whose larvae feed on their petals in bud throughout the mass-flowering event. In this way, the fecundity levels of these tree species are interdependent and maintained at a rough equilibrium. Such a system may therefore actually provide niche space for *additional* co-occurring tree species, by shortening the flowering period of individual species populations yet lengthening the overall flowering season of a clade to a maximum determined by the evolutionary plasticity of the rate of inflorescence and flower morphogenesis. The six most abundant *Shorea* (sect. *Mutica*) species at Pasoh are all late successional light hardwoods, with similar light responses as juveniles,[152] and appear ecologically complementary in other respects. Thus, these species manifest convergent evolution – or, more likely, conservation of an ancestral trait – in those floral characters which allow them to attract a common pollinator; yet simultaneous divergent evolution in flowering times. The fact that widespread species co-occurring in distant regions in mixture with local congeners flower in consistent sequence implies that this trait evolved early in their co-evolution.

2.5.2. Community reproductive phenology

Thus, extended seasons of flowering and fruiting with minimal year-to-year variation in reproductive phenology should in principle offer the maximum opportunity for niche specialisation among pollinators and seed dispersers. Such conditions should also provide opportunities for niche partitioning – through sequential flowering – of tree species that share the same pollinator. Given these expectations, it is therefore extraordinary that:

1. the most speciose forests of the Asian tropics, MDF, are distinguished by supra-annual mass flowering and tightly synchronised mast fruiting;

2. the ecotone from MDF to seasonal evergreen dipterocarp forest – with its extended annual, though fluctuating, seasons of flowering and reproduction – is correlated with a c.30% *reduction* in tree species diversity;

3. the diversity of MDF is not as great in the subcanopy as in the canopy, notwithstanding the longer period of flowering and fruiting occurring in the subcanopy, and the presence of annual reproductive seasons among many of its species.[153]

I infer that these constraints on tree species diversification in MDF are overridden by other forces favouring them. Compared with the Neotropics in particular, the diversity of pollinators and seed dispersers is itself relatively low.[154] Thus, the diversity-promoting role of co-occurring congeneric series of tree species sharing the same pollinator(s) – such a powerful influence on tropical rainforests in Asia – appears to have no analogue on the other tropical continents where co-occurring congeneric series of upper canopy species appear not to occur on the same scale.[155]

2.5.3. Pathogens and herbivores

Plants defend themselves against herbivores, seed predators and pathogenic micro-organisms by evolving a battery of physical defences such as thorns and leathery leaves, and particularly by synthesising a panoply of toxic or indigestible chemicals (Chapter 5). This great diversity of chemical defences has fostered a comparable diversity among the attackers of strategies to circumvent them. Because the energetic costs of defence and circumvention are each high, however, a given plant species and its enemies can remain competitive only by restricting their arsenals to combinations of several strategies. Over evolutionary time, the result has been a vast array of weapons and defences at the species level, among both plant defenders and attackers, and is the fundamental cause for the evolution of species diversity. Dan Janzen[156] and Joe Connell[157] hypothesised that tree species diversity should be enhanced by species-specific seed predators and tree pathogens killing conspecifics in high-density populations, thus freeing up space for individuals of other species to invade. These invading species could in principle differ *solely* in their degree of susceptibility to the same enemies.

However, seed predators, herbivores and saprophytic fungi usually show more generalised host specificity than pathogens, which are confronted directly by the host's immune system.[158] Bassett & Novotny found that seed predators and herbivores are no more host-specific in tropical than in temperate forests,[159] and they show low β-diversity at both local and regional scales.[160] Because their general level of specificity appears no greater in biodiverse tropical forests than in species-poor temperate forests, host-herbivore co-evolution cannot explain the vastly greater species richness of tropical trees, but could provide one explanation for their vastly greater *generic* representation (although evidence for pathogen diversification has been presented as a means of sustaining multiple species coexistence of the nitrogen-rich Neotropical leguminous genus *Inga*)[161] (see section 2.4.1 above). Herbivore or pathogen pressure, by simply reducing numbers, may thereby still provide opportunities for otherwise ecologically complementary species in different higher taxa to co-exist.

The influence of species-specific pathogens on tree demographics has hardly been examined in Asian tropical forests (Chapter 5), having as yet been the topic of only a single in-depth study.[162] Whereas a canker-forming fungus was found to be host-specific, saprophytic Polyporaceae were generalists, increasing with stand density but not stand diversity.[163]

A number of studies, including in Asia[164] (see also Chapter 5), have shown that juvenile (particularly seedling) mortality is positively correlated with the spatial density of a species juvenile population. This is most parsimoniously explained by the relatively greater ease with which pathogens can spread when the hosts are close to one another.[165] This phenomenon, if indeed universal in speciose rainforests, could by itself explain how so many apparently ecologically similar species can co-exist. It would do so by providing a further axis in niche space – in that respect contravening neutral theory.

Mortality among tree species of Sunda MDF has been found to increase when their populations exceed 80 individuals

> 1 cm dbh ha⁻¹.[166] Campbell Webb and colleagues found that survivorship among juveniles in MDF at Gunung Palung National Park (West Kalimantan) increased in relation to their communities' species diversity.[167] The fact that in MDF, communities of similar generic composition, consequently similar seed dispersal and clumping patterns, but very different species richness and stand densities share similar peaks of species abundance, provides supporting evidence (Fig. 7.8). The peak shifts towards lower population densities in both the less species-rich communities of the seasonal tropics and perhumid Sri Lanka, while the number of the most abundant species hardly increases. That can only partially be explained by the decline in density among the smallest size classes, and suggests that populations may share similar densities at which survivorship is maximal: that is, their optimal densities might be mediated by pathogen spore dispersal distances. Could the shift in the peak in seasonal climates be due in part to the greater distances across which pathogen spores may disperse where there are prevailing winds during the growing season? *Much more needs to be known, and investigations will be fraught with complexity*.[168] Alternating Lotka-Volterra cycles of abundance between particular pathogens and their tree host species may be anticipated, as in the well-known successive local cycles between epidemics of the gypsy moth and its *Bacillus thuringiensis* pathogen in North-East American woodlands. Because host-specific predator or pathogen interactions are positively related to host density, they are unlikely by themselves to cause extinction.[169] This again adds to the possibility that equable perhumid climates may accumulate species populations which can survive in very low population numbers by escape from pathogens, provided they overcome Allee effects (see section 2.3.2 above), when they would in effect be ecologically complementary. All that can be presently said of their niche specificity is that field observations convince me that they are no less habitat-specific than commoner species (see section 2.3.2 above)

Seed and seedling predation, in contrast, is unlikely to promote enhanced species co-existence because few predators are host-specific. In Asia, as in the Neotropics, ants play major roles in defending trees against leaf and seed herbivores, and are attracted to extrafloral nectaries (Chapter 5).[170] These are most frequent among pioneer tree species, comprising the *least* diverse of the dynamic guilds, but also universal among the seedlings of dipterocarps, which comprise among the *most*. These symbioses are not species-, or even genus-specific. Nevertheless, extrafloral nectaries are best represented and diverse in the most species-rich forest communities. A group of *Macaranga* species that attract and house ant colonies by hollow stems as well as by nutritious lipid bodies is confined to perhumid Malesia.[171] Perhaps this region alone provides an environment sufficiently equable for such complex symbioses

(see above).

There could exist a tension between the density that a species population can sustain in the presence of a pathogen with a given mean dispersal distance, and the likelihood of extinction through Allee effects (Chapter 5). Such an equilibrium, in a particular historic biogeographic region and climate, could govern the putative asymptote in α species richness that exists in the continuous and widespread habitat of lowland clay loam soils in Sundaland. This could involve a Red Queen model[172] in which local extinction of rare species provides the opportunity for the competitive success of an immigrant, and perhaps a change in rank orders of abundance.

In summary, the influence of highly species-specific plant pathogens, rather than that of predators, may be a leading cause of the accumulation and maintenance of hyperdiversity in some rainforests, *but much more research into specific instances is necessary, particularly in Asia*. Among all these examples, though, much of the difference in diversity is due to variance in the number of species that naturally exist in extremely low population densities. Those *variances* cannot convincingly be explained by the impact of species-specific and density-dependent pathogenicity.

2.5.4. *Symbiotic associations as promoters of species diversity*

Fungal associations with plant species also extend to the symbiotic relationships of root mycorrhizae (Chapter 5). Ectotrophic mycorrhizae are notably characteristic of the two tree families that dominate Asian rainforests – Dipterocarpaceae in the lowlands and Fagaceae in lower montane forests of the Himalaya and East Asia. Mycorrhizae form complex associations both with particular soils and with the particular dipterocarp species that grow in them.[173] At Lambir, mycorrhizae are most abundant in MDF on deep humult sands where α-diversity is also greatest.[174]

In the rainforests of other tropical continents, individual species belonging to the third ectotrophic tropical tree taxon, Fabaceae subfamily Caesalpinioideae, become monodominant.[175] It remains unclear why the ectotrophic association in Asia with Dipterocarpoideae has led instead to *family* dominance and hyper-speciation in some genera. One possibility I mentioned is the abundance and rapid mobility of resins in the intercellular canals that are unique to Asian dipterocarps. It is thought that dipterocarp seedlings gain early competitive advantage from infection with their specific mycorrhizal assemblage from parent roots. This may however be offset by enhanced negative density-dependent exposure to pathogens. *The reconciliation of these opposing forces offers an interesting challenge for future research*. There is some evidence that individual trees may grow more slowly when adjacent to conspecifics.[176] Of the few examples of monodominance among Asian dipterocarp species, none need to be explained by their ectotrophic mycorrhizal association alone, as simpler explanations already exist (see section 2.3.1 above).

Because these symbiotic interactions between plants, microorganisms and animals change continuously over evolutionary time, high genetic diversity among the competing populations is at a premium.

2.6. *Genetic variability as evidence for evolution of niche specificity*

The selective advantages of sexual reproduction remain controversial, although Charles Darwin's conclusion – that variation in heritable traits within populations is essential for their evolution and for their long-term survival in competitive environments – remains undisputed. Sexual reproduction also offers the opportunity to rid populations of harmful alleles. But asexual populations generally also include some genetic variability.[177] This can be maintained in long-lived modular organisms such as trees, for instance, at least in part by somatic mutations whose effects will be transferred to seeds in species that reproduce by adventive embryony (Chapter 5). Besides, adventive embryony is rarely, if ever, obligate. Sexual reproduction is necessary, though, for generating new allelic combinations and for sharing of advantageous genes, and thereby defines populations and ultimately species.

But sexual reproduction, thanks to the cost of successfully achieving cross-pollination between individuals, is expensive in both energy and nutrients. This is especially true in the perhumid tropics, where biodiversity – and distance between reproductive individuals of each species – is greatest, while soils are among the least fertile in the world (Chapters 1, 5). Successful pollination is particularly challenging in the leafy subcanopy. Pollen and seeds are among the plant parts richest in amino-acids and proteins, therefore nitrogen. The necessity of producing separate male and female organs and gametes, and in many species individuals of separate sex, halves the number of potentially successful new progeny.

Actually, sexual reproduction among plants is almost invariably achieved by the production of vastly more ovules than an individual has the resources to bring to maturity as ripe seed. This loss is further exacerbated – whatever the means of dispersal – by the vast pollen production needed to increase the likelihood of even a minute proportion successfully germinating on an appropriate stigma. I estimate, from Chan's observations at Pasoh of the number of flowers presented in a single flowering event,[178] that an emergent *Shorea leprosula* might produce, over a lifetime during which it experiences 30 flowering years at 5-year intervals, 21 million ovules and 7.5 billion pollen grains – in order to ensure that it is replaced by even a single individual. Sexual reproduction is therefore a cause of immense loss of nutritionally valuable (albeit low

biomass) material to the soil – where the ions released in decomposition may potentially be reabsorbed by a competitor.

Yet tropical forest trees are, overwhelmingly, outbreeders.[179] However, asexual seed production – that is, apomixis through adventive embryony – also appears widespread in Asian rainforests (Chapter 5).[180] Apomictic populations do sustain some genetic variability, either by partial sexual reproduction or through somatic mutation.[181] The level of this variability at population and larger scale must inevitably be lower than among purely sexual species. It is unknown whether the proportion of apomictic tree taxa varies consistently across habitats, and therefore whether it can contribute to the accumulation of co-occurring species, especially given the degree of spatial complexity of physical habitat in tropical forests. From what is currently known, apomixis appears particularly frequent among species belonging to the co-occurring congeneric series that contribute so much to tree species diversity in the rainforests of the perhumid tropics, and among those that possess morphological subspecies separated by physical barriers such as river valleys (Chapter 5). *Rigorous demonstration of apomixis entails the painstaking collection and microscopic examination of embryogenesis, and has been inadequately studied; but new microscopic technologies are easing the difficulty.*

The predominance of outcrossing and genetic variability among tree species in biodiverse rainforests confirms that the variability necessary for continued evolution of niche specificity – in particular for rapid ongoing evolution in response to pathogen pressure – can be maintained. Within-population genetic diversity would be an unnecessary and expensive luxury were species ecologically equivalent, especially in their pathogens. But outbreeding cannot be indisputably presented as evidence for selective niche specialisation, because it may also be necessary for the elimination of deleterious mitotic mutations, and therefore for cellular health. Some mechanism such as natural selection is nevertheless required to control what would otherwise be an explosive geometric expansion of genetic variation over successive generations. As it turns out, levels of variability within rainforest tree populations appear comparable to those found in the species of relatively depauperate temperate forests. This does imply that such a control on variability exists. That control is most likely natural selection.

But what is the selective advantage of apomixis in that minority of species in which it is found? Does the presence of apomictic species in hyper-diverse MDF constitute the exception that proves the rule? If so, why are they generally no more abundant than their sister species which lack apomixis? These issues are further discussed in Chapter 5.

Further support for selection is implied by the rarity of reproductively mature hybrid individuals among co-occurring sister species populations (Chapter 5).[182] Hybridisation results in a surge of variability from new combinations as yet unrestrained by selection. Molecular evidence reveals that hybridisation has played a significant part in speciation among dipterocarps.[183] First generation (F_1) hybrids comprise a substantial fraction of populations of *Shorea curtisii* and its sister species in Bukit Timah reserve (Singapore). Yet F_2 hybrid individuals are absent or rare (Chapter 6): is that the consequence of natural selection, or some genetic or morphological malfunction? Likewise, seedlings of *S. megistophylla* in Sinharaja forest resulting from selfing experience higher mortality than those resulting from cross-pollination.[184]

3. The origins of tree species diversity in tropical rainforests

The maximum age of tree species endemic to a rainforest region can be inferred from the age of its climate. For perhumid western Malesia, this appears to be c.23 million years. Some widespread species there with limited seed dispersal may indeed have arisen that long ago, but most endemic species will be more recent (Chapter 6). Low regional endemism within widespread habitats, such as the MDF of lowland Sumatra compared to Peninsular Malaysia or Borneo, implies that such forests are younger, in Sumatra's case possibly since the return of perhumid conditions in the Holocene, c.10,000 years ago. High endemism in species diverse habitat-islands, such as Sarawak's Lambir Hills, and the sharing of these endemics between these isolated islands implies a longer period of climatic stability, therefore opportunities for both immigration and speciation. Koichi Kamiya found genetic evidence of diversification among three sister species co-occurring at Lambir to have originated in the late Miocene or Pliocene, coincident with the uplift and emergence of Lambir's Neogene sediments (Chapter 6).[185]

I have inferred that rates of tree speciation, especially in widespread and relatively continuous habitats, may be slowing down. But I have also hypothesised that extinction rates may be exceptionally slow in the equable perhumid climate of Sundaland, thereby offsetting low speciation rates in setting regional species richness and also community species diversity. The relative rarity of point endemics, even in ecological islands, implies the former. Co-evolution of tree taxa, and of the more mobile organisms with which some are specifically interdependent, results from the evolution of specific chemical signals by tree taxa. These signals are variously specific, from within species metapopulations as well as to higher phylogenetic levels. They point to the central role of phylogeny – both of plants and their dependent organisms – in the evolution and structure of rainforest biodiversity.[186] *As yet, this role has hardly been explored.* However, recent studies of co-evolution among figs and their pollinators,[187] and of *Crematogaster* ants

and the *Macaranga* species they protect,[188] indicate either that co-evolution is *not* precise, or that the evolution of the shorter-lived dependent organisms is periodically interrupted by transfers to more distantly related hosts (Chapter 5).

It has been argued that tropical biodiversity could be explained by faster rates of genetic processes in the tropics, thanks to the influence of continuous high temperatures.[189] *Whether this is so remains to be confirmed*, but the rates of speciation I infer from the evidence presented here appear unexceptional. Furthermore, the species poverty of many tropical deciduous woodlands would then need to be explained. Nevertheless, ancient interdependencies between higher taxa do survive. Examples such as the relationships between the archaic tree families Annonaceae and Myristicaceae and archaic insects – particularly beetles – imply a bedrock conservatism in plant-insect interdependencies.[190] Such conservatism could restrain rates of diversification, and in part explain the low species richness in regions long deprived of sources for immigration, such as the evergreen forest regions of South Asia.

4. The role of community drift: a last laugh for neutral theory?

The unified neutral theory predicts that community species richness will eventually reach an asymptote within any uniform habitat in a region sharing a common history, but that there will be a drift in species composition over distance. This drift will be due to constraints imposed by limited seed dispersal, combined with random processes of immigration and extinction taking place over scales from local to regional.[191] Compositional drift would also be enhanced by speciation, assumed in the neutral model to occur through point mutation. Overall species abundances would thus be distributed according to R.A. Fisher's log series. Species' turnover rates would be most rapid at the local scale, and among species with small populations. They would be vastly slower in sessile long-lived organisms such as trees – so slow that one wonders whether other events, such as climate change or changes in dispersal opportunities, might not override them.

In Sarawak, in 105 MDF plots with a range of soils and at distances sometimes more than 500 km apart, we found (as might be expected) that the most abundant species had the highest likelihood of being constantly so in their chosen habitats[192] (section 2.3 above): when in their preferred habitat, species tend to be either consistently abundant or rare, irrespective of the distance between their communities where physical barriers do not isolate them. Although detailed replicated censuses are generally lacking, it seems that rare species are frequently absent from apparently suitable localities. Yet there appears to be no significant difference between the geographic ranges of low- and high-density species, either at Sinharaja, Pasoh or Lambir. At Lambir especially, local endemics such as *Dipterocarpus globosus* and *Shorea laxa* are among the most abundant. On the other hand, many consistently low-density species – such as the large-fruited *Mangifera* and *Durio* (sect. *Durio*), *Neesia* and *Coelostegia*, all with large animal-dispersed seeds – have unexceptional rates of endemicity (20% of 15 species at Lambir, 38% at Pasoh). In these cases, relative abundance seems to be determined by reproductive biology and by overall fecundity (section 2.3.2 above and Chapters 3, 5). All but the most abundant species do appear, on casual observation, to fluctuate somewhat in relative abundance at local to landscape scales, in accordance with neutral community drift theory – but that appears to have no bearing on their geographical range.

The prevalence of clumping of individuals within rainforest tree species populations is indisputable.[193] The extent to which clumping, especially at larger scales, is due to constraints other than habitat heterogeneity is difficult to evaluate, however. When clumping is taken into account, the shape of the curve relating species to area occupied follows the same mathematical law, whatever the forest type.[194] Most remarkable, all our CTFS lowland evergreen and semi-evergreen forest plots succeed in capturing about a quarter of the known tree flora within the land mass they occupy: 207 at Sinharaja (out of the 800 species of Sri Lanka's perhumid zone), 817 at Pasoh (out of 3,100 for Peninsular Malaysia), 1,260 at Lambir (out of 4,500 in Malaysian Borneo and Brunei). If other climates (north-west Peninsular Malaysia) and habitats (podsols, peats, mountains and so on) are excluded, this would approach half. For preferred habitats in local landscapes, it would exceed half (c.1,500 known lowland tree species in the well-collected Lambir National Park and 81 of the 86 dipterocarp species known from that region from all habitats!).

Dispersal barriers therefore seem to have been substantially overcome, over thousands rather than millions of years, when less than say 20 km wide, presumably by sweepstakes dispersal. The low rate of point endemism in the Sunda tree flora at all altitudes is striking, too, except in rare habitats whose physical constraints are unique and strong, such as ultramafics or in Kinabalu's alpine zone. Rather, as in the example of the coastal hills of north-west Borneo, species occur in islands of shared habitat along more than 400 km of coast, and often across to Peninsular Malaysia or the Riau Archipelago.[195]

I hypothesise that local extinction and immigration events do occur among tropical tree populations according to the neutral theory. This could happen in the following way: though community composition and diversity within stands and small areas of uniform habitat (especially among species of lower density) may fluctuate at random, community diversity within larger areas and plant geographic provinces may remain more constant, especially among more common species.

However, if extinction events are rare over long time periods, due in part to reinvasion at larger (regional) spatial scales, regional floras will continue to accumulate until diminished by changes in physical opportunity. Each floristic province of area exceeding, say, c.10,000 km² would thus accumulate species through lower extinction than immigration rates. This pattern of accumulation would be combined with a series of internal speciation episodes, beginning with the fresh opportunities provided by a prior extinction event and declining over time. Tree species within such a biogeographic region might eventually still approach an asymptote in species richness,[196] but would rarely achieve it before ecosystems are reset by changes in climate or geomorphology.

5. Is tree diversity correlated with that of other life forms?

We can expect that tree species diversity will be more closely correlated with that of organisms specifically dependent on tree species (especially – but not only – those that depend on individual tree taxa for nutrition and can exploit them while paying the energetic cost entailed in breaking down the tree's chemical defences). Examples are Lepidopteran larvae, phloem-sucking scale insects, plant-specific pollinators, and –especially – pathogenic and symbiotic fungi. Co-variation with habitat or with species richness in the specificity of organisms dependent on host tree taxa has yet to be established between any group of organisms and tree species in species-rich rainforest tree communities (Chapter 5).[197] Correlations with Lepidoptera have proven elusive. Vojtech Novotny and his colleagues[198] have confirmed that most insect herbivores interact with hosts at super-specific, that is, generic or higher, levels: the vast richness of rainforest insect faunas is thus only loosely correlated with tree community richness. Correlations and specificity between tree species, their pathogens and ectotrophic mycorrhizae appear likely, but are yet to be demonstrated.

Tom Givnish[199] interpreted the high species richness of fleshy-fruited lobelioid (Campanulaceae) shrubs in the understorey of Hawaiian forests as due to limited dispersal by resident locally endemic birds and the opportunities they provide for the isolation of small founder populations. Species diversity and point endemism among fleshy-fruited subcanopy trees in MDF and other Asian rainforest types is unexceptional, perhaps because of the dominance of juvenile and subcanopy dipterocarps.

Leachate from the leaf litter of the tree flora itself influences variation in rates of topsoil decomposition, and therefore nutrient availability. The strength of these influences on soil nutrients should correlate with soil organismal diversity.[200] Although such diversity of litter chemicals would be unlikely to show much species-specificity, it should nevertheless be greatest at the interface between humult and udult soils. That does indeed coincide with the peak in tree species diversity. Peay and colleagues[201] found that chemically identified ectomycorrhizal taxon diversity was richer on the more species-rich nutrient-poor acid humult than clay mull soils in Lambir MDF.

The optimal frequency and intensity of canopy catastrophe, through drought or other causes, for the maintenance of tree species diversity does not necessarily coincide with that for other plant life forms, or animals. Rhett Harrison, for instance, has documented local drought-induced extinction of the wasps specific to certain local continuously reproducing fig species.[202] Similarly, canopy drought may affect the subcanopy light climate and influence the early establishment and survivorship of semi-epiphytic fig juveniles – thereby contributing to niche differentiation among figs.[203]

Less specific interdependencies show little correlation with tree species diversity. The MacArthurs[204] showed that the species richness of nesting birds in eastern North America is correlated with forest structural complexity. There is presently some limited evidence from at least one region that this is also the case in Asian rainforests.[205] However, the pattern is obscured by the concentration of birds – along with other arboreal and terrestrial vertebrates – where food is most available and diverse, on and near riverbanks where

forest structure is relatively simple (Fig. 7.9). These habitats are relatively rich in nutrients, eutrophic water, and light (due to disturbance), but their forests are poorer in tree species than adjacent uplands. In Gunung Palung National Park (West Kalimantan), frugivorous species have been observed to fan out from riverine sites over adjacent MDF during mast fruiting years.[206] This reflects the dependence of the bird- and mammal-dispersed tree species of less-fertile sites, such as *kerangas* and peat swamp, on adjacent frugiferous ecosystems for sustainment of their mobile links.[207]

The change in vertebrate faunal species diversity occurring in the north of the Isthmus of Kra is closely correlated with the major phenological and structural ecotone from evergreen to semi-evergreen and deciduous forest. In contrast, we see very little change in faunal diversity at the major floristic and tree species diversity ecotone across the Kangar–Pattani Line where there is a major change in tree species composition and community diversity, but not forest structure or leafing phenology[208] (Chapter 6). There is a gradual reduction of vertebrate species numbers with increasing altitude in the perhumid tropics, with a particularly distinct change at the lower-to-upper montane ecotone on the great equatorial mountains of Kinabalu and New Guinea.[209] The montane grasslands of South Asia and the alpine grasslands of New Guinea harbour their own distinct flora and fauna. These include (in South Asia) palaearctic migrant birds during the winter months;[210] but these migrants do not compensate for the loss of lowland species. The distinct vertebrate fauna of the wet evergreen forests of the eastern Himalaya and northernmost Indo-Burma change very little at the tropical–warm temperate frontier. A more distinct – albeit still gradual – transition in species composition occurs across the great range of altitudes up the high mountains of this region but, here again, no clear correlation between forest diversity and animal species richness has been documented.

Species richness of other plant life forms is similarly uncorrelated with that of trees. A striking case is that of ferns and epiphytes, which reach their greatest species richness in lower montane forests, though they are most abundant in foggy upper montane forest (Chapter 4).[211] Indeed, on their account, lower montane forests are the richest in plant species overall.[212] Their tree crown habitat is both relatively uniform and relatively unstable, as conditions within a site change over the lifetime of the tree and sites are frequently lost through branch break, promoted by the weight of the epiphytes themselves. Despite its relative uniformity, there are significant habitat gradients within a single tree crown, including those of light and associated humidity levels and, in some cases, nutrient availability from leachates and bark surface topography. It would seem that the number of orchid species on a single host tree can far exceed what competition

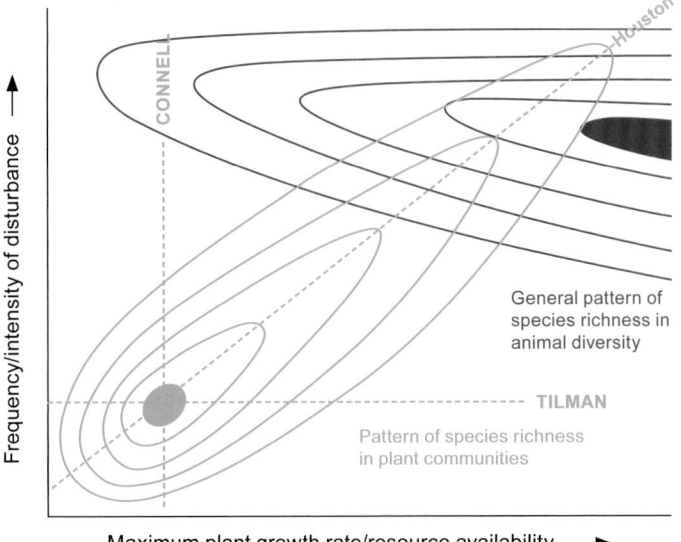

Fig. 7.9 Vertebrates and trees have differing patterns of species diversity in relation to habitat: whereas tree species richness is greatest in forests where rates of canopy disturbance, and maximum tree growth rates, are moderately low, for vertebrates it is the opposite. (From P.S. Ashton & LaFrankie 1999)

for niches along these gradients can reasonably accommodate – although orchidologist Philip Cribb[213] asserts that related species are indeed spatially niche-specific within tree crowns, possibly owing to their dependence over their entire life histories on a succession of species-specific mycorrhizal associations. Orchid seed is minute, wind-dispersed and uniform, but flower morphology and scent chemistry is unequalled in its diversity and is associated with individual pollinator species. Population numbers may therefore be limited by the influence of pollinator numbers on fecundity, but community composition here is likely overwhelmingly determined by seed dispersal limitation and chance.

6. Conclusions

Neutral theory should serve as a first rigorous null hypothesis against which to test evidence for niche differentiation (although it has also been claimed that the patterns of diversity predicted by neutral theory are, in fact, also consistent with those arising from niche differentiation[214]).

Studies in tropical Asia, discussed in Chapters 3–6 and synthesised in this chapter, have revealed the following patterns:

6.1. *Patterns of species richness and diversity among regional floras*

1. Floras long isolated and with differing histories, yet occupying ostensibly similar climate and soil, differ in their tree species richness and community diversity. Geological history, particularly the time since the last change in climate and opportunities for migration during the Pleistocene, is the major determinant of the size of a tree flora within a currently isolated land mass in a climatically uniform region – more important than is its current land area. Historical biogeographic elements especially, such as the Riau Pocket flora of the Sunda Shelf – best represented on humic leached soils occupying ecological 'islands' – do vary in richness through their region. But throughout the tectonically unitary Sunda region, communities occupying those zonal habitats with the most complete spatial continuity – namely the MDF on clay and sandy-clay loams – seem to share similar species diversity within similar physical habitats, notwithstanding differences in the size of their floras available to those habitats, and the differing area of the major land masses. Therefore, *communities* may approach an asymptote of α-diversity within specific habitats, which to some extent overcomes differences in the available species pool.

2. A species range varies with the range of habitats it is capable of occupying – so ranges are limited when a species habitat is limited. Ranges consistently appear most restricted by (i) geographical barriers, including river basins, and (ii) changes in geology and in soils. In the lowlands, for instance among the Sunda lands, these constraints may be as significant as relatively young sea barriers.

3. We can therefore infer the age of a species from the relationship between its habitat's geological and geomorphological history and its geographical distribution, together with phylogenetic evidence. On this basis, most modern species appear to be at least half a million years old, and many over three million years – making them pre-Pleistocene in origin. Clinal variation in morphology is as yet unknown. I infer that there has been ample time for most tropical Asian tree species to colonise all of their appropriate habitats, and/or that there have been sufficient (if intermittent) opportunity for genetic exchange, wherever habitats are near-continuous within land masses.

4. I conclude, overall, that the constraints on species ranges imposed by limited seed dispersal are overwhelmingly local and temporary within any continuous habitat. Neutral forces mediated by seed dispersal alone can therefore play no more than a relatively small part in determining overall community composition and diversity.

6.2. *Patterns of species diversity within and between communities at regional scale*

Community species diversity is influenced by a variety of limiting factors, which operate differently in different habitats and at different spatial scales. The overriding environmental determinant of variation in α-diversity (and richness) is climate. Community diversity is broadly correlated with seasonality, specifically the length of the dry season. As the dry season increases in length, its influence as a diversity, therefore richness, limiting factor increases.

1. Increasing length of the dry season is also associated with increasing *variability* in its length. Variability in the date of its onset increasingly limits species lacking prolonged seed dormancy.

2. In nature, *fire* becomes an important limiting factor in climates where the mean length of dry season exceeds five months. Fire reaches its peak of influence on diversity in destructive crown fires occurring in tall and short mixed deciduous forest, where the dry season is six to eight months long. It then declines with increasing length of dry season, and is absent in natural semi-arid and arid zone vegetation.

3. *Browsing* can eliminate some palatable species by inhibiting regeneration, thereby reducing diversity locally. It can be pervasive, but is most intense where food varies seasonally in abundance. Those conditions are found in deciduous forests with abundant seasonal herbaceous vegetation, particularly grasses.

6.3. *Variation in community α-diversity among tree species communities at landscape and larger scale within uniform climates*

There is no doubt that tree seed mostly falls beneath the parent crown, that densities of fallen seed decline rapidly away from it, and that the slope of this decline is determined by the means of dispersal. Reproductively mature individuals thus remain clumped in most (though probably not all) species. Community composition, including rank order of species abundance, is influenced by dispersal strategies irrespective of other physical and biotic influences; but this influence declines with increasing spatial scale.

1. Tree community species diversity is correlated with *community structural complexity* and *forest stature*.

    i. Diversity is greatest in tall evergreen forests with canopies of three strata. Such structure results from intermittent levels of canopy disturbance, whereby canopy gaps of high variability in area and intermediate frequency sustain a relatively rich pioneer and successional flora without loss of the mature phase, while emergent canopy mortality provides imtervening space for a rich main canopy flora. Variable subcanopy light is promoted with intermittent minor canopy disturbance, thereby enhancing the number of co-occurring species, as juveniles or as mature subcanopy individuals, through the diverse opportunities provided for response in growth rate and shade tolerance. Such conditions are found on upland soils where rooting is moderately shallow, in perhumid climates and in sheltered coves in moist seasonal climates.

    ii. Whereas infrequent drought-induced mortality enhances forest dynamics and diversity of canopy structure, thereby enhancing species diversity (see above), frequent – intermittent or seasonal – soil water stress reduces the potential height of the forest canopy. Intermittent catastrophic drought reduces or eliminates the emergent stratum. Both reduce species diversity.

    iii. Occasional cyclonic and other windstorms result in massive wind-throws. These may increase or decrease tree species diversity. If they occur at a high but erratic frequency they result in increased richness of pioneer and successional species, but restrict the opportunity for late successional and climax species to establish and reproduce, thereby reducing species diversity and richness at landscape scale. If such storms are occasional and occur at intervals at least as long as the period required for late successional and mature phase climax species to reach reproductive maturity, they may increase the area of successional as well as mature stands, leading to accumulation of both pioneer and climax species, therefore raising overall diversity at the landscape scale. But where such windstorms persist with frequency and regularity within the lifespan of a tree, forest structure and composition eventually becomes adapted to them. These adaptations include persistence of a climax flora in all structural guilds, and reduced stature with a single, smooth stratum of emergent species (or, under increasingly frequent storms, main canopy species). This is accompanied by a decline in species richness and diversity of both pioneer and climax species.

    iv. Where climate is calm and soils permit deep rooting, with enough water storage capacity to overcome the most prolonged droughts, the emergent canopy becomes a tall but continuous blanket with only occasional (and usually small) gaps. The light climate becomes more uniform, and species richness is reduced in all structural guilds. Such dense continuous emergent canopies can develop at stand level either where large windthrows occur, where they are dominated by pioneer or successional climax species; or can result from gradual succession of increasingly shade tolerant species.

2. Tree species diversity is also correlated with *soil nutrient* levels:

    i. When the availability of soil nutrients is strongly constrained, species richness is low because only a few species can survive in a competitive environment.

    ii. Species richness – and diversity – rises to a maximum at relatively low soil nutrient levels, at similar levels on both humult yellow sands and leached yellow silty clays. Both soil types can carry forests of great stature, with three well-developed strata and high spatial diversity of subcanopy light climates. They sustain great species diversity on that account also, but dynamics and structure are themselves mediated in large part by soils nutrient levels and water economy.

495

iii. On low-nutrient humult yellow sandy soils, high species diversity is also associated with diffuse but variable mature phase canopy foliage, and predominantly small gaps associated with secure rooting and single canopy tree mortality. These lend competitive advantage to species with high survivorship in moderate shade.

iv. On the often somewhat higher-nutrient but less securely rooted leached but udult silty clay soils, it is the dense mature phase canopy with more frequent gaps of variable area that provides variable but often high insolation within the canopy. This gives competitive advantage to a diversity of successional species.

v. High-nutrient upland soils are more friable clay loams with high water-retaining capacity and deep rooting. Their dense emergent canopy and paucity of gaps, occasioned by resistance to all but the strongest winds, is associated with depressed species diversity. High-nutrient floodplain soils are anoxic at depth, promoting shallow rooting. High rates of canopy disturbance there are associated with broken canopies, pre-eminence of early successional species, and variable α-diversity. In all cases, the species that dominate the canopy on fertile soils are those that attain both by rapid growth in canopy gaps.

3. The relative influence of soil nutrients on species richness declines as rainfall seasonality and length of dry season increase. These climatic factors come to override nutrients in influencing forest structure and phenology.

4. Whereas variation in canopy structure and dynamics is a major proximal mediator of species *diversity*, where soil nutrients vary across the landscape at critical levels they play the major part in determining interspecific competitive success, therefore species *composition* (Chapter 3).

5. Decline in mean annual *temperature* on mountains gradually limits the number of tree species that can maintain a positive carbon balance; but competitive relations change in a stepped manner correlated with the influences of temperature on soil organic matter decomposition rate, cloud shadow, fog and sunlight which influence photosynthesis and transpiration, and ultimately water stress and frost (Chapter 4).

6. Competitive success and niche breadth cannot be defined along single environmental gradients, but are manifested through combinations of life history traits. Furthermore, differing competitive attributes exist at different phylogenetic levels from interspecific to intergeneric and interfamilial. Phylogenetic relationships therefore mediate the outcome of competition and the biological diversity that results.[215]

6.4. *Summary and overall conclusions*

Correlates of tree species richness and diversity in Asian tropical forests, and where they are yet to be detected:

1. *The relative influence of historical and environmental factors.* The two major regions of Asian tropical forests, South Asia and the Far East, differ in the richness of their floras, and diversity of their tree communities, in a manner that is most likely explained by their differing histories of Pleistocene climate change and extinction. But the influences of historical geography and seed dispersal constraints on the differences of species diversity in their communities are subordinate to those of physical-environmental factors, at least among their more abundant tree species. That has become more apparent as it has become clear that the outcome of interspecific competition depends on a panoply of species life-history attributes. It is consistent with views reached from my early field observations and initial field analyses.[216] Nevertheless, the role of neutral factors in determining composition and diversity may increase with a decline in population densities of species, a possibility that will be difficult to test.

2. *Incident radiation.* Tree species richness and diversity is not correlated with regional gradients in above-canopy incident radiation, except in the case of their continuous decline with altitude on wet tropical mountains, and above the summer cloud base on mountains where seasonally dry lowland climates prevail. (Yet this species richness, and diversity, continues to decline in subalpine woodlands where sunny periods again increase.)

3. *The latitudinal gradient.* There is no consistent global decline in tree species diversity with latitude. The least-diverse lowland tropical communities are as poor as their boreal counterparts. Although seasonal evergreen dipterocarp forests at the northern margin tropical Asia are somewhat less diverse than those in similar rainfall regimes but lower latitudes in the Far East, it is unclear whether that is also due to other factors such as opportunities for continuing immigration.

4. *Niche breadth versus other factors.* High tree species diversity is partly correlated with niche breadth, but partly also with intermediate levels of disturbance and therefore with the

number and range of microhabitats available. Local species richness (but not necessarily community α-diversity) is also partly correlated with the age of the current climate regime, and depends on historical opportunities for migration. But it is reduced in relation to area only when habitat islands are small and isolated.

5. *Primary productivity.* Species richness and primary productivity are not closely correlated in Asia, either at the regional scale or within climatically uniform yet heterogeneous landscapes. Primary productivity appears highest where the greatest amount of sunshine is annually available to evergreen leaves that continue photosynthesising through sunny dry seasons, and in habitats where soils do not suffer prolonged water deficits. High productivity is also associated with fertile well-watered soils, as long as occasional severe canopy damage maintains fast-growing successional stages. Forests displaying this combination of characteristics are evergreen forests on the banks and alluvia of major rivers, and some in the wet seasonal tropics subject to occasional cyclones. Such climatic conditions and vegetation do not exist on a large scale in perhumid Asia, but may do so in the volcanic regions of the partially perhumid hurricane belt of Central America.

6. Given that the maximum species diversity attainable by any particular habitat or forest type (irrespective of the β-richness of the type) is achieved at intermediate levels of canopy disturbance, maximum sustained productivity levels for a given physical habitat may thus also be attained, because greater species diversity tends to reduce mortality from pathogen-caused epidemics (see *pathogen loads* below).

7. *Disturbance patterns.* The most species-diverse forests occur where climate is most equable, and in that sense stable; but some degree of instability in the form of regular yet moderate canopy disturbance is associated with highest species diversity levels. The species richness of these evergreen rainforests is highly sensitive to changes in their regime of canopy perturbation: Because of the lack of seed dormancy, limited dispersal distances, and demanding requirements for successful establishment and regeneration, these forests are easily impoverished, suffering local or regional extinction (for instance, through human activity). In that sense, these forests of equable climates are themselves unstable.

The most resilient, that is stable in structure and composition, of tropical forests are the species-poor woodlands of the semi-arid zone, which experience

predictable annual catastrophic drought and, in some cases, wind and fire at frequencies shorter than the lifespan of their trees. Although unpredictable disturbance at that scale will reduce species diversity even in this vegetation, the more species-poor a forest, the more resilient it is to predictable disturbance. Species diversity in this respect appears at first negatively correlated with ecosystem stability in tropical forest systems within a given rainfall regime.

8. *Pathogen loads.* If further research confirms that species-specific pathogenicity and positive density-dependent mortality are pervasive, it will imply that the species richness of speciose communities is mainly generated by the evolution of niche partitioning and its maintenance via natural selection. These interdependencies may endow speciose lowland tropical rainforests with considerable resilience against epidemic predation or pathogenicity – qualities notably absent from the relatively depauperate temperate forests as they currently lose species to exotic predators and pathogens.

Although other mediators of competition play a larger role than neutral phenomena in the sustainment of commoner species populations over time scales of several generations, it remains unclear, on account of the diversity of life history traits displayed by speciose communities, in what manner many of the species in consistently low populations are sustained.

9. *Climate history.* The most species-rich Asian forest floras – those of the Sunda Shelf and in particular north-west Borneo – appear to be those associated with the longest continuous persistence of the present perhumid climate regime. This is true even though these forests are located in moderately low-nutrient habitats and confined to ecological islands. Diversification through speciation also appears to have been ongoing since the establishment of the perhumid climate regime on extensive lowlands (c.23 Ma). The peak in speciation rates seems to have occurred well before the Pleistocene. There is no evidence for different rates of speciation from other regions or climates (Chapters 5, 6).

I thus conclude that patterns of species richness among Asian tropical forests broadly concur with the predictions of Givnish.[217] My observations differ with regard to the influence of soil nutrients on the rate of diversity decline with altitude, and with regard to the role of vertebrate seed dispersers among subcanopy species.

References

[1]Basset et al. 1996. [2]Condit et al. 2000. [3]P.S. Ashton 1969 . [4]Hubbell 2001. [5]MacArthur & Wilson 1967; Hubbell 1979, 2001. [6]Chave 2004. [7]Fisher et al. 1943. [8]Leigh 2004.' [9]MacArthur & Wilson 1967. [10]Francis & Currie 2003; Begon et al. 2006. [11]ter Steege et al. 2003; P.S. Ashton & CTFS Working Group 2004. [12]P.S. Ashton & CTFS Working Group 2004. [13]P.S. Ashton & CTFS Working Group 2004; Condit et al. 2004; Davidar et al. 2004, 2005. [14]Yoshifuji et al. 2006. [15]Fine et al. 2008; Ricklefs et al. 1999. [16]Cannon et al. 2009. [17]P.S. Ashton 1972. [18]S.L. Lee et al. 2000b. [19]Condit et al. 2002. [20]Hardy & Sonké 2004. [21]Leigh et al. 2004. [22]Okuda et al. 2004. [23]P.S. Ashton 2005. [24]Richardson et al. 2001; Kamiya et al., unpubl. ms. [25]Leigh et al. 2004. [26]Clark 1998. [27]Slik et al. 2003.. [28]Raes 2009; Raes et al. 2009. [29]Morley 1999. [30]Kartawinata 1980. [31]E. Soepadmo, pers. comm. 2009. [32]Ridley 1930 [33]P.S. Ashton 1972, 1982, 2004a, 2011. [34]P.S. Ashton 1982, 2004a, 2011. [35]Biffin et al. 2010. [36]Clarke 1998. [37]P.S. Ashton & CTFS Working Group 2004. [38]Pascal & Pélissier 1996. [39]Poulsen et al. 1996. [40]Rennolls & Laumonier 2000; Turner 2001b. [41]Sukumar et al. 2004. [42]Kanzaki et al. 1999. [43]Plotkin et al. 2002. [44]Potts et al. 2004. [45]Cao & Zhang 1997. [46]Ediriweera et al. 2008. [47]Swamy et al. 2000. [48]Takyu et al. 2002. [49]Givnish 1999. [50]Beaman & Beaman 1989; Beaman et al. 2001, 2004; P.S. Ashton 2003a. [51]Gunatilleke et al. 2004a, c, 2006. [52]Gunatilleke et al. 2004a. [53]Kohyama 1993, 1996, 1999; Kohyama et al. 1999, 2003; Kohyama 1994. [54]King et al. 2006a, c. [55]Thomas 1996a, b. [56]Davies 1996, 2001b; Davies & Ashton 1999; Davies et al. 1998. [57]Terborgh 1985; Terborgh et al. 2004. [58]P.S. Ashton 1982. [59]Paoli 2006. [60]Gale 1997, 2000; Gale & Hall 2001. [61]Valencia et al. 2004. [62]LaFrankie et al. 2003; LaFrankie 2005. [63]Gale & Barfod 1998; Gale 2000; Gale & Hall 2001. [64]Sheil & Burslem 2003. [65]Sheil 2001; Potts et al. 2005. [66]Brokaw 1987; Denslow 1987; Denslow et al. 1990, 1998. [67]Zipperlen & Press 1996; Hubbell et al. 1999; Bungard et al. 2000; Bebber et al. 2002b. [68]N.D. Brown & Whitmore 1992; Barker et al. 1997; Brown et al. 1999; Bungard et al. 2000. [69]Tanner et al. 2005. [70]Whitmore 1984a. [71]P.S. Ashton & Hall 1992. [72]Potts et al. 2002. [73]P.M.S. Ashton 1990b,1995; P.M.S. Ashton & Berlyn 1992; P.M.S. Ashton et al. 1992, 1995, 2006; Gamage et al. 2003; Singhakumara et al. 2003. [74]Whitmore 1989a; Burslem & Whitmore 1999. [75]Baker & Wilson 2003; Baker et al. 2005. [76]Chesson & Warner 1981. [77]Connell 1978; Connell et al. 1984. [78]Connell 1978; Connell et al. 1984. [79]Tuomisto et al. 2003a, b; Duivenvoorden et al. 2006. [80]P.S. Ashton 1977. [81]Huston 1980. [82]Tilman 1982, 1985, 1986a, b, 1987. [83]Tilman 1985, 1986a, b. [84]P.S. Ashton 1988, 1992b. [85]Bungard et al. 2000. [86]Miller et al. 2005. [87]Coomes & Grubb 1998. [88]Crane 2005. [89]Schmida & Wilson 1985. [90]Pulliam 1988. [91]Pulliam 1988. [92]P.S. Ashton 1984. [93]Yamada et al. 2006. [94]Russo et al. 2005, 2008, 2010. [95]Chesson & Warner 1981. [96]Russo et al. 2005, 2008; Yamada et al. 1997, 2007. [97]Potts et al. 2002; Potts 2003 . [98]Potts et al. 2002. [99]Suresh et al. 2011. [100]P.S. Ashton 1963. [101]Brearley 2012. [102]W. Smits, pers. comm. 1983. [103]Mukerji & Kapoor 1986. [104]Itoh et al. 1995a, 2004. [105]Wyatt-Smith 1963; Newbery et al. 1988c; Hart et al. 1989; Hart 1990, 1995; Itoh et al. 1995a; Newbery 2005. [106]Whitmore 1984a. [107]Itoh et al. 1995a.

[108]Itoh et al. 1995a. [109]Moad 1992. [110]Itoh et al. 1999. [111]Zipperlen & Press 1996, 1997. [112]Moad 1992; Itoh et al. 1999. [113]Itoh et al. 1997. [114]van Nieuwstadt & Shiel 2005. [115]Wyatt-Smith 1963; Burgess 1975. [116]Janzen 1974. [117]Suresh et al. 2011. [118]Rabinowitz 1981. [119]He et al. 1996. [120]Potts 2003. [121]Baker et al. 2005. [122]Faridah-Hanum et al., undated. [123]P.S. Ashton 1998a. [124]Comita et al. 2010. [125]P.S. Ashton 1972. [126]Volkov et al. 2003; Ricklefs 2004. [127]Fedorov 1966. [128]Givnish 1999. [129]P.M.S. Ashton & Berlyn 1992; P.M.S. Ashton et al. 1995; P.M.S. Ashton 1995; Singhakumara et al. 2003; P.M.S. Ashton et al. 2006. [130]Yamada et al. 2006. [131]Yamada et al. 1997, 2000a, b. [132]Yamada et al. 2000b. [133]Shanahan & Compton 2001. [134]Laman 1996c. [135]Rogstad 1990. [136]Rogstad 1994. [137]Debski et al. 2002. [138]Davies et al. 1998; Davies & Ashton 1999. [139]Cavender-Bares et al. 2004. [140]P.S. Ashton 1984. [141]Kursar et al. 2009. [142]Tilman 1999. [143]Grubb 1977a. [144]P.M.S. Ashton & Berlyn 1992; P.M.S. Ashton 1995; P.M.S. Ashton et al. 1995, 2006. [145]Davies 1996, 2001b; Davies et al. 1998. [146]Bell 2005; Comita et al. 2007. [147]Russo et al. 2005, 2008. [148]Latham 1992. [149]P.S. Ashton et al. 1988; Appanah 1993. [150]Kamiya et al. 2011. [151]Parkinson 1932; Pooma 2002. [152]Tanner et al. 2005. [153]Sakai et al. 1999. [154]Corlett & Primack 2011. [155]Gentry 1992. [156]Janzen 1970. [157]Connell 1971, 1978. [158]Basset & Novotny 1999; Novotny & Basset 2005; Novotny et al. 2006. [159]Novotny et al. 2002, 2006. [160]Novotny et al. 2007. [161]1Kursar et al. 2009. [162]Gilbert et al. 2002. [163]Takeuchi & Nakashizuka 2007. [164]Okuda et al. 1997; Harms et al. 2000; Webb & Peart 2000; John et al. 2002; John & Sukumar 2004; Wills et al. 2004. [165]Gilbert et al. 1994, 2002. [166]Wills et al. 2004. [167]Webb 2000; Webb et al. 2006. [168]Carson et al. 2008. [169]Ricklefs 2004. [170]Fiala & Linsenmair 1995. [171]Quek et al. 2007. [172]Maynard Smith 1982. [173]P.S. Ashton 1984; Alexander et al. 1992; Smits 1994. [174]Peay et al. 2009. [175]Hart et al. 1989; Hart 1990, 1995; Newbery 2005. [176]Stoll & Newbery 2005. [177]Wickneswari et al. 1994. [178]Chan 1977. [179]P.S. Ashton 1998c; Bawa 1974; Gan et al. 1977. [180]Kaur 1977; Kaur et al. 1978, 1986. [181]Wickneswari et al. 1994. [182]P.S. Ashton 1982. [183]Kamiya et al., unpubl. ms. [184]Murawski et al. 1994a. [185]Kamiya et al., unpubl. ms. [186]Ackerley 2000; Amarasekere 2000; Webb 2000; Webb et al. 2006. [187]Weiblen 2001, 2004; Rønsted et al. 2008. [188]Quek et al. 2007. [189]Wright et al. 2006. [190]Farrell & Mitter 1993; Farrell 1998. [191]Hubbell 1979, 2001. [192]Potts et al. 2002. [193]P.S. Ashton 1969; Condit et al. 2000. [194]May & Stumpf 2001. [195]P.S. Ashton 1984, 1992b, 2001, 2005. [196]Simpson 1953. [197]Basset et al. 1996; Basset & Novotny 1999. [198]Novotny et al. 2006. [199]Givnish 1998, 1999. [200]Giller 1996. [201]Peay et al. 2009. [202]Harrison 2001, 2006; Harrison et al. 2003. [203]Laman 1995, 1996a, c, Harrison et al. 2003. [204]MacArthur & Macarthur 1961. [205]M. Leighton, pers. comm. 1999; D. Wells, pers. comm. 2008. [206]M. Leighton pers. comm. 2001; T. Laman, pers. comm. 2010. [207]Waterman et al. 1988; Marshall & Leighton 2006; Marshall 2009. [208]Smythies 1960; King et al. 1998. [209]Smythies 1960; Medway 1977. [210]Henry 1971. [211]Henderson et al. 1991. [212]Takeuchi & Golman 2001. [213]P. Cribb, pers. comm. Feb. 2011. [214]Chave et al. 2002; Chave 2004, 2008. [215]Webb 2000; Silvertown et al. 2001. [216]P.S. Ashton 1964b, 1969. [217]Givnish 1999.

Helicopter view of recently logged mixed
dipterocarp forest, E. Kalimantan (W. Smits)

Chapter 8

PEOPLE AND THE FOREST:
A TIGHTLY INTERWOVEN TAPESTRY HAS FRAYED

With Reinmar Seidler

We have seen in previous chapters that tropical forests display at least as much variety as temperate forests. Particularly in Asia, their variability of structure and composition imposes characteristic patterns on the local as well as the regional landscape. Seen from an aircraft nowadays, much of Asia presents a predominantly agricultural landscape, with forest reduced to patches. Yet Peninsular Malaysia and much of Borneo and New Guinea still appear dark and green, at least over the hills. Why is there such a contrast?

1. The influence of ancient traditions

Uniquely among the tropical regions of the world, Asian climates consist either of an annual cycle of single wet and dry seasons, or of aseasonal perhumidity (with local exceptions, see Chapter 1). These distinctive climatic patterns have contributed to distinctive histories of land use. Land use patterns have in turn shaped wider aspects of social organisation. Until the advent of agriculture, the Asian tropics seem to have been almost completely covered by closed forest, except in floodplains where major rivers issue from the Himalaya, and on high mountain summits. Thus the land did not encourage major nomadic herding cultures. The lowland forests of monsoon Asia, though much reduced by farming since early times, were probably less disturbed until the colonial period than those of Africa or the Neotropics. Furthermore – and in contrast to the Neotropics – the ancient, complex civilisations of Asia were not swept away wholesale by European colonisation.

1.1. *Lands and peoples of the seasonal tropics*

Humans and their immediate predecessors have lived in tropical Asia for almost two million years. For most of that time, they are likely to have used fire in the deciduous forests as a means of driving game. Regions with two or more dependably dry months each year have supported irrigation-based civilisations since ancient times: irrigated wet rice cultivation had been developed by about 7,000 BP in the plains of temperate China, where it spurred increases in agricultural productivity and cultural complexity. Our first records of wet rice cultivation in the moist tropics come from north-east Indo-Burma, at

6,500 BP. The techniques appear to have become widespread in monsoon tropical Asia by 2,500 BP[1].

Between 6,000 and 4,500 BP, at the far northwestern edge of the Asian tropics, the valley of the Indus developed into the centre of a remarkably uniform 'nation' of small communalistic city-states. This civilisation was very different from the contemporary Mesopotamian civilisations to the north and west of them, but nevertheless depended on the same temperate cereals rather than on rice. Their sophisticated flood control and city drainage systems, inter-city communications and water storage tanks all imply a long familiarity with control of water, if not crop irrigation. The domestic ox was certainly known to them.

Many gods of the later Hindu pantheon were already being venerated by this early Indus Valley civilisation, but it remains unclear exactly who these people were. Were they related to the 'indigenous' Dravidians of ancient South Asia – who have inherited the oldest extant written language, and were clearly advanced for their time – or were they Semitic? South Asia apparently early evolved a distinct civilisation in which the buffalo was the primary beast of burden.

The Indus Valley civilisation endured for a millennium and a half. It collapsed either shortly before or perhaps as a consequence of the migration of Indo-European 'Aryan' pastoralist peoples moving down the valleys of Afghanistan from Central Asia. The Indo-European peoples brought horses, which they had domesticated by 6,400 BP, and horse-drawn chariots. On arrival they learned to cultivate rice, which has remained the dominant cereal in the Gangetic plain to the present day. These Indo-Europeans spread slowly southwards and eastwards into the lowlands of modern northern India and Pakistan. By 2,800 BP, forest clearance was widespread and cities were arising, implying that much of the plain was being irrigated. Irrigation spread with the expansion of Buddhism, arriving in Lanka c.2,400 BP. It was apparently well established in southern India by the time trade with the Roman Empire was established in the late third millennium before the present. The first cities on the Deccan Plateau date from 2,200 BP.

It is apparent that the Indo-Europeans venerated mountains and forests, the springs that issue from forested mountains, particular tree species, and mother-goddesses of abundance and fertility. They believed in a human descent into a watery after-life. They were sophisticated in the use of horses and of chariots. Many aspects of what we can discern of their world-view are those that might be expected of a pastoralist people originating from the dry steppes and isolated forested mountain ranges of west-central Asia.[2] Many of these aspects have been incorporated into modern Hinduism (Plates 8.1, 8.2).

Plate 8.1 (below) The ancient spiritual respect for forests. **a,** The ancient Indo-European tradition of venerating forests, mountains and springs extended from eastern Asia to western Europe. The first Indo-Europeans to reach North-West Europe, the Bronze Age people and Iron Age Celts, made Glastonbury tor a religious site. Said to then have been forested with oak, sacred to their Druids, and ash, the invading Saxons also accepted the tradition. But it is said that zealous Norman missionaries, attempting to eradicate 'superstitious anemism' cut the forest and crowned the hill with a church. The hill is still bare; **b,** but in Thailand many hills and their forests still remain venerated: in this case a new shrine has recently been built – funded by a petroleum company. (**a** Google; **b** H.H.)

Plate 8.2 (above) Veneration of water and trees. **a,** Venerating the water at a sacred spring, Tampaksiring, Bali; **b,** Offerings at the foot of a sacred strangling fig below the sacred bathing pools of Tampaksiring, Bali. Strangler figs are sacred throughout the tropics, and are found on occasion in Muslim cemeteries along with other plants venerated in the Hindu-Buddhist tradition, including *Mesua ferrea* and also frangipani, which was introduced from the Americas by the Portuguese. (**a,** Mary Ashton; **b,** P.A.)

Millennia before the spread of the Abrahamic religions, variants of this basic world-view had extended over much of Eurasia from eastern Indonesia as far north and west as the British islands: in all the places where forests, trees and springs have been venerated as elements of place-based nature worship. This still survives in the wetter regions of South Asia and on Bali in particular, and in the surviving tradition of the sacred groves, some of which are still strictly protected against harvesting.[3] Is the ancient ubiquity of sacred groves evidence for the spread of a common Indo-European culture, carried southwards, eastwards and westwards out of western Asia by waves of emigrants from 4,000 BP? Or for the migration of compelling ideas along ancient trade routes? In East Asia it was partially the latter (Plate 8.1).

The ancient deities were specific to a locality and its community, a 'village republic' whose territory was encompassed by a catchment. They oversaw the farming hazards of particular landscapes and climates. The ancient gods of South Asia were often demons (and their continued veneration in Sri Lanka is now known as devil worship). Their influence was ameliorated by Indo-European nature goddesses (*shakti*). Throughout tropical Asia, indeed throughout the global tropics, the strangler fig was universally venerated, sometimes feared. In the cities and cultivated plains, its extraordinary form (branches adorned with ghoulish descending roots) and its life-habit (as a killer parasite of noble forest giants) may command a certain awe (Plate 8.2). But forest people also know that strangler figs preside over the most fertile soils, and that their fruit are the famine food of forest wildlife (Chapter 5).

Between 1,500–1,200 BP Indian traders and *sadhus*, innocent of imperial design, spread their essential world-view eastwards throughout Indo-Burma, south-east to Sumatra but especially through the fertile volcanic and seasonally dry lowlands of East Java to Bali, and on to the Lesser Sunda Islands. Ancient animistic and location-specific understandings of the interdependence between man and nature, pervasive in humid climates north and east to Tibet, China, Korea and Japan, were assimilated. Later, as Buddhism spread, it was incorporated into this world-view or, as with the Shinto tradition of Japan, prospered alongside it. In all these regions, the landscape still bears its stamp.

1.1.1. *Irrigation: a fundamental influence*

The monsoon lands – particularly those with a shorter dry season, or where perennial streams descended from the mountains – permitted the control of water for agriculture. This ancient, yet fundamental, technological innovation of terraced rice (paddy) irrigation apparently approached its present form in northern Indo-Burma and what is now South China around 2,500 BP. It spread west and eventually south to the Sunda Islands. It permitted farmers to gradually enrich

the nutrient supply in their fields by mulching, and eased weed control. In wetter seasonal climates, up to three crops of rice can be harvested annually in perpetuity, without cleaning or restorative crops intervening, as long as the water supply can be controlled. The diversion of water from rivers, and its equitable distribution among peasant farmers, requires a scale of communal cooperation unknown elsewhere in rural societies. This astonishingly intensive system of agriculture, combined with relatively easy transportation by cart or by water, has in turn fostered the most sophisticated agrarian civilisations our world has known. The culture and technology of South Asia later spread rapidly through trade and settlement into those regions of the Far East that shared a similar climate.

Remarkably, the necessary combination of a spiritual unity with the natural landscape as a whole, and an agricultural system of extraordinary sociological and technological subtlety, was successfully transferred to indigenous societies. What a contrast with the lumpen efforts of contemporary 'foreign experts' (one of which I sadly admit to being!). Our western traditions have led us to focus over-heavily on technological solutions to individual problems, with insufficient consideration of the whole – and sometimes with dire consequences.

Irrigation became widespread wherever there was sufficient controllable water. Forest in the seasonal Asian tropics on irrigable land had been almost entirely converted to human use long before the arrival of global commodification and industrialisation. Where, in continental Asia, reliable water was lacking and forests survived on shallow soils, cattle were kept for milk, and these had long ago begun competing successfully with indigenous grazing and browsing mammals. Such forests have also served as continuous sources of fuelwood and as vital sources of famine food for humans in drier regions with least predictable monsoons. As such, forests have always been of utmost value to rural communities, and continue to be so today. Forests near communities were enshrined as sacred groves. Many of these were subject to locally and informally regulated patterns of use, and were set within increasingly deforested agricultural landscapes.[4] Meanwhile, more distant forested uplands unsuitable for permanent agriculture became the perquisite of the ruler, who was often traditionally invested with divine powers. These were his hunting grounds, but he also had a duty to protect the forest for his subjects as well. Rulers of traditional societies expected fealty from smallholders and assistance in times of war, in return upholding certain land rights. Grazing access for commoners was limited in years of average weather, but could be opened for the collection of fodder and famine foods in times of hardship. The system worked adequately until internecine warfare and infant mortality began to decline in the 19th Century, and rural populations started to increase beyond the capacity of the land to yield using traditional technologies alone (Plate 8.3).

Plate 8.3 When British forces overran the Deccan (Karnataka) 200 years ago they were mortified to find, not a tyranny of oppressed peasants, but a green and pleasant land where roads were paved in a concrete, and lined with shade-bearing figs (*Ficus bengalensis*) and other trees, while vast forests were conserved for game. Everywhere, the country bore the stamp of an enlightened ruler, Maharaja Krishna Mahendra Vodiyar of Mysore, whose benificence has survived to the present day. (H.H.)

Recent scholarship has, to some extent, picked apart the impression of smoothly functioning traditional systems of resource-use regulation in pre-colonial Asia, and the question of whether such systems can meaningfully be related to 'conservation' is contentious.[5] Nevertheless, the central point here is that, as human populations slowly expanded in Asia, there was a need for organisation of resource-productive landscapes. This led to the development of a variety of systems of land 'ownership', or control that were quite well-adapted to local conditions and rather robust over time.

1.1.2. *The later Mughal influence in South Asia*

Taxation on a large scale was not imposed in India until the Mughal Empire, when imperial taxes began to be collected as rents by *zamindars*. The word *zamindar* derives from the Persian, meaning approximately 'land owner'. The *zamindars* were not technically proprietors of land, but they effectively controlled the uses of land over swathes of territory, sometimes enormous, and were also given some of the judicial authority of a rural magistrate over disputes. The *zamindari* system thus gave a hereditary aristocracy practically unlimited powers over the lives of peasants, recodifying and reinforcing the rigid caste system throughout much of rural India. This widely hated feudalistic system endured into the colonial era, and was unfortunately co-

opted as a practical measure by the British Raj. Forests under *zamindari* authority were often degraded, presenting difficult negotiations for the young Indian Forest Service.[6] They set up a central conflict – still animating many of the conservation disputes in tropical Asia – between varied, locally adaptive patterns of control over land and forest, versus larger-scale, centralised systems established in the name of administrative efficiency but often privileging the few at the cost of the many.

The situation in colonial-era Neotropical regions was perhaps not so very different in practical terms, but it contained an important formal contrast. There, legally codified European conventions of land ownership were rapidly extended to encompass private dominion over vast estates. These imposed an artificial and anomalous system of outright private ownership of land – above and beyond control of productive uses of the land – that still proves resistant to land reform. This system also provided a historical model for legal land titling which has smoothed the entry into Latin America of today's growing emphasis on commercial incentives for conservation on private lands[7] (Chapter 9).

The security afforded by clearly defined ownership of the land by those who work it – when underwritten by a fair price for produce at local and national markets – is in recent times often regarded as fundamental to good management. Yet the Asian tropics encompass a hugely varied range of ownership arrangements, both historically and currently. Many of these have centred on customary *communal* control of land, often in combination with privately held plots. Where such arrangements have been part of the fabric of traditional society, such as in many Adivasi ('tribal') areas of India, there has been bitter conflict between highly centralised government agencies and the more flexible community structures. The latter have traditionally focused on tribal usufruct rights in forest land, but are finding a new emphasis in India on land titling with the passage of the Forest Rights Act of 2006. Such moves toward legal enshrinement of land titling restore security of ownership in principle, but they remain slow and halting. The process is often open to charges of neoliberal domination by markets and co-optation by commercial interests as well as by the state agencies (Chapter 9).[8]

In Sri Lanka, where private land titling has long been the norm, and where traditional respect for nature is also strongly engrained, a populist conservation movement among rural and urban groups has achieved legislation of policies for sustainable use and conservation of indigenous forests rivalling any in the world. In much of tropical Asia, however, virtually all forest land is still controlled and managed in principle by government agencies, with all the ambiguous administrative fall-out that status implies. (This is true whether or not 'forest' is defined as including tree cover. In India – as it also was in medieval Britain – 'forest' remains largely an administrative category: forest land is largely that *not* administered by the Revenue Department.)

Especially in the drier Asian climate regions, population pressures have led to overexploitation of forest for fuelwood and grazing, for hunting, and for hardscrabble dryland agriculture. Social scientists have emphasised the need to involve communities in forest management in these instances. Starting in the 1920s in northern India, *panchayat* forests were legislated for in some states to ensure the needs of village communities for forest goods.[9] These moves were consolidated nationally in the 1927 Indian Forest Rights Act and the 1931 *Van Panchayat* Act. But it was perhaps unrealistic to expect that village franchise and village leadership alone could implement and maintain the long-term consistency of policy necessary to sustain the growth of hardwood trees and the survival of forests. This early attempt at community management could be successful only where earlier traditions of continuous forest use were maintained, since the laws enshrining the reserved timber production forests remained in force, deterring alternative uses. Less-restricted state lands and community-managed forests meanwhile became degraded.[10]

South Asia, West Java and Bali support the densest rural communities in the seasonal wet tropics of Asia. In the latter regions, fertile base-rich volcanic soils have supported irrigated rice throughout the lowlands. The only forest stands remaining in the lowlands of West Java are in the Udjung Kulon National Park. This area now receives protection as the last refuge of the Javan rhinoceros, but it has also survived as forest because it rests on infertile acid volcanic substrate. The flora does not, therefore, represent the original West Javan lowland seasonal evergreen dipterocarp forest. This has long since gone, excepting patches and relict individuals of a few species.

1.2. *Agriculture in the perhumid Far East*

Control of water has always proved difficult in perhumid regions prone to flash floods, especially where sandy soils over siliceous rocks prevent the construction of stable field bunds and banks. High and continuous rainfall prevents accumulation of humus and leaches nutrients, while absence of a clearly defined dry season reduces the reliability and volume of rice harvests. Despite all this, evidence for settled agricultural activity in New Guinea stretches back 26,000 years.[11] *Homo sapiens* seems to have reached here by at least 50,000 BP. About that time, Australia also became inhabited, but that continent did not know complex agriculture before European settlement. Rising sea levels only separated New Guinea definitively from the Australian mainland around 6,000 BP. Therefore, either cultivation must have arrived very late to Australia from the north, or more likely, there were other reasons why simple agriculture was expedient there. Evidence of an increase in fire appears at 17,800 BP in the northern Sumatran mountains,

implying that swiddening immigrants may have arrived at that time;[12] but recent research on charcoal records indicates that bushfires on the Australian continent may have reached a low just at this time (c.18,000 BP). If so, fire there may have been responding to climate changes more than to human activity.[13]

Until the colonial era, in any case, power in Malesia lay along the coast where extensive sedimentation from the rain-washed hills supports mangrove and productive fisheries, and where the sea facilitates communications and trade. Nations developed around settlements in the upper tidewaters of the major rivers, thereby exerting control over both the rich coastal fisheries and the hinterlands while retaining access to plentiful fresh water. Well-placed communities dominated the regional trade in natural resources, including rice. Nevertheless, control of the forested hinterlands was more intermittent, less pervasive, and the upland peoples fewer and more widely dispersed.

1.3. *Throughout tropical Asia, a common tradition and history of land use*

1.3.1. *The organisation of agricultural landscapes*

Throughout the well-watered regions of South Asia and the Far East wherever land tenure is secure, rural inhabitants live in villages and towns strategically placed on the lower slopes of the hills (Fig. 8.1, Plate 8.4). Around their houses they cultivate miniature groves or 'home gardens', each containing a rich mixture of useful herbs including vegetables, medicinals, cosmetics and fibre sources, as well as fruit and timber trees. The wetter the climate, the more diverse are the available species and the richer the mixture.[14] These groves still incorporate mostly native species, but they also include abundant evidence of the ancient exchange of seed throughout tropical Asia (Table 9.1). They even contain exotic species lacking seed dormancy, which must have been transported under sail as seedlings. These include fruit trees native to the Western Ghats and North-East India now in Java and the Far East, such as jack (*Artocarpus heterophyllus*), mango (*Mangifera indica*), jambu (*Syzygium aqueum*), the *Averrhoa* starfruits, ornamental flowering trees (Chapter 6), probably the sugar palm (*Borassus flabellifer*) – extolled in the *vedas* for its 800 uses – and even the riverbank tree *Terminalia arjuna*, said to reduce the hardness of water. Some of these South Asian species already appear in frescos at the great 9th Century Buddhist shrine at Borobudur in East Java. The trade was a dynamic of mutual exchange: *Syzygium jambos*, *S. samarangense*, mangosteen (*Garcinia mangostana*), rambutan (*Nephelium lappaceum*), bananas (*Musa* spp.), durian (*Durio zibethinus*) and coconut (*Cocos nucifera*) each arrived in southern India and Sri Lanka from the opposite direction. In the Far East, forests have themselves been modified by selectively managing the useful species.[15]

Fig. 8.1 The traditional land use in perhumid and moist seasonal south-west Sri Lanka. More productive annual crops occupy the floodplain and lower slope, the predominantly woody home garden the mid-slopes, merging with those of neighbours to form a continuous forest band. Above are steeper slopes used for swiddening if the paddy harvest fails; while aboriginal forest along the ridge provides fuelwood and NTFP. (Expanded from P.M.S. Ashton *et al.* 1997)

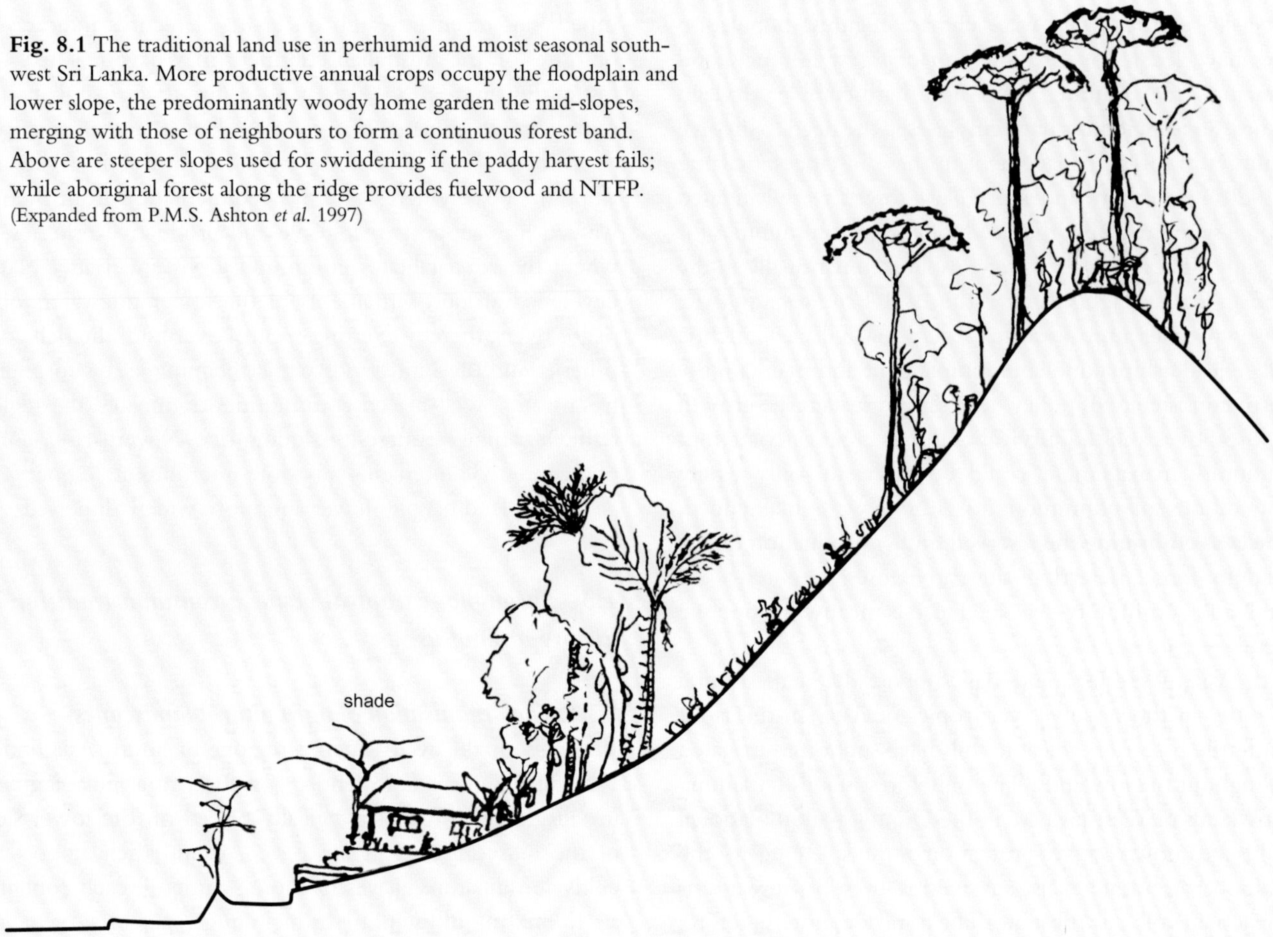

shade

Plate 8.4 The intelligently devised and picturesque traditional landscape of the moist regions of Asia, north to Japan, in which every site achieves its sustainable potential. **a,** A village landscape in south-west Sri Lanka: paddies where irrigation is feasible; the road between the villages, and homes with gardens on lower slopes; swidden above, and patches of forest along the ridge. A scene of traditional rural Asia throughout the region of tropical seasonal wet evergreen, semi-evergreen, and tall deciduous forests, and into the moist warm temperate Far East: wherever clay loam soils prevail; **b,** Neem, *Azadirachta indica*, a taller species of short deciduous and semi-arid woodlands, is one of the most valued trees in what is hard-scrabble country: it flourishes on poor siliceous soils better than on the fertile soils over basalt. Neem is commonly planted, as here, in villages and along field margins. It is the major source of fuelwood, and of fodder; it is unpalatable to cows, who cannot digest it, but relished by goats and sheep. Women and children scale the trees to shred the branches, as they did alders in medieval Europe (illustrated in Hobbema's classic painting *The Avenue at Middelharnis*). The bark is rich in antibacterials and the twigs are popular for brushing teeth. An extraction of the leaves, too, is taken as a decoction, which is drunk as a tonic. (P.A.)

The upper slopes – especially above 1,200 m where few lowland crops bear fruit (Table 4.11) – were generally left under forest (New Guinea excepted) and respected, even venerated, as the home of benevolent deities and the source of the all-important reliable water supply. Nearer to habitations some forest would be cut for swidden, originally in famine years but recently more frequently as families have expanded. Even on poor soils, a patch of extra swidden rice can be grown up to about 1,500 m.

The landscape was thereby organised as a whole, rather than by individual parts and separate uses. In this, its use is analogous with the concept of *satoyama*[16]. That Japanese tradition does not include man-made mixed-species and multi-purpose groves, nor does it incorporate exotic species; but the patterns of use across these landscapes are remarkably similar, being imposed by the landscapes' shared capacities and constraints.

In other places, the ebb and flow of armed conflict has led to the abandonment of regions by the vanquished. In the West Himalayan foothills, Banj oak (*Quercus oblongata*) forests have arisen from the deserted farms of the formerly powerful Kumaon nation. In Pegu (Bago, Burma), much of the finest teak has established on and near the abandoned farms of Chinese settlers evicted in an 18th Century conflict. Yet great tracts of aboriginal forest still survive in South Asia, even now, on lands that were formerly often under the protection of a ruler imbued with divine authority, who hunted in it with the court yet tolerated the forest tribes.

1.3.2. *Upland peoples*

1.3.2.1. *Graziers.* It is unclear when domestic cattle entered the Asian tropics, but they would certainly have been present in India after the Indo-European immigration. Cattle do not prosper in the lowland wet evergreen forests, but graziers have long left cattle to roam in the deciduous forests and the lower montane evergreen forests of the Himalaya and Indo-Burma, where remnants of an annual transhumance cycle of extraordinary complexity still exist in some places (Table 8.1; Plate 8.5).

Table 8.1 The main categories of cattle of the Eastern Himalaya. Herders with *mitun*, and those with other types of cattle on the hills above, move up into summer pastures, while descending to spend the winter in or near the tropical zone. In autumn and spring the roads are clogged with transhuming families and their animals.

Principal habitat	Animal
Summer: alpine pastures; winter: temperate forests	Yak, *Poephagus grunniens*
As above, but more often lower	*Zho* (yak x domestic cattle)
Temperate forests and meadows	Domestic cattle
Tropical lower montane forests	*Mitun* (domestic cattle x gaur)
Tropical lowland and lower montane forests; not domesticated	*Gaur*, wild Indian ox (*Bos gaurus*)

Plate 8.5 Grazing traditions that successfully persist. **a,** The Mitun of the Eastern Himalayan lower montane and warm temperate forests and pastures, a cross between the gaur and the domestic cow; **b,** Hay ricks of a smallholder in a dry district, 30 km north of Udaipur, Rajasthan, March 2013. Private landholdings allow creation of security against stray cattle, enabling hay production impossible in state forest reserves perceived by the populace to be common land, and difficult in communal forests. (**a** H.H., **b** P.A.)

1.3.2.2. *Farmers*. Elsewhere, especially in the perhumid tropics, and large hill tracts of the drier tropics where ridges are high enough to attract rain, the climate or topography have been too intractable for irrigation cultures to master. Minority cultures have survived in such areas by means of *slash and burn* or shifting agriculture – *swidden* in the Old English terminology. Dry-rice fields are left as woody fallow to regenerate by coppice or seed, eventually to be cut and burnt again, thereby releasing nutrients before replanting. Woody fallow periods were maintained traditionally for up to 15 years in the perhumid tropics, or up to 5 years in seasonal regions where soil nutrient levels are generally higher[17] (Chapter 1). There is considerable debate about the long-term effects, positive and negative, of swidden on the ecology of tropical forests. There is certainly wide variation in these effects, as there is in the surrounding forest type and in swidden techniques. But judgments about effects depend heavily on points of comparison. Thus, traditional swidden-based subsistence systems in South-East Asia were found to be far less damaging to soils and hydrology than almost any more intensive form of cultivation, particularly since the latter commonly involve the construction of roads and infrastructure;[18] while swidden under short-fallow rotation times results in poorer carbon storage and soil quality than swidden under long-fallow regimes.[19] Government agencies over the last decades have tended to focus on the negatives associated with intensified, short-rotation swidden and the cultivation of opium as a cash crop, both of which have increased rapidly along with increasing populations of invading landless lowland colonists.

The Western Ghats of southern India and perhumid south-west Sri Lanka are less-fertile metamorphosed granite. Human populations only colonised them densely in the last few centuries, and at first only on trap basalts and more seasonal districts to the north. In India, the lowlands of the Carnatic coast and the foothills abound in sacred groves, while the upper slopes and hills remained until recently under primary forest cover,[20] but for the expansion of upper montane grasslands under cattle grazing. Though spices such as pepper and cardamom have been exported from the Malabar Coast on a relatively small scale for millennia, it was the introduction of large-scale export commodity crops during the colonial period, here coffee and tea, that has been responsible for most forest conversion. Where no profitable commodity crop has transformed the local economy, and where local land ownership and customary rights have been maintained, the rural economy remained substantially at subsistence level. Paddy cultivation provides the excess crops which have ensured national food security.

In the seasonal mountain landscape of northern Luzon, people with irrigation technologies of great subtlety invaded several thousand years ago. These wet tropical irrigation-based cultures, originating in south-east China or north-east Indo-Burma, evolved the most dramatic paddy agricultural landscapes ever devised. They are still to be seen in tropical and warm temperate Asia, from Luzon and Indo-China north-east as far as Honshu. In the high hills and narrow valleys, even the steepest slopes (sometimes exceeding 45° of inclination) have been painstakingly stepped, walled and bunded, creating tier upon tier of intensively-managed, meandering paddies and a landscape of unsurpassed beauty and productivity. But such crops are seasonally dependent on water captured from fog by the ridge forests high above.

In Indo-Burma also, as Han populations to the north expanded, hill tribes such as the Hmong were pushed south and brought with them the warm-temperate traditions of terraced but non-irrigated vegetable cultivation for the market. As a consequence, in northern Laos and Vietnam in particular, very little lower montane forest remains today.

1.3.3. *Division of land use leads to inequality of power*

Ubiquitous lowland irrigation had its drawbacks. It produced a fundamental division within societies, between those who occupied the plains and lower slopes amenable to irrigation, and those in the hills who had to rely on the rains directly.[21] Up to 1,200 m, most slopes are amenable to swidden. In perhumid climates, though, only the less steep lower slopes can be cultivated continuously (unless the soils are exceptionally stable and fertile, as on and near the volcanic Barisan range of Sumatra). Thus, on the one hand, the ox or buffalo and plough combined with naturally high irrigation-based productivity and good communications provided the means to political consolidation and power in the lowlands. It also fostered specialisation of work and social hierarchy. On the other hand, the people of the high hills were restricted to the dibble stick, extensive farming practices and diffuse populations. Difficult communications encouraged the separation of societies and the diversification of languages. Political power among these indigenous groups was limited and local, lying ultimately with the 'foreigners' of the plains who controlled, through their higher demand and better communications, the trade in natural resources. Even with the extraordinary opportunities now available for communication, this division continues hard to bridge, let alone eliminate. In Burma, and to a lesser extent elsewhere in the Far East, the expansion of lowlanders or their commercial interests into the traditional lands of the uplanders has led to civil wars.

1.3.4. *Swidden cultivation*

The timing of the first agriculture in the perhumid tropics of Asia is unknown, although taro (*Colocasia esculenta*) appears to have come into cultivation in New Guinea a remarkable 10,000 BP.[22] Swidden (Plate 8.6) has apparently been guided sustainably by communal traditions of considerable sophistication. In some areas it continues to be so – providing human population numbers remain stable. But even swidden cycles could only be sustained over time on the more fertile substrates and the deeper clay loam soils of lower slopes and alluvium. Swiddeners in perhumid regions learned that burning, essential for release of nutrients from felled woody biomass, also consumed the greater part of the soil organic matter in less fertile humult soils; the raw humus which takes generations to regenerate. In sharp contrast to Africa and the Neotropics therefore, most of the Sunda Hills and much of the Philippines and New Guinea remained under primary forest until European colonisation. Irrigated agriculture reached into the less fertile areas relatively recently, and generally only under one of two conditions: when local populations flourished (as when Javanese and other Hindu-Buddhist satellite nation-states expanded, such as Sri Vijaya in Sumatra, at Samarinda in Kutei (East Kalimantan), and at Ligor in peninsular Thailand-northern Peninsular Malaysia), spill-over populations moved into marginal areas; or, on the other hand, when alien nations invaded and inhabitants retreated onto land they had previously avoided as happened, for instance, in south-west Sri Lanka and in parts of Burma.

Plate 8.6 Murut swidden, Ulu Senuko, Batu Apoi, Brunei, June 1957. The clearing is no bigger than a large windthrow, a feature of these clay loam soils and their MDF where primary forest species re-invade after abandonment. (P.A.)

Upland swidden cultivators developed deep knowledge of their demanding and variable terrain and, in times of stable population numbers (often reduced by periodic warfare), managed to sustain productivity by accepting traditional rules of constraint by mutual assent[23] (Chapter 9). It would be misleading, however, to claim that traditional swiddening tribes always lived in sustainable harmony with the land upon which they depended. Some, like the Iban of north-west Borneo, were habitually expansionist, moving onto and taking over the lands of other tribes after their own land was invaded with perennial weeds, especially the deep-rooted *Imperata cylindrica* grass, as can be seen over the hills of Sarawak's Saribas district. The seafaring Bajau (a generic term for seafaring peoples of south Philippine and north-east Bornean origin who have spread over much of Indonesia) have habitually settled near the coast, cut nearby forests to cultivate, then moved on in search of new land rather than adopting permanent swidden or other agriculture. The Hmong of Laos and Thailand intensively exploited hill land until it was exhausted, then moved on, often invading the lands of more sedentary tribes. The *Imperata* grasslands of the Plaine des Jarres and Plaine des Bolovens, plateaux on the Annamite range, are testimony to the failure of hill tribe cultivation to be sustainable in the past. Now, such 'green deserts' survive, and even expand owing to burning by hunters.

1.3.5. *Village governance and the state*

Before the arrival of European colonialism, the political concept of the nation-state, as currently understood, did not exist. The disposition of land, especially uncultivated forest land, was broadly under the authority of the local ruler; but in reality it was determined by local communities, although hunting was often more strictly under the ruler's control. The establishment of reserved forests for timber production by colonial governments, and their regulation of forest use, thus transferred much power at a stroke from the village to the metropolis and from upland minorities to the lowland majority. This wholesale transfer – and the later transfer of power from colonial to national administrations – created a series of harsh dissonances between people's needs and experiences and the legal-administrative frameworks available to accommodate them. In many places these dissonances have yet to be adequately resolved.

2. The modern world of global trade

2.1. *A long history*

Trade has existed between the Asian tropics and temperate Eurasia at least since Roman times. For over a thousand years, the states of Indo-Burma and around the China

Sea have been sources for China of raw materials from the rainforests, including timber, rattan, and birds' nests. High-value medicinals such as the antidote bezoar stone, still extracted from the guts of infected monkeys for export, have been exchanged with China for ceramics and other products of its already advanced economy. Chinese merchants came to do business and to reside in these 'tributary states', becoming influential in the courts on account of their commercial skills which provided access to exotic luxury items such as ceramics and silk. They often achieved a dominant influence as intermediaries between rulers and their people in trade, finance and sometimes governance. From the 17th Century, some European adventurers, including functionaries of the British East India Company such as Samuel 'Siamese' White and Elihu Yale, attempted to intervene in such relationships, often to their own disastrous undoing.[24] As early as the 14th Century, Arab traders discovered the spices and aromatics of South Asian and Far Eastern rain forests, and lively trade ensued in nutmeg, cinnamon, camphor (the aromatic resin of *Dryobalanops*) and eaglewood (the scented wood, infected by a fungus, of *Aquilaria* and *Aetoxylom*). Portuguese coastal trading colonies spread as far as Japan in the 15th and 16th Centuries. They left a major mark on the cultures of tropical Asia, in the *baila*, *lagu*, and *kroncong* folk songs of coastal Sri Lanka, Malaya and Java, but also in the many trees and shrubs brought from Brazil: papaya (*Carica papaya*), sapodilla (*Achras sapota*), avocado (*Persea americana*), custard apple (*Annona* spp.), and – amazingly – chili (*Capsicum*), now grown in home gardens throughout the wetter parts of the region. Who could today imagine a curry spiced, as formerly, solely by the indigenous black pepper (*Piper nigrum*)?!

2.2. *Commodity plantations arrive*

In contrast to the Portuguese, the later Dutch and British brought neither plants for home gardens nor folk music to enliven a languid evening. Instead, they focused on their commercial interests: exchange of commodity crop germplasm. Botanical gardens were established, at Calcutta in 1787, Buitenzorg (Bogor) in 1817 (Plate 8.7), Peradeniya in 1821, and Singapore in 1859, for the purposes of cultivating

and improving plants of commercial potential and exchanging them between the continents (Plate 8.7). By their means, coffee and tea were introduced into the hills, and latterly Brazilian *Hevea* rubber and West African *Elaeis* oil palm in the lowlands. Having wrested control of the indigenous forests from the rulers and constrained the activities of the swiddeners, they then proceeded to clear residual wet upland forests, which they regarded as wastelands. In 1848 in Sri Lanka, Kandyan farmers rebelled against the colonial government of the 7th Viscount Torrington who, with extraordinary insensitivity, allowed coffee planters to expropriate swidden fallow *chena* fields, and had even imposed a tax on village dogs![25] Only in India were rulers of local states sometimes able to withstand this quest for prosperity for the few.

Plate 8.7 The Indonesian National Botanic Gardens at Bogor (Buitenzorg), now an ornamental park and botanical research collection but founded in the early 19th Century by the Dutch, who regarded introduction and evaluation of new plants of commercial potential a major priority. The monument is a memorial to Mariamne, first wife of Stamford Raffles, founder of Singapore, who died in Indonesia when Raffles acted as Governor of the East Indies while the Netherlands was under the sway of Napoleon. Raffles was himself a keen naturalist and, accompanied by botanists, made an important early botanical collection in West Sumatra. (P.A.)

First, it was the lower montane forests on less-steep slopes in wet seasonal climates, with their humus-rich loams, that were found to be ideal for woody tropical commodity crops. In the mid-19th Century, Arabian coffee (*Coffea arabica*) was introduced, but the dense coffee monocultures were early eliminated by a fungal pathogen, *Hemileia vasta*. Coffee was largely replaced by tea (*Camellia sinensis*), a species indigenous to warm temperate East Asia and to the lower montane forests of Indo-Burma (Chapter 4). Powerful defensive phenolic metabolites have allowed tea to withstand epidemic pathogens. But the lower montane oak-laurel, *shola* and *rata dun* forests of South Asia were thereby widely eliminated on all but the steepest slopes.

Then in the late 19th Century, a Euphorbiaceous, late-successional light hardwood from Amazonia, *Hevea brasiliensis* (hevea rubber), was found to be far more productive and dependable than the native gutta percha (*Palaquium gutta*), which had to be felled to extract its latex. Hence rubber – introduced without its pathogens – was planted: the first lowland commodity crop suitable for wet climates and soils of unexceptional fertility. Over the last 40 years, rubber has been exceeded in profitability only by oil palm (*Elaeis guineensis*), though rubber remains competitive. Malaysia hosts the world's largest condom manufacturers. These two commodity crops can yield a family with four hectares of land an annual net pre-tax net income exceeding US$4,000, more than many (unsubsidised) cereal farmers can expect in temperate climates. Together, they are all but eliminating MDF outside of protected areas. Protected areas have in any case been declared almost entirely on land too steep, too rocky or too infertile for agriculture, where the MDF is itself suboptimal. The Pasoh Research Forest is one rare, accidental exception.

Cardamom (*Elettaria* and *Amomum* spp. of the ginger family) is an indigenous crop of lower montane forest, the former mostly in the Western Ghats and Sri Lanka, while *Amomum* has also been a significant commercial crop for smallholders of the Eastern Himalaya.[26] Cardamom is grown in the cleared understorey of the aboriginal canopy, and has been expanding until recently in response to commercial demand. Cardamom cultivation seems to have little short-term impact on carbon assimilation, erosion or water quality, but of course it prevents regeneration of the canopy species above it.

What nation would have done otherwise than convert the forest? From the air, the land still appears forested, but on the road, one observes mile after mile and rank upon monotonous rank of a single palm or tree species, just tall enough to obscure the exquisite landscape of hills behind. Alternatives to this simplified and fragile landscape, so inviting to an invasive pathogen, must realistically be driven by economic considerations, and some of these will be discussed in the next chapter.

2.3. Declining forests induce efforts to sustain them

Western trading companies in the East were interested in a limited number of forest products, amongst which timber became pre-eminent by the end of the 18th Century. The companies commandeered the power of government to stabilise and improve their terms of trade. They were uninterested in sustainably managing the resources that yielded goods of value until those resources began to be depleted. European forestry practice in the tropics, focused on sustaining timber yields, intervened eventually but developed differently under the varying climatic conditions (and hence silvicultural ecology) of the forests in question. In India and Burma, where it developed first, tropical timber producing forests are predominantly deciduous; in Malesia they are evergreen rainforest.

2.3.1. European forestry practice arrives in Asia

In India a single species, teak, became indispensable to the British navy for decking and other ship furniture. By the early 19th Century, the market value of teak was rising in response to the gradual exhaustion of accessible sources along the western coast of peninsular India. Forest degradation accelerated soil erosion and the siltation of harbours, so that vessels of international trade were forced to anchor off-shore. The British East India Company had appointed a 'Conservator of Forests' for the teak-bearing lands of Malabar and Travancore (now Kerala) in 1806. But Britain had no scientific tradition in forestry, and its efforts proved disastrous. The first commercial teak plantation was established at the initiative of the Company's district administrator, or Collector (of taxes, nominally on behalf of the Sultan of Delhi). This was Henry Conolly, also based at Malabar.[27] Conolly was concerned that the practice of buying timber directly from forest owners, the 'proprietors', gave them no incentive to regenerate the teak. Several rulers, including the Raja of Nilambur and Sultan Tippu of Mysore, were found by the British to be managing their forests and lands with care, but many other forests had become degraded. In Jan. 1840, Conolly wrote to the Company government in Madras (Chennai), suggesting that active management of regeneration should be enshrined in law. His recommendations constitute the basic principles of forest management that continue in India to the present day. Their aims were, and are, to obtain 'a complete knowledge' of the quantity and quality of timber in a prospective forest; to forestall any kind of depredation being committed on forests, whether government-owned or leased; and to improve the forests by new plantings and by unremittingly fostering the growth of young trees.

Conolly first negotiated to replant a logged teak forest at Nilambur. He tried more than once to place management in the hands of young Englishmen, none of whom rose to the challenge of inducing teak seed germination. He finally appointed Chatter Menon who, based on his own prior experience and by seeking advice, succeeded by soaking the seed in water and planting under light soil cover. Thus the first commercial timber plantation in the tropics was established, and the first forester in the Western 'scientific' tradition appointed. It says much for the spirit of India that the trees of Conolly and Menon became so magnificent that the plantation has become a conservation area, its trees revered as manifestations of the divine.[28]

By mid-century, commercial deforestation in India had become serious – so much so that at its annual conference in 1852, the British Association for the Advancement of Science sponsored a session on deforestation in the tropics! A committee was appointed to 'consider the probable effects in an economic and physical point of view of the destruction of tropical forests'. The committee recognized the necessity of converting accessible forest land to agriculture to feed and employ a population that was increasing under stable rule, but emphasised that much land was too steep, or its soils too infertile, for continuous agriculture, and should therefore be maintained under managed forest. The continuing concern engendered by this topic today, after one and one half centuries, is a measure of its intractability. (Conditions necessary to its solution will be presented in our final Chapter.) Recommendations resulting from this meeting eventually led to the founding of the Indian Forest Service, as an early action of the British government that had taken over the governing of India from the humiliated and disbanded chartered company following the great rebellion of 1857.

In 1855, Governor-General Dalhousie had issued a proclamation on the rapid commercial degradation of the Pegu (Bago) forests of lower Burma, which became known as the 'Charter of the Indian Forests'. In 1864, Dietrich Brandis was appointed Inspector-General of the new Indian (then Imperial) Forest Department, which became the Imperial Forest Service (IFS) in 1866. He was responsible for the establishment of an institution of pre-eminent reputation and *esprit de corps*.[29] No facilities existed in British universities for the training of professional foresters for another 40 years, and then only as a consequence of the growing demand from India. (The US Forest Service was effectively founded in 1881 partly in consultation with foresters in India, becoming the Bureau of Forestry in 1901. The British Forestry Commission was founded only in 1922.) France and Germany, then having the most distinguished record for 'scientific' – that is, investigative and systematic – forestry, were sought for professional foresters. The future development of scientific forestry, in India and in the tropics, was fortunate in the appointment of three outstanding practitioners: Dr (later Sir) Dietrich Brandis, father of Indian forestry; Sir William Schlich, who went on to found the first forestry department at a British university, which eventually

became the Commonwealth Forestry Institute at Oxford; and B. Ribbentrop.

Brandis was first posted to Pegu in order to administer the moist teak forests under British administration. There he proved himself tireless in the field and in science, in administration and in policy. However, he immediately came up against the leading challenge in tropical forest management, which in various forms continues to the present day: the chronic disharmony between commercial and conservation interests. Commercial timber interests, 'renting' rather than owning the forest, tend to favour profit maximisation from a single crop while disregarding damage to the forest's regenerative capacity; while the forester or conservator, representing the owner – namely, in principle, the nation – endeavours to sustain the value of the forest by preparing for its regeneration following harvest. British commercial interests had negotiated logging concessions for the magnificent teak from local rulers in Pegu that were favourable to both parties but disastrous for the forest. Brandis saw it as his task to amend that, and he took early steps to institute two practices fundamental to sustainable forestry in the tropics.

First, Brandis introduced strict management protocols. In Burma, he soon ran into administrative opposition when the appointed Inspector of Forests – a Mr Leeds who lacked formal forestry training – backed commercial interests against the new regulations. Even then, and commendably, the colonial government backed Brandis. The track record among British colonial governments, lest we forget, remained admirable in this respect.

Second, Brandis took early steps toward a kind of community involvement in forest management. In contrast to India, where forests were in danger of degradation by cattle browsing and fuelwood collection near villages, in Burma the upland people widely practiced a form of swidden there called *taungya*. At the same time, landless lowland Burmese with little knowledge of upland forests or agricultural practice were invading accessible forest areas. Teak had proved difficult both to plant and to regenerate naturally, owing to the difficulty of insuring seed germination. Yet Brandis observed that regeneration was often present in the disturbed semi-evergreen forest near villages. Brandis at first advocated demarcation of remaining good stands as strict forest reserves for teak, within which *taungya* should be strictly proscribed. But he also advocated reservation and sustainable management of scrub forests in the Irrawaddy plains and elsewhere. Communities depended on such forests for fuel, browse and other products, and they were in danger of degradation. Meanwhile, one of Brandis' conservators, Seaton, had persuaded Karen upland swiddeners to plant teak before they abandoned their fields to woody fallow. After some decades of mostly passive resistance by at least some elements of the Karen, and as populations continued to expand (particularly from immigration), this approach gradually became accepted. It has been adopted worldwide in a hybrid system known as *taungya* forestry, by which forest villages are permitted to plant crops for several years in reserved timber production forests in return for their labour in clearing and planting seedlings of timber species.[30] When the seedlings grow large enough to shade out the crops, the villages are moved on to new areas. The system was designed to simultaneously satisfy both local subsistence and external commercial demands, but it has been strongly criticised on grounds of political economy, as a method of extracting a 'stealth' economic surplus (in the form of labour) from villagers even while building the state's political control over them and gradually restricting their ability to practice traditional cultivation.[31]

The specific Burmese system aside, *taungya* forestry certainly displays at least two severe disadvantages. It removes traditional forms of land security while replacing them with little else: they pay no tax, so why should they merit services? And the need for forest labour is never great enough to absorb increasing human populations.

2.3.2. *The search for an equitable sharing of forest resources*

Brandis wisely came to see opportunity in gaining the participation of forest villagers in the management of the teak rather than their exclusion which, in so many parts of India, was advocated by European foresters trained in the tidy management traditions of temperate Europe. Nevertheless, successful negotiation of collaborative forest management with forest-dependent groups increasingly eluded colonial foresters. Reconciling commercial and community interests under equitable government oversight remains an elusive goal to this day. A consequence throughout Asia has been the perception that the forested landscapes, inhabited from time immemorial by swiddening minorities, have been usurped by 'government'. Even where communities cultivated only the more fertile lower slopes, they often assumed the upper forested ridges and slopes to be their heritage, depending on them for water, game, and non-timber goods, and seeing them as the abode of their gods. This was sensed by Rajah Vyner Brook of Sarawak, who for many years resisted advice to set limits on the expansive swiddening activities of the invading Iban by creating reserves. Though he conceded eventually, attempts by his government to legalise landholdings largely failed.[32]

Under ideal circumstances, the government should represent the nation, and uplanders should receive services such as schools and medical facilities in exchange for their contributions to the natural economy, and agreed limits to their activities. Eventually, timber harvesting brings roads to the uplands, and with them the opportunity for uplanders to share more equally in the benefits of the national economy. However, improved social and economic equity remains the

remotest of ideals in most places. Even the relatively benign rule of the kings of Bhutan has been criticised from an equity standpoint. In the worst cases – and not infrequently – the benefits of logging have failed to trickle down to anyone beyond the concessionaires and their political patrons. Even where they have, urban and lowland populations have benefitted at the cost of the forest uplanders. In short, there has been a massive transfer of wealth, but from the poor to the rich. Tragically, forest reservation is legislated from the capital without negotiation or consultation with traditional keepers, in effect abrogating all authority from forest peoples to metropolitan power centres. This has exacerbated the already deep divisions between the lowland owners of the best land and communications, and the impoverished upland forest peoples who are the keepers of accumulated site-specific forest knowledge. That knowledge is now itself endangered.

2.3.3. *Environmental consequences of forest conversion*

The introduction of tea cultivation in the second half of the 19th Century did not by itself lead to the intrusion of commodity agriculture into the traditional lands and forests of the upland tribes. Their lands rarely extended as high as the lower montane forests, where tea was planted but where cool climates yield few lowland tropical crops (Table 4.11). Rather, the environmental consequences to mountain forests became an increasing concern to Western biologists because they affected both water supplies and wildlife, including plant communities[33]. J.D. Hooker, Director of the Royal Botanic Gardens, Kew (1855–1885), used his influence against deforestation in British dependencies, and to campaign for the establishment of formal Forest Services.[34] As early as 1874, he initiated an enquiry into evidence for changes in rainfall patterns as a consequence of montane forest clearance. His recommendation that forest conversion should be confined below 5,000 ft (1,600 m), and slopes of greater than 45% avoided, in India and Sri Lanka was heeded. It continues there in some measure to the present day, and in Peninsular Malaysia and elsewhere where attempts are made to protect slopes greater than 30% from conversion.

2.4. *The quest for sustainable forest management*

This book is not a treatise on forest management or silviculture, though the physiological and ecological ground rules (as I understand them) are presented. For former British India – a vast swathe of tropical Asia from Pakistan to Burma – and also for Peninsular Malaysia (to which we return below), we are fortunate to have four key masterly publications: E.P. Stebbing's *The Forests of India*, which recounts the history of colonial forestry up to its heyday in 1923; R.S. Troup's *The Silviculture of Indian Trees* (1922–1926), an encyclopaedic survey of the silviculture and botany of the leading species, exhaustively revised by a team under H.B. Joshi at the Indian Forest Research Institute; H.G. Champion's 1938 *General Silviculture for India*, which reviews experience in silvicultural management for former British India, and which in 1968 was revised by S.K. Seth for the modern State of India; and J. Wyatt-Smith's 1963 *Manual of Silviculture for Malayan Inland Forests*.[35]

2.4.1. *The deciduous forests: teak and sal*

Colonial forestry focused at first on commercial timber production for their metropolitan home markets, but by the late 19th Century was increasingly serving local markets. Attempts at management of tropical forests started with the less species-rich deciduous forests:

> In forest working, it is commonly the position that even of the top (6–10) canopy species only one (e.g. teak) or few are workable at a profit, having a sale value in excess of cost of felling and extraction to market. Management aims at increasing the proportion of such profitable species as far as silvicultural considerations indicate to be desirable – the limit has sometimes been overstepped, as locally with teak and sandal. Many forests are workable for teak only, where this species may form 10 percent or less of the crop.
>
> Stebbing, 1922, *The Forests of India*, p.221

European foresters had been trained to manage their forests, which are unusually poor in species, even for the temperate world. In India they had gravitated first, and unsurprisingly, to the temperate western Himalaya where most forests were dominated by a single conifer species, fortunately also of commercial interest. Their interest in tropical India was overwhelmingly in the sustainable management of the two commercially important durable hardwood species, teak (*Tectona grandis*) and sal (*Shorea robusta*). Each of these frequently dominates the forests in which it occurs. Teak was in international demand; sal was formerly used for railway sleepers, and is still nationally important in construction. Both species are deciduous, and their populations prosper only in relatively species-poor deciduous forests. Both are restricted to lowland forests, but over a wide range of rainfall climates, from five to nine dry months in the case of teak and up to eight for sal. The two species rarely co-occur in nature, being separated both geographically and by soil preference (Chapter 3). Of the two, teak alone occurs in Indo-Burma, and it may be indigenous to East Java (Chapter 6).

2.4.1.1. *Fire management*. The first priorities for management, following European forestry tradition, were considered the exclusion of fire, and the reduction or elimination of browsing and trampling by domestic cattle (which alienated local farming communities early on). Fire was observed to kill young seedlings and to reduce older seedlings to root stumps, though most did successfully resprout. It was soon found that fire damage was insignificant after a few years in the drier forests, where the monsoon flush of the sparse grassy field layer presented no serious competition to tree coppice shoots (Chapter 2). However, fire protection seemed to inhibit regeneration in the higher-quality forests in climates with shorter dry seasons. Besides, teak seed was proving impossible to germinate at an acceptable rate. Even burying the seeds in cow dung was tried, with some but limited success! Curiously, seed germination rates were observed to be high in unmanaged forests where fire persisted. One reason appeared to be the reduction of leaf litter which, becoming waterlogged and anoxic during the monsoon, was found to impede the survival of seedlings. Critically, in climates with six or fewer dry months – and in the absence of fire – the moist forests were invaded by fire-sensitive evergreen species flourishing along water courses and in swales. These shaded out the deciduous species of choice (Chapter 2). It was Herbert Slade who, at the end of the 19th Century, broke the impasse by experimenting with fire as a management tool in Burmese tall teak forest.[36] He discovered not only that fire eliminated the evergreen species competing with teak, but that fire was a requirement for teak germination. It is said that he himself was in turn fired for his heterodoxy – but landed on his feet when King Rama V, Chulalongkorn of Siam, hired him to found forestry on Indian principles in that country.

The *Indian Forester* is among the earliest of forestry journals worldwide. For 50 years the journal was dense with articles recounting experience with attempts to regenerate and sustain the production of teak and especially of sal. Sal occurs mostly on siliceous soils of moderate to low fertility; these are highly prone to trampling, especially in drier climates (Chapters 2, 3). The resulting management protocols came to vary regionally and locally, according to climate, soil, and the pressures of the human economy, particularly domestic cattle.[37] Always, the aim was to increase stocking rates of one of these two tree species. Success in that goal almost inevitably came at the cost of other species (although sal could sometimes be restocked from existing advance regeneration without removing other species). But many of these competing species were (and are) important in the rural economy.

2.4.1.2. *Pest management*. Inevitably also, communicability of host-specific pests and pathogens increased (Chapter 5). Notorious among these is the longicorn beetle of sal, *Hoplocerambyx spinicornis*. This pest is widespread and also attacks other trees elsewhere in its range.[38] Similar problems arose in teak. The best management results, both for sal and teak, were often achieved in mixed stands where the species of interest was not strongly dominant. Plantation was increasingly resorted to wherever natural regeneration proved refractory. Today, probably no untreated primary teak or sal forest survives in India. (A few good teak stands, albeit selectively logged many years ago, do survive in northern Thailand and in the more inaccessible forests of Burma.) That notwithstanding, the regenerative capacities of the flora of these forests are such that it is unlikely that extirpations have resulted from these management methods (Chapters 2, 3).

2.4.2. *The evergreen rain forests: shelterwood systems*

Serious attempts to manage mixed evergreen forests came late, and were at first similarly focused on individual commercial species or genera. Increasing seasonality is accompanied by increasing periodicity and intensity of the natural catastrophes, which (as I argue in Chapter 7) are principal mediators of regional patterns of tree species diversity. The lower species diversity of seasonal evergreen forests, combined with greater innate capacity to regenerate after catastrophe, makes these forests easier to manage than the MDF of western Malesia. Notable experiments in seasonal evergreen dipterocarp forests included those of Chengappa[39] in the Andamans (1934–1944), and of Kadambi (1939–1949)[40] in the Western Ghats. Both these men focused on dipterocarp regeneration, especially the *Dipterocarpus* species used for railway sleepers.

Where natural regeneration of focal species was relatively reliable, *shelterwood* methods of silvicultural management were attempted. Shelterwood systems vary in pattern and timing of canopy removal, but all have several common characteristics. Firstly, the shelterwood regeneration method is designed to emulate an episodic natural disturbance, such that there is always one dominant cohort of recruits that develops, matures and is regenerated again. Secondly, the 'disturbance' and canopy removal is carried out over a large area of similar floristics and structure (that is, a forester's 'stand', but see Chapter 2) in an effort to homogenise stand development and thereby create a more economically manageable unit of timber, 'compartment'. Thus, forester's stands are inevitably larger than natural stands, other than those occasionally caused by hurricanes.[41] Thirdly, success of the shelterwood system depends on the existence of established regeneration at the time of final felling; this may require prior subcanopy removal and site treatments to facilitate the initial germination and establishment of canopy tree seedlings. All silvicultural manipulations aimed at releasing a well-stocked, fast-growing future stand of trees take place during or shortly after final timber harvest of the canopy trees. It involves partial or total removal of unwanted species and competitors to well-spaced superior individuals of the species desired. (Pre-felling, canopy opening, and understorey clearing were each attempted, but with mixed success.) Where it could

be adopted, enrichment planting proved most consistently successful;[42] it became indispensable in the Andamans. Only in North-East India was there sufficient success with natural regeneration in seasonal evergreen dipterocarp forests to incorporate research results into practice.

2.4.3. *The Malayan experience of sustainable forest management*

The most consistent and comprehensive attempts to manage natural mixed evergreen forest for sustained timber production have been in Peninsular Malaysia. The emergent canopy of mixed dipterocarp forest in the perhumid Sunda climate is rich in red meranti (*Shorea* spp.) and keruing (*Dipterocarpus* spp.). These yield general-purpose light hardwoods with favourable properties including ease of planing and resistance to warping.[43] By the 1930s, timber was being sold on local markets and exports were restricted by law to ensure a sustainable supply. Though they are light-demanding and fast-growing as juveniles, these species have the reproductive characteristics of climax species, and therefore occupy a late successional stage in the forest cycle (Chapter 2). On other continents, individual species such as the mahoganies show similar silvicultural characteristics, but widespread canopy dominance of this guild is unique to the perhumid tropics of the Far East, and to south-west Sri Lanka where the light hardwood *tiniya* group of *Shorea* (sect. *Doona*) can also codominate.

The silvicultural programme has been exhaustively documented by John Wyatt-Smith in his monumental *Manual of Malayan Silviculture for Inland Forest* of 1963, while Jeya Kathirithambi-Wells' 2005 monograph, *Nature and Nation*, provides a unique and enthralling account of the socio-political context of Peninsular Malaysian forestry. These histories reflect the many challenges to achieving the goal of timber harvests which are at once economically optimal, commercially attractive to implement, and sustainable. Indeed, one lesson may be that economic optimisation, and still less commercial maximisation, may alone not be viable objectives if sustainability over multiple tree generations is taken as the organising principle.

From an ecological perspective, the attempts to establish commercial forestry on a sustainable basis in Malaysian MDF must ultimately be viewed as a succession of responses to rapidly changing markets. Market developments have induced changes in emphasis from single species to many, and from climax species to successional – and all these changes have taken place *within* the life-span of a commercial light hardwood individual!

2.4.3.1. *Social and economic background.* Colonial administration in Malaya (now Peninsular Malaysia) led to the cessation of local hostilities, stabilisation of government and to the introduction of formal medical services heralding the control of malaria. These changes (combined with immigration from denser populations, especially in China) brought about a population growth spurt, which continues today. As towns grew and commodity plantations became widely established – initially almost entirely in the western half of the Peninsula – populations migrated inland. Early commercial interests and forest product export markets were in gutta percha (several species of Sapotaceae, but mostly *Palaquium gutta*). Stocks were rapidly depleted, and there was limited success in planting within the forest following canopy thinning. Highest demand within the country at first was for heavy durable hardwoods, particularly cengal (the dipterocarp *Neobalanocarpus heimii*) and the rosewood merbau (*Intsia palembanica*). Both of these species are shade tolerant, slow growing and produce abundant seedlings, but neither responded well to initial pre-felling canopy opening. It was observed that while opening was necessary to stimulate height growth in heavy hardwood juveniles, there was nevertheless an optimal canopy gap diameter beyond which growth rates were lower.

By the early 1920s, growing population in Malaya led to a rapidly expanding demand for fuelwood, creating a market for 'improvement fellings' of the pole-sized trees removed to enhance growth of the shade-tolerant heavy hardwoods then in demand. This eased the problem of high management costs incurred early in the felling cycle, which chronically affects the economic viability of sustainable management (especially when discount rates are accounted for, which they were not at that time, presumably because inflation rates were then negligible). Shelterwood silvicultural systems were consequently adopted, following Indian practice. The canopy was manipulated by intermediate fellings to encourage regeneration prior to the main cut. But by the mid-1930s, kerosene and other cheap petroleum-based fuels led to declines in the fuelwood and charcoal markets in an increasingly prosperous economy. Improvement fellings continued at a reduced rate, but removal by girdling was started, and poisons experimented with. When lowland forest silvicultural operations ended in the 1970s, the use of the poison sodium arsenite for removal of undesirable pole-sized trees had become universal, but was eventually legislated against owing to its toxicity.

In the 1930s, the main felling was still carried out with axes and saws and the logs were extracted by buffalo, causing minimal damage to established regeneration. It was observed that the main felling itself often produced a satisfactory increase in the survival and growth of regeneration. Pre-felling thinning was therefore finally abandoned, though thinning continued in primary forest and post-logging.

Economically viable sustainable management, which had become difficult during the depression of the 1930s, returned later in that decade, this time thanks to the mechanisation of milling. Hand pit-sawing was giving way to small saw mills which sprang up wherever markets were expanding, and this led to demand for a wider range of timbers. One million acres (400,000 ha) within reserved forests were slated for intensive management, of which two thirds were for the light hardwoods coming into demand. But this plan was interrupted by the war. During those years, management basically ceased and great destruction was wrought both by unmanaged exploitation for timber and by clearing forest for food production. Much of the forest that had been silviculturally treated, both prior to and following logging, was lost. Following the war, labour costs increased rapidly. Before the forest service was able to thoroughly assess the condition of the forest estate and regain full control, the communist-inspired civil war, or Emergency, had started. This lasted for over a decade, during which time much of the forest was inaccessible to all but the military. At the same time, world timber prices greatly increased as did local demand, and research focused increasingly on intensive management of hardwoods. Results of this research were applied where possible.

2.4.3.2. The Malayan Uniform System. This difficult period nevertheless led to the only successful method yet devised for sustaining timber production in a biodiverse tropical mixed hardwood rainforest: the Malayan Uniform System (MUS). MUS aimed at sustained production of light hardwood red and yellow merantis (*Shorea* spp.). In Sundaland, these are generally abundant in the lowland MDF on zonal soils. Treatment of timber by preservatives under pressure had added value to light hardwoods, and a wider range of species could now be marketed.

MUS is a simple shelterwood system, relying as always upon the presence of regeneration established beneath the canopy.[44] The removal of the overstorey in a single cutting was thought to be enough to release the regeneration. A felling cycle of 70 years was predicted: the forest would simply be demarcated through area regulation such that 1/70 of the production forest would be logged and regenerated each year into perpetuity. These coups would in effect still simulate large natural canopy gaps, and the resulting single-aged cohorts (or 'stands') would constitute the basic silvicultural management unit of the management system.

Under MUS, the first logging was intended to be heavy, and individuals of non-commercial species were poisoned, resulting in open conditions. Freed from competition and with their crowns in full sun, the early regeneration would respond vigorously as if in plantation – as long as it survived the logging event. Mechanised logging had by mid-century become general, so careful planning and oversight of the operation was critical. Where seedling densities were judged inadequate, it was assumed to be due to the absence of a recent mast year (Chapter 5), and felling was to be postponed. Post-felling regeneration monitoring was to be carried out at five-year intervals.

MDF that had been felled and treated in this manner (or by the earlier shelterwood system of the 1930s, which had provided similar conditions for regeneration) produced spectacular stands of light red meranti whose volume exceeded that in the original forest. However, MUS did reduce tree species richness drastically where poisoning of unwanted juveniles was practised, and it would likely lead to gradual elimination of shade-tolerant species over successive felling cycles because many would fail to reach reproductive age within a cycle. It may seem surprising that Wyatt-Smith, a keen naturalist and a founder of the Malayan Nature Society, is not known to have expressed concern about this; but the prevailing view at the time was that MDF was floristically uniform, and that therefore the few large forests conserved in the national park and wildlife sanctuaries would be sufficient.

2.4.3.3. Selective logging systems, and the demise of the dipterocarp forests.

The forests of the equable perhumid climates, subject to relatively small scale natural canopy disturbance (being outside the typhoon belt) and with high species diversity, have proven the most vulnerable to overexploitation.

MUS had scarcely been implemented before peace returned in 1959, also the year when the nation became independent. It nevertheless proved highly successful in achieving its primary purpose for managing MDF: increasing, even doubling, the yield of light dipterocarp hardwoods, the red and yellow meranti *Shorea*, on a 70-year rotation.[45] Girdling and poisoning of competing subcanopy species was abandoned as labour costs increased. Even then, basal areas of meranti were already equalling those of the original crop after 40 years.[46] Tragically, the final introduction of MUS in Peninsular Malaysia coincided with a policy decision to convert all lowland forest on flat or undulating land – except the peat swamps – to commodity plantation.

Selective systems of management had been recommended in the Philippines, where an almost continuous emergent canopy of light red meranti, diffuse on account of periodic partial defoliation by typhoons (Chapter 2), provided favourable light for dense pole-sized meranti regeneration[47]; but these had generally been sparse in the more densely canopied MDF of relatively windless western Malesia. In true selection systems, only canopy individuals within the mature stands (comprising perhaps one-tenth of the area of the floristic association under management) are harvested. The residual stand is then further modified to create conditions for optimal growth of promising pole-sized juveniles of the required species (here merantis) smaller than a defined minimum felling diameter,

recognized by their superior bole and crown form and marked. Growth rates are subsequently enhanced by periodic reduction of competition by 'regeneration thinnings'.[48] It is therefore essential to maintain a permanent road system for access. Felling cycles must be of 20-30(40) years to be economic. Correctly practiced, selection systems require intense book-keeping in species-diverse tropical forests, to monitor individuals of varying growth rates. They are therefore demanding in both labour and silvicultural skills.

Selective logging was later optimistically introduced in Malaysia and Indonesia. The seeming advantage of selective logging is financial, in that it promises to shorten the felling cycle by relying on advance regeneration at or above a prescribed minimum felling diameter. It was originally expected that trees of merchantable size could be logged on a cutting cycle as short as 15 years – on the assumption that they would grow at near-maximum rates. This would reduce the negative impact of high discount rates on profitability. In practice though, these so-called 'selection systems' relied solely on enforcement of the minimum felling diameter without the supervision of skilled technicians, and removal of canopy individuals was often complete, leaving only a few inferior individuals as seed trees to ensure seedling establishment and enhanced growth in the gaps left by felling.[49] Optimistic forecasts were made on the basis of short-term measurements, which captured the initial resurgence of growth following crown exposure of previously suppressed trees.[50] Primack *et al.*[51] were soon able to infer, from established experimental plots in Sarawak (unfortunately lacking a control), that in the absence of real selection and marking of promising individuals, mean growth rates of the residual stock were not maintained after an initial surge.

The major disadvantage of such selective logging systems, as practised in Indonesia and Malaysia, is ecological. Medium-sized individuals of a diversity of economic species are indeed often found in considerable densities (albeit few of the desired merantis); but most of these individuals are in fact suppressed, dying slowly. Owing to their malformed crowns, they cannot recover high growth rates even when released.[52] Individuals of good crown form may indeed maintain diameter growth at c.1 cm y^{-1},[53] but in forests of calm climates such trees are confined to natural gaps occupying less than 20% of the forest area. This problem can only be overcome by the pre- and post-harvest liberation thinnings that are fundamental to true selection systems – but such treatments require skilled, salaried field workers representing the owners, and were universally dispensed with.

By 1985, Thang[54] predicted that Peninsular Malaysian *domestic* timber demand would exceed supply by 1995 or, in the case of rubber wood, by 2000. The rush was on to log these hill forests as the lowland forests receded. A so-called selection system appeared to be the only management system practicable which was profitable,[55] yet no long-term monitoring seems to have been initiated in the 1980s, nor were control plots set aside. As expected, however, in the few cases of lowland MDF that have been monitored over several years, growth returned to low pre-felling rates (below 5 mm y^{-1}) following canopy closure.[56] Successive felling cycles would leave swathes of late-successional stands created by previous logging events, but would such logged areas retain sufficient soil with its early regeneration, untrammelled by machinery, however sparse? And what proportion of trees in the residual stands would achieve economically viable growth rates?

It is an indictment of the regional forest services that no-one established carefully monitored long-term plots to rigorously examine the response of stands to selection systems. Protected areas had usefully been established as controls for monitoring silvicultural systems in India early in the 20th Century,[57] and in Peninsular Malaysia in the 1950s, the Virgin Jungle Reserves (VJRs). But these were never extended as different forest types came under the chain saw.

The post-war timber boom was accompanied by the introduction of crawler tractors, which could be used both for road building and for timber extraction. This reduced costs to the concessionaire but greatly increased soil compression and scarification, causing loss of porosity and killing advance regeneration[58] (Table 8.2; Plate 8.8). Tractor damage to soil in lowland forest on undulating land was reported variously at 20–80% of the logged area. Tractor and felling damage to the residual stands, by debarking, branch and crown snap, was estimated at 40% to more than 50%.[59] Damage was correlated with logging intensity. The highest density and basal area of timber in the hill forests has been along the ridges where,

without consideration of subsequent crops, the extraction roads are most easily and therefore frequently located. High-lead logging led to even greater damage than tractor logging, both to seedlings and crowns in the residual stands on slopes.[60] Even in forest managed by MUS, and where dipterocarp regeneration remained abundant, the merantis performed poorly after tractor logging owing to a dense canopy of pioneers, which quickly established on soil surfaces scarified by machinery and thus shorn of dipterocarp regeneration.[61] Similar observations have been made elsewhere in the tropics.[62] Nevertheless, when it is not destroyed by machinery, red meranti regeneration often survives the first cut.[63] Survivorship and performance following a selection felling can then be enhanced by canopy opening 6–12 years *after* logging,[64] thereby avoiding pioneer competition.

2.4.3.4. *Reduced impact logging systems.* Attempts have been made to reduce damage, for instance by climber cutting,[65] and to restore by enrichment with planted timber saplings. The latter has generally failed through inadequate tending, rather than because of any innate property of the plants. Several protocols have been suggested for reduced impact logging (RIL), aiming to reduce damage to soil and stand by lowering logging intensity and maintaining partial canopy continuity. In Sabah, Pinard[66] found that RIL reduced soil damage in lowland MDF from 17% to 6% of total area. Pioneers still regenerated on scarified soil, though, whereas dipterocarps and other climax species failed.[67] Differences in species richness persisted after 18 years. A large-scale experiment by J.-G Bertault and P. Sist in Berau (East Kalimantan), in MDF on undulating land, recorded 48% damage to the residual trees >10 cm dbh following conventional logging practices.[68] Crown damage was predominant, and half the damaged trees died within five years.[69] RIL here reduced damage by 30%; careful planning of the direction of felling proved essential.

Frederickson and Putz[70] argued that RIL does not provide sufficient canopy opening to achieve high growth rates among meranti growing stock. Sist and Brown[71] replied that management need no longer be aimed at maximising yield of certain light hardwoods, because the market will now accept a wide range of species; and besides, forest ecosystem services, including biodiversity conservation, are enhanced by RIL and should be included in economic assessments. A similar argument has been made for selection systems in general, because the minimum cutting diameter protects the residual stock below it irrespective of species[72] (but see Chapter 9). There is

Plate 8.8 Lack of dipterocarp regeneration along an abandoned logging road, Danum, Sabah. Little has grown where vehicles have churned and compressed the soil. (H.H.)

some evidence nevertheless that species of low population density (often rare or endemic species) are particularly at risk in logged forest, whatever management system is adopted.[73] These observations and prescriptions have been extrapolated to the diversity of floristic communities represented in MDF throughout its geographical and edaphic range. In South Asia, where the greatly reduced forest estate is being re-evaluated, due consideration of the differing light and moisture responses of dipterocarps and other canopy trees in relation to local site conditions is now being promoted.[74]

In the Far East, such 'counsels of perfection' have yet to reach policy makers, and for obvious reasons: their interest has been in maximising financial returns rather than achieving sustainable economic value of a renewable resource. A 35-year felling cycle appears to promise higher and earlier profits, and perhaps even increased timber yield, but the evidence is entirely to the contrary. Forest departments have not been funded to meet the need for oversight of larger and more remote areas under exploitation. Oversight has in reality been abandoned, leaving the fate of the forest in the hands of the concessionaire. But the prevalence of licenses of shorter duration than any felling cycle deprives that same concessionaire of any incentive to protect the residual stand – while it motivates the politician, acting as rentier, to take maximum and most frequent rent.

2.4.3.5. *What could be sustained, and which system could best sustain it?* The MUS met its objective of sustainably yielding a maximum volume of light hardwood. Its drawbacks were three: (1) the necessity for careful management and a well-trained management team; (2) the long interval of more than half a century between harvests; and now increasingly (3) the focus on a single product from an extraordinarily diverse resource. Natural MDF includes many species important to local communities that might have suffered from strict suppression of non-timber species; though the use of arsenic compounds in landscapes where they would inevitably leach into aquifers did not survive eventual legal challenge. The MUS favoured regeneration of the light hardwood merantis in a MDF forest type – 'red meranti-keruing' – prone to the large squall-induced canopy gaps favouring meranti regeneration in nature.

It can be seen from Chapter 2, in which forest structure and dynamics are considered from an ecological perspective, that MUS simulates the regeneration characteristics of the quite large predominant windfall gaps that prevail on the udult clay and sandy clays soils in Peninsular Malaysia, and are the principal lowland soils of lowland Sundaland; whereas true selection systems, and RIL in principle, more closely simulate the small gaps that prevail on freely draining, deeper rooting but less fertile yellow soils of MDF on high ridges and sandstone (Chapters 2, 7).

Either system, *if strictly implemented*, can be efficacious in its appropriate forest type. Each management system can be expected to optimise conservation of the tree flora and dependent organisms to the extent possible when practised in the appropriate habitats and their forest communities, bearing in mind the inevitable impact of consequent increased size and relative area of canopy gaps. Selection systems in particular must rely on the commitment, skill and experience of the forester in charge, who must identify candidate trees, evaluate their potential and mark them for liberation. In practice, reliance on the mere goodwill of concessionaires and loggers has been naïve at best, and has most often led to the triumph of personal greed over the public interest. Poorly planned or unplanned logging with widespread soil compaction from heavy machinery has had more detrimental effects on forest regeneration – especially along ridges – than any putative management system. Where felling is carefully executed, a substantial part of the original forest structure can be maintained. Where the understorey is left unthinned, much of the biodiversity including wildlife can survive. But the relative population densities of tree species are bound to change in response to altered subcanopy light conditions (Chapters 2, 7). In the mid- to long term, this inevitably leads to local or regional extinctions (Chapter 9). Reliance for later crops on subcanopy juveniles – most of which are in slow decline through overshading – is based on poor ecological understanding (Chapter 2).

RIL is but a crude version of a selection system. It constitutes an attempt to provide a general method of forest management that would maintain some canopy continuity, thereby conserving it for arboreal animals. Like a selection system, it proposes a method in which the unit of management is the whole forest, rather than the individual stand. But does not a forest that is divided into a number of management units that simulate natural stands, equal to the number of years in the felling cycle, each at a different stage of regeneration and with an initially continuous canopy early developing, if initially of short stature (that is, MUS), not equally well or better achieve the RIL objective? That question is becoming irrelevant as suitable site conditions for MUS are those also suitable for oil palm.

Without knowledge of species on the part of the forest manager, it will be hard indeed to regenerate timber stands following unsupervised logging of forests on those soils for which true selection systems are most appropriate. Nor are their leachable and erodible soils, or the steep lands that are increasingly all that is left under forest, amenable to plantation in wet tropical climates. We advocate their exclusion from production forests as the only economically viable option when, if protected against fire, these remnants may eventually become recreation areas with some residual value for soil, and perhaps biodiversity, conservation.

2.4.3.6. *Oil palm arrives*. Perhaps these issues hardly matter. The years following independence led to the early demise of the MUS in Peninsular Malaysia, the only country in the Far East which had the skilled forest service necessary to implement it. Senior colonial officers, including researchers, were early replaced, either by highly experienced but less well-trained nationals, or by young graduates from overseas universities with little experience of local forest conditions and protocols. More seriously, the early five-year development plans in Peninsular

Plate 8.9 The stamp of mercantilism on a landscape once carrying unique and hyperdiverse forest ecosystems. **a,** MDF cleared and burned for plantation; **b,** Clearing the peat swamp forest for oil palm plantation. Oil palm sequesters but a fraction of the carbon within the lofty forest it replaces, while drainage oxidises the peat, and fertiliser, hydrocarbon-intensive to manufacture, and lime application accelerate decomposition. Scale: 1:30,000. (**a** W. Smits; **b** Google Earth)

Malaysia took note of the aspirations of an electorate increasing in both numbers and prosperity, and envisaged the eventual conversion of virtually all lowland forest on slopes of less than 30% to commodity plantation (Plate 8.9).

Oil palm overtook *Hevea* rubber in profitability by the 1980s as the global prepared-foods market exploded[75]. The remaining lowland MDF, and more recently even the peat swamp forest, is being converted to oil palm plantation, leaving only a few small patches of research forest and some National Parks.[76] The same process has almost run its course in Sumatra, and is rapidly progressing in Borneo.[77] Thus, during the 55 years of my career, a forest formation with origins 23 million years ago (Chapter 6) has been reduced from perhaps 75% of its original area to less than 10%. Almost the whole of this remnant outside the modest conserved areas has itself been logged and has an uncertain future. Most countries do still hold to a prohibition of agriculture on steep slopes, >30% in Peninsular Malaysia and Sri Lanka. This still leaves, in principle, a permanent forest estate in excess of 40% of the land in Malesian nations (map inside back cover).

2.4.3.7. *A dwindling forest estate*. In effect, the permanent forest estate has retreated to the steep lands and the mountains. While the tree species composition of the slopes differs little from that of the undulating lowlands, species relative abundances do differ. Stocking rates are lower on the steep slopes, and established regeneration is sparse and patchy. MDF on steep slopes generally lacks reliable light meranti advance regeneration. Timber stocking was always greatest on the convex surfaces of ridges and spurs below c.600 m, and was especially impressive in Peninsular Malaysia where the heavy dark red meranti, seraya (*Shorea curtisii*), is widely dominant. Beneath seraya though, grow extensive patches of the stemless palm, bertam (*Eugeissona tristis*). Bertam, which also responds to canopy opening, impedes regeneration of light-demanding species by its dense litter and shade, though seraya is relatively shade-tolerant (Box 2.1).

Fine regenerating stands of seraya-dominated hill forest on drought-prone slopes and ridges had emerged following a shelterwood logging system in the 1930s at Sinaling Inas Forest Reserve (Negeri Sembilan). But, as might have been anticipated, by the 1970s occasional severe droughts started killing some of the tallest regenerating individuals, candidates for the next harvest (Chapter 2).[78] By the 1980s Peter Burgess[79] had become concerned both by the weak regeneration of seraya and by the bertam problem. He recommended a selection system for dipterocarp forests, and this might have succeeded had it been carried out by the orthodox procedures. But minimum diameter cutting limits were the primary guideline for concessionaires, and there was no 'regeneration thinning' either before or after logging. In days when labour was cheap,

Unlogged: Bukit Lagong, Selangor[a]			Logged: Jengai Forest Reserve, Terengganu[b]		
Tree family	No. trees/ha	%	Tree family	No. trees/ha	%
Dipterocarpaceae	57	12	Myrtaceae	156	28
Flacourtiaceae	47	10	Lauraceae	31	6
Euphorbiaceae	36	7	Anacardiaceae (*Swintonia*)	28	5
Rhizophoraceae	26	5	Ebenaceae	28	5
Tiliaceae	22	4.5	Myristicaceae	27	5
Sterculiaceae	21	4	Burseraceae	24	4
Lauraceae	19	4	Sapotaceae	23	4
Anacardiaceae	19	4	Lecythidaceae	21	4
Myristicaceae	18	3.5	Clusiaceae (*Calophyllum*)	16	3
Burseraceae	16	3	Dipterocarpaceae (*Vatica*)	12	2
Other families	213	44		173	32
Total	494		Total	55	

[a] After Manokaran & Swaine 1994   [b] After Wan Mohd Shukri *et al.* 2005

Table 8.2 Comparison of an unlogged and a logged area after 12 years, of hill dipterocarp *Shorea curtisii* (seraya) dominated forests in Peninsular Malaysia, trees >10 cm dbh. (After Manokaran & Swaine 1994, Shukri *et al.* undated, c.2005)

physical clearing of bertam had been resorted to on the seraya-dominated coastal hills – but even then results were mixed. In Andulau Forest Reserve (Brunei), MDF on drought-prone, low-nutrient yellow sandy soils had been harvested since the 1930s. It was managed from the first by a shelterwood system and later by MUS. At first, regeneration was as dense as at Sinaling Inas, but in the absence of liberation thinnings it has stagnated.

The greatly increased demand for timber occasioned by relaxation of export quotas in the 1970s was substantially satisfied – as was the intention – by conversion of both unreserved and reserved lowland forests to agriculture; but it spurred logging in the hills as well. Insufficient field staff led to the abandonment of silvicultural thinning or even monitoring. Oversight of logging was neglected by the owner's representative, the Forest Service. Little research was sponsored, or logging experience analysed, upon which a management system to sustain timber yield in the hill forests might have been proposed and tested, before much of the damage had been done (Table 8.2).

In 1988, Duncan Poore[80] estimated that worldwide, hardly one million hectares of tropical rainforest were being sustainably managed for timber. A 2005 CIFOR[81] report nevertheless claimed that 7% of global tropical rainforests (c.25 million ha) are sustainably managed, of which 40% are in India or Malaysia. Can we believe that?

There are a few crumbs of comfort. For instance, Sabah state (Malaysia) has instigated a universal 100-year license policy for its remaining forest estate, thereby in effect handing over forest ownership to the concessionaire. The implications of this will be considered in Chapter 9. Several countries in the seasonal tropics, but including Sri Lanka, now have moratoria on logging in their indigenous evergreen forest residues.

The truth is that the days of the commercial light or medium hardwood dipterocarp are numbered. Dipterocarps are being replaced by pioneer and successional species which have more consistent annual seed production and the ability to establish on bare ground. Only now, when 80% of the hill forest has been logged and the time has arrived to initiate the second cycle, has an investigation (again without a control) been undertaken.[82] Results indicate that the high-quality dipterocarp species have been subordinated by *Syzygium*, *Elateriospermum*, Myristicaceae and other canopy trees of inferior timber quality (Table 8.2). Growth rates, even for these species, have failed to meet targets. In short, the forest has been trashed. And what can be expected in subsequent cycles? Technological advances now permit the profitable harvest of a wide range of species. But will the heterogeneous post-harvest canopy conditions stimulate the development of dense stands of fast-growing and competing individuals, such as occur in primary forest gaps – and which were created by MUS? These are the conditions leading to the tall, straight, barely-tapering boles that determine the high volumes of the best primary stands. Successive crops will lack emergent individuals, producing shorter heights and bole lengths and, with the lighter timber of successional species dominating, perhaps no more than half the capacity of the original forest to store carbon (Chapter 9).

So-called selection systems have become universal in tropical rainforest designated for timber production, but for financial, not silvicultural – nor even economic – reasons. In Sarawak and west Kalimantan, MDF seldom carried the standing volumes of timber species found in the Philippines and eastern Borneo. In colonial times, imposition and collection of relatively high royalties, combined by strict oversight of logging operations, deterred all but a handful of logging concessionaires, except in the peat swamps where the abundance of a few species and flat terrain increased profitability. Current customary logging – there and wherever MDF still exists – continues unsupervised by governments as owners. The results are soil scarification by tractor extraction, skid trails and roads, and unsupervised felling which is damaging the residual stands. The best stands, on well-watered high-nutrient clay loam ultisols, are invaded by large-leaved *Merremia* and other fast-growing vines (Chapter 2) (Plate 9.2b). These throw a monolayer blanket over tree regeneration and persist for decades, reducing both stand densities and height growth rates, and largely eliminating light demanding light hardwoods. But in any case, unless they are designated as wildlife sanctuaries or national parks, these forests are destined for conversion to oil palm.[83]

3. The indigenous forest still shrinks

Little surviving lowland rainforest remains unlogged in Malesia today. Logging roads provide opportunities for landless settlers to invade the forest, often supported by local politicians. In India and north-west Thailand, for example, teak forests have been particularly vulnerable as they occur on arable land and their timber is favoured in house construction. Up to 45% have been lost in some areas.[84] In Thailand, swiddeners already living within reserved forests were given legal tenure in the 1970s. Swidden has now been almost replaced by permanent cultivation, but farmers still continue to nibble at the forest margins. Thai forest law (following the example of India) forbids cutting live trees, so trees are first illegally bark-girdled. Once dead, they are felled, their roots are removed, and cultivation begins. In due course, the new land is legally recognized thanks to political support. In neither country do the rural poor pay tax, but they do receive minimal government services here, including primary schools and mobile clinics. The Forest Service, unable to secure prosecutions, becomes demoralised. Some small farmers had favoured commodity crops, but government credit was often misused for domestic

appliances and other luxuries. Many ended in bankruptcy. In Thailand, smallholders are increasingly selling their land to local businessmen in order to gain release from loans taken at extortionate rates following bad harvests. Concerned with these trends, the King – along with the king of Bhutan, the last still-venerated rulers in the ancient Hindu-Buddhist tradition – is advocating a 'sufficiency economy'. Excessive wealth is discouraged, and the aim is for farmers to have enough land for their well-being, with a fish pond and orchard. But, so far, the forest still recedes, albeit more slowly.[85]

In India, a national forest policy has existed since 1894, but the princely states followed it with varying degrees of attention.[86] Rajasthan, on the desert fringe, was long noted for the poor state of its short deciduous forest and semi-arid *Acacia* woodlands. A new national forest policy was initiated in 1952, recognising social needs as well as timber production as priorities. The forests protected under the princely rulers were nationalised and added to the reserved forest estate in 1956, when the rulers lost their legally protected rights. Certain forest reserves, such as Buxa in West Bengal (where there are still tigers) were transferred into park status at that time.

Although the area of forest in India is now said to be expanding again through plantation, the indigenous forest still shrinks. One example is the Similipahar massif in Odisha, which rises to 1,158 m at Meghasena and enjoys the wettest climate in the region. This is the foremost island of deciduous lowland and lower montane forest remaining between Meghalaya and the Western Ghats. The hills are also of outstanding conservation interest for an intact terrestrial fauna, including tigers, elephants and several ungulate species. Formerly under the dominion of the Maharaja of Mayurbhanj, this predominantly tall sal forest was declared the Similipal National Park in 1957.[87] The area had been licensed for timber since 1906, and the park today harbours several tribes, including the nomadic foraging Aharia, Ujun and Birhor. The tribes have their own local divinities, and also still venerate their Hindu maharajah – although his father would have hanged any forest malefactors amongst them! Today, their numbers are rapidly increasing: families with five to seven children are common. 'Eco-development committees' and some of the earliest joint forest management (JFM) programmes were established at nearby Baripada, 'to protect from fire and harvest edible potatoes and mushrooms'. Initially very promising, problems mounted over time.

Researchers (including myself) find that few forest communities retain strong traditions of resource conservation, beyond recognising family ownership of certain valued trees (see section 7.2 below). At Similipal as elsewhere in deciduous forests, firewood is almost universally harvested by illegally ring-barking live trees and collecting the wood after they die. Cattle are increasing in number. Fields and pasture prevail around villages, the few remaining sal standing gaunt following shredding of their lateral branches for forage and fuel. In 2008, photographer Hans Hazebroek and I waited by a flowering *Butea monosperma* vine in forest near a village within Similipal National Park, to photograph its bird pollinators. There were none: they had been eaten. In 2006, the federal government legislated land tenure for tribal farmers throughout India, a long-overdue decision in principle (Chapter 9). In Similipal, though, there is presently no clear tribal leadership and no adherence to forest law. Traders, including criminal elements, have negotiated contracts with the tribals to collect orchids, sal resin, the medicinal vine *Bauhinia vahlii* upon which elephants also depend, *Curcuma* (wild turmeric), *Madhuca* (palu) flowers, honey, and now even tiger carcasses. The junior forest staff are too few to be effective and lack the demanding cultural and social skills necessary to mediate. They are demoralised; the law has been made an ass. In 2009, the elegant old hunting lodges of the Maharaja, which had been adapted for tourists, were burned down by the Maoist Naxalite faction. As if with prior knowledge, criminal gangs, to which some at least of the tribal communities are in thrall, moved in at once, hunting down tigers and logging the best stands.

4. The triumph of mercantilism

> Is it the maintenance and improvement of the Forest Estate, making it at the same time economically (not necessarily financially) worth the while of the nation, now and forever, that we want? Or is it mere 'exploitation' in the baser sense of the term that we have in view? In other words is the idea to work the forests in accordance with the accepted principles of silviculture and national forestry? Or is it merely to make a profit by cutting them down? If the former there is nothing new in it, but it is the same old idea which, I believe, we have been endeavoring all these years to carry out!
>
> P. Govinda Menon, Conservator at Nilambur, 1925, *Indian Forester*.

The end of the Second World War was accompanied by the end of colonialism. Much of the world was economically devastated. Enormous new political emphasis was put on universal and rapid economic development. The engines were cheap energy, cheap primary resources and the technological advances achieved by military-industrial research. Stocks of physical resources were once again assumed to be effectively limitless – especially when supplemented by the flexibility of the new plastics – and the world's forests seemed to exceed any predicted demand. Sustainable management was seen as technically feasible provided timber prices remained

sufficiently high to ensure its economic viability. Large-scale logging started in the rainforests of the Far East in the 1960s. Within four decades, indigenous commercial forest resources were exhausted in several countries.

At first glance, it seems extraordinary that newly independent countries exploited and converted their forests at a rate never approached by former colonial regimes. The observer from the ex-colonising world is often quick to identify political corruption and an expedient short-term vision of development, egged on by self-serving commercial interests, as the cause. To discount these factors altogether would be naïve. But how conscientiously did those same colonialists, heirs to ancient traditions of public franchise, participate in preparing their erstwhile subjects for commitment to national over personal interests and the long-term over the short-term view? How well established were the foundations for a reasonably equitable distribution of wealth – upon which a stable national economy living within the productive capacity of its assets must rest – even in the most prosperous of the new or re-established countries? The first priority had to be to provide work and better livelihoods for rapidly increasing populations, and to develop the necessary infrastructure and sources of investment for industrialisation and economic development. What values did the forested lowlands have to a developing economy, other than as timber and land surface ideal for commodity plantation? The hydrological value of forests is in the hills, at the water sources and the adjacent slopes. Oil palm and rubber are demonstrably as capable of conserving soil as production forest, unless on steep slopes where the risk of severe erosion between successive crops is prohibitive.[88] Much – though by no means all – wildlife prospers on the young growth of logged forest, and does not discriminate between gentle and steep slopes.

It would be equally naïve to discount the extraordinarily powerful effect, in itself, of the growing availability of a vast forest-clearing technology – the crawler tractors, yarders and loaders, even cable cranes and feller-bunchers, along with the cheap fuel to run them – and the cheap credit extended to logging companies, especially the international conglomerates. Over the last half-century, tropical commercial logging has been able to exploit a renewable resource as if it were a non-renewable mineral, to great financial profit for the few and immense long-term cost to the majority. The remorseless demands of free global trade have been blamed, but were this the root problem, the selling price of timber should have increased as the commodity declined in availability, providing an incentive for reinvestment in forest management. Instead, owing to perverse national policies including under-taxation (low royalties) and increasingly intercontinental raw materials sourcing, the appropriate market signals have been neither sent nor received[89].

4.1. *Mercantilism in the forests of Malaysia, as one example*

4.1.1. *Peninsular Malaysia*

The social and political history of this period for Peninsular Malaysia has been even-handedly recounted in scholarly detail by Jeya Kathirithamby-Wells in *Nature and Nation* (2005), and need not be repeated here. Peninsular Malaysia is unique among the nations of the perhumid Far East in its long history of commercial forestry and research, and its early well-developed home market. One consequence, in part also owing to control of forests and licenses by individual state governments, was that timber companies remained local and focused on the home market. Following the abandonment of pre- and post-harvest silvicultural treatments, sustainability must rest entirely on the care with which timber extraction is planned and executed. Nonetheless, concessions have too often been awarded for areas too small and over periods too limited to generate any realistic incentive for sustainable management by concessionnaires. Dr Wells' excellent account shows how far short of the management ideal even a successful economy such as that of Peninsular Malaysia has fallen.

Taxes on timber have often failed to increase with inflation, and in some cases have decreased; they have often been evaded. Timber exploitation with weak government intervention has long been irresistible to commercial enterprises and their political patrons, witness the 19th Century depredation of the forests of Upper Michigan.

4.1.2. *Sarawak (Malaysian Borneo)*

In Sarawak, the Forest Department originally prescribed a selection felling system, cautiously employing a 70-year cycle owing to the steepness of the inland hills and patchiness of advance regeneration. But concessions were increasingly distributed under political pressure, and often for disastrously short tenure, overriding stated Forest Department policy. Local leaders were bought off.[90] The International Tropical Timber Organisation (ITTO), in a commissioned review as recently as 1990, concluded that Sarawak's export levels were in substantial excess of the productive capacity of the forest, and recommended a reduction in logging and greater emphasis on in-country processing if production were to be sustained.[91] Although the reduction was imposed at first, the government later mandated a 35-year cycle selection system to increase timber production for export, and to meet increasing demand from local industry. But timber out-turn continued to decline. Forests are re-entered at ever shorter intervals, pole-sized trees are now exploited and the forests degraded, even in areas proposed – but not yet legislated – as wildlife reserves. Richard Primack[92] recounts the melancholy story of Yong Khaw and his sons, who logged forests in Sarawak and Brunei for the

oilfields in colonial days, successfully sustaining 3 cutting cycles between 1938 and 1990; but gave up on account of the insecurity wrought when licences began to be given as political favours and squatters entered the concessions, unhindered by the authorities, in anticipation that the land would eventually be re-zoned for housing and its value increase.

Land ownership among many ethnicities had been locally established in Sarawak by the planting of useful trees, or the claiming of harvest rights over individual trees in the forest. These included tengkawang (engkabang) (see section 7.2.2 below). Traditionally, high quality timber trees were planted when a child was born, and cut when he or she married. The Forest Department photogrammetrically mapped the forest under state control for management planning, then disregarded traditional claims not legally enshrined.

Today, the concessionaire's agent may visit communities with promises of schools, clinics, roads and employment. Once the logging is over, promises are met to varying degrees and the land, once partly owned or at least controlled by villagers, is converted to company-owned oil palm plantation. Fire is used both to clear and to lay claim to disputed land.[93] After this, continuing rent is paid only for legally-recognized private land incorporated into commodity plantation. The designated fate of the upland farmer, imbued with generations of rural knowledge, has been to become a labourer on what was once his own land. This he has often declined. The companies have had to import foreign labour to harvest their estates. In contrast, in those fewer long settled communities with fruit gardens and permanent agriculture, as in west Sarawak, land has been secured by law. A century ago, across the frontier in West Kalimantan, the Sultan of Landak planted a private forest as species-rich as the justly celebrated Sri Lankan home garden.[94]

Rivers throughout Borneo have suffered drastic siltation, and degradation of the stabilising tree-associations characteristic of their banks below the flood line. Important protein sources such as fish, molluscs and crustaceans have greatly declined, especially where pesticides are being intensively applied in oil palm plantations. Nevertheless, although stream water turbidity increases at the time of logging, it appears to decline rapidly once logging activity ceases, where forest succession is allowed to re-initiate.[95] But the same is not true of commodity plantation on steep land.

With timber all but exhausted, the Sarawak Forest Department became partly privatised as a corporation charged with management, but dependent on the declining revenue of a degraded resource. Little primary unlogged forest remains outside the National Parks. Some of these too have been logged by criminal elements, enforcement having broken down. The forests in much of the Sematan Wildlife Sanctuary and the Pelagus and Similajau National Parks have essentially been trashed; likewise the lowland forest, once so rich in endemics. Maludam Forest, a peat swamp National Park, is being destroyed by uncontrolled illegal incursion. Proclaimed new parks, mostly already once logged, have yet to be legislated. And still, barges can be seen descending the rivers, loaded with trunks too small for their trees to have reached reproductive size. In west Sarawak, there is a lucrative trade in logs illegally imported from adjacent Indonesia, which nevertheless receive the Forest Department stamp implying tax paid (Plate 8.10). Recently, the failure of the intended reform of forest administration was recognized, and the corporation was returned to government ownership.

Behind this tale of forest conversion and degradation lies the fact that Sarawak, with neighbouring West Kalimantan and Brunei, held the most biodiverse forests of the Old World, as judged by numbers and endemism of tree species (Chapters 3, 6). Where have the profits from conversion gone? In part, they have been invested in oil palm on the deforested

Plate 8.10 Log stealing from Gunung Palung National Park, West Kalimantan. (T. Laman)

land, thereby ensuring at least a living wage for many. But how much has been invested elsewhere, in more lucrative opportunities overseas, or in private houses in Canada, New Zealand or Britain? In Indonesia, and most particularly in Sarawak, logging companies of global ambition have arisen through the enormous profits extracted from concessions at home. To their new contracts in Brazil and New Guinea they bring their own workers, among them the Iban Dayaks who are prized for their courage and ability with large machinery and felling in dangerous terrain. (These are the youths who traditionally went on a great personal expedition, *bejalai*, to qualify for manhood.)

Profits on this vast scale can only be made by draconian savings in capital investments, in labour for management, or by tax evasion or under-taxation by governments. Thus, the same crawler-tractors accomplish both road-building and timber extraction, with disastrous consequences for advance regeneration and soil compaction. There remains a dearth of careful documentation of the impact of this unmanaged exploitation on subsequent succession, and especially on the regeneration of late successional and climax quality timber species.

The hill forest on steep land, which is all that now remains of the permanent production forest estate, will in my estimation not recover its capacity to produce *quality* hardwood timber for several centuries unless enriched by planting. There is still profit to be made, for now almost any species can be used in pulp and as filler, but the quality has plummeted and the

owner (the public) is the loser. Discussions are arising now in Peninsular Malaysia and elsewhere on the commercial viability of forsaking the dipterocarps for those early successional and other species that do regenerate in degraded forest.

4.1.3. *Some general conclusions*
In the rainforest-holding countries of the Far East, the privileging of concessionaires over national interest (that is, of financial over economic priorities) has resulted neither in a means of sustaining timber production, nor in the conservation of other goods and services. Biodiversity, in particular, must decline over subsequent felling cycles (Chapter 9). Blame rests with those who, from positions of political power, claimed to represent the people.

It is also true that the greatly increasing cost of labour has imposed severe limitations on the commercial viability of sustainable timber production. On the one hand, it has nullified the economic advantages of pre- and post-harvest silvicultural treatment, so that the future of the forest now depends purely on the skills and motivation of the loggers. On the other, rising labour cost, in absence of commensurate raising of royalty taxes, has been one factor in the failure of forest services to expand field staff sufficiently to maintain oversight of forest operations. But that cost could have represented a more equitable distribution of profits among the rural population who could have supplied the labour for management, than the once-in-a-generation opportunity to fell.

For thirty years, dipterocarp timber was pre-eminent

in international trade. The dominant importing countries were in North-East Asia; first Japan and now China. Much of that imported timber was re-exported as finished goods. Notably, both countries have instituted strict and successful conservation policies for the protection and maintenance of their own indigenous forests.

India, Bangladesh, Sri Lanka and Thailand, all countries with well-developed government services but with greatly reduced remaining indigenous wet evergreen production forests, have imposed moratoria on logging to allow their forests to regenerate. Adoption of commodity crops there – notably tea and cardamom – by smallholders has increased rural prosperity but also increased the incentive for illegal entry into ever-shrinking forest fragments. They are actively researching optimal means to conserve their representative biodiversity, and legislating new conservation areas because earlier legislation focused almost exclusively on the large terrestrial mammal concentrations in the deciduous and dry evergreen forests (Chapter 9).

4.2. Why have lowland forests survived better in the seasonal tropics?

The forests and woodlands of seasonal Asia, adapted to climates prone to large-scale climatic catastrophes – drought and hurricane – are also better adapted to regenerate following human intervention than those of the perhumid regions. With increasing seasonality, annual flowering becomes more reliable, more species regenerate from coppice, and the proportion with seed dormancy increase.

4.2.1. Where lowland forests survive

It is still possible to see the various forest types and landscape patterns described in these essays in those nations that have legislated and actively protected networks of forest preserves for nature conservation. Within each climatic region, geology and soils impose distinctive patterns of canopy crown size and heterogeneity on the topography (Chapter 2). The physical nature of the geological substrate, of its rocks and their inclination, together with the Holocene history of sea-level change and retreat over the last 5,200 years, constitute the dominant influences on agricultural opportunities and forest structure at the local level. Superb protected forested landscapes remain in India, Sri Lanka and New Guinea; some of them are well maintained. Forested landscapes survive in a more limited way in Indo-Burma and West Malesia. Nevertheless, many of the forests we see in the seasonal tropics are in recovery from ancient natural or man-made catastrophes (Chapter 2). The fact that any considerable amount of seasonal lowland forest survives at all – though often in degraded condition – is due to the persistent dependence of rural communities on the

forest, combined with the difficulty of growing any of the major exotic commodity crops in the more seasonal lowlands.

4.2.2. Seasonal forests of India

Independence brought the taste of political freedom in societies where the sanctity of forests had been compromised, then forgotten. The loss of hereditary power suffered by traditional rulers in India following independence, and of will by the forest service under relentless criticism from both rural communities and the politically motivated public, resulted in a perception that government administered Forest Reserves were in effect common land: under-supervised and illegal logging, the cutting of fuelwood, and ubiquitous invasion of domestic cattle has prevailed. Forests suffered degradation as common lands everywhere have done, as population and consequent demand on their resources has increased.

As the population continued to increase in India, illegal cropping for forage and fuel became pervasive. Catchments were deforested, especially in the drier, western Siwalik Himalayan foothills.[96] Four out of five farmers tilled less than 2 ha, the average being 0.64 ha. Between 1951 and 1991 there had been a 20% increase in cultivated land and 43% reduction in 'fallow'. Yet in 2002, only 47% of the national land area was being cultivated according to some estimates,[97] much of the rest having been abandoned. In Madhya Pradesh alone, it was predicted in 1978 that there would be a fuelwood deficit of 8 million m^3 y^{-1} by 2000, and that plantation of 850 km^2 y^{-1} would be necessary, at INR 128 million. Fortunately, this turned out to be pessimistic owing to conversion to fossil fuels and the adoption of more efficient stoves. However, although the legislated forest area was increased, specifically for fuelwood plantation, forest degradation also continued to increase. The rate of degradation has been repeatedly underestimated: in Gujarat, satellite imagery revealed that only 42% of the state was still forested in 2001 – 66% less than had been reported.[98] Estimated forest cover in Gujarat has increased from 9,500 km^2 in the mid-1970s to 12,578 km^2 in 2001, but 'dense' (that is, un-degraded) forest declined by 5,000 km^2 over the same period. In the rainforested northeast, expansion of *jhum*, together with the illegal logging associated with rampant corruption, has decimated the forest estate.[99] Much forest degraded by illicit felling became commercially uneconomic.[100]

In response, the timely policy of JFM was introduced. Meanwhile, 5-year forest development plans had been initiated in 1952; the eighth, 1992–1997, included biodiversity legislation.[101] Shortly after, a national moratorium on logging was proposed, and continues to the present. In short, it is not the logging that has degraded Indian deciduous forests, so much as the increasing demands of an ever-expanding population, rural and urban, the supply of whose needs are not as yet being optimised through reform of forest policy and management.

But a major challenge remains: cattle. The cow was venerated historically, and is an important source of protein in the South Asian diet, in the form of milk and from it curd, cheese and other products. In many areas of short deciduous forest and thorn woodland, such as Rajasthan, cattle are restricted during the monsoon, enabling the cutting and storage of hay for fodder during the long dry season (Plate 8.5b). With the creation of government forest reserves, combined with the loss of control over forests by community authorities, many forests are treated as open-access commons. The grossly overstocked cattle starve during the long dry season; their browsing eliminates both seedling and coppice regeneration, except among species poisonous to them. Even these may be foraged by the goats and sheep preferred by graziers in semi-arid regions. Trampling, especially during the monsoon, hardens the soil surface to the detriment of worm and microbial decomposers, preventing the penetration of seedling taproots. It also increases storm run-off, reducing water storage in the aquifer. Removal of grass by cattle before it dies in the early dry season does reduce fire frequency and intensity, but this too alters the natural environment for tree regeneration. Thus, much of India's former deciduous forest estate is currently termed *wasteland*.

5. The social and political environment for forests in post-colonial Asia

In the 1950s, I was very fortunate to witness the primeval forests of Malaya and of Borneo in their magnificence, more or less as Beccari and Wallace had seen them. Now, the lowland primary forests have all but gone, and the hills are ravaged by uncontrolled logging.

5.1. *The colonial legacy*

At the end of the colonial period, half a century ago, there remained three categories of still-forested land in Asia:

1. *Upland forests*. Flood-free undulating uplands are widespread in the Asian tropics, and generally form continuous and well-defined landscapes structured by their geomorphological history. In the dry seasonal tropics their deciduous forests had long been inhabited by pastoralists and upland rice and millet cultivators. In the wet seasonal tropics, and especially in the perhumid tropics where malaria was rife, this land had remained widely under primeval forest until the late 19th Century. There is record of widespread settlement and conversion only near the great medieval centres of civilisation, in regions of fertile soil, and around the coastal capitals of the sultanates. The uplands have now been almost entirely converted, first to rubber and now increasingly to oil palm.

2. *Montane forests*. Early on, the steep slopes of major mountain ranges were hardly inhabited above c.1,000 m. Exceptions were found in Indo-China, where plateaux permitted the cultivation of hill rice; and New Guinea where yams have been cultivated for thousands years, and notably in northern Sumatra, the Kelabit Highlands and the Kinabalu Massif of Borneo, and in Sulawesi; and where pastoralists invaded the subalpine grasslands in South Asia. In seasonal regions, the slopes were the terrain of diverse swiddening hill tribes, often in intermittent warfare; but their populations were sparse and mostly confined to the lower slopes in perhumid Peninsular Malaysia and Borneo. Some of these, like the Batak in Sumatra, and Hmong in Laos and north-west Thailand, have become successful commercial farmers, producing fruit, temperate vegetables and garlic for local and regional markets. Several of the hill tribes of northern Indo-Burma had emigrated from warm temperate South China, where they had learned to terrace without irrigation. The trade in opium still continues among the tribes of north-east Burma. The majority of tribes have been less adept in trade, often labouring for the more successful tribes.

In South Asia and North-East India, coffee then tea cultivation had led to the almost total conversion of lower montane forest by the end of the 19th Century, except on the steepest land and in sanctuaries.

3. *Floodplain, swamp and peat forests*. About a fifth of continental tropical Asia and the Sunda continental islands consist of floodplain or swamp. In the seasonal tropics, flooding along the smaller rivers was controlled early, and more or less flood-free settlements were established by adapting irrigated agriculture to the flood season. This led to widespread deforestation many centuries ago, and woody vegetation survived only along unstable river courses and in sacred groves. In contrast, the peat swamp forests of perhumid regions, of the great plains of coastal Sumatra and Borneo, and the less extensive plains of New Guinea and Peninsular Malaysia, remained pristine well into the 20th Century. Now they are undergoing a process of complete deforestation, owing to the opportunistic – and from the perspective of carbon sequestration disastrous – practice of planting oil palm on deep peat (Chapter 9).

It has been estimated that, worldwide, 70–90% of the c.100,000 km² of tropical forest lost each year to other uses is in the lowlands, where biomass and biodiversity is greatest.[102] Asia is no exception.

5.2. *Post-colonial populations: Thailand as an example*

The end of the colonial experience following the Second World War released pent-up expectations among the populations of every country in tropical Asia. Forest-poor countries, already with dense and impoverished rural populations, were focused on enhancing their forests' socio-economic value.[103] Governments of countries still well endowed with forests, therefore with as yet relatively small populations, wished to log and convert many to commodity crops, not so much to provide land for the poor as to raise capital, ostensibly for industrial development and infrastructure – which in turn increased local demand for timber.[104] In reality and as we have seen, much of the private profit has been invested elsewhere, and lost to the forest-based economy.

There was a precedent in the feudal Kingdom of Siam, where in 1896 King Chulalongkorn, concerned by the increasing exploitation of his northern teak by European companies who had negotiated cosy agreements with local chiefs, hired European foresters to establish a forest service under principles similar to those established by Brandis in India. Felling cycles and strict limits set to ensure sustainable yields in the teak forests achieved a steady level of exports, contrary to some claims.[105] But in 1934, a military coup led to the establishment of a constitutional monarchy and the devolution of political power to an elite in which the military were prominent, which has continued in various forms to the present day. The European forestry leadership was terminated. It was not long before concessionaires found means to negotiate with the new leadership, to mutual advantage, for a more exploitative and financially profitable approach to logging. Villagers saw no reason why they, too, should not take what they needed. Settlement followed the logging roads.

In seasonal regions such as Thailand, the indigenous grazing and browsing fauna and their predators have suffered more than the forest flora. The tall deciduous forests are relatively low in endemic and rare plant species, though some do occur in unusual habitats. The trees are adapted to withstand fire when mature. Nearly all species readily sprout suckers from surviving rootstocks, thereby rapidly responding to occasional runs of wet, fire-free years (Chapter 2). Thus, these forests suffered the loss of essential resources for the rural economy and the loss of large fauna, rather than loss of plant species. Tall teak forests are also being degraded by invasions of two Neotropical weeds – *Chromolaena odorata* (Asteraceae), a coarse perennial herb, and *Lantana camara* (Lamiaceae), a scandent shrub – both of which form dense thickets, excluding seedling regeneration and increasing the incidence of hot destructive crown fires. As yet, no economically viable means for their large-scale control has been developed.

5.3. *Colonial and post-colonial administrations*

Tropical Asia differs from the unitary continental regions of Africa and South America in its ease of communication. No forest is far from the coast or from a port. The most fundamental change brought about by Western colonists, especially British and American, was the promotion of global trade, especially in natural resources. Colonial administrations had neither the skill nor the interest in preparing citizens for effective democratic government, for which open discussion and interest in public affairs, a healthy scepticism, and the confidence to challenge political leadership is fundamental. Colonial administrations were increasingly concerned at the growing populations in the forested hill regions, yet seemed queasy to introduce family planning. The 'white rajahs' of Sarawak had followed tradition in recognising the inland tribes as prime inheritors of the forest, but the third rajah, seeing increasing competition for land as populations swelled, established a forest service in 1920 – whose primary purpose, ahead of its time, was to adjudicate equitable distribution of access to forest resources among the forest people.[106]

The colonial epoch left forests in the hands of individuals who had no *personal* stake in their fate: politicians and civil servants of the forest service. Provincial politicians gained the power to influence forestry officials to relax the management rules necessary for sustaining the forests for timber. Licences were used to persuade local leaders to join the party in power. They were increased, and issued over vast tracts without commensurate increase in forest staff or training. This was to the immediate advantage of commercial interests – and often also to personal advantage of certain senior forest staff – but at the eventual cost of forest communities.[107]

The fate of forests was further compromised in almost every country, either by the dislocations of wars of independence or by militant movements opposing those governments helped into place by outgoing colonial administrations.[108] Forests harboured the contestants and became inaccessible to both management and concessionaires, often for years at a time. Foresters continued to support former rules and boundaries, but were often forced to become deskmen; the discipline of long weeks in the field, and the attendant knowledge, was lost.

6. Forests as an arena of conflict between the rural population and urban interests

Growing rural communities have contributed to forest degradation and deforestation, via fuelwood collection and cattle in dry regions, shortened swidden fallow in wet. But the increase of rural populations has everywhere gone hand in hand with rising expectations and the intrusions of broad-based market forces into fragile rural economies. Throughout the tropics, powerful interests have been the principal and proximate cause of misuse, especially through road-building allowing illegal access, conversion of commodity crops, subsidised credit and exploitation of weak tenure laws.[109] Together – and in the absence of oversight by silviculturists, investment in technological innovation and fencing (costly but permitting a rotation system of management), or any equitable distribution of benefits – these trends put intolerable pressure on the land and forest resources. Faced with serious hardship, traditional constraints on overexploitation are forsaken.[110] Future options are discussed in the next chapter.[111] These trends early became apparent in regions that had long been densely populated, in the seasonal tropics.

Demands on forests came to exceed their capacity to yield. Populations in regions amenable to irrigation, which had already become dense in classical times, continue to expand, degrading their forests and its potential to yield the goods and services they desire. Especially in India where cattle are a dominant component of subsistence agriculture, early attempts to exclude cattle from forests reserved for timber production alienated the forest service from rural communities and set the stage for confrontation:

> After fire, grazing is probably the most influential factor controlling the composition of the forest over most of the country, though not usually in the wet forests. Commonly associated with grazing damage is lopping for fodder and fuel, and many other injuries inflicted by the grazier.

Stebbing, 1922, *The Forests of India*, p. 225.

Excessive browsing of tree regeneration, and grazing and trampling by domestic cattle can eventually degrade the forest understorey to grass, leading to less intense, dry season fires. Toxic weeds then invade: *Rhus*, *Aegle*, various Apocynaceae, thorny *Randia*, *Acacia* (in the absence of goats and camels), *Holarrhena*, *Calotropis* and sour-smelling *Holoptelea*. Too much grazing and branch lopping, like over-intensive logging but with more immediate effect, reduce productivity for the users – the rural poor themselves. The number of tree species then declines,[112] although it is unclear as to how much grazing and lopping each contribute to this. In the short forest of the Gir National Park, Gujarat, last retreat of the Eurasian lion, the grazier Maldhari tribe were partially transferred elsewhere to reduce forest degradation. But the population that remained increased and started commercial milk production, introducing more cattle and degrading the area around their villages.[113] That is but one of the several threats to the Gir ecosystem; others include incompetent park personnel, illegal felling, mining, and over-use by religious pilgrims.

Although many are now under considerable pressure, India still has the highest concentration of sacred forests in the world – as many as 100,000 to 150,000.[114] Some are serving important conservation functions: for instance, sacred forests on the south-east coast of India are virtually the sole remaining remnants of semi-evergreen notophyll forest habitat there, while in northern Kerala, ecological interaction between reserved forest, sacred grove and shade coffee plantation strengthened the biological value of all three landscape elements.[115] Some sacred groves remain strictly inviolate and, though often small, retain a surprisingly intact flora and a reduced but still substantial fauna. More often, groves receive some use for grazing, fuelwood collection and licit or illicit logging even where use has been restricted by traditional custom.[116] New sacred groves are reportedly also being designated in some areas, through community concern at loss of forest quality and independent of any government agency. These may provide seasonal or occasional common-property economic benefits to the community.[117]

Beginning only in the 19th Century, conversion of the hill tribes to the Abrahamic religions – particularly the introduction of Christianity in North-East India and in northern Indo-Burma – was at times accompanied by denigration of the animistic traditions upon which the common internalised respect for nature was founded. Christianity brought education and a certain empowerment, though hardly increased communal opportunities for political power. Although local empowerment has been achieved in parts of Indonesia, eastern Malaysia and the Philippines, where tribal ethnicities are in the majority, these regions now harbour some of the most deforested landscapes.

6.1. *Joint forest management in India, and other participatory management*

> [Indian village communities are] little republics having nearly everything they want within themselves, and almost independent of any foreign relations.

Sir Charles Metcalfe, 1830, in J.W. Kaye, 1850, *The Life and Correspondence of Charles Lord Metcalfe.*

This statement was made from the 'bird's-eye' view of an official who would soon be appointed Governor General of Bengal (1835–1836). But it recalls the village republics of the ancient Bali-Hindu tradition, still partially surviving in Kintamani and described by Miguel Covarrubias as late as the 1930s;[118] also the longhouse communities of the Bornean Iban.[119] If village life was ever quite so independent from the point of view of the villagers themselves,[120] it is today far from the case. Even the most isolated of rural economies now finds itself enmeshed in a web of relations with wider economies, with administrative agencies, and with surrounding ecological patterns of change, all of which both create incentives and constrain actions. Under former conditions of low and fluctuating population densities, long-standing patterns of resource use could be maintained in some places, such as in the biodiverse forests of West Kalimantan. There, community forests, enriched with tengkawang fruit trees (*Shorea* spp.), hardly differed in their composition in commoner species from the nearby primary forest.[121] The same applied elsewhere in tribal societies where traditions of local decision-making prevailed.[122] Now, though, denser human populations call for increasingly intensive land use. It is unsurprising that the most promising examples of participatory forest management in general have been reported from the less-diverse deciduous forests, where uncontrolled grazing had compromised forest regeneration.[123] These systems have sometimes involved the introduction of exotic tree species.

6.1.1. *Joint forest management history in India*

The history of JFM in India, where it has been most experimented with and developed, has been excellently reviewed by Saxena, and Ravindranath and Sudha.[124] During and since the colonial period, the Government of India continued to enshrine certain traditional native rights to forest products in the law, but has allowed state forest departments to so encumber them with regulation that they often hardly existed in practise. The increases in forest area achieved through national policy since 1952 were overwhelmingly through plantation, often with exotic species, which did nothing to satisfy local needs. During the early 1970s Ajit Kumar Banerjee, a young officer in the West Bengal Forest Department, finding his silvicultural trials in sal forest in Arabari Forest Range hampered due to cattle browsing, resorted in desperation to direct negotiations with villagers. There was success, and the idea spread. In 1988 a national policy was initiated to manage appropriate forests in cooperation with the local communities, and JFM was born.[125] At the same time, the value of forest environmental services was re-emphasised. Their role in the rural economy of India has been comprehensively and elegantly reviewed by Saxena.[126]

This initiative, so clearly necessary when viewed from afar, has achieved some successes but is fraught with difficulties, some of them deeply historical and sociological in nature[127] (Chapter 9). Success of the JFM programmes has been greatest in regions where analogous traditions had persisted, where rights to certain products from community forests were recognized and regulated by traditional understandings and assent.[128] In some short deciduous forests, village participation in the rehabilitation of forests degraded by overgrazing and fuelwood cutting has secured results superior to those achieved by the Forest Department independently.[129] Women, who have always been subordinate yet are responsible for most of the land work, have been empowered in some cases, facilitating social

change. There are plenty of heart-warming instances, such as Dalit women of Ganashpura (Rajasthan) joining a tree nursery scheme, increasing their personal income and quality of life.[130]

But JFM has been at least as notable for its failures as for its successes. The major critiques point to flaws not only in implementation, but also in the basic design of the programmes, which largely fail to cede practical decision-making powers to forest users, especially the Adivasis.[131] Sometimes owing to divisions of caste and social status from those they serve, sometimes owing to simple overwork, Forest Department staff have found it difficult to relax authority where that is essential for success.[132] Even in West Bengal, where the success of JFM has been touted by international NGOs and agencies, it has been a mixed success.[133] Authoritarian attitudes and political influence are still common among *panchayat* leaderships. In some cases there have been significant problems with the system of electing Forest Protection Committees, where village elites have gained power over decision-making, especially for areas larger than a single village or community. Such social-political problems are partly rooted in lack of education and capacity among precisely those marginalised groups that stand to gain or lose the most under JFM, because they depend most directly on forest resources. In other cases, village groups have managed to appropriate benefits without making the corresponding investments in self-regulation and resource protection, and it is often difficult to impose sanctions on such behaviour.[134] In yet others, Forest Department officials have dominated and skewed what is intended to be a partnership with communities. At least one study exploring the views of Forest Department officials found that the rigid structure of the IFS, by severely limiting feedback within the institution, makes the task of responding to local conditions nearly impossible for lower-level personnel.[135] JFM thus remains a continuing struggle (see also Chapter 9).

In West Bengal, natural tall sal forests have often produced better overall results under JFM than have plantations, owing to the higher availability in them of the non-timber forest products (NTFPs) which contribute most of the community income. Net income varies much among years, and many opportunities to increase it through silvicultural intervention have been missed. It often remains uncertain whether the benefits provide sufficient incentive for the necessary sustained management effort.[136] Elsewhere, JFM has led to overexploitation of indigenous species uncompensated for by replanting.[137] It has been most successful where the land has the greatest capacity to respond resiliently, showing less success on ferricrete and other marginal sites where it is often most needed.[138]

The traditional values of medicinal and other products cannot be completely expressed monetarily.[139] NTFPs are the major source of forest-derived income in many parts of India – accounting in large part for the generally much higher value to villagers of natural forest over plantation. Nevertheless, even in West Bengal, with a history of relatively sympathetic and consistent NTFP policies by the Communist Party of India, JFM income per member has at best been a useful supplement to basic employment and subsistence income, rather than a ticket out of poverty.

6.1.2. Participatory forest management in other countries

Nepal has maintained a long-standing engagement with PFM policies since the 1970s, although the Maoist insurgency has had adverse consequences for community-based programmes in some rural areas.[140] Bhutan has also made considerable progress in combining strengths of state-centred and community-based development approaches.[141] Outside of South Asia, some similar initiatives have been adopted in dry northeastern Thailand.[142] In the Philippines, a 1986 amendment to the Constitution created autonomous regional authorities with the brief to devolve power over land to communities, but implementation has been slow.[143] In most places there has been a tendency for a 'widening gap' to open up 'between the written policies of the national government and the reality at the local government and community levels'.[144] This gap centres partly on the difference between those forests that provide sustenance value to local communities but have little commercial value, and those that retain potential significant value in timber or for other commodity production. Control over the latter tends not to devolve easily into community hands, reflected in the virtual absence of JFM in regions formerly clothed in commercially profitable MDF. In contrast, policies to promote the adoption of agroforestry methods to sustain village needs for fuel wood and other minor forest products are fairly widespread.[145]

6.2. The regional trade in forest resources

> Burma is a Buddhist country. Burma's leaders meditate. They say the country lives in peace.
> We have three neighbours: Laos, Cambodia and Burma. We use their resources, all three of them. If we have this great relationship, why should we pick on them?
> Westerners have a saying, 'look at both sides of the coin', but Westerners only look at one side.

Thai Prime Minister Samak Sundarajev, 17 March 2008, reported in the *Bangkok Post*.

6.2.1. A trend toward timber imports and its immediate consequences

Nations that have imposed logging moratoria must import on a large scale to meet home demand. These include Thailand and India; but Sri Lanka is a laudable exception, with plantations and extensive forests remaining in its double-monsoon northern and eastern sectors. These imports are now mostly coming from countries which either never had strong forestry oversight, or in which the overseeing agencies have been overridden and demoralised by tyrannical regimes. Nowhere has this been more tragic than in Burma. There, the world's finest teak forests, excellently managed for over a century, have been rapidly decimated for timber exports to China, India and even Thailand, in exchange for foreign currency to prop up the military junta.[146] The upland minorities are effectively reduced to servitude, often without legal title to land and cowed by persisting *corvée* labour. The *taungya* system of forest management, which was formalised by the British there, has turned thousands of uplanders and lowland Burmese migrants into temporary labourers, farming government forest lands in exchange for replanting teak and other species, but without land of their own and often without schools or health services.[147] Laos and Cambodia are suffering similar albeit less socially harsh exploitation. Illegal logging of rosewood (*Dalbergia*) for the Chinese market persists in Thailand, a moratorium notwithstanding.

The expansion of China's mercantile interests in South-East Asia has a long albeit intermittent history.[148] In the post-colonial era, aided by modern communications, the overseas Chinese community in the Far East has been able to reassert its long-established influence in trade. Between 1997 and 2007 the value of wood imports into China tripled, making it the world's leading importer. Wood imports to China are predicted to double again in the coming decade. Asian and Pacific countries, including Russia, have supplied c.70% of demand, so far. At the same time, China has become the major exporter of finished wood products.[149] It makes a third of the world's furniture, with exports rising from 91 million individual pieces in 2000 to 248 million in 2006.[150] China has also become the second largest producer of paper and cardboard. Between 1997 and 2005, exports of wood products from China to Western nations rose 700%. Clearly, the West can point no accusing finger!

Entrepreneurs, mostly of Chinese ancestry, amassed capital from the exploitation of Sarawak's forests and leveraged this to gain control of forest-mining conglomerates of great power. These concerns are presently reducing the last great tropical forest landscapes on Earth – from South America to New Guinea – to the natural equivalent of smoking rubble. In South-East Asia, there have been declines in the rate of loss of indigenous forest cover in Thailand, Malaysia, Laos and the Philippines, as there is little left on accessible topography to convert. Elsewhere, the rate of destruction continues.[151]

6.2.2. A new resource imperialism?

In effect, a new *commercial* imperialism has arisen. Burma has been becoming a virtual economic vassal of China, recalling the earlier conquest of inland southern China by the Han.[152]

Besides timber, products illegal in international trade, including tiger parts and ivory, are being exported to China and Vietnam.[153] Were the outflow of capital from tropical Asia into the Chinese economy fully accounted for with the outflow of goods, an increasing vassal status for the whole region could reasonably be suggested. It is hard to understand the apparent lack of serious attempts, on the part of Chinese planners, to encourage trade policies that might *sustain* that flow of goods and capital emanating from the tropical Far East. Such exploitation of neighbour states represents an extreme case of the financial interests of mercantilism being allowed to override the longer-term economic interest of nations. In the overwhelming quest for industrial raw materials, the tropical forests of southern Yunnan are themselves being converted to rubber plantations at a rate of 14,000 ha y^{-1}, up to the putative frost line at 1,200 m, leaving but 3.6% of Xishuangbanna province now under indigenous forest, almost all of it warm temperate forest located at higher altitudes.[154]

Yet, this seemingly irresistible vacuuming up of forest-based products into the insatiable maw of growing and industrialising economies is clearly just one element in the overall pattern of long-armed national and international resource exploitation presently being facilitated by the combination of cheap transport, rapid communications, hyper-mobile capital and half-hearted oversight that we know as 'globalisation'. Modern globalisation is itself the logical outcome of previous cycles of economic imperialism, including the industrialisation of the West in the 19th Century. But among the Western European nations, many boats 'rode up on the tide' together, benefiting from the products of scientific inquiry, relatively well-functioning institutions, and a culture that rewarded entrepreneurship, though also cheap labour, and cheap raw materials (especially timber) sourced from economically subservient colonies.

The pattern of exploitation by commercial interests serving one community against the interests of another (usually a forest minority) happens within individual national economies too. The social and economic consequences of such exploitation have often been disastrous. A poignant example is the Mountain Province of Luzon. The Ifugao, who live in narrow rain-shadow valleys there, patiently carved and sculpted their near-perpendicular slopes to produce irrigated rice. They were able to cultivate rice throughout the year, thanks to perennial springs dependent on of precipitation from fog in mossy upper montane forest along the high ridges during the dry northeast monsoon. But farmers from the lowlands, discovering that the cool moist climate and peaty soils of the ridges are ideal for potato and temperate vegetable cultivation, converted many. The streams dried up during the dry season as never before, and some Ifugao farmers, losing one rice crop, became destitute.

Another aspect of the exploitation of vulnerable forest minorities by lowland and metropolitan interests has been the plunder of natural resources made possible by the boom wages offered to locals to participate in exploiting their own forest resources. The costs to rural economies come later. Meat and fish are often severely depleted, as the number of guns skyrockets, and as the forest streams are choked by siltation and abandoned logs. Along the extended network of logging roads come an uncontrolled influx of lowlanders with antagonistic ambitions and little sense of the land's agricultural constraints. But the benefits of development, such as health services, schools and a more equitable role in wider commercial markets, have been acquired by upland minorities in only a few formerly forested Far Eastern regions.

6.2.3. *Large-scale population movements*

Migration of the landless and the enterprising from areas of high population into the forested hinterlands, whether organised in settlement schemes or illegal, has been a major proximate cause of deforestation. Migrants are seldom familiar with the intricacies of efficiently working the terrain and the soils on which they settle, or with the peoples whose traditional forest lands they usurp.[155] In many cases, these lands had remained forested over time because they are agriculturally marginal. Although the introduction of commodity crops on gentle topography has hardly increased soil erosion on udult clay soils, infertile humult soils have been more or less severely degraded.

Today's estimated rates of tropical deforestation vary, in part because the estimates represent different things. Importantly, Myers[156] estimated in 1986 that 73–91% of tropical deforestation was then occurring in the lowlands. There appear to be no reliable estimates of the residual unlogged forest area disaggregated by major forest formation. If any deciduous forests unaffected by human activity exist in Asia, they will be on rocky crags and insignificant in area. Nearly all have passed through a phase, or are now experiencing one, of management for timber. Nearly all deciduous forests are affected by some level of cattle browsing and fuelwood collection. Lowland *primary* evergreen and tall deciduous forests in Asia have almost certainly been reduced to less than 5% (probably closer to 2%) of their original area; moist montane forests to perhaps 10–15%.

Yet access to information has become almost universal in the electronic age, and the longing for betterment, even the luxury of a suburban existence, is irresistible. In one successful policy meriting replication elsewhere, the Malaysian Federal Land Development Authority (FELDA) successfully promoted the conversion of logged lowland wet evergreen forests to rubber then oil palm on a massive scale. This provided the entryway for many poor farmers to enter the cash economy on a permanent and secure basis, either as a labourer, or better, as a smallholder with legal tenure.

6.2.4. *An economic perspective*

Stumpage value, which is the surplus value remaining after the cost of the logging operation has been deducted from the log sale value, ought to provide the financial incentive for good forest management. But Vincent[157] showed that stumpage values have actually *decreased* over time, thereby suppressing the increase in value that should accrue with increasing scarcity of the commodity – a classic 'market failure'. He advocates providing incentives to the agent at the 'cutting edge', so to speak, of forest management – namely the concessionaires – by vesting in them rights and responsibilities comparable to those of ownership (since the representatives of the true owners – who are now the general public – regularly fail to act in their interest). This policy has been instigated by the far-sighted government of Sabah, Malaysian Borneo. In this way, long-term employment could be assured in the forest sector. But stumpage values would only increase if such policies were *widely* adopted.

It has been argued that profitable and sustainable forest management for timber is difficult to achieve when distances from markets are great[158] – or in the presence of cheap energy for transport of *un*sustainably sourced logs. This is particularly true when the extra costs of planning and overseeing RIL operations, with their attendant reduced yields, but even correctly executed selection systems of management, are factored in.[159] Profitability of sustained timber production is now increasingly uncertain since the most productive forests, those of the arable lowlands, are now almost entirely converted to commodity crops. When transportation once again becomes a truly significant cost factor, this trend will increase.

6.3. *A historical pattern and two exceptions*

In several Far Eastern nations, the resource capital is now almost exhausted. Where did it go? What has been achieved in those forest-holding countries, in terms of economic development, since the serious draw-down of forest resources began in the 1970s and 1980s? Increasing population is among the ultimate causes of deforestation, but is not in itself the primary cause.[160] Today, only two avenues seem open to the landless poor of the seasonal uplands: to move to the city in search of work, or to ignore traditional and legal constraints and trade in those forest plants and animal products whose value – because they are rare and endangered – has soared.

Impressive examples of long-term planning and policy do persist, nevertheless. There are two small Asian nations in which the greater part of the forest estate remains pristine, or altered only by the ancient practices of the inhabitants. In the Sultanate of Brunei, oil had deferred the exploitation of forest capital, but now formal legislation of a conservation policy and conserved forest areas is being planned and legislated

Plate 8.11 Brunei seen from space: an emerald surviving in a sea of desolation. The frontier of Brunei's two enclaves: grey line. (Google Earth)

(Plate 8.11). In the Kingdom of Bhutan, the annual estimate of Gross National Happiness is preferred to Gross Domestic Product.[161] So far, the Bhutanese royal government sees long-term economic value in combining intensified agriculture on fertile sites with conservation of ancient forest landscapes for their existence value, tourism and export of hydropower. The King is now a constitutional monarch, but has wisely retained ultimate authority over land. The country is steep and rocky, much of it inaccessible. 72% of the land area is forested, of which only 6–7% has been open for timber exploitation. Bhutan's rural population has yet to outgrow its arable land. Swidden is still practiced in areas of lowest population – on a sustainable 10–15-year rotation long unheard of elsewhere in South Asia. Bhutan's farmers have legal title to their land. Forests near villages have been given legal community ownership. Management is jointly planned with the forest service, which provides oversight for timber harvesting. Cattle browse throughout the forest, providing the greatest value from the forest to the community. Remote forests are claimed by the state, managed and exploited by the State Timber Corporation. Tropical lower montane forests are managed here on a modified shelterwood system. Trees >10 cm diameter are selectively cut on a 10-year cycle, where the understorey is first cleared in 6 ha patches. The trees are then replanted at the end of an 80-year cycle. Only about seven indigenous timber species are planted; besides these, only bird-dispersed species return. Understorey brush is collected by the community and used for a variety of purposes. But the 'Achilles heel' in this system remains the cattle. In their presence, the forest can only be regenerated by fencing and replanting. The cost of fencing materials is relatively high. Meanwhile, the human population increases. If cattle also increase, unpalatable exotic weeds will invade, and accessible forest will degrade to scrub. Will the population accept new restrictions on cattle and their freedom to roam?

At the scale of the village and the district, plenty of other examples exist elsewhere of rational resolution of the conflicts between individual and community interests in resource use.[162] The problem seems to be partly one of scale: large numbers, whether of population size or area, afford a degree of anonymity that tends to erode the sense of accountability and hence responsibility. Hence, the details of governance systems that forge secure links between levels of accountability become important. This is discussed further in Chapter 9.

7. More than timber has been lost

Long ago, the advanced economies of the older advanced nations reduced their forests to less than 15% of their total land area – in many cases much less. Japan is the outstanding exception, being still more than half forested. Aside from its sacred forest fragments, Japan's natural forest estate is confined to the earthquake-prone slopes of its precipitous mountains, too unstable for irrigation or settlement. Is it reasonable for us in the already-industrialised countries to lobby as we are doing, without offer of compensation, for younger nations to forego what we did before we well understood the consequences?

7.1. *Loss of traditional forest knowledge*

Those directly dependent, culturally and economically, on diminishing stocks of natural resources are still subsidising those whose *indirect* dependence on those same resources has been obscured by technology. The degrading resources include people's capacities as well as forests. Once radio and then TV became available to everyone in tropical Asia, the media and the schools together bore a responsibility to imbue pride in traditional knowledge and skills, but they have seldom risen to the task. Now it is nearly too late. A vast corpus of indigenous knowledge, much of it profoundly place-specific, is fading away. In China, forestry students from its many minority cultures attend a university in the capital, where their experience and cultural understanding becomes homogenised within the national Han cultural traditions. The ancient understandings embodied in animistic religions expressed a close spiritual unity between man and nature. Their remaining practitioners are ashamed to speak out as we shun their fundamental tenets.

7.2. *Non-timber forest products: still critically important in many Asian forests*

7.2.1. *Non-timber forest products in South Asian forests*
NTFP remain a critical component in the household economies of the poorest regions which, with some exceptions, are those

with seasonally dry climates (Plate 8.12). The diversity of species yielding NTFP may be higher in drier climates, but their net value may be lower.[163] Harvesting levels are increasing with population, but some also with growing expectations and increasing pressure from national urban and even international markets.[164] Overexploitation has been encouraged by the near-universal growth of trade through middlemen. The Soligas of the Biligiri Rangan Hills of southern Karnataka, until recently hunter-gatherers, have harvested more than 85% of the fruit crop of amla from the small trees *Phyllanthus emblica* and *P. indofischeri*. This has led to lopping of up to 15% of the above-ground biomass and reduction of next year's harvest, calling for an agreed harvesting limit.[165] The Forest Department has, since 2006, banned all NTFP harvesting within the forest. The putative harm to the crop is nevertheless contentious, and it is claimed by some that lopping is a traditional practice that reduces mistletoe infestation.[166] But the issue is now moot as the federal government has declared the forest a tiger reserve, therefore excluding further research! Such is one background to discussions of policy reform aimed at sustainably ameliorating the livelihoods of minorities, and at the same time conserving biological resources, in several Asian countries.

The seasonal tropics of Asia are rich in NTFP derived from two monocotyledonous plant families: palm[167] and bamboo.[168] Among palms, the palmyra (*Borassus flabellifer*) has been widely grown among rice paddies from India to Rote in the Lesser Sundas.[169] The discipline required for the successful harvesting of its bounty is recorded in a Tamil Nadu sutra, which documents and extols its 800 uses. These uses spread with it during the eastward expansion of Indian culture. But now in South India, the young of the Nadar caste who once specialised in palmyra culture are uninterested in the discipline, and the trees are being felled for timber. Such is the case with NTFP in many regions with fast-developing economies.

Plate 8.12 (right) Non-timber forest products. **a,** New Guinea villagers prepared for an occasion. Their elaborate clothing and ornamentation illustrates the bounty of the forest. But the forest cannot respond to a continuously increasing demand; **b,** Bamboo, the traditional raw material for hundreds of artefacts in seasonal tropical Asia. Doi Chiang Dao, Thailand, 2008; **c,** Palmyra, *Borassus flabellifer*, a palm once having 800 uses, central Thailand, 1958; **d,** Tapping the kitul palm, *Caryota urens*, for syrup, Sinharaja forest; **e,** Rattan collectors, Kudawe, south-west Sri Lanka; **f,** The medicinal vine, wenewel geta, *Coscinium fenestratum*, collected from forest gaps at Sinharaja, is even exported to India; **g,** Petai, the pods of the tree *Parkia speciosa* yields large garlic-tasting beans that are popular in curries; **h,** A honey collectors ladder up a *Tetrameles* tree, festooned with nests of the giant honey bee; **i,** The 1987 Brunei telephone book: a symphony to the home garden, with cultivars of indigenous rambutan (*Nephelium lappaceum*), cempadak (*Artocarpus integer*), and the yellow durian (*Durio graveolens*). (**a** T. Laman; **b,h** H.H.; **c, i** P.A.; **d–f** N. Gunatilleke; **g** C. Lobato)

In other places, however, NTFP continue as a very big business in the aggregate, and some have reached international markets.

Some products other than timber increase mightily following logging, fire or other disturbance. The most influential on the future trajectory of the forest is bamboo (Chapters 2, 3). Bamboos are more diverse in species and uses in Asia than anywhere else. Released by the decline of elephant populations which forage on the young shoots, bamboo has invaded old swidden fields, and is penetrating remaining lowland and lower montane moist deciduous and evergreen forests of the seasonal tropics as far south as Perak in Peninsular Malaysia. Bamboos are encouraged in woody fallows where they accumulate soil nutrients, besides providing primary materials for a diversity of NTFP.[170] But bamboo reduces forest regeneration by over-shading, and may impede it for decades (Chapter 2).

In low-wage economies such as those in drier rural India, other NTFP have been successfully commercialised[171]. Notable among them is tendu, the leaves of the dry deciduous forest tree *Diospyros melanoxylon*, which are used in a cheroot.[172] By 1991, Madhya Pradesh was producing 60% of India's crop, netting a formidable INR 1,500–2,000 million annually and employing over two million workers, after the state cut out middlemen by organising the trade into co-operative societies.[173] Productivity has increased through increased labour rather than mechanisation or increased efficiency, which puts into question long-term competitiveness if demand continues to increase.

In many parts of the moist seasonal Asian tropics, forests have been set aside at least in part to yield NTFP on a sustainable basis. These may include forests regarded as sacred.[174] The Soppina Betta forests of the Western Ghats are natural patches set aside by hamlets specialising in betel (the palm *Areca catechu*, indigenous to the Ghats). They harvest forest foliage and litter for composting their betel plantations. Poor and rich hamlets use similar methods, though the poor harvest most intensively: 'neither the poor fail in the upkeep of their forests, nor the rich succeed in managing them splendidly'.[175] In many locations, those sacred groves also used for harvesting products are becoming overexploited.[176]

7.2.2. Non-timber forest products in South-East Asian forests
Formerly, NTFP were often common property, but harvesting by a community within its watershed was constrained by private ownership of particular trees or other implicit limits. Especially in wetter climates, forest minorities forage for NTFP in the farming off-season, and sell to increase their income.[177] International demand, in particular, can lead to overexploitation to the point of endangerment. Tribesmen in the northern Burmese Adung valley were digging *Fritillaria cirrhosa* bulbs to sell to traders from Yunnan as long as 80 years ago.[178] *Tengkawang* (Indonesian) or engkabang (Iban and Sarawak), the fruit of

several *Shorea*, especially *S. macrophylla* and *S. stenoptera*,[179] yield a cocoa butter with exceptionally high melting point, which was collected from MDF and home gardens in mast fruiting years, for export as well as domestic use (Fig. 8.2). A local refinery was established in Pontianak. But growing in nature on the very soils sought for oil palm, the trees have been logged and the industry has collapsed, and the refinery's machinery exported to West Africa.[180] Gaharu, or eaglewood (*Aquilaria* spp., also *Aetoxylon sympetalum* endemic to north-west Borneo), yields a perfumed, deformed wood that can only be discovered by felling infected individuals. These populations are currently in steep decline in the Far Eastern forests where they were once common, due to almost total removal of reproductive individuals,[181] even in the Pasoh 50 ha plot of the CTFS network.[182] By contrast, in stable middle-income economies with secure land ownership and a strong tradition of resource conservation, many NTFP and timber species are planted in home gardens,[183] while NTFP such as rattan may be planted in mixed plantations using indigenous species.[184] These are ancient traditions and, even where conservation had been relatively lax, the rulers often sanctioned them.[185]

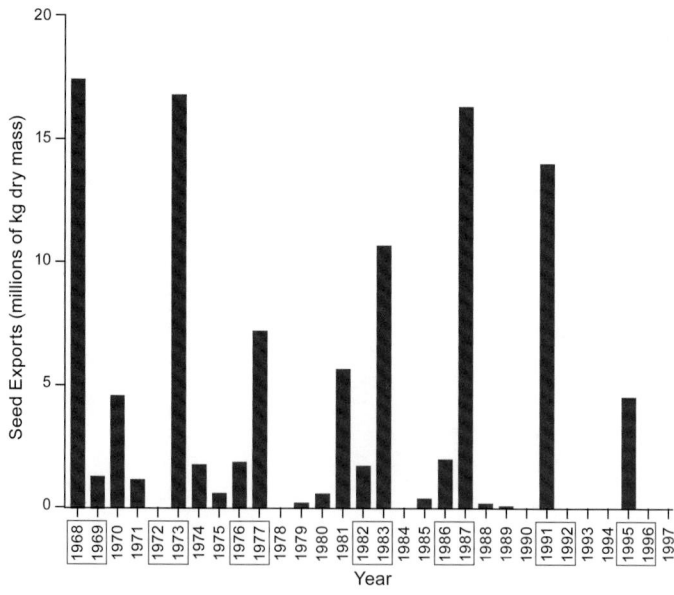

Fig. 8.2 The frequency and size of illipe exports over a 30-year period, and their relation to ENSO events (dates in boxes). (After Curran & Leighton 2000)

Whereas the rural poor of the less-developed rural economies continue to depend on forests for grass and browse, game, fuelwood, timber and medicinal herbs, advancing economies (notably that of Peninsular Malaysia) are experiencing massive changes in forest use. Prized products of former times such as Borneo camphor, the aromatic crystalline resin of *Dryobalanops aromatica*, or the oleaginous resin of Indo-Burmese *Dipterocarpus* traditionally used for varnish and caulking are all but neglected, though the latter are still tapped for caulking and as a paint additive in the poorest regions. The latex gutta percha was an early victim to the destructive exploitation of Far Eastern forests by Western mercantile interests. Early in the 19th Century, it had been found ideal for coating electric cables, and before the introduction of *Hevea* rubber it was collected for the international market from forests throughout Sundaland. Jelutong, another latex from *Dyera* species, was until recently exploited, though not destructively, for chewing gum manufacture. (The manufacturers were perhaps unaware that *jelutong* was coagulated in the field by the simple expedient of peeing into the collecting vessel.) The logging tide has depleted these resources.

The upland Temuan minority living near Pasoh Research Forest (Negeri Sembilan) still use the forest far more than do the Malay oil-palm cultivating settler population, and they still cherish it. Up until the time of forest clearance in the 1970s, they regularly collected c.270 different forest products, but with increased access to shops, this number has declined to c.70.[186] Now, they still use the forest, but increasingly for recreation including hunting.[187] Edible frogs can no longer be found in the forested valleys around Kuala Lumpur; so high a price did they command that only juveniles survive in most forests.[188] Everywhere, game has been reduced by overhunting, including inside many conservation areas. Animals yielding medicinal products, especially in the Chinese tradition such as tiger, bear and some reptiles, are under intense pressure, incurring vastly increased costs of protection. In parts of Indo-Burma and Sarawak, primates, hornbills, even squirrels have been hunted to local extinction and pervasive rarity. Their extirpation is having an impact on seed dispersal and germination success.[189]

The Sinharaja UNESCO World Heritage forest (southwest Sri Lanka) is surrounded by villages, some with close access to, others far from, local markets. Ruben Lubowsky[190] studied four of each. He found that villagers further from markets, the landless, the old and the poor, and women most often entered the forest to collect. Men undertook expeditions for more distant resources. That is reminiscent of Borneo, where young Iban men would spend months away in search of *Agathis* copal, or hunting the feckless rhinoceros.

The Peninsular Malaysian experience has shown that rural people, like all of us, given the opportunity to secure increased income, and with secure land holdings, forfeit traditional forest products for the goods of the market. Marketable NTFPs are exploited to local or regional extinction. Products of primarily local value may be sustainable in rural economies where opportunity for advancement are limited, but here too overexploitation becomes inevitable if human infant mortality is reduced without corresponding reduction of fecundity.

7.2.3. *Medicinal plants*

While plant collecting in the Limbang headquarters, I observed my local Murut associate searching out particular herbs. I asked him to instruct me concerning his traditions in herbal medicine. 'Alas', he replied, 'for humans we only know about poisons; but to increase the sense of smell of our hunting dogs, there we know everything.'

Tropical Asia is home to an unparalleled diversity of sophisticated medicinal traditions. The compendia of Watt for British India,[191] Burkill for Peninsular Malaysia,[192] and Heyne for Indonesia[193] (now further elaborated by the masterly PROSEA project[194]) attest to this richness. Broadly, plants yielding high concentrations of pharmaceutically active compounds, like culinary herbs, tend to concentrate in moderately dry seasonal climates and on fertile or calcium-rich soils, as they do in temperate regions. The karst country of warm temperate China, for instance, is exceptionally rich in them.

The two most highly developed traditional medicinal systems of Asia are the Hindu-Buddhist ayurvedic tradition and the Chinese Confucian system. The former employs herbal medicines derived from the flora of monsoon India, while the latter uses herbs overwhelmingly from warm temperate China (see the monograph by Needham[195]). The Javanese pharmacopoeia includes many ayurvedic herbs, either native or naturalised in seasonal Indonesia; but the rainforests are also a rich indigenous source.[196] The Malay pharmacopoeia is less rich, based predominantly on indigenous species of the perhumid climate, though many ayurvedic and Chinese medicinals have also been adopted. The upland tribes vary greatly in the complexity of their medicinal knowledge, and the plant medicinals known to them vary with the indigenous flora from which they are derived. Importantly, lowland populations have long depended on the uplanders for many medicinals, who have collected from the forest and bartered them since before recorded history. As a result, many medicinal herbs are local, even rare, in nature, being confined to specific soils. Traditional folk medicine in the advanced economy of Peninsular Malaysia has been reduced to a limited demand for a small number of popular remedies. A few are now commercially marketed, like the restorative, blood-clotter and putative aphrodisiac tongkat ali (*Eurycoma longifolia*; I have tried it – but with disappointing results) – which has become locally extinct.

7.2.4. *Summary*

As with logging itself, NTFP appear to be most sustainably used in the less-prosperous rural economies where there are stable smallholder communities. This is especially true where there are sophisticated traditions of home gardening, with

Name	Family	Vernacular name	Provenance
Achras zapota	Sapotaceae	*Ciku*	Mexico
Annona muricata	Annonaceae	*Durian belanda*	Neotropics
Antidesma bunius	Euphorbiaceae	*Buni, berunai*	Asian tropics
Artocarpus altilis	Moraceae	*Dukun*	Melanesia
A. odoratissimus		*Terap*	Philippines[a]
A. heterophyllus		*Nangka*	S. Asia
A. integer		*Cempedak*	Far East
Averrhoa bilimbi	Oxalidaceae	*Belimbing*	Uncertain
A. carambola		*Carambola*	Uncertain
Baccaurea costulata	Euphorbiaceae	*Tampoi silau*	Borneo
B. lanceolata		*Limpaoung*	Sunda
B. motleyana		*Rambai*	Sunda
Bouea oppositifolia	Anacardiaceae	*Rambaniah*	Far East
Canarium latistipulatum	Burseraceae	*Mitus*	Regional endemic
C. odontophyllum		*Kembayau*	Sumatra-Philippines
Carica papaya	Caricaceae	*Papaya*	Neotropics
Castanopsis cf. *ovoformis*	Fagaceae	*Berangan petak kemudi*	Borneo
Citrus grandis	Rutaceae	*Limau besar*	Far East
C. reticulata		*Limau kupas*	Far East
C. mitis		*Limau kesturi*	Far East
C. aurantiifolia		*Limau asam*	Far East
Dacryodes rostrata	Burseraceae	*Pinanasan*	Sunda
Dialium cochinchinense	Leguminosae	*Keranji*	Far East
Durio dulcis	Malvaceae-Bombacoideae	*Durian isa*	Borneo
D. graveolens		*Durian kuning*	Sunda
D. macrophyllus		*Durian kura*	Sunda
D. oxleyanus		*Durian sukang, sukang*	Sunda
D. zibethinus		*Durian biasa*	?Sumatra (probably extinct in wild)
Flacourtia rukam	Salicaceae	*Rukam*	Asian tropics
Garcinia mangostana	Guttiferae	*Manggis*	Sunda
Garcinia sp.		*Kandis*	
Gnetum gnemon	Gnetaceae	*Meninjau*	Malesia
Lansium domesticum	Meliaceae	*Langsat*	Malesia
Mangifera caesia vars.	Anacardiaceae	*Binjai, belunu, ukong*	Sunda
M. decandra		*Binjai nasi*	Borneo, Sumatra
M. foetida		*Membangan*	Sunda
M. griffithii		*Rancah rancah*	Sunda
M. indica		*Empelam, mangga*	Seasonal Asia
M. longipes		*Mangga air*	Sunda
M. odorata		*Kwini*	Unknown wild
M. pentandra		*Asam damaran*	Peninsular Malaysia
M. quadrifida		*Asam damaran*	Malesia
M. torquenda		*Kemantan, ma(n)tan*	Borneo
M. pajang		*Membangan*	Borneo
Musa acuminata	Musaceae	*Pisang* (various vars)	Sunda
M. paradisiaca		*Pisang* (cooking vars)	Far East
Nephelium cuspidatum	Sapindaceae		
v. subv. *cuspidatum*		*Pari*	Borneo
v. *ophioides* subv. *beccarianum*		*Ketidahan*	Borneo
N. daedaleum		*Kalas*	Borneo
N. lappaceum			
v. *lappaceum*		*Rambutan*	Far East
N. ramboutan-ake		*Meritam*	Far East
N. uncinatum		*Rambutan hutan*	Sunda
Pangium edule	Salicaceae	*Kepayang*	Malesia
Parkia speciosa	Leguminosae	*Petai*	Far East
Archidendrom jiringa	Leguminosae	*Jaring*	Sunda
Psidium guajava	Myrtaceae	*Jambu biabas*	Neotropics
Sandoricum koetjape	Meliaceae	*Sentul*	Malesia
Syzygium aqueum	Myrtaceae	*Jambu air*	S. Asia
S. jambos		*Jambu mawar*	SE Asia
S. malaccense		*Jambu merah*	?Sunda
S. samarangense		*Jambu puteh*	?Sunda

(Trees with other uses, including timber, are also grown)   [a] The species of MDF on sandy soils commonly identified as *A. odoratissimus* is not that.

Table 8.3 (left) Fruit tree species noted in Kedayan and Tutong home gardens (*dusun*), Brunei Darussalam. Kedayans, who trace their origin to Majapahit Java, are the only ethnicity in north-west Borneo to plant wet rice, and to follow the South Asian tradition of home gardens on the rising ground above paddies. Tutongs are swiddening neighbours. Their home gardens, the richest known to us, provide fresh fruit for the sophisticated citizenry of the modern capital. Many of those fruit trees have their ancestry in local forests.

a diversity of tree as well as herbaceous crops grown. The Kedayan communities around Bandar Seri Begawan (Brunei) and along the north-east Bornean coast formerly cultivated irrigated rice and maintained home gardens of immense complexity. These involved a diversity of indigenous fruit tree species unmatched elsewhere in Asia (Table 8.3). Now, high wage levels in a petroleum economy have undermined their worth. The tradition remains strong, though, in the perhumid south-west of Sri Lanka. Here, smallholders still occasionally plant the slow-growing but valuable endemic ebony (*Diospyros quaesita*), as well as *Artocarpus*, as a future dowry. This reflects a similar tradition among some Japanese villagers who still, in one of the world's largest economies, plant tsugi (*Cryptomeria japonica*) and hinoki (*Chamaecyparis obtusa*) with the same expectation.

7.3. *Carbon sequestration*

Forests are major repositories of carbon sequestered from the atmosphere. Some temperate forests sequester carbon through accumulation of organic matter in soils, but soil carbon is low and rapidly oxidised in most lowland rainforest soils (Chapter 1). For comparison, soil under Indian seasonal evergreen dipterocarp forest was estimated to sequester a mere 0.82 t C ha^{-1}, while temperate Himalayan spruce forest sequesters 386 t C ha^{-1}.[197] The obvious exceptions in the humid tropics are montane forest soils, *kerangas* and peatlands. But it has also been estimated that below-ground biomass and soil may contribute more than half the carbon stock in the deciduous forests of India.[198] Another study calculated that with improved management of agricultural and forestry soils (including degraded soil restoration, erosion control and agricultural practices such as no-till farming, cover crops, manuring and sludge application, water conservation and harvesting, and agroforestry systems), the total potential of soil carbon sequestration in India could reach 39–49 (44 ±5) Tg C y^{-1}.[199] Many of these improvements, however, would entail substantial cultural and micro-economic changes in farm energy budgeting and household practices in India – for instance, in the absence of trees or woody shrubs, crop residues and manure are traditionally removed from the fields to be used as household fuel. In any case, India's total carbon-flux from soil erosion (4.8–7.2 Tg C y^{-1}) compares rather favourably with other large countries (USA: 15 Tg C y^{-1}; Brazil: 60 Tg C y^{-1}).

Since 1860, South-East Asian forests have released an estimated 8,000–19,000 Tg of carbon. In the year 1980 *alone*, 150–430 Tg of carbon were released, mostly from forest conversion and swidden.[200] Much of this was due to felling, burning of lowland peat swamp forests, and draining and burning (therefore oxidation) of their peat soils on a vast scale. The carbon content of peat is constant, and peat accumulation rates are highest in the lowland perhumid tropics: c.1.1 mm y^{-1}, which is double the rate at higher latitudes. Neutzil estimated the average natural carbon flux from the atmosphere to tropical lowland peats at 61–145 g m^{-2} y^{-1}, on average perhaps six times as much as that in temperate peats (10–50 g m^{-2} y^{-1}). Therefore, although tropical peat swamps constitute only 10% of global peat by area, they have been responsible for 20–25% of global peat carbon storage.[201]

Worldwide, forest destruction is estimated to be releasing c.20% of the carbon currently entering the atmosphere. This is occurring mainly in the tropics and largely through burning.[202]

8. Other forest pressures are gathering momentum

8.1. *Fire*

Fire is a natural phenomenon in the seasonal tropics, and its frequency and intensity determines the ecotone between evergreen and deciduous forest in the lowlands (Chapter 2). Since humans first arrived on the scene, they have been using fire to attract wild ungulates, and latterly to encourage grass for cattle.[203] Swiddeners, originally hill tribals but now including landless migrants,[204] quite rightly regard fire as a fundamental instrument in their method of cultivating marginal land,[205] so inducements for them to seek alternative employment on commodity crop estates substantially reduces carbon release.[206] This is achieved, though, at the cost of the carbon released by one-time conversion of forest to commodity crop, whereas during the fallow period, woody swidden fixes as much carbon as it releases during the burning phase. (Such conversion, when on gentle topography, apparently does not necessarily increase soil erosion or impair water yield or quality.[207])

Fire frequency has increased markedly in many places over the last half-century, including in National Parks. At Mudumalai Wildlife Sanctuary, Karnataka fire occurred in the moist mixed deciduous forests 3.3 times more often during 1989–2002 as during 1909–1921.[208] Recently, fires have been occurring at intervals of less than 7 years, with an average return time of 3.3 years. Pervasive increase of fire frequency is a threat to both wildlife and vegetation.[209] More frequent fire lends competitive advantage to deciduous tree species, hence tending to convert semi-evergreen forest to tall deciduous forest, while late dry season hot fires in tall dead grass or shrubs can reach the canopy and lead to replacement of tall by short deciduous forest (Chapter 2). In the latter especially, fire facilitates the invasion of the unpalatable shrubs and weed species of thorn forest.[210]

During the last 50 years, fire has even been recorded in wet seasonal and perhumid climates, albeit in or close to logged forests, in association with intense ENSO-induced supra-annual droughts, for instance in 1982–1983 and 1996–1997.[211]

The drought of 1982–1983 is reported to have killed 16–20% of understorey trees and up to 70% of canopy trees in the primary forest of Kutei National Park (East Kalimantan). The addition of fire in some areas, however, resulted in 90% understorey mortality, seriously endangering forest recovery (Table 8.4).[212] Logging may heighten the intensity of droughts at the local scale, causing canopy mortality.[213] By increasing fallen fuel, it greatly enhances proneness to fire. A fire in Sabah killed 19–27% of canopy trees in an unlogged forest tract, 38–94% in a logged tract; juveniles and understorey trees in both tracts experienced over 80% mortality.[214] Recovery then proceeds with sharply reduced species richness. Dipterocarps are notably patchy or absent in the regeneration, which is at first saturated with *Macaranga* and other pioneers.[215] Successive episodes of burning eventually lead to the replacement of *Macaranga* by *Imperata*, producing the true 'green desert' (Plate 9.2e).

Site condition	Vines >4 cm dbh Percentage (N)	Trees 2–4 cm dbh Percentage (N)	Trees >50 cm dbh Percentage (N)
Moist, 70% burned	n.a.	67.8 (242)	6.2 (16)
Well drained, 100% burned	94.8 (83)	87.6 (680)	21.5 (77)

Table 8.4 Percentage mortality among woody plants in burned forest, one year after the 1982 ENSO event, Kutei National Park, Borneo. (From Leighton & Wirawan 1986)

In eastern Borneo and south-east Sumatra, many logged forests were effectively destroyed during the drought of 1982–1983. Particularly worrying has been the conversion of peat swamp forests throughout the Far East into oil palm plantations.[216] Not only is a unique ecosystem being destroyed wholesale,[217] but the burning of the cut forest and the peat – to a depth increased by the construction of drainage canals – contributes massively to atmospheric carbon. Regional peat swamps have been estimated to have sequestered 70 times the annual global carbon emissions from fossil fuels.[218] During the 1982–1983 drought, I estimated the particulate and gaseous carbon releases, at 0.5Gt,[219] to be comparable to that emitted by all industry in the industrialised nations over the same period. The resulting pervasive regional haze may have reduced rates of photosynthesis by as much as half.[220] In Indonesia alone, peat swamps occur over 16.5–27 million ha, and are said to be capable of *sequestering* 0.01–0.03 Gt C y^{-1} [221] (see Chapter 9).

Severe droughts, extending over decades, have occurred in the historic past and may have caused the demise of the medieval Khmer civilisation at Angkor.[222] It has been claimed that forest fire is an ancient natural phenomenon even in perhumid Borneo,[223] but there is no irrefutable evidence for that over historic time, and it seems unlikely because fire-tolerant indigenous woody species are absent.[224]

Swiddeners, even in perhumid climates, cultivated by felling a patch, burning its felled biomass, then planting annual crops for a few seasons, before abandoning the patch to regenerate by coppice and seed. But the burning may lead to irreversible acidification of the surface, and leaching, in freely draining humult soils (Chapters 2, 3), until recently shunned by swidders. Both Shono and colleagues in Singapore, and Slik and his colleagues in south-east Kalimantan[225] failed to find any re-establishment of primary forest species by seed, even several years following burning and even adjacent to the surviving forest's margin which had suffered little damage in this windless climate.

Fire also converts upper montane forest to grassland. The sharp boundary thus created between forest and grassland is enhanced by periodic frosts, killing tree seedlings which establish in the grassland[226] (Chapter 4). Forest burning severely affects the water yield of upper montane forests. When healthy – their surface area much amplified and enhanced by cryptogamic epiphytes – these physically rake water droplets from the foggy monsoon winds.[227]

8.2. *Forest change and wildlife*

People have lived with wildlife and shared its resources for millennia in the deciduous forest regions. This continues today, especially in the wildlife parks. Thus, along the foothill *terai* of the Himalaya, villagers live adjacent to or within the preserves and collect grass for thatch and fodder. They may also, illegally, harvest fuelwood. In one study, villagers recognized the value of wildlife protection, but valued the grass and fuelwood higher.[228] In rainforest regions, certain mammals persist and even increase in numbers in logged and disturbed primary forest, in secondary forest and close to villages where hunting has not intensified.[229] These include certain ungulates and a variety of terrestrial species (Chapter 5). Ickes[230] reported a startling increase in wild boar (*Sus scrofa*) entering oil palm plantations from adjacent Pasoh forest to feed on fallen fruit. They build their crèches twice a year at about two per hectare, using hundreds of saplings, some species of which are avoided and others preferred (Chapter 5). This may alter relative tree species densities, even precipitating local extinctions.

Any increasing trend among some animal populations in logged areas, however, does not take into account the pervasive hunting for some of these same species. This often extirpates the less fecund species, including large vertebrates such as wild cattle and elephants. Fragmented forests are prone to local extinctions, exacerbated both by hunting and by the resource 'booms and busts' associated with the ENSO cycle; extinctions of some frugivores are already influencing tree establishment patterns in some forests[231] (Chapter 5). Arboreal vertebrates eventually return to regenerating selectively logged forest as the canopy closes, and may even increase where dipterocarps have declined.[232] However, where they are deprived of canopy corridors and fruit trees, crucial behavioural patterns such as pairing and mating may be inhibited, and some species decline over the longer term.[233] Small mammal and bird species diversity of unlogged primary forest, often including local endemics, declines in logged forest even where overall richness may increase owing to the intrusion of forest margin and open country species.[234] Specialist species dependent on resources that decline, such as woodpeckers reliant on hollow old and dead canopy trees, are particularly affected.[235]

The extirpation of predators can also be expected to lead to massive changes in forest composition where hunting has not replaced them in limiting the abundance of browsing mammals. On the other hand, hunting licences for game are now becoming mandatory in those Peninsular Malaysian states where tigers and other predators survive, as prey has become so rare that starving tigers enter villages in search of meat. Mankind now wholly determines the fate of the forest. There has been some reduction in the capture and export of wildlife following the ratification of the Convention on International Trade for Endangered Species (CITES), but real reduction only comes when governments seriously decide to implement it[236].

Most biodiversity, though, is comprised of insects and microorganisms. Although one study of bees found that inherent local variations in abundance, composition and richness exceeded the differences between unlogged and logged forest,[237] other studies have been less sanguine. Butterfly richness decreases after logging owing to the loss of specialists[238] but, as with birds, may increase above primary levels in logged forest when forest margin and secondary vegetation species enter in dry seasons.[239] Termite communities can recover completely in returning forest within 50 years if there is adjacent primary forest, but may need a millennium without it.[240]

Many pollinating insects in the tropics are specialists. Their loss may be expected to reduce the fecundity of their plant hosts, especially where the hosts themselves also experience lowered densities of reproductives[241] (Chapter 9).

8.3. *Fragmentation*

The remaining primary lowland forests in Asia are now fragments. In some, such as the Bukit Timah Sanctuary within the City of Singapore, primary forest may cover less than 100 ha. Others may cover a few thousand, such as the 6,800 ha Lambir National Park (Sarawak), or the Danum Forest area in Sabah. Some larger patches of primary forest of uncertain area survive within larger conserved areas – and these are often substantially mountainous. Examples include Gunung Palung and Kutei National Parks in Indonesia, Gunung Gading, Gunung Mulu and Taman Negara in Malaysia. Most other patches of true primary forest persist as small or tiny relict fragments, containing just a few remnant individuals of endangered tree species (Chapter 9). All deciduous forest, and nearly all seasonal evergreen forest (perhaps excepting what little may remain in Peninsular Burma), have been selectively logged at least once.

However, for biological reasons (explored in Chapters 2, 9), the impact of logging is generally less severe on endangered populations in deciduous forest than it is in rainforest. Johns found that, provided they are large enough, fragments of unlogged primary forest embedded in logged secondary forest can play critical roles in the survival of some species[242] (Chapter 9). Habitat area has been found to influence the species composition of leaf-litter ant communities in Sabah.[243] Area and degree of isolation influences butterfly communities, no endemic species being recorded in fragments smaller that 4,000 ha.[244] Butterfly species richness was nevertheless found to be greater among than within small sites, strengthening the case for strict preservation of remaining small fragments.

In South America, much research has shown the influence of edges on the interior of rainforest fragments, bringing drier climate, increased light, and invasions of open-country vertebrates into the understorey.[245] The effect is most severe in smaller fragments with longer edge relative to overall area, and declines from the edge inwards. Such effects have not been documented in Asia, and appear to be milder. This may be for a combination of anthropogenic and natural reasons: Asian forests have rarely been *directly* converted to open pasture as they are in the Neotropics, while MDF do not experience prevailing dry winds (where there is a windy period, it is associated with the wet monsoon). Nevertheless, an increase in light-demanding pioneer and other species has been recorded in seasonal evergreen forest fragments in southern Yunnan, where the endangerment of species of low population density was feared.[246]

The case for conservation of fragments is particularly strong for plants, being sessile organisms (Chapter 9). According to the theory of island biogeography, 50% of species should eventually be lost when a true island is reduced by 90%. Of course, this will be less in an ecological 'island' because some species also survive in the surrounding vegetation. The organisms most immediately affected are the large, specialised, or sparsely distributed ones, and those with short life cycles.[247] Several studies have shown that Singapore Island, its reserves of primary vegetation now confined to a minute 0.25% of its total 616 km², has suffered 34–87% extinction of different animal groups. However, though 62% of epiphytic plants appear to be lost, only 22% of tree species appear to have become extinct as yet. Re-census of plots within the astonishing primary coastal MDF fragments within the city (admirably conserved within Bukit Timah and much visited) gives evidence of decline in population size for very few tree species. The exceptions are a slight decline in wild nutmegs (Myristicaceae), whose fruit were dispersed by now-absent large pigeons and hornbills (recent citings confirm that the latter are visiting again); and reduction of some pioneers, which suggests increased stability under current management.[248] Plants in fragments may, however, experience reduced fecundity if seed or seedling predators, such as wild pigs, invade.[249] Conversely, where forest fragments are large enough to harbour herbivorous mammals, human settlements near the margins frequently experience crop depredation, and will lobby for their control.[250]

Valuable adjuncts to conservation in India have been the Preservation Plots; over 300 of which have been initiated since 1905.[251] Generally covering only a few hectares, their purpose has variously been to act as controls for silvicultural treatments, to serve as refuges for biota threatened by logging operations, and to represent pristine examples of major forest types for staff training. *It is unclear how many survive today.*

Following India in 1959, the Forest Department of Peninsular Malaysia instituted a policy to set aside Virgin Jungle Reserves (VJR) throughout the production forest reserve system – examples of primary forest to act as controls for monitoring silvicultural interventions, for research and staff training, and for conservation. Remarkably, most of these stands survive imbedded in logged forest, but sometimes even where they are now isolated in a sea of oil palm. They were intended as patches of forest equivalent in area to a management compartment, generally about 40–200 ha. Where some logged forest remained around them, it has served to conserve most

wildlife. It is a tragedy that their implementation virtually ceased following independence.

8.4. *Invasive plant species*

Of the vast array of plant species that have been introduced among different regions of the tropics,[252] few have become invasive in unmodified indigenous forests.[253] Noxious weeds have nevertheless become pervasive in deciduous and seasonal evergreen forests of Asia, where fire and other disturbances, a periodically open canopy, and low species diversity provide conditions where competition is reduced. Most arrived from parts of the Neotropics. Notable among them are *Chromolaena odorata*[254] and *Parthenium hysterophorus*[255] in North-East India, both tall gregarious perennial herbs invading logged and burned areas in seasonal evergreen forests; and *Lantana camara*, a deep-rooted dominant understorey shrub of the Neotropics, which has invaded deciduous forests in Peninsular India.[256] Other weeds include the vine *Mikania micrantha*, growing in seasonal evergreen forest and mixed dipterocarp forest on fertile loams,[257] and *Mucuna bracteata*, a perennial creeping and climbing legume of disturbed habitats. *Mucuna* was purposely introduced from North-East India into Malaysia and Indonesia as a cover crop for *Hevea* rubber and oil palm plantations during the 1970s, and has since become a pest, growing at a rate of up to 10–15 cm/day. Certain plantation trees may succeed in severely degraded vegetation, notably the Australian *Acacia auriculiformis* and, locally, *Falcataria moluccana* (native to Wallacea).[258] The Neotropical pioneer trees *Muntingia calabura* and *Piper aduncum*, like the shrub *Cassia alata*, invade wasteland and forest margins. Very few weeds, however, are able to invade primary perhumid rainforest. One species that does invade is the fecund berry-bearing Melastomaceous shrub *Clidemia hirta*, again from the Neotropics and now widespread in Asia.[259] Perhumid south-west Sri Lanka – an ecological island within a true island – has attracted several more species, all successional and more or less confined to regenerating stands, including the Malesian pioneer canopy tree *Alstonia macrophylla*.

8.5. *The neglect of silvicultural research*

The last fifty years have seen an enormous increase in research into theoretical aspects of biodiverse plant communities, notably into the generation and maintenance of species diversity in hyper-diverse rainforests. There is a growing literature on the conservation biology of animal species, and a steady flow on other conservation-related subjects including the impacts of fragmentation. At the same time there has been a steady decline in peer-reviewed publications addressing the applied management of rainforests, for whatever purpose. Research emphasis has also shifted from the forests of tropical Asia to those of the Neotropics. The silvicultural and primary forest control plots set up in India and Malaysia seem to have gone almost unused for their intended purpose in the last half century.[260] Though there are outstanding researchers, they often operate in isolation and their work is rarely transferred into policy or practice. Forestry research must be driven by hypotheses, built on empirical field observation and tested by experimental manipulation of forest stands. Today, even descriptive data are only sporadically collected, and there is a critical shortage of long-term monitoring. The fundamental methods of modern science, based on sceptical enquiry, have been widely abandoned by tropical forestry.

In India, there has been a surge of socio-economic policy and some ecological research on sustainable yield of the forest products and services essential to dense rural communities; but this research rarely includes rigorous testing of sustainability through continuous monitoring, and rarely considers the benefits (including biodiversity) that forests provide to cities and the nation, let alone the world. The Ashoka Trust for Research in Ecology and the Environment (ATREE) programme, based in Bangalore, represents an impressive exception that can serve as a model. Despite the efforts of such groups to bridge the disciplinary divide, a split has developed in India between urban wildlife conservation movements and some biologists on the one hand, and rural social and economic researchers on the other. This split has made it difficult to find common ground on basic questions such as the causes of forest degradation, rights and responsibilities of community groups, and the potential (but inadequately explored) role of the Forest Department in fostering aspects of sustainable rural development.

These changes in research emphasis reflect the fact that the forests are themselves a declining resource, and the recognition that active management is increasingly expensive wherever labour costs are rising. In the Far Eastern perhumid tropics, there seems to be a worrying lack of political interest in forest values other than timber. Many non-timber values are 'services to the nation'; difficult or impossible to commercialise. As a consequence, career opportunities in forest science are limited in many nations of the Asian tropics, and are seldom attracting the best minds. There is also a hint, in the virtual absence of rigorous monitoring of post-harvest forest regeneration, that forest departments everywhere – including those which formerly conducted the lion's share of silvicultural research – no longer care to look too closely at the consequences of their lax oversight. Where survey data are collected, they are often kept unavailable to independent researchers, to the serious detriment of a vigorous indigenous research community. Strategic environmental assessment, planning and monitoring require multidisciplinary approaches, and can only achieve sustainable management where research data, particularly those arising from periodic long-term assays, are freely accessible, and

the process is open to public debate.

Hope rests, nevertheless, in the existence of outstanding individual research initiatives directed at critical conservation problems. As one example among several, Lilian Chua and her colleague Saw Leng Guan are leading a project sponsored by the Forest Research Institute Malaysia (FRIM), using the distribution of threatened dipterocarp species – well documented on account of their economic importance – to identify sites of exceptional conservation value.[261] An IUCN Red Data List has been compiled for the Peninsular Dipterocarpaceae, in which estimates for the species in Malaysian Borneo are included.[262] A search for the eleven dipterocarp species considered endangered has so far rediscovered nine of them. Recently the only known populations of the point endemic *Hopea subalata* were examined and means explored for long-term monitoring. This site is on quartzite, is threatened by realignment of the north-south superhighway, and also harbours the only known population in the west of the peninsula of *Dryobalanops aromatica*, as well as a nearby site of the widespread but rare tree *Rhodoleia championii* (Hamamelidaceae). Negotiations eventually led to redirecting the highway!

Dipterocarpus semivestitus has only ever been recorded from two widely separated populations. One was in the lower Perak River freshwater swamp, the other, last seen early in the 20th Century, in a similar habitat in South Kalimantan. The species was recently rediscovered at the Perak site, along with two other rare dipterocarps, *Shorea macrantha* and *Shorea hemsleyana*, both species of Borneo but represented here by distinct subspecies. The habitat is one of the most threatened in the region, but may have been more widespread during Pleistocene low sea levels. The site belongs to the local university and was scheduled for buildings, but agreement was reached to build elsewhere. The same approach has led to discovery of a new dipterocarp species in the same Virgin Jungle Reserve as the first record of the Bornean *Dryobalanops beccarii* in Johor. A parallel programme has started in Sarawak under the impressive leadership of Julia Sang, and by Yayan Kusuma of the Botanic Gardens of Indonesia. Palms, Begoniaceae and well-known representatives of other life forms are now being investigated similarly in Peninsular Malaysia.

Research has unfortunately not always been helped by CITES and other national and international regulations. These are intended to prevent theft of national heritage by malefactors or foreign interests, but have also led to draconian restrictions on collection for *bona fide* research purposes. In the case of Sri Lanka, these have even prevented its own nationals from pursuing research which would yield results essential to the understanding and better management of its endemic species! We are witnessing the suppression of research into biodiversity dynamics at the very moment when extinction rates are accelerating. The issue admittedly is complex. Policy makers forget that crops, especially the long-lived woody crops that are the principal revenue earners of their countries, are and will always be exotic. This is no surprise to the ecologist, who is coming to realise that tree species diversity – greatest in the climates best-suited to the multiplication of pathogens – appears to be maintained by the minimum distance that must separate conspecifics if they are to avoid host-specific pathogen infection, thereby providing space for others (Chapters 5, 7). That realisation poses another policy challenge. Can political support be gained for investments in the conservation of resources that may be of commercial benefit only to others, in the trust that they will reciprocate? Is *Homo* truly *sapiens* (wise) enough to meet that challenge?

Meanwhile, the global partnership of the Center for Tropical Forest Science began in 1985 to establish a network of control plots – the most extensive in Asia and now globally – for research into the fundamental functional principles of biodiverse forest ecosystems,[263] and how their diversity is generated and maintained. CTFS is thus providing an indispensable tool for generating a more robust theoretical base for and a renaissance of applied research into the sustainable management of tropical forests, as well as documented controls by which to evaluate future attempts to resuscitate indigenous forests.

8.6. *Neglect of the gene pool*

Whereas farmers in ancient China and Japan, and later Europe, improved the quality of fruits, and other products of trees introduced from the forests, by generations of more or less methodical selection, such a tradition hardly developed in those places where the opportunity was and still remains greatest – the lands of the hyperdiverse MDF. Instead, that richest of genetic resources in Asia has all but completely been replaced by exotic oil palm monoculture! How can a rich and diverse rural economy be established for an advanced economy on such a basis?

Yet the opportunity for novel introductions, for selecting novel strains and for enhancing by hybridisation, still remains; and the demand will increase if development results in prosperous rural communities, cultivating diverse produce and exporting to urban markets. Neglect of opportunity results from failure to appreciate the importance of experimental projects whose end results are uncertain, but such experiment was the very foundation upon which western advances in science, and their subsequent application, have rested. An example of such a failure is the sad story of the 'Borneo Illipe nut' of international trade: Buah Tengkawang, Kawang, Engkbang. The nut is the fruit of certain Shorea species, mostly in sect. Pachycarpae. The large fleshy cotyledons yield a vegetable fat of higher melting point than cocoa butter, but similar chemistry albeit pale colour. Mixed with cocoa butter, a chocolate is produced that does not melt at tropical temperatures; there are other uses,

in polishes, paints and more. Illipe has the major commercial disadvantage of mast fruiting (Fig. 8.2), resulting in irregular crops of unpredictable size. Formerly collected from the forest when bonanzas were achieved, mast trees have been logged and the trade has ceased. Although plantation was attempted in Java,[264] no research into floral primordium initiation has been attempted. Even an obvious test, whether planting in seasonal regions of Indo-Burma or the Philippines would yield annual crops, has yet to be attempted.

9. Stirrings of a conservation movement?

9.1. *Halting progress toward sustainable forest management and conservation of forest biota*

The movement to legally conserve areas for the wildlife of tropical Asia originated among European colonial officials in the late 19th Century. It arose from concern over excessive trophy hunting and killing of indigenous birds for their feathers, then fashionable in hats; awareness of the consequences of the increasing pace of forest conversion was only a secondary motivation. Formal protection started in India with the first Wildlife Sanctuary in 1898.[265] Wildlife reserves began to be widely inaugurated in the interwar years, culminating in the first National Parks in then Dutch East Indies. After prolonged debate and dissent, Taman Negara ('National Park') was established in central Peninsular Malaysia.[266] This park includes the Peninsula's highest mountain, but relatively little of the astonishingly diverse zonal lowland MDF. Parks then began to be established in South Asia, often fashioned from ex-game reserves. Following the Second World War and cessation of local insurrections, the gazetting of National Parks accelerated in several countries, notably Thailand, Indonesia and the Philippines, and there were further increases in India and Sri Lanka. The objectives of these parks are, in most cases, to protect extensive examples of natural landscapes, together with their native vegetation and wildlife, uninhabited wherever possible although that was only the case in rainforest areas. Parks have nevertheless remained most abundant in deciduous forest regions, and are mostly confined to areas where at least one charismatic 'flagship' vertebrate has survived.

There is only a single park system in the Asian tropics that was actually planned and implemented with the aim of conserving examples of the main landforms and tree species associations, and which is thought to be big enough for this purpose. This is in Sarawak. It was initiated with the gazetting of Bako National Park in 1957. Running along the coast to the east of the Sarawak estuary and its associated mangrove, Bako National Park protects a landscape of unique forest associations; its coastline was an obvious tourist attraction. In succeeding years, a series of parks including the large Mulu

National Park were inaugurated. These are mostly situated in or near lowland areas now experiencing increasing accessibility, settlement and conversion of surrounding forest. Elizabeth Bennett of Wildlife Conservation International spent several years reviewing and advising on Sarawak's conservation system. She rightly recognized that most of these parks, by then often isolated and lacking corridors to larger logged forests, were too small individually to conserve populations of larger vertebrates. She therefore advocated the creation of a second tier of fewer but larger inland sanctuaries, partly or wholly logged, to be delineated *without* special consideration for vegetation criteria. Several of these still await legislation, but they include the major Lanjak-Entimau/Betung-Kerihun Transnational Park, the first in the region and at 8,010 km² the largest park in Borneo. Meanwhile, the smaller, older parks – so essential for the conservation of regional and local biodiversity (see Chapter 9) – are being intermittently degraded by illegal logging (Plate 8.10). Thailand's park system was not planned primarily with botanical goals in mind, but has approached botanical comprehensiveness

Fig. 8.3 Remaining indigenous forest in perhumid south-west Sri Lanka. All but small fragments in the lowlands have been selectively logged. Black areas are national parks and other reserves, nearly all of which are montane; but a moratorium on logging now holds on all wet zone forests.

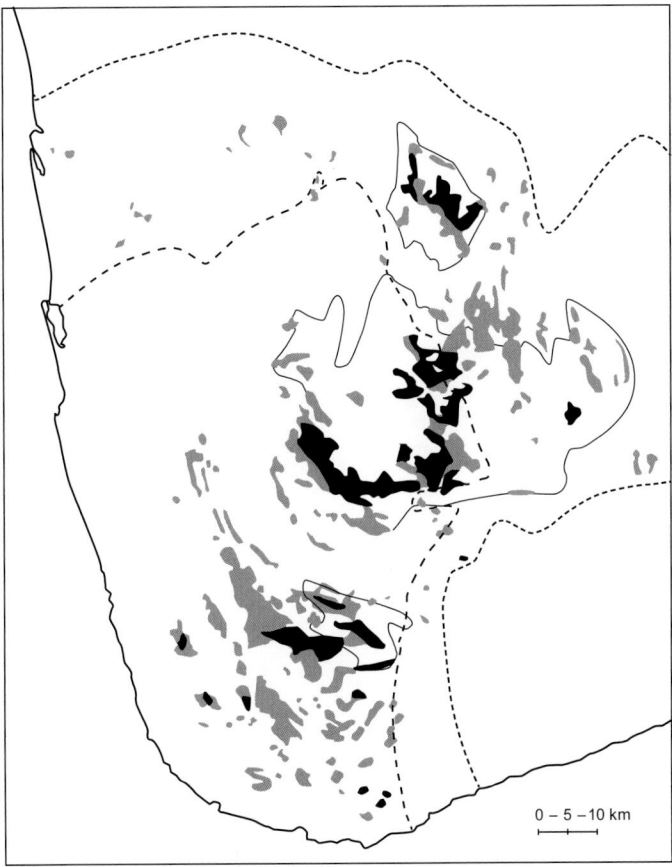

0 – 5 – 10 km

by incorporating the principal landscape features and geology of this climatically and geologically diverse country. The richest lowland rainforests, on undulating arable land are, as always, poorly represented.

Other countries, with Sri Lanka setting a particularly noteworthy example (Fig. 8.3), have recently been expanding their conservation systems to achieve more complete representation of overall species diversity. A foreign consultancy company formulated a draft national forest management plan. When it became clear that the plan envisaged continuing (though 'sustainable') logging in the remaining lowland rainforests, it raised a public outcry and was thrown out of parliament. It was instead decided that continued logging was contrary to the nation's social objectives. Only in Sri Lanka can it be said that there is truly 'grassroots' support for conservation: Sarath Kotagama's movement, the March for Conservation, has aroused true commitment, not only from the middle class but also from rural smallholders who are secure in the ownership of their land and rights.

In the Philippines, originally said to have been clothed entirely in rainforest, a National Parks system was legally established earlier than elsewhere in Asia, but where is it now? Little lowland MDF, the richest of the Philippine terrestrial ecosystems, remains in any legally conserved forest area.[267] Primary forest is reduced to less than 800,000 ha, and it has been estimated that between 24–56% of the flora has already been extirpated. These islands are rich in island-specific endemics, especially plants with shorter life cycles. The shrub genus *Psychotria* is comprised of 112 species there, all but three of them endemic to the archipelago.[268] Most *Psychotria* species are confined to undisturbed indigenous forest. Ten now appear extinct, 34 critically endangered, 19 endangered, 20 vulnerable, 21 threatened, and a mere eight 'of least concern'.[269]

What remains of primary MDF in Peninsular Malaysia is almost confined to the National Park system, part of which has been illegally logged,[270] and in those of Indonesia. In Indonesia, though, the great Kutei National Park has been largely destroyed by fire, illegal logging and settlement,[271] and the remaining forest is inaccessible to all but the visitor with time and assistance with transportation. Illegal logging is pervasive in Indonesia's national parks, Java excepted (Plate 8.10).

A particularly grievous result of the lack – indeed the discouragement – of public participation through discussion in policy making in Indo-Burma and Malesia is the virtual absence of management plans, or indeed any active management, of protected areas. Illegal hunting and the collection of NTFP of commercial value have become all but ubiquitous in parks. These activities originate with traditional rights, but as human populations increase they become impossible to manage sustainably without close collaboration between local communities and park managers.

a

b

ci

cii

d

PLIS TOK SAVE LONG tuPELA OL MAN NA MERI NA PIKNINI
TAMBU TRU LONG KULKIM GRAS LONG ROT. PLIS tu LUKIM
NA BIHANIM SAPOS tu iNo BIHANIM EM BAI tu BAIM
50k PAIN EM TASOL TOK SAVE BILONG MI

e

มูลนิธิศึกษาธรรมเพื่อสิ่งแวดล้อม

DHAMMA RESEARCH FOR ENVIRONMENT FOUND.

Plate 8.13 States are coming to foster conservation. **a,** A sign in Karnataka; **b,** The sign at the entrance to the Sinharaja World Heritage forest ensures that everybody knows about it. Nimal Gunatilleke , left, CTFS plot manager and long-time loyal aide to the Gunatillekes Gunadasa, right; **ci,** Thai villagers, supported by a local conservation movement, tie pleas for the retention of a majestic *Dipterocarpus turbinatus* grove threatened by road widening, Chiangdao, northern Thailand; **cii,** One of the *D. turbinatus* trees, bearing the scar of former resin tapping; **d,** A similar protest is proclaimed by Nuginian villagers; **e,** The Buddhist church is bringing traditional respect for nature into the modern world of ethical conviction and political action. Mae Sariang, north-west Thailand. (**a, b** P.A.; **ci, cii, e** H.H.; **d** S. Davies)

547

9.2. *The tide could yet turn*

It has taken decades for the citizenry to raise its voice against forest degradation, let alone conversion. Even the well-informed remain mute in the less democratic states. All will suffer when the resource is exhausted. Yet India and especially Sri Lanka are nations whose ancient respect for the natural universe remains pervasive in both countryside and in the corridors of power, and in both of which there are valiant voices speaking out. In these countries I sense the tide turning in favour of forest conservation. The professional middle classes are raising their voice, while the ancient respect for nature remains strong among the rural poor.[272] In Peninsular Malaysia and especially Thailand, the growing urban middle class and professionals are gaining in influence at least on central governments for sound resource management and conservation of forest resources (Plate 8.13).

The Malaysian Nature Society (MNS), for example, is now among the largest associations for amateur naturalists in the tropics. Backed by research at FRIM, the MNS has taken up the challenge of identifying and evaluating tree species under putative threat. Using IUCN criteria, recommendations arising from this research are being taken seriously, because certification of sustainably managed forests and timber exports from them depends on meeting these strictures. FRIM, the Botanic Gardens of Indonesia and Sarawak Forestry Corporation have each initiated a programme to relocate surviving populations of endangered tree species and negotiate their conservation (see section 8.5 above and Chapter 9). We may yet witness the start of a new era where government is sound and answerable to the people, and when the multiple values of indigenous forests can be sustained – if not by traditional perceptions and usages, then by the increasingly universal attitudes and values of urbanised and industrialised society.

Peninsular Malaysia is a middle-income nation in which the rural agrarian poor are being replaced by a new generation made prosperous by oil palm. Thailand, too, is on the way. But for India, Burma, and the Indo-Chinese nations, rural poverty remains chronic. Meanwhile the urban elite, despite some growing concerns for the state of their natural heritage, have yet to reconcile the concept of conservation with the aspirations of the rural poor, still dependent on the eroding forest resources. The greatest hope for the future of the forests lies in the growth of indigenous conservation movements in which rural poor and prosperous urban groups can unite. Strongest at present in India and Sri Lanka,[273] dynamic movements are gaining political clout also in Thailand, Malaysia, the Philippines, Indonesia and elsewhere. They have arisen within an urban middle class educated in traditions of sceptical inquiry, articulate and themselves only indirectly dependent on the land.[274] Through their influence, potentially outrageous examples of forest abuse

have at times been prevented. New conservation areas have been promulgated, old ones revived.

Unfortunately, it is also true that at present in India, for example, urban-based conservation spokespeople are looked on with suspicion by advocates of rural development and civil rights. In the short term, their interests often appear divergent; so the Forest Rights Act of 2006, for instance, has created fault-lines between those who see a chance for improved community involvement in forest protection and the benefits of secure tenure, and those who see only destruction of the remaining forest retreats of the tiger.[275] In the long term, the interests of conservation-minded urbanites and those who promote livelihoods of forest villagers converge on a vision of healthy sustainable forests. How to get there is the question.

In Thailand, Seub Nakhasathien became Director of Thung Yai Wildlife Sanctuary and adjacent Huai Kha Khaeng Wildlife Sanctuary – now UNESCO World Heritage Sites – in the 1980s, when tigers were being brazenly poached from these parks under the patronage of influential people, while eaglewood was being illegally and destructively exploited (Plate 8.14). Guards had been murdered in 1978 and 1979, yet Seub marshalled those remaining to courageously pursue malefactors. One of the latter was shot in the back, as a result of which Seub was temporarily imprisoned and deprived of his pension. Back in charge, but increasingly frustrated by lack of support either from his service or the politicians, Seub took his own life. Only then did the media awaken. Seub had truly given his life for wildlife, for his death provoked tens of thousands of young Thais to come onto the streets of Bangkok. The King intimated his support. Now, Thailand's parks have become increasingly popular with its public for recreation. Interest in natural history and nature conservation is increasing. Tiger and elephant populations are said to be growing again in Huai Kha Khaeng and Thung Yai. Even still, powerful interests threaten forests harbouring assets of value.[276]

Although the Philippines led the way in commercial export of light hardwood meranti (lauan) timber, and now has the lowest proportion of forest among Far Eastern countries, its several conservation NGOs are highly articulate and are meeting with increasing success in that turbulent democracy, while government policies, devolving power to the provinces, are promoting reforestation by private interests as well as on public land.

9.3. *The growth of ecotourism*

With the growth of parks systems and the decline of sporting hunting in the tropics has come the growth of 'ecotourism', aided by the popularity of wildlife programmes on television. Starting with a trickle of avid birdwatchers and other naturalists, often wealthy and preferring long stays in the tranquillity of remote forest bungalows (Plate 9.5b), it is vastly expanding as a mass market develops both for local and international middle-income tourists. These are primarily attracted to picturesque landscapes, exotic means of transport and accommodation which allows them to experience adventure without danger or discomfort, to hike and to swim, and to see for themselves

Plate 8.14 A few courageous individuals set the pace. (See also Prologue Plate 1d) **a,** Memorial statue of park director Seub Nakhasathien, who took his own life when politicians failed to act as poaching raged in his park, Huai Kha Khaeng Wildlife Sanctuary; he thereby aroused public protest, which succeeded; **b,** Gravestones of rangers murdered in the course of duty by tiger and ivory hunters, Huai Kha Khaeng, west Thailand. (H.H.)

what television has promised them: an abundant and easily accessible exotic flora and fauna. This latter perception is, of course, a myth, and has led to much disappointment. In reality, most rainforest sanctuaries only continue to attract the mass market if they include reliably observable vertebrates such as the orang utans at Sepilok (Sabah), where they have been re-introduced; or the astonishing, lasso-like nightly exodus of bats from Deer Cave at Mulu. Many reserves also fail because they are poorly marketed, or are provided with facilities unsatisfactory to an urban or foreign visitor. Heavy visitation may adversely affect the quality of the environment conserved, particularly in fragile ecosystems such as those of the Mahableshwar Plateau and Koyna watershed on the northern summits of the Western Ghats, beloved of visitors from nearby conurbations. Their montane forests are an important rainwater trap, yet are precariously adapted to a short but heavy wet monsoon and an extended dry season. Sitting on shallow lateritic soils, they are becoming severely degraded. In short, as is recognized in Thailand and in India, rarely do the objectives of biodiversity conservation and ecotourism closely coincide. In some countries, two categories of parks have been instituted in recognition of these facts. Can the biodiversity sanctuary be economically justified without the presence of ecotourism? We will address one possible solution in the our final chapter.

In some parts of India, community involvement in ecotourism is being advocated to financially underpin the conservation of nature reserves.[277] It has been initiated in Kerala, at the Periyar Tiger Reserve.[278] There, community groups manage the Anthirapilly and Chai waterfalls and the Varhachal picnic spot. Decision-making, though, remains firmly in the hands of the Forest Department, and the areas suffer from over-use. In many other parts of India, small-scale ecotourism programmes focused on village home-stays and guided treks are being promoted by social-environmental NGOs.[279] Success has been variable, pointing up among other things the unresolved vagueness of the distinction between 'eco'-tourism and 'regular' tourism.[280] This topic is also further discussed in Chapter 9.

9.4. Global-scale carbon commercialisation efforts

The world community is now finally talking about enacting measures to compensate nations that dedicate land to carbon sequestration. Discussions about means of addressing conservation of biodiversity at the international level are also in full swing. Who are the potential beneficiaries, and who should shoulder the economic burden? On the one hand, 'Reduced Emissions from Deforestation and Forest Degradation' (REDD) payments could provide a mechanism for the transfer of resources from the industrialised and polluting parts of the world to those with tropical forests left to protect. REDD+ programmes go farther, specifying additional targets for biodiversity conservation, community participation and benefits, sustainable management of native forests, and enhanced forest carbon stocks. Published estimates of the potential financial flows from economies of the 'North' to those of the 'South' under REDD+ reach as high as US$30 billion annually. In principle, that seems to represent an ethically defensible plan to sustain forests and sequester carbon simultaneously – a 'win-win solution'. Implementing it, though, is fraught with extraordinary complexity.

To begin with, tropical nations are far from unitary in their interests. Who would receive and distribute REDD funds? To whom would they be distributed – to the wealthy owner of a timber company, in exchange for an agreement not to harvest a forest tract? Or to small agriculturalists struggling to make a marginal living on the fringes of that forest? The former may have the power to make a perceptible difference on global scale, the latter perhaps not, and yet there is a natural reluctance to pay someone to give up the very activities from which he has been profiting at considerable social cost. Clearly, rich opportunities for social blackmail and elite corruption lie in wait. Yet even within the legal framework posited as a precondition for REDD programmes, forest-dependent and indigenous communities seem unlikely to benefit. Many of the forest resources they need (fruits, leaves, poles, roots, and the particular suites of species that grow them) may not be compatible with optimal forest-wide carbon sequestration capabilities. If strong incentives are put in place to maximise carbon storage function, with international investment backing them up, it seems unlikely, to say the least, that use and control of the forest's bounty could long remain in the hands of communities.

Secondly, there are staggering difficulties in the more technical aspects of any REDD programme. There is the problem of ensuring 'additionality': not all forests are equally likely to be deforested in the near future, and it would make little sense to award funding for the protection of forests that are not in any case endangered. Yet how do authorities establish reliably which areas are most in danger? Then there is the need to prevent 'leakage': if a particular tract were, in fact, somehow protected in perpetuity, who would guarantee that timber companies, or displaced swiddeners, did not simply move to a neighbouring tract and begin again?

And even before these considerations, lie the challenges inherent in estimating, with any degree of accuracy, first how much carbon a forest might be likely to store over any time period, and secondly whether it is in fact doing so. As these essays have demonstrated, forests are each unique in various ways, and they cannot be straightforwardly categorised according to their carbon-storage capacities. Assessing capacity and monitoring compliance are huge challenges, the solutions to which must be partly technical and partly social and political.

References

[1]Higham 1984. [2]Ohsawa 2006b. [3]Chandrakanth et al. 2004. [4]Dove et al. 2011. [5]Greenough & Tsing 2003; Dove et al. 2011. [6]Stebbing 1922–1923. [7]Pattanayak et al. 2010; Vincent 2010. [8]Anon. 2010a. [9]Stebbing 1922–1923. [10]Pandey 2001. [11]Haberle et al. 1991. [12]Maloney 1985; Flenley 1988. [13]Mooney et al. 2011. [14]Michon et al. 1986; Jacob & Alles 1987. [15]Michon et al. 1983, 1986. [16]Iwatsuki 2008. [17]Nye & Greenland 1960. [18]Ziegler et al. 2009. [19]Bruun et al. 2009. [20]Pélissier 1997; Bhagwat et al. 2005; P.S. Ashton 2007. [21]P.S. Ashton 2007. [22]Denham et al. 2003. [23]Conklin 1957; Kleinman et al. 1996. [24]Collis 1936; Church 2006. [25]Pethiyagoda 2007. [26]Caldecott 2006. [27]Stebbing 1922–1923. [28]Nouhare 2001. [29]Stebbing 1922–1923; Liese 1986; Church 2006. [30]Boonkird et al. 1984; Raintree & Warner 1986; Bryant 1994, 1997. [31]Bryant 1994, 1997, 2007; Dove et al. 2011. [32]Brünig 1996. [33]e.g. Bradshaw et al. 2007; Ziegler et al. 2009. [34]Stebbing 1922–1923. [35]Stebbing 1922–1923; Champion 1936, 1975; Wyatt-Smith 1963; Burkill 1966. [36]Slade 1896. [37]Troup 1921. [38]Beeson 1941. [39]Chengappa 1934, 1937, 1944. [40]Kadambi 1941, 1942. [41]Oliver & Larson 1990; P.M.S. Ashton et al. 1993a; Smith et al. 1996. [42]Gupta 1939; Thangam 1982; Rai 1989, 1990; Chana 1991; Kumara et al. 2003. [43]Wyatt-Smith 1963. [44]Wyatt-Smith 1963. [45]Wyatt-Smith 1963; Dawkins & Philip 1998. [46]Manokaran 1998; Okuda et al. 2003b. [47]Whitmore 1984; Anon. 1991. [48]Oliver & Larson 1990; Smith et al. 1996. [49]P.M.S. Ashton 2003; P.M.S. Ashton & Peters 1999. [50]Muller 1981. [51]Primack et al. 1987b. [52]Dawkins 1963; Dawkins & Philip 1996; Vincent et al. 2002. [53]Primack et al. 1987a, b. [54]Thang 1985. [55]Appanah 1985; Appanah & Weinland 1990; Thang 1987. [56]Sist et al. 1998a, b. [57]Rodgers 1991; Gupta & Totey 1994. [58]Cannon et al. 1994. [59]Abdulhadi et al. 1981; Malmer & Grip 1990; Dykstra 1995; Elias 1996; Ahmad & Rahim 1998; Manokaran 1998; Pinard et al. 2000. [60]Mohamad et al. 1987. [61]Kuusipalo et al. 1996; Sist & Nguyen-Thé 2002; Bischoff et al. 2005. [62]Poore et al. 1988; da Silva et al. 1996. [63]Primack et al. 1987a, b. [64]Kuusipalo et al. 1997. [65]Forshed et al. 2005. [66]Pinard et al. 2000. [67]Pinard 1996. [68]Sist et al. 1998a, b. [69]Bertault & Sist 1997. [70]Frederickson & Putz 2003. [71]Sist & Brown 2004. [72]Jennings et al. 2001. [73]Abhilash et al. 2005. [74]P.M.S. Ashton et al. 2001a; Giriraj et al. 2008. [75]Hartley 1967; Corley & Tinker 2008. [76]Liew 1974. [77]Mittermeier et al. 2004. [78]Tang & Chong 1979. [79]Burgess 1969–1975. [80]Poore et al. 1989. [81]Spilsbury 2007. [82]FRIM staff, pers. comm. 2008. [83]Marsh & Greer 1992. [84]Boonkird et al. 1984; Hafner 1989. [85]P. Wangda, pers. comm. 2008. [86]Stebbing 1922–1923; Rawat 1998. [87]Srivastava & Singh 2002. [88]Hartley 1967; Corley & Tinker 2008. [89]Callister 1992; Yu 1994. [90]Brünig 1996. [91]Brünig 1996. [92]Primack & Tieh 1994; Borhan Mohamad 1993. [93]Dennis et al. 2005. [94]Suzuki et al. 1997a. [95]Yusop et al. 1987. [96]Gorrie 1947. [97]Bon 2002. [98]Dixit et al. 2001. [99]Rajkhowa 1986; J.N. Singh 2004. [100]Bowonder 1982. [101]Rawat 1998. [102]Myers 1986. [103]Lélé 2000, 2003. [104]Guritno & Murao 2001. [105]de Ath 1992. [106]Brünig 1996. [107]Lélé 1998; R.R. Pandey 1998; Larsen & Ribot 2007; Gill et al. 2009. [108]P.S. Ashton 1986; Peluso 1989. [109]Laurance & Peres 2006. [110]Jusoff & Majid 1995. [111]Bawa et al. 2010. [112]Kumar et al. 1994. [113]Shah 2007. [114]Gadgil 1989; Chandrakanth et al. 1990; Chandrashekara & Sankar 1998; Khumbongmayum et al. 2005; Malhotra et al. 2007. [115]Bhagwat et al. 2005; Bhagwat & Rutte 2006; Ormsby & Bhagwat 2010. [116]Sukumaran et al. 2005. [117]Dhaila-Adhikari & Adhikari 2009. [118]Covarrubias 1937. [119]Freeman 1960. [120]Kessinger 1974. [121]Marjokorpi & Ruokalainan 2003. [122]Hafner 1989; Tripathi 2005. [123]Payne 1985; Gill & Roy 2001; Kumara et al. 2003; Bala et al. 2007. [124]Saxena 1997; Ravindranath & Sudha 2004. [125]Wood 2001; Bon 2002. [126]Saxena 1997. [127]Saxena 1997. [128]Ramakrishnan et al. 1998; Nayak et al. 2000; Nagendra 2007. [129]Dasgupta 2001; Gill & Roy 2001. [130]Verma 1989; Batliwala & Dhanraj 2009. [131]Khare et al. 2000; Lélé 2000, 2003; Ghate 2003; Sarin et al. 2003; Caruso & Modi 2004; Nayak & Berkes 2008. [132]Sarkar 1994; Dasgupta 2004. [133]Pandey 1993;

Sarkar & Chattopadhyay 2002; Mukerji 2006; Sarkar 2006. [134]Sarkar 2006. [135]Matta *et al.* 2005. [136]Sarkar & Chattopadhyay 2002. [137]Pande 2005; Sunderlin *et al.* 2005; Saravanan 2009; Sarker 2011. [138]Sarkar 1994. [139]Brown 1994. [140]Matta *et al.* 2005. [141]Menon *et al.* 2007. [142]Fujita *et al.* 2002. [143]Sajise & Omegan 1989. [144]Guiang *et al.* 2008. [145]Kumara *et al.* 2003. [146]Anon. 2003. [147]Tani 2000. [148]Wang 1958. [149]Katsigris *et al.* 2004. [150]Katsigris *et al.* 2004. [151]Koh 2007. [152]Elvin 2004. [153]Hemley 1994. [154]Liu *et al.* 2008; Qiu 2009. [155]Chopa & Gulati 2001. [156]Myers 1986. [157]Vincent 1992. [158]Anon. 1991; Rice *et al.* 1997. [159]Pierce *et al.* 2003. [160]Geist & Lambin 2002. [161]http://www.grossnationalhappiness. com [accessed 2013]. [162]Agrawal 2001; Ostrom *et al.* 2002. [163]Narendran 2001; Murthy *et al.* 2005. [164]Chitwadgi 2001; Larsen & Olsen 2007. [165]Hegde *et al.* 1996; Murali *et al.* 1996; Shankar *et al.* 1996; Rist *et al.* 2008. [166]Rist *et al.* 2008, 2010. [167]Dransfield *et al.* 2008. [168]de Padua *et al.* 1999–2003. [169]Fox 1977. [170]Mailly *et al.* 1997. [171]Mahapatra & Mitchell 1997. [172]Dhar *et al.* 1989. [173]Lal & Daka 1991. [174]Basu 2000; Ramakrishnan *et al.* 1998. [175]Nayak *et al.* 2000. [176]Jamir & Pandey 2002. [177]Sinha 1979. [178]Kingdon-Ward 1937. [179]Seibert 1996. [180]H.S. Barlow, pers. comm. 26 Feb. 2007. [181]Chakrabarty *et al.* 1994; Soehartono & Newton 2000, 2001. [182]Chakrabarty *et al.* 1994; Soehartono & Newton 2001, 2001. [183]Pethiyagoda 2007; P.M.S. Ashton *et al.* 1993b. [184]Kathriarachchi 2002. [185]Suzuki *et al.* 1997a. [186]T. Okuda, pers. comm. 2006. [187]T. Okuda, pers. comm. 2006. [188]Private discussion in market, 2008. [189]Harrison 2011; Harrison *et al.* 2013. [190]Lubowsky 1998; Kathriarachchi 2002. [191]Watt, 1889–1893, rev. as Anon 2001, 2013–; Ambasta 1986. [192]Burkill 1966. [193]Heyne 1950. [194]de Padua *et al.* 1999–2003. [195]Needham & Lu 1986. [196]Heyne 1950. [197]Jha *et al.* 2003. [198]Ravindranath *et al.* 1997. [199]Lal 2004a, b. [200]Palm *et al.* 1986. [201]Neutzil 1997; Hirano *et al.* 2009. [202]Hao *et al.* 1990; Karl & Trenbarth 2003; Humphreys 2007. [203]Johnsingh 1986. [204]Stolle *et al.* 2003; Brühl *et al.* 2003. [205]Ketterings *et al.* 1999. [206]Tomich *et al.* 1998, 2001. [207]Kleinman *et al.* 1996. [208]Kodandapani *et al.* 2009. [209]Narendran 2001, 2004. [210]Johnsingh 1986. [211]Prance 1986; Goldammer 1990; Goldammer & Seibert 1990; Dennis & Colfer 2006; Dennis *et al.* 2005. [212]Leighton & Wirawan 1986. [213]Curran *et al.* 1999. [214]Woods 1989. [215]Nykvist 1996; Slik *et al.* 2002; Cleary & Piadjati 2005. [216]Rieley & Page 1997; Yeager *et al.* 2004; Saharjo *et al.* 2006. [217]Wosten *et al.* 2008. [218]Hooijer *et al.* 2006. [219]P.S. Ashton 1991, unpubl. calculation. [220]Takahashi 1999; Davies & Unam 1999. [221]Sorensen 1993. [222]Buckley *et al.* 2010. [223]Goldammer 1990. [224]Leighton & Wirawan 1986; Taylor *et al.* 1999. [225]Slik *et al.* 2002. [226]Srivastava *et al.* 1999; Srivastava & Singh 2002. [227]Srivastava *et al.* 1999; Srivastava & Singh 2002. [228]Lehmkuhl *et al.* 1988. [229]Wharton 1968; Davies *et al.* 2001. [230]Ickes *et al.* 2001. [231]Curran *et al.* 2004; Wong *et al.* 2005; Harrison *et al.* 2013. [232]Shukri *et al.*, undated ms. [233]van Schaik & Rao 1997; Bawa & Seidler 1998; Wong *et al.* 2005; Marshall & Leighton 2006. [234]Bawa & Seidler 1998; Johns 1992; Wijesinghe & Brooke 2005; Shahabuddin & Kumar 2007. [235]Styring & Ickes 2001. [236]Hemley 1994. [237]Liow *et al.* 2001. [238]Ghazoul 2002. [239]Hamer *et al.* 2005. [240]Gathorne-Hardy *et al.* 2002a. [241]Samananthan & Borges 2000. [242]Johns 1996. [243]Brühl *et al.* 2003. [244]Benedick *et al.* 2006. [245]Bierregaard *et al.* 1992; Laurance *et al.* 1997, 2012. [246]Zhu *et al.* 2004. [247]Turner *et al.* 1993. [248]LaFrankie *et al.* 2005. [249]Ickes *et al.* 2001; Blundell & Peart 2004. [250]Lehmkuhl *et al.* 1988; Veeramani *et al.* 2004. [251]Ghosh & Kaul 1977; Khullar 1992. [252]MacMillan *et al.* 1991. [253]Whitmore 1991; Whitmore & Sayer 1994. [254]Prasad *et al.* 1996. [255]Naithani 1987. [256]Thakur *et al.* 1992; Carrion-Tacuri *et al.* 2011. [257]Swamy & Ramakrishnan 1987a, b. [258]Whitmore 1991; Singhakumara *et al.* 2000. [259]Peters 2001. [260]Seth 1951; Kaul *et al.* 1975; Rodgers 1991. [261]Seth 1951; Kaul *et al.* 1975; Rodgers 1991. [262]Anon. 2009; Saw & Sam 2000a, b; Saw *et al.* 2009. [263]Anon. 2009. [264]Tantra 1977. [265]Khullar 1992. [266]Kathirithamby-Wells 2005. [267]Roth 1983. [268]Sohmer 2001. [269]Davis & Sohmer 2004; Sohmer & Davis 2004. [270]Kathirithamby-Wells 2005. [271]W.T.M. Smits, pers. comm. 1988. [272]Bahuguna 1986. [273]Wickramaratne *et al.* 1997; Nayak *et al.* 2000. [274]Bahuguna 1986. [275]Sekhsaria 2007. [276]Maxwell 1999. [277]Pandey 1993; Mukerji 2006. [278]Thampi 2005. [279]e.g. http://www.aboriginal-ecotourism.org/spip.php?article161; http://www.sikkimhomestay.com/ [280]Butcher 2005.

Photo M. Ashton

Chapter 9

THE FUTURE: CAN WE RETAIN FOREST OPTIONS PROFITABLY?

With Reinmar Seidler

In the aftermath of the invasions of Goths and Slavs, the single event that most emphatically set back opportunity for an early return of learning to Europe was the burning of the great library at Alexandria in the 7th Century. Repository of the collective memory of ancient Greece and Rome, each book was unique and irreplaceable at that time, six centuries before the printing press. Although tropical rainforests obey the same dynamic rules as temperate forests, the genetic information enshrined in their vastly greater array of species is, like those books, unique and irreplaceable. Like the books, each species may yield important information or insight no more than once a century, yet future generations will not forgive their loss.

It has fallen to the generations straddling the turn of the 21st Century to decide the fate of much of the global biodiversity library. Each of us, wherever we live, bears some responsibility for the outcome. But what is the nature of that responsibility? Does it derive from history? Or does it derive from a globalised economy? Does it vary with geographical proximity, or with our personal or professional position in society? How can our collective and individual responsibilities toward tropical biodiversity be first defined, and then met? These are some of the thorniest questions presently facing humanity. This chapter cannot hope to fully answer them, but it will try to highlight some of the central issues, from the viewpoint of one who has witnessed many of the vast and often astounding changes that have posed these questions in such powerful and increasingly urgent ways.

1. Social and economic challenges for sustainable use of natural resources in tropical Asia

We humans continue trying to extract more from nature than she is capable of yielding. We add drainage, nutrients, water, pesticides, but they all cost and, in inherently unproductive land, their costs may not eventually justify the expense. The tendency then is to take all, and leave others to clear up the mess. Those others here are the forest communities who, although in poverty, knew the way to eke out a continuing living; and it is our children. Rural communities depended on the forest. And even the forest on the arable land, so little of which now remains in Asia, harbours unique resources that,

in the longer term, we cannot lose without consequence.

I, a Briton, can hardly reproach the feckless inheritors of Asia's tropical forest heritage, when my own ancestors reduced their own primeval woods to a few scrawny patches in order to build warships to fight our neighbours! Forest management brings out all the familiar and eternal problems of immediate satisfaction versus long-term investment, private interest versus collective security and the temptations of the rich versus the needs of the poor. But my own British forest endowment is depauperate in biodiversity compared with that of tropical Asia, which even now remains on a par with that of the American tropics. In many parts of the industrialised world, reason in forestry appears to have prevailed for the moment, and remaining forests are becoming increasingly well managed. In China, too, and in the other metropolitan nations of the region, remarkable progress is now being made in spite of set-backs. But 'leakage' from these powerful Asian economies still drives forward the depredation of forests in the less-developed nations now harbouring the lion's share of regional biodiversity. In tiny Sri Lanka, only 5% of indigenous rainforests remain but ancient respect for nature persists, and the logging moratorium is holding. In vast India, the same is true in many areas, but most forests continue to be nibbled away by economic appetites large and small. Bhutan and Brunei (the latter with the richest and most diverse tropical lowland rainforests remaining in Asia) are now the standard bearers. They alone are in a position to succour the best of tropical Asia's remaining biodiversity for future generations. But these nations will not succeed in doing so unless the wider world acknowledges the value of this biodiversity by finding ways to contribute to its support.

1.1. *Our present predicament*

Forests in the tropics are in crisis. In Asia, ancient traditions of governance that once organised forest use have themselves been degraded, along with profound sensitivities to the dependence of mankind upon nature. Social systems that appear to have guarded against resource depredation have broken down nearly everywhere; and this has happened over less than the lifetime of an old-growth tree. Nevertheless, we must also guard against romanticising the past. Asian governance systems were location-specific and varied so widely as to make any generalisation suspect. Many enduring traditional systems of resource use may have been outcomes less of intuitively or spiritually 'conservationist' impulses than of the simple inability of pre-industrial civilisations to thoroughly dominate nature. This has changed principally through the 'magic' of fossil fuels, permitting a single man to do the work of hundreds. As long as our economies remain founded on cheap energy from fossil fuels, we will need to devise original solutions to over-use dilemmas; traditional templates may provide pointers, road signs and clues, but will not suffice wholesale.

Economic globalisation, of course, began well before the reign of cheap oil. Eager 18th Century Western trading companies, and then their governments, saw tropical timber as a source of wealth first to themselves and then to their subjects. Within two centuries, the primary forests of tropical Asia had been thoroughly logged over. At the same time, commercial interests converted forests to commodity farms, first coffee then tea, on unheard-of spatial and temporal scales.[1] They changed the rules of forest use to favour production of a single commodity, altering at a stroke the fundamental relationship of forest-dependent societies with the forests they inhabited. This was true even where the colonial powers retained those limited traditional systems of administration and taxation which appeared to meet their needs.[2] The colonial dynamic thus drained resource wealth from tropical nations at inequitable rates of exchange, while accelerating the concentration of wealth among elite groups within the client states. Rural populations burgeoned with the benefit of reduced infant mortality rates, but without commensurate escape from poverty. As colonial administrations subsequently departed, the new leaders who assumed the reins of power had typically come of age in a polity lacking in open debate and deprived of an articulate well-informed public franchise.

The impacts of European philosophies on forests were not unremittingly negative. The spirit of sceptical enquiry was reawakened in India, where it had always been strong,[3] and was applied by Western researchers throughout tropical Asia in a search for the means to produce timber sustainably from indigenous forests (Chapter 8). The introduction of Western enlightenment concepts of impartial justice, and the education of new elites including forest scientists trained in the spirit of open enquiry, appeared to set the stage for a promising future.

At the same time, colonial administrations applied these concepts and opportunities unevenly and arbitrarily. Principle was often contradicted in practice, presenting conflicting models. The introduction, or reinforcement, of rationalist and utilitarian ways of thinking about forest use was also a two-edged sword. Smallholders today are very hard-headed about their margins of profit, having abandoned old strictures of tradition, community or religious scruple. Has contact with Western values, and with Western science, influenced people to behave more like *Homo economicus*? Scholarship remains divided about the extent of conservationist or sustainable habits among pre-colonial societies; much variation probably always prevailed. Colonial governance certainly provided crass examples of mercantile treatment of natural resources.

One great disservice of colonial administrations was their constraint of free interplay between the emerging educated elites and the people they were to serve. Unfettered debate of policy and political process is essential to the evolution of institutions in which citizens can have confidence and a sense of ownership. It provides the least risky foundation for the long-term vision upon which successful environmental policy must rest. This social and political evolution often (though not always) runs parallel to the curve of successful economic development. Authoritarian, 'top-down' political philosophies may appear more successful or less risky, under certain conditions and in the short term, as seen in Singapore and in the impressive wave of current environmental reforms in China; but they are inherently riskier in the long term because they are less profoundly rooted in public understanding, participation, and therefore conviction. The foundations of post-colonial polity have been disastrously undermined by persistent lack of open dialogue with rural populations in policy reform, and by reliance — perhaps unavoidable under the circumstances — on indigenous mercantile elites to implement forest policy. Consequently the forest became a 'cash cow'. Vital investments in forest management and regeneration, even of timber species, were not made. Exploitation of every product profitable on national or international markets (including medicinal plants and wildlife) or needed by rural communities (such as meat) proceeded without control and increasingly to exhaustion. When you next visit some national parks in Thailand, for instance, ask yourself why there are no squirrels!

We recognize that there are often vast differences in the understandings of forest function and dynamics between an academic ecologist and a subsistence farmer at the forest edge. Meanwhile, the logger and the landless immigrant farmer heeds neither. The citizen of a rich country is mostly concerned that imported goods be affordable, whether or not their low price is at the cost of local well-being in the country of origin, or results in the degradation of resources and opportunities. The list of ingredients of many processed foods includes palm oil among the fats that are currently consumed far in excess of nutritional requirements; the demand for these sustains a price which guarantees pressure for further rainforest conversion. From one perspective, sustainability is the exercise of mediating among these competing interests.

1.2. *Traditional high value of forests*

And yet, to the rural communities directly dependent on them, forests have historically held the highest of values. In Asia, farmers long ago learned through empirical experience that mountain forests ensure a reliable source of dry-season water. This truth became embedded in myth and in spirituality, and the strength of myth can outbalance the weight of secular law. Nevertheless, forests by themselves rarely yield enough of the carbohydrates on which humanity depends. The presence of nomadic forest cultures reflects the presence of carbohydrate-yielding forest plants, mostly yams in the seasonal tropics and sago palms (*Eugeissona utilis* and *Metroxylon sagu*) in Borneo and New Guinea respectively. To subsist on these natural carbohydrate resources, each family group needs access to a forest area exceeding 50 km² (Plate 9.1). If there no longer is extensive forest in which to forage, people must cut forest to establish the conditions for growing essential cereals. Subsequently, irrigation technology has allowed for agricultural intensification and the easier communication necessary for urbanisation and the specialisation of skills, the foundation of complex civilisations. But urbanisation produces a growing and powerful minority that loses touch with the old interdependence between people and forest. Townspeople need timber and other forest products for building and fuel, but may no longer understand the forest and the value of its services. Some such sequence of social changes has been repeated many times since civilisations first arose, and has left its mark on ancient landscapes such as those of the Mediterranean

and England, where no primeval forest remains. Temperate eastern North America and much of forested Australia and New Zealand were almost entirely denuded of forest cover during the 18th and 19th Centuries (though many areas have since reforested with secondary growth – itself now experiencing a new wave of deforestation and fragmentation under novel pressures of sub- and exurbanisation, or 'sprawl'.[4]

1.3. *What case for retaining land under rainforest for timber, let alone biodiversity, when commodity crops beckon?*

It is the deciduous forests and woodlands of the dry tropics that continue to sustain the densest — and poorest — rural populations; these forests are the most resilient, but the least rich in overall biodiversity. Here, the rocky hills and other marginal lands, former hunting preserves of the rulers, largely remain as sanctuaries for the larger vertebrates beloved of tourists. On the other hand, the evergreen rainforests, whose wildlife is so difficult for the casual visitor to find, are the threatened home of so much of the world's biodiversity.

The demise of tropical forests at the hands of loggers has often been blamed on the liberalisation of global trade. This has created an apparently insatiable demand which concessionaires have found profit in satisfying.[5] Some have blamed rapid increases in the demand for raw materials within forest-holding developing economies.[6] Repetto[7] and others instead stress that governments of rainforest-holding nations are to blame, failing the public that they are meant to represent. Taxes have not been increased to keep up with the growing value of the commodity, nor with the exploding profits going to concessionaires and their political patrons operating in rentier capacity.

1.3.1. *Commodity export crops*

The present dilemma has been most drastically exposed first by rubber and more recently by oil palm. The most biodiverse ecosystems in the region (and among the most speciose in the world), namely the mixed dipterocarp forest (MDF), coincide with the low hills most favourable for the cultivation of these exotic commodity tree species. Without much sharper investment in wastage reduction, it is unlikely that customary selective logging methods, or even reduced impact logging (RIL), can achieve their sustained rate of return as high as 16%. Indeed, it is unclear whether

Plate 9.1 Camp of nomadic Punan, in Mulu National Park, Sarawak in 1958, now excluded from the forest by logging interests. (P.A.)

ecologically sustainable management of rainforest for timber alone can ever be profitable as long as discount rates exceed 5%, as they consistently have in developing countries.

But, when the red meranti-keruing forest of the lowlands is converted to smallholdings of commodity crops in Peninsular Malaysia, a nation providing strong infrastructural support to markets, the small farmer can expect a net pre-tax profit of c.$500–800 ha^{-1} y^{-1} (and these estimates continue to increase with demand for the oil).[8] An individual smallholder can reap an acceptable family income from three hectares, the land thus supporting up to 20 times as many farmers and their families per unit area as in the temperate West.

It is therefore hardly surprising that almost all of this lowland MDF, the most biodiverse terrestrial ecosystem in the Old World, has gone. In the current economic environment, rainforests only present much of a case for retention where they are on land difficult or dangerous to cultivate, or where they yield immediate essential services such as reliable and clean water supply or protection from soil siltation and erosion. Would that suffice to conserve their several undervalued services, such as their biodiversity?

The cost *to the owner* of strictly conserving a forest — the 'opportunity cost' — is generally the difference between its value for sustained timber production (plus the associated forest services, or at least those services accruing directly to the owner) and the value of the land under commodity plantations. The opportunity cost for Asia's most biodiverse forests stands at more than $300 ha^{-1} y^{-1}. The last major conserved area of hyper-diverse lowland MDF, at Danum (Sabah, 300 km^2), therefore foregoes roughly $9 million y^{-1}, a not inconsiderable sum in any developing economy[9].

1.3.2. Ecological services

Strict conservation of some amount of area is clearly required if the biodiversity of a region is to be retained. Total biodiversity, though, is a complex abstraction. Its value (relative or absolute) is extraordinarily difficult to estimate objectively, though particular components of it may have demonstrable utility to particular groups. In the final analysis, biodiversity itself may be of greater benefit to the global community, mainly to the rich countries, than to rural parts of a nation in the earlier stages of development (Chapter 8). Agricultural economies tend to be preoccupied with fertility of the soil, availability of water and defending growing crops against disease, bad weather and hungry wildlife. These are some of the most visible interfaces between natural forces and human ingenuity; they have only recently, and mostly in the scientific community, come to be considered 'ecosystem services' (and 'disservices').[10] Along with other less easily observed functions such as crop pollination and natural pest control, these service values have always been taken more or less for granted, and are at present conveniently

accepted *gratis* by their beneficiaries. Forest service values may accrue to non-local beneficiaries, perhaps humanity at large, who have benefited from the stabilisation of climate through carbon sequestration, and from access to biodiversity conserved in distant forests; or national governments, which gain from forest hydrological function and weather amelioration. In an emphatically mercantile environment favouring conversion to palm oil plantations, indigenous forest can only survive if the difference in realised values can be substantially made up by those forest goods and services which plantations are unable to yield. Most such benefits are gained in the long term as much by abutting populations as by the state. They must come to be seen as politically as well as economically advantageous.

1.4. Changing values

1.4.1. The Kuznets Curve

The economic value of forests therefore changes during the course of national development, following what economists call a *Kuznets Curve*[11] (Fig. 9.1). The high value assigned to forest by subsistence societies plummets in comparison with the access to cash-based markets and the cash employment offered by timber concessionaires. These emoluments prove irresistible. Technologies, including hunting guns and chainsaws, make life so much easier that future costs due to the draw-down of forest capital are ignored. However, if the capital thereby released as profits is attracted to the same national and local economy, rather than invested worldwide wherever the returns promise to be greater, more and better employment options will result. Such reinvestment depends on political stability, land security, and some means by which the workforce can acquire new skills. In some countries in

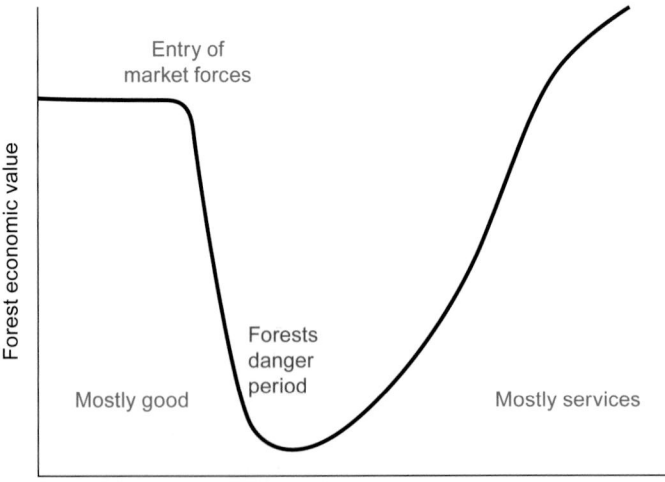

Fig. 9.1 The changing value of forests during economic development follows a Kuznets Curve.

Traditional agrarian economies	Early commercial access	With successful development

Benefits

To forest dwellers, hill farmers *Goods* • Game, fruit vegetables, medicinals, fibre; construction, roofing, matting, basket-making materials *Services* • 'Home ground', heritage • Water • Spiritual values **To lowland farmers** *Goods* • Fish and game • Goods acquired by trade *Services* • Water	**To commercial interests** *Goods* • Timber, minerals, other marketable products *Services* • Recreational hunting **To local communities (upland and lowland)** *Services* • Direct access to markets • Increased income through wage labour • Improved access to schools, medical facilities **To government** • Increased tax revenue **To the nation** • Increased wealth from profit distribution and improved government services	**To forest abutters** *Goods* • As in traditional economy but with formal regulation *Services* • Return and maintenance of traditional services • Secure employment in the service sector • Continued access to markets and government services **To the nation** *Services* • Water, weather amelioration, heritage • Recreation: hiking, nature, hunting • International ecotourism *Goods* • Continued access to timber, and other marketable forest products **To nation and government** • Increased wealth and revenue **To the world** *Services* • Carbon sequestration and potential assimilation • Climate amelioration (continental_ forests) • Biodiversity conservation • Ecotourism *Goods* • As the nation

Costs

(Insignificant)	**To commercial interests** • Cost of access and exploitation **To local communities** • Local communities gain only while exploitation continues, thereafter -short to long term loss of overexploited forest products • Increased access to forest resources by outsiders • Loss of security, difficulty of retaining traditional property rights • Water pollution	**To everyone** • Active management, research, policy reform • Facilities development • Education and marketing

Summary

Sustainable economy focused on goods, but beneficiaries local	*Almost invariably unsustainable owing to management breakdown; can lead to loss of resources that cannot be recovered. Downstream beneficiaries gain at expense of forest communities*	*Successful development is accompanied by restoration of values to forests, but now service values increasingly outweigh those of goods*

Table 9.1 Benefits and costs of forests during the stages of economic development.

the Far Eastern tropics, reinvestment has initially been in commodity plantation agriculture. The profits from this, when retained in-country, have in turn been invested in industry, which is providing better-paid employment and demanding a skilled workforce. But it is also drawing people off the land into the cities, and degrading traditional experience-based forest knowledge. It is impoverishing what were mixed rural economies, which themselves had promise of further diversification and prosperity, thereby turning proud freeholders into estate labourers.

Further incentives for sustainable management can be secured if the values other than timber are accounted for. Ecosystem benefits accrue to a wide range of stakeholders, from forest-fringe villages to the global community.[12] Vital goods and services also derive directly from forests, from medicinal plants and fuelwood to reliable clean water, crop pollination, pest control and weather amelioration. Unfortunately, these services are difficult to value;[13] and compensatory payments, because they must be directed to those individuals who actually have the power to make a difference in forest fates, could end up subsidising loggers or speculative elites rather than the villagers living sustainably with forest.[14]

A growing, financially secure and educated citizenry comes to appreciate that forests yield services essential to a stable prosperous economy, including dependable high quality water, equable weather and amenity.[15] Through their influence, policy can be reformed to secure a sustainable forest estate. Meanwhile, if rural communities come to share in a successfully diversifying economy, excessive demand for traditional uses by an increasing rural population will ameliorate. Thus, the remaining forests may become revalued even higher than they originally were.[16] Their benefits are then recognized more for the services than the material products that they can yield (Fig. 9.1; Table 9.1); and many of these accrue services to all society.[17] Thus, security with prosperity – the final achievement of a stable 'developed' society, in the modern sense – depends on sustaining forest services, including biodiversity (see Fig. 9.1) and water supplies, through the earlier, exploitative economic stage.

The early industrialising economies of Western Europe had more than a century to adjust to the changes wrought by development. In the current modernising economies of the tropics, similar changes are happening in less than half that time and with rapidly growing populations.

1.4.2. A case of sustainable public-private forest management in the West

Today, virtually everyone is influenced by the rules of the global economy. Yale University owns a second-growth forest managed by the School of Forestry and Environmental Studies in the state of Connecticut, in the eastern USA. The management goals for the forest are: (1) to provide a hands-on working (managed) forest laboratory for teaching; (2) to maintain a permanent, fully owned site for scientific research, especially research that extends over several decades; (3) to provide a financially sustainable asset to the School; and (4) to maintain and/or increase the overall integrity and health of the forest ecosystem. Some 2,100 ha of this forest is therefore managed for timber production, mostly red oak, with the objective of sustaining or increasing age-class, structural and species diversity across the forest while obtaining income. The forest is Forest Stewardship Council certified for sustainable forest management. If the whole forest were managed primarily for profit (as is the adjacent land managed by Hull Forest Products timber company), the net profits derived chiefly from timber stumpage in 2005, before the market crash of 2007, would come to approximately $90 ha^{-1} y^{-1}.[18] That would represent an internal rate of return of 6–8%. Net income per hectare is similar to that achieved by small farmers in the eastern USA from dairy herds or cereals on similar undulating land of moderate to low fertility.

However, 15% of the 3,270 ha total forest area consists of wetlands and other categories protected from any destructive use by law, and a further 15% is set aside by Yale for research and for conservation of representative forest types. The net income therefore amounts to only $75 ha^{-1}. To professionally run this forest requires the employment of one forester part-time to prepare the silvicultural prescriptions and oversee the felling operation. Obviously, the university does not do this. Instead, in accordance with its educational mission, a recent graduate is employed as a full-time forest manager who is also responsible for training 10–12 students in an internship program, who undertake tree marking and other silvicultural operations; and two part-time managerial assistants. As in a contemporary tropical rainforest, the logging operation itself serves as the major silvicultural intervention, simply by opening the canopy. Careful operation of the logging machinery determines the fate and quality of future crops. An equivalent area of small farms, in contrast, might currently employ 100 people.

Yale Forest's timber is sold largely on the home market, as tends to be the case where forest ownership and management are secure, irrespective of the extent of the resource remaining. Yet, as a symptom of the pervasiveness of the global economy, Yale Forest's veneer logs sometimes go to China or Europe for peeling, and its white pine logs are often milled in Canada.

Yale is recognized by government as a tax-free charitable institution. However, as a reflection of its educational value it chooses to pay taxes as a demonstration of sustainable private land management, and forgoes a portion of its timber profits to support forest services, including wildlife conservation and the protection of rare species. The state benefits directly by issuing hunting licenses for a fee, a right gifted to it. Ordinarily a private landowner, such as Hull Forest Products, would lease hunting rights to a hunting club for a significant amount of money ($25–35 ha^{-1} y^{-1}). Hunting simultaneously serves management by controlling browsing of the regenerating stock by deer, in the absence of the original indigenous human population and the wolves that were hunted to extinction by the early colonists. The state also gains hydrological benefits from the conservation of headwaters as a drinking water supply for cities downstream, and by the tourism attracted to this 'quiet corner' – the last extensive forest in its region. The whole world profits from the carbon captured in its biomass.

The Yale example is only possible where there is sufficient public support to forego maximisation of current financial return, and where the public has power to influence policy. Besides, the forest is on land unable to yield substantially higher agricultural income. Thus, the Kuznets curve exposes a critical gap in the progress of reliable husbandry during economic development, an era of extraordinary peril that forests must endure if they are to serve the more prosperous and stable economies of the future. Once the semi-divine authority of the ruler wanes, the forests have to survive independently until the people, in self-interest enlightened by scientific understanding, and encouraged by a fair return on global forest services from the world community, have the means to ensure that policies for their sustainment are implemented by a representative government.

Crucially, there is no good evidence for a Kuznets curve in biodiversity. That is, there is no clear tendency for the value of biodiversity to decline rather than increase, or the numbers of endangered species to increase with intermediate levels of development and then decrease with greater prosperity,[19] although much depends on the specific criteria used and at least one study did show a weak relationship.[20] That study indicated a possible turning point in number of bird and mammal species under threat when per capita income reaches around $10,000–15,000. Other factors such as political rights and stability, civil liberties, and legal institutions were co-correlated with species threats. It is a priori clear there can be no Kuznets Curve in extinction.

1.4.3. *Sustainable forest management challenges in tropical Asia*

Unlike those in Africa and South America, no rainforest in Asia is far from a port. Trees in sustainably managed tropical lowland MDF on equivalent land area might show no higher diameter growth than the temperate Yale Forest. The Yale oaks grow an *average* 5mm diameter per annum, comparable to light hardwood dipterocarps in a successfully regenerating red-meranti-keruing MDF. However, the length of convertible bole among the Yale oaks is only 10 m, whereas a well-managed red meranti-keruing forest should yield boles twice that length – though shorter than from the original primary forest (Chapter 8). Annualised stumpage values of hill forest in Peninsular Malaysia have been estimated at c. $50 ha^{-1}.[21] Assuming the timber is similarly used for veneer and lumber and has the same commercial value per unit, net profit from a future well-managed red meranti-keruing forest (lowland MDF) could be conservatively estimated at about twice that of Yale, provided discount rates are similar. Employment needs would be similar to the Yale example. Well-managed hill dipterocarp forests situated below 1,000 m and on land too steep for plantation could likely yield timber comparable in value to the Yale Forest.

Yet sustainable management of such a forest faces a host of additional challenges (Plate 9.2). Not the least of these is the presence of rural communities. The needs of these communities for the goods and services of the forest, and their claims to the land it occupies, are diverse, urgent, and not easy to reconcile with the requirements of a sustainable timber market.

Plate 9.2a Forests and their wildlife can only survive with enlightened policies, with committed and qualified public servants to carry them out, and an enlightened public to support them. Logging allows outsiders to enter and hunt. In this image logging company employees, replanting during the day, return at night to hunt. They start fires to flush game and regenerate grass for wild ungulates, destroying the benefits of their daily work. Ma Da Forest, Dong Nai Province, Vietnam. 1982. (P.A)

Plate 9.2b–e Forests can only survive with enlightened policies and committed and with qualified public servants to carry them out. **b,** MDF does not necessarily immediately regenerate following logging into productive forest: vines smother early regeneration on clay loams a decade after logging. Danum, north-east Sabah; **c,** Major roads through forests lead to illegal collection and sale of fuelwood by village women to truckdrivers, who sell in the city; Ranchi Plateau, India, 1983; **d,** As populations increase and forests decrease, conflict increases between people and wildlife. Guardposts of Nepalese immigrants protect crops against bears. Buxa, West Bengal, northern India; **e,** Population expansion, exploitative traditions, or migrant inexperience leads to shortening of swidden fallow, and eventually invasion of the deep-rooted fire resistant grass *Imperata cylindrica*. Land of the Sungei people, south-east Kalimantan. (**c** P.A.; **b** C.-E. Lee; **d** H.H.; **e** W. Smits)

1.4.4. *An unrealised value of indigenous forest.*

Several countries in the Asian tropics are building industrial economies. Will demand for non-exploitative products of forest diversity expand? As forest goods such as long-lived hardwoods and wild fruit trees — or tigers — become rare through overexploitation, their value inevitably increases. In theory, this *scarcity value* should provide incentive for their conservation, but that can only happen where ownership of the product is clear and secure.

The early industrial revolution in the West was accompanied by the greatest era of experimentation with natural forest products. Exploration led to introduction of many ancient fruit tree cultivars from temperate Asia and elsewhere, and growing scientific understanding led to hundreds of new varieties through hybridisation. Now, many command high value as regional delicacies, while others provide the genetic base for new mass-marketed varieties. Would such advances have been made had countries in those days been unwilling to exchange their endemic germplasm? Where is this research advancing in the tropics of Asia? In India, there is continued research on the mango, but none, as far as I know, incorporating genes from the many wild species of the Far East – some of which are delicious.

Other than in India, and now in China and Singapore, indigenous researchers in tropical Asia have been hesitant to publish. Particularly, they have found it challenging to derive and test hypotheses from carefully designed and documented data which, from the 14th Century, has been the means by which science has advanced. There were once some experiments in fruit improvement of durians by the Malaysian Department of Agriculture, and one of the resulting durian varieties is still prized; but does the research continue? Only in Brunei do I know of truly original advances. Brunei, a wealthy country, manages its forest estate solely for the home market and remains 70% forested. Jumaat Alim, of the Department of Agriculture, initiated hybridisation experiments between species of durian and of *Artocarpus* in widespread cultivation. Local markets there still offer a range of indigenous species that produce edible (if at times less tasty) fruits but are well adapted to the infertile sandy soils of Borneo (Table 8.3).

2. Present and future potentials of the indigenous forest

Successful economic development depends on retaining the full range of options for future use of the aboriginal forest estate: an asset many of whose species are truly unique to each country and therefore irreplaceable national heritage. Retention of options must include protecting the genetic library for which Asian and American tropical lowland rainforest far exceeds any other terrestrial ecosystem on Earth. Such forests must survive the stage of economic development when forests decline to their lowest perceived values (Fig. 9.1).

From where might future capital investment — for environmental services in particular — be generated? Urban populations have grown, and feel better off, but have lost awareness of the extent to which their wealth has arisen from their forests; while those left behind in the inland valleys and hills have been disenfranchised. Modern tropical deforestation in Asia is in many ways a new manifestation of an age-old pattern: exploitation of isolated and poorly organised upland communities by lowlanders with access to information, resources or political patronage. This pattern is enormously strengthened by modern communications and technologies.[22] To insure that enough (and sufficiently representative) primeval forests be left intact and strictly protected, we must all be willing to contribute to the cost of sustaining them.

As I hope these essays make starkly clear, we are only just beginning to understand how species-rich forests are maintained in nature, and therefore how and at what rate timber harvest, fragmentation and other modifications influence the future of a forest. Practical management will benefit from the emerging general theory of the evolution and dynamics of highly biodiverse natural ecosystems that is being developed via the international collaborations of the Center for Tropical Forest Science at their twelve forest sites in tropical Asia. Equally important is a deeper understanding of the conditions under which forest minorities protect or imperil the resources upon which they depend. Such research is being advanced by initiatives such as the Ashoka Trust for Research in Ecology and the Environment (ATREE) in India, the Centre for International Forestry Research in Bogor, Indonesia, and the International Forestry Resources and Institutions network (IFRI), based in the USA but with participants in more than ten countries. Meanwhile, some predictions can be tentatively made from current knowledge.

2.1. *What future for forest minorities?*

Commercial interests and governments tend to favour uses that maximise financial returns to them. Governments may derive taxes and other emoluments from commercial activities, while subsistence farmers generally pay no tax (even though, under more enlightened regimes, they are entitled to educational and health services). Governments have thus allowed the hill tribes' main asset, the forest, to be taken from them while failing to provide a secure alternative living other than as unskilled estate labourers downriver.[23] Logging offers only sporadic employment. Young uplanders, faced with declining resources and opportunities, and witness to the fantasy suburban life portrayed on television, willingly forsake traditional constraints on exploitation in their attempts to escape poverty. The consequences are often dire. Investment in active management for sustaining forest products and services can provide continuous rural employment and higher incomes, which would then become taxable. The cost of management can thereby serve to more equitably distribute profits. Ecological and social goals could thus be brought into harmony with economic ones.

Current governments have generally followed colonial administrations in blaming swiddening traditions as the greatest threat to the forest estate, while ignoring the devastating impact of vastly increased and unsupervised logging. Swidden had earlier remained in near-harmonious equilibrium with the forest in North-East India and elsewhere in the northern tropics, where the lower montane forests regenerate relatively easily[24] (Chapter 4). Swidden has increased with population growth, but in many places this increase is attributable to immigration of inexperienced landless lowlanders from densely populated regions long dependent on irrigation. In Sarawak, and less overtly in several other regions, government policies encourage swiddening forest minorities to migrate to the lowlands. These have been converted to massive oil palm plantations with widespread labour needs. This is a drastic strategy, erasing ancient cultures and their deep understanding of the land. In the short term it has often failed (in Sarawak, for instance) because swiddeners have tended to shun the monotonous discipline and loss of freedom of a plantation worker's life. Such resettlement policies are often driven by commercial interests, but they may turn out to be inevitable if they can do anything to reduce the poverty and suffering which results from the *laissez-faire* approach so widely seen wherever government infrastructure is weak. Just occasionally, as in the Peninsular Malaysian Federal Land Development Authority (FELDA) schemes, government has overseen oil palm plantation by smallholders who come to own their land, with new villages and services established and the product refined, therefore quality-assured; and the hill forest has remained conserved on the steep land, often nearby. But the best mix of government and private investment has, not surprisingly, proven difficult to achieve.

Of course, such policies are only possible where new land and new crops are available to absorb the growing population or immigrant labour. Policy should reduce overhunting, small-scale illegal logging and incursions into sanctuaries by minorities, although criminal elements from elsewhere will persist where easy access is maintained. Far more destructive are the swiddening attempts of immigrants inexperienced to the lowland forests who, lacking credit or security, have no alternative but to slash, burn and move on.[25]

Nevertheless, we should heed the timeless anger of the Highland Scots, subsistence farmers now dispersed worldwide following clearance from their ancestral lands two centuries ago. The consequence to their homeland has been a derelict landscape, eaten out by sheep. Forest minorities have little opportunity to disperse beyond their countries. They have reason for anger, and it will have to be redressed if a nation is to move toward reconciliation.[26] This is one more reason why, to create incentives for those nations that possess the world's major biodiverse resources to manage them judiciously, the industrialised nations as beneficiaries must recognize the need to make a fair and sustained contribution.

Many readers, both in industrialised and less developed nations, may bridle at this suggestion. These essays are not the place to digress on the knotty but critical political problems involved. But the facts must speak. The costs of forest loss are being shared by all humanity, as will be the consequences of economic failure. Tropical forests and their biodiversity are the proximate heritage, with few exceptions, of less developed and often less prosperous nations; but sustaining this patrimony benefits us all. We cannot afford to ignore our collective responsibility towards it.

2.2. *Challenges of multiple use*

In developed and urbanised economies natural forests remain important sources of quality hardwoods and other products, but they are even more important for their ecosystem services, as predicted by Kuznets[27] (Fig. 9.1). Besides those cited in section 1.3.2 above, these services include clean and dependable water for industry and consumers, soil conservation and reduction of siltation and, most apparent to the citizenry, places to relax and regenerate from the pressures of urban life. The degree to which forested hinterlands mitigate major flooding downriver is under debate,[28] but it is clear that by reducing erosion of steep uplands, they do minimise soil loss and river siltation, stabilising river systems over the medium and long term. Forested riverine wetlands, such as the khair-sissoo (*Acacia catechu–Dalbergia sissoo*) groves that line the torrents issuing from the Himalaya, mitigate moderate seasonal flooding.[29] Local weather patterns are modified by the presence of forests: deforested areas experience heavier rainfall, therefore more loss and erosion by run-off, and at longer intervals. In industrialised economies, forests have become important sources of recreation and income generation for local communities, including commercial hunting, hiking, and ecotourism. 'Ecotourism' is now said to comprise 20% of international travel.[30]

On a larger scale, at least some tropical forests are thought to be major carbon banks. Research at Pasoh (Malaysia) has shown that they may be net sinks of atmospheric carbon, some possibly going into the soil and aquifer (see also section 4.2 below). The lower standing biomass of deciduous forests is partially offset by the higher carbon content of their soils.

But in many parts of Asia, particularly in the drier seasonal regions, rural populations continue to live in subsistence economies, that is, near the starting point of the Kuznets curve (Fig. 9.1). There, the tree species are adapted to regenerate by coppicing following fire and intense browsing. A number of Indian authors[31] have pointed out that traditional land uses have often maintained a widespread equilibrium with the forest resource, notwithstanding high rural population densities. This has contrasted starkly with the depredations of some commercial interests. In some places, traditional uses remain influential despite the potential immediate financial benefits of overexploitation. But population increase has led to harvesting of some resources beyond the capacity of the forest to produce, in particular livestock forage, medicinal plants and other non-timber forest products (NTFPs).[32] Wood remains the principal energy source in poorer agricultural communities, although improvements in stove technology and conversion to alternative energy sources have reduced urban demand.[33] Overexploitation of forests for fuelwood has been accelerated by the ease in which these same communities can earn extra income through sale, usually illegal, to truckers who market it in the city; conversion to charcoal for urban use continues to increase (Plate 9.2c).[34] Importantly for the search for an equitable management model, rural communities tend to benefit most by keeping forest stands at an early successional stage, thereby coming into conflict with the requirements of timber production.[35] In short, the economies of different beneficiaries of the same forest resource are at different points along the Kuznets curve: How can their varying demands be realised?

Social researchers[36] have questioned whether current exploitation levels are degrading the deciduous forest resource. Although these forests are highly resilient, the adverse impact of soil trampling by domestic cattle, and lack of regeneration owing to browsing and more frequent fire, must cast doubt on these assertions. *Critically, the dialogue is being conducted in near-total absence of data from rigorous monitoring, of both the forest resource and patterns of its use.* In the short term, socio-economic reforms such as legal guarantees of fair prices for harvests may somewhat reduce poverty.[37] Research into crop ecology can result in harvesting and management improvements, thereby potentially increasing sustainable yield levels. But in the long term, it is the complex interrelation of human population size, demand growth and market arrangements that determines the local pressure on a natural resource. Resource use can only be sustainable when all these factors can be controlled, so that demand for the resource remains within its productive capacity. Alternative employment, including a wide range of options from agricultural intensification to urban migration, may also form part of a solution; indeed, these are already measurable trends in many places.[38] At present, though, birth rates remain high among many marginalised communities: those upcountry within the forests, and beyond the regular reach of health services.

It must be recognized that many of the options for economic development, including commodity plantation or urban industry, are restricted to appropriate lands or to accessible or mineral-rich regions. Vast regions therefore must be expected to remain tied to traditional subsistence-based agriculture for the foreseeable future. Such communities will remain dependent on well-maintained forests for fuel and forage especially, but also for building materials and much more. But how and to what extent can the wider market continue to gain from these same forests, in production of sustainably managed timber for instance, or can their service values alone be realised?

The fate of humanity is to be played out in the next one or two centuries. Deforestation continues in much of the Asian tropics. Although it has been arrested in a few countries, little unlogged primary forest remains in them. Extinction rates have started to accelerate, and though the trend is predicted to continue, it need not be so. Our objective must be to get through these centuries, representing less than the lifespan of one tree. After this, conditions can again improve – or they could get very much worse, if corrective steps are not taken now.

2.3. *Sustainable forest management*

2.3.1. *Deciduous forests: fire, fuel and browsing pressures*

At present, exhortations by social scientists for more equitable and sustainable distribution of the products and services of tropical deciduous forests have rarely or never been calibrated by rigorous estimates of the forests' capacity to yield. The primary determinants of succession in these forests are the frequency and timing therefore intensity of fire, biomass (fuelwood) removal, and browsing and trampling by domestic cattle. The impact of these includes the reduction of combustible materials and the consequent deflection of forest succession from moister, that is richer and taller, floristic associations toward drier (Chapter 2). Semi-evergreen and tall deciduous forests are most susceptible. On the one hand, regular but infrequent (supra-annual) ground fires can sustain tall deciduous and semi-evergreen forests with valuable deciduous rosewoods (*Dalbergia, Pterocarpus, Afzelia, Xylia*), teak and sal by excluding evergreen species of little timber value. On the other hand, more frequent or intense burning of tall deciduous forests, by selectively eliminating regeneration even of some deciduous species, directs succession toward a shorter deciduous forest type, and reduction of such valuable species in numbers and form. The optimal fire regime is difficult to maintain, varying both across sites and according to management objectives.

Foresters long ago came to appreciate the critical role of fire in sustaining tall deciduous forests (Chapter 2), but *long-term monitoring of experimental regimes remains vital.*

Grazing and browsing pressures can be ameliorated by establishing grazing rotations, which require close management and, ideally, fencing.[39] This allows soils to rejuvenate from trampling and grasses to achieve their maximum productivity during the monsoon. The resultant excess hay can then be harvested at peak nutritional value and stored for the dry season when cattle carrying-capacity is harshly determined, potentially reducing their pressure on wildlife (Plate 8.5b). Fencing also offers potential for set-asides, which can permit tree regeneration even from seed if fire is controlled. Fencing is expensive, but current conditions may often make it economically feasible. *Again, and for the same reasons, experimentation is required.* Fuelwood rotation should also enhance its productivity. The necessary control will be easier to execute on private than public land. But we cannot think of a better investment for government subsidy aimed at rural poverty amelioration.

2.3.2. Perhumid tropical rainforests: focus on sustainable timber production

Timber harvesting in indigenous forest can be achieved carefully in a manner which maximises sustainable out-turn of logs. Treated well, a logged forest may continue to provide many natural services, but this depends entirely on the details of how it is managed. (See Chapter 8 for a fuller discussion of silvicultural management systems). In developing economies, where discount rates often remain comparatively high, it can be tough to make a commercial case for growing heavy hardwoods that may take 120 years to reach 75 cm diameter on lower nutrient freely draining sandy soils and ridges. The Yale forest does grow veneer sugar maple on that cycle, and veneer oak on 80–100-year rotations by harvesting competing trees of the same cohorts in two previous crown thinnings for sawtimber and firewood. The Far Eastern balau (*Shorea* sect. *Shorea*) and merbau (*Intsia*, Fabaceae-Caesalpinioideae), of which so much garden furniture currently on sale in the West is made, are unlikely to be regenerated commercially if additional values cannot be realised by the forest owner. In the Sunda MDF, the yellow and especially the red light hardwood merantis (*Shorea* species in several sections) could have formed the backbone of a sustainable timber industry. Their juveniles have pyramidal crowns with dominant leaders like temperate conifers, and leaf area index comparable to birch (Chapter 2). On the best sites they can grow at 1cm y^{-1} in diameter, attaining 40–50 cm over 50 years with a straight bole length of c.30 m.

There is no escaping the necessity for a skilled workforce, and active participation of experienced professional silviculturists in field operations, if the remaining rainforests reserved for timber, which are the hill forests on steep land, are to continue to produce at commercially attractive levels. These forests require skill-demanding management by strict adherence to a true selection system, and knowledge of the species (Chapter 8). Regenerated stands will never reach the stature, clear bole length or diameter of old growth, but they will be comparable to the best stands in temperate and tropical deciduous forests. The future, smaller, logs will be extractable with lighter machinery, with less damage to soil and regeneration.

A major reason for the continuing (comparative) success of natural forest management in Peninsular Malaysia has been administrative. Responsibility for control of licensing and oversight of the logging operations lies with local forest offices and the states, rather than with the federal service headquarters. This, and the strength of the home market from early on, has fostered relatively small enterprises and a processing industry. Recent devolment of government to the provinces in Indonesia and the Philippines provides opportunity to gain the same advantages. These offer the best hope of a continuing rainforest-based timber industry recalling the successful forestry traditions of Japan and Germany.

2.3.3. Selective logging

A prime objective of Far Eastern selective logging management systems has been to shorten the felling cycle in order to yield timber more frequently (see Chapter 8 for a description of methods and their silvicultural outcomes). These 'selection' systems, as have been practiced in the Asian tropics, aim to secure sustained timber yield but have not considered economic optimisation of yield, or net present value (NPV) specifically. Mohamad Shawahid[40] presented the concept of *financial maturity*, observing that as trees grow the NPV of timber harvests increases, but that it eventually declines when the growth rate falls below the prevailing discount rate. This discounted value of *conversion surplus* should therefore be maximised within the silvicultural constraints imposed by the need to sustain production. The forest owner (which in Asia is usually the state as representative of the population and nation) only incurs the costs of a pre-felling inventory, volume assessment, tree marking and logging oversight, so that for them the conversion surplus approaches stumpage prices in value. Because the owner concedes the rights of stumpage to the concessionaire, it is the conversion surplus that should be used as the basis for setting royalty rates. That was not even closely approached in the early post-colonial decades; but more recently some Peninsular Malaysian states have introduced auctions for allocating harvesting rights, thereby introducing a means to restore harvesting rights to the state.

Marco Boscolo and Jeff Vincent concluded, from a

simulation analysis, that concession licences over longer periods provided little incentive for loggers to adopt logging methods with reduced impact on the residual stand, owing to the increasing impact over time of discount rates on profitability.[41] They also found that performance-based licence renewal conditions, in the form of performance bonds, would provide a strong incentive both to adopt a regime with reduced impact and to adhere to minimum diameter cutting limits, whatever the licence period. There would still remain an incentive to return to the stand early, though, because remaining trees would have reached financial maturity. Loggers, in summary, prefer intensive harvesting at longer intervals, as in the Malayan Uniform System (MUS) (see Chapter 8, and section 2.3.5 below), but even then could not achieve the rate of return they aspire to if royalty rates were based on the conversion surplus. However, the scenarios explored in these simulations take no account of the impact that operations on the ground, especially on steep land, have on the established regeneration. It is upon this regeneration that a new crop of hardwood meranti must entirely depend. In short, although rainforests are claimed by governments as owners on behalf of their nations as a whole, it is moot whether a *sustainable* return can be made to the owner by timber harvesting alone: for that, other values must be taken into account.

In Sri Lanka, where lowland tea can achieve NPV of $23,000 ha^{-1}y^{-1}, one financial analysis estimated that comparable values could be achieved by a uniform shelterwood system if cut stands were enriched with commercially marketable NTFPs, and providing discount rates do not exceed 5%.[42] This study was performed in the context of what was at that time a comparatively low-wage economy, but continuing demand for these products in that diversified economy has ensured that the case remains valid today.

By maintaining low stumpage values the state forest owner, in nations holding large forests, instead encourages the rapid exploitation of the timber resource at minimum profit to the citizenry (though maximum profit to the concessionaires and their political patrons). Exceeding local industrial capacity, most timber is exported. This muffles the expected market signals of increasing scarcity, preventing the rise in timber price that should occur as the resource declines, and is the central cause of the massive fluctuations in world timber market prices.[43] Consequently, although 70% of forestry sector jobs remain in developing countries, less than a quarter of the value added through processing ever returns to their workers.[44]

2.3.4. *Shelterwood systems* (see also Chapter 8)

The MUS of management for MDF, especially those forests dominated by light hardwood meranti *Shorea*, and *Dipterocarpus*, resulted in magnificent stands (Chapter 8). Others, established following felling according to a similar shelterwood system in the 1930s, could still be seen in Peninsular Malaysia in the 1980s. When these were felled they yielded more timber than had the original primary forest. But everything depended on careful management of machinery and road-building to ensure survival of the established regeneration for the new crop, and on the initial poisoning and thinning to reduce competition among the regenerating cohort, thus serving as an investment for the succeeding crop. These treatments became increasingly uneconomic as wages increased with the nation's prosperity, and ignored as machinery increased in size and use. Besides, the site requirements of light hardwood 'red meranti-keruing' MDF economically favoured oil palm.

2.3.5. *So what management system is preferable now?*

Both MUS and other shelterwood systems can sustain timber production if well managed.[45] In most lowland Sunda MDF on clay loams, pole-sized juvenile light hardwood merantis are in low density, but advance light demanding seedling and sapling regeneration is abundant, albeit often patchy on the ground. Where this is true, reduced logging damage due to the 70-year logging interval in MUS lends it the advantage on silvicultural grounds.[46] Historically, MUS maximised timber production in indigenous wet evergreen forest on clay and sandy clay soils.

MUS would have yielded greater volume of timber in subsequent cuts, and arguably more revenue by reducing the road-building required for each coupe and concentrating management costs to a smaller felling area each year. But its success depends, like any other system, on careful management of the felling process and in particular road-building and use of tractor crawlers and other machinery which destroy established regeneration. Most of all therefore, it depends on an agreement between the owner and concessionaire which lasts long enough to provide incentive for the concessionaire to tend for at least one subsequent crop.

MUS protected soil and water. It also increased carbon sequestration, because the product, sawn logs and veneer, was retained in structures and furniture. It would, by promoting regeneration of an early, more continuous, canopy have provided optimal physical habitat for canopy vertebrates and other wildlife, while early post-harvest regeneration would attract browsing fauna. It did not directly harm terrestrial wildlife and, because logging proceeded on a rotational system stand by stand, even arboreal animal populations could be sustained though many of their food trees were reduced by increased dipterocarp densities, especially the strangling figs which are the forest's blessing but which were once – and sometimes still are – considered the forester's bane. But once poisoning of juveniles was abandoned, that cause of tree species depletion was avoided.

MUS is the most productive system of management for

timber in MDF dominated by light hardwoods, but the tree species diversity itself, and that large part of biodiversity which is interdependent with tree species diversity, is increasingly and unavoidably lost. Species yielding traditional non-timber products may also be reduced or eventually eliminated if their regeneration is removed, though that is no longer current practice. Any form of logging inevitably increases the mean gap size and, more importantly, the frequency of canopy gap formation. Because tree species diversity of a forest is largely determined by its natural regime of gap size and frequency, therefore forest vertical structure (see Chapter 7), any change in that regime will lead to diversity loss (gain is only possible over millennia, through speciation or immigration); and that loss will increase in each successive felling cycle. Even less rigorous forms of shelterwood silvicultural systems (Chapter 8) will therefore eventually lead to loss of biodiversity. Some unlogged forest must therefore be set aside for these purposes if they are to survive (see sections 2.4 and 3.1 below).

Further, recreation value of successional forests is limited, because the early regeneration is impenetrable, and felling takes place before the majesty of old growth is attained.

But in reality, MUS is rarely feasible now, because the lowland MDF for which it was designed has almost entirely been converted to commodity crops but for patches surviving in conservation forests. The residual forest estate is, in effect, now confined to lands too steep to merit clearing and planting, even terracing for tree plantation crops without severe risk of erosion. There, timber stands are densest in nature on the drought-prone ridges, and require a skill-demanding true selection system of management (see sections 2.3.3 and 2.3.5 below). But unmanaged logging, and road construction along the ridges which held the highest timber volumes, will necessitate enrichment of regeneration by planting. Ridges may merit planting of an initial 'nurse' crop, for which exotics such as *Acacia* and *Albizia*, with diffuse crowns, may be preferred. Such species only invade in severely degraded lands.

The great majority of surviving regional timber production forest will be on land too steep for commodity crops ('hill forest' in the Peninsular Malaysian terminology) or on low-nutrient, erodible sandy soils. A true selection system of management was recommended by foresters Burgess[47] for Peninsular Malaysian hill dipterocarp *seraya* forests and by Brünig for Sarawak *kerangas*,[48] both of which are dominated by slow-growing shade-tolerant species. However, the requirement of re-entry for the liberation of regeneration from competition is disastrous unless permanent road systems are maintained. We have seen that hill forests vary throughout the region in species composition and in dynamics, and therefore in silvicultural potential[49] (Chapters 2, 3). The principal axis of variation is along the catena: the lower slopes, with deeper, moister soil, originally supported MDF, rich in dense-crowned light hardwoods, especially red and yellow meranti *Shorea*. The stands on the steeper hill slopes were lower in basal area and had more gaps of greater size than those on the more stable surfaces of the undulating lowlands now given over to commodity crops. Ridges, by contrast, supported high densities of emergent diffuse-crowned canopy species; natural gaps were fewer and smaller due to the predominant pattern of single-tree mortality (Chapter 2). In forests below 600 m, the best timber there is thus found on broad ridges with relatively deep soil, while the narrower high ridges and shallow soil carry shorter trees.[50] But shade tolerant medium to heavy hardwoods dominated: even in succession following logging, felling cycles will prove substantially longer than predicted from experience in 'red meranti-keruing' MDF.

Selective logging based solely on minimum diameter cutting limits are a far cry from any form of RIL aimed at conserving tree species and hence overall biodiversity. M. Ashton and B.M.P. Singhakumara have shown, in south-west Sri Lanka, how the canopy species components across the catena from ridge to valley rely on different canopy density and gap sizes for their successful regeneration, as they compete with species with different light requirements.[51] They concluded that all such forest types can be optimally managed both silviculturally[52] and economically[53] by differing stages and intensities of canopy removal in relation to topographic position and floristic asssociation. All would be characterised as different kinds of shelterwood, though that for the upper slopes and ridges approaches a true selection system: for the shade tolerant associations on ridges and freely draining sandy humult soil this requires the initial removal of lesser species in the sub-canopy. In the Peninsular Malaysian seraya (*Shorea curtisii*) ridge forests this would include systematic removal of the stemless bertam palm, *Eugeissona tristis*. Like mountain laurel in New England woods, this palm is a clump-forming, shade- and drought-tolerant under-shrub which outshades tree regeneration and greatly increases when the canopy is opened by felling.

The shelterwood method therefore differs fundamentally from *current* selective logging systems. The primary objective would be to maintain or enhance the growth and production of *chosen* canopy species and purposely establish and release their advance growth beneath the canopy. Consequently, the selection of individuals in the canopy to be felled is not determined by a minimum felling diameter. Instead, it relies on careful, stand-by-stand planning, choosing those to be cut with the aim of increasing the number of canopy gaps without changing their mean size. This promotes the sub-canopy most favourable for regeneration of the chosen species. Unfortunately, most of these forests in the Far East are now trashed, and liberation treatments, often followed by enrichment planting, will initially be necessary, if stagnation in stand growth is to be avoided. Even then, the forest cannot be more productive than the potential offered by its site conditions. Achievement of sustainable felling cycles of less than fifty years will require skilful and stringent, management. Appropriate choice of species for planting is essential. In short, the forest cannot recover unless returned to management by well-trained, experienced and dedicated forest services staff: servants — and heroes — of the nation.[54]

2.3.6. *Reduced Impact Logging*
Conservation ecologists and silviculturists have promoted the concept of RIL[55] (Chapter 8). Less a protocol than an attitude, RIL has developed as a reaction to the degradation caused by unsupervised logging masquerading as a selection system: in lowland forests that would have best been managed by MUS when coups ('stands') of the full range of stature and canopy closure are well represented in the forest as a whole. Its aim is to extract timber with minimal damage to canopy and soil, retaining, as far as possible, the continuous canopy corridors in which arboreal vertebrates can forage. When carried out strictly 'by the book', RIL logging excludes forest on steep slopes and near watercourses. Taken together, such areas might incorporate enough mature-phase stands to sustain even the most threatened climax tree species, were the rules strictly enforced. Currently that is rare, however, and the reasons are primarily financial. RIL rarely maximises profits and requires a highly skilled workforce to plan operations and oversee execution. Additional necessary investments include separate equipment for road building and timber extraction. Some studies predict that RIL can be at best financially marginal to the concessionaires, who have never enthusiastically embraced it. Other studies have shown that, when the conditions of RIL are met, damage both to the residual stand and to regeneration can be reduced; although it also appears to be true that in well-stocked forests (such as were once common in Peninsular Malaysia), harvesting intensities tend to be high enough that even using RIL techniques faithfully, damage to the residual stand can exceed 50%.[56] On the other hand, largely due to better preliminary planning, the wastage of salable timber may be much reduced under RIL: in some cases actually counterbalancing increased costs.[57] In the end, results seem to depend on forest conditions, terrain, timber markets, skills of the logging crews, methods of compensating them, among other factors.[58] Critically, RIL underplays the need to manage forests differentially according to soil and topography if production is to be sustained.

It has been estimated that even some destructively logged forests can regenerate economically, but discount rates must be low[59] (Chapter 8). In one simulation model based on data from an East Kalimantan concession in high yielding MDF on clay loams over undulating land,[60] yields were compared between logging using the standard Indonesian selection system versus RIL. Felling cycles of 25, 35 and 45 years were compared over successive cycles of an Indonesian selection system with minimum diameter cutting limit. Average yield of 80 m^3 ha^{-1} at first harvest was halved thereafter. With RIL, a 35-year cutting cycle with maximum allowable yield of 50 m^3 ha^{-1}, or 45 years with 60 m^3 ha^{-1}, were predicted. However, neither could achieve a 16% rate of financial return unless damage was strictly controlled. Further, diameter growth rates were not followed long enough to gain realistic regeneration rates. Carefully managing RIL on a 45-year cycle, while increasing timber conversion efficiency, was therefore advocated in this case.

Even where timber production remains the primary objective, enhanced protection of wildlife requires management changes that almost certainly constrain profitability. Management becomes a questionable investment unless forest ecosystem service values are taken into account by the forest owner. That may appear a luxury feasible only in fully industrialised economies; but failure to conserve biodiversity will seriously limit options for future development of the rural agricultural economy as well. Again, RIL requires careful silvicultural management and is relatively labour-intensive. Most productive light hardwood-dominated MDF is on the very land sought for oil palm plantation.

We conclude that sustainable production forestry may still be possible in the now hilly remaining West Malesian forest estate, but only if focus is returned to optimisation of timber production, and only if skills are reacquired to implement a strict classical selection system of management, concentrating on those most productive ridge stands where regeneration still survives. This would best be achieved by small local enterprises, with local silvicultural experience and part-time employment from local communities. Even then, the economic case will likely on be made if service and other values of the forest are taken into account. Further, we suspect that careful and appropriate silvicultural management is at least as effective at conserving wildlife as RIL: it is lack of management that degrades it, and the forests productiveness as well.

2.3.7. *An ecologist's perspective*
Viewing the problem as ecologists, we see no solution other than to return to Govinda Menon's (Chapter 8) plea to re-establish as a first priority the maintenance and improvement of the forest estate *as a whole* for the good of the nation, that is for *all* its people; and to value long-term economic benefit over initial financial profits. There is no alternative method but to focus management of individual forests for a single overriding use, or on the improvement or restoration of *individual forest stands*, which vary in their silvicultural properties and potential with site conditions. If poisoning of non-commercial species is foregone, non-timber products will continue to regenerate,

and those yielded by pioneer species may even increase. Such a focus, if for sustainable timber production, will result in forests which retain much albeit not all vertebrate diversity, and most service values. But much tree species diversity, and animals and micro-organisms dependent on it, will eventually be lost. It must therefore be conserved in other forests, or in strictly conserved refuges within the production forest, strictly set aside for that purpose.

2.4. *Protecting indigenous forest biodiversity*

The forests richest in biodiversity, which are the lowland MDF of Malesia, have fluctuated in area during periods of climate change, including under exceptionally high sea levels as at present, when perhumid Sunda is fragmented by the sea. This has occurred several times during the Pleistocene, reducing the Sunda continent to islands albeit over relatively short geological periods.[61] But the fragmentation now occurring due to expansion of agriculture and urbanisation is beyond any that occurred in the last 23 million years, which has been the era in which our forest flora and fauna has speciated. Our task must be to sustain as much as possible of that biodiversity, as an integral requirement in our quest for a sustainable future for ourselves.

2.4.1. *Forests are genetic libraries*

To return to the metaphor with which this essay began: biodiversity is genetic information, a living library of gene sequences contained in populations, accruing over evolutionary (that is geological) time. Most of us have been happy to allocate some portion of our tax obligations to the protection of information gathered through human endeavour, that is, to our public libraries. Biodiversity maintains itself, as long as breeding populations are sufficiently numerous to reproduce the level of genetic variability of one generation in the next, and to persist in their natural environments. Unlike the information in most modern books, however, when genetic information is lost it cannot be recreated. The extraordinary chemical diversity of plants and microorganisms continues to yield novel complex compounds, any one of which could prove important in medicine or industry and yet is highly unlikely to be developed in the laboratory. Chemical diversity also sustains wild genotypes possessing resistance to particular pathogens (Chapters 5, 7). This diversity confers stability on ecosystems, by reducing the incidence of epidemic disease and mortality through complex mechanisms that remain subjects of active research. Importantly, too, genes from these localised varieties can potentially be introduced into many cultivars currently susceptible to the panoply of introduced diseases increasing with worldwide trade.

More than half of the world's biodiversity is thought to be stored in the vast library of the lowland tropical rainforests. Within that biome, diversity is not distributed at random, but is overwhelmingly concentrated below 400 m altitude and in those large, continuously-wet equatorial land masses which did not experience (or are adjacent to regions that did not experience) rainfall seasonality during the Pleistocene ice ages. The Sunda lands and New Guinea are two such major land masses. South-west Sri Lanka is smaller but important; distinctly different gene pools are stored in each.

2.4.2. *Varying patterns of diversity within forest habitats: tree species as indicators*

Living diversity is overwhelmingly concentrated among the arthropods (especially insects) and microorganisms (including the fungi). The diversity of these groups directly or indirectly depends on the diversity of plants as primary producers. It also reflects diversification of plants in response to the diversity of plant predators and pathogens (Chapters 3, 7).[62] In the evolutionary record, insects and flowering plants go hand in hand; the diversity in each has both stimulated and responded to the diversity of the other. As sessile organisms, trees and other plants are particularly habitat-specific within the forests where they are richest in species. At the regional scale, their diversity is much enhanced by diversity of local habitats (especially of soils, see Chapters 3, 7).[63] This is a markedly different pattern from the vertebrates,[64] and the conservation priorities for plants are thus distinctly different from those for vertebrates alone.

Because trees comprise the predominant biomass of forests, maintenance of their diversity must be a leading conservation objective. Any reduction in tree diversity or change in relative species abundance will bring subsequent changes in overall biodiversity. The species richness of trees is only imperfectly correlated with that of other organisms in a particular area (Chapter 7), but tree species do nevertheless represent the most accessible measure of overall biodiversity. The tree species of Asian forests are now mostly well known and, unlike most arthropods and micro-organisms, are possible to identify with experience and simple field tools. Tree species richness thus serves as a valid and convenient proxy for genetic diversity in biodiverse tropical forests. Documenting these patterns is a major function of the primary forest tree demography plots of CTFS. Monitoring changes in tree species can alert us to unnoticed local and potentially wider changes.

2.4.3. *Different organisms pose different conservation challenges*

Biodiversity varies within any forest along several axes. Soil fertility (Chapter 7) is an important one: the greatest richness of tree species is consistently found in areas of moderately low soil fertility. But such areas tend to lack dense populations of vertebrates, including the seed and pollen dispersers upon which so much depends (Chapter 5). Such populations require lands of higher fertility that are more productive of nutritious foods. And indeed, the rainforests richest in vertebrate diversity and biomass grow on nutrient-rich soils that experience high disturbance levels. Such forests often support a distinct flora, but are not among the richest in tree species (Chapter 7).

Most of the tree species of Asian rainforests are represented in their chosen communities by less than one reproductive individual per hectare. However, CTFS plots indicate that 500 ha are sufficient to conserve at least 200 mature individuals of a third of the species in a rich MDF community (Chapter 6). Several patches of suitable habitat for commoner tree species of at least 200 ha, and for some species a few of at least 10,000 ha, are therefore necessary for a region. A few large conservation areas might therefore appear optimal for conservation of both trees and animals. But many plants are confined in nature to habitat islands, some of which have always been small. These include karst limestone and ultramafic rock exposures. Gunung Silam in Sabah and the Knuckles Range montane forest in Sri Lanka are each less than 10,000 ha, but each nevertheless supports several endemic tree species that may have evolved and survived in these small areas for more than one million years. Amazingly, 105 plots set out in Sarawak MDF at 13 judiciously chosen locations and totalling a mere 63 ha overall, captured 83% of the 164 dipterocarp species known to occur in MDF there (of individuals over 10 cm dbh), representative of other tree families also.[65]

Examples like these imply that full protection of plant diversity will require many small conservation areas in addition to the few large ones.[66] Conservation of animals, especially large vertebrates, requires large protected areas, within which some disturbances such as careful selective logging may even enhance numbers of certain taxa. Plant conservation requires, at least for the foreseeable future, a network of preserves representative of the climate, geology and landforms of their region. These may be quite small, but they must be strictly protected from canopy disturbance therefore logging, and they are best set in a matrix of other forest. The model is the sacred grove so strong in many Asian traditions, and the Virgin Jungle Reserve (VJR) system of Peninsular Malaysia, whose origins are in Indian forestry. Joann and colleagues[67] of the Malaysian Forest Research Institute found, surprisingly, that the number of insectivorous bats remained constant in VJRs and adjacent regenerating logged forest, but declined more than 600 m away, from which they inferred that recolonisation of the forest can be achieved over at least that distance. Remarkably, they found 20 bat species in the smallest of six VJRs examined, whereas the largest, 1,834 ha, revealed only 17; but larger VJRs included more true forest foliage-roosting species. Other primary forest-dependent vertebrates, such as woodpeckers, which need dead

standing trees in which to nest, may show similar results. But such small isolated preserves cannot survive more than one tree generation if their dispersers are lost, or if browsing intensifies. Many island forests are now being encroached: Bukit Timah in Singapore by urbanisation and roads,[68] Lambir National Park in Sarawak by swidden, highways and suburban development, Sinharaja and many small surviving forests in Sri Lanka's wet south-west by smallholders whose several children lack adequate landholdings. Forest margins are gradually eaten away by illegal tree felling. In the perhumid tropics, newly exposed then abandoned forest margins are soon colonised by dense pioneer vegetation, which acts as a buffer to microclimate changes in the fragmented understorey. In more seasonal regions, however, prevailing dry monsoon winds may desiccate the forest margin and impede rapid succession,[69] and fire is a more frequent hazard (Chapters 2, 8).

2.4.4. *A botanist's view of SLOSS*

A debate raged in the pages of the journal *Conservation Biology* in its early days, about the relative merits of protecting 'Single Large or Several Small' (SLOSS) areas. The controversy has largely died down, less, perhaps, by being resolved than because it gradually became clear that scientists rarely enjoyed the luxury of making such a choice. Protected area establishment in the 'real world' has turned out to be largely a question of political and financial opportunity.

Nevertheless, it is true that many in the early conservation movement concentrated on safeguarding the appealing vertebrate fauna. Fauna are essential to the integrity of forest systems, since some serve as seed dispersers (Chapter 5), others in the control of invasive plants such as bamboo (Chapters 2, 5), while others reduce the depredations of browsers (Chapter 2) and seed predators (Chapter 5). But vertebrates are mobile, and are overwhelmingly generalists of the physical habitat. Even vertebrates highly specialised on trees rarely depend on particular tree species: woodpeckers, for example, depend on the persistence of rotting trunks.[70] Many conservation biologists therefore advocated for the creation of a few large preserves to contain these mobile populations.[71] Yet far-ranging vertebrates may require more space than is available even in exceptionally large protected areas. Solutions to the challenge of maintaining options must thus be sought at the scale of landscapes and regions.[72] These are some of the reasons why conservation biology finds itself having to 'think outside the fence' and consider carefully what is happening in production forest, mixed-use agricultural and human-dominated lands throughout the developing world. Much, though not all, plant diversity (including trees) can be conserved by setting aside forests on agriculturally marginal land, though the costs of protection will remain substantial. Much animal diversity can be maintained by retaining forest corridors between smaller conservation areas. River bank stabilisation by woodland corridors can often serve this purpose, as can windbreaks and hedgerows in agricultural lands. Forest fragments and groves are important elements of a landscape-level conservation strategy.

The International Union for the Conservation of Nature (IUCN) has codified criteria for endangerment categories. Critically endangered species are those that, among other criteria, have experienced population size reductions of 80% over three generations, endangered species 50%, and vulnerable species 20%, estimated either via habitat area loss or via population density.[73] These criteria must surely be derived from mobile organisms, particularly birds and large vertebrates, for almost all climax rainforest tree species would thereby qualify as endangered, which, correct or not when considered from a population genetic standpoint, cannot be regarded as practical policy advice! In the long term, though, we must accept these criteria as applicable to trees as well. The reduction of primary forest to isolated fragments must radically reduce genetic exchange within meta-populations, indeed generally eliminate it.

2.4.5. *Some conclusions in light of current reality*

We are 'five minutes before midnight' for the conservation of biodiversity in Asia's rainforests, while the warnings of ancient times lie quiescent: only three decades back, President Soeharto heeded the goddess of Gunung Meru's prophetic warning that a blight threatening the Javanese rice harvest would herald the end of a regime; but is she silent on deforestation now? Today, unlogged lowland MDF are almost gone throughout the region, except in those pathetically few cases that have been set aside for conservation or research. Examples of MUS-regenerated MDF are even rarer. The less diverse hill and upper dipterocarp forests have greater hope of long-term survival, because they persist on land too steep for prudent agriculture: periodic clearing for planting leads to massive soil erosion and river siltation. In Peninsular Malaysia, Sri Lanka and some other countries, legislation forbids commodity plantation and other cultivation on steep slopes, and logging on mountains above 1,000 m: rules which have been ignored in parts of Borneo. These upper dipterocarp forests also therefore serve as protection forests, while upper montane forests on ridges penetrating the cloud base actually increase water supply, by physically raking water droplets from the cloud.

To have left representative areas the size of Bukit Timah Reserve (Singapore) in each reserved forest would have had no impact on the national economy, and only a little on the companies and leading individuals benefitting from logging. But it would have kept options open for alternative land uses based on indigenous species, and it would have provided lasting educational and heritage benefits. This is still possible

in some regions. Success will depend on securing local support and commitment, active protection against overhunting, conservation of fauna crucial for pollination and seed dispersal, and predators to control herbivores; and, essentially, on international subsidy for biodiversity conservation.

These are predictions, urgently in need of rigorous testing. Opportunities do not at present exist to test the impact of successive logging cycles on biodiversity, since this would require samples in which the impact of all previous logging had been recorded. *Current research must therefore establish and document permanent field samples capable of providing such information in the future. In the meantime, comparative data from samples documented before and after first felling must be used to develop predictive models, which can serve as initial tests of the hypothesis that successive logging cycles impair biodiversity.*

2.5. *How long might it take to regenerate a rainforest to its original condition?*

Nobody knows the answer to this question with accuracy. Of course, the less severe the modification, the less time is required to regenerate.

The reforestation of eastern North America following the abandonment of large-scale agriculture a century ago, and the more recent natural reforestation of parts of Western Europe, may give some the impression that tropical forests will recover with equal rapidity. They will not, for many reasons, but especially because so few rainforest species possess dormant seeds, median seed dispersal distances of most are so short, and fewer species, once past their sapling stage, can recover by coppicing. Furthermore, surface soil organic matter, which is the main external source of readily available nutrients, is easily lost, especially on acid humult soils which, following deforestation and cultivation, effectively become permanently impoverished of nutrients (Chapters 3, 6, 8).

Asian tropical deciduous forests and the Himalayan lower montane forests, with prevalence of coppicing species and lower species richness than the rainforests, have survived the impacts of cattle grazing and periodic fire for centuries. They appear amenable to restoration within relatively short periods.[74] In these respects they may be comparable to their temperate analogues. But here, too, active intercession requires detailed knowledge of reproductive characteristics.[75]

The Phnom Kulen Plateau, north of the great medieval city of Angkor Thom, was deforested more than 1,000 years ago in order to dam the tributaries of its river for annual release during the dry season, thereby ensuring continuous water to the city and its great fish tanks (*baray*). Angkor was abandoned six centuries ago. Now, seasonal evergreen dipterocarp forest, to all appearances primary but for the lychee trees (*Litchi chinensis*) persisting in it, clothes the plateau; only the scattered brick *prasat* (towers) betray its past. Soedarsono Riswan and colleagues,[76] modelling from growth data gathered in East Kalimantan MDF, predicted that it would take up to 70 years for pioneer and successional species to fully re-establish from bare ground and up to 150 years for climax species to reach full height; but as much as 250 years for the forest to regain its original canopy structure and dynamics. These authors did not consider constraints imposed by limited seed dispersal which, at an *average* rate of spread of perhaps 1 m y⁻¹, would add further centuries to the total. A rainforest carefully and selectively logged, with minimal disturbance to soil and to established regeneration, may achieve sufficient timber stocking to be logged successively and sustainably every 50 years. However, it may not return to anything close to its original composition and species abundances until the canopy gaps have regained their original size distributions, and until reproductively mature individuals of the culled species have returned to their original numbers and population structure. This may take at least two centuries. Repeated logging at shorter intervals is predicted to lead eventually to extinction of shade-tolerant species, starting with those with low densities of reproductive individuals in nature. If the forest has been degraded, reproductive individuals of canopy species locally eliminated, or the established regeneration reduced to islands, full recovery as here defined could take at least a millennium. Recovery of the forest's capacity for sequestering carbon and producing quality hardwood, which in upland rainforest is concentrated overwhelmingly in the emergent trees, would then require enrichment by planting.

Current knowledge does provide some guide to the length of time to restore a rainforest, other things – such as vertebrate extinctions – being equal. In more seasonal climates, where composition and structure is most simple, recovery will take a shorter time than in perhumid regions. Recovery on udult soils, more resistant to leaching and especially where subject to surface erosion in nature, will be most rapid, whereas recovery on freely draining humult soils may be permanently deflected through soil impoverishment following forest removal. The floristic composition and orders of abundance of species in small deforested patches, as those occasioned by traditional swidden, may fully recover on udult soils within a few centuries, even in perhumid regions; but areas more than a few hundreds of metres from the edge of surviving forest may take millennia. Even their physical structure will take many centuries, dependent on the occasional chance distant dispersal of the seed of an emergent dipterocarp or tualang (tapang, *Koompassia*).

Few studies of the reinvasion of climax forest tree species into abandoned fields have been made in the wet tropics. I have stood in what appeared to be primary MDF on clay loam in Brunei, with a Murut swiddener who informed me that I was standing in his grandfather's small *ladang* of half a century earlier, cut as an island no bigger than a large windthrow within the forest (Plate 8.6). The forest had recovered much of its structure, though emergents were only just erupting above the canopy, but its species composition must still have been undergoing rapid change. On the other hand, extensive hill slopes on leached sandy clay soils in Sarawak's Rejang valley between Belaga and the Pelagus rapids, when I saw them in the 1970s, still bore forest devoid of emergents, or indeed any dipterocarps. They were dominated instead by *Syzygium* and other successional trees with well-dispersed seeds. This had been the land of Kayan Dayaks, forced out in wars with the Iban around the turn of the 18th and 19th Centuries. Here, deforestation had occurred over an area sufficiently large to preclude rapid reinvasion of those many MDF tree species with limited seed dispersal (Chapter 5). Soil had likely been depleted by previous repeated swiddening, perhaps beyond the tolerance of some species. Savitri Gunatilleke, as a student, found the same absence of reinvasion in a Sri Lankan swidden.[77] Soil fauna, mycorrhizae and other microorganisms may be slow to re-establish (Plate 9.3).

Plate 9.3 MDF forest at edge of an oil palm plantation. The forest is not degraded in this windless climate, but neither does it reinvade where soils have been degraded and leached. (T. Laman)

Such depleted lands now extend for thousands of square kilometres in the Saribas region of Sarawak, south-east Indonesian Borneo and elsewhere. Abandonment of oil palm or rubber plantations would produce similar conditions. The return of hyper-diverse rainforest to such areas cannot be expected to occur for thousands of years, if ever. Once lost, the aboriginal primary forest diversity cannot be reconstructed. Mixed indigenous species plantations may achieve similar or indeed enhanced timber productivity and carbon storage to aboriginal rainforests, and can accommodate non-timber producers including medicinal, rattan and sugar palm,[78] but they cannot accumulate its biodiversity. Even in mixed deciduous forests, cultivation including *taungya* leads to declines in tree species diversity.

On the low granite hills of Singapore, forested lands around reservoirs were converted to intense vegetable and cereal cultivation during the Second World War. Abandoned thereafter, the depleted soil now supports a deflected successional vegetation dominated by pioneer species with multi-layer crowns more often seen in regenerating *kerangas*. The community consisted initially of *Trema*, which typically invades dry depleted sites after burning, then *Adinandra dumosa*.[79] The soil has accumulated some acid raw humus and shows evidence of accelerated leaching. Species richness continues to increase, but irrespective of the proximity of seed sources, climax MDF species have hardly expanded beyond the tiny fragments that survived the destruction in Bukit Timah Reserve and the MacRitchie Reservoir.[80] Succession has instead favoured species whose litter accelerates leaching. Forest succession is thereby being deflected, and the trend may be irreversible. Similar soil impoverishment also seems to be occurring widely on sandstone substrates in Borneo (Chapter 3). It may be that MDF could only return, and then only very slowly, if a good seed source were adjacent and if the soils were rejuvenated by surface erosion.

Such marginal lands include the deep peats in swamp centres; and the lowland podsols and ultramafics, and humult yellow sands overlying late Cenozoic sediments on the hills uplifted from coastal basins in north-west Borneo, which are intensely erodable following logging (Chapter 1). These soils are infertile and vulnerable to drought; tree growth rates are slow (Chapter 2). Sustainable felling cycles of less than 70 years can be dismissed, so timber production will probably never prove economic. Once the forest is fragmented, fire becomes a constant hazard. These lands are uneconomic for permanent agriculture, but harbour unique flora and fauna (though few unique vertebrates). Nevertheless, they incur high protection costs. Throughout the region, their derelict remains lie testimony to one of the greatest challenges to sustainable land-use in the wet tropics of Asia.

Repeated burning of secondary growth often leads to invasion of the deep-rooted perennial grass *Imperata cylindrica*, the mono-dominant grass of the vast *cogoñales* of the Philippines and of 'green desert' of the Sungei people's lands in south-east Kalimantan. Under such conditions, trees may only persist along river banks. Natural reinvasion may occur from these edges, but only if fire is excluded.[81] That requires strong public and political will, and strong infrastructure to enforce it.[82] Even then, little of the original biodiversity survives in these narrow degraded gallery woods. Ironically, the biologically uninteresting *Imperata* grasslands have at first been avoided by companies seeking to plant oil palm. This is simply because without timber, there is no profit to cover the cost of plantation.

The question, how long would it take for a rainforest to reassemble, is therefore one of immense complexity, which needs first to be considered in the next section.

3. Meeting the hidden ecological costs of sustainable timber production

The theory of island biogeography predicts that reduction of an area of continuous and uniform habitat by 90% will lead to extinction from that area of half of its species, even in the absence of hunting[83]. The theory has been tested and confirmed for mammals in North American and African conservation areas. One consequence of forest conversion is fragmentation of the remaining forest.[84] For animals with large territories such as top predators (tigers and eagles) and large herbivores (serow and hornbills) on the one hand, or short life cycles (frogs and squirrels) on the other, the increase in extinction rate is almost immediate and the predicted new equilibrium may be reached within decades. Generalists who can survive in logged, even degraded forest are less affected.[85] Thus, extinction rates for vertebrates can reliably be forecast for the remaining forest fragments in Asia under 'business-as-usual' scenarios (though it may be possible to forestall them by active management for selected key species).

But extinction rates in forest fragments may be very slow for long-lived trees with limited pollen and seed dispersal (Chapter 5), and new population equilibria may even take thousands of years: provided that habitat requirements including light for juveniles continue, dispersers and pollinators survive in sufficient numbers, and there is no hunting. Although usually poor in species relative to primary forest, well-conserved sacred groves have therefore provided effective sanctuaries for many trees, including some rare and endemic species.[86]

With strictly conserved areas set aside, management of the rest of the forest can be directed toward the maintenance of any or all the goods and services produced by forests. Nowadays, active management is even required to sustain unlogged forests, for these have been influenced without exception by human activity, especially by reductions in area and the extinction of top predators. Sustainable management, nevertheless, is still usually discussed solely in the context of timber production.

3.1. *Multiple use forests, or multiple use forested landscapes?*

Is it better to concentrate on developing methods of forest management that promote multiple uses throughout a given area, or to designate distinct areas for the maximisation of specific products or services? And if, as we have argued, production forest management should concentrate on optimising the sustainable yield of timber, a goal incompatible with biodiversity conservation, what role would set-asides within production forests play in biodiversity conservation, compared with whole forests designated for that purpose? This is the forest land's version of the 'land sparing versus land sharing' controversy more familiar in the context of agricultural and human-dominated landscapes.[87] As in those contexts, the question comes down in part to how the necessary planning to meet various societal needs takes place, and at what landscape scale. Answers must take into account local societal conditions, including governance, land tenure regimes, and history of land use.

Full restoration of biodiversity in tropical rainforest is unachievable in a human lifetime, and may take centuries or longer if there has been local extinction (section 2.5 above). However, plantations of species of economic including commercial value, incorporating many tree species and a variety of life forms, are feasible, though biodiversity cannot thus be restored. This too requires skill, especially knowledge of species' ecological requirements. Pre-establishment of a suitable nurse crop can include exotics such as *Pinus caribaea* or *Acacia* spp., even alongside surviving forest, if opportunities for enrichment of degraded forest do not exist.[88]

3.1.1. *Facts to bear in mind*

3.1.1.1. *Vertebrates: some can thrive.* Provided hunting is controlled, many terrestrial vertebrates can survive in logged and secondary forest, though at first in reduced numbers.[89] Some browsers such as pigs, deer and elephants, along with certain predators, may increase following logging and may need judicious control.[90] These animals include important tree seed dispersers. Arboreal vertebrates, in contrast, may be negatively affected when the canopy is opened by logging; by reduction of their food sources and fragmentation of arboreal pathways.[91] There have been suggestions that carefully managed logging, in particular selective logging with high minimum harvesting diameters, may have a significant positive impact on local (or α-) diversity,[92] while removal of strangler figs eliminates food for famine years, and removal of dead trees takes away the nesting sites of many. This notion is supported by the undoubted increase of some wildlife (where hunting has not also increased), consequent on the higher productivity and nutritional value of the many pioneer and successional plant species which increase following logging; also the possibility that some fruit trees, *Artocarpus* and *Ficus*, on fertile soils especially, *Syzygium* on less fertile, may replace dipterocarps in secondary forest canopies. However, it rarely if ever has a positive effect on diversity at the regional or higher level. Where humans are living and active in the forest, concentrations of wildlife

may lead to increases in human-wildlife conflict. In extreme cases this may drastically tilt the popular balance of opinion against conservation.[93]

In summary, forest stands that are managed to optimise timber production do serve a secondary function in wildlife conservation, but are insufficient as they do not conserve biodiversity overall.

3.1.1.2. *Does tree species diversity increase following logging?* The claim is sometimes made that logging can also lead to *increased* tree species diversity. In reality, species richness of a tree community cannot be increased by modifying its canopy disturbance regime. Apparent post-logging jumps in species numbers are based on a sampling artefact: because recent canopy gaps occupy less than 15% by area in most primary rainforests,[94] their pioneer species are distributed patchily and will often evade pre-logging sampling. After logging, canopy gaps occupy more area than the surviving mature phase (old growth), so pioneer species are more likely to be sampled, resulting in an increase in species richness at the scale of the sample. But the species numbers of the forest as a whole will not have changed. Increases in species richness must depend on dispersal rates, and competitive success, of immigrant species from adjacent forest: these processes would take many centuries to reach the centre of even a moderate-sized forest compartment. Speciation processes, though they may also contribute to increases in diversity, are a good deal slower, taking place among trees on a geological time scale.

The only true additions to forest plant communities in historic times have been due to invasions of exotic weedy species (Chapter 8). They outshade and exclude tree regeneration, which may lead to local or regional tree species extinctions and eventual species impoverishment. Few exotics have succeeded, though, in invading unmolested hyper-diverse rainforest. The shrub *Clidemia hirta* is the one exception in Asia. With high fecundity and small bird-dispersed berries, it is rapidly dispersed. *Establishing in natural forest gaps and persisting long after canopy closure, its effect on the species diversity of regeneration remains unknown.*

3.1.1.3. *Soils and tree species composition and diversity* (a recapitulation from Chapter 7). Different surface geology yields soils with different nutrient regimes, permitting different rooting depths, and undergo differing degrees of water deficit during drought. In complex interactions with a panoply of biotic and historical factors, such abiotic factors bear more or less distinct tree floras, and therefore biodiversity at large (Chapters 2, 3, 7).

Soils in floodplains, and those dominated by short-lattice clays with minute pore spaces, are recurrently or permanently anoxic at shallow depths. This constrains a tree's rooting depth, so that frequent windthrow is ubiquitous in floodplains and shallow peat swamps (Chapter 2). Steep slopes experience more frequent landslips, and therefore often include large areas of successional stands. Different geomorphologies provide varying levels of resistance to windthrows, and are again differentially prone to landslips. These conditions affect the relative proportions of individual versus multiple tree deaths, and create a range of frequencies and intensities of canopy opening in natural forests. This in turn accounts for some of the variation in the proportions of pioneer, successional and climax species in their canopies. Although canopy openings occur episodically and are unpredictable within the scale of our lifetime or that of a tree, they differ consistently in both frequency and intensity among habitats over the millennia required to reach stable species numbers. Forests on different land forms and soils therefore differ substantially, not only in the inherent species richness and composition, but in the time required for full restoration of their species and primary structure following human modification.

3.1.1.4. *At what disturbance level does tree species diversity peak?* In Sarawak,[95] we found that species diversity is greatest at intermediate levels of canopy disturbance (Chapter 7). There, emergents are scattered and often clumped, with space for a well-developed and species-rich main canopy of varying height in between (Lambir, in Fig. 7.3ab,ae). Where landscape-scale catastrophe occasionally occurred, severe drought in the case observed by us (Bako in Table 2.6), emergents suffered most. Nevertheless, at Bako the drought in combination with low soil fertility, apparently affected tree genera and species overall and therefore reduced biodiversity more generally. In *kerangas* forests, which are confined to drought-prone podsols and which lack a distinct emergent canopy, the predominant pattern is single-tree mortality, and canopy gaps are small and few (Chapter 2). These forests are relatively low in tree species richness, especially among pioneers, though their tree floras are distinct and merit conservation on their own account (Chapters 2, 3, 7). Unexpectedly, we also found the extreme contrary case: in sheltered, well-watered sites on fertile soils (such as Mersing in Table 2.6), disturbance to the emergent canopy is rare, though gaps when they occur may be large, so that extensive stands of mature emergent individuals form a continuous closed canopy. Their crowns are dense, and impart deep shade beneath. The main canopy is patchy or absent, the subcanopy sparse, and the tree species diversity remains low, even among the emergents and pioneers (see Chapter 7). Each of these disturbance regimes and physical habitats support their own distinctive flora, worthy of conservation. Data from Sri Lanka are consistent with these observations,[96] albeit different in detail (see Chapter 7). Over the long term,

each of these forest types supports different proportions, by area, of stands at the various successional stages.

3.1.1.5. *A hypothesis on the impact of logging on tree species diversity.* We hypothesised that the observed variation in species richness among forests subject to differing long-term disturbance regimes is ultimately determined in large part by the relative area occupied, respectively, by the emergent and main canopy strata, and the heterogeneity of light climate beneath the canopy (Chapter 7). Because the species mixtures specialised to each stand development phase — gap, successional, and mature — cover different relative areas in different habitats, the balance between rates of immigration and extinction varies among them.

Tree species richness, not surprisingly, is also overwhelmingly greatest among the climax species, tolerant of shade as juveniles, in primary forests because late successional and mature stands occupy most of their area; and among forests that experience intermediate levels of canopy disturbance in nature (Chapter 7). Richness is relatively low among pioneers, which comprise less than one tenth of the species in old growth stands. But there appears to be no consistent difference in levels of endemism between climax and pioneer species assemblages, especially among species-rich genera (Chapter 6).

The skilled silviculturist, aiming to achieve maximum levels of sustainable timber production, must take preferred species' different light requirements into account in planning felling regimes and subsequent regeneration conditions therefore must inevitably do so at the cost of local and wider extinctions among species of those successional phases he reduces.

Logging produces different *relative* areas, and frequencies of recurrence, of the gap, successional and mature phases of the forest. Initially, the proportion of gap phase is increased, later that of the successional phase. However, the area of mature phase canopy, the phase richest in species under natural conditions, is reduced, as is the area of the subcanopy light regime where its species' juveniles have their greatest competitive advantage. The stands become increasingly dominated by late successional species at the time of each subsequent felling cycle. The island biogeography of the forests' dynamic phases is reversed, with the expected effects: the gap and successional phases become the matrix, while the species-rich mature phase, which is the most susceptible to perturbation, is reduced to an archipelago of fragments which become the targets of successive felling cycles. Local extinctions inevitably increase, widening in area following successive logging cycles.

Slow-growing climax species rarely reproduce quickly enough to sustain their population numbers through repeated iterations of a felling regime, and are often cut before reaching full reproductive size. Provided the soil and established regeneration is strictly protected during timber extraction, the first logging event may critically reduce populations of only a few tree species, namely slow growing climax species that occur naturally in low population numbers in the mature phase stands. Examples of this group include genera of high utilitarian conservation value, such as some wild *Mangifera* and the heavy hardwood emergent legumes (*merbau* and rosewoods) and *balau* (*Shorea* sect. *Shorea*). Even in selective logging systems, though, when near-mature individuals of climax species are left for subsequent harvest, slower-growing species will be the first to suffer population decline over successive logging cycles. Declining numbers of interbreeding individuals inevitably reduces genetic variability within these small populations, compromising their genetic fitness.[97] Extinctions may then ensue, due either to the failure of establishment conditions to return, or to catastrophic increases in the mortality rates of residual reproductive individuals. In effect, forest dynamics will come to resemble that of forests in seasonal rainfall climates, whose lower species diversity (as I argue in Chapter 7) is due both to failure of early regeneration owing to changes in habitat conditions following large-scale disturbance, and to periodic mortality of reproductives caused by such catastrophes.

Conservation of biodiversity in forest stands managed for timber, let alone plantations, may therefore prove unachievably demanding.

For these reasons, as well as the *relative* paucity of coppicing species in MDF (Chapter 2), the tradition of community forests and sacred groves where medicinal plants and fuelwood are collected works less well in more humid climates also. (Nevertheless, *strictly* conserved sacred groves, even when isolated from forest tracts, have conserved plant diversity to a remarkable extent.[98])

3.1.2. *Testing the value of primary forest set-asides within timber production forest*

Using data from Pasoh, Potts and Vincent used simulations to compare the effectiveness of tree species conservation via logging-free sanctuaries (VJRs) incorporated *within* conventional timber concessions versus RIL.[99] They found that, when the clumping of conspecific tree individuals was taken into account, the sanctuaries model did better. The effect was stronger with heavier logging damage, and it was stronger in selective logging as currently practiced, than in uniform shelterwood management systems.[100] Their analysis confirmed that logging, however carefully executed, suddenly and fundamentally alters the canopy gap regime.

3.1.3. *Ecological impacts of logging and forest fragmentation on species diversity: a summary*

Logging affects not only particular groups of organisms or populations, but also the 'services' — or ecological functions — naturally provided by forests. There are a number of ways in which logging *consistently* impacts biodiversity and other ecological services in those forests where biodiversity is greatest, in perhumid Malesia and Sri Lanka:

1. These species-rich lowland forests are the least well-adapted to changes in canopy damage such as occurs during timber harvesting (Chapters 2, 7). MDF species richness is correlated with at least two significant elements of forest canopy structure: the ratio of emergent to main canopy crowns, and the density of sub-canopy populations. These populations are greatest where there is variable including high diffuse light penetration through the canopy above.

2. Logging fundamentally alters the ratio of canopy to gap, and the subcanopy light climate regime. It unavoidably creates relatively large open gaps, favouring species-poor pioneer assemblages over the rich old-growth guilds. Subsequent harvesting interventions will, it is predicted, accelerate local, and eventually regional, elimination of climax species and therefore forest biodiversity, beginning with the many species whose reproductive individuals naturally persist in low densities.

3. Changes in the size and frequency of canopy gaps lead to gradual but accelerating changes in the relative abundance of tree species with differing light requirements for regeneration (Chapters 3, 7). In particular, the shade-tolerant heavy hardwoods, especially understorey species, in many cases exceptionally rare in nature, become increasingly prone to local and regional extinction. Careless road construction and timber extraction reduce even those species capable of coppice. Epiphytes of emergent crowns suffer early and steep decline.

4. Pioneer species of the humid tropics, which dominate the disturbed soils (especially those of moist clay and spaces opened by logging), are selected to quickly develop dense crowns of a single overlapping layer of leaves (Chapter 2). This creates a dense uniform shade inimical to height growth beneath, especially of juveniles of light-demanding climax species surviving or subsequently seeding.

5. At the same time, the low leaf area index (LAI) of pioneers limits the growth rate of their woody mass. This means that forest degradation through overexploitation, by curtailing succession at the pioneer and early successional stages, cannot enhance primary productivity even of low value pulp woods. It consequently reduces the forest's capacity to store carbon (Chapter 2).

6. Logging initially causes loss of crown connectivity in the canopy and increased openness in the sub-canopy. These changes often lead to reduction and even local extirpation of arboreal vertebrates including some primates, and sub-canopy specialists especially birds (Chapter 5). Trees and other plant species most tolerant of shade experience competitive disadvantage.

7. Tropical climax species, which produce the most valued timbers and other products, generally lack seed dormancy, while canopy species in MDF mostly fruit at several-year intervals (Chapters 2, 5). Removal of seed trees over successive logging cycles therefore serves to enhance the dominance of pioneer species that leave a seed bank in the soil, and their usurpation of more light-demanding climax species. This trend is further emphasised when heavy machinery scarifies the soil surface along roads and skid trails, removing the established regeneration of climax species that is the essential precursor of indigenous moist hardwood stands worldwide.

8. Sustaining populations of tree species depends in many cases on sustaining adequate populations of seed and pollen dispersers, which may themselves require large areas of forest. Asian examples include hornbills, gibbons, and the nectarivorous bat *Eonycteris spelaea* (Chapter 5).

9. The soils of the forests richest in tree species are poorer in nutrients (Chapter 7). What nutrients are stored in the soils are concentrated in the organic matter at or near the surface (Chapters 1, 3). Mechanical damage or exposure to sun accelerates decomposition of this crucial organic nutrient store, also releasing carbon dioxide by oxidation. The rate of soil restoration may be slower than the rate of forest stand restoration (Chapter 1).

10. Control of population densities of sapling browsers such as wild pig and deer may require conservation of their predators. In some places these include animals also dangerous to man, such as the tiger. Where adequate populations of such predators can no longer be supported, the only alternative is strictly managed and licensed hunting. These can also be a significant source of income.

11. Forested landscapes in the perhumid tropics support uneven densities of vertebrates; this is true especially of the birds and larger mammals. Riverine and bottomland forests can support relatively high vertebrate species diversity and density. Large browsing mammals disturb the forest locally, increasing the growth of those successional plants lower in toxic chemical defences and often richer in nutrients (Chapters 2, 5). Vertebrate species find their most reliable food sources, such as dense populations of figs, successional stands and herbaceous vegetation, in these fertile habitats. Only in mast fruiting years do vertebrates forage intensively in upland forests.[101] Upland and peat swamp forests, notably in nutrient-poor regions of Borneo, therefore depend for dispersers and predators on the survival of adjacent valley-bottom and riverine forests; yet these fertile valleys and plains are the very habitats most sought out by farmers and commodity plantation investors.

12. Riverine forests often act as corridors between more extensive areas. Although apparently less important for the conservation of sessile organisms dependent on trees, their survival and accessibility for wide ranging and migratory animals is crucial as it radically restricts the dispersal of many species. The loss of such corridors can prove disastrous for wildlife and mankind alike.[102]

It must be concluded that comprehensive conservation of tropical tree species diversity and the diversity of tree-dependent organisms can thus *only* be achieved in the presence of sizeable tracts of inviolate primary forest. Projects such as the Heart of Borneo[103], which focus on logged forests, will not permanently conserve biodiversity; nor, by focusing on forests with high vertebrate diversity, would those many the forests richest in unique *overall* biodiversity that are confined to infertile habitats be included. Exactly how slippery concepts such as 'comprehensive', 'sizeable' and 'inviolate' should be defined in different contexts has been the subject of debate. But it is clear that even a single episode of felling within national parks and other conservation areas will inevitably have an impact on biodiversity that is unpredictable and may last centuries.[104] Areas experiencing such interventions without detailed documentation also lose much of their usefulness for research or for serious educational ventures.

It is, of course, too late for complete conservation of Asian tropical forests, but set-asides — 'virgin jungle reserves' (VJRs) — established by foresters can serve as vital refugia for many species within production forests.[105] These refugia include the VJRs within production forests in Peninsular Malaysia, and the research plot network in India. Even now, not many tree species have been extirpated from surviving production forest reserves, though the genetic diversity of rarer species will inevitably have been reduced. It is not too late to search out suitable set-asides, and to legislate and manage their conservation. This is not to argue against the strict conservation, including exclusion of logging and regulation of hunting and gathering, of those larger areas of primary forest already legislated for as national parks and nature reserves. These serve the larger objective of conservation at whole landscape scale and beyond, unachievable in isolated set-asides.

In the medium term, the inevitable extinction of larger animals from small and isolated forest patches will affect the demography of many tree species through loss of dispersers. This can be ameliorated if tree and strangler figs are left, even encouraged, in the surrounding landscape.[106] The loss of specialist dispersers may at first lead to their replacement by generalists such as the long-tailed macaques (*Macaca fascicularis*), omnipresent in the Far East. Uncontrolled hunting, though, is proving disastrous in small isolated parks such as Lambir in Sarawak (Chapter 8).[107]

3.1.4. *Overall conclusions*

1. Forest policy and management must be first executed at the scale of whole landscapes, because optimisation for more than one objective, notably timber production and biodiversity conservation, is unachievable within the same stand or forest.

2. Conservation areas such as national parks and nature reserves must *never* be logged.

3. Inviolate reserves are essential to overall biodiversity conservation, even when they are too small to support many vertebrates.

4. Setting aside representative tracts of primary forest within timber production forests serve two essential purposes: as refuges into which motile organisms can retreat, and survive, and from which they can reinvade logged stands as they regenerate; and as control samples for evaluation of management methods. But they can never serve the landscape-scale conservation function of national parks and nature reserves.

5. Conservation of species with broad ecological tolerances, which include many vertebrates, can best be achieved by sustaining habitat connectivity in a landscape divided into fragments with differing uses and management; but it must be borne in mind that generalists represent a mere fraction of biodiversity.

6. The necessity for managing hunting becomes ever more urgent as forests become more accessible.

3.2. *How can priority areas for conservation of biodiversity be identified?*

The method is straightforward, though not necessarily easily realised:

1. Identify areas which represent (i) geographical regions with distinct climate or climatic history (if suspected); (ii) every kind of surface geology, especially unusual rocks such as limestone and ultramafics, and landforms and soils such as podsols and peat lands.

2. Search for residual unlogged (or lightly logged if absent) forest fragments in each, by satellite imagery and local enquiry.

3. Explore candidate areas, to assess their flora. Where they occur, dipterocarps and Fagaceae, being rich in species and specific to habitats, and dipterocarps, being easily identifiable from fallen leaves alone, are excellent indicators of specific floristic associations.

4. Prioritise candidate areas according to the level of endangerment of their habitats and associations, their species (as far as known) and their endemism.

5. Negotiate conservation of a comprehensive but parsimonious series of reserves.

6. Seek to maximise the presence of forested corridors, whether logged, secondary or otherwise, upland or riparian, between reserves so designated.

7. Collaborate in gaining community and political interest and support, and legal protection.

3.3. *What of timber plantations?*

Under favourable conditions, timber plantations of a single or few species are capable of rapid and uniform production. To that extent, they can be a useful means to intensify wood production on limited land area. Plantations of indigenous climax species such as dipterocarps, if well tended initially, may sustain average growth rates two to three times those in primary forest where the majority of juveniles are in competitive decline yet occupying growing space[108] (Chapter 2). In the wet tropics, the only lands still available to timber and pulp plantations are steep and erodible, or those left derelict by mining or war. Like any other commodity crop, trees can only be grown there if the soil is protected by careful retention of surface layers, litter and woody regeneration during harvesting and replanting. Enrichment is therefore recommended over complete replanting.

On the downside, plantations often represent the final and complete transfer of the forest away from local inhabitants who may still depend on its resources.[109] When plantations achieve high timber productivity, it is generally at the cost of eliminating any other products that benefit rural populations, although some game animals may prosper. Like commodity plantations, timber plantations in the humid tropics are biodiversity deserts.[110] Use of indigenous species does not appreciably increase their biodiversity conservation value either, unless they yield important food sources for vertebrates; but that may only be possible at the cost of wood yield.

Timber plantations require planting and active management, incurring greater labour costs than natural forests. Like commodity plantations, they also require periodic clear-cutting and replanting. Short cycles are commercially preferred, which may result in frequent phases of high soil erosion and river

Plate 9.4 Reintroduction of useful indigenous MDF trees, palms and herbs into restorative plantation, using the diffuse-crowned *Pinus caribaea* as nurse. In the buffer zone of the Sinharaja World Heritage forest. Gunadasa as scale (M.A.)

siltation. Nevertheless, sal and teak forests in India are now widely managed by enrichment planting in which a mixture of timber species has often been used. Even in monospecific plantations, deciduous tree species return more readily than in evergreen forest, both from stump sprouts and, if nearby sources are available, from seed. In Sri Lanka, mixed indigenous climax species and NTFP have been successfully re-established through enrichment planting using an existing *Pinus caribaea* plantation as nurse (Plate 9.4). North-south oriented lines of pine were felled to increase understorey light. Re-invasion of other indigenous species from adjacent forest was also noted under the conditions created, and could be prolific depending upon distance from the rainforest seed source; but almost all species were late successional site generalists or pioneers.[111]

Recently, plantations of indigenous species have been advocated in Thailand. The Royal Forest Department receives subsidies for planting, which makes it preferable to management by natural regeneration. In the peninsula and at Sakhaerat Environmental Research Station, plantations of the indigenous dipterocarps *Dipterocarpus alatus* and *Hopea odorata* have proven successful if weeded at first, with a mean diameter increment of 1 cm y^{-1} and height increment of 1 m y^{-1}. *D. alatus* at Kampaeng Phet reached 25 m height in 45 years.[112] In the perhumid tropics, plantations of light hardwood dipterocarps at Forest Research Institute Malaysia (FRIM) have increased in estimated wood volume by 6–8 m^3 ha^{-1}, far exceeding the 1–2 m^3 ha^{-1} estimated in logged forest. A possible solution to the dismal state of regeneration in Peninsular Malaysian hill forests, proposed by FRIM silvicultural ecologists, is to enrich with appropriate dipterocarp species in cleared patches, thereby forming artificial stands with the growth attributes of plantations.[113]

Experimental results in east Sabah and elsewhere, where diameter growth rates of more than 2 cm y^{-1} have been reported, have engendered optimism concerning the potential of indigenous plantations of fast-growing, long-fibre species in the humid tropics. This is unexpected because stands of these rainforest pioneers, such as *Neolamarckia cadamba* and *Octomeles sumatrana*, rarely have an LAI exceeding 4, so they compare unfavourably in photosynthetic area with exotic fast-growing species like Australasian *Acacia mangium* and eucalypts. These exotics are evergreens adapted to the relatively dry and windy climates from which they originate, and they produce timber of higher specific gravity than many native species.[114] Their rates of biomass production and carbon assimilation are more consistent and often higher than native pioneer plantations.[115] Many of these indigenous pioneer species are confined in nature to relatively fertile soils. It is unrealistic to expect comparable growth rates on the generally infertile soils available to wood plantations in the wet tropics (Chapter 2). They will likely prove particularly prone to insect attack and disease.

Throughout Malesia, notably in the Philippines and south-east Borneo, we see thousands of square kilometres of coarse *Imperata* grasslands. Prior to the boom in oil palm, investment in their conversion to agriculture was rare. That is unsurprising, as the initial cost has been estimated as at least $840 ha^{-1}. Even short-rotation pulpwood trees could only yield a real rate of return of 3.3% at a pulpwood price of $20 m^{-3} and mean annual increment of 25 m^3 ha^{-1}, although the *economic* rate of return might exceed 20%.[116]

The carbon stored in plantations of fast-growing species in short rotation cannot be compared with that in naturally regenerating hardwood forest. Besides, the wood thereby produced is mostly used for pulp, which is in turn used for ephemeral products. Such plantations can be highly mechanised, employing few to achieve their single product. In sum, they are socio-economically the very opposite of biodiverse indigenous forest, which in the past brought so many benefits to the rural poor.

The increased incidence of insect- and pathogen-induced mortality in teak and sal which resulted from silviculture aimed at creating pure stands from mixed species forests (Chapters 5, 8) should not be forgotten, nor the current devastating impact of accidentally introduced exotic pests on many tree species in temperate North American and East Asian deciduous forests.[117] We therefore strongly advocate mixed species plantations. These permit planting in a manner that exploits differences in branch architecture and leaf arrangement, which enhances overall LAI. Most importantly, because rainforest tree species appear to be maintained at low population densities in nature by pathogen contagion probabilities (Chapter 5), plantations of *single* indigenous species entail a high and recurrent risk of epidemic wipe-out. While it is impossible to put enough distance between conspecifics to completely eliminate the risk of pathogen or folivore infection, mixed species plantations would spread and reduce this risk by a factor related to the number of species utilised.

3.4. *Can ecotourism help protect biodiversity?*

Public interest worldwide has been attracted to the wonder of tropical forests and wildlife partly by programmes of extraordinary beauty and authority offered by several television networks, particularly the BBC. Unfortunately, these also encourage the unrealistic expectation that charismatic animals can easily be seen at close quarters in rainforests, and that they can be viewed without the discomfort of sticky climates and biting insects.[118] The mass market is mostly attracted to parks where large animals can reliably be seen; commercial success is often determined by the attractions of the location rather than by its conservation value.[119] Habitat degradation in such parks is a continuing threat, which must be ameliorated by careful management in combination with community involvement.[120] Mass tourism also results in social problems and dislocation, often including the loss of hunting, fishing or grazing rights. These effects can be disruptive to local communities.[121] To fight this trend, community-based tourism programmes, such as opportunities for home-stays and participation in the life of remote villages, have multiplied in recent years (Chapter 8) (Plate 9.5). Many such programmes are run by locally based NGOs, often with external funding, but aiming to retain

Plate 9.5 Nature tourism beckons. **a,** Discriminating high-end tourists love nothing more than to immerse themselves in the realities of forest life. Approaching camp on a tributary of the Tinjar, Mulu National Park, Sarawak; **b,** A guesthouse in the former forest domain of the Maharaja of Mayurbhanj, now Similipal National Park. Since this photograph was taken by Hans Hazebroek, in 2008, this triumph of vernacular art, and other tourist buildings, have been torched by Naxalites, in league with timber thieves. (**a** Mary Ashton; **b** H.H.)

proceeds within the local economy. It is probably too early to judge their impact in general, but we can be sure that there will be a variety of unexpected outcomes, both social and economic, from the juxtaposition of cultural attitudes and assumptions inherent in all such programmes. A study from Costa Rica, where experience of the many varieties of nature-based tourism is perhaps strongest, showed very mixed results in terms of community participation and benefits, and called for renewed commitment to environmental education not only of local communities but especially of the tourist pool.[122] There is plenty of rhetoric aimed at 'empowering' communities to benefit from ecotourism, but much depends on people's ability to manage a rather complex business with national and international links and, potentially, substantial funding flows.[123] Legal frameworks and recognition of tenure rights are fundamental, yet are still weak in many rural areas.

High-end ecotourism, becoming familiar in those parts of Africa associated with the big game trade, has not penetrated far into tropical Asia. The international acumen (not to speak of the capital) required to design, implement and maintain such high-end facilities can only be found among private companies with an international operations-base; as a result, profits may flow out of the local area as fast as they flow in. Studies have not shown this model to be beneficial to local communities or to the environment. 'Tourism is a fickle industry, one that is especially vulnerable to changes in the global economy and the impact of distant political events.'[124] When viewed from the perspective of their broader social and economic impacts, tourism and ecotourism have some commonality: *all* tourism tends to promote and deepen economic 'liberalisation'.[125] Perhaps the most widely-read book of case studies on the ecotourism quandary is now in a second edition.[126] Though it still contains no case study from Asia, it shows that the global picture is exceedingly complex and dynamic, depending much on the good will (spontaneous or compulsory) of involved individuals. An analysis of the peculiarities of the ecotourism industry in Asian contexts diagnoses fundamental differences between Asian and Western attitudes toward biodiversity conservation and protected areas. These are said to strongly influence the development of 'ecotourism', recognising that the very term is used as a form of 'communicative staging' to deflect attention from environmental damage for commercial gain.[127] It is uncertain whether that is a specifically 'Asian' characteristic, or just another opportunity for profitable green-washing in an ecologically and environmentally illiterate market-place.

Ecotourism enterprises can probably only strengthen conservation long term where there is strong technical and regulatory oversight by public-minded government agencies or by resourceful, ethically committed and politically adroit NGOs; for in practice, commercial interests are too often at variance with conservation principles. In short, ecotourism is least likely to lead to negative impacts, on nature and on communities, in societies that have already achieved an advanced level of education and economic sophistication.

4. Climate change and tropical forests

> The primary influence of anthropogenic climate change on ecosystems is expected to be through changes in the rate and magnitude of climate means and extremes....rapid relative to the speed at which ecosystems can adapt and re-establish themselves – and through the direct effects of increased CO_2 concentrations.
>
> Intergovernmental Panel on Climate Change (IPCC), 1997.

4.1. *Impacts of changes in weather patterns*

Climate change is acting both globally and locally, by differentially altering weather patterns. Increases in greenhouse gases (particularly carbon dioxide and methane) in the atmosphere are responsible at least in substantial part for increasing mean global temperatures.[128] Regional and local means vary about the global mean. In the tropics, the greatest increases in temperature are expected in continental, temperate and seasonal regions.[129] The humid tropics are expected to experience less temperature increase, perhaps less than 2°C, but likely sufficient to alter mean and maximum tree growth rates and consequently competitive relationships.

Initially, increasing mean global temperatures are expected to be associated with increasing year-to-year fluctuations in climate, and with increasing frequency and intensity of catastrophic weather events: drought, hurricane, flood, fire. In the absence of a tropical continental landmass in tropical Asia on the scale of Africa or South America, the steep declines in annual rainfall inland such as those predicted for sub-Saharan Africa and the Amazon basin are also thought unlikely, though northern India is already beginning to experience failure of monsoon rains with increasing frequency.[130] More widespread in the Asian tropics is a likely increase in rainfall seasonality, including penetration of seasonality into currently perhumid regions, and an increase in the frequency and severity of severe cyclones (typhoons) and droughts, some of which are already being experienced[131] (Chapter 1).

In Asia, severe typhoons, droughts, and killing frosts in the mountains, already bear testimony to the coming changes (Chapters 2, 4, 8). In the year 2008 Asia experienced persistent high pressure over China, resulting in unusually low temperatures and low altitude frost events (Chapter 6), combined with the highest dry-season rainfall over the Deccan Plateau since 1875. The years 2011 and 2012 witnessed weak

Indian monsoons and failed crops in drier regions, while southern China suffered catastrophic flooding.

4.1.1. *Change in rainfall seasonality*
In seasonal continental Asia, semi-evergreen dipterocarp and other lowland forests are replaced by less species-rich and less biologically productive tall deciduous forests as rainfall seasonality, and with it fire frequency increases (Chapters 2, 3). Increases in rainfall seasonality can therefore be expected to exert powerful influence on the future distribution of species and ecosystems (Chapters 2, 3, 6). Forests of currently seasonal regions would shift towards drier types, but the impact of any intrusion of seasonal climates into the main islands of Sunda, where the tree flora of seasonal region is mostly absent, is unforeseeable. Increasing seasonality, or perhaps aridity, will therefore shrink the range of the most biodiverse forest ecosystems and the species confined to them. Forest fragmentation has severely constrained the opportunities for species to disperse into new habitats. Currently predicted rates of climate change may well contribute to increases in the incidence of fire, including catastrophic fire, in semi-evergreen and logged seasonal evergreen forests.

Increased frequency and intensity of ENSO-related droughts over the last century have been well documented, especially in perhumid Sundaland (Chapter 1). Their impact has been intensified by frequent human-induced fires, facilitating dominance of the deep-rooted perennial coarse grass *Imperata cylindrica* (*lalang*, *alang-alang*, *cogon*). Through human agency, a green desert can thus be created from a majestic forest. The reductions in transpiration effected by these changes are also expected to reduce atmospheric humidity, and hence rainfall, over the continent during droughts and dry seasons. El Niño droughts and La Niña floods have increased in severity over the last half century, but we cannot as yet be certain whether there is any connection with greenhouse gas-induced climate change (Chapters 1, 8).

So far, ENSO-correlated droughts and floods seem to have had limited influence on primary forests in perhumid Asia. Occasional moderate droughts may even assist in sustaining rare species populations[132] (Chapter 7). But their impact on degraded forests, through human-induced fire, has been devastating and can be expected to increase.[133] Whether any reduction in mean rainfall occurs, and the extent of rainfall seasonality spread into currently perhumid regions, remains uncertain. Whereas increased seasonality, judged by present vegetation patterns, would impose dramatic changes in competitive relationships among species[134] (Chapter 3), lowered mean rainfall alone would hardly affect regions currently experiencing excess, unless mean monthly rainfall consistently declines seasonally below evapotranspiration rates. Measuring and interpreting the influence of climate change

on rainforests is difficult, particularly because rainforest trees seldom manifest growth rings, and because stands of trees in primary forests are never in dynamic equilibrium.[135]

There is evidence that deforestation and urbanisation can alter local rainfall patterns. Total annual rainfall may not change, but heavier storms tend to occur at longer intervals. A higher proportion of the rain therefore falls on already-saturated soil, and is consequently lost into drainage systems.

The recent, apparently novel, mortality of emergent or otherwise exposed canopy trees near the mountain treeline appears to be caused by occasional intense drought. Decreasing atmospheric humidity also leads to increase in the adiabatic lapse rate of temperature decline with altitude. Whether drought or freezing has caused the mortality is unknown, but likely both (Chapter 4). A decline in local atmospheric humidity levels may be associated with lowland deforestation, which would also generate the increase in adiabatic lapse rate. The mossy upper montane and subalpine forests are major mediators of permanent streams in seasonal regions by virtue of precipitation out of wind-blown fog. If drought continues to increase in such areas, a further compelling case – this one at the local scale – exists for the immediate reforestation of derelict lands.

4.1.2. *Changing temperatures*
Increased temperatures lead to increased metabolic rates. Although increased temperature, by changing relative growth rates of co-occurring species, may alter their competitive relationships at stand level, the impact of increased temperature on shortened life-cycles, therefore evolutionary rates, among arthropod herbivores and predators, and pathogenic micro-organisms, could prove more dramatic. This, combined with increased costs of energy, fertilisers and pesticides, means that pathogens can be predicted to have devastating impact on woody monocultures, such as oil palm and rubber, in the equable humid tropics. The case for conservation of alternative germplasm, and for research into multiple species plantation, has never been stronger.

Any decline in the incidence of killing frosts at the north-east margin of the wet tropics is less likely, in reality, to lead to northerly expansion of tropical biota[136] than to an expansion of the land area available for tropical crops, especially rubber in the Far East.

4.2. *Impacts of increasing atmospheric carbon*

4.2.1. *Change in forest growth rates*
The growth of trees is continuous, while their death is sudden and may occasion major loss in stand biomass. Stands are permanently in a state of recovery from periodic natural

catastrophe; recovery follows a predominant successional path distinct to each forest type (Chapter 2). An individual-based, spatially explicit model for simulating forest dynamics, based on growth and mortality in the Pasoh CTFS plot over one 5-year census interval, predicted a near doubling of above-ground volume over two centuries.[137] This was most obviously explained by the absence of a major canopy mortality event during the period on which the model was based. Major surges in mortality usually take place at long intervals in biodiverse equatorial rainforests. Predictions of globally increasing tropical rainforest biomass have consequently been brought into question (see Chapter 8), and were negated after a major drought and accompanying mortality was experienced in Amazonia.[138] Suggestions in both the scientific literature and the media that the Amazon forests might be becoming consistent carbon emitters also appear premature.[139] In any case, the evidence for and against carbon-driven growth rate changes appears inconsistent at this stage. Local and regional trends may be driven by other factors. To date the most persuasive evidence points to a *decline* in mean growth rates with rising temperature, detected in data from the Pasoh 50 ha plot (Peninsular Malaysia) and from the La Selva research forest (Costa Rica).[140] This may be caused by increased respiration rates associated with a 1°C increase in nocturnal temperatures at both forests. Correlations have been found between occasional lower seasonal night temperatures and late fruiting in the Neotropics,[141] but these require further substantiation.

4.2.2. *Deforestation as a source of atmospheric carbon*

Deforestation, which for 40 years has overwhelmingly been concentrated in the tropics, is currently estimated to be releasing carbon into the atmosphere at a rate equivalent to the world's automobile emissions, about 20% of all emissions.[142] Reduction of deforestation rates, and plantation or natural regeneration, are crucial parts of the solution to a problem that otherwise threatens starvation and anarchy when large sections of the world's best arable land, let alone major cities, begin to sink below the sea. In the 140 years between 1850 and 1990, the total net flux of carbon from tropical Asia has been estimated at 38.6 Pg, compared with 30.5 Pg for the Neotropics and <15 Pg carbon for the tropics elsewhere.[143] The average annual flux over Asia is estimated to have been about twice that in the Neotropics during that period. In recent decades the problem has been exacerbated in the Far East by the felling and clearing of the peat swamps, and subsequent drainage schemes aimed at converting ombrogenous peat to commodity production. The carbon locked in the peat, up to 15 m deep in some domes (Chapter 2), may exceed that in the standing biomass above. The burning of peat during the 1995–1996 ENSO-related drought released a quantity of carbon comparable to that released by the world's power plants and industry over

the same period (Chapter 8). During the months of pervasive haze, and in spite of elevated carbon dioxide levels, stomatal conductance and rates of photosynthesis were reduced in species examined.[144]

Stands of tall old trees contain the greatest carbon in their biomass, by virtue of their stature and volume of heartwood, and more particularly their great diameter to height ratio.[145] It is a myth that dense young plantations accumulate greater biomass than old growth stands.[146] The logged evergreen dipterocarp forests will not approach their original biomass, therefore carbon content, until their emergent canopy re-establishes equilibrium, which would take centuries (see section 2.5 above). This is unlikely to be allowed to happen! Besides, regenerating dipterocarps in heavily logged forests will establish mature crowns and full height, as exemplified by the free-standing individuals in the dipterocarp arboretum at Kepong (Selangor), at perhaps no more than two thirds of their former stature, and therefore with commensurately less sequestered carbon. Some primary unlogged temperate forests are carbon sinks by virtue of continuous accumulation of organic carbon in their soils.[147] Upland rainforest soils store less carbon than do temperate (especially cool temperate) conifer forests, while organic carbon accumulation quite rapidly comes into equilibrium with its decomposition at tropical high temperatures and humidity (Chapter 1). Nevertheless, transfer of dissolved carbon through the aquifer into the drainage is possible. This carbon in solution finds its way into rivers where it is re-oxidised, particularly when it mixes with marine salt water, but some, which is assimilated into aquatic organisms, may eventually be deposited on the sea bed. *Much research is still needed into the sequestration pathways followed by atmospheric carbon in forest systems.*

Evergreen forests therefore constitute an important carbon bank. The sequestered carbon of Malaysian forests has been valued at $US330–760 billion.[148] If they do prove to act as sinks, it must be due to solution of carbon as humic acids and their leaching through the soil. In any case, carbon offsets will only provide incentives for rainforest conservation where opportunity costs are low or where other values, including service values, are sufficient to justify conservation.[149]

In stark contrast, the tropical lowland peat swamps of the Far East (Chapter 3) are accumulating carbon in their peat at a more rapid rate than anywhere else, worldwide (Chapters 3, 8). Their continued destruction is a catastrophe. Compensation for their protection on behalf of all humanity must be given, among all forests, the highest priority. Tropical deciduous forests have greater below- than above-ground biomass, in part owing to their deep root systems. Their soils contain higher organic carbon (Chapters 1, 8). Indian forests, which are predominantly deciduous and subject to periodic burning, are estimated to sequester 56% of their carbon in the soils.[150] This may partly represent charcoal sequestered as a consequence of centuries, even millennia, of burning, so that there is a possibility of continued sequestration. Carbon *incremental* value of Indian forests has been estimated at $450 ha^{-1} y^{-1}, but that estimate may have been enhanced by enhanced growth rates following logging prior to the nationwide ban, and other periodic destruction.

One of the paradoxes resulting from the recent availability of cheap hydrocarbon-derived energy has been the almost complete change from timber to concrete residential houses in forest-rich tropical countries, in contrast to Scandinavia and much of North America. This is understandable in view of the overwhelming conversion to row housing, with consequent fire hazards; but current energy prices, which favour cement manufacture, are unlikely to continue into the medium and long term. A return to timber will both increase the value of the residual forest and the incentive to actively manage production forests. It will increase these forests' capacity for carbon sequestration, by yielding timber that will be retained in semi-permanent structures.

At present, compensation to forest owners for conserving carbon in trees does not extend to unlogged primary forest. A United Nations initiative, Reduced Emissions from Deforestation and Degradation (REDD), is intended to correct that. Vast problems, however, lurk in obtaining reliable data on qualifications, and yet more on identifying the many opportunities for cheating that will arise (Chapter 8).[151]

4.3. *Sea level rise*

The peripheral water tables of the coastal peat swamps of Borneo, Sumatra and New Guinea, are finely balanced. The extraordinary inability of the seedlings of *Shorea albida*, the dominant species in the peat swamps of north-west Borneo, to survive in the coastal peripheral swamp phasic community was first observed in the 1950s. In 2001, I observed saplings surviving and growing robustly on the upper banks of a drainage canal at Seria (Brunei), as they also do in communities on peat domes in the swamp interiors, and in some *kerangas* (Chapter 3). I infer that a rising base water table could explain their absence at the swamp periphery.

Current predictions vary in their estimates of the rate of future eustatic rise in sea level due to global climate change. More pessimistic rates, approaching 1 m over the coming century, would pose severe challenges even for mangrove to migrate and re-establish without extinctions.[152]

Whatever the rate, increasing eustatic sea levels will erode the majority, perhaps all, of the coastal peat swamps of the Far East, releasing vast additional quantities of atmospheric carbon as they degrade. Most tree species will survive on those upland plateau peat swamps and *kerangas* that survive

plantation agriculture and fire. But the major impact will be the inundation and increase in salinity of the great rice bowls of the Asian coastal and riverine plains: Bangladesh, lower Burma, Thailand, Vietnam, China and parts of Cambodia and East Java. Many farmers made landless will likely migrate into the residual forests of the hills.

5. We can meet these varying challenges of sustainable use

What we have said can be summarised in the commonplace current example, where forest conversion to commodity crops brings quick returns to several groups:

1. The developer gains the windfall profit from the felled forest timber.

2. The owner — the national population, represented by government — gains early tax revenues.

3. The local population may gain a temporary rise in employment during logging and the option thereafter to join the estate labour force.

On the other hand, such conversion leads to dereliction and poverty where soils are unsuitable, as they are in many parts of Borneo. Given the level of initial investment required, the continuing lack of enthusiasm among developers for re-converting derelict *Imperata* grasslands is understandable.

Yet forests on these same poor soils could have sustainably yielded timber, with other products of value to the community and those ecosystem services that plantations cannot provide. The most favourable way, indeed probably the only way, to optimise the variety of goods and services yielded by forests in the modern economy is to plan and negotiate forest use at the level of the landscape and the whole region. Forest land area should be optimised at the regional level, while yields of individual goods and services should be planned and allocated at the local or individual forest level. By the same reasoning, policy should favour *economic* over *financial* advantage: that is, the sustainable long-term return to both forest owner and concessionaire.[153]

The goal of comprehensive conservation of regional biodiversity will also require retention of representative examples of forests on arable soils. The economic viability of such areas can only be assured by recognition of, and compensation for, the full economic value of their biodiversity. That can only be realised by *international* economic commitment, which requires political recognition of the service values of tropical forests, especially for biodiversity conservation, and for carbon sequestration.

6. Reconciling conflicting demands on forests: toward equitable solutions

Today, some of the uncertainties and disputes around forest use simmer, others flare, and still others lurk just below the daily experience of forest communities and the wider public. Their resolution will require a combination of technical, ethical and jurisprudential wisdom beyond the reach of any single observer, however committed. In what remains, we put forward some thoughts and observations, based on experience of changes in Asian forests and societies over the last couple of generations.

6.1. *Who should be the forests' keepers in our current world?*

We have claimed that policy should provide incentives for the logger and forest owner to manage forest areas set aside specifically for *optimal and sustainable* production of timber, that is by optimising the economic value of their timber yield. Forest area and site conditions must be sufficient to achieve that objective irrespective of the impact this may have on other goods and services, except in the case of plant species essential to vertebrates, especially figs and species bearing large fruits, which must be retained; while conservation of biodiversity additionally requires *other* areas to be set aside, inviolate. As rural populations expand and gain secure land tenure, forest products other than timber would increasingly be grown in home gardens. In practice, we predict that impact of forest use on most ecosystem-scale services would then be minimal.

A prerequisite to this favourable scenario would be an incentive structure that would either provide security of long-term returns to a timber concessionaire directly responsible for management, or a strong enough assurance of long-term returns to the forest owners to motivate them to oversee management assiduously. Crucial is the management of the logging operation itself, which serves as the principal (and usually the only) silvicultural intervention. Government policies, and those responsible for implementing them, have changed several times over the lifetime of a tree. Given that fact, what level of security guarantees can governments be expected to offer, either to the forest or to the concessionaire? Meanwhile, the interests of forest-dwelling communities have been appallingly neglected by authoritarian or paternalistic governments or their ministries; these are already a major cause of conflict in India,[154] Burma,[155] Sarawak,[156] Sabah,[157] Thailand,[158] northern Australia,[159] Papua New Guinea,[160] and indeed in most tropical forest-retaining nations worldwide.

The most intact forested landscapes left in the tropical Asian lowlands are found near the Equator in Brunei, and at the tropical margins in Bhutan (though in the latter, lowland forests have recently been intruded upon by revolutionaries seeking refuge from across the Indian frontier). What can be learned from these two countries? Is it an accident that they have both been ruled by ancient monarchies, without the services of an elected parliament? In both, therefore, the rulers have had the incentive to retain a proprietorial interest in the future of their forests. They have other traits in common too: for one thing, their areas and populations are relatively small. This fosters local participation and accountability, and renders centralised administrative control easier while also facilitating decentralisation of some functions. Both countries aim at a high *quality* of life for the citizenry, to some extent independent of individual means. Both have enjoyed low rural population density, universal public education, land allocation (more or less at the discretion of the monarch) with secure tenure to the small farmer, and a strong ethic of social responsibility in part embedded in religion. Both have maintained a significant degree of economic isolation from global trends. Can these factors be adapted to contemporary realities elsewhere, or even to changing realities within the countries themselves?[161] Universal public education and family planning opportunities are possible, in principle, anywhere. For the other factors, a more reliable alternative to the benign autocracy model will need to be found. The most promising one will be open public franchise among an educated and articulate public informed by an unfettered media.

6.2. *Some forms of community forest management*

Colonial forestry's central objective was to co-opt the productivity of the forests for wider markets within the state and beyond. That diversion of production has yet to be reconciled with the claims of the traditional local economy. Attempts to return the management of state forests to community leadership under the *panchayat* system began as early as the 1920s in India (Chapter 8). Experience with reconciling local with wider demand had been decidedly mixed in Europe, and in India the Forest Department was generally left with power of oversight. Nevertheless, few of the forests whose management was founded on established tradition have survived.[162]

As rural communities grow, and their needs approach the capacity of the forest to yield, the case for their participation in forest policy and management becomes compelling. It has to be recognized though that the track record of community management has been decidedly mixed. Nevertheless, enormous unexplored opportunities exist, not only to increase opportunities for women and enhance family planning, but to establish secure community access to resources and fair participation in markets. Again, Bhutan serves as one example, where forest products remain in use for a wide variety of products. The remarkable ATREE initiative, based in Bangalore but with projects in several parts of the subcontinent, is exploring a wide range of livelihood alternatives for rural Indian communities. In their flagship programme in the Biligiri Rangan Hills of southern Karnataka, tribal groups collect amla (*Phyllanthus emblica*) fruits and other NTFPs.[163] Widespread throughout the deciduous forests of continental Asia but most abundant in short deciduous forests, amla fruit are used for pickles and ayurvedic medicines (as a restorative, and for colds and fevers) and shampoo. The tribes-people, as is usual, had failed to gain a fair price from middlemen. With ATREE intervention, they have been enabled to short-circuit the line of trading intermediaries, thereby gaining better prices. They have started their own modest processing industries, including for wild honey and for furniture manufactured from the noxious shrubby exotic forest weed *Lantana camara*. Similar programmes are now being quite widely undertaken in India. As examples, a further project promoted by ATREE includes making crafts using an invasive bamboo in Darjeeling; blending water hyacinth and other natural fibres with cotton in handloom-produced textiles sponsored by the North Eastern Development Finance Corporation[164] and the Resource Centre for Natural Fibre Craft[165] in Assam; and other lantana (*Lantana camara*) craft enterprises run by The Shola Trust[166] and by Osai[167] in villages around the Mudumalai and Corbett Tiger Reserves.

Throughout the tropics of Asia, the traditional respect for and spiritual value of forests persists amongst many of the inhabitants of forested landscapes, as witness the remarkable levels of conservation achieved in sacred groves. However, community and family-owned indigenous forest patches — whether set aside for the harvest of particular products, such as the Soppina Betta forests of the Western Ghats,[168] or for their sacred status[169] — vary widely, both in management intensity and in level of degradation. In short, the variation in outcomes from traditional community management systems has been too great for them unassisted to provide a long-term, large-scale solution for forest biodiversity conservation.

It must be the goal of national policy to achieve a harmonious collaboration, whereby the technical and legal oversight of sound government provides a framework within which communities can freely manage their forest resources without overexploitation, yet with due respect for goods and services yielded to regional and wider entities. Policies towards traditional forest rights and joint forest management (JFM) have themselves been inconsistent, and have provided a weak sociopolitical foundation for the level of planning required even to sustain production of goods essential to communities.[170] Too often though, policies conceived at headquarters have been imposed without consideration or respect for local forest and social conditions, resulting in loss rather than gain in community rights, both in perception and reality.[171] Institutional reform within some state forest departments is seen as an urgent prerequisite for a successful JFM programme.

However, perhaps the most common complaint about JFM in its current manifestation is that it has not managed to clearly define legal property or resource access rights over communally managed forests. This fatally weakens the developing institutions responsible for community management of forests.[172] A related criticism is that decision-making powers, though officially meant to reside partly with democratic community institutions, are in practice too rarely relinquished by forest departments operating at the state or local level. Thus the 'jointness' of JFM has been strongly questioned.[173]

One partial solution is to reduce local demand on the forest by offering more attractive employment opportunities in commodity plantations or developing urban economies. This may now be emerging in South Asia, but too late to save most forests. Some regional economies may have great difficulty in ever offering such options, and the effects on some vulnerable communities may be severely disruptive.

Optimistic early reports on the progress of JFM identified innovative means, especially plantations of species that can support village industry, by which forest productivity can be sustained and diversified.[174] Much can be accomplished when villagers, especially women who have traditionally been the gatherers and processors of forest products in many places, gain incentives and a sense of co-ownership.[175] Even timber stocks can be regenerated with their participation.[176]

More recent analyses have identified essential conditions without which long-term success cannot be achieved.

1. Identification of sustainable forest resources, and objectives for their management, must be clearly defined and explicitly communicated.

2. There must be an equitable distribution of power among participants. Concentration of power among a village elite or a single leader invites favouritism and corruption, and guarantees variability in the quality of leadership as personalities change over time.[177] Particularly flagrant examples are the rash of destructive logging in Papua New Guinea where forests have remained in tribal ownership, but where the power of chiefs has remained pre-eminent, and in Sarawak where, though tribal traditions have often been democratic, the leadership has often become the dominant beneficiary of timber licenses. Women, who have played such a pivotal role in the harvesting of forest products in rural societies, must be given the authority that this role requires.

3. Participation by all social classes (castes) is a prerequisite, thereby dispersing power to several centres; participants must all be fully involved and benefit equitably through their efforts; 'sleeping partners' must be disqualified.

4. A continuous and sufficient flow of incentives must be established, including early benefits for the poor (to offset perceived high discount rates), with responsibility for all community members to contribute and to forego immediate gain when longer term gain is greater.

5. Above all, forests must be recognized as ecologically diverse, therefore diverse in their potential to yield and in their requirements for sustainable management. Equally, forest communities must be recognized as diverse in their traditions and in their knowledge, also their needs. As a general principle it must be accepted that it is these communities who have the deepest knowledge of the forest's potentials and its limitations.[178] Traditional practices must be understood and heeded before government officials or NGOs attempt to reform present practices.[179] Nevertheless, forest communities will often be unaware of the advances in understanding achieved by science.[180] Technical and other necessary knowledge must be made accessible, through the forest departments as well as NGOs.[181] *Rigorous monitoring and analysis of both socio-economic outcomes and forest productivity is essential.*[182]

6. Beyond the technical challenges, forest governance remains the most difficult issue. Devolution in Asia remains strictly 'a work in progress'.[183] Perhaps nowhere has an ideal balance been achieved between local community and state power. Both are necessary, since in principle they represent different interests. An organic entity such as a forest cannot be well managed piecemeal, as a collection of independent local responsibilities; yet for historical and ethical reasons, local commitments must have a voice and a certain privilege. An important part of the challenge is a lack on both sides of capacity, including training and indeed education, that would permit taking a wider view. In many nations a perceived social divide between civil servants and villagers remains widespread, and it does not help that a predominance of under-supervised, under-qualified and underpaid male forestry field staff serve a largely female village constituency.[184]

For all these reasons, forestry departments in many tropical Asian countries will themselves need to undergo processes of reform in order for any participatory forest management to meet its highest expectations. There have been trenchant observations of these facts for at least 30 years, for instance those of Ponna Wignaraja, writing on Sri Lanka in 1984:[185]

> A truly participatory development process cannot be generated spontaneously given the existing power relations at all levels and the deep-rooted dependency relationships.

> It requires a catalyst or initiator who can break this vicious circle, who identifies with the interests of the poor and who has faith in people…

Just as local communities often need some help and stimulation in becoming involved in a participatory process, so the forest department field staff also need to learn how to engage collaboratively. This change in thinking is unlikely to be accomplished through conventional 'retraining', but will also involve restructuring the more hierarchical forest service bureaucracies to allow for, and promote, greater flexibility and initiative on the part of the field staff. Wignaraja again:

> Conventional training methodology — delivering a pre-packaged basket of knowledge or skills through lecture and instruction — cannot be used for the purpose of sensitizing people for participatory development work.

Rather, a process of 'reorientation' to a different set of roles and responsibilities is needed. In some cases, this may involve substantial change in value systems and attitudes;[186] in others, forestry personnel should focus on technical assistance to management and initiating monitoring systems, rather than being drawn into the complexities of adjudication and local arbitration. Each case must to some extent be considered separately, as each rests on merits unique to itself. These conditions can be difficult and demanding, but they can be successfully confronted.[187]

It need hardly be added that, where forests extensive enough to provide opportunity for sustainable management for timber survive, the case for revival of local logging enterprises serving the home market and industry, with staff experienced with local forest conditions and species, over big national and international enterprises is irrefutable.

The potential to enhance income through improved management exists, though the costs to the state and the community of participatory engagement remain uncertain as does, therefore, net profitability. Whatever success is achieved, the limits to the biological productivity of forests will always eventually be reached, as many likely already have, and with them the limits to the densities of human populations that can directly depend on them. *This must be estimated and planned for.*

By passing the forest to large logging enterprises that cut at provincial or larger scales, often in several countries, governments not only degraded a resource unequalled in its local diversity but devalued centuries of accumulated local knowledge upon which optimal forest management must depend. Such local forest diversity and knowledge has proven an important source of local products in fully developed rural societies, but prolonged confinement of rural employment opportunities to unskilled harvesting of global crops will surely eliminate these unique and rich resources for niche markets.

6.3. *Wider horizons of forest dependency and responsibility*

In our present globally linked economy, forests benefit not just the adjacent rural communities, but all inhabitants of the watershed, the nation, and the wider world. However, different sets of benefits accrue to each group, and benefits are not directly comparable. Values of some services, especially biodiversity maintenance, are experienced over the long term and are therefore difficult to quantify. The economic value of carbon held in Malaysian forests has been estimated at a staggering US$330–760 billion,[188] whereas the money required to offset carbon released by forest conversion in India, based on an annual release valued at $433 ha^{-1}, could be $277 billion.[189] All forest beneficiaries should therefore have some influence over forests' fate, but only if they contribute to them in proportion to what they gain. That is extraordinarily difficult to measure and to manage, but it must be aimed for.

The increasing concern and sense of urgency to 'save the rainforest' among the urban middle classes of the advanced temperate economies cannot *by itself* greatly influence the future fate of the forest (although their past influence can hardly be ignored and their future financial subsidy should be expected). Solutions must come from the owners. In real terms, this means that community or local private ownership can only sustain a forest resource if legislation is enacted to protect forest from conversion; and those laws must be effectively administered by strong provincial and federal forestry bureaucracies.[190] Policy must be defined which provides the necessary incentives for this to happen. These ideals will never be achieved unless the citizen, and especially the upland forest dweller, can speak up against malefactors without fear of retribution, either from criminal elements or from self-interested representatives of government.[191]

The mission of international NGOs is always to address forest conservation in pursuit of one or a limited range of objectives. Their role will thus always be limited, and has often been at variance with those of major local stakeholders.[192] Strong and uncorrupted national institutions and the free exchange of ideas and political franchise within the citizenry are necessary for a swift transition to responsible and reliable forest management. Unfortunately, in many polities secrecy has even entered into the realm of research. Forest services are often better funded than university researchers or even government research institutes, but invaluable management data gathered by forest services are often withheld from researchers. Could it be that there is something to hide? Strategic planning and implementation for a sustainable forest

environment is impossible without free access to data and open public debate of the issues.

The initial changes accompanying development take place locally, and at varying paces. Thus, nations in process of economic development often include regions and groupings at all stages of development. The earliest and most vulnerable stages and the far-advanced may coexist within a single country. It is crucial that the citizenry of the more advanced groups have a positive influence on the process and outcome of sectors further back.

The rich world has played an ambiguous role in this regard, for it has often been responsible for offering economic incentives for overexploitation. In future, in addition to carbon sequestration,[193] strictly protected areas for wildlife and biodiversity will need to be subsidised by the global community. In 2008, a United Nations collaborative programme on Reducing Emissions from Deforestation and Degradation (REDD) in the less-developed countries was negotiated. It focuses on protecting high-biodiversity areas – while leaving unspecified how that might be achieved.[194] Two years later at Nagoya (Japan), representatives of more than half the world's nations agreed to a Convention on Biological Diversity. This set an objective of permanently protecting 17% of remaining forest land. (However, it did not define 'forest land', or articulate whether timber harvests should be permissible.) New economic incentives to conserve what is left must be created, and ways must be found to incorporate and justify appropriate oversight[195] (see also sections 6.4.3 and 6.4.5 below). But none of this will come cheap!

6.4. *What, then, should be the future role of the state?*

> Silence is a powerful enemy of social justice.
>
> Amartya Sen, 2006, *The Argumentative Indian. Writings on Indian History, Culture and Identity.*

> Currently, costs and benefits from conventional logging in tropical rainforest are neither transparent nor equally shared.
>
> Eberhard Brünig, 1996, *Conservation and Management of Tropical Rain Forests. An integrated approach to sustainability*

Continued ownership and protection of forest by the state on behalf of the people has been intermittently successful in modern times, and only where an articulate and involved public, informed by a vibrant and unbridled media and supported by the law, has had the political power to rapidly and effectively intervene in policy. It is a fact that the future of the forests, especially in regions of high population, rests most immediately in local hands. They represent the widest range of demands on the forest and have in many cases accumulated deep knowledge of its requirements for sustainable use. The fate of many forests depends significantly on local decisions addressing local forest use. Social scientists have advocated the transfer of forest management to local communities. Too often, though, those decisions have been over-ridden by vested interests, corporate actors, and self-serving government representatives or agencies.

But it is unrealistic to expect that village franchise and village leadership will always maintain the long-term view and consistency of policy necessary to sustain the forests without outside intervention, either. It is also true that decisions which appear optimal at local level are frequently against the national or wider interest (Chapter 8).

6.4.1. *The age of the internet*
We have reached a time when technologies facilitate instant, worldwide communication. Urban and rural values can be compared as never before. The poorest swiddener can see how the urban middle class live (or wish to live) in the fantasy of television soap opera. Few young people want to be confined to the hard traditional existence, and many aspire to enter the burgeoning market economy by trading forest products whose rarity has raised their value. At the same time, an increasing number of citizens are sufficiently wealthy to resist placing immediate priorities over the long-term requirements of society. But many are coming to realise that their descendants will need indigenous forests if climate, and with it local weather patterns, are not to be irremediably damaged, and if their rural cultural heritage is not to be lost along with the social, educational and economic benefits of forest recreation. The opportunity for reconciliation between communities uniting in a common endeavour has never been better, thanks to the transparency that is being forced by the internet.

6.4.2. *Top-down forest management*
Several types of 'top-down' political systems for forest control have provided relatively effective, if not equitable, protection of indigenous forest assets. Among these are traditional hereditary ownership by the ruler, and more recently private ownership under taxation regimes that allow families to retain authority over forests for generations. These may also guarantee some degree of equitability in cases where (1) the law restricts the land to forest uses, and (2) the traditional rights of other dependent parties are accommodated. Research by social scientists has brought to light plenty of examples of more informal, community-based systems that have endured over generations and furnished effective management.[196] The exact extent of these systems' relevance under present conditions is as yet uncertain: these include growing rural and urban populations, change in relative value to service and global values, technological change, and pervasive market involvement.

Restoration of authority usurped by politicians to the forestry service will be essential if forest goods and services are to be optimised and sustained. Preconditions are that *local* rights and franchise are respected and the training and development of forest service staff is both appropriate and adequate.

It is a striking indictment of human venality where political leadership forsakes its responsibility that only two tropical Asian nations retain more than half of their area under indigenous forest. These are therefore the best endowed with opportunities in a fully developed future. Again, they are Bhutan and Brunei Darussalam. In Japan, outside the tropics, a long-standing and successful tradition of village cooperatives was once supported by the shogunate. Such systems survive where there is dedicated and enlightened leadership, and functional support from the law.

6.4.3. *Incentives to protect*
Experience in advanced economies has shown that conservation of resources, and indeed their sustainable use, is often most successful in the hands of individuals and organisations with long-term interests, and therefore incentives, to protect, *as long as these incentives come with good information, good communication, and clearly-defined powers of decision-making within a legal framework*. Every tropical forest has a multitude of stakeholders. These range from the local inhabitants who harvest products directly, to those whose health[197] and welfare depends on ecosystem services, and to the global community which benefits from forests' amelioration of climate, carbon sequestration and the biodiversity unique to each forest. In most tropical forest lands today, there is a clear need for institutional reform, so that all stakeholders may participate in planning the future of the forest[198] and understand and accept that, in this role, they assume concrete responsibilities. Even then, options for changes in land-use must be defined by the law, overseen by an enlightened forest service and mediated by an independent judiciary.

In the drier regions, especially in South Asia, ongoing degradation of deciduous forests is rightly being met by efforts to increase the decision-making authority of local communities, and to more equitably distribute forest benefits. This trend toward community management or co-management is being led by individual Indian states, and perhaps most progressively by Nepal.[199] But too little has yet been invested in long-term research to gauge optimal approaches to forest production, so local management solutions are often based partly on guesswork. Strong arguments are being made for the integration of traditional and local knowledge into scientific analysis, and there are plenty of cases where these sources can offer guidance toward sustainable resource use.[200] But the caveat remains that traditional systems were not developed under conditions of continuing population growth and industrialisation; the main forest pressures remain the multiplying needs of growing populations juxtaposed upon the vast appetites, stimulated by increasingly under-regulated trade, of industry for raw materials.

The major challenge to actively manage cattle both to optimise their numbers and health, and to restore the forest, remains. We have often been told by graziers that any attempt to fence common land, even private land where fencing has not been practiced, will be abused by neighbours with cattle but no landholding. This recalls the 'tragedy of the commons' in medieval England, where common land was equally degraded through overgrazing by the cattle of both rich farmers and peasants during the summer growing season. But the Enclosures Act of 1808, which permitted division and legal tenure of formerly common lands among those graziers who had traditional rights over them, quickly led to stall feeding and hay harvesting for winter among smallholders willing to protect their interests. And so it is in Rajasthan today (Plate 8.5b). In England then, as in India now, those without land can seek employment in a rapidly developing urban economy.

6.4.4. Bottom-up structures

The era of the large timber corporation, logging with minimum reinvestment in the forest resource, is over in Asia. Their activities have left us with the necessity, especially where the forest has been degraded, for highly skilled staff and site-specific management. The added costs will fall on those governments that had earlier forgone investment in the management of first logging operations; these costs will now be passed on to consumers and taxpayers. Greater local authority over the fate of the forest will need to be exercised by smaller, local enterprises, involved on the scale that had been successfully managed in India and Malaysia.

Public trusts, such as The Nature Conservancy in the USA and the Malaysian Nature Society, have negotiated control over natural landscapes with the objective of managing their resources, including biodiversity, sustainably (as well as profitably) for the common good. This model might provide a mechanism for the reformation of public forest management, were politicians and the forest service to accept boards of trustees, representing the variety of public interests at both local and national scales, and including both traditional and technical scientific expertise, to maintain oversight with the power to sanction. That, however, would require drastic institutional reform in many forest services.

Private ownership of forest lands is one of the alternatives, but it has proved extraordinarily complex to implement in such a way that the many potential beneficiaries share equitably. Most large owners or investors demand returns on investment at least equivalent to the financially most profitable alternative use. Legislation to prevent conversion to non-forest use is

essential. Long-term timber licenses, such as those now being introduced in Sabah, in effect give ownership analogous to the 100-year tenancies offered by private estates in London and Hawaii. In several Western European economies, governments are now offering subsidies for land farmed at below-maximum crop productivity. This gives a certain advantage to service values, including biodiversity conservation. Owners in less-developed economies should qualify for equivalent funding from international sources. These kinds of arrangements, however, will require careful scrutiny to implement under conditions of fragmented or contested ownership, a common pattern in many parts of Asia where traditional tenure systems have morphed only gradually into more formally documented ones.

Asia is also rich with important traditions of commonly held lands, but many such traditions were interrupted by colonial and post-colonial government action and have been difficult to re-establish.[201] The halting progress of forest tenure reform under the Indian Forest Rights Act of 2006 shows how difficult even limited reform can be under conditions of insecure governance and uneven capacity.[202] Retention of customary rights to forest products will also be difficult to assure and to regulate, since demand inevitably increases as rights-holding populations grow and markets expand, while national and wider interests must also be incorporated. Nevertheless, locally based forms and institutions must be systematically explored as models for new frameworks appropriate to present conditions.[203]

6.4.5. The monetisation of service values and its translation into policy

The difficulty of valuing services notwithstanding, there is no easy alternative to doing so as a means to weigh their importance relative to that of goods yielded from a common resource. Economists have developed several more or less indirect ways of valuing, in international or local currency, such services as weather amelioration, water conservation, carbon sequestration, and – most difficult – biodiversity conservation. Their attempts are invaluable, but open to misuse by policy-makers.

There is a broader critique, prevalent among social activists and theorists of social justice especially in India, that regards with some suspicion the panoply of 'market-based solutions' that are now being promulgated. These range from local NTFP commercialisation schemes at the smallest scale to Payments for Ecosystem Services (PES), Clean Development Mechanism (CDM) and REDD+ at the national and international levels. Such projects are often promoted as strategies to 'harness the power of markets' in the service of conservation objectives, promising the possibility of 'win-win' outcomes in which poverty is reduced and social justice increased even as nature is protected[204]. The social justice critique sees such projects as founded upon a broad push toward transnational 'marketisation' of the land and of its natural productivity potentials, and possibly as further steps in the direction of an already globally-pervasive neoliberal ideology and agenda. This ideology, though couched in terms of freedom and enterprise, has often had the effect of enriching the few at the cost of the many, and has been used to justify excessive consumption of the Earth's resources. Over the course of little more than a single generation this economic vision has become so pervasive that it is often perceived as an inevitable process, rather than as the outcome of social, political and economic choices made by individuals of influence with personal interests or convictions. But what might be a more equitable, yet practical, alternative?

As sovereign states have exhausted their primary forest resource, they have cut forest agency budgets and privatised services over the last three decades, often in response to demands from international lending and aid agencies. But significantly, this is being implemented at a moment when the resource capital, from whose revenue services could have been funded, has been exhausted on expenditures unrelated to its sustainable management or the good of forest-dependent communities. The status and influence of forest departments within governments was already weakened. Boundaries between institutions of the state, the private sector and the non-profit sector have become increasingly porous, not only in the forestry and natural resource sectors.[205] Many functions previously served by government agencies, as representatives of the national polity, have been taken over by private groups, thereby avoiding commitment of tax revenues on essential rural and forest services, and to ensure that the resource and its dependent community, both impoverished by underinvestment and exploitation, somehow pays its way.

The philosophy dominant between the 1950s and 1980s, that investment in the development of rural resources would lead to a positive economic return, has been widely abandoned as governments have allowed, even encouraged, unsustainable exploitation by commercial interests. Failure to stem population growth in already overpopulated agrarian societies, itself a consequence of underinvestment, has also contributed an incentive. Governments in smaller countries have thus come to perceive the large and powerful international conservation NGOs as one more lobbying group among transnational firms seeking to help set agendas for resource conservation, research and policy. The most widely influential NGOs all maintain close dependence on the private sector for their funding and even oversight,[206] including multinational resource extractive corporations whose activities they monitor and often deplore.

It is difficult for organisations wielding relatively vast wealth and influence to deal flexibly and creatively with the small details of life 'on the ground' in rural economies with very little financial through-flow. Thus it may not be possible for them to serve the interests of small and unorganised groups adequately, even with the best of intentions. A well-established international carbon trading system to mitigate climate change, for instance, might end by benefitting powerful private interests: it will always be easier and cheaper to pay for reforestation schemes of uncertain outcome than to develop new energy technologies, let alone to move toward a new economics of sobriety in the industrialised nations. Carbon trading must be designed so that it benefits vulnerable forest ecosystems and the rural people whose economic future depends on them, as well as their investors. Despite optimistic scenarios,[207] there is as yet little information and no consensus on how this can be guaranteed.[208] One likely element in a solution will be building up the capacities of 'indigenous' environmental organisations (whether governmental or non-governmental) to act on an international stage, and increasing the reliance of large transnational conservation organisations on their in-country services.

7. All may not be lost

7.1. *Public action is being aroused*

The primary aim of conservation must always be the survival and sustainable use of renewable resources. This means protecting the capacity of landscapes to produce goods and services, and ensuring that resource benefits are shared equitably among stakeholders. Among these benefits is biodiversity, which once lost cannot be recreated. The protection of biodiversity will be critical at all levels, from genetic through ecosystem: both for genetic improvements in agriculture, including the protection of plantations, and to stabilise the fragmented forests of future landscapes.

In every country in the Asian tropics, there is a stirring of interest and pride in indigenous wildlife and in its conservation (Plate 9.6), but so far this is happening mostly among the urban educated elite. In 2010, Liang Congjie died in China at 78: a courageous man who exposed corrupt officials and led in achieving a ban on the conversion of indigenous forest. But China, like India, still *imports* wood, much of it illegally logged from the remnant virgin forests of tropical neighbours. Urban conservation movements in the tropics have been stimulated in part by the media. Just as in countries outside the tropics, interest is mostly expressed as sentiment and political opinion, although a lively growth of interest in natural history is also contributing important advances in knowledge.

That, however, is in tragic contrast with rural trends. Fifty years ago, I learned the botany of Bornean forests from young Iban men in their twenties, extraordinarily knowledgeable about forest plants and animals, proud and confident in their own spiritual and utilitarian traditions about the forest though

Plate 9.6 Villagers of the Santal tribes, living in Similipal National Park, Odisha, set up their own conservation corps with the encouragement of the park director. (H.H.)

they lacked formal schooling. These essays are dedicated to them in friendship and respect. Few of today's generation have their interest or their level of knowledge. This is in large part because the intended benefit of universal education has ignored the *particular*, which is the land on which each country person lives and on which each depends. It has thereby devalued traditional knowledge, the basis for understanding our forests, and undermined the confidence of those in the countryside who have the greatest potential for learning and intellectual achievement. It is those young people whose enthusiasm has been kindled by the details of their local natural history, who can form bridges between the villager and the university biologist. University students must be given the opportunity to become steeped in their natural heritage by fieldwork, and to learn to study it. Education must include an understanding of our dependence on nature, the renewability of natural resources but also the limits to their potential sustained productivity, and the basic rules for maintaining their well-being and bounty. Biology must be reinstated on a par with arithmetic, writing and language, and taught in rural primary schools by teachers with natural history as a vocation.

Yet, how many publications on ecology and conservation biology are emanating from the universities in forest-holding countries? Indeed, how many active professional ecologists are there left on their faculties? It is therefore just as important to resolve the growing gulf, in mutual understanding as well as in socio-economic position, between the growing urban middle class and those left behind in the countryside. This gulf has to be addressed by reinvigorated contact and dialogue, and a shared understanding and vision. An intellectual concern for the forest will achieve nothing unless translated into actions supportive of those whose decisions determine its fate. Alone, these urban movements are no more capable of inducing reform than their well-meaning brethren in the industrialised nations.

7.2. *A vision for the landscape*

Biodiversity cannot be conserved in zoos and botanical gardens. The cost would be astronomical and besides, in the medium to long term, diversity relies on the web of interactions between species, their populations and individuals, which are the motor of natural ecosystems. Biodiversity conservation requires, above all, adequate and representative pristine areas to be set aside for strict protection. It also requires some degree of connectivity between those areas, which can be provided by production forest and continuous forest strips along riverbanks.[209] Increasingly sophisticated methods are being formulated to achieve optimal design solutions,[210] taking into account differences in dispersal ranges and population densities of component organisms. Above all, management planning must start, and remain coordinated, at regional and whole-landscape scale. Regional distribution patterns are now quite well known[211] (Chapters 3, 6). I visualise the economically optimal landscape of the wet Asian tropics approximately as follows:

1. The lowlands would have been converted to intensive commodity agriculture and food production. In fertile areas, irrigation would advance up the lower, less steep slopes. Degraded, mostly steep sites with poor soils would support plantations, initially of exotic fast-growing timber and long-fibre wood species, which can act as nurse for restoration of indigenous species for both timber and other products of value.[212] Throughout, rivers would each retain a strip of native riverine tree species, such as have been conserved along Sabah's Kinabatangan (Plate 9.7), but ideally with occasional upland forest fragments located behind their floodplains. Where still possible, 5% of this lowland landscape would be set aside as forest, half of which would be mixed species plantation (exotic or native) for fuelwood and half (or all that now remains) would be strictly conserved indigenous forest. This portion would be physically separate but connected to riverine forest belts acting as corridors to the forested hills. Above all, existing legislated parks and nature reserves would

Plate 9.7 Sabah has adopted the policy of retaining primary forest along the banks of the Kinabatangan and other major rivers, but it is a narrow band, representing only riparian forest and habitat. (T. Laman)

be rigorously maintained and protected with the active participation of local communities in management. Villages and towns would be located primarily in the foothills, as was the case in traditional times in the seasonal wet tropics. Gardens would grow tree and other plant species yielding products for household use, but would also shelter wildlife. Remarkably, such lowland landscapes still actually exist in Sri Lanka, in the western coastal lowlands of India and in parts of north-west Borneo.

2. Hilly landscapes below 900 m in the wet tropics, where most slopes exceed 40 degrees would be set aside as permanent indigenous forest estate, and restored to this condition wherever feasible. All riverbank forest would be strictly protected. The rest of the forest would be carefully managed to optimise sustained timber production, and would be certified for that purpose, yet species yielding other valued goods, many of which are successional, would be encouraged. Active management would ensure continuous employment opportunities and distribution of profits back to forest people.

3. Forest within the cloud base would be totally protected for soil protection and water yield, including upper montane forest with its high point endemism. It would also serve as the principal conservation area for top predators and other large animals, and the mobile links (Chapter 5). The Heart of Borneo project, coordinated by the World Wide Fund for Nature, serves as an example. These areas should be connected by tracts of timber production forest to other conservation sites.

4. Above all, large reserved areas can only be protected with the support and participation of communities within or adjacent to them. Unless these communities gain clear benefits, and understand and accept clear rights and obligations, especially as regards limits on hunting and harvesting, such parks will prove illusory, and no amount of policy legislation will succeed.

5. Smaller but strict conservation areas, of unlogged primary forest where still existing, throughout the lowlands of each region would include sufficiently large examples (minimum 5 km² area where still possible, but smaller areas are also valuable), representative of all habitats. Particularly high priority must be given to include (i) habitats rich in rare and endemic species, and fragile habitats including podsolised raised beaches and summits, ultramafic exposures, and limestone karst, and (ii) the last few remaining forest patches surviving on arable land but which are not yet strictly legally conserved.

6. Landscapes of exceptional richness or endemism must also be targeted (the Tigapuluh Hill landscape of southern Riau province in Sumatra is an outstanding example[213]). These small areas of exceptional biodiversity, so critical to conserve, are almost always easy to locate, being identifiable from maps and satellite images owing to their unusual geology or topography: coastal hills, volcanic or other igneous exposures in regions of sedimentary rock, limestone, lowland podsols, seasonally flooded alluvium, mountain summits above 900 m. Ground truthing is essential. Wherever possible, conservation forest fragments would be connected to larger ones by riparian corridors wide enough to permit vertebrates to migrate and thereby subsidise their naturally dwindling populations (Plate 9.7).

7. Each management plan would necessarily be specific to its area, because each must aim at realising the ecological and economic potentials within its particular climate, soil (as characterised by topography and geology), and surviving biota. In all cases, hunting must be regulated, and regulations adhered to through benefits gained and support thereby given by local communities.

8. Land use and forest rehabilitation and conservation would be monitored by satellite imagery and by periodic field inspection including documentation of changes by use of controls.

9. But, above all, the rural population — both the old farming communities and the immigrant farmers and estate employees — would be settled in villages and small towns, logically set within a landscape each of whose physical components carries the forest or crop that optimises its ecological and social benefits. Communities would be set among these component areas so as to most benefit from and contribute to all.

These ideals are already compromised in regions where high population density and extensive deforestation already exists, including most of South Asia and Vietnam, but even here they can and must be aimed for. In Sri Lanka, the flora was comprehensively documented in the mid-19th Century. Tree endemism is almost entirely confined to the perhumid south-west, where the forest that remains is mostly montane. A national botanical survey in 1998,[214] and exploration by naturalists, has revealed that all but a very few plant species (most of them over-collected orchids) still survive. However, many exist only as a few individuals, and are unlikely to sustain populations without intervention. Here, the approach of Lilian Chua and her colleagues in Peninsular Malaysia provides a possible solution (Chapter 8), although it would be prudent

wherever possible to back up tiny sanctuaries by ex situ plantings in botanical gardens and arboreta.

All is not lost though. The conservation ethic and the recent history of conservation legislation have been strong in these same regions (Chapter 8). Where else in the world but in India would giant timber trees, surviving in managed production forests, be honoured with the names of divinities?[215] Where else but in Sri Lanka would smallholders, secure on their land, plant slow-growing calamander (*Diospyros quaesita*) trees, sources of the finest ebony, whose benefits will only be reaped by their great-grandchildren? And although forest area in Sri Lanka is reduced to 22% of the country's total area, and to a critical 5% in the perhumid south-west, and though small scale depredations (such as illegal felling by villagers and gem-mining) continue,[216] all but a very few plant species still survive, if only as a few individuals.

8. Epilogue

The future of the rainforest depends on a shared understanding of its multiple values: to communities, to forest-holding nations, and to the international community and trade. Unless timber-importing nations take action to ensure that those imports come from well-managed forests, the forest will cease to exist as a productive resource. The trade in endangered species must be brought under control as a matter of priority. The persistence of medieval medicinal traditions in the face of contemporary research showing them to be superstition is perhaps not unexpected in remote agrarian societies, but among the wealthy of advanced countries it is unacceptable. The collection and consumption of endangered plants and animals for medicinal or restorative purposes and as a spurious mark of prestige, is surely a pathetic admission of barbarity, and commits future generations to an impoverished patrimony. In the same way, can a state claim to be civilised while it allows the rampant exploitation of the resources of a less powerful neighbour, harvesting their renewable assets to destruction? Investors take no account of the cost of mitigation following failure, an issue becoming increasingly obvious in mangrove forest conversion to aquaculture beds, where the high overall revenue and employment gained also by the dependence on mangrove of the onshore fishing industry is seldom considered.[217] Again, over-exploitation of NTFP impacts the poor most.[218] Primary rainforests bear a greater diversity of NTFP, though not necessarily greater *quantity*, than do logged or deciduous forests. This is true whether they are silviculturally managed or degraded (though that need not be so true of deciduous forests).[219] In India, products including tendu leaves (from *Diospyros melanoxylon* – used for bidi cigarettes)[220] and lac (from *Butea monosperma*)[221] are becoming industrialised through better management aimed at production via intensifying labour rather than adding machinery[222] (Chapter 8). Over-exploitation of NTFP harvested for commercial gain from the wild and from community forests is likely in the absence of consistent and participatory monitoring.[223]

I therefore conclude that the following concepts and goals must be integrated into forest policies if adequate examples of indigenous forests, particularly those rich in biodiversity and endemism, are to survive the coming bottleneck:

1. No individual forest can simultaneously meet all the demands placed on it by modern economies. Sustainable yields of goods and services are optimised by assigning different objectives (and consequently different management protocols) to different forests. In many cases (and specifically the conservation of biodiversity and timber production) the means for optimally managing one forest for several uses will prove incompatible.

2. Sustainable timber production can only achieve acceptable rates of return if appropriate machinery and extraction methods are adopted, minimising disturbance to soil and established regeneration, and if conversion wastage is greatly reduced from present practice.

3. Oversight must be shared by *all* who benefit from forest biodiversity, and therefore wish for its maintenance (the 'stakeholders'). These include local communities who gain from goods, services and employment, commercial interests that gain mostly from timber extraction in production forest that can also conserve wildlife, a well-trained forest service, and governments who gain revenue; but also the world community, which stands to benefit both from carbon sequestration and from the protection of unique biodiversity. This implies that Trusts would provide the optimal solution to forest oversight, with local representation pre-eminent.

4. Forest conservation is impossible unless local communities support and foster it; therefore they must benefit from conservation in as many ways as possible, including and beyond wage labour. They should then be brought within the taxation system, thereby honourably gaining the medical and educational benefits offered to the population at large.

5. Government forestry services must reform to focus on working jointly with those in whose hands the future of the forest will rest: the local communities that continue to depend on the forest for both goods and services. But they can only do so if their authority, undermined by politicians and their commercial allies, is fully restored and supported by the judiciary, unencumbered by short-term political concerns.

6. Forests' yield in goods and services are optimised by taking the variation, according to their site conditions, into account. Their sustainable management consequently demands a high level of skill. Policy and management require oversight from well-trained and experienced professionals, who must have the power to veto destructive policies and actions. Success requires the care and subtlety borne of deep local knowledge of people and their forests.

The optimal distribution of benefits, we all know, is never achieved. This is a major reason why natural resources are so rarely managed sustainably for the common good. To take one example, the rate of national compliance with the Kyoto Protocol is still so incomplete, and the Protocol is itself so far from what is required, that one wonders whether humanity is capable of ensuring its own survival. Our species, during the millennia when we were subject to natural selection, lived in extended family groups, then ethnicities, then nation-states which were the antithesis of an interdependent global economy. Our traditional religious ethics reflect this. Are our religious and ethical mentors guiding us wisely in the context of present reality? That ancient recognition of our interdependence with nature, which achieved an unequalled sophistication of understanding in Asia, still persists but urgently needs reinvigoration and communication worldwide.[224] Some religious leaders are at last beginning to recognise, once again, the moral urgency behind the conservation of nature, and in a case in Thailand (Plate 8.13e), are participating in practical efforts. Without such reinvigoration, the collective memory of our ancient forest heritage will surely fade among our grandchildren. Although they may never realise it, it will have mattered, and it is you and I who will have caused them to suffer for it.

Here we are, in the English-speaking West particularly, being swept along lemming-like by our immediate concerns, while the vortex of changing climate draws ever closer. Many of us who can afford it move to the countryside in retirement, seeking in a diffident way that ancient spiritual communion with the natural world that our Abrahamic tide swept away so long ago. My dear friend, the late Tamiji Inoue, once explained to me that, in Japanese tradition, one returns to farm as a means to prepare for death – consistent with the deep spiritual understanding, embodied in *satoyama*, that reached its apogee in South and East Asia. That spiritual reverence survives, but is in decline. Those ancient Japanese villages are half-empty, in decay. How many urban South or East Asians do actually retire to their rural landscapes these days? The Indian writer Pankaj Mishra has challenged Asia to create a political and moral philosophy that can replace the still dominant Western, which he claims is approaching bankruptcy.[225] Certainly, it is from Asia alone that a deep political philosophy that marries urban and rural perceptions, fired by spiritual as well as mechanistic understanding of the interdependence of man and nature, must arise. Without it we surely cannot build the commitment needed to avert catastrophe.

Now that Asia is setting the economic pace, can it set new priorities on these ancient values? Our immediate task is to get biodiversity through the impending, and in many cases ongoing, population bottleneck. We need to act ethically and in our own interests at the large scale, in space, time and policy. Yet, some fear that our species — which has presumptuously named itself *sapiens* (wise) — lacks the genetic predisposition and wherewithal to do so. If that proves to be so, why were we created in this manner? That may turn out to be the *last* question to be addressed by theologians.

References

[1]Eden 1976. [2]Poffenberger 2000. [3]Sen 2006. [4]Forman *et al.* 2003. [5]Katsigiris *et al.* 2004; Humphreys 2007. [6]Guritno & Murao 2001. [7]Repetto 1990. [8]Nantha & Tisdell 2009. [9]Abd. Rahim & Shahwahid 2009. [10]Daily 1997; Engel *et al.* 2008; Norgaard 2010; Pattanayak *et al.* 2010; Vincent 2010. [11]Panayotou & Ashton 1992. [12]Anon. 2010b. [13]Narendran *et al.* 2011. [14]Wunder 2007. [15]Turner 2001c. [16]Anon. 2010 b. [17]Turner 2001c. [18]P.M.S. Ashton, pers. comm. Nov. 2012. [19]Naidoo & Adamowicz 2001. [20]McPherson & Nieswiadomy 2005. [21]Mohd Shawahid 1985. [22]Tewari 1983. [23]Gill *et al.* 2009. [24]Ramakrishan 2002a. [25]Tomich *et al.* 1998. [26]Anon. 1982; Parnwell 1988. [27]Dasgupta 1990; Panayotou & Ashton 1992, 1995. [28]Anon. 2005; Bradshaw *et al.* 2007; van Dijk *et al.* 2009. [29]Sharma *et al.* undated. [30]World Tourism Organisation 1998. [31]Sarin *et al.* 2003; Caruso & Modi 2004. [32]Sinha & Bawa 2002; Ganesan & Setty 2004. [33]Pandey 2001. [34]Arnold *et al.* 2003. [35]Caruso & Modi 2004. [36]Arnold *et al.* 2003. [37]Bawa *et al.* 2010. [38]Overseas Development Institute 2007. [39]Fischer 1918; Sagreiya 1939; Mohan 1943; Kumar 1946. [40]Mohd. Shawahid 1985; Mohd. Shawahid & Noor 1999. [41]Boscolo & Vincent 1998. [42]P.M.S. Ashton *et al.* 2001b. [43]Vincent 1992. [44]Sunderlin *et al.* 2005. [45]P.M.S. Ashton, in press. [46]P.S. Ashton, in press. [47]Burgess 1975. [48]Brünig 1974, 1996. [49]Brünig 1996. [50]Wyatt-Smith 1960; P.S. Ashton 1964b. [51]P.M.S. Ashton, pers. comm. 2010. [52]P.M.S. Ashton *et al.*1993b; P.M.S. Ashton & Peters 1999. [53]P.M.S. Ashton *et al.* 2001b. [54]Brünig 1996. [55]Sist *et al.* 1999; Sist & Brown 2004. [56]Putz *et al.* 2008. [57]Healey *et al.* 2000. [58]Dykstra 2004. [59]Korpelainen *et al.* 1995. [60]Gardingen *et al.* 2003. [61]Cannon *et al.* 2009. [62]Basset & Novony 1999; Novotny *et al.* 2006. [63]Puri *et al.* 1983. [64]Xie *et al.* 2004. [65]P.S. Ashton, unpubl. data. [66]S. Gupta 1996; P.S. Ashton 2008. [67]Joann *et al.* 2013. [68]Corlett 1988a. [69]Laurance *et al.* 1997; Laurance & Peres 2006. [70]Styring & Hussin 2004. [71]Pimm & Raven 2000; Pimm *et al.* 2001; Laurance & Peres 2006. [72]Primack 1993; Primack & Lovejoy 1995. [73]http://www.iucn.org/ [Accessed 2005]. [74]Fang & Peng 1997; Hardwick *et al.* 1997; Elliott *et al.* 2003. [75]Gariguata & Pinard 1998; Elliott *et al.* 2003. [76]Riswan 1982; Riswan & Kartawinata 1991. [77]Peeris 1975. [78]P.M.S. Ashton *et al.* 1993b, 1997, 1998. [79]Sim *et al.* 1992. [80]Cohen *et al.* 1996; Shono *et al.* 2006. [81]Srivastava *et al.* 1994. [82]Dennis *et al.* 2005; Dennis & Colfer 2006. [83]MacArthur & Wilson 1967. [84]Newmark 1986; Primack & Lovejoy 1995; Narendran 2004; Corlett 2009a. [85]Sekercioglu *et al.* 2002; Harrison & Rasplus 2006. [86]Gadgil & Huga1992; Amirlingam 2004; Jameer & Pandey 2002; H. Li *et al.* 2002; Ramanujan & Cyril 2003; Sukumaran *et al.* 2005; Pandit & Bhakat 2007. [87]Fischer *et al.* 2011; Phalan *et al.* 2011. [88]P.M.S. Ashton *et al.* 1997, 1998, 2001a. [89]A.D. Johns 1986; A.G. Johns 1996; Nakagawa *et al.* 2006; Anon. 2006b. [90]Mohd Aslan & Abd. Gulam Asad 2006. [91]Gupta 1996; P.S. Ashton 2008. [92]P.S. Ashton 2007. [93]Treves *et al.* 2006. [94]Whitmore 1984a; Nicholson *et al.* 1987. [95]P.S. Ashton & Hall 1992. [96]Gunatilleke *et al.* 2004a, 2006. [97]Tsumura 2002. [98]Amirlingam 2004; Bhagwat *et al.* 2005; Bhagwat & Rutte 2006. [99]Potts & Vincent 2008b. [100]Potts & Vincent 2008a, b. [101]Marshall & Leighton 2006; M. Leighton, pers. comm. 2007. [102]Swain 2001a, b. [103]Anon. 2006b. [104]Bawa & Seidler 1998. [105]Van Schaik & Terborgh 1993; Turner &

Corlett 1996. [106]Maudsley *et al.* 1998. [107]Harrison 2011; Harrison *et al.* 2013. [108]Tan *et al.* 1987. [109]Sayer *et al.* 2004. [110]Chandrasekar-Rao & Sunquist 1996; Webb & Sah 2003. [111]P.M.S. Ashton *et al.* 1993b, 1997, 1998. [112]B. Sarayudh, pers. comm. Oct. 2008. [113]Abd. Rahman Kassim, pers. comm. July 2008. [114]Anon. 2011a. [115]Cheah 1995. [116]Kosonen *et al.* 1997. [117]Carrington 2011. [118]Kiss 2004. [119]Wunder 2000. [120]Bhattacharya & Bannerjee 2003. [121]Mitchell & Ashley 2007; Stronza & Gordillo 2008. [122]Stem *et al.* 2003. [123]Zeppel 2007. [124]Igoe & Brockington 2007. [125]Duffy & Moore 2011. [126]Honey 1999, 2008. [127]Cochrane 2007. [128]Karl & Trenbarth 2003. [129]Solbrig *et al.* 1994; Vitousek 1994. [130]Salati *et al.* 1986. [131]Walsh 1996a, b. [132]Potts *et al.* 2002. [133]Leighton & Wirawan 1986; Goldammer 1990. [134]Russo *et al.* 2005, 2008, 2010; Balzer *et al.* 2007, 2008. [135]Clark & Clark 2011. [136]Ohsawa 2006a. [137]Liu & Ashton 1998. [138]Nelson *et al.* 1994. [139]Philips 1998; Philips *et al.* 1998. [140]Feeley *et al.* 2007c; Clark & Clark 2011. [141]Chapman *et al.* 2005. [142]Stern 2006. [143]Palm *et al.* 1986. [144]Davies & Unam 1999. [145]Luyssaert *et al.* 2008. [146]Luyssaert *et al.* 2008. [147]Luyssaert *et al.* 2008. [148]Morita 2002. [149]Kremen *et al.* 2000. [150]Yoda *et al.* 1995; Ravindranath *et al.* 1997; Rai & Sharma 2003. [151]Sharma *et al.* undated. [152]Fujimoto 1997. [153]Panayotou & Ashton 1992. [154]Lélé 1998; Pandey 1998; Sarin *et al.* 2003; Sunderlin *et al.* 2005; Saravanan 2009; Sarker 2011. [155]Tani 2000. [156]Kaur 1998. [157]Doolittle 2011. [158]Buergin & Kessler 2000. [159]Adams 2004. [160]Walton & Barnett 2008. [161]Wangdi 2010. [162]Sarin *et al.* 2003. [163]Morita 2002. [164]http://www.nedfi.net/. [165]http://www.fibreresource.org/. [166]http://www.thesholatrust.org/. [167]http://www.greenosai.org/ [168]Nayak *et al.* 2000. [169]Chandrashekara & Sankar 1998. [170]Jattan & Pratima 2001. [171]Sarin *et al.* 2003. [172]Ghate 2003. [173]Lélé 1998. [174]A.K. Singh *et al.* 1985. [175]Gupta *et al.* 2004; Rijsoort & Zhang 2005; Batliwala & Dhanraj 2009. [176]Dasgupta 2001. [177]Ghate 2003. [178]Ghate 2003. [179]Sarin *et al.* 2003. [180]Ghate 2003. [181]Lélé 1998, 2000, 2003. [182]Mahapatra & Mitchell 1997. [183]Colfer *et al.* 2008. [184]Ostrom *et al.* 2002. [185]Wignaraja 1984. [186]Gronow & Shrestha 1990. [187]Thompson 1995. [188]Morita 2002. [189]Lal & Singh 2003. [190]Batliwala & Dhanraj 2009. [191]Sen 2006. [192]Brown & Ashman 1996. [193]Engel *et al.* 2008. [194]Anon. 2010b. [195]Santili *et al.* 2005; P.S. Ashton 2007; Chomnitz & Kumari 2008. [196]Ostrom *et al.* 2002. [197]Pattanayak & Wendland 2007; Vincent 2010. [198]Ostrom *et al.* 2002. [199]Nagendra *et al.* 2008; Ram Dahal & Chapagain 2008. [200]Becker & Ghimire 2003; Brook & McLachlan 2005. [201]Edmunds & Wollenberg 2001; Guha 2001. [202]Saravanan 2009; Bose 2010; Sarker 2011. [203]Dove *et al.* 2011. [204]Yu 1994. [205]Corson 2010. [206]Chapin 2004. [207]Boyd *et al.* 2005. [208]Corbera *et al.* 2007; Putz & Redford 2009. [209]Given 1998; Anon. 2006b, 2008; P.S. Ashton 2008, 2013. [210]Solbrig *et al.* 1994; Heywood & Dulloo 2005. [211]Davis *et al.* 1995; P.S. Ashton 2010. [212]P.M.S. Ashton *et al.* 1993b, 1997, 1998. [213]Danielson & Schumacher 1997; Laurance & Peres 2006. [214]Gunatilleke & Gunatilleke 1991; Green & Gunawardena 1993; Jayasuriya 1995. [215]Nouhare 2001. [216]Wickramaratne *et al.* 1997. [217]Bennet & Reynolds 1993. [218]Nasi *et al.* 2008. [219]Mohd. Shawahid & Noor 1999. [220]Baghel & Gupta 2002. [221]Sequeira & Bezkorowanjnyj 1998. [222]Mahapatra & Mitchell 1997. [223]Chitwadgi 2001. [224]Biodiversity & Conservation Special Issues 1995 (4(8)), 2000 (9(8)). [225]Misra 2012.

Plate 9.8 The future of the forest – and of ourselves – will depend on bringing our own population into balance. (Mary Ashton)

Glossary

Adivasi. Indian tribal minority.

Adventive embryony. Embryogenesis directly from maternal tissue; that is, without male gamete and prior zygote. Often, several are formed from the same tissue (see **apomixis**).

AFLP. Amplified fragment length polymorphism, a method of genetic mapping, DNA fingerprinting.

Agamospermy. Development of seeds without zygote formation, from maternal tissue.

Ali Baba. Used in reference to politicians who co-opt public resources for their personal or political gain (referring to the tale of *Ali Baba and the Forty Thieves*).

Allee effect (in the present context). Decline in population due to decline in density below that required for sustainment of fecundity through cross-pollination.

Allele. One of a pair that comprise the genes that occupy specific locations (loci) on the chromosome.

Allopatric. Originating following isolation of populations and consequent genetic or phenotypic divergence.

Allozyme. Structurally differing enzymes formed by different alleles sharing the same gene locus (see **isozyme**).

Alluvium. Sediment that settles during the flooding of rivers.

Androdioecious. Species and their populations that are divided into functionally male and hermaphrodite (bisexual) individuals. Considered rare in nature.

Angiosperm. A flowering plant: in which the seed is entirely enclosed, within the ovule.

Anion. A negatively charged atom or molecule (see **cation**).

Anoxic. Devoid of oxygen.

Apomixis. Reproduction without fertilisation, asexual reproduction.

Arbuscule. Tree-shaped, that is ramifying, fungal hyphae which penetrate the cytoplasm of host higher plant roots, and defines the *Vesicular arbuscular mycorrhiza*. They are the site of exchange of water, phosphorus and perhaps other ions in solution from fungus to plant, and of carbon from plant to fungus.

Archaean. The period (eon) on Earth before life began, c.3.8–2.5 billion years ago.

Argillic. Rich in clay minerals.

Asymptotic height. Trees, like fish, continue to grow until they die. Their height, however, is limited by environmental constraints such as drought, so that a maximum, specific for a given environment, is reached that is rarely exceeded.

Autogamy. Self-fertilisation.

Available. Nutrient ions that are in solution and accessible to plant roots (see **exchangeable**).

Axillary. Arising between the twig and petiole.

Back arc. Basin caused by roll-back of the subduction of one continental plate edge onto the edge of the continent being subducted, combined with the collapse of the latter.

Basal area. Summed cross-sectional area, usually at c1.3 m above ground ('breast height'), of trees in a sample; providing a misleading estimate of volume or biomass, as variation with tree height and forest stature is considerable.

Basement rocks. The ancient crystalline or metamorphic rocks of the Earth; the rocks that underlie the sedimentary rocks that originated from them.

Benioff zone. Zone of greatest tectonic instability, that is, high earthquake frequency, within a subduction zone where one tectonic plate is being subducted beneath another (see **subduction**).

Biodiversity. The diversity of life, fundamentally expressed as its genetotypic diversity.

Biota. The total array of life in a site or region of specific area, either in space or over time.

Biotype. Organism of identical genetic composition.

Boundary layer (of leaves). The layer of air immediately adjacent to the surface whose movement is reduced by resulting viscosity (affected by roughness, hairs, etc.)

BP. Before present.

Brown earths. The predominant soils of the moist temperate northern regions and similar to those of moist tropical lower montane forests: humic mull soils, rich in earthworms.

Bryophyte. Non-vascular plants: mosses and liverworts.

Bunga (Malay). A flower.

Building phase. The stage of forest stand regeneration between plant establishment and attainment of full height (see **gap phase**, **mature phase**).

Canopy. The upper layer of foliage of a forest.

Capitate. With a distinct head.

Carpophore (of fungi). The structure that bears the spores, a mushroom.

Cataphyll. A modified leaf functioning as a protective bud scale.

Catena. The soil gradient across the topography, from ridge to valley.

Cation. A positively charged atom or molecule (see **anion**).

Cauliflory. Flowers borne on the trunk.

Cavitation (of cells, vessels etc.). Implosion of a cavity arising when the tension of the liquid in them is sufficient to cause a vacuum bubble (embolism).

Chatterjee's Partition. The belt of evergreen forest mountains that separate the South Asian deciduous forest floristic province from the Indo-Burmese.

Clade. A group of species sharing a common ancestor.

Clastic. Sedimentary (rocks).

Climax (species). Those long-lived species that persist, whether in the canopy or within it, after a stand has reached full height.

Cohort. A group of individuals that established at the same time.

Colluvium. Sediment that has shifted downslope and there accumulated.

Comminution. Reduction to minute particles.

Competitive exclusion. According to Gause, species cannot survive together if identically dependent on resources: one will exclude the others by more successfully competing for them.

Coriolis force. Winds are created by two forces: (1) incident solar energy which, greatest at the Equator and over high dark land surfaces such as Tibet, cause warmed air to rise thereby dragging cold air to replace it and; (2) the Coriolis force, that is the distortion in their direction, reversed at the Equator, which is caused by the Earth's rotation.

Corvée. Labour exacted by public authorities *in lieu* of taxes.

Craton. Larger pieces of pre-Cambrian continental crust (younger deposits can overlay) when manifested as a single coherent unit.

Crown shyness. Crowns in an even canopy in which branches and crowns are nevertheless spatially separate, most easily seen when crown foliage is dense. The distance of separation increases with the length of the branch. Apparently caused by abrasion during twig expansion.

Crown. The part of a tree that bears the leaves.

Cuesta. A ridge representing the broken upper edge of sloping rock strata (see **dip slope** and **scarp**).

Cumulonimbus. Clouds of great height, which result in thunderstorms.

Cyclone. Encircling winds, counterclockwise in the Northern Hemisphere, ubiquitous in regions with prevailing winds centred on a zone of high atmospheric pressure in the tropics, low in temperate regions. Tropical cyclones may result in hurricanes (typhoons).

Dalit. The lowest rank in the Indian caste system, the formerly so-called 'untouchables'.

dbh. Diameter at breast height; diameter of a trunk or bole at c.1.3 m above ground.

Determinism. A condition where both environmental and genetic constraints are such that they can only lead to a single result.

Devarakadu. A sacred grove.

Dichogamy. A hermaphrodite state in which male and female functions are expressed sequentually and separately, resulting in sequential unisexuality. Originally applied to individual flowers, but may be applied to whole plants.

Dioecy. Separation of male and female floral functions onto different plants.

Dip slope. The backslope of a cuesta that follows the slope of the rock strata.

Distichous. Arranged in two opposite rows on an axis.

Diversity. A measure of the total range of variation in an attribute. Biological diversity may be measured, for example, as genetic, phylogenetic, morphological or structural (see also **species diversity**).

Dormant bud. A bud that has reached maturity but has yet to start expanding into a shoot.

Ecotone. Vegetation of intermediate structure and composition between larger areas of differing but relatively uniform physical environment, and consequently vegetation.

Ectomycorrhiza. Fungi whose mycelia (fungal soil nets) form symbiotic associations with host plant roots without penetrating host plant cells (ectotrophic). Those that infect trees are Basidiomycetes (mushrooms). Restricted to Dipterocarps, Fagaceae and Caesalpinioid legumes in the Asian tropics.

Ectotrophic. See **Ectomycorrhiza**.

Edaphic. Qualities pertaining to soil.

Embolism. See **cavitation**.

Emergent crown. Tree crowns whose main branches arise in or above the principal outer forest canopy.

Endemic. Unique to a specific area.

ENSO. El Niño Southern Ocillation.

Epicontinental. Over a continental tectonic land mass (plate), as opposed to an oceanic.

Epicormic leader. A new leading shoot of epicormic origin.

Epicormic. Developing from a nascent bud arising from the cambium, beneath the bark.

Epigeal. Seeds in which the cotyledons burst through at germination and photosynthesise; possible only when the seed rests on, not in, the soil.

Epiphyte. A plant that lives on another plant.

Eustatic (sea level). In response to a worldwide change caused by melting or freezing of land-borne ice caps.

Evapotranspiration. The combination of rainwater evaporation off leaves, and loss from their tissues as transpiration.

Exchangeable. Nutrients when in solution as ions.

Extrusive. Rocks which are extruded with lava onto the Earth's surface, where they cool and solidify.

Ferricrete. An encrusted soil horizon of iron and/or aluminium sesquioxides.

Fisher's α. A measure of species richness in a vegetation sample that has the advantage of being independent of sample area or density. But it is not independent of intrinsic non-random population spatial distributions, such as clumping due to limited seed dispersal. Adopted here because, with the exception of deciduous-evergreen or altitudinal zone comparisons, forests share similar generic composition throughout the region, therefore similar seed dispersal patterns.

Flysch. 'Syntectonics'. Sediments resulting from mountain building, for example, landslips and larger phenomena. Recognized by their sharp-angled rocks.

Foredeep, or foreland basin, is the basin caused by sinking of lands adjacent to rising mountains, caused by the immense weight of the accumulating mountain rocks.

Friable. Crumbly.

FRIM. Forest Research Institute Malaysia.

FSC. Forest Stewardship Council.

Gamete. Male or female, that is unisexual, reproductive cell.

Gap phase. The stage of plant establishment within a forest stand (see building, mature phases).

gbh. Girth at breast height (1.3 m above ground).

Geitonogamy. Self-compatible only by cross-pollination among different flowers on the same plant.

Gene frequency. The frequency of a specific allele relative to all other alleles in a population that share the same gene locus.

Gene sequence. The arrangement of genes, that is of nucleotide bases, along a DNA molecule.

Genome. The complete genetic, that is heritable, content of a species.

Genotype. The genetic composition of an individual organism.

Geomorphology. The history and nature of land forms.

Ghat (India). A mountain wall or scarp and its crest.

Gigatonne (Gt). 10^9 tonnes.

Granite. A crystalline igneous rock consisting mainly of quartz, mica and feldspar, generally yielding soils of moderate to moderately low nutrient content.

Granitoid. Granite and related rocks, generally related to volcanic activity at subduction zones.

Granodiorite. A siliceous igneous rock, similar to granite but in which most of its feldspar consists of plagioclase (one of the silicate minerals).

Group III elements. In this context, iron and alaminium which, in welldrained lowland tropical soils, occur as sesquioxides and impart to them their yellow- to rust-brown colour. Their increase in concentration with depth is a measure of leaching intensity.

Guild. A group of species sharing similar ecological roles within a community.

Gunung (Malay). Mountain, hill.

Gymnosperm. Plants (all woody) in which the ovule and seed is not enclosed within an ovary, therefore exposed; including conifers, *Gnetum* and cycads.

Hadley cell. A circumglobal wind circulation system, of which there are three: latitudinally separated each side of the Earth's climatic equator.

Halophyte. A plant with roots tolerant to living in saline water, therefore generally confined there.

Haplotype. A sequence of genes along a chromosome that are inherited together, therefore stay together in the gametes following meiosis.

Heterosis. Hybrid vigour, including that gained by cross-pollination between genetically different individuals within a population.

Homozygous. Bearing identical alleles at a gene locus.

Horse latitudes. The calm, high pressure zone, generally at a latitude of 30–35° N, out of which the winds of the tropical and temperate Hadley Cells originate.

Horst and graben topography. Landscapes either side of faults; horst being squeezed upwards, graben being pushed downwards by the fault.

Hortonian runoff. Surface run-off caused by the inability of soil to infiltrate intense rainfall sufficiently to absorb it all, notwithstanding that the soil is not saturated.

Humidity Index. An index of climatic humidity, devised by C.W. Thornthwaite.

Humult. Soils in which conditions favour accumulation of acid-reacting organic matter at the surface.

Humus (raw, acid, etc.). Decomposed litter, comminuted organic material within the soil.

Hypogeal (seeds). Seeds in which cotyledons remain inside the seed shell and on or in the soil, not photosynthesising but acting as food reserves for the seedling.

Hypsophyll. Any form of leaf modified into a floral or protective, that is not principally photosynthetic, function (see **cataphyll**).

IFS. Indian Forest Service.

Illite. Pulverised mica with the consistency, but not the expanding property, of clay forming a compact mineral soil prone to waterlogging and loss of oxygen.

Indo-Burma. The unitary historical biogeographic area which comprises the tropics from China and Tibet south to the Isthmus of Kra (the Kangar-Pattani Line), and from the moist tropics of the eastern Himalaya and the eastern Indian states east to the South China Sea.

Inselberg. An isolated hill arising from a plain.

Intertropical Convergence Zone. The calm area around the climatic equator (that is, moving north and south with the zenith of the sun), in which the trade winds cease and their air moves upwards to condense its moisture, therefore the wet season; they then return, dry, at high altitude to descend at the horse latitudes.

Introgressive hybridisation. Crossing between two genetically distinct but similar populations or species in which one is smaller, therefore becomes genetically unstable.

Intrusive. Rocks that cool and form within volcanic magma.

Ion. An atom in which the number of protons are not equal to that of electrons, resulting in a negative (less protons: anion) or positive (less electrons: cation) charge.

Isozyme, isoenzyme. Chemically distinct forms of enzymes that have similar or identical catalytic although different kinetic properties and are therefore regarded as selectively neutral, and are encoded by different alleles and at different loci (see **allozyme**).

Isostatic (sea level). Local or regional changes due to crustal movement rather than the volume of water in the oceans.

Joint Forest Management (JFM). Management in which forest-dependent communities participate with the forest service.

Jhum (North-East India). Swidden.

Ka. Thousand years ago.

Kapok. The fine silky fibre derived from the small seeds of the kapok tree, *Ceiba pentandra,* which serves to assist their dispersal by wind.

Karst. Precipitous limestone topography caused by solution of the rock by rain, with caverns and underground rivers.

Kerangas (Iban). Upland upon which rice cannot be grown. In ecology, a forest type defined by its even, generally comparatively short, canopy and distinct flora.

Kerapa (Iban and Borneo Malay). Short dense peat swamp forest, either on the top of a peat dome or on poorly drained humic podsols.

Khan (Kannada). Patches of semi-evergreen or tall deciduous forest on the Deccan Plateau, Central India.

Landas (Sarawak Malay). The north-east wet monsoon in Malaysia, which can dump more than a metre of rain a month.

Lapse rate. The decline in mean ambient temperature with increasing altitude.

Lateral light. Light penetrating beneath the forest canopy from its side, on hill slopes.

Laterite. A hardpan of iron and aluminium oxides, deposited in seasonally alternatively waterlogged and dry soil. Now technically named **ferricrete**.

Leader. The tallest vertical shoot of a tree.

Leaf area density (LAD). The (upper) surface area of foliage in a standard volume of air, expressed as m^2 surface/m^3 air.

Leaf area index (LAI). The surface area of foliage over a standard area of ground expressed as m^2/m^2.

Leaf osmotic potential. The degree to which a leaf is able to adjust to water stress.

Leaf size classes. The standard leaf areas originally proposed by Raunkaeir, and modified for trees by Webb: leptophyll <25 mm^2, nanophyll: 26–225 mm^2, microphyll: 226–2,025 mm^2, notophyll: 2,025–4,500 mm^2, mesophyll: 4,501–18,225 mm^2, macrophyll: 18,226–164,025 mm^2, megaphyll: >164,025 mm^2.

Leptophyll. See **leaf size classes**.

Lignotuber. A root thickened into tuber-like proportions.

Light compensation point. The intensity of photosynthetically active radiation at which the rate of respiration equals that of photosynthesis.

Loam. Freely draining soils with a substantial clay content, imbuing them with fertility and conditions favourable to rapid breakdown of small litter.

m. Metres

Ma. Million years ago.

Macrophyll. See **leaf size classes**.

Malayan Uniform System (MUS). The (shelterwood) silvicultural management system devised for the light hardwood dipterocarp forests of Peninsular Malaysia.

Malesia. The equatorial region between the Kangar-Pattani Line in the southern Isthmus of Kra (see Chapter 6) and the Louisiades Archipelago east of New Guinea, including the Philippines.

Mantle. The layer of hot and semi-plastic but recognisably crystalline rock between the Earth's crust and liquid magmatic core.

Massart's model. The tree architectural form in which a monopodial orthotropic leading shoot intermittently radiates plagiotropic branches.

Mast fruiting. Supra-annual mass fruiting of a tree species population, or a whole community, generally concentrated in the canopy.

Mature phase. The dynamic stage of a stand following attainment of full height (see **gap phase** and **building phase**).

MDF. Mixed dipterocarp forest.

Megaphyll. See **leaf size classes**.

Mélange. Mixed rock fragments associated with a subduction zone.

Merishdar. Land proprietor.

Meristem. Tissue comprised of undifferentiated cells capable of division (mitosis).

Mesic. A freely draining habitat with moist soil.

Mesophyll. See **leaf size classes**.

Meta-. Sharing.

Metapopulation. All populations of a species that, though currently more or less isolated, continue to exchange genes over evolutionary time.

Metamorphic. Rocks that have been transformed by heat and pressure.

Meteoric. Atmospheric.

Microphyll. See **leaf size classes**.

Microsatellite. Repeating sequences of 2–6 pairs of DNA.

Microtherm. Used by van Steenis to describe frost-hardy species of high tropical mountains, generally herbs, in genera that are otherwise temperate.

Moder (in the present context). Well drained yellow-red forest soils of intermediate nutrient status and acidity with a thin surface horizon of partially raw humus (mor).

Molasse. Deposits originating during quiet periods between uplifts; on landscape scale. Recognized by their rounded rocks.

Monoecy. In which the male and female functions are served by separate flowers of the same plant.

Monophyletic. Sharing a common origin.

Monopodial. Trees in which the trunk and leader remains derived from the same apical shoot.

Monotypic. Containing a single taxon.

Monsoon. A season in the dominant single dry, single wet climate of tropical and warm temperate Asia; referring to the wet season when unqualified. (From Arabic: *musim* = season.)

Montmorillonite. A mineral, generally blackish, composed of clay crystals that form a very large lattice, thereby imbuing it with sponginess and capacity to contract on losing water.

Mor. Raw humus (see **raw humus**).

Morphology, -ical. Form, shape.

Morphospecies. A plant recognized as a separate taxonomic entity solely on the basis of vegetative characters, as yet without formal identity.

Mull. Humus formed under non-acid conditions; clay loam soils with well developed crumb structure.

MUS. Malayan Uniform System of forest management for timber.

Mycelium (plural: **mycelia**). The weft of branching fungal threads (hyphae) in the soil or elsewhere; the asexual tissue of fungi.

Mycorrhiza(e). Fungal mycelia that form symbiotic association with the mesophyll in plant roots.

Nanophyll. See leaf size classes.

Nastic. Movements of plants in response to environmental stimuli that are non-directional, unlike for example, phototropism.

Neotropical. Tropics of the New World, or Americas.

Neritic. Sea over the continental shelf.

Niche. The response of a successful individual or taxon to competition and available resources.

Noah's Ark. Used to describe a tectonic plate that breaks off from a larger, more static plate and is rafted with its biota to collide with another.

Notophyll. See **leaf size classes**.

NPV. Net present value The difference between the present value of inflows and that of outflows. It is a means of comparing the present value of something to its future value, when future returns and inflation have been taken into account **NTFP**. Non-timber forest products. Any natural forest resource, other than timber, that enters local or wider economies.

O, A, B, C horizon (of soils). The distinct layers into which soils are divided from the surface down. O = surface litter; A = decomposed organic matter; B = mineral soil, sometimes with a distinct indurated layer towards its base; C = the layer of decomposing substrate, where present.

Obduction. The tectonic collision of two continental or oceanic land masses, that is masses of similar density (cf. **subduction**).

Oligotrophic. Lacking in nutrients, though well-oxygenated: for example, soils whose sole external source of nutrients is rainwater.

Ontogeny, -etic. The genetically determined sequential development of an organism.

Ophiolite. Volcanic rock arising from magma extruded along the suture between two collided continental blocks.

Ordination. Any quantitative method for organising ecological data, such as soils or plant community data collected in plots, in such a way as their interrelationships can be evaluated.

Orographic. Induced by the presence of mountains.

Orthotropic (of branches). Ascending and with branching and leaf arrangement identical to that of the leader.

Osmotic potential. The potential of water to move from a less to more concentrated solution of salts across a semi-permeable membrane.

Pa. Pascal, the unit of vapour pressure, a measure of perpendicular force per unit area, equal to 1 kilogram per second per second.

p.a. (per annum). In or for each year.

Padang (bahasa Indonesia, Melayu). An open area; in forests often dense with short trees.

Padi. Unhusked rice, or the field in which rice is grown.

Pagoda tree. Term created by Corner for monopodial trees with plagiotropic branches, giving the appearance of a pagoda.

Palimpsest. A surface (of, for example, a text or landscape) in which early patterns (such as writing, vegetation formations) have been partly or largely effaced by later patterns.

Panchayat. Indian village leadership.

Panmixis. A population in which every individual has an equal chance of mating with any other.

Pantropical. Throughout the tropics.

Photosynthetically active radiation (PAR). The quantity of solar radiation (energy) that is of the wave lengths utilized in photosynthesis

Paya (Malay). Swamp.

Peneplain. The surface resulting from the complete erosion of a previously lifted one.

Peneplanation. The process of peneplain formation.

Perhumid. Constantly wet. In the present context: climates in which mean monthly rainfall always exceeds expected mean monthly evapotranspiration of a forest stand.

Perhumidity Index. See **Humidity Index**.

Participatory forest management (PFM). The generic term for forest management policies and methods in which rural communities share responsibility with government forest agencies.

Phenology. Seasonal or otherwise periodic changes in the presentation of plant parts, such as leaves, flowers, fruit.

Photoinhibition. Reduced photosynthesis caused by excessive radiation.

Photon flux density (PFD). Photons/unit area/unit time. The amount of photosynthetically active light in an area.

Phylogeny. The evolutionary history of an organism or group of organisms.

Physiognomy. Surface appearance.

Pioneer (species). Plant whose seed requires direct sunlight to germinate (see Chapter 2).

Plagiotropic. Growing at an angle approaching the horizontal.

Plasticity (phenotypic, evolutionary etc.). The genetically inherited ability to change in response to environment.

Pluton. Intrusive igneous rock raised from beneath the Earth's crust with lava, and slowly solidified (crystallised) within it.

Podsol. An oligotrophic soil in which the mineral soil is leached of nutrients, and in which iron and/or humic acids have been leached and reprecipitated in a pan, and acid organic raw humus has accumulated on the surface.

Primary (forest). Forest free of human interference; often more broadly applied to forest in which no trees have been felled by humans.

Primordium. The cells from which an organ will develop.

Probabilistic. Of or pertaining to probability, likelihood without certainty (see **determinism**).

Prolepsis. Expansion of an axial shoot in advance of laterals.

Protogyny. Maturity of female

before male functions.

Pseudogamy. Flowering plant reproduction in which pollination is a necessary stimulus for embryogenesis, but in which no male gametes, therefore no zygotes, are involved.

Random (genetic) drift. The variation of gene frequencies due to random deviation as a result of sampling accidents from one generation to the next.

RAPD. Rapid amplificaton of polymorphic DNA. A method of analysis of genetic differences between closely related organisms.

Rauh's model. The tree architectural form in which a monopodial orthotropic leader continuously gives rise to plagiotropic branches.

Raw humus. Acid-reacting cominnuted organic matter at the soils surface in which decomposition is slow and mediated by fungi.

REDD. The United Nations programme for Reduced Emissions from Deforestation and forest Degradation.

Reiteration. Restorative shooting which repeats the architecture of a previous form, specifically the regrowth of a vertical leading shoot from a lateral.

Rejuvenation (of soil). Redevelopment of a soil profile following its truncation.

Relative growth rate. Growth rate relative to others in its sample.

Rentier. A person receiving an annual income effectively from a government security that yields it.

'Reserve'. A measure of soil fertility used by the Sarawak Department of Agriculture, obtained through the inexpensive method of nutrient extraction by concentrated hydrochloric acid.

Rhyolite. An extrusive volcanic rock rich in silica, yielding soil of low nutrient content and acid reaction.

Richness. See **species richness**.

RIL. Reduced impact logging. A set of management practices which aim to minimise impact on forest stands and soils.

Rotundate. Globe shaped.

Sarcotesta. A fleshy seed coast, such as the flesh of the rambutan.

Scarp. The slope of a cuesta, generally the steeper, beneath the broken edge of the rock strata which defines the ridge.

Sclerenchyma. Cells whose walls are lignified, thereby losing wall permeability and dying.

Sclerophylly. Stiffening of leaves through sclenchyma development.

Secondary forest. Forest recovering from large scale tree felling, usually through swidden or logging intervention, but sometimes thus termed following destruction by hurricanes.

Sedimentary. Rocks derived from the cementing of deposited inorganic materials, usually from pulverisation of previous rocks.

Seed dormancy. Capacity of mature seed to survive while awaiting conditions suitable for germination.

Selection system. A method of forest management for timber whereby trees meeting certain economic and/or silvicultural criteria are felled, leaving trees, mostly of intermediate size, to yield the following crop (in contrast to shelterwood systems which rely on juveniles and therefore longer intervals between harvesting individual stands.

Semelparous. Capable of reproducing once only, followed by death.

Semi-scandent. Insufficiently strengthened to support its own weight.

Sesquioixides. Oxides of Group IV elements (which include iron and aluminium) in which two of their atoms are bound as cations with three anions of

oxygen.

Shear. Stress or deformation due to parallel surfaces passing one another.

Shelterwood. A method of forest management for timber, whereby all trees large enough to produce timber, therefore the whole forest canopy in a coup, is removed by one intense logging comprising a clear cut. The residual understorey is selectively thinned, as necessary, to free juveniles of preferred species from competition. To be contrasted with the **selection system**.

Shola. A forest patch in the Western Ghats, mostly referring to montane forests.

Shrub. A woody plant that branches from the ground as several shoots.

Sib, half-sib. An individual with two or one parents in common with another, respectively.

Sibumasu. One of three granite blocks (terranes) which, rafted from the north-east edge of Gondwana as the Tethys Sea opened during the Permian geological period. Now sandwiched between the other two in an S-shape, it comprises the granites of north Sumatra, the west of the Malaysian Peninsula, and the frontier mountains that divide Burma and Indo-china north to South China.

Siliceous. Rich in silica, sandy.

Slash. To cut woody vegetation, also used for the cut product.

South Asia. The nations of Indian Gondwana: Bangladesh, India, Pakistan.

Species diversity. The common measure of biodiversity. Here referring (unless stated otherwise) to tree species. α-diversity: a measure of the diversity within a community occupying a homogeneous physical environment. β-diversity: the diversity expressed across a physical environmental gradient, usually local and therefore of soils or topography.

601

Species richness. Used here to express the number of species in a given area (usually a region) irrespective of the variability of its physical environment (see also **diversity**)

Specific leaf area. The ratio of a leaf's area to its dry weight.

Stand. A patch of vegetation in which the conditions for its regeneration arose simultaneously. In forestry, a patch of forest in which the trees share the same management history; that is, a combination of development age, site conditions and species composition.

Stochasticity. Randomness, lack of predictability.

Stomatal conductance. The rate at which carbon doxide or water vapour can diffuse through the stomata.

Stratum (plural **strata**). Layer, especially detectable layering of sedimentary rock.

Subcanopy. Beneath the forest canopy.

Subduction. The process by which a heavier, generally oceanic tectonic plate descends beneath a lighter, generally continental plate.

Succession, successional. The process whereby a stand (q.v.) is regenerated to its full height, by competitive succession of species.

Sundaland, Sunda Shelf. The lands on the continental plates of western Malesia: the Malay peninsula south of the Kangar-Pattani Line, Sumatra, Borneo, Palawan, Java and intervening islands.

Sungei (Malay). River.

Sunlight utilisation efficiency. The efficiency with which photosynthesis uses solar radiative energy to synthesise sugars.

Surface truncation. Physical cutting off of surface layers.

Suture. The fusion line between two separate bodies, such as tectonic plates or valves of an ovary.

Sweepstakes dispersal. Dispersal by chance.

Swidden. Crop cultivation in which forest is cut and burned in preparation for planting a cereal, the resulting field abandoned once weeds invade and fertility declines, and the subsequent woody fallow left for several years before re-clearance (slash-and-burn agriculture).

Syllepsis. Simultaneous expansion of axial and lateral shoots.

Symbiosis. Physiological interdependence between different species of organism.

Sympodial (branching). Branching in which the apical bud aborts and growth is continued from lateral buds.

Taluk. Tenure.

Taungya. (Karen, Burma) Swidden. Now widely adopted as the term for planting trees in woody fallow by communities following swidden.

Taxon (plural **taxa**). A group of similar plants which, being presumed to be genetically distinct, are provided a formal name. Taxonomy involves the grouping of the smallest, most homogeneous taxa into higher and higher taxa, based on their morphological and, increasingly, genetic similarities.

Tectonic. The movement of the crustal plates of which Earth's surface is composed.

Terai. The raised terraces that front the length of the Himalaya.

Terrane. A fragment of continental plate.

Tessellated. Forming a surface from similar small pieces, such as tiles or leaves, with no, or minimum, overlap.

Tg. Teragram = 1 megatonne = 1 million metric tons = 10^{12} grams.

Trade winds. The predominantly, Equator directed, winds of the northern and southern seasonal tropical Hadley cells.

Transpiration. Loss of water, as vapour, from leaf tissues through stomata and cuticle.

Trapliner. An animal that visits its prey (including plant food) on a more or less consistent repeating circuit, like a hunter visiting traps.

Triploid. Having chromosomes in triples instead of pairs (diploid). Triploidy precludes sexual reproduction; triploid plans are therefore apomictic.

Troll's model. A tree architectural form in which an orthotropic leader with radial leaf arrangement grows to the horizontal with alternate leaf arrangement; after which a new leader arises from an axillary bud and repeats the process.

Turbidite. Sediments that have been deposited on the slopes of the continental shelf, where local channels and flow water lead to undulating strata and local folding. Predominantly fine-grained (clay).

Udult. Soils in which the rate of litter breakdown is greater than the rate of litter fall.

Ultisol (as used here) Yellow-red soils in which the colour of the mineral horizons intensify with depth, as evidence for leaching of clays and sesquioxides.

Ultramafic (= ultrabasic). Igneous rocks, notably ophiolites, that are low in silica but high in oxides of magnesium, iron, chromium and cobalt, often weathering to soils toxic to plants.

Understorey. The woody vegetation that is growing beneath the forest canopy.

Unit leaf rate. The gain in plant growth expressed by a standard area and time period.

Upland. Used here for any dry land, the mountains excepted.

Vesicular arbuscular. Symbiotic associations between fungi and plant roots wherein fungal hyphae enter the host cell and develop organs of exchange, the vesicles, within the host cytoplasm.

VJR. Virgin Jungle Reserve. Small areas of primary forest, established in Peninsular Malaysia following Indian Forest Service practice, serving to act as controls for silvicultural trials.

Wallacea. The biogeographic province of islands between the continental shelves of Asia (the Sunda Shelf) and Australasia (the Sahul Shelf).

Wallace's Line. The biogeographical divide, first recognized by naturalist Alfred Russel Wallace, which follows the eastern edge of tectonic continental Asia within Malesia, and includes the Philippine Islands and Bali to its west.

West Malesia. The plant geographic region west of Wallace's Line, and south of the Kangar-Pattani Line (see Chapter 6).

Whorl. Plant organs attached round an axis at the same linear position.

Windpruning. Leaf loss and abortion of young leaves in tree crowns due to abrasion and desiccation by wind, thereby smoothing crowns and canopies.

Wyall (Kannada). Grassy floodplains within forest.

Xenogamous. Self incompatible.

Xeric. Tending to be dry.

Xylem. The principal tissue of wood, comprising the tubular cells that conduct the water.

Zonal. Of an ecological region.

Zygote. The diploid cell resulting from the marriage of two haploid gametes; the fundamental process of sexual reproduction.

References

ABDUL RAHIM, A.S. & MOHD SHAWAHID, H.O. 2009. Short- and long-run effects of sustainable forest management practices on West Malaysian log supply: an ARDL approach. *J. Trop. Forest Sci.* 21: 369–376.

ABDUL RAHMAN, K., NIIYAMA, K., AZIZI, R., APPANAH, S. & IIDA, S. 2002. Species assembly and site preference of tree species in a primary seraya-ridge forest of Peninsular Malaysia. *J. Trop. Forest Sci.* 14: 287–303.

ABDULHADI, D.T., KATAWINATA, K., & SUKARDJO, S. 1981. Effects of mechanized logging in the lowland dipterocarp forest at Lempaka, East Kalimantan. *Malayan Forester* 44: 407–448.

ABE, T. & HIGASHI, M. 1991. Cellulose centered perspective on terrestrial community structure. *Oikos* 60: 127–133.

ABHILASH, E.S., MENON, A.R.R. & BALASUBRAMANIAN, K. 2005. Regeneration status in natural habitats of *Nageia wallichiana* (Presl.) O. Ktz., Goodrical Reserved Forest of western Ghats of India. *Indian Forester* 131: 183–200.

ACKERLY, D.D. 1996. Canopy structure and dynamics: Integration of growth processes in tropical pioneer trees. In: *Tropical Plant Ecophysiology*, eds S.S. Mulkey, R.L. Chazdon & A.P. Smith, pp. 619–658. Chapman & Hall, London & New York.

ACKERLY, D.D. 1997. Plant life histories: a meeting of phylogeny and ecology. *Trends Ecol. Evol.* 12: 7–9.

ACKERLY, D.D. 2000. Taxon sampling, correlated evolution, and independent contrasts. *Evolution* 54: 1480–1492.

ACKERLY, D.D. 2003. Community assembly, niche conservatism, and adaptive evolution in changing environments. *Int. J. Pl. Sci.* 144 (3 Suppl.) S165–184.

ACKERLY, D.D. & BAZZAZ, F.A. 1995a. Seedling crown orientation and interruption of diffuse radiation in tropical forest gaps. *Ecology* 76: 1134–1146.

ACKERLY, D.D. & BAZZAZ, F.A. 1995b. Leaf dynamics, self-shading and carbon gain in seedlings of a tropical pioneer tree. *Oecologia* 101: 289–298.

ACKERLY, D.D. & DONOGHUE, M.J. 1998. Leaf size, sapling allometry, and Corner's rules: phylogeny and correlated evolution in maples (*Acer*). *Amer. Naturalist* 152: 767–791.

ACKERLEY, D.D. & RAICH, P.B. 1999. Convergence and correlations among leaf size and function in seed plants: A comparative test using independent contrasts. *Amer. J. Bot.* 86: 1272–1281.

ACKERLY, D.D., DUDLEY, S.A., SULTAN, S.E., SCHMITT, J., COLEMAN, J., LINDER, C.R., SANDQUIST, D.R., GEBER, M.A., EVANS, A.S., DAWSON T.E. & LECHOWICZ, M.J. 2000. The evolution of plant ecophysiological traits: recent advances and future direction. *BioScience* 50: 979–995.

ADAMS, M. 2004. Negotiating nature: collaboration and conflict between Aboriginal and conservation interests in New South Wales, Australia. *Austral. J. Environ. Ed.* 20 (1): 3–12.

ADZMI, Y., SUHAIMI, W.C., AMIR HUSNI, M.S., MOHD GHAZALI, H., AMIR, S.K. & BAILLIE, I. 2009. Heterogeneity of soil morphology and hydrology on the 50 ha long term ecological research plot at Pasoh, West Malaysia. *J. Trop. Forest Sci.* 22: 21–35.

AGRAWAL, A. 2001. Common Property Institutions and Sustainable Governance of Resources. *World Devel.* 29: 1649–1672

AHAMEDULLAH, M. & NAYAR, M.P. 1986. *Endemic Plants of the Indian Region*. Botanical Survey of India. Calcutta.

AHMAD, W.Y.W. & RAHIM, A.R.A. 1998. Status of enrichment planting in peninsular Malaysia. *Malayan Forester* 61: 196–212.

AIBA, M. & NAKASHIZUKA, T. 2007. Variation in juvenile survival and related physiological traits among dipterocarp species coexisting in a Bornean forest. *J. Veg. Sci.* 18: 379–388.

AIBA, S.-I. & KITAYAMA, K. 1999. Structure, composition and species diversity in an altitude-substrate matrix of rainforest tree communities on Mount Kinabalu, Borneo. *Plant Ecol.* 140: 139–157.

AIBA, S.-I. & KITAYAMA, K. 2002. Effects of the 1997–98 El Niño drought on rain forests of Mount Kinabalu, Borneo. *J. Trop. Ecol.* 18: 215–230.

AIBA, S.-I. & KOHYAMA, T. 1996. Tree species stratification in relation to allometry and demography in a warm-temperate rain forest. *J. Ecol.* 84: 207–218.

AIBA, S.-I., & KOHYAMA, T. 1997. Crown architecture and life-history traits of 14 tree species in a warm-temperate rain forest: significance of spatial heterogeneity. *J. Ecol.* 85: 611–624.

AIBA, S. I., KITAYAMA, K. & TAKYU, M. 2004. Habitat associations with topography and canopy structure of tree species in a tropical montane forest on Mount Kinabalu, Borneo. *Plant Ecol.* 174: 147–161.

AIBA, S.- I., TAKYU, M. & KITAYAMA, K. 2005. Dynamics, productivity, and species richness of tropical rainforests along elevational and edaphic gradients on Mount Kinabalu, Borneo. *Ecol. Res.* 20: 279–286.

AIBA, S.-I., SUZUKI, E. & KITAYAMA, K. 2006. Structural and floristic variation among small replicate plots of a tropical montane forest on Mount Kinabalu, Sabah, Malaysia. *Tropics* 15: 219–236.

AIRY-SHAW, H.K. 1975. *The Euphorbiaceae of Borneo*. H.M. Stationary Office for Royal Botanic Gardens, Kew, London.

AIYAR, T.V.V. 1932. The sholas of the Palghat Division. A study of the ecology and silviculture of the tropical rain forests of Western Ghats. *Indian Forester* 58: 414–432.

ALEXANDER, I. 1989. Mycorrhizas in tropical forests. In: *Mineral Nutrients in Tropical Forests and Savanna Ecosystems*, ed. J. Proctor, pp. 169–188. Blackwell, Oxford.

ALEXANDER, I.J. & LEE, S.S. 2005. Mycorrhizas and ecosystem processes in tropical rain forests. In: *Biotic Interactions in the Tropics: their Role in the Maintenance of Species Diversity*, eds D.R.F.P. Burslem, M. Pinard & S. Hartley, pp. 204–225. Cambridge University Press.

ALEXANDER, I.J., AHMAD, N. & LEE, S.S. 1992. The role of mycorrhizas in the regeneration of some Malaysian forest trees. *Philos. Trans., Ser. B* 335: 379–388.

ALLEE, W.C. 1931. *Animal Aggregations. A Study in General Sociology*. University of Chicago Press, Chicago.

ALVERSON, W.S., KAROL, K.G., BAUM, D.A., CHASE, M.W., SWENSON, S.M., MCCOURT, R. & SYTSMA, K.J. 1998. Circumscription of the Malvales and relationships with other Rosidae: evidence from *rbcL* sequence data. *Amer. J. Bot.* 85: 876–887.

ALVERSON, W.S., WHITLOCK, B.A. NYFFELER, R., BAYER, C. & BAUM, D.A. 1999. Phylogeny of the core Malvales: evidence from *ndhF* sedquence data. *Amer. J. Bot.* 86: 1474–1486.

AMARASEKERE, P. 2000. The geometry of coexistence. *Biol. J. Linn. Soc.* 71: 1–31.

AMBASTA, S.P., ed. 1986. *The Useful Plants of India*. Council of Scientific and Industrial Research, New Delhi.

AMIRLINGAM, M. 2004. The sacred groves of Tamil Nadu. *Indian Forester* 130: 1279–1285.

ANDERSON, J.A.R. 1958. Précis of studies of fresh-water swamp forests of Sarawak and Brunei. *Study of Tropical Vegetation: Proceedings of the Kandy Symposium*, pp. 195–196. UNESCO, Paris and Government of Ceylon.

ANDERSON, J.A.R. 1961. The destruction of *Shorea albida* forest by an unidentified insect. *Empire Forest. Rev.* 40 (103): 19–30.

ANDERSON, J.A.R.1963. The flora of the peat swamp forests of Sarawak and Brunei, including a catalogue of all recorded species of flowering plants, ferns and fern allies. *Gard. Bull. Singapore* 20: 131–228.

ANDERSON, J.A.R. 1964. The structure and development of the peat swamps of Sarawak and Brunei. *J. Trop. Geogr.* 18: 7–16.

ANDERSON, J.A.R. 1965. Limestone habitat in Sarawak. *Proceedings of the Symposium on Ecological Research in Humid Tropics Vegetation*, pp. 49–57. Government of Sarawak & UNESCO Science Co-operation Office for South-East Asia.

ANDERSON, J.A.R. 1976. Observations on the ecology of five peat-swamp forests in Sumatra and Kalimantan. In: *Peat and Podsolic Soils in Indonesia*. Soil Research Institute Bogor, Bulletin 3: 45–55.

ANDERSON, J.A.R. 1980. *A Checklist of the Trees of Sarawak*. Forest Department, Sarawak.

ANDERSON, J.A.R. & MULLER, J. 1975. Palynological study of a Holocene peat and a Miocene coal deposit from NW Borneo. *Rev. Palaeobot. Palynol.* 19: 291–351.

ANDERSON, J.A.R., JERMY, A.C. & CRANBROOK, LORD. 1982. *Gunong Mulu National Park: A Management and Development Plan*. Royal Geographical Society, London.

ANDERSON, J.M., PROCTOR, J. & VALLAK, H.W. 1983. Ecological studies in tour contrasting lowland forests in Gunung Mulu National Park, Sarawak. III. Decomposition processes and nutrient losses from leaf litter. *J. Ecol.* 71: 503–527.

ANDRIESSE, J.P. 1969. A study of the environment and characteristics of tropical podzols in Sarawak East Malaysia. *Geoderma* 2: 210–226.

ANGIOSPERM PHYLOGENY GROUP. 2009. An update of the Angiosperm Phylogeny Group classification for the orders and families of flowering plants: APG III. *Bot. J. Linn. Soc.* 161: 105–121.

ANON. (E.G.C.) 1892. Fire caused by sparks. *Indian Forester* 18: 153.

ANON. (E.M.B.) 1892. The Andamans. *Indian Forester* 20: 155.

ANON. (L.M.) 1888. Forest set on fire by lightning. *Indian Forester* 14: 264.

ANON. 1891. The question of village forests. *Indian Forester* 17: 132–133.

ANON. 1953. Bearded pig swim again. *Malayan Nat. J.* 8: 53–56.

ANON. 1958a. *Evaporation and Evapotranspiration*. World Meteorological Union, Geneva.

ANON. 1958b. *Study of Tropical Vegetation: Proceedings of the Kandy Symposium on Humid Tropics Vegetation*. UNESCO, Paris and Government of Ceylon.

ANON. 1963. *Proceedings of the UNESCO Symposium on Humid Tropics Vegetation, Kuching*. Government of Sarawak & UNESCO Science Cooperation Office, Jakarta.

ANON. c.1964. *Sarawak Hydrological Year Book for the Water Year 1962–63*. Public Works Department, Sarawak.

ANON. 1982. *Report on Committee on Forest and Tribals in India*. Ministry Of Home Affairs Tribal Development Division, Government of India.

ANON. 1988. *Natural Forest Resources of the Philippines*. Philippine-German Forest Resources Inventory Project. Department of Forest Management, Department of Environment and Natural Resources, Manila.

ANON. 1990. *The Promotion of Sustainable Forest Management; a Case Study in Sarawak, Malaysia*. International Tropical Timber Organization. Unpubl. report.

ANON. 1991. *Sustainable Management of Natural Forests in the Philippines: Possibility or Illusion*. Philippine-German Dipterocarp Forest Management Project PN 88.2047.4

ANON. 1995a. Special issue on ecology and ethics. *Biodivers. & Conservation* 4 (8).

ANON. 1995b. *Sri Lanka Forestry Sector Master Plan*. Forestry Planning Unit, Ministry of Agriculture, Lands and Forestry. Battaramulla, Sri Lanka.

ANON. 1997a. *IPCC. An Assessment of Vulnerability*. Report, Working Group 2, Intergovernmental Panel on Climate Change. The United Nations.

ANON. 1997b. *Designing an Optimum Protected Areas System for Sri Lanka's Natural Forests*. Vol. 1. IUCN, WCMC, FAO.

ANON. 1999. Changes and disturbances in tropical rain forest in South-east Asia. (Symposium). *Philos. Trans., Ser. B* 354: 1391.

ANON. 2001. *Census of India*. (Electronic Resource). Office of the Registrar General, Government of India, New Delhi.

ANON. 2003. A conflict of interests, the uncertain future of Burma's forests. *A Briefing Document by Global Witness*, London.

ANON. 2004. Issue based on the proceedings of symposium 'Tropical Forests and Global Atmospheric Change'. *Philos. Trans., Ser. B* 359: 1443.

ANON. 2005. *Forests and Flood: Drowning in Fiction or Thriving on Facts?* RAP Publication 2005/3, Forest Perspectives 2, CIFOR & FAO.

ANON. 2006a– *The Wealth of India. A Dictionary of India's Raw Material and Industrial Products. Resources*. CSIR National Institute of Science Communication and Information Resources, New Delhi. URL: www.niscair.res.in

ANON. 2006b. *Heart of Borneo. Three Countries One Conservation Vision*. World Wildlife Fund Indonesia. Jakarta.

ANON. 2008. *A Common Vision on Biodiversity in Government and the Development Process. 1. Synthesis, 2. Reference Document; for Planners, Decision-Makers and Practitioners*. Ministry of Natural Resources and Environment, Putrajaya, Malaysia.

ANON. 2009. *Red Data List of Threatened and Endangered Peninsular Malaysia*. Forest Research Institute Malaysia (FRIM), Kepong.

ANON. 2010a. Council for Social Development. 2010. *Summary Report on Implementation of the Forest Rights Act*. Council for Social Development, New Delhi. URL: http://www.forestrightsact.com/component/k2/item/15

ANON. 2010b. Money can grow on trees. *The Economist*, 25th Sept.: 4–14.

ANON. 2011. *Gross National Happiness*. Centre for Bhutan Studies, Thimphu. URL: http://grossnationalhappiness.com

APPANAH, S. 1979. *The Ecology of Insect Pollination of Some Tropical Rain Forest Trees*. PhD thesis, University of Malaya.

APPANAH, S. 1982. Pollination of androdioecious *Xerospermum intermedium* Radlk. (Sapindaceae) in a rain forest. *Biol. J. Linn. Soc.* 18: 11–34.

APPANAH, S. 1985. General flowering in the climax rain forests of South-East Asia. *J. Trop. Forest Sci.* 1: 225–240.

APPANAH, S. 1990. Plant-pollinator interactions in Malaysian rain forests. In: *Reproductive Ecology of Tropical Forest Plants*, eds K.S. Bawa, & M. Hadley, pp. 85–101. UNESCO, Paris and the Parthenon Publishing Group, Carnforth.

APPANAH, S. 1993. Mass flowering of dipterocarp forests in the aseasonal tropics. In: *Diversity and Flexibility of Biotic Communities in Fluctuating Environments*, ed. M.K. Chandrashekaran, *J. Biosciences* 18 (4) special issue, pp. 457–474.

APPANAH, S. & CHAN, H.T. 1981. Thrips; the pollinators of some dipterocarps. *Malaysian Forester* 44: 234–252.

APPANAH, S. & TURNBULL, J.M. 1998. *A Review of Dipterocarps: Taxonomy, Ecology and Silviculture*. CIFOR, Bogor.

APPANAH, S. & WEINLAND, G. 1990. Will the management systems for hill forests stand up? *J. Trop. Forest Sci.* 3: 140–158.

APPANAH, S., WEINLAND, G., BOSSEL, H. & KRIEGER, H. 1989. Are tropical rain forests non-renewable? An enquiry through modeling. *J. Trop. Forest Sci.* 2: 331–348.

ARENTZ, F. 1983. *Nothofagus* dieback on Mt. Giluwe, Papua New Guinea. *Pac. Sci.* 37: 453–464.

ARMITAGE, F.B. & BURLEY, J., eds. 1980. *Pinus kesiya* Royle ex Gordon (syn. *P. insularis* Endl.). *Comm. For. Inst. Trop. For. Paper* 9. Oxford.

ARMSTRONG, J.E. & IRVINE, A.K. 1982. Flowering, sex ratios, pollen-ovule ratios, fruit set, and reproductive effort of a dioecioius tree, *Mysitica insipida*, in two different rain forest communities. *Amer. J. Bot.* 76: 74–85.

ARNOLD, M., KÖHLIN, G., PERSSON, R. & SHEPHERD, G. 2003. *Fuelwood Revisited: What has Changed in the Last Decade?* CIFOR Occasional Paper 39.

ARUNACHALAM, A. & ARUNACHALAM, K. 2002. Evaluation of bamboos in eco-restoration of 'jhum' fallows in Arunachal Pradesh: ground vegetation, soil and microbial biomass. *Forest Ecol. Managem.* 159: 231–239.

ARUNACHALAM, A. PANDEY, H.N., TRIPATHI, R.S. & MAITHANI, K. 1996. Biomass and production of fine and coarse roots during regrowth of a disturbed subtropical humid forest in north-east India. *Vegetatio* 123: 73–80.

ASH, J. 1988. *Nothofagus* (Fagaceae) forest on Mt. Giluwe, New Guinea. *New Zealand J. Bot.* 26: 245–258.

ASHTON, P.M.S. 1990a. Method for the evaluation of advanced regeneration in forest types of south and southeast Asia. *Forest Ecol. Managem.* 36: 163–175.

ASHTON, P.M.S. 1990b. *Seedling Response of Shorea Species Across Moisture and Light Regimes in a Sri Lankan Rain Forest*. PhD thesis, Yale University.

ASHTON, P.M.S. 1992. Some measurements of the microclimate within a Sri Lankan tropical rainforest. *Agr. Forest Meteorol.* 59: 217–235.

ASHTON, P.M.S. 1995. Seedling growth of co-occurring *Shorea* species in the simulated light environments of a rain forest. *Forest Ecol. Managem.* 72: 1–12.

ASHTON, P.M.S. 2003. Regeneration methods for dipterocarp forests of wet tropical Asia. *For. Chron.* 79: 263–267.

ASHTON, P.M.S. & BERLYN, G.P. 1992. Leaf adaptations of some *Shorea* species to sun and shade. *New Phytol.* 121: 587–596.

ASHTON, P.M.S. & PETERS, C.M. 1999. Even-aged silviculture in mixed moist tropical forest, with special reference to Asia. Lessons learned and myths perpetrated. *J. Forest. (Washington)* 97: 14–19.

ASHTON, P.M.S. & DE ZOYSA, N.D. 1990. Performance of *Shorea trapezifolia* (Thwaites) Ashton seedlings growing in different light regimes. *J. Trop. Forest Sci.* 1: 356–364.

ASHTON, P.M.S., GUNATLLEKE, C.V.S. & GUNATILLEKE, I.A.U.N. 1993a. A shelterwood method of regeneration for sustained timber production in *Mesua-Shorea* forest of Southwest Sri Lanka. In: *Ecology and Landscape Management in Sri Lanka, Proceedings from the International and Interdisciplinary Symposium, Colombo, Sri Lanka, 12–26 March 1990*, eds W. Erdelen, C. Preu, N. Ishwaran & C.M. Madduma Bandara pp. 255–274. V.J. Markgraf, Weikersheim.

ASHTON, P.M.S., GUNATILLEKE, C.V.S. & GUNATILLEKE, I.A.U.N. 1993b. A case for the evaluation and development of mixed-species even-aged plantation in Sri Lanka's lowland wet zone. In: *Ecology and Landscape Management in Sri Lanka, Proceedings from the International and Interdisciplinary Symposium, Colombo, Sri Lanka, 12–26 March 1990*, eds W. Erdelen, C. Preu, N. Ishwaran & C.M. Madduma Bandara, pp. 275–288. V.J. Markgraf, Weikersheim.

ASHTON, P.M.S., GUNATILLEKE, C.V.S. & GUNATILLEKE, I.A.U.N. 1995. Seedling survival and growth of four *Shorea* species in a Sri Lankan rain forest. *J. Trop. Ecol.* 11: 263–279.

ASHTON, P.M.S., GAMAGE, S., GUNATILLEKE, I.A.U.N. & GUNATILLEKE, C.V.S. 1997. Restoration of a Sri Lankan rain forest: using Caribbean pine *Pinus caribaea* as a nurse for establishing late successional tree species. *J. Appl. Ecol.* 34: 915–925.

ASHTON, P.M.S., GAMAGE, S., GUNATLLEKE, I.A.U.N. & GUNATILLEKE, C.V.S. 1998. Using Caribbean pine to establish mixed plantations: testing effects of pine canopy removal on plantings of rain forest tree species. *Forest Ecol. Managem.* 106: 211–222.

ASHTON, P.M.S., GUNATILLEKE, C.V.S., SINGHAKUMARA, B.M.P. & GUNATILLEKE, I.A.U.N. 2001a. Restoration pathways for rain forest in southwest Sri Lanka: a review of concepts and models. *Forest Ecol. Managem.* 154: 409–430.

ASHTON, P.M.S., MENDELSOHN, R., SINGHAKUMARA, B.M.P., GUNATILLEKE, C.V.S, GUNATILLEKE, I.A.U.N. & EVANS, A. 2001b. A financial analysis of rain forest silviculture in southwestern Sri Lanka. *Forest Ecol. Managem.* 154: 431–441.

ASHTON, P.M.S., SINGHAKUMARA, B.M.P. & GAMAGE, H.K. 2006. Interaction between light and drought affect performance of Asian tropical tree species that have differing topographic affinities. *Forest Ecol. Managem.* 221: 42–45.

ASHTON, P.S. 1958a. Notes on the primary vegetation and soils of Brunei. In: *Proceedings of the Symposium on Humid Tropics Vegetation, Tjiawi, Indonesia*, pp. 149–154. UNESCO Science Cooperation Office for SE Asia, Jakarta.

ASHTON, P.S. 1958b. *Summary Report on Expedition to Gunung Pagon Periok Area, Brunei, March–May 1958*. Unpubl. ms.

ASHTON, P.S. 1963. Notes on the formation of a rational classification within the mixed dipterocarp forest of Sarawak and Brunei, for forestry and land use planning. In: *Proceedings of the UNESCO Symposium on Humid Tropics Vegetation, Kuching*, Anon, pp. 185–198. Government of Sarawak & UNESCO Science Cooperation Office, Jakarta.

ASHTON, P.S. 1964a. *Manual of the Dipterocarp Trees of Brunei State*. Oxford University Press.

ASHTON, P.S. 1964b. Ecological Studies in the Mixed Dipterocarp forests of Brunei State. *Oxford For. Mem.* 25.

ASHTON, P.S. 1969. Speciation among tropical forest trees: Some deductions in the light of recent evidence. In: *Speciation in Tropical Environments*, ed. R.H. Lowe-McConnell, pp. 155–196. Academic Press, London.

ASHTON, P.S. 1970. The plants and vegetation of Bako National Park. *Malayan Naturalist J.* 24: 151–162.

ASHTON, P.S. 1972. The quaternary geomorphological history of western Malesia and lowland forest phytogeography. In: *The Quaternary Era in Malesia. Transactions of the Second Aberdeen Hull Symposium on Malesian Ecology*, eds P.S. Ashton & H.M. Ashton, pp. 35–49. Hull Geog. Dept. Misc. Ser. 13.

ASHTON, P.S. 1976a. Mixed dipterocarp forest and its variation with habitat in the Malayan lowlands; a re-evaluation at Pasoh. *Malaysian Forester* 39: 56–72.

ASHTON, P.S. 1976b. An approach to the study of breeding systems, population structure and taxonomy of tropical trees. In: *Tropical Trees: Variation, Breeding and Conservation*, eds J. Burley & B.T. Styles, pp. 35–42. Academic Press, Carnforth.

ASHTON, P.S. 1977. A contribution of rain forest research to evolutionary theory. *Ann. Missouri Bot. Gard.* 64: 694–705.

ASHTON, P.S. 1978. Crown characteristics of tropical trees. In: *Tropical Trees as Living Systems. Proceedings of the Fourth Cabot Symposium*, eds P.B. Tomlinson & M.H. Zimmermann, pp. 591–615. Cambridge University Press.

ASHTON, P.S. 1980a. Some geographic trends in morphological variation in the Asian tropics, and their possible significance. In: *Tropical Botany*, eds K. Larson & L.B. Holm-Nielson, pp. 35–48. New York: Academic Press.

ASHTON, P.S. 1980b. Dipterocarpaceae. In: *A Revised Handbook to the Flora of Ceylon*, eds M.D. Dassanayake & F.R. Fosberg, pp. 364–423. Amerind, New Delhi, for the Smithsonian Institution.

ASHTON, P.S. 1980c. The biological and ecological basis for the utilization of dipterocarps. *BioIndonesia* 7: 43–54.

ASHTON, P.S. 1981. The need for information regarding tree age and growth in tropical forests. In: *Age and Growth Rate of Tropical Trees: New Directions for Research*, eds F.H. Bormann & G. Berlyn, pp. 3–6. Yale University School of Forestry and Environmental Studies Bulletin 94.

ASHTON, P.S. 1982. Dipterocarpaceae. In: *Flora Malesiana* 1(9), ed. C.G.G.J. van

Steenis, pp. 237–552. Martinus Nijhoff Publishers, The Hague.

ASHTON, P.S. 1984. Biosystematics of tropical forest plants: a problem of rare species. In: *Plant Biosystematics*, ed. W.F. Grant, pp. 497–518. Academis Press, Toronto.

ASHTON, P.S. 1986. La regénération dans les forêts de plaine de l'interieur du Sud Vietnam, 10 ans suive pulverization aerienne de l'agent Orange comme défoliant. *Bois Forets Trop.* 211: 3–34.

ASHTON, P.S. 1988. Dipterocarp biology as a window to the understanding of tropical forest structure. *Annual Rev. Ecol. Syst.* 19: 347–370.

ASHTON, P.S. 1989. Dipterocarp reproductive biology. In: *Tropical Rain Forest Ecosystems. Biogeographical and Ecological Studies, Ecosystems of the World* 14B, eds H. Lieth & M.J.A. Werger, pp. 219–240. Elsevier, Amsterdam.

ASHTON, P.S. 1991. Towards a regional classification of the humid tropics of Asia. *Tropics* 1: 1–12.

ASHTON, P.S. 1992a. Plant conservation in the Malaysian region. In: *Harmony with Nature, Proceedings of the International Conference on Conservation of Tropical Biodiversity*, eds S.K. Yap & S.W. Lee, pp. 86–93. Malayan Nature Society, Kuala Lumpur.

ASHTON, P.S. 1992b. The structure and dynamics of tropical rain forest in relation to tree species richness. In: *The Ecology and Silviculture of Mixed Species Forests*, eds M.J. Kelty, B.C. Larson & C.D. Oliver, pp. 53–64. Kluwer, Boston.

ASHTON, P.S. 1993. The community ecology of Asian rain forests in relation to catastrophic events. *J. Biosciences* 18: 501–514.

ASHTON, P.S. 1995a. Towards a regional forest classification of the humid tropics of Asia. In: *Vegetation Science in Forestry*, eds E.O. Box, R.K. Peet, T. Masuzawa, I. Yamada, K. Fujiwara and P.F. Maycock, pp. 453–464. Dordrecht: Kluwer Handbooks of Vegetation Science Series, H. Lieth.

ASHTON, P.S. 1995b. Biogeography and ecology. In: *Tree Flora of Sabah and Sarawak* 1, eds E. Soepadmo & K.M. Wong, pp. xliii–li. Forest Research Institute Malaysia (FRIM), Kepong.

ASHTON, P.S. 1997. Before the memory fades: some notes on the indigenous forests of the Philippines. *Sandakania* 9: 1–19.

ASHTON, P.S. 1998a. Niche specificity among tropical trees: a question of scales. In: *Dynamics of Tropical Communities. 37th Symposium of British Ecological Society, Cambridge University*, eds D.M. Newbery, H.H. Prins & N.D. Brown, pp. 491–514. Blackwell Science, Oxford.

ASHTON, P.S. 1998b. A global network of plots for understanding tree species diversity in tropical forests. In: *Forest Biodiversity Research, Monitoring and Modeling. Conceptual Background and Old World Case Studies*, eds F. Dallmeier & J.A. Comiskey, pp. 47–62. UNESCO, Paris & Parthenon, Carnforth.

ASHTON, P.S. 1998c. Plant diversity in tropical forests. In: *Modern Trends in Ecology and the Environment*, ed. R.S.D. Ambasht, pp. 241–252. Backhuys, Leiden.

ASHTON, P.S. 2001. Patterns of species

variation in West and East Malaysia. *Malaysian Nat. J.* 55: 181–186.

ASHTON, P.S. 2002. *List of Species Seen at Kelani F.R., Malabodda (W. of 7 Virgins, Sri Lanka)*. Unpubl. ms.

ASHTON, P.S. 2003a. Floristic zonation on wet tropical mountains revisited. *Perspect. Pl. Ecol.* 6, 2: 87–104.

ASHTON, P.S. 2003b. *A Field Guide to the Forest Trees of Brunei Darussalam.* 1. University of Brunei Darussalam.

ASHTON, P.S. 2004a. Dipterocarpaceae. In: *Tree Flora of Sabah and Sarawak* 5, eds E. Soepadmo, L.G. Saw & R.C.K. Chung, pp. 63–388. Forest Research Institute Malaysia (FRIM), Kepong.

ASHTON, P.S. 2004b. Soils in the tropics. In: *Tropical Forest Diversity and Dynamism: Findings from a Large-Scale Plot Network*, eds E.C. Losos & E.G. Leigh jr., pp. 56–68. University of Chicago Press, Chicago.

ASHTON, P.S. 2005. Lambir's forest: the world's most diverse known tree assemblage? In: *Pollination Ecology and the Rain Forest*, eds D.J. Roubik, S. Sakai & A.A. Hamid., pp. 191–216. Springer, New York.

ASHTON, P.S. 2007. Asia's tropics are the most intensively used: contrasting conservation strategies between south and east. *Curr. Sci.* 93: 1538–1543.

ASHTON, P.S. 2008. The changing values of Malaysian forests: the challenge of biodiversity and its sustainable management. *J. Trop. Forest Sci.* 20: 282–291.

ASHTON, P.S. 2010. Conservation of Borneo biodiversity: do small lowland parks have a role, or are big inland sanctuaries sufficient? Brunei as an example. *Biodivers. & Conservation* 19: 343–356.

ASHTON, P.S. 2011. Myrtaceae. In: *Tree Flora of Sabah and Sarawak* 7, eds E. Soepadmo, L.G. Saw, R.C.K. Chung, & R. Kiew, pp. 87–330. Forest Research Institute Malaysia (FRIM), Kepong.

ASHTON, P.S. 2013. A view from the trees. *Raffles Bull. Zoology* Suppl. 29: 41–47.

ASHTON, P.S. & ASHTON, H.M.,(eds). 1972. *The Quaternary Era in Malesia. Transactions of the Second Aberdeen Hull Symposium on Malesian Ecology.* Hull Geog. Dept. Misc. Ser. 13.

ASHTON, P.S. & CTFS WORKING GROUP. 2004. Floristics and vegetation of the forest dynamics plots. In: *Tropical Forest Diversity and Dynamism: Findings from a Large-Scale Plot Network*, eds E.C. Losos & E.G. Leigh Jr., pp. 90–102. University of Chicago Press, Chicago.

ASHTON, P.S. & GUNATILLEKE, C.V.S. 1987. New light on the plant geography of Ceylon; I, Historical plant geography. *J. Biogeogr.* 14: 249–285.

ASHTON, P.S. & HALL, P. 1992. Comparisons of structure among mixed diperocarp forests of north-western Borneo. *J. Ecol.* 80: 459–481.

ASHTON, P.S. & LAFRANKIE, J.V. 1999. Patterns of tree species diversity among tropical rain forests. In: *The Biology of Biodiversity*, ed. M. Kato, pp. 161–177. Springer, Tokyo.

ASHTON, P.S., HOPKINS, M.J., WEBB, L.J., WILLIAMS, W.T. & PALMER, J. 1978a. Description, functioning and evolution of tropical forest ecosystems; the natural

forest: plant biology, regeneration and tree growth. In: Anon., *Tropical Forests Ecosystems*, pp. 180–214. Natural Resources Research 16. UNESCO-UNEP-FAO.

ASHTON, P.S., RISWAN, S. & KENWORTHY, J.B. 1978b. *Rainforests of the Far East and Latin America: Some comparisons*. Unpubl. ms.

ASHTON, P.S., GAN, Y.Y. & ROBERTSON, F.W. 1984. Electrophoretic and morphological comparisons in ten rain forest species of *Shorea* (Dipterocarpaceae). *Bot. J. Linn. Soc.* 89: 293–304.

ASHTON, P.S., GIVNISH, T.J. & APPANAH, S. 1988. Staggered flowering in the Dipterocarpaceae: new insights into floral induction and the evolution of mast fruiting in the aseasonal tropics. *Amer. Naturalist* 132: 44–66.

ASHTON, P.S., APPANAH, S. & CHAN, H.T. 1995. Tree prosthesis for crown access. *Selbyana* 16: 174–178.

ASHTON, P.S., BOSCOLO, M., LIU, J. & LaFRANKIE, J.V. 1999. A global programme in interdisciplinary forest research: the CTFS perspective. *J. Trop. Forest Sci.* 11: 180–204.

ASKEW, G.P. 1964. The mountain soils of the east ridge of Mt. Kinabalu. Royal Society Expedition to North Borneo 1961: reports. *Proc. Linn. Soc. Lond.* 175: 32–35.

ATH, C. DE 1992. A history of timber exports from Thailand with emphasis on the 1870–1937 period. *Nat. Hist. Bull. Siam. Soc.* 40: 49–66.

AUBRÉVILLE, A. & LEROY, J.-F. eds 1960-. *Flore du Cambodge, du Laos et du Vietnam*. Muséum National d'Histoire Naturelle, Paris.

AUDLEY-CHARLES, M.G. & HALLAM, A., eds 1988. *Gondwana and Tethys*. Geological Society Special Publication 37. Oxford University Press.

AUGSPURGER, C.K. 1983a. Seed dispersal of the tropical tree *Platypodium elegans*, and the escape of its seedlings from fungal infection. *J. Ecol.* 71: 759–771.

AUGSPURGER, C.K. 1983b. Offspring recruitment around tropical trees: changes in cohort distance with time. *Oikos* 20: 189–196.

AUGSPURGER, C.K. 1984. Seedling survival of tropical tree species: interactions of dispersal distance, light gaps and pathogens. *Ecology* 65: 1705–1712.

AUGSPURGER, C.K. & WILKINSON, H.T. 2007. Host specificity of pathogenic *Pythium* species: implications for tropical tree species diversity. *Biotropica* 39: 702–788.

AUSTIN, M.P., ASHTON, P.S. & GREIG-SMITH, P. 1972. The application of quantitative methods to vegetation survey. III. A re-examination of rain forest data from Brunei. *J. Ecol.* 60: 305–324.

AVERYANOV, L. 2003. Introduction. In: *Slipper Orchids of Vietnam, with an Introduction to the Flora of Vietnam*, eds L. Averyanov, P. Cribb, P.K. Loc, P.K. Hiep & N.T. Hiep, pp. 1–53. Royal Botanic Gardens, Kew.

AYYAPPAN, N. & PARTHASARATHY, N. 1999. Biodiversity inventory of trees in a large-scale permanent plot of tropical evergreen forest at Varagaliar, Anamalais, Western Ghats, India. *Biodivers. & Conservation* 8: 1533–1554.

AZUMA, H., GARCIA-FRANCO, J.-G., RICO-GREY, V. & THIEN, L.B. 2001. Molecular phylogeny of Magnoliaceae: the biogeography of tropical and temperate disjunctions. *Amer. J. Bot.* 88: 2275–2285.

BAAS, P. & VETTER, R.E. 1989. Growth rings in tropical trees. *Int. Assoc. Wood Anat. News Bull.* 10, n. ser.: 95–174.

BAAS, P., KALKMAN, K. & GEESINK, R. eds 1989. *The Plant Diversity of Malesia*. Kluwer, Dordrecht.

BABIK, W., BUTLIN, R.K., BAKER, W.J., PAPADOPULOS A.S.T., BOULESTEIX, M., ANSTETT M.-C., LEXER, C., HUTTON I. & SAVOLAINEN, V. 2009. How sympatric is speciation in the *Howea* palms of Lord Howe island? *Molec. Ecol.* 18: 3629–3638.

BACKER, P.F., DAVIES, S.J., MOKSIN, M. ISMAIL, M.Z. & SIMANJUNTAK, P.M. 1999. Leaf size distribution of understory plants in mixed dipterocarp and heath forests in Brunei. *J. Trop. Ecol.* 15: 123–128.

BAGCHI, S., GOYAL, S.P. & SANKAR, K. 2004. Herbivore density and biomass in a semi-arid dry deciduous forest of western India. *J. Trop. Ecol.* 20: 475–478.

BAGHEL, L.M.S. & GUPTA, A. 2002. Econometric analysis of production functions and technical change for bidi industry of Madhya Pradesh. *Indian Forester* 128: 551–561.

BAHUGUNA, V.G. 1986. Survey of public opinion for wildlife – a case study. *Indian Forester* 112: 874–880.

BAILEY, J.M. 1967. Chemical changes in a Sarawak soil after fertilisation and crop growth. *Plant Soil* 27: 33–52.

BAILLIE, I.C. 1972. *Further Studies on the Occurrence of Drought in Sarawak*. Sarawak Forest Department Soil Survey, Research Section F.7.

BAILLIE, I.C. 1975. Subsoil charcoal in the Upper Balleh. *Sar. Mus. J.* 23 (N.S.) 271–273.

BAILLIE, I.C. 1982. *Studies of Site-Forest Relationships, Sarawak*. UK Ministry of Overseas Development. Unpubl. ms.

BAILLIE, I.C. 1996. Soils of the humid tropics. In: *The Tropical Rain Forest*, 2nd ed., ed. P.W. Richards, pp. 256–286. Cambridge University Press.

BAILLIE, I.C. & ASHTON, P.S. 1983. Some soil aspects of the nutrient cycle in Mixed Dipterocarp Forests in Sarawak. In: *Tropical Rain Forest: Ecology and Management*, eds S.L. Sutton, T.C. Whitmore & A.C. Chadwick, p. 347–356. Blackwell, Oxford.

BAILLIE, I.C., ASHTON, P.S., COURT, M.N., ANDERSON, J.A.R., FITZPATRICK, E.A. & TINSLEY, J. 1987. Site characteristics and the distribution of tree species in Mixed Dipterocarp Forest on Tertiary sediments in central Sarawak, Malaysia. *J. Trop. Ecol.* 3: 201–220.

BAILLIE, I.C., EVANGELISTA, P.M. & INCIONG, N.B. 2000. Differentiation of upland soils on the Palawan ophiolitic complex, Philippines. *Catena* 39: 283–299.

BAILLIE, I.C., INCIONG, N.B. & EVANGELISTA, P.M. 2001. Soils and land use on lithologically diverse ophiolitic alluvia on the coastal plain of Palawan, Philippines. *J. Trop. Geogr.* 22: 1–14.

BAILLIE, I.C., TSHERING, K., DORJI, T., TAMANG, H.B., DORJI, T., NORBU, C., HUTCHEON, A.A. & BÄUMLER, R. 2004. Regolith and soils in Bhutan, Eastern Himalayas. *Eur. J. Soil. Sci.* 55: 9–27.

BAILLIE, I.C., BUNYAVEJCHEWIN, S. & MANOP, K. 2005. *Soils of Mae Ping Long Term Forest Ecological Research Plot, Northern Thailand.* Unpubl. ms.

BAILLIE, I.C., ASHTON, P.S., CHIN, S.P., DAVIES, S.J., PALMIOTTO, P.A., RUSSO, S.E. & TAN, S. 2006a. Spatial associations of humus, nutrients and soils in mixed dipterocarp forest at Lambir, Sarawak, Malaysian Borneo. *J. Trop. Ecol.* 22: 543–553.

BAILLIE, I.C., GUNATILLEKE, I.A.U.N., SENEVIRATNE, G., GUNATILLEKE, C.V.S. & ASHTON, P.S. 2006b. Preliminary characterisation of the physiography and soils of the 25 ha long-term ecological research plot at Sinharaja, south-western Sri Lanka. *Sri Lankan Forester* 29: 23–45.

BAKER, A.J.M., PROCTOR, J., VAN BALGOOY, M.M.J. & REEVES, R.D. 1992. Hyperaccumulation of nickel by the flora of the ultramafics of Palawan, Republic of the Philippines. In: *The Vegetation of Ultramafic (Serpentine) Soils, Proceedings of the First International Conference on Serpentine Ecology,* eds A.J.M. Baker, J. Proctor & R.D. Reeves, pp. 291–304. Intercept, Andover.

BAKER, P. 1985. Research item: catastrophic mortality of *Shorea leprosula* juveniles in a small gap. *Malaysian Forester* 48: 263–265.

BAKER, P.J. 1997. Seedling establishment and growth across forest types in an evergreen/deciduous forest mosaic in western Thailand. *Nat. Hist. Bull. Siam Soc.* 45: 17–41.

BAKER, P.J. & BUNYAVEJCHEWIN, S. 2006a. Bark thickness and the influence of forest fire on tree population structure in a seasonal evergreen tropical forest. *Nat. Hist. Bull. Siam Soc.* 54: 215–225.

BAKER, P.J. & BUNYAVEJCHEWIN, S. 2006b. Suppression, release and canopy recruitment in five tree species from a seasonal tropical forest in western Thailand. *J. Trop. Ecol.* 22: 521–529.

BAKER, P.J. & BUNYAVEJCHEWIN, S. 2009. Fire behavior and fire effects across the forest landscape of continental Southeast Asia. In: *Tropical Fire Ecology. Climate Change, Land Use and Ecosystem Dynamics,* ed. M.A. Cochrane, pp. 311–334. Springer & Paxis, Chichester.

BAKER, P.J. & WILSON, J.S. 2003. Coexistence and relative abundance in forest trees. *Nature* 422: 581–582.

BAKER, P.J., BUNYAVEJCHEWIN, S., OLIVER, C.D. & ASHTON, P.S. 2005. Disturbance history and historical stand dynamics of a seasonal tropical forest in western Thailand. *Ecol. Monogr.* 75: 317–343.

BAKER, P.J., BUNYAVEJCHEWIN, S. & ROBINSON, A.P. 2008. The impacts of large-scale, low-intensity fires on the forests of continental south-east Asia. *Int. J. Wildland Fire* 17: 782–792.

BAKER, T.R., BURSLEM, F.R.P. & SWAINE, M.D. 2003. Association between tree growth, soil fertility and water availability at local and regional scales in Ghanaian tropical rain forest. *J. Trop. Ecol.* 19: 109–125.

BAKER, T.R., PHILLIPS, O.L., MALHI, Y., ALMEIDA, S., ARROYO, L., DI FIORE, A.,

ERWIN, T., HIGUCHI, N., KILLEEN,T.J., LAURANCE, S.G., LAURANCE, W.F., LEWIS, S.L., MONTEAGUDO, A., NEILL, D.A., VARGAS, P.N., PITMAN, N.C., SILVA, J.N., MARTÍNEZ, R.V. 2004. Increasing biomass in Amazonian forest plots. *Philos. Trans., Ser. B* 359: 353–365.

BAKER, W.J., COODE, M.J.E., DRANSFIELD, J., DRANSFIELD, M.M., HARLEY, P., HOFFMANN, P. & JOHNS, R.J. 1998. Patterns of distribution of Malesian vascular plants. In: *Biogeography and Geological Evolution of S.E. Asia,* eds R. Hall & J.D. Holloway, pp. 243–258. Backhuys, Leiden.

BAKEWELL, S. 2010. *How to Live, or a Life of Montaigne. In One Question, and Twenty Attempts at an Answer.* Chatto & Windus & Other Press, New York.

BALA, N., KUMAR, P., SINGH, G. & LIMBA, N.K. 2007. Poverty alleviation and natural resource restoration through community participation: a case study in north-western Rajasthan. *Indian Forester* 133: 351–359.

BALAN MENON, P.K. 1956. *Siliceous Timbers of Malaya.* Malayan Forest Records 19. Forest Research Institute Malaysia (FRIM), Kepong.

BALASUBRAMANYAM, S., RATNAYAKE, S. & WHITE, R. 1993. The montane forests of the Horton Plains Nature Reserve. In: *Ecology and Landscape Management in Sri Lanka, Proceedings from the International and Interdisciplinary Symposium, Colombo, Sri Lanka, 12–26 March 1990,* eds W. Erdelen, C. Preu, N. Ishwaran & C.M. Madduma Bandara, pp. 95–108. V.J. Markgraf, Weikersheim.

BALL M.C., EGERTON, J.J.G., LEUNING, R., CUNNINGHAM, R.B. & DURING, P. 1997. Microclimate and grass adversely affects spring growth of seedling snow gum (*Eucalyptus pauciflora*). *Pl. Cell Environ.* 20: 155–166.

BALZER, J.L., DAVIES, S.J., NOOR, N.S.M., KASSIM, A.R. & LAFRANKIE, J.L. 2007. Geographical distributions in tropical trees: can geographical range predict performance and habitat association in co-occurring tree species? *J. Biogeogr.* 34: 1916–1926.

BALZER, J.L., DAVIES, S.J., BUNYAVEJCHEWIN, S. & NOOR, N.S.M. 2008. The role of desiccation tolerance in determining tree species distributions along the Malay-Thai Peninsula. *Funct. Ecol.* 22: 221–231.

BALZER, J.L., GREGOIRE, D.M., BUNYAVEJCHEWIN, S., NOOR, N.S.M. & DAVIES, S.J. 2009. Coordination of foliar and wood anatomical traits contributes to tropical tree distributions and productivity along the Malay-Thai Peninsula. *Amer. J. Bot.* 96: 2214–2223.

BANCROFT, H. 1933. A contribution to the geological history of the Dipteocarpaceae. *Geol. Fören. Förhandl.* 55 (1): 59–100.

BÄNFER, G., MOOG, U., FIALA, B., MOHAMED, M., WEISING, K. & BLATTNER, F.R. 2006. A chloroplast genealogy of myrmecophytic *Macaranga* (Euphorbiaceae) species in Southeast Asia reveals hybridization, vicariance, and long-distance dispersal. *Molec. Ecol.* 15: 4409–4424.

BARKER, M.G., PRESS, M.C. & BROWN, N.C. 1997. Photosynthetic characteristics of dipterocarp seedlings in three tropical rain forest light environments: a basis

for niche partitioning? *Oecologia* 112: 453–463.

BARLEY, K.P. 1964. Earthworms and the decay of plant litter and dung. *Proc. Austral. Soc. Anim. Prod.* 5: 236–240.

BARNETT, T.P., DUMENIL, L. & SCHLESE, U., ROECHNER, E. & LATIF, M. 1989. The effect of Eurasian snow cover on regional and global climate variations. *J. Atmos. Sci.* 46: 661–685.

BARRINGTON, A.H.M. 1931. Forest soil and vegetation in the Hlaing Forest Circle, Burma. *Burma Forest Bull., Ecol. Ser.* 25. Government Printing and Stationery, Rangoon.

BARRY, R.G. 1980. Mountain climates of New Guinea. In: *The Alpine Flora of New Guinea* 1, ed. P. van Royen, pp. 75–109. Cramer Verlag, Vaduz.

BASSET, Y. & NOVOTNY, V. 1999. Species richness of insect herbivore communities on *Ficus* in Papua New Guinea. *Biol. J. Linn. Soc.* 67: 477–499.

BASSET, Y., SAMUELSON, G.A., ALLISON, A. & MULLER, S.E. 1996. How many species of host-specific insects feed on a species of tropical tree? *Biol. J. Linn. Soc.* 59: 201–216.

BASSET, Y., CIZEK, L., CUÉNOUD, P., DIDHAM R.K., GUILHAUMON, F., MISSA, O., NOVOTNY, V., ØDEGAARD, F., ROSLIN, T., SCHMIDL, J., TISHECHKIN, A.K., WINCHESTER, N.N., ROUBIK, D.W., ABERLANC, H.-P., BAIL, J., BARRIOS, H., BRIDLE, J.R., CASTAÑO-MENESES, G., CORBARA, B., CURLETTI, G., DA ROCHA, W.D., DE BAKKER, D., DELABIE, J.H.C., DEJEAN, A., FAGAN, L.L., FLOREN, A., KITCHING, R.L., MEDIANERO, E., MILLER, S.E., DE OLIVEIRA, E.G., ORIVEL, J., POLLET, M., RAPP, M., RIBEIRO, S.P., ROISIN, Y., SCHMIDT, J.B., SØRENSEN, L. & LEPONCE, M. 2012. Arthropod diversity in a tropical forest. *Science* 338 (6113): 1481–1484.

BASU, R. 2000. Studies on sacred groves and taboos in Purulia District of West Bengal. *Indian Forester* 126: 1309–1318.

BATEMAN, A.J. 1947. Number of S-alleles in a population. *Nature* 160: 337.

BATEMEN, A.J. 1952. Self-incompatibility systems in angiosperms. 1. Theory. *Heredity* 6: 285–310.

BATLIWALA, S. & DHANRAJ, D. 2009. Gender myths that instrumentalise women: a view from the Indian frontline. *IDS Bulletin* 35: 11–18.

BAWA, K.S. 1974. Breeding systems of tree species of a lowland tropical community. *Evolution* 28: 85–92.

BAWA, K.S. 1990. Evolution of dioecy in flowering plants. *Annual Rev. Ecol. Syst.* 11: 15–39.

BAWA, K.S. 1994. *Conservation of Genetic Resources in the Dipterocarpaceae.* Unpubl. ms. for CIFOR.

BAWA, K.S. & HADLEY, M., eds. 1990. *Reproductive Ecology of Tropical Forest Plants.* UNESCO, Paris and the Parthenon Publishing Group, Carnforth.

BAWA, K.S. & MENON, S. 1997. Biodiversity monitoring: the missing ingredients. *Trends Ecol. Evol.* 12: 42.

BAWA, K.S. & OPLER, P.A. 1975. Dioecy in tropical forest trees. *Evolution* 29: 167–179.

BAWA, K.S. & SEIDLER, R. 1998. Natural Forest Management and Conservation

of Biodiversity in Tropical Forests. *Conservation Biol.* 12: 46–55.

BAWA, K.S., PERRY, D.R. & BEACH, J.H. 1982. Reproductive biology of tropical lowland rain forest trees. I. Sexual systems and incompatability mechanisms. *Amer. J. Bot.* 72: 331–345.

BAWA, K.S., ASHTON, P.S. & MD. NOR, SALLEH. 1990. Reproductive ecology of tropical forest plants: management issues. In: *Reproductive Ecology of Tropical Forest Plants,* eds K.S. Bawa & M. Hadley, pp. 3–13. UNESCO, Paris and the Parthenon Publishing Group, Carnforth.

BAWA, K.S., SEIDLER, R. & RAVEN, P.H. 2004. Reconciling conservation paradigms. *Conservation Biol.* 18: 859–860.

BAWA, K.S., PRIMACK, R.B. & OMMEEN, M. 2010. *Conservation Biology: a Primer for South Asia.* Universities Press, Hyderabad.

BAZZAZ, F.A. 1984. Demographic consequences of plant physiological traits. In: *Perspectives on Plant Population Ecology,* eds R. Dirzo & J. Sarukhan, pp. 324–346. J. Sinauer.

BAZZAZ, F.A. 1990. The response of natural ecosystems to the rising global CO_2 levels. *Annual Rev. Ecol. Syst.* 21: 167–196.

BAZZAZ, F.A. 1996. *Plants in Changing Environments: Linking Physiological, Population, and Community Ecology.* Cambridge University Press.

BAZZAZ, F.A. & PICKETT, S.T.A. 1980. Physiological ecology of tropical succession: a comparative review. *Annual Rev. Ecol. Syst.* 11: 287–311.

BEALES, E.W. 1973. Ordination: mathematical elegance and ecological naivité. *J. Ecol.* 61: 23–25.

BEAMAN, J.H. 1998. Preliminary enumeration of the summit flora, Mount Murud, Kelabit highlands, Sarawak. In: *A Scientific Journey Through Borneo. Bario. The Kelabit Highlands of Sarawak,* eds G. Ismail, G. & L. Din, pp. 51–82. Pelanduk Press, Selangor.

BEAMAN, J.H. & BEAMAN, R.S. 1989. Diversity and distribution patterns in the flora of Mount Kinabalu. In: *The Plant Diversity of Malesia,* eds P. Baas, K. Klakman & R. Geesink, pp. 147–160. Kluwer, Dordrecht.

BEAMAN, J.H. & BEAMAN, R.S. 1998. *The Plants of Mount Kinabalu 3. Gymnosperms and Non-Orchid Monocotyledons.* Natural History Publications (Borneo), Kota Kinabalu & Royal Botanic Gardens, Kew.

BEAMAN, J.H. & EDWARDS, P.J. 2007. *Ferns of Kinabalu: an Introduction.* Natural History Publications (Borneo). Kota Kinabalu, Sabah.

BEAMAN, J.H., ANDERSON, C. & BEAMAN, R.S. 2001. *The Plants of Mount Kinabalu 4. Dicotyledon Families Acanthaceaed to Lythraceae.* Natural History Publications (Borneo), Kota Kinabalu, Sabah.

BEAMAN, J.H., ANDERSON, C. & BEAMAN, R.S. 2004. *The Plants of Mount Kinabalu 5. Magnoliacedae to Winteraceae.* Natural History Publications (Borneo), Kota Kinabalu, Sabah.

BEARDSLEY, T. 1998. In the heat of the night. *Sci. Amer.* Oct.: 20.

BEBBER, D., BROWN, N. & SPEIGHT, M.

2002a. Drought & root herbivory in understorey *Parashorea* Kurz (Dipterocarpaceae) seedlings in Borneo. *J. Trop. Ecol.* 18: 795–804.

BEBBER, D., BROWN, N., SPEIGHT, M., MOURA-COSTA, P. & YAP, S.W. 2002b. Spatial structure of light and dipterocarp seedling growth in a tropical secondary forest. *Forest Ecol. Managem.* 157: 65–75.

BEBBER, D.P., BROWN, N.D., & SPEIGHT, M.R. 2003. Dipterocarp seedling population dynamics in Bornean primary lowland forest during the 1997–8 El Niño oscillation. *J. Trop. Ecol.* 19: 11–19.

BECCARI, O. 1904. *Wanderings in the Great Forests of Borneo*. Constable, London.

BECKER, C.D. & GHIMIRE, K. 2003. Synergy between traditional ecological knowledge and conservation science supports forest preservation in Ecuador. *Conserv. Ecol.* 8(1): 1. [online] URL: www.consecol.org/vol8/iss1/art1

BECKER, P. 1996. Sap flow in Bornean heath and dipterocarp forest trees during wet and dry periods. *Tree Physiol.* 16: 295–299.

BECKER, P. & WONG, M. 1993. Drought-induced mortality in tropical heath forest. *J. Trop. Forest Sci.* 5: 416–419.

BECKER, P., LEIGHTON, M. & PAYNE, J.B. 1985. Why tropical squirrels carry seeds out of source crowns. *J. Trop. Ecol.* 1: 183–186.

BECKER, P., RABENOLD, P.E., IDOL, J.R. & SMITH, A.P. 1988. Water potential gradients for gaps and slopes in a Panamanian tropical moist forest's dry season. *J. Trop. Ecol.* 4: 173–184.

BECKER, P., TYREE, M.T. & TSUDA, M. 1999. Hydraulic conductances of angiosperm versus conifers: similar transport sufficiency at the whole plant level. *Tree Physiol.* 19: 445–452.

BEDARD, P.W. 1958. Reconnaissance classification and mapping of Philippine forests. *Study of Tropical Vegetation: Proceedings of the Kandy Symposium on Humid Tropics Vegetation*, pp. 49–52. UNESCO, Paris and Government of Ceylon.

BEESON, C.F.C. 1941. *The Ecology and Control of the Forest Insects of India and the Neighboring Countries*. Vasant Press, Dehra Dun.

BEGON, M., TOWNSEND, C.R. & HARPER, J.L. 2006. *Ecology: from Individuals to Ecosystems*. Blackwell, Oxford.

BEHERA, B., GIRI, S., DASH, N.C., SAHN, J & SENAPATI, B.K. 1999. Earthworm bioindication of forest land use pattern. *Indian Forester* 125: 272–281.

BEHERA, M.D. & KUSHWAHA, S.P.S. 2007. An analysis of tree species in Subansiri district, Eastern Himalaya. *Biodivers. & Conservation* 16: 1851–1865.

BEHERA, N., JOSHI, S.K. & PATI, D.P. 1990. Root contribution to total soil metabolism in a tropical forest soil from Orissa, India. *Forest Ecol. Managem.* 36: 125–134.

BELL, G. 2001. Neutral macroecology. *Science* 293: 2413–2418.

BELL, G. 2005. The co-distribution of species in relation to the neutral theory of community ecology. *Ecology* 86: 1757–1770.

BENDER, F. 1983. *Geology of Burma*. Borntraeger, Berlin & Stuttgart.

BENEDICK, S., HILL, S.K., MUSTAFFA, N., CHEY, V.A., MARYATI, M., SEARLE, J.B., SCHILTHUIZEN, M. & HAMER, K.C. 2006. Impacts of rain forest fragmentation on butterflies in northern Borneo: species richness, turnover and the value of small fragments. *J. Ecol.* 43: 967–977.

BENNET, E.L. & REYNOLDS, C.J. 1993. The value of a mangrove area in Sarawak. *Biol. Conservation* 2: 359–375.

BENTLEY, B.L. 1979. Longevity of individual leaves in a tropical rain forest under-storey. *Ann. Bot. London* 43: 119–121.

BERGMAN, S.C., COFFIELD, D.Q., TALBOT, J.P. & GARRARD, R.A. 1996. Tertiary tectonic and magmatic evolution of western Sulawesi and the Makassar Strait, Indonesia: evidence for a Miocene continent-continent collision. In: *Tectonic Evolution of Southeast Asia*, eds R. Hall & D. Blundell. Geological Society Special Publication 106: 391–429. Geological Society, London.

BERMINGHAM, E. & DICK, C.W. 2005. Overview: The history and ecology of tropical rainforest communities. In: *Tropical Rainforests. Past, Present and Future*, eds E. Bermingham, C.W. Dick & C. Moritz, pp. 2–15. University of Chicago Press, Chicago.

BERRY, R.J. 1983. A steady and devout lover of natural history. *Biol. J. Linn. Soc.* 20: 1–2.

BERTAULT, J.-G. & SYST, P. 1997. An experimental comparison of different harvesting intensities with reduced-impact and conventional logging in East Kalimantan, Indonesia. *Forest Ecol. Managem.* 94: 209–218.

BHAGWAT, S. & RUTTE, C. 2006. Sacred groves: potential for biodiversity management. *Front. Ecol. Environ.* 4: 519–524.

BHAGWAT, S., KUSHALAPPA, C., WILLIAMS, P., BROWN, N. 2005. The role of informal protected areas in maintaining biodiversity in the Western Ghats of India. *Ecology and Society* 10 (1): 8. URL: www.ecologyandsociety.org/vol10/iss1/art8

BHAKAT, R.K. & PANDIT, P.K. 2003. Role of sacred groves in conservation of medicinal plants. *Indian Forester* 129: 224–232.

BHARGAVA, O.P. 1958. Tropical evergreen virgin forests of Andaman Islands. *Indian Forester* 84: 20–29.

BHATTACHARYA, A.K. & BANNERJEE, S. 2003. Relevance of carrying capacity and eco-development linkages for sustainable tourism. *Indian Forester* 129: 332–340.

BHUYAN, P., KHAN, M.L. & TRIPATHI, R.S. 2003. Tree diversity and population structure in undisturbed and human-impacted stands of tropical evergreen forest in Arunachal Pradesh, Eastern Himalaya, India. *Biodivers. & Conservation* 12: 1753–1773.

BIÊN, C. & THÌ, N.N. 1992. *Updated List of Cuc Phuong Flora*. B Lâm Nghi p, Hanoi. [in Vietnamese]

BIERREGAARD, R.O. JR, LOVEJOY, T.E., KAPOS, V., AUGUSTO DOS SANTOS, A. & HUTCHINGS, R.W. 1992. The biological dynamics of tropical rainforest fragments. *BioScience* 42: 859–866.

BIFFIN, E., CRAVEN, L.A., CRISP, M.D., & GADEK, P.A. 2006. Molecular systematics of *Syzygium* and allied genera (Myrtaceae): evidence from the chloroplast genome. *Taxon* 55: 79–94.

BIFFIN, E., HARRINGTON, M.G., CRISP, M.D., CRAVEN, L.A. & GADEK, P.A. 2007. Structural partitioning, paired-sites models and evolution of the ITS transcript in *Syzygium* and Myrtaceae. *Molec. Phylogenet. Evol.* 43: 124–139.

BIFFIN, E., LUCAS, E., CRAVEN, L.A., RIBEIRO DA COSTA, I., HARRINGTON, M.G. & CRISP, M.D. 2010. Evolution of exceptional species richness among lineages of fleshy-fruited Myrtaceae. *Ann. Bot. London* 106: 79–93.

BILLINGS, W.D. 1966. *Plants and the Ecosystem*. Wadsworth. Belmont, California.

BINOY, C.F. & MATTHEW, G. 2001. Butterflies visiting flower heads of *Terminalia paniculata* Roth. in Kerala, India. *Indian Forester* 127: 1185–1187.

BIRD, M.I., TAYLOR, D. & HUNT, C. 2005. Palaeoenvironments of insular Southeast Asia during the last glacial Period: a savanna corridor in Sundaland? *Quat. Sci. Rev.* 24: 2228–2242.

BIRD, M.I., BOOBYER, E.M., BRYANT, C., LEWIS, H.A., PAZ, V. & STEPHENS, W.E. 2007. A long record of environmental change from bat guano deposits in Makangit Cave, Palawan, Philippines. *TRSE Earth* 98: 59–69.

BISCHOFF, W., NEWBERY, D.M., LINGENFELDER, M., SCHNAECKEL, R., PETOL, G.H., MADANI, L. & RIDSDALE, C.E. 2005. Secondary succession and dipterocarp recruitment in Bornean rain forest after logging. *Forest Ecol. Managem.* 218: 174–192.

BISWAS, S.R. & MISBAHUZZAMAN, K. 2006. Structural composition of trees based on diameter class distribution in a dipterocarp forest of Bangladesh. *Indian Forester* 132: 1083–1089.

BLAK, M. Undated. *The Botanical Diversity in the Ayawasi arwea, Irian Jaya, Indonesia*. Unpubl. ms.

BLAKER, P. 1980. The alpine soils on the New Guinea mountains. In: *The Alpine Flora of New Guinea. Volume 1*, ed. P. van Royen, pp. 59–74. Cramer Verlag, Vaduz.

BLANC, L., MAURY-LECHON, G. & PASCAL, J.-P. 2000. Structure, floristic composition and natural regeneration of the forests of Cat Tien National Park, Vietnam: an analysis of the successional trends. *J. Biogeogr.* 27: 141–158.

BLANFORD, H.R. 1915. Some notes on the regeneration of In and Kanyin in Upper Burma division. *Indian Forester* 41: 78.

BLASCO, F. 1971. *Montagnes du Sud de l'Inde: Forêts, Savanes, Écologie*. Institut Français de Pondichéry 10.

BLASCO, F., BELLAN, M.F. & AIZPURU, M. 1996. A vegetation map of tropical continental Asia at scale 1:5 million. *J. Veg. Sci.* 7: 623–634.

BLASCO, F., WHITMORE, T.C. & GERS, C. 2000. A framework for the worldwide comparison of tropical woody vegetation types. *Biol. Conservation* 95: 175–189.

BLATTER, E. 1929. The flowering of bamboos, I. *J. Bombay Nat. Hist. Soc.* 33: 899–921.

BLATTER, E. 1930. The flowering of bamboos, II. *J. Bombay Nat. Hist. Soc.* 34: 135–141.

BLEIWEISS, R. 1998. Tempo and mode of hummingbird evolution. *Biol. J. Linn. Soc.* 65: 63–76.

BLOOMFIELD, C. 1954. A study of podzolisation. III. The mobilization of iron and aluminium by a species of rimu (*Dacrydium cupressinum*). *J. Soil. Sci.* 5: 39–45.

BLOOR, M.G. & GRUBB, P.G. 2003. Growth and mortality in high and low light: trends among shade-tolerant tropical rain forest species. *J. Ecol.* 91: 77–85.

BLUNDELL, A.G. & PEART, D.R. 2004. Seedling recruitment failure following dipterocarp mast fruiting. *J. Trop. Ecol.* 20: 229–231.

BODEN KLOSS, C. 1920. Quoted by H.N. Ridley, in Peninsular Siamese plants. *J. Fed. Malay States Mus.* 10: 65–126.

BON, E. 2002. Common property resources at the interface of agriculture and forestry policies: before and after 1988. In: *Traditional Ecological Knowledge, Conservation of Biodiversity and Sustainable Development*, eds D. Depommier & P.S. Ramakrishnan, pp. 190–209. Institut Français de Pondichéry.

BOND, W.J. 1989. The tortoise and the hare: ecology of angiosperm dominance and gymnosperm persistence. *Biol. J. Linn. Soc.* 36: 227–249.

BOOJH, R. & RAMAKRISHNAN, P.S. 1982a. Growth strategy of trees related to successional status. I. Architecture and extension growth. *Forest Ecol. Managem.* 4: 359–374.

BOOJH, R. & RAMAKRISHNAN, P.S. 1982b. Growth strategy of trees related to successional status. II. Leaf dynamics. *Forest Ecol. Managem.* 4: 375–386.

BOONKIRD, S.A., FERNANDES, E.C.M. & NAIR, P.K.R. 1984. Forest villages: an agroforestry approach to rehabilitating forest land degraded by shifting cultivation in Thailand. *Agrofor. Systems* 2: 87–102.

BOOSE, E.R., FOSTER, D.R., & FLUET, M. 1994. Hurricane impacts to tropical and temperate landscapes. *Ecol. Monogr.* 64: 309–400.

BOR, N.L. 1942. Relic vegetation of the Shillong plateau, Assam. *Indian Forest Rec. (Bot.)* 3: 153–195.

BORCHERT, R. 1983. Phenology and control of flowering in tropical trees. *Biotropica* 15: 81–89.

BORHAN MOHAMAD. 1993. *Some Aspects on the Demography and Growth of the Tropical Rain Forests of Brunei Darussalam*. Unpubl. ms., Universiti Brunei Darussallam/Royal Geographical Society conference on tropical rain forest research. Bandar Seri Begawan, 9–17 April 1993.

BOSCOLO, M & VINCENT, J.R. 1998. *Promoting Better Logging Practices in Tropical Forests. A Simulation Analysis of Alternative Regulations*. Harvard Institute for International Development Discussion Paper 652.

BOSE, I. 2010. *How Did the Indian Forest Rights Act, 2006, Emerge?* IPPG Discussion Paper Series No. 39, IPPG Programme Office, IDPM, School of Environment & Development, University of Manchester.

BOURDILLON, T.F. 1894. The forests of Travancore. *Indian Forester* 26: 458–470.

BOWONDER, B. 1982. Deforestation in India. *Int. J. Environ. Stud.* 18: 223–236.

BOX, E.O., PEET R.K., MASUZAWA T., YAMADA, I., FUJIWARA, K. & MAYCOCK, P.F., eds. 1995. *Vegetation Science in Forestry*. Kluwer, Dordrecht.

BOYD, E., GUTIERREZ, M., & CHANG, M. 2005. Adapting small-scale CDM sinks projects to low-income communities. *Tyndall Centre Working Paper* 71. Tyndall Centre for Climate Change Research, University of East Anglia.

BOYLE, W.A., GANONG, C.N., CLARK, D.B. & HAST, M.A. 2008. Density, distribution and attributes of tree cavities in an old growth tropical rain forest. *Biotropica* 38: 241–245.

BRADLEY, J.W. 1914. Flowering of kyauthaung bamboo, (*Bambusa polymorpha*) in the Prome Division, Burma. *Indian Forester* 40: 526–529.

BRADSHAW, C.J.A., SODHI, N., PEH, K.S. & BROOK, B.W. 2007. Global evidence that deforestation amplifies flood risk and severity in the developing world. *Global Change Biol.* 13: 2379–2395.

BRADY, N. & WEIL, R.R. 1999. *The Nature and Properties of Soils*. Prentice Hall, Upper Saddle River, New Jersey.

BRANDIS, D. 1897. Biological notes on Indian bamboos. *Indian Forester* 25: 1–25.

BRANDIS, D. 1903. Teak in evergreen forest. *Indian Forester* 29: 187–189.

BRANDIS, D. 1907. *Indian Trees: an Account of the Trees, Shrubs, Woody Climbers, Bamboos and Palms Indigenous or Commonly Cultivated in the British Indian Empire*. A. Constable, London. Constable, London.

BRASS, L.J. 1941. The 1938–39 expedition to the Snow Mountains, Netherlands New Guinea. *J. Arnold Arbor.* 22: 271–342.

BRAY, J.R. & CURTIS, J.T. 1957. The ordination of upland forest communities in southern Wisconsin. *Ecol. Monogr.* 27: 325–349.

BREARLEY, F. 2006. Project Barito Ulu Rekut Camp, Central Kalimantan, Indonesia. *Tropinet* 17 (1): 7.

BREARLEY, F.Q. 2012. Ectomycorrhizal associations of the Dipterocarpaceae. *Biotropica* 44: 637–648.

BREARLEY, F.Q., PROCTOR, J., SURIANTATA, NAGY, L., DALRYMPLE, G. & VOYSEY, B.C. 2007. Reproductive phenology over a 10-year period in a lowland evergreen rain forest in Central Borneo. *J. Ecol.* 98: 828–839.

BREMER, K. 1979. Taxonomy of *Memecylon* (Melastomataceae) in Ceylon. *Opera Bot.* 50: 1–32.

BREMER, K. 1983. Taxonomy of *Memecylon* (Melastomataceae) in Borneo. *Opera Bot.* 69: 5–57.

BRODRIBB, T.J. & FIELD, T.S. 2008. Evolutionary significance of a flat-leaved *Pinus* in Vietnamese rainforest. *New Phytol.* 178: 201–209.

BROKAW, N.V.L. 1987. Gap phase regeneration of the pioneer species in a tropical rain forest. *J. Ecol.* 75: 9–19.

BROOK, B.W., SODHI, N.S. & NG, P.K.L. 2003. Catastrophic extinctions follow deforestation in Singapore. *Nature* 424: 420–423.

BROOK, R.K. & McLACHLAN S.M. 2005. On using expert-based science to 'test' local ecological knowledge. *Ecology and Society* 10 (2): r3. [online] URL: www.ecologyandsociety.org/vol10/iss2/resp3/

BROOKFIELD, H.C. 1964. The ecology of highland settlement. *Amer. Anthropol.* 66: ii, 20–308.

BROWN, K. 1994. Approaches to valuing plant medicines: the economics of culture and the culture of economics. *Biodivers. & Conservation* 3: 734–750.

BROWN, K. 1997. Plain tales from the grasslands: extraction, volume and utilization of biomass in Royal Bardia National Park, Nepal. *Biodivers. & Conservation* 6: 59–74.

BROWN, L.D. & ASHMAN, D. 1996. Participation, social capital, and intersectoral problem solving: African and Asian cases. *World Developm.* 24: 1467–1479.

BROWN, M. & POWELL, J.M. 1974. Frost and drought in the highlands of Papua New Guinea. *J. Trop. Geogr.* 38: 1–6.

BROWN, M.R. & WHITMORE, T.C. 1980. *Agathis. Commonw. Forest Rev.* 59: 307–310.

BROWN, N., PRESS, M. & BEBBER, D. 1999. Growth and survivorship of dipterocarp seedlings: differences in shade persistence create a special case of dispersal limitation. *Philos. Trans., Ser. B* 354: 1847–1855.

BROWN, N.D. 1993. The implications of climate and gap microclimate for seedling growth conditions in a Bornean lowland forest. *J. Trop. Ecol.* 9: 153–168.

BROWN, N.D. & WHITMORE, T.C. 1992. Do dipterocarp seedlings really partition rain forest gaps? *Philos. Trans., Ser. B* 335: 369–378.

BROWN, W.H. 1917. The rate of growth of *Podocarpus imbricatus* at the top of Mount Banahao, Luzon, Philippine Islands. *Philipp. J. Sci., C* 12: 317–328.

BROWN, W.H. 1919. *Vegetation of the Philippine Mountains*. Bureau of Printing, Manila.

BROWN, W.H. & MATTHEWS, D.M. 1914. Philippine dipterocarp forests. *Philipp. J. Sci., A* 9: 413–491.

BROWNE, F.G. 1955. *Forest Trees of Sarawak and Brunei, and their Products*. Government Printing Offics, Kuching, Sarawak.

BROWNE, F.G. 1966. The biology of Malayan Scolytidae and Platypodidae. *Malayan Forest Rec.* 22.

BROWNE, F.G. 1968. *Pests and Deseases of Plantation Trees: an Annotated List of the Principal Species Occurring in the British Commonwealth*. Clarendon Press, Oxford.

BRÜHL, C.A., ELTZ, T. & LINSENMAIR, K.E. 2003. Size does matter – effects of tropical rainforest fragmentation on the leaf litter ant community in Sabah, Malaysia. *Biodivers. & Conservation* 12: 1371–1389.

BRUIJNZEEL, L.A. 1990. *Hydrology of Moist Tropical Forests and Effects of Conversion: a State of Knowledge Review*. Free University of Amsterdam.

BRUIJNZEEL, L.A. & PROCTOR, J. 1993. Hydrology and biogeochemistry of tropical montane forests: what do we really know? In: *Tropical Montane Forest*, eds L.S. Hamilton, J.O. Juvick & F.N. Scatena, pp. 25–46. USDA Forest Service, Puerto Rico.

BRUIJNZEEL, L.A. & VENEKLAAS, E.J. 1998. Climatic conditions and tropical montane forest productivity: the fog has not lifted yet. *Ecology* 79: 3–9.

BRUIJNZEEL, L.A., WATERLOO, M.J., PROCTOR, J., KUITERS, A.T. & KOTTERINK, B. 1993. Hydrological observations in montane rain forests on Gunung Silam, Sabah, Malaysia with special reference to the 'massenerhebung' effect. *J. Ecol.* 81: 145–167.

BRUIJNZEEL, L.A., SCATENA, F.N. & HAMILTON, L.S., eds. 2011. *Tropical Montane Cloud Forests. Science for Conservation and Management*. Cambridge University Press.

BRÜNIG, E.F. 1969. On the seasonality of drought in the lowlands of Sarawak (Borneo). *Erdkunde* 23: 127–133.

BRÜNIG, E.F. 1970. Stand structure, physiognomy and environmental factors in some lowland forests in Sarawak. *Trop. Ecol.* 11: 26–43.

BRÜNIG, E.F. 1971. On the ecological significance of drought in the equatorial wet evergreen (rain) forest of Sarawak, Borneo. *Transactions of the First Aberdeen-Hull Symposium on Malesian Ecology*, ed. J.R. Flenley, pp. 66–88.

BRÜNIG, E.F. 1974. *Ecological Studies in the Kerangas Forests of Sarawak and Brunei*. Borneo Literature Bureau, Kuching, Sarawak.

BRÜNIG, E.F. 1996. *Conservation and Management of Tropical Rain Forests: an Integrated Approach to Sustainability*. CABI, Wallingford.

BRUUN, T.B., DE NEERGAARD, A., LAWRENCE, D., & ZIEGLER, A.D. 2009. Environmewntal consequences of the demise of swidden cultivation in Southeast Asia: carbon storage and soil quality. *For. Ecol. Manag.* 255: 2347–2361.

BRYAN, M.B. 1981. *Studies of Timber Growth and Mortality in the Mixed Dipterocarp Forests of Sarawak*. Consulting Report to FAO Project MAL/76/008, Sarawak Forest Department, Malaysia.

BRYANT, R.L. 1994. Shifting the cultivator: the politics of teak regeneration in colonial Burma. *Modern Asian Studies* 28: 225–250.

BRYANT, R.L. 1997. *The Political Ecology of Forestry in Burma*. C. Hurst, London.

BRYANT, R.L. 2007. *Consuming Burmese Teak: Anatomy of a Violent Resource*. Paper 23. Department of Geography, Kings College, London.

BUCHANAN HAMILTON, F. 1807. *Journey from Madras through Mysore, Canara and Malabar: For the Express Purpose of Investigating the State of Agriculture, Arts and Commerce, the Religion, Manners, and Customs, the History Natural and Civil, and Antiquities*. Reprinted Asian Educational Services, New Delhi 1988.

BUCKLEY, B.M., ANCHUKAITIS, K.J., PENNY, D., FLETCHER, R., COOK, E.R., SANO M., NAM, LE C., WICHIENKEEO, A., MINH, T.T. & HONG, T.M. 2010. Climate as a contributing factor in the demise of Angkor, Cambodia. *Proc. Natl. Acad. Sci. U.S.A.*, B, 107: 6748–6752.

BUERGIN, R. & KESSLER, C. 2000. Intrusions and exclusions: democratization in Thailand in the context of environmental discourses and resource conflicts. *Geojournal* 52: 71–80.

BULLOCK, S.H., MOONEY, H.A. & MEDINA, E., eds. 1995. *Seasonally Dry Tropical Forests*. Cambridge University Press.

BUNGARD, R.A., PRESS, M.C. & SCHOLES, J.D. 2000. The influence of nitrogen on rain forest dipterocarp seedlings exposed to a large increase in radiance. *Pl. Cell Environ.* 23: 1183–1194.

BUNGARD, R.A., ZIPPERLIN, S.A., PRESS, M.C. & SCHOLES, J.D. 2002. The influence of nutrients on growth and photosynthesis of seedlings of two rainforest dipterocarp trees. *Funct. Pl. Biol.* 29: 505–515.

BUNYAVEJCHEWIN, S. 1983a. Analysis of the tropical dry deciduous forest of Thailand. I Characteristics of the dominance types. *Nat. Hist. Bull. Siam Soc.* 31: 109–122.

BUNYAVEJCHEWIN, S. 1983b. Canopy structure of the dry dipterocarp forest of Thailand. *Thai Forest Bull., Bot.* 14: 1–132.

BUNYAVEJCHEWIN, S. 1985. Analysis of the tropical dry deciduous forest of Thailand. II. Vegetation in relation to topographic and soils gradients. *Nat. Hist. Bull. Siam. Soc.* 33: 3–20.

BUNYAVEJCHEWIN, S. 1986. Ecological studies of tropical semi-evergreen rain forest at Sakhaerat, Nakhorn Ratchasima, north-east Thailand. I. Vegetation patterns. *Nat. Hist. Bull. Siam. Soc.* 34: 35–57.

BUNYAVEJCHEWIN, S. 1999. Structure and dynamics in seasonal dry evergreen forest in north-east Thailand. *J. Veg. Sci.* 10: 787–792.

BUNYAVEJCHEWIN, S., BAKER, P.J., LaFRANKIE, J.V., & ASHTON, P.S. 2001. Stand structure of a seasonal dry evergreen forest at Huai Kha Khaeng Wildlife Sanctuary, western Thailand. *Nat. Hist. Bull. Siam. Soc.* 49: 89–106.

BUNYAVEJCHEWIN, S., BAKER, P.J., LaFRANKIE, J.V. & ASHTON, P.S. 2002. Floristic composition of a seasonal dry evergreen forest at Huai Kha Khaeng Wildlife Sanctuary, western Thailand. *Nat. Hist. Bull. Siam. Soc.* 50: 125–134.

BUNYAVEJCHEWIN, S., LaFRANKIE, J.V., BAKER, P.J., KANZAKI, ASHTON, P.S. & YAMAKURA, T. 2003. Spatial distribution patterns of the dominant canopy dipterocarp species in a seasonal dry evergreen forest in western Thailand. *Forest Ecol. Managem.* 175: 87–101.

BUNYAVEJCHEWIN, S., BAKER, P.J., LaFRANKIE, J.V. & ASHTON, P.S. 2004. Structure, history and rarity in a seasonal evergreen forest in western Thailand. In: *Tropical Forest Diversity and Dynamism: Findings from a Large-Scale Plot Network*, eds E.C. Losos & E.G. Leigh Jr., pp.145–158. University of Chicago Press, Chicago.

BUNYAVEJCHEWIN, S., TAN, S., GUNATILLEKE, I.A.U.N., CHIN S. P., SENEVIRATNE, G. & BAILLIE, I. 2005. *Soils of some CTFS Plots in Asia*. International Forest Biology Field Course, Center for Tropical Forest Science, Khao Chong.

BUNYAVEJCHEWIN, S., LaFRANKIE, J.V., BAKER, P.J., DAVIES, S.J. & ASHTON, P.S. 2009. *Forest Trees of Huai Kha Khaeng Wildlife Sanctuary, Thailand. Data from the 50-Hectare Forest Dynamics Plot*. National Parks, Wildlife and Plant Conservation Department, Thailand.

BUNYAVEJCHEWIN, S., BAKER, P.J. & DAVIES, S.J. 2011. Seasonally dry forests in Continental Southeast Asia: structure, composition and dynamics. In: *The Ecology and Conservation of Seasonally Dry Forest in Asia*, eds W.J. McShea, S.J. Davies & N. Bhumpakphan, pp. 9–35. Rowman & Littlefield, Smithsonian Institution Scholarly Press, Washington, DC.

Buot, I.E. Jr & Okitsu, S. 1998. Vertical distribution and structure of the tree vegetation in the montane forest of Mt. Pulog, Cordillera mountain range, the highest mountain in Luzon Is., Philippines. *Veg. Sci.* 15: 19–32.

Buot, I.E. Jr & Okitsu, S. 1999. Leaf size zonation pattern of woody species along an altitudinal gradient on Mt Pulog, Philippines. *Pl. Ecol.* 145: 197–208.

Burgess, P.F. 1969. Ecological factors in the hill and mountain forests of the States of Malaya. *Malayan Nat. J.* 22: 119–128.

Burgess, P.F. 1970. An approach towards a silvicultural system for the hill forests of the Malay Peninsula. *Malayan Forester* 33: 126–134.

Burgess, P.F. 1971. Effect of logging on hill dipterocarp forest. *Malayan Nat. J.* 24: 231–237.

Burgess, P.F. 1975. Silviculture in the hill forests of the Malay Peninsula. *Malaysian Forest Dept. Res. Pamphl.* 52.

Burghouts, T.B.A., van Straalen, N.M. & Bruinzeel, L.A. 1998. Spatial heterogeneity of elements and litter in a Bornean rainforest. *J. Trop. Ecol.* 14: 477–506.

Burkill, I.H. 1935 (reprinted 1966). *A Dictionary of the Economic Products of the Malay Peninsula*. 2 vols. Crown Agents for the Colonies, Oxford.

Burley, J. & Styles, B.T., eds 1976. *Tropical Trees: Variation, Breeding and Conservation*. Academic Press, Carnforth.

Burslem, D.F.R.P. 1996. Differential responses to nutrients, shade and drought among tree seedlings of lowland tropical forest in Singapore. In: *The Ecology of Tropical Forest Tree Seedlings*, ed. M.D. Swaine, pp. 211–245. MAB/Parthenon.

Burslem, D.F.R.P. 2010. Phenology and pollination biology of co-occurring dipterocarps in Sabah. ATB Bali conference. Unpubl. ms.

Burslem, D.F.R.P. & Whitmore, T.C. 1999. Species diversity, susceptibility to disturbance and tree population dynamics in tropical rain forest. *J. Veg. Sci.* 10: 767–776.

Burslem, D.F.R.P., Turner, I.M. & Grubb, P.J. 1994. Mineral nutrient status of coastal hill dipterocarp forest and *Adinandra belukar* in Singapore: bioassays of nutrient limitation. *J. Trop. Ecol.* 10: 579–599.

Burslem, D.F.R.P., Grubb, P.J. & Turner, I.M. 1996. Responses to simulated drought and elevated nutrient supply among shade-tolerant tree seedlings of lowland tropical forest in Singapore. *Biotropica* 28: 636–648.

Burslem, D.F.R.P., Whitmore, T.C. & Denmark, N. 1998. A thirty-year record of forest dynamics from Kolombangara, Solomon Islands. In: *Forest Biodiversity Research, Monitoring and Modeling: Conceptual Background and Old World Case Studies*, eds F. Dallmeier & J.A. Comiskey, pp. 633–648. UNESCO, Paris & Parthenon, Carnforth.

Burslem, D.F.R.P., Whitmore, T.C.

& Brown, G.C. 2000. Short-term effects of cyclone impact and long-term recovery of tropical rain forest on Kolombangara, Solomon Islands. *J. Ecol.* 88: 1063–1088.

Burslem, D.R.F.P., Pinard, M. & Hartley, S., eds 2005. *Biotic Interactions in the Tropics: their Role in the Maintenance of Species Diversity*. Cambridge University Press.

Burtt Davy, J. 1938. The classification of tropical woody vegetation types. *Imp. For. Inst. Paper* 13.

Bush, M.B. & Whittaker, R.J. 1991. Krakatau: colonization patterns and hierarchies. *J. Biogeogr.* 18: 341–356.

Butcher, J. 2005. The moral authority of ecotourism: a critique. *Curr. Iss. Tour.* 8: 114–124.

Bystriakova, N., Kapos, V., Lysenko, I & Stapleton, C.M.A. 2003. Distribution and conservation status of forest bamboo biodiversity in the Asia Pacific region. *Biodivers. & Conservation* 12: 1833–1841.

Cai, Z.-Q. & Bongers, F. 2007. Contrasting nitrogen and phosphorus resorption efficiencies in trees and lianes from a tropical montane rain forest in Xishuangbanna South-west China. *J. Trop. Ecol.* 23: 115–118.

Cai, Z-Q., Schnitzer, S.A., Wen, B., Chen, Y.-J. & Bongers, F. 2009. Liana communities in three forest types in Xishuangbanna, south-west China. *J. Trop. Forest Sci.* 21: 252–264.

Caldecott, J. & Caldecott, S. 1985. A horde of pork. *New Sci.* 15, August: 32–35.

Caldecott, T. 2006. *The Divine Science of Life*. Elsevier/Mosby.

Callister, D.J. 1992. *Illegal Tropical Timber Trade: Asia-Pacific*. TRAFFIC International, Cambridge.

Campbell, E.J.F. & Newbery, D. Mc. 1993. Ecological relationships between lianes and trees in lowland rain forest in Sabah, East Malaysia. *J. Trop. Ecol.* 9: 469–490.

Caner, L., Seen, D.L., Gunnell Y., Ramesh B.R. & Bourgeon G. 2007. Spatial heterogeneity of land cover response to climatic change in the Nilgiri highlands (Southern India) since the last glacial maximum. *Holocene* 17: 195–205.

Canham, C.D. & Burbank, D.H. 1994. Causes and consequences of resource heterogeneity in forests: interspecific variation in light transmission by canopy trees. *Canad. J. Forest Res.* 24: 337–348.

Cannon, C.H. & Manos, P.S. 2003. Phylogeography of the Southeast Asian stone oaks (*Lithocarpus*). *J. Biogeogr.* 30: 211–226.

Cannon, C.H., Peart, D.R., Leighton, M., & Kartawinata, K. 1994. The structure of lowland rainforest after selective logging in West Kalimantan, Indonesia. *Forest Ecol. Managem.* 67: 49–68.

Cannon, C.H., Oh, S-H., Harting, J.R. & Manos, P.S. c.2005. *Diversification of Southeast Asian Rain Forests Ancient Origins and Current Disequilibrium*. Unpubl. ms.

Cannon, C.H., Morley, R.J. & Bush, A.B.G. 2009. The current refugial rainforests of Sundaland are unrepresentative of their biogeographic past and highly sensitive to disturbance. *Proc. Natl. Acad. Sci. U.S.A.* 106:

11188–11193.

Cao, C.P., Finkeldey, R., Siregar, I.Z., Siregar, U.J., & Gailing, O. 2006. Genetic diversity within and among populations of *Shorea leprosula* Miq. and *Shorea parvifolia* Dyer (Dipterocarpaceae) in Indonesia detected by AFLPs. *Tree Genet. Genomes* 2: 225–239.

Cao, K.-F. 2000. Water relations and gas exchange of tropical saplings during a prolonged drought in a Bornean heath forest, with reference to root architecture. *J. Trop. Ecol.* 16: 101–116.

Cao, M. & Zhang, J. 1997. Tree species diversity of tropical forest vegetation in Xishuangbanna, SW China. *Biodivers. & Conservation* 6: 995–1006.

Cao, M., Zhang, J., Feng, Z., Deng, J. & Deng, X. 1996. The species composition of a seasonal rain forest in Xishuangbanna, Southeast China. *Trop. Ecol.* 37: 183–192.

Cao, M., Zhou, H., Wang, H., Lan, G.Y., Hu, Y.H., Zhou, S.S., Deng, X.B. & Cui, J.Y. 2008. *Xishuangbanna Tropical Seasonal Rainforest Dynamics Plot: Tree Distribution Maps, Diameter Tables and Species Documentation*. Yunnan Science and Technology Press, Kunming.

Carrington, D. 2011. Scientists warn Amazon drought may mean it switches to emitting carbon. *The Guardian*, 4 Feb.: 26.

Carrión-Tacuri, A.E., Rubio-Casal, A.E., Figueroa, M.E. & Castillo, J.M. 2011. *Lantana camara* L.: a weed with great light-acclimation capacity. *Photosynthetica* 49: 321–329.

Carson, W.P. & Schnitzer, S.A., eds. 2008. *Tropical Forest Community Ecology*. Wiley-Blackwell, Chichester.

Carson, W.P., Anderson, J.T., Leigh, E.G. Jr, Schnitzer, S.A. 2008. Challenges associated with testing – falsifying the Janzen-Connell hypothesis: a review and critique. In: *Tropical Forest Community Ecology*, eds W.P. Carson & S.A. Schnitzer, pp. 210–241. Wiley-Blackwell, Chichester.

Caruso, E. & Modi, A. 2004. *Village Forest Protection Committees in Madhya Pradesh: an Update and Critical Evaluation*. Forest Peoples Programme. URL: www.forestpeoples.org/region/india/publication/2010/village–forest–protection–committees–madhya–pradesh–update–and–critica

Cassou, J. & Depommier, D. 2002. Management and dynamics of Palmyra (*Borassus flabellifer* L.) stands in Tamil Nadu: the end of a traditional agroforestry system? In: *Traditional Ecological Knowledge, Conservation of Biodiversity and Sustainable Development*, eds D. Depommier & P.S. Ramakrishnan, pp. 210–223. Institut Français de Pondichéry.

Cavalier, J. & Goldstein, G. 1989. Mist and fog interception in elfin cloud forests in Colombia and Venezuela. *J. Trop. Ecol.* 5: 309–322.

Cavender-Bares, J., Kitajima, K. & Bazzaz, F.A. 2004. Multiple trait associations in relation to habitat differentiation among 17 Floridian oak species. *Ecol. Monogr.* 74: 635–662.

Chabot, B.F. & Hicks, D.J. 1982. The ecology of leaf life spans. *Annual Rev. Ecol. Syst.* 13: 229–259.

Chadwick, O.A., Derry, L.A., Vitousek, P.M., Huebert, B.J. & Hedin, L.O. 1999. Changing sources of nutrients during four million years of ecosystem development. *Nature* 397: 491–497.

Chai, E.O.K. 1988. *Regression Models for Shorea Species in Three Primary Hill Mixed Dipterocarp Forests in Sarawak*. Forest Department, Sarawak.

Chakrabarty, K. Kumar, A. & Menon, V. 1994. *Trade in Agarwood*. TRAFFIC-India Publications, New Delhi.

Chambers, J.Q., Higuchi, N. & Schimel, J.P. 1998. Ancient trees in Amazonia. *Nature* 391: 135–136.

Champion, H.G. 1920. Letter to the *Indian Forester* 46: 152–154.

Champion, H.G. 1936. A preliminary survey of the forest types of India. *Ind. Forest Rec. N.S., Silviculure* 1 (1).

Champion, H.G. 1938. *The Manual of Indian Silviculture. General Silviculture for India*. Publication Branch, Dept. of Printing and Stationery, Government of India, Delhi.

Champion, H.G. 1975. Indian silviculture and research over the century. *Indian Forester* 101: 3–8.

Champion, H.G. & Seth, S.K. 1968a. *A Revised Survey of the Forest Types of India*. Govt. of India Press, Nasik.

Champion, H.G. & Seth, S.K. 1968b. *General Silvicuture for India*. Government of India, Publication Branch, Delhi.

Chan, H.T. 1977. *Reproductive Biology of some Malaysian Dipterocarps*. PhD thesis, University of Aberdeen.

Chan, H.T. 1981. Reproductive biology of some Malaysian dipterocarps. III. Breeding systems. *Malaysian Forester* 44: 28–36.

Chan, H.T. & Appanah, S. 1980. Reproductive biology of some Malayan dipterocarps. 1. Flowering biology. *Malaysian Forester* 43: 132–143.

Chana, S.S. 1991. Hollong-Mikai forests of Arunachal Pradesh, their management and regeneration. *Indian Forester* 117: 162–167.

Chanderbali, A.S., Werff, H. & van der Renner, S.S. 2001. Phylogeny and historical biogeography of Lauraceae: Evidence from the chloroplast and nuclear genomes. *Ann. Missouri Bot. Gard.* 88: 104–134.

Chandrakanth, M.G., Gilless, J.K., Gowramma, V. & Nagaraja, M.G. 1990. Temple forests in India's forest development. *Agrofor. Systems* 11: 199–211.

Chandrakanth, M.G., Mahadev G. B. & Accavva, M.S. 2004. Socio economic changes and sacred groves in South India: protecting a community based resource management institution. *Natural Resources Forum* 28 (2). Blackwell Publishing.

Chandrasekar-Rao, A. & Sunquist, M.E. 1996. Ecology of small mammals in tropical forest habitats of southern India. *J. Trop. Ecol.* 12: 561–571.

Chandrasekhara, U.M. & Ramakrishnan, P.S. 1993. Germinable soil seed bank dynamics during the gap phase of a humid topical forest in the Western Ghats of Kerala, India. *J. Trop. Ecol.* 9: 455–461.

609

CHANDRASHEKARA, U.M. & SANKAR, S. 1998. Ecology and management of sacred groves in Kerala, India. *Forest Ecol. Managem.* 112: 165–177.

CHANDRASEKHARA, U.M., MUALEEDHARAN, P.K. & SIBICHAM, V. 2001. Disturbed shola forests of Kerala and strategies for its conservation and management. In: *Shola Forests of Kerala: Environment and Biodiversity*, eds K.K.N. Nair, S.K. Khanduri & K. Balasubramanyan, pp. 395–437. Kerala Forest Research Institute.

CHANDRASEKHARAN, C. 1962a. A general note on the vegetation of Kerala State. *Indian Forester* 88: 440–441.

CHANDRASEKHARAN, C. 1962b. Ecological study of the forests of Kerala state. *Indian Forester* 88: 473–480.

CHANDRASEKHARAN, C. 1962c. Forest types of Kerala state. *Indian Forester* 88: 660–674, 731–742, 837–847.

CHANDRASHEKARAN, M.K., ed. 1993. Diversity and flexibility of biotic communities in fluctuating environments. *J. Biosciences* 18(4), special issue.

CHANG-FU, H., ZUENG-SANG, C., YUEH-MEI, H., KUOH-CHIANG, Y. & TSUNG-HSIN, H. 1998. Altitudinal zonation of evergreen broad-leaved forest on Mount Lopei, Taiwan. *J. Veg. Sci.* 9: 201–212.

CHAPIN, M. 2004. A challenge to conservationists. *Worldwatch Magazine*, Worldwatch Institute.

CHAPMAN, C.A., CHAPMAN, L.J., STRUHSAKER, T.T., ZANNE, A.E., CLARKE, C.J. & POULSEN, J.R. 2005. A long-term evaluation of fruiting phenology: importance of climate change. *J. Trop. Ecol.* 21: 31–45.

CHAPMAN, G.P. 1997, ed. *The Bamboos*. Academic Press.

CHARLESWORTH, B. 1984. Androdioecy and the evolution of dioecy. *Biol. J. Linn. Soc.* 22: 333–348.

CHARLESWORTH, B. 2007. Why bother? The evolutionary function of sex. *Daedalus* 136: 37–46.

CHATTERJEE, D. 1939. Studies on the endemic flora of India and Burma. *J. Roy. Asiat. Soc. Bengal (Sci.)* 5: 19–67.

CHATURVEDI, A.N. 1992. History of forests in India. *Indian Forester* 118: 729–734.

CHAUHAN, B.S. 1991. Wildlife management in Himachal Pradesh. *Indian Forester* 117: 896–900.

CHAVARRÍA-VILLASEÑOR, G. 1996. *Systematics and Behavior of the Neotropical Bumble Bees (Hymenoptera: Apidae: Bombus)*. PhD thesis, Harvard University.

CHAVE, J. 2004. Neutral theory and community ecology. 2004. *Ecol. Lett.* 7: 241–253.

CHAVE, J. 2008. Spatial variation in tree species composition across tropical forests. In: *Tropical Forest Community Ecology*, eds W.P. Carson & S.A. Schnitzer, pp. 11–30. Wiley-Blackwell, Chichester. [Quotes R.M. May, 1973.]

CHAVE, J., MULLER-LANDAU, H.C. & LEVIN, S.A. 2002. Comparing classical community models: theoretical consequences for patterns of diversity. *Amer. Naturalist* 159: 1–23.

CHAZDON, R.L. 1986. Light variation and carbon gain in rainforest understorey

palms. *J. Ecol.* 74: 995–1012.

CHAZDON, R.L. & FETCHER, N. 1984. Photosynthetic light environments in a lowland tropical rain forest in Costa Rica. *J. Ecol.* 72: 553–564.

CHAZDON, R.L. & FIELD, C.B. 1987. Determinants of photosynthetic capacity in six rainforest *Piper* species. *Oecologia* 73: 222–230.

CHAZDON, R.L., PEARCY, R., LEE, D. & FETCHER, N. 1996. Photosynthesis response of tropical forest plants to contrasting light environments. In: *Tropical Plant Forest Ecophysiology*, eds S.S. Mulkey, R.L. Chazdon & A.P. Smith, pp. 5–55. Chapman & Hall, London & New York.

CHEAH, L.C. 1995. *Pioneer Species for Fast Growing Tree Plantations in Malaysia – and Evaluation*. FRIM Technical Information 53.

CHENGAPPA, B.S. 1934. Andaman forests and their reproduction. *Indian Forester* 60: 117–119, 185–198.

CHENGAPPA, B.S. 1937. Reproduction of the Andaman Forests. *Indian Forester* 63: 16–29.

CHENGAPPA, B.S. 1944. The Andaman forests and their regeneration. *Indian Forester* 70: 297–304, 339–351, 380–385, 421–431.

CHESSON, P.L. & WARNER, R.R. 1981. Environmental variability promotes coexistence in lottery competitive systems. *Amer. Naturalist* 117: 923–943.

CHESTER, E.G. 1892. Effects of the cyclone of November 2nd 1891 on the forests of the Andamans. *Indian Forester* 18: 156–157.

CHHANGANI, A.K. 2004a. Flowering and fruiting. *Indian Forester* 131: 771–784.

CHHANGANI, A.K. 2004b. Leaf phenology of plants of a dry deciduous forest in the Aravalli Hills of Rajasthan, India. *Indian Forester* 131: 1095–1104.

CHIARUGI, A. 1933. Legni fossile della Somalia Italiana, paleontologia della Somalia. *Paleontog. Ital.* 32 (suppl. 1): 97–167.

CHICHILNISKY, G. 1994. North-south trade in the global environment. *Amer. Econ. Rev.* 84: 851–874.

CHIEN, P.D. 2006. *Demography of Threatened Tree Species in Vietnam*. Published PhD thesis, University of Utrecht.

CHIN, S.C. 1977. The limestone flora of Malaya. I. *Gard. Bull. Sing.* 30: 165–219.

CHIN, S.C. 1979. The limestone flora of Malaya. II. *Gard. Bull. Sing.* 32: 64–203.

CHITWADGI, S.S. 2001. Sustainable harvest of non-wood forest produce. *Indian Forester* 127: 1420–1421.

CHIVERS, D.J. 1971. The Malayan siamang. *Malayan Nat. J.* 24: 78–86.

CHIVERS, D.J. 1972. The siamang and the gibbon in the Malay Peninsula. In: *Gibbon and Siamang* 1, ed. D.M. Rumbaugh, pp. 103–135. Karger, Basel.

CHIVERS, D.J. 1973. Introduction to the socio-ecology of Malayan forest primates. In: *Comparative Ecology and Behaviour of Primates*, eds R.P. Michael & J.H. Crook, pp. 101–146. Academic Press, London.

CHIVERS, D.J. 1974a. The Siamang in Malaya. A field study of a primate in tropical rain forest. *Contrib. Primatol.* 4: 1–333. Kargedr, Basel.

CHIVERS, D.J. 1974b. Daily patterns of ranging and feeding in siamang. *IBP Synthesis Meeting, Kuala Lumpur, 12–18 August, 1974*.

CHOMNITZ, K.M. & KUMARI, K. 1998. The domestic benefits of tropical forests: a critical review. *World Bank Res. Obs.* 13: 13–35.

CHOPA, K. & GULATI, S.C. 2001. *Migration, Common Property Resources and Environmental Degradation*. SAGE Publications.

CHOWDHURY, M.A.M., HUDA, M.K. & ISLAM, A.S.M.T. 2000. Phytodiversity of *Dipterocarpus turbinatus* Gaertn. F. (Gurjan) undergrowth at Dulahazara gurjan forest, Cox's Bazar, Bangladesh. *Indian Forester* 126: 674–684.

CHRISTENSEN, J. 2005. Are we consuming too much? *Conservation & Practice* 4: 15.

CHURCH, P. 2006. *A Short History of South-East Asia*. Wiley, Singapore.

CLARKE, C.B. 1898. On the subsubareas of British India, illustrated by the distribution of the Cyperaceae. *J. Linn. Soc.* 34: 1–146.

CLARK, D.A. & CLARK, D.B. 1994. Climate-induced variation in canopy tree growth in a Costa Rican tropical rain forest. *J. Ecol.* 82: 865–872.

CLARK, D.A. & CLARK, D.B. 2011. Assessing tropical forests' climatic sensitivities with long-term data. *Biotropica* 43: 31–40.

CLARK, D.A., PIPER, S.C., KEELING C.D., & CLARK, D.B. 2003. Tropical rain forest tree growth and atmospheric carbon dynamics linked to international temperature variation during 1984–2000. *Proc. Natl. Acad. Sci. U.S.A.* 100: 5852–5857.

CLARK, J.S. 1998. Why trees migrate so fast: confronting theory with dispersal biology and the paleorecord. *Amer. Naturalist* 152: 204–225.

CLEARY, D. & PIADJATI, A. 2005. Vegetation response to burning in a rain forest in Borneo. *Pl. Ecol.* 177: 145–163.

CLEMENTS, F.E. 1936. Nature and structure of the climax. *J. Ecol.* 24: 252–284.

CLENNELL, B. 1996. Far-field and gravity tectonics in Miocene basins of Sabah, Malaysia. In: *Tectonic Evolution of Southeast Asia*, eds R. Hall & D. Blundell, pp. 307–320. Geological Society Special Publication 106. Geological Society, London.

CLIFT. P.D. & PLUMB, R.A. 2008. *The Asian monsoon: Causes, History and Effects*. Cambridge University Press.

CLIMAP PROJECT MEMBERS. 1976. The surface of the ice-age Earth. *Science* 191, 4232: 1131–1137.

CO, L., LAGUNZAD, D.A., LaFRANKIE, J.V., BARTOLOME, N.A., MOLINA, J.E., YAP, S.L., GARCIA, H.G., BAUTISTA, J.P., GUMPAL, E.C., ARANOAVIES, S.J. 2004. Palanan forest dynamics plot, Philippines. In: *Tropical Forest Diversity and Dynamism: Findings from a Large-Scale Plot Network*, eds E.C. Losos & E.G. Leigh Jr., pp. 574–584. University of Chicago Press, Chicago.

CO, L., LaFRANKIE, J.V., LAGUNZAD, D.A., PASION, K.A.C., CONSUNJI, H.T., BARTOLOME, N.A., YAP, S.L., MOLINA, J.E., TONGCO, M.D.C., FERRERAS, U.F., DAVIES, S.J. & ASHTON, P.S. 2006. Forest trees of Palanan, Philippines. A Study in

Population Ecology. Center for Integrative and Development Studies, University of the Philippines, Diliman.

COCHRANE, J. 2007. Ecotourism and biodiversity conservation in Asia: institutional challenges and opportunities. In: *Critical Issues in Ecotourism: Understanding a Complex Tourism Phenomenon*, ed. J. Higham, pp. 294–305. Elsevier Butterworth-Heinemann, Oxford.

COCHRANE, M.A. 2009. *Tropical Fire Ecology. Climate Change, Land Use, and Ecosystem Dynamics*. Springer & Praxis, Chichester.

COCKBURN, P.F. 1972. Fagaceae. In: *Tree Flora of Malaya*, ed. T.C. Whitmore, pp. 196–232. Longmans.

COCKBURN, P.F. 1974. Phenology of dipterocarp trees in Sabah. IBP Synthesis Meeting. Unpubl. ms.

COHEN, A.L., SINGHAKUMARA, B.M.P. & ASHTON, P.M.S. 1996. Releasing rain forest succession: a case study in the *Dicranopteris linearis* fernlands of Sri Lanka. *Restoration Ecol.* 3: 261–270.

COLENETTE, P. 1958. *The Geology and Mineral Resources of the Jesselton-Kinabalu area, North Borneo*. Mem. 6, Geological Survey Deptartment, British Territories Borneo.

COLENETTE, P. 1964. A short account of the geology and geolocial history of Mt. Kinabalu. In: *A Discussion on the Results of the Royal Society Expedition to North Borneo, 1961*, ed. E.J.H. Corner. *Proc. Roy. Soc. London, Ser. B., Biol. Sci.* 161: 56–63.

COLEY, P.D. 1983. Herbivory and defensive characteristics of tree species in a lowland tropical forest. *Ecol. Monogr.* 53: 209–233.

COLEY, P.D. & KURSAR, T. 1996. Anti-herbivore defenses of young tropical leaves: physiological constraints and ecological trade-offs. In: *Tropical Plant Forest Ecophysiology*, eds S.S. Mulkey, R.L. Chazdon & A.P. Smith, pp. 305–336. Chapman & Hall, London & New York.

COLEY, P.D., BRYANT, J.P. & CHAPIN, F.S. III. 1985. Resource availability and plant antiherbivore defence. *Science* 230: 895–899.

COLFER, C.J.P., DAHA, G.R., & CAPISTRANO, D., eds. 2008. *Lessons from Forest Decentralization: Money, Justice and the Quest for Good Governance in Asia-Pacific*. Earthscan, London.

COLLINS, N.M. 1980. The distribution of soil macrofauna on the West Ridge of Gunung (Mount) Mulu, Sarawak. *Oecologia* 44: 263–275.

COLLINS, N.M. 1983. Termite populations and their role in litter removal in Malaysian rain forests. In: *Tropical Rain Forest: Ecology and Management*, eds S.L. Sutton, T.C. Whitmore & A.C. Chadwick, pp. 311–325. Blackwell, Oxford.

COLLIS, M. 1936. *Siamese White*. Faber & Faber, London.

COMITA, L.S., CONDIT, R. & HUBBELL, S.P. 2007. Developmental stages in habitat associations in tropical trees. *J. Ecol.* 95: 482–492.

COMITA, L.S., MULLER-LANDAU, H.C., AGUILAR, S. & HUBBELL, S.P. 2010.

Asymmetric density dependence shapes species abundances in a tropical tree community. *Science* 329: 330–332.

CONDIT, R. 1998. *Tropical Forest Census Plots: Methods and Results from Barro Colorado Island, Panama and comparison with other Plots*. Springer, Berlin.

CONDIT, R., HUBBELL, S.P., LAFRANKIE, J.V., SUKUMAR, R., MANOKARAN, N., FOSTER, R.B. & ASHTON, P.S. 1996. Species-area and species-individual relationships for tropical trees: a comparison of three 50 ha plots. *J. Ecol.* 84: 549–562.

CONDIT, R., ASHTON, P.S., MANOKARAN, N., LAFRANKIE, J.V., HUBBELL, S.P. & FOSTER, R.B. 1999. Dynamics of the forest communities at Pasoh and Barro Colorado: comparing two 50 ha plots. *Philos. Trans., Ser. B* 354: 1739–1748.

CONDIT, R., ASHTON, P.S., BAKER, P., BUNYAVEJCHEWIN, S., GUNATILLEKE, S., GUNATILLEKE, N., HUBBELL, S.P., FOSTER, R.B., ITOH, A., LAFRANKIE, J.V., LEE, H.S., LOSOS, E., MANOKARAN, N., SUKUMAR, R. & YAMAKURA, T. 2000. Spatial patterns in the distribution of tropical tree species. *Science* 288: 1414–1418.

CONDIT, R., PITMAN, N., LEIGH, E.G. JR., CHAVE, J., TERBORGH, J., FOSTER, R.B., NÚÑEZ, P., AGUILAR, S., VALENCIA, R., VILLA, G., MULLER-LANDAU, H.C., LOSOS, E. & HUBBELL, S.P. 2002. Beta diversity in tropical forest trees. *Science* 295: 666–669.

CONDIT, R., LEIGH, E.G. JR., LOO DE LAO, S. & CTFS WORKING GROUP. 2004. Species-area relationships and diversity measures in the forest dynamics plots. In: *Tropical Forest Diversity and Dynamism: Findings from a Large-Scale Plot Network*, eds E.C. Losos & E.G. Leigh Jr., pp. 79–89. University of Chicago Press, Chicago.

CONDIT, R., ASHTON, P., BALSLEV, H., BROKAW, N., BUNYAVEJCHEWIN, S., CHUYONG, G., CO, L., DATTARAJA, H.S., DAVIES, S., ESUFALI, S., EWANGO, C.E.N., FOSTER, R., GUNATILLEKE, N., GUNATILLEKE, S., HERNANDEZ, C., HUBBELL, S., JOHN, R., KENFACK, D., KIRATIPRAYOON, S., HALL, P., HART, T., ITOH, A., LAFRANKIE, J., LIENGOLA, I., LAGUNZAD, D., LAO, S., LOSOS, E., MAGARD, E., MAKANA, J., MANOKARAN, R., NAVARETTE, H., MOHAMMED NUR, S., OKHUBO, T., PÉREZ, R., SAMPER, C., HUA SENG, L., SUKUMAR, R., SVENNING, J.C., TAN, S., THOMAS, D., THOMPSON, J., VALLEJO, M., VILLA MUÑOZ, G., VALENCIA, R., YAMAKURA, T. & ZIMMERMAN, J. 2005a. Tropical tree α-diversity: results from a worldwide network of large plots. *Biol. Skr.* 55: 565–582.

CONDIT, R., PÉREZ, R., LAO, S., AGUILAR, S., & SOMOZA, A. 2005b. Geographic ranges and b-diversity: discovering how many tree species there are and where. *Biol. Skr.* 55: 57–71.

CONGDON, G. 1982. The vegetation of Tarutao National Park. *Nat. Hist Bull. Siam Soc.* 30: 135–198.

CONKLIN, H.C. 1957. *Hanunóo Agriculture. A Report on an Integral System of Shifting Agriculture in the Philippines*. FAO, Rome.

CONNELL, J.H. 1971. On the role of natural enemies in preventing competitive exclusion in some marine animals and rain forest trees. In: *Dynamics*

of Populations, eds P.J. den Boer, & G.P. Gradwell, pp. 298–312. Centre for Agricultural Publication and Documentation, Wageningen.

CONNELL, J.H. 1978. Diversity in tropical rain forests and coral reefs. *Science* 199: 1302–1310.

CONNELL, J.H. 1979. Tropical rain forests and coral reefs as open non-equilibrium systems. In: *Population Dynamics*, eds R.M. Anderson, L.R. Taylor, & B.D. Turner, pp. 141–163. Blackwell, Oxford.

CONNELL, J.H. & LOWMAN, M.D. 1989. Low-diversity tropical rain forests: some possible mechanisms for their existence. *Amer. Naturalist* 134: 88–119.

CONNELL, J.H., TRACEY, J.G. & WEBB, L.J. 1984. Compensatory recruitment, growth, and mortality as factors maintaining rain forest tree diversity. *Ecol. Monogr.* 54: 141–164.

CONSTANZA, R., D'ARGE, R., GROOT, R. DE, FARBER, S., GRASSO, M., HANNON, B., LIMBURG, K., NAEEM, S., O'NEILL, R.V., PARUELO, J., RASKIN, R.G., SUTTON, P. & BELT, M. VAN DEN. 1997. The value of the world's ecosystem services and natural capital. *Nature* 387: 253–260.

CONTI, E., ERIKKSON, T., SCHÖNENBERGER, J., SYTSMA, K.J. & BAUM, D.A. 2002. Early Tertiary out-of-India dispersal of the Crypteroniaceae: evidence from phylogeny and molecular dating. *Evolution* 56: 1931–1942.

COOMES, D.A. & GRUBB, P.J. 1998. Responses of juvenile trees to above- and below-ground competition in nutrient-starved Amazonian rain forest. *Ecology* 79: 768–782.

COORAY, P.G. 1967. *An Introduction to the Geology of Ceylon*. National Museums of Ceylon, Colombo.

CORBERA, E., BROWN, K. & ADGER, W.N. 2007. The equity and legitimacy of markets for ecosystem services. *Developm. Change* 38: 587–613.

CORLETT, R.T. 1984. Human impact on the subalpine vegetation of Mt. Wilhelm, Papua New Guinea. *J. Ecol.* 72: 841–854.

CORLETT, R.T. 1988a. Bukit Timah: the history and significance of a small rain-forest reserve. *Environm. Conservation* 17: 37–44.

CORLETT, R.T. 1988b. The naturalized flora of Singapore. *J. Biogeogr.* 15: 657–665.

CORLETT, R.T. 1990. Flora and reproductive phenology of the rain forest of Bukit Timah, Singapore. *J. Trop. Ecol.* 6: 55.

CORLETT, R.T. 1992. The ecological transformation of Singapore, 1819–1919. *J. Biogeogr.* 19: 411–420.

CORLETT, R.T. 1998. Frugivory and seed dispersal by vertebrates in the Oriental (Indomalayan) region. *Biol. Rev.* 73: 413–448.

CORLETT, R.T. 2002. Frugivory and seed dispersal in degraded tropical East Asian landscapes. In: *Seed Dispersal and Frugivory: Ecology, Evolution and Conservation*, eds D.J. Levey, W.W.R Silva & M. Galetti, pp. 451–465. CABI, New York.

CORLETT, R.T. 2004. Flower visitors and pollination in the Oriental (Indo-Malayan) Region. *Biol. Rev.* 79: 492–532.

CORLETT, R.T. 2009a. *The Ecology of Tropical East Asia*. Oxford University Press.

CORLETT, R.T. 2009b. Seed dispersal distances and plant migration potential in tropical East Asia. *Biotropica* 41: 592–598.

CORLETT, R.T. & PRIMACK, R.B. 2011. *Tropical Rain Forests: an Ecological and Biogeographical Comparison*. 2nd ed. Oxford University Press.

CORLEY, R.H.V. & TINKER, P.B.H. 2008. *The Oil Palm*. Blackwell Scientific, Oxford.

CORNER, E.J.H. Undated. *The Malayan Flora*. Unpubl. ms.

CORNER, E.J.H. 1949. The Durian Theory or the origin of the modern tree. *Ann. Bot.* 13 (4): 367–414.

CORNER, E.J.H. 1954. The evolution of tropical forest. In: *Evolution as a Process*, eds J.S. Huxley, A.C. Hardy & E.B. Ford, pp. 34–46. Allen & Unwin, London.

CORNER, E.J.H. 1964a. A discussion on the results of the Royal Society Expedition to North Borneo, 1961. *Proc. Roy. Soc. B., Biol. Sci.* 161: 1–191.

CORNER, E.J.H. 1964b. [and subsequent authors] Royal Society Expedition to North Borneo 1961: reports. *Proc. Linn. Soc. Lond.* 175: 9–56.

CORNER, E.J.H. 1965. *Check-List of Ficus in Asia and Australasia with Keys to Identification*. Government Printing Office, Singapore.

CORNER, E.J.H. 1978. *The fresh-water Swamp Forest of South Johore and Singapore*. Botanic Gardens, Park and Recreation Dept., Singapore.

CORNER, E.J.H. 1988. *The Wayside Trees of Malaya*. 1st ed., 1940, 2nd ed. 1942, Government Printer, Singapore, 3rd ed., Malayan Nature Society.

CORNER, E.J.H. 1990. On *Trigonobalanus* (Fagaceae). *Bot. J. Linn. Soc.* 102: 219–223.

CORSON, C. 2010. Shifting environmental governance in a neoliberal world: US AID for conservation. *Antipode* 42: 576.

COULTER, J.M. & CHAMBERLAIN, C.J. 1903. *The Morphology of Angiosperms*. D. Appleton, New York.

COUVREUR, T.L.P., PIRIE, M.D., CHATROU, L.W., SANDERS, R.M.K., SU, Y.C.F., RICHARDSON, J.E. & ERKENS, R.J.H. 2011. Early evolutionary history of the flowering plant family Annonaceae: steady diversification and boreotropical geodispersal. *J. Biogeogr.* 38: 664–680.

COVARRUBIAS, M. 1937. *Island of Bali*. Alfred A. Knopf, New York.

COX, P.A. 1984. Chiropterophily and ornithophily in *Freycinetia* (Pandanaceae) in Samoa. *Pl. Syst. Evol.* 144: 277–290.

CRANBROOK, EARL OF. 1962. Notes on the habits and vertical distribution of some insectivores from the Burma-Tibetan frontier. *Proc. Linn. Soc. Lond.* 173: 121–127.

CRANBROOK, EARL OF. 2010. Late Quaternary turnover of mammals in Borneo: the zooarchaeological record. *Biodivers. & Conservation* 19: 373–391.

CRANBROOK, EARL OF & EDWARDS, D.S. 1994. *Belalong. A Tropical Rain Forest*. Royal Geographical Society, London & Sun Tree Publishing, Singapore.

CRANBROOK, EARL OF, & PIPER, P.J. 2009.

Borneo records of Malay tapir, *Tapirus indicus* Desmarest: a zooarchaeological and historical review. *Int. J. Osteoarchaeol.* 19: 491–507.

CRANE, J.M. 2005. Essay review: Reconciling plant strategy theories of Grime and Tilman. *J. Ecol.* 93: 1041–1052.

CRAWLEY, M.J., JOHNSTON, A.E., SILVERTOWN, J., DODD, M., DE MAZANCOURT, C., HEARD, M.S., HENMAN, D.F. & EDWARDS, G.R. 2005. Determinants of species richness in the park grass experiment. *Amer. Naturalist* 165: 179–192.

CROWTHER, J. 1982. Ecological observations in a tropical karst terrain, West Malaysia. I. Variations in topography, soils and vegetation. *J. Biogeogr.* 9: 65–78. II. Rainfall interception, litterfall and nutrient cycling. *ibid.* 14: 145–155; III. Dynamics of the vegetation-soil-bedrock system. *ibid.* 14: 157–164.

CRUDEN, R.W. 1977. Pollen-ovule ratios: a conservative indication of breeding systems in flowering plants. *Evolution* 31: 32–46.

CRUZ, H. 1973. Nature conservation in Sri Lanka (Ceylon). *Biol. Conservation* 5: 199–208.

CURRAN, L.M. 1994. *The Ecology and Evolution of Mast-Fruiting in Bornean Dipterocarpaceae: a General Ectomycorrhizal Theory*. PhD thesis, Princeton University.

CURRAN, L.M. & LEIGHTON, M. 2000. Vertebrate responses to spatiotemporal variation in seed production of mast-fruiting Dipterocarpaceae. *Ecol. Monogr.* 70: 101–128.

CURRAN, L.M. & WEBB, C.O. 2000. Experimental tests of the spatiotemporal scale of seed predation in mast-fruiting Dipterocarpaceae. *Ecol. Monogr.* 70: 129–148.

CURRAN, L.M., CANIAGO, I., PAOLI, G.D., ASTIANTI, D., KUSNETI, M., LEIGHTON, M., NIRARITA. 1999. Impact of El Nino and logging on canopy tree recruitment in Borneo. *Science* 286: 2184–2188.

CURRAN, L.M., TRIGG, S.N., McDONALD, A.K., ASTIANTI, D., HARDIONO, Y.M., SIRIGAR, P., CANIAGO, I. & KASISCHKE, E. 2004. Lowland forest loss in protected areas of Indonesian Borneo. *Science* 303: 1000–1003.

DAILY, G.C., ed. 1997. *Nature's Services: Societal Dependence on Natural Systems*. Island Press, Washington, DC.

DALBIR SINGH A. (née AWTAR KAUR). 1977. *Embryological and Cytological Studies of Some Members of the Dipterocarpaceae*. PhD thesis, University of Aberdeen.

DALLING, J.W. & JOHN, R. 2008. Seed limitation and the coexistence of pioneer tree species. In: *Tropical Forest Community Ecology*, eds W.P. Carson & S.A. Schnitzer, pp. 242–253. Wiley-Blackwell, Chichester.

DALLING, J.W., HUBBELL, S.P. & SILVERA, K. 1996. Seed dispersal, seedling establishment and gap partitioning among tropical pioneer species. *J. Ecol.* 86: 674–689.

DALLMEIER, F. & COMISKEY, J.A. 1998. *Forest Biodiversity Research, Monitoring and Modeling. Conceptual Background and Old World Case Studies*. UNESCO, Paris & Parthenon, Carnforth.

DALY, H.E., CZECH, B., TRAUGER, D.L., REES, W.E., GROVER, M., DOBSON, T. & TROMBULAK, S.C. 2007. Are we consuming too much – for what? *Conservation Biol.* 21: 1359–1362.

DAM, R.A.C. 2001. Quaternary environmental changes in the Indonesian region. *Palaeogeogr. Palaeoclimatol. Palaeoecol.* 11: 91–95.

DAM, R.A.C. & KARS, S. VAN DER, eds. 2001. Quaternary Environmental Change in the Indonesian Region. *Palaeogeogr. Palaeoclimatol. Palaeoecol.* 171: 3–4.

DAMES, T.W.G. 1955. The Soils of East-Central Java. *Contributions of the General Agricultural Stations, Bogor* 141.

DANIELS, L.D. & VEBLEN, T.D. 2004. Spatiotemporal influences of climate on altitudinal tree line in northern Patagonia. *Ecology* 85: 1284–1296.

DANIELSON, F. & SCHUMACHER, T. 1997. The importance of Tigapuluh Hills, southern Riau, Indonesia to biodiversity conservation. *Trop. Biodivers.* 4: 129–159.

DARGIE, T.C.D. & EPITAWATTA, D.S. 1988. Environment, succession and land-use potential in the semi-evergreen forests of southern Sri Lanka. *J. Biogeogr.* 15: 209–220.

DARWIN, C. 1882. *The Formation of Vegetable Mould, through the Action of Worms, with Observations on their Habits.* John Murray, London.

DAS, B.K. 2005. Growth of ethnic groups in forest villages of Buxa Tiger Reserve, West Bengal. *Indian Forester* 131: 504–518.

DASANAYAKE, M.D. *et al.*, eds. 1980–. *Revised Handbook to the Flora of Ceylon.* 15 vols. Various publishers.

DASAPPA & SWAMINATH, M.H. 2000. A new species of *Semicarpus* L. from the *Myristica* swamps of Western Ghats of North Canara, Karnataka, India. *Indian Forester* 126: 78–82.

DASGUPTA, P. 1990. The environment as a commodity. *Oxf. Rev. Econ. Policy* 6: 51–67.

DASGUPTA, S. 2001. A note on improvement in regeneration status in forests of Madhya Pradesh under Joint Forest Management. *Indian Forester* 127: 823–826.

DASGUPTA, S. 2004. Community forest management in Madhya Pradesh – the future challenge? *Indian Forester* 130: 1381–1389.

DATTA, K., RATHORE, L.S., SANJEEVA RAO, P. & SASEENDRAN, M. 1998. Climate-related stress conditions in semi-arid to humid (dry) tropics: historical perspective. In: *Red and Lateritic Soils, 1, Red and Lateritic Soils of the World,* eds J. Sehgal, W.E. Blum & K.S. Gajbhiye, pp. 93–111. Balkema, Rotterdam & Brookfield.

DATTARAJA, H.S., SURESH, H.S. & SUKUMAR, R. Undated. *Impact of Forest Fires on Species Diversity and Community Structure of a Tropical Dry Deciduous Forest, Mudumalai Wildlife Sanctuary, Southern India.* Unpubl. ms. a.

DATTARAJA, H.S., SURESH, H.S. & SUKUMAR, R. Undated. *Vegetation Structure along a Precipitation Gradient in Mudumalai Wildlife Sanctuary, Southern India.* Unpubl. ms. b.

DAVIDAR, P., DICK, C.W., PUYREVAUD, J.P., TERBORGH, J., TER STEEGE, H., & WRIGHT, S.J. 2004. Why do some tropical forests have so many species of trees? *Biotropica* 36: 447–473.

DAVIDAR, P., PUYRAVAUD, J-P. & LEIGH, E.G. JR. 2005. Changes in rain forest tree diversity, dominance and rarity across a seasonality gradient in the Western Ghats, India. *J. Biogeogr.* 32: 493–501.

DAVIDAR, P., MOHANDASS, D. & VIJAYAN, S.L. 2007. Floristic inventory of woody plants in a tropical montane (shola) forest in the Palni hills of the Western Ghats, India. *Trop. Ecol.* 48: 15–25.

DAVIDSON, J. 2000. *Ecology and Management of Broadleaved Forests of Eastern Bhutan.* Field Document FD00:7. Forestry Development Project, Department of Forestry Servides, Royal Government of Bhutan.

DAVIDSON, J., THO, Y.P. & BIJLEVELD, M., eds. 1985. *The Future of Tropical rain Forests in South East Asia.* International Union for Conservation of Nature and Natural Resources, Morges.

DAVIES, G., HEYDON, M., LEADER-WILLIAMS, N., MACKINNON, J. & NEWING, H. 2001. The effects of logging on tropical forest ungulates. In: *The Cutting Edge: Conserving Wildlife in Tropical Forest,* eds R.A. Fimbel, A. Grajal & J.G. Robinson, pp. 91–124. Columbia University Press, New York.

DAVIES, R.G., EGGLETON, P., JONES, D.T., GATHORNE-HARDY, F.J. & HERNANDEZ, L.M. 2003. Evolution of termite functional diversity: analysis and synthesis of local species richness. *J. Biogeogr.* 30: 847–877.

DAVIES, S.J. 1996. *The Comparative Ecology of Macaranga (Euphorbiaceae).* PhD thesis, Harvard University.

DAVIES, S.J. 2001a. Systematics of *Macaranga* sects. *Pachystemon* and *Pruinosae* (Euphorbiaceae). *Harvard Pap. Bot.* 6: 371–448.

DAVIES, S.J. 2001b. Tree mortality and growth in 11 sympatric *Macaranga* species in Borneo. *Ecology* 82: 920–932.

DAVIES, S.J. & ASHTON, P.S. 1999. Phenology and fecundity of 11 sympatric pioneers species of *Macaranga* (Euphorbiaceae) in Borneo. *Amer. J. Bot.* 86: 1786–1795.

DAVIES, S.J. & BECKER, P. 1996. Floristic composition and stand structure of mixed dipterocarp and heath forests in Brunei Darussalam. *J. Trop. Forest Sci.* 8: 542–569.

DAVIES, S.J. & SEMUI, H. 2006. Competitive dominance in a secondary successional rain-forest community in Borneo. *J. Trop. Ecol.* 22: 53–64.

DAVIES, S.J. & UNAM, L. 1999. Smoke-haze from the 1997 Indonesian forest fires: effects on pollution levels, local climate and atmospheric CO_2 concentrations. *Forest Ecol. Managem.* 124: 137–144.

DAVIES, S.J., PALMIOTTO, P.A., ASHTON, P.S., LEE, H.S. & LAFRANKIE, J.V. 1998. Comparative ecology of 11 sympatric species of *Macaranga* in Borneo: tree distribution in relation to horizontal and vertical resource heterogeneity. *J. Ecol.* 86: 662–673.

DAVIES, S.J., TAN, S.T., LAFRANKIE, J.V. & POTTS, M.D. 2005. Soil-related floristic variation in a hyperdiverse dipterocarp forest. In: *Pollination Ecology and the Rain Forest,* eds D.J. Roubik, S. Sakai & A.A. Hamid, pp. 22–334. Springer, New York.

DAVIS, A.P. & SOHMER, S.H. 2004. Current status of *Psychotria* in the Philippines. *AGRIS Records.* FAO, Rome.

DAVIS, C.C., BELL, C.D., MATHEWS, S., & DONOGHUE, M.J. 2002. Laurasian migration explains Gondwanan disjunctions: evidence from Malpighiaceae. *Proc. Natl. Acad. Sci. U.S.A.* 99: 6833–6837.

DAVIS, C.C., FRITSCH, P.W., BELL, C.D. & MATHEWS, S. 2004. High latitude Tertiary migrations of an exclusively tropical clade: evidence from Malpighiaceae. *Int. J. Pl. Sci.* 165 (4 suppl.): S107–S121.

DAVIS, C.C., WEBB, C.O., WURDACK, K.J., JARAMILLO, C.A. & DONOGHUE, M.J. 2005. Explosive radiation of Malpighiales supports a mid-Cretaceous origin of modern tropical rain forests. *Amer. Naturalist* 165: E36–65.

DAVIS, S.S., HEYWOOD, V.H. & HAMILTON, A.C. 1995. *Centres of Plant Diversity. A Guide and Strategy for their Conservation.* Vol 2. Asia Australia and the Pacific. WWF & IUCN, Cambridge.

DAWKINS, H.C. 1963. Crown diameters: the relation to bole diameters in tropical forest trees. *Commonw. Forest. Rev.* 42 (114): 318–333.

DAWKINS, H.C. & PHILIP, M.S. 1998. *Tropical Moist Forest Silviculture and Management. A History of Success and Failure.* CABI, Wallingford.

DAYANANDA, D.T., DE SILVA, M.P., INOMATA, N., YAMAZAKI, T. & SZMIDT, A.E. 2006. Comprehensive molecular phylogeny of the sub-family Dipterocarpoideae (Dipterocarpaceae) based on chloroplast DNA sequences. *Genes Gen. Syst.* 81 (1): 1–12.

DAYANANDAN, S., ATTYGALLA, D.N.C., ABEYGUNASEKERA, A.W.W. L., GUNATILLEKE, I.A.U.N. & GUNATILLEKE, C.V.S. 1990. Phenological and floral morphology in relation to pollination of some Sri Lankan dipterocarps. In: *Reproductive Ecology of Tropical Forest Plants,* eds K.S. Bawa & M. Hadley, pp. 103–133. UNESCO, Paris and the Parthenon Publishing Group, Carnforth.

DAYANANDAN, S., ASHTON, P.S., WILLIAMS, S.M. & PRIMACK, R.B. 1999. Phylogeny of the tropical tree family Dipterocarpaceae based on nucleotide sequences of the chloroplast *rbcL* gene. *Amer. J. Bot.* 86: 1182–1190.

DE, R.N. 1923. Assam to Burma across the hills. *Indian Forester* 429: 529–537.

DEBRAL, B.G., BALI, H.K. & BHALLA, H.K.L. 1964a. Dew studies under forest plantations at New Forest Dehra Dun. *Indian Forester* 90: 169.

DEBRAL, B.G., SACHIDANAND & SWARUP, R. 1964b. Dew at New Forest, Dehra Dun. *Indian Forester* 90: 24.

DEBSKI, I., BURSLEM, D.F.R.P., PALMIOTTO, P.A., LAFRANKIE, J.V., LEE, H.S. & MANOKARAN, N. 2002. Habitat preferences of *Aporosa* in two Malaysian forests: implications for abundance and coexistence. *Ecology* 83: 2005–2018.

DELISSIO, L.J. & PRIMACK, R.B. 2003. The impact of drought on the population dynamics of canopy-tree seedlings in an aseasonal Malaysian rain forest. *J. Trop. Ecol.* 19: 489–500.

DELISSIO, L.J., PRIMACK, R.B., HALL, P. & LEE, H.S. 2002. A decade of canopy tree seedling survival and growth in two Bornean rain-forests: persistence aand recovery from suppression. *J. Trop. Ecol.* 18: 645–658.

DENHAM, T., HABERLE, S.G., LENTFER, C., FULLAGER, R., FIELD, J., THERIN, M., PORCH, N. & WINSBROUGH, B. 2003. Origins of agriculture in Kuk Swamp in the Highlands of New Guinea. *Science* 301, 11 July: 189–193.

DENNIS, R.A., MAYER, J., APPLEGATE, G. & CHOKKALINGAM, U. 2005. Fire, people and pixels: linking social science and remote sensing to understand underlying causes and impacts of fires in Indonesia. *Human Ecol.* 33: 465–504, and *Singapore J. Trop. Geogr.* 27: 37–48.

DENNIS, R.A. & COLFER, C.P. 2006. Impacts of land use and fire on the loss and degradation of lowland forest in 1983–2000 in East Kutai District, East Kalimantan, Indonesia.

DENSLOW, J.S. 1987. Tropical rainforest gaps and tree species diversity. *Annual Rev. Ecol. Syst.* 18: 431–451.

DENSLOW, J.S., SCHULTZ, J.C., VITOUSEK, P.M. & STRAIN, B.R. 1990. Growth response of tropical shrubs to treefall gap environments. *Ecology* 71: 165–179.

DENSLOW, J.S., ELLISON, A.M. & SANFORD, R.E. 1998. Treefall gap size effects on above- and below-ground processes in tropical wet forest. *J. Ecol.* 86: 597–609.

DENT, D.H. & BURSLEM, D.F.R.P. 2009. Performance trade-offs driven by morphological plasticity contribute to habitat specialization of Bornean tree species. *Biotropica* 41: 424–434.

DEPOMMIER, D. & RAMAKRISHNAN, P.S., eds. 2002. *Traditional Ecological Knowledge, Conservation of Biodiversity and Sustainable Development.* Institut Français de Pondichéry.

DERANIYAGALA, P.E.P. 1958. The Pleistocene of Ceylon. *Ceylon Nat. Hist. Mus. Nat. Hist. Ser.* Colombo.

DEWALT, S.J., ICKES, K., NILUS, R., HARMS, K.E. & BURSLEM, D.F.R.P. 2006. Liana habitat associations and community structure in a Bornean lowland rain forest. *Pl. Ecol.* 186: 203–216.

DHAILA-ADHIKARI, S. & ADHIKARI, B.S. 2009. Conserving biodiversity through faith and beliefs in Kumaun Region, Uttarakhand. Unpubl. ms., Mountain Action Research Group, Nainital. [online] URL: http://oldwww.wii.gov.in/publications/researchpapers/2009/2009_Dhaila–Adhikari_Conserving.pdf

DHAR, S., MAHENDRA, A.K. & ANSARI, M.Y. 1989. Tendu leaves – their collection and trade for the benefit of rural people. *Indian Forester* 115: 296–302.

DICK, C.W. 2002. Effect of pollinator composition on the breeding structure of tropical timber trees. In: *Modeling and Experimental Research on Genetic Processes in Tropical and Temperate Trees,* eds B. Degen, M. Loveless, & A. Kremer, pp. 140–152. EMBRAPA, Belém.

DICK, C.W., ABDUL-SALIM, K. & BERMINGHAM, E. 2003. Molecular systematic analysis reveals cryptic diversification of a widespread tropical rain forest tree. *Amer. Naturalist* 162: 691–703.

DICK, C.W., ROUBIK, D.W., GRUBER, K.F. & BERMINGHAM, E. 2004. Long-distance gene flow and cross-Andean dispersal of lowland rainforest bees (Apidae: Euglossiniae) revealed by comparative mitochondrial DNA phylogeography. *Molec. Ecol.* 13: 3775–3785.

DICK, C.W., BERMINGHAM, E., LEMES, M.R. & GRIBEL, R. 2007. Extreme long-distance dispersal of the lowland tropical rainforest tree *Ceiba pentandra* L. (Malvaceae) in Africa and the Neotropics. *Molec. Ecol.* 16: 3039–3049.

DICKINSON, J.A. & TANNER, E.J.V. 1976. Exploitation of hollow trunks by tropical trees. *Biotropica* 10: 231–233.

DIETZ, S. & ADGER, W.N. 2003. Economic growth, biodiversity loss, and conservation effort. *J. Environ. Managem.* 68: 23–35.

DIJK, A.J.M. van, NOORDWIJK, M. van, CALDER, I.R., BRUIJNZEEL, S.L.A., SCHELLEKENS, J. & CHAPPELL, N.A. 2009. Forest-flood relation still tenuous – comment on 'Global evidence that deforestation amplifies flood risk and severity in the developing world' by C.J.A. Bradshaw, N. Sodhi, K.S. Peh & B.W. Brook. *Global Change Biol.* 15: 110–115.

DIJK, P.P. van, TORDOFF, A.W., FELLOWES, J., LAU, M. & MA, J. 2004. Indo-Burma. In: *Hotspots Revisited: Earth's Biologically Richest and Most Endangered Terrestrial Ecoregions*, eds R. Mitteremeier, P.R. Gil, M. Hoffman, J. Pilgrim, T. Brooks, C.G. Mittermeier, J. Lamoreux, G.A.B. da Fonceca, P.A. Seligmann & H. Ford, pp. 323–332. Cemex, Mexico City.

DINERSTEIN, E. 1979. An ecological survey of the Royal Karneli-Bardia Wildlife Reserve, Nepal. I. Vegetation, modifying factors and successional relationships. *Biol. Conservation* 15: 127–150.

DINERSTEIN, E. 1987. Deer, plant phenology, and succession in South Asian lowland forests. In: *Biology and Management of the Cervidae*, ed. C. Wemmer, pp. 272–288. Smithsonian Institution Press, Washington, DC.

DINERSTEIN, E. 1992. Effects of *Rhinoceros unicornis* on riverine forest structure in lowland Nepal. *Ecology* 73: 701–704.

DINERSTEIN, E. & WEMMER, C. 1988. Fruits rhinoceros eat: dispersal of *Trewia nudiflora* (Euphorbiaceae) in lowland Nepal. *Ecology* 69: 1768–1774.

DIXIT, A.M., GEEVAN, C.P. & SILORI, C.S. 2001. Status of natural terrestrial vegetation in Gujarat – a reassessment. *Indian Forester* 127: 533–546.

DOOLITTLE, A.A. 2011. Redefining native customary law: struggles over property rights between native peoples and colonial rulers in Sabah, Malaysia, 1950–1996. In: *Beyond the Sacred Forest: Complicating Conservation in Southeast Asia*, eds M.R. Dove, P.E. Sajise & A.A. Doolittle, pp. 151–179. Duke University Press.

DOUGLAS, I. 1992. Hydrological investigation of forest disturbance and land cover impacts in South-east Asia: a review. *Philos. Trans., Ser. B* 335: 1739–1748.

DOVE, M.R., SAJISE, P.E. & DOOLITTLE, A.A., eds. 2011. *Beyond the Sacred Forest: Complicating Conservation in Southeast*

Asia. Duke University Press.

DOYLE, J.A., SAUQUET, H., SCHARASCHKIN, T. & LE THOMAS, A. 2004. Phylogeny, molecular and fossil dating, and biogeographic history of Annonaceae and Myristicaceae (Magnoliales). *Int. J. Pl. Sci.* 165 (4 suppl.): S55–67.

DOYLE, J.A., MANCHESTER, S.R. & SAUQUET, H. 2008. A seed related to Myristicaceae in the early Eocene of southern England. *Syst. Bot.* 33: 636–646.

DRANSFIELD, J., UHL, N.W., ASMUSSEN, C.B., BAKER, W.J., HARLEY, M.M. & LEWIS, C.E. 2008. *Genera Palmarum: the Evolution and Classification of Palms.* Kew Publishing, London.

DRANSFIELD, S. & WIDJAJA, E. eds 1995. *Plant Resources of South-East Asia (PROSEA). No. 7 Bamboos.* Backhuys, Leiden.

DRESSLER, S. 1999. Revision of the genus *Cleistanthus* (Euphorbiaceae) in the Philippines. *Blumea* 44: 109–148.

DUCOUSSO, M., BÉNA, G., BOURGEOIS, C., BUYCK, B., EYSSARTIER, G., VINCELETTE, M., RABEVOHITRA, R., RANDRIHASIPARA, L., DREYFUS, B. & PRIN, Y. 2004. The last common ancestor of Sarcolaenaceae and Asian dipterocarp trees was ectomycorrhizal before the India-Madagascar separation, about 88 million years ago. *Molec. Ecol.* 13: 231–236.

DUDGEON, D. & CORLETT, R.T. 2004. *The Ecology and Biodiversity of Hong Kong.* Friends of the Country Parks, Hong Kong.

DUFFY, R. & MOORE, L. 2011. Neoliberalising nature? Elephant-back tourism in Thailand and Botswana. In: *Capitalism and Conservation*, eds D. Brockington and R. Duffy, pp. XX–XX. Wiley-Blackwell, Oxford.

DUIVENVOORDEN, J.F. 1995. Tree species composition and rain forest-environment relationships in the middle Caquetá area, Colombia, NW Amazonia. *Vegetatio* 120: 91–113.

DUIVENVOORDEN, J.F. 1996. Patterns of species richness in rain forests of the middle Caquetá area, Colombia, NW Amazonia. *Biotropica* 28: 142–158.

DUIVENVOORDEN, J.F., SVENNING, J.-C. & WRIGHT, S.J. 2006. Beta diversity in tropical forests. *Science* 295: 636–637.

DUPÉRON-LOUDOUENRIX, M. 1991. Importance of fossil woods (conifers and gymnosperms) discovered in continental Mesozoic sediments of northwestern Equatorial Africa. *J. Afr. Earth Sci.* 12: 391–396.

DY PHON, P. 1970. La Végétation du sud-ouest du Cambodge. *Ann. Fac. Sci. Phnom Penh* 3: 1–136; *ibid.* 4: 1–59.

DYKES, A.P. 1994a. Landform processes. In: *Belalong. A Tropical Rain Forest*, eds Earl of Cranbrook & D.S. Edwards, pp. 31–49. Royal Geographical Society, London & Sun Tree Publishing, Singapore.

DYKES, A.P. 1994b. Landslides in the Batu Apoi Forest Reserve. *Brunei Mus. J.* 71–87.

DYKES, A.P. 1996. Analysis of factors contributing to the stability of steep hillslopes in the tropical rainforest of Temburong, Brunei Darussalam. In: *Tropical Rainforest Research – Current Issues*, eds D.S. Edwards, W.E. Booth & S.C. Choy, pp. 387–409. Kluwer, Dordrecht.

DYKES, A.P. 2000. Climatic patterns in a tropical rainforest in Brunei. *Geogr. J.* 166: 63–80.

DYKES, A.P. 2002. Weathering-limited rainfall-triggered shallow mass movements in undisturbed steepland tropical rainforest. *Geomorphology* 46: 73–93.

DYKSTRA, D.P. 1996. *Forest Operations for Sustainable Forestry in the Tropics. Proceedings of a Symposium.* Organised by IUFRO subject group S3.05–00, Forest Operations in the Tropics, at XX IUFRO World Congress, 6–12 Aug. 1995, Jakarta.

DYKSTRA, D.P. 2004. RILSIM 2.0. US Forest Service, International Programs Office, Washington, DC. URL: www.blueoxforestry.com/RILSIM/

EDEN, T. 1976. *Tea.* Longmans, London.

EDIRIWEERA, S., SINGHAKUMARA, B.M.P. & ASHTON, P.M.S. 2008. Variation in canopy structure, light and soil nutrition across elevation of a Sri Lankan tropical rain forest. *Forest Ecol. Managem.* 256: 1339–1349.

EDITORIAL COMMITTEE 1994–2000. (see also Hoang, W.-L.) *Flora of Taiwan,* 2nd ed. Editorial Committee of the Flora of Taiwan, Taipei.

EDMUNDS, D. & WOLLENBERG, E. 2001. Historical perspectives on forest policy change in Asia: an introduction. *Environmental History* 6:190–212.

EDWARDS, D.S., BOOTH, W.E. & CHOY, S.C., eds. 1996. *Tropical Rainforest Research – Current Issues.* Kluwer, Dordrecht.

EDWARDS, I.D., MACDONALD, A.A.Q. & PROCTOR, J. 1993. *Natural History of Ceram, Maluku, Indonesia.* Raleigh and Intercept, Andover.

EDWARDS, J.P. 1930. Growth of Malayan forest trees as shown by sample plot records. *Malayan Forest Rec.* 9.

EGUCHI, T. 1991. Moisture conditions of midland dry valleys in Bhutan. In: *Life Zone Ecology of the Bhutan Himalaya II*, ed. M. Ohsawa, pp. 21–40. Laboratory of Ecology, Chiba University.

EHRLICH, P.R. & GOULDER, L.H. 2007. Is current consumption excessive? A general framework and some indications for the United States. *Conserv. Biol.* 21: 1145–1154.

EICHHORN, M.P., COMPTON, S.G. & HARTLEY, S.E. 2008. The influence of soil type on rain forest insect herbivore communities. *Biotropica* 40: 707–713.

ELIAS, I. 1996. A case study on forest harvesting damages, structure and composition dynamic changes in the residual stand stand for dipterocarp forest in East Kalimantan, Indonesia. In: *Forest Operations for Sustainable Forestry in the Tropics.* ed. D.P. Dylstra, pp. 13–27. IUCN/CIFOR.

ELKUNCHWAR, S., SAVANT, P.V. & RAI, S.N. 1997. Status of natural regeneration in tropical forests of the Andaman Islands. *Indian Forester* 123: 1091–1106.

ELLEN, R.F. 1985. Patterns of indigenous timber extraction from Moluccan rain forest fringes. *J. Biogeogr.* 12: 559–587.

ELLIOTT, S. 1994. Can community forestry save biodiversity? *Nat. Hist. Bull. Siam. Soc.* 42: 150–152.

ELLIOTT, S., MAXWELL, J.F. & BEAVER, O.P.

1989. A transect survey of monsoon forest in Doi Suthep-Pui National Park. *Nat. Hist. Bull. Siam. Soc.* 37: 131–171.

ELLIOTT, S., PATTIPONG, N., CHERDAK, K., SUDARAT, Z., VILAIWAN, A. & BLAKESLEY, D. 2003. Selecting framework tree species for restoring seasonally dry tropical forests in northern Thailand based on field performance. *Forest Ecol. Managem.* 184: 177–191.

ELTZ, T. 2004. Spatio-temporal variation of Apine bee attraction to honey-baits in Bornean forests. *J. Trop. Ecol.* 20: 317–324.

ELOUARD, C., HOULLIER, F., PASCAL, J.-P., PÉLLISIER, R. & RAMESH, B.R. 1997. *Dynamics of the Dense Moist Evergreen Forests. Long term Monitoring of an Experimental Station in Kodagu District (Karnataka, India).* Institut Français de Pondichéry.

ELVIN, M. 2004. *The Retreat of the Elephants.* Yale University Press.

ENDRESS, P.K., HYLAND, B.P.M. & TRACEY, J.G. 1985. *Noahdendron,* a new Australian genus of the Hamamelidaceae. *Bot. Jahrb. Syst.* 107: 369–378.

ENGEL, S., PAGIOLA, S. & WUNDER, S. 2008. Designing payments for environmental services in theory and practice: an overview of the issues. *Ecol. Econ.* 65: 663–674.

ENGELBRECHT, B.M.J., COMITA, L.S., CONDIT, R., KURSAR, T.A., TYREE, M.T., TURNER, B.L. & HUBBELL, S.P. 2007. Drought sensitivity shapes species distribution patterns in tropical forests. *Nature* 447: 80–83.

ERCELAWN, A.C. 1997. *Forest Dynamics in a 2 ha Long-Term Plot at Bukit Timah Nature Reserve, Singapore.* Masters dissertation, National Institute of Education, Namyang Technological University, Singapore.

ERDELEN, W. 1984. The genus *Calotes* (Sauria, Agamidae) in Sri Lanka: distribution patterns. *J. Biogeogr.* 11: 515–525.

ERDELEN, W., PREU, C., ISHWARAN, N. & MADDUMA BANDARA, C.M., eds. 1993. *Ecology and Landscape Management in Sri Lanka. Proceedings from the International and Interdisciplinary Symposium, Colombo, Sri Lanka, 12–26 March 1990.* V.J. Markgraf, Weikersheim.

ETIENNE, R.S. & OLFF, H. 2004. A novel genealogical approach to neutral biodiversity theory. *Ecol. Lett.* 7: 170–175.

EVERETT, H.D. & WHITFORD, H.N. 1914. A preliminary working plan for the public forest tract of the Insular Lumber Company, Occidental Negros, P.I. *Bull. P.I. Bureau of Forests.*

EWEL, J.J., CHAI, P. & LIM, M.T. 1983. Biomass and floristics of three young second-growth forests in Sarawak. *Malayan Forester* 46: 347–364.

EYDE, R.H & XIANG, Q.Y. 1990. Fossil Mastixioid (Cornaceae) alive in eastern Asia. *Amer. J. Bot.* 77: 689–692.

FAEGRI, K & VAN DER PIJL, L. 1971. *The Principles of Pollination Ecology.* 2nd ed. Oxford University Press.

FAEGRI, K. & VAN DER PIJL, L. 1979. *The Principles of Pollination Ecology.* 3rd rev. ed. Pergamon, Oxford.

FANG, W. & PENG, S.L. 1997. Development of species diversity in the restoration process of establishing a tropical man-made forest ecosystem in China. *Forest Ecol. Managem.* 99: 185–196.

FARIDAH-HANUM, I., LINATOC, A.C. & LATIFF, A. 1999. *Dynamics of Mangifera L. in Relation to Soil and Topographic Factors in Pasoh 50 ha plot, Malaysia.* Unpubl. ms., University Pertanian Malaysia.

FARJON, A. 2008. *A Natural History of Conifers.* Timber Press, Portland, Oregon.

FARJON, A. 2010. *A Handbook of the World's Conifers.* Brill, Leiden.

FARJON, A. & HIEP, N.J. 2004. *Vietnam Conifers: Conservation Status Review.* Flora & Fauna International Vietnam Programme, Hanoi.

FARJON, A., THOMAS, P. & LUU, N.D.T. 2004. *Conifer Conservation in Vietnam: Three Potential Flagship Species.* Unpubl. ms.

FARRELL, B.D. 1998. 'Inordinate fondness' explained: why are there so many beetles? *Science* 281: 555–559.

FARRELL, B.D. & MITTER, C. 1993. Phylogenetic determinants of insect/plant community diversity. In: *Species Diversity in Ecological Communities. Historical and Geographical Perspectives,* eds R.E. Ricklefs & D. Schluter, pp. 253–266. University of Chicago Press.

FARRELL, B.D. & MITTER, C. 1998. The timing of insect/plant diversification: might *Tetraopes* (Coleoptera: Cerambicidae) and *Asclepias* (Asclepiadaceae) have co-evolved? *Biol. J. Linn. Soc.* 63: 553–577.

FEDOROV, A.A. 1966. The structure of the tropical rain forest and speciation in the humid tropics. *J. Ecol.* 54: 1–11.

FEDOROV, A.A. 1969. *Chromosome Numbers of Flowering Plants* [title in Russian]. Komarov Botanical Institute, 'Nauca', Leningrad.

FEELEY, K.J., DAVIES, S.J., ASHTON, P.S., BUNYAVEJCHEWIN, S., NUR SUPARDI, M.N., KASSIM, A.R., TAN, S. & CHAVE, J. 2007a. The role of gap phase processes in the biomass dynamics of tropical forests. *Proc. Roy. Soc. London, Ser. B, Biol. Sci.* 274: 2857–2864.

FEELEY, K.J., DAVIES, S.J., NUR SUPARDI, M.N., RAHMAN KASSIM, A. & TAN, S. 2007b. Do current stem size distributions predict future population changes? An empirical test of intraspecific patterns in tropical trees at two spatial scales. *J. Trop. Ecol.* 23: 1–8.

FEELEY, K.J., WRIGHT, S.J., NUR SUPARDI, M.N., KASSIM, A.R. & DAVIES, S.J. 2007c. Decelerating growth in tropical forest trees. *Ecol. Lett.* 10: 461–469.

FELKER, P. 1981. Use of tree legumes in semi-arid regions. *Econ. Bot.* 35: 174–186.

FENNER, M., LEE, W.G. & BASTOW WILSON, J. 1997. A comparative study of the distribution of genus size in twenty angiosperm floras. *Biol. J. Linn. Soc.* 62: 225–237.

FERGUSON, J. 1868, reprinted 1971. *Tree and Serpent Worship: Illustrations of Mythology and Art in India in the First and Fourth Centuries after Christ. From the Sculptures of the Buddhist Topes at Sanchi and Amravati.* Prepared under the authority of the Secretary of State for India in Council. London, India Museum. W.H. Allen.

FERNANDO, S.S. & RANASINGHE, P.N. 1997. Diversity of family Orchidaceae in different forest formations of Adam's Peak in the Peak Wilderness Sanctuary. *Proc. Part 1, Abstracts, Sri Lanka Association for the Advancement of Science, 53rd Annual Session,* p. 254.

FIALA, B. & LINSENMAIR, K.E. 1995. Distribution and abundance of plants with extrafloral nectaries in the woody flora of a lowland primary forest in Malaysia. *Biodivers. & Conservation* 4: 165–182.

FIALA, B. & MASCHWITZ, U. 1991. Extrafloral nectaries in the genus *Macaranga* (Euphorbiaceae) in Malaysia: comparative studies of their possible significance as predispositions for myrmecophytism. *Biol. J. Linn. Soc.* 44: 287–305.

FIALA, B. & MASCHWITZ, U. 1992. Food bodies and their significance for obligate ant-association in the tree genus *Macaranga* (Euphorbiaceae). *Bot. J. Linn. Soc.* 110: 61–75.

FIALA, B., JAKOB, A., MASCHWITZ, U. & LINSENMAIR, K.E. 1999. Diversity, evolutionary specialization and geographic distribution of a mutualistic ant-plant complex: *Macaranga* and *Crematogaster* in South East Asia. *Biol. J. Linn. Soc.* 66: 305–331.

FIALA, B., MEYER, U., HASHIM, R. & MASCHWITZ, U. 2011. Pollination systems in pioneer trees of the genus *Macaranga* (Euphorbiaceae) in Malaysian rainforests. *Biol. J. Linn. Soc.* 103: 935–953.

FIEDLER, K. 1998. Diet breadth and host plant specificity of tropical versus temperate-zone herbivores: South-East Asian and West Palaearctic butterflies as a special case. *Ecol. Entomol.* 23: 285–297.

FIELD, C.B. 1988. On the role of photosynthetic responses in constraining the habitat distribution of tropical trees. *Austral. J. Pl. Physiol.* 15: 343–358.

FIELD, T.S., ZWIENECKI, M.A., DOYLE, M.J., & HOLLBROOK, N.M. 1998. Structural plugs of *Drimys winteri* (Winteraceae) protect leaves from mist not drought. *Proc. Natl. Acad. Sc. U.S.A.* 95: 14256–14259.

FIMBEL, R.A., GRAJAL, A., & ROBINSON, J.G., eds. 2001. *The Cutting Edge: Conserving Wildlife in Logged Tropical Forest.* Columbia University Press.

FINE, P.V.A., MESONES, I. & COLEY, P.D. 2004. Herbivores promote habitat specialization by trees in Amazonian forests. *Science* 305: 663–665.

FINE, P.V.A., DALY, D.C., MUÑOZ, G.V., MESONES, I. & CAMERON, K.C. 2005. The contribution of edaphic heterogeneity to the evolution and diversity of Burseraceae trees in the western Amazon. *Evolution* 59: 1464–1478.

FINE, P.V.A., REE, R.H. & BURNHAM, R.J. 2008. The disparity in tree species richness among tropical, temperate, and boreal biomes: the geographical area and age hypothesis. In: *Tropical Forest Community Ecology,* eds W.P. Carson & S.A. Schnitzer, pp. 31–45. Wiley-Blackwell, Chichester.

FISCHER, E.C. 1918. Forest grazing of the Nellore 'kancha' system. *Indian Forester* 44: 531–547.

FISCHER, J., BATÁRY, P., BAWA, K.S., BRUSSARD, L., CHAPPELL, M.J., CLOUGH, Y., DAILY, G.C., DORROUGH, J., HARTEL, T., JACKSON, L.E., KLEIN, A., KREMEN, C., KUEMMERLE, T., LINDENMAYER, D.B., MOONEY, H.A., PERFECTO, I., PHILPOTT, S.M., TSCHARNTKE, T., VANDERMEER, J., WANGER, T.C. & WEHDREN, H. VON. 2011. Conservation: limits of land sparing. *Science* 334: 593.

FISHER, R.A., CORBET, A.S. & WILLIAM, C.B. 1943. The relation between the number of species and the number of individuals in a random sample of an animal population. *J. Anim. Ecol.* 12: 42–58.

FLENLEY, J.R. 1979. *The Equatorial Rainforest: a Geological History.* Butterworth, London.

FLENLEY, J.R. 1988. Palynological evidence for land use changes in South-East Asia. *J. Biogeog.* 15: 185–197.

FLENLEY, J.R. 1995. Cloud forest, the Massenerhebung effect, and ultraviolet insulation. In: *Tropical Montane Cloud Forests,* eds L.S. Hamilton, J.O. Juvick & F.N. Scatena, pp. 150–155. Springer.

FLORIN, R. 1963. The distribution of conifer and taxad genera in time and space. *Acta Horti Berg.* 20: 121–312.

FONTANEL, J. & CHANTEFORT, A. 1978. *Microclimats du Monde Indonesien.* Institut Français de Pondichéry 16.

FORMAN, R.T.T., SPERLING, D., BISSONETTE, J.A., CLEVENGER, A.P., CUTSHALL, C.D., DALE, V.H., LENORE, F., FRANCE, R.L., GOLDMAN, C.R., HEANUE, K., JONES, J., SWANSON, F., TURRENTINE, T. & WINTER, T.C. 2003. *Road Ecology: Science and Solutions.* Island Press, Washington DC.

FORSHED, O., UDARBE, T., KARLSON, A. & FALCK, J. 2005. Initial impact of supervised logging climber cutting compared with conventional logging in a dipterocarp rainforest in Sabah, Malaysia. *Forest Ecol. Manag.* 221: 233–240.

FOSTER, D.M. & BOOSE, E.R. 1995. Hurricane disturbance regions in temperate and tropical forest ecosystems. In: *Wind & Trees,* eds M.P. Coutts & J.J. Grace, pp. 305–339. Cambridge University Press.

FOX, J.J. 1977. *Harvest of the Palm: Ecological Change in Eastern Indonesia.* Harvard University Press.

FRAHM, J.-P. & GRADSTEIN, S.R. 1991. An altitudinal zonation of rain forests using bryophytes. *J. Biogeogr.* 18: 669–678.

FRANCE-LANORD, C., DERRY, L. & MICHARD, A. 1993. Evolution of the Himalaya since Miocene time: isotopic and sedimentological evidence from the Bengal Fan. In: *Himalayan Tectonics,* eds P.J. Treloar & M.P. Searle. Geological Society Special Publication 74: 603–621. Geological Society, London.

FRANCIS, A.P. & CURRIE, D.J. 2003. Globally consistent richness climate relationship for angiosperms. *Amer. Naturalist* 161: 523–536.

FRANCIS, C.M. 1994. Vertical stratification of fruit bats (Pteropodidae) in lowland dipterocarp forest in Malaysia. *J. Trop. Ecol.* 10: 523–530.

FREDERICKSON, T.S. & PUTZ, F.E. 2003. Silvicultural intensification for tropical forest conservation. *Biodivers. & Conservation* 12: 1445–1453.

FREEMAN, J.D. 1960. The Iban of Western Borneo. In: *Social Structure in Southeast Asia,* ed G.P. Murdock, pp. 65–87. Quadrangle Books, Chicago.

FUJIMOTO, K. 1997. Mangrove habitat evolution related to Holocene sea-level changes in Pacific islands. *Tropics* 6: 203–213.

FUJITA, W., SUKWONG, S. & OGINO, K. 2002. Why do they plant trees? Wan Kaset agroforestry practice in eastern Thailand and people's strategy. *Tropics* 11: 169–185.

FULLER, D.O. & CHOWDHURY, R.R. 2006. Monitoring and modeling tropical deforestation: introduction to the special issue. *Singapore J. Trop. Geogr.* 27: 1–3.

FURUKAWA, A., TOMA, T., MARUYAMA, Y., MATSUMOTO, Y., UEMURA, A., ABDULLAH, A.M. & AWANG, M. 2001. Photosynthetic rates of four tree species in the upper canopy of a tropical rain forest at Pasoh Forest reserve in Peninsular Malaysia. *Tropics* 10: 519–527.

FURUKAWA, H. 1997. Geomorphological processes of the coastal wetlands in Indonesian archipelagoes. *Tropics* 6: 163–188.

FYSON, P.F. 1915. *The Flora of the Nilgiri and Pulney Hill-Tops.* 3 vols. Government Press, Madras.

FYSON, P.F. 1932. *The Flora of the South Indian Hill Stations.* 1. Government Press, Madras.

GADGIL, M. 1989. The Indian heritage of a conservation ethic. In: *Conservation of the Indian Heritage,* eds B. Allchin, E.R. Allchin & B.K. Thapar, pp. 13–21. Cosmo Publications, New Delhi.

GADGIL, M. & GUHA, R. 1992. *This Fissured Earth. An Ecological History of India.* Oxford University Press, Delhi.

GADGIL, M. & PRASAD, S.N. 1984. Ecological determinants of life history evolution of two Indian bamboo species. *Biotropica* 16: 161–172.

GADGIL, M. & VARTAK, V.D. 1976. The sacred groves of Western Ghats of India. *Econ. Bot.* 30: 152–160.

GADRINAB, & BELIN, M. 1981. Biology of the green spots in the leaves of some dipterocarps. *Malaysian Forester* 44: 253–266.

GALE, N. 1997. *Modes of Tree Death in Four Tropical Forests.* PhD thesis, University of Aarhus.

GALE, N. 2000. The relationship between canopy gaps and topography in a Western Ecuadorian rain forest. *Biotropica* 32: 653–661.

GALE, N. & BARFOD, A.S. 1998. Canopy tree mode of death in a western Ecuadorian forest. *J. Trop. Ecol.* 15: 415–436.

GALE, N. & HALL, P. 2001. Factors determining the modes of tree death in three Bornean forests. *J. Veg. Sci.* 12: 337–346.

GAMAGE, D.T., DE SILVA, M.P., INOMATA, N., YAMAZAKI, T. & SZMIDT, A.E. 2006. Comprehensive molecular phylogeny of the sub-family Dipteocarpoideae (Dipterocarpaceae) based on chloroplast DNA sequences. *Genes Genet. Syst.* 81: 1–12.

GAMAGE, H.K., ASHTON, M.S. &

SINGHAKUMARA, B.M.P. 2003. Leaf structure of Syzygium spp.(Myrtaceae) in relation to site affinity within a tropical rain forest. Bot. J. Linn. Soc. 141: 365–377.

GAMAGE, H.K., SINGHAKUMARA, B.M.P. & ASHTON, P.M.S. 2004. Effects of light and fertilization on arbuscular mycorrhizal colonization and growth of tropical rain-forest Syzygium tree seedlings. J. Trop. Ecol. 20: 525–534.

GAMBLE, J.S. 1896. The Bambusae of British India. Bengal Secretariat Press, Calcutta.

GAN, Y.Y. 1976. (née YAP, Y.Y.: see there).

GAN, Y.Y., ROBERTSON F.W., ASHTON, P.S., F.W., SOEPADMO, E. & LEE, D.W. 1977. Genetic variation in wild populations of rain forest trees. Nature 269: 323–325.

GANESAN, R. 2002. Evergreen forest swamps and their plant species diversity in Kalakad-Mundathunai Tiger Reserve, south western Ghats, India. Indian Forester 128: 1351–1359.

GANESAN, R. & SETTY, R.S. 2004. Regeneration of Amla, an important non-timber forest product from southern India. Conservation & Soc. 2: 365–375.

GANESH, T. & DAVIDAR, P. 1997. Flowering phenology and flower predation of Cullenia exarillata (Bombacaceae) by arboreal vertebrates in Western Ghats, India. J. Trop. Ecol. 13: 359–368.

GANESH, T. & DAVIDAR, P. 1999. Fruit biomass and relative abundance of frugivores in a rain forest in the Western Ghats, India. J. Trop. Ecol. 15: 399–413.

GANSSER, A. 1983. The Geology of Bhutan. Birkhäuser, Basel.

GARDNER, S., SIDISUNTHORN, P. & ANUSARNSUNTHORN, V. 2000. A Field Guide to Forest Trees of Northern Thailand. Kobfai Publishing Project, Bangkok.

GARDINGEN, P.R. van, MCLEISH, M.J., PHILLIPS, P.D. FADILAH, D., TYRIE, G. & YASMAN, I. 2003. Financial and ecological analysis of management options for logged-over forests in Indonesian Borneo. Forest Ecol. Managem.183: 1–29.

GARIGUATA, M.R. & PINARD, M.A. 1998. Ecological knowledge of regeneration from seed in neotropical forest trees: Implications for natural forest management. Forest Ecol. Managem.112: 87–99.

GATES, D.M. 1962. Energy Exchange in the Biosphere. Biological Monographs. Harper & Row, New York.

GATHORNE-HARDY, F.J., JONES, D.T. & SYAKANI. 2002a. A regional perspective on the effects of human disturbance on the termites of Sundaland. Biodivers. & Conservation 11: 1991–2000.

GATHORNE-HARDY, F.J., SYAUKANI, DAVIES, R.E., EGGLETON, F. & JONES, D.T. 2002b. Quaternary rain forest refugia in South-east Asia: using termites (Isoptera) as indicators. Biol. J. Linn. Soc. 75: 453–466.

GAUSSEN, H. 1955. Expression des milieux par les formules écologiques; leur représentation cartographique. Coll. Int. Cent. Nat. Rech. Sci. 59: 257–269.

GAUSSEN, H. 1959. The vegetation maps. Inst. Fr. Pondichéry, Trav. Sect. Sci. Tech. 1, 4: 155–179.

GAUSSEN, F. 1978. Les Cartes Internationales du Tapis Végétale et des Conditions Ecologiques.

Institut Français de Pondichéry.

GAUSSEN, H., LEGRIS, P. & VIART, M. 1961a. a. Notes on the sheet Cape Comorin. Inst. Fr. Pondichéry, Trav. Sect. Sci. Tech. 1: 108.

GAUSSEN, H., LEGRIS, P. & VIART, M. 1961b. Sheet Cape Comorin. Inst. Fr. Pondichéry, Trav. Sect. Sci. Tech. 1a.

GAUSSEN, H., LEGRIS, P., VIART, M., MEHER-HOMJI, V. & LABROUE, L. 1965a. Sheet Bombay. Inst. Fr. Pondichéry, Trav. Sect. Sci. Tech. 9a.

GAUSSEN, H., LeGRIS, P., LABROUE, L., MEHER-HOMJI, V. & VIART, M. 1965b. Carte Internationale Du Tapis Végétale. Notice explicative de la feuille Mysore. Inst. Fr. Pondichery, Trav. Sect. Sci. Tech. 7.

GAUSSEN, H., LEGRIS, P., VIART, M., MEHER-HOMJI, V. & LABROUE, L. 1966. b. Explanatory booklet on the sheet Bombay. Inst. Fr. Pondichéry, Trav. Sect. Sci. Tech. 9.

GAUSSEN, H., LEGRIS, P. & BLASCO, P. 1967. Bioclimats du Sud-Est Asiatique. Institut Français de Pondichéry.

GAUSSEN, H., LEGRIS, P., MEHER-HOMJI, V.M., FONTANEL, J. & PASCAL, J.P. 1992. Notes on the sheet Orissa. Inst. Fr. Pondichéry, Int. Map. Veg., Publ. Dép. d'Écol.

GEAGH, M.A., KUDRASS, H.R. & STREIF, H. 1979. Sea-level changes during the late Pleistocene and Holocene in the Straits of Malacca. Nature 278: 441–443.

GEIGER, R., ARON, R.H. & TODHUNTER, P. 2003. The Climate Near the Ground. 6th ed. Rowman & Littlefield, Lanham, Maryland.

GEIST, H. & LAMBIN, E. 2002. Proximate causes and underlying foreces of tropical deforestation. Bioscience 52: 143–150.

GENTRY, A.H. 1988. Changes in plant community diversity and floristic composition on environmental and geographic gradients. Ann. Missouri Bot. Gard. 75: 1–34.

GENTRY, A.H. 1992. Tropical forest biodiversity: distribution patterns. Oikos 63: 19–26.

GERA, M., BISHT, N.S. & RANA, A.K. 2003. Market information system for sustainable managfement of medicinal plants. Indian Forester 129: 102–108.

GHATE, R. 2003. Ensuring 'Collective Action' in Participatory Forest Management. South Asian Network for Development and Environmental Economics (SANDEE) Working Paper No. 3–03, Kathmandu, Nepal.

GHAZOUL, J. 2002. Impact of logging on the richness and diversity of forest butterflies in a tropical dry forest in Thailand. Biodivers. & Conservation 11:521–541.

GHAZOUL, J., LISTON, K.A. & BOYLE, T.J.B. 1998. Disturbance-induced density-dependent seed set in Shorea siamensis (Dipterocarpaceae), a tropical tree. J. Ecol. 86: 462–473.

GHOSH, R.C. & KAUL, O.N. 1977. Nature reserves in India. Indian Forester 103: 497–512.

GIBBONS, J.M. & NEWBERY, D.M. 2002. Drought avoidance and the effect of local topography on trees in the understory of Bornean lowland rain forest. Pl. Ecol. 164: 1–18.

GILBERT, D.G., HUBBELL, S.P. & FOSTER, R.B. 1994. Density and distance-to-adult effects of a canker disease of trees in a moist tropical forest. Oecologia 98: 100–108.

GILBERT, G.S., FERRER, A. & CARRANZA, J. 2002. Polypore fungal diversity and host density in a moist tropical forest. Biodivers. & Conservation 11: 947–957.

GILL, A.S. & ROY, M.M. 2001. Improved silvopastures for joint forest management in India. Indian Forester 127: 303–315.

GILL, S.K., ROSS, W.H. & PANYA, O. 2009. Moving beyond rhetoric: the need for participatory forest management with the Jakun of South-East Pahang, Malaysia. J. Trop. Forest. Sci. 21: 123–138.

GILLER, P.S. 1996. The diversity of soil communities, 'poor man's tropical rain forest'. Biodivers. & Conservation 5: 135–168.

GILLIS, M. 1980. Fiscal and Financial Issues in Tropical Hardwood Concessions. Development Discussion Paper 10, Harvard Institute for International Development, Cambridge, Massachusetts.

GILLISON, A.N. 1970. Structure and floristics of a montane grassland/forest transition, Doma Peaks region, Papua. Blumea 18: 71–86.

GIMLETTE, J.D. 1929. Malay Poisons and Charm Cures. J. & A. Churchill, London.

GIRIRAJ, A., MURTHY, M.S.R. & RAMESH, B.R. 2008. Vegetation composition, structure and diversity pattern: a case study from tropical wet evergreen forests of Western Ghats, India. Edinb. J. Bot. 65: 447–468.

GIVEN, D.H. 1998. Conserving landscapes and species in Sarawak. Sarawak Gazette 125, 1538: 5–14.

GIVNISH, T. 1980. Ecological constraints on the evolution of breeding systems in seed plants: dioecy and dispersal in gymnosperms. Evolution 34: 959–972.

GIVNISH, T.J. 1982. Outcrossing versus ecological constraints in th evolution of dioecy. Amer. Naturalist 119: 849–863.

GIVNISH, T.J. 1987. Comparative studies of leaf form: assessing the relative roles of selective pressures and phylogenetic constraints. New Phytol. 106: 131–160.

GIVNISH, T.J. 1988. Adaptation to sun and shade: a whole-plant perspective. Austral. J. Pl. Physiol. 15: 63–92.

GIVNISH, T.J. 1998. Adaptive plant evolution on islands: classical patterns, molecular data, new insights. In: Evolution on Islands, ed. P.R. Grant, pp. 28–304. Oxford University Press.

GIVNISH, T.J. 1999. On the causes of gradients in tropical tree diversity. J. Ecol. 87: 193–210.

GIVNISH, T.J. & SYTSMA, K.J. 1997. Molecular Evolution and Adaptive Radiation. Cambridge University Press.

GIVNISH, T.J. & VERMEIJ, G. 1976. The sizes and shapes of liane leaves. Amer. Naturalist 110: 743–778.

GOBBETT, D.J. & HUTCHISON, C.S. 1973. The Geology of the Malay Peninsula. Wiley, London.

GOLDAMMER, J.G. 1988. Rural land-use and wildfires in the tropics. Agrofor. Syst. 6: 235–252.

GOLDAMMER, J.G., ed. 1990. Fire in the

Tropical Biota: Ecosystem Processes and Global Challenges. Springer.

GOLDAMMER, J.G., ed. 1992. Tropical Forests in Transition: Ecology of Natural and Anthropogenic Disturbance Processes. Birkhäuser, Basel.

GOLDAMMER, J.G. & PEÑAFIEL, S.R. 1990. Fire in the pine-grassland biomes of tropical and subtropical Asia. In: Fire in the Tropical Biota: Ecosystem Processes and Global Challenges, ed. J.G. Goldammer, pp. 45–62. Springer.

GOLDAMMER, J.G. & SEIBERT, B. 1989. Natural rain forest fires in Eastern Borneo during the Pleistocene and Holocene. Naturwissenschaften 76: 518–520.

GOLDAMMER, J.G. & SEIBERT, B. 1990. The impact of drought and forest fires on tropical lowland rain forests of East Kalimantan. In: Fire in the Tropical Biota: Ecosystem Processes and Global Challenges, ed. J.G. Goldammer, pp. 11–31. Springer.

GOLTAPEH, E.M., DANESH, Y.R., PRASAD, R. & VARMA, A. 2008. Mycorrhizal fungi: what do we know and what should we know? In: Mycorrhiza, 3rd ed., ed. A. Varma, pp. 3–27. Springer, Berlin & Heidelberg.

GOOD, R. 1974. The Geography of the Flowering Plants. 4th ed. Longmans, London.

GONG, W.K. 1981. Studies in the natural regeneration of a hill dipterocarp species, Hopea pedicellata. Malayan Forester 44: 357–369.

GONG, W.K. 1982. Leaf litter fall, decomposition and nutrient element release in a lowland dipterocarp forest. Malayan Forester 45: 367–378.

GONG, W.K. & ONG, J.E. 1983. Litter production and decomposition in a coastal hill dipterocarp forest. In: Tropical Rain Forest: Ecology and Management, eds S.L Sutton, T.C. Whitmore & A.C. Chadwick, pp. 275–285. Blackwell, Oxford.

GORE, A.J.P. ed., 1983. Mires: Swamp, Bog, Fen and Moor. Ecosystems of the World, 4A & B. Elsevier, Amsterdam.

GORRIE, R.M. 1947. Soil erosion survey and conservation working plans. Indian Forester 73: 65–70

GOULD, E. 1977. Foraging behavior of Pterocarpus vampinus on the flowers of Durio zibethinus. Malayan Nat. J. 30: 53–57.

GOULD, E. 1978. Foraging behavior of Malaysian nectar-feeding bats. Biotropica 10: 184–193.

GOULD, S.J. & LEWONTIN, R.C. 1979. The spandrels of San Marco and the Panglossian paradigm: a critique of the adaptationist paradigm. Proc. Roy. Soc. London, Ser. B., Biol. Sci.: 581–598.

GOYAL, A.K. 2001. 'Polluters to pay' – managing the ecotourism zone and ensuring conservation of forests through participatory forest management – Kerala experience. Indian Forester 127: 1416–1417.

GRADSTEIN, F., OGG, J. & SMITH, A. 2004. A Geological Time Scale. Cambridge University Press.

GRAHAM, E.A., MULKEY, S.S., KITAJIMA, K., PHILLIPS, N.G. & WRIGHT, S.J. 2003. Cloud cover limits net CO_2 uptake and growth of a rain forest tree during tropical rainy seasons. *Proc. Natl. Acad. Sci. U.S.A.* 100: 572–576.

GRANT, V. 1981. *Plant Speciation*. Columbia University Press.

GRAY, A. 1846. Analogy between the flora of Japan and that of the United States. *Am. J. Sci. Arts 2* (2): 175–176.

GREEN, M.J.B. & GUNAWARDENA, E.R.N. 1993. *Conservation Evaluation of some Natural Forest in Sri Lanka*. UNDP, FAO & IUCN. Unpubl. ms.

GREEN, P.T. & JUNIPER, P.A. 2004. Seed-seedling allometry in tropical rain forest trees: seed mass-related paths of resource allocation and the 'reserve effect'. *J. Ecol.* 92: 397–408.

GREENOUGH, P. & TSING, A.L., eds. 2003. Nature in the global south: environmental projects in South and Southeast Asia. Duke University Press.

GREIG-SMITH, P., AUSTIN, M.P. & WHITMORE, T.C. 1967. The application of quantitative methods to vegetation survey. I. Association analysis and principal component ordination of rain forest. *J. Ecol.* 55: 483–505.

GRELLER, A.M. & BALASUBRAMANYAM, S. 1980. A preliminary floristic climatic classification of the forests of Sri Lanka. *Sri Lankan Forester* 4: 163–169.

GRELLER, A.M. & BALASUBRAMANYAM, S. 1988. Vegetational composition, leaf size, and climatic warmth in an altitudinal sequence of evergreen forests in Sri Lanka (Ceylon). *Trop. Ecol.* 29: 121–145.

GRELLER, A.M. & BALASUBRAMANYAM, S. 1993. Physiognomic, floristic and bioclimatological characterization of the major forest types of Sri Lanka. In: *Ecology and Landscape Management in Sri Lanka. Proceedings from the International and Interdisciplinary Symposium, Colombo, Sri Lanka, 12–26 March 1990*, eds W. Erdelen, C. Preu, N. Ishwaran & C.M. Madduma Bandara, pp. 55–77. V.J. Markgraf, Weikersheim.

GRELLER, A.M., GUNATILLEKE, I.A.U.N., JAYASURIYA, A.H.M., GUNATILLEKE, C.V.S., BALASUBRAMANIAM, S. & DASSANAYAKE, M.D. 1987. *Stemonoporus* (Dipterocarpaceae)-dominated montane forests in the Adam's Peak Wilderness, Sri Lanka. *J. Trop. Ecol.* 3: 243–253.

GRESSITT, J.L., ed. 1982. *Biogeography and Ecology of New Guinea*. W. Junk, The Hague.

GRIERSON, A.J.C. & LONG, D.G. 1983–2001. *Flora of Bhutan: Including a Record of Plants from Sikkim and Darjeeling*. 3 Vols. Royal Botanic Gardens, Edinburgh & Government of Bhutan.

GRIFFITH, A.L. 1945. Snowfall in Dehra Dun. *Indian Forester* 71: 117.

GRONOW, J. & SHRESTHA, N.K. 1990. From policing to participation: reorientation of Forest Department field staff in Nepal. *Research Report Series* 11. HMG Ministry of Agriculture – Winrock International, Kathmandu.

GRUBB, P.J. 1971. Interpretation of the 'Massenerhebung' effect on tropical mountains. *Nature* 229, 5279: 44–45.

GRUBB, P.J. 1973. Factors controlling the distribution of forest-types on tropical mountains: new facts and a new perspective. In: *Transactions of the Third Aberdeen-Hull Symposium on Malesian*

Ecology, ed. J.R. Flenley, pp. 13–46. Hull Geog. Dept. Misc. Ser. 13.

GRUBB, P.J. 1977a. The maintenance of species richness in plant communities: the importance of the regeneration niche. *Biol. Rev.* 52: 107–145.

GRUBB, P.J. 1977b. Control of forest growth and distribution on wet tropical mountains: with special reference to mineral nutrition. *Annual Rev. Ecol. Syst.* 8: 83–107.

GRUBB, P.J. 1998. Seeds and fruits of tropical rainforest plants: interpretation of the range in seed size, degree of defense and flesh/seed quotients. In: *Dynamics of Tropical Communities. 37th Symposium of the British Ecological Society, Cambridge University*, eds D.M. Newbery, H.H.T. Prins & N.D. Brown, pp.1–24. Blackwell Science, Oxford.

GRUBB, P.J. & STEVENS, P.F. 1985. *The Forests of the Fatima Basin and Mt. Kerigomna, Papua New Guinea with a Review of Montane and Subalpine Rainforets in Papuasia*. Australian National University.

GRUBB, P.J., LLOYD, J.R., PENNINGTON, T.D. & WHITMORE, T.C. 1963. A comparison of montane and lowland forest in Ecuador. I. The forest structure, physiognomy and floristics. *J. Ecol.* 51: 567–601.

GRUBB, P.J., TURNER, I.M. & BURSLEM, D.F.R.P. 1994. Mineral nutrient status of coastal hill dipterocarp forest and *Adinandra* belukar in Singapore. *J. Ecol.* 82: 559–577.

GRUBB, P.J., LEE, W.G., KOLLMANN, J. & WILSON, J.B. 1996. Interraction of irradiance and soil nutrient supply on growth of seedlings of ten European tall-shrub species and *Fagus sylvatica*. *J. Ecol.* 84: 827–840.

GRUBB, P.J., COOMBES, D.A. & METCALFE, D.J. 2005. Comment on 'A Brief History of Seed Size'. *Science* 310: 783.

GUHA, R. 2001. The prehistory of community forestry in India. *Environm. Hist.* 6: 213–238.

GUIANG, E.S., ESGUERRA, F. & BACALLA, D. 2008. Devolved and decentralized forest management in the Philippines: triggers and constraints in investments. In: *Lessons from Forest Decentralization: Money, Justice and the Quest for Good Governance in Asia–Pacific*, eds C.J.P. Colfer, G.R. Dahal & D. Capistrano, pp. 161–182. Earthscan, London.

GUNASEKERA, N., ASHTON, P. & DAYANANDAN, S. Undated. Molecular phylogeny of the Dipterocarpaceae as a window to understand evolution of tropical tree diversity. Unpubl. ms.

GUNATILLEKE, C.V.S. (as née, PEERIS). 1975. *The Ecology of the Endemic Tree Species of Sri Lanka in Relation to their Conservation*. PhD thesis, University of Aberdeen.

GUNATILLEKE, C.V.S. & ASHTON, P.S. 1987. New light on the plant geography of Ceylon. II. The ecological biogeography of the lowland endemic flora. *J. Biogeogr.* 4: 295–327.

GUNATILLEKE, C.V.S. & GUNATILLEKE, I.A.U.N., ASHTON, P.M.S. & ASHTON, P.S. 1996. An overview of seed and seedling ecology of *Shorea* (section *Doona*) Dipterocarpaceae. *DIWPA* 1: 81–102.

GUNATILLEKE, C.V.S, GUNATILLEKE, I.A.U.N., PERERA, G.D., BURSLEM,

D.F.R.P., ASHTON, P.M.S. & ASHTON, P.S. 1997. Responses to nutrient addition among seedlings of eight closely related species of *Shorea* in Sri Lanka. *J. Ecol.* 85: 301–312.

GUNATILLEKE, C.V.S. & GUNATILLEKE, I.A.U.N., ASHTON, P.M.S. & ASHTON, P.S. 1998. Seedling growth of *Shorea* (Dipterocarpaceae) across an elevational range in Southwest Sri Lanka. *J. Trop. Ecol.* 14: 231–245.

GUNATILLEKE, C.V.S. & GUNATILLEKE, I.A.U.N., ETHUGALA, A.U.K. & ESUFALI, S. 2004a. *Ecology of Sinharaja Rain Forest and the Forest Dynamics Plot*. WHT publications, Colombo.

GUNATILLEKE, C.V.S., GUNATILLEKE, I.A.U.N., ETHUGALA, A.U.K., WEERASEKARA, N.S., ESUFALI, S., ASHTON, P.S., ASHTON, M.S. & WIJESUNDARA, S.A. 2004b. Community ecology in an everwet forest in Sri Lanka. In: *Tropical Forest Diversity and Dynamism: Findings from a Large-Scale Plot Network*, eds E.C. Losos & E.G. Leigh jr., pp. 119–144. University of Chicago Press, Chicago.

GUNATILLEKE, C.V.S. & GUNATILLEKE, I.A.U.N., ASHTON, P.S., ETHUGALA, A.U.K., WEERASEKARA, N.S. & ESUFALI, S. 2004c. Sinharaja forest dynamics plot, Sri Lanka. In: *Tropical Forest Diversity and Dynamism: Findings from a Large-Scale Plot Network*, eds E.C. Losos & E.G. Leigh Jr., pp. 599–608. University of Chicago Press, Chicago.

GUNATILLEKE, C.V.S., GUNATILLEKE, I.A.U.N., ESUFALI, S., HARMS, K.E., ASHTON M.S., BURSLEM, D.F.R.P. & ASHTON, P.S. 2006. Species-habitat associations in a Sri Lankan dipterocarp forest. *J. Trop. Ecol.* 22: 1–14.

GUNATILLEKE, I.A.U.N. 1999. *Reproductive Genetics of some Selected Rain Forest Plant Species of Sri Lanka: Implications for Conservation*. Presidential address, section d, 1998. Proceedings 54th annual session, II, 14–19 Dec. 1998, pp. 95–122. Sri Lanka Association for the Advancement of Science.

GUNATILLEKE, I.A.U.N. & GUNATILLEKE, C.V.S. 1991. Threatened woody endemics of the wet lowlands of Sri Lanka and their conservation. *Biol. Conservation* 55: 17–36.

GUNATILLEKE, I.A.U.N., GUNATILLEKE, C.V.S., UMANTHA, S.K.N., BAWA, K.S., DAYANANDAN. S., DAYANANDAN, B., MURAWSKI, D. & ASHTON, P.S. Undated, c.1999. *Reproductive Biology and Genetics of some* Shorea *(section Doona) Species in Sri Lankan Rain Forests: a Case for Long-Term Research*. Unpubl. ms.

GUPTA, B.N. & TOTEY, N.G. 1994. An account of preservation plots in central India. *Indian Forester* 120: 844–859.

GUPTA, J.N. 1939. *Dipterocarpus* (gurjun) forests in India and their regeneration. *Indian Forester Rec.* 3: 61–164.

GUPTA, R., SRIVASTAVA, S.K., MAHENDRA, A.K., PUNDIR, I. & KUMAR, I. 2004. Impact of participatory forest management on socio-economic development of rural people: A cased study in Kodsi and Talaichittor villages of Dehra Dun District. *Indian Forester* 130: 243–252.

GUPTA, S. 1996. Patterns of rarity and endemism in flowering plants – a new approach in conservation planning.

Indian Forester 122: 804–807.

GUPTA, V. 2005. Community resource governance in the tribal societies of Arunachal Pradesh. *Indian Forester* 131: 634–646.

GURITNO, A.D. & MURAO, K. 2001. Deforestation issues related to continued forest industrialization in Indonesia – with special reference to supply and demand of raw materials. *Tropics* 10: 609–623.

HA, C.O., SANDS, V.E., SOEPADMO, E. & JONG, K. 1988a. Reproductive patterns of selected understorey trees in the Malaysian rain forest: the sexual species. *Bot. J. Linn. Soc.* 97: 295–316.

HA, C.O., SANDS, V.E., SOEPADMO, E. & JONG, K. 1988b. Reproductive patterns of selected understorey trees in the Malaysian rain forest: the apomictic species. *Bot. J. Linn. Soc.* 97: 317–331.

HAANTJENS, H.A. *et al.* 1964–1973. Papua-New Guinea Land Reports. *CSIRO Land Res. Ser.* 10, 12, 14, 15, 17, 20, 22, 23, 27, 29–32.

HAANTJENS, H.A., McALPINE, J.R., REINER, E., ROBBINS, R.G. & SAUNDERS, J.C. 1970. Lands of the Goroka-Mount Hagen Area, Territory of Papua and New Guinea. *CSIRO Land Res. Ser.* 27.

HABERLE, S.G., HOPE, G.S. & DeFRETES, Y. 1991. Environmental change in the Baliem Valley, montane Irian Jaya, Republic of Indonesia. *J. Biogeogr.* 18: 25–40.

HADFIELD, W. 1968. Leaf temperature, leaf pose and productivity of the tea bush. *Nature* 219: 282–284.

HADFIELD, W. 1974. Shade in north-east Indian tea plantations. II. Foliar illumination and canopy characteristics. *J. Appl. Ecol.* 11: 179–199.

HAFNER, J.A. 1989. Forces and policy issues affecting forest use in Northeast Thailand. In: *Keepers of the Forest: Land Management Alternatives in Southeast Asia*, ed. M. Poffenberger, pp. 69–94. Kumarian Press, West Hartford, Conneticut.

HAGERUP, E. & HAGERUP, O. 1953. Thrips pollination in *Erica tetralix*. *New Phytol.* 52: 1–5.

HAGERUP, O. 1950. Thrips pollination in *Calluna*. *Kongel. Danske Vidensk.-Selsk. Biol. Medd.* 8(4): 1–16.

HAILE, N.S. 1968. The Quaternary geomorphological history of Sarawak: discussion. *Sar. Mus. J.* 16 (N.S.): 277–281.

HAILE, N.S. 1969. Quaternary deposits and geomorphology of the Sunda shelf off Malaysian shores. *Proc. VIII Congrés INQUA, Paris*: 161–164.

HAILE, N.S. 1971. Quaternary shorelines in West Malaysia and adjacent parts of the Sunda Shelf. *Quaternaria* 15: 333–343.

HAILE, N.S. 1973. The geomorphology and geology of the northern part of the Sunda Shelf and its place in the Sunda orogene. *Pacific Geology* 6: 73–90.

HALL, P. 1991. *Structure, Stand Dynamics and Species Compositional Change in Three Mixed Dipterocarp Forests of Northwest Borneo*. PhD thesis, Boston University.

HALL, P. & BAWA, K. 1993. Methods to assess the impact of extraction of non-timber tropical forest products on plant populations. *Econ. Bot.* 47: 234–247.

HALL, R. 1996. Reconstructing Cenozoic S.E. Asia. In: *Tectonic Evolution of Southeast Asia,* eds R. Hall & D. Blundell. Geological Society Special Publication 106: 153–184. Geological Society, London.

HALL, R. & BLUNDELL, D., eds. 1996. *Tectonic Evolution of Southeast Asia.* Geological Society Special Publication 106. Geological Society, London.

HALL, R. & HOLLOWAY, J.D. eds., 1998. *Biogeography and Geological Evolution of S.E. Asia.* Backhuys, Leiden.

HALLÉ, F. & NG, F.S.P. 1981. Crown construction of mature dipterocarp trees. *Malaysian Forester* 44: 222–233.

HALLÉ, F. & OLDEMAN, R.A.A. 1970. *Essai sur L'architecture et la Dynamique de Croissance des Arbres Tropicaux.* Masson, Paris.

HALLÉ, F., & OLDEMAN, R.A.A. 1975. *An Essay on the Architecture and Dynamics of Growth of Tropical Trees.* Transl. B.C. Stone. University of Malaya Press.

HALLÉ, F., OLDEMAN, R.A.A. & TOMLINSON, P.B. 1978. *Tropical Trees and Forests: an Architectural Analysis.* Springer, New York and Berlin.

HAMANN, A. 2004. Flowering and fruiting phenology of Philippines submontane forest: climatic factors as proximate and ultimate causes. *J. Ecol.* 92: 24–31.

HAMANN, A. & CURIO, E. 1999. Interactions among frugivores and fleshy fruited trees in a Philippine submontane rain forest. *Conservation Biol.* 13: 766–733.

HAMANN, A., BARBON, E.B., CURIO, E. & MADULID, D.A. 1999. A botanical inventory of a submontane tropical rainforest on Negros Island, Philippines. *Biodivers. & Conservation* 8: 1017–1031.

HAMER, K.C., HILL, J.K., MUSTAFFA, N., BENEDICK, S., SHERRATT, T.N., CHEY, V.K. & MARIATI, M.J. 2005. Temporal variation in abundance and diversity of butterflies in Bornean rain forests: opposite impacts of logging recorded in different seasons. *J. Trop. Ecol.* 21: 417–425.

HAMILTON, J.D. 1930. Teak-bearing rocks. *Indian Forester* 56: 147–156.

HAMILTON, L.S., JUVICK, J.O. & SCATENA, F.N., eds. 1995. *Tropical Montane Cloud Forests.* Springer.

HAMILTON, W. 1979. *Tectonics of the Indonesian Region.* U.S. Government Printing Office, Washington, DC.

HAMRICK, J.L. & LOVELESS, M.D. 1986. Isozyme variation in tropical trees. Techniques and preliminary results. *Biotropica* 18: 201–207.

HAMRICK, J.L. & MURAWSKI, D.A. 1990. The breeding structure of tropical tree populations. *Pl. Spec. Biol.* 5: 157–165.

HANEBUTH, T., STATTEGGER, K. & GROOTES, P.M. 2000. Rapid flooding of the Sunda Shelf: a late-glacial sea-level record. *Science* 288: 1033–1035.

HAO, W.M., LIU, M.-H. & CRUTZEN, P.J. 1990. Estimates of annual and regional releases of CO_2 and other trace gases to the atmosphere from fires in the tropics, based on FAO statistics for the period 1975–1980. In: *Fire in the Tropical Biota: Ecosystem Processes and Global Challenges,* ed. J.G. Goldammer, pp. 440–462. Springer.

HARA, M. 1991a. Forest structure of several broad-leaved forests of Bhutan. In: *Life Zone Ecology of the Bhutan Himalaya II,* ed. M. Ohsawa, pp. 157–170. Laboratory of Ecology, Chiba University.

HARA, M. 1991b. Comparison of seedling stages in species of Fagaceae in Bhutan. In: *Life Zone Ecology of the Bhutan Himalaya II,* ed. M. Ohsawa, pp. 171–188. Laboratory of Ecology, Chiba University.

HARA, M., KANZAKI, M., MIZUNO, T., NOGUCHI, H., SRI NGERNYUANG, K., TEEJUNTUK, S., SUNGPALEE, C., OHKUBO, T., YAMAKURA, T., SAHUNALU, P., DHANMANONDA, P. & BUNYAVEJCHEWIN, S. 2002. The floristic composition of tropical montane forest in Doi Inthanon National Park, Northern Thailand, with special reference to its phytogeographical relation with montane forests in tropical Asia. *Nat. Hist. Res. (Chiba)* 7: 1–17.

HARAGUCHI, A. & YABE, K. 2001. Vertical and horizontal distribution of redox potential of soil in wetland forests in Lahei, Central Kalimantan, Indonesia. *Tropics* 10: 91–100.

HARATA, T., NANAMI, S., YAMAKURA, T., MATSUYAMA, S., CHONG, L., DIWAY, B.M., TAN, S. & ITOH, A. 2012. Fine scale spatial genetic structure of ten dipterocarp tree species in a Bornean rain forest. *Biotropica* 44: 586–594.

HÄRDTER, R., WOO, W.C. & OOI, S.H. 1997. Intensive plantation cropping, a source of sustainable food and energy production in the tropical rain forest areas in southeast Asia. *Forest Ecol. Managem.* 91: 93–102.

HARDWICK, K., HEALEY, J., ELLIOTT, S., GARWOOD, N. & VILAIWAN, A. 1997. Understanding and assisting natural regeneration processes in degraded evergreen seasonal forests in northern Thailand. *Forest Ecol. Managem.* 99: 203–214.

HARDY, O.J. & SONKÉ, B. 2004. Spatial pattern analysis of tree species didtributions in a tropical rain forest of Cameroon: assessing the role of limited dispersal and niche differentiation. *Forest Ecol. Managem.* 192: 191–202.

HARMS, K.E., WRIGHT, S.J., CALDERÓN, O., HERNANDEZ, A. & HERRE, E.A. 2000. Pervasive density-dependent recruitment enhances seedling diversity in a tropical forest. *Nature* 404: 493–495.

HARRISON, R.D. 2000. Repercussions of El Niño: drought causes extinction and breakdown of mutualism in Borneo. *Proc. Roy. Soc. London, Ser. B, Biol. Sci.* 267: 911–915.

HARRISON, R.D. 2001. Drought and the consequences of of El Niño in Borneo: a case study of figs. *Popul. Ecol.* 43: 63–75.

HARRISON, R.D. 2003. Fig wasp dispersal and the stability of a keystone plant resource in Borneo. *Proc. Roy. Soc. London, Ser. B, Biol. Sci.* 270: S76–79.

HARRISON, R.D. 2005a. A severe drought in Lambir Hills National Park. In: *Pollination Ecology and the Rain Forest,* eds D.J. Roubik, S. Sakai & A.A. Hamid, pp. 51–64. Springer, New York.

HARRISON, R.D. 2005b. Figs and the diversity of tropical rainforests. *Bioscience* 55: 1053–1064.

HARRISON, R.D. 2006. Mortality and recruitment of hemi-epiphytic figs in the canopy of a Bornean rain forest. *J. Trop. Ecol.* 22: 477–480.

HARRISON, R.D. 2011. Emptying the forest. Hunting and the extirpation of wildlife from tropical nature reserves. *Bioscience* 61: 919–924.

HARRISON, R.D. & RASPLUS, J.-Y. 2006. Dispersal of fig pollinators in Asian tropical forests. *J. Trop. Ecol.* 22: 631–639.

HARRISON, R.D., HAMID, A.A., KENTA, T., LAFRANKIE, J., LEE, H.S., NAGAMASU, H., NAKASHIZUKA, T. & PALMIOTTO, P. 2003. The diversity of hemi-epiphytic figs (*Ficus; Moraceae*) in a Bornean lowland rain forest. *Biol. J. Linn. Soc.* 73: 439–455.

HARRISON, R.D., RNSTED & PENG, Y.-Q. 2008. Fig and fig-wasp biology: a perspective from the East. *Symbiosis* 45: 1–8.

HARRISON, R.D., TAN, S., PLOTKIN J., SLIK F., DETTO M., BRENES T., ITOH, A. & DAVIES, S.J. 2013. Consequences of defaunation for a tropical tree community. *Ecol. Lett.* 16: 687–694.

HART, T.B. 1990. Monospecific dominance in tropical rain forests. *Trends Ecol. Evol.* 5: 6–11.

HART, T.B. 1995. Seed, seedlings and subcanopy survival in monodominant and mixed forests of the Ituri forest, Africa. *J. Trop. Ecol.* 11: 443–459.

HART, T.B., HART, J.A. & MURPHY, P.G. 1989. Humid tropics: causes for their co-occurrence. *Amer. Naturalist* 133: 613–633.

HARTLEY, C.W.S. 1967. *The Oil Palm.* CABI, Wallingford.

HARTSHORN, G.S. 1988. Tropical and subtropical vegetation of Meso-America. In: *North American Terrestrial Vegetation,* eds M.G. Barbour & W.D. Billings, pp. 365–390. Cambridge University Press.

HASAN, P. & ASHTON, P.S. 1963. *A Checklist of Brunei Trees.* Borneo Bulletin Press, Brunei Darussalam.

HASTINGS, A. 1987. Can competition be detected using species co-occurrence data? *Ecology* 68: 117–123.

HAZEBROEK, H.P. & TAN, D.N.K. 1993. Tertiary tectonic evolution of the NW Borneo continental margin. *Bull. Geol. Soc. Malaysia* 33: 195–210.

HAZEBROEK, H.P., ADLIN, T.Z. & SINUN, W. 2004. *Maliau Basin: Sabah's Lost World.* Natural History Publications (Borneo), Kota Kinabalu.

HE, F., LEGENDRE, P. & LAFRANKIE, J.V. 1996. Distribution patterns of tree species in a Malaysian tropical rain forest. *J. Veg. Sci.* 8: 105–114.

HE, S.-Y., ed. 1982. Ding Hu Shan Ecosystem Station, Academia Sinica In: *Tropical and Subtropical Forest Ecosystems,* I. Popular Science Press, Guangzhou Branch.

HEADS, M. 2009. Inferring biogeographic history from molecular phylogenies. *Biol. J. Linn. Soc.* 98: 757–774.

HEALEY, J.R., PRICE, C., TAY, J. 2000. The cost of carbon retention by reduced impact logging. *Forest Ecol. Managem.* 139: 237–255.

HEANEY, L.R. 1984. Mammalian species richness on islands on the Sunda Shelf, Southeast, Asia. *Oecologia* 61: 11–17.

HEANEY, L.R. 1991. A synopsis of climatic and vegetational change in Southeast Asia. *Climatic Change* 19: 53–61.

HEANEY, L.R. 1999. Historical biogeography of SE Asia: integrating paradigm and refining the details. *J. Biogeogr.* 26: 435–438.

HECKROTH, H.P., FIALA, B., GULLAN, P.J., IDRIS, A.H. & MASCHWITZ, U. 1998. The soft-scale (Coccidae) associates of Malaysian ant-plants. *J. Trop. Ecol.* 14: 427–443.

HEGDE, E.R., SURYAPRAKASH, S., ACOTH, L. & BAWA, K.S. 1996. Extraction of non-timber forest products in the forests of Biligiri Rangan Hills, India. 1. Contribution to rural income. *Econ. Bot.* 50: 243–251.

HEIDEMAN, P.D. 1989. Temporal and spatial variation in the phenology of flowering and fruiting in a tropical rain forest. *J. Ecol.* 77: 1059–1079.

HEIL, M., FIALA, B., LINSENMAIR, K.E., ZOTZ, G., MENKE, P. & MASCHWITZ, U. 1997. Food body production in *Macaranga triloba* (Euphorbiaceae): a plant investment in anti-herbivore defense via symbiotic ant partners. *J. Ecol.* 85: 847–861.

HEILBUTH, J.C. 2000. Lower species richness in dioecious clades. *Amer. Naturalist* 156: 221–241.

HEMLEY, G. 1994. *International Wildlife Trade: A CITES Sourcebook.* World Wildlife Fund.

HENDERSON, A., CHURCHILL, S.P. & LUTEYN, J.L. 1991 Neotropical plant diversity: are the northern Andes richer than the Amazon Basin? *Nature* 351: 21–22.

HENDERSON, M.R. 1939. The flora of the limestone hills of the Maly Peninsula. *J. Malayan Br. Roy. Asiatic. Soc.* 17: 13–87.

HENRY, G.M. 1971. *A Guide to the Birds of Ceylon.* De Silva, Kandy.

HERRE, E.A., MEJA, L.C., KYLLO, D.A., ROJAS, E., MAYNARD, Z., BUTLER, A. & VAN BAEL, S.A. 2007. Ecological implications of anti-pathogen effects of tropical fungal endophytes and mycorrhizae. *Ecology* 88: 550–558.

HERRE, E.A., JANDÉR, K.C. & MACHADO, C.A. 2008. Evolutionary ecology of figs and their associates: recent progress and outstanding puzzles. *Annual Rev. Ecol. Syst.* 39: 439–458.

HEYNE, K. 1950. De Nuttige Planten van Nederlandsch-Indië. 2 vols, 2nd ed. Van Hoeve, The Hague.

HEYWOOD, C.A., HEYWOOD, V.H. & JACKSON, P.W. 1990. *International Directory of Botanical Gardens,* 5th ed. Koeltz Scientific Books, Königstein.

HEYWOOD, V.H. & DULLOO, M.H. 2005. In situ conservation of wild plant species: a critical global review of best practices. *International Plant Genetic Resources Institute Technical Bulletin.*

HIGHAM, C.F.W. 1984. Prehistoric rice cultivation in Southeast Asia. *Sci. Amer.* 250: 138–146.

HIGHAM, T.F.G., BARTON, H., TURNEY, C.S.M., BARKER, G., BRONK RAMSEY, C., BROCK, F. 2009. Radiocarbon dating of charcoal from tropical sequences: results from the Niah Great Cave, Sarawak, and their broader implications. *J. Quaternary Sci.* 24 (2): 189–197.

617

HILL, R.D. 1972. Soil moisture under forest, Bukit Timah Nature Reserve, Singapore. *Gard. Bull. Singapore* 26: 85–93.

HIRAI, H., MATSUMURA, H., HIROTIMI, H., SAKURAI, K., OGINO, K. & LEE, H.S. 1997. Soils and the distribution of *Dryobalanops aromatica* and *D. lanceolata* in mixed dipterocarp forest – a case study at Lambir Hills National Park, Sarawak, Malaysia. *Tropics* 7: 21–33.

HIRANO, T., JAUHIAINEN, J., INONES, T. & TAKAHASHI, H. 2009. Controls on the carbon balance of tropical peatlands. *Ecosystems* 12: 873–877.

HIROSE, T. & TERASHIMA, I. 2005. Preface: structure and function of plant canopies. *Ann. Bot.* 95: 481–570.

HISLOP, J.A. 1949. Some field notes on the bearded pig. *Malayan Nat. J.* 4: 62–65.

HO, C.C., NEWBERY, M.C. & POORE, M.E.D. 1987. Forest comparisons and inferred dynamics in Jengka Forest Reserve, Malaya. *J. Trop. Ecol.* 3: 25–56.

HO, W.S., WICKNESWARI, R. & MAHANI, M.C. 2001. Changes in population structure and genetic diversity of *Shorea curtisii* Dyer *ex* King due to selective logging. *Pl. Syst. Biodivers. Malaysia 2010* 2: 5–6.

HOANG, W.L. 1993. *Vegetation of Taiwan.* Beijing, Chinese Environmental Press. [in Chinese]

HOCKINGS, P., ed. 1989. *Blue Mountains: The Ethnography and Biogeography of a South Indian Region.* Oxford University Press, New Delhi.

HODGKISON, R., BALDING, S.T., ZUBAID, A. & KUNZ, T.H. 2004. Habitat structure, wing morphology, and the vertical stratification of Malaysian fruit bats (Megachiroptera-Megapodidae). *J. Trop. Ecol.* 20: 667–673.

HOLDRIDGE, L.R. 1947. Determination of world plant formations from simple climatic data. *Science* 105: 367–368.

HOLDRIDGE, L.R., GRENKE, W.C., HATHEWAY, W.H., LIANG, T. & TOSI, J.A. 1971. *Forest Environments in Tropical Life Zones: a Pilot Study.* Pergamon, Oxford.

HOLMES, C.H. 1956. The broad pattern of climate and vegetation distribution in Ceylon. *Ceylon Forester* 2, 4: 209–225.

HOLMES, C.H. 1957. The natural regeneration of the wet and dry evergreen forests of Ceylon. *Ceylon Forester* 3: 15–41.

HÖLSCHER, D., LEUSCHNER, C., BOHMAN, K., JUHRBANDT, J. & TJITROSEMITO, S. 2004. Photosynthetic characteristics in relation to leaf traits in eight co-existing pioneer tree species in Central Sulawesi, Indonesia. *J. Trop. Ecol.* 20: 157–167.

HOLTTUM, R.E. 1954. *Adinandra* belukar. A succession from bare ground on Singapore island. *Malaysian J. Trop. Geogr.* 3: 27–32.

HONEY, M. 1999, 2008. *Ecotourism and Sustainable Development: Who Owns Paradise?* 1st and 2nd eds. Island Press, Washington, DC.

HONG, D. 2008. *Hevea,* Euphorbiaceae. In: *Flora of China* 11, *(Oxalidaceae through Aceraceae),* eds Z.Y. Wu, P.H. Raven & D.Y. Hong, pp. 265–266. Science Press, Bejing & Missouri Botanical Garden Press, St. Louis.

HOOIJER, A., SILVIUS, M., WÖSTEN, H. & PAGE, S. 2006. *PEAT–CO₂: Assessment of CO₂ Emissions from Drained Peatlands in SE Asia.* Technical Report Q3943. Delft Hydraulics.

HOOKER, J.D. 1848. Observations when following the Grand Trunk Road across the hills of Upper Bengal, Paras Nath in the Soane valley; and on the Kymaon branch of the Vindhya Hills. *J. Asiat. Soc. Bengal* 17(2): 355–411.

HOOKER, J.D. 1907. Botany. *Imp. Gaz. India* 1: 157–212.

HOPE, C.W. 1885. How lightning strikes trees. *Indian Forester* 11: 269–270.

HOPE, G.S. 1976. The vegetational history of Mt. Wilhelm, Papua New Guinea. *J. Ecol.* 64: 627–663.

HOPE, G.S. 1980. New Guinea mountain vegetation communities. In: *The Alpine Flora of New Guinea* 1, ed. P. van Royen, pp. 153–222. Cramer Verlag, Vaduz.

HOPE, G.S., GOLSON, J. & ALLEN, J. 1983. Palaeoecology and prehistory in New Guinea. *J. Hum. Evol.* 12: 37–60.

HOPE, G.S., GILLIESON, D. & HEAD, J. 1988. A comparison of sedimentation and environmental change in New Guinea shallow lakes. *J. Biogeogr.* 15: 603–618.

HOPKINS, G.M. 1924. Frost as a cause of unsoundness. *Indian Forester* 50: 243.

HOPKINS, H.C. 1983. The taxonomy, reproductive biology and economic potential of *Parkia* (Leguminosae: Mimosoideae) in Africa and Madagascar. *Bot. J. Linn. Soc.* 87: 135–167.

HORN, H. 1971. *The Adaptive Geometry of Trees.* Princeton University Press.

HORTON, T.R. & BRUNS, T.D. 2001. The molecular revolution in ectomycorrhizal ecology: peeking into the black box. *Molec. Ecol.* 10: 1855–1871.

HOU, H.-Y., ed. 1979. *Vegetation Map of China.* Cartographic Publishing House, Peking.

HOU, H.-Y., 1983. Vegetation of China with reference to its geographical distribution. *Ann. Missouri Bot. Gard.* 70: 509–548.

HOU, H.-Y., SUN, S.C., CHANG, J.W., HO, M.K., WANG, Y.F., KONG, D.C. & WANG, S.C. 1979. *The Legend to The Vegetation Map of China.* Chinese Academy of Sciences Institute of Botany. Mao Publisher, P.R.C.

HU, S.-S., 1979. The phytocoenological features of evergreen broad-leaf forest in Guangxi. *Acta Bot. Sin.* 21: 362–370.

HUANG, C., ZHANG, Y. & BARTHOLOEMEW, B. 1999. Fagaceae, in Wu *et al.,* Flora of China 4: 314.

HUBBELL, S.P. 1979. Tree dispersion, abundance, and diversity in a tropical dry forest. *Science* 203: 1299–1309.

HUBBELL, S.P. 2001. *The Unified Neutral Theory of Biodiversity and Biogeography.* Monographs in Population Biology 32. Princeton University Press.

HUBBELL, S.P., FOSTER, R.B., O'BRIEN, S.T., HARMS, K.E., CONDIT, R., WECHSLER, B., WRIGHT, S.J. & LOO DE LAO, S. 1999. Light-gap disturbances, recruitment limitation, and tree diversity in a neotropical forest. *Science* 283: 554–557.

HUGHES, J.B., ROUND, P.D. & WOODRUFF, D.S. 2003. The Indo-Chinese-Sundaic faunal transition at the Isthmus of Kra: an analysis of resident forest bird species distributions. *J. Biogeography* 30: 569–580.

HUISMAN, J. & WEISSING, F.J. 2001. Fundamental unpredictability of multispecies competition. *Amer. Naturalist* 157: 488–494.

HULME, M., WIGLEY, T., JIANG, T., ZHAO, Z., WANG, F., DING, Y., LEEMANS, R. & MARKHAM, A. 1992. *Climate Change due to Greenhouse Effect and its Implications for China.* World Wildlife Fund for Nature, Gland.

HUMPHREYS, D. 2007. *Logjam: Deforestation and the Crisis of Global Governance.* Earthscan.

HUSTON, M. 1979. A general hypothesis of species diversity. *Amer. Naturalist* 113: 81–101.

HUSTON, M. 1980. Soil nutrients and tree species richness in Costa Rican forests. *J. Biogeogr.* 7: 147–157.

HUSTON, M.A. 1994. *Biological Diversity: the Coexistence of Species in Changing Landscapes.* Cambridge University Press.

HUSTON, M.A. 1997. Hidden treatments in ecological experiments: re-evaluating the ecosystem function of biodiversity. *Oecologia* 110: 449–460.

HUSTON, M.A. & GILBERT, L. 1996. Consumer diversity and secondary production. In: *Biodiversity and Ecosystem Processes in Tropical Forests,* eds G. Orians, R. Dirzo & J.H. Cushman, pp. 33–47. Springer, New York.

HUSTON, M.A. & SMITH, T.M. 1987. Plant succession: life history and competition. *Amer. Naturalist* 130: 168–198.

HUTCHISON, C.S. 1989. *Geological Evolution of South-east Asia.* Oxford Monographs of Geology and Geophysics 13. Clarendon Press, Oxford.

HUTCHISON, C.S. 1996. The 'Rejang accretionary prism' and 'Lupar Line' problem of Borneo. In: *Tectonic Evolution of Southeast Asia,* eds R. Hall & D. Blundell, Geological Society Special Publication 106: 247–261. Geological Society, London.

HUTCHISON, C.S., BERGMAN, S.C., SWANGER, D.A. & GRAVES, J.E. 2000. A Miocene collisional belt in north Borneo: uplift mechanism and isostatic adjustment quantified by thermochronology. *J. Geol. Soc. Lond.* 157: 783–793.

HYLAND, B.P.M., WHIFFEN, T., CHRISTOPHEL, D.C., GREY, B., ELICK, R.W. & FORD, A.J. 1999. *Australian Tropical Rain Forest Trees and Shrubs. User Guide.* CSIRO Publishing, Victoria.

HYNDEMAN, D.C. & MENZIES, J.I. 1990. Rain forests of the Ok Tedi headwaters, New Guinea: an ecological analysis. *J. Biogeogr.* 17: 241–273.

HYNES, R.A. 1970. Nothofagus *Forest in New Guinea – an Ecological Introduction.* Unpubl. ms.

ICHIE, T., HIROMI, T., YONEDA, R., KAMIYA, K., KOHIRA, M., NINOMIYA, I., & OGINO, K. 2004. Short-term drought causes synchronous leaf shedding and flushing in a lowland mixed dipterocarp forest, Sarawak, Malaysia. *J. Trop. Ecol.* 20: 1–5.

ICHIE, T., KENTA, T., NAKAGAWA, M., SATO, K. & NAKASHIZUKA, T. 2005. Resource allocation to reproductive organs during masting in the tropical emergent tree, *Dipterocarpus tempehes.* *J. Trop. Ecol.* 21: 237–241.

ICKES, K., DEWALT, S. & APPANAH, S. 2001. Effects of native pigs (*Sus scrofa*) on woody understorey vegetation in a Malaysian lowland rain forest. *J. Trop. Ecol.* 17: 191–206.

ICKES, K., DEWALT, S.J. & THOMAS, S.C. 2003. Resprouting of woody saplings following stem snap by wild pigs in a Malaysian forest. *J. Ecol.* 91: 222–233.

IGOE, J. & BROCKINGTON, D. 2007. Neoliberal conservation: a brief introduction. *Conservation & Soc.* 5 (4): 432–449.

ILORKAR, V.M. & KHATRI, P.K. 2003. Phytosociological study of Navegaon National Park (Maharashtra). *Indian Forester* 129: 377–387.

IMAI, N., KITAYAMA, K. & TITIN, J. 2010. Distribution of phosphorus in an above-to-below-ground profile in a Bornean tropical rain forest. *J. Trop. Ecol.* 26: 627–636.

INAMATI, S.S., DEVAR, K.V. & KRISHNA, A. 2007. Floristic composition along altitudinal gradation in Devimane, Western Ghats, Karnataka. *Indian Forester* 133: 679–689.

INDERJIT & MALLIK, A.U. 1997. Effect of phenolic compounds on selected soil properties. *Forest Ecol. Managem.* 92: 11–18.

INGLEBY, K., MUNRO, R.C., NOOR, M., MASON, P.A. & CLEARWATER, M.J. 1998. Ectomycorrhizal populations and growth of *Shorea parvifolia* (Dipterocarpaceae) seedlings regenerating under three different forest canopies following logging. *Forest Ecol. Managem.* 111: 171–179.

INOUE, M. & IGIN, B. 1991. Changes in economic life of the hunters and gatherers: the Kelay Punan of East Kalimantan. *Tropics* 1: 143–153.

INOUE, T. & ABDUL HAMID, A., eds. 1994. *Plant Reproductive Systems and Animal Seasonal Dynamics.* Center for Ecological Research, Kyoto University.

INOUE, T. & ABDUL HAMID, A., eds. 1997. *General Flowering of Tropical Rainforests in Sarawak.* Center for Ecological Research, Kyoto University.

INTERGOVERNMENTAL PANEL ON CLIMATE CHANGE (IPCC). 1997. *An Assessment of Vulnerability.* Report, Working Group 2. IPCC, The United Nations.

INTERNATIONAL UNION FOR THE CONSERVATION OF NATURE (IUCN). 2013. *Red List of Threatened Species.* IUCN, Morges.

ISHIDA, A., TOMA, T., MATSUMOTO, Y., YAP, S.K. & MARUYAMA, Y. 1996. Diurnal change in leaf gas exchange characteristics in the uppermost canopy of a rain forest tree, *Dryobalanops aromatica* Gaertn. f. *Tree Physiol.* 16: 779–785.

ISHIGAWA, H., INOMATA, N., YAMAYAKI, N. SHUKOR, AB. & SZMIDT, A.E. 2008. Demographic history and interspecific hybridization of four *Shorea* species (Dipterocarpaceae) from Peninsular Malaysia inferred from nucleotide polymorphism in nucleargene regions. *Can. J. Forest Res.* 38: 996–1007.

ISKANDAR, H., SNOOK, L.K., TOMA, T., MACDICKEN, K. & KANNINEN, M. 2006. A comparison of damage due to logging under different forms of resource access in East Kalimantan, Indonesia. *Forest*

Ecol. Managem. 237: 83–93.

ISMAIL, G. & DIN, L., eds. 1998. *A Scientific Journey Through Borneo. Bario. The Kelabit Highlands of Sarawak*. Pelanduk Press, Selangor, Malaysia.

ITIOKA, T. & YAMAUTI, M. 2004. Severe drought, leafing phenology, leaf damage and Lepidoptera abundance in the canopy of a Bornean aseasonal tropical forest. *J. Trop. Ecol.* 20: 479–482.

ITIOKA, T., NOMURA, M., INUI, Y., ITINO, T. & INOUE, T. 2000. Difference in intensity of ant defense among three species of *Macaranga* myrmecophytes in a Southeast Asian dipterocarp forest. *Biotropica* 32: 318–326.

ITIOKA, T., INOUE, T., KALIANG, H., KATO, M., NAGAMITSU, T., MOMOSE, K., SAKAI, S., YUMOTO, T., MOHAMAD, S.U., HAMID, A.A. & YAMANE, S. 2001. Six-year population fluctuation of the giant honey bee *Apis dorsata* (Hymenoptera: Apidae) in a tropical lowland dipterocarp forest in Sarawak. *Ann. Entomol. Soc. Am.* 94: 545–549.

ITOH, A. 1995. Effects of forest floor environment on germination and seedling establishment of two Bornean rainforest emergent species. *J. Trop. Ecol.* 11: 517–527.

ITOH, A., YAMAKURA, T., OGINO, K. & LEE, H.S. 1995a. Survivorship and growth of seedlings of four dipterocarp species in a tropical rainforest of Sarawak, East Malaysia. *Ecol. Res.* 10: 327–338.

ITOH, A., YAMAKURA, T., OGINO, K., LEE, H.S. & ASHTON, P.S. 1995b. Population structure and canopy dominance of two emergent dipterocarp species in a tropical rain forest of Sarawak, Malaysia. *Tropics* 4: 133–141.

ITOH, A., YAMAKURA, T., OGINO, K., LEE, H.S. & ASHTON, P.S. 1997. Spatial distribution patterns of two predominant emergent trees in a tropical rainforest in Sarawak, Malaysia. *Pl. Ecol.* 132: 121–136.

ITOH, A., YAMAKURA, T. & LEE, H.S. 1999. Effects of light intensity on the growth and allometry of two Bornean *Dryobalanops* (Dipterocarpaceae) seedlings. *J. Trop. Forest Sci.* 11: 610–618.

ITOH, A., YAMAKURA, T., KANZAKI, M., OHKUBO, T., PALMIOTTO, P.A., LAFRANKIE, J.V., KENDAWANG, J.J. & LEE, H.S. 2002. Rooting ability of cutting relates to phylogeny, habitat preference and growth characteristics of tropical rainforest trees. *Forest Ecol. Managem.* 168: 275–287.

ITOH, A., YAMAKURA, T., OHKUBO, T., KANZAKI, M., PALMIOTTO, P.A., TAN, S. & LEE, H.S. 2003a. Spatially aggregated fruiting in an emergent Bornean tree. *J. Trop. Ecol.* 19: 531–538.

ITOH, A., YAMAKURA, T., OHKUBO, T, KANZAKI, M., PALMIOTTO, P.A., LAFRANKIE, J.V., ASHTON, P.S. & LEE, H.S. 2003b. Importance of topography and soil texture in spatial distribution of two sympatric dipterocarp trees in a Bornean rain forest. *Ecol. Res.* 18: 307–320.

ITOH, A., NANAMI, S., HARATA, T., OHKUBO, T., TAN, S., CHONG, L., DAVIES, S.J. & YAMAKURA, T. 2012. The effect of habitat association and edaphic conditions on tree mortality during El Niño-induced drought in a Bornean

dipterocarp forest. *Biotropica* 44: 606–617.

IWATA, S. & MUJAMOTO, S. 1996. History of deforestation in the Himalayas. *Tropics* 5: 243–262.

IWATSUKI, K. 2008. Harmonious coexistence between nature and mankind: an ideal lifestyle for sustainability carried out in the traditional Japanese spirit. *Humans & Nature* 19: 1–18.

JACOB, V.J. & ALLES, W.S. 1987. Kandyan gardens of Sri Lanka. *Agroforestry Systems* 5: 123–127.

JACOBS, M. 1958. Contribution to the botany of Mount Kerinci and adjacent area in West Sumatera. *Ann. Bogor.* 3: 45–104.

JACOBS, M. 1960. Juglandaceae. In: *Flora Malesiana* 1(6), ed. C.G.G.J. van Steenis, pp. 143–154. Goningen.

JACOBS, M. 1961. Plant life on mount Kinabalu. *Mal. Nat. J.* 15: 134–140.

JACOBS, M. 1962. *Pometia* (Sapindaceae), a study in variability. *Reinwardtia* 6: 109–144.

JACOBS, M. 1972. *The Plant World on Luzon's Highest Mountains*. Rijksherbarium, Leiden.

JAIN, S.K., ed. 1981. *Glimpses of Indian Ethnobotany*. Oxford, New Delhi.

JAMEER, S.A. & PANDEY, H.N. 2002. Status of biodiversity in the sacred groves of Jaintia Hills, Meghalaya. *Indian Forester* 128: 738–744.

JAMIR, S.A. & PANDEY, H.N. 2003. Vascular plant diversity in the sacred groves of Jaintia Hills in northeast India. *Biodivers. & Conservation* 12: 1497–1510.

JANKOWSKA-BLASZCZUK, M. & GRUBB, P.J. 2006. Changing perspectives on the role of the soil seed bank in northern temperate forests and in tropical lowland rain forests: parallels and contrasts. *Perspect. Pl. Ecol. Evol. Syst.* 8: 3–21.

JANZEN, D.H. 1967. Why mountain passes are higher in the tropics. *Amer. Naturalist* 101: 233–249.

JANZEN, D.H. 1970. Herbivores and the number of tree species in tropical forests. *Amer. Naturalist* 104: 501–518.

JANZEN, D.H. 1971. Euglossine bees as long-distance pollinators of tropical plants. *Science* 171: 203–205.

JANZEN, D.H. 1974. Tropical blackwater rivers, animals, and mast fruiting by the Dipterocarpaceae. *Biotropica* 6: 69–103.

JANZEN, D.H. 1976a. Why bamboos take so long to flower. *Annual Rev. Ecol. Syst.* 7:347–391.

JANZEN, D.H. 1976b. Why tropical trees have rotten cores. *Biotropica* 8: 110.

JANZEN, D.H. 1978. Seeding patterns in tropical trees. In: *Tropical Trees as Living Systems. Proceedings of the Fourth Cabot Symposium*, eds P.B. Tomlinson & M.H. Zimmermann, pp. 83–128. Cambridge University Press.

JANZEN, D.H. 1979. How to be a fig. *Annual Rev. Ecol. Syst.* 10: 13–51.

JANZEN, D.H. & WATERMAN, P.G. 1984. A seasonal census of phenolics, fibre and alkaloids in foliage of forest trees in Costa Rica: some factors influencing their distribution and relation to host selection by Sphingidae and Saturnidae. *Biol. J. Linn. Soc.* 21: 439–454.

JATTAN, S.S. & PRATIMA. 2001. Participatory

forest management objectives in India. *Indian Forester* 127: 505–511.

JAYASINGAM, T. & VIVEKANANTHARAJA, S. 1994. Vegetation survey of the Wasgomuwa Oya forest. *Vegetatio* 113: 1–8.

JAYASURIYA, A.H.M. 1995. National conservation review: The discovery of extinct plants in Sri Lanka. *Ambio* 24: 313–316.

JAYASURIYA, A.H.M. & PEMADASA, M.A. 1983. Factors affecting the distribution of tree species in a dry zone mature forest in Sri Lanka. *J. Ecol.* 71: 571–583.

JENNINGS, S.B., BROWN, N.D., BOSHIER, D.H., WHITMORE, T.C. & LOPES, J. DO C.A. 2001. Ecology provides a pragmatic solution to the maintenance of genetic diversity in sustainably managed tropical forests. *Forest Ecol. Managem.* 154: 1–10.

JESSUP, T.C. & PELUSO, N.L. 1986. Minor forest products as common property resources in East Kalimantan, Indonesia. *Proceedings of the Conference on Common Property Resources Management, April 21–26, 1985*, pp. 503–531. National Academy Press, Washington, DC.

JEWITT, S. 1995. Voluntary and 'official' forest protection committees in Bihar: solutions to India's deforestation? *J. Biogeogr.* 22: 1003–1021.

JHA, M.N., GUPTA, M.K., SAXENA, A. & KUMAR, R. 2003. Soil organic carbon store in different forests in India. *Indian Forester* 129: 714–724.

JOANN, C.L., FLETCHER, C. & ABD. RAHMAN, K. 2013. Spatial effects of virgin jungle reserves (VJR) on the community of insectivorous bats in Peninsular Malaysia. *J. Trop. Forest Sci.* 25: 118–130.

JOHAN, L.C., VAN VALKENBERG, H. & KETNER, P. 1994. Vegetation changes following human disturbance in mid-montane forest in the Wau area, Papua-New Guinea. *J. Trop. Ecol.* 10: 41–45.

JOHN, R. & SUKUMAR, R. 2004. Distance- and density-related effects in a tropical dry deciduous forest tree community at Mudumalai, southern India. In: *Tropical Forest Diversity and Dynamism: Findings from a Large-Scale Plot Network*, eds E.C. Losos & E.G. Leigh Jr., pp. 363–383. University of Chicago Press, Chicago.

JOHN, R., DATTARAJA, H.S., SURESH, H.S. & SUKUMAR, R. 2002. Density-dependence in common tree species in a tropical dry forest in Musumalai, southern India. *J. Veg. Sci.* 13: 45–56.

JOHNS, A.D. 1986. Effects of selective logging on the behavioral ecology of West Malaysian primates. *Ecology* 67: 684–694.

JOHNS, A.D. 1992. Vertebrate responses to selective logging: implications for the design of logging systems. *Philos. Trans., Ser. B* 335: 437–442.

JOHNS, A.G. 1996. Bird population persistence in Sabahan logging concessions. *Biol. Conservation* 75: 3–10.

JOHNS, R.J. 1982. Plant zonation. In: *Biogeography and Ecology of New Guinea*, ed. J.L. Gressitt, pp. 309–330. W. Junk, The Hague.

JOHNS, R.J. 1986. The instability of the tropical ecosystems of New Guinea. *Blumea* 31: 341–371.

JOHNSINGH, A.J.T. 1986. Impact of fire on wildlife ecology in two dry deciduous

forests in South India. *Indian Forester* 112: 933–944.

JOHNSON, L.A.S. & BRIGGS, B.G. 1975. On the Proteaceae – the evolution and classification of a southern family. *Bot. J. Linn. Soc.* 70: 83–183.

JOHNSON, L.A. & DEARDEN, P. 2009. Fire ecology and management of seasonal evergreen forests in mainland Southeast Asia. In: *Tropical Fire Ecology. Climate Change, Land Use and Ecosystem Dynamics*, ed. M.A. Cochrane, pp. 290–310. Springer & Paxis, Chichester.

JONES, A.W. & KENNEDY, R.S. 2008. Evolution of a tropical archipelago: comparative phylogeography of Philippine flora and fauna reveals complex patterns of colonization and diversification. *Biol. J. Linn. Soc.* 95: 620–639.

JONG, K. 1982. Cytotaxonomy. In: Dipterocarpaceae, ed P.S. Ashton, *Flora Malesiana* 1 (9), ed. C.G.G.J. van Steenis, pp. 268–273. Martinus Nijhoff, The Hague.

JOSHI, H.B. 1975–. *Troup's Silviculture of Indian Trees, Revised and Enlarged*. Controller of Publications, Delhi.

JU, J. & SLINGO, J. 1995. The Asian summer monsoon and ENSO. *Quart. J. Roy. Meteor. Soc.* 121: 1133–1168.

JUSOFF, K. & MAJID, N.M. 1995. Integrating needs of the local community to conserve forest biodiversity in the state of Kelantan. *Biodivers. & Conservation* 4: 108–114.

KAARS, W.A. VAN DER & DAM, M.A.C. 1995. A 135,000 year record of vegetational and climatic change from the Bandung area, West Java, Indonesia. *Palaeogeogr. Palaeoclimatol. Palaeoecol.* 117: 55–72.

KADAMBI, K. 1939. The montane forest. *Indian Forester* 65: 189–201.

KADAMBI, K. 1941. The evergreen Ghat rain forests; Agumbe Kilandur zone (a study in the tropical rain forest of Western Ghats of Mysore). *Indian Forester* 67: 187–203.

KADAMBI, K. 1942. The evergreen Ghat rain forest of the Tunga and Bhadra river sources – Kadur District – Mysore State. *Indian Forester* 68: 233–240; 305–312.

KADAMBI, K. 1949. On the ecology and silviculture of *Dendrocalamus strictus* in the bamboo forests of Bhadravathi Division, Mysore State and comparative notes on the species *Bambusa arundinacea, Ochlandra travancorica, Oxytenanthera monostigma* and *O. stocksii*. *Indian Forester* 75: 289–299.

KADAMBI, K. 1955. *Terminalia myriocarpa* Huerck & Muell. Arg. *Indian Forester* 81: 32–42.

KADAVUL, K. & PARTHASARATHY, N. 1999. Plant biodiversity and conservation of tropical semi-evergreen forest in the Shervaryan hills of Eastern Ghats, India. *Biodivers. & Conservation* 8: 421–439.

KAJITA, T., KAMIYA, K., NAKAMURA, K., TACHIDA, H., WICKNEWARI, R., TSUMURA, Y., YOSHIMARU, H. & YAMAZAKI, T. 1998. Molecular phylogeny of Dipterocarpaceae in Southeast Asia based on nucleotide sequences of *matK, trn*L intron, and the *trn*L–*trn*F intergenic spacer region in chloroplast DNA. *Molec. Phylogen. Evol.* 10: 202–209.

KAKU, S., IWAYA, M & JEON, K.B. 1981. Supercooling ability, water content and hardiness of *Rhododendron* flower buds during cold acclimation and deacclimation. *Pl. Cell Physiol.* 22: 1561–1569.

KALA, J.C. 1992. Common lands: management perspective and emerging trends. *Indian Forester* 118: 379–383.

KALKMAN, C. & VINK, W. 1970. Botanical exploration of the Doma Peak region, New Guinea. *Blumea* 18: 87–135.

KAMBEJ, R.D., SINGH, M. & RAVAL, B.R. 1997. Analysis of threats to Gir ecosystem. *Indian Forester* 123: 964–972.

KAMIYA, K., HARADA, K., TACHIDA, H., ASHTON, P.S. 2005. Phylogeny of *PgiC* gene in *Shorea* and its closely related genera (Dipterocarpaceae), the dominant trees in Southeast Asian tropical rain forests. *Amer. J. Bot.* 92, 775–788.

KAMIYA, K., GAN, Y.Y., LUM, K.Y., KHOO, M.S., CHUA, S.C. & FAIZU, N.H. 2011. Morphological and molecular evidence of natural hybridization in *Shorea* (Dipterocarpaceae). *Tree Genet. Genomes* 7: 297–306.

KAMIYA, K. NANAMI, S., KENZO, T., YONEDA, R., DIWAY, B., CHONG, L., AZANI, M.A., MAJID, N.M., LUM, S.K.Y., WONG, K.-M. & HARADA, K. 2012. Demographic history of *Shorea curtisii* (Dipterocarpaceae) inferred from chloroplast DNA sequence variations. *Biotropica* 44: 577–585.

KAMIYA, K., TACHIDA, H. & ICHIE, T. Undated. *Testing 'Isolation with Migration Model' among Closely Related Species of Shorea (Dipterocarpaceae), a Species-Rich Genus in Asian Tropical Forests.* Unpubl. ms.

KAMMESCHEIDT, L., BERHAMAN, A., TAY, J., ABDULLAH, G. & AZWAL, M. 2009. Liana abundance, diversity and tree infestation in the Imbak Canyon Conservation Area, Sabah, Malaysia. *J. Trop. Forest Sci.* 21: 265–272.

KANE, R.P. 1998. Extremes of the ENSO phenomenon and Indian summer monsoon rainfall. *Int. J. Climatol.* 18: 775–791.

KANZAKI, M., WATANABE, M., KUWAHARA, J., KENDAWANG, J.J., LEE, H.S., SERIZAWA, S. & YAMAKURA, T. 1996. Genetic structure of *Shorea macrophylla* in Sarawak, Malaysia. *Tropics* 6: 153–160.

KANZAKI, M., YAP, S.K., OKAUCHI, Y. & YAMAKURA, T. 1997. Survival and germination of buried seeds of non-dipterocarp species in a tropical rain forest at Pasoh, West Malaysia. *Tropics* 7: 9–20.

KANZAKI, M., SRI-NGERNYUANG, K., FUJI, N., MIZUNO, T., NOGUCHI, H., YAMAKURA, T., HARA, M., OHKUBO, T., SAHUNALU, P., DHANMANOND, P., BUNYAVEJCHEWIN, S. & MATHAVARARUG, A. 1999. Comparative community ecology of tropical montane and temperate lucidophyll forests. Comparison of niche differentiation pattern. *Tropics* 9: 211–228.

KANZAKI, M., HARA, M., YAMAKURA, T., OHKUBO, T., TAMURA, M.N., SRI-NGEMYUANG, K., SAHUNALU, P., TEEJUNTUK, S. & BUNYAVEJCHEWIN, S. 2004. Doi Inthanon Forest Dynamics Plot, Thailand. In: *Tropical Forest Diversity and Dynamism: Findings from a Large-Scale Plot Network,* eds E.C.

Losos & E.G. Leigh Jr., pp. 474–481. University of Chicago Press, Chicago.

KAO, D. & IIDA, S. 2006. Structural characteristics of logged evergreen forests in Preah Vihear, Cambodia, 3 years after logging. *Forest Ecol. Managem.* 225: 62–73.

KAPELLE, M. 2008. *Biodiversity of the Oak Forests of Tropical America.* INBIO. Sto. Domindo de Heredia, Costa Rica.

KAREWKROM, P., GAJASENI, J., JORDAN, C.F. & GAJASENI, N. 2005. Floristic composition in 5 types of teak plantations in Thailand. *Forest Ecol. Managem.* 210: 351–361.

KARL, T.R. & TRENBARTH, K.E. 2003. Modern global climate change. *Science* 302: 1719–1723.

KARTAWINATA, K. 1980. A note on kerangas (Heath) forest at Sebulu, East Kalimantan. *Reinwardtia* 9: 429–447.

KARTAWINATA, K., ABDULHADI, R. & PARTIMIHARDJO, T. 1981. Composition and structure of a lowland dipterocarp forest at Wanariset, East Kalimantan. *Malayan Forester* 44: 397–408.

KATHIRITHAMBY-WELLS, J. 2005. *Nature and Nation: Forests and Development in Peninsular Malaysia.* Nias, Copenhagen.

KATHRIARACHCHI, H. 2002. *Performance of Selected Forest Species Providing Non-Timber Forest Products in Sinharaja MAB Reserve, Sri Lanka.* Masters dissertation, University of Peradeniya.

KATO, M. & INOUE, T. 1994. Origin of insect pollination. *Nature* 368: 159.

KATO, M., ITINO, T., & NAGAMITSU, T. 1993. Melittophily and ornithophily of long-tubed flowers in Zingiberaceae and Gesneriaceae in West Sumatra. *Tropics* 2: 129–142.

KATO, R., TADAKI, Y. & OGAWA, H. 1974. Plant biomass and growth increment studies in Pasoh forest. *Malayan Nat. J.* 30: 211–224.

KATSIGIRIS, E., BULL G.Q., WHITE, A., BARR, C., BARNEY, K., BUN, Y., KAHRL, F., KING, T., LANKIN, A., LEBEDEV, A., SHEARMAN, P., SHEINGAUZ, A., SU, Y. & WEYERHAEUSER, H. 2004. The China forest products trade: overview of Asia-Pacific supplying countries, impacts and implications. *Int. Forest. Rev.* 6: 254–266.

KAUL, O.N., GUPTA, A.C. & SHARMA, D.C. 1975. Preservation plots in India. *Indian Forester Bull.* 271.

KAUR, A. (as MRS DALBIR SINGH). 1977. *Embryological and Cytological Studies on some Members of the Dipterocarpaceae.* PhD thesis, University of Aberdeen.

KAUR, A. 1998. A history of forestry in Sarawak. *Mod. Asian Stud.* 32: 117–147.

KAUR, A., HA, C.O., JONG, K., SANDS, V.E., CHAN, H.T., SOEPADMO, E. & ASHTON, P.S. 1978. Apomixis may be widespread among trees of the climax rain forest. *Nature* 271: 440–442.

KAUR, A., JONG, K., SANDS, V.E. & SOEPADMO, E. 1986. Cytoembryology of some Malaysian dipterocarps, with some evidence of apomixis. *Bot. J. Linn. Soc.* 92: 75–88.

KAYE, J.W. 1854. *The Life and Correspondence of Charles Lord Metcalfe.* Richard Bentley, London (Google e-books).

KEELEY, J.E. & BOND, W.J. 1999. Mast flowering and semelparity in bamboos:

the bamboo fire cycle hypothesis. *Amer. Naturalist* 154: 383–391.

KELLY, D. 1994. The evolutionary ecology of mast seeding. *Trends Ecol. Evol.* 9:465–467.

KELLY, D. & SORK, L. 2002. Mast seeding in perennial plants: why, how, where? *Annual Rev. Ecol. Syst.* 23: 427–447.

KELTY, M.J., LARSON, B.C. & OLIVER, C.D., eds. 1992. *The Ecology and Silviculture of Mixed Species Forests.* Kluwer, Boston.

KENNEDY, D.N. & SWAINE, M.D. 1992. Germination and growth of colonizing species in artrificial gaps of different sizes in dipterocarp rain forest. *Philos. Trans., Ser. B* 335: 357–361.

KENTA, T., SHIMIZU, K.K., NAKAGAWA, M., OKADA, K., HAMID, A.A. & NAKASHIZUKA, T. 2002. Multiple factors contribute to outcrossing in a tropical emergent *Dipterocarpus tempehes,* including a new pollen-tube guidance mechanism for self-incompatibility. *Amer. J. Bot.* 89: 60–66.

KENTA, T., ICHIE, T., NINOMIYA, J. & KIOKE, T. 2003. Photosynthetic activity in seed wings of Dipterocarpaceae in a masting year: does wing photosynthesis contribute to reproduction? *Photosynthetica* 41: 551–557.

KENTA, T., ISAGI, M. & NAKAGAWA, M., YAMASHITA, M. & NAKASHIZUKA, T. 2004. Variation in pollen dispersal between years with different pollination conditions in a tropical emergent tree. *Molec. Ecol.* 13: 3375–3384.

KENZO, T., ICHIE, T., WATANABE, Y., & HIROMI, T. 2007. Ecological distribution of homobaric and heterobaric leaves in tree species of Malaysian lowland tropical rainforest. *Amer. J. Bot.* 94: 764–775.

KERFOOT, O. 1968. Mist precipitation on vegetation. *Forestry Abstr.* 29: 8–20.

KERR, A.F.G. 1928. Kaw Tao, its physical features and vegetation. *J. Siam. Soc.* 7:1–149.

KERR, A.F.G. 1933. A trip to Phu Bia in French Laos. *J. Siam. Soc.* 9: 193–223.

KERR, R.A. 1997. Moderating a tropical climate. *Science* 276: 1793.

KEBLER, M., KEBLER, P.J.A., GRADSTEIN, S.R., BACH, K., SCHMULL, M.A. & PITOPONG, R. 2005. Tree diversity in primary forest and different land use systems in Central Sulawesi, Indonesia. *Biodivers. & Conservation* 14: 547–560.

KESSINGER, T.G. 1974. *Vilyatpur, 1848–1968: Social and Economic Change in a North Indian Village.* University of California, Berkeley.

KETTERINGS, Q.M., WIBOWO, T.T., NOORDWIJK, M. VAN & PENOT, E. 1999. Farmer's perspectives on slash-and-burn as a land clearing method for small-scale rubbere producers in Sepunggor, Jambi Province, Sumatra, Indonesia. *Forest Ecol. Managem.* 170: 157–169.

KETTLE, C.J. 2012. Flower size and differential pollination success in tropical forest trees. ATB Bali conference paper. Unpubl.

KEVAN, P.G. & LACK, A.J. 1985. Pollination in a cryptically dioecious plant *Decaspermum parviflorum* (Lam.) A.J. Scott (Myrtaceae) by pollen-collecting bees in Sulawesi, Indonesia. *Biol. J. Linn. Soc.* 25: 319–330.

KHAMYONG, S., LYKKE, A.M.,

SERAMETHAKUN, D. & BARFOD, A.S. 2004. Species composition and vegetation structure of an upper montane forest at the summit of Mt. Doi Inthanon, Thailand. *Nordic J. Bot.* 23:83–97.

KHAN, M., & OLIVIER, R.C.D. Undated c.1975. *The Asian Elephant (*Elephas maximus indicus, *Cuvier) in West Malaysia. Its Past, Present and Future.* Unpubl. ms.

KHAN, M.L. & TRIPATHI, R.S. 1986. The regeneration in a disturbed sub-tropical wet hill forest of north-east India: effect of stump diameter and height on sprouting of four tree species. *Forest Ecol. Managem.* 17: 199–209.

KHAN, M.L. & TRIPATHI, R.S. 1989a. Survival and growth of transplanted nursery seedlings of three sub-tropical trees at burnt and unburnt sites in dense and sparse forest stands. *Trop. Ecol.* 30: 20–30.

KHAN, M.L. & TRIPATHI, R.S. 1989b. Effects of stump diameter, stump height and sprout density on the sprout growth of four tree species in burnt and unburnt forest plots. *Acta Oecol. (Oecol. Appl.)* 10: 303–316.

KHAN, M.L. & TRIPATHI, R.S. 1991. Seedling survival and growth of early and late successional tree species as affected by insect herbivory and pathogen attack in sub-tropical humid forest stands of north-east India. *Acta Oecol.* 12: 569–579.

KHAN, M.L., RAI, J.P.N. & TRIPATHI, R.S. 1986. Regeneration and survival of tree seedlings and sprouts in tropical deciduous and subtropical forests of Meghalaya, India. *Forest Ecol. Managem.* 14: 293–304.

KHAN, M.L., RAI, J.P.N. & TRIPATHI, R.S. 1987. Population structure of some tree species in disturbed and protected subtropical forests of north-east India. *Acta Oecol. (Oecol. Appl.)* 8: 247–255.

KHAN, M.L., MENON, S. & BAWA, K.S. 1997. Effectiveness of the protected area network of biodiversity conservation: a case study of Meghalaya state. *Biodivers. & Conservation* 8: 853–868.

KHANNA, P. 1994. Research agenda for joint management. *Indian Forester* 120: 564–569.

KHANNA, P. & SAMA, R.K. 2001. People, panchayat and forests. *Indian Forester* 127: 1326–1332.

KHARE, A., SARIN, M., SAXENA, N.C., PALIT, S., BATHLA, S., VANIA, F. & SATYANARAYANA, M. 2000. *Joint Forest Management: Policy, Practice and Prospects.* Policy that Works for Forests and People, Series No. 3. WWF-India, New Delhi and IIED, London.

KHATRI, P.K., TOKY, N.G. & PANDEY, R.K. 2004. Altitudinal variation in structural composition of vegetation in Satpura National Park. *Indian Forester* 130: 1141–1159.

KHEMACHALERM, W. 1988. Forest agriculture: New options for Thai Agriculture. In: *Turning Point of Thai Farmers. Thai Institute for Rural Management, Culture and Development Series 2,* eds S. Phongphit & R. Bennoun, pp. 147–156. Thai Institute for Rural Development.

KHEMNARK, U. 1970. Natural regeneration of the deciduous forests in Thailand.

Trop. Agr. Res. Ser. 12: 31–43.

KHUANG, X. Y. & CORLETT, R.T. 1997. Forest and forest succession in Hong Kong, China. *J. Trop. Ecol.* 14: 857–866.

KHULLAR, P. 1992. Conservation of biodiversity in natural forests through preservation plots. *Indian Forester* 118: 327–338.

KHUMBONGMAYUM, A.D., KHAN, M.L. & TRIPATHI, R.S. 2005. Sacred groves of Manipur, north-east India. Biodiversity value, status and strategies for their conservation. *Biodivers. & Conservation* 14: 1541–1582.

KHURANA, E. & SINGH, J.S. 2004. Germination and seedling growth of five tree species from tropical dry forest in relation to water stress: impact of seed size. *J. Trop. Ecol.* 20: 385–396.

KHURANA, E. & SINGH, J.S. 2006. Impact of life-history traits on response of seedlings of five tree species of tropical dry forest to shade. *J. Trop. Ecol.* 22: 653–661.

KIEW, R. 1992. Germination and seedling survival of kemenyan, *Styrax benzoin*. *Malayan Forester* 45: 69–80.

KILADIS, G.N. & SINHA, S.K. 1991. ENSO, monsoon and drought in India. In: *Teleconnections Linking Worldwide Climate Anomalies.* eds M.H. Glantz, R.W. Katz & N. Nicholls, pp. 431–458. Cambridge University Press.

KIMURA, K., YUMOTO, T. & KIKUZAWA, K. 2001. Fruiting phenology of fleshy-fruited plants and seasonal dynamics of frugivorous birds in four vegetation zones on Mt. Kinabalu, Borneo. *J. Trop. Ecol.* 17: 833–858.

KING, B., WOODCOCK, M. & DICKINSON, E.C. 1998. *Birds of Southeast Asia.* Periplus, Hong Kong.

KING, D.A. 1990. Allometry of saplings and understorey trees in a Panamanian forest. *Funct. Ecol.* 4: 27–32.

KING, D.A. 1995. Allometry and life history of tropical trees. *J. Trop. Ecol.* 12: 25–44.

KING, D.A. 1997. Relationship between branch spacing, growth rate and light in tropical tree saplings. *Funct. Ecol.* 11: 627–635.

KING, D.A. 1998a. Relationship between crown architecture and branch orientation in rain forest trees. *Ann. Bot. London* 82: 1–7.

KING, D.A. 1998b. Influence of leaf size on tree architecture: first branch height and crown dimensions in tropical rain forest trees. *Trees* 12: 438–445.

KING, D.A. & MAINDONALD, J.H. 1999. The architecture in relation to leaf dimensions and tree stature in temperate and tropical rain forests. *J. Ecol.* 87: 1012–1024.

KING, D.A., LEIGH, E.G. JR., CONDIT, R., FOSTER, R.B., & HUBBELL, S.P. 1997. Relationships between branch spacing, growth rate, and light in tropical forest saplings. *Funct. Ecol.* 11: 627–635.

KING, D.A., DAVIES, S.J., NUR SUPARDI M.N. & TAN, S. 2005. Tree growth is related to light interception and wood density in two mixed dipterocarp forests of Malaysia. *Funct. Ecol.* 19: 445–453.

KING, D.A., DAVIES, S.J. & NUR SUPARDI M.N. 2006a. Growth and mortality are related to adult tree size in a Malaysian mixed dipterocarp forest. *Forest Ecol. Managem.* 223: 152–158.

KING, D.A., DAVIES, S.J., TAN, S. & NUR SUPARDI, M.N. 2006b. The role of wood density and stem support costs in the growth and mortality of tropical trees. *J. Ecol.* 94: 670–680.

KING, D.A., WRIGHT, S.J. & CONNELL, J.H. 2006c. The contribution of interspecific variation in maximum tree height to tropical and temperate diversity. *J. Trop. Ecol.* 22: 11–24.

KINGDON-WARD, F. 1930. *Plant Hunting on the Edge of the World.* V. Gollanz, London.

KINGDON-WARD, F. 1937. *Plant Hunter's Paradise.* Jonathan Cape, London.

KINGDON-WARD, F. 1944. A sketch of the botany and geography of North Burma. II. *J. Bombay Nat. Hist Soc.* 45: 16–30.

KINGDON-WARD, F. 1949. *Burma's Icy Mountains.* Jonathan Cape, London.

KIRA, T. 1977. A climatological interpretation of Japanese vegetation zones. In: *Vegetation Science and Environmental Protection*, eds A. Mayawaki & R. Tüxen, pp. 21–30. Maruzen, Tokyo.

KIRA, T. 1978. Community architecture and organic matter dynamics in tropical lowland rainforests of Southeast Asia with special reference to Pasoh Forest, West Malaysia. In: *Tropical Trees as Living Systems. Proceedings of the Fourth Cabot Symposium,* eds P.B. Tomlinson and M.H. Zimmermann, pp. 561–590. Cambridge University Press.

KIRA, T. & YODA, K. 1989. Vertical stratification in microclimate. In: *Tropical Rain Forest Ecosystems,* eds H. Lieth & M.J.A. Werger, pp. 55–71. Elsevier, Amsterdam.

KIRK, W.D.J. 1988. Thrips and pollination biology. In: *Dynamics of Insect-Plant Interaction.* T.N. Ananthakrishnan & A. Raman, pp 129–135. Oxford & IBH, New Delhi.

KISHIMOTO-YAMADA, K. & ITIOKA, T. 2008. Survival of flower-visiting Chrysomelids during non-general-flowering periods in Bornean dipterocarp forests. *Biotropica* 40: 600–606.

KISS, A. 2004. Is community-based ecotourism a good use of biodiversity conservation funds? *Trends Ecol. Evol.* 19: 232–237.

KITAJIMA, K. & POORTER, L. 2008. Functional basis for resource niche partitionaing by tropical trees. In: *Tropical Forest Community Ecology,* eds W.P. Carson & S.A. Schnitzer, pp. 160–181.Wiley-Blackwell, Chichester.

KITAJIMA, K., MULKEY, S.S. & WRIGHT, S.J. 2005. Variation in crown light utilization characteristics among tropical canopy trees. *Ann. Bot. Lond.* 95: 535–547.

KITAMURA, S., SUZUKI, S., YUMOTO, T., POONSWAD, P., CHUAILUA, P., PLONGMAI, K., NOMA, N., MURUHASHI, T. & SUCKASAM, C. 2004a. Dispersal of *Aglaia spectabilis,* a large-seeded tree species in a moist evergreen forest in Thailand. *J. Trop. Ecol.* 20: 421–427.

KITAMURA, S., YUMOTO, T., POONSWAD, P., NOMA, N., CHUAILUA, P., PLONGMAI, K., MARUHASHI, T. & SUCKASAM, C. 2004b. Pattern and impact of hornbill seed dispersal at nest trees in a moist forest in Thailand. *J. Trop. Ecol.* 20: 545–553.

KITAMURA, S., SUZUKI, S., YUMOTO, T., CHUAILUA, P., PLONGMAI, K.,

POONSWAD, P., NOMA, N., MARUHASHI, T. & SUCKASAM, C. 2005. A botanical inventory of a tropical seasonal forest in Khao Yai National Park, Thailand: implications for fruit-frugivore interactions. *Biodivers. & Conservation* 14: 1241–1262.

KITAMURA, S., SUZUKI, S., YUMOTO, T., POONSWAD, P., CHUAILUA, P., PLONGMAI, K., MURUHASHI, K., NOMA, M. & SUCKASAM, C. 2006. Dispersal of *Canarium euphyllum* (Burseraceae), a large-seeded tree species, in a moist evergreen forest in Thailand. *J. Trop. Ecol.* 22: 137–146.

KITAYAMA, K. 1991. Vegetation of Mount Kinabalu Park, Sabah, Malaysia. Unpubl. project paper.

KITAYAMA, K. 1992. An altitudinal transect study of the vegetation on Mt. Kinabalu, Borneo. *Vegetatio* 102: 149–171.

KITAYAMA, K. 1995. Biophysical conditions of the montane cloud forests of Mount Kinabalu, Sabah, Malaysia. In: *Tropical Montane Cloud Forests,* eds L.S. Hamilton, J.O. Juvick & F.N. Scatena, pp. 183–197. USDA Forest Service, Puerto Rico.

KITAYAMA, K. 1996. Climate of the summit region of Kinabalu (Borneo) in 1992, an El Niño year. *Mountain Res. Dev.* 16: 65–75.

KITAYAMA, K. & AIBA, S.-I. 2002. Ecosystem structure and productivity of tropical rain forests along altitudinal gradients with contrasting soil phosphorus pools on Mount Kinabalu, Borneo. *J. Ecol.* 90: 37–51.

KITAYAMA, K., SCHUUR, E.A.G., DRAKE, D.R. & MUELLER-DOMBOIS, D. 1997. Fate of a wet montane forest during soil ageing in Hawaii. *J. Ecol.* 85: 669–679.

KITAYAMA, K., MAJALAP-LEE, N. & AIBA, S.-I. 2000. Soil phosphorus fractionation and phosphorus-use efficiencies of tropical rainforests along altitudinal gradients of Mount Kinabalu, Borneo. *Oecologia* 123: 342–349.

KITAYAMA, K. AIBA, S.-I., TAKYU, M., MAJALAP, N. & WAGAI, R. 2004. Soil phosphorus fractionation and phosphorus-use efficiency of a Bornean tropical montane rain forest during soil aging with podzolization. *Ecosystems* 7: 259–274.

KITAYAMA, K., AIBA, S.-I., USHIO, M., SEINO, T. & FUJIKI, Y. 2012. The ecology of podocarps in tropical montane forests of Borneo: distribution, population dynamics, and soil nutrient acquisition. *Smithsonian Contr. Bot.* 95: 101–117.

KLEINMAN, P.J.A., PIMENTEL & BRYANT, R.B. 1996. The ecological sustainability of slash-and burn agriculture. *Agric. Eco-syst. Environm.* 52: 235–249.

KNOTT, C.D. 1998. Changes in orangutan caloric intake, energy balance, and ketones in response to fluctuating fruit availability. *Int. J. Primatol.* 19: 1061–1079.

KNUDSEN, J.T. & TOLLSTEN, L. 1993. Trends in floral scent chemistry in pollination syndromes: floral scent composition in moth-pollinated taxa. *Bot. J. Linn. Soc.* 113: 263–284.

KNUDSEN, J.T. & TOLLSTEN, L. 1995. Floral scent in bat-pollinated plants: a case of convergent evolution. *Bot. J. Linn. Soc.* 119: 45–57.

KOBE, R. K. 1999. Light gradient

partitioning among tropical tree species through differential seedling mortality and growth. *Ecology* 80: 187–201.

KOCHUMMEN, K.M. 1966. Natural plant succession after farming in Sg. Kroh. *Malayan Forester* 29: 170–181.

KOCHUMMEN, K.M. 1980. The occurrence of *Agathis borneensis* Warburg (Damar minyak) in heath forest in Peninsular Malaysia. *Malayan Forester* 43: 117–123.

KOCHUMMEN, K.M. 1982. *Effects of Elevation on Vegetation on Gunung Jerai, Kedah.* Forest Research Institute Research Pamphlet 87. Forest Research Institute Malaysia (FRIM), Kepong.

KOCHUMMEN, K.M. 1997. *The Flora of Pasoh Forest.* Malayan Forest Records 44. Forest Research Institute Malaysia (FRIM), Kepong.

KOCHUMMEN, K.M. & GO, R. 2000. Moraceae. In: *Tree Flora of Sabah and Sarawak 3,* eds E. Soepadmo & L.G. Saw, pp. 181–334. Forest Research Institute Malaysia (FRIM).

KOCHUMMEN, K.M. & NG, F.S.P. 1977. Natural plant succession after farming in Kepong. *Malayan Forester* 40: 61–78.

KODANDAPANI, N., COCHRANE, M.A. & SUKUMAR, R. 2009. Forest fire regimes and their ecological effects in seasonally dry tropical systems in the Western Ghats, India. In: *Tropical Fire Ecology. Climate Change, Land Use and Ecosystem Dynamics,* ed. M.A. Cochrane, pp. 335–353. Springer & Paxis, Chichester.

KOELMEYER, K.O. 1957. Climate classification and distribution of vegetation in Ceylon. I. *Ceylon Forester* 3: 144–163.

KOELMEYER, K.O. 1959. Climatic classification and the distribution of vegetation in Ceylon. II. *Ceylon Forester* 3: 265–293.

KOELMEYER, K.O. 1960. The periodicity of leaf change and flowering in the principal forest communities of Ceylon. II. *Ceylon Forester* 4: 308–364.

KOH, L.P. 2007. Impending disaster or sliver of hope for South-East Asian forests? The devil may lie in the details. *Biodivers. & Conservation* 16: 3935–3938.

KOHYAMA, K. 1991. Simulating stationary size distribution of trees in rain forests. *Ann. Bot. Lond.* 68: 173–180.

KOHYAMA, K. 1994. Size-structure-based models of forest dynamics to interpret population- and community-level mechanisms. *J. Pl. Res.* 107: 107–116.

KOHYAMA, T. 1993. Size-structured tree populations in gap-dynamic forest – the forest architecture hypothesis for the stable coexistence of species. *J. Ecol.* 81: 131–143.

KOHYAMA, T. 1996. The role of architecture in enhancing plant species diversity. In: *Biodiversity: an Ecological Perspective,* eds T. Abe, S.A. Levin & M. Higashi, M., pp. 21–33 in Springer, New York.

KOHYAMA, T. 1999. Functional differentiation and positive feedback enhancing plant biodiversity. In: *The Biology of Biodiversity,* ed. M. Kato, pp. 179–191. Springer, Tokyo.

KOHYAMA, T. 2003. Tree species differentiation of growth, recruitment and allometry in relation to maximum height in a Bornean mixed dipterocarp forest. *J. Ecol.* 91: 797–806.

KOHYAMA, T. & HOTTA, M. 1990. Significance of allometry in tropical saplings. *Funct. Ecol.* 4: 515–521.

KOHYAMA, T., SUZUKI, E., AIBA, S.-I. & SEINO, T. 1999. Functional differentiation and positive feedback enhancing plant biodiversity. In: *The Biology of Biodiversity,* ed. M. Kato, pp. 179–191. Springer, Tokyo.

KOHYAMA, T., SUZUKI, E., PARTOMIHARDJO, T., YAMADA, T. & KUBO, T. 2003. Tree species differentiation in growth, recruitment and allometry in relation to maximum height in a Bornean mixed dipterocarp forest. *J. Ecol.* 91: 797–806.

KONG, G.H., HUANG, Z.L., ZHANG, Q.M., LIU, S.Z., MO, J.M. & HE, D.Q. 1997. Type, structure, dynamics, and management of the lower subtropical evergreen broad-leaved forest in the Dinghushan Biosphere Reserve of China. *Tropics* 6: 335–350.

KONG, G.H., DALLMEIER, F., COMISKEY, J.A., HUANG, H. Z., WEI, P., MO, J.M., HE, D.Q., ZHANG, Q. M., & WANG, Y. J. 1998. Structure, composition, and dynamics of an evergreen broadleaf forest in Dinghushan Biosphere Reserve, China. In: *Forest Biodiversity Research, Monitoring and Modeling: Conceptual Background and Old World Case Studies,* eds F. Dallmeier & J.A. Comiskey, pp. 533–551. UNESCO, Paris & Parthenon, Carnforth.

KONUMA, A., TSUMURA, Y., LEE, C.T., LEE, S.L. & OKUDA, T. 2000. Estimation of gene flow in the tropical rainforest tree *Neobalanocarpus heimii* (Dipterocarpaceae) inferred from paternity analysis. *Molec. Ecol.* 9: 1843–1852.

KÖPPEN, W. & GEIGER, W. 1936. *Handbuch der Klimatologie.* 1, C. Gebruder Berntrager, Berlin.

KORIBA, K. 1958. On the periodicity of tree-growth in the tropics, with reference to the mode of branching, the leaf-fall, and the formation of the resting bud. *Gard. Bull. Singapore* 17: 11–81.

KÖRNER, C. 1989. The nutritional status of plants from high altitudes. A worldwide comparison. *Oecologia* 81: 379–391.

KÖRNER, C. 1998. A reassessment of high-elevation treeline positions and their explanation. *Oecologia* 115: 445–459.

KÖRNER, C., ALLISON, A. & HILSCHER, H. 1983. Altitudinal variation of leaf conductance and leaf anatomy in heliophytes of montane New Guinea and their interrelation with microclimate. *Flora* 174: 91–135.

KORPELAINEN, H., ÅDJERS, G., KUUSIPALO, J. NURYANTO, K. & OTSUNO, A. 1995. Profitability of rehabilitation of overlogged dipterocarp forest: a case study form South Kalimantan, Indonesia. *Forest Ecol. Managem.* 79: 207–215.

KOSKELA, J., KUUSIPALO, J. & SIRIKUL, W. 1995. Natural regeneration dynamics of *Pinus merkusii* in northern Thailand. *Forest Ecol. Managem.* 77: 169–179.

KOSONEN, M., OTSAMO, A. & KUUSIPALO, J. 1997. Financial, economic and environmental profitability of reforestation of *Imperata* grasslands in Indonesia. *Forest Ecol. Managem.* 99: 247–259.

KOSTERMANS, A.J.G.H. 1992. *A Handbook of the Dipterocarpaceae of Sr Lanka.* Wildlife Heritage Trust of Sri Lanka, Colombo.

KOYAMA, H. 1981. Photosynthetic rates of lowland rain forest trees of peninsular Malaysia. *Jap. J. Ecol.* 31: 361–369.

KREMEN, C., NILES, J.O., DALTON, M.G., DAILY, G.C., EHRLICH, P.R., FAY, J.P., GREWAL, D. & GUILLERY, R.P. 2000. Economic incentives for rain forest conservation across scales. *Science* 288: 1828–1832.

KRESS, W.J. & BEACH, J.H., eds. 1994. Flowering plant reproductive systems. In: *La Selva: Ecology and Natural History of a Neotopical Rain Forest,* eds W.J. Kress & J.H. Beach, pp. 161–169. University of Chicago.

KRESS, W.J., DE FILIPPS, R.A., FARR, E. & KIY, D.Y.Y. 2003. *A Checklist of the Trees, Shrubs, Herbs and Climbers of Myanmar.* Department of Systematic Biology, National Museum of Natural History, Washington, DC.

KRIANGSAK, S.-N., KWANCHAI, C.-U., KANZAKI, M., OHKUBO, T. & YAMAKURA, T. 2003. Survival and germination of an experimental seed bank population of two species of Lauraceae in a tropical montane forest in Thailand. *J. Forest Res.* 8: 311–316.

KRISHNA MOORTHY, K. 1960. *Mysistica* swamps in the evergreen forests of Travancore. *Indian Forester* 86: 314–315.

KRISHNAN, M.S. 1968. *Geology of India and Burma.* Higginbothams, Madras.

KRISHNAN, R.M. & DAVIDAR, P. 1996. The sholas of the Western Ghats (South India), floristics and structure. *J. Biogeogr.* 23: 783–789.

KRISHNASWAMY, K. 1939. The montane evergreen forest, Bisale region. *Indian Forester* 65: 189–201.

KUDO, G. 2001. Unpubl. meteorological data.

KUDO, G. & SUZUKI, S. 2004. Flowering phenology of tropical-alpine dwarf trees on Mount Kinabalu, Borneo. *J. Trop. Ecol.* 20: 563–571.

KUHLE, M. 1998. New findings on the inland glaciations of Tibet from South and Central West Tibet with evidences for its importance as an Ice Age trigger. *Himal. Geol.* 19: 3–22.

KUMAGAI, T., KATUL, G.G., SAITOH, T.M., SATO, Y., MANFROI, O.J., MOROOKA, T., ICHIE, T., KURAJI, K., SUZUKI, M. & PORPORATO, A. 2001. Water cycling in a Bornean tropical rain forest under current and projected precipitation scenarios. *Water Resour. Res.* 40, WO1104, doi: 10.1029/2003WR002226.

KUMAR, B.H. & DEEPU, J.K. 1992. Litter production and decomposition dynamics in moist deciduous forests of the western Ghats of peninsular India. *Forest Ecol. Managem.* 50: 181–201.

KUMAR, L.S.S. 1946. Better utilization of forest grasslands in Bombay Province. *Indian Forester* 72: 162–164.

KUMAR, M., SHARMA, C.M. & RAJWAR, G.S. 2004. A study on community structure and diversity of a sub-tropical forest of Garhwal Himalayas. *Indian Forester* 130: 207–212.

KUMAR, R. SINGH, A.K. & ABBAS, S.G. 1994. Changes in population structure of some dominant tree species of dry peninsular sal forest. *Indian Forester* 120: 343–348.

KUMARA, H.J., UDAYA, R. & HAYASHI, S. 2003. Sustainable agroforestry systems in dry zone of Sri Lanka: a case study. *Tropics* 12: 103–114.

KUME, T., TAKIZAWA, H., YOSHIFUJI, N., TANAKA, K., TANTASIRIN, C., TANAKA, N. & SUZUKI, M. 2007. Impact of soil drought on sap flow and water status of evergreen trees in a tropical monsoon forest in northern Thailand. *Forest Ecol. Managem.* 238: 220–230.

KUME, T., TANAKA, N., KURAJI, K., KOMATSU, H., YOSHIFUJI, N., SAITOH, T.M., SUZUKI, M. & KUMAGAI, T. 2011. Ten year evapotranspiration estimates in a Bornean tropical rainforest. *Agr. Forest. Meteorol.* 151: 1183–1192.

KUNTH, R. 1906. *Handbook of Flower Pollination: Based on Hermann Müller's Work 'The Fertilisation of Flowers by Insects'.* Clarendon Press, Oxford.

KURIYAMA, K. 2002. The values of tropical forests for local residents. *Tropics* 11: 205–212. [In Japanese]

KUROKAWA, H., YOSHIDA, T. NAKAMURA, T., LAI, L. & NAKASHIZUKA, T. 2003. The age of a tropical rain forest canopy species, Borneo Ironwood (*Eusideroxylon zwageri* Teijsm. & Binnend.) determined by C[14] dating. *J. Trop. Ecol.* 19: 1–8.

KURSAR, T.A., DEXTER, K.G., LOKVAM, J., PENNINGTON, R.T., RICHARDSON, J.E., WEBER, M.G., MURAKAMI, E.T., DRAKE, C., McGREGOR, R., & COLEY, P.D. 2009. The evolution of antiherbivore defenses and their contribution to species coexistence in the tropical tree genus *Inga.* *Proc. Natl. Acad. Sci. U.S.A.* 106: 18073–18078. URL: http://www.pnas.org/cgi/doi/10.1073/pnas.0904786106

KURUVILLA, P.K. 1989. Pollination biology, seed setting and fruit setting in *Madhuca indica* (Sapotaceae). *Indian Forester* 115: 22–28.

KURZ, S. 1875. *Preliminary Report on the Forest and other Vegetation of Pegu.* Baptist Mission Press, Calcutta.

KURZ, S. 1876. Bamboo and its uses. *Indian Forester* 1: 219–269; 335–355.

KUSHALAPPA, K.A. 1988. Silvicultural systems in tropical rain forest of Karnataka. *Indian Forester* 114: 372–378.

KUSHWAHA, C.P. & SINGH, K.P. 2005. Diversity of leaf phenology in a tropical deciduous forest in India. *J. Trop. Ecol.* 21: 47–56.

KUUSIPALO, J., JAFARSIDIK, Y., ÅDJERS, G. & TUOMELA, K. 1996. Population dynamics of tree seedlings in a mixed dipterocarp rainforest before and after crown logging and crown liberation. *Forest Ecol. Managem.* 81: 85–94.

KUUSIPALO, J., HADENGAMMAN, S., ÅDJERS, G. & SAGALA, A.P. 1997. Effect of gap liberation on the performance and growth of dipterocarp trees on logged over forest. *Forest Ecol. Managem.* 92: 209–219.

KWE-TU-WET-TU. 1903. The flowering of *Bambusa polymorpha. Indian Forester* 29: 513–516.

LABROUE, L., LEGRIS, P. VIART, M. 1965. Bioclimat du sous-continent indien. *Inst. Fr. Pondichéry, Trav. Sect. Sci. Tech.* 3.

LACK, A.J. & KEVAN, P.G. 1987. The reproductive biology of a distylous tree, *Sarcotheca celebica* (Oxalidaceae) in Sulawesi, Indonesia. *Bot. J. Linn. Soc.* 95: 1–8.

LaFRANKIE, J.V. 2005. Lowland tropical rain forests of Asia and America: parallels, convergence and divergence. In: *Pollination Ecology and the Rain Forest,* eds D.J. Roubik, S. Sakai & A.A. Hamid, pp. 178–190. Springer, New York.

LaFRANKIE, J.V. & SAW, L.G. 2005. The understorey palm *Licuala* (Arecaceae) suppresses tree regeneration in a lowland forest in Asia. *J. Trop. Ecol.* 21: 703–706.

LaFRANKIE, J.V., LEE, H.-S., NAGAMASU, H., NAKASHIZUKA, T. & PALMIOTTO, P. 2003. The diversity of hemi-epiphytic figs in a Bornean lowland rain forest. *Biol. J. Linn. Soc.* 78: 456.

LaFRANKIE, J.V., DAVIES, S.J., WANG, L.K., LEE, S.K. & LUM, S.K.Y. 2005. Forest Trees of Bukit Timah. Population Ecology in a Tropical Forest. Simply Green, Singapore.

LaFRANKIE, J.V., ASHTON, P.A., CHUYONG, G.B., CO, L., CONDIT, R., DAVIES, S.J., FOSTER, R., HUBBELL, S.P., KENFACK, D., LAGUNZAD, D., LOSOS, E.C., NOR, N.S.M., TAN, S., THOMAS, D.W., VALENCIA, R. & VILLA, G. 2006. Contrasting structure and composition of the understory in species-rich tropical rain forests. *Ecology* 87: 2298–2305.

LAL, A.B. 1942. Significance of teak-sal mixture from the standpoint of plant succession, with note by T.V. Dent. *Indian Forester* 68: 181.

LAL, A.K. & SINGH, P.P. 2003. Economic worth of carbon stored in aboveground biomass of India's forests. *Indian Forester* 129: 874–879.

LAL, C.B., ANNAPURNA, C., RAGUBANSHI, A.S. & SINGH, J.S. 2001. Foliar demand and resource economy of nutrients in dry tropical forest species. *J. Veg. Sci.* 12: 5–14.

LAL, J.B. & DAKA, R.K. 1991. Tendu leaves trade in Madhya Pradesh. A big cooperative venture. *Indian Forester* 117: 728–732.

LAL, R. 2004a. Soil carbon sequestration impacts on global climate change and food security. *Science* 304: 1623.

LAL, R. 2004b. Soil carbon sequestration in India. *Climate Change* 65: 277–296.

LAMAN, T.G. 1994. *The Ecology of Strangler Figs (Hemiepiphytic* Ficus *spp.) in the Rain Forest Canopy of Borneo.* PhD thesis, Harvard University.

LAMAN, T.G. 1995. The ecology of strangler fig seedling establishment. *Selbyana* 16: 223–229.

LAMAN, T.G. 1996a. The impact of seed harvesting ants (*Pheidole* sp. nov.) on *Ficus* establishment in the canopy. *Biotropica* 28: 777–781.

LAMAN, T.G. 1996b. *Ficus* seed shadows in a Bornean forest. *Oecologia* 107: 347–355.

LAMAN, T.G. 1996c. Specialization for canopy position by hemi-epiphytic *Ficus* species in a Bornean rain forest. *J. Trop. Ecol.* 12: 789–803.

LAMBERT, F.R. & MARSHALL, A.G. 1991. Keystone characteristics of bird-dispersed *Ficus* in a Malaysian lowland rain forest. *J. Ecol.* 79: 793–809.

LANGENHEIM, J. H., OSMOND, C.B., BROOKS, A. & FERRAR, P.J. 1984. Photosynthetic responses to light in seedlings of selected Amazonian and Australian rainforest tree species. *Oecologia* 63: 215–224.

LARSEN, A.M. & RIBOT, J.C. 2007. The

poverty of forestry policy: double standards on an uneven playing field. *Sust. Sci.* 2: 189–204.

LARSEN, H.O. & OLSEN, S.S. 2007. Unsustainable collection and unfair trade? Uncovering and assessing assumptions regarding Central Himalayan medicinal plant conservation. *Biodivers. & Conservation* 16: 1679–1697.

LATHAM, R.E. 1992. Co-occurring tree species changed rank in seedling performance with resources varied experimentally. *Ecology* 73: 2129–2144.

LAUMONIER, Y. 1990. Search for phytogeographic provinces in Sumatra. In: *The Plant Diversity of Malesia,* eds P. Baas, K. Kalkman & R. Geesink, pp. 193–211. Kluwer, Dordrecht.

LAUMONIER, Y. 1994. The vegetation and tree flora of Kerinci-Seblat National Park, Sumatera. *Trop. Biod.* 1 (2): 232–251.

LAUMONIER, Y. 1997. The vegetation and physiography of Sumatra. *Geobotany* 22. Kluwer, Dordrecht.

LAURANCE, W.F. & PERES, C.A. 2006. *Emerging Threats to Tropical Forests.* University of Chicago Press.

LAURANCE, W.F., LAURANCE, S.G., FERREIRA, L.V., RANKIN-DE MERONA, J.M., GASCON, C. & LOVEJOY, T.E. 1997. Biomass in Amazonian forest fragments. *Science* 278: 1117–1118.

LAURANCE, W.F., NASCIEMENTO, H.E.M., LAURANCE, S.G., ANDRADE, A., EWERS, R.M., HARMS, K.E., LUIZÃO, R.C.C. & RIBEIRO, J.E. 2007. Habitat fragmentation, variable edge effects, and the landscape-divergence hypothesis. *PLoS ONE* 2 (10): e1017.

LAURANCE, W.F., USECHE, D.C., RENDEIRO, J., KALKA, M., BRADSHAW, C.J.A., SLOAN, S.P., LAURANCE, S.G. *ET AL.* 2012. Averting biodiversity collapse in tropical forest protected areas. *Nature,* 489: 290–294.

LAVELLE, P. 2002. Functional domains in soils. *Ecological Research* 17: 441–450.

LAVELLE, P., BLANCHART, E., MARTIN, A., MARTIN, S., BAROIS, I., TOUTAIN, F., SPAIN, A. & SCHAEFER, R. 1993. A hierarchical model for decomposition in terrestrial ecosystems. Application to soils in the humid tropics. *Biotropica* 25: 130–150.

LAVELLE, P., BRUSSARD, L. & HENDRIX, P., eds. 1999. *Earthworm Management in Tropical Agroecosystems.* CABI, Wallingford.

LAVELLE, P. & SPAIN, A.V. 2001. *Soil Ecology.* Kluwer Scientific Publications, Amsterdam.

LAWRENCE, D. 2001. Nitrogen and phosphorus enhance growth and luxury consumption of four secondary forest tree species *Melicope glabra, Macaranga gigantea, Persea ramosa, Peronema canescens*. *J. Trop. Ecol.* 17: 859–869.

LAWTON, R.O. & PUTZ, F.E. 1988. Natural disturbance and gap-phase regeneration in a wind-exposed tropical cloud forest. *Ecology* 69: 764–777.

LECOMTE, H., GAGNEPAIN, F. & HUMBERT, H., eds. 1907-1951. *Flore Générale de l'Indochine.* Masson, Paris.

LEE, C.T., LEE, S.L., NG, K.K.S., FARIDAH, Q.Z., SIRAJ, S.S. & NORWATI, M. 2011. Estimation of outcrossing rates

in *Koompassia malaccensis* from open-pollinated population in Peninsular Malaysia using microsatellite markers. *J. Trop. Forest Sci.* 23: 410–416.

LEE, D.W. 2007. *Nature's Palette: The Science of Plant Color.* University of Chicago Press.

LEE, D.W. & GRAHAM, R. 1986. Leaf optical properties of rainforest sun and extreme shade plants. *Amer. J. Bot.* 73: 1100–1108.

LEE, D.W. & LOWRY, J.B. 1980a. Solar ultraviolet on tropical mountains: can it affect plant speciation? *Amer. Naturalist* 115: 880–883.

LEE, D.W. & LOWRY, J.B. 1980b. Plant speciation on tropical mountains: *Leptospermum* (Myrtaceae) on Mount Kinabalu, Borneo. *Bot. J. Linn. Soc.* 80: 223–242.

LEE, D.W., LOWRY, J.B. & STONE, B.C. 1979. Abaxial anthocyanin layer in leaves of tropical rain forest plants: enhancer of light capture in deep shade. *Biotropica* 11: 70–77.

LEE, D.W., BOONE, R.A., TARSIS, S.L. & STORCH, D. 1990. Correlates of leaf optimal properties in tropical forest sun and extreme-shade plants. *Amer. J. Bot.* 77: 370–380.

LEE, D.W., BASKARAN, K., MANSOR, M., MOHAMAD, H. & YAP, S.K. 1996. Irradiance and spectral quality affect Asian tropical rain forest tree seedling development. *Ecology* 77: 568–580.

LEE, H.L., SODHI, N.S. & ELMQVIST, T. 2001. Bee diversity along a disturbance gradient in tropical lowland forest of south-east Asia. *J. Appl. Ecol.* 38: 180–192.

LEE, H.S., ASHTON, P.A., YAMAKURA, T., TAN, S., DAVIES, S.J., ITOH, A., CHAI, E.O.K., OKHUBO, T. & LAFRANKIE, J.V. 2002a. *The 52-hectare Forest Research Plot at Lambir Hills, Sarawak, Malaysia: Tree Distribution Maps, Diameter Tables and Species Documentation.* Sarawak Forest Department, Arnold Arboretum of Harvard University & Smithsonian Tropical Research Institute.

LEE, H.S., DAVIES, S.J., LAFRANKIE, J.V., TAN, S., YAMAKURA, T., ITOH, A., OHKUBO, T. & ASHTON, P.S. 2002b. Floristic and structural diversity of mixed dipterocarp forest in Lambir Hills National Park, Sarawak, Malaysia. *J. Trop. Forest Sci.* 14: 379–400.

LEE, H.S., TAN, S., DAVIES, S.J., LAFRANKIE, J.V., YAMAKURA, T., ITOH, A., OKHUBO, T. & HARRISON, R. 2004. Lambir Forest Dynamics plot, Sarawak, Malaysia. In: *Tropical Forest Diversity and Dynamism: Findings from a Large-Scale Plot Network,* eds E.C. Losos & E.G. Leigh Jr., pp. 527–539. University of Chicago Press, Chicago.

LEE, S.L., WICKNESWARI, R., MAHANI, M.C. & ZAKRI, A.H. 2000a. Genetic diversity of a tropical tree species, *Shorea leprosula* Miq. (Dipterocarpaceae), in Malaysia: implications for conservation of genetic resources and tree improvement. *Biotropica* 32: 213–224.

LEE, S.L., WICKNESWARI, R., MAHANI, M.C. & ZAKRI, A.H. 2000b. Mating system parameters in a tropical tree species, *Shorea leprosula* Miq. (Dipterocarpaceae), from Malaysian lowland dipterocarp forest. *Biotropica* 32: 693–702.

LEE, S.L., WICKNESWARI, R., MAHANI, M.C. & ZAKRI, A.H. 2001. Comparative genetic diversity studies of *Shorea leprosula* Miq. (Dipterocarpaceae) using RAPD and allozyme markers. *J. Trop. Forest Sci.* 13: 202–215.

LEE, S.L., SAW, L.G., NORWATI, A., SITI SALWANA H., LEE, C.T. & NORWATI, M. 2002. Population genetics of *Intsia palembanica* (Leguminosae) and genetic conservation of Virgin Jungle Reserves (VJRs) in peninsular Malaysia. *Amer. J. Bot.* 89: 447–459.

LEE, S.L., NG, K.K.S, SAW, L.G., LEE, C.T., NORWATI, M., TANI, N., TSUNUMA, Y. & KOSKELA, J. 2006. Linking the gaps between conservation research and conservation management of rare dipterocarps: a case study of *Shorea lumutensis*. *Biol. Conservation* 131: 72–92.

LEE, S.S. 1998. Root symbiosis and nutrition. In: *A Review of Dipterocarps: Taxonomy, Ecology and Silviculture,* eds S. Appanah & J.M. Turnbull, pp. 99–114. CIFOR, Bogor, Indonesia.

LEE, S.S. & ALEXANDER, I.J. 1994. The response of seedlings of two dipterocarp species to nutrient additions and ectomycorrhizal infections. *Pl. & Soil* 163: 299–306.

LEENHOUTS, P.W. 1967. A conspectus of the genus *Allophylus* (Sapindaceae). The problem of the complex species. *Blumea* 15: 301–358.

LEENHOUTS, P.W. 1986. A taxonomic revision of *Nephelium* (Sapindaceae). *Blumea* 31: 373–436.

LEGRIS, P. 1963. La végétation de l'Inde. *Inst. Fr. Pondichéry, Trav. Sec. Sci. Tech.* 6.

LEGRIS, P. & BLASCO, F. 1969. Variabilité des facteurs du climat: cas des montagnes du sud de l'Inde et de Ceylan. *Inst. Fr. Pondichéry, Trav. Sec. Sci. Tech.* 8: 1–95.

LEGRIS, P. & VIART, M. 1961. Bioclimates of South India and Ceylon. *Inst. Fr. Pondichéry, Trav. Sec. Sci. Tech.* 3.

LEHMKUHL, J.F. 1994. A classification of subtropical riverine grassland and forest in Chitwan National Park, Nepal. *Vegetatio* 111: 29–43.

LEHMKUHL, J.F., UPRETI, R.K. & SHARMA, U.R. 1988. National parks and local development: grasses and people in Royal Chitwan National Park, Nepal. *Env. Conserv.* 15: 143–148.

LEIGH, E.G., JR. 1990. Tree shape and leaf arrangement: a quantitative comparison of montane forests, with emphasis on Malaysia and South India. In: *Conservation in Developing Countries: Problems and Prospects.* eds J.C. Daniel & J.S. Serrao, pp. 119–174. Bombay Natural History Society, Oxford University Press, Bombay.

LEIGH, E.G., JR. 2004. The neutral theory of forest ecology. In: *Tropical Forest Diversity and Dynamism: Findings from a Large-Scale Plot Network,* eds E.C. Losos & E.G. Leigh Jr., pp. 244–263. University of Chicago Press, Chicago.

LEIGH, E.G., DAVIDAR, P., DICK, C., PUYRAVAUD, J.P., TERBORGH, J., STEEGE, H.T. & WRIGHT, S.J. 2004. Why do some tropical forests have so many kinds of trees? *Biotropica* 36: 447–473.

LEIGHTON, M. & WIRAWAN, N. 1986. Catastrophic drought and fire in Borneo tropical rain forest associated with the 1982–1983 El Niño Southern Oscillation

event. In: *Tropical Rain Forests and the World Atmosphere,* ed. G.T. Prance, pp. 75–102. Westview Press, Colorado.

LÉLÉ, S. 1998. Why, who, and how of jointness in joint forest management: theoretical considerations and empirical insights from the Western Ghats of Karnataka. *Publication 11, Project on Linking Ecology, Economics, and Institutions of Forest Use in the Western Ghats,* Institute for Social and Economic Change and Pacific Institute for Studies in Development, Environment, and Security, Bangalore.

LÉLÉ, S. 2000. Godsend, sleight of hand, or just muddling through: joint water and forest management in India. *Natural Resource Perspectives* 53. Overseas Development Institute.

LÉLÉ, S. 2003. Participatory Forest Management in Karnataka at the crossroads. *Community Forestry* 2: 4.

LÉLÉ, S. Undated. *Institutional Issues in (J) FM(R).* Unpubl. ms.

LÉLÉ, S. 1998. *Indian Forest Policy, Forest Law, and Forest Rights Settlement: a Serious Mismatch.* Paper presented at international workshop on capacity building in environmental governance for sustainable development, I Gandhi Institute for Developmental Research, Mumbai, Dec. 8–10 1998.

LEPŠ, J., NOVOTN, V. & BASSET, Y. 2001. Habitat and successional status of plants in relation to the communities of leaf-chewing herbivores in Papua-New Guinea. *J. Ecol.* 89: 186–199.

LEWIS, S.L., MALHI, Y., PHILLIPS, O.L. 2004a. Fingerprinting the impacts of global change on tropical forests. *Philos. Trans., Ser. B* 359: 437–462.

LEWIS, S.L., PHILLIPS, O.L., BAKER, T.R., LLOYD, J., MALHI, Y., ALMEIDA, S., HIGUCHI, N., LAURANCE, W.F., NEILL, D.A., SILVA, J.N.M., TERBORGH, J., TORRES LEZAMA, A., VÁSQUEZ MARTÍNEZ, R., BROWN, S., CHAVE, J., KUEBLER, C., NÚÑEZ VARGAS, P. & VINCETI, P. 2004b. Concerted changes in tropical forest structure and dynamics: evidence from 50 South American long-term plots. *Phil. Trans. R. Soc. London, Ser. B* 359: 421–436.

LEWIS, S.L., PHILLIPS, O.L., SHEIL, D., VINCETI, B., BAKER, T.R., BROWN, S., GRAHAM, A.W., HIGUCHI, N., HILBERT, D.W., LAURANCE, W.F., LEJOLY, J., MALHI, Y., MONTEAGUDO, A., VARGAS, P.N., SONKÉ, B., SUPARDI, N., M.N., TERBORGH, J.W. & MARTÍNEZ, R.V. 2004c. Tropical forest tree mortality, recruitment and turnover rates: calculation, interpretation and comparison when census intervals vary. *J. Ecol.* 92: 929–944.

LEYSSAERT, S., SCHULZE, E.-D., BÖRNER, A., KNOHL, A., HESSENMÖLLER, D., LAW, B.E., CIAIS, P. & GRACE, J. 2008. Old-growth forests as global carbon sinks. *Nature* 455: 213–215.

LI, H., XU, Z., XU, Y. & WANG, J. 2002. Practise of conserving plant diversity through traditional beliefs: a case study in Xishuangbanna, southwest China. *Biodivers. & Conservation* 11: 705–713.

LI, H., AIDE, M., MA, X., LIU, W. & CAO, M. 2007. Demand for rubber is causing the loss of high diversity rain forest in SW China. *Biodivers. & Conservation* 16: 1731–1745.

LI, H.-L. & KENG, H. 1950. Phytogeographical affinities of Southern Vietnam. *Taiwania* 1: 103–128.

LI, J.H., BOGLE, A.L. & KLEIN, A.S. 1999a. Phylogenetic relationships in the Hamamelidaceae. Evidence from the nucleotide sequences of the plastic gene *matK*. *Pl. Syst. Evol.* 218: 205–219.

LI, J.H., BOGLE, A.L. & KLEIN, A.S. 1999b. Phylogenetic relationships in the Hamamelidaceae: inferred from sequences of internal transcribed spacer (ITS) of nuclear ribosomal DNA. *Amer. J. Bot.* 86: 1027–1037.

LI, X. & WALKER, D. 1986. The plant geography of Yunnan Province, southwest China. *J. Biogeogr.* 13: 367–397.

LI, Z., ZHANG, Y., WANG, S., YUAN, G., YANG, Y. & CAO, M. 2010. Evapotranspiration of a tropical forest at Xishuangbanna, southeast China. *Hydrol. Process* 24: 2405–2416.

LIEBERMAN, D., LIEBERMAN, M., PERALTA, R. & HARTSHORN, G.S. 1996. Tropical forest structure and composition on a large scale altitudinal gradient in Costa Rica. *J. Ecol.* 84: 137–152.

LIEBERMAN, M. & LIEBERMAN, D. 1989. Forests are not just Swiss cheese: canopy stereo geometry of non-gaps in tropical forests. *Ecology* 70: 550–552.

LIECHTI, P. 1960. *The Geology of Sarawak, Brunei and the Western Part of North Borneo*. Geological Survey Department Bulletin 3, British Territories in Borneo.

LIECHTI, P., ROE, F.W. & HAILE, N.S. 1960. *The Geology of Sarawak, Brunei, and the Western Part of North Borneo*. Government Printing Office, Kuching.

LIENGSIRI, C., YEH, F.C. & BOYLE, T.J.B. 1995. Isozyme analysis of a tropical forest tree, *Pterocarpus macrocarpus* Kurz in Thailand. *Forest Ecol. Managem.* 74: 13–22.

LIESE, W. 1986. To the memory of Sir Dietrich Brandis. *Indian Forester* 112: 640–645.

LIEW, T.C. 1974. Forest manipulation and regeneration in Sabah. Cyclostyled, PT/PP/10. IBP Synthesis Meeting, Kuala Lumpur.

LIM, C.-L., SITI-MUNIRAH, M.Y. & KIEW, R. 2009. Botany of the Berembun Forest Reserve, Negeri Sembilan, Malaysia. *Malayan Nat. J.* 61: 143–189.

LIN, S., QIAO, Y. & WALKER, D. 1986. Late Pleistocene and Holocene vegetation history at Xi Hu, Er Yuan, Yunnan Province, southwest China. *J. Biogeogr.* 13: 419–440.

LIOW, L.H., SODHI, N.S. & ELMQVIST, T. 2001. Bee diversity along a disturbance gradient in tropical lowland forests of south-east Asia. *J. Appl. Ecol.* 38: 190–192.

LIU, J., TANG, L., QIAO, Y., HEAD, M.J. & WALKER, D. 1986. Late Quaternary vegetation history at Menghai, Yunnan. *J. Biogeogr.* 13: 399–418.

LIU, J. & ASHTON, P.S. 1998. FORMOSAIC: an individual-based spatially explicit model for simulating forest dynamics in landscape mosaics. *Ecol. Model.* 106: 177–200.

LIU, W., MENG, F.R., ZHANG, Y., LIU, K. & LI, H. 2004. Water input from fog drip in the tropical seasonal rain forest of Xishuangbanna, south-west China. *J. Trop. Ecol.* 20: 517–524.

LIU, W.J., WANG, P.Y., LI, J.T., LI, P.J. & LIU, W.Y. 2008. The importance of radiation fog in the tropical seasonal forest of Xishuangbanna, south-west China. *Hydrol. Res.* 39: 79–87.

LODHIYAL, N. & LODHIYAL, L.S. 2003. Aspects of nutrient cycling and nutrient use pattern of Bhabar Shisham forests in central Himalaya, India. *Forest Ecol. Managem.* 176: 237–252.

LÖFFLER, E. 1972. Pleistocene glaciations in Papua and New Guinea. *Z. Geomorphol. N.F. Suppl. Bd.* 13: 32–58.

LÖFFLER, E. 1980. Geology and geomorphology of the New Guinea high mountains. In: *The Alpine Flora of New Guinea* 1, ed. P. van Royen, pp. 29–57. Cramer Verlag, Vaduz.

LONG, C.L. & ZHOU, Y. 2001. Indigenous community forest management if Jinuo people's swidden agroecosystem in southeast China. *Biodivers. & Conservation* 10: 753–767.

LOSOS, E.C. 2004. Habitat specialization and species rarity in forest dynamic plots. Introduction. In: *Tropical Forest Diversity and Dynamism: Findings from a Large-Scale Plot Network*, eds E.C. Losos & E.G. Leigh Jr., pp. 103–106. University of Chicago Press, Chicago.

LOSOS, E.C. & LEIGH, E.G. JR. 2004. *Tropical Forest Diversity and Dynamism: Findings from a Large-Scale Plot Network*. University of Chicago Press, Chicago.

LOTKA, A.J. 1925. *Elements of Physical Biology*. Williams & Wilkins, Baltimore.

LOUYS, J., CURNOE, D. & TONG, H. 2007. Characteristics of Pleistocene megafauna extinctions in Southeast Asia. *Palaeogeogr. Palaeoclimatol. Palaeoecol.* 243: 152–173.

LOVEGROVE, W.H. 1900. Notes on *Dendrocalamus strictus*. *Indian Forester* 20: 433–442.

LOWMAN, M.D. & NADKARNI, M., eds 1995. *Forest Canopies*. Academic Press, San Diego.

LOWMAN, M.D. & RINKER, H.B., eds 2004. *Forest Canopies*. Elsevier Academic Press, Amsterdam & Boston.

LOWRIE, A.E. 1891. An authentic case of a forest fire caused by lightning. *Indian Forester* 17: 196.

LOWRIE, A.G. 1903. The effects of the late drought in the Chanda District. *Indian Forester* 26: 503–506.

LOWRY, D.B. 2012. Ecotypes and the controversy over stages in the formation of new species. *Biol. J. Linn. Soc.* 106: 241–257.

LOWRY, J.B., LEE, D.W. & STONE, B.C. 1973. Effects of drought on Mount Kinabalu. *Symposium on Ecological Research in Humid Tropics Vegetation, Kuching*, pp. 325–364. UNESCO & Government of Sarawak.

LÜ, X.T., YIN, J.X. & TANG, J.W. 2010. Structure, tree species diversity and composition of tropical seasonal rainforest in Xishuangbanna, south-west China. *J. Trop. Forest Sci.* 22: 260–270.

LUBOWSKY, R. 1998. *The Importance of Rain Forest Products in Rural Household Income: a Case Study of Two Villlages in the Sinharaja Forest of Sri Lanka*. Unpubl. report.

LUC, D. 1992. *Update List of Cucphuong Flora*. Bo Lam Nghiep. Vuon Quoc Gia Cuc Phuong. [In Vietnamese except for scientific names]

LUM, S.K.Y., LEE, S.K. & LaFRANKIE, J.V. 2004. Bukit Timah Forest Dynamics plot, Singapore. In: *Tropical Forest Diversity and Dynamism: Findings from a Large-Scale Plot Network*, eds E.C. Losos & E.G. Leigh Jr., pp. 464–473. University of Chicago Press, Chicago.

LUNA, A.C., OSUMI, K., GASCON, A.F., LASCO, R.D., PALIJON, A.M. & CASTILLO, M.L. 1999. The community structure of a logged-over tropical rain forest in Mt. Maquiling forest reserve, Philippines. *J. Trop. Forest Sci.* 11: 446–458.

LUSK, C. 2008. Constraints on the evolution and geographical range of *Pinus*. *New Phytol.* 178: 1–3.

LUYSSAERT, S., INGLIMA, I., JUNG, M., RICHARDSON, A.D., REICHSTEIN, M., PAPALE, D., PIAO, S.L., SCHULZE, E.-D., WINGATE, L., MATTEUCCI, G., ARAGAO, L., AUBINET, M., BEER, C., BERNHOFER, C., BLACK, K.G., BONAL, D., BONNEFOND, J.-M, CHAMBERS, J., CIAIS, P., COOK, B., DAVIS, K.J., DOLMAN, A.J., GIELEN, B., GOULDEN, M., GRACE, J., GRANIER, A., GRELLE, A., FRIFFIS, T., GRÜNWALD, T., GUIDOLOTTI, G., HANSON, P.J., HARDING, R., HOLLINGER, D.Y., HUTYRA, L.R., KOLARI, P., KRUIJT, B., KUTSCH, W., LAGERGREN, F., LAURILA, T., LAW, B.E., LE MAIRE, H., LINDROTH, A., LOUSTAU, D., MALHI, Y., MATEUS, J., MIGLIAVACCA, M., MISSON, L., MONTAGNANI, L., MONCRIEFF, J., MOORS, E., MUNGER, J.W., NIKINMAA, E., OLLINGER, S.V., PITA, G., REBMANN, C., ROUPSARD, O., SAIGUSA, N., SANZ, M.J., SEUFERT, G., SIERRA, C., SMITH, M.-L., TANG, J., VALENTINI, R., VESALA, T. & JANSSENS, A. 2007. CO_2 balance of boreal, temperate and tropical forests derived from a global data base. *Global Change Biol.* 13: 2509–2537.

LUYSSAERT, S., SCHULZE, E.-D., BÖRNER, A., KNOHL, A., HESSENMÖLLER, D., LAW, B.E., CIAIS, P. & JOHN, G. 2008. Old-growth forest as global carbon sinks. *Nature* 455: 213–215.

MA, K.P., CHEN, B., MI, X.C., FANG, T., CHEN, L. & REN, H.B. 2009. *Gutianshan Forest Dynamic Plot. Tree Species and Their Distribution Patterns*. China Forestry Publishing House, Beijing.

MacARTHUR, R.H. & MacARTHUR, J.W. 1961. On bird species diversity. *Ecology* 42: 594–598.

MacARTHUR, R.H. & WILSON, E.O. 1967. *The Theory of Island Biogeography*. Monographs in Population Biology. Princeton University Press.

MACK, A.L. 1997. Spatial distribution of fruit production and seed removal of a rare, dioecious canopy tree species (*Aglaia aff. flavida* Merr. & Perry) in Papua New Guinea. *J. Trop. Ecol.* 13: 305–316.

MACKINNON, J.R. 1996. *A Biodiversity Review of China*. World Wide Fund for Nature. WWT International, China Programme, Hong Kong.

MacMILLAN, H.F. 1910. *Tropical Planting and Gardening*. Macmillan Co., London.

MacMILLAN, H.F., BARLOW, H.S., ENOCH, L. & RUSSELL, R.A., eds 1991. *Tropical Planting and Gardening*. 6th rev. ed. Oxford University Press.

MAGURRAN, A.E. 2005. Species abundance distributions: pattern or process? *Funct. Ecol.* 19: 177–181.

MAGURRAN, A.E. & HENDERSON, P.A. 2003. Explaining the excess of rare species in natural species abundance distributions. *Nature* 422: 714–716.

MAHAPATRA, A. & MITCHELL, C.P. 1997. Sustainable development of non-timber forest products: implications for forest management in India. *Forest Ecol. Managem.* 94: 15–29.

MAILLY, D., CHRISTANTY, L. & KIMMINS, J.P. 1997. 'Without bamboo, the land dies': mutrient cycling and biogeochemistry of a Javanase bamboo talon-kebun system. *Forest Ecol. Managem.* 91: 155–173.

MALHOTRA, K.C., GOKHALE, Y., CHATTERJEE, S. & SRIVASTAVA, S. 2007. *Sacred Groves in India*. Aryan Books International, New Delhi.

MALLET, J. 2008. Mayr's view of Darwin: was Darwin wrong about speciation? *Biol. J. Linn. Soc.* 95: 3–36.

MALMER, A. & GRIP, H. 1990. Soil disturbance and loss of infiltrability caused by mechanized and manual extraction of tropical rainforest in Sabah, Malaysia. *Forest Ecol. Managem.* 38: 1–12.

MALONEY, B.K. 1985. Man's impact on the rainforests of West Malesia: the palynological record. *J. Biogeogr.* 12: 537–558.

MANCHESTER, S.R., CHEN, Z.-D., LU, A.-M. & UEMURA, K. 2009. Eastern Asiatic seed plant genera and other paleogeographic history throughout the Northern Hemisphere. *J. Syst. Evol.* 47:1–41.

MANEN, J.-F., BOULTER, M.C. & YACIRI-GRAVEN, Y. 2002. The complex history of the genus *Ilex* L. (Aquifoliaceae): evidence from the comparison of plastid and nuclear DNA sequences and from fossil data. *Pl. Syst. Evol.* 235: 79–98.

MANEN, J.-F., BARRIERA, G., LOIZEAU, P.-A. & NACIN, Y. 2010. The history of extant *Ilex* species (Aquifoliaceae): evidence of hybridization within a Miocene radiation. *Molec. Phylogenet. Evol.* 57: 961–977.

MANGEN, J.-M. 1993. *Ecology and Vegetation of Mt. Trikora New Guinea (Irian Jaya, Indonesia)*. Travaux Scientifiques Musee National d'Histoire Naturelle de Luxembourg. Ministere des Affaires Culturelles.

MANI, M.S. 1974. *Ecology and Biogeography in India*. W. Junk, The Hague.

MANOKARAN, N. 1998. Effect, 34 years after, of selective logging in the lowland dipterocarp forest at Pasoh, Peninsular Malaysia, and implications on present-day logging in the hill forests. In: *Conservation, Management and Development of Forest Resources. Proceedings of the Malaysia-UK Programme Workshop 21–24 1996, Kuala Lumpur*, eds S.S. Lee, Y.M. Dan, I.D. Gauls & J. Bishop, pp. 41–60. Forest Research Institute Malaysia, Kepong.

MANOKARAN, N. & SWAINE, M.D. 1994. *Population Dynamics of Trees in Dipterocarp Forests of Peninsular Malaysia*. Malayan Forest Records 40. Forest Research Institute Malaysia (FRIM), Kepong.

MANOKARAN, N., LaFRANKIE, J.V., KOCHUMMEN, K.M., QUAH, E.S., KLAHN, J.E., ASHTON, P.S. & HUBBELL, S.P. 1990. *Methodology for the Fifty Hectare*

Research Plot at Pasoh Forest Reserve. Forest Research Institute Malaysia Research Pamphlet 104.

MANOKARAN, N., LAFRANKIE, J.V., KOCHUMMEN, K.M., QUAH, E.S., KLAHN, J.E., ASHTON, P.S. & HUBBELL, S.P. 1992. *Stand Tables and Distributions of Species in the 50 ha Research Plot at Pasoh Forest Reserve*. Forest Research Institute Malaysia Research Data Series 1. Forest Research Institute Malaysia, Kepong. Kepong.

MANOKARAN, N., SENG, Q.E., ASHTON, P.S., LAFRANKIE, J.V., NOOR, N.S.M., AHMAD, W.M.S.W. & OKUDA, T. 2004. Pasoh forest dynamics plot, Peninsular Malaysia. In: *Tropical Forest Diversity and Dynamism: Findings from a Large-Scale Plot Network*, eds E.C. Losos & E.G. Leigh Jr., pp. 585–598. University of Chicago Press, Chicago.

MANOS, P.S. & STANFORD, A.M. 2001. The historical biogeography of Fagaceae: tracking the Tertiary history of temperate and subtropical forests of the northern hemisphere. *Int. J. Pl. Sci.* 162 (6 suppl.): S77–93.

MARJOKORPI, A. & RUOKOLAINEN, K. 2003. The role of traditional forest gardens in the conservation of tree species in W. Kalimantan. *Biodivers. & Conservation* 12: 799–822.

MAROD, D., KUTINTARA, U., YARWADHI, C., TANAKA, H. & NAKASHIZUKA, T. 1999. Structural dynamics of a natural mixed deciduous forest in western Thailand. *J. Veg. Sci.* 10: 763–860.

MAROD, D., KUTINTARA, U., TANAKA, H. & NALASHIZUKA, T. 2002. The effects of drought and fire on seed and seedling dynamics in a tropical seasonal forest in Thailand. *Pl. Ecol.* 161: 41–57.

MARSH, C.W. & GREER, A.G. 1992. Forest land use in Malaysia: an introduction to Danum valley. *Philos. Trans., Ser. B* 347: 331–339.

MARSHALL, A.G. 1983. Bats, flowers and fruit: evolutionary relationships in the Old World. *Biol. J. Linn. Soc.* 20: 115–135.

MARSHALL, A.G. 1985. Old world phytophagous bats (Megachiroptera) and their food plants: a survey. *J. Linn. Soc.* 83: 351–369.

MARSHALL, A.J. 2009. Are montane forests demographic sinks for Bornean white-bearded gibbons *Hylobates albibarbis*? *Biotropica* 41: 257–267.

MARSHALL, A.J. & LEIGHTON, M. 2006. How does food availability limit the population density of agile gibbons? In: *Feeding Ecology in Apes and Other Primates,* eds G. Hohmann, M.M. Robbins & C. Boesch, pp. 313–335. Cambridge University Press.

MARTIN, H.A. 1977. The history of *Ilex* (Aquifoliaceae) with special reference to Australia: evidence from pollen. *Austral. J. Bot.* 25: 655–673.

MATSUMOTO, T. & ABE, T. 1979. The role of termites in an equatorial rain forest ecosystem of West Malaysia. II. Leaf litter consumption on the forest floor. *Bioscience* 38: 261–274.

MATSUMOTO, Y., MARUYAMA, Y. & ANG, L.H. 2000. Maximum gas exchange rates and osmotic potential in sun leaves of tropical tree species. *Tropics* 9: 195–209.

MATTA, J., ALAVALAPATI, J. KERR, J. &

MERCER, E. 2005. Agency perspectives on transition to participatory forest management: a case study from Tamil Nadu, India. *Soc. Nat. Res.*18: 859–870.

MATTHEW, K.M. 1971. The flowering of the Strobilanth (Acanthaceae). *J. Bombay Nat. Hist. Soc.* 67: 502–506.

MATTHEW, K.M. 1999. *The Flora of the Palni Hills*. I. Polypetalae. II. Gamopetalae. The Rapirat Herbarium, St. Joseph's College, Tiruchirapalli, India.

MAUDSLEY, N.A., COMPTON, S.G. & WHITTAKER, R.J. 1998. Population persistence, pollination mutualisms and figs in a fragmented landscape. *Conservation Biol.* 12: 1416–1420.

MAURIN, O., CHASE, M.W., JORDAAN, M. & VAN DER BANK, M. 2010. Phylogenetic relationships of Combretaceae inferred from nuclear and plastid DNA sequence data: implications for generic classification. *Bot. J. Linn. Soc.* 162: 453–376.

MAURY-LECHON, G. & CURTET, L. 1998. Biogeography and evolutionary systematics of Dipterocarpaceae. In: *A Review of Dipterocarps: Taxonomy, Ecology and Silviculture,* eds S. Appanah & J.M. Turnbull, pp. 5–44. CIFOR, Bogor.

MAXWELL, J.F. 1988. The vegetation of Doi Suthep-Pui National Park, Chiang Mai Province, Thailand. *Tigerpaper* 15: 6–14.

MAXWELL, J.F. 1995. Vegetation and vascular flora of the Ban Saneh Pawng area, Lai Wo subdistrict, Sangkaburi District, Kanchanaburi, Thailand. *Nat. Hist. Bull. Siam. Soc.* 43: 131–170.

MAXWELL, J.F. 1999. Mae Yom National Park: a precious national botanical treasure. *Nat. Hist. Bull. Siam Soc.* 47: 7–11.

MAXWELL, J.F. 2001. Vegetation and vascular flora along the Yetayun-Yadana gas pipeline, Taninthayi (Tennaserim) Division, Myanmar. *Nat. Hist. Bull. Siam. Soc.* 49: 29–59.

MAXWELL, J.F. 2004. A synopsis of the vegetation of Thailand. *Nat. Hist. J., Chulalongkorn Univ.* 4: 19–29.

MAXWELL, J.F. & ELLIOTT, S. 2001. Vegetation and vascular flora of Doi Sutep-Pui National Park. Northern Thailand. *Thai Studies in Biodiversity* 5. Biodiversity Research & Training Program, Bangkok.

MAXWELL, J.F., ELLIOTT, S., PALEE, P. & ANUSARNSUNTHORN, V. 1995. The vegetation of Doi Khuntan National Park, Lamphun-Lampang Provinces, Thailand. *Nat. Hist. Bull. Siam. Soc.* 43: 185–205.

MAXWELL, J.F., ELLIOTT, S. & ANUSARNSUNTHORN, V. 1997. The vegetation of the Jae Sawn National Park, Lampang Province, Thailand. *Nat. Hist. Bull. Siam. Soc.* 45: 35–52.

MAY, R.M. 1973. *Stability and Complexity in Model Ecosystems*. Princeton University Press, New Jersey.

MAY, R.M. & STUMPF, P.H. 2001. Species-area relations in tropical forests. *Science* 290: 2084–2086.

MAYCOCK, C.R., THEWLIS, R.N., GHAZOUL, J., NILUS, R. & BURSLEM, D.F.R.P. 2005. Reproduction of dipterocarps during low intensity masting events in a Bornean rain forest. *J. Veg. Sci.* 16: 635–646.

MAYCOCK, C.R., KETTLE, C.J., KHOO, E., PEREIRA, J.T., SUGAU, J.B., NILUS, R., ONG, R.C., AMALUDIN, N.A., NEWMAN, M.F. & BURSLEM, D.F.R.P. 2012. A revised conservation assessment of dipterocarps in Sabah. *Biotropica* 44: 649–657.

MAYNARD SMITH, J. 1982. *Evolution and the Theory of Games*. Cambridge University Press.

MAYR, E. 1969. The biological meaning of species. *Biol. J. Linn. Soc.* 1: 311–320.

McA., M.E. 1885. Forest fire caused by lightning. *Indian Forester* 11: 58.

McA., M.E. 1888. Authentic case of forest fire caused by lightning. *Indian Forester* 14: 318.

McCLURE, H.E. 1975. Effect of cool weather upon a rain forest and its inhabitants. *Nat. Hist. Bull. Siam. Soc.* 26: 35–70.

McCONKEY, K.R. 2005. The influence of gibbon primary seed shadows on post-dispersal seed fate in a lowland dipterocarp forest in Central Borneo. *J. Trop. Ecol.* 21: 255–262.

McGUIRE, K.L. 2007. Common ectomycorrhizal networks may maintain monodominance in a tropical rain forest. *Ecology* 88: 567–574.

McPHERSON, M.A. & NIESWIADOMY, M.L. 2005. Environmental Kuznets curve: threatened species and spatial effects. *Ecol. Econ.* 55: 395– 407.

McSHEA, W.J. & DAVIES, S.J. 2011. Introduction. In: *The Ecology and Conservation of Seasonally Dry Forest in Asia*, eds W.J. McShea, S.J. Davies & N. Bhumpakphan, pp. 1–8. Rowman & Littlefield, Smithsonian Institution Scholarly Press, Washington, DC.

McSHEA, W.J., DAVIES, S.J. & BHUMPAKPHAN, N., eds. 2011. *The Ecology and Conservation of Seasonally Dry Forest in Asia*. Rowman & Littlefield, Smithsonian Institution Scholarly Press, Washington, DC.

MEDINA, E., SOBRADO, M. & HERRERA, R. 1978. Significance of leaf orientation for leaf temperature in an Amazonian sclerophyll vegetation. *Rad. Environm. Biophys.* 15: 131–140.

MEDWAY, LORD. 1972. Phenology of a tropical rain forest in Malaya. *Biol. J. Linn. Soc.* 4: 117–146.

MEDWAY, LORD. 1977. *Mammals of Borneo: Field Keys and an Annotated Checklist*. Monographs of the Malaysian Branch of the Royal Asiatic Society, Kuala Lumpur.

MEDWAY, LORD & McCLURE, H.E. 1966. Flowering, fruiting, and animals in the canopy of a tropical rain forest. *Malayan Forester* 29: 182–203.

MEEGASKUMBURA, M. Undated. *Wildlife and Amphibian Biogeography, Restoration, and Fragmentation of Forests of SW Sri Lanka*. Unpubl. ms.

MEER, P.J., VAN DER, KUNNE, P.L.B., BRUNSTING, A.M.H., DIBOR, L.A. & JANSEN, P.A. 2008. Evidence for scatter-hoarding in a tropical peat swamp forest in Malaysia. *J. Trop. Forest Sci.* 20: 340–343.

MEHER-HOMJI, V.M. 1970. Notes on some peculiar cases of phytogeographic distributions. *J. Bombay Nat. Hist. Soc.* 67: 1–6.

MEHER-HOMJI, V.M. 1973. A phytosociological study of the *Albizia amara* Boiv. Community of India. *Phytocoenologia* 1: 114–129.

MEHER-HOMJI, V.M. 1977. Vegetation-climate parallelism from Pondicherry-Mysore-Murkal transect, South India. *Phytocoenologica* 4: 206–217.

MEHER-HOMJI, V.M. 1978. *The Term Subtropical in Phytogeography: Facts and Fallacies*. Prof. U.N. Chatterji Commemoration Volume.

MEHER-HOMJI, V.M. 2001. *Bioclimatology and Plant Geography of Peninsular India*. Scientific Publishers (India), Jodhpur.

MEHER-HOMJI, V.M., & MISRA, K.C. 1973. Phytogeography of the Indian subcontinent. In: *Progress of Plant Ecology in India* 1, eds R. Misra, B. Gopal, K.P. Singh & J.S. Singh, pp. 9–89. Today & Tomorrow's Printers & Publishers.

MEIJAARD, E. & GROVES, C.P. 2006. The geography of mammals and rivers in mainland Southeast Asia. In: *Primate Biogeography. Progress and Prospects,* eds S.M. Lehman & J.G. Fleagle, pp. 305–330. Springer, New York.

MEIJAARD, E. & SHIEL, D. 2007. Is wildlife research useful for wildlife conservation in the tropics? A review for Borneo with global implications. *Biodivers. & Conservation* 11: 3053–3065.

MEIJER, W. 1959a. Plant sociological analysis of montane forest near Tjibodas, West Java. *Act. Bot. Neerl.* 8: 277–291.

MEIJER, W. 1959b. On the flora of Mt. Sago nearf Pajakumbuh, Central Sumatra. *Penggamar Alam* 38: 1–12.

MEINZER, F.C., GOLDSTEIN, G., JACKSON, P., HOLBROOK, N.M., GUTIÉRREZ, M.V. & CAVALIER, J. 1995. Environmental and physiological regulation of transpiration in tropical forest gap species: the influence of boundary layer and hydraulic properties. *Oecologia* 101: 514–522.

MENON, A., SINGH P., SHAH, E., LÉLÉ S., PARANJAPE S. & JOY, K. 2007. Community-based natural resource management in Bhutan: the case of the Lingmuteychhu watershed. In: *Community-Based Natural Resource Management: Issues and Cases from South Asia*, pp.159–195. SAGE Publications India, New Delhi.

MENON, V. 2009. *Mammals of India*. Princeton University Press.

MENZEL, R., VENTURA, D.F., WERNER, A., JOAQUIN, L.C.M. & BACKHAUS, W. 1989. Spectral sensitivity of single receptors and colour vision in the stingless bee *Melipona quadrifasciata*. *J. Comp. Physiol. A*, 166: 151–164.

MERGEN, F., BURLEY, J. & FURNIVAL. G.M. 1974. Provenance-temperature interactions in four coniferous species. *Silvae Genet.* 23: 167–226.

MERRILL, E.D. 1907. The flora of Mount Halcon, Mindoro. *Philipp. J. Sc.* 2: 251–309.

MERRILL, E.D. 1926. *An Enumeration of Philippine Flowering Plants*. Bureau of Printing, Manila.

MERTZ, O., RAVNBORG, H.M., LÖVEI, G.L., NIELSEN, I. & KONIJNDIJL, C.C. 2007. Ecosystem services and biodiversity in developing countries. *Biodivers. & Conservation* 16: 2729–2737.

METCALFE, C.R. & CHALKE, L. 1979– Anatomy of the Dicotyledons. 2nd ed. Clarendon, Oxford.

METCALFE, D.J. & GRUBB, P.J. 1995. Seed mass and light requirements for regeneration in South East Asian rain forest. Canad. J. Bot. 73: 817–826.

METCALFE, D.J. & GRUBB, P.J. 1997. The responses to shade of seedlings of very small-seeded tree and shrub species from a tropical rain forest in Singapore. Funct. Ecol. 11: 215–221.

METCALFE, D.J., GRUBB, P.J. & TURNER, I.M. 1998. The ecology of very small-seeded shade-tolerant trees and shrubs in lowland forest in Singapore. Pl. Ecol. 134: 131–149.

METCALFE, I. 1988. Origin and assembly of south-east Asian continental terranes. In: Gondwana and Tethys, eds M.G. Audley-Charles & A. Hallam, Geological Society Special Publication 37: 101–118. Oxford University Press.

METCALFE, I. 1996. Pre-Cretaceous evolution of S.E. Asian terraines. In: Tectonic Evolution of Southeast Asia, eds R. Hall & D. Blundell, Geological Society Special Publication 106: 97–122. Geological Society, London.

METCALFE, I., REN, J.-S., CHARVET, J. & HADA, S. 1999. Gondwana Dispersion and Asian Accretion. Balkema, Rotterdam & Brookfield.

MEYER DE SCHAUENSEE, R, & PHELPS, W.H. JR. 1978. A Guide to the Birds of Venezuela. Princeton University Press.

MICHAEL, R.P. & CROOK, J.H. 1973. Comparative Ecology and Behaviour of Primates: Proceedings of a Conference. Academic Press, London and New York.

MICHAUX, B. 2010. Biogeology of Wallacea: geotectonic models, areas of endemism and natural biogeographical units. Biol. J. Linn. Soc. 101: 193–212.

MICHON, G., BOMPARD, J., HECKETSWEILER, P. & DUCATILLON, C. 1983. Tropical forest architectural analysis as applied to agroforests in the humid tropics: the example of traditional village-forests in Java. Agrofor. Systems 1: 117–129.

MICHON, G., BOMPARD, M. F. & BOMPARD, J. 1986. Multistoried agroforestry garden system in West Sumatra, Indonesia. Agrofor. Systems 4: 315–338.

MILLER, T.E., BURNS, J.H., MUNGUIA, P., WALTERS, E.L., KNEITEL, J.M., RICHARDS, P.M., MOUQUET, N. & BUCKLEY, H.L. 2005. A critical review of twenty years' use of the resource-ratio theory. Amer. Naturalist 165: 439–448.

MILLET, J., PASCAL, J.P. & KIET, L.C. 2010. Effects of disturbance over 60 years on a lowland forest in southern Vietnam. J. Trop. Forest Sci. 22: 237–246.

MILLIKEN, W. & PROCTOR, J. 1999. Montane forest in Dumoga Bone National Park, Sulawesi. Edinburgh J. Bot. 56:449–458.

MILROY, J.W. 1930. The relationship between sal forests and fire. Indian Forester 56: 442–447.

MIRMANTO, E., TSUYZAKI, S. & KOHYAMA, T. 2003. Investigation of the effects of distance from river and peat depth on tropical wetland forest communities. Tropics 10: 287–294.

MISHRA, B.P., TRIPATHI, O.P., TRIPATHI, R.S., & PANDEY, H.V. 2004. Effects of anthropogenic disturbance on plant diversity and community structure of a sacred grove in Meghalaya, northeast India. Biodivers. & Conservation 13: 421–426.

MISRA, P. 2012. From the Ruins of Empire: the Revolt Against the West and the Remaking of Asia. Allen Lane, London.

MITCHELL, J. & ASHLEY, C. 2007. Can Tourism Offer Pro-Poor Pathways to Prosperity? Overseas Development Institute, London.

MITTERMEIER, R.A., GIL, P.R., HOFFMANN, M., PILGRIM, J., BROOKS, T., MITTERMEIER, P.G., LAMOREUX, J. & DA FONSECA, G.A.B. 2004. Hotspots Revisited. Cemex, Mexico City.

MIYAMOTO, M. 1998. Natural Regeneration of Dipterocarps in Tropical Lowland Forest in Malaysia. Masters thesis, Hiroshima University.

MOAD, A.S. 1992. Dipterocarp Juvenile Growth and Understory Light Availability in Malaysian Tropical Forest. PhD thesis, Harvard University.

MOFFETT, M.W. 1993. The High Frontier. Exploring the Tropical Rain Forest Canopy. Harvard University Press.

MOHAMAD, B., BAHARUDIN, J. & QUAH, E.-S. 1987. Studies on logging damage due to different methods and intensities of harvesting in a hill dipterocarp forest in Peninsular Malaysia. Malayan Forester 50: 135–147.

MOHAN, N.P. 1943. How to increase fodder in the Low Hills, Punjab. Indian Forester 69: 182–186.

MOHAN JAIN, S. & PRIYADARSHAN, P.M., eds. 2009. Breeding Plantation Tree Crops. Tropical Species. Springer Science & Business Media, New York.

MOHD. ASLAN, J. & ABD. GULAM AZAD M.A.J. 2006. Mammal diversity in a secondary forest in Peninsular Malaysia. Biodivers. & Conservation 15: 1013–1025.

MOHD. SHAWAHID, H.O. 1985. Determining an economic cutting regime for the tropical rainforest. Malayan Forester 48: 57–74.

MOHD. SHAWAHID, H.O. & AWANG NOOR, A.G. 1999. Estimating the non-timber values of forest: beginning with natural stands of rattan. Trop. Biodivers. 6: 161–178.

MOHR, E.C.J. & VAN BAREN, F.A. 1959. Tropical Soils: a Critical Study of Soils Genesis as Related to Climate, Rock and Vegetation. Uitgeverij & v. Hoeve, The Hague.

MOLENGRAAFF, G.A.F & WEBER, M. 1921. On the relation between the Pleistocene glacial period and the origin of the Sunda Sea (Java- and South China-Sea), and its influence of the distribution of coralreefs and on the land- and freshwater fauna. Proc. Sect. Sci. 23: 395–439.

MOMOSE, K. & HAMID KARIM, ABG. A. 2005. The plant pollinator community in a lowland dipterocarp forest. In: Pollination Ecology and the Rain Forest, eds D.J. Roubik, S. Sakai & A.A. Hamid, pp. 65–72. Springer, New York.

MOMOSE, K., NAGAMITSU, T. & INOUE, T. 1996. The reproductive ecology of an emergent dipterocarp in a lowland rain forest in Sarawak. Pl. Sci. Biol. 11: 189–198.

MOMOSE, K., YUMOTO, T., NAGAMITSU, T., KATO, M., NAGAMASU, H., SAKAI, S.,

HARRISON, R.D., ITIOKA, T., HAMID, A.A. & INOUE, T. 1998. Pollination biology in a lowland dipterocarp forest in Sarawak, Malaysia. I. Characteristics of the plant-pollinator community in a lowland dipterocarp forest. Amer. J. Bot. 85: 1477–1501.

MONTGOMERY, R.A. & CHAZDON, R.L. 2002. Light gradient partitioning by tropical tree seedlings in the absence of canopy gaps. Oecologia 131: 165–174.

MOOG, U., FIALA, B., FEDERLE, W. & MASCHWITZ, U. 2002. Thrips pollination of the dioecious ant plant Macaranga hullettii (Euphorbiaceae) in Southeast Asia. Amer. J. Bot. 89: 50–59.

MOONEY, H.F. 1938. A synecological study of the forests of western Singhbhum. Indian Forester Rec. 2: 259–355.

MOONEY, S.D., HARRISON, S.P., BARTLEIN, P.J., DANIAU, A.-L., STEVENSON, J., BROWNLIE, K.C., BUCKMAN, S., CUPPER, M., LULY, J., BLACK, M., COLHOUN, E., D'COSTA, D., DODSON, J., HABERLE, S., HOPE, G.S., KERSHAW, P., KENYON, C., MCKENZIE, M. & WILLIAMS, N. 2011. Late Quaternary fire regimes of Australasia. Quat. Sci. Rev. 30: 28–46.

MOORE, T.A. & SHEARER, J.C. 1997. Evidence for aerobic degradation of Palangka Raya peat and implications for its sustainability. In: Biodiversity and Sustainability of Tropical Peatlands, eds J.O. Rieley & S.E. Page, pp. 157–167. Samara, Cardigan.

MOORMAN, F.R. & PANABOKKE, C.R. 1961. Soils of Ceylon. Trop. Agric. 117: 4–65.

MORITA, T. 2002. Economic value of tropical forest from the carbon sequestration viewpoint. Tropics 11: 213–219.

MORLEY, R.J. 1981. Development and vegetation dynamics of a lowland ombrogenous peat swamp in Kalimantan Tengah, Indonesia. J. Biogeogr. 8: 383–404.

MORLEY, R.J. 1982. A palaeoecological interpretation of a 10,000 year pollen record from Danau Padang, Central Sumatra, Indonesia. J. Biogeogr. 9: 151–190.

MORLEY, R.J. 1998. Palynological evidence for Tertiary plant dispersals in the SE Asian region in relation to plate tectonics and climate. In: Biogeography and Geological Evolution of S.E. Asia, eds R. Hall, & J.D. Holloway, pp. 211–234. Backhuys, Leiden.

MORLEY, R.J. 1999. Origin and Evolution of Tropical Rain Forests. Wiley, Chichester.

MORLEY, R.J. 2011. Cretaceous and Tertiary climate change and the past distribution of megathermal rainforests. In: Tropical Rainforest Responses to Climatic Change, eds M. Bush, J. Flenley & W. Gosling, W, pp. 1–34. Springer Praxis.

MORLEY, R.J. 2012. A review of the Cenozoic palaeoclimate history of Southeast Asia. In: Biotic Evolution and Environmental Change in Southeast Asia, eds D.J. Gower, K.G. Johnson, J.E. Richardson. B.R. Rosen, L. Rüber & S. Williams. Systematics Association Special Volume 82: 79–114. Cambridge University Press.

MORLEY, R.J. & FLENLEY J.R. 1987. Late Cainozoic vegetational and environmental changes in the Malay Archipelago. In: Biogeographical Evolution of the Malay Archipelago, ed.

T.C. Whitmore, pp. 50–59. Oxford Monographs on Biogeography 4. Oxford Scientific Publications.

MORLEY, R.J., MORLEY, H.P. & JAIS, J. 2013. Oligocene to Pliocene palaeovegetation maps for Sundaland. Conference paper at Southeast Asian Gateway Evolution: SAGE2013, Museum fuer Naturkunde, Berlin.

MORTON, C.M., DAYANANDAN, S. & DASSANAYAKE, D. 1999. Phylogeny and biosystematics of Pseudomonotes (Dipterocarpaceae) based on molecular and morphological data. Pl. Syst. Evol. 216: 197–205.

MOYERSOEN, B., BAKER, P. & ALEXANDER, I.J. 2001. Are ectomycorrhizas more abundant than arbuscular mycorrhizas in tropical heath forests? New Phytol. 150: 591–599.

MUELLER-DOMBOIS, D. 1968. Ecogeographic analysis of a climate map of Ceylon with particular reference to vegetation. Ceylon Forester 8: 39–58.

MUELLER-DOMBOIS, D. 1988. Forest decline and die-back – a global ecological problem, Trends Ecol. Evol. 3: 310–312.

MUKERJI, A.K. 2006. Conservation and management of biodiversity in Indian forests – a new eco-development approach. Indian Forester 132: 921–930.

MUKERJI, K.G. & KAPOOR, A. 1986. Occurrence and importance of vesicular-arbuscular mycorrhizal fungi in semiarid regions of India. Forest Ecol. Managem. 16: 117–126.

MUKHTAR, E., SUZUKI, E. KOHYAMA, T. & RAHMAN, R. 1991. Regeneration process of a climax species Calophyllum cf. soulattri in tropical rain forest of West Sumatra. Tropics 1: 1–12.

MUKHTAR, E., YONEDA, T. & RAHMAN, R. 1998. Population dynamics of a cohort from mast fruiting in 1981: regeneration process of a climax species Calophyllum cf. soulattrei in a tropical rain forest of West Sumatra. Tropics 7: 183–194.

MULKEY, S.S., CHAZDON, R.L. & SMITH, A.P. 1996a. Tropical Forest Plant Ecophysiology. Chapman & Hall, London & New York.

MULKEY, S.S., KITAJIMA, K. & WRIGHT, S.J. 1996b. Plant physiological ecology of tropical forest canopies. Trends Ecol. Evol. 11: 408–412.

MULLER, J. 1968. Palynology of the Pedawan and Plateau Sandstone formations (Cretaceous-Eocene) in Sarawak, Malaysia. Micropaleontology 14: 1–37.

MULLER, J. 1972. Palynological evidence for change in geomorphology, climate, and vegetation in the Mio-Pliocene of Malesia. In: The Quaternary Era in Malesia. Transactions of the Second Aberdeen Hull Symposium on Malesian Ecology, eds P.S. Ashton & H.M. Ashton, pp. 6–16. Hull Geog. Dept. Misc. Ser. 13.

MULLER, J. 1978. New observations on pollen morphology and fossil distribution of genus Sonneratia (Sonneratiaceae). Rev. Palaeobot. Palynol. 26: 277–300.

MULLER, T.B. 1981. Growth and yield of logged-over mixed dipterocarp forest in East Kalimantan. Malayan Forester 44: 419–431.

MULLER-LANDAU, H.C., WRIGHT, S.J. & CALDERON, CONDIT, R. & HUBBELL, S.P.

2008. Interspecific variation in primary seed dispersal in a tropical forest. *J. Ecol.* 96: 653–667.

MURALI, K.S. & SUKUMAR, R. 1993. Leaf flushing phenology and herbivory in a tropical dry deciduous forest, southern India. *Oecologia* 94: 114–119.

MURALI, K.S. & SUKUMAR, R. 1994. Reproductive phenology of a tropical dry forest in Mudumalai, southern India. *J. Ecol.* 82: 759–767.

MURALI, K.S., SHANKAR, U., SHAANKER, R.U., GANESHAIAH, K.N. & BAWA, K.S. 1996. Extraction of non-timber forest products in the forests of Biligiri Rangan Hills, India. *Econ. Bot.* 50: 252–269.

MURAWSKI, D.A. 1995. Reproductive biology and genetics of tropical trees from a canopy perspective. In: *Forest Canopies,* eds M.D. Lowman M.D. & M. Nadkarni, pp. 457–493. Academic in Press, San Diego.

MURAWSKI, D.A. & BAWA, K.S. 1994. Genetic structure and mating system of *Stemonoporus oblongifolius* (Dipterocarpaceae) in Sri Lanka. *Amer. J. Bot.* 81: 155–160.

MURAWSKI, D.A. & HAMRICK, J.L. 1991. The effects of density of flowering individuals on the mating systems of nine tropical tree species. *Heredity* 67: 167–174.

MURAWSKI, D.A., DAYANANDAN, B. & BAWA, K.S. 1994a. Outcrossing rates of two endemic *Shorea* species from Sri Lankan tropical rain forest. *Biotropica* 26: 23–29.

MURAWSKI, D.A., GUNATILLEKE, I.A.U.N., & BAWA, K.S. 1994b. The effects of selective logging on inbreeding in *Shorea megistophylla* (Dipterocarpaceae) from Sri Lanka. *Conservation Biol.* 8: 997–1002.

MURDIYARSO, D., VAN NOORDWIJK, M., WASRIN, U.R., TOMICH, T.P. & GULLISON, A.N. 2002. Environmental benefits and sustainable land-use options in the Jambi transect, Sumatra. *J. Veg. Sci.* 13: 429–438.

MURTHY, I.K., BHAT, P.R., RAVINDRANATH, N.H. & SUKUMAR, R. 2005. Financial valuation of non-timber forest product flows in Uttara Kannada district, western Ghats, India. *Current Science* 88: 1573–1579.

MUTHURAMKUMAR, S. & PARTHASARATHY, N. 2000. Alpha diversity of lianas in a tropical evergreen forest in the Anamalais, Western Ghats, India. *Diversity Distrib.* 6: 1–14.

MYERS, N. 1986. Tropical forests: patterns of depletion. In: *Tropical Rain Forests and the World Atmosphere. American Association for the Advancement of Science Symposium* 101, ed. G.T. Prance, pp. 9–22. Westview Press, Colorado.

MYERS, R.J., ROBICHAUX, R.H., UNWIN, G.L. & CRAIG, I.E. 1987. Leaf water relations and anatomy of a tropical rainforest tree species vary with crown position. *Oecologia* 74: 81–85.

NAGENDRA, H. 2007. Drivers of reforestation in human-dominated forests. *Proc. Natl. Acad. Sci. U.S.A.* 104: 15218–15223.

NAGENDRA, H., PAREETH, S., SHARMA, B., SCHWEIK, C. & ADHIKARI, K.R. 2008. Forest fragmentation and regrowth in an institutional mosaic of community, government and private ownership in Nepal. *Landscape Ecol.* 23: 41–54.

NAGY, L. & PROCTOR, J. 2000. Leaf δ13C signatures in heath and lowland

evergreen rain forest species from Borneo. *J. Trop. Ecol.* 16: 757–761.

NAIDOO, R. & ADAMOWICZ, W. L. 2001. Effects of economic prosperity on numbers of threatened species. *Conservation Biol.* 15: 1021–1029.

NAIDU, P.D. 1995. *High-Resolution Studies of Asian Quaternary Monsoon Climate and Carbonate Records from the Equatorial Indian Ocean.* Earth Sciences Centre, Göteborg.

NAIR, K.K.N., KHANDURI, S.K. & BALASUBRAMANYAN, K., eds. 2001. *Shola Forests of Kerala: Environment and Biodiversity.* Kerala Forest Research Institute.

NAITHANI, H.B. 1987. *Parthenium histerophorus* – a pernicious weed in Arunachal Pradesh and Nagaland. *Indian Forester* 113: 709–710.

NAITHANI, H.B., PAL, R.C. & SRIVASTAVA, R.K. 2006. Vegetation analysis of the Tirumala Hills, Andhra Pradesh. *Indian Forester* 132: 1110–1130.

NAJMAN, Y., CLIFT, P., JOHNSON, M.R.W. & ROBERTSON, A.H.F. 1993. Early stages of foreland basin evolution in the Lesser Himalaya, N. India. In: *Himalayan Tectonics,* eds P.J. Treloar & M.P. Searle. Geological Society Special Publication 74: 541–558. Geological Society, London.

NAKAGAWA, M., TANAKA, K., NAKASHIZUKA, T., OHKUBO, T., KATO, T., MAEDA, T., SATO, K., MIGUCHI, H., NAGAMASU, H., OGINO, K., TEO, S., HAMID, A.A. & SENG L.H. 2000. Impact of severe drought associated with the 1997–1998 El Niño in a tropical forest in Sarawak. *J. Trop. Ecol.* 16: 355–367.

NAKAGAWA, M., ITIOKA, T., MOMOSE, K., YUMOTO, T., KOMAI, F., MORIMOTO, K., JORDAL, B.H., KATO, M., KALIANG, H., HAMID, A.A., INOUE, T. & NAKASHIZUKA, T. 2003. Resource use of insect seed predators during general flowering and seeding events in Bornean dipterocarp rain forest. *Bull. Entomol. Res.* 93: 455–466.

NAKAGAWA, M., YAYOI, T., TANAKA, K. & TOHRU, N. 2005. Predispersal seed predation by insects vs. vertebrates in six dipterocarp species in Malaysia. *Biotropica* 37: 389–396.

NAKAGAWA, M., MIGUCHI, H. & NAKASHIZUKA, T. 2006. The effects of various forest uses on small mammal communities in Sarawak, Malaysia. *Forest Ecol. Managem.* 231: 55–62.

NANDI, O.I. 1998. Ovule and seed anatomy of Cistaceae and related Malvanae. *Pl. Syst. Evol.* 209: 239–264.

NANTHA, H.S. & TISDELL, C. 2009. The orang utan-palm oil conflict: economic constraints and opportunities for conservation. *Biodivers. & Conservation* 18: 487–502.

NARENDRAN, K. 2001. NTFP in Nilgiris. *Econ. Bot.* 55: 528–538.

NARENDRAN, K. 2004. Fire as a conservation threat. *Conservation Biol.* 18: 1553–1561.

NARENDRAN, K., MURTHY, I. K. SURESH, H. S., DATTARAJA, H.S., RAVINDRANATH, N.H., & SUKUMAR. R. 2001. Nontimber forest product extraction, utilization and valuation: a case study from the Nilgiri Biosphere Reserve, Southern India. *Econ. Bot.* 55: 528–538.

NASI, R., BROWN, D. & WILKIE, D. 2008.

Conservation and Use of Wildlife-Based Resources: the Bushmeat Crisis. Secretariat of the Convention on Biological Diversity, Montreal.

NATH, P.C., ARUNACHALAM, A., KHAN, M.L., ARUNACHALAM, K. & BARBHUIYA, A.R. 2005. Vegetation analysis and tree population structure of tropical wet evergreen forests in and around Namdampha National Park, northeast India. *Biodivers. & Conservation* 14: 2109–2136.

NAUTIYAL, B. 1934. Damage by frost at New Forest, Dehra Dun, during the period 1930 to 1934. *Forest Bulletin* 91. Manager of Publications, Delhi.

NAUTIYAL, S. & KAECHELE, H. 2007. Conserving the Himalayan forests: approaches and implications of different conservation regimes. *Biodivers. & Conservation* 16: 3737–3754.

NAYAK, P.K. & BERKES, F. 2008. Politics of co-optation: community forest management versus joint forest management in Orissa, India. *Environ. Managem.* 41: 707–718.

NAYAK, S.N.V., SWAMY, H.R., NAGARAJA, B.C., RAO, U. & CHANDRASHEKARA, U.M. 2000. Farmers' attitude towards sustainable management of Soppina Betta forests of Sringeri area of the western Ghats, India. *Forest Ecol. Managem.* 132: 223–241.

NAYAR, M.O. 1996. *Hotspots of Endemic Plants of India, Nepal, and Bhutan.* Tropical Botanical Garden and Research Institute, Thiruvananthapuram.

NCHANJI, A.C. & PLUMPTRE, A.J. 2003. Seed germination and early seedling establishment of some elephant-dispersed species in Banyang-Mbo Wildlife Sanctuary, south-western Cameroon. *J. Trop. Ecol.* 19: 229–237.

NEEDHAM, J., & LU, G.-D. 1986. *Science and Civilisation in China 6, Biology and Biological Technology,* 1: *Botany.* Cambridge University Press.

NELSON, B.W., KAPOS, V., ADAMS, J.B., OLIVEIRA, W.J., & BRAUN, O.P.G. 1994. Forest disturbance by large blowdowns in the Brazilian Amazon. *Ecology* 75: 853–858.

NEPSTAD, D.C., CARVALHO, C.R. DE, DAVIDSON, E.A., JIPP, P.H., LEFEBVRE, P.A., NEGREIROS, G.H., SILVA, E.D. DA, STONE, T., TRUMBORE, S.E. & VIEIRA, S. 1994. The role of deep roots in the hydrological and carbon cycles of Amazonian forests and pastures. *Nature* 372: 666–669.

NESS, J.S., BRONSTEIN, J.L., ANDERSEN, A.N. & HOLLAND, J.N. 2004. Ant body size predicts the dispersal distance of ant-adapted seeds: implications of small-ant invasions. *Ecology* 85: 1244–1250.

NETO, V., AHMAD AINUDDIN, N., WONG, M.Y. & TING, H.L. 2012. Contributions of forest biomass and organic matter to above - and below - ground carbon contents at Ayer Hitam Forest Reserve, Malaysia. *J. Trop. Forest Sci.* 24: 217–230.

NEUTZIL, S.G. 1997. Onset and rate of peat and carbon accumulation in four domed ombrogenous peat deposits, Indonesia. In: *Biodiversity and Sustainability of Tropical Peatlands,* eds J.O. Rieley & S.E. Page, pp. 55–72. Samara, Cardigan.

NEWBERY, D.M. 2005. Ectomycorrhizas and mast fruiting in trees: linked by climate-

driven tree resources? *New Phytol.* 167: 324–326.

NEWBERY, D.M. & LINGENFELDER, M. 2004. Resistance of a lowland rain forest to increasing drought intensity in Sabah, Borneo. *J. Trop. Ecol.* 20: 613–624.

NEWBERY, D. M., CAMPBELL, E.J.F., PROCTOR, J. & STILL, M.J. 1996. Primary lowland dipterocarp forest at Danum Valley, Sabah, Malaysia. *Vegetatio* 122: 193–220.

NEWBERY, D.M., PRINS, H.H.T. & BROWN, N.D. 1998a. *Dynamics of Tropical Communities.* 37th Symposium of British Ecological Society, Cambridge University. Blackwell Science, Oxford.

NEWBERY, D.M., SONGWE, N.C. & CHUYONG, G.B. 1998b. Phenology and dynamics of an African rainforest at Korup, Cameroon. In: *Dynamics of Tropical Communities. 37th Symposium of the British Ecological Society, Cambridge University,* eds D.M. Newbery, H.H.T. Prins & N.D. Brown, pp. 267–308. Blackwell Science, Oxford.

NEWBERY, D.M., ALEXANDER, I.J., THOMAS, D.W. & GARTLAN, J.S. 1988c. Ectomycorrhizal rain - forest legumes and soil phosphorus in Korup National Park, Cameroon. *New Phytol.* 109: 433–450.

NEWBERY, D.M., KENNEDY, D.N., PETOL, G.H., MADANI, L. & RIDSDALE, C.E. 1999. Primary forest dynamics in a lowland dipterocarp forest at Danum Valley, Sabah, Malaysia, and the role of the understorey. *Philos. Trans., Ser. B* 354: 1763–1782.

NEWBERY, D.M., CLUTTON-BROCK, T.H. & PRANCE, G.T., eds. 2000. *Changes and Disturbance in Tropical Rainforest in South-East Asia.* The Royal Society and Imperial College Press, London. Also *Philos. Trans., Ser. B* 354: 1869–1883.

NEWBERY, D. MC C. 1991. Floristic variation within *kerangas* (heath) forest: re-evaluation of data from Sarawak and Brunei. *Vegetatio* 96: 43–86.

NEWBERY, D. MC C. & PROCTOR, J. 1984. Ecological studies in four contrasting lowland forests in Gunung Mulu National Park, Sarawak. IV. Association between tree distribution and soils factors. *J. Ecol.* 72: 475–493.

NEWBERY, D.MC C., RENSHAW, E. & BRÜNIG, E.F. 1986. Spatial pattern of trees in kerangas forest, Sarawak. *Vegetatio* 65: 77–89.

NEWBERY, D.MC C., CAMPBELL, E.J.F., LEE, Y.F., RIDSDALE, C.E. & STILL, M.J. 1992. Primary lowland dipterocarp forest at Danum Valley, Sabah, Malaysia: structure, relative abundance and family composition. *Philos. Trans., Ser. B* 335: 341–356.

NEWMARK, W.D. 1986. Species-area relationship and its determinants for mammals in western North American national parks. *Biol. J. Linn. Soc.* 28: 83–98.

NEWSOME, J. & FLENLEY, J.R. 1988. Late Quaternary vegetational history of the Central Highlands of Sumatra. II. Palaeopalynology and vegetational history. *J. Biogeogr.* 15: 555–578.

NG, F.S.P. 1974. The life of leaves in a humid tropical environment with special reference to Malaysian trees. *IBP Synthesis Meeting,* Kuala Lumpur.

NG, F.S.P. 1975. Germination and establishment of fresh seeds of Malaysian trees II. *Malayan Forester* 38: 171–176.

NG, F.S.P. 1976. The fruits, seeds and seedlings of Malaysian trees xii–xv. *Malayan Forester* 39: 110–146.

NG, F.S.P. 1977. Germination of fresh seeds of Malaysian trees III. *Malayan Forester* 40: 160–163.

NG, F.S.P. 1978. Strategies for establishment in Malaysian Forest trees. In: *Tropical Trees and Forests: an Architectural Analysis*, eds F. Hallé, R.A.A. Oldeman & P.B. Tomlinson, pp. 129–162. Springer, New York & Berlin.

NG, F.S.P. 1980. Germination ecology of Malaysian woody plants. *Malayan Forester* 43: 406–437.

NG, F.S.P. 1981. Vegetative and reproductive phenology of dipterocarps. *Malayan Forester* 44: 197–221.

NG, F.S.P. 1986. Tropical sapwood trees. *Naturalia Monspeliensis. Colloque International sur l'Arbre*: 61–67.

NG, F.S.P. 1991. *Manual of Forest Fruits, Seeds, and Seedlings*. Malaysian Forest Records 34. Forest Research Institute Malaysia, Kepong.

NG, F.S.P. 2010. Guest editorial: making and maintaining impact in science. *J. Trop. Forest Sci.* 22: vii–viii.

NG, F.S.P. 2013. Age of trees in tropical rainforests estimated by timing of wood decay. *J. Trop. Forest Sci.* 434–441.

NG, F.S.P. & LOH, H.S. 1974. Flowering-to-fruiting periods of Malaysian trees. *Malayan Forester* 37: 127–132.

NICHOLSON, D.I., HENRY, N.B. & RUDDER, J. 1987. Stand changes in North Queensland forests. *Proc. Ecol. Soc. Austral.* 15: 61–80.

NICHOLSON, J.W. 1922. Note on the distribution and habit of *Dendrocalamus strictus* and *Bambusa arundinacea* in Orissa. *Indian Forester* 48: 425–428.

NICODEMUS, A., JACOB, J.P., & NAGARAJAN, B. 2005. Pollination by nectarivorous birds in teak clonal seed orchards. *Indian Forester* 131: 1613–1616.

NIE, Z.-L., WEN, J., AZUMA, H., QIU, Y.L., SUN, H., MENG, Y., SUN, W.B. & ZIMMER, E.A. 2008. Phylogeny and biogeographic complexity of Magnoliaceae in the Northern Hemisphere inferred from three nuclear data sets. *Molec. Phylogenet. Evol.* 48: 1027–1040.

NIEUWSTADT, M.G.L. VAN & SHIEL, D. 2005. Drought, fire and tree survival in a Borneo rain forest, East Kalimantan, Borneo. *J. Ecol.* 93: 191–201.

NIIYAMA, K., RAHMAN, K. ABD., IIDA, S., KIMURA, K., AZIZI, R. & APPANAH, S. 1999a. Spatial patterns of common tree species relating to topography, canopy gaps and understorey vegetation in a hill dipterocarp forest at Semangkok Forest Reserve, Peninsular Malaysia. *J. Trop. Forest Sci.* 11: 731–745.

NIIYAMA, K., KASSUM, A.R., KIMURA, K., TANGE, T., IIDA, S., QUAH, E.S., CHAN, Y.C., RIPIN, A. & APPANAH, S. 1999b. *Design and Methods for the Study on Tree Demography in a Hill Dipterocarp Forest at Semangkok Forest Reserve, Peninsular Malaysia*. Forest Institute Malaysia, Research Pamphlet 123. Forest Research Institute Malaysia (FRIM), Kepong.

NIIYAMA, K., RAHMAN, K. ABD., IIDA, S., KIMURA, K., AZIZI, R. & APPANAH, S. 2003. Regeneration of a clear-cut plot in a lowland dipterocarp forest in Pasoh Forest Reserve, Peninsular Malaysia. In: *Pasoh. Ecology of a Lowland Rain Forest in Southeast Asia*, eds T. Okuda, N. Manokaran, Y. Matsumoto, K. Niiyama, S.C. Thomas & P.S. Ashton, pp. 559–568. Springer.

NIK, A.R. 1987. Impact of forest conversion on water yield in Peninsular Malaysia. *Malayan Forester* 50: 258–273.

NILSEN, E.T. 1992. Thermonastic leaf movements: a synthesis of research with *Rhododendron*. *Bot. J. Linn. Soc.* 110: 205–233.

NINOMIYA, I. 1995. Photosynthesis of canopy tree species in a Sarawak rain forest. *Tropics* 4: 297–305.

NISHIMURA, S., YONEDA, T., FUJII, S., MUKHTAR, E. & KANZAKI, M. 2008. Spatial patterns and habitat associations of Fagaceae in a hill dipterocarp forest in West Sumatra. *J. Trop. Ecol.* 24: 535–550.

NISHIMURA, S., ed. 1980. *Physical Geology of Indonesian Island Arcs*. Kyoto University.

NISHIMURA, T.B. & SUZUKI, K. 2001. Allometric differentiation among tropical tree seedlings in heath and peat swamp forests. *J. Trop. Ecol.* 17: 661–681.

NISHIMURA, T.B., SUZUKI, E., KOHYAMA, T. & TSUYUZAKI, S. 2007. Mortality and growth of trees in peat-swamp and heath forests in Cenral Kalimantan after severe drought. *Pl. Ecol.* 188: 165–177.

NITTA, I. & OHSAWA, M. 1998. Bud structure and shoot architecture of canopy and understorey evergreen broad-leaved trees at their northern limit in East Asia. *Ann. Bot. Lond.* 81: 115–129.

NITTA, I. & OHSAWA, M. 2001. Geographical transition of sylleptic/proleptic branching in three *Cinnamomum* species with different bud types. *Ann. Bot. Lond.* 87: 35–45.

NIYOMDHAM, C. 1986. A list of the flowering plants in the swamp area of Peninsular Thailand. *Thai Forest Bull., Bot.* 16: 211–229.

NIYOMDHAM, C. 1988. Some important characters and a special note on the flora of peat swamp forest in Thaland. *Thai Forest Bull., Bot.* 17: 106–115.

NIYOMDHAM, C. 2002. An account of *Dalbergia* (Leguminosae-Papilionoideae) in Thailand. *Thai Forest Bull., Bot.* 30: 124–166.

NOCK, W. 1899. The effects of the recent frost near and at Nuwara Eliya. *Trop. Agric.* 18: 7–3.

NOGUCHI, H., ITOH, A., MIZUNO, T., SRI-NGERNYUANG, K., KANZAKI, M., TEEJUNTUK, S., SUNGPALEE, W., HARA, M., OHKUBO, T., SAHUNALU, P., DHANMMANONDA, P. & YAMAKURA, T. 2007. Habitat divergence in sympatric Fagaceae tree species of a tropical montane forest in northern Thailand. *J. Trop. Ecol.* 23: 549–558.

NOLTIE, H.J. 2000. *Flora of Bhutan: Including a Record of Plants from Sikkim and Darjeeling* 3. Royal Botanic Gardens, Edinburgh & Government of Bhutan.

NOMURA, M., ITIOKA, T. & ITINO, T. 2000. Variations in abiotic defense within myrmecophytic and non-myrmecophytic species of *Macaranga* in a Bornean

dipterocarp forest. *Ecol. Res.* 15: 1–11.

NOOTEBOOM, H.P. 1993. *Magnoliaceae*. In: *The Families and Genera of Vascular Plants 2*, ed. K. Kubitzki, pp.391–401. Springer, Heidelberg.

NORGAARD, R.B. 2010. Ecosystem services: from eye-opening metaphor to complexity blinder. *Ecol. Econ.* 69: 1219–1227.

NORISADA, M. & KOJIMA, K. 2005. Nitrogen form preference of six dipterocarp species. *Forest Ecol. Managem.* 216: 175–186.

NORTON, I. & SCLATER, J.G. 1979. A model for the evolution of the Indian Ocean and the breakup of Gondwanaland. *J. Geophys. Res.* 84: 6803–6830.

NORUP, M.V., DRANSFIELD, J., CHASE, M.W., BARFOD, A.S., FERNANDO, E.S. & BAKER, W.J. 2006. Homoplasious character combinations and generic delimitation: a case study from the Indo-Pacific Arecoid palms (Arecaceae: Areceae). *Amer. J. Bot.* 93: 1065–1080.

NOSIL, P. 2008. Ernst Mayr and the integration of geographic and ecological factors in speciation. *Biol. J. Linn. Soc.* 94: 26–46.

NOUHARE, B.P. 2001. Letter to editor: four giant trees of teak, *Tectona grandis* Linn. (Sagon) – Rama, Laxman, Bharat and Shatrugham. *Indian Forester* 127: 596–597.

NOVICK, R.S., DICK, C.W., LEMES, M., NAVARRO, C., CACCONE, A. & BERMINGHAM, E. 2003. Genetic structure of Mesoamerican populations of big-leaf mahogany (*Swietenia macrophylla*) inferred by microsatellite analysis. *Molec. Ecol.* 12: 2885–2893.

NOVOTNY, V. & BASSET, Y. 2005. Host-specificity of insect herbivores in tropical forests. *Proc. Roy. Soc. London, Ser. B, Biol. Sci.* 272: 1083–1090.

NOVOTNY, V., BASSET, Y., MILLER, S.E., DROZD, P. & CIZEK, L. 2002. Host specialization of leaf-chewing insects in a New Guinea rainforest. *J. Anim. Ecol.* 71: 400–412.

NOVOTNY, V., CLARKE, A.R., DREW, R.A., BALAGAWI, S. & CLIFFORD, B. 2005. Host specialization and species richness of fruit flies (Diptera: Tephritidae) in a New Guinea rain forest. *J. Trop. Ecol.* 21: 67–77.

NOVOTNY, V., DROZD, P., MILLER, S.E., KULFAN, M., JANDA, M., BASSET, Y. & WEIBLEN, G.D. 2006. Why are there so many species of herbivorous insects in tropical forests? *Science* 313: 1115–1118.

NOVOTNY, V., MILLER, S.E., HULCR, J., DREW, R.A.I., BASSET, Y., JANDA, M., SETLIFF, G.P., DARROW, K., STEWART, A.J.A., AUGA, J., ISUA, B., MOLEM, K., MANUMBOR, M., TAMTIAI, T., MOGIA, M. & WEIBLEN, G.D. 2007. Low beta diversity of herbivorous insects in tropical forests. *Nature* 448: 692–697.

NUMATA, M., ed. 1981. *Ecological Studies in the Arun Valley, East Nepal and Mountaineering of Mount Baruntse*. Chiba University.

NUMATA, S., KACHI, N., OKUDA, T. & MANOKARAN, N. 1999. Chemical defenses of fruits and mast-fruiting of dipterocarps. *J. Trop. Ecol.* 15: 695–700.

NUMATA, S., KACHI, N., OKUDA, T. & MANOKARAN, N. 2000. Leaf damage and traits of dipterocarp seedlings in

a lowland rain forest in Peninsular Malaysia. *Tropics* 9: 237–243.

NUMATA, S., YASUDA, M., OKUDA, T., KACHI, N. & NUR SUPARDI, M.N. 2003. Temporal and spatial patterns of mass flowerings on the Malay Peninsula. *Amer. J. Bot.* 90: 1025–1031.

NUMATA, S., KACHI, N., OKUDA, T. & MANOKARAN, N. 2004. Delayed greening, leaf expansion, and damage of sympatric *Shorea* species in a lowland rain forest. *J. Pl. Res.* 117: 19–25.

NUSSBAUM, R., ANDERSON, J. & SPENCER, T. 1995. Factors limiting the growth of indigenous tree seedlings planted in degraded rainforest soils in Sabah, Malaysia. *Forest Ecol. Managem.* 74: 149–159.

NYE, P.H. 1955. Some soil-forming processes in the humid tropics IV. The action of soil fauna. *J. Soil. Sci.* 6: 73–83.

NYE, P.H. & GREENLAND, D.J. 1960. *The Soil under Shifting Cultivation*. CABI, Farnham Royal.

NYKVIST, N. 1996. Regrowth of secondary vegetation after the 'Borneo fire' of 1982–83. *J. Trop. Ecol.* 12: 307–312.

OBERPRIELER, C., MEISTER, J., SCHNEIDER, C. & KILIAN, N. 2009. Genetic structure of *Anogeissus dhofarica* (Combretaceae) populations endemic to the monsoonal fog oases of the southern Arabian Peninsula. *Biol. J. Linn. Soc.* 97: 40–51.

OGAWA, H. 1978. Litter production and carbon cycling in Pasoh Forest. *Malayan Nat. J.* 30: 367–373.

OGAWA, H., YODA, K. & KIRA, Y. 1961. A preliminary survey on the vegetation of Thailand. In: *Nature and Life in Southeast Asia.* I, eds T. Kira & T. Umesao, pp. 21–157. Fauna and Flora Research Society, Kyoto.

OGHUSHI, T. 1997. Plant-mediated interactions between herbivorous insects. In: *Biodiversity, an Ecological Perspective*, eds T. Abe, S.A. Levin & M. Higashi, pp. 115–130. Springer, New York.

OHKUBO, T. & ASHTON, P.S. 2002. Floristic and structural diversity of mixed dipterocarp forest in Lambir Hills National Park, Sarawak, Malaysia. *J. Trop. Forest Sci.* 14: 379–400.

OHSAWA, M. 1983. Distribution, structure and regeneration of forest communities in eastern Nepal. In: *Structure and Dynamics of Vegetation in Eastern Nepal*, ed. M. Numata, pp. 89–120. Chiba University.

OHSAWA, M., ed. 1987. *Life Zone Ecology of the Bhutan Himalaya*. Laboratory of Ecology, Chiba University.

OHSAWA, M. 1990. An interpretation of latitudinal patterns of floristic limits in south and east Asian mountains. *J. Ecol.* 78: 326–339.

OHSAWA, M., ed. 1991a. *Life Zone Ecology of the Bhutan Himalaya II*. Laboratory of Ecology, Chiba University.

OHSAWA, M. 1991b. Montane evergreen broad-leaved forests in the Bhutan Himalaya. In: *Life Zone Ecology of the Bhutan Himalaya II*, ed. M. Ohsawa, pp. 89–156. Laboratory of Ecology, Chiba University.

OHSAWA, M. 1991c. Structural comparison of tropical montane rain forests along latitudinal and altitudinal gradients in south and east Asia. *Vegetatio* 97: 1–10.

OHSAWA, M. 1993. Latitudinal pattern of mountain vegetation zonation in southern and eastern Asia. *J. Veg. Sci.* 4: 13–18.

OHSAWA, M. 1995a. The montane cloud forest and its gradational changes in Southeast Asia. In: *Tropical Montane Cloud Forests*, eds L.S. Hamilton, J.O. Juvick & F.N. Scatena, pp. 254–265. Springer.

OHSAWA, M. 1995b. Latitudinal comparison of altitudinal change in forest structure, leaf type, and species richness in humid monsoon Asia. *Vegetatio* 121: 3–10.

OHSAWA, M. 1996. An interpretation of latitudinal patterns of floristic limits in south and east Asian mountains. *J. Ecol.* 78: 326–339.

OHSAWA, M. 2006a. Climate change impacts on vegetation in humid Asian mountains. *Global Env. Res.* 10: 13–20.

OHSAWA, M. 2006b. Sacred mountains and landscape supporting biodiversity and human life: lessons from Mount Fuji and the Himalyas. *International Symposium, United Nations Universityu, Tokyo: Conserving Cultural and Biological Diversity: The Role of Sacred Natural Sites and Cultural Landscapes. Section 1*, pp. 44–49. UNESCO Div. of Ecological and Earth Sciences, Paris.

OHSAWA, M. & NITTA, I. 1997. Patterning of subtropical/warm-temperate evergreen broad-leaved forests in East Asian mountains with special reference to shoot phenology. *Tropics* 6: 317–334.

OHSAWA, M. & NITTA, I. 2002. Forest zonation and morphological tree-traits along latitudinal and altitudinal gradients in humid monsoon Asia. *Global Env. Res.* 6: 41–52.

OHSAWA, M., NAINGGOLAN, P.H.J., TANAKA, N. & ANWAR, C. 1985. Altitudinal zonation of forest vegetation on Mount Kerinci, Sumatra: with comparisons of zonation in temperate regions of East Asia. *J. Trop. Ecol.* 1: 193–216.

OHTSUKA, T. 1999. Early stages of secondary succession on abandoned cropland in north-east Borneo island. *Ecol. Res.* 14: 281–290.

OI, J. 1953. *Flora of Japan*. Shibundo, Tokyo [in Japanese]. 1957, English transl., rev. by F.G. Meyer & E.H. Walker. Smithsonian Institution, Washington, DC.

OKUDA, T., KACHI, N., YAP, S.K. & MANOKARAN, N. 1997. The distribution pattern and fate of juveniles in a lowland tropical rain forest – implications for regeneration and maintenance of species diversity. *Pl. Ecol.* 131: 155–171.

OKUDA, T., MANOKARAN N., MATSUMOTO, Y., NIIYAMA, K. THOMAS, S.C. & ASHTON, P.S., eds 2003a. *Pasoh. Ecology of a Lowland Rain Forest in Southeast Asia*. Springer.

OKUDA, T., SUZUKI, E., ADACHI, N., QUAH, E.S., HUSSEIN, N.A. & MANOKARAN, N. 2003b. Effect of selective logging on canopy and stand structure and tree species composition in a lowland dipterocarp forest in Peninsular Malaysia. *Forest Ecol. Managem.* 175: 297–320.

OKUDA, T., NOR AZMAN, H., MANOKARAN, N., SAW, L.Q., AMIR, H.M.S. & ASHTON, P.S. 2004. Local variation of canopy structure in relation to soils and topography and the implications for

tree species diversity in a rainforest of Peninsular Malaysia. In: *Tropical Forest Diversity and Dynamism: Findings from a Large-Scale Plot Network*, eds E.C. Losos & E.G. Leigh jr., pp. 221–239. University of Chicago Press, Chicago.

OLIVER, C.D. & LARSON, B.C. 1990. *Forest Stand Dynamics*. McGraw-Hill, New York.

OLIVIER, R.C.D. 1978. *On the Ecology of the Asian Elephant*. PhD thesis, University of Cambridge. Quoted in Sukumar, R., 1989. Ecology of the Asian elephant in southern India. I. Movement and habitat. *J. Trop. Ecol.* 5: 1–18.

ONGUENE, N.A. & KUYPER, T.W. 2001. Mycorrhizal associations in the rain forest of South Cameroon. *Forest Ecol. Managem.*140: 277–287.

OOMEN, M.A. & SHANKAR, K. 2005. Elevational species richness patterns emerge from multiple local mechanisms in Himalayan woody plants. *Ecology* 86: 3039–3047.

ORMSBY, A.A. & BHAGWAT, S.A. 2010. Sacred forests of India: a strong tradition of community-based natural resource management. *Environ. Conservation* 37: 320–326.

OSADA, N., TAKEDA, H., FURUKAWA, A. & AWANG, M. 2001a. Fruit dispersal of two dipterocarp species in a Malaysian rain forest. *J. Trop. Ecol.* 17: 911–917.

OSADA, N., TAKEDA, H., FURUKAWA, A. & AWANG, M. 2001b. Leaf dynamics and maintenance of tree crowns in a Malaysian rainforest stand. *J. Ecol.* 89: 774–782.

OSADA, N., TAKEDA, H., FURUKAWA, A. & AWANG, M. 2002. Ontogenetic changes in leaf phenology of a canopy species, *Elateriospermum tapos* (Euphorbiaceae) in a Malaysian rain forest. *J. Trop. Ecol.* 18: 91–105.

OSADA, N., TAKEDA, H., KAWAGUCHI, H., FURUKAWA, A. & AWANG, M. 2003. Estimation of crown characters and leaf biomass from leaf litter in a Malaysian canopy species, *Elateriospermum tapos* (Euphorbiaceae). *Forest Ecol. Managem.* 177: 379–386.

OSMASTON, A.E. 1927. Hill forets and sal regeneration in the United Provinces. *Indian Forester* 53: 432.

OSTROM, E. 1990. *Governing the Commons: The Evolution of Institutions for Collective Action*. Cambridge University Press.

OSTROM, E., DIETZ, T., DOLSAK, N., STERN, P. C., STONICH, S., & WEBER, E.U., eds. 2002. *The Drama of the Commons*. Committee on the Human Dimensions of Global Change, National Research Council. National Academy Press, Washington, DC.

OVERPECK, J., ANDERSON, D., TRUMBORE, S. & PRELL, W. 1996. The southwest Indian monsoon over the last 18,000 years. *Clim. Dynam.* 12: 213–225.

OVERSEAS DEVELOPMENT INSTITUTE. 2007. Rural employment and migration: in search of decent work. *Briefing Paper* 27. Overseas Development Institute, London.

PADUA, L.S. DE, BUNYAPRAPHATSARA, N., LEMMENS, R.H.M.J. & VALKENBURG, J.L.C.H. VAN. 1999–2003. *Plant Resources of South-East Asia (PROSEA)*12: *Medicinal and Poisonous Plants*. Backhuys,

Leiden.

PAGE, S., HOSCILO, A., LANGNER, A., TANSEY, K., SIEGERT, F., LIMIN, S., & RIELEY, J. 2009. Tropical peatland fires in Southeast Asia. In: *Tropical Fire Ecology. Climate Change, Land Use and Ecosystem Dynamics*, ed. M.A. Cochrane, pp. 263–287. Springer & Paxis, Chichester.

PAGE, S.E., RIELEY, J.O., SHOTYK, O.W. & WEISS, D. 1999. Interdependence of peat and vegetation in a tropical peat swamp forest. *Philos. Trans., Ser. B* 354: 1885–1897.

PAIJMANS, K.J. 1970. An analysis of four tropical rain forest sites in New Guinea. *J. Ecol.* 58: 77–101.

PAIJMANS, K., ed. 1976. *New Guinea Vegetation*. ANU Press, Canberra & Elsevier Scientific Publishing, Amsterdam.

PAIJMANS, K. 1976. Vegetation. In: *New Guinea Vegetation*, ed. K. Paijmans, pp. 23–105. ANU Press, Canberra & Elsevier Scientific Publishing, Amsterdam.

PALIT, S. 1990. Tribals and conservation of forests. *Indian Forester* 116: 93–98.

PALM, C.A., HOUGHTON, R.A., & MELILLO, J.M. 1986. Atmospheric carbon dioxide from deforestation in Southeast Asia. *Biotropica* 18: 177–188.

PALMER, E. J. 1948. Hybrid oaks of North America. *J. Arnold Arbor.* 29: 1–48.

PALMIOTTO, P.A., DAVIES, S.J., VOGT, K.A., ASHTON, M.S., VOGT, D.J. & ASHTON, P.S. 2004a. Soil-related habitat specificialization in dipterocarp rain forest species in Borneo. *J. Ecol.* 92: 609–624.

PALMIOTTO, P.A., VOGT, K.V., ASHTON, P.M.S., ASHTON, P.S., VOGT, D.J., SEMUI, H. & SENG, L.W. 2004b. Linking canopy gaps, topographic position, and edaphic variation in a tropical rainforest: implications for species diversity. In: *Tropical Forest Diversity and Dynamism: Findings from a Large-Scale Plot Network*, eds E.C. Losos & E.G. Leigh jr., pp. 195–220. University of Chicago Press, Chicago.

PANAYOTOU, T. & ASHTON, P.S. 1992. *Not by Timber Alone. Economic and Ecology for Sustaining Tropical Forests*. Island Press, Washington, DC.

PANAYOTOU, T. & ASHTON, P.S. 1995. Sustainable use of tropical forests in Asia. In: *Biodiversity Conservation*, eds C.A. Perrings, K.-G. Mäler, C. Folke, C.S. Holling & B.-O. Jansson, pp. 257–277. Kluwer Academic, Dordrecht.

PANDE, P.K. 2005. Ecological assessment of vegetation in JFM adopted village-forests in Satpura plateau, Madhya Pradesh. *Indian Forester* 131: 97–114.

PANDEY, A.K. 2001. Letter to the editor: where are the village forests? *Indian Forester* 127: 372.

PANDEY, C.B. & SINGH, J.S. 1992. Rainfall and grazing effects on net productivity in a tropical savanna, India. *Ecology* 73: 2007–2021.

PANDEY, D. 2002. *Fuelwood Studies in India: Myth and Reality*. CIFOR.

PANDEY, D.N. 1993. Wildlife, national parks and people. *Indian Forester* 119: 521–527.

PANDEY, R.R. 1998. Social forestry: an approach that is demanding more

objectivity. *Indian Forester* 124: 224–231.

PANDEY, S.K. & SHUKLA, R.P. 2003. Plant diversity in managed sal (*Shorea robusta* Gaertn.) forests of Gorakhpur, India: species composition, regeneration and conservation. *Biodivers. & Conservation* 12: 2295–2319.

PANDIT, P.K. & BHAKAT, R.K. 2007. Conservation of biodiversity and ethnic culture through sacred groves in Midnapur District, West Bengal, India. *Indian Forester* 133: 323–344.

PANDITHARATHNA, P.A.K., SINGHAKUMARA, B.M.P., GRISCOM, H.P. & ASHTON, P.M.S. 2008. Changes in leaf structure in relation to crown position and size class for tree species within a Sri Lankan tropical rain forest. *Canad. J. Botany* 86: 633–640.

PAOLI, G.D. 2006. Divergent leaf traits among congeneric tropical trees with contrasting habitat associations on Borneo. *J. Trop. Ecol.* 22: 397–408.

PAOLI, G.D., CURRAN, L.M. & ZAK, D.R. 2005. Phosphorus efficiency of Bornean rain forest productivity: evidence against the unimodal efficiency hypothesis. *Ecology* 86: 1548–1561.

PAOLI, G.D., CURRAN, L.M. & ZAK, D.R. 2006. Soil nutrients and beta diversity in the Bornean Dipterocarpaceae. *J. Ecol.* 94: 157–170.

PAOLI, G.D., CURRAN, L.M. & SLIK, J.W.F. 2008. Soil nutrients affect spatial patterns of aboveground biomass and emergent tree density in southwestern Borneo. *Oecologia* 155: 287–299.

PARK, S.-Y. & FURUKAWA, A. 1999. Photosynthetic and stomatal responses of two tropical and two temperate trees to atmospheric humidity. *Photosynthetica* 36: 181–186.

PARKER, G.G. & BROWN, M.J. 2000. Forest canopy stratification – is it useful? *Amer. Naturalist* 155: 473–484.

PARKINSON, C.E. 1932. A note on the Burmese species of the genus *Dipterocarpus. Burma Forest Bull.* 27.

PARNWELL, M.J.G. 1988. Rural poverty, development and the environment: the case of North-East Thailand. *J. Biogeogr.* 15: 199–208.

PARRIS, B.S., BEAMAN, R.S. & BEAMAN, J.H. 1992. *The Plants of Mount Kinabalu*. Royal Botanic Gardens, Kew.

PARTHASARATHY, N. 1999. Tree diversity and distribution in undisturbed and human-impacted sites of tropical wet evergreen forest in southern Western Ghats, India. *Biodivers. & Conservation* 8: 1365–1381.

PARTHASARATHY, N. & KARTHIKEYAN, R. 1997. Plant biodiversity inventory and conservation of two tropical dry evergreen forests on the Coromandel coast, south India. *Biodivers. & Conservation* 6: 1063–1083.

PARTHASARATHY, N., MUTHURAMKUMAR, S. & SRIDHAR REDDY, M. 2004. Patterns of liana diversity in tropical evergreen forests of peninsular India. *Forest Ecol. Managem.* 190: 15–31.

PASCAL, J.-P. 1982. Bioclimates of the Western Ghats. *Map Publ. Dep. Ecol.* 17. Institut Français de Pondichéry.

PASCAL, J.-P. 1986. *Explanatory booklet on the Forest Map of South India*. Institut Français de Pondichéry.

PASCAL, J.-P. 1988. Wet evergreen forests of the Western Ghats of India. *Map Publ. Dep. Ecol.* 18. Institut Français de Pondichéry.

PASCAL, J.-P. & PÉLISSIER, R. 1996. Structure and floristic composition of a tropical evergreen forest in south-west India. *J. Trop. Ecol.* 12: 191–214.

PASCAL, J.-P. & RAMESH, B.R. 1995. Forest map of South India – note on the sheet Bangalore-Salem. *Map Publ. Dep. Ecol.* 21. Institut Français de Pondichéry.

PASCAL, J.-P., RAMESH, B.R. & BOURGEON, G. 1988. The 'Khan' forests of the Karnataka Plateau (India): Structure and floristic composition, trends in changes due to their exploitation. *Trop. Ecol.* 29: 9–23.

PASCAL, J.-P., PÉLISSIER, R., LOFFEIER, M.E., & RAMESH, B.R. 1998. Floristic composition, structure, diversity, and dynamics of two evergreen forest plots in Karnataka State, India. In: *Forest Biodiversity Research, Monitoring and Modeling: Conceptual Background and Old World Case Studies*, eds F. Dallmeier & J.A. Comiskey, pp. 507–520. UNESCO, Paris & Parthenon, Carnforth.

PATEL, A. 1996. Variation in a mutualism: phenology and the maintenance of gynodioecy in two Indian figs. *J. Ecol.* 84: 667–680.

PATI, B.P. 2000. Impact of livelihood practices of Maldhari tribe on wildlife habitat of Gir protected area. *Indian Forester* 126: 1120–1127.

PATNAIK, S. 2002. Sustainable non-timber forest products management: challenges and opportunities. In: *Traditional Ecological Knowledge for Managing Biosphere Reserves in South and Central Asia*, P.S. Ramakrishnan, R.K. Rai, R.P.S. Katawal & S. Mehndiratta, pp. 97–127. UNESCO, Oxford & IBH, New Delhi.

PATRA, A.K., PANDE, P.K. & SAHU, T.R. 2007. Vegetation analysis of open and closed sal and miscellaneous forests of Satpura Plateau, India. *Indian Forester* 133: 611–627.

PATTANAYAK, S.K. & WENDLAND, K.J. 2007. Nature's care: diarrhea, watershed protection, and biodiversity conservation in Flores, Indonesia. *Biodivers. & Conservation* 16: 2801–2819.

PATTANAYAK, S.K., WUNDER, S. & FERRARO, P.J. 2010. Show me the money: do payments supply environmental services in developing countries. *Rev. Environ. Econ. Pol.* 4: 254–274.

PAYNE, J., FRANCIS, C.M. & PHILLIPS, K. 1985. *A Field Guide to the Mammals of Borneo*. The Sabah Society, Kota Kinabalu & World Wildlife Fund Malaysia, Kuala Lumpur.

PAYNE, N.J.A. 1985. A review of the possibilities for integrating cattle and tree crop production systems in the tropics. *Forest Ecol. Managem.* 12: 1–36.

PEACE, W.J.H. & GRUBB, P.J. 1982. Interaction of light and mineral nutrient supply in the growth of *Impatiens parviflora*. *New Phytol.* 90: 127–150.

PEACE, W.J.H. & MACDONALD, F.D. 1981. An investigation of the leaf anatomy, foliar mineral levels, and water relations of a Sarawak forest. *Biotropica* 13: 100–109.

PEARSON, T.R.H., BURSLEM, D.R.F.P., LULLINGS, C.E. & DALLING, J.W. 2002. Germination ecology of neotropical pioneers: interacting effects of environmental conditions and seed size. *Ecology* 83: 2798–2807.

PEAY, K.G., KENNEDY, P.G., DAVIES, S.J., TAN, S. & BRUNS, T.D. 2009. Potential link between plant and fungal distribution in a dipterocarp rainforest: community and phylogenetic structure of tropical ectomycorrhizal fungi across a plant soil ecotone. *New Phytol.* 185: 529–542.

PEERIS, C.V.S. (See Gunatilleke, C.V.S.)

PEITSCH, D., FIETZ, A., HERTEL, H., DE SOUZA, J., VENTURA, D.F. & MENZEL, R. 1992. The spectral input system of hymenopteran insects and their receptor-based colour vision. *J. Comp. Physiol. A* 170: 23.

PÉLISSIER, R. 1997. Hétérogénéité spatiale et dynamique d'une forêt dense humide dans les Ghats Occidentaux de l'Inde. *Publ. Dép. d'Ecol.* 37. Institut Français de Pondichéry.

PÉLISSIER, R. 1998. Tree spatial patterns in three contrasting plots of a southern Indian tropical moist evergreen forest. *J. Trop. Ecol.* 14: 1–16.

PELUSO, N.L. 1989. A history of state forest management in Java. In: *Keepers of the Forest: Land Management Alternatives in Southeast Asia*, ed. M. Poffenberger, pp. 27–55. Kumarian Press, West Hartford, Conneticut.

PEÑAFIEL, S.R. 1982. Forest fire costs and losses in the Upper Agno River Basin over a four year period. *Philippine Lumberman* 28: 20–21.

PENDRY, C.A. & PROCTOR, J. 1996. The causes of altitudinal zonation of rain forests on Bukit Belalong, Brunei. *J. Ecol.* 84: 407–418.

PENMAN, H.L. 1948. Natural evaporation from open water, bare soil and grass. *Proc. Roy. Soc. Lond. A* 193: 120–145.

PENNINGTON, R.T. & DICK, C.W. 2004. The role of immigrants in the assembly of the South American rainforest tree flora. *Philos. Trans., Ser. B* 359: 1611–1622.

PENNINGTON, R.T., LAVIN, M., PRADO, D.E., PENDRY, C.A., PELL, S.K. & BUTTERWORTH, C.A. 2004. Historical climate change and speciation: neotropical seasonally dry forest plants show patterns of both tertiary and quaternary diversification. *Philos. Trans., Ser. B* 359: 515–537.

PENNY, D. 2001. A 40,000 year palynological record from north-east Thailand; implications for biogeography and palaeo-environmental reconstruction. *Palaeogeogr. Palaeoclimatol. Palaeoecol.* 171: 97–128.

PERERA, A.H. & RAJAPAKSE, R.M.M. 1991. A baseline study of Kandyan forest gardens of Sri Lanka: structure, composition and utilization. *Forest Ecol. Managem.* 45: 269–280.

PERERA, N.P. 1975. A physiognomic vegetation map of Sri Lanka (Ceylon). *J. Trop. Ecol.* 2: 185–203.

PERERA, W.R.H. 1971–1972. A study of the protective benefits of the wet zone forestry reserves of Sri Lanka. *Ceylon Forester* 3 & 4: 87–101.

PERERA, W.R.H. 1978. Thotupolakanda – an environmental disaster? *Sri Lankan Forester* 13: 53–55.

PETERS, H.A. 2001. *Clidemia hirta* invasion at the Pasoh Forest Reserve: an unexpected plant invasion in an undisturbed tropical forest. *Biotropica* 33: 60–68.

PETERS, H.A. 2003. Neighborhood-regulated mortality: the influence of positive and negative density dependence on tree populations in species-rich tropical forests. *Ecol. Lett.* 6: 757–765.

PETHIYAGODA, R. 2007. *Pearls, Spices and Green Gold: an Illustrated History of Biodiversity Exploration in Sri Lanka*. WHT Publications, Colombo.

PETIT, R.J., HU, F.S. & DICK, C.W. 2008. Forests of the past: a window to future change. *Science* 320: 1450–1451.

PFEFFER, P. 1939. Biologie et migrations du sanglier de Bornéo. *Mammalia* 23: 277–303.

PFEIFFER, M., NAIS, J. & LINSENMAIR, K.E. 2006. Worker size and seed size selection in seed-collecting ant assemblages (Hymenoptera: Formicidae) in primary forests in Borneo. *J. Trop. Ecol.* 22: 685–693.

PHALAN, B., ONIAL, M., BALMFORD, A. & GREEN, R.E. 2011. Reconciling food production and biodiversity conservation: land sharing and land sparing compared. *Science* 333: 1289.

PHENGKLAI, C. 1989. Peat swamp forest of Thailand. *Thai Forest Bull.* 18: 1–41.

PHILIPSON, C. 2009. *Plant Growth Analysis of Bornean Dipterocarp Seedlings*. PhD thesis, University of Zurich.

PHILIPSON, W.R. 1986. Monimiaceae. In *Flora Malesiana* 1 (10), ed. C.G.G.J. van Steenis, pp. 225–236. Martinus Nijhoff, The Hague.

PHILLIPS, O.L. 1998. Increasing turnover in tropical forests as measured in permanent plots. In: *Forest Biodiversity Research, Monitoring and Modeling: Conceptual Background and Old World Case Studies*, eds F. Dallmeier & J.A. Comiskey, pp. 221–245. UNESCO, Paris & Parthenon, Carnforth.

PHILLIPS, O.L. & SHIEL, D. 1997. Forest turnover, diversity and CO_2. *Trends Ecol. Evol.* 12: 404.

PHILLIPS, O.L., HALL, P., GENTRY, A.H., SAWYER, S.A. & VASQUEZ, R. 1994. Dynamics and species richness of tropical rain forests. *Proc. Natl. Acad. Sci. U.S.A.* 91: 2805–2809.

PHILLIPS, O.L., MALHI, Y., HIGUCHI, N., LAURANCE, W.F., NUNEZ, P.V., VASQUEZ, M., LAURANCE, S.G., FERREIRA, L.V., STERN, M., BROWN, S. & GRACE, J. 1998. Changes in the carbon balance of tropical forests: evidence from long-term plots. *Science* 282: 439–442.

PHONGPHIT, S. & BENNOUN, R. 1988. *Turning Point of Thai Farms*. Institute of Rural Development. Moo Bem Press, Bangkok.

PIERCE, D., PUTZ, F.E. & VANCLAY, J.K. 2003. Sustainable forestry in the tropics: panacea or folly? *Forest Ecol. Managem.* 172: 229–247.

PIJL, L. VAN DER 1934. ber die polyembryonie bei *Eugenia*. *Recueil Trav. Bot. Néerl.* 3: 113–186.

PIJL, L. VAN DER 1953. On the floral biololgy of some plants from Java. *Ann. Bogor.* 1 (2): 77–99.

PIJL, L. VAN DER 1954. *Xylocopa* and flowers in the tropics. *Proc. Kon. Ned. Akad. Wetensch. C* 57 (3): 413–423, 552–562.

PIJL, L. VAN DER. 1956. Remarks on the pollination by bats in the genera *Freycinetia, Duabanga* and *Haplophragma*. *Acta Bot. Neerl.* 5: 135–144.

PIJL, L. VAN DER. 1957. The dispersal of plants by bats (chiropterochory). *Acta Bot. Neerl.* 6: 291–315.

PIJL, L. VAN DER. 1969, 2nd ed. 1972, 3rd ed. 1982. *Principles of Dispersal in Higher Plants*. Springer, Berlin & New York.

PIMM, S.L. & RAVEN, P.H. 2000. Extinction by numbers. *Nature* 403: 843–844.

PIMM, S.L., AYRES, M., BLAMFORD, A., BRANCH, G., BRANDON, K., BROOKS, T., BUSTAMANTE, R., COSTANZA, R., COWLING, R., CURRAN, L.M., DOBSON, A., FARBER, S., DA FONSECA, G.A.B., GASCON, C., KITCHING, R., MCNEELY, J., LOVEJOY, T., MITTERMEIER, R.A., MYSERS, N., PATZ, J.A., RAFFLE, B., RAPPORT, D., RAVEN, P., ROBERTS, C., RODRÍGUEZ, J.P., RYLANDS, A.B., TUCKER, C., SAFINA, C., SAMPER, C., STIASSNY, M.L.J., SUPRIATNA, J., WALL, D.H. & WILCOVE D. 2001. Can we defy nature's end? *Science* 293: 2207–2208.

PINARD, M. 1996. Site conditions limit pioneer tree recruitment after logging of dipterocarp forests in Sabah, Malaysia. *Biotropica* 28: 2–12.

PINARD, M., BARKER, M.G. & TAY, J. 2000. Soil disturbance and post-logging forest recovery on bulldozer paths in Sabah, Malaysia. *Forest Ecol. Managem.* 130: 213–225.

PIPER, P.J., OCHOA, J., PAZ, V., LEWIS, H. & RONQUILLO, W.P. 2008. The first evidence for the past presence of the tiger *Panthera tigris* (L.) on the island of Palawan, Philippines: extinction in an island population. *Palaeogeogr. Palaeoclimatol. Palaeoecol.* 264: 123–127.

PIPOLI, J.J. III & MADULID, D.A. 1997. The vegetation of a submontane forest on Mt. Kinasalapi, Kitangland Range, Mindanao, Philippines. In: *Plant Diversity in Malesia III, Proceedings of the Third International Flora Malesian Symposium 1995*, eds J. Dransfield, M.J.E. Coode & D.A. Simpson, pp. 177–185. Royal Botanic Gardens, Kew.

PITMAN, N.C.A., TERBORGH, J.W., SILMAN, M.R., NÚÑEZ, P.V., NEILL, D.A., CERÓN, C.E., PALACIOS, W.A. & AULESTIA, M. 2001. Dominance and distribution of tree species in upper Amazonian terra firme forests. *Ecology* 82: 2101–2117.

PLOTKIN, J.B., POTTS, M.D., LESLIE, N., MANOKARAN, N., LAFRANKIE, J.V. & ASHTON, P.S. 2000a. Species-area curves, spatial aggregation, and habitat specialization in tropical forests. *J. Theor. Biol.* 207: 81–99.

PLOTKIN, J.B., POTTS, M.D., YU, D.W., BUNYAVEJCHEWIN, S., CONDIT, R., FOSTER, R., HUBBELL, S., LAFRANKIE, J., MANOKARAN, N., SENG, L.H., SUKUMAR, R., NOWAK, M.A. & ASHTON, P.S. 2000b. Predicting species diversity in tropical forests. *Proc. Natl. Acad. Sci. U.S.A.* 97: 10850–10854.

PLOTKIN, J.B., CHAVE, J. & ASHTON, P.S. 2002. Cluster analysis of spatial patterns in Malaysian tree species. *Amer. Naturalist* 160: 629–644.

POFFENBERGER, M., ed. 1989a. *Keepers of*

the Forest: Land Management Alternatives in Southeast Asia. Kumarian Press, West Hartford, Conneticut.

POFFENBERGER, M. 1989b. Introduction. In: Keepers of the Forest: Land Management Alternatives in Southeast Asia, ed. M. Poffenberger, pp. 27–55. Kumarian Press, West Hartford, Conneticut.

POFFENBERGER, M. 1995. India's forest keepers. Wasteland News (Aug–Oct.): 65–79.

POFFENBERGER, M. 2000. Communities and Forest Management in South Asia. Working Group on Community Involvement in Forest Management.

POOMA, R. 2002. Further notes on Thai Dipterocarpaceae. Thai Forest Bull. 30: 7–27.

POORE, D., BURGESS, P., PALMER, J., RIETBERGEN, S. & SYNNOTT, T. 1989. No Timber Without Trees. Sustainability in the Tropical Forest. Earthscan, London.

POORE, M.E.D., BURGESS, P., PALMER, J., RIETBERGEN, S. & SYNOTT, T. 1988. No Timber Without Trees. ITTO, Yokahama.

POTTS, M.D. 2003. Drought in a Bornean everwet rain forest. J. Ecol. 91: 467–474.

POTTS, M.D. & VINCENT, J.R. 2008a. Spatial distribution of species populations, relative economic values, and the optimal size and numbers of reserves. Environ. Resource Econ. 39: 91–112.

POTTS, M.D. & VINCENT, J.R. 2008b. Harvest and extinction in multi-species ecosystems. Ecol. Econ. 65: 336–347.

POTTS, M.D., ASHTON, P.S., KAUFMAN, L.S. & PLOTKIN, J.B. 2002. Habitat patterns in tropical rain forests: a comparison of 105 plots in northwest Borneo. Ecology 83: 2782–2797.

POTTS, M.D., DAVIES, S.J., BOSSERT, W.H., TAN, S. & NUR SUPARDI, M.N. 2004. Habitat heterogeneity and niche structure of trees in two tropical rain forests. Oecologia 139: 446–453.

POTTS, M.D., KASSIM, A.R., SUPARDI, M.N., TAN, S. & BOSSERT, W.H. 2005. Sampling tree diversity in Malaysian forests: an evaluation of a pre–felling inventory. Forest Ecol. Managem. 205: 385–395.

POULSEN, A.D. 1996. Species richness and density of ground herbs within a plot of lowland rainforest in north–west Borneo. J. Trop. Ecol. 12: 177–190.

POULSEN, A.D. & PENDRY, C.A. 1995. Inventories of ground herbs at three altitudes on Bukit Belalong, Brunei, Borneo. Biodivers. & Conservation 4: 745–757.

POULSEN, A., NIELSEN, I., TAN, S. & BALSLEV, H. 1996. A quantitative inventory of trees in one hectare of mixed dipterocarp forest in Temburong, Brunei Darussalam. In: Tropical Rainforest Research – Current Issues, eds D.S. Edwards, W.E. Booth & S.C. Choy, pp. 139–150. Kluwer, Dordrecht.

PRADHAN, R. 2001. 1. Trumsingla National Park Vegetation Survey Report, Oct. 2000. Unpubl. ms.

PRAKASH, I. 1994. Biodiversity conservation in the Thar desert. Indian Forester 120: 873–879.

PRANCE, G.T., ed. 1986. Tropical Rain Forests and the World Atmosphere. AAAS Symposium 101. Westview Press, Colorado.

PRASAD, R. & JAMALUDDIN. 1985. Preliminary observations on sal mortality in Madhya Pradesh. Indian Forester 111: 250–271.

PRASAD, S. & SUKUMAR, R. 2010. Context-dependency of a complex fruit-frugivore mutualism: temporal variation in crop size and neighborhood effects. Oikos 119: 514–523.

PRASAD, S., PITTET, A. & SUKUMAR, R. 2010. Who really ate the fruit? A novel approach to camera trapping for quantifying frugivory by ruminants. Ecol. Res. 25: 225–231.

PRASAD, U.K., MUNIAPPAN, R., FERRAR, P., AESCHLIMANN, J.P. & DE FORESTA, H., eds. 1996. Distribution, ecology, and management of Chromolaena odorata. Proceedings of the Third International Chromolaena Workshop, Abidjan, Cote d'Ivoire, 1993. Agricultural Experimental Station, University of Guam, Publication 202.

PRASAD, V., FAROOQUI, A., TRIPATHI, S.K.M., GARG, R. & THAKUR, B. 2009. Evidence of late Palaeocene-Eocene equatorial rain forest refugia in southern Western Ghats, India. J. Bioscience 34: 777–797.

PRELL, W.L. 1984. Monsoon climates of the Arabian Sea during the late Quaternary: a response to changing solar radiation. In: Milankovich and Climate, eds A. Berger, J. Lubrie, J. Hays, J. Kukla & B. Saltzman, pp. 349–336. Reidel, Dordrecht.

PRIMACK, R.B. 1993. Essentials of Conservation Biology. Sinauer, Sunderland, Massachusetts.

PRIMACK, R.B. & LEE, H.S. 1991. Population dynamics of pioneer (Macaranga) and understorey (Mallotus) trees (Euphorbiaceae) in primary and selectively logged Bornean rain forests. J. Trop. Ecol. 7: 439–458.

PRIMACK, R.B. & LOVEJOY, T.E., eds. 1995. Ecology, Conservation and Management of Southeast Asian Rainforests. Yale University Press, Connecticut.

PRIMACK, R.B. & TIEH, F. 1994. Long-term timber harvesting in Bornean forests: the Yong Khow case. J. Trop. For. Sci. 7: 262–279.

PRIMACK, R.B., ASHTON, P.S., CHAI, P. & LEE, H.S. 1985. Growth rates and population structure of Moraceae trees in Sarawak, East Malaysia. Ecology 66: 577–588.

PRIMACK, R.B., CHAI, E.O.K., TAN, S.S. & LEE, H.S. 1987a. The silviculture of dipterocarp trees in Sarawak, Malaysia. I. Introduction to the series and performance in primary forest. Malayan Forester 50: 29–42.

PRIMACK, R.B., CHAI, E.O.K., TAN, S.S. & LEE, H.S. 1987b. The silviculture of dipterocarp trees in Sarawak, Malaysia. II. Improvement felling in primary forest stands. Malayan Forester 50: 43–61.

PRIMACK, R.B., HALL, P. & LEE, H.S. 1987c. The silviculture of dipterocarp trees in Sarawak, Malaysia. IV. Seedling establishment and adult regeneration in a selectively logged forest and three primary forests. Malayan Forester 50: 162–178.

PROCTOR, J. 1999. Heath forests and acid soils. Bot. J. Scotland 51: 1–14.

PROCTOR, J., ANDERSON, J.M., CHAI, P. &

VALLACK, H.W. 1983a. Ecological studies in four contrasting lowland forests in Gunung Mulu National Park, Sarawak. I. Forest environment, structure and floristics. J. Ecol. 71: 261–283.

PROCTOR, J., ANDERSON, J.M., FOGDEN, S.C.L. & VALLACK, H.W. 1983b. Ecological studies in four contrasting lowland forests in Gunung Mulu National Park, Sarawak. II. Litterfall, litter standing crop and preliminary observations on herbivory. J. Ecol. 71: 261–283.

PROCTOR, J., LEE, Y.F., LANGLEY, A.M., MUNRO, W.R.C. & NELSON, T. 1988. Ecological studies on Gunung Silam, a small ultrabasic mountain in Sabah, Malaysia. I. Environment, forest structure and floristics. J. Ecol. 76: 320–340.

PROCTOR, J., PHILLIPS, C., DUFF, G., HEANEY, A. & ROBERTSON, F.M. 1989. Ecological studies on Gunung Silam, a small ultrabasic mountain in Sabah, Malaysia. II. Some forest processes. J. Ecol. 77: 317–331.

PROCTOR, J., HARIDASAN, K. & SMITH, G.W. 1998. How far north does lowland evergreen tropical rain forest go? Global Ecol. Biogeog. 7: 141–146.

PROCTOR, J., BREARLEY, F.Q., DUNLOP, H., PROCTOR, K. SUPRAMONO & TAYLOR, D. 2001. Local wind damage in Barito Ulu, Central Kalimantan: a rare but essential event in a lowland dipterocarp forest. J. Trop. Ecol. 17: 473–475.

PROCTOR, M., YEO, P. & LACK, A. 1996. The Natural History of Pollination. Harper Collins, London.

PROMDEJ, C., CHAEMCHUMROON, V. SATAPORN, P. & DEJPOPHRA, S. Undated. A Preliminary Study of the Vegetation of the Peninsular Botanic Garden (Thung Khai.).Unpubl. ms.

PUBELLIER, M., QUEBRAL, R., AURELIO, M. & RANGIN, C. 1996. Docking and post-docking escape tectonics in the southern Philippines. In: Tectonic Evolution of Southeast Asia, eds R. Hall & D. Blundell. Geological Society Special Publication 106: 511–523. Geological Society, London.

PUBELLIER, M., GIRARDEAU, J. & TJASHURI, T. 1999. Accretion history of Borneo inferred from the polyphase structural features in the Meratus Mountains. In: Gondwana Dispersion and Asian Accretion, eds I. Metcalfe, J.-S. Ren, J. Charvet, & S. Hada, pp. 141–160. A.A. Balkema, Rotterdam.

PULLIAM, H.R. 1988. Sources, sinks and population regulation. Amer. Naturalist 132: 652–661.

PURI, G.S. & MAHAJAN S.D. 1960. Study of evergreen vegetation of Mahableshwar. Bull. Bot. Surv. India 2: 109–137.

PURI, G.S., GUPTA, R.K., MEHER–HOMJI, V.M. & PURI S. 1983. Forest Ecology. Vol. 2. Plant Form, Diversity, Communities and Succession. 2nd ed. Oxford & IBH, New Delhi.

PURI, G.S., MEHER–HOMJI, V.M., GUPTA, R.K. & PURI, S. 1989. Forest Ecology. Vol. 1. Phytogeography and Forest Conversion. 2nd ed. Oxford & IBH, New Delhi.

PUTZ, F.E. & CHAI, P. 1987. Ecological studies of lianas in Lambir National Park, Sarawak, Malaysia. J. Ecol. 75: 523–531.

PUTZ, F.E. & REDFORD K.H. 2009. Dangers of carbon-based conservation. Global Environ. Change 19: 400–401.

PUTZ, F.E. & VIANA, V. 1996. Biological challenges for certification of tropical timber. Biotropica 28: 323–330.

PUTZ, F.E., SIST, P., FREDERICKSEN, T., DYKSTRA, D. 2008. Reduced-impact logging: challenges and opportunities. Forest Ecol. Managem. 256: 1427–1433.

PYNE, S.J. 1990. Fire conservancy: the origins of wildland fire protection in British India, America and Australia. In: Fire in the Tropical Biota: Ecosystem Processes and Global Challenges, ed. J.G. Goldammer, pp. 319–336. Springer.

QIAN, H. 2002. Floristic relationships between eastern Asia and North America: test of Grey's hypothesis. Amer. Naturalist 160: 317–332.

QIAN, H. & RICKLEFS, R.E. 1999. A comparison of the taxonomic richness of vascular plants in China and The United States. Amer. Naturalist 154: 160–181.

QIAN, H. & RICKLEFS, R.E. 2000. Large-scale processes and the Asian bias in species diversity in temperate plants. Nature 407: 180–182.

QIAN, H. & RICKLEFS, R.E. 2004. Taxonomic richness and climate in angiosperms: is there a globally consistent relationship which precludes regional effects? Amer. Naturalist 163: 773–779.

QIU, J. 2009. Where the rubber meets the garden. Nature 457, Special Report 246–247.

QUADE, J., CERLING, T.E. & BOWMAN, J.R. 1989. Development of Asian monsoon revealed by marked ecological shift during the latest Miocene in northern Pakistan. Nature 342: 163–165.

QUEK, S.-P., DAVIES, S.J., ASHTON, P.S., ITINO, T. & PIERCE, N.E. 2007. The geography of diversification in mutualistic ants: a gene's-eye view into the Neogene history of Sundaland rain forests. Molec. Ecol. 16: 2045–2062.

QUINN, W. 1992. A study of Southern Oscillation–related climatic activity for AD 622–1900 incorporating Nile River flood data. In: El Niño. Historical and Paleoclimatic Aspects of the Southern Oscillation, eds H.F. Diaz & V. Markgraf, pp. 119–149. Cambridge University Press.

QURESHI, I.M., SUBBA RAO, B.K. & DABRAL, B.G. 1966. Frost at New Forest, Dehra Dun during 1963–64. Indian Forester 92: 253.

RABINOWITZ, D. 1981. Seven forms of rarity. In: The Biological Aspects of Rare Plant Conservation, ed. H. Synge, pp. 207–217. Wiley, Chichester.

RACEY, A., GOODALL, J.G.S., LOVE, M.A., POLACHAN, S. & JONES, P.D. 1994. New age data for the Mesozoic Khorat Group of Northeast Thailand. Proceedings of the International Symposium on Stratigraphic Correlation of Southeast Asia, Bangkok, November 1994: 245–252.

RAES, N. 2009. Borneo: a Quantitative Analysis of Botanical Richness, Endemicity and Floristic Regions Based on Herbarium records. National Herbarium of the Netherlands, Leiden.

RAES, N., ROOS, M.C., SLIK, J.W.F., VANLOON, E.E. & TER STEEGE, H. 2009. Botanical richness and endemicity pattern of Borneo derived from species distribution models. *Ecography* 32: 180–192.

RAI, S.C. & SHARMA, P. 2003. Carbon sequestration with forestry and land use/cover change: an overview. *Indian Forester* 129: 776–786.

RAI, S.N. 1989. Tropical forests of India – their management and regeneration. *Indian Forester* 115: 82–88.

RAI, S.N. 1990. Restoration of degraded tropical rain forests of Western Ghats. *Indian Forester* 116: 179–188.

RAI, S.N. 2000. *Productivity of Tropical Rain Forests of Karnataka*. Punarvasu Publisher, Dharwad, Karnataka.

RAI, S.N. & PROCTOR, J. 1986. Ecological studies on four rain forests in Karnataka, India. I. Environment, structure, floristics and biomass. *J. Ecol.* 74: 439–454. II. Litterfall. *Ibid.*: 455–463.

RAI, S.N. & SAXENA, A. 1997. The extent of forest fires, grazing and regeneration status of inventoried forest areas of India. *Indian Forester* 123: 689–702.

RAICH, J.W. 1989. Seasonal and spatial variation in the light environment in a tropical dipterocarp forest and gaps. *Biotropica* 21: 299–302.

RAICH, J.W. & GONG, W.K. 1990. Effects of canopy tree opening on tree seed germination in a Malaysian dipterocarp forest. *J. Trop. Ecol.* 6: 203–217.

RAINA, A.K. & JHA, M.N. 2005. Inter-relationship between geology, soil and vegetation in Raipur range of Mussoorie Forest Division, Urranchal. *Indian Forester* 131: 1201–1211.

RAINTREE, J.B. & WARNER, K. 1986. Agroforestry pathways for the intensification of shifting cultivation. *Agrofor. Systems* 4: 39–54.

RAJ, A. 1979. *Onset of Effective Monsoon and Critical Dry Spells: a Computer Based Forecasting Technique*. Water Technology Centre, Indian Agricultural Research Institute, New Delhi.

RAJAKARUNA, N. & BAKER, A.J.M. 2004. Serpentine: a model habitat for botanical research in Sri Lanka. *Ceyl. J. Sci., Biol. Sci.* 32: 1–19.

RAJKHOWA, S. 1986. Problems of forestry in Assam. *Indian Forester* 112: 616–620.

RAM, N. & MAZUMDAR, S.N. 2006. Microclimate change in the Darjeeling Himalayas. *Indian Forester* 132: 1673–1688.

RAM DAHAL, G. & CHAPAGAIN, A. 2008. Community forestry in Nepal: decentralized forest governance. In: *Lessons from Forest Decentralization: Money, Justice and the Quest for Good Governance in Asia-Pacific*, eds C.J.P. Colfer, G.R. Dahal & D. Capistrano, pp. 67–82. Earthscan, London.

RAMAKRISHNAN, P.S., ed. 1991. *Ecology of Biological Invasions in the Tropics*. International Scientific Publications, New Delhi.

RAMAKRISHNAN, P.S. 1992. *Shifting Agriculture and Sustainable Development: an Interdisciplinary Study from North-Eastern India*. UNESCO-MAB. Academic Press, Paris & Carnforth.

RAMAKRISHNAN, P.S. 1998. Ecology, economics and ethics: some key issues relevant to natural resource management in developing countries. *Int. J. Soc. Econ.* 25: 207–227.

RAMAKRISHNAN, P.S. 2002a. Linking natural resource management with sustainable development of traditional mountain societies. In: *Traditional Ecological Knowledge, Conservation of Biodiversity and Sustainable Development*, eds D. Depommier & P.S. Ramakrishnan, pp. 9–17. Institut Français de Pondichéry.

RAMAKRISHNAN, P.S. 2002b. *Traditional Ecological Knowledge for Managing Biosphere Reserves in South and Central Asia*. Oxford & IBH Publishing Co., New Delhi.

RAMAKRISHNAN, P.S. 2007. Traditional forest knowledge and sustainable forestry: a north-east India perspective. *Forest Ecol. Managem.* 249: 91–99.

RAMAKRISHNAN, P.S., SAXENA, K.G. & CHANDRA SEKARA, U.M., eds. 1998. *Conserving the Sacred for Biodiversity Management*. Oxford & IBH Publishing Co., New Delhi.

RAMAKRISHNAN, P.S., RAI, R.K., KATAWAL, R.P.S. & MEHNDIRATTA, S. 2002. *Traditional Ecological Knowledge for Managing Biosphere Reserves in South and Central Asia*. Oxford & IBH Publishing Co., New Delhi.

RAMANUJAN, M.P. & CYRIL, K.P.K. 2003. Woody species diversity of four sacred groves in the Pondicherry region of South India. *Biodivers. & Conservation* 12: 289–299.

RAMESH, B.R. & PASCAL, J.P. 1997. *Atlas of Endemics of the Western Ghats (India)*. Institut Français de Pondichéry.

RAMESH, B.R., PASCAL, J.P. & DE FRANCESCHI, D. 1991. Distribution of endemic, arborescent evergreen species in the Western Ghats. *Proceedings of Rare, Endangered, Endemic Species of the Western Ghats*, pp. 20–29. Kerala Forest Department, Trivandrum.

RAMESH, B.R., VENUGOPAL, P.D., PÉLISSIER, R., PATIL, S.V., SWAMINATH, M.H. & COUTERON, P. 2010. Mesoscale patterns in the floristic composition of forests in the Central Western Ghats of Karnataka, India. *Biotropica* 42: 435–443.

RANA, B.S., SINGH, S.P. & SINGH, R.P. 1988. Biomass and productivity of Central Himalayan sal (*Shorea robusta*) forest. *Trop. Ecol.* 29: 1–5.

RANGAIAH, K., SOLOMON RAJU, A.J. & RAO, S.P. 2004. Passerine bird-pollination in the Indian coral tree, *Erythrina variegata* var. *orientalis* (Fabaceae). *Curr. Sci.* 87: 736–739.

RAO, K.S. & RAMAKRISHNAN, P.S. 1987. Comparative analysis of the population dynamics of two bamboo species, *Dendrocalamus hamiltonii* and *Neohouzeua dulloa*, in a successional environment. *Forest Ecol. Managem.* 21: 177–189.

RATNAYAKE, R.M.C., GUNATILLEKE, I.A.U.N., WIJESUNDERE, D.S.A. & SAUNDERS, R.M.K. 2006. Reproductive biology of two sympatric species of *Polyalthia* (Annonaceae) in Sri Lanka. I. Pollination by curculionid weevils, *Int. J. Pl. Sci.* 167: 483–493; II. Breeding systems and population genetic structure. *Ibid.*: 495–502.

RAUNKIAER, C. 1934. *The Life Forms of Plants and Statistical Plant Geography*. Oxford University Press.

RAUTIAINEN, O. & SUOHEIMO, J. 1997. Natural generation potential and early development of *Shorea robusta* Gaertn. f. forest after regeneration felling in the Bhabar-Terai zone of Nepal. *Forest Ecol. Managem.* 92: 243–251.

RAVEN, P.H. 1964. Catastrophic selection and edphic endemism. *Evolution* 18: 336–338.

RAVEN, P.H. & AXELROD, D.I. 1974. Angiosperm biogeography and past continental movements. *Ann. Missouri Bot. Gard.* 61: 539–673.

RAVINDRANATH, N.H. & SUDHA, P. 2004. *Joint Forest Management in India: Spread, Performance and Impact*. Universities Press (India), Hyderabad.

RAVINDRANATH, N.H., SOMASHEKAR, B.S. & GADGIL, M. 1997. Carbon flow in Indian forests. *Climate Change* 35: 297–320.

RAWAT, G.S. 1998. 50 years of forests and forestry in India after independence. *Indian Forester* 124: 367–380.

READ, J., HOPE, G. & HILL, R. 1990. The dynamics of some *Nothofagus*-dominated forests in Papua New Guinea. *J. Biogeogr.* 17: 185–204.

REDDY, M.S. & PARTHASARATHY, N. 2003. Liana diversity and distribution in four tropical dry evergreen forests on the Coromandel Coast of South India. *Biodivers. & Conservation* 12: 1609–1627.

REEKIE, E.G. & BAZZAZ, F., eds. 2005. *Reproductive Allocation in Plants*. Elsevier.

REICH, P.B. & BORCHERT, R. 1982. Phenology and ecophysiology of the tropical tree *Tabebuia neochrysantha* (Bignoniaceae). *Ecology* 63: 294–299.

RENNER, S.S. & RICKLEFS, R.E. 1995. Dioecy and its correlates in the flowering plants. *Amer. J. Bot.* 82: 596–606.

RENNER, S.S. & SCHAEFER, H. 2010. The evolution and loss of oil-offering flowers – new insights from dated phylogenies for plants and bees. *Philos. Trans., Ser. B* 365: 423–435.

RENNER, S.S., STRIJK, J.S., STRASBERG, D. & THÉBAUD, C. 2010. Biogeography of the Monimiaceae (Laurales): a role for East Gondwana and long-distance dispersal, but not West Gondwana. *J. Biogeogr.* 37: 1227–1238.

RENNOLLS, K. & LAUMONIER, Y. 2000. Species diversity structure at two sites in the tropical rain forest of Sumatra. *J. Trop. Ecol.* 16: 253–270.

RENNOLLS, K. & LAUMONIER, Y. 2006. A new estimator of regional species diversity in terms of 'shadow species', with a case study from Sumatra. *J. Trop. Ecol.* 22: 313–320.

REPETTO, R. 1990. Deforestation in the tropics. *Sci. Amer.* 262 (4): 36–42.

RICE, R., GULLISON, R.E. & REID, J.W. 1997. Can sustainable management save tropical forests? *Sci. Amer.* 276 (4): 44–49.

RICHARDS, A.J. 1990a. Studies in *Garcinia*, dioecious tropical forest trees: agamospermy. *Bot. J. Linn. Soc.* 103: 233–250.

RICHARDS, A.J. 1990b. Studies in *Garcinia*, dioecious tropical forest trees: the phenology, pollination biology and fertilization of *G. hombroniana* Pierre. *Bot. J. Linn. Soc.* 103: 251–261.

RICHARDS, A.J. 1990c. Studies in *Garcinia*, dioecious tropical forest trees: the origin of the mangosteen (*Garcinia mangostana* L.). *Bot. J. Linn. Soc.* 103: 301–308.

RICHARDS, A.J., KIRSCHNER, J., ŠTEPÁNIK, J. & MARHOLD, K., eds. 1996. *Apomixis and Taxonomy: Special Features in Biosystematics and Biodiversity*. No. 1. Institute of Botany. Opulus Press.

RICHARDS, P.W. 1958, 2nd ed. 1996. *The Tropical Rain Forest. An Ecological Study*. Cambridge University Press.

RICHARDSON, J.E., PENNINGTON, R.T., PENNINGTON, T.D. & HOLLINGSWORTH, P.M. 2001. Rapid diversification of a species-rich genus of neotropical rain forest trees. *Science* 293: 2242–2243.

RICKLEFS, R.E. 2004. A comprehensive framework for global patterns in biodiversity. *Ecol. Lett.* 7:1–5.

RICKLEFS, R.E., LATHAM, R.E. & QIAN, H. 1999. Global patterns of tree species richness in moist forests: distinguishing ecological influences and historical contingency. *Oikos* 86: 369–373.

RICKLEFS, R.E. & SCHLUTER, D., eds. 1993a. *Species Diversity in Ecological Communities. Historical and Geographical Perspectives*. University of Chicago Press.

RICKLEFS, R.E. & SCHLUTER, D. 1993b. Species diversity: regional and historical influences. In: *Species Diversity in Ecological Communities. Historical and Geographical Perspectives*, eds R.E. Ricklefs & D. Schluter, pp. 350–363. University of Chicago Press.

RIDLEY, H.N. 1930. *The Dispersal of Plants Throughout the World*. Reeve & Co., Kent.

RIELEY, J.O. & PAGE, S.E., eds. 1997. *Biodiversity and Sustainability of Tropical Peatlands*. Samara, Cardigan.

RIELEY, J.O., PAGE, S.E., LIMIN, S.H. & WINARTI, S. 1997 The peatland resource of Indonesia and the Kalimantan peat swamp forest project. In: *Biodiversity and Sustainability of Tropical Peatlands*, eds J.O. Rieley & S.E. Page, pp. 37–44. Samara, Cardigan.

RIGBY, J. F. 1998. Upper Palaeozoic floras of S.E. Asia. In: *Biogeography and Geological Evolution of S.E. Asia*, eds R. Hall & J.D. Holloway, pp. 73–82. Backhuys, Leiden.

RIJSOORT, J.V. & ZHANG, J. 2005. Participatory resource monitoring as a means for promoting social change in Yunnan, China. *Biodivers. & Conservation* 14: 2543–2573.

RIST, L., UMA SHAANKER, R., MILNER-GULLAND, E.J. & GHAZOUL, J. 2008. Managing mistletoes: the value of local practices for a non-timber forest resource. *Forest Ecol. Managem.* 255: 1684–1691.

RIST, L., UMA SHAANKER, R., MILNER-GULLAND, E.J. & GHAZOUL, J. 2010. The use of traditional ecological knowledge in forest management: an example from India. *Ecology & Society* 15 (1).

RISWAN, S. 1982. *Ecological Studies on Primary, Secondary and Experimentally Cleared Mixed Dipterocarp and Kerangas Forest in East Kalimantan, Indonesia*. PhD thesis, University of Aberdeen.

RISWAN, S. & KARTAWINATA, K. 1991. Species strategy in early stages of secondary succession associated with soil properties status in a lowland mixed dipterocarp forest and kerangas forest in East Kalimantan. *Tropics* 1: 13–34.

ROBBINS, R.G. 1960. Montane formations in the Central Highlands of New Guinea. *Proceedings of the Symposium on Vegetation in the Tropics: Tjiawi (Indonesia)*: 176–195.

UNESCO Science Cooperation Office for Southeast Asia.

ROBBINS, R.G. & KAOSA-ARD, A. 1981. Teak (*Tectona grandis* Linn. *f.*). Its natural distribution and related factors. *Nat. Hist. Bull. Siam. Soc.* 29: 55–74.

ROBBINS, R.G. & SMITINAND, T. 1966. A botanical ascent of Doi Inthanond. *J. Siam. Soc.* 21: 205–233.

ROBERT, A. & MORAVIE, M.-A. 2003. Topographic variation and stand heterogeneity in a wet evergreen forest of India. *J. Trop. Ecol.* 19: 697–707.

ROBERTSON, I., FROYD, C.A., WALSH, R.P.D., NEWBERY, D.M., WOODBORNE, S. & ONG, R.C. 2004. The dating of dipterocarp tree rings: establishing a record of carbon cycling and climatic change in the tropics. *J. Quaternary Sci.* 19: 6567–6664.

ROBINSON, M.E. 1935. The flowering of strobilanths in 1934. *J. Bombay Nat. Hist. Soc.* 38: 117–122.

RODGERS, W.A. 1986. The role of fire in the management of wildlife habitats: a review. *Indian Forester* 112: 846–857.

RODGERS, W.A. 1991. Forest preservation plots in India. I. Management status and value. *Indian Forester* 117: 425–433.

ROGSTAD, S.H. 1986. *A Biosystematic Investigation of the* Polyalthia hypoleuca *Complex (Annonaceae) of Malesia*. PhD thesis, Harvard University.

ROGSTAD, S.H. 1990. The biosystematics and evolution of the *Polyalthia hypoleuca* species complex (Annonaceae) of Malesia. II. Comparative distributional ecology. *J. Trop. Ecol.* 6: 387–408.

ROGSTAD, S.H. 1994. The biosystematics and evolution of the *Polyalthia hypoleuca* complex (Annonaceae) in Malesia. III. Floral ontogeny and breeding systems. *Amer. J. Bot.* 81: 145–154.

ROMAN, L., SCATENA, F.N. & BRUIJNZEEL, L.A. 2010. Global and local variation in tropical montane forest soils. In: *Tropical Montane Cloud Forest. Science for Conservation and Management*, eds L.A. Bruijnzeel, F.N. Scatena & L.S. Hamilton, pp. 77–89. Cambridge University Press.

ROMM, J. 1980a. Towards a research agenda for social forestry. *Indian Forester* 106: 164–188.

ROMM, J. 1980b. Assessing the benefits and costs of social forestry projects. *Indian Forester* 106: 445–455.

ROMM, J. 1981. The uncultivated half of India. *Indian Forester* 107: 1–23, 69–86.

RONSTED, N., WEIBLEN, G.D., CLEMENT, W.L., ZEREGA, N.J.C. & SAVOLAINEN, V. 2008. Reconstructing the phylogeny of figs (*Ficus*, Moraceae) to reveal the history of the fig pollinator mutualism. *Symbiosis* 45: 45–56.

ROOS, M.C., KESSLER, P.J.A., GRADSTEIN, S.R. & BAASA, P. 2004. Species diversity and endemism of five major Malesian islands; diversity-area relationships. *J. Biogeogr.* 31: 1893–1908.

ROPELEWSKI, C.F. & HALPERT, M.S. 1989. Precipitation patterns associated with the high index phase of the Southern Oscillation. *J. Climate* 2: 268–284.

ROSAYRO, R.A. DE. 1950. Ecological conceptions and vegetational types with special reference to Ceylon. *Trop. Agric. (Ceylon)* 56: 108–121.

ROSAYRO, R.A. DE. 1955. The silviculture and management of tropical rain forest with special reference to Ceylon. *Ceylon Forester* 2: 5–26.

ROSAYRO, R.A. DE. 1956. Tropical ecological studies in Ceylon. *Ceylon Forester* 2: 168–175.

ROSENZWEIG, M.C. & ABRAMSKY, Z. 1993. How are diversity and productivity related? In: *Species Diversity in Ecological Communities. Historical and Geographical Perspectives*, eds R.E. Ricklefs & D. Schluter, pp. 52–65. University of Chicago Press.

ROSS, S.M. & DYKES, A. 1996. Soil conditions, erosion and nutrient loss on steep slopes under mixed dipterocarp forest in Brunei Darussalam. In: *Tropical Rainforest Research – Current Issues*, eds D.S. Edwards, W.E. Booth & S.C. Choy, pp. 259–270. Kluwer, Dordrecht.

ROTH, D. 1983. Philippines forests and forestry 1565–1920. In: *Global Deforestation and the Nineteenth Century World Economy*, eds R. Tucker & J.F. Richards, pp. 30–49. Duke University Press, Durham, N.C.

ROTHSCHILD, M. 1987. Changing conditions and conservation at Ashton Wold, the birthplace of the SPNR. *Biol. J. Linn. Soc.* 32: 161–170.

ROUBIK, D.W. 1993. Direct cost of forest reproduction, bee-cycling and the efficiency of pollination modes. *J. Biosci.* 18: 537–552.

ROUBIK, D.W., INOUE, T. & HAMID, A.A. 1995. Canopy foraging by tropical honeybees. Bee height fidelity and tree genetic neighborhoods. *Tropics* 5: 81–93.

ROUBIK, D.W., SAKAI, S. & HAMID, A.A., eds. 2005. *Pollination Ecology and the Rain Forest*. Springer, New York.

ROYEN, P. VAN. 1967. Some observations on the alpine vegetation of Mount Biota (Papua). *Acta Bot. Neerl.* 15: 530–534.

ROYEN, P. VAN, ed. 1980a. *The Alpine Flora of New Guinea*. Cramer Verlag, Vaduz.

ROYEN, P. VAN. 1980b. General Part. Definition and descriptions of high altitude vegetation types. In: *The Alpine Flora of New Guinea* 1, ed. P. van Royen, pp. 11–14. Cramer Verlag, Vaduz.

ROZARIA, S.A. 1995. Association between mites and leaf domatia: evidence form Bangladesh, South Asia. *J. Trop. Ecol.* 11: 99–108.

RUANGPANIT, N. 1995. Tropical seasonal forests in monsoon Asia: with special emphasis on continental southeast Asia. *Vegetatio* 121: 31–40.

RUNDEL, P.W. 1999. *Forest Habitats and Flora in Lao PDR, Cambodia and Vietnam. Conservation Priorities in Indochina* – WWF Desk Study. WWF, Indochina Programme Office, Hanoi.

RUNDEL, P.W., PATTERSON, M.T. & BOONPRAYOB, K. 2004. Photosynthetic responses to light and the ecological dominance of *Hopea ferrea* (Dipterocarpaceae) in a semi-evergreen forest of northern Thailand. *Nat. Hist. Bull. Siam Soc.* 52: 55–70.

RUSSELL, A.E. & VITOUSEK, P.M. 1997. Decomposition and potential nitrogen fixation in *Dicranopteris linearis* litter in Mauna Loa, Hawai'i. *J. Trop. Ecol.* 13: 579–594.

RUSSELL, E., RAICH, J.A. & VITOUSEK, P.M.

1998. The ecology of the climbing fern *Dicranopteris linearis* on windward Mona Loa, Hawaii. *J. Ecol.* 86: 765–779.

RUSSO, S.E. 2003. Responses of dispersal agents to tree and fruit traits in *Virola calophylla* (Myristicaceae): implications for selection. *Oecologia* 136: 80–87.

RUSSO, S.E. & AUGSPURGER, C.K. 2004. Aggregated seed dispersal by spider monkeys limits recruitment to clumped patterns in *Virola calophylla*. *Ecol. Lett.* 7: 1058–1067.

RUSSO, S.E., DAVIES, S.J., KING, D.A. & TAN, S. 2005. Soil-related performance variation and distribution of tree species in a Bornean rain forest. *J. Ecol.* 93: 879–889.

RUSSO, S.E., BROWN, P., TAN, S. & DAVIES, S.J. 2008. Interspecific demographic trade-offs and soil-related habitat associations of tree species along resource gradients. *J. Ecol.* 96: 192–203.

RUSSO, S.E., CANNON, W.L., BLOWSKY, C., TAN, S. & DAVIES, S.J. 2010. Variation in leaf structural traits of 28 tree species in relation to gas exchange along an edaphic gradient in a Bornean rain forest. *Amer. J. Bot.* 97: 1109–1120.

RUSSO, S.E., LEGGE, R., WEBER, K.A., BRODIE, J., GOLDFARB, K.C., BENSON, A.K. & TAN, S. 2012. Bacterial community structure of contrasting soils underlying Bornean rain forests: inferences from microarray and next-generation sequencing methods. *Soil Biol. Biochem.* 55: 48–59.

RUST, J., SINGH, H., RANA, R.S., McCANN, T., SINGH, L., ANDERSON, K., SARKAR, N., NASCIMBENE, P.C., STEBNER, F., THOMAS, J.C., KRAEMER, M.S., WILLIAMS, C.J., ENGEL, M.S., SAHNI, A. & GRIMALDI, D. 2010. Biogeographic and evolutionary implications of a diverse paleobiota in amber from the early Eocene of India. *Proc. Natl. Acad. Sci. U.S.A.* 107: 18360–18365.

SACK, L. & GRUBB, P.J. 2001. Why do species of woody species change rank in relative growth rate between low and high irradiance? *Funct. Ecol.* 15: 145–154.

SAGREIYA, K.P. 1939. Making better grazing available in open pasture forests. *Indian Forester* 65: 425–433.

SAHARJO, B.H., SUDO, S., YONEMURA, S. & TSURUTA, H. 2006. Greenhouse gases produced during burning in the land preparation area using fire in peat area belong to the community. *Forest Ecol. Managem.* 234 (1 suppl.): S247.

SAJISE, P.E. & OMEGAN, E.A. 1989. The changing upland landscape of the northern Philippines. In: *Keepers of the Forest: Land Management Alternatives in Southeast Asia*, ed. M. Poffenberger, pp. 56–68. Kumarian Press, West Hartford, Conneticut.

SAKAI, S. 2000. Reproductive phenology of gingers in a lowland mixed dipterocarp forest in Borneo. *J. Trop. Ecol.* 16: 337–354.

SAKAI, S., MOMOSE, K., YUMOTO, T., NAGAMASU, H., HAMID, A.A., NAKASHIZUKA, T. & INOUE, T. 1999. Plant reproductive biology over four years including an episode of general flowering in a lowland dipterocarp forest, Sarawak, Malaysia. *Amer. J. Bot.* 86: 1414–1436.

SAKAI, S., KATO, M. & NAGAMASU, H. 2000. *Artocarpus* (Moraceae) – gall midge pollination mutualism mediated by a male-flower parasitic fungus. *Amer. J. Bot.* 87: 440–445.

SAKAI, S., MOMOSE, K., YUMOTO, T., NAGAMITSU, T., NAGAMASU, H., HAMID, A.A., NAKASHIZUKA, T. & INOUE, T. 2005. Plant reproductive phenology and general flowering in a mixed dipterocarp forest. In: *Pollination Ecology and the Rain Forest*, eds D.J. Roubik, S. Sakai & A.A. Hamid, pp. 35–50. Springer, New York.

SAKURAI, K., TANAKA, S., ISHIZUKA, S. & KANZAKI, M. 1998. Differences in soil properties of dry evergreen and dry deciduous forests in the Sakhaerat Experimental Research Station. *Tropics* 8: 61–80.

SALATI, E., VOSE, P.B. & LOVEJOY, T.E. 1986. Amazon rainfall, potential effects of deforestation, and plans for future research. In: *Tropical Rain Forests and the World Atmosphere*, AAAS Symposium 101, ed. G.T. Prance, pp. 61–74. Westview Press, Colorado.

SAMANANTHAN, H. & BORGES, R.M. 2000. Influence of exploitation on population structure, spatial distributon and reproductive success of dioecious species in a fragmented forest. *Biol. Conservation* 94: 243–256.

SANTILI, M., MOUTINHO, P., SCHWARTZMAN, S., NEPSTAD, D. CURRAN, L. & NOBRE, C. 2005. Tropical deforestation and the Kyoto protocol. *Climate Change* 71: 267–276.

SANTISUK, T. 1988. *An Account of the Vegetation of Northern Thailand*. Steiner, Stuttgart.

SANTISUK, T. 1997. Geographical and ecological distribution of the two tropical pines, *Pinus kesiya* and *Pinus merkusii*, in South-east Asia. *Thai Forest Bull.* 25: 102–123.

SARAVANAN, V. 2009. Political economy of the recognition of Forest Rights Act, 2006: conflict between environment and tribal development. *S. Asia Res.* 29: 199–221.

SARAYUDH, B. (See Bunyavejchewin, S.)

SARIN, M., SINGH, N. M., SUNDAR, N. & BHOGAL, R. K. 2003. *Devolution as a Threat to Democratic Decision-making in Forestry? Findings from Three States in India*. Working Paper 197, Overseas Development Institute, London.

SARKAR, S.K. 1994. The strengths and weaknesses of the joint forest management in West Bengal. *Indian Forester* 120: 579–584.

SARKAR, S.K. 2006. Present status and future prospects of joint forest management in West Bengal. *Indian Forester* 132: 11–18.

SARKAR, S.K. & CHATTOPADHYAY, R.N. 2002. The extent of economic viability of institutions formed for JFM – case study from Midnapore District of West Bengal. *Indian Forester* 128: 3–18.

SARKER, D. 2011.The implementation of the Forest Rights Act in India: critical issues. *Econ. Affairs* 31: 25–29.

SASAKI, S. & MORI, T. 1980. *Growth Responses of Dipterocarp Seedlings to Light*. International Working Group in Dipterocarpaceae, second round table conference. Unpubl. ms.

SASAOKA, M. 2003. Customary forest esource management in Seram Island, Central

Maluku: The 'Seli Keitahu' system. *Tropics* 12: 247–260.

SAVOLAINEN, V., ANSTETT, M.-C., LEXER, C., HUTTON, I., CLARKSON, J.J., NORUP, M.V., POWELL, M.P., SPRINGATE, D., SALAMIN, N. & BAKER, W.J. 2006. Sympatric speciation in palms on an oceanic island. *Nature* 441: 2010–2013.

SAW, L.G. & SAM, Y.Y. 2000a *Red Data List of Threatened and Endangered Peninsular Malaysia*. Forest Research Institute Malaysia (FRIM), Kepong.

SAW, L.G. & SAM, Y.Y. 2000b. Conservation of Dipterocarpaceae in Peninsular Malaysia. *J. Trop. Forest. Ecol.* 12: 593–615.

SAW, L.G., LAFRANKIE, J.V., KOCHUMMEN, K.M. & YAP, S.K. 1991. Fruit trees in a Malaysian rainforest. *Econ. Bot.* 45: 120–136.

SAW, L.G., CHUA, L.S.L. & ABDUL RAHIM, N. 2009. *Malaysia. National Strategy for Plant Conservation*. Ministry of Natural Resources and Environment & Forest Research Institute Malaysia (FRIM).

SAWYER, J.O. & CHERMSIRIVATHANA, C. 1969. A flora of Doi Suthep, Doi Pui, Chiang Mai, Thailand. *Nat. Hist. Bull. Siam. Soc.* 26: 99–132.

SAXENA, H.O. & BRAHMAM, M. 1989. *The Flora of Similipahar (Similipal), Orissa with Special Reference to the Potential Economic Plants*. Regional Research Laboratory, Bhubaneswar.

SAXENA, K.G. 1997. *The Saga of Participatory Forest Management in India*. CIFOR Special Publication, Jakarta, Indonesia.

SAYER, J., CHOKKALINGAM, U. & POULSEN, J. 2004. The restoration of forest biodiversity and ecological values. *Forest Ecol. Managem.* 201: 3–11.

SAYER, J., MCNEELY, J., MAGINNIS, S., BOEDIHARTONO, I., SHEPHERD, G. & FISHER, B. 2008. *Local Rights and Tenure for Forests: Opportunity or Threat for Conservation?* Rights and Resources Initiative, Washington, DC.

SCHAIK, C.P. VAN. 1986. Phenological changes in a Sumatran rain forest. *J. Trop. Ecol.* 2: 327–347.

SCHAIK, C.P. VAN & RAO, M. 1997. The behavioural ecology of Sumatran orangutans in logged and unlogged forest. *Trop. Biodivers.* 4: 173–185.

SCHAIK, C.P. VAN & TERBORGH, J. 1993. Production forests and protected forests: the potential for mutualism in the tropics. *Trop. Biodivers.* 1: 183–194.

SCHAIK, C.P. VAN, TERBORGH, J.W. & WRIGHT, S.J. 1993. The phenology of tropical forests: adaptive significance and consequences for primary consumers. *Annual Rev. Ecol. Syst.* 24: 353–379.

SCHIMPER, A.F.W. 1903. *Plant Geography Upon a Physiological Basis*. Transl. eds W.R. Fisher, P. Groom & I.B. Balfour. Oxford University Press.

SCHLUTER, D. & RICKLEFS, R.E. 1993a. Species diversity: an introduction to the problem. In: *Species Diversity in Ecological Communities. Historical and Geographical Perspectives*, eds R.E. Ricklefs & D. Schluter, pp. 1–10. University of Chicago Press.

SCHLUTER, D. & RICKLEFS, R.E. 1993b. Convergence and the regional component of species diversity. In: *Species Diversity in Ecological Communities.*

Historical and Geographical Perspectives, eds R.E. Ricklefs & D. Schluter, pp. 230–242. University of Chicago Press.

SCHMIDA, A. & WILSON, M.V. 1985. Biological determinants of species diversity. *J. Biogeogr.* 12: 1–20.

SCHMIDT, F.H. & FERGUSON, J.H.A. 1951. Rainfall types based on wet and dry period ratios for Indonesia with western New Guinea. *Verh. Djawatan Met. Geofisik. Djakarta*: 42.

SCHNITZER, S.A., DALLING, J.W. & CARSON, W.P. 2000. The impact of lianas on tree regeneration in tropical forest canopy gaps: evidence for an alternative pathway of gap phase regeneration. *J. Ecol.* 88: 655–666.

SCHÖNENBERGER, J., FRIIS, E.M., MATTHEWS, M.L. & ENDRESS, P.K. 2001. Cunoniaceae in the Cretaceous of Europe: evidence from fossil flowers. *Ann. Bot. Lond.* 88: 423–427.

SCHULTE, A. & SCHÖNE, D. 1996. *Dipterocarp Forest Ecosystems. Towards Sustainable Management*. World Scientific Publishing, Singapore.

SCHWARZWÄLLER, W., CHAI, F.Y.C. & HAHN-SCHILLING, B. 1999. Growth characteristics and response to illumination of some *Shorea* species in the logged-over mixed dipterocarp forest of Sarawak, Malaysia. *J. Trop. Forest Sci.* 11: 554–569.

SCHWEINFURTH, U. 1957. Die horizontale und vertikale Verbreitung der Vegetation im Himalaya. *Bonner Geogr. Abh.* Heft 20.

SCHWEINFURTH, U. 1988. *Pinus in Southeast Asia. Beitr. Biol. Pflanzen* 63: 253–269.

SCHWEITHELM, J., WIRAWAN, N., ELLIOTT, J. & KHAN, A. 1992. *Sulawesi Parks Program. Land Use and Socio-Economic Survey, Lore Linu National Park and Morowali Nature Reserve*. Ministry of Forestry, Indonesia & US Nature Conservancy.

SEARLE, M.P. & TRELOAR, P.J. 1993. Introduction. In: *Himalayan Tectonics*, Geological Society Special Publication 74, eds P.J. Treloar & M.P. Searle, pp. 1–7. Geological Society, London.

SEDDON, G. 1974. Xerophytes, xeromorphs, and sclerophylls: the history of some concepts in ecology. *Biol. J. Linn. Soc.* 6: 65–87.

SEETHALAKSHMI, K.K. & MUKTESH KUMAR, M.S., SANKARA PILLAI, M.S. & SAROJAM, N. 1998. *Bamboos of India. A Compendium*. Bamboo Information Centre, India & International Network for Bamboo & Rattan, Beijing. Eindhoven, New Delhi.

SEIBERT, B. 1996. Utilization of the tengkawang species group for nut and fat production. In: *Dipterocarp Forest Ecosystems. Towards Sustainable Management*, eds A. Schulte & D. Schöne, pp. 616–623. World Scientific Publishing, Singapore.

SEIFRIZ, W. 1923. The altitudinal distribution of plants on Mt. Gedeh, Java. *Bull. Torrey Bot. Club* 50: 283–305.

SEKERCIOGLU, H., EHRLICH, P.R., DAILY, G.C., AYGEN, D., GOEHRING, D. & SANDI, R.F. 2002. Disappearance of insectivorous birds from tropical forest fragments. *Proc. Natl. Acad. Sci. U.S.A.* 99: 263–267.

SEKHSARIA, P. 2007. Conservation in India

and the need to think beyond 'Tiger vs. Tribal'. *Biotropica* 39: 575–577.

SELWYN, M.A. & PARTHASARATHY, N. 2006. Reproductive traits and phenology of plants in tropical dry evergreen forest on the Coromandel coast of India. *Biodivers. & Conservation* 15: 3207–3234.

SEN, A. 2006. *The Argumentative Indian: Writings on Indian History, Culture, and Identity*. Picador/Farrar, Strausss & Giroux, New York.

SENANAYAKE, F.R., SOULÉ, M. & SENNER, J.W. 1977. Habitat values and endemicity in the vanishing rain forests of Sri Lanka. *Nature* 265: 351–354.

SENARATNA, L.K. 2001. *A Check List of the Flowering Plants of Sri Lanka*. National Science Foundation of Sri Lanka, Colombo.

SEN GUPTA, B.C. 1917. Sal at its extreme eastern limit. *Indian Forester* 43: 225.

SENTHILKUMAR, N., BARTHAKUR, N.D. & SINGH, A.N. 2009. Record of seed insect pests of *Dipterocarpus retusus* in Hollongapar Reserve Forests, Assam. *J. Trop. Forest Sci.* 21: 8–12.

SEQUEIRA, V. & BEZKOROWAJNYJ, P.J. 1998. Improved management of *Butea monosperma* (Lam.) Taub. for lac production in India. *Forest Ecol. Managem.* 102: 225–234.

SETH, S.K. 1951. *Manual of Instructions on Experiments, Preservation Plots, Protected Trees, Statistical Sample Plots, Linear Increment Plots, and Annual Research Reports*. Superintendent of Printing and Stationery, Allahabad.

SETH, S.K. & DABRAL, S.N. 1957. The effect of forest cover on air and soil temperatures during frost period. *Indian Forester* 83: 91.

SETH, S.K. & YADAV, J.S.B. 1959. Teak soils. *Indian Forester* 85: 1–16.

SHAH, A. 2007. Management of protected areas: exploring an alternative at Gir. *Econ. Polit. Weekly* 42: 2923–2930.

SHAHABUDDIN, G. & KUMAR, R. 2007. Effects of extractive disturbances on bird assemblages, vegetation structure and floristics in tropical scrub forest, Sariska Tiger Reserve, India. *Forest Ecol. Managem.* 246: 175–185.

SHANAHAN, M. & COMPTON, C. 2000. Fig eating by Bornean tree shrews (*Tupaia* spp.): evidence for a role as seed dispersers. *Biotropica* 32: 759–763.

SHANAHAN, M. & COMPTON, S.G. 2001. Vertical stratification of figs and fig-eaters in a Bornean lowland rain forest: how is the canopy different? *Pl. Ecol.* 153: 1–2.

SHANKAR, U., MURALI, K.S., SHAANKER, U., GANESHAIAH, K.N. & BAWA, K.S. 1996. Extraction of non-timber forest products in the forests of Biligiri Rangan Hills, India. 3. Productivity, extraction and prospects of sustainable harvest of amla, *Phyllanthus emblica* (Euphorbiaceae). *Econ. Bot.* 50: 271–279.

SHARMA, M.K., SINGAL, R.M. & POKHRIYAL, T.C. Undated. *Dalbergia sissoo* in India. FAO Forest Department Corporate Document Depository. Unpubl. ms.

SHEBBEARE, E.O. 1928. *Fire Protection and Fire Control in India*. Third British Empire Forestry Conference. Government Printing Office, Canberra.

SHEIL, D. 2001. Long-term observations of rain forest succession, tree diversity and

responses to disturbance. *Pl. Ecol.* 155: 183–199.

SHEIL, D. & BURSLEM, D.F.R.P. 2003. Disturbing hypotheses in tropical forests. *Trends Ecol. Evol.* 18: 18–26.

SHI, S., HUANG, Y., ZHONG, Y., DU, Y., ZHANG, Q., CHANG, H. & BOUFFORD, D.E. 2001. Phylogeny of the Altingiaceae based on cpDNA mat K, PY-IGS and nrDNA ITS sequences. *Pl. Syst. Evol.* 230: 13–24.

SHIMMIELD, G.B., MOWBRAY, S.R. & WEEDON, G.P. 1990. A 350 ka history of the Indian southwest monsoon – evidence from deep sea cores, northwest Arabian Sea. *Trans. Roy Soc. Edinburgh Earth* 81: 289–299.

SHONO, K., DAVIES, S.J. & CHUA, Y.K. 2006. Regeneration of native plant species in restored forests on degraded lands in Singapore. *Forest Ecol. Managem.* 237: 574–582.

SHONO, K., DAVIES, S.J. & KHENG, C.Y. 2007. Performance of 45 native tree species on degraded lands in Singapore. *J. Trop. Forest Sci.* 19: 25–34.

SHRIVASTAVA, R.K. 2002. Forest fire and biotic interference – a great threat to Nilgiri Biosphere Reserve. *Indian Forester* 128: 667–673.

SHUKLA, R.P. & RAMAKRISHNAN, P.S. 1981. The nature and ecological significance of heterophylly in *Artocarpus chaplasha* Roxb. *Proc. Ind. Nat. Sci. Acad.* B, 47: 551–556.

SHUKLA, R.P. & RAMAKRISHNAN, P.S. 1984. Biomass allocation strategies and productivity of tropical trees related to successional status. *Forest Ecol. Managem.* 9: 315–324.

SHUKLA, R.P. & RAMAKRISHNAN, P.S. 1986. Architecture and growth strategies of tropical trees in relation to successional status. *J. Ecol.* 74: 33–46.

SHUKRI, W.A., WAN MOHD, ISMAIL, H., ABD. RAHMAN, K. & NUR HAJAR, Z.S. Undated, c. 2005. *Status of Logged-Over Forest in Compartment 53 & 114 Jengai F.R., Terengganu*. Unpubl. ms.

SIDLE R.C., TANI M. & ZIEGLER A.D. 2006. Catchment processes in Southeast Asia. *Forest Ecol. Managem.*, special issue 224.

SIEBERET, S.F. 2005. The abundance and distribution of rattan over an elevation gradient in Sulawesi, Indonesia. *Forest Ecol. Managem.* 210: 143–158.

SIEFFERMANN, G.R. 1990. Origin of iron carbonate layers in Tertiary coastal sediments of Central Kalimantan Province (Borneo), Indonesia. *Spec. Publ. Int.* 11: 139–146.

SILORI, C.S. & MISHRA, B.K. 2001. Assessment of livestock grazing presence in and around the elephant corridors in Mudumalai Wildlife Sanctuary, South India. *Biodivers. & Conservation* 10: 2181–2195.

SILVA, J.M.C. DA, UHL, C. & MURRAY, G. 1996. Plant succession, landscape management, and the ecology of frugivorous birds in abandoned Amazonian pastures. *Conservation Biol.* 10: 491–503.

SILVERTOWN, J.W. 1980. The evolutionary ecology of mast seeding in trees. *Biol. J. Linn. Soc.* 14: 235–250.

SILVERTOWN, T., DODD, M. & GOWING, D. 2001. Phylogeny and the niche structure of meadow plant communities. *J. Ecol.* 89: 428–435.

SIM, J.W.S., TAN, H.T.W. & TURNER, I.M. 1992. *Adinandra* belukar: an anthropogenetic heath forest in Singapore. *Vegetatio* 102: 125–137.

SIMMONDS, W.T.F. 1980. Monocarpy, calendars and flowering cycles in Angiosperms. *Kew Bull.* 35: 235–245.

SIMONS, W.J.F., SOCQUET, A., VIGNY, C., AMBROSIUS, B.A.C., ABU, S.H., PROMTHING, C., SUBARYA, C., SARSITO, D.A., MATHEUSSEN, S., MORGAN, P. & SPAKMAN, W. 2007. A decade of GPS in Southeast Asia: resolving Sundaland motion and boundaries. *J. Geophys. Res.* 112, B06420.

SIMPSON, G.G. 1944. *Tempo and Mode in Evolution*. Columbia University Press.

SIMPSON, G.G. 1953. *The Major Features of Evolution*. Simon & Schuster, New York.

SIMS, R.W. & EASTON, E.G. 1972. A numerical revision of the earthworm genus *Pheretima* auct. (Megascolecidae: Oligochaeta) with the recognition of new genera and an appendix on the earthworms collected by the Royal Society North Borneo Expedition. *Biol. J. Linn. Soc.* 4: 169–268.

SINGH, A., REDDY, V.S. & SINGH, J.S. 1995. Analysis of woody vegetation of Corbett National Park, India. *Vegetatio* 120: 69–79.

SINGH, A.K., SINGH, M.K. & MASCARENHAS, O.A.J. 1985. Community forestry for revitalizing rural ecosystems: a case study. *Forest Ecol. Managem.* 10: 209–232.

SINGH, D.R.R. 2002. Studies on management of natural bamboo (*Dendrocalamus strictus* Nees) in Shiwalik Hills of Haryana state, India. *Indian Forester* 128: 257–265.

SINGH, J.N. 2004. Edaphic characteristics of Nokrek Biosphere Reserve, Meghalaya and their relationship with its flora. *Indian Forester* 130: 921–938.

SINGH, K. DALJEET-D. 1974. Pests of some dipterocarps. *Malayan Forester* 37: 24–31.

SINGH, K.P. & SINGH, G. 2003. Insect visitors of mango inflorescence. *Indian Forester* 129: 1289–1292.

SINGH, M. & NANTUJAL, S. 1993. *Grewia oppositifolia* – a tree for fodder, fuel and fibre. *Indian Forester* 119: 674.

SINGH, R.S. 1992. Effect of winter fire on primary productivity and nutrient concentrations of a dry tropical savanna. *Vegetatio* 106: 63–71.

SINGH, S.P., RAWAT, Y.S., BANA, B.S. & NEGI, G.C.S. 1990. Effects of unusually large seed crop on litterfall and nitrogen translocation in Himalayan oaks. *J. Trop. Forest Sci.* 32: 79–86.

SINGHAKUMARA, B.M.P. 1995. *Floristic survey of Adam's Peak Wilderness*. Ministry of Agriculture, Lands, and Forestry. Forestry Sector Development project. Unpubl. ms.

SINGHAKUMARA, B.M.P., UDUPORUWA, R.S.J.P. & ASHTON, P.M.S. 2000. Soil seed banks in relation to light and topographic position of a hill dipterocarp forest in Sri Lanka. *Biotropica* 32: 190–196.

SINGHAKUMARA, B.M.P., GAMAGE, H.K. & ASHTON, P.M.S. 2003. Comparative growth of four *Syzygium* species within simulated shade environments of a Sri

Lankan rain forest. *Forest Ecol. Managem.* 174: 511–520.

SINHA, A. & BAWA, K.S. 2002. Harvesting techniques, hemiparasites and fruit production in two non-timber forest tree species in South India. *Forest Ecol. Managem.* 168: 289–300.

SINHA, J.N. 1979. Serve the tribal to save the forest. *Indian Forester* 105: 290–292.

SIST, P. & BROWN, N. 2004. Silvicultural intensification for tropical forest conservation: a response to Frederickson and Putz. *Biodivers. & Conservation* 13: 2381–2385.

SIST, P. & NGUYEN-THÉ, N. 2002. Logging damage and the subsequent damage of a dipterocarp forest in East Kalimantan (1990–1996). *Forest Ecol. Managem.* 165: 85–103.

SIST, P., DYKSTRA, D. & FIMBEL, R. 1998a. *Reduced-impact Logging Guidelines for Lowland and Hill Dipterocarp Forests in Indonesia*. CIFOR Occasional Paper 15. CIFOR, Bogor.

SIST, P., NOLAN, T., BERTAULT, J.-G. & DYKSTRA, D. 1998b. Harvesting intensity versus sustainability in Indonesia. *Forest Ecol. Managem.* 108: 251–260.

SIST, P., SABOGAL, C. & BYRON, Y. 1999. *Management of Secondary and Logged-Over Forests in Indonesia: Selected Proceedings of an International Workshop, 17–19 November 1997*. CIFOR, Bogor.

SIVARAJ, N. & KRISHNAMURTHY, K.V. 1988. Flowering phenology in the vegetation of the Shavaroys, South India. *Vegetatio* 79: 85–88.

SLADE, H. 1896. Too much fire protection in Burna. *Indian Forester* 22: 176.

SLEUMER, H. 1966. Ericaceae. In: *Flora Malesiana* 1 (6), ed. C.G.G.J. van Steenis, pp. 469–914. P. Noordhoff, Groningen.

SLIK, J.W.F., VERBERG, R.W. & KESSLER, P.J.A. 2002. Effects of fire and selective logging on the tree species composition of lowland dipterocarp forest in East Kalimantan, Indonesia. *Biodivers. & Conservation* 11: 85–98.

SLIK, J.W.F., POULSEN, A.D., ASHTON, P.S., CANNON, C.H., EICHHORN, K.A.O., KARTAWINATA, K., LANNIARI, I., NAGAMASU, H., NAKAGAWA, M., VAN NIEUWSTADT, M.G.L., PAYNE, J., PURWANINGSIH, SARIDAN, A., SIDIYASA, K.D., VERBURG, R.W. & WEBB, C.O. 2003. A floristic analysis of the lowland dipterocarp forests of Borneo. *J. Biogeogr.* 30: 1517–1531.

SLIK, J.W.F., ALBA, S.-I., BASTIAN, M., BREARLET, F.Q., CANNON, C.H., EICHHORN, K.A.O., FREDRIKSSON, G., KARTAWINATA, K., LAUMONIER, Y., MANSOR, A., MARJOKORPI, A., MEIJAARD, E., MORLEY, R.J., NAGAMASU, H., NILUS, R., NURTJAHYA, E., PAYNE, J., PERMANA, A., POULSEN, A.D., RAES, N., SIRWAN, S., CAN SCHAIK, C.P., SHEIL, D., SIDIYASA, K., SUZUKI, E., CAN VALKENBURG, L.C.H., WEBB, C.O., WICH, S., YONEDA, T., ZAKARIA, R. & ZWEIFEL, N. 2011. Soils on exposed Sunda Shelf shaped biogeographic patterns in the equatorial forests of Southeast Asia. *Proc. Natl. Acad. Sci. U.S.A.* 108: 12343–12347.

SMITH, A.G. & BRIDEN, J.C. 1977. *Mesozoic and Cenozoic Palaeocontinental Maps*. Cambridge University Press.

SMITH, D.M., LARSON, B.C., KELTY, J. &

ASHTON, P.M.S. 1996. *The Practice of Silviculture*, 6th ed. Wiley, Chichester.

SMITH, J.M.B. 1970. Herbaceous communities in the summit zone of Mount Kinabalu. *Malayan Nat. J.* 24: 16–29.

SMITH, J.M.B. 1977. Origins and ecology of the tropicalpine flora of Mt. Wilhelm, New Guinea. *Biol. J. Linn. Soc.* 9: 87–131.

SMITINAND, T. 1966. The vegetation of Doi Chiengdao, a limestone massif in Chiengmai, North Thailand. *Nat. Hist. Bull. Siam. Soc.* 21: 93–128.

SMITINAND, T. 1968. The vegetation of Khao Yai National Park. *Nat. Hist. Bull. Siam. Soc.* 22: 289–305.

SMITINAND, T. 1969–1970. The distribution of the Dipterocarpaceae in Thailand. *Bull. Siam. Soc.* 23: 167–175.

SMITINAND, T. 2001. *Thai Plant Names*. 2nd ed. Thailand Forest Herbarium, Royal Forest Department, Bangkok.

SMITINAND, T., LARSON, K. & SANTISUK, T., eds. 1970-. *Flora of Thailand*. Forest Herbarium, Royal Forest Department, Bangkok.

SMITINAND, T., SANTISUK, T. & PHENGKHLAI, C. 1979. The manual of Dipterocarpaceae of South-East Asia. *Thai Forest Bull., Bot.* 12: 1–110.

SMITINAND, T., VIDAL, J.E. & HO, P.H. 1990. Dipterocarpacées. In: *Flore du Cambodge, du Laos et du Viêtnam*, 25, ed. P. Morat. Museum National d'Histoire Naturelle, Paris.

SMITS, W.T.M. 1992. Mycorrhizal studies in dipterocarp forests in Indonesia. In: *Mycorrhizas in Ecosystems*, eds D.J. Read, D.H. Leevis, A.H. Fitter & I.J. Alexander, pp. 283–292. CABI, Wallingford.

SMITS, W.T.M. 1994. *Dipterocarpaceae: Mycorrhiza and Regeneration*. Tropenbos Foundation, Wageningen.

SMYTHIES, A. 1900. Drought and forests in C.P. *Indian Forester* 20: 336–341.

SMYTHIES, B.E. 1960. *The Birds of Borneo*. Oliver & Boyd, Edinburgh.

SMYTHIES, B.E. 1964. The birds of Mt Kinabalu and their zoogeographical relationships. *Proc. Roy. Soc. London, Ser. B., Biol. Sci.* 161: 75–80.

SMYTHIES, E.A. 1919. Geology and forest distribution. *Indian Forester* 45: 240–243.

SOEHARTONO, T. & MARDIASTUTI, A. 2001. Kutai National Park: where to go? *Trop. Biodivers.* 7: 83–101.

SOEHARTONO, T. & NEWTON, A.C. 2000. Conservation and sustainable use of tropical trees in the genus *Aquilaria*. I. Status and distribution in Indonesia. *Biol. Conservation* 96: 83–94.

SOEHARTONO, T. & NEWTON, A.C. 2001. Conservation and sustainable use of tropical trees in the genus *Aquilaria*. II. The impact of gaharu harvesting in Indonesia. *Biol. Conservation* 97: 29–41.

SOEPADMO, E. 1976. Fagaceae. In: *Flora Malesiana*, ed. C.G.G.J. van Steenis, ed., pp. 265–403. Noordhoff, Leiden.

SOEPADMO, E. 1979. Genetic resources of Malaysian fruit trees. *Malaysian Appl. Biol.* 8, 1: 33–42.

SOEPADMO, E., JULIA, S. & GO, R. 2000. Fagaceae. In: *Tree Flora of Sabah and Sarawak 3*, eds E. Soepadmo & L.G. Saw,

pp. 1–117. Forest Research Institute Malaysia (FRIM), Kepong.

SOEPADMO, E., WONG, K.M. & SAW, L.G., EDS. 1995- . *Tree Flora of Sabah and Sarawak*. Forest Research Institute Malaysia & Sabah & Sarawak Forest Departments.

SOHMER, S.H. 2001. Conservation lessons from studies in Philippine *Psychotria*. *Malaysian Nat. J.* 55: 43–47.

SOHMER, S.H. & DAVIS, A. 2004. Current status of *Psychotria* (Rubiaceae) in the Philippines. *Proceedings of the 6th International Flora Malesiana Symposium* Los Baños, Luzon, Philippines. Programme & Abstracts, pp 34–35.

SOLBRIG, O.T., EMDEN, H.M. VAN & OORDT, P.G.W.J. VAN. 1994. *Biodiversity and Global Change*. CABI, Wallingford.

SOLLINS, P. 1998. Factors influencing species composition in tropical lowland rain forest: does soil matter? *Ecology* 79: 23–30.

SOLOMON RAJU, A.J., RAO, S.P. & EZRADANAM, V. 2004. Bird pollination in *Sterculia colorata* Roxb, (Sterculiaceae), a rare tree species in the eastern Ghats of Viasakhapatnam and East Godavari Districts of Andhra Pradesh. *Curr. Sci.* 87: 28–31.

SORENSEN, K.W. 1993. Indonesian peat swamp forests and their role as a carbon sink. *Chemosphere* 27: 1065–1082.

SPALIK, K. 1991. On evolution of andromonoecy and 'over-production' of flowers: a resource allocation model. *Biol. J. Linn. Soc.* 42: 325–336.

SPILSBURY, M.J. 2007. Forest management: assessing the impact of CIFOR criteria and indicators research. In: *International Research on Natural Resource Management: Advances in Impact*, eds H. Waibel & D. Zilberman, pp. 217–233. CABI, Wallingford.

SRIKOSAMATARA, S. 1993. Density and biomass of large herbivores and other mammals in a dry tropical forest of western Thailand. *J. Trop. Ecol.* 9: 33–43.

SRI-NGERNYANG, K., KANZAKI, M., MIZUNO, T., NOGUCHI, H., TEEJUNTUK, S., SUNGPALEE, C., HARA, M., YAMAKURA, T., SAHUNALU, P., DHANMANONDA, P. & BUNYAVEJCHEWIN, S. 2003. Doi Inthanon. Habitat differentiation of Lauraceae species in a tropical lower montane forest in northern Thailand. *Ecol. Res.* 18: 1–4.

SRIVASTAVA, K.K., ZACHARIAS, V.J., BHARDWAJ, A.K., JONY, A. & SHIRLEY, J. 1994. A preliminary study of the grasslands of Periyar Tiger Reserve, Kerala. *Indian Forester* 120: 898–907.

SRIVASTAVA, R.K. 1994. Re-establishment of sholas in grassland. *Indian Forester* 120: 868–870.

SRIVASTAVA, R.K. 2002a. Biotic interferences and other problems faced by the elephants in the most crucial corridors of Eastern Ghats. *Indian Forester* 128: 169–178.

SRIVASTAVA, R.K. 2002b. Forest fire and biotic interference – a great threat to Nilgiri Biosphere Reserve. *Indian Forester* 128: 667–673.

SRIVASTAVA, R.K., CHIDAMBARAM, K. & KUMARAVELU, G. 1999. Impact of forest fire and biotic interference on the biodiversity of the Eastern Ghats. *Indian Forester* 125: 439–443.

SRIVASTAVA, S. & SINGH, L.K. 2002. Similipal Biosphere Reserve. In: *Traditional Ecological Knowledge for Managing Biosphere Reserves in South and Central Asia*, eds P.S. Ramakrishnan, R.K. Rai, R.P.S. Katawal & S. Mehndiratta, pp. 485–491. UNESCO, Oxford University Press & IBH New Delhi.

ST. JOHN, S. 1863. *Life in the Forests of the Far East, or, Travels in Northern Borneo*. Smith, Elder. London.

STACY, E.A. 2001a. *Relatedness-Dependent Cross-fertility in Tropical Rain Forest Trees and the Nature of Species' Boundaries*. PhD thesis, Boston University.

STACY, E.A. 2001b. Cross-fertility in two tropical tree species: evidence of inbreeding depression within populations and genetic divergence among populations. *Amer. J. Bot.* 88: 1041–1051.

STACY, E.A., GUNATILLEKE, I.A.U.N., GUNATILLEKE, C.V.S., DAYANANDAN, S., SNEIDER, C.J. & KHASA, P.D. 2001. Crossing barriers within rain forest tree species can arise over small spatial scales. In: *Proceedings of the International Conference on Tropical Ecosystems: Structure, Diversity & Human Welfare*, eds K.N. Ganeshaiah, R. Uma Shaanker & K.S. Bawa, pp. 581–585. Oxford & IBH Publishing Co., New Delhi.

STADTMÜLLER, T. 1987. *Cloud Forests in the Humid Tropics*. The United Nations University, Tokyo.

STAINTON, J.D.A. 1972. *Forests of Nepal*. John Murray, London.

STAMP, L.D. 1924. *The Vegetation of Burma from an Ecological Standpoint*. University of Rangoon.

STAPF, O. 1894. Of the flora of Mt. Kinabalu in North Borneo. *Trans. Linn. Soc., Bot.*, ser. ii, 4: 69–263.

START, A.N. 1974. *The Feeding Biology in Relation to Food Sources of Nectarivorous Bats (Chiroptera: Macroglossinae) in Malaysia*. PhD thesis, Aberdeen University.

STAUBWASSER, M. 1999. *Early Holocene Variability of the Indian Monsoon and Arabian Sea Thermocline Ventilation*. PhD thesis, Christian-Albrechts-Universität, Kiel.

STAUFFER, P.H., NISHIMURA, S. & BATCHELOR, B.C. 1980. Volcanic ash in Malaya from the catastrophic eruption of Toba, Sumatra, 30,000 years ago. In: *Physical Geology of Indonesian Island Arcs*, ed. S. Nishimura, pp. 156–164. Kyoto University.

STEBBING, E.P. 1922–1923. *The Forests of India*. 2 vols. The Bodley Head, London.

STEBBINS, G.L. 1957. Self-fertilization and population variability in higher plants. *Amer. Naturalist* 91: 337–354.

STEBBINS, G.L. 1977. *Processes of Organic Evolution*. 3rd ed. Prentice-Hall, Englewood Cliffs, New Jersey.

STEEGE, H. TER, PITMAN, N., SABATIER, D., CASTELLANOS, H., VAN DER HOUT, P., DALY, D.C., SILVEIRA, M., PHILLIPS, O., VASQUEZ, R., VAN ANDEL, T., DUIVENVOORDEN, J., DE OLIVEIRA, A.A., EK, R., LILWAH, R., THOMAS, R., VAN ESSEN, J., BAIDER, C., MAAS, P., MORI, S., TERBORGH, J., VARGAS, N., MOGOLLÓN, H. & MORAWETZ, W. 2003. Spatial model of tree α-diversity and tree density for the Amazon. *Biodivers. & Conservation* 12: 2255–2277.

STEENIS, C.G.G.J. VAN. 1933. Report of a botanical trip to the Ranau region, South Sumatra. *Bull. Jard. Bot. Buitenzorg* 3 (13): 53.

STEENIS, C.G.G.J. VAN. 1935. On the origin of the Malaysian mountain flora. II. *Bull. Jard. Bot. Buitenzorg* 3 (13): 289–417.

STEENIS, C.G.G.J. VAN. 1957. Specific and infraspecific delimination. *Flora Malesiana* 1(5): clxviii–ccxxxiv.

STEENIS, C.G.G. J. VAN. 1968. Frost in the tropics. *Proceedings of the Symposium on Recent Advances in Tropical Ecology*. Banaras Hindu University, Varanasi, pp. 154–167.

STEENIS, C.G.G.J. VAN. 1972a. *The Mountain Flora of Java*. Brill, Leiden.

STEENIS, C.G.G. J. VAN. 1972b. The mountain flora of the Malaysian tropics. *Endeavour* 21: 183–193.

STEENIS, C.G.G.J. VAN. 1979. Plant-geography of East Malesia. *Bot. J. Linn. Soc.* 79: 97–178.

STEENIS, C.G.G.J. VAN 1981. *Rheophytes of the World. An Account of the Flood-Resistant Flowering Plants and Ferns of the World and the Theory of Autonomous Evolution*. Sijthoff & Noordhoff, Alphen, Netherlands and Rockville, Maryland.

STEENIS, C.G.G.J. VAN. 1984. Floristic altitudinal zones in Malesia. *Bot. J. Linn. Soc.* 89: 289–292.

STEENIS, C.G.G.J. VAN & SCHIPPERS-LAMMERTSE, A.F. 1965. Concise plant-geography of Java. In: *Flora of Java*, eds C.A. Backer, & R.C. Bakhuizen van den Brink, Jr., pp. 1–72. Noordhof, Groningen.

STEM, C. J., LASSOIE, J.P., LEE, D.R., DESHLER, D.D. & SCHELHAS, J.W. 2003. Community participation in ecotourism benefits: the link to conservation practices and perspectives. *Soc. Nat. Res.* 16: 387–413.

STERN, N. 2006. *Report on the Economics of Climate Change*. H.M. Treasury, London.

STERCK, F.J. & BONGERS, F. 1998. Ontogenetic changes in size, allometry, and mechanical design of tropical rain forest trees. *Amer. J. Bot.* 85: 266–272.

STERCK, F.J. & BONGERS, F. 2001. Crown development in tropical rain forest trees: patterns with tree height and light availability. *J. Ecol.* 89: 1–13.

STERCK, F.J., BONGERS, F. & NEWBERY, D.M. 2001. Tree architecture in a Bornean lowland forest: intraspecific and interspecific patterns. *Pl. Ecol.* 153: 279–292.

STERLING, E.J., HURLEY, M.M. & MINH, L.D. 2006. *Vietnam: a Natural History*. Yale University Press.

STEVENS, P.F. 1976. The altitudinal and geographical distributions of flower types in *Rhododendron* section *Vireya*, especially in the Papuasian species, and their significance. *Bot. J. Linn. Soc.* 72: 1–33.

STEVENS, P.F. 1980. A revision of the Old World species of Calophyllum (Guttiferae). *J. Arnold Arbor.* 61: 117–424.

STEVENS, P.F. 2001–. Angiosperm Phylogeny Website. URL: www.mobot.org/MOBOT/research/APweb/

STOLL, P. & NEWBERY, D.M. 2005. Evidence of species specific neighboring effects in the Dipterocarpaceae of a Bornean rain forest. *Ecology* 86: 3048–3063.

STOLLE, F., CHOMITZ, K.M., LAMBIN, E.F. & TOMICH, T.P. 2003. Land use and vegetation fires in Jambi Province, Sumatra, Indonesia. *Forest Ecol. Managem.* 179: 277–292.

STOTT, P.A. 1976. Recent trends in the classification and mapping of dry deciduous dipterocarp forest in Thailand. In: *The Classification and Mapping of South-East Asian Vegetation. Transactions of the Fourth Aberdeen-Hull Symposium on Malesian Ecology*, eds P. Ashton & M. Ashton, pp. 22–56, Hull Geog. Dept. Misc. Ser. 17.

STOTT, P.A. 1986. The spatial pattern of dry season fires in the savanna forests of Thailand. *J. Biogeogr.* 13: 345–358.

STOTT, P.A. 1988. The forest as phoenix: towards a biogeography of fire in mainland South-east Asia. *Geogr. J.* 154: 337–350.

STOTT, P.A., GOLDAMMER, J.G. & WERNER, W.L. 1990. The role of fire in the tropical lowland deciduous forests of Asia. In: *Fire in the Tropical Biota: Ecosystem Processes and Global Challenges*, ed. J.G. Goldammer, pp. 32–44. Springer.

STRONZA, A. & GORDILLO, J. 2008. Community views of ecotourism. *Ann. Tourism Res.* 35 (2): 448–468.

STUIJTS, I., NEWSOME, J.C. & FLENLEY, J.R. 1988. Evidence for late Quaternary vegetational change in the Sumatran and Javan highlands. *Rev. Palaeobot. Palynol.* 55: 207–216.

STYRING, A.R. & HUSSIN, M.Z. 2004. Effects of logging on woodpeckers in a Malaysian rain forest: the relationship between resource availability and woodpecker abundance. *J. Trop. Ecol.* 20: 495–504.

STYRING, A.R. & ICKES, K. 2001. Woodpecker abundance in a logged (40 years ago) versus unlogged lowland dipterocarp forest in Peninsular Malaysia. *J. Trop. Ecol.* 17: 261–268.

SU, H.J. 1984. Studies on the climate and vegetation types of the natural forests in Taiwan. 2. Altitudinal vegetation zones in relation to temperature gradient. *Quart. J. Chin. Forest.* 17: 57–73.

SU, S.-H., CHANG-YANG, C.H., LU, C.L., TSUI, C.C., LIN, T.T, LIN, C.L., CHIOU, W.L., KUAN, L.H., CHEN, Z.S. & HSIEH, C.F. 2007. *Fushan Subtropical Forest Dynamics Plot: Tree Species Characteristics and Distribution Patterns*. Taiwan Forestry Research Institute, Taipei.

SUGDEN, A.M. 1985. Leaf anatomy in a Venezuelan montane forest. *Bot. J. Linn. Soc.* 90: 231–241.

SUKRI, R.S., WAHAB, R.A., SALIM, K.A. & BURSLEM, D.F.R.P. 2012. Habitat associations and community structure of dipterocarps in response to environment and soil conditions in Brunei Darussalam, Northwest Borneo. *Biotropica* 44: 595–605.

SUKUMAR, R. 2011. *The Story of Asia's Elephants*. Radhika Sabavala for the Marg Foundation, Mumbai.

SUKUMAR, R., DATTARAJA, H.S., SURESH, H.S., RADHAKRISHNAN, J., VASUDEVA, S., NIRMALA, S. & JOSHI, N.V. 1992. Long-term monitoring of vegetation in a tropical deciduous forest in Mudumalai, southern India. *Curr. Sci.* 62: 606–616.

SUKUMAR, R., RAMESH, R., PANT, R.K & RAJAGOPALAN, G. 1993. A ^{13}C record of late Quaternary climate change from tropical peats in southern India. *Nature* 364: 703–706.

SUKUMAR, R., SURESH, H.S., DATTARAJA, H.S. & JOHN, R. 1998. Dynamics of a tropical deciduous forest: population changes (1988 though 1993) in a 50 ha plot at Mudumalai, Southern India. In: *Forest Biodiversity Research, Monitoring and Modeling: Conceptual Background and Old World Case Studies*, eds F. Dallmeier & J.A. Comiskey, pp. 495–648. UNESCO, Paris & Parthenon, Carnforth.

SUKUMAR, R., SURESH, H.S., DATTARAJA, H.S., JOHN, R. & JOSHI, N.V. 2004. Mudumalai Forest Dynamics Plot, India. In: *Tropical Forest Diversity and Dynamism: Findings from a Large-Scale Plot Network*, eds E.C. Losos & E.G. Leigh Jr., pp. 551–563. University of Chicago Press, Chicago.

SUKUMARAN, S., REJINI BALASINGH, G.B., KAVITHA, A. & RAJ, A.D.S. 2005. The floristic composition of sacred groves – a functional tool to analyse the miniforest ecosystem. *Indian Forester* 131: 773–785.

SUN, G., DILCHER, D.L., ZHENG, S. & ZHOU, Z. 1998. In search of the first flower: a Jurassic angiosperm, *Archaefructus*, from northwest China. *Science* 282 (5394): 1692–1695.

SUN, G., DILCHER, D.L., ZHENG, S., ZHOU, Z., NIXON, K.C. & WANG, X. 2002. *Archaefructaceae*, a new basal angiosperm family. *Science* 296 (5569): 899–904.

SUN, I.F. & HSIEH, C.F. 2004. Nanjenshan forest dynamics plot, Taiwan. In: *Tropical Forest Diversity and Dynamism: Findings from a Large-Scale Plot Network*, eds E.C. Losos & E.G. Leigh Jr., pp. 564–573. University of Chicago Press, Chicago.

SUN, I.F., HSIEH, C.F. & HUBBELL, S.P. 1998. Structure and species composition of a sub-tropical rain forest in southern Taiwan on a wind-stress gradient. In: *Forest Biodiversity Research, Monitoring and Modeling: Conceptual Background and Old World Case Studies*, eds F. Dallmeier & J.A. Comiskey, pp. 563–590. UNESCO, Paris & Parthenon, Carnforth.

SUN, X. & WANG, T.P. 2005. How old is trhe Asian monsoon system? Palaeobotanical records from China. *Palaeogeogr. Palaeoclimatol. Palaeoecol.* 222: 181–222.

SUN, X., WU, Y., QIAO, Y. & WALKER, D. 1986. Late Pleistocene and Holocene vegetation history at Kunming, Yunnan Province, southwest China. *J. Biogeogr.* 13: 441–476.

SUNDERLIN, W.D., ANGELSEN, A., BELCHER, B. & BURGERS, P. 2005. Livelihoods, forests, and conservation in developing countries: an overview. *World Dev.* 33: 1383–1402.

SUNDERLIN, W.D., HATCHER, J. & LIDDLE, M. 2008. *From Exclusion to Ownership? Challenges and Opportunities in Advancing Forest Tenure System*. Rights and Resources Initiative, Washington, DC.

SURESH, H.S., DATTARAJA, H.S. & SUKUMAR, R. 1996. Tree flora of Mudumalai Sanctuary, Tamil Nadu, southern India. *Indian Forester* 122: 507–519.

SURESH, H.S., DATTARAJA, H.S., MONDAL, N. & SUKUMAR, R. 2011. Seasonally dry tropical forests in southern India: an analysis of floristic composition, structure and dynamics. In: *The Ecology and Conservation of Seasonally Dry Forest in Asia*, eds W.J. McShea, S.J. Davies & N. Bhumpakphan, pp. 37–58. Rowman & Littlefield, Smithsonian Institution Scholarly Press, Washington, DC.

SUTTON, S.L., WHITMORE, T.C. & CHADWICK, A.C. 1983. *Tropical Rain Forest: Ecology and Management*. Blackwell, Oxford.

SUZUKI, E. & ASHTON, P.S. 1996. Sepal and nut size ratio of fruits of Asian Dipterocarpaceae and its implications for dispersal. *J. Trop. Ecol.* 12: 853–870.

SUZUKI, E. & KOHYAMA, T. 1991. Spatial distributions of wind-dispersed fruits and trees of *Swintonia schwenkii* (Anacardiaceae) in a tropical forest in West Sumatra. *Tropics* 1: 131–142.

SUZUKI, E., HOTTA, M., PARTMONIHARDJO, T., SILE, A., KIOKE, F., NOMA, N., YAMADA, T. & KAJI, M. 1997a. Ecology of tengkawang forests under varying degrees of management in West Kalimantan. *Tropics* 7: 35–53.

SUZUKI, E., KOHYAMA, T., SIMBOLON, H., HARAGUCHI, A., TSUZAKI, S. & NISHIMURA, T. 1997b. *Vegetation of Kerangas and Peat Swamp Forets in Lahei, Central Kalimantan*. Proceedings of the International Workshop 6–9 August 1997, Palangka Raya, Central Kalimantan. Unpubl. ms.

SUZUKI, K. & NIYOMDHAM, C. 1992. Phytosociological studies on tropical peat swamps. 1. Classification of vegetation at Narathiwat, Thailand. *Tropics* 2: 49–65.

SWAIN, D. 2001a. Letter to editor: elephant menace in Kendujhar District: causes and remedial measures. *Indian Forester* 127: 254–255.

SWAIN, D. 2001b. Man and wild elephant conflict in Orissa. *Indian Forester* 127: 1134–1142.

SWAIN, D. & NANDA, F. 1997. Study of plant diversity in a newly established preservation plot inside Similipal National Park, Orissa. In: *Similipal: a Natural Habitat of Unique Biodiversity*, ed. Anon., pp. 46–59. Orissa Environmental Society.

SWAINE, M.D. & WHITMORE, T.C. 1988. On the definition of ecological species groups in tropical rain forests. *Vegetatio* 75: 81–86.

SWAMY, P.S. & RAMAKRISHNAN, P.S. 1987a. Contribution of *Mikania micrantha* during secondary succession following slash-and-burn agriculture (jhum) in north-east India. I. Biomass, litterfall and productivity. *Forest Ecol. Managem.* 22: 229–238.

SWAMY, P.S. & RAMAKRISHNAN, P.S. 1987b. Contribution of *Mikania micrantha* during secondary succession following slash-and-burn agriculture (jhum) in north-east India. II. Nutrient cycling. *Forest Ecol. Managem.* 22: 239–250.

SWAMY, P.S., SUNDARAPANDIAN, S.M., CHANDRASEKAR, P. & CHANDRASEKARAN, S. 2000. Plant species diversity and tree population structure of a humid tropical forest in Tamil Nadu, India. *Biodivers. & Conservation* 9: 1643–1669.

SYMINGTON, C.F. 1943, revised ASHTON, P.S. & APPANAH, S. 2004. *Foresters' Manual of Dipterocarps*. Malayan Forest Records

16. Forest Research Institute Malaysia (FRIM) and Malayan Nature Society.

SZECHOWYCZ, R.W. 1959. The study of tropical vegetation. *Ceylon Forester* 4: 65–79.

TAGAWA, H. 1995. Distribution of lucidophyll oak-laurel forest formation in Asia and other areas. *Tropics* 5: 1–40.

TAGAWA, H. 1997. Worldwide distribution of evergreen lucidophyll oak-laurel forests. *Tropics* 6: 295–316.

TAKADA, M. 1991. Landform and quaternary geohistory of the Bhutan Himalaya. In: *Life Zone Ecology of the Bhutan Himalaya* II, ed. M. Ohsawa, pp. 41–88. Laboratory of Ecology, Chiba University.

TAKAHASHI, H. 1999. Hydrological and meteorological environments of inland peat swamp forest in Central Kalimantan, Indonesia with special reference to the effects of forest fire. *Tropics* 9: 17–25.

TAKEDA, T. 1995. Evolution of semelparous and iteroparous perennial plants: comparison between density-independent and density-dependent dynamics. *J. Theor. Biol.* 173: 51–60.

TAKEUCHI, W. 2003. Botanical survey of a lowland utrabasic flora in Papua New Guinea. *Sida* 20: 1491–1559.

TAKEUCHI, W. & GOLMAN, M. 2001. The richest floristic interval in Papuasia is the premontane. *Sida* 19: 445–468.

TAKEUCHI, Y. & NAKASHIZUKA, T. 2007. Effect of distance and density on seed/ seedling fate of two dipterocarp species. *Forest Ecol. Managem.* 247: 167–174.

TAKEUCHI, Y., KENTA, T. & NAKASHIZUKA, T. 2005. Comparison of sapling demography of four dipterocarp species with different seed-dispersal strategies. *Forest Ecol. Managem.* 208: 237–248.

TAKHTAJAN, A. 1969. *Flowering Plants: Origin and Dispersal*. Transl. C. Jeffrey. Oliver & Boyd, Edinburgh.

TAKYU, M., AIBA, S.-I. & KITAYAMA, K. 2002. Effects of topography on tropical lower montane forests under different geological conditions on Mount Kinabalu, Borneo. *Pl. Ecol.* 159: 35–49.

TAKYU, M., AIBA, S.-I. & KITAYAMA, K. 2003. Changes in biomass, productivity and decomposition along topographical gradients under different geological conditions in tropical lower montane forests on Mount Kinabalu, Borneo. *Oecologia* 134: 397–404.

TAN, K.H., ZUBAID, A. & KUNZ, T.H. 1998. Food habits of *Cynopteris brachyotis* (Muller) (Chiroptera: Pteripodidae) in Peninsular Malaysia. *J. Trop. Ecol.* 14: 299–307.

TAN, S., YAMAKURA, T., TANI, M., PALMIOTTO, P., MAMIT, J. D., CHIN, S.-P., DAVIES, S., ASHTON, P. & BAILLIE, I. 2009. Mini Review – review of soils on the 52 ha long term ecological research plot in mixed dipterocarp forest at Lambir, Sarawak, Malaysian Borneo. *Tropics* 18: 61–86.

TAN, S.S., PRIMACK, R.B., CHAI, E.O.K. & LEE, H.S. 1987. The silviculture of dipterocarp trees in Sarawak, Malaysia. III. Plantation forest. *Malayan Forester* 50: 148–161.

TANG, H.T. & CHONG, P.E. 1979. Sudden mortality in a regenerated stand of *Shorea curtisii* in Senaling Inas Forest Reserve,

Negri Sembilan. *Malayan Forester* 42: 240–254.

TANGE, T., YAGI, H., SASAKI, S., NIIYAMA, K. & ABDUL RAHMAN, K. 1998. Relationship between topography and soil properties in a hill dipterocarp forest dominated by *Shorea curtisii* at Semengkok Forest Reserve, Peninsular Malaysia. *J. Trop. For. Sci.* 10: 398–409.

TANI, Y. 2000. Ecological factor affecting taungya farmers behavior in teak plantation projects: a case study in Bago Range, Union of Myanmar. *Tropics* 10: 273–286.

TANNER, E.V.J., VITOUSEK, P.M. & CUEVAS, P.M. 1998. Experimental investigation of nutrient limitation of forest growth on wet tropical mountains. *Ecology* 79: 10–22.

TANNER, E.V.J., TEO, V.K., COOMES, D.A. & MIDGLEY, J.J. 2005. Pairwise competition trials amongst seedlings of ten dipterocarp species; the role of initial height, growth rate and leaf attributes. *J. Trop. Ecol.* 21: 317–328.

TANSLEY, A.G. 1939. *The British Islands and their Vegetation*. Cambridge University Press.

TANTRA, I.G.M. 1977. The establishment of tengkawang (*Shorea* spp.) plantations in Indonesia. In: *Biotropica Special Publication* 4: 232–241. Proceedings of the Symposium on Management of Forest Production in South-East Asia. SEAMEO Regional Centre of Tropical Biology, Bogor.

TARLING, N. 1988. Gondwana and the evolution of the Indian Ocean. In: *Gondwana and Tethys*, eds M.G. Audley-Charles & A. Hallam. Geological Society Special Publication 37: 61–77. Oxford University Press.

TAYASU, I., SUGIMOTO, A., WADA, E. & ABE, T. 1994. Xylophylous termites depending on atmospheric nitrogen. *Naturwissenschaften* 81: 229–231.

TAYLOR, D., SAKSENA, P.G., SANDERSON, P.G. & KUCERA, K. 1999. Environmental change and rain forests on the Sunda shelf of Southeast Asia: drought, fire and the biological cooling of biodiversity hotspots. *Biodivers. & Conservation* 8:1159–1177.

TAYLOR, G.F. 1982. The development of forest policy in India – the forgotten policy: Sir Herbert Howard's post-war policy of 1944. *Indian Forester* 108: 196–201.

TEEJUNTUK, S., SAHUNALU, K., SAKURAI, K. & SUNGPALEE, W. 2003. Forest structure and tree species diversity along an altitudinal gradient in Doi Inthanon National Park, Thailand. *Tropics* 12: 85–102.

TEEUW, R.M., RHODES, E,J, & N.K. PERKINS. 1999. Dating of Quaternary sediments from western Borneo, using optically stimulated luminescence. *J. Trop. Geogr.* 20: 181–192.

TENNAKOON, M.M.D, GUNATILLEKE, I.A.U.N., HAFEEL, K.M., SENEVIRATNE, G., GUNATILLEKE, C.V.S. & ASHTON, P.M.S. 2005. Ectomycorrhizal colonization and seedling growth of *Shorea* (Dipterocarpaceae) seedlings in simulated shade environmentsof a Sri Lankan rain forest. *Forest Ecol. Managem.* 208: 399–405.

TERBORGH, J. 1985. The vertical component of plant species diversity in temperate and tropical forests. *Amer. Naturalist* 126: 760–776.

TERBORGH, J. 1986. Keystone plant resources in the tropical forest. In: *Conservation Biology: the Science of Scarcity and Diversity*, ed. M.E. Soulé, pp. 330–344. Sinauer Associates, Sunderland, Massachusetts.

TERBORGH, J., TER STEEGE, H. & WRIGHT, S.J. 2004. Why do some tropical forests have so many species of trees? *Biotropica* 36: 447–473.

TEWARI, D.N. 1983. Development strategy for forests, tribals and rural poor. *Indian Forester* 109: 795–803.

TEWARI, J.C., RIKHARI, H.C. & SINGH, S.P. 1989. Compositional and structural features of certain tree stands along an elevational gradient in Central Himalaya. *Vegetatio* 85: 117–120.

TEWARI, R. & KALA, R.P. 2007. Constraints on marketing Hoplo-affected sal timber. *Indian Forester* 133: 822–830.

THADANI, R. & ASHTON, M.S. 1995. Regeneration of banj oak (*Quercus leucotrichophora* A. Camus) in the Central Himalaya. *Forest Ecol. Managem.* 78: 217–224.

THAKUR, A.S. & KHARE, P.K. 2006. Disappearing *Boswellia serrata* Roxb. in Sagar District, Madhya Pradesh. *Indian Forester* 132: 889–893.

THAKUR, M.L., AHMAD, M. & THAKUR, R.K. 1992. Lantana weed (*Lantana camara* var. *aculeata* Linn.) and its possible management through insect pests in India. *Indian Forester* 118: 466–488.

THAMPI, S.P. 2005. Ecotourism in Kerala, India: lessons from the ecodevelopment project in Periyar Tiger Reserve. *ECOCLUB. Co. E-paper* series 13.

THANG, H.C. 1985. Timber supply and domestic demand in Peninsular Malaysia. *Malayan Forester* 48: 87–100.

THANG, H.C. 1987. Forest management systems for tropical high forest, with special reference to peninsular Malaysia. *Forest Ecol. Managem.* 21: 3–20.

THANGAM, E.S. 1982. Regeneration methods of *Dipterocarpus* species in India. *Indian Forester* 108: 637–647.

THOMAS, M.F. 2006. Lessons from the tropics for a global geomorphology. *J. Trop. Geogr.* 27: 111–127.

THOMAS, S.C. 1993. *Interspecific Allometry in Malaysian Forest Trees*. PhD thesis, Harvard University.

THOMAS, S.C. 1996a. Asymptotic height as a predictor of growth and allometric characteristics in Malaysian rain forest trees. *Amer. J. Bot.* 83: 556–566.

THOMAS, S.C. 1996b. Relative size at onset of maturity in rain forest trees: a comparative analysis of 37 Malaysian species. *Oikos* 76: 145–154.

THOMAS, S.C. 1996c. Reproductive allometry in Malaysian rain forest trees: biomechanics versus optimal allocation. *Evol. Ecol.* 10: 517–530.

THOMAS, S.C. 1997. Geographic parthenogenesis in a tropical forest tree. *Amer. J. Bot.* 84: 1012–1015.

THOMAS, S.C. & ICKES, K. 1995. Ontogenetic changes in leaf size in Malaysian forest trees. *Biotropica* 27: 427–434.

THOMAS, S.C. & LAFRANKIE, J.V. 1993. Sex, size, and interyear variation in flowering among dioecious trees of the Malayan rain forest. *Ecology* 74: 1529–1537.

THOMPSON, J. 1995. Participatory approaches in government bureaucracies: facilitating the process of institutional change. *World Developm.* 23 (9): 1521–1554.

THORNTON, I.W.B. 1996. *Krakatau, the Destruction and Reassembly of an Island Ecosystem*. Harvard University Press.

TIBBETTS, R.A. & SMITH, J.A.C. 1992. Vacuolar accumulation of calcium and its interaction with magnesium. In: *The Vegetation of Ultramafic (Serpentine) Soils*, eds A.J.M. Baker, J. Proctor & R.D. Reeves, pp. 367–373. Intercept, Andover, New Hampshire.

TILMAN, D. 1982. *Resource Competition and Community Structure*. Princeton Monographs in Population Biology 7.

TILMAN, D. 1985. The resource-ratio hypothesis of plant succession. *Amer. Nat.* 125: 827–852.

TILMAN, D. 1986a. Evolution and differentiation in terrestrial plant communities: the importance of the soil resource: light gradient. In: *Community Ecology*, eds J. Diamond & T.J. Case, pp. 359–380. Harper & Rowe, New York.

TILMAN, D. 1986b. Nitrogen-limited growth in plants from different successional stages. *Ecology* 67: 555–563.

TILMAN, D. 1987. Secondary succession and the pattern of plant dominance along experimental nitrogen gradients. *Ecol. Monog.* 57: 189–214.

TILMAN, D. 1999. The ecological consequences of changes in biodiversity: a search for general principles. *Ecology* 80: 1455–1474.

TIWARY, N.K. 1926. On the occurrence of polyembryony in the genus *Eugenia*. *J. Indian Bot. Soc.* 5: 124–136.

TOBE, H. & RAVEN, P.H. 1987a. The embryology and relationships of *Crypteronia* (Crypteroniaceae). *Bot. Gaz.* 148: 96–102.

TOBE, H. & RAVEN, P.H. 1987b. The embryology and relationships of *Dactylocladus* (Crypteroniaceae) and a discussion of the family. *Bot. Gaz.* 148: 103–111.

TOKY, O.P. & RAMAKRISHNAN, P.S. 1983. Secondary succession following slash and burn agriculture in north-eastern India. II. Biomass, litterfall and productivity. *J. Ecol.* 71: 735–747.

TOMA, T. 1999. Exceptional droughts and forest fires in eastern part of Borneo island. *Tropics*, 9: 55–72.

TOMAR, M.S. & JOSHI, S.C. 1978. Madhya Pradesh forests and people's demands – a situation by 2000 A.D. *Indian Forester* 104: 661–675.

TOMICH, T.P., VAN NOORDWIJK, M., BUDIDARSONO, S., GILLISON, A., KUSUMANTO, T., MURDIYARSO, D., STOLLE, F. & FAGI, A.M., eds. 1998. *Alternatives to Slash-and-Burn in Indonesia*. Summary Report and Synthesis of Phase II. ICRAF SE Asia Bogor.

TOMICH, T.P., VAN NOORDWIJK, M., BUIDARSONO, S., GILLISON, A., KUSUMANOT, T., MURDIYARSO, D., STOLLE, F. & FAGI, A.M. 2001. Agricultural intensification, deforestation and the environment: assessing tradeoffs

in Sumatra, Indonesia. In: *Trade-offs or Synergies? Agricultural Intensification, Economic Development and the Environment*, eds D.R. Lee & C.B. Barrett, pp. 221–244. CABI, Wallingford.

TOMLINSON, P.B. 1979. Systematics and ecology of the Palmae. *Annual Rev. Ecol. Syst.* 10: 85–107.

TOMLINSON, P.B. 1986. *The Botany of Mangroves*. Cambridge University Press.

TOMLINSON, P.B. 2006. The uniqueness of palms. *Bot. J. Linn. Soc.* 151: 5–14.

TOMLINSON, P.B. & ZIMMERMAN, M.H., eds. 1978. *Tropical Trees as Living Systems. Proceedings of the Fourth Cabot Symposium.* Cambridge University Press.

TONGKUB, F. 1991. Tectonic evolution of Sabah, Malaysia. *J. S.E. Asian Earth Sciences* 6: 393–406.

TOUGARD, C. 2001. Biogeography and migration routes of large mammal faunas in South-East Asia during the late middle Pleistocene: focus on the fossil and extant faunas from Thailand. *Palaeogeogr. Palaeoclimatol. Palaeoecol.* 168: 337–358.

TOUGARD, C. & MONTUIRE, S. 2006. Pleistocene palaeoenvironmental reconstructions and mammalian evolution in South-East Asia: focus on fossil faunas from Thailand. *Quatern. Sci. Rev.* 25: 126–141.

TOURANONT, R. 2000. *Effects of the White-Handed Gibbon (Hylobates lar) on Habitat, Spatial Distribution,and Seed Dispersal of Trees in Khao Yai National Park, Thailand*. Masters dissertation, Harvard University.

TOYE, R.J., MARSHALL, A.G. & THO, Y.P. 1992. Fruiting phenology and the survival of insect fruit predators: a case study from the South-East Asian Dipterocarpaceae. *Philos. Trans., Ser. B* 335: 417–423.

TRELOAR, P.J. & SEARLE, M.P., eds. 1993. *Himalayan Tectonics*. Geological Society Special Publication 74. Geological Society, London.

TREUB, M. 1911. Le sac embryonnaire et l'embryon dans les Angiospermes. *Ann. Jard. Bot. Buitenzorg* 24: 1–27.

TREVES, A., WALLACE, R.B., NAUGHTON-TREVES, L. & MORALES, A. 2006. Comanaging human-wildlife conflicts: a review. *Hum. Dimens. Wildl.* 11: 383–396.

TRIPATHI, K.P., MEHOTRA, S. & PUSHPANGADAN, P. 2006a. Vegetation characteristics of tropical forests of Andaman Islands. *Indian Forester* 132: 165–180.

TRIPATHI, K.P., PANDEY, H.N. & TRIPATHI, R.S. 2006b. Fragmentation and plant diversity status of major forest types in Meghalaya, North-East India. *Indian Forester* 132: 1598–1608.

TRIPATHI, R.S. & KHAN, M.L. 1990. Effects of seed weight and microsite characteristics on seedling fitness in two species of *Quercus* in a subtropical wet hill forest. *Oikos* 57: 289–296.

TRIPATHI, R.S. & KHAN, M.L. 1992. Regeneration pattern and population structure of tropical trees in sub-tropical forests of North-East India. In: *Tropical Ecosystems. Ecology and Management*, eds K.P. Singh & J.S. Singh, pp. 431–441. Wiley Eastern, New Delhi.

TRIPATHI, S.P. 2005. Status of joint forest management in Nagaland. *Indian Forester* 131: 1158–1171.

TROUP, R.S. 1913. A note on the causes and effects of the drought of 1907 and 1908 on the *sal* forests of the United Provinces. *Indian Forest Bull.* 22: 1–17.

TROUP, R.S. 1921. The Silviculture of Indian Trees. Oxford University Press.

TSUMURA, Y. 2002. Genetic diversity and conservation of tropical forest. *Tropics* 11: 241–247.

TUOMELA, K., KUUSIPALO, J., VESA, L., NURYANTO, K., SAGALA, A.P.S. & ÅDJERS, G. 1996. Growth of dipterocarp seedlings in artificial gaps: an experiment in a logged-over rainforest in South Kalimantan, Indonesia. *Forest Ecol. Managem.* 81: 95–100.

TUOMISTO, H., RUOKALAINAN, K., AQUILAR, M. & SARMIENTO, A. 2003a. Floristic patterns along a 43 km long transect in an Amazonian rain forest. *J. Ecol.* 91: 743–756.

TUOMISTO, H., RUOKALAINAN, K. & YLI-HALLA, M. 2003b. Dispersal, environment, and floristic variation of western Amazonian forests. *Science* 299: 241–244.

TURNER, H., HOVENKAMP, P. & VAN WELZEN, P.S. 2001. Biogeography of Southeast Asian and West Pacific. *J. Biogeogr.* 28: 217–230.

TURNER, I.M. 1989. An enumeration of one hectare of Pantai Aceh Forest Reserve, Penang. *Gard. Bull. Singapore* 42: 29–44.

TURNER, I.M. 1990. The seedling survivorship and growth of three *Shorea* species in a Malaysian tropical rain forest. *J. Trop. Ecol.* 6: 469–478.

TURNER, I.M. 1994a. Sclerophylly: primarily protective? *Funct. Ecol.* 8: 669–675.

TURNER, I.M. 1994b. The taxonomy and ecology of the vascular plant flora of Singapore: a statistical analysis. *Bot. J. Linn. Soc.* 114: 215–227.

TURNER, I.M. 1996. Species loss in fragments of tropical rain forest: a review of the evidence. *J. Appl. Ecol.* 33: 200–209.

TURNER, I.M. 2001a. *The Ecology of Trees in the Tropical Rain Forest*. Cambridge University Press.

TURNER, I.M. 2001b. An overview of the plant diversity of South-East Asia. *Asian J. Trop. Biol.* 4: 1–16.

TURNER, I.M. 2001c. Rainforest, ecosystems, plant diversity. In: *Encyclopoedia of Biodiversity* 5, ed. S.A. Levin, pp. 13–23. Academic Press.

TURNER, I.M. & CORLETT, R.T. 1996. The conservation value of small isolated fragments of lowland tropical rainforest, *Trends Ecol. Evol.* 11: 330–333.

TURNER, I.M., RAICH, J.W., GONG, W.K. ONG, J.E. & WHITMORE, T.C. 1991. The dynamics of Pantai Acheh Forest Reserve: a synthesis of recent research. *Malayan Nat. J.* 45: 166–174. [Also in: *In Harmony with Nature. Proceedings of the International Conference on Conservation of Tropical Biodiversity*, eds S.K. Yap & S.W. Lee, pp. 166–174. Malaysian Nature Society.]

TURNER, I.M., TAN, H.T.W., WEE, W.C., IBRAHIM, A., CHEW, P.T. & CORLETT, R.T. 1993. A study of plant species extinction in Singapore: lessons for the conservation of tropical biodiversity. *Conservation Biol.* 8: 705–712.

TURNER, I.M., WONG, W.K., CHEW, P.T.

& IBRAHIM, A.B. 1997a. Tree species richness in primary and old secondary forest in Singapore. *Biodivers. & Conservation* 6: 537–544.

TURNER, I.M., YONG, J.W.H., ISMAIL, A.Z., & LATIFF, A. 1997b. The botany of the islands of Mersing District, Johore, Peninsular Malaysia. 1. The plants and vegetation of Pulau Tinggi. *Gard. Bull. Singapore* 49: 119–141.

TURNER, I.M., LUCAS, P.W., BECKER, P., WONG, S.C., WONG, J.W.H., CHOONG, M.F. & TYREE, M.T. 2000. Tree leaf form in Brunei: a heath forest and mixed dipterocarp forest compared. *Biotropica* 32: 53–61.

TYREE, M. & EWERS, F. 1996. Hydraulic architecture of woody tropical plants. In: *Tropical Plant Forest Ecophysiology*, eds S.S. Mulkey, R.L. Chazdon & A.P. Smith, pp. 217–243. Chapman & Hall, London & New York.

TYREE, M.T., PATIÑO, S. & BECKER, P. 1998. Vulnerability to drought-induced embolism of Bornean heath and dipterocarp forest trees. *Tree Physiol.* 18: 583–588.

UPADHAYU, K., PANDEY, H.N., LAW, P.S. & TRIPATHI, R.S. 2003. Tree diversity in sacred groves of Jaintia Hills in Meghalaya, northeast India. *Biodivers. & Conservation* 12: 583–597.

UTKARSH, G., JOSHI, N.V. & GADGIL, M. 1998. On the patterns of tree diversity in the Western Ghats of India. *Curr. Sci.* 75: 594–603.

VALENCIA, R., CONDIT, R., FOSTER, R.B., ROMOLEROUX, K., MUÑOZ, G.V., SVENNING, J.-C., MAGÅRD, E., BASS, M., LOSOS, E.C. & BALSLEV, H. 2004. Yasuní forest dynamics plot, Ecuador. In: *Tropical Forest Diversity and Dynamism: Findings from a Large-Scale Plot Network*, eds E.C. Losos & E.G. Leigh Jr., pp. 609–620. University of Chicago Press, Chicago.

VARGHESE, A.O. & MENON, A.R.R. 1999. Floristic composition, dynamics and diversity of *Myristica* swamp forests of south-western Ghats, Kerala. *Indian Forester* 125: 775–783.

VASANTHY G. 1988. Pollen analysis of late Quaternary sediments: evolution of upland savanna in Sandynallah (Nilgiris, South India). *Rev. Palaeobot. Palynol.* 55: 175–192.

VÁSQUEZ, G.J.A & GIVNISH, T.J. 1998. Altitudinal gradients in tropical forest composition, structure and diversity in the Sierre de Manantlán. *J. Ecol.* 86: 999–1020.

VÁSQUEZ-YÁNES, C. & SMITH, H. 1982. Phytochrome controls of seed germination in the tropical rain forest pioneer trees *Cecropia obtusifolia* and *Piper auritum* and its ecological significance. *New Phytol.* 92: 477–485.

VÁSQUEZ-YÁNES, C., OROZCO-SEGOVIA, A., RINCÓN, E., SÁNCHEZ-CORONADO, M.E., HUANTE, P., TOLEDO, J.R. & BARRADOS, V.L. 1990. Light beneath the litter in a tropical forest: effect on seed predation. *Ecology* 71: 1952–1958.

VEERAMANI, A., EASA, P.S. & JAYSON, E.A. 2004. Socio-economic status of cultivators and their interface with wild animals: a case study of Marayur forest range, Kerala. *Indian Forester* 130: 513–520.

VENKATASWARAN, R. & PARTHASARATHY, N. 2005. Tree population changes in a tropical dry evergreen forest of South India over a decade (1992–2002). *Biodivers. & Conservation* 14: 1335–1344.

VERMA, D.P.S. 1989. Women and forestry (tale of six harijan women of Ganashpura). *Indian Forester* 115: 780–788.

VIA, S. 2002. The ecological genetics of speciation. *Amer. Naturalist* 159: S1–7.

VIDAL, J. 1956. *La Végétation dul Laos. I. Le Milieu;* 1960. *II. Groupements Végétaux et Flore.* Douladoure, Toulouse.

VINCENT, G. 2001. Leaf photosynthetic capacity and nitrogen content adjustment to canopy openness in tropical forest tree seedlings. *J. Trop. Ecol.* 17: 495–509.

VINCENT, G., FORESTA, H. DE & MULIA, R. 2002. Predictions of tree growth in a dipterocarp-based agroforest: a critical assessment. *Forest Ecol. Managem.* 161: 39–52.

VINCENT, J.R. 1992. The tropical timber trade and sustainable development. *Science* 256: 1651–1654.

VINCENT, J.R. 2010. Microeconomic analysis of innovative environmental programs in developing countries. *Rev. Environm. Econ. Policy* 4: 221–233.

VINK, W. 1970. The Winteraceae of the Old World. I. *Pseudowintera* and *Drimys*. *Blumea* 18: 225–354.

VIR, S. & JINDAL, S.K. 1994. Fruit infestation of *Acacia tortilis* (Forsk.) Hyne by *Bruchidius andrewesii* Pic. (Coleoptera, Bruchidae) in the Thar desert. *Forest Ecol. Managem.* 70: 349–352.

VITANAGE, P.W. 1972. Post-precambrian uplifts and regional neotectonic movements in Ceylon. *Proceedings of the 24th International Geological Congress,* Montreal, Sect. 3: 641–654.

VITOUSEK, P.M. 1994. Beyond global warming: ecology and global change. *Ecology* 75: 1861–1876.

VITOUSEK, P.M. 1998. The structure and functioning of montane tropical forests: control by climate, soils and disturbance. *Ecology* 79: 1–2.

VOLKOV, I., BANAVAR, J.R., HUBBELL, S.P. & MARITAN, A. 2003. Neutral theory and relative species abundance in ecology. *Nature* 424: 1035–1037.

VOLTERRA, V. 1926. Variations and fluctuations of the number of individuals in animal species living together. Reprinted, 1931, in: *Animal Ecology,* ed. R.N. Chapman. McGraw-Hill, New York.

VORIS, H. 2000. Maps of putative Pleistocene sea levels in Southeast Asia: shorelines, river systems and time durations. *J. Biogeogr.* 27: 1153–1167.

WADE, L.K. & MCVEAN, D.N. 1969. *The Alpine and Subalpine Vegetation. Mt. Wilhelm Studies* 1. The Australian National University Press, Canberra.

WAIDE, R.B., ZIMMERMAN, J.K. & SCATENA, F.N. 1998. Controls of primary productivity: lessons from the liquillo mountains in Puerto Rico. *Ecology* 79: 31–37.

WALKER, D. 1970. The changing vegetation of the montane tropics. *Search* 1: 217–221.

WALKER, D. 1972. Vegetation of the Lake Ipea region, New Guinea Highlands. II. Kayamanda Swamp. *J. Ecol.* 60: 479–504.

WALKER, D. 1986. Late Pleistocene–early Holocene vegetational and climatic changes in Yunnan Province, southwest China. *J. Biogeogr.* 13: 477–486.

WALLACE, A.R. 1891. *Natural Selection and Tropical Nature; Essays on Descriptive and Theoretical Biology.* Macmillan, London.

WALSH, R.P.D. 1992. Representation and classification of tropical climates for ecological purposes using the perhumidity index. *Swansea Geogr.* 24: 109–129.

WALSH, R.P.D. 1996a. Drought frequency changes in Sabah and adjacent parts of northern Borneo since the late nineteenth century and possible implications for tropical rain forest dynamics. *J. Trop. Ecol.* 12: 385–407.

WALSH, R.P.D. 1996b. Climate. In: *The Tropical Rain Forest,* 2nd ed, ed. P.W. Richards, pp. 159–205. Cambridge University Press.

WALSH, R.P.D. & NEWBERY, D.M. 1999. The ecoclimatology of Danum, Sabah, in the context of the World's rainforest regions, with particular reference to dry periods and their impact. *Philos. Trans., Ser. B* 354: 1869–1883.

WALSH, R.P.D. & NEWBERY, D.M. 2000. The ecoclimatology of Danum, Sabah, in the context of the World's rainforest regions, with particular reference to dry periods and their impact. In: *Changes and Disturbance in Tropical Rainforest in South-East Asia,* eds D.M. Newbery, T.H. Clutton-Brock & G.T. Prance, pp. 145–160. Royal Society and Imperial College Press, London.

WALTER, H. & LIETH H. 1960. *Klimadiagramm Weltatlas.* G. Fischer, Jena.

WALTER, H., HARNICKELL, E. & MUELLER-DOMBOIS, D. 1975. *Climate-Diagram Maps of the Ecological Climatic Regions of the Earth.* Springer, Berlin.

WALTON, G. & BARNETT, J. 2008. The ambiguities of 'environmental' conflict: insights from the Tolukuma gold mine, Papua New Guinea. *Soc. Nat. Res.* 21 (1): 1–16.

WANANDORN, P.W. 1934. Death of forest trees and animals through drought. *J. Siam. Soc.* 9 Suppl.: 326.

WAN JULIANA, W.A. 2001. *Habitat Specialization of Tree Species in a Tropical Rain Forest.* PhD thesis, University of Aberdeen.

WAN JULIANA, W.A., BURSLEM, D.F.R.P. & SWAINE, M.D. 2009. Nutrient limitation of seedling growth on contrasting soils from Pasoh Forest Reserve, Peninsular Malaysia. *J. Trop. Forest Sci.* 21: 316–327.

WANG, B., CLEMENS, S.C. & LIU, P.I. 2003. Contrasting the Indian and East Asian monsoons. *Mar. Ecol.* 201: 5–21.

WANG, C. 1961. *The Forests of China.* Maria Moors Cabot Foundation Publication 5, Harvard University.

WANG, G. 1958. The Nanthai trade. A study of the early history of Chinese trade in the South China Sea. *Monogr. Malaysian Branch Roy. Asiat. Soc.* 31: 1–135.

WANG, L., SARNTHEIN, M., ERLENKEUSER, H., GRIMALT, J., GROOTES, P., HEILIG, S., IVANOVA, E., KIENAST, M., PELEJERO, C. & PFLAUMANN, U. 1999. East Asian monsoon climate during the late Pleistocene: high resolution sediment records from the South China Sea. *Mar. Geol.* 156: 245–284.

WANGDA, P. & OHSAWA, M. 2006a. Gradational forest changes along the climatically dry valley slopes of Bhutan in the midst of humid eastern Himalaya. *Pl. Ecol.* 186: 109–128.

WANGDA, P. & OHSAWA, M. 2006b. Structure and regeneration dynamics of dominant tree species along altitudinal gradient in a dry valley slopes of the Bhutan Himalaya. *Forest Ecol. Managem.* 230: 136–150.

WANGDI, K. 2010. To join or not to join WTO: a study on its negative impacts. *J. Bhutan Stud.* 23: 55–117.

WATANABE, H. & SAWAENG, R. 1984. Cast production of the Megascoleid earthworm *Pheretima* sp. in Northeastern Thailand. *Pedobiologia* 26: 37–44.

WATANABE, T. & OSAKI, M. 2001. Influence of aluminium and phosphorus on growth and xylem sap composition in *Melastoma malabathricum* L. *Pl. & Soil* 237: 63–70.

WATANABE, T. & OSAKI, M. 2002. Role of organic acids in aluminium accumulation and plant growth in *Melastoma malabathricum. Tree Physiol.* 22: 785–792.

WATERMAN, P.G., ROSS, J.A., BENNETT, E.L. & DAVIES, A.G. 1988. A comparison of the floristics and leaf chemistry of the tree flora in two Malaysian rain forests and the influence of leaf chemistry on populations of colobine monkeys in the Old World. *Biol. J. Linn. Soc.* 34: 1–32.

WATSON, J.G. 1928. The mangrove swamps of the Malay peninsula. *Malayan Forest Rec.* 6.

WATT, G., ed. 1899–1993. *A Dictionary of the Economic Products of India.* Superintendent of Government Printing, Calcutta.

WEBB, C.O. 2000. The phylogenetic structure of ecological communities. An example for rain forest trees. *Amer. Naturalist* 158: 145–155.

WEBB, C.O. & PEART, D.R. 2000. Habitat associations of trees and seedlings in a Bornean rain forest. *J. Ecol.* 88: 464–478.

WEBB, C.O., ACKERLY, D.A., MCPEEK, M.A. & DONOGHUE, M.J. 2002. Phylogenies and community ecology. *Annual Rev. Ecol. Syst.* 33: 475–505.

WEBB, C.O., GILBERT, G.S. & DONOGHUE, M.J. 2006. Phylogenetic-dependent seedling mortality, size structure, and disease in a Bornean rain forest. *Ecology* 87 (7 suppl.): S123–S131.

WEBB, C.O., CANNON, C.H. & DAVIES, S.J. 2008. Ecological organization, biogeography, and the phylogenetic structure of tropical forest tree communities. In: *Tropical Forest Community Ecology,* eds W.P. Carson & S.A. Schnitzer, pp. 79–97. Wiley-Blackwell, Chichester.

WEBB, E.L. & SAH, R.N. 2003. Structure and diversity of natural and managed sal (*Shorea robusta* Gaertn. f.) forest in the terai of Nepal. *Forest Ecol. Managem.* 176: 337–353.

WEBB, L.J. 1959. A physiognomic classification of Australian rain forest. *J. Ecol.* 47: 551–570.

WEBB, L.J. 1986. *Report on a visit to the South China Institute of Botany, Academia Sinica, Guangzhou, 13–27 October 1986.* Rainforest Conservation Society of Queensland. Unpubl. ms. [*per* P.J. Grubb]

WEBB, L.J., TRACEY, J.G. & WILLIAMS, W.T. 1984. A floristic framework of Australian rainforests. *Austral. J. Ecol.* 9: 169–198.

WEBBER, M.L. 1934. Fruit dispersal. *Malayan Forester* 3: 18–19.

WEERASINGHE H.A.S. & IQBAL, M.C.M. 2011. Plant diversity and soil characteristics of the Ussangoda serpentine site. *J. Natl. Sci. Council Sri Lanka* 39: 355–363.

WEIBLEN, G.D. 1999. *Phylogeny and Ecology of Dioecious Fig Pollination.* PhD thesis, Harvard University.

WEIBLEN, G.D. 2001. Phylogenetic relationships of fig wasps pollinating functionally dioecious *Ficus* based on mitochondrial DNA sequences and morphology. *Syst. Biol.* 50: 243–267.

WEIBLEN, G.D. 2004. Correlated evolution in fig pollination. *Syst. Biol.* 53: 128–139.

WELLS, D.R. 1999. *The Birds of the Thai-Malay Peninsula. I. Nonpasserines.* Introduction: xvi–xxxvi. Academic Press.

WELZEN, P.C. VAN, PARNELL, J.A.N. & SLIK, J.W.F. 2011. Wallace's Line and plate tectonics: two or three phytogeographical areas and where to group Java? *Biol. J. Linn. Soc.* 103: 531–545.

WERNER, W.L. 1984. Die Hohen- und Nebel walder auf der Insel Ceylon (Sri Lanka). *Trop. Subtrop. Pflanzenwelt* 46, Wiesbaden.

WERNER, W.L. 1988. Canopy die-back in the upper montane rain forests of Sri Lanka. *Geojournal* 17: 245–248.

WERNER, W.L. 1993. *Pinus in Thailand. Geoecological Research* 7. F. Steiner, Stuttgart.

WERNER, W.L. 1997. Pines and other conifers in Thailand – a Quaternary relic? *J. Quaternary Sci.* 12: 451–454.

WERNER, W. 2001. *Sri Lanka's Magnificent Cloud Forests.* World Heritage Trust Publications, Colombo.

WHARTON, C.H. 1968. Man, fire and wild cattle in Southeast Asia. *Proceedings of the Eighth Annual Tall Timbers Fire Ecology Conference,* pp. 107–167. Tall Timbers Research Station, Tallahassee, Florida.

WHEELWRIGHT, N.T. 1985a. Fruit size, gape width, and the diets of fruit-eating birds. *Ecology* 66: 808–818.

WHEELWRIGHT, N.T. 1985b. Competition for dispersers, and the timing of flowering and fruiting in a guild of tropical trees. *Oikos* 44: 465–477.

WHITE, F. 1965. The savanna woodlands of the Zambezian and Sudanian domains. *Webbia* 19: 651–671.

WHITFORD, H.N. 1906. The vegetation of the Lamão Forest Reserve. *Philipp. J. Sc.* 1: 373–431, 637–682.

WHITFORD, H.N. 1909. *Studies in the Vegetation of the Philippines. I. The Composition and Volume of the Dipterocarp Forest.* Bureau of Printing, Manila.

WHITFORD, H.N. 1911. *The Forests of the Philippines. I. Forest types and products.* Bulletin 10, Department of the Interior Bureau of Forestry. Bureau of Printing, Manila.

WHITMORE, T.C. 1973. Datiscaceae. In: *Tree Flora of Malaya,* ed. T.C. Whitmore, pp. 29–31. Longmans, Kuala Lumpur.

WHITMORE, T.C. 1975. *Tropical Rain Forests of the Far East.* 1st ed. Clarendon Press, Oxford.

WHITMORE, T.C., ed. 1981. *Wallace's Line and Plate Tectonics.* Clarendon Press, Oxford.

WHITMORE, T.C. 1982. Wallace's Line: a result of plate tectonics. *Ann. Missouri Bot. Gard.* 69: 668–675.

WHITMORE, T.C. 1984a. *Tropical Rain Forests of the Far East.* 2nd ed. Clarendon Press, Oxford.

WHITMORE, T.C. 1984b. A vegetation map of Malesia at scaled 1: 5 million. *J. Biogeogr.* 11: 461–473.

WHITMORE, T.C., ed. 1987. *Biogeographical Evolution of the Malay Archipelago.* Clarendon Press, Oxford.

WHITMORE, T.C. 1988. Phytogeography of the eastern end of Tethys. In: *Gondwana and Tethys,* Geological Society Special Publication 37, eds M.G. Audley-Charles & A. Hallam, pp. 307–311. Oxford University Press.

WHITMORE, T.C. 1989a. Changes over twenty-one years in the Kolombangara rain forests. *J. Ecol.* 77: 469–483.

WHITMORE, T.C. 1989b. Canopy gaps and the two major groups of forest trees. *Ecology* 70: 536–538.

WHITMORE, T.C. 1989c. South-East Asian Tropical Forests. In: *Tropical Rain Forest Ecosystems: Biogeographical and Ecological Studies,* eds H. Lieth & M.J.A. Werger, pp. 195–218. Elsevier, Amsterdam.

WHITMORE, T.C. 1990. *An Introduction to Tropical Rain Forests.* Clarendon Press, Oxford.

WHITMORE, T.C. 1991. Invasive woody plants in perhumid tropical climates. In: *Ecology of Biological Invasions in the Tropics,* ed. P. Ramakrishnan, pp. 35–40. International Scientific, New Delhi.

WHITMORE, T.C. 1998a. *An Introduction to Tropical Rain Forests,* 2nd ed. Oxford University Press, Oxford.

WHITMORE, T.C. 1998b. *Palms of Malaya,* 2nd ed. White Lotus Press, Thailand.

WHITMORE, T.C. 2008. *The Genus Macaranga: a Prodromus.* Royal Botanic Gardens, Kew.

WHITMORE, T.C. & BROWN, N.D. 1996. Dipterocarp seedling growth in rain forest canopy gaps during six and a half years. *Philos. Trans., Ser. B* 351: 1195–1203.

WHITMORE, T.C. & BURNHAM, C.P. 1969. The altitudinal sequence of forests and soils on granite near Kuala Lumpur. *Malayan Nat. J.* 22: 99–118.

WHITMORE, T.C. & BURSLEM, D.F.R.P. 1998. Major disturbance in tropical forests. In: *Dynamics of Tropical Communities. 37th Symposium of the British Ecological Society, Cambridge University,* eds D.M. Newbery, H.H.T. Prins & N.D. Brown, pp. 549–565. Blackwell Science, Oxford.

WHITMORE, T.C. & NG, F.S.P., eds. 1972–1989. *Tree Flora of Malaya.* 4 vols. Longman, Malaysia.

WHITMORE, T.C. & SAYER, J.A. eds. 1994. *Tropical Deforestation and Species Extinction.* IUCN Forest Conservation Programme. Chapman & Hall.

WHITMORE, T.C. & SIDIYASA, K. 1986. Composition and structure of a lowland rain forest in Toraut, northern Sulawesi. *Kew Bull.* 41: 747–757.

WHITMORE, T.C., SIDIYASA, K. & WHITMORE, T.J. 1987. Tree species enumeration of 0.5 ha on Halmahera. *Gard. Bull. Singapore* 40: 31–34.

WHITTAKER, R.J. 1989. Plant recolonization and vegetation succession in the Krakatau Islands, Indonesia. *Ecol. Monog.* 59: 60–163.

WHITTEN, T., SOERIAATMADJA, R.E. & AFIFF, S.A. 1996. *Java and Bali.* Periplus Editions, Hong Kong.

WICKNESWARI, R., ISMAIL, Z., LEE, S.L. & MUHAMMAD, N. 1994. Genetic diversity in remnant and planted populations of *Hopea odorata* in Peninsular Malaysia. In: *Proceedings of the International Workshop, BIO-REFOR, Kangar, Malaysia,* pp. 72–76. BIO-REFOR, Tokyo.

WICKRAMARATNE, S.N., HAYASHI, S. & HERATH, S. 1997. The current status of conserving tropical forests in Sri Lanka. *Tropics* 6: 129–137.

WIEBES, J.T. 1979. Coevolution of figs and their insect pollinators. *Annual Rev. Ecol. Syst.* 10: 1–12.

WIGNARAJA, P. 1984. Towards a theory and practice of rural development. *Development* 2: 3–11.

WIDJAJA-ADHI, I.P.G. 1997. Developing tropical peatlands for agriculture. In: *Biodiversity and Sustainability of Tropical Peatlands,* eds J.O. Rieley & S.E. Page, pp. 293–300. Samara, Cardigan.

WIJESINGHE, M.R. & BROOKE, M. DE L. 2005. Impact of habitat disturbance on the distribution of endemic species of mammals and birds in a tropical rain forest in Sri Lanka. *J. Trop. Ecol.* 21: 661–668.

WIJESINGHE, Y. 1994. *Checklist of Woody Perennial Plants of Sri Lanka.* NARESA & Forest Department, Battaramulla, Sri Lanka.

WIJESOORIYA, W.A.D.A & GUNATILLEKE, C.V.S. 2003. Buffer zone of the Sinharaja Biosphere Reserve in Sri Lanka and its management strategies. *J. Natl. Sci. Found. Sri Lanka* 31: 57–71.

WILFORD, G.E. 1961. The geology and mineral Resources of Brunei and adjacent parts of Sarawak, with descriptions of the Seria and Miri oilfields. *British Borneo Geological Survey, Memoir* 10.

WILLI, Y., VAN BUSKIRK, J. & HOFFMANN, A.A. 2006. Limits to the adaptive potential of small populations. *Annual Rev. Ecol. Syst.* 37: 433–458.

WILLIAMS, L.J., BUNYAVEJCHEWIN, S. & BAKER, P.J. 2008. Deciduousness in a seasonal tropical forest in western Thailand: interannual and interspecific variation in timing, duration and environmental cues. *Oecologia* 55: 571–582.

WILLS, C. & CONDIT, R. 1999. Similar non-random processes maintain diversity in two tropical rainforests. *Proc. Roy. Soc. London, Ser. B., Biol. Sci.* 266: 1445–1452.

WILLS, C., CONDIT, R., FOSTER, R.B. & HUBBELL, S.P. 1997. Strong density- and diversity-related effects help to maintain tree species diversity in tropical forest. *Proc. Natl. Acad. Sci. U.S.A.* 94: 1252–1257.

WILLS, C., CONDIT, R., HUBBELL, S.P. & FOSTER, R.B. & MANOKARAN, N. 2004. Comparable nonrandom forces act to maintain diversity in both a New World and an Old World rainforest plot. In: *Tropical Forest Diversity and Dynamism: Findings from a Large-Scale Plot Network,* eds E.C. Losos & E.G. Leigh Jr., pp. 384–407. University of Chicago Press, Chicago.

WILSON, K.C. & CLIFFORD, M.N. 1992. *Tea. Cultivation and Consumption.* Chapman & Hall, London & New York.

WILSON, P.G., O'BRIEN, M.M., HESLEWOOD, M.M. & QUINN, C.J. 2005. Relationships within Myrtaceae *sensu lato* based on a *matK* phylogeny. *Pl. Syst. Evol.* 251: 3–19.

WIN, U.N. 1951. A note on Kyathaung (*Bambusa polymorpha*) flowering in Pyinmana Forest Division. *Burmese Forester* 1 (2): 51–56.

WONG, K.M. 2011. A biogeographic history of Southeast Asian rainforests. In: *Managing the Future of Southeast Asia's Valuable Tropical Rainforests,* Advances in Asian Human-Environmental Research 2, eds R. Wickneswari & C. Cannon, pp. 21–55. Springer.

WONG, K.M. & PHILLIPS, A. 1996. *Kinabalu, Summit of Borneo.* The Sabah Society.

WONG, K.M., SUGUMARAN, K.M., LEE, D.K.P. & ZAHID, M.S. 2010. Ecological aspects of endemic plant populations on Klang Gates quartz ridge, a habitat island in Peninsular Malaysia. *Biodivers. & Conservation* 19: 435–447.

WONG, S.T., SERVHEEN, C., AMBU, L. & NORHAYATI, A. 2005. Impact of fruit production cycles on Malayan sun bears and bearded pigs in lowland tropical forest of Sabah, Malaysian Borneo. *J. Trop Ecol.* 21: 627–639.

WONG, Y.K. & WHITMORE, T.C. 1970. The influence of soil properties on species distribution in a Malayan lowland forest. *Malayan Forester* 33: 42–54.

WOOD, G.H.S. 1956. The dipterocarp flowering season in North Borneo, 1955. *Malayan Forester* 19: 193–201.

WOOD, G.H.S. & MEIJER, W. 1964. *Dipterocarps of Sabah (North Borneo).* Sabah Forest Department, Malaysia.

WOOD, P.J. 2001. Forests, democracy and the forest service. *Indian Forester* 127: 615–630.

WOOD, T.J., BEAMAN, R.S. & BEAMAN, J.S. 1993. *The Plants of Mount Kinabalu 2: Orchids.* Natural History Publications (Borneo) & Royal Botanic Gardens, Kew.

WOODRUFF, D.S. 2003. Neogene marine transgressions, palaeogeography and biogeographic transitions in the Thai-Malay Peninsula. *J. Biogeogr.* 30: 551–568.

WOODRUFF, D.S. & TURNER, L.M. 2009. The Indo-Chinese-Sundaic biogeographic transition: a description and analysis of terrestrial mammal distributions. *J. Biogeogr.* 36: 803–821.

WOODS, P. 1989. Effects of logging, drought, and fire on structure and composition of tropical forests in Sabah, Malaysia. *Biotropica* 21: 290–298.

WORBES, M. & JUNK, W.J. 1999. How old are tropical trees? The persistence of a myth. *IAWA J.* 20: 255–260.

WORLD TOURISM ORGANISATION. 1998. Ecotourism now one-fifth of market. *World Tour. Org. News* 1: 6.

WÖSTEN, J.H.M., CLYMANS, E., PAGE, S.E., RIELEY, J.O. & LIMIN, S.H. 2008. Peat-water interrelationships in a tropical peatland ecosystem in Southeast Asia. *Catena* 73: 212–224.

WRIGHT, D.H., CURRIE, D.J. & MAURER, B.A. 1993. Energy supply and patterns of species' richness on local and regional scales. In: *Species Diversity in Ecological Communities. Historical and Geographical Perspectives,* eds R.E. Ricklefs & D. Schluter, pp. 66–74. University of Chicago Press.

WRIGHT, S. 1965. The interpretation of population structure by F statistics with special regard to systems of mating. *Evolution* 19: 395–420.

WRIGHT, S., KEELING, J. & GILMAN, L. 2006. The road from Santa Rosalis: a faster tempo of evolution in tropical climates. *Proc. Ntl. Acad. Sci. U.S.A.* 103: 7718–7722.

WRIGHT, S.J. & CORNEJO, F.H. 1990. Seasonal drought and the timing of flowering and leaf fall in a neotropical forest. In: *Reproductive Ecology of Tropical Forest Plants,* eds K.S. Bawa & M. Hadley, pp. 103–133. UNESCO, Paris and the Parthenon Publishing Group, Carnforth.

WRIGHT, S.J. & SCHAIK, C.P. VAN. 1994. Light and the phenology of tropical trees. *Amer. Naturalist* 143: 192–199.

WRIGHT, S.J., CARRASCO, C., CALDERON, O. & PATTON, S. 1999. The El Niño oscillation, variable fruit production, and famine in a tropical forest. *Ecology* 80: 1632–1647.

WU, C.-W. 1961. *The Forests of China.* Harvard University Press.

WU, Z., RAVEN, P.H., & HONG, D., eds. 1994–. *Flora of China.* Science Press, Beijing & Missouri Botanical Garden Press, St Louis.

WUNDER, S. 2000. Ecotourism and economic incentives – an empirical approach. *Ecol. Econ.* 32: 465–479.

WUNDER, S. 2007. The efficiency of payments for environmental services in tropical conservation. *Conserv. Biol.* 21:48–58.

WURSTER, C.M., BIRD, M.I., BULL, I.D., CREED, F., BRYANT, C., DUNGAIT, J.A.J. & PAZ, V. 2010. Forest contraction in north equatorial Southeast Asia during the last glacial period. *Proc. Ntl. Acad. Sci. U.S.A.* 107: 15508–15511.

WYATT-SMITH, J. 1949. Natural plant succession. *Malayan Forester* 12: 148–152.

WYATT-SMITH, J. 1950. Virgin jungle reserves. *Malayan Forester* 23: 40–45.

WYATT-SMITH, J. 1954a. Storm forest in Kelantan. *Malayan Forester* 17: 5–11.

WYATT-SMITH, J. 1954b. Forest memories of the Philippines. *Malayan Forester* 17: 135.

WYATT-SMITH, J. 1955. Changes in composition in early natural plant succession. *Malayan Forester* 18: 44–49.

WYATT-SMITH, J. 1960. Stems per acre and topography. *Malayan Forester* 23: 57–59.

WYATT-SMITH, J. 1963. *Manual of Malayan Silviculture for Inland Forests*. Malayan Forest Records 23.

WYATT-SMITH, J. 1966. *Ecological studies on Malayan Forests*. 1. Malayan Forest Department Research Pamphlet 52.

XI, Z., RUHFEL, B.R., SCHAEFER, H., AMORIM, A.M., SUGUMARAN, M., WURDACK, K.J., ENDRESS, P.K., MATTHEWS, M.L., STEVENS, P.F., MATHEWS, S. & DAVIS, C.C. 2012. Phylogenomics and a posteriori data partitioning resolve the Cretaceous angiosperm radiation Malpighiales. *Proc. Natl. Acad. Sci. U.S.A.* 109: 17519–17524.

XIE, S.-C. 1997. Ecosystem perspectives and conservation of middle-mountain moist evergreen broad-leaved forests in Yunnan, China and its preservation. *Tropics* 6: 351–359.

XIE, Y., MacKINNON, J. & LI, D. 2004. Study on biogeographical divisions of China. *Biodivers. & Conservation* 1391–1417.

YAACOB, A., SUHAIMI, W.C., AMIR, H., AMIR, S., AHMAD, S., ARI, Z., SHANZAN & BAILLIE, I. 2007. *Soil Heterogeneity in the 50 ha Long Term Tropical Forest Ecological Research Plot at Pasoh, West Malaysia*. Unpubl. ms.

YADAV, J.S.P. 1963. Site and soil characteristics of bamboo forests. *Indian Forester* 89: 177–193.

YADAVA, P.S. 1990. Savannas of north-east India. *J. Biogeogr.* 17: 385–394.

YAMADA, I. 1975. Forest ecological studies of the montane forest of Mt. Pangrango, West Java. I. Stratification and floristic composition of the montane rain forest near Cibodas. *Tonan Ajia Kenkyu* (The Southeast Asian Studies) 13 (3).

YAMADA, I. 1976a. Forest ecological studies of the montane forest of Mt. Pangrango, West Java. II. The stratification and floristic composition of the forest vegetation of the higher part of Mt. Pangrango. *Tonan Ajia Kenkyu* (The Southeast Asian Studies) 13 (4).

YAMADA, I. 1976b. Forest ecological studies of the montane forest of Mt. Pangrango, West Java. III. Litterfall of the tropical montane forest near Cibodas. *Tonan Ajia Kenkyu* (The Southeast Asian Studies) 14 (2).

YAMADA, I. 1977. Forest ecological studies of the montane forest of Mt. Pangrango, West Java. IV. Floristic composition along the altitude. *Tonan Ajia Kenkyu* (The Southeast Asian Studies) 15 (2).

YAMADA, T. & SUZUKI, E. 1999. Comparative morphology and allometry of winged diaspores among Asian Sterculiaceae. *J. Trop. Ecol.* 15: 619–636.

YAMADA, T. & SUZUKI, E. 2004. Ecological role of vegetative sprouting in the regeneration of *Dryobalanops rappa*, an emergent species in a Bornean tropical wetland forest. *J. Trop. Ecol.* 20: 377–384.

YAMADA, T., YAMAKURA, T., KANZAKI, M., ITOH, A., OHKUBO, T., OGINO, K., LEE, H.S. & ASHTON, P.S. 1997. Topography-dependent spatial pattern and habitat segregation of sympatric *Scaphium* species in a tropical rain forest at Lambir, Sarawak. *Tropics* 7: 57–66.

YAMADA, T., ITOH, A., KANZAKI, M., YAMAKURA, T., SUZUKI, E. & ASHTON,

P.S. 2000a Local and geographical distributions for a tropical genus, *Scaphium* (Sterculiaceae) in the Far East. *Pl. Ecol.* 148: 23–30.

YAMADA, T., YAMAKURA, T & LEE, H.S. 2000b. Architectural and allometric differences among *Scaphium* species are related to microhabitat differences. *Funct. Ecol.* 14: 731–737.

YAMADA, T., SUZUKI, E., YAMAKURA, T. & TAN, S. 2005. Tap-root depth of tropical seedlings in relation to species-specific edaphic preferences. *J. Trop. Ecol.* 21: 155–160.

YAMADA, T., TOMETA, A., ITOH, A., YAMAKURA, T., OHKUBO, T., KANZAKI, M., TAN, S. & ASHTON, P.S. 2006. Habitat associations of Sterculiaceae trees in a Bornean rain forest. *J. Veg. Sci.* 17: 559–566.

YAMADA, T., ZUIDEMA, P.A., ITOH, A., YAMAKURA, T., OHKUBO, T. KANZAKI, M., TAN, S. & ASHTON, P.S. 2007. Strong habitat preference of a tropical rain forest tree does not imply large differences in population dynamics across habitats. *J. Ecol.* 95: 332–342.

YAMAKURA, T., YAMADA, I., INOUE, T. & OGINO, K. 1995. A long-term and large-scale research of the Lambir rainforest in Sarawak: progress and conceptual background of Japanese activities. *Tropics* 4: 259–276.

YAMAKURA, T., KANZAKI, M., ITOH, A., OHKUBO, T., OGINO, T., CHAI, E.O.K, LEE, H.S. & ASHTON, P.S. 1996. Forest structure of a tropical rain forest at Lambir, Sarawak with special reference to the dependency of its physiognomic dimensions on topography. *Tropics* 6: 1–18.

YAMASHITA, S, FUKUDA, K. & UGAWA, S. 2007. Ectomycorrhizal communities on tree roots and in soil propagule banks along a secondary successional vegetation gradient. *Forest Sci.* 53: 635–644.

YAMASHITA, S., HIROSE, D. & NAKASHIZUKA, T. 2012. Effects of local rainfall exclosure on the ectomycorrhizal fungi of a canopy tree in a tropical forest. *J. Trop. Forest Sci.*24: 322–331.

YANEDA, T., OKUDA, T., ABDULLAH, M., AWANG, M. & FURUKAWA, A. 2000. The leaf development process and its significance for reducing self-shading of a tropical pioneer tree species. *Oecologia* 125: 476–482.

YAP, S.K. 1976. *The Reproductive Biology of Some Understorey Fruit Tree Species in the Lowland Dipterocarp Forest of West Malaysia*. PhD thesis, University of Malaya.

YAP, S.K. 1982. The phenology of some fruit tree species in a lowland dipterocarp forest. *Malayan Forester* 45: 21–35.

YAP, S.K. 1989. Sapindaceae. In: *Tree Flora of Malaya 4*, ed. F.S.P. Ng, pp.434–461. Longman, Malaysia.

YAP, S.K. & CHAN, H.Y. 1990. Phenological behaviour of some *Shorea* species in Peninsular Malaysia. In: *Reproductive Ecology of Tropical Forest Plants*, eds K.S. Bawa & M. Hadley, pp. 21–35. UNESCO, Paris and the Parthenon Publishing Group, Carnforth.

YAP, Y.Y. 1976. *Population and Phylogenetic Studies on Species of Malaysian Rainforest*

Trees. PhD thesis, University of Aberdeen.

YASUDA, M., MATSUMOTO, J., OSADA, N., ICHIKAWA, S., KACHI, N., TANI, M., OKUDA, T., FURUKAWA, A., NIK, A.R. & MANOKARAN, N. 1999. The mechanism of general flowering in Dipterocarpaceae in the Malay Peninsula. *J. Trop. Ecol.* 15: 437–449.

YEAGER, C.P., MARSHALL, A.J., STICKLER, C.M. & CHAPMAN, C.A. 2004. Effects of fires on peat swamp and lowland dipterocarp forests in Kalimantan, Indonesia. *Tropical Biodiversity* 8: 121–138.

YODA, K. 1974. *Three-dimensional Distribution of Light Intensity in a Tropical Rain Forest of West Malaysia*. IBP Synthesis Meeting, Kuala Lumpur, Unpubl. ms.

YODA, K, SAHUNALU, P. & KANZAKI, M., eds 1995. *Elucidation of the Missing Link in the Global Carbon Cycling – Focussing on the Dynamics of Tropical Seasonal Forests*. Osaka City University.

YOSHIFUJI, N., KUMAGAI, T., TANAKA, K., TANAKA, N., KOMATSU, H., SUZUKI, M. & TANTASIRIN, C. 2006. Inter-annual variation in growing season length of a tropical seasonal forest in northern Thailand. *Forest Ecol. Managem.* 229: 333–339.

YOUNG, T.P. & HUBBELL, S.P. 1991. Crown asymmetry, tree falls, and repeat disturbance of broad-leaved forest gaps. *Ecology* 72: 1464–1471.

YU, D. 1994. Free trade is green, protectionism is not. *Conservation Biol.* 8: 989–996.

YUMOTO, T. 2000. Bird pollination of three *Durio* species (Bombacaceae) in a tropical rainforest in Sarawak, Malaysia. *Amer. J. Bot.* 87: 1181–1188.

YUMOTO, T., ITINO, T. & NAGAMASU, H. 1997. Pollination of hemiparasites (Loranthaceae) by spider hunters (Nectariniidae) in the canopy of a Bornean tropical rain forest. *Selbyana* 18: 51–60.

YUMOTO, T., NOMA, N. & MARUHASHI, T. 1998. Cheek-pouch dispersal of seeds by Japanese monkeys (*Macaca fascata yakui*) on Yakushima Island, Japan. *Primates* 39: 325–338.

YUSOP, Z., SUKI, A. & ZAKARIA, M.P. 1987. Effects of selective logging on physical streamwater quality in hill tropical rain forest. *Malayan Forester* 50: 281–310.

ZEPPEL, H. 2007. Indigenous ecotourism: conservation and resource rights. In: *Critical Issues in Ecotourism: Understanding a Complex Tourism Phenomenon*, ed. J. Highman, pp. 308–348. Elsevier Butterworth-Heinemann, Oxford.

ZHANG, J. & CAO, M. 1995. Tropical forest vegetation of Xishuangbanna, SW China and its secondary changes, with special reference to some problems in local nature conservation. *Biol. Conservation* 73: 229–238.

ZHOU, Z.Y., CREPET, W. & DIXON, K. 2001. The earliest fossil evidence of the Hamamelidaceae: late Cretaceous (Turonian) inflorescence and fruits of Altingioideae. *Amer. J. Bot.* 88: 753–766.

ZHU, H. 1997. Ecological and biogeographical studies in the tropical rain forest of south Yunnan, SW China, with special reference to its relation with

rain forests of tropical Asia. *J. Biogeogr.* 24: 647–662.

ZHU, H. 2006. Forest vegetation of Xishuangbanna, South China. *For. Stud. China* 8: 1–58.

ZHU, H. 2008. Advances in biogeography of the tropical rain forest of southern Yunnan, Southwestern China. *Trop. Conserv. Sci.* 1: 34–42.

ZHU, H. & CAI, L. 2004. Vegetation of the Upper Mekong Valley. *Mankind and Nature Suppl.* 7. Yunnan Education Publishing House, Kunming.

ZHU, H., XU, Z.F., WANG, H. & LI, B.G. 2004. Tropical forest fragmentation and its ecological and species diversity changes in southern Yunnan. *Biodivers. & Conservation* 13: 1355–1372.

ZHU, H., SHI, J.P. & ZHAO, C.J. 2005. Species composition, physiognomy and plant diversity of the tropical montane evergreen broad-leaved forest in southern Yunnan. *Biodivers. & Conservation* 14: 2855–2870.

ZHU, H., WANG, H. & LI, B.-G. 2006. Species composition and biogeography of tropical montane rain forest in southern Yunnan of China. *Gard. Bull. Singapore* 58: 81–132.

ZHUANG, X.Y. & CORLETT, R.T. 1997. Forest and forest succession in Hong Kong, China. *J. Trop. Ecol.* 14: 857–866.

ZIEGLER, A.D., BRUUN, T.B., & GUARDIOLA-CLARAMONTE, M. 2009. Environmental consequences of the demise of swidden cultivation in montane mainland Southeast Asia: hydrology and geomorphology. *For. Ecol. Manag.* 251: 1–9.

ZIPPERLEN, S.W. & PRESS, M.C. 1996. Photosynthesis in relation to growth and seedling ecology of two dipterocarp rain forest tree species. *J. Ecol.* 84: 863–876.

ZIPPERLEN, S.W. & PRESS, M.C. 1997. Photosynthesis induction and stomatal oscillations in relation to the light environment of two dipterocarp rain forest tree species. *J. Ecol.* 85: 491–503.

ZOTZ, G. 2004. How prevalent is crassulacean acid metabolism among vascular epiphytes? *Oecologia* 138: 184–192.

ZOTZ, G. & WINTER, K. 1996. Diel pattern of CO_2 exchange in rain forest canopy plants. In: *Tropical Plant Forest Ecophysiology*, eds S.S. Mulkey, R.L. Chazdon & A.P. Smith, pp. 89–113. Chapman & Hall, London & New York.

Index

Scientific and common names

Place names

Some will dislike my frequent use of currently discarded place names, but I have tried to use those that I found to be most frequent in the literature.

The place names in bold text are those shown on the 'Rainfall seasonality in tropical Asia, and the line of occasional lowland frost' map at the **front** of the book and the numbers in square brackets relate to the numbers on that map.

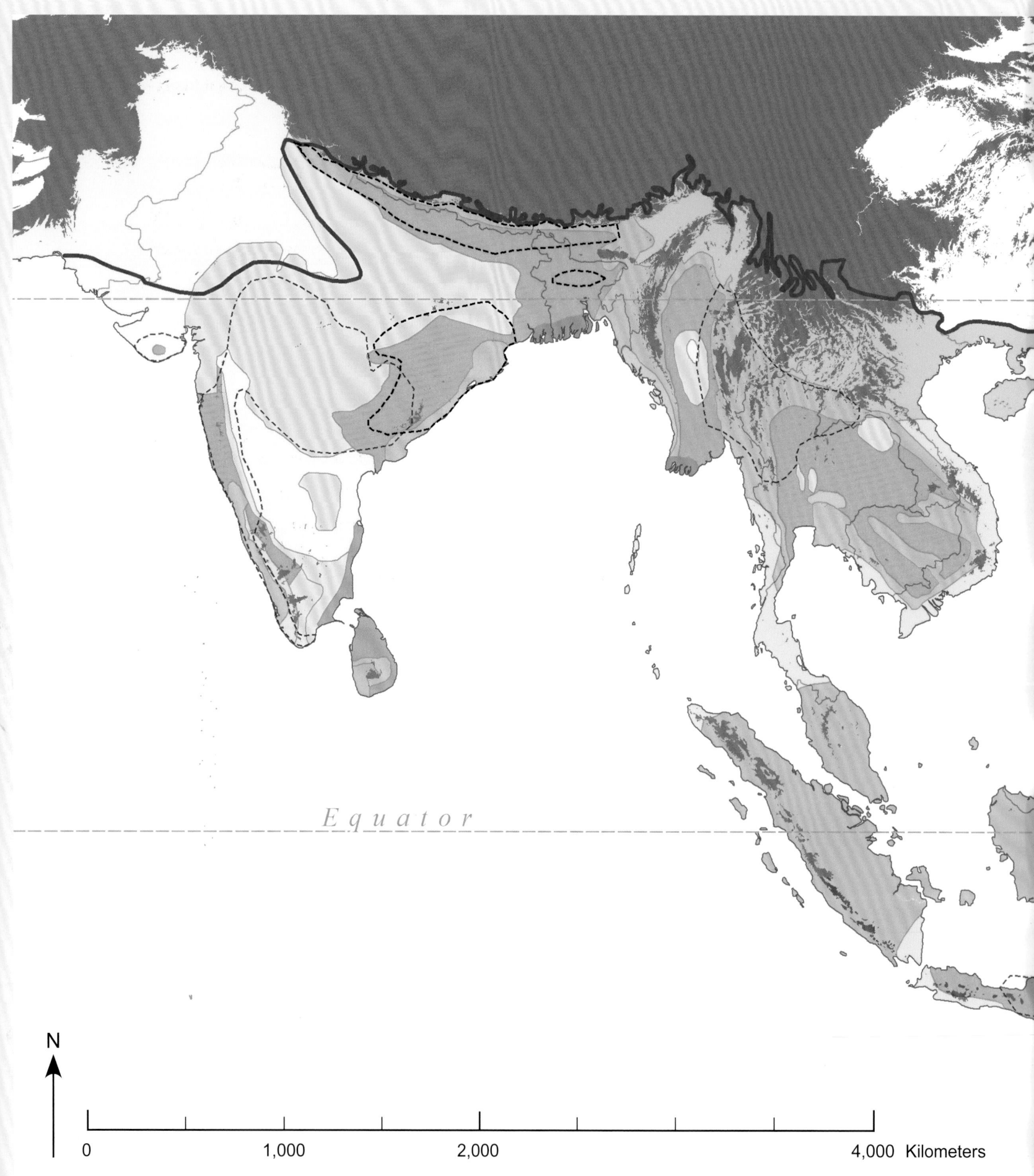

N

0	1,000	2,000		4,000	Kilometers

The major zonal forest formations of lowland tropical Asia. Boundaries
approximate, on the basis of herbarium specimens and personal experience.